Nanostructured Materials and Nanotechnology

Nanostructured Materials and Nanotechnology

Concise Edition

Edited by

Hari Singh Nalwa, M.Sc., Ph.D.
Hitachi Research Laboratory
Hitachi Ltd., Ibaraki, Japan

ACADEMIC PRESS

A Harcourt Science and Technology Company

San Diego San Francisco New York Boston London Sydney Tokyo

This book is printed on acid-free paper. ♾

Academic Press
A Harcourt Science and Technology Company
525 B Street, Suite 1900, San Diego, California 92101-4495, USA
http://www.academicpress.com

Academic Press
Harcourt Place, 32 Jamestown Road, London NW1 7BY, UK
http://www.academicpress.com

Library of Congress Catalog Card Number: 2001089144

International Standard Book Number: 0-12-513920-9

PRINTED IN THE UNITED STATES OF AMERICA
01 02 03 04 05 06 IP 9 8 7 6 5 4 3 2 1

To my children,
Surya, Ravina and Eric

Preface

Nanotechnology is the science and engineering of creating materials, functional structures, and devices on a nanometer scale. In scientific terminology "nano" means 10^{-9}: one nanometer is equivalent to one thousandth of a micrometer, one millionth of a millimeter, and one billionth of a meter. Nanostructured inorganic, organic, and biological materials may have existed in nature since the evolution of life on Earth. Some evident examples are microorganisms, fine-grained minerals in rocks, nanoparticles in bacteria, and smoke. From a biological viewpoint, the DNA double helix has a diameter of about 2 nm, whereas ribosomes have a diameter of 25 nm. Atoms have a size of 1 to 4 angstroms. Therefore, nanostructured materials could hold tens of thousands of atoms all together. On a micrometer scale, an excellent example is a human hair, which has a diameter of 50 to 100 μm.

Why are we so fascinated with downsizing materials to a nanoscale? The fundamental physical, chemical, and biological properties of materials are surprisingly altered as their constituent grains are decreased to a nanometer scale owing to their size, shape, surface chemistry, and topology. For example, 6-nm copper grains show five times more hardness than conventional copper. Cadmium selenide (CdSe) can be made to yield any color in the spectrum simply by controlling the size of its constituent grains. Therefore, nanotechnology has generated much interest in the scientific community, and it has become a very active area of research. There are several major government programs on research and development of nanostructured materials and nanotechnology in the United States, Europe, and Asia. A new vision of molecular nanotechnology will develop in the coming years, and we may see many technological breakthroughs in creating materials atom by atom. The resulting new inventions will have a widespread impact on the fields of science, engineering, and medicine.

The five-volume *Handbook of Nanostructured Materials and Nanotechnology* published in October 2000 by the Editor created much interest in the field of nanotechnology, and many of our colleagues looked for a short version of this major reference work. Therefore, the current condensed version of this handbook contains 16 selected chapters particularly focused on the synthesis and fabrication aspects and the electrical and optical properties of nanoscale materials. The chapters on synthesis include Chemical Synthesis of Nanostructured Metals, Metals Alloys, and Semiconductors, by K. E. Gonsalves, S. P. Rangarajan, and J. Wang; Nanocomposites Prepared by Sol-Gel Methods: Synthesis and Characterization, by K. C. Kwiatkowski and C. M. Lukehart; Low-Temperature Compaction of Nanosize Powders, by E. J. Gonzalez and G. J. Piermarini; Semiconductor Nanoparticles, by P. V. Kamat, K. Murakoshi, Y. Wada, and S. Yanagida; Colloidal Quantum Dots of III-V Semiconductors, by O. I. Micic and A. J. Nozik; Strained-Layer Heteroepitaxy to Fabricate Self-Assembled Semiconductor Islands, by W. H. Weinberg, C. M. Reaves, B. Z. Nosho, R. I. Pelzel, and S. P. Denbaars; Hybrid Magnetic–Semiconductor Nanostructures, by F. M. Peeters and J. De Boeck; Carbon Nanotubes, by P. M. Ajayan; and Encapsulation and Crystallization Behavior of Materials Inside Carbon Nanotubes, by J. Sloan and M. L. H. Green. Two chapters provide details of the electronic properties of nanostructured materials: Silicon-Based Nanostructures, by T. P. Sidiki and C. M. Sotomayor Torres; and Electronic Transport Properties of Quantum Dots, by M. A. Reed, J. W. Sleight, and M. R. Deshpande.

Because nanoscale semiconductor materials exhibit very interesting optical properties, we have included two chapters on this topic: Photorefractive Semiconductor Nanostructures, by D. D. Nolte, M. R. Melloch, Y. Ding, M. Dinu, K. M. Kwolek, and I. Lahiri; and Linear and Nonlinear Optical Spectroscopy of Semiconductor Nanocrystals, by V. I. Klimov. Organic materials offer tremendous possibilities of modifying structures and their functions on a nanometer scale, and we have included two chapters on supramolecular nanostructures: Molecular and Supramolecular Nanomachines, by M. Gomez-Lopez and J. Fraser Stoddart; and Functional Nanostructures Incorporating Responsive Modules, by A. C. Benniston and P. R. Mackie. Finally, we have included a chapter on the biological aspects of nanotechnology: Structure, Behavior, and Manipulation of Nanoscale Biological Assemblies, by T. Bayburt, J. Carlson, B. Godfrey, M. Shank-Retzlaff, and S. G. Sligar.

I hope this condensed version of the handbook will be useful for scientists and students working in chemistry, physics, materials science, electrical engineering, polymer science, surface science, spectroscopy, and biotechnology. This handbook provides state-of-the-art information on many aspects of nanostructured materials and emerging nanotechnology.

I owe my deepest appreciation for the cooperation and patience of my wife, Dr. Beena Singh Nalwa, the moral support of my parents, Sri Kadam Singh and Srimati Sukh Devi, and the love of my children, Surya Singh, Ravina Singh, and Eric Singh, in this exciting enterprise.

Hari Singh Nalwa

Contents

Chapter 7. HYBRID MAGNETIC–SEMICONDUCTOR NANOSTRUCTURES

François M. Peeters, Jo De Boeck

About the Editor

Dr. Hari Singh Nalwa is the Managing Director of the Stanford Scientific Corporation in Los Angeles, California. Previously, he was Head of Department and R&D Manager at the Ciba Specialty Chemicals Corporation in Los Angeles (1999–2000) and a staff scientist at the Hitachi Research Laboratory, Hitachi Ltd., Japan (1990–1999). He has authored over 150 scientific articles in journals and books. He has 18 patents, either issued or applied for, on electronic and photonic materials and devices based on them.

He has published 43 books including this single volume: *Ferroelectric Polymers* (Marcel Dekker, 1995), *Nonlinear Optics of Organic Molecules and Polymers* (CRC Press, 1997), *Organic Electroluminescent Materials and Devices* (Gordon & Breach, 1997), *Handbook of Organic Conductive Molecules and Polymers*, Vols. 1–4 (John Wiley & Sons, 1997), *Handbook of Low and High Dielectric Constant Materials and Their Applications*, Vols. 1–2 (Academic Press, 1999), *Handbook of Nanostructured Materials and Nanotechnology*, Vols. 1–5 (Academic Press, 2000), *Handbook of Advanced Electronic and Photonic Materials and Devices*, Vols. 1–10 (Academic Press, 2001), *Advanced Functional Molecules and Polymers*, Vols. 1–4 (Gordon & Breach, 2001), *Photodetectors and Fiber Optics* (Academic Press, 2001), *Silicon-Based Materials and Devices*, Vols. 1–2 (Academic Press, 2001), *Supramolecular Photosensitive and Electroactive Materials* (Academic Press, 2001), *Handbook of Nanostructured Materials and Nanotechnology*–Concise Edition (Academic Press, 2001), *Handbook of Surfaces and Interfaces of Materials*, Vols. 1–5 (Academic Press, 2001), *Handbook of Thin Film Materials*, Vols. 1–5 (Academic Press, 2001). The *Handbook of Nanostructured Materials and Nanotechnology* edited by him received the 1999 Award of Excellence in Engineering Handbooks from the Association of American Publishers.

Dr. Nalwa is the founder and Editor-in-Chief of the *Journal of Nanoscience and Nanotechnology* (2001–). He also was the founder and Editor-in-Chief of the *Journal of Porphyrins and Phthalocyanines* published by John Wiley & Sons (1997–2000) and serves or has served on the editorial boards of *Journal of Macromolecular Science-Physics* (1994–), *Applied Organometallic Chemistry* (1993–1999), *International Journal of Photoenergy* (1998–) and *Photonics Science News* (1995–). He has been a referee for many international journals including *Journal of American Chemical Society*, *Journal of Physical Chemistry*, *Applied Physics Letters*, *Journal of Applied Physics*, *Chemistry of Materials*, *Journal of Materials Science*, *Coordination Chemistry Reviews*, *Applied Organometallic Chemistry*, *Journal of Porphyrins and Phthalocyanines*, *Journal of Macromolecular Science-Physics*, *Applied Physics*, *Materials Research Bulletin*, and *Optical Communications*.

Dr. Nalwa helped organize the First International Symposium on the Crystal Growth of Organic Materials (Tokyo, 1989) and the Second International Symposium on Phthalocyanines (Edinburgh, 1998) under the auspices of the Royal Society of Chemistry. He also proposed a conference on porphyrins and phthalocyanies to the scientific community that, in part, was intended to promote public awareness of the *Journal of Porphyrins and Phthalocyanines*, which he founded in 1996. As a member of the organizing committee, he help effectuate the First International Conference on Porphyrins and Phthalocyanines, which was held in Dijon, France in 2000. Currently he is on the organizing committee of the BioMEMS and Smart Nanostructures, (December 17–19, 2001, Adelaide, Australia) and the World Congress on Biomimetics and Artificial Muscles (December 9–11, 2002, Albuquerque, USA).

Dr. Nalwa has been cited in the *Dictionary of International Biography*, *Who's Who in Science and Engineering*, *Who's Who in America*, and *Who's Who in the World*. He is a member of the American Chemical Society (ACS), the American Physical Society (APS), the Materials Research Society (MRS), the Electrochemical Society and the American Association for the Advancement of Science (AAAS). He has been awarded a number of prestigious fellowships including a National Merit Scholarship, an Indian Space Research Organization (ISRO) Fellowship, a Council of Scientific and Industrial Research (CSIR) Senior Fellowship, a NEC Fellowship, and Japanese Government Science & Technology Agency (STA) Fellowship. He was an Honorary Visiting Professor at the Indian Institute of Technology in New Delhi.

Dr. Nalwa received a B.Sc. degree in biosciences from Meerut University in 1974, a M.Sc. degree in organic chemistry from University of Roorkee in 1977, and a Ph.D. degree in polymer science from Indian Institute of Technology in New Delhi in 1983. His thesis research focused on the electrical properties of macromolecules. Since then, his research activities and professional career have been devoted to studies of electronic and photonic organic and polymeric materials. His endeavors include molecular design, chemical synthesis, spectroscopic characterization, structure-property relationships, and evaluation of novel high performance materials for electronic and photonic applications. He was a guest scientist at Hahn-Meitner Institute in Berlin, Germany (1983) and research associate at University of Southern California in Los Angeles (1984–1987) and State University of New York at Buffalo (1987–1988). In 1988, he moved to the Tokyo University of Agriculture and Technology, Japan as a lecturer (1988–1990), where he taught and conducted research on electronic and photonic materials. His research activities include studies of ferroelectric polymers, nonlinear optical materials for integrated optics, low and high dielectric constant materials for microelectronics packaging, electrically conducting polymers, electroluminescent materials, nanocrystalline and nanostructured materials, photocuring polymers, polymer electrets, organic semiconductors, Langmuir-Blodgett films, high temperature-resistant polymer composites, water-soluble polymers, rapid modeling, and stereolithography.

List of Contributors

Numbers in parenthesis indicate the pages on which the author's contribution begins.

P. M. AJAYAN (329)
Department of Materials Science and Engineering, Rensselaer Polytechnic Institute, Troy, New York, USA

TIMOTHY BAYBURT (749)
Beckman Institute, Urbana, Illinois, USA

ANDREW C. BENNISTON (693)
Chemistry Department, Glasgow University, Glasgow G12 8QQ, Scotland

JOSEPH CARLSON (749)
Beckman Institute, Urbana, Illinois, USA

JO DE BOECK (247)
IMEC, Leuven, Belgium

S. P. DENBAARS (207)
Departments of Chemical Engineering and Materials, University of California, Santa Barbara, California, USA

M. R. DESHPANDE (445)
Departments of Physics, Applied Physics, and Electrical Engineering, Yale University, New Haven, Connecticut, USA

Y. DING (483)
Department of Physics, Purdue University, 1396 Physics Building, West Lafayette, Indiana, USA

M. DINU (483)
Department of Physics, Purdue University, 1396 Physics Building, West Lafayette, Indiana, USA

BRUCE GODFREY (749)
Beckman Institute, Urbana, Illinois, USA

MARCOS GÓMEZ-LÓPEZ (641)
Department of Chemistry and Biochemistry, University of California at Los Angeles, California, USA

K. E. GONSALVES (1)
Department of Chemistry and Polymer Program, Institute of Materials Science U-136, University of Connecticut, Storrs, Connecticut, USA

E. J. GONZALEZ (93)
Ceramics Division, National Institute of Standards and Technology, Gaithersburg, Maryland, USA

M. L. H. GREEN (361)
Inorganic Chemistry Laboratory, University of Oxford, Oxford, UK

PRASHANT V. KAMAT (129)
Notre Dame Radiation Laboratory, Notre Dame, Indiana, USA

VICTOR I. KLIMOV (563)
Chemical Sciences and Technology Division, CST-6, MS-J585,
Los Alamos National Laboratory, Los Alamos, New Mexico, USA

KRZYSZTOF C. KWIATKOWSKI (57)
Department of Chemistry, Vanderbilt University, Nashville, Tennessee, USA

K. M. KWOLEK (483)
Department of Physics, Purdue University, 1396 Physics Building, West Lafayette, Indiana, USA

I. LAHIRI (483)
Department of Physics, Purdue University, 1396 Physics Building, West Lafayette, Indiana, USA

CHARLES M. LUKEHART (57)
Department of Chemistry, Vanderbilt University, Nashville, Tennessee, USA

PHILIP R. MACKIE (693)
Chemistry Department, Glasgow University, Glasgow G12 8QQ, Scotland

M. R. MELLOCH (483)
School of Electrical and Computer Engineering, Purdue University,
1396 Physics Building, West Lafayette, Indiana, USA

OLGA I. MIĆIĆ (183)
Center for Basic Sciences, National Renewable Energy Laboratory, Golden, Colorado, USA

KEI MURAKOSHI (129)
Chemical Process Engineering, Faculty of Engineering, Osaka University, Suita, Osaka, Japan

D. D. NOLTE (483)
Department of Physics, Purdue University, 1396 Physics Building, West Lafayette, Indiana, USA

B. Z. NOSHO (207)
Departments of Chemical Engineering and Materials, University of California, Santa Barbara, California, USA

ARTHUR J. NOZIK (183)
Center for Basic Sciences, National Renewable Energy Laboratory, Golden, Colorado, USA

FRANÇOIS M. PEETERS (247)
Departement Natuurkunde, Universiteit Antwerpen, Antwerpen, Belgium

R. I. PELZEL (207)
Departments of Chemical Engineering and Materials, University of California, Santa Barbara, California, USA

G. J. PIERMARINI (93)
Ceramics Division, National Institute of Standards and Technology, Gaithersburg, Maryland, USA

S. P. RANGARAJAN (1)
Department of Chemistry and Polymer Program, Institute of Materials Science U-136, University of Connecticut, Storrs, Connecticut, USA

C. M. Reaves (207)
Departments of Chemical Engineering and Materials, University of California, Santa Barbara, California, USA

M. A. Reed (445)
Departments of Physics, Applied Physics, and Electrical Engineering, Yale University, New Haven, Connecticut, USA

Mary Shank-Retzlaff (749)
Beckman Institute, Urbana, Illinois, USA

Tamim P. Sidiki (387)
Institute of Materials Science and Department of Electrical Engineering, University of Wuppertal, 42097 Wuppertal, Germany

J. W. Sleight (445)
Departments of Physics, Applied Physics, and Electrical Engineering, Yale University, New Haven, Connecticut, USA

Stephen G. Sligar (749)
Beckman Institute, Urbana, Illinois, USA

J. Sloan (361)
Inorganic Chemistry Laboratory and Department of Materials, University of Oxford, Oxford, UK

Clivia M. Sotomayor Torres (387)
Institute of Materials Science and Department of Electrical Engineering, University of Wuppertal, 42097 Wuppertal, Germany

J. Fraser Stoddart (641)
Department of Chemistry and Biochemistry, University of California at Los Angeles, California, USA

Yuji Wada (129)
Chemical Process Engineering, Faculty of Engineering, Osaka University, Suita, Osaka, Japan

J. Wang (1)
Department of Chemistry and Polymer Program, Institute of Materials Science U-136, University of Connecticut, Storrs, Connecticut, USA

W. H. Weinberg (207)
Departments of Chemical Engineering and Materials, University of California, Santa Barbara, California, USA

Shizo Yanagida (129)
Chemical Process Engineering, Faculty of Engineering, Osaka University, Suita, Osaka, Japan

Chapter 1

CHEMICAL SYNTHESIS OF NANOSTRUCTURED METALS, METAL ALLOYS, AND SEMICONDUCTORS

K. E. Gonsalves, S. P. Rangarajan, J. Wang

Department of Chemistry and Polymer Program, Institute of Materials Science U-136, University of Connecticut, Storrs, Connecticut, USA

Contents

1. INTRODUCTION

Ultrafine microstructures having an average phase or grain size on the order of a nanometer (10^{-9} m) are classified as nanostructured materials (NSMs) [1]. Currently, in a wider meaning of the term, any material that contains grains or clusters below 100 nm, or layers or filaments of that dimension, can be considered to be nanostructured [2]. The interest in these materials has been stimulated by the fact that, owing to the small size of the building blocks (particle, grain, or phase) and the high surface-to-volume ratio, these materials are expected to demonstrate unique mechanical, optical, electronic, and magnetic properties [3]. The properties of NSMs depend on the following four common microstructural features: (1) fine grain size and size distribution (<100 nm); (2) the chemical composition of the constituent phases; (3) the presence of interfaces, more specifically, grain boundaries, heterophase interfaces, or the free surface; and (4) interactions between the constituent domains. The presence and interplay of these four features largely determine the unique properties of NSMs.

Handbook of Nanostructured Materials and Nanotechnology, edited by H.S. Nalwa
Volume 1: Synthesis and Processing

In nanophase materials, a variety of size-related effects can be incorporated by controlling the sizes of the constituent components [4]. For example, nanostructured metals and ceramics can have improved mechanical properties compared to conventional materials as a result of the ultrafine microstructure. In addition, NSMs have the capability to be sintered at much lower temperatures than conventional powders, enabling the full densification of these materials at relatively lower temperatures. Semiconductor NSMs are currently also considered to have technological applications in optoelectronic devices such as semiconductor quantum dots and photodiodes, owing to the phenomenon of "quantum size effects" caused by the spatial confinement of delocalized electrons in confined grain sizes [5]. Magnetic applications of NSMs include fabrication of devices with giant magnetoresistance (GMR) effects, the property used by magnetic heads to read data on computer hard drives, as well as the development of magnetic refrigerators that use solid magnets as refrigerants rather than compressed ozone-destroying chlorofluorocarbons [6]. In addition, nanostructured metals and ceramics seem to be candidates for new catalytic applications [7].

The development of semiconductor nanoclusters is an area of intense research efforts. These nanoclusters are often referred to as quantum dots, nanocrystals, and Q-particles [8]. In the nanometer size regime, electron–hole confinement in nanosized spherical semiconductor particles results in three-dimensional size quantization. Band gap engineering by size and dimension quantization is important because it leads to electrical, optical, magnetic, optoelectronic, and magnetooptical properties substantially different from those observed for the bulk material [5a, 9]. As an example, quantum dots can be developed to emit and absorb a desired wavelength of light by changing the particle diameters. This feature allows the construction of a finely tunable and efficient semiconductor laser.

2. SYNTHESIS OF NANOSTRUCTURED MATERIALS

The synthesis of NSMs from atomic or molecular sources depends on the control of a variety of "nanoscale" attributes desired in the final product. In general, the following four methods have been used to make nanophase materials:

1. The first technique involves the production of isolated, ultrafine crystallites having uncontaminated free surfaces followed by a consolidation process either at room or at elevated temperatures. The specific processes used to isolate the NSMs are, for example, inert-gas condensation [10], decomposition of the starting chemicals or the precursors, and precipitation from solutions.
2. Chemical vapor deposition (CVD), physical vapor deposition (PVD) [11], and some electrochemical methods [12] have been used to deposit atoms or molecules of materials on suitable substrates. Nanocomposites can be produced by depositing chemically different molecules simultaneously or consecutively.
3. By introducing defects in a formerly perfect crystal such as dislocations or grain boundaries, new classes of NSMs can be synthesized. Such deformations may be brought about by subjecting the materials to high energy by either ball milling, extrusion, shear, or high-energy irradiation [13].
4. The final approach used to make NSMs is based on crystallization or precipitation from unstable states of condensed matter such as crystallization from glasses or precipitation from supersaturated solid or liquid solutions [14].

Although these are the general methodologies employed in the synthesis of nanostructured materials, several variants of these processes have been developed to generate compounds or alloys with specific compositions and properties and also for optimized production.

There are basically two broad areas of synthetic techniques for NSMs, namely, (1) physical methods [15] and (2) chemical methods [16].

2.1. Physical Methods

Several different physical methods are currently in use for the synthesis and commercial production of NSMs. The first and the most widely used technique involves the synthesis of single-phase metals and ceramic oxides by the inert-gas evaporation technique [17]. The generation of atom clusters by gas phase condensation proceeds by evaporating a precursor material, either a single metal or a compound, in a gas maintained at a low pressure, usually below 1 atm, in an apparatus similar to that shown in Figure 1. The evaporated atoms or molecules undergo a homogeneous condensation to form atom clusters via collisions with gas atoms or molecules in the vicinity of a cold-powder collection surface. The clusters once formed must be removed from the region of deposition to prevent further aggregation and coalescence of the clusters. These clusters are readily removed from the gas condensation chamber either by natural convection of the gas or by forced gas flow. Sputtering is another technique used to produce NSMs clusters as well as a variety of thin films. This method involves the ejection of atoms or clusters of designated materials by subjecting them to an accelerated and highly focused beam of inert gas such as argon or helium. The third physical method involves generation of NSMs via severe mechanical deformation [18]. In this method, NSMs are produced not by cluster assembly but rather by structural degradation of coarser-grained structures induced by the application of high

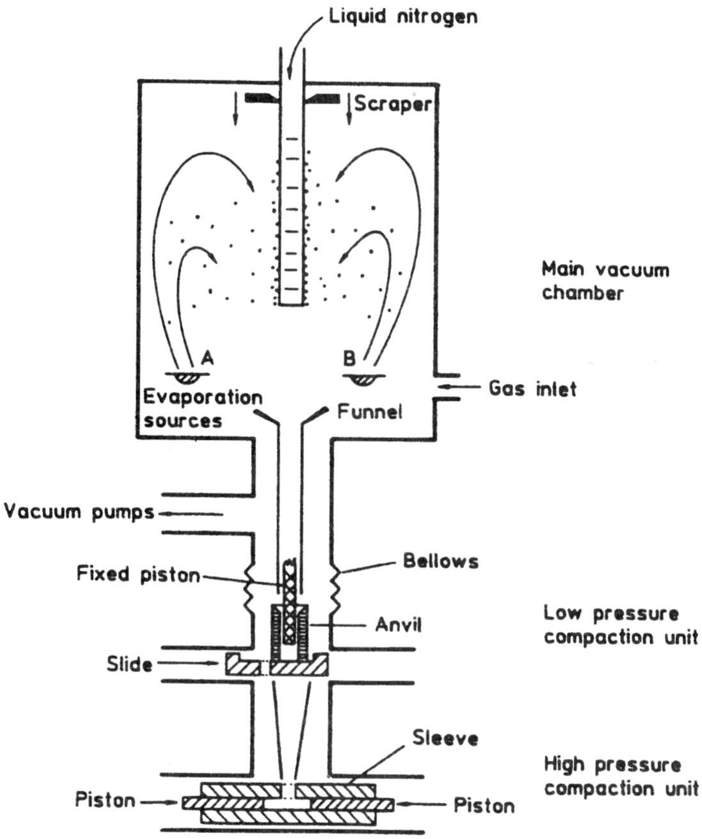

Fig. 1. Schematic drawing of a gas condensation chamber for the synthesis of nanophase materials. Precursor material evaporated from sources A and/or B condenses in the gas and is transported via convection to the liquid-nitrogen-filled cold finger. The clusters are then scraped from the cold finger, collected via the funnel, and consolidated first in the low-pressure compaction unit and then in the high-pressure compaction unit, all in vacuum. Reprinted from *Mater. Sci. Eng. A*, Siegel et al., 168, 189 (© 1993), with kind permission from Elsevier Science Ltd., The Boulevard, Langford Lane, Kidlington 0X5 1GB, UK.

mechanical energy. The nanometer-sized grains nucleate within the shear bands of the deformed materials converting a coarse-grained structure to an ultrafine powder. The heavy deformation of the coarser materials is effected by means of a high-energy ball mill or a high-energy shear process. Although this method is very useful in generating commercial quantities of the material, it suffers from the disadvantage of contamination problems resulting from the sources of the grinding media. Further details regarding the synthesis of NSMs by physical methods have been elaborated in other relevant chapters.

2.2. Chemical Methods

Chemistry has played a major role in developing new materials with novel and technologically important properties [19]. The advantage of chemical synthesis is its versatility in designing and synthesizing new materials that can be refined into the final product. The primary advantage that chemical processes offer over other methods is good chemical homogeneity, as chemical synthesis offers mixing at the molecular level. Molecular chemistry can be designed to prepare new materials by understanding how matter is assembled on an atomic and molecular level and the consequent effects on the desired material macroscopic properties. A basic understanding of the principles of crystal chemistry, thermodynamics, phase equilibrium, and reaction kinetics is important to take advantage of the many benefits that chemical processing has to offer [20].

However, there are certain difficulties in chemical processing. In some preparations, the chemistry is complex and hazardous. Contamination can also result from the byproducts being generated or side reactions in the chemical process. This should be minimized or avoided to obtain desirable properties in the final product. Agglomeration can also be a major cause of concern at any stage in a synthetic process and it can dramatically alter the properties of the materials. As an example, agglomeration frequently makes it more difficult to consolidate nanoparticles to a fully dense compact. Finally, although many chemical processes are scalable for economical production, it is not always straightforward for all systems.

Solution chemistry is used sometimes to prepare the precursor, which is subsequently converted to the nanophase particles by nonliquid phase chemical reactions. Precipitation of a solid from a solution is a common technique for the synthesis of fine particles. The general procedure involves reactions in aqueous or nonaqueous solutions containing the soluble or suspended salts. Once the solution becomes supersaturated with the product, the precipitate is formed by either homogeneous or heterogeneous nucleation. The formation of a stable material with or without the presence of a foreign species is referred to as heterogeneous or homogeneous nucleation [21]. The growth of the nuclei after formation usually proceeds by diffusion, in which case concentration gradients and reaction temperatures are very important in determining the growth rate of the particles, for example, to form monodispersed particles. For instance, to prepare unagglomerated particles with a very narrow size distribution, all the nuclei must form at nearly the same time and subsequent growth must occur without further nucleation or agglomeration of the particles.

In general, the particle size and particle size distribution, the physical properties such as crystallinity and crystal structure, and the degree of dispersion can be affected by reaction kinetics. In addition, the concentration of reactants, the reaction temperature, the pH, and the order of addition of reactants to the solution are also important. Even though a multielement material is often made by coprecipitation of batched ions, it is not always easy to coprecipitate all the desired ions simultaneously because different species may only precipitate at different pH. Thus, control of chemical homogeneity and stoichiometry requires a very careful control of reaction conditions. The problem of agglomeration may be avoided in liquid phase reactions by common methods such as spray drying and freeze drying [22, 23].

NSMs are also prepared by chemical vapor deposition (CVD) or chemical vapor condensation (CVC) [24]. In these processes, a chemical precursor is converted to the gas phase and it then undergoes decomposition at either low or atmospheric pressure to generate the nanostructured particles. These products are then subjected to transport in a carrier gas and collected on a cold substrate, from where they are scraped and collected. The CVC method may be used to produce a variety of powders and fibers of metals, compounds, or composites. The CVD method has been employed to synthesize several ceramic metals, intermetallics, and composite materials. For example, nanophase Si–N–C-containing ceramic particles were obtained by the thermal decomposition of liquid silazane precursors having the general formula $[CH_3SiHNH]_x$, $x = 3$ or 4, with 80% of the cyclic being $x = 4$. It is believed that in the pyrolysis reaction the –SiH–NH– groups were responsible for the extensive crosslinking and the nucleophilic displacements on the neighboring Si atoms, resulting in a three-dimensional network [25].

Semiconductor clusters have traditionally been prepared by use of colloids, micelles, polymers, crystalline hosts, and glasses [26]. The clusters prepared by these methods have poorly defined surfaces and a broad size distribution, which is detrimental to the properties of the semiconductor material. The synthesis of monodisperse clusters with very well defined surfaces is still a challenge to synthetic chemists. However, some of the recent approaches used to overcome these problems are: (1) the synthesis of the clusters within a porous host lattice (such as a zeolite) acting as a template and (2) the controlled fusion of clusters. In this chapter, an overview of the various synthetic methods used to make semiconductor nanoclusters is presented.

3. SYNTHESIS OF METALS, INTERMETALLICS, AND SEMICONDUCTORS

Metals and intermetallics are made by employing either aqueous or nonaqueous methods. Fine metal powders have applications in electronic and magnetic materials, explosives, catalysts, pharmaceuticals, and powder metallurgy [27]. The two main routes of chemical synthesis for the metals and intermetallics are: (1) thermal or ultrasonic decomposition of organometallic precursors to yield the respective elements or alloys and (2) reduction of inorganic or organometallic precursors by reducing agents. Semiconductor nanoclusters are made by incorporating the particles in micelles/colloids, polymers, glasses, or zeolites or by controlled cluster fusion.

Because this chapter deals with the chemical synthesis of nanophase metals and metal alloys and semiconductors, we will be describing here the various methods to prepare metals, for example, palladium, gold, chromium, molybdenum, and copper, as well as the synthesis of certain alloys such as Fe–Co, Fe–Ni, Ti–Al, and a four-component alloy of Fe, commercially known as M50 steel. The development of semiconductor clusters such as ZnO, CdS, ZnTe, and GaN by different synthetic methodologies is also described.

3.1. Chemical Synthesis of Metals

3.1.1. Thermal and Ultrasound Decomposition Methods

The most common example of thermal decomposition is the synthesis of colloidal iron dispersions by the decomposition of iron pentacarbonyl, $Fe(CO)_5$, in a high boiling solvent such as decalin. In a variation of the aforementioned procedure, dispersions of colloidal iron are produced by the decomposition of $Fe(CO)_5$ in two kinds of polymer solutions: (1) "active" polymers such as polymers having a nitrogen nucleophile and (2) "passive" polymers such as polymers having an alkenyl or benzylic functionality. Examples of the polymers used include: polybutadiene, poly(styrene-*co*-butadiene), poly[styrene-*co*-4-vinylpyridine], and poly[styrene-*co*-*N*-vinylpyrrolidone] [28].

The role of the polymeric catalysts in particle nucleation is their action as "dispersants," which stabilize metal particles by adsorption of a thick layer of polymer on the surface of the particles. Thus, the thermal decomposition of $Fe(CO)_5$ has been examined whereby $Fe(CO)_5$ undergoes facile valence disproportionation reactions with nitrogen nucleophiles. If analogous reactions were to occur between $Fe(CO)_5$ and nucleophilic residues [29] on a macromolecule, intermediate ligand–metal cluster compounds would be generated in the polymeric domain. It is postulated that such ligand clusters are more thermally labile than $Fe(CO)_5$ molecules in bulk solutions and thus, are precursors to colloidal iron particles. In the absence of a polymer, the mechanism and kinetics of the decomposition of $Fe(CO)_5$ in an "inert" hydrocarbon media such as decalin are quite complex [30]. Although it is not very well understood, it is certain, however, that the conversion is a sequential stepwise process in which increasingly larger clusters are formed as molecules of CO and $Fe(CO)_5$ are split out (Scheme I). Nitrogen copolymer systems have been termed "active" because the initial and overall rate of decomposition of $Fe(CO)_5$, as determined by the rate of CO production, was much faster in the presence of active polymers than in the solvent alone. Liganded polymer–metal carbonyl compounds can also be generated in passive functional substrates, that is, in molecules having alkenyl or benzylic (allylic) functionality [31]. These types of materials have been termed "passive" because they react only after the loss of a CO ligand to yield the reactive intermediate, $[Fe(CO)_4]$. The rate of decomposition of $Fe(CO)_5$ in the presence of the "passive" polymers—polystyrene, polybutadiene, and poly(Styrene-*co*-butadiene)—is initially similar to that in solvent alone. However, with the formation of the intermediate polymer–bound metal carbonyl compounds, the rate of evolution of CO increases over the rate in solvent alone.

Smith and Wychick [28] carried out the synthesis of Fe nanoparticles in the presence of butadiene-containing polymers. Uniform colloidal dispersions of approximately 70–80-Å Fe^0 particles, which are physically very stable, were obtained. In the initial phase of the thermolysis, it was felt that the major CO-evolving reaction is that resulting in the formation of $Fe_2(CO)_9$, just as is the case in the presence of decalin alone. The "catalyzed" reaction differs in that intermediate $[Fe(CO)_4]$ molecules can react with the isolated alkenyl residues on the polymer backbone. Subsequently, isomerization of the double bonds along the chain occurs, generating butadienyl–iron tricarbonyl residues in the polymer (Scheme II). Particles in the dispersion that were less than 100 Å were superparamagnetic (125 emu/g of Fe at 10,000 Oe), and particles in the 100–200-Å range

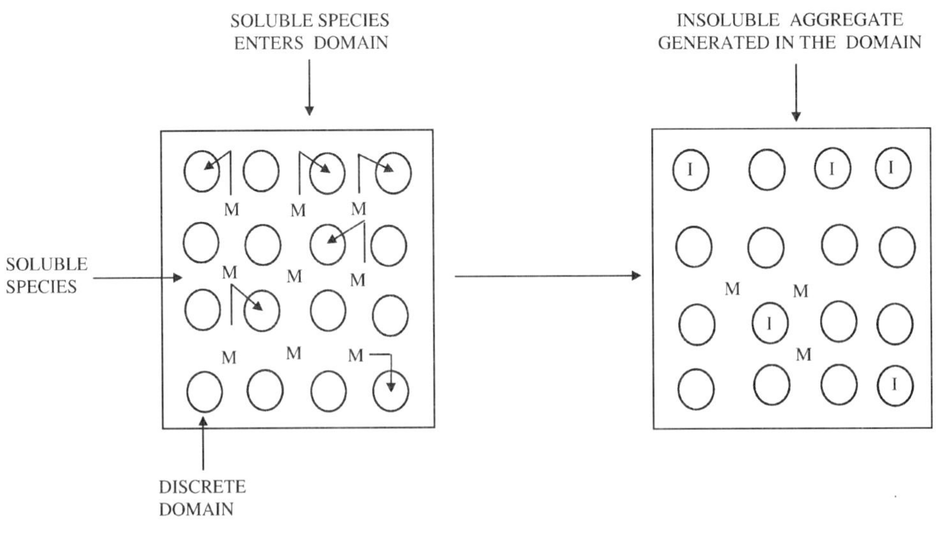

Scheme I. (Adapted from J. Phys. Chem. 84, 1621–29 (1980), Smith et al.)

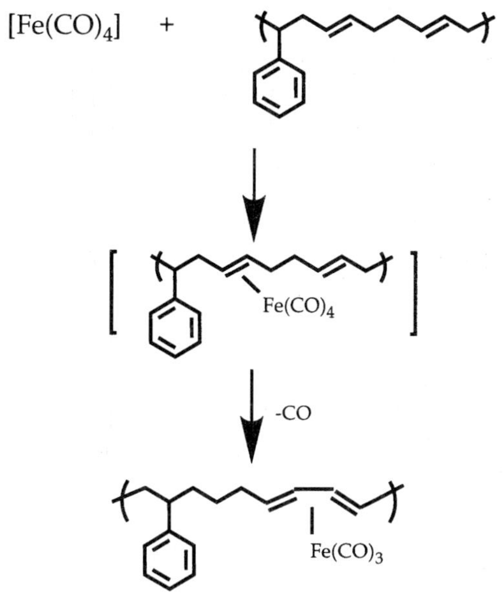

Scheme II. Reaction of $Fe(CO)_4$ with Alkenyl Residues. (Adapted from J. Phys. Chem. 84, 1621–29 (1980), Smith et al.)

had a time-dependent hysteresis. On exposure to the atmosphere, an approximately 30-Å-thick γ-Fe_2O_3 oxide film was produced on the surface of the particles. This is the "passive oxide film" detected previously by a number of techniques [32]. As water was adsorbed from the atmosphere, the chlorinated solvent-based dispersions reacted further to give β-FeOOH. This reaction was, in fact, promoted when the decompositions were carried out in a chlorinated solvent such as chlorobenzene. The magnetic moment decayed with oxidation roughly in proportion to the quantities of Fe, γ-Fe_2O_3, and β-FeOOH present.

In a slight variation of this process, Nakatani et al. [33] have developed a new metallic magnetic fluid with ε-Fe_3N fine particles dispersed in kerosene by the vapor-liquid chemical reaction between iron carbonyl and NH_3. A surfactant amine (polybutenylsuccinpolyamine) was added into the reaction to produce fine colloidal dispersions. The particles showed electron diffraction patterns indexed by the ε-Fe_3N structure. The ratio of amine to $Fe(CO)_5$ has a remarkable influence on the dispersion of the iron-nitride fine particles. Varying concentrations of the surfactant yielded magnetic fluids with different agglomerations (Fig. 2). The particles, in general, were highly uniform in size and well dispersed without agglomeration.

Iron magnetic fluids have high saturation magnetic flux densities up to 2330 G and with high relative initial permeabilities up to 160 or 180, depending on the particle diameter and particle number density as measured by the B–H loop tracer.

In another experiment, Wonterghem et al. [34] prepared a magnetic glass, or in other terms a ferrofluid [35], by the thermal decomposition of iron pentacarbonyl in decalin along with a surfactant (Sarkosyl-O). Usually, metallic glasses are prepared by the liquid quench or the vapor deposition techniques in which hot atoms are condensed onto a substrate kept at temperatures well below the glass transition temperature (T_g). In the case of glass formation by the carbonyl decomposition, the condensation process is presumably a vapor–solid transition. The condition for glass formation is that condensed atoms are prevented from diffusing more than one atomic distance at the surface before they are fixed in position by the arrival of additional atoms. It is noteworthy that the molecules forming the metallic glass particles are at a temperature that does not exceed the boiling point of the liquid

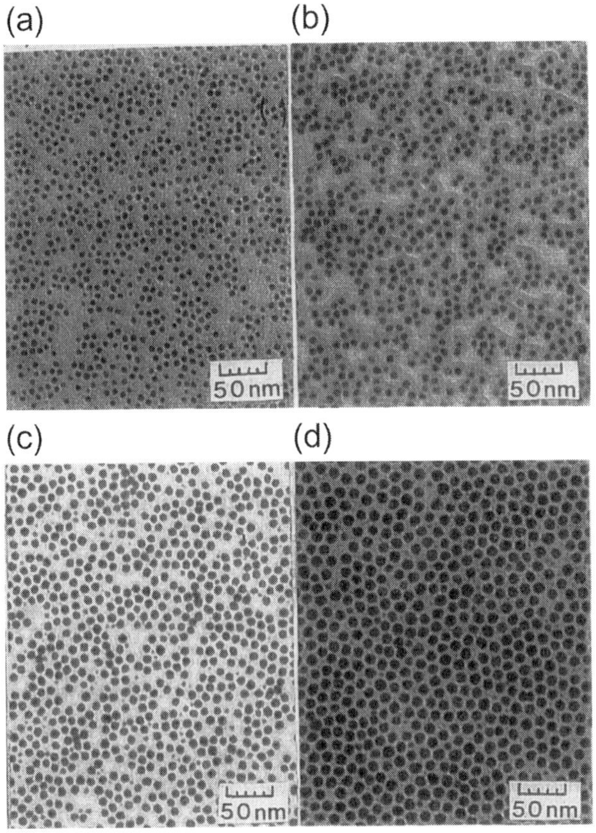

Fig. 2. Electron micrographs of iron-nitride magnetic fluids synthesized from solutions with various amounts of $Fe(CO)_5$ in 50.1 g kerosene: (a) synthesized from solution with 80 g $Fe(CO)_5$, (b) 120 g $Fe(CO)_5$, (c) 150 g $Fe(CO)_5$, and (d) 200 g $Fe(CO)_5$. Reprinted from *J. Magn. Magn. Mater.,* I. Nakatani et al., 122, 10 (© 1993), with kind permission from Elsevier Science Ltd., The Boulevard, Langford Lane, Kidlington 0X5 1GB, UK.

($\sim$460 K). Therefore, this carbonyl decomposition method is not based on rapid cooling from a high temperature, but, rather, the mechanism is based on the growth of alloy particles in a system that is kept at a temperature below T_g. Mössbauer spectra taken at room temperature, 80 K, and 166 K suggest that the particles in the colloid are not pure α-Fe but an amorphous iron–carbon alloy with 5–10 at% carbon. This type of metallic glass crystallizes into a mixture of α-Fe and iron carbides at 523 K. The Mössbauer studies [36] show that the decomposition of $Fe(CO)_5$ in the fluid results in the formation of metallic glass particles that crystallize into α-Fe and iron carbide upon heating. Figure 2 shows the room temperature Mössbauer spectrum of the particles after heating in hydrogen at 523 K. The predominant component was unambiguously identified as the spectrum of α-Fe. The remaining weak absorption lines indicate the presence of other magnetic phases with smaller magnetic hyperfine fields. The best computer fit (Fig. 3) was obtained by including three additional six-line components of low intensity. The parameters of these components are in accordance with the published values of χ-Fe_5C_2 [37]. The χ-carbide component constitutes about 8% of the spectral area corresponding to about 11 at% carbon in the particles.

Apart from thermal decomposition methods, iron nanoparticles have also been made by ultrasonic irradiation or by laser pyrolysis of iron pentacarbonyl. A dull powder was obtained in experiments performed by Cao et al. [38] on the sonochemical decomposition of $Fe(CO)_5$ in decane with varying solution concentrations. They obtained particles of varying sizes (59–243 nm), depending on the solution concentrations, with smaller particles being

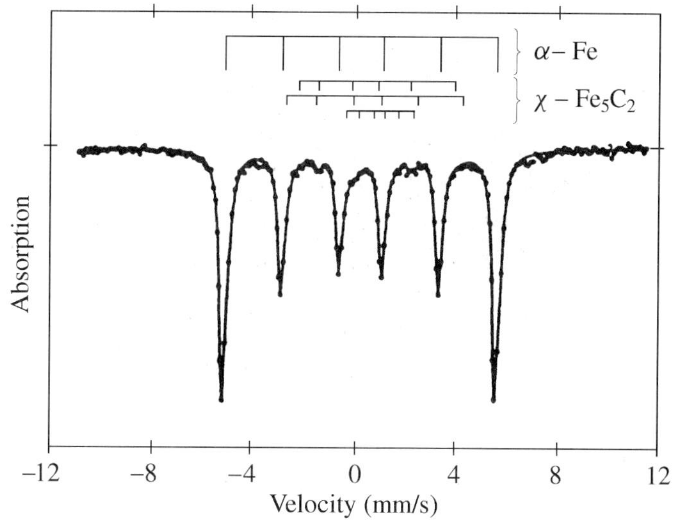

Fig. 3. Room temperature Mössbauer spectrum of the particles after heating in hydrogen at 523 K. The full line indicates a computer fit of the spectrum with a six-line component due to α-Fe and three six-line components due to χ-Fe_5C_2, as shown by the bar diagrams. Reprinted with permission from J. Wonterghem et al., *Phys. Rev. Lett.*, 55, 410 (© 1985 American Physical Society).

obtained for more dilute solutions. The nanophase iron powders do not exhibit saturation at magnetic fields up to 50 kG. The magnetic moment at 50 kG varies between 42.5 emu/g for a pure $Fe(CO)_5$ solution to 12 emu/g for the most dilute solution. This result is explained on the basis that in bulk ferromagnetic materials the local magnetic moments are organized in domains with a certain characteristic size. This yields a reduction in the magnetic energy of the demagnetizing field surrounding the sample. The total energy of the domain boundaries grows, thoroughly as a square of the effective radius of a sample, whereas the magnetic energy of a demagnetizing field is proportional to third degree. At some particle size, the domain formation is not profitable anymore and particles become single domain (all spins are oriented in a certain direction). In this case, the material becomes superparamagnetic and does not exhibit saturation. From Figure 4, it can be seen that the pure sample is the transition state from ferromagnetism to superparamagnetism because it still shows a quite flat region at high fields. Electron spin resonance (ESR) measurements on the sample showed strong signals, and it is certain that these come from the internal magnetic nature of the sample and not from the impurity contained in the amorphous iron. Interestingly, the ESR signals become stronger and sharper for more dilute solutions, implying that the magnetic exchange is becoming weaker and weaker.

By differential scanning calorimetry (DSC) measurements, it was confirmed that the crystallization temperature of iron nanoparticles obtained from pure iron pentacarbonyl was much lower than those of the other samples; that is, particles with a finer grain size had a lower crystallization temperature. This occurs because the powders produced by a decomposition of pure $Fe(CO)_5$ consist of denser particles and it is easier for denser particles to crystallize.

In our laboratory, we have synthesized nanostructured α-Fe by sonochemical as well as thermal decomposition of $Fe(CO)_5$ in decalin [39]. In a typical reaction, for the sonochemical procedure, a dispersion of 15 g (0.076 mol) of $Fe(CO)_5$ in dry decalin (200 mL) was sonicated at 50% amplitude using a Sonic and Materials VC-600 ultrasonic probe (20 kHz, 100 W cm^{-2}) for approximately 6 h at room temperature. In the thermal method, the same amounts of precursors were refluxed in decalin until the completion of the reaction. In either case, on completion of the reaction the formation of shiny metallic particles

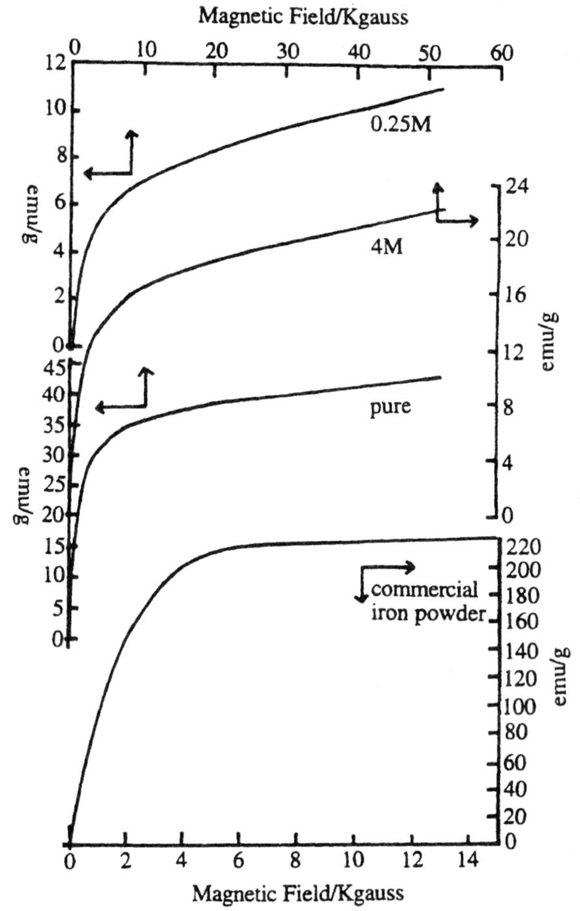

Fig. 4. Magnetization curves of commercial iron powder and amorphous iron at 100 K. Reprinted with permission from X. Cao et al., *J. Mater. Res.*, 10, 2952 (© 1995 Materials Research Society).

was observed on the walls of the reaction vessel. Decalin was removed by decantation and the resulting black powders were isolated and dried with heating under vacuum. These iron powders were then consolidated by vacuum hot pressing at 275 MPa at 700 °C for 1 h.

The as-synthesized iron powders were amorphous by X-ray diffraction (XRD). The major peaks in the X-ray spectrum were assigned to the α-Fe phase and the average crystallite size was calculated to be approximately 40 nm as determined by line broadening analysis [40]. The morphology under scanning electron microscopy (SEM) was found to be porous coral like (Fig. 5). The consolidated iron pellet had a smooth and homogeneous microstructure as confirmed by SEM (Fig. 6). This sample had a high Rockwell C (RC) hardness of 37 as compared to the hardness of conventional iron (4–5 RC). Carbon and oxygen concentrations were 0.05% and 1.1%, respectively. Details of the sonochemical synthesis of nanostructured metals and alloys are also provided by Suslick.

A pulsed laser pyrolysis technique [41] has been used to study the gas phase thermal decomposition of iron pentacarbonyl and chromium, tungsten, and molybdenum hexacarbonyl. In essence, a pulsed infrared CO_2 laser is used to heat an absorbing gas (SF_6), which then, by collision, transfers its energy to the reactive substrate and bath gas (N_2). Laser pyrolysis offers several advantages over other techniques. Gas phase measurements avoid the possible problems created by solvent or matrix effects on the molecules. Laser heating provides a wide temperature range and well-controlled reaction times. Lewis et al. [41] have

Fig. 5. SEM micrograph of the iron powder. Reprinted with permission from K. E. Gonsalves et al., *J. Mater. Sci. Lett.*, 15, 1261 (© 1996 Kluwer Academic Publishers).

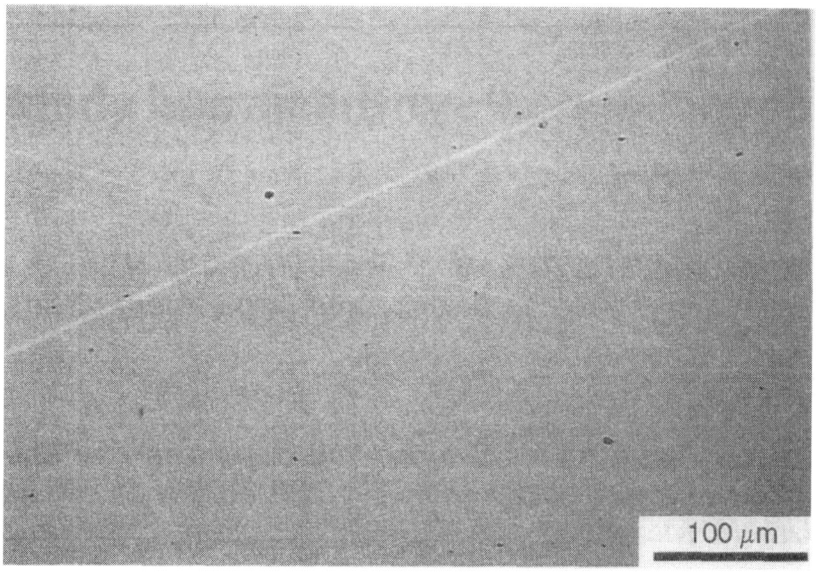

Fig. 6. SEM micrograph of the consolidated iron sample. Reprinted with permission from K. E. Gonsalves et al., *J. Mater. Sci. Lett.*, 15, 1261 (© 1996 Kluwer Academic Publishers).

measured the bond dissociation energies of various transition metal carbonyls by use of the laser pyrolysis method.

Stabilized colloidal cobalt nanoparticles were prepared by Thomas [42] by the thermal decomposition of $Co_2(CO)_8$ in a hydrocarbon solvent containing a suitable polymeric material (methyl methacrylate–ethyl acrylate–vinylpyrrolidone terpolymer of mole proportions 33:66:1) and about 3% methyl isobutyl ketone. It should be noted that the average particle size could be easily varied from about 20 Å to about 300 Å by variations in the reagent concentrations, the temperature, and the composition of the polymeric material.

Polymers with a relatively large percentage of highly polar groups promoted the growth of smaller particles. It was also found that copolymers of reasonably high molecular weight (on the order of 10^4 and greater) were unique in furnishing a high degree of stability to such colloids. The magnetic properties of the Co particles were measured on dried films formed by evaporation of the solvent with the polymeric material acting as a binder. In general, the magnetic properties were those to be expected from single-domain particles [43]. Two unique features were observed in these nanoparticles: (1) The particles were arranged more or less in the form of a chain, and (2) they exhibited a continuous range of coercive force between the limits of 200 and 900 Oe, while retaining constant remanence-to-saturation ratios of 0.5–0.6 for randomly oriented samples and 0.8–0.9 for oriented samples.

3.1.2. Reduction Methods

Several types of reducing agents have been employed to produce fine metal particles from inorganic salts. In the following discussion, examples of various metals produced by different reducing agents are listed and brief descriptions of some property characterizations are also discussed.

Reducing agents such as $NaBEt_3H$, $LiBEt_3H$, and $NaBH_4$ have been commonly used to yield metal nanopowders. The group 6 metal chlorides $CrCl_3$, $MoCl_3$, $MoCl_4$, and WCl_4 were reduced in toluene solution with $NaBEt_3H$ at room temperature to form the corresponding metal colloids in high yield [44]. When the same metal chlorides were reduced in tetrahydrofuran (THF) solution with $LiBEt_3H$ and $NaBEt_3H$, metal carbides (M_2C) were formed in approximately 95% yield (Scheme III). The metal and metal carbide colloids were shown to comprise 1–5-nm-sized particles by transmission electron microscopy (TEM). In general, the powders were isolated as agglomerates of these primary crystallites with dimensions of approximately 400–500 nm as determined by SEM. These 400–500-nm agglomerates were, in turn, composed of the primary 1–5-nm-sized crystallites. X-ray powder diffraction studies of the black powders exhibited broad peaks for the as-synthesized powders at room temperature where the crystallite size estimated from the broadening analysis generally corresponded to the primary particle size as determined by TEM. Initially, the powders were characterized with the NaCl byproduct intact. However, when the as-precipitated black powders were washed with the deoxygenated water, the NaCl could be removed without observable oxidation.

The proposed equation for the reduction is as follows, generalized for $M = Cr$ and Mo, $x = 3$, and $M = W$, $x = 4$:

$$MCl_x + xNaBEt_3H \rightarrow M + xNaCl + xBEt_3 + (x/2)H_2 \tag{1}$$

This reaction may result from direct hydride transfer to form a metal hydride intermediate, which subsequently reductively eliminates H_2, or may be reduced directly by an

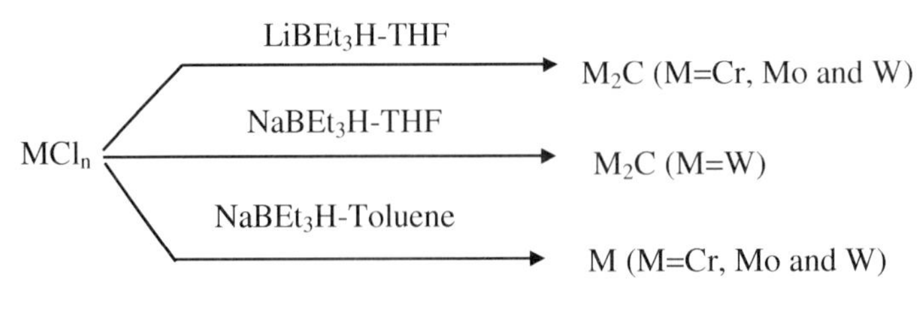

[a] M=Cr or Mo, n=3; M=W, n=4

Scheme III. (Adapted from Chem. Mater. 5, 689 (1993), Zeng et al.)

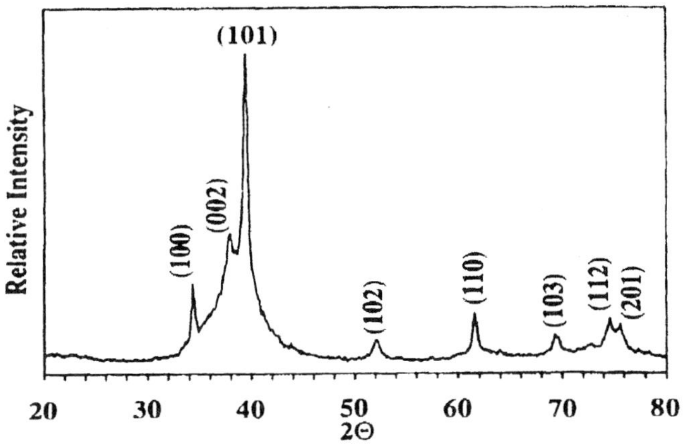

Fig. 7. X-ray powder diffraction data obtained from powder isolated from reduction of $MoCl_3(THF)_3$ in THF after heating to 500 °C for 4 h *in vacuo*. Reprinted with permission from D. Zeng and M. J. Hampden-Smith, *Chem. Mater.*, 5, 681 (© 1993 American Chemical Society).

electron transfer mechanism [45]. Reduction of $MoCl_4(THF)_2$ or $MoCl_3(THF)_3$ in THF at room temperature with a slight excess of stoichiometric $LiBEt_3H$ resulted in the formation of a black colloid accompanied by H_2 gas evolution. The reaction mixture turned black immediately on addition of $LiBEt_3H$. After stirring overnight at room temperature, a black colloid was formed that has been shown to comprise 2–4-nm-sized Mo_2C crystallites. X-ray powder diffraction of the room temperature product showed a broad diffraction peak centered at a d spacing of approximately 2.4 Å. On heating this powder to 500 °C, the XRD pattern sharpened and corresponded to that of Mo_2C (Fig. 7). No evidence of Mo was observed. Elemental analysis confirmed the carbon content to be 6.3%, close to the calculated carbon percentage, 5.9%, for Mo_2C.

W_2C was isolated when WCl_4 was reduced in THF at room temperature with a slight excess of $LiBEt_3H$. Both SEM and TEM data revealed that the grain size of the black powder was approximately 2–4 μm and was composed of agglomerated 1-nm-sized primary particles by TEM.

On the basis of the X-ray and electron diffraction data, it appears that the peaks present are attributable to the diffraction maxima of the metal carbide phase. The origin of the carbide is still unclear. It was proposed that the formation of the carbide may have been due to transfer of an ethyl anion rather than hydride from the triethylborohydride reducing reagent to the metal center. Ethyl groups normally rearrange to liberate ethylene and result in the formation of a metal hydride [46]. Therefore, it seems likely that the common factor in the formation of the metal carbide rather than the metal is the solvent and not the reagents. Hydrogen abstraction from a metal-bonded carbon would lead to retention and strengthening of the metal carbon bond.

Buhro et al. [47] have similarly prepared copper nanopowders by the use of $NaBH_4$ or $Li/NaBEt_3H$ reducing agents in THF solvents. The byproducts were removed by solvent and water washing and the product powders had average sizes ranging from 10–30 nm, depending on the reaction conditions. Copper powders stored in air for extended periods did not exhibit CuO or Cu_2O reflections in the XRD patterns. These powders were then hot pressed at 750 °C and 120 MPa for 2 h. Density measurements of 92% (Archimedes method) of the theoretical value were obtained. XRD and SEM analyses established that grain growth to approximately 38 nm had occurred. The compact had a yield strength of 401 MPa by three-point bending and compression tests. The same group also synthesized air-sensitive 40-nm-sized aluminum powders by the reduction of $AlCl_3$ with lithium aluminum hydride.

A series of fine metal powders has also been synthesized by Bönnemann et al. [48] using metal tetrahydroborates, $M[BH_4]$ (M = Li, Na). However, because all of the four hydrogen atoms in the BH_4^- anion can take part in the reducing process, the resulting metal powders are contaminated with borides [49]. Alkaline hydrotriorganoborates, $M[BR_3H]$, are also used as reducing agents [50]. Here, in contrast, the organoboron part has no reducing properties, but exclusively functions as a complexing agent to generate very soluble metal hydrides in organic media. This allows for the preparation of boride-free metal nanopowders by reducing metal salts with alkali or alkaline earth metal hydrides in organic phases. The hydrides are solubilized in organic media using BR_3 or $BR_n(OR')_{3-n}$ (R, R' = alkyl or aryl, $n = 0, 1, 2$) as complexing agents to form hydrotriorganoborates of the general formula $M'H_u{\cdot}(BR_3)_u$ or $M'H_u[BR_n(OR')_{3-n}]_u$ (M' = alkali or alkaline earth metal, $u = 1, 2$). Analogously, organogallium compounds, $GaR_n(OR')_{3-n}$ ($n = 0$, 1, 2, 3) may be used as complexing agents to give the corresponding hydrotriorganogallates in organic solvents. Tables I and II summarize the different preparation conditions for the synthesis of nanocrystalline metal powders in THF by different reducing agents. Table III summarizes the grain sizes of selected metal powders prepared by the hydrotriorganoborate reduction method, observed by TEM.

The reduction of metal halides in organic phases can also be achieved by adding only a catalytic amount of BR_3 to suspended $M'H_u$ (M' = alkali or alkaline earth metal, $u = 1, 2$). The organoboron complexing agents liberated according to Eq. (2) react with further metal hydrides in suspension to regenerate the hydrotriorganoborate complexes *in situ* [Eq. (3)]:

$$uMX_v + vM'(BR_3H)_u \rightarrow uM\downarrow + vM'Xu + uvBR_3 + uv/2H_2\uparrow \tag{2}$$

$$uMXv + vM'Hu \rightarrow uM\downarrow + vM'Xu + uv/2H_2\uparrow \tag{3}$$

Tetrabutylammonium hydrotriorganoborates [45], which are readily accessible from ammonium halides and alkali metal hydrotriorganoborates, may advantageously be used

Table I. Preparation of Nanocrystalline Metal Powders in THF with Formation of Soluble Borates

			Condition		Product	
Number	Metal salt	Reducing agent	t (h)	T (°C)	Metal content (%)	Boron content (%)
1	$Fe(OEt)_2$	$NaBEt_3H$	16	65	96.8	0.16
2	$Co(OH)_2$	$NaBEt_3H$	2	23	94.5	0.40
3	$Co(CN)_2$	$NaBEt_3H$	16	65	96.5	0.20
4a	$Ni(OH)_2$	$NaBEt_3H$	2	23	94.7	0.13
5	$Ni(OEt)_2$	$NaBEt_3H$	16	65	91.4	0.58
6	CuCN	$LiBEt_3H$	2	23	97.3	0.0
7	CuSCN	$NaBEt_3H$	16	65	95.0	0.23
8	$Pd(CN)_2$	$NaBEt_3H$	16	65	95.5	1.38
9	AgCN	$Ca(BEt_3H)_2$ [a]	2	23	89.6	0.20
10	$Cd(OH)_2$	$NaBEt_3H$	2	23	97.9	0.22
11	$Pt(CN)_2$	$NaBEt_3H$	16	65	87.5	0.93
12	AuCN	$NaBEt_3H$	2	23	97.5	0.0

Source: H. Bönnemann et al., *J. Mol. Catal.* 86, 129 (1994).

[a] Solvent: diglyme.

Table II. Preparation of Nanocrystalline Metal Powders in THF

			Condition		Product	
Number	Metal salt	Reducing agent	*t* (h)	*T* (°C)	Metal content (%)	Boron content (%)
1	$CrCl_3$	$NaBEt_3H$	2	23	93.3	0.3
2	$MnCl_2$	$LiBEt_3H$	1	23	94.0	0.42
3	$FeCl_3$	$LiBEt_3H$	2	23	97.1	0.36
4	CoF_2	$NaBEt_3H$	16	65	96.9	0.0
5	$CoCl_2$	$NaBEt_3H$	16	65	95.1	0.0
6	$CoCl_2$	LiH+10% BEt_3	16	65	95.8	0.0
7	$CoBr_2$	$LiBEt_3H$	2	23	86.6	0.0
8	$NiCl_2$	$NaBEt_3H$	16	65	96.9	0.0
9	$CuBr_2$	$LiBEt_3H$	2	23	94.9	0.0
10	$CuCl_2$	$Na(Et_2BOMe)H$	2	23	94.7	0.1
11	$ZnCl_2$	$LiBEt_3H$	12	65	97.8	0.0
12	$RuCl_3$	$NaBEt_3H$	16	65	95.2	0.52
13	$RhCl_3 \cdot 3H_2O$	$NaBEt_3H$	2	23	98.1	0.1
14	$RhCl_3$	$LiBEt_3H$	2	23	96.1	0.66
15	$PdCl_2$	$NaBEt_3H$	16	65	98.0	0.29
16	AgF	$NaB(OMe)_3H$	2	23	94.1	0.05
17	AgI	$NaBEt_3H$	2	23	95.3	0.02
18	$CdCl_2$	$LiBEt_3H$	2	23	99.4	0.0
19	$ReCl_3$	$LiBEt_3H$	2	23	95.4	0.0
20	$OsCl_3$	$NaBEt_3H$	2	23	95.8	0.0
21	$IrCl_3 \cdot 4H_2O$	$NaBEt_3H$	216	23	77.1	0.16
22	$IrCl_3$	$KBPr_3H$	2	65	94.7	0.08
23	$PtCl_2$	$NaBEt_3H$	5	23	98.2	0.21
24	$PtCl_2$	LiH+10% BEt_3	12	65	98.8	0.0
25	$PtCl_2$	$LiBEt_3H$	4	65	99.0	0.0
26	$PtCl_2$	$LiBEt_3H$	2	0	99.0	0.0
27	$SnCl_2$	$LiBEt_3H$	2	23	96.7	0.0
28	$SnBr_2$	$LiBEt_3H$	2	23	87.1	0.0
29	$PdCl_2$	$Na(GaEt_2OEt)H$	2	40	92.7	Ga:0.25
30	$Pt(NH_3)_2Cl_2$	$NaBEt_3H$	2	23	97.1	0.32
31	$Pt(Py)_2Cl_2$	$LiBEt_3H$	2	23	97.1	0.02
32	$Pt(Py)_4Cl_2$	$LiBEt_3H$	2	23	97.5	0.01
33	$CODPtCl_2$	$NaBEt_3H$	2	60	97.9	0.58

Source: H. Bönnemann et al., *J. Mol. Catal.* 86, 129 (1994).

Py, pyridine; COD, 1,5-cyclooctadiene.

Table III. Selected Grain Sizes of Metal Powders Measured by TEM

Number	Metal	Starting materials	Preparation	Grain size (nm)
1	Platinum	$PtCl_2$ $LiBEt_3H$	Table II, No. 25	2–5
2	Palladium	$PdCl_2$ $LiBEt_3H$	Similar to Table II, No. 15	12–28
3	Rhodium	$RhCl_3$ $LiBEt_3H$	Table II, No. 14	1–4
4	Copper	$CuCl_2$ $LiBEt_3H$	Similar to Table II, No. 9	25–90
5	Nickel	$Ni(OH)_2$ $NaBEt_3H$	Table I, No. 4	5–15
6	Cobalt	$CoCl_2$ $NaBEt_3H$	Table II, No. 5	3–5

Source: Bönnemann et al., *J. Mol. Catal.* 86, 129 (1994).

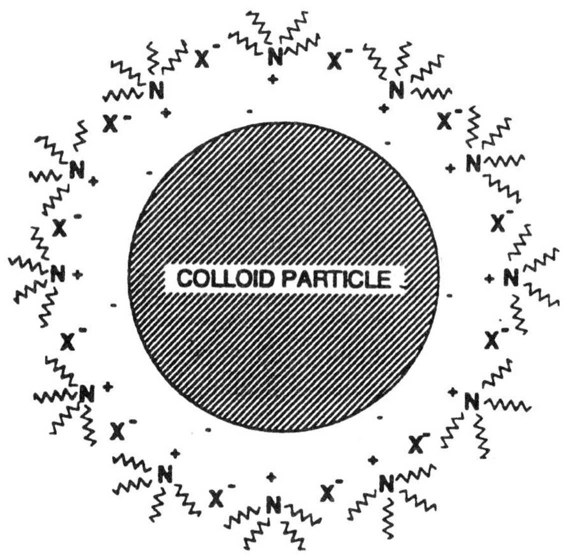

Fig. 8. Stabilization of the metal core with NR_4X. Reprinted from *J. Mol. Catal.*, H. Bönnemann et al., 86, 129 (© 1994), with kind permission of Elsevier Science-NL, Sara Burgerhartstraat 25, 105b KV Amsterdam, The Netherlands.

for the reduction of metal halides in THF. Because ammonium halides formed as the byproduct of the reaction [Eq. (4)] remain completely dissolved after the reaction, the precipitated metal powders can be easily isolated in pure form:

$$MX\nu + \nu NR_4(BEt_3H) \rightarrow M\downarrow + \nu NR_4X + \nu BEt_3 + \nu/2H_2\uparrow \quad (4)$$

where M is a metal of groups 6–11, X is a halogen; R is *n*-butyl, and $\nu = 1, 2, 3$.

Bönnemann et al. have prepared metal colloids of elements of groups 6–11 in the organic phase. The metal salts are suspended in an organic solvent such as THF and treated with tetraalkylammonium-hydrotriorganoborates, which are readily accessible from ammonium halides and alkali–metal hydrorganoborates. Brown–red metal colloid solutions are generated with hydrogen evolution, from which only a small portion of the reduced metal precipitates. The ammonium halide salt that forms in the process functions as a protective atmosphere for the metal particles (see Fig. 8) obviating the need for external stabilizers. The elemental analysis of the isolated metal colloid and results from mass spectrometry indicate that NR_4X is present. The results hitherto suggested that tetraalkylammonium ions surround the presumably negatively charged metal nucleus. The screening of the metal nucleus by large lipophilic alkyl groups explains the remarkable solubility of the metal colloids in organic solvents and their extraordinary stability. Table IV summarizes the reaction conditions and particle properties of various metal colloids obtained by using the tetraalkylammonium salts as reducing agents.

The metal colloids can be adsorbed on the surface of supports. The metal aggregates do not coalesce during the process. TEM investigations showed that the particle size remained the same and a very uniform distribution of the metal particles on the support was achieved. The supported metal colloids are effective catalysts for the hydrogenation of unsaturated compounds such as CO, C–C, C–O, and C–N multiple-bond systems, as well as the hydrogenation of naturally occurring products and mixtures such as soya bean oil [51].

Duteil et al. [52] have prepared Ni colloids by the reaction of $Ni(acac)_2$ (acac = acetylacetonate) and PPh_3 (Ph = phenyl) in diethyl ether with the reducing agent Et_2AlH at $-40\,^{\circ}C$. These colloids can be isolated in the solid state and redispersed in any concentration in polar solvents such as pyridine, owing to their ligand shell mainly consisting not of PPh_3 molecules but, instead, of PPh moieties that are generated from PPh_3 during the reaction with Et_2AlH with the formation of free benzene:

$$Et_2AlH + Ni(acac)_2 \rightarrow Et_2Al(acac) + Ni + 2H \tag{5}$$

$$PPh_3 + 2H \rightarrow C_6H_6 + PPh \tag{6}$$

High-resolution transmission electron microscopy (HRTEM) investigations of the nickel colloid, deposited on a grid from a dark-brown pyridine solution, indicate very narrow particle size distribution with an average diameter of 4 nm. The exact size of some of these particles correspond to diameters between 3.9 and 4.3 nm. Preliminary studies of the catalytic properties of the nickel colloids show very low activities in hydrogenation reactions. For instance, the turnover frequency for the hydrogenation of hex-2-yne to cis-hex-2-ene in a heterogeneous reaction is only 2 $mol_{prod}\ mol_{Ni}^{-1}\ h^{-1}$, compared with values of some hundreds using common hydrogenation catalysts of nickel or palladium. The lack of catalytic activity is due to the perfect protecting ligand shell covering the surface nickel atoms.

Another commonly used reduction procedure is called the polyol process [53]. In this method, liquid polyols such as ethylene glycol or diethylene glycol are used both as a solvent and as a reducing agent for the chemical preparation of metallic powders from various inorganic precursors. The basic reaction scheme for the synthesis of these metal powders by the polyol process involves the dissolution of the solid precursor, the reduction of the dissolved metallic species by the polyol itself, nucleation of the metallic phase, and growth of the nuclei. To obtain metal powders with a narrow size distribution, two conditions must be fulfilled: (1) A complete separation of the nucleation and growth steps is required and (2) the aggregation of metal particles must be avoided during the nucleation and growth steps.

Fievet et al. [54] have successfully used the polyol process for the synthesis of fine, highly pure, monodisperse, nonagglomerated particles of Cu. The precursor copper(II) oxide is dispersed in a given volume of ethylene glycol and the suspension is stirred at 300 rpm and heated at a rate of $6\,^{\circ}C\,min^{-1}$ up to the reaction temperature ranging from 150 to 195 °C. Typical reaction times were 30 min at 195 °C or 2 h at 175 °C. To prevent particle sintering, different organic protecting agents have been tested. The best result was obtained with a solution of D-sorbitol in ethylene glycol as a reaction medium. The size of the copper particles could be strictly controlled by dissolving various amounts of sodium hydroxide in the D-sorbitol/ethylene glycol solution and then suspending CuO in this mixture. Addition of this strong base enhances the solubility of the precursor CuO and intermediate Cu_2O [55].

The XRD pattern of the solid phases present during the reduction of Cu(II) oxide in ethylene glycol is shown in Figure 9. Highly crystallized Cu powders are produced as evidenced by the sharp XRD line of Figure 9c. It is inferred from the XRD analysis of the

Table IV. Preparation of Metal Colloids in THF Solution or as an Isolated Powder

Number	Metal salt	Reducing agent	Condition		Product colloid solution color	Workup solvent	Solvent added for precipitation	Metal content in isolated colloid (%)	Mean particle size (nm)
			t (h)	T (°C)					
1	MnI_2	$N(octyl)_4BEt_3H$	1	23	Dark brown Mn completely dissolved				
2[a]	$FeBr_2$	$N(octyl)_4BEt_3H$	18	90	Dark brown to black Fe almost completely dissolved	Ethanol	Ether	11.34	3.0
3	$RuCl_3$	$N(octyl)_4BEt_3H$	2	50	Dark reddishbrown to black Ru almost completely dissolved	Ethanol	Pentane	68.72	1.3
4	$OsCl_3$	NBu_4BEt_3H	1	23	Deep red to black				
5	$CoBr_2$	$N(octyl)_4BEt_3H$	16	23	Dark brown to black Co completely dissolved	Ethanol	Ether	37.45	2.8
6	$RhCl_3$	$N(octyl)_4BEt_3H$	3	40	Deep red to black Rh completely dissolved	Ether	Ethanol	73.40	2.1
7	$IrCl_3$	$N(octyl)_4BEt_3H$	1	50	Dark red to black Ir almost completely dissolved	Ethanol	Ether	65.55	1.5
8	$NiBr_2$	$N(octyl)_4BEt_3H$	16	23	Dark red to black Ni completely dissolved	Ethanol	Ether	66.13	2.8
9	$NiBr_2$	$N(octyl)_3MeBEt_3H$	16	23	Dark red to Ni completely dissolved	Ethanol	Ether	68.29	2.8
10	$PdCl_2$	$N(octyl)_4BEt_3H$	1	23	Dark brown to black Pd completely dissolved	Ether	Ethanol	83.62	2.5
11	$PtCl_2$	$N(hexyl)_4BEt_3H$	2	23	Dark brown to black Pt up to 80% dissolved				
12	$PtCl_2$	$N(octyl)_4BEt_3H$	18	23	Dark brown to black Pt completely dissolved	Ether	Ethanol	85.13	2.8
13	$PtCl_2$	$N(hexyl)_4BEt_3H$	2	23	Dark brown to black Pt up to 80% dissolved				
14	$CuCl_2$	$N(octyl)_4BEt_3H$	2	23	Deep red to black Cu almost completely dissolved	Ether	Ethanol	77.04	
15	$CuBr_2$	$N(octyl)_4BEt_3H$	2	23	Deep red to black Cu completely dissolved	Toluene	Ethanol	52.15	

Source: H. Bönnemann et al., *J. Mol. Catal.* 86, 129 (1994).

[a] Solvent: toluene.

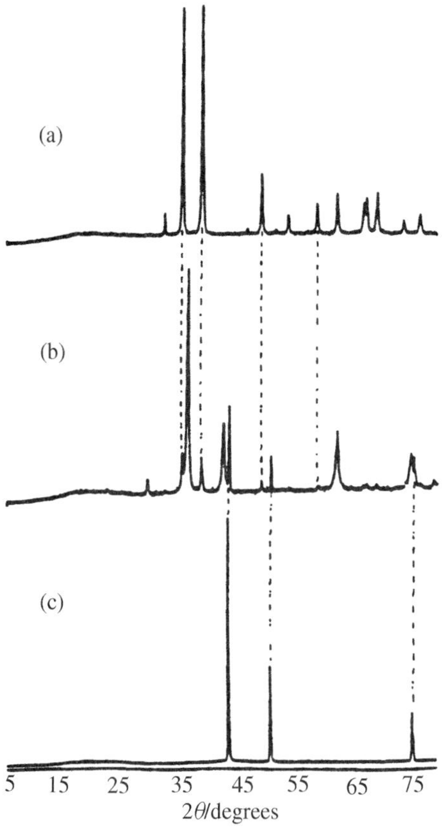

Fig. 9. XRD pattern of the solid phases present during the reduction of copper(II) oxide in ethylene glycol: (a) CuO precursor, (b) CuO + the intermediate Cu_2O + Cu, and (c) Cu obtained at the end of the reduction. Reprinted with permission from F. Fievet et al., *J. Mater. Chem.*, 3, 627 (© 1993 The Royal Society of Chemistry).

solid phases present during the course of the reaction that Cu_2O always exists as an intermediate crystalline phase. As a general rule, copper powders are made up of equiaxed particles whose size lies within the micrometer range. Some of these particles are single entities, but many of them are sintered with no definite shape. By similar reactions, the same group isolated cobalt and nickel in ethylene glycol by the reduction of $Co(OH)_2$ and $Ni(OH)_2$. The metals are recovered as a finely divided powder that appears in SEM as made up of particles with an isotropic shape and a size in the micrometer range. Almost perfectly spherical particles (Fig. 10a) can even be obtained in a mixture of diethylene glycol and ethylene glycol. The particles are usually homogeneous in size as shown in Figure 10b. The isotropically shaped particles have a narrow size distribution, owing to an effective separation between homogeneous nucleation and growth steps. Submicrometer-sized particles can be obtained either by raising the reaction temperature or by seeding the reactive medium with foreign metal nuclei, in order to induce a heterogeneous nucleation. The latter method is more convenient because the increase in temperature leads to degradation of the polyol and the heterogeneous nucleation allows, to some extent, control of the particle size in the submicrometer range.

A relatively new method for the synthesis of nanoscale metal/oxidized metal particles is described by Tsai and Dye [56] that utilizes homogeneous reduction of metal salts by dissolved alkalides or electrides in an aprotic solvent such as dimethyl ether or THF. Alkalides and electrides are crystalline ionic salts that contain either alkali–metal anions or trapped electrons. These compounds crystallize from solution to yield shiny bronze-

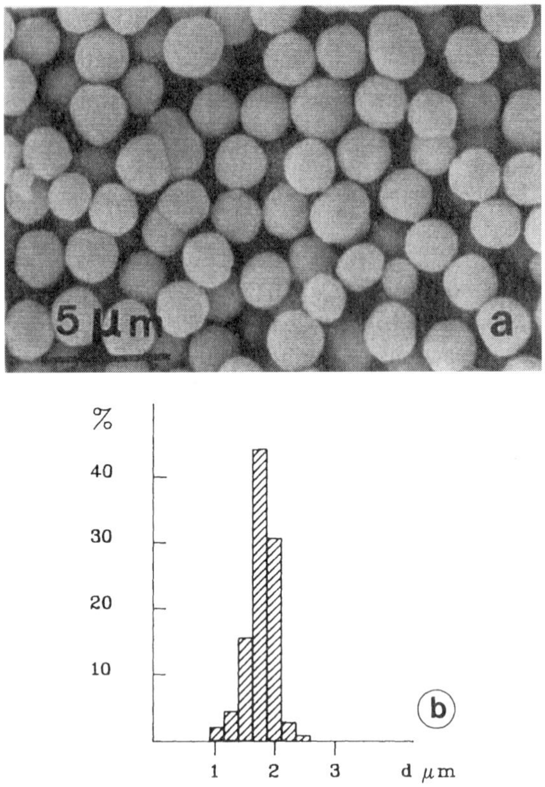

Fig. 10. (a) Cobalt powder obtained from $Co(OH)_2$ in a mixture of diethylene glycol and ethylene glycol. (b) Particle size distribution of the particles ($d_m = 1.75\ \mu m$, $\sigma = 0.24\ \mu m$). Reprinted from *Solid State Ionics*, F. Fievet et al., 32/33, 198 (© 1989), with kind permission of Elsevier Science-NL, Sara Burgerhartstraat 25, 105b KV Amsterdam, The Netherlands.

colored crystals (alkalides) or black crystals (electrides), which are all reactive toward air and moisture and thermally unstable at room temperature. Alkalides and electrides form M^- and e^-_{solv} when dissolved in a nonreducible solvent. These species are the strongest reducing agents that can exist in solution [57].

Soluble compounds of transition metals and post-transition metals in dimethyl ether or THF were rapidly reduced at −50 °C by dissolved alkalides or electrides to produce metal particles with crystallite sizes from less than 3 to 15 nm. Salts of Au, Cu, Te, and Pt formed metallic particles with little or no oxidation even when washed with degassed methanol. The reduction of salts of Ni, Zn, Ga, Mo, Sn, and Sb yielded surface oxidation over a metallic core. Stoichiometric amounts of the alkalide or electride were used and these were prepared either separately or *in situ*.

A typical reduction reaction follows the scheme [58]:

$$2AuCl_3 + 3K^+(15C5)_2e^- \rightarrow 2Au + 3K^+(15C5)_2Cl^- + 3KCl \qquad (7)$$

or

$$AuCl_3 + 3K^+(15C5)_2e^- \rightarrow Au + 3K^+(15C5)_2Cl^- \qquad (8)$$

where C5 = 15-crown-5 ether.

Only metallic gold peaks were detected by XRD from the precipitates after washing away the byproducts, $K^+(15C5)_2Cl^-$ and KCl, with water or methanol. The micrograph in Figure 11 agrees well with the particle size of about 100 Å as measured by the X-ray line broadening of gold produced in this reaction. The selected area diffraction (SAD) pattern

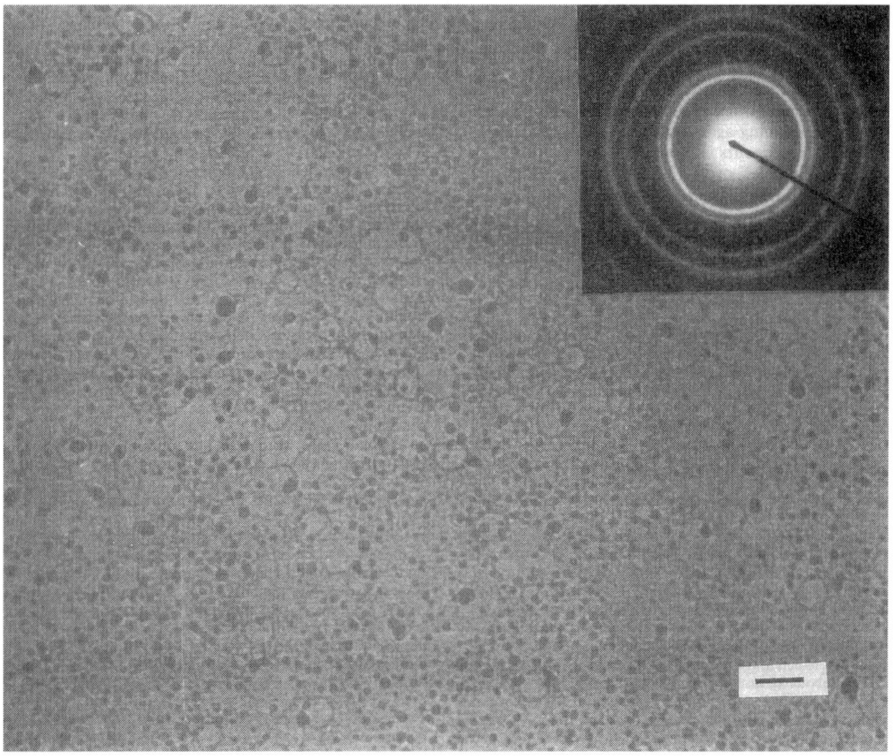

Fig. 11. Electron micrograph of Au particles (bar = 330 Å). The SAD pattern in the upper right corner shows the cubic structure of gold. Reprinted with permission from K. L. Tsai and J. L. Dye, *Chem. Mater.*, 5, 540 (© 1993 American Chemical Society).

was also obtained from the same area and is shown at the top of the micrograph. The ring pattern confirms all of the *d* spacings of gold. In most cases, with all metals, a colloidal suspension is first produced as indicated by light scattering and color, followed by slow aggregation of the colloid to a precipitate that can be separated by centrifugation. The average particle sizes obtained were in most cases less than 3 nm.

Whereas the noble metals and some others such as Cu and Te can be separated from the byproducts, K^+X^- and $K^+(15C5)_2Cl^-$ ($X = Cl, Br, I$), by washing with methanol, more oxophilic metals were oxidized by this procedure. Also the inclusion of organic residues on the surface of the highly reactive metals could be removed in some metals by heating the samples to 100–150 °C under vacuum. Such procedures coupled with washing should improve the purity. With highly oxophilic metals such as Ti, it might not be possible to prevent oxidation.

A major advantage of this reduction method is that these reactions occur rapidly with homogeneous solutions. This has permitted the formation of intermetallic compounds or alloys when two metal salts are reduced simultaneously [59]. The strong reducing power (~ -3 V) of alkalides and electrides means that practically any soluble metal salt can be reduced to the metallic state.

Certain noble metals such as palladium and rhodium [60] have been prepared by the reduction of their respective inorganic salts in the presence of a polymer such as poy(vinylpyrrolidone) (PVP) or poly(vinyl alcohol) (PVA) and methanol as a reducing agent. The polymers were added because they can act as a protective agent to prevent the coagulation and precipitation of metallic particles [61]. More specifically, palladium acetate and rhodium(III) chloride were used as starting materials. Formaldehyde is produced quantitatively with the reduction of rhodium(III) chloride to metallic rhodium. The rhodium particles in the colloidal dispersion are found to be of two types, about 8 and 40 Å in diameter, by electron microscopy. The number of small particles, which form the large majority of particles at the early stage of refluxing, gradually decreases; concurrently, the number of large particles increases on prolonged refluxing. An absorption peak [by ultraviolet–visible (UV–vis) spectroscopy] appears at 260 nm at the early stage of refluxing and indicates that the coordination of poly(vinyl alcohol) to the rhodium(III) ion is indispensable for the formation of a homogeneous colloidal dispersion of rhodium. With the reduction of the rhodium(III) chloride to rhodium(0), methanol is oxidized to formaldehyde [62].

Similarly, when palladium acetate was refluxed with PVP in methanol, Bradley and coworkers produced colloidal palladium particles in the form of well-shaped microcrystallites with a mean diameter of 70 Å. This colloid shows no tendency to precipitate on standing for at least several weeks. X-ray diffraction shows the presence of fcc palladium, with line widths consistent with the particle size shown by TEM. CO adsorbs readily on the colloid in methanol, occupying only bridging sites (ν_{CO} 1944 cm^{-1}), similarly to CO on Pd (111) surfaces, and consistent with the crystalline nature of the colloid particles [63].

3.2. Synthesis of Intermetallics

Intermetallics are defined as solid solutions of two or more metals in varying proportions. The properties of the intermetallics are unique [64]. The general chemical synthesis methods used to make intermetallics are very similar to those used to produce individual metals. Therefore, this section will focus on the respective techniques, characterizations, and properties of the intermetallic systems. The synthesis and microstructural study of a four-component intermetallic, known as M50 steel [65] in our labs, will be discussed in detail.

3.2.1. Two-Component Intermetallics

Most commonly, the intermetallics are prepared by reduction reactions. For example, Buhro et al. [47] synthesized nanocrystalline powders of TiAl, $TiAl_3$, NiAl, and Ni_3Al by the reductions of $TiCl_3$ or $NiCl_2$ with $LiAlH_4$ in a mesitylene slurry followed by heating in the solid state ($\leqslant 550\,°C$). In the course of the reactions, Al and Ti or Al and Ni are initially precipitated in a segregated component phase, which, on subsequent heating, undergo an exothermic reactive-sintering process to give the nanocrystalline intermetallic. The microstructure of the air-sensitive aluminide powders consists of porous sintered aggregates of nanocrystallites, in which the primary nanocrystalline sizes are in the range of 25–35 nm. $MoSi_2$ was also prepared by the same group by sonochemical coreduction as described in the following reaction:

$$MoCl_5 + 2SiCl_4 + (13/2)NaK \rightarrow MoSi_2 + (13/2)NaCl + (13/2)KCl \qquad (9)$$

The ultrasonic irradiation formed a fine emulsion of the liquid NaK alloy, which maximized interfacial contact of the heterogeneous reactant phases and afforded a rapid, uniform coreduction reaction.

Co–Cu and Fe–Cu powders with varying metal concentrations have been prepared by the conventional method using sodium borohydride reduction of metallic salts [66]. As-synthesized Co–Cu powders showed a face-centered cubic (fcc) structure and an amorphous phase. The amount of amorphous phase was found to increase with the ratio of Co/Cu and grain sizes of the alloys were typically 300 Å. SEM and TEM revealed that the Fe–Cu powders were apparently agglomerated, as is frequently observed in ultrafine powders, and the grain sizes varied between 300 and 400 Å. Magnetic measurements confirmed that both the alloys and the composites of Fe–Cu were ferromagnetic at room temperature. The magnetometry curves showed that the metastable alloy possesses a low coercivity (ranging from 10–40 Oe).

Various other intermetallics such as TiB_2, Ni_2B, WC–Co, Co–B, Fe–B, Ni–B, and Pd–B have been prepared by reduction reactions using $NaBH_4$ as a reducing agent on different inorganic precursors such as $TiCl_4$, $NiCl_2{\cdot}6H_2O$, etc. The residual byproducts were either sublimed out of the mixture at elevated temperatures or washed away by organic solvents or water. Particles of varying sizes and morphologies were obtained, depending on the reaction conditions and the type of intermetallic being synthesized [47].

Some novel methods to prepare Ni and Pd boride colloids with core diameters of 1.4 and 1.6 nm have been described by Schmid and co-workers [67]. The particles are formed by the reaction of the metal complexes $Cl_2M(PR_3)_2$ ($M = Ni, Pd$, $PR_3 = PPr_3, PBu_3$) with B_2H_6 in toluene at room temperature with 40–70% yield. From the elemental analyses of the four colloids, the lowest formula units could be concluded as $Ni_6B_{10}Cl_{1.5}(PPr_3)$, $Ni_6B_{10}Cl_{1.5}(PBu_3)$, $Pd_4B_6Cl(PPr_3)$, and $Pd_4B_6Cl(PBu_3)$. Particle sizes of 1.6 nm were obtained by HRTEM (Fig. 12). X-ray powder investigations support the assumption that the particles are amorphous. ^{31}P nuclear magnetic resonance (NMR) investigations of the four colloids in solution resulted in singlets of free PR_3 molecules at -33 ppm and broad multiplets at ≈ -5 ppm indicating coordinated phosphines. This means that, as could be expected, at least parts of the ligands are dissociated in solution.

A thermochemical processing method for preparing high-surface-area powders starting from homogeneous precursor compounds has been reported. The method has been applied successfully to the synthesis of nanophase WC–Co powders [68]. The method is known as "spray conversion processing" and consists of three sequential steps: (1) preparation and mixing of aqueous solutions of the precursor compounds to fix the composition of the starting solution, (2) spray drying of the starting solution to form a chemically homogeneous precursor powder, and (3) thermochemical conversion of the precursor powder to the desired nanostructured end-product powder (Scheme IV). In a typical reaction, WC–Co powders were formed by thermochemical processing of a single chemical precursor compound, cobalt tris(ethylenediamine)tungstate $[Co(en)_3WO_4]$. The precursor is crystallized from solution and reduced in flowing argon/hydrogen to yield nanoporous/nanophase W–Co. This high-surface-area reactive intermediate was then converted directly to nanostructured WC–Co powder by gas phase carburization in flowing CO/CO_2. The resulting powder particles had the same morphology as the original $Co(en)_3WO_4$ particles, but the size of each particle was reduced by about 50%. In this process, the scale of the powder particle structure may be controlled from nanometer up to micrometer dimensions by adjusting the temperature of the carburization reaction, the residence time at a particular temperature, and the carbon activity of the gas phase. Alternatively, precursors such as $(NH_4)_6(H_2W_{12}O_{40}){\cdot}4H_2O$ and $CoCl_2$, $Co(NO_3)_2$, or $Co(CH_3COO)_2$ can be used to produce the desired nanopowders by the spray technique.

The alkalide and electride method of reduction used earlier for the synthesis of single metals can also be used for the preparation of alloys or compounds by the simultaneous reduction of two or more metal salts [56]. All systems tested (Au–Zn, Au–Cu, Cu–Te, and Zn–Te) yielded binary alloys or compounds. The Au–Cu system has been the subject of many investigations, and its phase diagram has been extensively studied. The most ideal stoichiometry of Au–Cu consists of alternate layers of Au and Cu atoms parallel to a cube face. The symmetry is slightly distorted to tetragonal with $c/a = 0.93$. The structure transforms to orthorhombic with $b/a = 10.03$ at about 380 °C [69]. The particles produced by the reduction of stoichiometric mixtures of $AuCl_3$ and $CuCl_2$ at -50 °C have an SAD pattern that corresponds to a simple cubic pattern with each Au atom at the center of a cube of Cu atoms and vice versa.

The polyol reduction method has been used by Chow et al. [70] to produce nanocrystalline Co_xCu_{100-x} ($4 \leqslant x \leqslant 49$ at%) powders by refluxing metal acetates in a polyol. Typically, the powders were produced by suspending different proportions of Co(II) acetate tetrahydrate $[Co(O_2CCH_3)_2{\cdot}4H_2O]$ with Cu(II) acetate hydrate $[Cu(O_2CCH_3)_2{\cdot}4H_2O]$ in

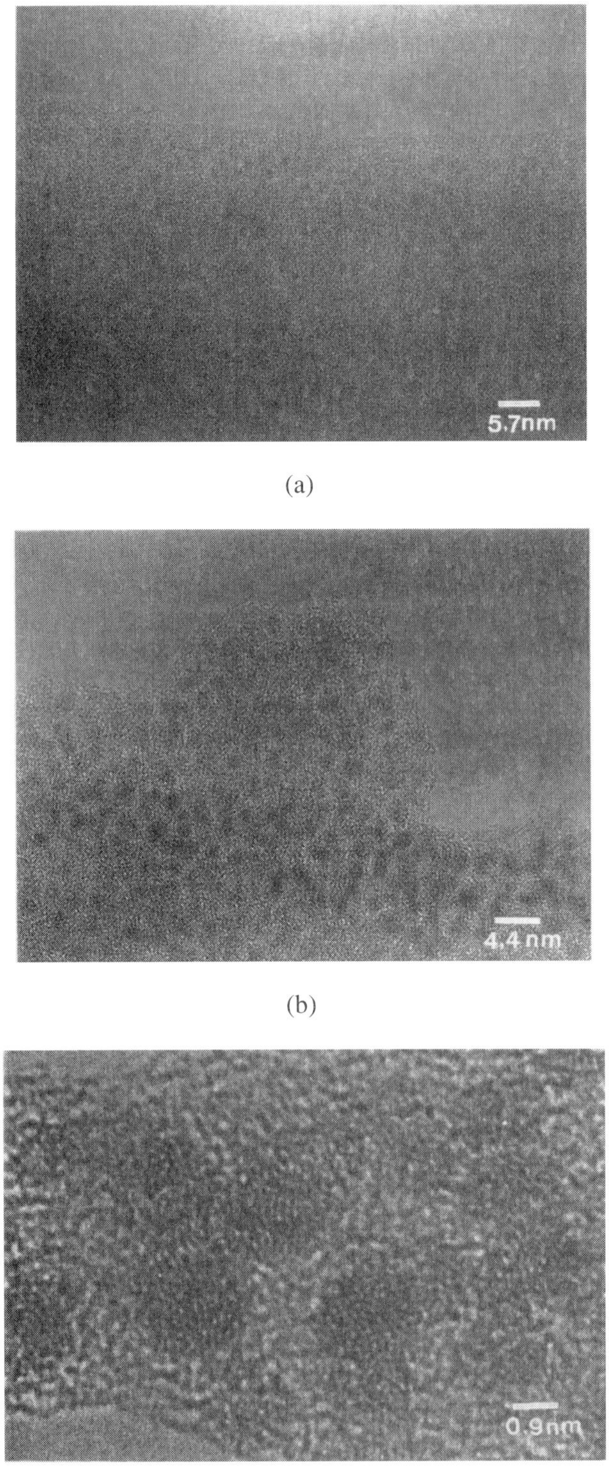

(c)

Fig. 12. (a) Larger area covered with mainly uniform nickel boride particles **1**. (b) Magnified image of the same particles showing a size distribution of $\approx 1.6 \pm 0.3$ nm. (c) High-resolution image of a few particles of **1** indicating the amorphous structure. Reprinted with permission from G. Schmid et al., *Z. Anorg. Allg. Chem.*, 620, 1170 (© 1994).

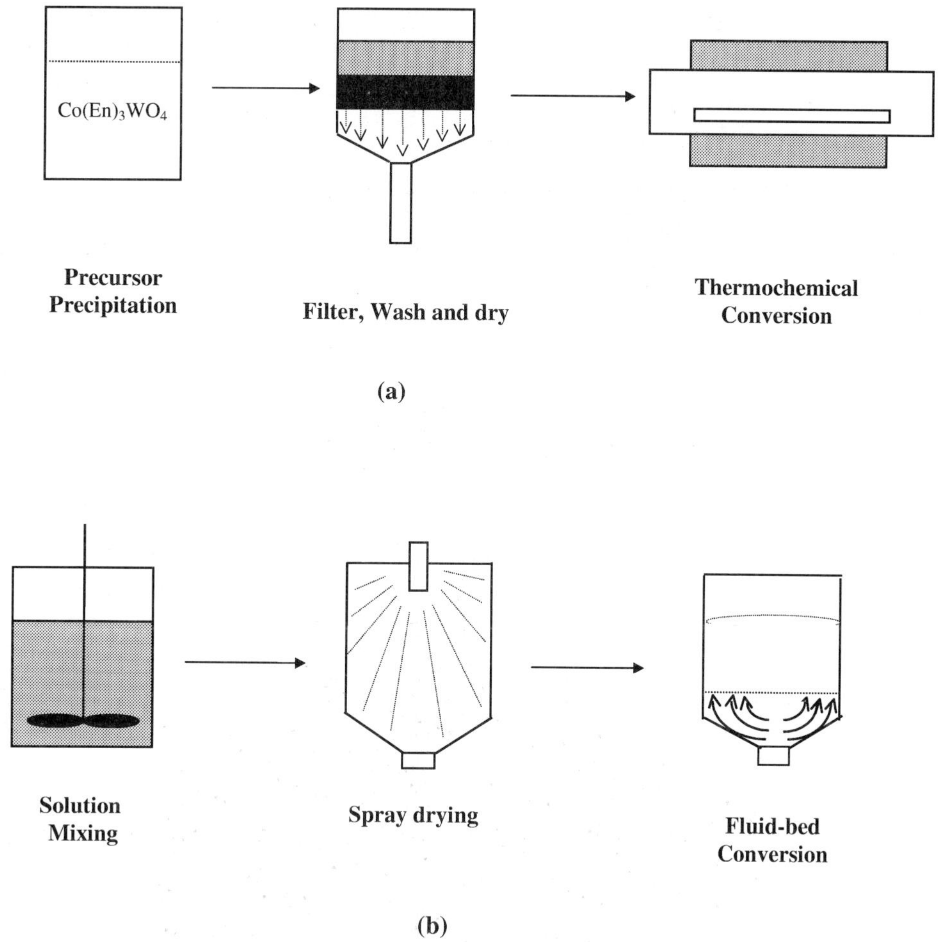

Scheme IV. Schematic diagrams of (a) laboratory scale process, (b) industrial scale process for the thermochemical processing of nanostructured powders, starting from aqueous solution mixtures. (Adapted from Nanostr. Mater. 3, 19–30 (1993), Kear et al.)

250 mL of ethylene glycol. The mixtures were refluxed at 180–190 °C for 2 h. During that time, Co–Cu particles precipitated out of solution. XRD data showed that the powders were crystalline. From the line broadening of the Cu(200) peaks, the crystallite size of Cu was in the range of 36 nm. The crystallite size of Co was estimated to be 18 nm from the Co(111) peaks for samples with $x \geqslant 30$. The morphology of the as-synthesized powders of Co_4Cu_{96} and $Co_{49}Cu_{51}$ is shown by TEM micrographs in Figure 13. All the powders appeared to be agglomerated, and the agglomerate size was typically 100 nm or larger. ^{59}Co spin-echo NMR spectra for the Co_xCu_{100-x} samples with $x = 4$, 19, and 49 have similar features in that there was a strong peak centered at 217 MHz and a broad shoulder in its high-frequency side with a small peak around 224–226 MHz. In addition, the spectra were very broad overall compared with the spectrum of pure Co metal. The strong 217-MHz peak and the absence of low-frequency resonances all indicated that the Co atoms existed as clusters rather than in a Co–Cu solid solution [71]. It is likely that the shoulder at the high-frequency side was due to either hexagonal closed-pack (hcp) Co grains or hcp stacking faults in fcc Co.

The magnetic properties of the samples were measured as a function of annealing temperature. The saturation magnetization (M_s) of as-synthesized powders with lower Co concentrations was much lower than those with higher Co content. Upon increasing the

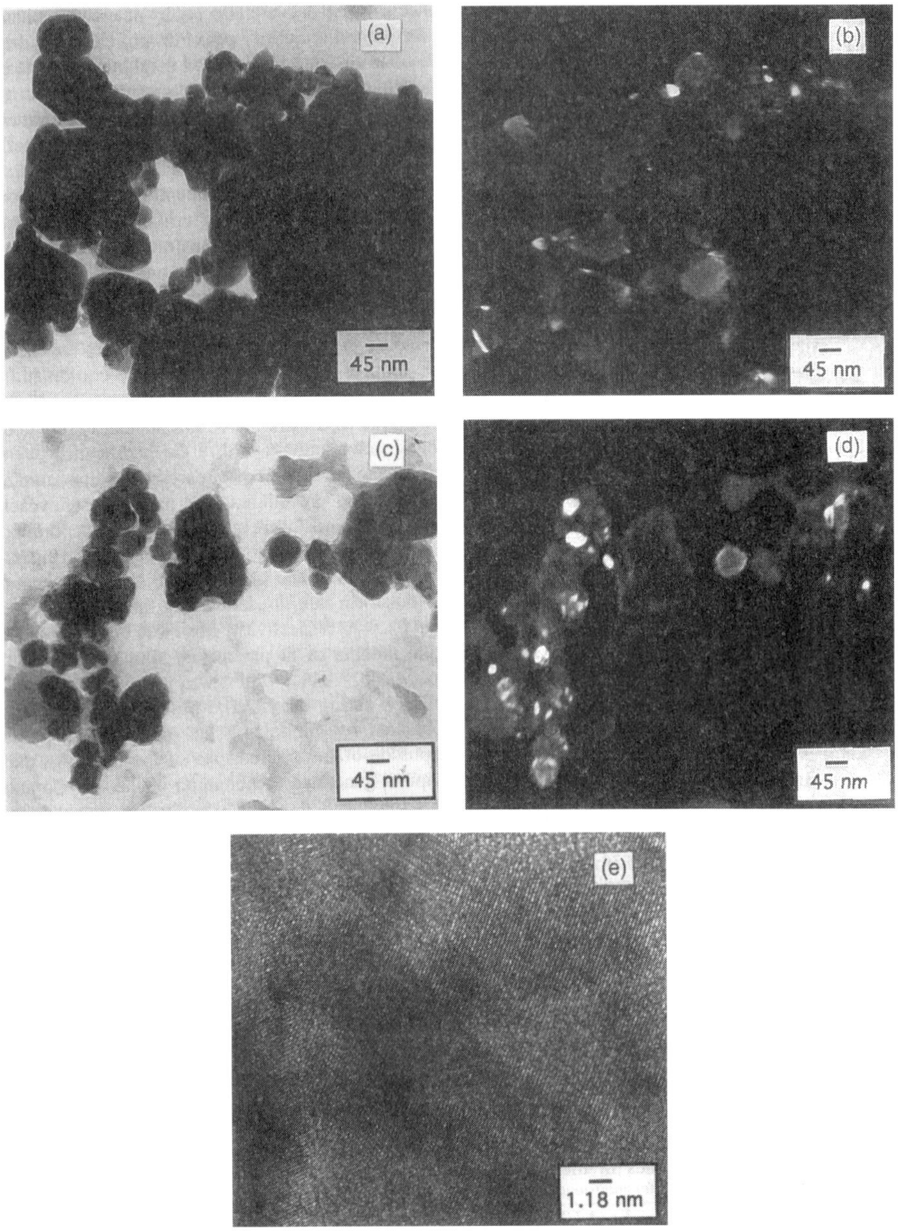

Fig. 13. TEM and HRTEM micrographs of powders: (a) bright-field image of Co_4Cu_{96}, (b) corresponding dark-field image of Co_4Cu_{96}, (c) bright-field image of $Co_{49}Cu_{51}$, (d) corresponding dark-field image of $Co_{49}Cu_{51}$, and (e) HRTEM showing the Co(111) lattice fringes in a $Co_{49}Cu_{51}$ sample. Reprinted with permission from G. M. Chow et al., *J. Mater. Res.*, 10, 1546 (© 1995 Materials Research Society).

annealing temperature, there was an increase in magnetization for samples with lower Co concentrations [72]. The increase in M_s could be due to precipitation of Co either from a metastable Co–Cu alloy or from the grain growth of superparamagnetic Co. Because HRTEM results of as-synthesized samples showed that alloying did not occur, the increase in M_s by annealing was attributed to the grain growth of Co clusters. It is notable that a coercivity as high as 370 Oe was observed for a nanocomposite with a low Co content (Co_4Cu_{96}) after annealing. $Co_{49}Cu_{51}$ showed a coercivity of 150 Oe and decreased to about 110 Oe at 650 °C.

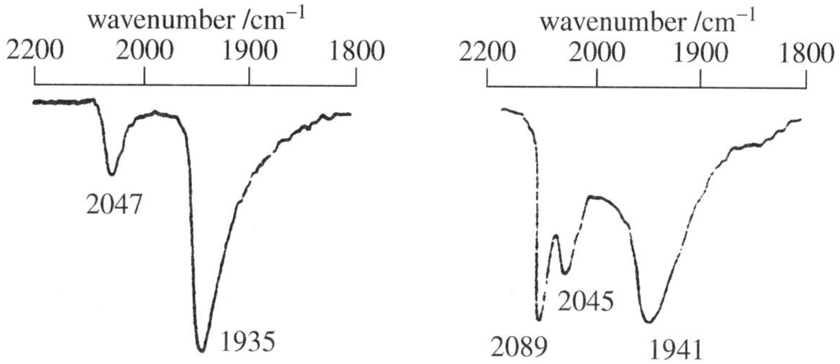

Fig. 14. Infrared spectra (CH_2Cl_2 solution) of carbon monoxide adsorbed on (a) PVP-stabilized colloidal palladium and (b) PVP-stabilized colloidal palladium–copper ($Pd_{37}Cu_{63}$). Reprinted with permission from J. S. Bradley et al., *Chem. Mater.*, 5, 254 (© 1993 American Chemical Society).

Bradley and co-workers [73] prepared Pd–Cu nanoscale colloids by heating mixtures of palladium acetate and copper acetate hydrate in 2-ethoxyethanol to reflux (135 °C) in the presence of poly(vinylpyrrolidone). The mean diameter of the particles varied in the range of 30–50 Å. EDAX analysis of the polymer films prepared with various PdCu ratios showed that the metal particles were bimetallic. Analysis of areas of the films between the colloid particles showed that no palladium(II) or copper(II) reamined unreduced. The observed electron diffraction rings were observed to be consistent with crystallinity, although the degree of crystallinity of the particles seemed to vary.

CO was adsorbed readily onto PdCu particles in dichloromethane at 25 °C, as shown by infrared (IR) spectroscopy (Fig. 14) on colloids with the composition $Cu_{63}Pd_{37}$ (particle size approximately 40 Å). On the bimetallic colloid, CO occupied both palladium and copper sites, as evidenced by the presence of IR bands at 2089, 2045, and 1941 cm^{-1}, demonstrating the presence of both metals at the surface of the particles [74].

Bönnemann and associates [48] have prepared several intermetallic colloids by the reduction of (1) two or more metal halides in the presence of hydrotriorganoborates, such as $LiBEt_3H$ or $NaBEt_3H$ (Table V summarizes the different products and some property characterizations of those powders), and (2) coreduction of inorganic salts such as metal halides in the presence of tetraalkylammonium-hydrotriorganoborates having alkyl groups preferentially of the chain lengths C_6–C_{20}. The NR_4^+ ions in these reactions stabilized the metal colloids by forming a protective shell around the metal ion and also helped in enabling the facile solvation of the colloids. The reactions occurring are similar to that described previously for the individual metal particles. A summary of the various intermetallics synthesized by this procedure is outlined in Table VI.

3.2.2. M50 Steel

M50 steel (4.0% Cr, 4.5% Mo, 1.0% V, and a balance of Fe) is widely used in the aircraft industry as the main-shaft bearings in gas turbine engines, because of its good resistance to tempering and wear and rolling contact fatigue. Conventional M50 steel consists of micrometer-sized carbide particles that can act as fatigue crack initiation sites in such bearing materials [75]. In contrast, an improvement in the mechanical properties of this material would be expected on reducing the grain size or, in other words, by making nanophase M50 steel.

Three different methods were used to synthesize the M50 nanopowders. The basic procedures are described in Scheme V and a detailed description of each procedure follows. After chemical synthesis, all the powders were subjected to an H_2 heat treatment at 420 °C to reduce the carbon and oxygen contents prior to compaction.

Table V. Preparation of Nanocrystalline Alloys by Coreduction of Metal Salts in THF

Number	Metal salt	Reducing agent	Condition		Product		DIF		Comments
			t (h)	T (°C)	Metal content (%)	Boron content (%)	2θ (°)	D (Å)	
1	$Co(OH)_2$ $Ni(OH)_2$	$NaBEt_3H$	7	65	Co: 48.3 Ni: 45.9	0.25	51.7	2.05	Single phase Nanocrystalline
2	$FeCl_3$ $CoCl_2$	LiH+10% BEt_3	6	65	Fe: 47.0 Co: 47.1	0.00	52.7	2.02	Single phase Grain size: 1–5 nm
3	$FeCl_3$ $CoCl_2$	$LiBEt_3H$	5	23	Fe: 54.8 Co: 24.5	0.00	52.5 99.9	2.02 1.17	Single phase Nanocrystalline
4	$CoCl_2$ $PtCl_2$	$LiBEt_3H$	7	65	Co: 21.6 Pt: 76.3	0.00	55.4 47.4	1.93 2.23	Single phase
5	$RhCl_3$ $PtCl_2$	$LiBEt_3H$	5	65	Rh: 26.5 Pt: 65.5	0.04	40.2 46.3	2.24 1.96	Single phase Grain size: 1–4 nm
6	$RhCl_3$ $IrCl_3$	$LiBEt_3H$	5	65	Rh: 33.5 Ir: 62.5	0.15	42.3	2.14	Single phase + Traces $IrCl_3$
7	$PdCl_2$ $PtCl_2$	$LiBEt_3H$	5	65	Pd: 33.6 Pt: 63.4	0.04	40.1 46.3	2.25 1.96	Single phase Grain size: 2–6 nm
8	$PtCl_2$ $IrCl_2$	$NaBEt_3H$	12	65	Pt: 50.2 Ir: 48.7	0.15	40.0 46.5	2.25 1.95	Single phase Nanocrystalline
9	$CuCl_2$ $SnCl_2$	$LiBEt_3H$	4	65	Cu: 49.6 Sn: 47.6	0.00	30.2 53.5	2.96 1.80	Cu_6Sn_5 + Cu + Sn
10	$FeCl_3$ $CoCl_2$	$LiBEt_3H$	1.5	23	Fe: 30.1 Co: 31.4	0.00	52.7 60.8	2.02 1.77	Single phase Nanocrystalline
	$NiCl_2$				Ni: 30.9		77.7 100.3	1.43 1.17	

Source: H. Bönnemann et al., *J. Mol. Catal.* 86, 129 (1994).

Table VI. Preparation of Colloidal Alloys

Number	Metal salt	Reducing agent	Condition		Colloidal alloy solution colour	Workup solvent	Solvent added for precipitation	Metal content in isolated colloid (%)	Mean particle size (nm)
			t (h)	*T* (°C)					
1	$RhCl_2$ $PtCl_2$	$N(octyl)_4BEt_3H$	18	50	Deep red to black Rh and Pt completely dissolved	Ether	Ethanol	Rh: 21.85 Pt: 45.96	2.3
2	$PdCl_2$ $PtCl_2$	$N(octyl)_4BEt_3H$	16	23	Deep brown to black Pd and Pt completely dissolved	Ether	Ethanol	Pd: 25.90 Pt: 33.60	2.8
3	$CuCl_2$ $PtCl_2$	$N(octyl)_4BEt_3H$	16	23	Deep red to black Cu and Pt completely dissolved	Ether	Ethanol	Cu: 15.60 Pt: 55.40	2.3
4	$PtCl_2$ $CoBr_2$	$N(octyl)_4BEt_3H$	18	23	Deep reddish brown to black Pt and Co completely dissolved	Toluene	Pentane/ethanol (25:1)	Pt: 25.40 Co: 6.47	—
5	$NiBr_2$ $CoBr_2$	$N(octyl)_4BEt_3H$	16	25	Dark red to black Ni and Co completely dissolved	ethanol	ether	Ni: 23.80 Co: 23.80	2.8
6	$FeBr_2$ $CoBr_2$	$N(octyl)_4BEt_3H$	18	50	Dark brown to black Fe and Co completely dissolved	ethanol	ether	Fe: 13.36 Co: 14.39	3.2

Source: H. Bönnemann et al., *J. Mol. Catal.* 86, 129 (1994).

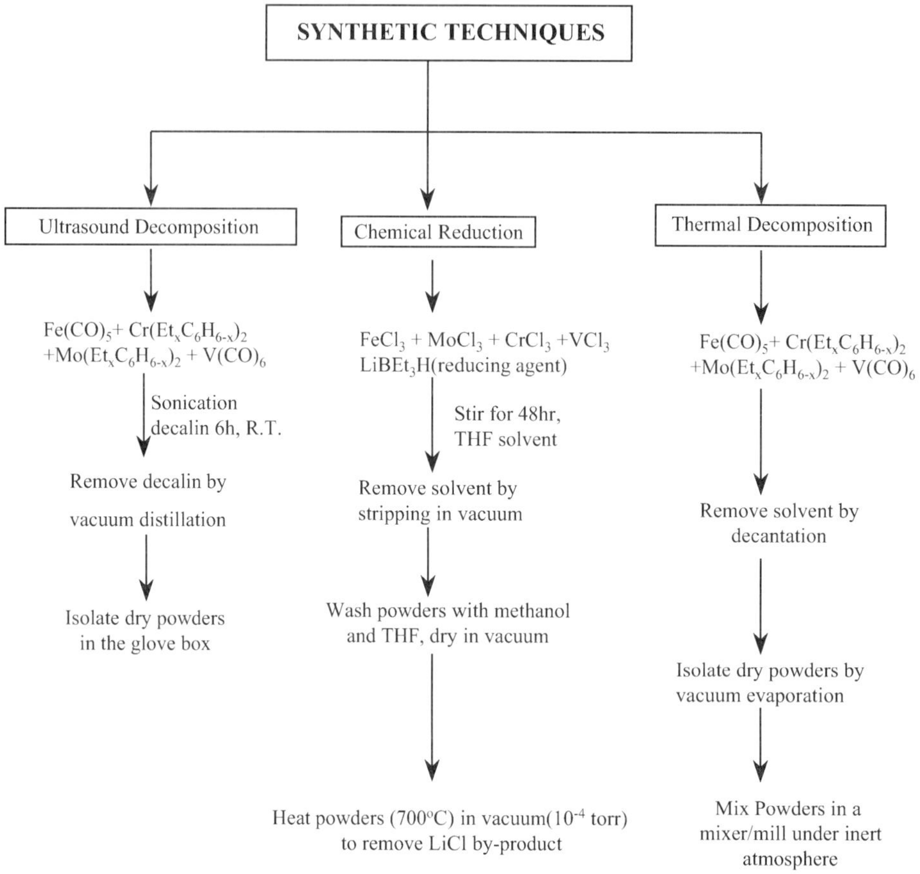

Scheme V. Schematic of synthetic procedures. (Adapted from Chemistry & Physics of Nanostructures and Related Non-Equilibrium Materials, TMS, p. 149, 1997, Gonsalves et al.)

3.2.2.1. Sonochemical Synthesis

A dispersion of 20 g $Fe(CO)_5$, 1.40 g $Cr(Et_xC_6H_{6-x})_2$, 1.03 g $Mo(Et_xC_6H_{6-x})_2$, and 0.063 g $V(CO)_6$ in dry decalin was sonicated at 50% of maximum vibration amplitude [sonochemical reactor horn (Sonics and Materials model VC-600, 20 kHz, 100 W cm^{-2})] for 6 h at room temperature in a sonochemical mercury bubbler. The color of the solution turned dark and then black within a few minutes and this reaction mixture was sonicated until the formation of shiny metallic particles was observed on the walls of the reaction vessel. The sonication was then stopped and the decalin solvent was removed from the reaction flask via vacuum distillation. Fine black powder (yield: 5.84 g) remained at the bottom of the reactor, which was then isolated and stored in a vial under nitrogen.

3.2.2.2. Coreduction Method

To a suspension of 10 g $FeCl_3$, 0.36 g $MoCl_3$, 0.46 g $CrCl_3$, and 0.11 g VCl_3 in 100 mL of THF, 202.98 mL of 1.0 M lithium triethyl borohydride in THF was added slowly while stirring at room temperature by a liquid addition funnel in the dry box. Slow effervescence was observed for a few minutes. After the reaction was stirred in the glove box for 48 h at room temperature, a black suspension was formed. The solvent THF was removed from the reaction flask by vacuum distillation and the black powders were washed with approximately100 mL of distilled degassed methanol until no further bubbling was observed. The fine black solid was washed with 50 mL of THF and dried under vacuum. The lithium chloride byproduct was removed from the preceding solid by vacuum sublimation in a tube furnace at 700 °C/10^{-4} torr. The yield of the powders after the sublimation was 3.38 g.

3.2.2.3. Thermal Decomposition

A dispersion of 20 g $Fe(CO)_5$, 1.40 g $Cr(Et_xC_6H_{6-x})_2$, 1.03 g $Mo(Et_xC_6H_{6-x})_2$, and 0.063 g of $V(CO)_6$ in dry decalin was refluxed for 6 h in a round-bottom flask fitted with a condenser and gas inlet and outlet tubes connected to a mercury bubbler. The color of the solution turned dark and then black within 3 h and this reaction mixture was left to reflux until the formation of shiny metallic particles was observed on the walls of the reaction vessel. The reaction was then stopped and the decalin solvent was decanted from the powders inside the glove box. A fine black powder (yield: 5.79 g) remained at the bottom of the flask, which was then thoroughly dried by gentle heating in vacuum, and then the dry powders were consequently mixed in a mixer/mill for 8 h under an inert atmosphere.

The morphology and the microstructure of the M50 steel powders produced by the three different techniques were examined by SEM, TEM, and XRD. The as-synthesized powders prepared by all the methods were amorphous and were agglomerated. Figure 15 shows the XRD spectra of a sonochemically synthesized M50 nanopowder. The broad peak [centered around 44–45° with full width at half-maximum (FWHM) of approximately 5° in 2θ units] is assigned to the main α-Fe reflection. The low processing temperatures involved in these methods prevented grain growth as well as full crystallization in the (Fig. 16) that they are highly agglomerated but consisted of smaller particles, approximately 4 nm in diameter.

M50 powders produced by the coreduction method, after workup of the reaction and washing with methanol and THF, were contaminated with the LiCl byproduct as shown in the EDAX spectrum (Fig. 17). The spectrum shows the presence of large amounts of LiCl as evidenced by the prominent chlorine peak. However, after sublimation of these powders at 700 °C/2 h and a vacuum of 10^{-4} torr in a tube furnace, all the chloride byproduct is removed (Fig. 18) and the pure M50 nanopowders are isolated.

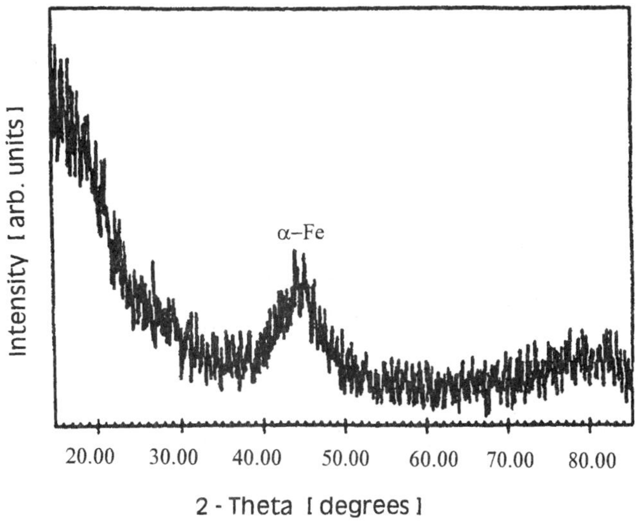

Fig. 15. XRD spectrum of as-synthesized steel powders (sonochemical method). Reprinted with permission from K. E. Gonsalves and S. P. Rangarajan, in "Chemistry and Physics of Nanostructures and Related Non-Equilibrium Materials" (E. Ma et al., eds.), p. 149. TMS Meeting, Orlando, FL (© 1997 Minerals, Metals & Materials Society).

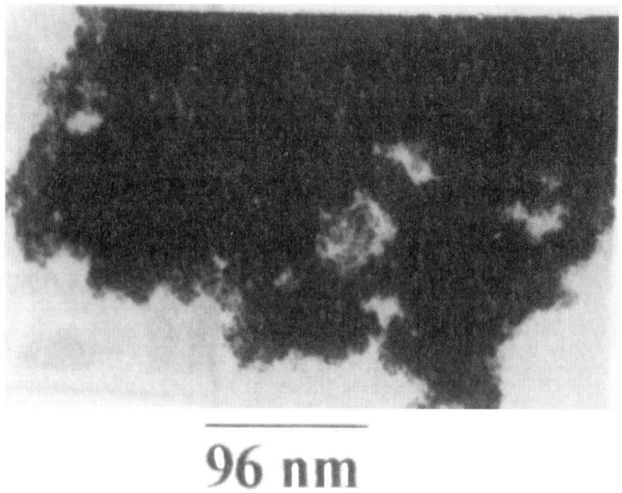

Fig. 16. TEM micrograph of M50 steel powders (sonochemical method). Reprinted with permission from K. E. Gonsalves and S. P. Rangarajan, in "Chemistry and Physics of Nanostructures and Related Non-Equilibrium Materials" (E. Ma et al., eds.), p. 149. TMS Meeting, Orlando, FL (© 1997 Minerals, Metals & Materials Society).

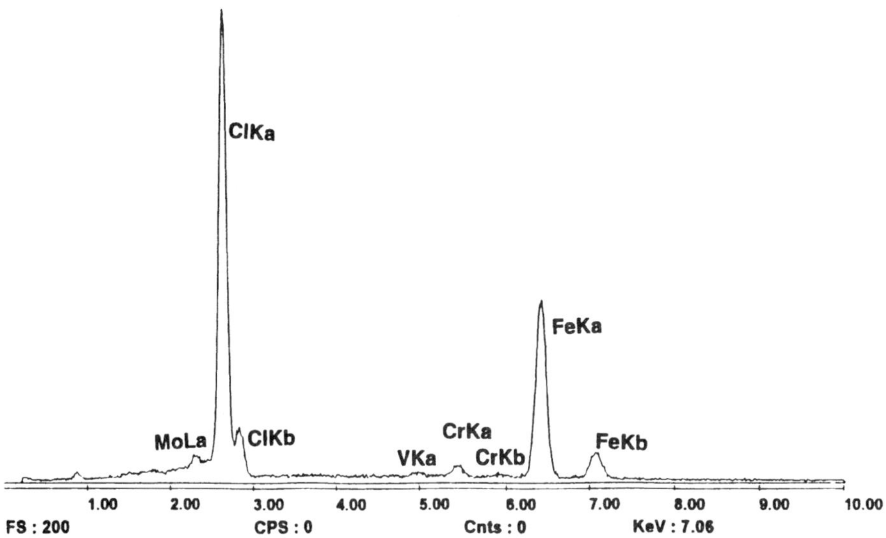

Fig. 17. EDAX spectrum of M50 steel powders (coreduction) before sublimation. Reprinted with permission from K. E. Gonsalves and S. P. Rangarajan, in "Chemistry and Physics of Nanostructures and Related Non-Equilibrium Materials" (E. Ma et al., eds.), p. 149. TMS Meeting, Orlando, FL (© 1997 Minerals, Metals & Materials Society).

The SEM micrograph of the coreduced powders taken after the heat treatment showed agglomeration and possibly sintering of the particles. This is indicated in Figure 19 by the presence of a chunky microstructure. Also the XRD spectrum of the powders taken after the heat treatment illustrated that the powders were crystallized and showed sharp peaks at the expected 2θ values of body-centered cubic (bcc) α-Fe, as well as the absence of any reflections from lithium chloride. This suggests that the sublimation process was effective in eliminating the powders of the chloride contaminant. The average grain size of these thermally treated coreduced powders as calculated from line broadening analysis was 34 nm ($\pm$3 nm).

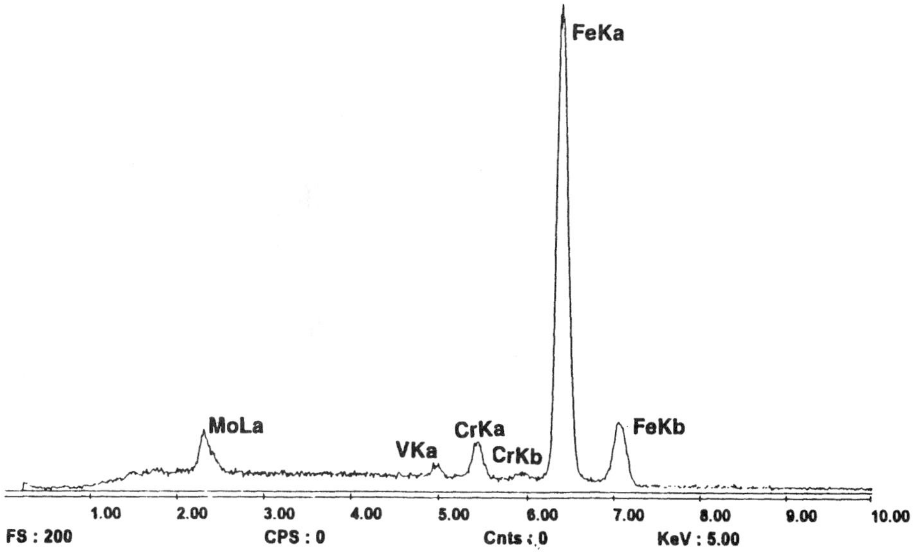

Fig. 18. EDAX spectrum of M50 steel powders (coreduction) after sublimation. Reprinted with permission from K. E. Gonsalves and S. P. Rangarajan, in "Chemistry and Physics of Nanostructures and Related Non-Equilibrium Materials" (E. Ma et al., eds.), p. 149. TMS Meeting, Orlando, FL (© 1997 Minerals, Metals & Materials Society).

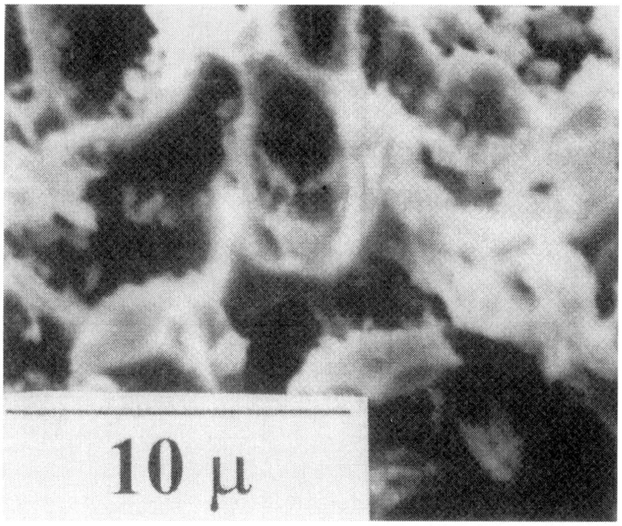

Fig. 19. SEM micrograph of powders (coreduction) taken after sublimation. Reprinted with permission from K. E. Gonsalves and S. P. Rangarajan, in "Chemistry and Physics of Nanostructures and Related Non-Equilibrium Materials" (E. Ma et al., eds.), p. 149. TMS Meeting, Orlando, FL (© 1997 Minerals, Metals & Materials Society).

The morphology of the powders produced by thermal decomposition was examined by SEM and their homogeneity was confirmed by EDAX analysis. These powders showed a porous coral-like morphology when observed in an SEM at 4400X as shown in Figure 20. Spot EDAX analysis at various locations on the micrograph (Fig. 21) showed the powders to be of homogeneous composition and of the expected stoichiometry for conventional M50 steel.

Fig. 20. SEM micrograph of powders (thermal decomposition). Reprinted with permission from K. E. Gonsalves and S. P. Rangarajan, in "Chemistry and Physics of Nanostructures and Related Non-Equilibrium Materials" (E. Ma et al., eds.), p. 149. TMS Meeting, Orlando, FL (© 1997 Minerals, Metals & Materials Society).

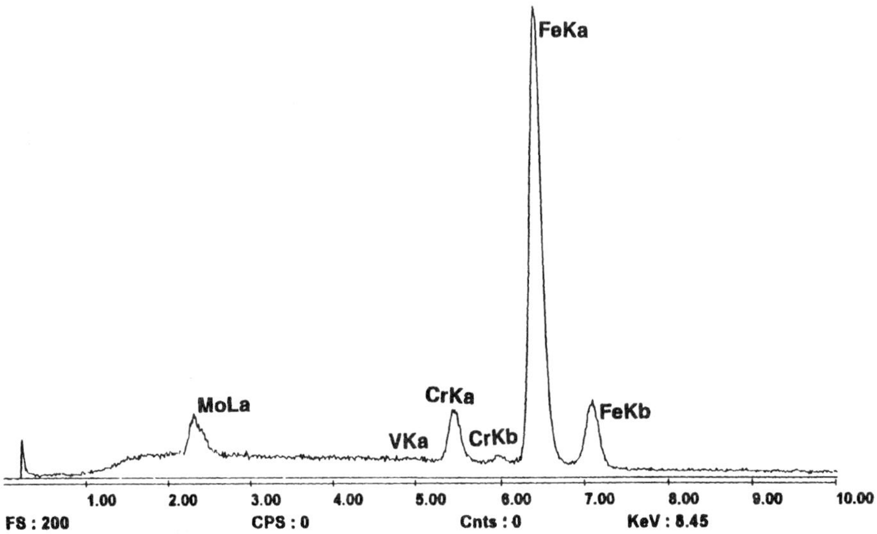

Fig. 21. EDAX spectrum of M50 steel powders (thermal decomposition). Reprinted with permission from K. E. Gonsalves and S. P. Rangarajan, in "Chemistry and Physics of Nanostructures and Related Non-Equilibrium Materials" (E. Ma et al., eds.), p. 149. TMS Meeting, Orlando, FL (© 1997 Minerals, Metals & Materials Society).

3.3. Synthesis of Semiconductors

The semiconductors being synthesized by chemical methods are mainly those of the group II–VI and III–V binary systems [76]. Frequently, these clusters are prepared in the form of dispersed colloids or can be trapped and stabilized within micelles, polymers, zeolites, or glasses [77]. The chemical synthesis is divided into categories according to the host material in which the semiconductor is created or embedded. In this section, we discuss the various methods that have been adopted to prepare single-size clusters, an important goal in this area.

3.3.1. Colloids/Micelles/Vesicles

There are several examples in the literature for the synthesis of semiconductor particles. The stabilization of a colloid in the small-cluster size regime requires an agent that can bind to the cluster surface and thereby prevent uncontrolled growth into larger particles. The simplest method to prepare these colloids involves using a solvent to act as a stabilizer of the small clusters; for example, base hydrolysis of a solution of a zinc salt results in the generation of ZnO in alcohol solvents [78]. On completion of this reaction, a transparent colloid is produced where the ZnO particles increase in size on standing. In a variation of this approach, that is, by a combined solvent/anion stabilization method followed by hydrolysis using LiOH, extremely high concentrations and very stable solutions of highly luminescent ZnO particles in ethanol have been produced.

Such colloids can also be made by the use of a polymeric surfactant/stabilizer that is added to a reaction designed to precipitate the bulk material. The polymer attaches to the surface of growing clusters and by either steric or electrostatic repulsion prevents the further growth of the nanoclusters. The most commonly used polymer is sodium polyphosphate (hexametaphosphate), and clusters of CdS, CdTe, and ZnTe with this surfactant have been studied [79].

The addition of anionic agents frequently called capping agents to the solutions of growing clusters prevents further growth of these materials by covalently binding to the cluster surface. Thiolates are the most commonly used capping agents and this method also forms the basis of synthesis of the monodispersed clusters described later on in this chapter. Figure 22 shows the synthetic strategies of producing clusters by this approach. This technique can be thought of as mimicking an organic polymerization reaction (initiation, propagation, and the termination phases) and is depicted for CdS in the Figure 22. Mixing the cadmium and sulfide ions initiates the polymerization, and the growth of the CdS clusters can be viewed as propagation steps that are sustained by the presence of additional cadmium and sulfide ions. The growth of the clusters can be terminated by providing a capping agent, in this case a thiophenolate ion that intercepts the growing clusters by binding to the cluster surface [80]. The average cluster size can then be controlled by adjusting the ratio of the sulfide to thiophenolate in the reaction solution. It has been found that thiophenolate-

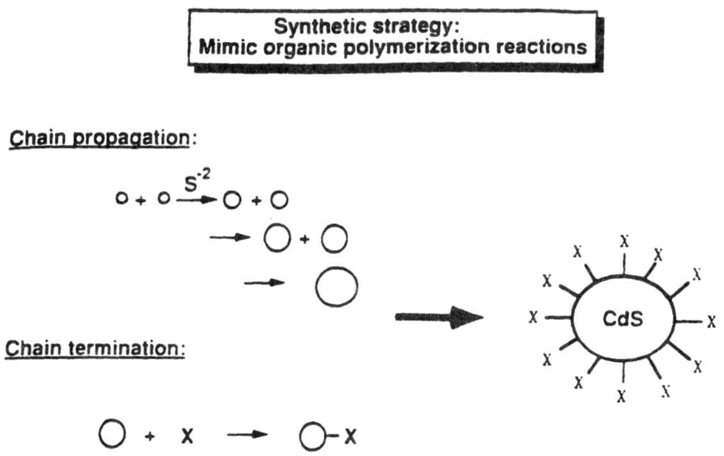

Fig. 22. Schematic diagram for the synthesis of the thiophenolate-capped CdS clusters (X represents the thiophenolate ion) drawing analogy to organic polymerizations. In the chain propagation step, the size of the circle represents the size of the CdS clusters. The growth of the CdS clusters is analogous to the growth of a polymer chain as long as the surfaces are not covered by the terminating agent, X. Reprinted with permission from N. Herron and Y. Wang, in "Nanomaterials: Synthesis, Properties and Applications" (A. S. Edelstein, ed.), p. 73 (© 1996 IOP Publishing, Bristol, U.K.).

capped CdS clusters act somewhat like living polymers that keep growing if fed on more sulfide ions [80].

Other capping agents that have been used are glutathione peptides produced by yeast for the capping of small CdS clusters. The generation of GaAs in THF solution or glycol ether may also be categorized as this type of a reaction [81]. The pentamethylcyclopentadienyl ligands on the Ga ions and the trimethylsilyl groups on the As ions regulate cluster growth and maintain GaAs colloids in solution. In another case, the acetylacetonate groups on the Ga ions and the trimethylsilyl groups on the As ions control cluster dimensions [82].

In similar approaches, micelle-forming reagents have been widely used as a method of controlling cluster growth. In this method, the semiconductor is precipitated in a small region of space defined by the micelle. In contrast to the colloidal approach, the micellar reagent acts as a physical boundary rather than a surface-capping agent. There are two approaches in the micellar technique: (1) normal micelles and (2) reverse micelles. Normal micelles using dihexadecyl phosphate or dioctadecyldimethylammonium chloride (DODAC) are generated in water to limit particle sizes between 150 and 300 nm [83]. The cadmium or zinc ions are dissolved in these clusters and precipitated with H_2S leading to clusters of up to 50 Å in diameter in these micelles. Reverse micelles using bis(2-ethylhexyl)sulfosuccinate salts (AOT) allow the formation of small water pools (<100 Å in radius) in heptane solvent, and again incorporation of metal ions followed by chalcogenide treatment can precipitate semiconductor clusters within these pools. Selenophenol-capped CdSe or CdSe/ZnS clusters have been developed by a combination of the reverse-micelle and the surface-capping approaches [84].

3.3.2. Polymers

The preceding methods have major problems of irreproducibility and colloidal instability, as well as the problem of good characterization. Also, it is often necessary to synthesize the semiconductor nanoclusters in solid thin-film form for use in practical applications. By using polymers as a matrix for the semiconductors, these problems can be alleviated. This method was first used with CdS in Nafion and PbS in Surlyn. These stable clusters were then identified by X-ray diffraction [85].

Composites of these clusters have been produced by several methods. The first is by ion exchange. Polymers such as Nafion or ethylene-methacrylic acid copolymer have cation exchange sites where Cd or Pb ions may be introduced into the polymer matrix (Fig. 23).

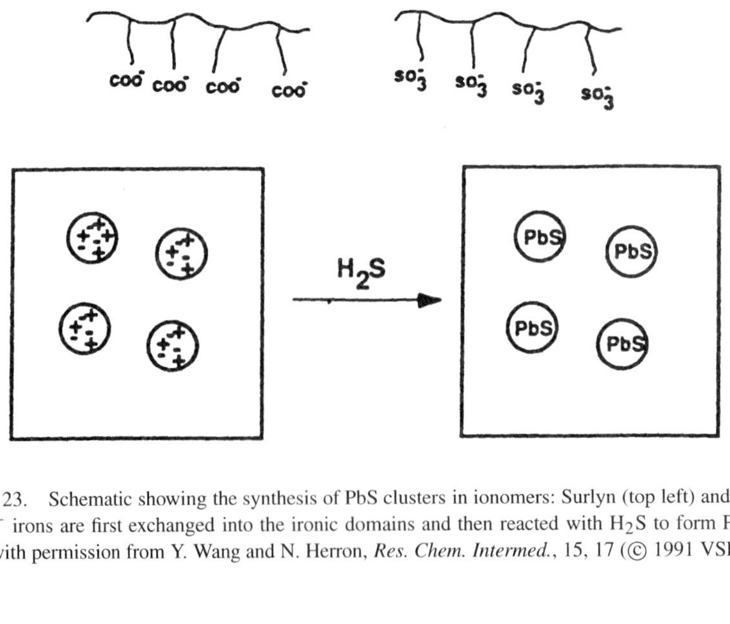

Fig. 23. Schematic showing the synthesis of PbS clusters in ionomers: Surlyn (top left) and Nafion (top right). Pb^{2+} irons are first exchanged into the ironic domains and then reacted with H_2S to form PbS clusters. Reprinted with permission from Y. Wang and N. Herron, *Res. Chem. Intermed.*, 15, 17 (© 1991 VSP BV).

Treatment of such ion-exchanged films with chalcogenide sources results in *in situ* precipitation of the compound semiconductor within the hydrophilic regions of the polymer. The ionic polymers can be considered as the solid-state analogs of micelles. The size of the trapped clusters is controlled by the appropriate control of the phase-separated hydrophilic and hydrophobic regions of the polymer [86]. Layered semiconductor clusters such as PbI_2, as well as magnetic particles such as Fe_2O_3, have been developed by this method [87].

Using ring opening methasesis polymerization (ROMP), Schrock et al. [88] incorporated the Pb^{2+} metal ion into monomer units that eventually became part of a norbornene-derived copolymer. In this case, the polymer is designed to phase separate into size-controlled regions of high and low hydrophobicity. The semiconductor is precipitated in these regions by gas phase treatment with H_2S.

Another approach that has been reported uses a combination of the polymer isolation and the surface-capped cluster approaches. A well-defined semiconductor cluster or colloid is prepared by the conventional capping technique described previously and is then dissolved in a solvent along with a soluble polymer. This mixed solution may then be spin coated onto a substrate and dried to produce a polymer film doped with the semiconductor particles. The CdS semiconductor has been incorporated into a photoconductive polymer such as polyvinylcarbazle using this method [89].

There are examples of the synthesis of CdS and CdSe nanoparticles stabilized by polymers in the literature. Yao et al. [90] first prepared CdS and CdSe particles by the reaction of $Cd(NO_3)_2{\cdot}4H_2O$ or $Cd(ClO_4)_2{\cdot}6H_2O$ with H_2S and H_2Se in the presence of PVP. In this case, PVP plays an essential role in preventing the flocculation of concentrated nanoparticles and controlling the particle size. The absorption spectra of CdS nanoparticles prepared with PVP (2.0 g/L) in acetonitrile showed a sharp peak at about 330 nm, and the particle diameters estimated from the absorption peak using tight binding approximation were about 2.0 nm. Similar blue-shifted absorption spectra of CdSe nanoparticles prepared in the presence of PVP in ethanol were also observed. The PVP-coated CdS and CdSe nanoparticles obtained were dispersed in acrylonitrile-styrene copolymer or in poly(2-hydroxyethyl methacrylate) (PHEMA) organic films, and the third-order nonlinear optical susceptibilities $\chi^{(3)}$ of these films were measured. A $\chi^{(3)}$ yield of 1.1×10^{-7} esu was obtained for a 4.0-vol% CdS-doped PHEMA film, and a value of 3.0×10^{-11} esu was obtained for a similar CdSe analog. CdS particles (2.5 nm in size) were also prepared by reactions of cadmium acetate, thiourea, and thioglycerol (which acts as a capping agent) by Chevreau et al. [91], who then dispersed them in polystyrene and cast films. The UV–vis spectrum of CdS thin films that were heated to 100 °C showed excitonic features at 375 nm. Luminescence spectra for the spin-coated films (Fig. 24) showed a maximum intensity around 540 nm. The transition of semiconductor chalcogenides from molecular CdX (X = Se, S) to bulk material has been observed in Nafion, a cation-exchanged membrane. The CdS and CdSe clusters were prepared by sonicating aqueous solutions of $CdCl_2{\cdot}4H_2O$ with the protonated Nafion. After drying, the samples were exposed to 1 atm of H_2S or H_2Se on a vacuum line for 10 h. TEM images indicate that CdS particle sizes varied between 2 and 4 nm. Exposure of the CdS and CdSe films to water caused a shift to longer wavelengths in the absorption spectra owing to Ostwald ripening and possible increase in grain size.

ZnO has many applications in luminescent devices, photocatalysis, and photoelectrochemistry [92]. A promising approach in nanocluster synthesis from a practical viewpoint is to use stabilizing agents to grow the particles in structured media. Highly stable, wurtzite quantum-sized ZnO colloids have been encapsulated in polymers such as PVP, PVA, and sodium hexa-meta phosphate (HMP) [93]. Freshly prepared uncapped ZnO clusters exhibited a blue shift in absorption, with the excitonic shoulder appearing at about 300 nm compared to the band edge for macrocrystalline ZnO at 365 nm. PVP-protected clusters showed an excitonic hump at about 310 nm. XRD revealed the particle sizes were 2.8 nm for the uncapped sample and 3.5 nm for the PVP-protected sample. It is, therefore, obvious

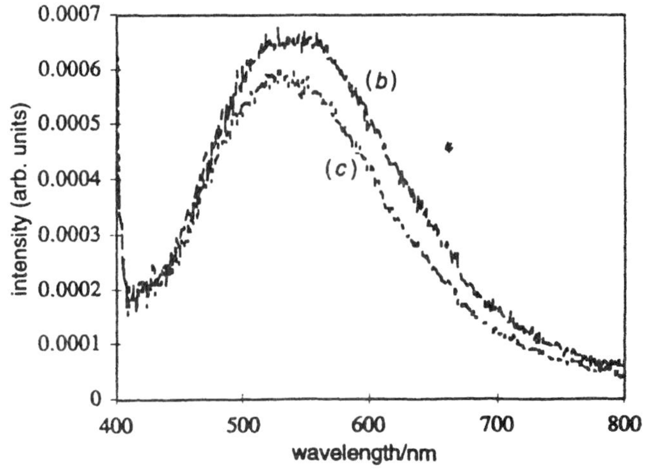

Fig. 24. Detail of luminescence spectra for 500 (b) and 1000 (c) rpm films showing the same emission linewidths. Reprinted with permission from A. Chevreau et al., *J. Mater. Chem.*, 6, 1643 (© 1996 The Royal Society of Chemistry).

that these clusters will exhibit quantum-sized effects as the Bohr exciton radius of ZnO is about 4.0 nm [94].

3.3.3. Glasses

Particles of CdS_xSe_{1-x} can be embedded in a borosilicate glass matrix and have traditionally been used as color filters [95]. These materials are prepared by traditional glass-making technology where Cd, S, and Se are added to a silicate or germanate glass melt at elevated temperatures. After casting, the glass is annealed at temperatures less than the melting point and the small semiconductors form within the dense glass matrix [96]. This technique is suitable for only a limited number of semiconductors as they must survive a very high temperature oxidizing environment during the glass-forming step. Thus, this method is not appropriate to make III–V semiconductor clusters of GaAs because of their thermal and oxidation sensitivity.

Rajh et al. [97] have reported size quantization effects in specially prepared silicate glasses containing small particles of CdS, CdSe, Bi_2S_3, PbS, HgSe, In_2Se_3, and AgI. The procedure involved mixing an aqueous colloidal solution of the semiconductor with tetramethoxysilane (TMOS) and accelerating the polymerization of SiO_2 with NH_4OH. In the starting colloidal solutions, the particle sizes were determined by electron microscopy and were in the 20–40-Å range. It was found that the luminescence intensity of colloidal semiconductor particles in silicate glasses is much higher compared to that of the same particles in aqueous solutions. The absorption and emission spectra for AgI (~100 Å) and CdS (~30 Å) are shown in Figure 25. The quantum yields for AgI and CdS in silicate glasses were 2% and 5%, respectively. The higher emission yields in silicate glasses could occur through passivation by the silicate polymer of surface states.

Minti et al. [98] have used a sol–gel method to prepare quantum dots of CdS in thin glass films, using tetraethoxysilane (TEOS), water, alcohol, and cadmium nitrate as precursors. Organically modified ceramic (ORMOCER) films were also prepared by the same group according to the procedures of Schmidt [99]. TEM micrographs of CdS particles in the glass films had sizes of 20–50 Å. Third-harmonic-generation (THG) signals from the CdS-doped thin glass and the ORMOCER films were easily observed. The strongest signals were about 1500–2000 times stronger than those generated by the glass substrate, and $\chi^{(3)}$ was measured to be 2×10^{-12} esu for the sol–gel CdS films and for the ORMOCER film it was approximately 10^{-12} esu. These values of $\chi^{(3)}$ are much lower than those of

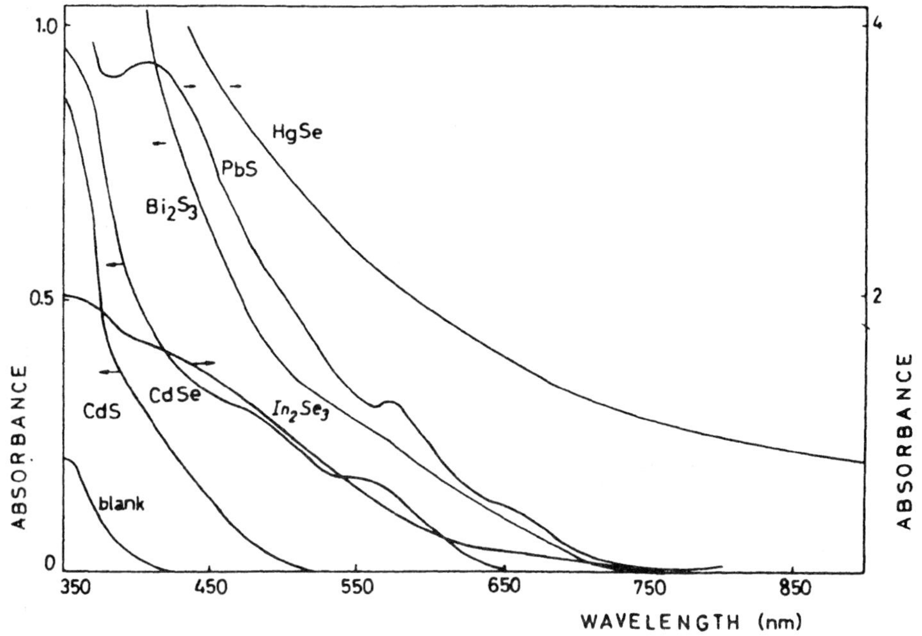

Fig. 25. Absorption spectra of various colloids incorporated in glasses. The monomer concentration of the starting colloids was 4×10^{-4} M and the stabilizing agents were: (a) 2×10^{-3} M polybrene for CdS and AgI, (b) 0.1% carbowax for CdSe and In_2Se_3, (c) 4×10^{-4} M hexametaphosphate for Bi_2S_3, (d) 0.1% PVA (MW 14,000) for PbS, and (e) 0.1% carbowax for HgSe. The absorption spectrum of the glass itself is also presented as blank. The optical path was 0.5 cm. Reprinted from *Chem. Phys. Lett.*, T. Rajh et al., 143, 305 (© 1988), with kind permission of Elsevier Science-NL, Sara Burgerhartstraat 25, 105b KV Amsterdam, The Netherlands.

pure CdS, but still quite high considering the low volume fraction of CdS particles in the film [100].

III–V semiconductors, because of their thermal and oxidation sensitivity, have been prepared by the selective acid leaching of borates from a borosilicate Vycor glass. The physical space restraint imposed by the pore sizes of the glass controls the size of the semiconductor that is precipitated *in situ* within the pores of this glass at low temperatures. Examples of II–VI and III–V semiconductors in the literature have been previously reported [101]. Sol–gel-derived porous silica glass has also been used for the production of these materials. In this case, the semiconductor particles were assembled within the pore structure of the host glass matrix and the residual porosity was then back filled with a polymer such as poly(methylmethacrylate) (PMMA) [102]. This construction strategy reduces light scattering, which is frequently observed from the empty pore network, and, therefore, makes the composite have more desirable optical properties.

A glassy thin film of TiO_2 nanoparticles has been reported by electrostatic spraying of an alkoxide solution. Radio frequency (rf) sputtering and gas evaporation techniques have also been used to produce small semiconductor clusters, which are sometimes doped in glasses. These methods are very useful for producing high-quality thin-film samples [103].

3.3.4. Crystalline and Zeolite Hosts

The use of a crystalline host lattice for the stabilization of semiconductor nanoclusters is an attempt to impose crystalline order and physical size constraints upon the included semiconductor particles. CuCl crystallites, ranging from 20 to 50 Å, have been prepared within single crystals of NaCl or KCl by a melt procedure and they exhibited exciton absorption features that were blue shifted from the bulk material [104]. This approach involves the use of a porous crystalline host, zeolite, into which semiconductor clusters are either directly

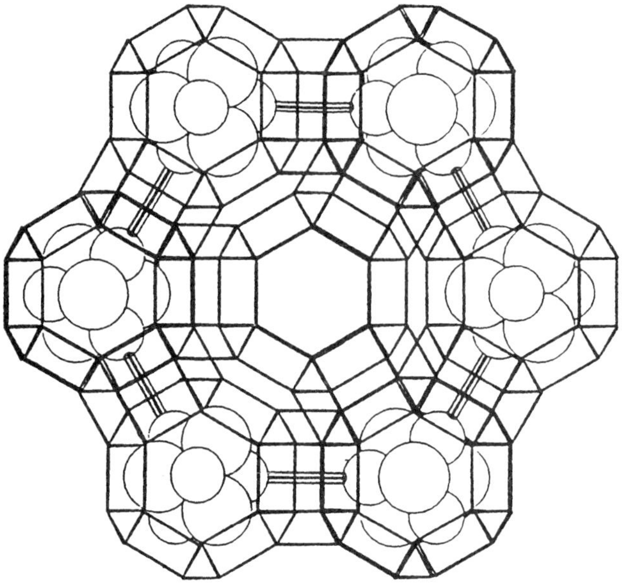

Fig. 26. Cd_4S_4 cluster array in the sodalite cages of zeolite Y. Circles represent CD_4S_4 clusters (smaller circles are Cd and larger circles are S). Six sodalite cages (occupied by Cd_4S_4) and one supercage (at the center, empty) are shown here. The sodalite cage has a diameter of ~5 Å and the supercage has a diameter of ~13 Å. Reprinted with permission from N. Herron and Y. Wang, in "Nanomaterials: Synthesis, Properties and Applications" (A. S. Edelstein, ed.), p. 73 (© 1996 IOP Publishing, Bristol, U.K.).

imbibed or created *in situ*. Examples of the former technique are the sublimation of Se or Te into aluminosilicate zeolites [105]. CdS and superclusters (cluster arrays) have been generated in zeolite Y by the second method, which is described next [106].

Zeolite Y occurs in nature as the mineral faujasite and consists of a porous network of aluminate and silicate tetrahedra linked through bridging oxygen atoms [107]. It has two types of cavities within the structure—the sodalite cavity of ~5 Å diameter with access through ~2.5-Å windows and the supercage of ~13 Å diameter with access through ~7.5-Å windows as shown in Figure 26. These well-defined and ordered cavities provide an ideal environment for the synthesis of single-sized clusters and cluster arrays, whereas the windows provide access for transporting reagents to the cavities. The ion-exchange properties of the zeolite are produced as a result of their chemical framework and are useful in producing the clusters. For example, CdS clusters were synthesized in zeolite Y by the ion-exchange method. Synchrotron X-ray, extended X-ray absorption fine structure (EXAFS), and optical absorption data reveal discrete $(CdS)_4$ cubes located within the small sodalite units of the structure (see Fig. 26). The Cd_4S_4 units are not isolated and they interact with each other via through-bond interactions to form a supercluster (or cluster array). The absorption spectrum shifts from approximately 290 nm for the isolated clusters to approximately 360 nm in the supercluster [108].

The zeolite confinement approach has been extended to include many other semiconductor guests and sodalite hosts [109]. AgI in zeolite mordenite has been shown to have unusual optical behavior in terms of its photosensitivity. Silver halides in the sodalites have been synthesized by Ozin et al. [110]. PbI_2 in X-, Y-, A-, and L-type zeolites shows evidence of a strong size effect on the exciton absorption. Cao et al. [111] investigated a variant of this approach and used layered crystalline hosts of the metal phosphonate series instead of zeolites.

The formation and the photophysical properties of CdS in zeolites with cages and channels such as the zeolite sodalites, A, X, chabazite, and offrertite, were investigated by Liu and Thomas [112]. The sizes of the clusters are constrained by the sizes of the cages. The results showed that the CdS clusters form in the biggest cages of the zeolites and in the main channel of the zeolites with channels. Absorption spectra of the CdS formed in these two types of systems show different features, indicating different particle sizes and distributions. The former covers a rather narrow range of wavelength and the latter covers a wider range, as the particles of CdS are smaller than and of a narrower size distribution in the former confined regions than in the latter.

Semiconductor superclusters of the preceding type represent a novel class of materials where the three-dimensional structure and electronic properties can be controlled by using different zeolites as the template [113]. The unique stability of the semiconductor clusters within these zeolite units is due to the coordination of Cd atoms with the framework oxygen atoms of the zeolites.

The zeolites offer the unique advantage of preparing three-dimensional arrays of mutually interacting clusters with geometric structures imposed by the zeolite internal pore structure. The electronic properties of the cluster array are controlled by the different spatial arrangements of the clusters, which, in turn, are controlled by using different zeolites as the template. However, imperfections still exist in this procedure, and the further success of this method will depend on the availability of high-quality zeolite single crystals as well as zeolites with larger pore sizes.

3.3.5. Synthesis of Single-Sized Clusters

It is imperative to produce clusters with a very narrow size distribution in order to realize their electrical and optical properties. Herron and Wang [26] have observed in their study of thiophenolate-capped CdS clusters that clusters can be grown simply by adding extra sulfide to the cluster solution. Based on this observation, it seems that if starting with a well-defined small molecular cluster, larger single-sized clusters could be obtained by the addition of reagents selected to cement the smaller units together. The $(NMe_4)_4Cd_{10}S_4SPh_{16}$ cluster synthesized by Dance and co-workers [114] provides an ideal candidate as a starting material. This compound belongs to a series of molecular clusters synthesized by Dance and co-workers, who produced crystalline, well-characterized semiconductor molecular clusters. The compounds $(NMe_4)_4Cd_{10}S_4SPh_{16}$ and $(NMe_4)_2Cd_{17}S_4SPh_{28}$ contain a crystalline core having the same atomic range as the bulk cubic phase of CdS and have particle dimensions in the 7–9-Å range [115]. These clusters are very soluble in polar organic solvents and have been used to synthesize single-sized clusters by two different approaches.

3.3.5.1. Controlled Cluster Fusion in Solution

In the work of Dance and associates [114], a molecular fragment of CdS containing a $Cd_{10}S_4$ core was capped by 16 thiophenolate groups. Although these syntheses are carried out in solution, the clusters can be collected as stable solids and redissolved. The reaction between the cadmium and the chalcogenide ions can be considered as an example of an inorganic polymerization with an initiation step forming a small nucleus, which then grows larger in a propagation step before precipitating in the termination step. It can be anticipated that the growth of the clusters can be controlled by the relative kinetics of attachment of the growing clusters by the propagation and termination steps, respectively.

The cluster fusion idea has been demonstrated by Dance et al., who synthesized the cluster $(Cd_{20}S_{13}(SPh)_{22})^{8-}$ just by the addition of extra sulfide ions to $(Cd_{10}S_4(SPh)_{16})^{4-}$:

$$2\left(Cd_{10}S_4(SPh)_{16}\right)^{4-} + 5S^{2-} \rightarrow \left(Cd_{20}S_{13}(SPh)_{22}\right)^{8-} + 10SPh^- \qquad (10)$$

The resultant 55-atom cluster showed a very sharp (FWHM $\sim$800 cm^{-1}) absorption peak at 351 nm (3.53 eV) and a weak shoulder at 330 nm (3.76 eV) at room temperature (Fig. 27). The clusters can be collected as stable solids and redissolved in solutions.

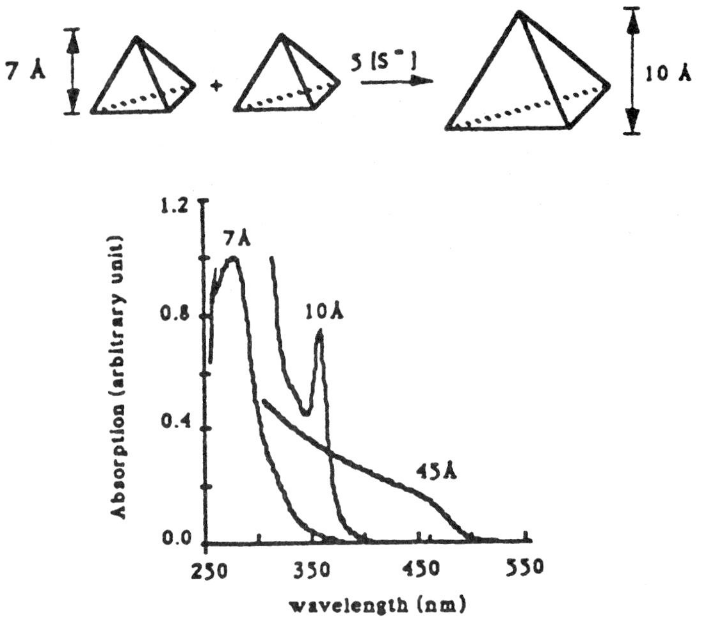

Fig. 27. The absorption of 7-Å $(Cd_{10}S_4(SPh)_{16})^{4-}$ clusters, 10-Å $(Cd_{20}S_{13}SPh_{22})^{8-}$ clusters, and 45-Å CdS clusters in Nafion film. Reprinted with permission from Y. Wang and N. Herron, *Res. Chem. Intermed.*, 15, 17 (© 1991 VSP BV).

By exercising care in controlling the mixing conditions (low concentrations and vigorous mixing) and component ratios during these types of syntheses, it is also possible to generate solutions of CdSe and CdTe clusters that are all capped by thiophenolate ion and that exhibit very sharp absorption features at the band edge in their electronic spectra. It is hoped that these controlled cluster fusion reactions coupled with various chemical separation techniques can eventually lead to the synthesis of single-sized clusters in the condensed phase.

3.3.5.2. Controlled Thermolysis

Thermolysis of the compound $(NMe_4)_4Cd_{10}S_4SPh_{16}$ in inert atmospheres leads to the loss of thiophenolate capping groups and the eventual formation of bulk CdS beginning at approximately 350 °C [116]. At an intermediate temperature of approximately 250 °C where a mass loss corresponding to 4 (NMe_4SPh) is observed, a material of stoichiometric $Cd_{10}S_4SPh_{12}$ can be isolated. This cluster has an 82-atom tetrahedral core of cubic phase CdS with an overall tetrahedral shape. This represents the largest crystallographically characterized semiconductor cluster and presents opportunities for the definitive assessment of optical and electronic properties for a monodispersed cluster size.

3.3.6. Nanostructured III–V Semiconductors

Binary III–V materials have recently attracted significant interest [117]. This is due to their wide band gap energy (3.1–3.8 eV), which makes them suitable for creating and processing blue light. Eventually, devices based on these materials will bring advances in the fields of communications and computing. However, most of the research on III–Vs has concentrated on vapor-deposited thin films.

Research has also been concentrated [118] on materials with large third-order nonlinear optical susceptibility ($\chi^{(3)}$). Third-order optical susceptibility ($\chi^{(3)}$) refers to the dependence of the polarizability on the third power of the optical electric vector (**E·E·E**).

One possible morphology for a $\chi^{(3)}$material is a nanostructured composite with a polymer matrix. Optical nonlinearity arises in nanostructured systems because of the quantum confinement of the particle's electrons. Gallium nitride is a material known [119] to exhibit a significant $\chi^{(3)}$.

Nanostructured GaN has been chemically synthesized by Gladfelter's group by the pyrolysis of cyclotrigallazane ($[H_2GaNH_2]_3$) at 600 °C in argon [120]. The XRD of the heat-treated samples showed a diffraction pattern corresponding to a mixture of hexagonal and cubic gallium nitride. On the basis of X-ray line broadening and TEM, the particle size was found to be 60 Å. This nanocrystalline material, however, converted into the known wurtzite phase at 900 °C. The same group also synthesized nanocrystalline wurtzite GaN from the thermal decomposition of Ga_2O_3 and NH_3 at 600 °C.

Janik et al. [121] have synthesized gallium nitride powders by the pyrolysis at 450–500 °C under NH_3 or vacuum of a polymeric gallium imide $\{Ga(NH)_{3/2}\}_n$. This precursor is, in turn, prepared by the ammonolysis of $[Ga(NMe_2)_3]_2$ and NH_3 at ambient temperatures. The gallium imide precursor was shown to yield, upon pyrolysis, a rare cubic/hexagonal variety of GaN. On the basis of the relative broadness of the XRD powder patterns (Scherrer's equation) and the TEM results, the solid from the pyrolysis at 500 °C under NH_3 consisted of larger and less disordered GaN crystallites (~7 nm) than the solid from the pyrolysis at 450 °C under vacuum (~2 nm).

Another thermal synthetic route to nanocrystalline GaN was demonstrated by Xie et al. [122], who synthesized hexagonal GaN by the reaction of Li_3N with $GaCl_3$ in benzene at 280 °C under pressure in an autoclave. The XRD pattern obtained could be indexed to the hexagonal cell of GaN with lattice constants $a = 3.188$ Å and $c = 5.176$ Å, which are near the reported values. Some minor reflections can also be indexed to cubic GaN in a rock salt structure with $a = 4.006$ Å. The GaN crystallites have an average size of 32 nm and display a uniform shape. The photoluminescence (PL) spectrum of these nanoparticles shows one broad emission feature at 370 nm, which is in agreement with that of bulk GaN. Hexagonal nanocrystalline GaN, ranging in size from 5 to 10 nm, was produced by the controlled thermal decomposition of $[Ga(N_3)_3]_\infty$.

In our laboratory, we have synthesized nanostructured gallium nitride (GaN) and GaN/PMMA composites and have characterized them by HRTEM, chemical analysis, XRD, and image analysis. The deagglomeration of the initial GaN product, and its dispersion into a polymer matrix, yielding a composite with 3–8-nm-diameter particles, is also described.

The dimeric precursor, $Ga_2[N(Me)_2]_6$, was synthesized by the reaction of gallium chloride, $GaCl_3$, with lithium dimethylamide, $LiNMe_2$ [123]. A slight excess (6.44 g) of $LiNMe_2$ was reacted with 7.17 g $GaCl_3$ at room temperature for 2 days in dried, distilled hexane. After stirring, the solution was filtered through Celite and the volatiles were removed *in vacuo* to yield a colorless solid. The pale-yellow pure dimer was isolated by vacuum sublimation at 110 °C. The LiCl byproduct remained on the frit and was discarded. The yield was 3.82 g (71%).

The precursor dimer was placed in an alumina boat and decomposed using a glass tube with an inside diameter (i.d.) of 1 in. placed in a Thermolyne 21100 furnace with a 12-in. heating zone. The material was heated for 4 h at 600 °C. NH_3 flow at ambient pressure was maintained during heat treatment and while the material cooled to room temperature over several hours. The product, a grayish powder, was handled only under an inert atmosphere to prevent oxidation.

To form the nanostructured composite, 60 mg of this powder and 1 mL of methyl methacrylate were charged to a flask under argon. The mixture was sonicated in a cleaning bath for 2 h. The liquid was then decanted and placed in a sealed vial with an initiator (azobisisobutylnitrile, ~5 mg). This mixture was polymerized thermally at 72 °C for 50 min. From the resulting solution, films were spin cast onto clean quartz substrates. The speed was 225 rpm and the time was 5 min. Film thicknesses, measured by profilometry, were approximately 10 μm.

Powder samples of GaN were examined by XRD under mineral oil to inhibit oxidation. The analysis was performed on a Scintag XDS-2000 diffractometer equipped with a Cu K_α source, $2\theta = 0$–$70°$ at a rate of 2°/min. Duplicate chemical analyses were performed (Galbraith Labs) to determine gallium and nitrogen. The results indicate that the material is somewhat nitrogen poor, with an empirical formula of $GaN_{0.86}$. In a separate analysis (Microlytics), carbon and hydrogen contamination were found to be 0.8 and 0.5 wt%, respectively.

The powder sample was examined using a JEOL-4000EX electron microscope with an accelerating voltage of 400 kV and a point-to-point resolution of approximately 1.7 Å. HRTEM images were obtained at optimum (Scherzer) defocus. To prevent any contamination, the sample was prepared for HRTEM by simply grinding the powder between two glass plates and bringing the fine powder into contact with a carbon-coated copper grid under nitrogen atmosphere. The composite sample was embedded in epoxy resin, microtomed to a thickness of approximately 80 nm, picked up on a carbon-coated copper grid, and examined with a Phillips EM300 electron microscope. Subsequently, this sample was also examined using the JEOL instrument described previously.

The gallium nitride/PMMA composite film was examined using a Perkin-Elmer Lambda 6 spectrophotometer. The sample was scanned at a rate of 2 nm/min and the slit width was 0.5 nm.

The X-ray spectrum (Fig. 28) exhibits three relatively broadened reflections at $2\theta =$ 35.5, 58, and 69°. The broadening of the reflections is probably due to the effect of small domain size. These reflections were assigned to the (111), (220), and (311) planes, respectively, of zincblende gallium nitride (unit cell parameter $a = 4.5$ Å). The asymmetry of the (111) peak can be attributed to a confined, low-intensity (200) peak ($2\theta = 40.5°$), which is not explicitly observed. The low intensity of this latter peak may be caused by some nitrogen vacancies (as the chemical analysis indicates) and/or a high density of stacking faults. XRD calculations showed that a good match to experimental data was only obtained by introducing a high density of stacking faults.

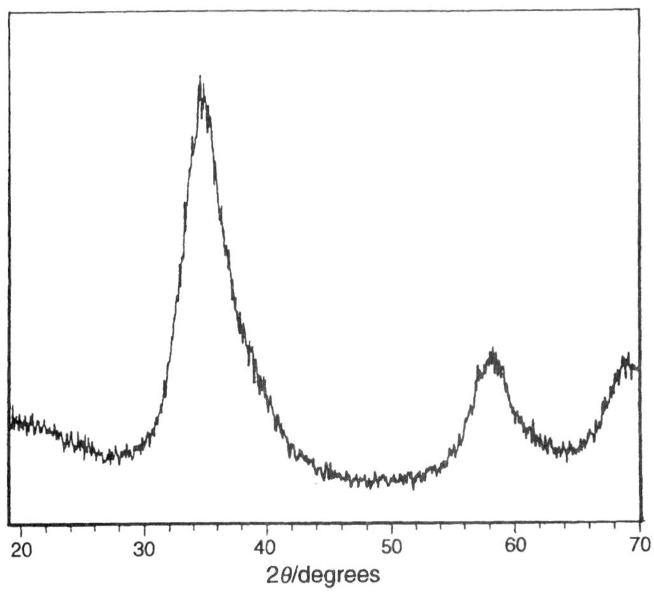

Fig. 28. X-Ray spectrum of GaN powder. Reprinted with permission from K. E. Gonsalves et al., *J. Mater. Chem.*, 8, 1451 (© 1996 The Royal Society of Chemistry).

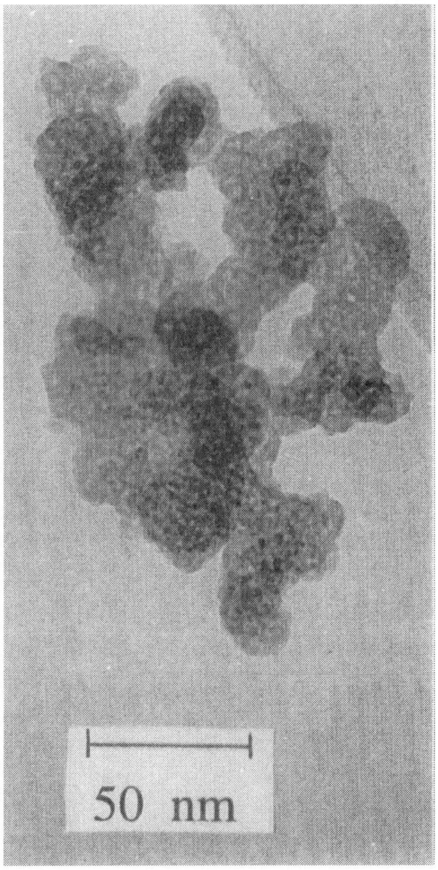

Fig. 29. Nanostructured GaN at high magnification. Reprinted with permission from K. E. Gonsalves et al., *J. Mater. Chem.*, 8, 1451 (© 1996 The Royal Society of Chemistry).

The low-magnification HRTEM image of the powdered GaN shows porous particles of relatively large size (around 50 nm major axis). Examination of the particles at higher magnification (Fig. 29) indicates that each of these large particles is an agglomeration of smaller particles, with nanostructured domains.

The mechanism of the particle deagglomeration is not fully characterized. Probably, the localized temperature and pressure transients introduced by the sonication broke the aggregates into their single-crystal domains, the amorphous material between the grains being weaker. However, there have been cases of stable nanoparticles produced by simple dissolution [124]. Similarly, the nature of the particle surface is unclear. The high curvature and roughness of the surface make it very reactive.

A close examination of a GaN particle (Fig. 30) shows evidence of a high density of stacking faults within the fcc structure. This particle is oriented along its $\langle 110 \rangle$ crystallographic axis. Usually, when a zincblende structure is oriented along this observation axis, the HRTEM image is composed of three atomic planes: two $\{111\}$ and a $\{002\}$. This gives rise to a pseudo-hexagonal symmetry appearance. When a stacking fault is introduced, one of the $\{111\}$ planes is displaced by a vector of $1/6\langle 112 \rangle$. In the case of our GaN nanoparticles, both $\{111\}$ planes underwent this displacement, resulting in a bidimensional disorder. This behavior explains the asymmetry observed in the first X-ray peak and the low contribution of the $\{002\}$ reflection.

The absorption of the composite film was examined using a Perkin-Elmer Lambda 6 spectrophotometer. The sample was scanned at a rate of 2 nm/s and the slit width was

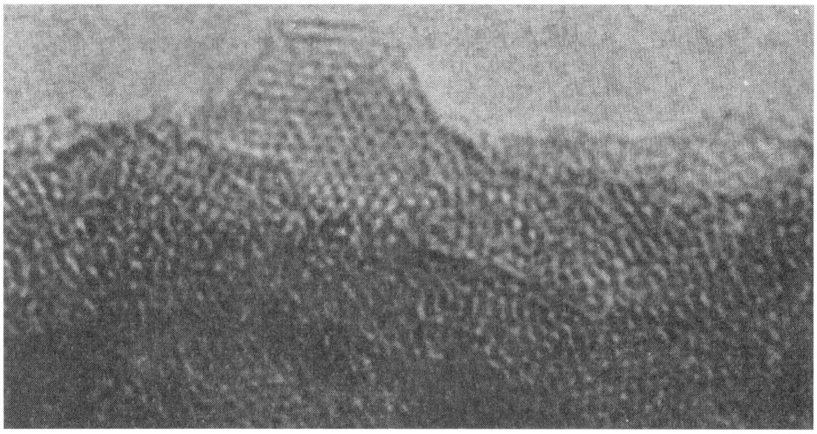

Fig. 30. HRTEM image of GaN. Reprinted with permission from K. E. Gonsalves et al., *J. Mater. Chem.*, 8, 1451 (© 1996 The Royal Society of Chemistry).

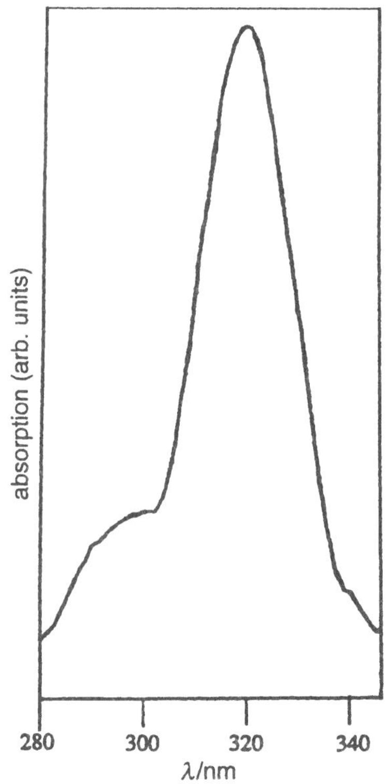

Fig. 31. Optical absorption spectrum. Reprinted with permission from K. E. Gonsalves et al., *J. Mater. Chem.*, 8, 1451 (© 1996 The Royal Society of Chemistry).

0.5 nm. The nanostructured GaN/poly(methyl methacrylate) composite exhibited a strong, symmetric optical absorption (Fig. 31) from 315 to 335 nm. This is probably the result of a surface-enhanced resonance that will lead to a large third-order nonlinear susceptibility ($\chi^{(3)}$) at this wavelength. The particle size distribution (Fig. 32), determined by analysis of TEM images of the composite, had a mean of 5.5 nm and a standard deviation of 2.6. High-resolution TEM of this composite (Fig. 33) confirms that the dispersed phase is zinc-blende GaN.

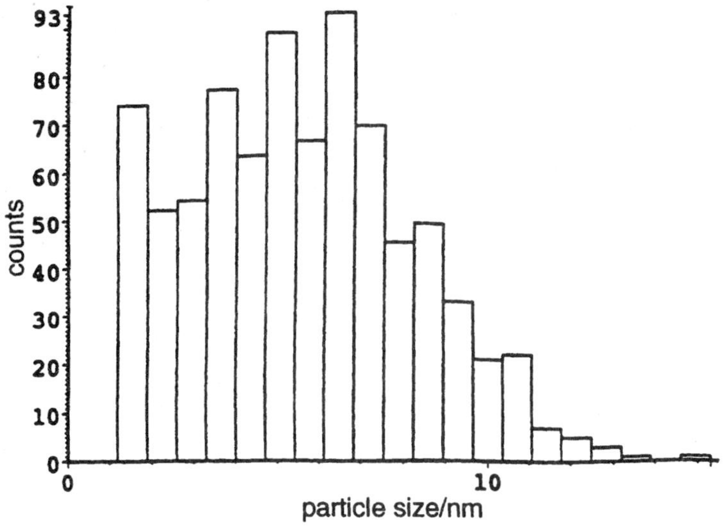

Fig. 32. Particle size distribution. Reprinted with permission from K. E. Gonsalves et al., *J. Mater. Chem.*, 8, 1451 (© 1996 The Royal Society of Chemistry).

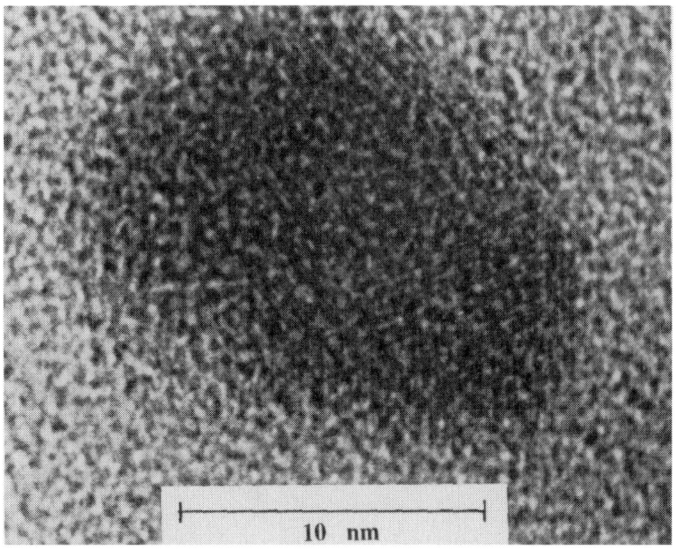

Fig. 33. HRTEM image of nanostructured GaN in PMMA. Reprinted with permission from K. E. Gonsalves et al., *J. Mater. Chem.*, 8, 1451 (© 1996 The Royal Society of Chemistry).

Photoluminescence was measured on a Perkin-Elmer LS 50B fluorescence spectrophotometer. The photoluminescence spectroscopy of GaN nanoparticles with a mean particle size of 5 nm has been carried out at room temperature in the UV excitation range. The excitation and emission features observed have not been reported in either epitaxially grown materials or nanoparticles reported elsewhere. The nanoparticles were enclosed in a quartz container that had been purged with nitrogen. During the measurement of photoluminescence, the sample compartment was continuously purged with nitrogen. The emission spectrum of the GaN nanoparticles was found to be extremely sensitive to the excitation wavelength. Figure 34a shows the excitation spectrum of GaN at room temperature when the emission was monitored at 350 nm. The quartz container has a cutoff around 200 nm

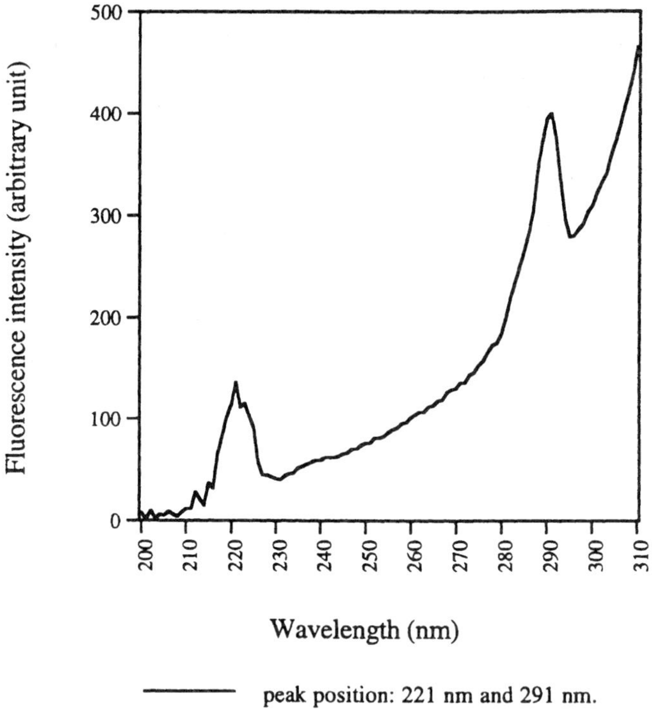

(a)

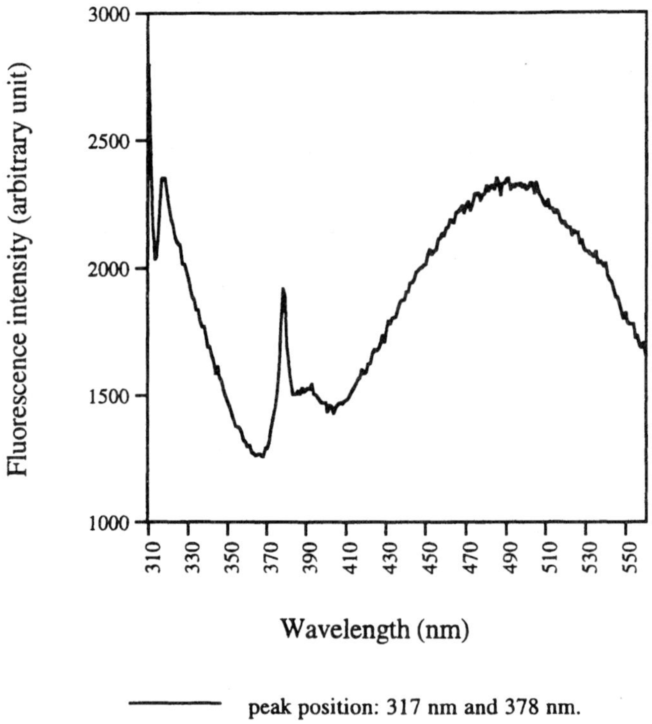

(b)

Fig. 34. (a) PL excitation spectrum of GaN powders (emission at 350 nm). (b) PL emission spectrum of GaN powders (emission at 293 nm).

and, therefore, no photoluminesence was observed at excitation below this wavelength. As the excitation wavelength is scanned to 300 nm, we see considerable structure in the strength of the fluorescence signal, suggesting the existence of certain resonances in the GaN nanoparticle electronic spectrum, possibly due to the quantum confinement effects. For a 5-nm particle size scale, we expect quantum confinement effects to modify the states in the conduction and valence bands considerably. Although the effective masses of electrons and holes in such small nanoparticles are not known, if we accept the bulk values we expect significant shifts to shorter wavelengths in the photoluminesence spectra. In addition, the large surface-to-volume ratio in such small particles may lead to the existence of surface states quite distinct from the states normally encountered in bulk or epitaxially grown GaN [125]. The photoluminesence emission spectrum obtained on excitation at 293 nm is shown in Figure 34b. There are several interesting features in the spectrum. A peak at 378 nm is observed, which is quite close to the 370-nm peak observed in 30-nm GaN particles and epitaxially grown material [122]. The observed 378-nm peak is very sharp, indicating that discrete energy states have taken the place of a continuous energy band. According to our proposed model, this is a transition from a discrete state situated at the bottom of the conduction band of bulk GaN to the discrete state situated at the top of the valence band of bulk GaN. The positions of these two discrete states are independent of the size and shape of the nanoparticle. Therefore, the distribution of the size and shape of the nanoparticle cannot appreciably broaden this peak, although it has broadened other peaks. According to the model, the allowed states are a series of sampling points on the energy band of bulk GaN. The transition between the discrete states situated on the conduction band of bulk GaN and those situated on the valence band should lead to the emission at wavelengths shorter than the band gap (378 nm) and this has been clearly observed by a peak at 317 nm. The excitation peaks at 221 and 291 (Fig. 34a) can be explained as the creation of electron–hole pairs in the discrete states (sampling points) with higher energy on the valence and conduction bands of bulk GaN. They are direct processes; that is, the created electron and hole have the same vector. Owing to the particle size and shape distribution, the emissions from various particle sizes would overlap and broaden the emission compared to a single particle size (except the 378-nm peak). A tighter control on the size can lead to narrower emissions. The emission at longer wavelengths is attributed to defect or impurity states. We also expect surface states to play an important role in the PL of these materials. PL in the visible range in GaN has been observed from epitaxial GaN as well and has been attributed to impurity and defect states [126].

The third-order nonlinear optical properties of nanoparticles in PMMA were studied using degenerate four-wave mixing (DFWM). The backward wave geometry was used, where the signal is a phase-conjugate replica of the probe, counterpropagating in the direction of the probe. The 532-nm, 30-ps pulses from a frequency-doubled, *Q*-switched neodymium-doped yttrium–aluminum–garnet (Nd:YAG) laser (Quantel) were used. The laser operated at a 10-Hz repetition rate. The average pulse energy was 25 mJ. A neutral density filter was placed in the probe beam to reduce its intensity to about 1% that of the pump beams. The phase-conjugate beam was separated from the signal using a beam splitter. The crossing angle was 6°. Fast silicon photodiodes calibrated against a laser energy meter were used to monitor the signal, the probe and the pump pulse energies. By comparison with a CS_2 standard, the $\chi^{(3)}$ was 2.6×10^{-11} esu, as compared to the $\chi^{(3)}$ value reported for gallium nitride films (1.5×10^{-11} esu) [119].

Several kinds of binary and ternary III–V semiconductor nanocrystals, such as GaAs or GaP, have been synthesized using dehalosilylation reactions of group III halides with $P[Si(CH_3)_3]_3$ or $As[Si(CH_3)_3]_3$ or in organic solvents. The byproduct, $(CH_3)_3SiCl$, is removed from the reaction mixture easily, as it is very soluble in commonly used organic solvents and can be isolated from the solid GaAs or GaP product [127].

3.3.7. Semiconductor Nanoparticle Films by Self-Assembly

The term self-assembly implies the spontaneous adsorption of molecules or nanoparticles onto a substrate. Self-assembled multilayer films are formed by the adsorption of subsequent monolayers of molecules or nanoparticles [128]. The evolution of self-assembled layers of molecularly nanostructured materials can be traced, from simple surfactant monolayers and multilayers, through their more complex particulate analogs to the self-assembly of simple molecules and larger particulates.

Self-assembly of nanoparticles to the oppositely charged substrate surface is governed by a delicate balance of the adsorption and desorption equilibria. The optimization of the self-assembly in terms of maximizing the adsorption of nanoparticles from their dispersions and minimizing their desorption on rinsing requires the judicious selection of stabilizers and the careful control of the kinetics of the process. The self-assembly of CdS and PbS nanoparticles was found to be most efficient, for example, if the semiconductor particles were coated by a 1:3 mixture of thiolactic acid and ethyl mercaptan [129]. The generation of size-quantized semiconductor nanoparticles as dispersed organoclay complexes and layered silicates illustrates the use of the nanophase reactors provided by adsorbed binary liquids. The binary liquid pairs, ethanol(1)–cyclohexane(2) and methanol(1)–cyclohexane(2), were selected because the polar component of the liquid mixture (1) preferentially adsorbed at the solid interface and the semiconductor precursors (Cd^{2+} and Zn^{2+}) were highly soluble in the liquid, which preferentially adsorbed at the interface (methanol and ethanol) but was insoluble in the bulk phase (predominantly cyclohexane). These conditions effectively limited the nucleation and growth of the semiconductors to the nanophase reactor provided by the adsorption layer at the solid interface. By varying the mole fraction of the polar liquid (1), it was possible to control the volume of the nanophase reactor and, hence, the size of the semiconductor particles grown therein.

3.3.8. Self-Assembly of Inorganic Nanoparticle Sandwich Films

3.3.8.1. Alternating Layers of Polyelectrolyte–Semiconductor Nanoparticles

The layer-by-layer self-assembly of polyelectrolyte–semiconductor nanoparticles onto substrates is simple. A cleaned substrate is primed by adsorbing a layer of surfactant or polyelectrolyte onto its surface. The primed substrate is then immersed into a dilute aqueous solution of a cationic polyelectrolyte, for adsorption of a monolayer, rinsed, and dried. Then the polyelectrolyte monolayer–covered substrate is immersed into a dilute dispersion of surfactant-coated negatively charged semiconductor nanoparticles, for a time optimized for adsorption of a monoparticulate layer, and subsequently rinsed and dried. These operations complete the self-assembly of a polyelectrolyte monolayer–monoparticulate layer of semiconductor nanoparticle sandwich onto the primed substrate. Subsequent sandwich units are deposited analogously. This method has been used to produce a self-assembly of a poly(diallylmethylammonium chloride), P, CdS nanoparticle onto different substrates (gold, silver, platinum, quartz, and Teflon) (Scheme VI) [129]. Similar methodologies were employed for the self-assembly of many other polyelectrolyte–semiconductor nanoparticle sandwich films.

3.3.8.2. Alternating Layers of Polyelectrolyte–Clay Platelet–Semiconductor Nanoparticles

The high lateral bond strength and aspect ratios have rendered clay organocomplexes such as Na^{+}-montmorillonite to be eminently suitable materials for nanoconstruction. Indeed, the nanocomposites prepared by mixing polymers and clay organocomplexes had superior mechanical properties.

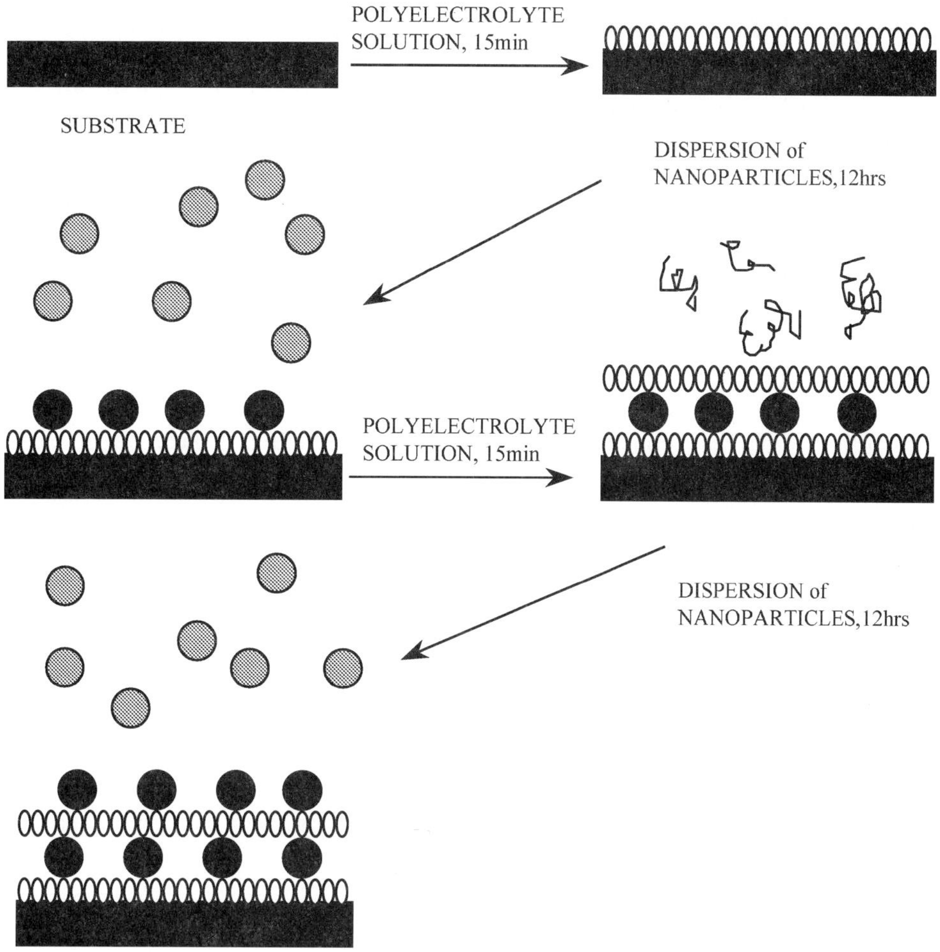

Scheme VI. Schematics of the self-assembly of an S-(P/CdS)$_n$ film. (Adapted from Chem. Mater. 8, 1616 (1996), Fendler et al.)

Alternating layers of polyelectrolyte(P)–clay platelet(M) sandwich films have self-assembled in a manner analogous to that shown for S-(P/CdS)$_n$ (Scheme VI). Significantly, the thickness of a given multilayer S-(P/M)$_n$ film was found to depend on the potential that was applied during the deposition(s) of M. Positive potentials increased the film thickness, whereas negative ones decreased it. A similar behavior has been noted in the electrophoretic deposition of metal particles onto conductive substrates.

3.3.8.3. Potential Applications of Self-Assembled Monolayers

Self-assembly has great advantages. This method is useful for the synthesis of composite films comprising different nanoparticles that can be layered in any desired order. Self-assembled films can function as membranes (barriers), with controllable levels of permeability, for gases, liquids, covalent molecules, ions, and, indeed, electrons (tunneling barriers). These properties have been exploited for the construction of insulators, passivators, sensors, and modified electrodes. Devices based on molecular recognition are candidates for synthesis by self-assembly. Supramolecular assemblies (nanoparticles) within a self-assembled layer can be aligned spontaneously or by changing the temperature, pressure, or pH, or by application of electric and magnetic fields. These properties permit the formation of superlattices with the desired symmetry, as well as the formation of a variety of photonic, electronic, magnetic, and nonlinear optical devices. Control of the sizes and agglomeration of the monodispersed nanoparticle within the self-assembled film can be exploited for construction of optical devices.

4. CONCLUSIONS

In this chapter, we have outlined several common and unique techniques for the synthesis of nanostructured metals, intermetallics, and semiconductors in powder and thin-film form. Considerable opportunities now exist for synthesizing nanophase materials with a variety of new architectures at nanometer length scales from atomic or molecular precursors via the assembly of atom clusters and several other chemical preparation methods. Solution chemistry can directly produce either the desired particles or the precursors that are further treated by various reaction methods to obtain the final products. It should be pointed out that detailed work on the characterization of these materials and their property evaluation is required. Concerns about the yield and the impurity content need to be critically addressed before a particular synthetic and processing method can be adopted for large-scale production. Although the problem of agglomeration of fine particles in solution can potentially be solved by controlling the interparticle forces, the subsequent processing must be designed carefully so that agglomeration can be minimized once the particles are removed from solution. The handling of fine particles after they are synthesized remains a serious problem. This problem is particularly acute for air-sensitive materials. The interesting and fast-developing area of membrane-mediated research, which addresses the stabilization and size control of nanoscale particles, is a potential practical approach to synthesize and process these materials.

The real future of nanophase materials, and other nanostructures as well, will depend on our ability to change significantly for the better, the properties of materials by structuring them artificially on nanometer scales and on developing economic and environmentally responsible methods for producing these materials in commercially viable quantities. Chemical synthesis of nanoparticles is a rapidly growing research area with a great potential to make technologically advanced and useful materials. The realization of this potential will require multidisciplinary interactions and collaborations between biologists, chemists, materials scientists, and engineers, in order to control and improve the properties of nanophase materials.

References

1. (a) H. Gleiter, *Adv. Mater.* 4, 474 (1992). (b) R. Birringer and H. Gleiter, in: "Encyclopedia of Materials Science and Engineering" (R. W. Cahn, ed.), Suppl. Vol. 1, p. 339. Pergamon Press, Oxford, U.K., 1988.
2. R. W. Seigel, *Nanostruct. Mater.* 3, 1 (1993).
3. R. W. Seigel, *Mater. Sci. Eng. B* 19, 37 (1993).
4. (a) R. W. Cahn, *Nature* 348, 389 (1990). (b) R. Dagani, *Chem. Eng. News.* 72, 18 (1992). (c) H. Gleiter, *Nanostruct. Mater.* 6, 3 (1995). (d) D. Chakravorty and A. K. Giri, "Chemistry for the 21st Century: Chemistry of Advanced Materials" (C. N. R. Rao, ed.), p. 217. Blackwell Scientific, London, 1993.
5. (a) Y. Wang and N. Herron, *J. Phys. Chem.* 95, 525 (1991). (b) J. P. Kuczynski, *J. Am. Chem. Soc.* 108, 2513 (1986). (c) E. E. Mendez and K. von Klitzing, *NATO Adv. Study Inst., Ser. B* 170, 1 (1989).
6. (a) Y. Yoshizawa, S. Oguma, and Y. Yamaguchi, *J. Appl. Phys.* 64, 6044 (1988). (b) H. J. de Witt, C. H. M. Wittmer, and F. W. A. Dirne, *Adv. Mater.* 3, 356 (1991).
7. (a) D. D. Beck and R. W. Seigel, *J. Mater. Res.* 7, 2840 (1992). (b) E. Boakye, *J. Colloid Interface Sci.* 163, 120 (1994).
8. (a) M. L. Steigerwald and L. E. Brus, *Acc. Chem. Res.* 23, 183 (1990). (b) A. I. Ekimov and A. A. Onushchenko, *Zh. Eskp. Teor. Fiz.* 40, 337 (1984). [English Translation, *JETP Lett.* 40, 1136 (1984)].
9. A. Henglein, *Top. Curr. Chem.* 143, 113 (1988).
10. (a) R. W. Seigel, *MRS Bull.* 15, 60 (1990). (b) H. Gleiter, *Prog. Mater. Sci.* 33, 223 (1990). (c) R. Herr, U. Birringer, and H. Gleiter, *Trans. Jpn. Inst. Met. Suppl.* 27, 43 (1986).

11. (a) B. W. Dodson, L. J. Schowalter, J. E. Cunningham, and F. E. Pollak, eds., *Mater. Res. Soc. Symp. Proc.* 160 (1989). (b) T. M. Bessman and B. M. Bessman, eds., *Mater. Res. Soc. Symp. Proc.* 168 (1990).
12. (a) L. M. Goldman, B. Blanpain, and F. Spaepen, *J. Appl. Phys.* 60, 1374 (1986). (b) D. S. Lashmore, R. Oberle, M. P. Dariel, L. H. Bennett, and L. Swartzendruber, *Mater. Res. Soc. Symp. Proc.* 132, 219 (1989).
13. (a) C. C. Koch, *Nanostruct. Mater.* 2, 109 (1993). (b) Materzzi et al., *Nanostruct. Mater.* 2, 217 (1993).
14. (a) G. Wasserman, in: "Proceedings of the 4th International Conference on Strength of Metals and Alloys," 1976, Vol. 3, p. 1343. (b) J. D. Embury, in: "Strengthening Methods in Crystals" (A. Kelly and R. B. Nicholson, eds.), p. 331. Applied Science Publishing, London, 1971.
15. (a) C. R. Aita, *Nanostruct. Mater.* 4, 257 (1994). (b) K. J. Balkus, Jr. et al., J. *Mater. Res. Soc. Symp. Proc.* 351, 437 (1994). (c) T. Kameyama et al., *J. Mater. Sci.* 25, 1058 (1994).
16. (a) J. Rivas et al., *J. Magn. Magn. Mater.* 122 (1993). (b) K. E. Gonsalves and T. D. Xiao, "Chemical Processing of Ceramics" (B. I. Lee and E. J. A. Pope, eds.), p. 359. Marcel Dekker, New York, 1994. (c) W. Chang et al., *Nanostruct. Mater.* 4, 345 (1994). (d) J. Phalippou, "Chemical Processing of Ceramics" (B. I. Lee and E. J. A. Pope, eds.), p. 265. Marcel Dekker, New York, 1994. (e) C. J. Brinker and J. Schener, "Sol–Gel Science, The Physics and Chemistry of Sol–Gel Processing." Academic Press, Boston, 1990.
17. R. W. Seigel, "Materials Science and Technology" (R. Cahn, ed.), Vol. 15, p. 583. VCH Publishers, Weinheim, Germany, 1991.
18. (a) S. K. Ganapathi and D. A. Rigney, *Scr. Metall. Mater.* 24, 1675 (1990). (b) H. J. Fecht, E. Hellstern, Z. Fu, and W. L. Johnson, *Adv. Powder Metall.* 1–2, 111 (1989). (c) E. Hellstern, H. J. Fecht, Z. Fu, and W. L. Johnson, *J. Appl. Phys.* 65, 305 (1989). (d) C. C. Kock, J. S. C. Jang, and S. S. Gross, *J. Mater. Res.* 4, 557 (1989).
19. P. A. Psaras and H. D. Langford, eds., "Advancing Materials Research, U.S. National Academy of Engineering and National Academy of Sciences," p. 203. National Academy Press, Washington, DC, 1987.
20. C. N. R. Rao, *Mater. Sci. Eng. B* 18, 1 (1993).
21. (a) V. K. LaMer and R. H. Dinegar, *J. Am. Chem. Soc.* 72, 4847 (1950). (b) J. T. G. Overbeek, *Adv. Colloid Interface Sci.* 15, 251 (1982).
22. F. Neilsen, *Manufacturing Chemist* 53, 38 (1982).
23. M. W. Real, *Proc. Br. Ceram. Soc.* 38, 59 (1986).
24. (a) W. Chang et al., *Nanostruct. Mater.* 4, 345 (1994). (b) H. Hahn and R. S. Averback, *J. Appl. Phys.* 67, 1113 (1990).
25. (a) K. E. Gonsalves et al., *J. Mater. Sci.* 27, 3231 (1992). (b) T. D. Xiao, et al., *J. Mater. Sci.* 28, 1334 (1993).
26. N. Herron and Y. Wang, in: "Nanomaterials: Synthesis, Properties and Applications" (A. S. Edelstein, ed.), p. 73. IOP Publishing, Bristol, U.K., 1996.
27. (a) N. Ichinose et al., "Superfine Particle Technology." Springer-Verlag, London, 1992. (b) Q. Xu and M. A. Anderson, *J. Am. Ceram. Soc.* 77, 1939 (1994). (c) K. E. Gonsalves et al., *Adv. Mater.* 6, 291 (1994). (d) B. H. Kear and L. F. McCandlish, *J. Adv. Mater.* 10, 11 (1993).
28. T. W. Smith and D. Wychick, *J. Phys. Chem.* 84, 1621 (1980).
29. (a) W. Heiber and R. Werner, *Chem. Ber.* 90, 286 (1957). (b) W. Heiber and N. Kahlen, *Chem. Ber.* 91, 2234 (1958).
30. (a) J. Dewar and H. O. Jones, *Proc. R. Soc. London, Ser. A* 76, 564 (1905). (b) H. G. Cutforth and P. W. Selwood, *J. Am. Chem. Soc.* 65, 2414 (1905).
31. (a) T. H. Whitesides and J. P. Neilan, *J. Am. Chem. Soc.* 95, 5811 (1973). (b) H. W. Whitlock, Jr. and Y. N. Chuah, *J. Am. Chem. Soc.* 87, 3605 (1965).
32. N. Smith, *J. Am. Chem. Soc.* 58, 173 (1936). (b) J. Kruger and J. P. Calvert, *J. Electrochem. Soc.* 114, 43 (1967).
33. I. Nakatani, M. Hijikata, and K. Ozawa, *J. Magn. Magn. Mater*. 122, 10 (1993).
34. J. Wonterghem, S. Morup, S. W. Charles, S. Wells, and J. Villadsen, *Phys. Rev. Lett.* 55, 410 (1985).
35. S. W. Charles and J. Popplewell, "Ferromagnetic Materials" (E. P. Wohlfarth, ed.), Vol. 2, p. 509. North-Holland, Amsterdam, 1980.
36. (a) B. S. Clausen, S. Morup, and H. Topsoe, *Surf. Sci.* 106, 438 (1981). (b) S. Morup, H. Topsoe, and B. S. Clausen, *Phys. Scr.* 25, 713 (1982).
37. (a) G. Le Caer, J. M. Dubois, M. Pijolat, V. Perrichon, and P. Bussiere, *J. Phys. Chem.* 86, 4799 (1982). (b) C. B. Ma, T. Ando, D. L. Williamson, and G. Krauss, *Metall. Trans. A* 14, 1033 (1983).
38. X. Cao, Y. Koltypin, G. Kataby, R. Prozorov, and A. Gedanken, *J. Mater. Res.* 10, 2952 (1995).
39. K. E. Gonsalves, S. P. Rangarajan, A. Garcia-Ruiz, and C. C. Law, *J. Mater. Sci. Lett.* 15, 1261 (1996).
40. J. Schneider, *Acta Crystallogr. Sect. A* 43, 295 (1987).
41. (a) D. M. McMillen, K. E. Lewis, G. P. Smith, and D. M. Golden, *J. Phys. Chem.* 8, 709 (1982). (b) W. M. Shaub and S. H. Bauer, *J. Chem. Kinet.* 7, 509 (1975).
42. J. R. Thomas, *Chem. Commun.* 2914 (1965).
43. F. E. Luborsky and T. O. Paine, *J. Appl. Phys.* 31, 68S (1960).
44. D. Zeng and M. J. Hampden-Smith, *Chem. Mater.* 5, 681 (1993).
45. H. Bönnemann, W. Brijoux, and T. Jousson, *Angew. Chem. Int. Ed. Engl.* 29, 273 (1990).

46. J. P. Collman, L. S. Hegedus, J. R. Norton, and R. G. Finke, "Principles and Applications of Organotransition Metal Chemistry." University Science Books, Berkeley, CA, 1987.
47. (a) W. E. Buhro, J. A. Haber, B. E. Waller, and T. J. Trentler, *Am. Chem. Soc. Symp. Ser.* 210, 20 (1995). (b) J. A. Haber, J. L. Crane, W. E. Buhro, C. A. Frey, S. M. L. Sastry, J. L. Balbach, and M. S. Conradi, *Adv. Mater.* 8, 163 (1996).
48. H. Bönnemann, W. Brijoux, R. Brinkmann, R. Fretzen, T. Joussen, R. Köpler, B. Korall, P. Neiteler, and J. Richter, *J. Mol. Catal.* 86, 129 (1994).
49. (a) A. Corrias, G. Ennas, G. Licheri, G. Marongin, and G. Paschina, *Chem. Mater.* 2, 363 (1990). (b) H. C. Brown and C. A. Brown, *J. Am. Chem. Soc.* 84, 1492 (1962).
50. H. Bönnemann, W. Brijoux, and T. Joussen, DE OS 3934351, Studiengesellschaft Kohle mbH, 1991.
51. G. Schmid, in: "Aspects of Homogeneous Catalysis" (R. Ugo, ed.), Vol. 7, p. 1. Kluwer Academic Publishers, Dordrecht, The Netherlands, 1990.
52. A. Duteil, G. Schmid, and W. Meyer-Zaika, *J. Chem. Soc., Chem. Commun.* 31 (1995).
53. (a) F. Fievet, J. P. Lagier, and M. Figlarz, *MRS Bull.* 14, 29 (1989). (b) M. Figlarz, F. Fievet, and J. P. Lagier, Eur. Patent, 0113281; U.S. Patent, 4,539,041.
54. (a) F. Fievet, F. Fievet-Vincent, J. P. Lagier, B. Dumont, and M. Figlarz, *J. Mater. Chem.* 3, 627 (1993). (b) F. Fievet, J. P. Lagier, B. Blin, B. Beaudoin, and M. Figlarz, *Solid State Ionics* 32/33, 198 (1989).
55. R. D. Nelson, Jr., "Dispersing Powders in Liquids," pp. 211. Elsevier, Amsterdam, 1988.
56. (a) K. L. Tsai and J. L. Dye, *J. Am. Chem. Soc.* 113, 1650 (1991). (b) K. L. Tsai and J. L. Dye, *Chem. Mater.* 5, 540 (1993).
57. (a) J. L. Dye, M. T. Lok, F. J. Tehan, R. B. Coolen, N. Papadakis, J. M. Ceraso, and M. G. DeBacker, *Ber. Bunsen-Ges. Phys. Chem.* 75, 3092 (1971). (b) J. L. Dye, M. G. DeBacker, and V. A. Nicely, *J. Am. Chem. Soc.* 92, 5226 (1970).
58. IUPAC name: 15-crown-5; 1,4,7,10,13-pentaoxacyclopentadecane.
59. J. L. Dye and K. L. Tsai, *Faraday Discuss. Chem. Soc.* 92, 45 (1991).
60. (a) J. S. Bradley, J. M. Millar, and E. W. Hill, *J. Am. Chem. Soc.* 113, 4016 (1991). (b) H. Hirai, Y. Nakao, and N. Toshima, *J. Macromol. Sci. Chem.* 12, 1117 (1978).
61. H. Thiele and H. S. von Levern, *J. Colloid Sci.* 3, 363 (1965).
62. J. H. Ross, *Anal. Chem.* 25, 1288 (1953).
63. A. M. Bradshaw and F. M. Hoffmann, *Surf. Sci.* 72, 513 (1978).
64. (a) M. J. Tracy and J. R. Groza, *Nanostruct. Mater.* 1, 369 (1992). (b) K. Higashi, T. Mukai, S. Tanimura, A. Inoue, K. Masumoto, K. Kita, K. Ohtera, and J. Nagahora, *Nanostruct. Mater.* 26, 191 (1992).
65. (a) K. E. Gonsalves, T. D. Xiao, G. M. Chow, and C. C. Law, *Nanostruct. Mater.* 4, 139 (1994). (b) K. E. Gonsalves, S. P. Rangarajan, C. C. Law, C. R. Feng, G.-M. Chow, and A. Garcia-Ruiz, *Am. Chem. Soc. Symp. Ser.* 622, 15. 220 (1996). (c) C. R. Feng, G. M. Chow, S. P. Rangarajan, X. Chen, K. E. Gonsalves, and C. C. Law, *Nanostruct. Mater.* 8, 45 (1997). (d) K. E. Gonsalves and S. P. Rangarajan, in: "Chemistry and Physics of Nanostructures and Related Non-Equilibrium Materials" (E. Ma, B. Fultz, R. Shull, J. Morral, and P. Nash, eds.), p. 149. TMS Meeting, Orlando, FL, 1997. (e) K. E. Gonsalves and S. P. Rangarajan, *J. Appl. Polym. Sci.* 64, 2667 (1997). (f) K. E. Gonsalves, U.S. Patent 389,778, 1996.
66. G. M. Chow, T. Ambrose, J. Xiao, F. Kaatz, and A. Ervin, *Nanostruct. Mater.* 2, 131 (1993).
67. G. Schmid, E. Schöps, J.-O. Malm, and J.-O. Bovin, *Z. Anorg. Allg. Chem.* 620, 1170 (1994).
68. (a) B. H. Kear and L. E. McCandlish, *Nanostruct. Mater.* 3, 19 (1993). (b) L. E. McCandlish and R. S. Polizzotti, *Solid State Ionics* 32/33, 795 (1989).
69. C. H. Johansson and J. O. Linde, *Ann. Phys.* 25, 1 (1936).
70. G. M. Chow, L. K. Kurihara, K. M. Kemner, P. E. Schoen, W. T. Elam, A. Ervin, S. Keller, Y. D. Zhang, J. Budnick, and T. Ambrose, *J. Mater. Res.* 10, 1546 (1995).
71. C. Meny, P. Panissod, and R. Loloee, *Phys. Rev. B* 45, 12269 (1992).
72. J. R. Childress, C. L. Chien, and M. Nathan, *J. Appl. Phys.* 70, 5885 (1991).
73. J. S. Bradley, E. W. Hill, C. Klein, B. Chaudret, and A. Duteil, *Chem. Mater.* 5, 254 (1993).
74. (a) H. A. C. M. Hendricks and V. Ponec, *Surf. Sci.* 192, 234 (1987). (b) J. S. Bradley, J. M. Millar, E. W. Hill, S. Behal, B. Chaudret, and A. Duteil, *Faraday Discuss. Chem. Soc.* 92, 255 (1991).
75. (a) F. Kayser and M. Cohen, *Metal. Progr.* 61, 79 (1952). (b) W. B. Pearson, "A Handbook of Lattice Spacing and Structures of Metals and Alloys," Vol. 1. Pergamon Press, Elmsford, NY, 1958.
76. (a) Al. L. Efros and A. L. Efros, *Fiz. Tekh. Poluprovodn.* 16, 1209 (1982) [English Translation, *Sov. Phys.-Semicond.* 16, 772 (1982)]. (b) M. L. Steigerwald and L. E. Brus, *Acc. Chem. Res.* 23, 183 (1990). (c) A. Henglein, *Top. Curr. Chem.* 143, 113 (1988). (d) Y. Wang, N. Herron, W. Mahler, and A. Suna, *J. Opt. Soc. Am. B* 6, 808 (1989).
77. N. Herron, J. C. Calabrese, W. E. Farneth, and Y. Wang, *Science* 259, 1426 (1993).
78. D. W. Bahnemann, C. Kormann, and M. R. Hoffmann, *J. Phys. Chem.* 91, 3789 (1987).
79. (a) L. Spanhel, M. Haase, H. Weller, and A. Henglein, *J. Am. Chem. Soc.* 109, 5649 (1987). (b) U. Resch, H. Weller, and A. Henglein, *Langmuir* 5, 1015 (1989).
80. N. Herron, Y. Wang, and H. Eckert, *J. Am. Chem. Soc.* 112, 1322 (1990).
81. C. T. Dameron, R. N. Reese, R. K. Mehra, A. R. Kortan, P. J. Carroll, M. L. Steigerwald, L. E. Brus, and D. R. Winge, *Nature* 338, 596 (1989).

82. H. Uchida, C. Curtis, P. V. Kamat, K. M. Jones, and A. J. Nozik, *J. Phys. Chem.* 96, 1156 (1992).
83. (a) Y. M. Tricot, A. Emeren, and J. H. Fendler, *J. Phys. Chem.* 89, 4721 (1985). (b) H.-C. Youn, S. Baral, and J. H. Fendler, *J. Phys. Chem.* 92, 6320 (1988).
84. (a) M. Meyer, C. Wallberg, K. Kurihara, and J. H. Fendler, *J. Chem. Soc. Chem. Commun.* 90 (1984). (b) C. Petit, P. Lixon, and M. P. Pileni, *J. Phys. Chem.* 94, 1598 (1990).
85. (a) Y. Wang, A. Suna, W. Mahler, and R. Kasowski, *J. Chem. Phys.* 87, 7315 (1987). (b) W. Mahler, *Inorg. Chem.* 27, 435 (1988). (c) E. Hilinski, P. Lucas, and Y. Wang, *J. Chem. Phys.* 89, 3435 (1988).
86. C. H. Fischer, H. Weler, L. Katsikas, and A. Henglein, *Langmuir* 5, 429 (1989).
87. (a) T. Goto, S. Saito, and M. Tanaka, *Solid State Commun.* 80, 331 (1991). (b) R. F. Ziolo, E. P. Giannelis, B. A. Weinstein, M. P. O'Horo, B. N. Ganguly, V. Mehrotra, M. W. Russell, and D. R. Huffman, *Science* 257, 219 (1992).
88. V. Sankaran, C. C. Cummins, R. R. Schrock, R. E. Cohen, and R. J. Silbey, *J. Am. Chem. Soc.* 112, 6858 (1990).
89. Y. Wang, N. Heron, M. Harmer, and A. Suna, "Proceedings of the MRS Spring Meeting," Pittsburgh, PA, 1992.
90. H. Yao, S. Takahara, H. Mizuma, T. Kozeki, and T. Hayashi, *Jpn. J. Appl. Phys.* 35, 4633 (1996).
91. A. Chevreau, B. Phillips, B. G. Higgins, and S. H. Risbud, *J. Mater. Chem.* 6, 1643 (1999).
92. (a) U. Koch, H. Fojtik, H. Weller, and A. Henglein, *Chem. Phys. Lett.* 122, 507 (1985). (b) P. Hoyer, R. Eichberger, and H. Weller, *Ber. Bunsen-Ges Phys. Chem.* 97, 630 (1993).
93. S. Mahamuni, B. S. Bendre, V. J. Leppert, C. A. Smith, D. Cooke, S. H. Risbud, and H. W. H. Lee, *Nanostruct. Mater.* 6, 659 (1996).
94. L. Spanhel and M. Anderson, *J. Am. Chem. Soc.* 113, 2826 (1990).
95. N. F. Borelli, D. W. Hall, H. J. Holland, and D. W. Smith, *J. Appl. Phys.* 61, 5399 (1987).
96. (a) B. G. Potter and J. H. Simmons, *Phys. Rev.* 37, 10838 (1988). (b) T. Arai, H. Fujumura, I. Umezu, T. Ogawa, and A. Fujii, *Jpn. J. Appl. Phys.* 28, 484 (1989).
97. T. Rajh, M. I. Vucemilovic, N. M. Dimitrijevic, and O. I. Micic, *Chem. Phys. Lett.* 143, 305 (1988).
98. H. Minti, M. Eyal, and R. Reisfeld, *Chem. Phys. Lett.* 183, 277 (1991).
99. H. Schmidt, "Proceedings of the Winter School on Glasses and Ceramics from Gels, Sol–Gel Science and Technology, Brazil" (M. A. Aegerter, M. Jafelicci, Jr., D. F. Souza, and E. D. Zanotto, eds.), p. 432. World Scientific, Singapore, 1989.
100. B. Buchalter and G. R. Meredith, *Appl. Opt.* 21, 3221 (1982).
101. (a) D. J. Scoberg, F. Greiser, and D. N. Furlong, *J. Chem. Soc., Chem. Commun.* 516 (1991). (b) J. C. Luong and N. F. Borelli, *Mater. Res. Soc. Symp. Proc.* 144, 695 (1989).
102. (a) Y. Wang and N. Herron, *Res. Chem. Intermed.* 15, 17 (1991). (b) R. Roy, S. Roy, and D. M. Komareni, *Mater. Res. Soc. Symp. Proc.* 32, 347 (1984). (c) Y. Kobayashi, S. Yamazaki, Y. Kurokawa, T. Miyakawa, and H. Kawaguchi, *J. Mater. Sci. Mater. Electron.* 2, 20 (1992).
103. (a) D. G. Park and J. M. Burlitch, *Chem. Mater.* 4, 500 (1992). (b) I. Tanahashi, A. Tsujimura, T. Mitsuyu, and A. Nishino, *Jpn. J. Appl. Phys.* 29, 2111 (1990). (c) M. Fujii, S. Hayashi, and K. Yamamoto, *Appl. Phys. Lett.* 57, 2962 (1990).
104. T. Itoh, Y. Iwabuchi, and M. Kataoka, *Phys. Status Solidi B* 145, 567 (1988).
105. (a) V. N. Bogomolov, V. V. Poborchii, and S. V. Kholodkevich, *JETP Lett.* 31, 434 (1980). (b) Y. Katayama, M. Yao, and Y. Ajiro, *J. Phys. Soc. Jpn.* 58, 1811 (1989). (c) J. B. Parise, J. E. MacDougall, N. Herron, R. Farlee, A. W. Sleight, Y. Wang, T. Bein, K. Moller, and L. M. Moroney, *Inorg. Chem.* 27, 210 (1988).
106. N. Herron, *Inorg. Chem.* 25, 4714 (1986).
107. (a) R. D. Stramel, T. Nakamura, and J. K. Thomas, *J. Chem. Soc., Faraday Trans. 1*, 84, 1287 (1988). (b) D. W. Breck, "Zeolite Molecular Sieves." Wiley, New York, 1974.
108. N. Herron, Y. Wang, M. Eddy, G. D. Stucky, D. E. Cox, T. Bein, and K. Moller, *J. Am. Chem. Soc.* 111, 530 (1989).
109. (a) T. Hirono, A. Kawana, and T. Yamada, *J. Appl. Phys.* 62, 1984 (1987). (b) Y. Nozue, Z. K. Tang, and T. Goto, *Solid State Commun.* 73, 31 (1990).
110. (a) G. A. Ozin, J. P. Godber, and A. Stein, U.S. Patent, August 1988. (b) A. Stein, Thesis, University of Toronto, 1988.
111. G. Cao, L. K. Rabenberg, C. M. Nunn, and T. E. Mallouk, *Chem. Mater.* 3, 149 (1991).
112. X. Liu and J. K. Thomas, *Langmuir* 5, 58 (1989).
113. V. Swayambunathan, D. Hayes, K. H. Schmidt, Y. X. Liao, and D. Meisel, *J. Am. Chem. Soc.* 112, 3831 (1990).
114. I. G. Dance, A. Choy, and M. L. Scudder, *J. Am. Chem. Soc.* 106, 6285 (1984).
115. G. S. H. Lee, D. C. Craig, I. Ma, M. L. Scudder, T. D. Bailey, and I. G. Dance, *J. Am. Chem. Soc.* 110, 4863 (1988).
116. W. E. Farneth, N. Herron, and Y. Wang, *Chem. Mater.* 4, 917 (1992).
117. (a) R. E. Treece, G. S. Macala, and R. B. Kaner, *Chem. Mater.* 4, 9 (1992). (b) B. K. Laurich, D. C. Smith, and M. D. Healy, *Mater. Res. Soc. Symp. Proc.* 351, 49 (1994). (c) R. L. Wells, S. R. Aubuchon, S. S. Kher, and M. S. Lube, *Chem. Mater.* 7, 793 (1995). (b) H. S. Nalwa and S. Miyata, eds., "Nonlinear Optics of Organic Molecules and Polymers." CRC Press, Boca Raton, FL, 1997.

118. (a) H. S. Nalwa and S. Miyata, eds., "Nonlinear Optics of Organic Molecules and Polymers," CRC Press, Boca Raton, FL, 1997. (b) J. W. Goodman, P. Chavel, and G. Roblin, eds., "Optical Computing," Vol. 963. SPIE, Washington, DC, 1989.
119. D. K. Wickenden, T. J. Kistenmacher, and J. Miragliotta, *J. Electron. Mater.* 23, 1209 (1994).
120. J. W. Hwang, J. P. Campbell, J. Kozubowski, S. A. Hanson, J. F. Evans, and W. L. Gladfelter, *Chem. Mater.* 7, 517 (1995).
121. J. F. Janik and R. L. Wells, *Chem. Mater.* 8, 2708 (1996).
122. Y. Xie, Y. Qian, W. Wang, S. Zhang, and Y. Zhang, *Science* 272, 1926 (1996).
123. H. Nöth and P. Konrad, *Z. Naturforsch. B: Chem. Sci.* 30, 681 (1975).
124. J. E. Turner, M. Hendewerk, J. Parmeter, D. Neiman, and G. A. Somorjai, *J. Electrochem. Soc.* 131, 1777 (1984).
125. H. P. Maruska and J. J. Tietjen, *Appl. Phys. Lett.* 15, 327 (1969).
126. S. Strite, *J. Vac. Sci. Technol. B* 9, 1924 (1991).
127. L. I. Halaoui, S. S. Kher, M. S. Lube, S. R. Aubuchon, C. R. S. Hagan, R. L. Wells, and L. A. Coury, *Am. Chem. Soc. Symp. Ser.* 622, 178 (1996).
128. J. H. Fendler, *Chem. Mater.* 8, 1616 (1996).
129. N. A. Kotov, I. Dekany, and J. H. Fendler, *J. Phys. Chem.* 99, 13065 (1995).

Chapter 2

NANOCOMPOSITES PREPARED BY SOL–GEL METHODS: SYNTHESIS AND CHARACTERIZATION

Krzysztof C. Kwiatkowski, Charles M. Lukehart

Department of Chemistry, Vanderbilt University, Nashville, Tennessee, USA

Contents

1. INTRODUCTION

In typical sol–gel syntheses, metal or main-group element compounds undergo hydrolysis and condensation reactions giving gel materials with extended three-dimensional structures [1–10]. As shown in the following equation for silicon, the addition of an acid or base catalyst to a solution of an alkoxysilane reagent, such as tetramethoxysilane (TMOS), water,

Handbook of Nanostructured Materials and Nanotechnology, edited by H.S. Nalwa
Volume 1: Synthesis and Processing

and methanol leads to hydrolysis of the Si–OMe bonds to form Si–OH functional groups:

$$\underset{\text{TMOS}}{Si(OMe)_4} + 4H_2O \xrightarrow[\text{catal.}]{\text{MeOH}} \text{“}Si(OH)_4\text{”} + 4MeOH \quad \xrightarrow{-2H_2O} \quad SiO_2 \text{ gel} \qquad (1)$$

Subsequent elimination of water from two such Si–OH groups gives eventually an extended silica gel matrix (known as a xerogel when dry). Because hydrolysis and condensation reactions occur concurrently, monomeric silanols proceed to xerogel through oligomeric and polymeric intermediates. As these reactions progress, the viscosity of sol–gel solutions increases and can reach a "spinable" stage at which point thin films or fibers can be produced. Otherwise, xerogel products are obtained as porous powders or as monoliths that assume the shape of their container. If desired, residual Si–OH groups remaining in the xerogel product can be removed at elevated temperature to give fully densified SiO_2.

A wide variety of elements exhibit sol–gel chemistry. Alkoxy, or related, compounds of any one such element undergo sol–gel conversions to produce a gel containing that element, as shown for the silica gel synthesis in Eq. (1). However, mixtures of precursor compounds for two or more elements can undergo heterocondensation to form mixed-element gels. Similarly, mixtures of different precursor compounds, such like a TMOS, and other reactants containing reactive groups, such as $Si(OR)_3$ substituents, likewise undergo sol–gel conversion to give gels of complex composition.

Sol–gel methods are commonly used to prepare nanocomposite materials because these conversions occur readily with a wide variety of precursors and can be conducted at or near room temperature. In addition, the gel products frequently have properties ideal for desired applications. Porous ceramic xerogels of high surface area can serve as supports for chemical catalysts [11–13], and thin-film deposition is useful for materials possessing desired optical or magnetic properties [14].

Reports appearing since January 1990 on the synthesis and characterization of nanocomposite materials prepared by sol–gel methods are reviewed. Nanocomposite materials are defined as materials consisting of a host matrix along with an identifiable, discrete nanoparticulate guest substance having a maximum dimension of less than 1000 nm. Either the host matrix, the guest substance, or both must be derived from sol–gel methods. Discussion of the nonlinear optical and magnetic properties of such nanocomposite materials is provided elsewhere in this book.

Important sol–gel materials not explicitly covered in this review include organic/inorganic hybrid gels lacking a nanoparticulate guest phase [15–21]; gels containing entrapped polymers [22–32]; organic [33–39] or inorganic [40, 41] molecules as gel dopants; gels doped with metal ions [42–46]; pure noncomposite materials, such as thin films or powders, prepared by sol–gel processing [47–68]; or mixed metal oxides lacking nanoparticulate phases [69–109].

2. NANOCOMPOSITES CONTAINING ELEMENTAL NANOPARTICULATES

Elemental nanoclusters are the least complex nanoparticulates by composition. Available review articles for nanocomposites of this type include (1) nano-Co/Mo, Cu, Fe, Ni, Pd, Pt, or Ru particles in Al_2O_3, SiO_2, TiO_2, or ZrO_2 gels [110]; (2) nano-C, Cu/Ni, Pd/Ni, or Pt in silica gel [111]; (3) nano-Ag, Ge, or Os in silica xerogel [112]; (4) nano-Ag, Cu, Fe, Mo, Os, Pd, Pt, Re, Ru, or PtSn in silica xerogel [113]; and (5) nano-Ge or Si composites [114].

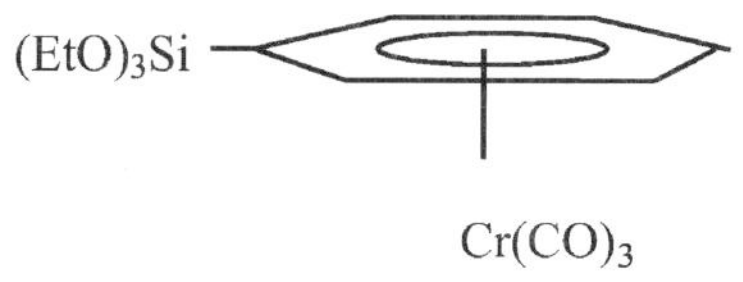

Structure I

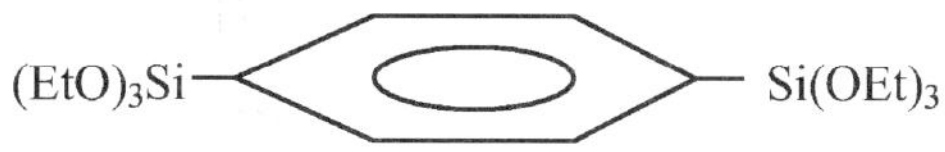

Structure II

2.1. Group VI Metal Nanocomposites

A 2:98 molar mixture of (alkoxy)silanes (Structures I and II) undergoes sol–gel conversion in aqueous tetrahydrofuran (THF) solution in the presence of an acid catalyst to give a poly(1,4-phenylene)-bridged silsesquioxane condensation product. The host matrix material is an organic/inorganic hybrid of typical sol–gel ceramic matrices in that bifunctional phenylene groups serve as spacers between extended silicate networks. The compound shown in Structure I is an η^6-arene complex of chromium. This complex serves as both an internal dopant and a precursor source of chromium. When the condensation product is thermally treated at 120 °C under high vacuum, decomposition of the $Cr(CO)_3$ complex results in the formation of irregularly shaped nanoclusters of cubic Cr metal approximately 10 nm in diameter [115, 116]. When using a low-valent Cr complex as a precursor, Cr nanoparticle formation occurs directly upon precursor decomposition without proceeding through a metal oxide intermediate.

2.2. Group VIII Metal Nanocomposites

Iron/silica xerogel nanocomposites are formed by dissolving Fe(III) nitrate or chloride salts in a conventional silica sol–gel formulation using tetraethylorthosilicate (TEOS) as a source of silica xerogel [117–125]. Ferric ions within the gel are reduced to metallic iron upon thermal treatment under hydrogen gas at 450–700 °C for up to 4 h. Iron particles with average diameters of 3–30 nm are produced depending on the weight percentage of Fe ion present in the xerogel and the temperature and period of thermal treatment. Higher Fe ion loading and more robust heating conditions yield larger average particle sizes. Fe/silica xerogel nanocomposites are usually pressed into pellets or prepared as thin films on glass slides for magnetic or electrical conductivity measurements. Experimental observations include the following: (1) Crystalline iron nanoclusters are usually formed; (2) FeO, Fe_3O_4, and Fe_2O_3 phases can be present also; (3) Fe_2SiO_4 is formed at elevated temperatures (700 °C) if the Fe ion content exceeds 16 wt%; (4) electrical resistivity decreases with increasing Fe nanocluster size, presumably because a greater number of charge percolation chains involve Fe particles; and (5) the effective magnetic anisotropy of Fe/silica gel nanocomposites can be greater than that of bulk iron.

Iron nanocomposites have been produced in an alumina matrix using Al(O*sec*-Bu)$_3$ as a sol–gel precursor [126]. Using (alkoxy)aluminum compounds in combination with TMOS affords a mullite matrix ($3Al_2O_3$–$2SiO_2$). Iron ion is introduced into these matrices using immersion infiltration of ferric nitrate solutions. Reduction under hydrogen at 600 °C produces crystalline Fe nanoclusters poorly dispersed throughout the solid matrix for Fe doping levels of 25 wt%. The Fe/alumina nanocomposite exhibits only weak saturation magnetization. Reaction of the Fe/mullite nanocomposite in air at 650 °C produces Fe_3O_4 particulates, whereas heating to 800 °C gives Fe_2O_3 (maghemite) particles.

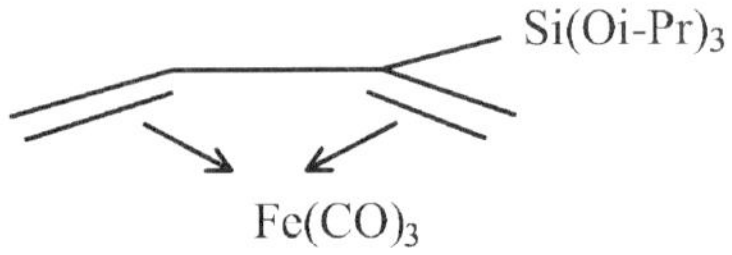

Structure III

$(EtO)_3Si - (CH_2)_6 - Si(OEt)_3$

Structure IV

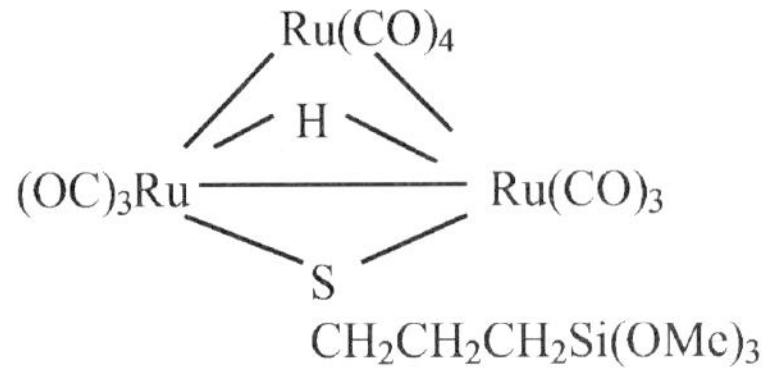

Structure V

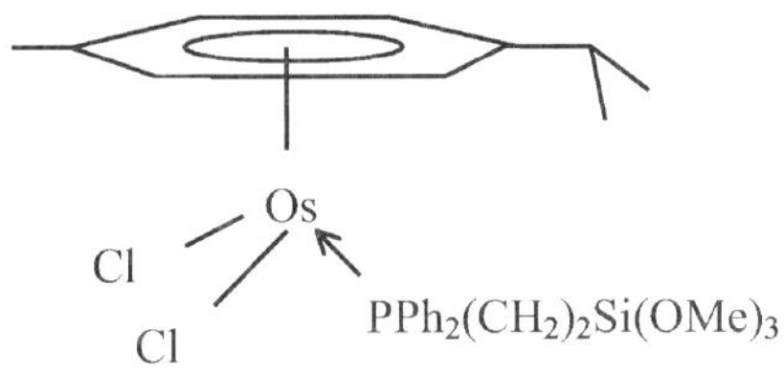

Structure VI

Sol–gel condensation in a 2:98 molar ratio of the $Fe(CO)_3$ complex shown in Structure III with either Structure II or IV gives inorganic/organic hybrid polysilsesquioxane gels containing Fe(III) as an internal dopant [127]. Irradiation of the resulting doped gel with ultraviolet (UV) light under vacuum leads to loss of CO and formation of Fe nanoclusters with an average diameter of 4 (0.5) nm. By using a low-valent iron precursor, Fe particles are formed without requiring calcination and subsequent reduction. The electrical conductivity of this poly(1,4-phenylene)-bridged silsesquioxane nanocomposite is in the semiconductor range.

Acid hydrolysis of ferric nitrate/cupric nitrate solutions affords inorganic gels that, upon thermal reduction, give nanocomposites consisting of Fe nanoclusters dispersed throughout a Cu metal matrix [124, 128]. Reduction by hydrogen at 425 °C gives Fe nanoclusters with an average diameter of 10 nm at a doping level of 29.6 wt% Fe. Mössbauer spectra confirm the presence of α-Fe particles, and magnetic measurements indicate that each Fe particle is a single magnetic domain.

Ru/silica xerogel nanocomposites have been prepared by conventional metal ion doping or by more unconventional covalent internal doping of silica xerogels. The addition of Ru(III) chloride to sol–gel solutions containing TEOS at the levels of 0.5, 1.0, and 2.0 wt% Ru leads to silica xerogels doped with Ru(III) ions. Thermal treatments up to 600 °C give Ru/silica xerogel nanocomposites that serve as hydrogenation catalysts [129]. Ru average particle diameters are less than 3 nm.

The addition of the triruthenium cluster complex, $Ru_3(CO)_{10}[\mu\text{-}S(CH_2)_3Si(OMe)_3][\mu\text{-}H]$ (Structure V), which contains a bifunctional thiolate ligand, to a TMOS silica sol–gel formulation gives covalent, internal incorporation of this Ru precursor complex into the silica xerogel matrix as it is being formed. Thermal treatment of this molecularly doped xerogel under hydrogen at 700 °C gives a Ru/silica xerogel nanocomposite [130]. The Ru nanoclusters exhibit near hexagonal shape and have average diameters of 3.1 nm [by transmission electron microscopy (TEM)] or 7.5 nm [by X-ray diffraction (XRD)]. Use of a low-valent Ru precursor complex gives Ru metal particles directly without proceeding through calcination thermal treatments that would produce highly toxic and volatile ruthenium oxides.

Using a similar strategy, covalent incorporation of the (η^6-arene)Os complex (Structure VI) into a silica xerogel using TMOS as a sol–gel precursor occurs through hydrolysis and condensation reactions involving the $Si(OMe)_3$ functional group of the bifunctional phosphine ligand. Thermal treatment of the resulting molecularly doped xerogel under hydrogen at 900 °C gives an Os/silica xerogel nanocomposite directly without calcination, thereby avoiding formation of the very volatile and toxic OsO_4 [131]. The measured average diameters of the Os nanoclusters are 2.7 nm (by TEM) or 5.5 nm (by XRD).

2.3. Group IX Metal Nanocomposites

Co/silica xerogels have been prepared by metal ion doping of conventional sol–gel formulations followed by reductive thermal treatment [118]. Similar Co nanocomposites have been prepared using ZrO_2-modified silica aerogels. Aerogels are formed using conventional sol–gel chemistry but are usually dried under supercritical conditions. Such composites have large surface areas, and these Co/aerogel nanocomposites are active Fischer–Tropsch catalysts [132].

In situ coordination of $Co(OAc)_2$ by the bifunctional diamine (Structure VII) in a sol–gel formulation based on TEOS leads to apparent covalent incorporation of Co ion into a silica xerogel matrix as it is being formed [133, 134]. Calcination of this molecularly doped xerogel in air at 500 °C affords CoO nanoclusters with an average diameter of 19.5 nm. Thermal reduction under hydrogen at 500 °C produces a Co/silica xerogel nanocomposite having crystalline Co nanoclusters with an average diameter of 17.4 nm. The particle size range for these Co nanoclusters is 11.0–24.9 nm.

Similarly, lysine (Structure VIII) reacts with $Ti(OEt)_4$ or $Zr(OPr)_4$ to form bifunctional complexes of the type $[(RO)_3M(\text{lysinate})]_n$, in which the butylamine functional group of lysine is chemically available to coordinate to metal ions. The addition of cobalt(II) acetate, $Co(OAc)_2$, to solutions of these precursor complexes leads to *in situ* complexation of Co ion. Subsequent sol–gel condensation gives gels containing Co ion as an internal dopant [135]. Calcination in air at 500 °C affords $CoTiO_3$/titania or CoO/zirconia nanocomposites. Thermal reduction under hydrogen at 500 °C gives the corresponding Co/titania or Co/zirconia gel nanocomposites. With a molar doping level of 20% Co, the average diameter of the Co nanoclusters is 19 nm (by XRD).

Rh/silica or Rh/zirconia-modified silica gel nanocomposites have been prepared by conventional internal metal ion doping of sol–gel formulations, as well as by using metal ion impregnation techniques [136]. The Rh nanoclusters produced by external doping are more active as hydrogenolysis catalysts, presumably because these metal nanoclusters have more exposed surface area.

2.4. Group X Metal Nanocomposites

Nanoscale electrodeposition of nickel in thin films of silica xerogel has been reported [137]. Thin-film nanocomposites of Ni in silica xerogel have also been prepared by doping TEOS sol–gel formulations with nickel nitrate followed by reduction under hydrogen at 650 °C [121]. In the latter films, the average diameter of the Ni particles increases from 5.2 to 10.5 nm as the period of reduction is increased from 15 to 35 min. The electrical conductivity of these films has been studied.

Bulk Ni/silica xerogel nanocomposites are prepared by doping silica sol–gel formulations with nickel salts, such as nickel(II) chloride or nitrate, followed by reduction under hydrogen at 500–650 °C [120, 138–140]. The average Ni particle diameters range from 5 to 13.4 nm, depending on the severity of thermal treatment. Bulk powder Ni/silica xerogels

$H_2NCH_2CH_2NH(CH_2)_3Si(OEt)_3$

Structure VII

$H_2NCH_2CH_2CH_2CH_2-CH(NH_2)-CO_2H$

Structure VIII

containing 7 wt% Ni and an average particle size of 5 nm show good catalytic activity in hydrogenation reactions [138]. Compacted Ni/silica xerogels [140] having Ni particles with diameters of 5–7 nm possess semiconductor electrical conductivity at 235–340 K [139], whereas similar compacts having Ni particles 9–13.4 nm in size are ferromagnetic [120].

In situ complexation of $Ni(OAc)_2$ by the bifunctional diamine shown in Structure VII gives molecularly doped silica xerogels following sol–gel processing [133, 134, 141]. The addition of TEOS to these sol formulations dilutes the doping level of nickel salt. Calcination in air at 500–525 °C and reduction under hydrogen at 500 °C or higher gives Ni/silica xerogel nanocomposites. Further sintering of these composites leads to growth in Ni nanocluster particle size. Sintering for 2 h at 500, 100, or 200 °C affords Ni nanoclusters of 30-, 100-, or 200-nm average diameter, respectively, as determined by XRD [141].

In situ complexation of $Ni(OAc)_2$ by bifunctional Ti or Zr complexes of the type $[(RO)_3M(\text{lysinate})]_n$, where R is Et or Pr, respectively, leads to Ni/titania or Ni/zirconia nanocomposites after sol–gel condensation and reduction under hydrogen [135]. At a 20-mol% loading of Ni, the average diameters of Ni nanoclusters in the titania or zirconia gel matrices are 11 and 7 nm, respectively. Ni/Y_2O_3–ZrO_2 gel nanocomposites are formed by *in situ* complexation of $Ni(OAc)_2$ to bifunctional amine ligands, such as glycine or triethanolamine, in the presence of $Zr(OPr)_4$ and yttrium nitrate [142, 143]. Calcination in air and reduction under hydrogen affords Ni nanoclusters having average sizes of 5–10 nm or 30–40 nm, depending on the conditions of thermal treatment and the doping level of Ni. These nanocomposites become electrically conductive upon hot pressing the powder when the Ni loading exceeds 30 mol% [143].

Ni/Al_2O_3 gel nanocomposites are formed by sol–gel condensation of aluminum alkoxides, such as $Al(OR)_3$, where R is isopropyl or *sec*-butyl, in the presence of nickel nitrate or formate salts followed by calcination and reduction [126, 144, 145]. Nickel loadings of 5–50 vol% give Ni nanoclusters with average diameters of 5–60 nm. The mechanical [144], magnetic [126], and catalytic properties [145] of these nanocomposites have been reported.

Nickel nanocomposites in $BaTiO_3$ or $Pb(Zr_{0.58}Ti_{0.42})O_3$ gel matrices are prepared by sol–gel condensation of a mixture of appropriate metal alkoxides along with added nickel salts [146, 147]. The ceramic matrix is formed upon calcination, and Ni nanoclusters with diameters of 2.0–7.6 nm are formed during subsequent thermal reduction under hydrogen. Such ferroelectric materials are of interest for capacitor fabrication. Compacted $Ni/BaTiO_3$ nanocomposite gels have a dielectric constant of 400–900 at 1 kHz [146], whereas the dielectric constant of thin films of the $Ni/Pb(Zr_{0.58}Ti_{0.42})O_3$ gel nanocomposites is 220–410 at 1 kHz at room temperature [147].

Nanocomposites of Pd and Pt are of interest primarily for use as chemical catalysts or as materials exhibiting interesting nonlinear optical (NLO) properties [148]. A variety of sol–gel preparative methods are available for fabricating such nanocomposites as either bulk powders or as thin films.

Palladium or Pt nanocomposites in γ-Al_2O_3 [149–152], TiO_2 [153–156], or Y_2O_3-stabilized ZrO_2 [157] gels are formed by doping conventional sol–gel formulations with metal salts or complexes. Sols are formed from metal alkoxides, such as TEOS, $Ti(Oi\text{-}Pr)_4$, or $M(OBu)_4$, where M is Ti or Zr, and Pd or Pt is introduced as palladium nitrate, $PdCl_2$, $PdCl_4^{2-}$, H_2PtCl_6, or as the acetylacetonate (acac) complexes. Ethylenediaminetetraacetic acid (EDTA) can also be added to the sol to effect *in situ* complex formation [149–151]. Thermal calcination and reduction of gels doped with Pd or Pt ions gives metal nanoclusters, although direct thermal reduction of these more noble metals can occur without using an explicit reducing agent. Palladium or Pt nanocluster average sizes typically range from 4 to 115 nm with narrow particle size distributions obtained through careful control of metal doping levels and the severity of thermal treatment.

Pd or Pt nanocomposites are also formed using organically modified gels or gel precursors, as shown in the following equation [158–163]:

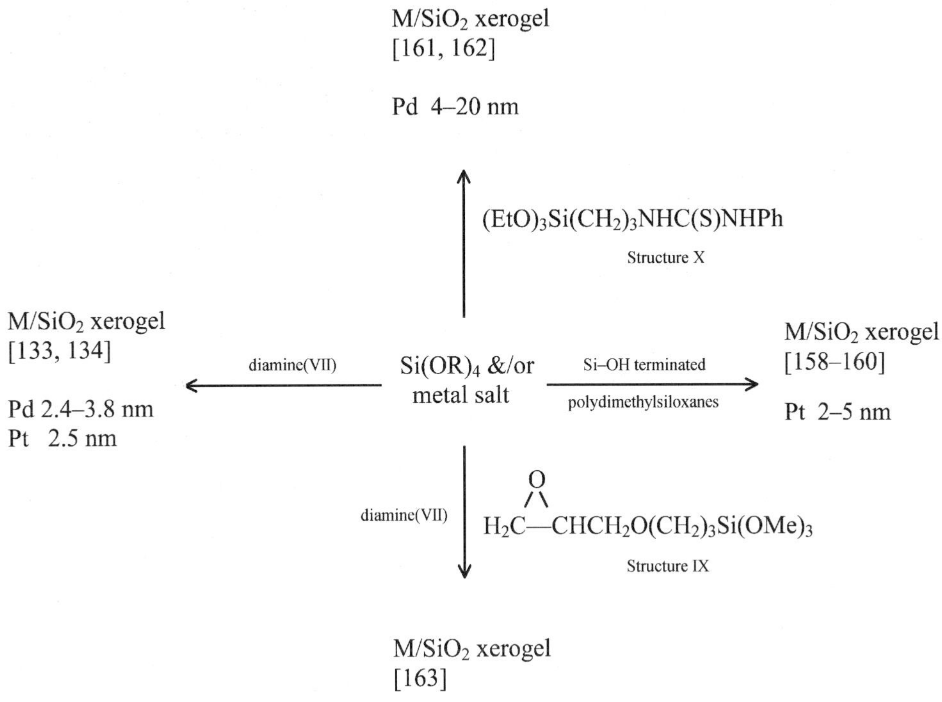

(2)

The bifunctional diamine (Structure VII) gives composites having average metal particle diameters of 2.4–3.8 nm [133, 134]. Pd particle sizes are essentially independent of the precursor, $Pd(acac)_2$, loading level. For the Pt nanocomposite, the Pt particle size distribution broadens and shifts to larger average diameters when treated thermally at higher temperatures (750–900 °C) [164]. Platinum silicide formation, such as Pt_3Si and $Pt_{12}Si_5$, also occurs at these conditions [165]. Use of the bifunctional diamine (Structure VII) in combination with the bifunctional epoxide (Structure IX) affords both *in situ* complexation of Pd ion and binding of the silica xerogel matrix to glass surfaces for casting and dip-coating applications [163]. *In situ* complexation of Pd salts to silica xerogel derivatized with the bifunctional benzoylthiourea (Structure X) gives a Pd/silica xerogel that acts as a hydrogenation catalyst [161, 162]. Either bulk or thin films of Pt/silica xerogel nanocomposites containing up to 5 wt% Pt are obtained from conventional sol–gel formulations using silanol-terminated polydimethylsiloxanes as additives [158–160]. Reduction of Pt ion using H_2PtCl_6 as a metal dopant is effected either thermally or photochemically. Other reports of Pd/silica xerogel nanocomposite formation using organic binders or modified sol–gel methods are available [166–168].

Addition of the bifunctional Pt complex, $PtCl_2[PPh_2(CH_2)_2Si(OMe)_3]_2$, to TMOS at a Si/Pt molar ratio of 100:1 under sol–gel conditions gives a silica xerogel containing this complex as a covalent, internal dopant. Subsequent thermal calcination in air and reduction under hydrogen at 700 °C gives a Pt/silica xerogel nanocomposite with Pt nanoclusters having an average diameter of 3.8 nm [131].

2.5. Group XI Metal Nanocomposites

A vast literature exists on nanocomposite materials containing nanoclusters of Cu, Ag, and Au. Principal interest in these materials is derived from their interesting optical properties. All three metals have surface plasmon resonances in the visible region for nanoclusters of typical size.

Cu/silica xerogel nanocomposites are formed as bulk powders [139, 140, 169] or as thin films [121, 122, 170, 171] by adding Cu(II) salts to conventional TEOS sol–gel formulations. Metal ion doped gels are usually calcined at 500–600 °C, and reduction to metal occurs upon heating from 400 to 700 °C in the presence of hydrogen. Average Cu particle sizes varying from 3 to 28 nm are obtained, depending on the level of metal doping and thermal conditions. Reduction at 400 °C for a gel containing 20 wt% Cu ion gives Cu nanoclusters of average diameter 6.2 nm with a monomodal size distribution. Thermal treatment at 700 °C gives a bimodal Cu particle size distribution with maxima at 8.5 and 28.0 nm [169].

Another route to Cu/silica gel nanocomposites is *in situ* complexation of Cu ion to organically modified gels followed by calcination and thermal reduction under hydrogen [158]. Use of the bifunctional diamine (Structure VII) permits a high doping level of Cu ion in a silica sol–gel formulation. The average diameter of the resulting Cu nanoclusters is 3.9 nm [133, 134]. A similar synthesis using a much lower concentration of Cu ion and thermal treatment at 900 °C gives a Cu/silica xerogel nanocomposite having Cu particles with an average diameter of 15 nm [172]. Silica sol–gel formulations containing the bifunctional diamine shown in Structure VII, 3-glycidoxypropyl triethoxysilane, TEOS, and Cu(II) ion give gels that coat glass through dip coating. Thermal reduction of these coatings gives a film of Cu/silica xerogel containing Cu nanoclusters 5–30 nm in diameter [173a]. Sol–gel condensation of the bifunctional titanium lysinate complex, $[Ti(OEt)_3(lysinate)]_2$, in the presence of $Cu(OAc)_2$ forms a TiO_2-based xerogel. Upon calcination and reduction, a bulk Cu/TiO_2 gel nanocomposite having Cu particles 20 nm in average diameter is formed [135]. A related preparation of a Cu/TiO_2 nanocomposite has appeared recently [173b].

Cocondensation of TMOS and the bifunctional complex, $CuCl[PPh_2(CH_2)_2Si(OEt)_3]_3$, in a 10:1 Si/Cu molar ratio under sol–gel conditions gives a silica xerogel covalently doped with this Cu precursor complex. Direct thermal reduction at 600 °C under hydrogen gives a Cu/silica xerogel nanocomposite containing Cu nanoclusters with an average diameter of 24 nm [131].

Nanocomposites of Ag and Au have received much interest primarily as materials exhibiting interesting optical properties. Because these metals are easily reduced, a variety of sol–gel preparative methods are available for fabricating Au or Ag nanocomposites as either bulk powders or thin films. Thin-film methods are particularly useful for optical studies to maintain adequate optical transmission.

Silver or Au nanocomposites in γ-Al_2O_3 [174, 175], SiO_2 [176–196], SiO_2/B_2O_3 [197], SiO_2/TiO_2 [198], TiO_2 [199–201], $BaTiO_3$ or PZT [202–205] gels, or micellular [206, 207] matrices are formed by direct metal ion doping of conventional sol–gel formulations with metal salts or complexes. Sols are formed from $AlCl_3$ or metal alkoxides, such as Al(O*sec*-Bu)$_3$, TEOS, H_3BO_3, Ti(O*i*-Pr)$_4$, $M(OBu)_4$, where M is Ti or Zr, and Ba or Pb acetates. The most common source of Ag is silver nitrate; gold is usually introduced as $HAuCl_4$ or $NaAuCl_4$ or as acetylacetonate complexes [177, 178]. Thermal calcination and hydrogen reduction of metal ion doped gels affords metal nanoclusters, although thermal reduction of these more noble metals, particularly Au, occurs without using an explicit reducing agent. Hydrazine hydrate has been used to reduce gold ion at reduced temperatures [175], and both Ag and Au ions can be reduced to metal by light [177–179] or by γ-irradiation [195, 196, 200].

Silver or Au nanocluster sizes typically fall in a range of 4 to 40 nm with narrow particle size distributions obtained by careful control of metal doping levels and the severity of thermal treatment. The growth of Ag particles as a function of annealing temperature and gaseous environment has been studied [167, 170, 194]. Ag/silica xerogel nanocomposites heated in air show some loss of silver at temperatures as low as 650 °C. Annealing such samples in inert or N_2/H_2 atmospheres from 200 to 600 °C leads to particle growth by Ostwald ripening growth processes, whereas particle growth at more elevated temperatures gives bimodal particle size distributions. Annealing further to 850 °C leads to a decrease in average particle size and to a sharpening of Ag particle size distributions as larger Ag clusters undergo fragmentation. The thermal bleaching effect observed for Ag/silica xerogel nanocomposites is attributed to thermal fragmentation of the larger Ag nanoclusters within the composite. The addition of Li_2O completely inhibits this bleaching effect [193].

Silver and Au nanocomposites exhibit properties useful for many applications including (1) resonant and nonlinear optical effects [171, 174–184, 193, 197, 199, 201, 202], (2) use as high-dielectric-strength media for capacitors [146, 187–189, 203], (3) enhanced electrical conductivity of ceramic media [204, 205], (4) pattern etching using HF [186], (5) augmentation of electrode response [185], and (5) as magnetic spin glasses for Ag nanocomposites [191].

Ag or Au/silica nanocomposites have been prepared using organically modified gels or gel precursors, as shown in the following equation:

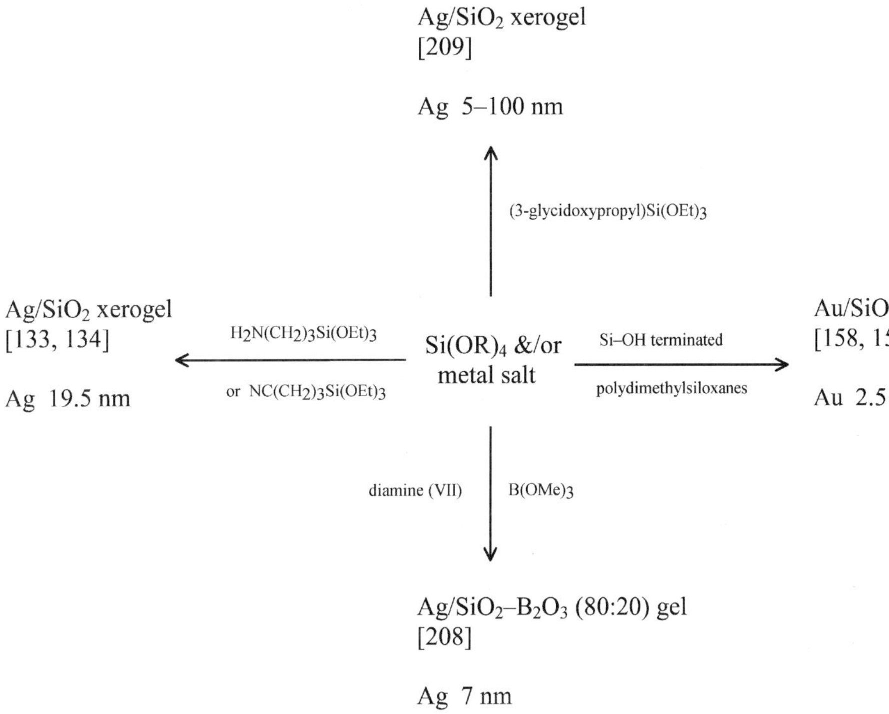

(3)

Bulk or thin-film products are obtained. The sources of the metals are $AgNO_3$, AgOAc, or $HAuCl_4$. Reduction to metal is accomplished either thermally [158, 159], by reaction with hydrogen at elevated temperature [133, 134, 209], or by photolysis [158, 159, 208]. The particle sizes shown in Eq. (3) represent either average particle size or a range of average particle sizes produced under controlled reduction conditions. Control over average metal particle size can be partially achieved by variation of reduction conditions. Peak absorption for the surface plasmon resonance of Ag nanoclusters in borosilicate gels ranges from 394 to 405 nm, depending on the metal particle size [208].

Sol–gel condensation of $Zr(Oi\text{-}Pr)_4$ in the presence of $HAuCl_4$, $CH_2=C(Me)CO_2(CH_2)_3Si(OMe)_3$, $CH_2=C(Me)CO_2H$, and diamine (Structure VII) leads to a zirconia gel doped with Au(III) ions. Photolysis initiates alkene polymerization and reduction of Au ion to form Au nanoclusters. Controlled irradiation gives nanocomposites having Au nanoclusters with average diameters of 10–40 nm [210].

Hydrogen reduction of $HAuCl_4$ by photolysis in the presence of poly(N-vinyl-2-pyrrolidone) (PVP) forms polymer-stabilized Au nanoclusters. The addition of this dispersion to $Ti(Oi\text{-}Pr)_4$ along with acetylacetone and water leads to sol–gel condensation and formation of a hybrid gel containing Au nanoclusters 2.8 nm in diameter [160]. The size variation of the Au particulates is accomplished by controlling the reducing conditions. Maxima of the surface plasmon resonance of the Au nanoclusters range typically from 530 to 620 nm, depending on the metal particle size. Analogous silica hybrid gels could not be produced by this procedure because of unfavorable precipitation upon mixing PVP with TEOS.

The addition of the bifunctional Ag mixed-ligand complex, $[AgS(CH_2)_2Si(OMe)_3]_7[AgCl]_3$, to a silica sol–gel formulation based on TMOS at a Si/Ag molar ratio of 100:1 gives a silica xerogel containing this complex as a covalent, internal dopant. Subsequent thermal calcination and reduction under hydrogen at 600 °C gives a Ag/silica xerogel nanocomposite containing Ag nanoclusters with an average diameter of 5.3 nm [131].

Au nanoclusters have also been prepared in silica materials using sol–gel conversions within inverse micelles as microreactors [211]. Poly(diethoxysilane) and $AuCl_3$ are dissolved in the presence of an inverse micelle. Reduction to Au metal is effected using $LiBH_4$ affording Au nanoclusters with diameters of 6.0 (1.5) nm. Control of metal particle size is complicated by the effects of gel precursor and sol–gel condensation products on the stability of the inverse micelle.

Silver or gold nanoclusters have also been prepared in silica/zirconia matrices using γ-radiation to initiate reduction [212].

2.6. Metal Alloy Nanocomposites

Four Cu-based alloy/silica xerogel nanocomposites have been prepared using sol–gel chemistry, as shown in the following equation:

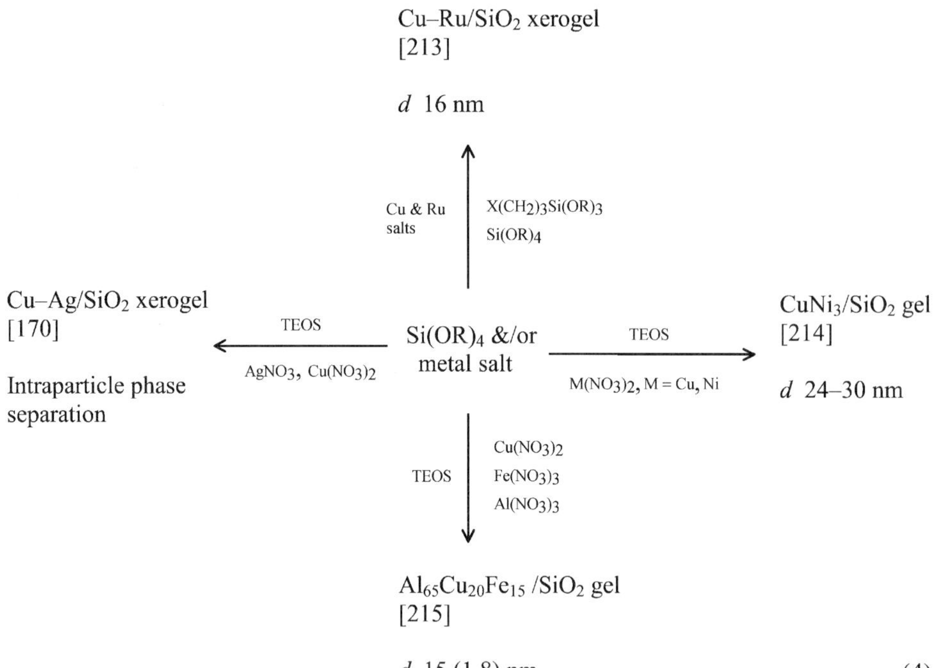

(4)

In situ complexation of Cu and Ru salts by bifunctional ligands gives a silica gel containing coordination complexes of these metals [213]. Subsequent calcination in air and thermal reduction under hydrogen affords a Cu/Ru nanocomposite containing particles with an average diameter of 16 nm. Alcohol solutions of copper and silver nitrate salts containing TEOS form, upon sol–gel conversion, a gel matrix doped with these metal ions [170]. Thermal treatment of the resulting mixed-metal-ion-doped xerogels under hydrogen gives Cu/Ag alloy particulates; however, phase separation of the metals occurs within each particle. Compaction at elevated temperatures of a physical mixture of silica xerogels containing either Cu or Ni metal nanoclusters at a Cu/Ni atomic ratio of 1:3 gives a $CuNi_3$ nanocomposite [214]. XRD data confirm a single-phase composition for these alloy particulates. The average diameter of the alloy particles ranges from 27 to 30 nm as the temperature of compaction increases from 450 to 650 °C. Sol–gel processing of solutions of Cu(II), Fe(III), and Al(III) nitrates with TEOS affords a mixed-metal-ion-doped silica gel [215]. Particles of the ternary alloy, $Al_{65}Cu_{20}Fe_{15}$, with an average diameters of 15 nm are formed in the gel matrix along with nanoparticulate Fe upon thermal treatment with hydrogen. XRD data indicate that the alloy particles are quasicrystalline, and Mössbauer spectra are consistent with these particle compositions.

Two nanocomposites containing binary alloys of Pt have been prepared using sol–gel methods. Gel formation using $Al(O\textit{sec}\text{-}Bu)_3$ and acetic acid in the presence of $Pt(acac)_2$ and $Re_2(CO)_{10}$ gives a $Pt\text{–}Re/Al_2O_3$ composite upon calcination at 400 °C [152]. At loadings of 1 wt% Pt and Re, nanoparticles with an average size of 3.5 nm are obtained. Although the presence of Re appears to inhibit Pt metal aggregation, direct evidence for the formation of alloy nanoclusters is lacking.

Sol–gel condensation of mixtures of TMOS and the bifunctional complex, $Pt(PPh_3)_2(Ph)[SnPh_2S(CH_2)_3Si(OMe)_3]$, leads to covalent incorporation of this binuclear complex into the resulting silica xerogel matrix [130, 131]. Thermal treatment of this molecularly doped xerogel under hydrogen at 700 °C gives a nanocomposite containing crystalline nanoclusters of niggliite, PtSn, with an average diameter of 6 nm. No other Pt/Sn phases are detected by either XRD or electron diffraction. The 1:1 Pt/Sn stoichiometry of the precursor complex apparently gives selective formation of niggliite nanocrystals.

An $Fe_{20}Ni_{80}$/silica xerogel nanocomposite has been prepared by sol–gel processing [119]. The alloy particles have an average diameter of 8(1) nm and face-centered cubic (fcc) structure. Magnetic measurements reveal an effective magnetic anisotropy for the composite that is larger than that of the bulk alloy.

2.7. Group XIV Nanocomposites

Fullerene nanocomposites have been prepared by (1) entrapment of C_{60}/C_{70} mixtures in silica aerogels through the addition of fullerenes to a conventional sol–gel mixture [216–220]; (2) absorption of C_{60} solutions into silica gel formed from TEOS, followed by impregnation and polymerization of an organic polymer giving a multiphasic inorganic/organic composite [221, 222]; or (3) derivatization of C_{60} with a bifunctional substituent, $H_xC_{60}[NH(CH_2)_3Si(OEt)_3]_x$, where $x = 1\text{–}3$, followed by sol–gel conversion using TEOS [223]. These aerogel composites are produced by supercritical fluid drying. Photoluminescence is observed at 2.26 eV when using laser excitation at 488 nm. Fullerene molecules are presumably entrapped in nanopores within the areogel matrix. For the multiphasic composites, the organic polymeric phase is formed from polymerization of methyl methacrylate or bisbenzothiazole 3,4-didecyloxythiophene or mixtures of the two. These composites are optically transparent and might be useful as materials having tailored optical responses. Composites prepared from covalently derivatized C_{60} absorb light at various wavelengths, depending on the C_{60} loading. Femtosecond excited-state absorption dynamics studies of C_{60} in inorganic sol–gel matrices reveal a crucial role of C_{60} in relaxation dynamics. This property may find application in optical-limiting devices [224, 225].

Pyrolysis of gaseous organic compounds in sol–gel aerogels forms carbon nanotubual composites [226]. The presence of carbon nanotubes reduces the infrared (IR) transmission of the aerogel and improves its thermal performance relative to that of the undoped aerogel.

Nanosilicon/silica xerogel composites can be formed by ultrasonic fracturing of porous silicon. The nanoscale Si fragments produced are mixed into conventional silica sol–gel formulations giving an n-Si/xerogel composite [227, 228]. These materials exhibit red (680 nm) and blue (415–460 nm) electroluminescence with decay times different from those observed from porous silicon.

Elemental Ge nanocomposites are accessible using conventional sol–gel methods. Reduction of intermediate germanium oxides under hydrogen and subsequent Ge particle growth occur at temperatures below 1000 °C. Ge/silica gel composites result from cocondensation of TEOS and $GeCl_4$ or $Ge(OEt)_4$ followed by thermal treatment [229–232]. Annealing temperatures above 800 °C usually give crystalline Ge nanoclusters of approximately 5-nm average diameter, although larger particle sizes are observed upon prolonged heating at elevated temperatures. Ge particles of 5 nm show an optical absorption edge onset near 2.2 eV and a broad photoluminescence peak at the same energy. Larger Ge particles do not exhibit photoluminescence. A decrease in Ge average particle size leads to an intensity increase in the observed photoluminescence peak along with a blue shift in peak position.

The addition of the bifunctional organogermane, $Me_3GeS(CH_2)_3Si(OMe)_3$, to a conventional TMOS sol–gel formulation leads to covalent incorporation of this molecular precursor into the silica xerogel matrix as it is being formed [131, 233]. Subsequent calcination in air at 600 °C followed by reduction in hydrogen at 600–900 °C gives a Ge/silica xerogel nanocomposite containing Ge particles with an average diameter of 6.8 nm. Optical absorption spectra of this composite are consistent with expected quantum confinement effects.

A Ge/alumina gel nanocomposite has also been prepared [234]. The host matrix is formed by sol–gel processing of $AlCl_3$.

3. NANOCOMPOSITES CONTAINING NANOPARTICULATE SUBSTANCES

In nanocomposites of chemical substances, two or more elements are present in the nanocluster, and, ideally, a single stoichiometry is desired. Although many such compositions are expected to show catalytic activity, such as supported metal oxides, there is much interest in preparing nanocomposites of semiconductor materials for property measurements or for device fabrication. The NLO and magnetic properties of such materials are reviewed elsewhere in this book.

3.1. Metal Carbide Nanocomposites

A mixed-metal gel is formed when $Ti(Oi\text{-}Pr)_4$ is added in a 70:30 molar ratio to a prehydrolyzed sol derived from $SiMe_2(OEt)_2$ [235]. Subsequent pyrolysis of this gel under Ar at 400–1600 °C gives eventual formation of cristobalite and crystalline TiC phases. XRD data indicate the appearance and growth of TiC particles between 1000 and 1600 °C, although particle size estimates have not been determined. Formation of a $TiC/SiO_2–TiO_2$ nanocomposite presumably occurs at intermediate temperatures.

The addition of the intensely purple tricobalt carbido cluster complex (Structure XI) to a TMOS sol–gel formulation leads to covalent incorporation of the complex to the silica xerogel matrix as it is being formed [130, 131, 236]:

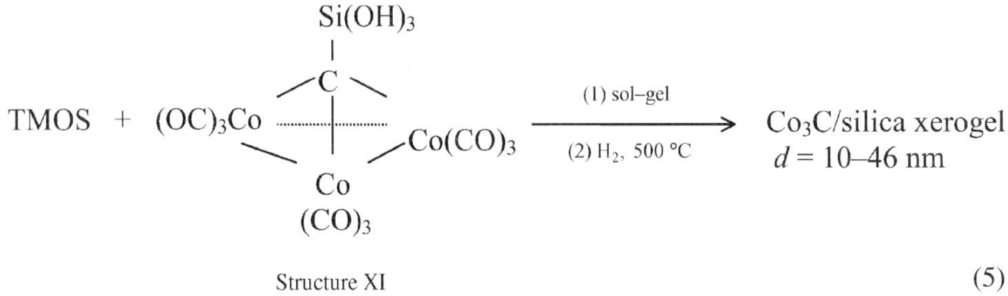

Structure XI (5)

Heterocondensation of the $Si(OH)_3$ functional group of the complex shown in Structure XI with hydrolysis products of TMOS assures a covalent molecular doping of the resulting silica xerogel, as confirmed by IR spectroscopy and by the inability to remove the complex (Structure XI) from the matrix by solvent extraction. A control reaction using a derivative of the complex shown in Structure XI in which the $Si(OH)_3$ group is replaced by a hydrogen atom substituent leads to a physically doped silica xerogel. Washing this latter xerogel with organic solvents leads to complete removal of the tricobalt cluster from the matrix. Thermal degradation of the covalently doped gel under hydrogen at 500 °C leads to a Co_3C/silica xerogel. Selected-area electron diffraction (SAED) and both powder and single-crystal XRD indicate that crystalline particles of Co_3C are formed. These nanoparticles have an average diameter of 25 nm and a stoichiometry identical to that of the precursor core. Phase separation into crystalline particles of other Co/C stoichiometries is not observed. Use of a more complex molecular precursor apparently provides control over the elemental stoichiometry of the resulting cobalt carbide nanoclusters.

Nanocomposites containing SiC particulates of either 30- or 600-nm average diameter have been prepared in an alumina host matrix by ultrasonic dispersion of preformed SiC nanoparticles in a sol of pseudo-boehmite, aluminum acetate, and acetic acid [237]. Gel formation is initiated by the addition of ammonia. Calcination in air at 900 °C and thermal treatment in vacuum at 1300 °C converts the host matrix to α-Al_2O_3. Equilibration of the resulting gel with solutions of ^{220}Ra and ^{224}Ra leads to a radioactive doping of the gel in the top 80 nm of the surface. Emanation thermal analysis and differential thermal analysis (DTA) studies indicate that the onset of structural change and grain growth of the alumina matrix shifts to a higher temperature when SiC nanoparticulates are present. This effect is greatest for smaller SiC particles. A related study has also been reported [238].

3.2. Metal Pnictide Nanocomposites

Hydrolysis and heterocondensation of TEOS and phenyl(trimethoxy)silane gives an organic functionalized silica gel. Thermal reaction of this gel with nitrogen gas at 1500 °C forms a nano-Si_3N_4/SiC composite. Changes in the composition of the resulting gel as a function of thermal treatment conditions have been studied [239].

Covalent incorporation of low-valent transition metal complexes containing bifunctional phosphine ligands into a silica gel matrix as it is being formed gives molecularly doped gels. Subsequent thermal treatment under hydrogen at 600–900 °C provides a convenient route to silica xerogel composites containing crystalline nanoclusters of the transition metal phosphides, Fe_2P, RuP, Co_2P, Rh_2P, Ni_2P, Pd_5P_2, or PtP_2 [130, 131, 240–243]:

$$
\begin{array}{lcl}
\mathrm{Fe(CO)_4[PPh_2CH_2CH_2Si(OMe)_3]} & & \mathrm{Fe_2P(4.7\ nm)/SiO_2\ xerogel} \\
\mathrm{RuCl_2(\eta^6\text{-cymene})[PPh_2(CH_2)_2Si(OEt)_3]} & & \mathrm{RuP(4.7\ nm)/SiO_2\ xerogel} \\
\mathrm{CO_2(CO)_6[PPh_2CH_2CH_2Si(OEt)_3]_2} & & \mathrm{Co_2P(5.0\ nm)/SiO_2\ xerogel} \\
\mathrm{RhCl[PEt_2CH_2CH_2Si(OEt)_3]_3} & \text{(1) TMOS sol–gel} & \mathrm{Rh_2P(2.0\ nm)/SiO_2\ xerogel} \\
\mathrm{Ni[PPh_2CH_2CH_2Si(OEt)_3]_4} & \xrightarrow{\hspace{3cm}} & \mathrm{Ni_2P(2.6\ nm)/SiO_2\ xerogel} \\
\mathrm{NiCl_2[PPh_2CH_2CH_2Si(OEt)_3]_2} & \text{(2) } \mathrm{H_2}, >600\,^{\circ}\mathrm{C} & \mathrm{Ni_2P(35.5\ nm)/SiO_2\ xerogel} \\
\mathrm{PdBr_2[PEt_2CH_2CH_2Si(OEt)_3]_2} & & \mathrm{Pd_5P_2(11.3\ nm)/SiO_2\ xerogel} \\
\mathrm{PtCl_2[PPh_2CH_2CH_2Si(OMe)_3]_2} & & \mathrm{PtP_2(4.0\ nm)/SiO_2\ xerogel}
\end{array} \quad (2)
$$

Each molecular complex serves as a single-source precursor for the specific metal phosphide phase shown. For the RuP and PtP_2 composites, the metal phosphide stoichiometry is the same as that of the molecular precursor. Selective formation of the other metal phosphides requires a relative loss of phosphorus from the molecular precursor during thermal treatment. In each case, the metal phosphide stoichiometry of the nanocluster is that of a congruently melting phase or that of a particularly stable composition. Equivalent precursor loadings of the Ni(0) and Ni(II) precursor complexes gives Ni_2P nanocrystals differing in average size by nearly 14-fold (by TEM). The origin of this effect is not known.

Methanolysis of the cadmium phosphide compound, $Cd[P(SiPh_3)_2]_2$, in a pyridine solution gives nanoclusters of Cd_3P_2 having diameters of 2–7 nm [244]. These particles are assumed to be surface passivated with OMe and $SiPh_3$ groups. This synthetic route is a variation of typical sol–gel procedures in which solvent evaporation from the reaction solution provides an initial composite of nano-Cd_3P_2 along with byproducts, such as $HP(SiPh)_2$ and $Si(OMe)Ph_3$.

Several processing techniques, including sol–gel methods, have been used to introduce 4-nm GaAs nanoclusters into amorphous matrices [245]. Sol–gel synthesis of GaAs/silica gels is one such technique. These bulk materials show quantum confinement effects as revealed by optical absorption spectroscopy.

3.3. Metal Oxide Nanocomposites

3.3.1. Al_2O_3 Gel Matrix

Alumina gel nanocomposites containing nanoclusters of CoO, FeO, TiO_2, or ZrO_2 have been prepared by sol–gel methods [246–250]. Common sources of the alumina matrix include aluminum trialkoxides or hydrated alumina powder. Titania and zirconia particles are usually derived from tetraalkoxide reagents, although metal chlorides are also used. The CoO and FeO composites test well as catalysts for selective reduction of NO by unsaturated hydrocarbons [246]. Individual TiO_2 particles are observed when the Ti/Al molar ratio is less than 17.5% and can be imaged by TEM when the particles have diameters greater than 1.2 nm [248]. Alumina-stabilized zirconia films are of interest for high-temperature protection of turbine blades [249].

3.3.2. SiO_2 Gel Matrix

Nanoparticulates of numerous metal oxides have been formed in silica gel matrices, as shown in Table I. Synthetic methods include *in situ* coordination of metal ions by a bifunctional ligand, such as $X(CH_2)_3Si(OEt)_3$, where X is NH_2, CN, or $NH(CH_2)_3NH_2$, in conventional sol–gel formulations, Method A; dissolution of metal salts in conventional sol–gel formulations, Method B; heterocondensation of metal alkoxide precursors and tetraalkoxide silicates, Method C; or direct sol–gel condensation of zirconium oxychloride octahydrate onto a silica–glass surface followed by thermal annealing, Method D. Drying-control chemical additives, such as formamide, are occasionally used in Methods B and C. Triethoxysilane [258], H_2SiO_3 [258], or organically modified silicates (Ormosils), such as silanol-terminated polydimethyl siloxane [259], can be used as sources of the host matrix. Ormosil composites are inorganic/organic hybrid matrices. The chemical composition of the resulting gel depends on the weight fraction of polymeric silanol used in the sol formulation and the severity of the resulting thermal treatment.

Metal oxide particle sizes are usually determined by electron microscopy. In some cases, however, metal oxide nanoparticulates are intermediates in the conversion of metal ions to metal nanoclusters, and their sizes are inferred from the sizes of the resulting metal nanoclusters if direct observation is not possible [133].

The properties of interest for metal oxide/silica gel nanocomposites include optical absorption in the visible region [271] and NLO effects (particularly for nanoparticulate ferroelectric materials, such as metal titanates and niobates). Nanoclusters of magnetic materials, such as maghemite (γ-Fe_2O_3), are stabilized through entrapment in silica gel matrices over a considerable temperature range [258]. These composites are superparamagnetic materials.

Table I. Metal Oxide/Silica Gel Nanocomposites

M_xO_y	Synthetic method (see text)	Particle size (nm)	Properties	Reference(s)
Ag_2O	A			[133]
CdO	A			[133]
CoO	B		Alternating-current electrical conductivity	[251]
Co_3O_4	B	8.9–9.7	Effect on optical transmittance	[181, 182]
Cr_2O_3	B	12		[181]
CuO	A			[133]
	B	5–8, 17–21	NLO	[181, 252]
FeO	B		Color variation	[253, 254]
Fe_2O_3	B	2–10, <300	Effect on optical transmittance Mössbauer, magnetism	[64, 181, 182, 254–258]
Mn_2O_3	B	8	IR filters	[181, 184]
NiO	A	1.4, 34.8		[133]
	B	7.5–8.1		[181, 260]
PdO	A			[133]
	B	10.7–27.1		[181]
PtO	A			[133]
TiO_2	C	20–50	NLO	[260, 261]
V_2O_5	C		Photoinitiated polymerization	[262]
$(Ni,Zn)Fe_2O_4$		13–35	Superparamagnetism	[263]
$(Li,K)NbO_3$	C		NLO	[259]
$PbTiO_3$	C	20–40	NLO	[264–266]
$Pb(Zr,Ti)O_3$	C	8–10	NLO	[259, 267]
$(In,Sn)O_2$ (ITO)	B	1–8.5	NLO	[268]
$LiMn_2O_4$	B	Submicrometer		[269]
ZrO_2	D	<15	Oxygen-deficient ZrO_2	[270]

3.3.3. MO_2 Gel Matrices, where M Is Ti, Zr, or Sn

Nanoclusters of several metal oxides have been prepared in TiO_2, ZrO_2, or SnO_2 gels, as shown in Table II. Synthetic methods include sol–gel conversion of metal halide or coordination complexes with alkoxide compounds of the second metal, Method A; hydrolysis and heterocondensation of homoleptic alkoxide compounds of both metals, Method B; sol–gel condensation of metal acetate salts with either $M(OBu)_4$ (Method C) or $M(OEt)_3$(lysinate) compounds, where M is Ti or Zr (Method D); sol–gel preparation followed by supercritical fluid drying, Method E; or hydrolysis and condensation of mixtures of metal halide or acetate salts, Method F. Metal oxide particle sizes are obtained from electron microscopy, XRD, or fluorescence measurements.

Table II. Metal Oxide/TiO_2, ZrO_2, or SnO_2 Nanocomposites

M_xO_y	Synthetic method (see text)	Particle size (nm)	Properties	Reference(s)
		TiO_2 Matrix		
Al_2O_3	A	<100	Membrane structure	[272]
CeO_2	A, B	1–5	Transparent, colorless films	[273–276]
CuO	C	8		[135]
	D	61		[135]
NiO				[277]
ZnO				[277]
$CoTiO_3$	C	28		[135]
	D	47		[135]
$NiTiO_3$	C	7		[135]
	D	22		[135]
		ZrO_2 Matrix		
CoO	C, D			[135]
CuO	C, D, E		Aerogel or xerogel	[135, 278, 279]
NiO	C, D			[135]
		SnO_2 Matrix		
CuO	F	20–23, 25–37		[280, 281]
$Sn(Eu)O_2$	F	2–3	Time-resolved fluorescence	[282]

3.3.4. Al or $MgAl_2O_4$ Matrices

Nanocomposites containing nanocrystals of mullite or ZrO_2 in an aluminum metal matrix have been reported; the metal oxide particles are formed using sol–gel processing [283]. The Vickers hardness values of these composites are substantially higher than that of aluminum.

Dispersion of commercial spinel powders in a sol solution of $Zr(OPr)_4$, acetylacetone, *n*-propanol, and water gives, after gelation and thermal treatment, nanocrystals of ZrO_2 in $MgAl_2O_4$ as a host matrix [284]. Intergranular ZrO_2 particles are 250 nm in size, whereas intragranular particles have diameters of only 50–100 nm. Orthorhombic ZrO_2 particles are observed.

3.3.5. Polymer Host Matrices

Metal oxide/organic polymer composites have been prepared through sol–gel processing under mild conditions. Polyacrylate composites containing nanoparticulate SiO_2 [285, 286], TiO_2 [287], ZrO_2 [288, 289], or $BaTiO_3$ [290] are formed by *in situ* sol–gel conversion of metal alkoxide precursors within a preformed polymer matrix or by cocondensation of metal alkoxides with derivatized monomers followed by initiated polymerization.

Hydrolysis and condensation of $Ti(O\textit{i}\text{-}Pr)_4$ in the presence of polyvinylacetate gives a polymer composite containing TiO_2 particles with an average diameter of 6 nm. Strong covalent cross-linking occurs between the organic and inorganic phases [287].

Prehydrolysis of TEOS affords colloidal silica gel particles 10.5 nm in size [285]. Nuclear magnetic resonance (NMR) data indicate that these particles possess ethoxy and hydroxyl groups at surface sites. Subsequent reaction of this colloid with the methacrylate monomer, $(MeO)_3Si(CH_2)_3O_2CC(Me){=}CH_2$, leads to sol–gel derivatization of the surface hydroxy groups and to formation of a 2-nm layer of methacrylate monomer on the silica gel particles. Copolymerization of these derivatized particles with excess methyl methacrylate under photochemical initiation gives the final inorganic/organic composite. At least 85% of the surface-bound methacrylate groups cross-link to the bulk polymer matrix.

Prehydrolysis of $Zr(OPr)_4$ in the presence of methacrylic acid and subsequent polymerization gives ZrO_2 particles with a diameter of 4 nm within a polymer matrix [288]. The presence of ZrO_2 alters the refractive index of the polymer matrix, and patterning techniques afford strip waveguides and other microoptical devices made from this material. Similar polymers containing ZrO_2 particles and an azobenzene chromophore formed under electrical poling exhibit second-harmonic NLO properties [289].

Sol–gel processing of methacryl(tri-isopropoxytitanium), barium alkoxide, and styrene forms $BaTiO_3$ particles of approximately 3 nm in diameter within a polymer matrix [290]. The dielectric properties of these inorganic/organic hybrid materials have been measured.

Several types of SiO_2 or SiO_2/TiO_2 nanocomposites in perfluorosulfonate ionomer matrices, such as Nafion, have been prepared [291–294]. Diffusion of TEOS into ionomer membranes occurs in aqueous alcohol solutions, and subsequent sol–gel hydrolysis and condensation affords particles of silica gel dispersed within the ionomer matrix. Particulate features have a dimension of approximately 5 nm. Postreaction with titanium alkoxides forms a TiO_2 gel layer as a shell on the silica particles and leads to interparticle binding. This percolative growth process occurs in the near-surface region of the membrane and controls the mechanical properties of the resulting inorganic/ionomer polymer.

Nanoparticulate SiO_2 or TiO_2 gels have been prepared in polymer or polymer composites of styrene-maleic anhydride copolymer [295, 296], poly(vinyl alcohol) [297], silicones [298,299], polyimides [300, 301], polyurethanes [302], or poly(hydroxypropylcellulose) [303]. The inorganic gel particulates are usually formed by *in situ* sol–gel processing of metal alkoxide precursors, such as TEOS. Gel particles are typically found in polymer pore structures with sizes smaller than 100 nm, although direct determination of ceramic particle sizes is lacking in many cases. These inorganic/organic composites are of interest because the structure and composition of the inorganic phase controls selected physical properties of the organic polymer. For example, a silica gel/poly(dimethylsilicone) nanocomposite has a butane/methane selectivity double that of the undoped polymer based on permeability measurements. Optical absorption spectra of films of a titania gel/poly(hydroxypropylcellulose) nanocomposite demonstrate that this composite is transparent in the visible region but acts as an efficient block for UV radiation.

3.4. Metal Chalcogenide (S, Se, or Te) Nanocomposites

3.4.1. CdS/Silica Gel Composites

Cadmium sulfide/silica gel nanocomposites are formed by a variety of synthetic methods, as shown in Figure 1. The most common procedure involves dissolving cadmium nitrate [148, 304–311] or cadmium acetate [312–319] in conventional silica sol–gel formulations, using TEOS, followed by reaction with H_2S (Method A). CdS particles with average diameters of 1.6–10 nm are produced, depending on the weight percentage of CdS present in the xerogel and the conditions of thermal annealing. The addition of particle stabilizers, such as poly(vinyl alcohol) or sodium hexametaphosphate (Method B) [320], or the bifunctional ligand, $H_2N(CH_2)_3Si(OEt)_3$, to solutions of a cadmium(II) salt (Method C) [321, 322] gives CdS particle sizes in the range of 2.3–4.2 nm. Sol–gel condensation of bifunctional $(EtO)_3SiX$ compounds, where X is NH_2, SH, or carboxylate, and cadmium

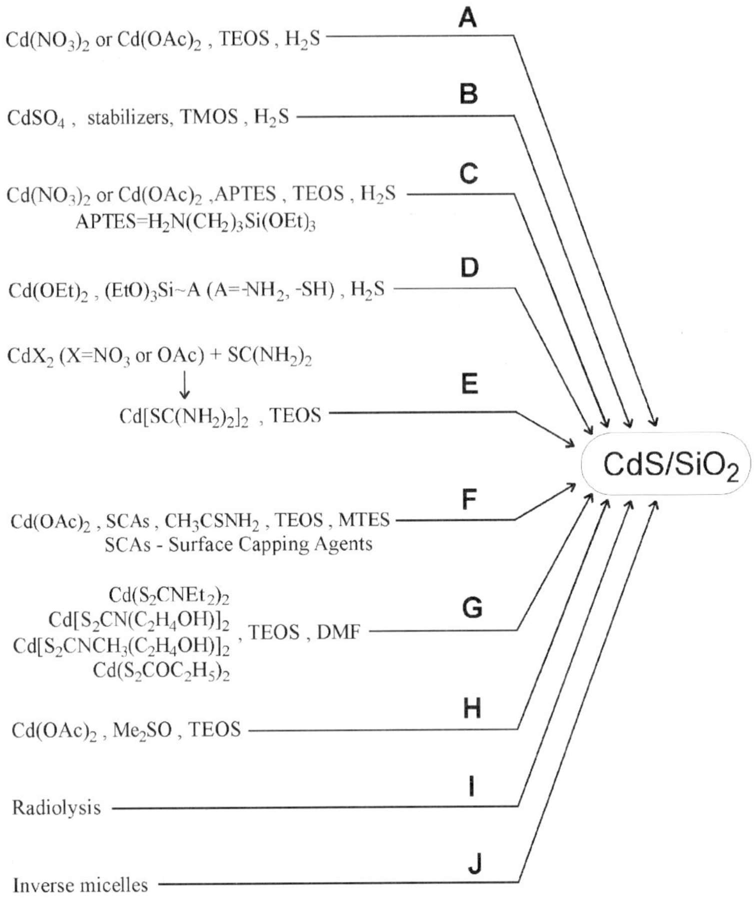

Fig. 1. Preparative routes to CdS/silica gel nanocomposites.

ethoxide gives a silica gel presumably doped with Cd(II) complexes [323]. Subsequent reaction with H_2S affords a nanocomposite containing CdS particles of less than 5 nm in diameter (Method D).

CdS/silica gel composites are also obtained by sol–gel condensation of TEOS, Cd(II) salts, and sources of sulfur other than H_2S. Using thiourea complexes of cadmium(II) nitrate [324–327] or acetate [328] together with TEOS gives CdS particles with average diameters of 2.1–8 nm (Method E). Thin-film CdS/silica gel nanocomposites have been prepared by mixing preformed CdS colloids with silica sols (Method F) [329]. CdS particles are also formed by the reaction of cadmium acetate with thioacetamide. Control of CdS particle size is achieved by using surface-capping agents as additives. Hydrolysis of TEOS along with dithiocarbamatocadmium complexes using dimethylformamide (DMF) as a cosolvent gives CdS particulates having average diameters of 2.7–7.2 nm (Method G) [330, 331]. Dimethylsulfoxide (DMSO) can serve as a source of sulfur and as solvent for the sol–gel conversion of TEOS and cadmium acetate to form a CdS/silica gel nanocomposite (Method H) [332]. The observed average CdS particle size is 2.7 nm.

Cadmium sulfide/silica gel composites can also be prepared using γ-radiolysis (Method I) [196, 333] and inverse micelle techniques (Method J) [196]. Other reports of CdS/silica gel nanocomposite formation are available [334–345].

Precipitation of CdS nanoparticles occurs in gel matrices at low temperatures, particularly when H_2S is the source of sulfur. More unreactive sources of sulfur, such as dithiocarbamate or thioacetamide, require mild heating for sulfur generation (usually as H_2S in an aquated sol–gel matrix). CdS/silica gel composites, once prepared, are frequently annealed at temperatures up to 1000 °C under nonoxidizing conditions. CdS particles are obtained as a distribution of sizes, although there is much interest in controlling this polydispersity [313, 317].

The spectral properties of CdS nanocomposites, in general, have been extensively studied because CdS nanoclusters behave as semiconductor quantum dots, and optical confinement effects are expected. CdS/silica gel nanocomposites have been studied by Raman spectroscopy [310, 311, 318, 335, 341, 343] and optical absorption spectroscopy [310, 312, 314, 316, 317, 319, 328, 337, 342, 346–348] including spectral hole burning [334]. Of particular interest are photoluminescence [307, 309, 323, 330, 331, 336, 340, 349] and NLO [148, 304–306, 314, 338, 339, 341, 342, 344, 345, 350] properties, as discussed elsewhere. Charge carrier dynamics has also been studied in CdS nanoparticles [320, 351].

Raman data reveal CdS vibrations at 300 and 600 cm^{-1} consistent with the bulk material. Exciton formation occurs by absorption of light with specific quantum confinement effects being observed depending on the relative size of the nanoparticle and the exciton radius. The silica xerogel represents a dielectric matrix surrounding the CdS particles, and significant third-order NLO effects are observed. Nanoparticulate CdS shows optical absorption that is blue-shifted relative to the bulk band gap of 512 nm at room temperature, and photoluminescence across the CdS band gap has been observed near 515 and 525 nm.

3.4.2. CdS/Sodium Borosilicate Gel Composites

Cadmium sulfide nanocomposites are formed in $Na_2O/B_2O_3/SiO_2$ matrices using sol–gel routes like those shown in Figure 2. Most commonly, a mixture of TMOS, $B(OEt)_3$, and aqueous $NaOAc/Cd(OAc)_2$ solution serves as a source of both the matrix and the CdS guest phases (Method A) [352–358]. In this process, $Cd(OAc)_2$ microcrystallites precipitate as a dopant in the wet gel. Thermal treatment in air converts the $Cd(OAc)_2$ to CdO, and subsequent exposure of the doped gel to H_2S transforms the CdO particles into CdS nanocrystals. The CdS-doped gel can then be densified further through drying procedures. The doping level and size distribution of the CdS nanoparticles, which affect the nonlinear optical properties of these composites, depends on specific experimental conditions. Typically, CdS average particle sizes range from 3.5 to 10 nm.

Nanosized CdS quantum dots with narrow size distribution can be fabricated by the addition of the bifunctional ligand, $H_2N(CH_2)_3Si(OEt)_3$, to a mixture of TEOS, triethylborate, NaOAc, and $Cd(OAc)_2$ (Method B) [359]. The amine ligand presumably binds to the Cd(II) ion through *in situ* coordination and anchors these ions homogeneously to the matrix through hydrolysis and condensation reactions of the triethoxysilyl group.

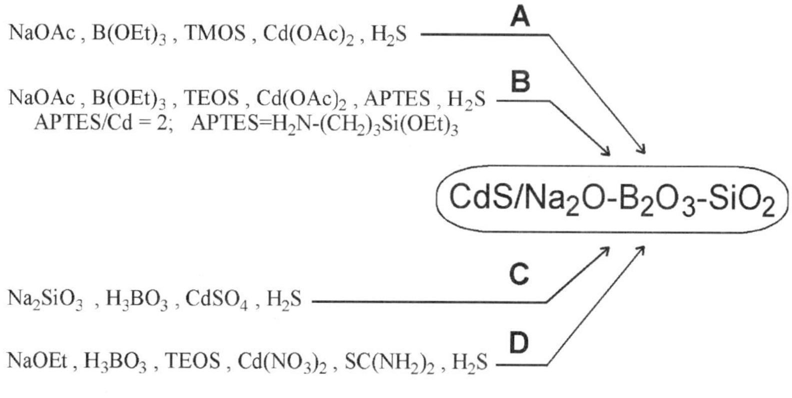

Fig. 2. Preparative routes to CdS/sodium borosilicate gel nanocomposites.

Sol–gel processing of sodium metasilicate, Na_2SiO_3, boric acid, H_3BO_3, with $CdSO_4$ as an additive gives a gel doped with Cd(II) ions. Following thermal treatment in air, these gels are heated to different temperatures while being exposed to H_2S gas (Method C) [360–363]. Careful control over the conditions of H_2S exposure affords nanocomposites containing CdS nanocrystals having average diameters of 2.15–9.2 nm.

Sodium borosilicate gels containing $Cd(NO_3)_2$ and $SC(NH_2)_2$ are formed using these compounds as additives and TEOS, boric acid, and sodium alkoxide as precursors of the host matrix. After heat treatment, the doped gels are exposed to H_2S gas to convert the formed CdO particles to CdS, giving average particle sizes of 8–10 nm (Method D) [364]. Other reports on the synthesis and properties of CdS/sodium borosilicate composites are available [365–367].

Because of the strong dependence of optical properties on semiconductor particle size, there is great interest in controlling the average size and particle size distributions in CdS nanocomposite materials. Regarding CdS/sodium borosilicate nanocomposites, studies of CdS particle size distributions on annealing conditions or exposure to H_2S have been reported [356, 360]. In general, CdS particle size increases with the temperature of reaction or annealing but is essentially independent of the period of exposure to H_2S. Narrow CdS particle size distributions [2.8 (0.9) nm in diameter] are obtained when $H_2N(CH_2)_3Si(OEt)_3$ is used as an additive [358, 359]. These materials have been successfully fabricated into channel waveguides.

Spectroscopic studies reported on CdS/sodium borosilicate gel nanocomposites include Raman spectroscopy [366], linear optical absorption spectroscopy [352, 354, 357, 359, 360, 362, 363], optical gain spectroscopy [354, 357], NLO properties [352, 353, 355, 367], and photoluminescence [358, 364]. The presence of CdS quantum dots is confirmed, and band edge luminescence near 495 nm is blue-shifted from the CdS bulk band gap energy of 512 nm with decreasing CdS particle size.

3.4.3. CdS/SiO_2–TiO_2 Gel Composites

Silica–titania ($30TiO_2$–$70SiO_2$) optical waveguides doped with CdS quantum dots have been prepared from TEOS, $Ti(OBu)_4$, and cadmium acetate [368]. Sulfide is obtained from thiourea, which is present as an additive. A more stable sol is achieved by adding acetylacetone to the cadmium acetate solution. The average CdS particle sizes obtained by this method range from 3.5 to 4.8 nm. Silica–titania (70:30) thin films containing CdS quantum dots form upon mixing CdS colloidal sols with a prehydrolyzed mixture of TEOS, methyltriethoxysilane (MTES), and $Ti(OBu)_4$ [329]. Cadmium sulfide particles are also obtained by the reaction of cadmium acetate with thioacetamide, and control of CdS particle size is achieved by using surface-capping agents (SCAs), such as acetylacetone, 3-aminopropyltriethoxysilane, 3-aminopropyltrimethoxysilane, or 3-mercaptopropyltrimethoxysilane. This procedure gives good control of CdS particle sizes in the 2- and 20-nm range.

In another approach, a silica–titania host layer 2.2 μm thick is deposited onto a Si substrate using sol–gel processing, and this layer is covered with a pure SiO_2 buffer layer 7.5 μm thick [369]. CdS crystallites precipitate in the porous matrix when this bilayer is soaked in CdF_2 solution and is subsequently treated with H_2S gas. The absorption spectrum of the resulting composite indicates that CdS crystallites having a maximum radius of approximately 2.6 nm are present.

A large blue shift (62 nm) of the CdS absorption edge is observed for SiO_2-TiO_2 films containing inclusions of CdS microcrystals [370]. This shift is attributed to quantum confinement effects in very small CdS quantum dots.

3.4.4. CdS Composites in Other Sol–Gel Matrices

Cadmium sulfide/alumina composites form when a mixture of cadmium acetate and $AlCl_3$ (which has been treated with NH_3 and peptized with acetic acid) is heated in a stream of

H_2S gas at temperatures between 100 and 600 °C [234, 371]. The NLO properties of these composites have been measured.

Stable confinement of CdS quantum dots in their lowest excitonic state has been achieved in a transparent ZrO_2 gel film. Optical absorption and fluorescent measurements of these films have been reported [372]. The addition of TiO_2 to the ZrO_2 gel matrix attenuates fluorescence emanating from impurities in the composite.

Polysilsequioxanes doped with CdS particles have been prepared by sol–gel condensation of either 1,4-bis(triethoxysilyl)benzene (Structure II) or 1,6-bis(triethoxysilyl)hexane (Structure IV) to give inorganic/organic gels. Soaking these xerogels for 2 days in aqueous $CdCl_2$ solution followed by treatment with a solution of Na_2S leads to precipitation of CdS [373, 374]. Because the host xerogels produced from the compound shown in Structure II or IV have different average pore sizes, the resulting CdS nanocomposites have average diameters of 5.8 and 9 nm, respectively.

Inorganic/organic composites containing nanoclusters of CdS and Co metal result when a Co nanocomposite prepared from 1,4-bis(triethoxysilyl)benzene (Structure II) and chromium tricarbonyl(triethoxysilyl)benzene (Structure I) is soaked with a solution of cadmium chloride. Sulfide is introduced by soaking the cadmium-doped gel with a solution of sodium sulfide, and CdS precipitation is induced by a subsequent heat treatment [116]. A mixture of phase-separated chromium and CdS nanocrystallites is obtained.

Optically transparent ZnO films on glass substrates (prepared by sol–gel dip processing) serve as a host matrix for CdS quantum dots [375]. In these CdS/ZnO composites, a rapid electron transfer from CdS particles to the ZnO matrix takes place rapidly following photoinduced electron–hole pair generation in the CdS nanoclusters. The stored electrons cause a blue shift of the absorption threshold of ZnO, an effect attributed to exciton polarization within the created excess electric field.

Silica glasses doped with $Cd_xZn_{1-x}S$ nanoclusters are formed by hydrolysis and polymerization of TEOS, cadmium acetate, and zinc acetate in DMSO solution [376]. Optical absorption measurements of the resulting composite indicate a greater shift of the optical absorption edge toward the blue than observed with related CdS/SiO_2 composites. The magnitude of this shift has been attributed to the possible formation of a solid solution in the glass matrix affording particularly small CdS clusters.

Sandwiched CdS–PbS composites are synthesized by the addition of a THF solution of lead methoxy ethanolate to a THF solution of $(CH_3)_3Si{-}S{-}Si(CH_3)_3$, cadmium ethoxide, and a bifunctional diamine that stabilizes the quantum dots [377]. PbS formation completely quenches the intense room temperature photoluminescence of the CdS clusters, indicating that PbS growth occurs at the surface of the CdS nanoparticles. PbS–CdS composites having cluster sizes less than 5 nm also exhibit third-order nonlinear optical properties.

CdS/Mn particles 1.2–2.4 nm in diameter have been prepared in silica xerogels using inverse micelle synthetic methods [336, 378]. Reagents include cadmium nitrate, manganese nitrate, sodium sulfide, pyridine (as a surface-capping agent), and $SiMe(OEt)_3$. Extensive studies (extended X-ray absorption fine structure, electron spin resonance, and electron–nuclear double resonance) reveal that incorporation of Mn(II) ions as a dopant in metal sulfide nanocomposites induces a bright photoluminescence (quantum yield of 7%) resulting from the $^4T_1 \rightarrow {}^6A_1$ transition centered on the Mn(II) ion.

3.4.5. CdS_xSe_{1-x}/Silica Gel Composites

Because of the difference in bulk band gaps between CdS (2.4 eV) and CdSe (1.7 eV), cadmium sulfoselenide/glass composites are commercially produced as color filters. Variation in the relative amounts of sulfur and selenium provides control over the transmission window of the glass filter. A desire to prepare color filters having cadmium chalcogenide particles of more uniform size has inspired preparations of CdS_xSe_{1-x}/silica gel composites [379–381]. Synthetic routes include mixing soluble sources of Cd(II) ion, sulfur, and selenium into conventional sol–gel formulations [379]; treatment of silica gels doped with

Cd(II) ions and selenide ions with gaseous H_2S [380]; successive diffusion of Cd(II) and chalcogenide solutions into preformed silica gels [381]; diffusion of chalcogenide ions into silica gels doped with Cd(II) ions [381]; or solid-couple diffusion of two separate gels, one containing Cd(II) ion and the other containing a source of chalcogenide ion [381]. $CdSeO_4$, NH_4SCN, K_2SeO_4, and elemental Se dissolved in nitric acid are common sources of chalcogenide ion. Ammonium thiocyanate is converted to thiourea in aqueous solution. Subsequent thermal treatment is the last step in these syntheses.

Variation of the S/Se atomic ratio is controlled through reactant stoichiometry. For those composites prepared by heating doped silica gels in an atmosphere of H_2S, the sulfide content increases with increasing temperature. XRD data confirm a single hexagonal nanoparticulate phase consistent with formation of a Cd/S/Se solid solution. The average sizes of the nanoparticles are usually less than 10 nm. Raman peaks at 300 and 200 cm^{-1} are assigned to the expected Cd–S and Cd–Se phonon bands, respectively. Optical absorption spectra reveal the expected blue shift in absorption onset.

3.4.6. CdSe/Gel Composites

CdSe nanocomposites have been prepared in silica gel [325, 338, 382–385], organically modified silica gel [386], sodium borosilicate gel [362, 363, 365, 387], or related matrices [337, 339, 346]. Synthetic methods parallel those used to prepare CdS/gel composites by sol–gel processing. Sources of selenide ion include KSeCN, H_2SeO_3, or Se dissolved in nitric acid. However, molecular precursors of Se, such as $Cd(SePh)_2$ or salts of $Cd[Se(NH_2)_2]_2^{2+}$, have also been used [325, 385]. The initially formed composites are usually treated thermally under nonoxidizing conditions to form CdSe quantum dots.

CdSe nanoparticle average sizes usually are 10 nm or less. Impregnation of silica gels doped with Cd(II) ions by solutions of KSeCN leads to bimodal CdSe particle size distributions [382]. Some control of CdSe average particle size is obtained through careful heat treatment [384]. As found with CdS nanocomposites, CdSe quantum dots provide both linear [337, 338, 384, 385, 387] and nonlinear [338, 339, 346] optical properties of interest.

3.4.7. CdTe/Gel Composites

CdTe nanocomposites in silica gel [388, 389], sodium borosilicate [141, 390, 391], or related matrices [346] are known. Sources of telluride ion include $CdTeO_4$, H_2TeO_4, or Te dissolved in nitric acid. The most common synthetic strategy involves adding soluble sources of both Cd(II) ion and tellurium to a conventional sol–gel formulation followed by thermal treatment. The CdTe average particle size is typically below 10 nm. Under controlled heat treatment, CdTe/silica gel composites having average CdTe particle sizes in the range of 4 to 9 nm have been prepared. Although the optical absorption edge is blue-shifted as CdTe nanocluster size decreases [388, 389], this shift is not always attributed to quantum confinement effects [390]. A mixed-metal telluride composite, $Cd_xHg_{1-x}Te$/silica gel, has been prepared by sol–gel methods using $Hg(OAc)_2$ as a soluble source of Hg [337, 338, 392]. The optical absorption spectrum and NLO properties of this composite have been measured. A blue shift in absorption edge is observed with decreasing nanocluster size [392].

3.4.8. ZnS/Gel Composites

Zinc sulfide is a semiconductor substance with an optical band gap of 3.6 eV (344 nm). As expected, quantum dots of ZnS have electronic structures between those of bulk and molecular materials, and quantum confinement effects should be evident by a blue shift of the absorption edge with decreasing particle size. As with cadmium chalcogenide quantum dots, sol–gel preparations of ZnS nanocomposites provide an entry into the fabrication of devices having tunable optical properties, including nonlinear optical effects.

ZnS nanocomposites have been prepared by sol–gel processing in silica gel [325, 326, 337–339, 393–396], aluminoborosilicate gel [397, 398], and related glassy matrices [333]. Synthetic routes include (1) the addition of zinc and sulfide sources, such as $Zn(OAc)_2$ and thiourea or a zinc thiourea complex, to a conventional sol–gel formulation [325, 326, 339]; (2) a two-step process in which Zn(II) ion is incorporated directly into a sol–gel matrix using zinc additives followed by the introduction of sulfide using H_2S thermal treatment [337, 338, 393–396]; (3) the preparation of a sol–gel matrix of well-defined pore structure followed by successive incorporation of Zn(II) by solution absorption and sulfide ion by H_2S thermal treatment [397, 398]; or (4) γ-radiolysis or membrane methods [333]. Thermal aging or annealing of the initially prepared nanocomposite induces ZnS particle growth.

ZnS average particle sizes are typically less than 10 nm. Synthetic control of ZnS particle size distribution by careful thermal treatment and regulation of doping level leads to size distributions of approximately 2–3 nm in breadth. Size distributions of approximately 6 nm in breadth are otherwise more common. ZnS particle size increases with increasing dopant level and annealing temperature. Careful preparations of ZnS/aluminoborosilicate composites indicate that the pore size of the host matrix constrains the size of ZnS particulates. Average pore sizes of 2.0–3.5 nm give ZnS average particle sizes of 2.3–4.6 nm. The larger ZnS particles are formed at higher annealing temperatures where pore size restraint begins to break down [397, 398].

ZnS quantum dots crystallize with the sphalerite structure [396], and Raman spectra reveal Zn–S vibrations at 283 and 544 cm^{-1} [395]. The optical absorption edges of ZnS quantum dots of average sizes 2.3, 4.0, or 4.6 nm appear at 280, 310, or 315 nm, respectively, showing the expected blue shift of absorption onset with decreasing dot size [398]. NLO properties of the ZnS nanocomposites have also been observed [338, 339, 393–395], including photoluminescence [394].

3.4.9. ZnSe or ZnTe/Gel Composites

Zinc chalcogenide/silica gel nanocomposites of the heavier chalcogens, Se and Te, have been prepared using synthetic routes analogous to those used for the preparation of ZnS composites [325, 338, 339]. Ascorbic acid can be used as an antioxidant additive [325]. Optical absorption spectra reveal band gaps of the ZnX quantum dots greater than that of the bulk substances, and NLO measurements have been reported [338, 339].

3.4.10. PbS/Gel Composites

Lead sulfide is a semiconductor with an optical band gap near 0.41 eV (3020 nm). Composite materials containing PbS quantum dots are of interest as possible optical devices based on expected quantum confinement effects.

PbS nanocomposites have been prepared via sol–gel processing in silica gel [329, 393, 399–402], aluminoborosilicate gel [397, 403], silica–titania gels [329, 368], and related glassy matrices using methods similar to those used to prepare ZnS nanocomposites [333, 339, 404]. Synthetic routes include (1) the addition of lead and sulfide sources, such as lead acetate or nitrate salts or lead thiourea complexes and thiourea or thioacetamide, to conventional sol–gel formulations [329, 368, 401]; (2) a two-step process in which Pb(II) ion is incorporated directly into a sol–gel matrix using lead additives followed by the introduction of sulfide using H_2S thermal treatment [339, 393, 399]; (3) the preparation of a sol–gel matrix of well-defined pore structure followed by successive incorporation of Pb(II) by solution absorption and sulfide ion by H_2S thermal treatment [397]; (4) mixing colloidal PbS dispersions into conventional sols [400]; or (5) γ-radiolysis and membrane methods [333, 404]. Thermal annealing of the formed nanocomposites is a common practice.

Average PbS particle sizes are typically less than 10 nm, although PbS particles with average diameters of 25–31 nm have also been reported [368]. Synthetic control of PbS particle size distribution by careful thermal treatment and regulation of doping level leads to distributions of approximately 0.8–3.5 nm in breadth. Size distributions of approximately 2–6 nm in breadth are otherwise more commonly observed. The PbS average particle size increases with increasing dopant level and annealing temperature [401]. Careful preparations of PbS/aluminoborosilicate composites reveal that the pore size of the host matrix constrains PbS particle size formation, but to a lesser degree than observed for ZnS particulate formation [397].

The optical absorption edge of bulk PbS near 3020 nm can be blue-shifted to near 650 nm through quantum confinement effects within PbS quantum dots 2–3 nm in size [401]. Intermediate absorption edge onsets can be tailored by control of the PbS average particle size. NLO properties of the PbS nanocomposites have also been observed [339, 400].

3.4.11. PbSe or PbTe/Gel Composites

PbSe/porous glass composites are obtained by soaking gels doped with Pb(II) ions with solutions containing selenide ion [339]. The resulting glassy composites show quantum confinement effects and third-order harmonic generation.

PbTe nanocomposites in silica or borosilicate gels are formed by conventional sol–gel methods using lead nitrate and H_6TeO_6 as additives [405]. The initial gel is treated in air at 450 °C followed by further annealing in nitrogen/hydrogen atmosphere. Annealing temperatures of 400–500 °C for periods of 10–120 min give crystalline PbTe particulates 2–15 nm in size. PbTe particle growth is more rapid in a borosilicate matrix. Although the band gap of bulk PbTe is in the infrared (0.31 eV, 4000 nm), the absorption edge onset is shifted into the near-visible region for these PbTe nanocomposites.

3.4.12. Other Metal Chalcogenide/Gel Nanocomposites

The addition of bismuth nitrate and thiourea to a conventional silica sol–gel formulation followed by thermal treatments of 350 °C in air and 400–500 °C in nitrogen gives a Bi_2S_3/silica gel nanocomposite [148, 406–408]. The Bi_2S_3 average particle size is less than 10 nm. Complex optical absorption spectra show absorption valleys ranging from 510 to 660 nm with a red shift being observed for gels subjected to longer heating. Thermal treatment for 2 h at 400 °C affords Bi_2S_3 particles with an average diameter of 2 nm. This composite emits luminescence ranging from 345 to 540 nm upon photoexcitation.

Sol–gel processing using the techniques mentioned previously has been used to prepare several metal chalcogenide nanocomposites for designed applications or property enhancements. Materials of these types include (1) WSe_3 nanocomposites as patternable silicate–copolymer films and monoliths for optical applications [386], (2) submicrometer silica gel particles suspended in styrene–butadiene latex rubber for strength reinforcement [409], (3) silica gel particulates on silicon for integrated optical devices [316], (4) SnO_2 deposition in the pores of porous silicon for gas sensor applications [410], (5) ZnO membrane formation from sol–gel-derived ZnO colloids as luminescent chemical sensors [411], (6) nanoparticles of Cu or CuO as coatings for coloration [412], or (7) boehmite coatings of Si_3N_4 particles to provide fracture-free green-state deformation of solid compacts [413].

3.5. Metal Halide Nanocomposites

3.5.1. Silver Halide/Gel Nanocomposites

Colloidal and surface-passivated submicrometer particles of a Ag halide have been prepared as a dispersion in a gel matrix using freeze–dry and sol–gel methods [414]. This nanocomposite exhibits large NLO properties because of quantum size effects.

3.5.2. Copper Halide/Gel Nanocomposites

Suspensions of Cu_2O or CuX (X = Cl or Br) in acetonitrile have been used in the preparation of glasses doped with quantum dots of CuCl and CuBr, as shown in Figure 3 [148, 304, 415, 416]. Complete dissolution of the copper reagent occurs in the sol.

Thin-film glass composites of CuCl or CuBr having waveguide properties are obtained by the addition of a Cu_2O/acetonitrile suspension to partially hydrolyzed TEOS [304]. Sol–gel processing is initiated by the addition of water, methanol, and either HCl or HBr. Copper halide average particle size is controlled (48.2–51.9 nm) through regulation of the Cu_2O concentration and by using montmorillonite as an entrapping support (23.9 nm). These nanocomposites exhibit NLO properties.

Silica composites containing cubic CuCl nanocrystals 3–6 nm in size are formed by sol–gel processing of solutions of TEOS and CuCl using HCl as a catalyst [415]. Optical absorption measurements indicate strong quantum confinement effects, and peaks observed at approximately 370 and 380 nm for samples heated above 700 °C are attributed, respectively, to excitations of the confined $Z_{1,2}$ and Z_3 excitons in CuCl microcrystals. As the size of the CuCl crystals decreases, exciton energies show a blue shift. A resonant third-order nonlinear susceptibility of 1.1×10^{-8} esu is observed at 77 K for this nanocomposite.

CuBr doped silica–alumina composites are obtained through sol–gel processing of solutions containing TEOS, $Al(OC_4H_9)_3$, and CuBr [416]. Heating the resulting gels to 900 °C in N_2 results in the formation of cubic CuBr nanocrystals 5–8 nm in diameter. Optical absorption spectra of samples heated above 700 °C show bands near 3.05 and 3.20 eV at 77 K. These peaks are attributed to the $Z_{1,2}$ and Z_3 excitons, respectively, in CuBr nanocrystals and show a blue shift as the size of the CuBr crystals decreases.

3.5.3. Magnesium Halide/Gel Nanocomposites

MgF_2/silica gel composites, containing MgF_2 particles with an average size of 5–10 nm, depending on the conditions of thermal treatment, are formed when a MgF_2 sol is mixed with silicate sols. The MgF_2 sol is formed by reacting HF with a sol prepared from methanolic H_2O_2 and $Mg(OCH_3)_2$, whereas the silicate sols are prepared from acid-catalyzed hydrolysis and condensation of TMOS [417].

3.5.4. Lead Halide/Gel Nanocomposites

PbI_2/silica gel nanocomposite films containing PbI_2 particles with an average diameter of 6 nm have been prepared by sol–gel methods [148, 418–420]. The Pb(II) ions are introduced into the sol as an additive, while iodide ions are added at a later stage. X-ray

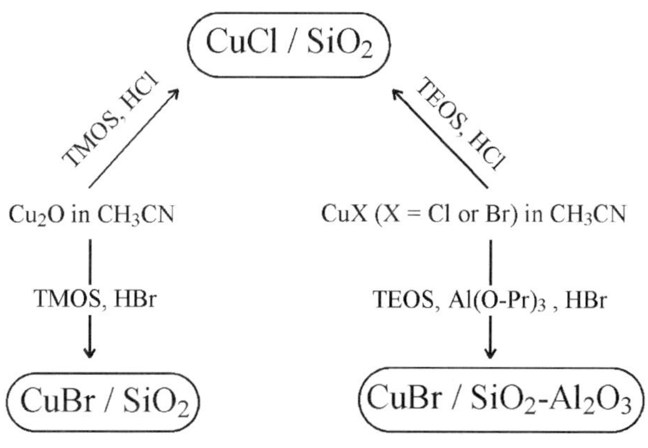

Fig. 3. Preparative routes to copper halide/gel nanocomposites.

diffraction from the PbI_2 particles is observed only when the average particle size is larger than 8 nm [418]. Low-temperature luminescence from the PbI_2 nanoparticles consists of a series of exciton lines near 2.5 eV and additional donor-acceptor combination luminescent bands at lower energies, centered at 2.44 eV (green) and 2.07 eV (red), respectively. A correlation between donor–acceptor recombination emission processes and relaxation by lattice imperfection was examined utilizing optically detected magnetic resonance (ODMR) spectroscopy.

4. SUMMARY

Sol–gel processing of nanocomposite materials has attracted much interest because of the generality of this synthetic method. A wide variety of host matrices can be formed under very mild conditions with relatively inexpensive reagents using sol–gel chemistry. Conversion of a sol to a gel using hydrolysis and intermolecular condensation reactions occurs, under proper conditions, even for complex reactant mixtures affording convenient syntheses of host matrices having complex compositions. This flexibility enables the tuning of the pore size, surface area, density, dielectric constant, refractive index, and chemical composition of a host matrix. Inorganic/organic matrices are also accessible, and sol–gel chemistry even occurs within inverse micelles.

Formation of the nanoparticulate phase is achieved most commonly through reduction processes for elemental particles or precipitation reactions for chemical substances. A variety of methods are available to initiate particle formation. Nanoparticle growth is usually promoted by thermal annealing, and, for many substances, thermal energy is required for single-crystal formation. This growth process typically affords nanoclusters of the guest phase having a distribution of particle sizes. Synthetic control of nanoparticle size has been, and remains, a challenge in sol–gel processing. Although control of host matrix pore size gives some degree of control of nanoparticle size, convenient routes to monodisperse nanocomposites are still needed. Although a high degree of monodispersity might not be required for some applications, it is highly desired for most optical applications.

Classical homogeneous nucleation theories, as applied to the growth of CdS or CdSe quantum dots in a glass matrix, predict (1) a nucleation stage in which nanocluster size distribution is Gaussian; (2) a normal growth stage where the average radius of the nanoclusters is proportional to the square root of the annealing time; and (3) a coalescence stage in which a supersaturated glass solid solution is established, and the average radius of the nanoclusters is proportional to the cube root of the annealing time [421]. Coalescence-like particle growth is observed for CdSe particles of average radius 2.5 (1.4) nm grown in a silicate-based glass matrix. The growths of Ag nanoclusters in a sputtered SiO_2 matrix and of ZnS nanoclusters in a sol–gel silica matrix also follow a cube root dependence on annealing time [396, 422]. Given the dependence of average nanoparticle size and particle size distribution on the time and temperature of thermal treatment, less energy intensive synthetic strategies are needed to achieve highly monodisperse nanocomposites.

Acknowledgments

C.M.L. thanks the U.S. Army Research Office under Grant DAAH04-95-1-0146 for support. We also thank Mr. Stan Griffin for assistance in preparing the manuscript.

References

1. C. J. Brinker and G. W. Scherer, "Sol–Gel Science." Academic Press, Boston, 1990.
2. A. K. Cheetham, C. J. Brinker, M. L. Mecartney, and C. Sanchez, eds., "Better Ceramics through Chemistry VI," Vol. 346. Materials Research Society, Pittsburgh, PA, 1994.

3. M. J. Hampden-Smith, W. G. Klemperer, and C. J. Brinker, eds., "Better Ceramics through Chemistry V," Vol. 271. Materials Research Society, Pittsburgh, PA, 1992.
4. L. L. Hench and J. K. West, *Chem. Rev.* 90, 33 (1990).
5. R. J. P. Corriu and D. Leclercq, *Angew. Chem., Int. Ed. Engl.* 35, 1420 (1996).
6. J. Zarzycki, *J. Sol.–Gel Sci. Technol.* 8, 17 (1997).
7. S. Sakka, Proc. SPIE-Int. *Soc. Opt. Eng.* 1758, 2 (1992).
8. A. M. Klonkiowski, *Wiad. Chem.* 47, 497 (1993).
9. J. D. Mackenzie and D. R. Ulrich, *Proc. SPIE-Int. Soc. Opt. Eng.* 1328, 2 (1990).
10. (a) R. M. Laine, *Proc. SPIE-Int. Soc. Opt. Eng.* 1328, 16 (1990). (b) J. Y. Ying, ed., "Sol–Gel Derived Materials," *Chem. Mater.* 9 (1997).
11. M. A. Cauqui and J. M. Rodriguez-Izquierdo, *J. Non-Cryst. Solids* 147–148, 724 (1992).
12. G. M. Pajonk, *Appl. Catal.* 72, 217 (1991).
13. U. Schubert, *New J. Chem.* 18, 1049 (1994).
14. D. R. Uhlmann, S. Motakef, T. Suratwala, J. Young, J. M. Boulton, B. J. J. Zelinski, Z. Gardlund, G. Teowee, and J. Cronin, *Proc. SPIE* 2288, 2 (1994).
15. B. K. Coltrain, C. Sanchez, D. W. Schaefer, and G. L. Wilkes, eds., "Better Ceramics through Chemistry VII," Vol. 435. Materials Research Society, Pittsburgh, PA, 1996.
16. U. Schubert, N. Husing, and A. Lorenz, *Chem. Mater.* 7, 2010 (1995).
17. B. Lebeau, S. Brasselet, J. Zyss, and C. Sanchez, *Chem. Mater.* 9, 1012 (1997).
18. L. Ukrainczyk, R. A. Bellman, and A. B. Anderson, *J. Phys. Chem. B* 101, 531 (1997).
19. J. C. Schrotter, M. Smaihi, and C. Guizard, *Mater. Res. Soc. Symp. Proc.* 435, 199 (1996).
20. G. M. Jamison, D. A. Loy, K. A. Opperman, J. V. Beach, and R. M. Waymouth, *Polym. Prepr.* 37, 297 (1996).
21. S. K. Young, Q. Deng, and K. A. Mauritz, *Polym. Mater. Sci. Eng.* 74, 309 (1996).
22. C. L. Beaudry and L. C. Klein, *ACS Symp. Ser.* 622, 382 (1996).
23. G. W. Jang, C. Chen, R. W. Gumbs, Y. Wei, and J. M. Yeh, *J. Electrochem. Soc.* 143, 2591 (1996).
24. F. Leroux, B. E. Koene, and L. F. Nazar, *J. Electrochem. Soc.* 143, L181 (1996).
25. C.G. Wu, D. C. DeGroot, H. O. Marcy, J. L. Schindler, C. R. Kannewurf, Y.-J. Liu, W. Hirpo, and M. G. Kanatzidis, *Chem. Mater.* 8, 1992 (1996).
26. Y.J. Liu, J. L. Schindler, D. C. DeGroot, C. R. Kannewurf, W. Hirpo, and M. G. Kanatzidis, *Chem. Mater.* 8, 525 (1996).
27. P. R. McDaniel and T. L. St Clair, *Mater. Res. Soc. Symp. Proc.* 405, 535 (1996).
28. M. A. Harmer, W. E. Farneth, and Q. Sun, *J. Am. Chem. Soc.* 118, 7708 (1996).
29. G. M. Kloster, J. A. Thomas, P. W. Brazis, C. R. Kannewurf, and D. F. Shriver, *Chem. Mater.* 8, 2418 (1996).
30. M. A. Harmer and Q. Sun, PCT Int. Appl. WO 9619288 A1 960627, 59 pp.
31. H. Krug, N. Merl, and H. Schmidt, *Teubner-Texte Phys.* 27, 192 (1993).
32. C. Sanchez, M. In, P. Toledano, and P. Griesmar, *Mater. Res. Soc. Symp. Proc.* 271, 669 (1992).
33. M. A. Meneses-Nava, S. Chavez-Cerda, and J. J. Sanchez-Mondragon, *Proc. SPIE-Int. Soc. Opt. Eng.* 2778, 395 (1996).
34. P. N. Prasad, R. Gvishi, G. S. He, J. D. Bhawalkar, N. D. Kumar, G. Ruland, C. F. Zhao, and B. A. Reinhardt, *Mater. Res. Soc. Symp. Proc.* 413, 203 (1996).
35. E. J. A. Pope, *Mater. Res. Soc. Symp. Proc.* 345, 331 (1994).
36. D. L'Esperance and E. L. Chronister, in "Act. Mater. Adapt. Struct., Proc. ADPA/AIAA/ASME/SPIE Conference" (J. G. Knowles, ed.), pp. 875–878. IOP Publishing, Bristol, UK 1992.
37. Y. Wei, D. Jin, and T. Ding, *J. Phys. Chem. B* 101, 3318 (1997).
38. S. Yano and M. Kodomari, *Nihon Reoroji Gakkaishi* 24, 15 (1996).
39. F. Bentivegna, M. Canva, A. Brun, F. Chaput, and J. P. Boilot, *J. Appl. Phys.* 80, 4655 (1996).
40. G. Qian and M. Wang, *J. Phys. Chem. Solids* 58, 375 (1997).
41. R. L. Matthews and E. T. Knobbe, *Chem. Mater.* 5, 1697 (1993).
42. Y. J. Lin and C. J. Wu, *Surf. Coat. Technol.* 88, 238 (1997).
43. R. Reisfeld, V. Chernyak, M. Eyal, and C. K. Jorgensen, in "Int. Sch. Excited States Transition Elem." (W. Strek, ed.), 2nd ed., Vol. 247. World Scientific, Singapore, 1992.
44. T. N. Vasilevskaya and R. I. Zakharchenya, *Fiz. Tverd. Tela* 38, 3129 (1996).
45. G. E. S. Brito, S. J. L. Ribeiro, V. Briois, J. Dexpert-Ghys, C. V. Santilli, and S. H. Pulcinelli, *J. Sol.–Gel Sci. Technol.* 8, 261 (1997).
46. S. P. Feofilov, A. A. Kaplyanskii, A. B. Kutsenko, T. N. Vasilevskaya, and R. I. Zakharchenya, *Mater. Sci. Forum* 239–241, 687 (1997).
47. K. Yao, K. Kong, L. Kong, L. Zhang, and X. Yao, *Xi'an Jiaotong Daxue Xuebao* 30, 40, 59 (1996).
48. M. C. Gust, L. A. Momoda, and M. L. Mecartney, *Ceram. Trans.* 55, 85 (1995).
49. Z. Xu, H. K. Chae, M. H. Frey, and D. A. Payne, *Mater. Res. Soc. Symp. Proc.* 271, 339 (1992).
50. Z. Zhong, Z. Hu, O. Yan, X. Fu, and H. Peng, *Hunan Shifan Daxue Ziran Kexue Xuebao* 19, 48 (1996).
51. Y. Wu, T. Yamaguchi, and C. Xianghao, *Nippon Kagaku Kaishi* 12, 1376 (1993).
52. S. Zhao, H. Cai, Z. Peng, C. Zhao, B. Xu, M. Zhao, and J. Xing, *Wuji Huaxue Xuebao* 10, 409 (1994).

53. Z. Zhong, L. Chen, Q. Yan, X. Fu, and J. Hong, *Stud. Surf. Sci. Catal.* 91, 647 (1995).
54. W. H. Shih and Q. Lu, *Ferroelectrics* 154, 1379 (1994).
55. W. L. Zhong, B. Jiang, P. L. Zhang, J. M. Ma, H. M. Cheng, Z. H. Yang, and L. X. Li, *J. Phys.: Condens. Matter.* 5, 2619 (1993).
56. S. Lu, L. Zhang, and X. Yao, *Proc. IEEE Int. Symp. Appl. Ferroelectr.* 8, 385 (1992).
57. Q. Zhang, Z. Zhang, D. Shen, W. Xue, H. Wang, and M. Zhao, *Proc. IEEE Int. Symp. Appl. Ferroelectr.* 8, 551 (1992).
58. Q. F. Zhou, Y. Han, L. Y. Zhang, and X. Yao, *Proc. IEEE Int. Symp. Appl. Ferroelectr.* 8, 621 (1992).
59. Q. Zhou, Y. Li, L. Zhang, and X. Yao, *Chin. Sci. Bull.* 37, 1402 (1992).
60. T. R. Narayanan Kutty and P. Padmini, *J. Mater. Chem.* 7, 521 (1997).
61. K. No, D. S. Yoon, and J. M. Kim, *J. Mater. Res.* 8, 245 (1993).
62. Y. A. Attia, D. K. Sengupta, and H. A. Hamza, in "Sol.–Gel Process. Appl." (Y. A. Attia, ed.), pp. 159–168. Plenum, New York, 1994.
63. A. R. Modak and S. K. Dey, *Integr. Ferroelectr.* 5, 321 (1994).
64. J. F. Meng, B. K. Rai, R. S. Katiyar, and G. T. Zou, *Phys. Lett. A* 229, 254 (1997).
65. A. K.Bhattacharya, K. K. Mallick, A. Hartridge, and J. L. Woodhead, *Mater. Lett.* 18, 247 (1994).
66. J. Mira, J. Rivas, D. Fiorani, R. Caciuffo, D. Rinaldi, C. Vazquez Vazquez, J. Mahia, M. A. Lopez Quintela, and S. B. Oseroff, *J. Magn. Magn. Mater.* 164, 241 (1996).
67. Y. K. Sun, I. H. Oh, and S. A. Hong, *J. Mater. Sci.* 31, 3617 (1996).
68. Y. Fan and D. Chen, *Wuji Huaxue Xuebao* 8, 194 (1992).
69. B. K. Kim, N. Mizuno, and I. Yasui, *Kongop Hwahak* 7, 1034 (1996).
70. R. M. Almeida, M. I. DeBarros Marques, and X. Orignac, *J. Sol.–Gel Sci. Technol.* 8, 293 (1997).
71. S. M. Koo, D. H. Lee, C. S. Ryu, and Y. E. Lee, *Kongop Hwahak* 8, 301 (1997).
72. G. Dagan, S. Sampath, and O. Lev, *Chem. Mater.* 7, 446 (1995).
73. I. M. Miranda Salvado, F. M. A.Margaca, and J. Teixeira, *J. Mol. Struct.* 383, 271 (1996).
74. A. Hanprasopwattana, S. Srinivasan, A. G. Sault, and A. K. Datye, *Langmuir* 12, 3173 (1996).
75. S. M. Melpolder, A. W. West, and C. L. Bauer, *Mater. Res. Soc. Symp. Proc.* 239, 371 (1992).
76. G. W. Koebrugge, L. Winnubst, and A. J. Burggraaf, *J. Mater. Chem.* 3, 1095 (1993).
77. N. Ozer, S. De Souza, and C. M. Lampert, *Proc. SPIE-Int. Soc. Opt. Eng.* 2531, 143 (1995).
78. K. Kameyama, S. Shohji, S. Onoue, K. Nishimura, K. Yahikozawa, and Y. Takasu, *J. Electrochem. Soc.* 140, 1034 (1993).
79. S. Onoue, K. Kameyama, K. Yahikozawa, and Y. Takasu, *Proc. Electrochem. Soc.* 93-14, 158 (1993).
80. K. E. Swider, C. I. Merzbacher, P. L. Hagans, and D. R. Rolison, *Chem. Mater.* 9, 1248 (1997).
81. J. Y. Jeong, H. M. Lee, and H. L. Lee, *Yoop Hakhoechi* 33, 1138 (1996).
82. H. L. Lee, J. Y. Jeong, and H. M. Lee, *Yoop Hakhoechi* 33, 1147 (1996).
83. F. Lu, Q. Xu, Y. Lu, R. Ye, and B. Song, *J. Univ. Sci. Technol. Beijing* 3, 1 (1996).
84. D. L. Bourell and K. W. Parimal, *J. Am. Ceram. Soc.* 76, 705 (1993).
85. G. B. Prabhu and D. L. Bourell, *Nanostruct. Mater.* 5, 727 (1995).
86. V. M. Sglavo, R. Dal. Maschio, G. D. Soraru, and G. Carturan, *J. Eur. Ceram. Soc.* 11, 439 (1993).
87. Y. Murakami, H. Ohkawauchi, M. Ito, K. Yahikozawa, and Y. Takasu, *Electrochim. Acta* 39, 2551 (1994).
88. J. Yuan, Z. Liu, Q. Fang, and X. Xu, *Huadong Ligong Daxue Xuebao* 22, 316 (1996).
89. M. Sando, A. Towata, and A. Tsuge, *Funtai Kogaku Kaishi* 29, 755 (1992).
90. L. Armelao, R. Bertoncello, L. Crociani, G. Depaoli, G. Granozzi, E. Tondello, and M. Bettinelli, *J. Mater. Chem.* 5, 79 (1995).
91. C. Suerig, K. A. Hempel, and C. Sauer, *J. Magn. Magn. Mater.* 157/158, 268 (1996).
92. A. K. Bhattacharya, A. Hartridge, K. K. Mallick, C. K. Majumdar, D. Das, and S. N. Chintalapudi, *J. Mater. Sci.* 32, 557 (1997).
93. T. Ishiwaki, H. Inoue, and A. Makishima, *J. Non-Cryst. Solids* 203, 43 (1996).
94. C. Liu, Y. Shen, and Q. Xi, *Yingyong Huaxue* 13, 72 (1996).
95. P. G. Harrison, *Catal. Today* 17, 483 (1993).
96. H. Wang and W. J. Thomson, *AIChE J.* 41, 1790 (1995).
97. M. Nofz, R. Stoesser, B. Unger, and W. Herrmann, *J. Non-Cryst. Solids* 149, 62 (1992).
98. M. Zhou, S. Zhou, and Y. Wang, *Huadong Huagong Xueyuan Xuebao* 18, 74 (1992).
99. Z. Liu, Y. Sun, X. Du, and J. Cheng, *Huadong Huagong Xueyuan Xuebao* 18, 1 (1992).
100. N. Satyanarayana, R. Patcheammalle, P. Muralidharan, and M. Venkateswarlu, *Bull. Electrochem.* 12, 658 (1996).
101. N. Satyanarayana, P. Muralidharan, R. Patcheammalle, M. Venkateswarlu, and G. V. Rama Rao, *Solid State Ionics* 86–88, 543 (1996).
102. S. Komarneni and L. Rani, *Proc. SPIE-Int. Soc. Opt. Eng.* 1622, 147 (1992).
103. M. Ishida, H. Jin, and T. Okamoto, *Energy Fuels* 10, 958 (1996).
104. N. N. Ghosh and P. Pramanik, *Br. Ceram. Trans.* 95, 209 (1996).
105. X. Li, H. Zhang, F. Chi, S. Li, B. Xu, and M. Zhao, *Mater. Sci. Eng., B* 18, 209 (1993).
106. W. Nie, G. Boulon, C. Mai, C. Esnouf, R. Xu, and J. Zarzycki, *J. Non-Cryst. Solids* 121, 282 (1990).
107. W. Nie, G. Boulon, C. Mai, C. Esnouf, R. Xu, and J. Zarzycki, *Chem. Mater.* 4, 216 (1992).

108. W. Nie, G. Boulon, C. Mai, C. Esnouf, R. Xu, and J. Zarzycki, *Mater. Chem. Phys.* 25, 105 (1990).
109. F. Keller-Besrest and S. Benazeth, *J. Phys. IV* 4, C9/117 (1994).
110. G. M. Pajonk, *Heterog. Chem. Rev.* 2, 129 (1995).
111. U. Schubert, F. Schwertfeger, and C. Doersmann, *ACS Symp. Ser.* 622, 366 (1996).
112. C. M. Lukehart, J. P. Carpenter, S. B. Milne, and K. J. Burnam, *Chem. Tech.* 23, 29 (1993).
113. J. P. Carpenter, C. M. Lukehart, S. B. Milne, S. R. Stock, J. E. Wittig, B. D. Jones, R. Glosser, D. O. Henderson, and R. Mu, *Int. SAMPE Tech. Conf.* 27, 549 (1995).
114. W. D. King, D. L. Boxall, and C. M. Lukehart, *J. Cluster Sci.* 8, 267 (1997).
115. K. M. Choi and K. J. Shea, *Mater. Res. Soc. Symp. Proc.* 346, 763 (1994).
116. K. M. Choi and K. J. Shea, *J. Am. Chem. Soc.* 116, 9052 (1994).
117. J. P. Wang and H. L. Luo, *J. Magn. Magn. Mater.* 131, 54 (1994).
118. A. Basumallick, K. Biswas, S. Mukherjee, and G. C. Das, *Mater. Lett.* 30, 363 (1997).
119. J. P. Wang, H. Han, H. L. Luo, B. Liu, S. B. Hu, and T. S. Low, *IEEE Trans. Magn.* 32, 4496 (1996).
120. S. Roy, D. Das, D. Chakravorty, and D. C. Agrawal, *J. Appl. Phys.* 74, 4746 (1993).
121. A. Chatterjee and D. Chakravorty, *J. Phys. D: Appl. Phys.* 23, 1097 (1990).
122. A. Chatterjee and D. Chakravorty, *J. Mater. Sci.* 27, 4115 (1992).
123. S. Roy and D. Chakravorty, *Jpn. J. Appl. Phys., Part 1* 32, 3515 (1993).
124. J. P. Wang, D. Han, H. L. Luo, Q. X. Lu, and Y. W. Sun, *Appl. Phys. A* 61, 407 (1995).
125. J. P. Wang, D. Han, H. L. Luo, Y. W. Sun, and Q. X. Lu, *Wuli Xuebao* 44, 963 (1995).
126. V. Vendange and P. Colomban, *Mater. Sci. Eng., A* 168, 199 (1993).
127. K. M. Choi and K. J. Shea, *J. Sol.–Gel Sci. Technol.* 5, 143 (1995).
128. A. Chatterjee, A. Datta, A. K. Giri, and D. Chakravorty, *J. Appl. Phys.* 72, 3832 (1992).
129. T. Lopez, R. Gomez, O. Novaro, A. R. Solis, E. S. Mora, S. Castillo, E. Poulain, and J. M. Magadan, *J. Catal.* 141, 114 (1993).
130. J. P. Carpenter, C. M. Lukehart, S. B. Milne, S. R. Stock, and J. E. Wittig, *Inorg. Chim. Acta* 251, 151 (1996).
131. K. J. Burnam, J. P. Carpenter, C. M. Lukehart, S. B. Milne, S. R Stock., B. D. Jones, R. Glosser, and J. E. Wittig, *Nanostruct. Mater.* 5, 155 (1995).
132. Y. Zhang, Q. Wang, and B. Zhong, *Ranliao Huaxue Xuebao* 24, 517 (1996).
133. B. Breitscheidel, J. Zieder, and U. Schubert, *Chem. Mater.* 3, 559 (1991).
134. C. Goersmann, B. Breitscheidel, and U. Schubert, in "Organosilicon Chemistry" (N. Auner and J. Weis, eds.), p. 319. VCH Publisher, Weinheim, Germany, 1994.
135. U. Schubert, S. Tewinkel, and R. Lamber, *Chem. Mater.* 8, 2047 (1996).
136. P. Reyes, M. Morales, and G. Pechhi, *Bol. Soc. Chil. Quim.* 41, 221 (1996).
137. T. J. Lee, K. G. Sheppard, A. Gamburg, and L. Klein, *Proc. Electrochem. Soc.* 94, 231 (1994).
138. R. Monaci, A. Musinu, G. Piccaluga, and G. Pinna, *Mater. Sci. Forum* 195, 1 (1995).
139. S. Roy and D. Chakravorty, *J. Phys.: Condens. Matter* 6, 8599 (1994).
140. S. Roy, A. Chatterjee, and D. Chakravorty, *J. Mater. Res.* 8, 689 (1993).
141. J. Petrullat, S. Ray, U. Schubert, G. Guldner, C. Egger, and B. Breitscheidel, *J. Non-Cryst. Solids* 147–148, 594 (1992).
142. D. Spron, J. Grossmann, A. Kaiser, R. Jahn, and A. Berger, *Nanostruct. Mater.* 6, 328 (1995).
143. J. Grossmann, K. Rose, and D. Sporn, *Ceram. Trans.* 51, 713 (1995).
144. E. D. Rodeghiero, O. K. Tse, B. S. Wolkenberg, and E. P. Giannelis, *Process. Fabr. Adv. Mater. IV, Proc. Symp., 4th*, 775 (1995). CA: 126:10481.
145. T. Okubo, M. Watanabe, K. Kusakabe, and S. Morooka, *Key Eng. Mater.* 61–62, 71 (1991).
146. T. K. Kundu and D. Chakravorty, *J. Phys. Soc. Jpn.* 65, 3357 (1996).
147. T. K. Kundu and D. Chakravorty, *Appl. Phys. Lett.* 66, 3576 (1995).
148. R. Reisfeld, *Struct. Bonding* 85, 99 (1996).
149. A. W. Li, G. X. Xiong, H. B. Zhao, J. H. Gu, S. S. Sheng, W. Cui, and L. B. Zheng, *Chin. Chem. Lett.* 7, 193 (1996).
150. C. He and R. Wang, *Cuihua Xuebao* 18, 93 (1997).
151. H. Zhao, A. Li, J. Gu, and G. Xiong, *Cuihua Xuebao* 17, 263 (1996).
152. S. Rezgui, R. Jentoft, and B. C. Gates, *J. Catal.* 163, 496 (1996).
153. S. Sakka, H. Kozuka, and G. Zhao, *Proc. SPIE-Int. Soc. Opt. Eng.* 2288, 108 (1994).
154. E. Sanchez, T. Lopez, R. Gomez, I. L. Bokhimi, A. Morales, and O. Novaro, *J. Solid State Chem.* 122, 309 (1996).
155. M. Boutonnet Kizling, C. Bigey, and R. Touroude, *Appl. Catal., A* 135, L13 (1996).
156. J. L. Bokhimi, J. L. Boldu, E. Munoz, O. Novaro, T. Lopez, and R. Gomez, *Mater. Res. Soc. Symp. Proc.* 405, 523 (1996).
157. H. Shiga, T. Okubo, and M. Sadakata, *Ind. Eng. Chem. Res.* 35, 4498 (1996).
158. C. Y. Li, J. Y. Tseng, C. Lechner, and J. D. Mackenzie, *Mater. Res. Soc. Symp. Proc.* 272, 133 (1992).
159. J. Y. Tseng, C. Y. Li, T. Takada, C. Lechner, and J. D. Mackenzie, *Proc. SPIE-Int. Soc. Opt. Eng.* 1758, 612 (1992).
160. M. Ohtaki, Y. Oshima, K. Eguchi, and H. Arai, *Chem. Lett.* 11, 2201 (1992).

161. C. Ferrari, G. Predieri, A. Tiripicchio, and M. Costa, *Chem. Mater.* 4, 243 (1992).
162. D. Cauzzi, G. Marzolini, G. Predieri, A. Tiripicchio, M. Costa, G. Salviati, A. Armigliato, L. Basini, and R. Zanoni, *J. Mater. Chem.* 5, 1375 (1995).
163. T. Burkhart, M. Mennig, H. Schmidt, and A. Licciulli, *Mater. Res. Soc. Symp. Proc.* 346, 779 (1994).
164. C. Goersmann, U. Schubert, J. Leyrer, and E. Lox, *Mater. Res. Soc. Symp. Proc.* 435, 625 (1996).
165. C. Hippe, R. Lamber, G. Schulz-Ekloff, and U. Schubert, *Catal. Lett.* 43, 195 (1997).
166. M. Catalano, E. Carlino, M. A. Tagliente, A. Licciulli, and L. Tapfer, *Microsc. Microanal. Microstruct.* 6, 611 (1995).
167. (a) M. Catalano, G. De, A. Licciulli, and L. Tapfer, *Mater. Sci. Forum* 195, 87 (1995). (b) U. Werner, M. Schmitt, and H. Schmidt, *Mater. Res. Soc. Symp. Proc.* 435, 637 (1996).
168. A. Julbe, C. Balzer, A. Larbot, C. Guizard, and L. Cot, *Recent Prog. Genie Proc.* 6, 127 (1992).
169. S. Szu, C. Y. Lin, and C. H. Lin, *J. Sol.–Gel Sci. Technol.* 2, 881 (1994).
170. G. De, M. Gusso, L. Tapfer, M. Catalano, F. Gonella, G. Mattei, P. Mazzoldi, and G. Battaglin, *J. Appl. Phys.* 80, 6734 (1996).
171. G. De, L. Tapfer, M. Catalano, G. Battaglin, F. Caccavale, F. Gonella, P. Mazzoldi, and R. F. Haglund, Jr., *Appl. Phys. Lett.* 68, 3820 (1996).
172. G. De, M. Epifani, and A. Licciulli, *J. Non-Cryst. Solids* 201, 250 (1996).
173. (a) M. Mennig, M. Schmitt, B. Kutsch, and H. Schmidt, *Proc. SPIE-Int. Soc. Opt. Eng.* 2288, 120 (1994). (b) X. Bokhimi, A. Morales, O. Novaro, T. Lopez, O. Chimal, M. Asomoza, and R. Gomez, *Chem. Mater.* 9, 2616 (1997).
174. Y. Hosoya, T. Suga, T. Yanagawa, and Y. Kurokawa, *J. Appl. Phys.* 81, 1475 (1997).
175. H. Kozuka, M. Okuno, and T. Yoko, *J. Ceram. Soc. Jpn.* 103, 1305 (1995).
176. M. Lee, T. S. Kim, and Y. S. Choi, *J. Non-Cryst. Solids* 211, 143 (1997).
177. F. Akbarian, B. S. Dunn, and J. I. Zink, *Proc. SPIE-Int. Soc. Opt. Eng.* 2288, 140 (1994).
178. F. Akbarian, B. S. Dunn, and J. I. Zink, *J. Raman Spectrosc.* 27, 775 (1996).
179. I. Tanahashi and T. Tohda, *J. Am. Ceram. Soc.* 79, 796 (1996).
180. H. Kozuka, T. Hashimoto, T. Uchino, and T. Yoko, *ICR Ann. Rep.* 2, 22 (1996).
181. K. Yasumoto, N. Koshizaki, and K. Suga, *Proc. SPIE-Int. Soc. Opt. Eng.* 1758, 604 (1992).
182. K. Yasumoto and N. Koshizaki, *Busshitsu Kogaku Kogyo Gijutsu Kenkyusho Hokoku* 4, 121 (1996).
183. A. Maegawa, *Rep. Ind. Res. Cent. Shiga Prefect.* 8, 7 (1994).
184. T. Miura, H. Sakai, and Y. Asahara, Jpn. Kokai Tokkyo Koho JP 02225342 A2 900907 Heisei, 11 pp.
185. L. Zhang, C. Y. Fan, and L. Jiang, *Chin. Chem. Lett.* 7, 77 (1996).
186. B. E. Baker, M. J. Natan, H. Zhu, and T. P. Beebe, Jr., *Supramol. Sci.* 4, 147 (1997).
187. G. C. Vezzoli and M. F. Chen, *Mater. Res. Soc. Symp. Proc.* 286, 271 (1993).
188. G. C. Vezzoli, M. F. Chen, and J. Caslavsky, *Nanostruct. Mater.* 4, 985 (1994).
189. G. C. Vezzoli, M. F. Chen, and J. Caslavsky, *Ceram. Int.* 23, 105 (1997).
190. G. De, A. Licciulli, C. Massaro, L. Tapfer, M. Catalano, G. Battaglin, C. Meneghini, P. Mazzoldi, *J. Non-Cryst. Solids* 194, 225 (1996).
191. S. K. Ma and J. T. Lue, *Solid State Commun.* 97, 979 (1996).
192. P. Innocenzi and H. Kozuka, *J. Sol.–Gel Sci. Technol.* 3, 229 (1994).
193. B. Ritzer, M. A. Villegas, and J. M. Fernandez Navarro, *J. Sol.–Gel Sci. Technol.* 8, 917 (1997).
194. A. Licciulli, G. De, P. Mazzoldi, M. Catalano, L. Mirenghi, and L. Tapfer, *Mater. Sci. Forum* 203, 59 (1996).
195. Y. Zhu, Y. Qian, M. Zhang, Z. Chen, and G. Zhou, *J. Mater. Chem.* 4, 1619 (1994).
196. T. Gacoin, F. Chaput, J. P. Boilot, M. Mostafavi, and M. O. Delcourt, *Eur. Mater. Res. Soc. Mon.* 5, 159 (1992).
197. W. Xiang, C. Wang, Z. Wang, Q. Yang, W. Zhao, Z. Ding, M. Wang, and J. Wang, *Guangxue Xuebao* 16, 967 (1996).
198. P. Innocenzi, G. Brusatin, A. Martucci, and K. Urabe, *Thin Solid Films* 279, 23 (1996).
199. (a) G. Zhao, H. Kozuka, and T. Yoko, *J. Ceram. Soc. Jpn.* 104, 164 (1996). (b) G. Zhao, H. Kozuka, and T. Yoko, *Thin Solid Films* 277, 147 (1996).
200. Y. Zhu, Y. Qian, H. Huang, M. Zhang, and S. Liu, *Mater. Lett.* 28, 259 (1996).
201. H. Kozuka, G. Zhao, and S. Sakka, *Bull. Inst. Chem. Res. Kyoto Univ.* 72, 209 (1994).
202. T. Kineri, E. Matano, and T. Tsuchiya, *Proc. SPIE* 2288, 145 (1994).
203. T. K. Kundu and D. Chakravorty, *Appl. Phys. Lett.* 67, 2732 (1995).
204. T. K. Kundu and D. Chakravorty, *J. Mater. Res.* 11, 200 (1996).
205. D. Chakravorty, *Ferroelectrics* 102, 33 (1990).
206. A. N. Krasovskii and A. I. Andreeva, *Zh. Nauchn. Prikl. Fotogr.* 41, 7 (1996).
207. A. N. Krasovskii and A. I. Andreeva, *Zh. Prikl. Khim.* 69, 834 (1996).
208. M. Mennig, J. Spanhel, H. Schmidt, and S. Betzholz, *J. Non-Cryst. Solids* 147–148, 326 (1992).
209. M. Mennig, M. Schmitt, and H. Schmidt, *J. Sol.–Gel Sci. Technol.* 8, 1035 (1997).
210. M. Mennig, U. Becker, M. Schmitt, and H. Schmit, in "Advanced Materials in Optics, Electro-Optics and Communication Technologies" (P. Vincenzini, ed.), p. 39. Techna Sri., 1995.
211. A. Martino, S. A. Yamanaka, J. S. Kawola, and D. A. Loy, *Chem. Mater.* 9, 423 (1997).
212. T. Gacoin, F. Chaput, J. P. Boilot, and G. Jaskierowicz, *Chem. Mater.* 5, 1150 (1993).

213. C. Goersmann, B. Breitscheidel, and U. Schubert, in "Organosilicon Chemistry" (N. Auner and J. Weis, eds.), p. 319. VCH Publishers, Weinheim, Germany, 1994.
214. S. K. Pradhan, A. Datta, M. Pal, and D. Chakravorty, *Metall. Mater. Trans. A* 27, 4213 (1996).
215. A. Chatterjee, A. K. Giri, D. Das, and D. Chakravorty, *J. Mater. Sci. Lett.* 11, 518 (1992).
216. L. Zhu, Y. Li, J. Wang, and J. Shen, *J. Appl. Phys.* 77, 2801 (1995).
217. L. Zhu, Y. Li, J. Wang, and J. Shen, *Chem. Phys. Lett.* 239, 393 (1995).
218. J. Shen, J. Wang, W. Jiang, B. Zhou, B. Zhang, and L. Zhu, *Gongneng Cailiao* 26, 168 (1995).
219. J. Shen, L. Zhu, J. Wang, Y. Li, and X. Wu, *Chin. Phys. Lett.* 12, 693 (1995).
220. J. Shen, L. Zhu, J. Wang, X. Wu, and Y. Li, *Wuji Cailiao Xuebao* 11, 371 (1996).
221. R. Gvishi, J. D. Bhawalkar, N. D. Kumar, G. Ruland, U. Narang, P. N. Prasad, and B. A. Reinhardt, *Chem. Mater.* 7, 2199 (1995).
222. P. N. Prasad, R. Gvishi, G. Ruland, N. D. Kumar, J. D. Bhawalkar, U. Narang, and B. A. Reinhardt, *Proc. SPIE-Int. Soc. Opt. Eng.* 2530, 128 (1995).
223. H. Peng, S. M. Leung, C. F. Au, X. Wu, N. T. Yu, and B. Z. Tang, *Polym. Mater. Sci. Eng.* 75, 247 (1996).
224. D. McBranch, V. Klimov, L. Smilowitz, M. Grigorova, J. M. Robinson, A. Koskelo, B. R. Mattes, H. Wang, and F. Wudl, *Proc. SPIE-Int. Soc. Opt. Eng.* 2854, 140 (1996).
225. D. McBranch, V. Klimov, L. Smilowitz, M. Grigorova, and B. R. Mattes, *Proc. Electrochem. Soc.* 96-10, 384 (1996).
226. W. Cao, X. Y. Song, and A. J. Hunt, *Mater. Res. Soc. Symp. Proc.* 349, 87 (1994).
227. G. R. Delgado, H. W. H. Lee, and K. Pakbaz, *Mater. Res. Soc. Symp. Proc.* 452, 669 (1997).
228. H. W. H. Lee, J. E. Davis, M. L. Olsen, S. M. Kauzlarich, R. A. Bley, S. H. Risbud, and D. J. Duval, *Mater. Res. Soc. Symp. Proc.* 351, 129 (1994).
229. M. Nogami and Y. Abe, *J. Sol.–Gel Sci. Technol.* 9, 139 (1997).
230. M. Nogami and Y. Abe, *Appl. Phys. Lett.* 65, 2545 (1994).
231. M. Nogami, T. Kasuga, and Y. Abe, *Proc. SPIE-Int. Soc. Opt. Eng.* 2288, 193 (1994).
232. Y. Maeda, Jpn. *J. Appl. Phys., Part 1* 34, 254 (1995).
233. J. P. Carpenter, C. M. Lukehart, D. O. Henderson, R. Mu, B. D. Jones, R. Glosser, S. R. Stock, J. E. Wittig, and J. G. Zhu, *Chem. Mater.* 8, 1268 (1996).
234. Y. Kurokawa and Y. Kobayashi, *Zairyo Gijutsu* 11, 197 (1993).
235. S. Dire and F. Babonneau, *J. Sol.–Gel Sci. Technol.* 2, 139 (1994).
236. J. P. Carpenter, C. M. Lukehart, S. R. Stock, and J. E. Wittig, *Chem. Mater.* 7, 201 (1995).
237. V. Balek, E. Klosova, M. Murat, and N. A. Camargo, *Am. Ceram. Soc. Bull.* 75, 73 (1996).
238. Y. Xu, A. Nakahira, and K. Niihara, *Ceram. Trans.* 44, 275 (1994).
239. J. Y. Choi, C. H. Kim, and D. K. Kim, *Ceram. Trans.* 74, 153 (1996).
240. C. M. Lukehart, S. B. Milne, S. R. Stock, R. D. Shull, and J. E. Wittig, *ACS Symp. Ser.* 622, 195 (1996).
241. C. M. Lukehart, S. B Milne, S. R. Stock, R. D. Shull, and J. E. Wittig, *Mater. Sci. Eng., A* 204, 176 (1995).
242. S. B. Milne, 176 pp. Available from Univ. Microfilms Int., Order No.: DA9541802. From: *Diss. Abstr. Int., B 1995* 56, 4309 (1995).
243. C. M. Lukehart, S. B. Milne, S. R. Stock, R. D. Shull, and J. E. Wittig, *Polym. Mater. Sci. Eng.* 73, 436 (1995).
244. M. A. Matchett, A. M. Viano, N. L. Adolphi, R. D. Stoddard, W. E. Buhro, M. S. Conradi, and P. C. Gibbons, *Chem. Mater.* 4, 508 (1992).
245. S. H. Risbud and H. B. Underwood, *J. Mater. Synth. Process.* 1, 225 (1993).
246. H. Hamada, Y. Kintaichi, M. Inaba, M. Tabata, T. Yoshinari, and H. Tsuchida, *Catal. Today* 29, 53 (1996).
247. A. Towata, M. Awano, M. Sando, and A. Tsuge, *Kagaku Kogaku Ronbunshu* 18, 315 (1992).
248. G. Pacheco-Malagon, A. Garcia-Borquez, D. Coster, A. Sklyarov, S. Petit, and J. J. Fripiat, *J. Mater. Res.* 10, 1264 (1995).
249. A. Nazeri and S. B. Qadri, *Surf. Coat. Technol.* 86–87, 166 (1996).
250. J. Kuo and D. L. Bourell, in "Synth. Process. Nanocryst. Powder, Proc. Symp." (D. L. Bourell, ed.), pp. 237–248. Minerals, Metals and Materials Society, Warrendale, PA, 1996.
251. S. Hazra, A. Ghosh, and D. Chakravorty, *J. Phys.: Condens. Matter* 8, 10279 (1996).
252. Q. Zhou, Q. Zhang, J. Zhang, L. Zhang, and X. Yao, *Chin. Sci. Bull.* 41, 555 (1996).
253. M. G. F. Da Silva and J. M. F. Navarro, *J. Sol.–Gel Sci. Technol.* 6, 169 (1996).
254. M.G. F. Da Silva and C. B. Koch, *J. Sol.–Gel Sci. Technol.* 8, 311 (1997).
255. D. Niznansky, N. Viart, and J. L. Rehspringer, *J. Sol.–Gel Sci. Technol.* 8, 615 (1997).
256. T. P. Sinha, M. Mukherjee, D. Chakravorty, and M. Bhattacharya, *Indian J. Phys., A* 70, 741 (1996).
257. F. del Monte, M. P. Morales, D. Levy, A. Fernandez, M. Ocana, A. Roig, E. Molins, K. O'Grady, and C. J. Serna, *Langmuir* 13, 3627 (1997).
258. C. Chaneac, E. Tronc, and J. P. Jolivet, *J. Mater. Chem.* 6, 1905 (1996).
259. C. H. Cheng, Y. Xu, J. D. Mackenzie, J. Chee, and J. M. Liu, *Proc. SPIE-Int. Soc. Opt. Eng.* 1758, 485 (1992).
260. K. Takeuchi, T. Isobe, and M. Senna, *J. Non-Cryst. Solids* 194, 58 (1996).
261. Q. F. Zhou, Q. Q. Zhang, J. X. Zhang, L. Y. Zhang, and X. Yao, *Mater. Lett.* 31, 39 (1997).

262. A. E. Stiegman, H. Eckert, G. Plett, S. S. Kim, M. Anderson, and A. Yavrouian, *Chem. Mater.* 5, 1591 (1993).
263. A. Chatterjee, D. Das, S. K. Pradhan, and D. Chakravorty, *J. Magn. Magn. Mater.* 127, 214 (1993).
264. Q. Zhou, S. Wang, Y. Han, L. Zhang, and X. Yao, *Ferroelectrics* 154, 1433 (1994).
265. Q. Zhou, J. Zhang, L. Zhang, and X. Yao, *Cailiao Yanjiu Xuebao* 9, 71 (1995).
266. K. Yao, L. Zhang, and X. Yao, *Ferroelectrics* 19, 113 (1995).
267. K. Yao, L. Zhang, and X. Yao, *Guisuanyan Xuebao* 24, 355 (1996).
268. T. Sei, H. Takeda, T. Tsuchiya, and T. Kineri, *Ann. Chim.* 18, 329 (1993).
269. B. Ammundsen, G. R. Burns, A. Amran, and S. E. Friberg, *J. Sol.–Gel Sci. Technol.* 2, 341 (1994).
270. S. Jana and P. K. Biswas, *Mater. Lett.* 30, 53 (1997).
271. H. Inoe, Jpn. Kokai Tokkyo Koho JP 08175823 A2 960709 Heisei, 5 pp.
272. S. Luo and L. Gui, *Mater. Res. Soc. Symp. Proc.* 371, 303 (1995).
273. D. Keomany, C. Poinsignon, and D. Deroo, *Sol. Energy Mater. Sol. Cells* 33, 429 (1994).
274. D. Keomany, J. P. Petit, and D. Deroo, *Proc. SPIE-Int. Soc. Opt. Eng.* 2255, 363 (1994).
275. D. Keomany, J. P. Petit, and D. Deroo, *Sol. Energy Mater. Sol. Cells* 36, 397 (1995).
276. D. Camino, D. Deroo, J. Salardenne, and N. Treuil, *Sol. Energy Mater. Sol. Cells* 39, 349 (1995).
277. G. Zhou, S. Liu, and D. Peng, *Huaxue Wuli Xuebao* 9, 54 (1996).
278. J. Shi, J. Liu, Q. Zhu, L. Zhang, and J. Zhang, *Cuihua Xuebao* 17, 277 (1996).
279. J. Shi, J. Liu, Q. Zhu, L. Zhang, and J. Zhang, *Cuihua Xuebao* 17, 323 (1996).
280. G. Fang, Z. Liu, Z. Zhang, Y. Hu, I. A. Ashur, and K. L. Yao, *Phys. Status Solidi A* 156, 15 (1996).
281. G. Fang, Z. Liu, Y. Hu, and K. Yao, *Wuji Cailiao Xuebao* 11, 537 (1996).
282. S. J. L. Ribeiro, S. H. Pulcinelli, and C. V. Santilli, *Chem. Phys. Lett.* 190, 64 (1992).
283. S. K. Pradhan, A. Datta, A. Chatterjee, M. De, and D. Chakravorty, *Bull. Mater. Sci.* 17, 849 (1994).
284. R. Guinebretiere, Z. Oudjedi, and A. Dauger, *Scr. Mater.* 34, 1039 (1996).
285. R. Joseph, S. Zhang, and W. T. Ford, *Macromolecules* 29, 1305 (1996).
286. M. Motomatsu, T. Takahashi, H. Y. Nie, W. Mizutani, and H. Tokumoto, *Polymer* 38, 177 (1997).
287. B. Lantelme, M. Dumon, C. Mai, and J. P. Pascault, *J. Non-Cryst. Solids* 194, 63 (1996).
288. H. Krug and H. Schmidt, *New J. Chem.* 18, 1125 (1994).
289. L. Kador, R. Fischer, D. Haarer, R. Kasemann, S. Brueck, H. Schmidt, and H. Duerr, *Adv. Mater.* 5, 270 (1993).
290. S. Hirano, T. Yogo, K. Kikuta, and S. Yamada, *Ceram. Trans.* 68, 131 (1996).
291. P. L. Shao, K. A. Mauritz, and R. B. Moore, *J. Polym. Sci., Part B: Polym. Phys.* 34, 873 (1996).
292. Q. Deng, K. M. Cable, R. B. Moore, and K. A. Mauritz, *J. Polym. Sci., Part B: Polym. Phys.* 34, 1917 (1996).
293. M. A. F. Robertson and K. A. Mauritz, *Polym. Prepr.* 37, 248 (1996).
294. P. L. Shao, K. A. Mauritz, and R. B. Moore, *Polym. Mater. Sci. Eng.* 73, 427 (1995).
295. Z. Zhao, Z. Gao, Y. Ou, Z. Qi, and F. Wang, *Gaofenzi Xuebao* 2, 228 (1996).
296. Z. Gao, Z. Zhao, Y. Ou, Z. Qi, and F. Wang, *Polym. Int.* 40, 187 (1996).
297. S. Yano, T. Furukawa, and M. Kodomari, *Kimio Kurita* 53, 218 (1996).
298. J. E. Mark, *Int. SAMPE Tech. Conf.* 27, 539 (1995).
299. S. P. Nunes, J. Schultz, and K. V. Peinemann, *J. Mater. Sci. Lett.* 15, 1139 (1996).
300. N. D. Kumar, G. Ruland, M. Yoshida, M. Lal, J. Bhawalkar, G. S. He, and P. N. Prasad, *Mater. Res. Soc. Symp. Proc.* 435, 535 (1996).
301. Y. Chen, X. Wang, Z. Gao, X. Zhu, Z. Qi, and C. L. Choy, *Gaofenzi Xuebao* 1, 73 (1997).
302. H. Kaddami, S. Cuney, J. P. Pascault, and J. F. Gerard, *AIP Conf. Proc.* 354, 522 (1996).
303. V. J. Nagpal, R. M. Davis, and S. B. Desu, *J. Mater. Res.* 10, 3068 (1995).
304. R. Reisfeld and H. Minti, *J. Sol.–Gel Sci. Technol.* 2, 641 (1994).
305. R. Reisfeld, *Proc. SPIE-Int. Soc. Opt. Eng.* 1758, 546 (1992).
306. H. Minti, M. Eyal, R. Reisfeld, and G. Berkovic, *Chem. Phys. Lett.* 183, 277 (1991).
307. R. Reisfeld, H. Minti, and M. Eyal, *Proc. SPIE-Int. Soc. Opt. Eng.* 1513, 360 (1991).
308. E. Blanco, R. Litran, M. Ramirez-del-Solar, N. De La Rosa-Fox, and L. Esquivias, *J. Mater. Res.* 9, 2873 (1994).
309. T. Fujii, Y. Hisakawa, E. J. Winder, and A. B. Ellis, *Bull. Chem. Soc. Jpn.* 68, 1559 (1995).
310. M. Pinero, R. Litran, C. Fernandez-Lorenzo, E. Blanco, M. Ramirez-Del-Solar, N. De La Rosa-Fox, L. Esquivias, A. Craievich, and J. Zarzycki, *J. Sol.–Gel Sci. Technol.* 2, 689 (1994).
311. R. Litran, R. Alcantara, E. Blanco, and M. Ramirez-Del-Solar, *J. Sol.–Gel Sci. Technol.* 8, 275 (1997).
312. M. Nogami, K. Nagasaka, and M. Takata, *J. Non-Cryst. Solids* 122, 101 (1990).
313. A. Othmani, J. C. Plenet, E. Bernstein, F. Paille, C. Bovier, J. Dumas, and C. Mai, *J. Mater. Sci.* 30, 2425 (1995).
314. A. Othmani, J. C. Plenet, E. Bernstein, C. Bovier, J. Dumas, P. Riblet, P. Gilliot, R. Levy, and J. B. Grun, *J. Cryst. Growth* 144, 141 (1994).
315. A. Othmani, C. Bovier, J. Dumas, and B. Champagnon, *J. Phys. IV* 2, C2-275 (1992).
316. E. M. Yeatman, M. Green, E. J. C. Dawnay, M. A. Fardad, and F. Horowitz, *J. Sol.–Gel Sci. Technol.* 2, 711 (1994).

317. M. Nogami, K. Nagaska, and E. Kato, *J. Am. Ceram. Soc.* 73, 2097 (1990).
318. J. Huang, Q. Dai, G. Guo, and L. Gui, *Wuli Huaxue Xuebao* 12, 621 (1996).
319. G. Tu, Z. Sui, and C. Wang, *J. Less-Common Met.* 175, 205 (1991).
320. T. Rajh, O. I. Micic, D. Lawless, and N. Serpone, *J. Phys. Chem.* 96, 4633 (1992).
321. Y. L. Chia, Y. Kao, K. Hayashi, T. Takada, J. D. Mackenzie, K. Kang, S. Lee, N. Peyghambarian, and M. Yamane, *Proc. SPIE-Int. Soc. Opt. Eng.* 2288, 151 (1994).
322. T. Takada, C. Li, J. Y. Tseng, and J. D. Mackenzie, *J. Sol.–Gel Sci. Technol.* 1, 123 (1994).
323. L. Spanhel, E. Arpac, and H. Schmidt, *J. Non-Cryst. Solids* 147–148, 657 (1992).
324. N. Tohge, M. Asuka, and T. Minami, *Proc. SPIE-Int. Soc. Opt. Eng.* 1328, 125 (1990).
325. N. Tohge and T. Minami, *Proc. SPIE-Int. Soc. Opt. Eng.* 1758, 587 (1992).
326. N. Tohge, M. Asuka, and T. Minami, *J. Non-Cryst. Solids* 147–148, 652 (1992).
327. T. A. King, D. West, D. L. Williams, C. Moussu, and M. Bradford, *Adv. Sci. Technol.* 11, 21 (1995).
328. N. Tohge, M. Asuka, and T. Minami, *Chem. Express* 5, 521 (1990).
329. M. Guglielmi, A. Martucci, E. Menegazzo, G. C. Righini, S. Pelli, J. Fick, and G. Vitrant, *J. Sol.–Gel Sci. Technol.* 8, 1017 (1997).
330. M. Tsutomu, I. Tomoaki, T. Kiyoharu, and T. Masahiro, *Proc. SPIE-Int. Soc. Opt. Eng.* 2288, 183 (1994).
331. T. Iwami, K. Tadanaga, M. Tatsumisago, T. Minami, and N. Tohge, *J. Am. Ceram. Soc.* 78, 1668 (1995).
332. E. Cordoncillo, P. Escribano, G. Monros, M. A. Tena, V. Orera, and J. Carda, *J. Solid State Chem.* 118, 1 (1995).
333. T. Gacoin, J. P. Boilot, F. Chaput, and A. Lecomte, *Mater. Res. Soc. Symp. Proc.* 272, 21 (1992).
334. K. Kang, A. D. Kepner, Y. Z. Hu, S. W. Koch, N. Peyghambarian, C. Y. Li, T. Takada, Y. Kao, and J. D. Mackenzie, *Appl. Phys. Lett.* 64, 1487 (1994).
335. A. Othmani, C. Bovier, J. C. Plenet, J. Dumas, B. Champagnon, and C. Mai, *Mater. Sci. Eng., A* 168, 263 (1993).
336. G. Counio, S. Esnouf, T. Gacoin, P. Barboux, A. Hofstaetter, and J.-P. Boilot, *Mater. Res. Soc. Symp. Proc.* 452, 311 (1997).
337. H. L. Liu, S. S. Wang, L. Y. Zhang, and X. Yao, *Ferroelectrics* 196, 245 (1997).
338. H. Liu, S. Wang, L. Zhang, and X. Yao, *Xi'an Jiaotong Daxue Xuebao* 29, 13 (1995).
339. N. Toge, T. Minami, I. Tanahashi, and T. Mitsuyu, Jpn. Kokai Tokkyo Koho JP 04270131 A2 920925 Heisei, 5 pp.
340. S. Lu, Q. Xie, L. Zhang, and X. Yao, *Chin. Sci. Bull.* 39, 96 (1994).
341. B. Honerlage, P. Gilliot, and R. Levy, *Nuovo Cimento Soc. Ital. Fis., D* 17, 1247 (1995).
342. J. C. Plenet, A. Othmani, F. Paille, J. Mugnier, E. Bernstein, and J. Dumas, *Opt. Mater.* 7, 129 (1997).
343. G. Yu, K. Wang, J. Huan, and B. Zhao, *Wuli Huaxue Xuebao* 13, 230 (1997).
344. S. Lu, H. Yu, L. Zhang, and X. Yao, *Guisuanyan Xuebao* 22, 71 (1994).
345. C. S. Mun, J. B. Kang, and K. M. Kim, *Yoop Hakhoechi* 33, 1353 (1996).
346. N. Herron, *ACS Symp. Ser.* 455, 582 (1991).
347. C. Y. Li, M. Wilson, N. Haegel, J. D. Mackenzie, E. T. Knobbe, C. Porter, and R. Reeves, *Mater. Res. Soc. Symp. Proc.* 272, 41 (1992).
348. H. Ye, D. He, and Z. Jiang, *Guisuanyan Xuebao* 22, 434 (1994).
349. J. Zhao, K. Dou, Y. Chen, C. Jin, L. Sun, S. Huang, J. Yu, W. Xiang, and Z. Ding, *J. Lumin.* 66-67, 332 (1995).
350. E. J. C. Dawnay, J. Fick, M. Green, M. Guglielmi, A. Martucci, S. Pelli, G. C. Righini, G. Vitrant, and E. M. Yeatman, *Adv. Sci. Technol.* 11, 15 (1995).
351. E. Vanagas, J. Moniatte, M. Mazilu, P. Riblet, B. Honerlage, S. Juodkazis, F. Paille, J. C. Plenet, J. G. Dumas, M. Petrauskas, and J. Vaitkus, *J. Appl. Phys.* 81, 3586 (1997).
352. T. Takada, T. Yano, A. Yasumori, M. Yamane, and J. D. Mackenzie, *J. Non-Cryst. Solids* 147–148, 631 (1992).
353. T. Takada, J. D. Mackenzie, M. Yamane, K. Kang, N. Peyghambarian, R. J. Reeves, E. T. Knobbe, and R. C. Powell, *J. Mater. Sci.* 31, 423 (1996).
354. J. Butty, N. Peyghambarian, Y. H. Kao, and J. D. Mackenzie, *Appl. Phys. Lett.* 69, 3224 (1996).
355. J. D. Mackenzie, *J. Ceram. Soc. Jpn.* 101, 1 (1993).
356. M. Yamane, T. Takada, J. D. Mackenzie, and C. Y. Li, *Proc. SPIE-Int. Soc. Opt. Eng.* 1758, 577 (1992).
357. J. Butty, Y. Z. Hu, N. Peyghambarian, Y. H. Kao, and J. D. Mackenzie, *Appl. Phys. Lett.* 67, 2672 (1995).
358. C. Li, Y. Kao, K. Hayashi, T. Takada, J. D. Mackenzie, K. Kang, S. Lee, N. Peyghambarian, and M. Yamane, *Proc. SPIE-Int. Soc. Opt. Eng.* 2288, 151 (1994).
359. Y. Kao, K. Hayashi, L. Yu, M. Yamane, and J. D. Mackenzie, *Proc. SPIE-Int. Soc. Opt. Eng.* 2288, 752 (1994).
360. H. Mathieu, T. Richard, J. Allegre, P. Lefebvre, G. Arnaud, W. Granier, L. Boudes, J. L. Marc, A. Pradel, and M. Ribes, *J. Appl. Phys.* 77, 287 (1995).
361. W. Granier, L. Boudes, A. Pradel, M. Ribes, J. Allegre, G. Arnaud, P. Lefebvre, and H. Mathieu, *J. Sol.–Gel Sci. Technol.* 2, 765 (1994).

362. T. Richard, J. Allegre, P. Lefebvre, H. Mathieu, C. Jouanin, L. Boudes, J. L. Marc, W. Granier, A. Pradel, and M. Ribes, in "Int. Conf. Phys. Semicond." (D. J. Lockwood, ed.), Vol. 3, p. 2003. World Scientific, Singapore, 1995.
363. P. Lefebvre, T. Richard, J. Allegre, H. Mathieu, A. Pradel, J. L. Marc, L. Boudes, W. Granier, and M. Ribes, *Proc. SPIE-Int. Soc. Opt. Eng.* 2288, 163 (1994).
364. J. Zhao, K. Dou, S. Lu, Y. Chen, S. Huang, J. Yu, W. Xiang, and Z. Ding, *J. Mater. Sci. Lett.* 15, 702 (1996).
365. P. Lefebvre, T. Richard, J. Allegre, H. Mathieu, A. Pradel, J. L. Marc, L. Boudes, W. Granier, and M. Ribes, *Superlattices Microstruct.* 15, 447 (1994).
366. W. Granier, L. Boudes, A. Pradel, M. Ribes, J. Allegre, G. Arnaud, P. Lefebvre, and H. Mathieu, *Mater. Sci. Forum* 152–153, 351 (1994).
367. J. D. Mackenzie and Y. Kao, *Proc. SPIE-Int. Soc. Opt. Eng.* 2145, 90 (1994).
368. M. Guglielmi, A. Martucci, G. C. Righini, and S. Pelli, *Proc. SPIE-Int. Soc. Opt. Eng.* 2288, 174 (1994).
369. M. A. Fardad, E. M. Yeatman, E. J. C. Dawnay, M. Green, J. Fick, M. Guntau, and G. Vitrant, *IEE Proc.: Optoelectron.* 143, 298 (1996).
370. J. Fick, G. Vitrant, A. Martucci, M. Guglielmi, S. Pelli, and G. C. Righini, *MCLC S&T, Sect. B: Nonlinear Opt.* 12, 203 (1995).
371. H. Kawaguchi, T. Miyakawa, N. Tanno, Y. Kobayashi, and Y. Kurokawa, *Jpn. J. Appl. Phys., Part 2* 30, L280 (1991).
372. A. K. Atta, P. K. Biswas, and D. Ganguli, *Polym. Other Adv. Mater.* 3, 645 (1995).
373. K. M. Choi and K. J. Shea, *J. Phys. Chem.* 98, 3207 (1994).
374. K. M. Choi and K. J. Shea, *Chem. Mater.* 5, 1067 (1993).
375. L. Spanhel, *Eur. Mater. Res. Soc. Monogr.* 5, 407 (1992).
376. E. Cordoncillo, J. B. Carda, M. A. Tena, G. Monros, and P. Escribano, *J. Sol.–Gel Sci. Technol.* 8, 1043 (1997).
377. L. Spanhel, H. Schmidt, A. Uhrig, and C. Klingshirn, *Mater. Res. Soc. Symp. Proc.* 272, 53 (1992).
378. G. Counio, S. Esnouf, T. Gacoin, and J. P. Boilot, *J. Phys. Chem.* 100, 20021 (1996).
379. M. Nogami and A. Kato, *J. Sol.–Gel Sci. Technol.* 2, 751 (1994).
380. M. Nogami, A. Kato, and Y. Tanaka, *J. Mater. Sci.* 28, 4129 (1993).
381. C. M. Bagnall and J. Karzycki, *Proc. SPIE-Int. Soc. Opt. Eng.* 1328, 108 (1990).
382. D. C. Hummel, I. L. Torriani, A. Y. Ramos, A. F. Craievich, N. Fox, and L. Esquivias, *Mater. Res. Soc. Symp. Proc.* 346, 673 (1994).
383. D. A. Hummel, I. L. Torriani, A. F. Craievich, N. De La Rosa Fox, A. Y. Ramos, O. Lyon, *J. Sol.–Gel Sci. Technol.* 8, 285 (1997).
384. M. Nogami, S. Suzuki, and K. Nagasaka, *J. Non-Cryst. Solids* 135, 182 (1991).
385. J. L. Coffer, G. Beauchamp, and T. W. Zerda, *J. Non-Cryst. Solids* 142, 208 (1992).
386. L. Spanhel, M. Popall, and G. Mueller, *Proc. Indian Acad. Sci., Chem. Sci.* 107, 637 (1995).
387. J. L. Marc, W. Granier, A. Pradel, M. Ribes, T. Richard, J. Allegre, and P. Lefebvre, *Mater. Res. Soc. Symp. Proc.* 346, 901 (1994).
388. M. Nogami, K. Nagasaka, and T. Suzuki, *J. Am. Ceram. Soc.* 75, 220 (1992).
389. M. Nogami, I. Kojima, and K. Nagasaka, *Proc. SPIE-Int. Soc. Opt. Eng.* 1758, 557 (1992).
390. P. Lefebvre, T. Richard, J. Allegre, H. Mathieu, A. Combette-Roos, and W. Granier, *Phys. Rev. B: Condens. Matter* 53, 15440 (1996).
391. P. Lefebvre, T. Richard, J. Allegre, H. Mathieu, A. Pradel, J. L. Marc, L. Boudes, W. Granier, and M. Ribes, *Superlattices Microstruct.* 15, 447 (1994).
392. H. Liu, S. Wang, L. Zhang, and X. Yao, *Ferroelectrics* 19, 83 (1995).
393. M. Nogami, M. Watabe, and K. Nagasaka, *Proc. SPIE-Int. Soc. Opt. Eng.* 1328, 119 (1990).
394. S. Lu, H. Liu, L. Zhang, and X. Yao, in "Proc. Int. Conf. Electron. Compon. Mater., Sens. Actuators," p. 230. International Academic Publishers, Beijing, 1995.
395. S. Lu, H. Liu, L. Zhang, and X. Yao, *Chin. Sci. Bull.* 41, 1923 (1996).
396. S. Lu, H. Liu, L. Zhang, and X. Yao, *Chin. Sci. Bull.* 41, 83 (1996).
397. Y. Zhang, J. K. Bailey, C. J. Brinker, N. Raman, R. M. Crooks, and C. S. Ashley, *Proc. SPIE-Int. Soc. Opt. Eng.* 1758, 596 (1992).
398. Y. Zhang, N. Raman, J. K. Bailey, C. J. Brinker, and R. M. Crooks, *J. Phys. Chem.* 96, 9098 (1992).
399. M. Nogami, K. Nagasaka, and K. Kotani, *J. Non-Cryst. Solids* 126, 87 (1990).
400. N. Pellegri, R. Trbojevich, O. De Sanctis, and K. J. Kadono, *J. Sol.–Gel Sci. Technol.* 8, 1023 (1997).
401. N. Narvathy, G. M. Pajonk, and A. V. Rao, *Mater. Res. Bull.* 32, 397 (1997).
402. T. K. Singh, S. K. Sharma, and C. L. Jain, *Def. Sci. J.* 46, 105 (1996).
403. T. Takada, T. Yano, A. Yasumori, and M. Yamane, *J. Ceram. Soc. Jpn.* 101, 73 (1993).
404. T. Gacoin, J. P. Boilot, M. Gandias, C. Ricolleau, and M. Chamarro, *Mater. Res. Soc. Symp. Proc.* 358, 247 (1995).
405. G. Li and M. Nogami, *J. Sol.–Gel Sci. Technol.* 1, 79 (1993).
406. Z. Jiang and Y. Hui, *Proc. SPIE-Int. Soc. Opt. Eng.* 2288, 200 (1994).
407. H. Ye, D. He, Z. Jiang, and Y. Ding, *J. Sol.–Gel Sci. Technol.* 3, 235 (1994).
408. H. Ye, D. He, and Z. Jiang, *Guangxue Xuebao* 14, 518 (1994).

409. K. Yoshikai, M. Yamaguchi, and K. Nishimura, *Nippon Gomu Kyokaishi* 69, 485 (1996).
410. C. Cobianu, C. Savaniu, O. Buiu, D. Dascalu, M. Zabarescu, C. Parlog, A. van den Berg, and B. Pecz, in "CAS 96 Proc., Int. Semicond. Conf.," Vol. 2, p. 633. IEEE, New York, 1996. CA: 126:53304.
411. S. Sakohara, L. D. Tickanen, and M. A. Anderson, *J. Phys. Chem.* 96, 11086 (1992).
412. N. Maliavski, O. Dushkin, E. Tchekounova, P. Innocenzi, G. Scarinci, and M. Guglielmi, *Chim. Chron.* 23, 181 (1994).
413. W. H. Shih, D. J. Farrell, and W. Y. Shih, *Ceram. Trans.* 62, 233 (1996).
414. T. Oozeki and S. Urabe, Jpn. Kokai Tokkyo Koho JP 05127200 A2 930525 Heisei, 34 pp.
415. M. Nogami, Y. Q. Zhu, Y. Tohyama, K. Nagasaka, T. Tokizaki, and A. Nakamura, *J. Am. Ceram. Soc.* 74, 238 (1991).
416. M. Nogami, Y. Q. Zhu, and K. Nagasaka, *J. Non-Cryst. Solids* 134, 71 (1991).
417. A. A. Rywak and J. M. Burlitch, *Chem. Mater.* 8, 60 (1996).
418. E. Lifshitz, M. Yassen, L. Bykov, I. Dag, and R. Chaim, *NATO ASI Ser., Ser. E* 260, 503 (1994).
419. E. Lifshitz, *Mater. Res. Soc. Symp. Proc.* 452, 377 (1997).
420. E. Lifshitz, L. Bykov, and M. Yassen, *Proc. Electrochem. Soc.* 95-25, 477 (1996).
421. L. C. Liu and S. H. Risbud, *J. Appl. Phys.* 68, 28 (1990).
422. I. Tanahashi, M. Yoshida, Y. Manabe, and T. Tohda, *J. Mater. Res.* 10, 361 (1995).

Chapter 3

LOW-TEMPERATURE COMPACTION OF NANOSIZE POWDERS

E. J. Gonzalez, G. J. Piermarini

Ceramics Division, National Institute of Standards and Technology, Gaithersburg, Maryland, USA

Contents

1. INTRODUCTION

Recently, there has been a significant increase in interest in fabricating ceramic materials from ultrafine powders that consist of nanosize primary particles ranging in mean diameter from 1 to 100 nm. Theoretical predictions by Frenkel [1] and Herring [2] clearly indicate that the rate of densification varies inversely as a function of particle size. Thus, based on this prediction, as particle size decreases from micrometers to nanometers, a substantial decrease in sintering time can be expected at a given temperature. Indeed, many experimental investigations support this theoretical prediction. For example, Rhodes [3] produced densely packed compacts of nanosize zirconia particles and observed sintering of the compacts to near theoretical density at much lower temperatures than are used for sintering coarse zirconia particles. Recently, Skandan et al. [4] sintered nanosize titania at 800 °C, well below the sintering temperature for conventional titania powders.

These results suggest that nanosize particles as starting materials might offer considerable advantages for fabricating ceramics, especially because the reduced sintering

Handbook of Nanostructured Materials and Nanotechnology, edited by H.S. Nalwa
Volume 1: Synthesis and Processing

temperatures and sintering times required may inhibit undesired grain growth, one of the most important microstructural parameters because it is directly related to materials properties. It has been substantiated in many polycrystalline metallic and nonmetallic materials that the finer the grain size the higher the yield strength [5]. The expected reduced grain size (possibly on a nanoscale dimension) in a ceramic fabricated from nanosize particles results in a significant increase in grain interfaces, so that the number of interfacial atoms becomes comparable to the number of lattice site atoms. Thus, the overall properties of the solid will be influenced strongly by the local atomic arrangements in the interfaces, which are quite different from the atoms in the ordered crystalline state. Under these conditions, where significant contributions are made by the interfacial atoms, novel and possibly improved physical properties in ceramic materials fabricated from nanosize particles are anticipated.

One of the more important challenges in the fabrication of nanostructured materials is how to achieve full densification of the powder while simultaneously retaining a nanoscale microstructure. At present, relatively high densities (>95%) can be achieved with 10–20-nm particles, but the high densities invariably are accompanied by significant grain growth (grain sizes >50 nm) [6]. To reduce grain growth during sintering (where most of the densification process occurs), a high-density (random close packed) homogeneous green body with minimum pore size is desired, because large-size pores begin to close only after substantial grain growth has occurred and most of the small pores have been eliminated. Thus, prior to sintering a body composed of nanosize powders, it is important to try to achieve high densities, that is, full random close packing, of the nanosize particles in the green body. One way to achieve this condition is by compaction of the powder at high pressures. However, it is now known that forming densely packed green compacts from nanosize ceramic particles is very difficult to achieve by the application of pressure, because strong aggregation forces, such as the van der Waals attraction, increase dramatically as particle size decreases. For nanosize particles, the van der Waals attraction forces can essentially prevent the particles from sliding by one another, thereby causing agglomeration by diffusion during compaction [7]. As a result of aggregation and subsequent agglomeration of primary particles, green compacts of nanosize particles usually have relatively low densities after room temperature (RT) pressing or compaction.

The low green density of nanosize particle compacts is due primarily to two factors: (1) the presence of large voids formed by agglomerates of irregular shapes that cannot be easily broken down using conventional compaction techniques because of the presence of very strong bridging forces and (2) inefficient packing of individual particles because of large interparticle (interagglomerate) friction forces [8]. To obtain the desired nanoscale microstructure in the green body, agglomeration of the particles during synthesis and subsequent processing must be minimized. Recently, Duran et al. [9] sintered 3 mol% Y_2O_3–ZrO_2 powder to full density at relatively low temperatures. The authors emphasized that it is important to avoid the formation of gradients in the green-body density, inhomogeneities, and agglomerations, to produce defect-free sintered specimens. Their results also indicated that a change in sintering temperature of only 70 °C can have a significant effect on the densification kinetics. Based on these results, it is evident that special criteria must be met when selecting the processing conditions for nanosize powders.

High-pressure compaction of nanosize particles has been used in an attempt to achieve the desired nanoscale microstructure and high density in the green body. Typical compaction pressures usually exceed 1.0 GPa and, therefore, require specialized equipment. For example, to attain high green density by uniaxial compaction of a nanosize powder, a high-strength tungsten carbide–cobalt (WC/Co) piston–cylinder die has been used, and, for even higher pressures, a diamond anvil-type pressure cell (DAC) has been utilized. Because of the demanding strength requirements of apparatus capable of producing pressures in the gigapascal range, the compaction equipment necessarily must be designed to produce relatively small samples. For these reasons, the DAC was one of the first devices to be utilized in compaction of nanosize particle experiments. Moreover, it is relatively easy to use and cost effective when compared with large press equipment.

Recently, it was discovered that the densification of nanosize powders can be improved in the green body by the use of lubricants that allow particles to slide easily over one another, permitting their rearrangement to produce a minimum interstice structure, that is, random close packing [10]. A good lubricant can improve significantly the packing properties of the nanosize particles. However, selection of a suitable lubricant is severely limited because of the existence of strongly reacting interfaces between nanosize particles, which the lubricant must overcome, as well as the presence of nanoscale voids in the structure formed by densely packed nanosize particles, which the lubricant must be able to penetrate to be effective. Thus, the lubricant must be necessarily a low-molecular-weight substance, small enough to fit into these nanoscale voids. High-density green-body compacts of nanosize particles are typically transparent, indicating that the voids contained within them are also of nanoscale dimensions, that is, less than the wavelengths associated with white light. Thus, only a few substances are commonly available that can be used as lubricants meeting this dimensional requirement.

Pechenik et al. [10] demonstrated the concept that simple molecular liquids can provide efficient lubrication to enhance the densification of nanosize particles when subjected to high pressures. They compacted nanosize silicon nitride particles at liquid nitrogen temperatures and also used LN_2 as a lubricant. The first compactions were performed in a DAC because that was the only pressure equipment conveniently available to them at that time. This cryogenic compaction process was successful in producing densely packed transparent (light brownish in color) compacts of nanosize particles that sintered close to full density at a much lower temperature (1200 °C) than that required for the conventional processing of silicon nitride. As a follow-up to this original research, Chen et al. [8] studied the cryogenic powder compaction process in much greater detail. To do this, these authors designed and constructed a special scaled-up WC/Co piston–cylinder die apparatus for fabricating much larger samples than those that were made in the DAC. The compaction equipment they developed is capable of producing 3-mm disk-shaped samples under vacuum or in a variety of controlled conditions, such as temperatures in the range from 77 to 1000 K and pressures up to 3 GPa. In addition, throughout the compaction process, continuous measurements of the sample volume, applied force, and frictional force between the sample and the die walls were performed. This novel system, because of its uniqueness and its applicability to understanding the rheology and compaction of nanosize powders, is described in great detail later in this chapter, along with a number of experimental studies on the compaction of nanosize silicon nitride and γ-alumina powders.

Unfortunately, published reports on the effects of lubricants on nanosize powder compaction processes are very limited. However, the authors are aware of unpublished current activity in private industry related to the use of lubrication techniques to achieve the high green densities necessary for low-temperature sintering of nanosize powders. Unfortunately, results related to such work have not been reported publicly by the different ceramic and metal processing industries because the information is regarded as being very exclusive and proprietary. In the following sections, we will describe the more important published techniques used to compact nanosize ceramic powders at low temperatures. The advantages and disadvantages of these techniques will be discussed also. Because of their prominence in this field, the high-pressure piston–cylinder compaction system developed by Chen et al. [8] and the diamond anvil method used by Pechenik et al. [10] will be described in detail in conjunction with the different lubrication experiments that were conducted with these systems.

2. LOW-TEMPERATURE–HIGH-PRESSURE POWDER COMPACTION

2.1. Diamond Anvil Pressure Cell

The diamond anvil pressure cell has been used for many years to investigate a variety of physical phenomena too numerous to list here. For the purposes of this discussion, it is sufficient to note that one of the more recent and less well known applications of the DAC has been in ceramic processing, particularly of nanosize powders at high pressures and moderate temperatures ($<600\,^\circ$C). Because the DAC is relatively easy to use and can typically achieve pressures in excess of 5 GPa, it is an attractive alternative for routinely compacting small samples of nanosize powders for exploratory research purposes. Because much of the early results in this area of ceramic processing were obtained by utilizing the DAC, we consider the device important enough to describe in detail.

A schematic diagram of the typical design of a DAC for use at room temperature is shown in Figure 1 [11]. For the sake of clarity, there are no features incorporated in this particular design for specifically heating this cell. However, cells designed for simple resistance heating have been developed that can be routinely heated to temperatures in the range of 500 °C with occasional excursions to 600 °C. The sample, located between two opposed diamond anvils shown in detail on the left side of Figure 1, is confined by a metallic (usually Inconel X750)* gasket with a small circular hole about 250 μm in diameter for confining the nanosize powder sample. The initial thickness of the gasket is also about 250 μm. Pressure is applied by rotation of the large screw, which compresses the Belleville spring washers, generating a load transmitted to the gasket/sample via the lever arms. The design has 180° optical access (see Fig. 1), which permits the ruby fluorescence method of pressure measurement as well as other optical and X-ray spectroscopic techniques to be used [12, 13]. The fluorescence pressure measurement method utilizes a small ruby sphere ($\approx$10 μm in diameter), which is included with the powder to be pressed and serves as the internal pressure sensor.

Heating an anvil-type pressure cell of this design is typically done with a cylindrical resistance coil heating furnace positioned either around the anvils themselves internally or around the anvil–piston assembly [13]. The latter method, shown in Figure 2, is a modification of the design shown in Figure 1. This static method of heating was the first developed for this type of pressure cell and is limited to routine temperatures of about 500 °C, because at higher temperatures the metal components of the pressure cell lose their required strength and hardness and, thus, may lead to anvil failure [10]. Utilizing the resistance coil heating method with the DAC is not useful for processing ceramic powders at typical sintering temperatures, usually well above 500 °C. The DAC is more readily adaptable to low-temperature or cryogenic compaction experiments, which is accomplished by total immersion of the pressure cell in LN_2 as pressure is applied to the sample.

The main disadvantages of using the DAC for ceramic processing experiments is that the sample's volume is very small, making characterization of the resulting sintered compact very difficult (a disk $\approx$0.2 mm in diameter and 0.2 mm in height). Thus, although the DAC has some serious limitations, it has enough advantages to justify its use as a device for relatively simple and rapid exploratory research. Its role in ceramic processing has evolved into a complementary tool to a more useful compaction technique, that is, the piston–cylinder die developed by Chen et al. [8], which is described in the following section.

* Certain trade names and company products are mentioned in the text or identified in illustrations in order to specify adequately the experimental procedure and equipment used. In no case does such identification imply recommendation or endorsement by the National Institute of Standards and Technology, nor does it imply that the products are necessarily the best available for the purpose.

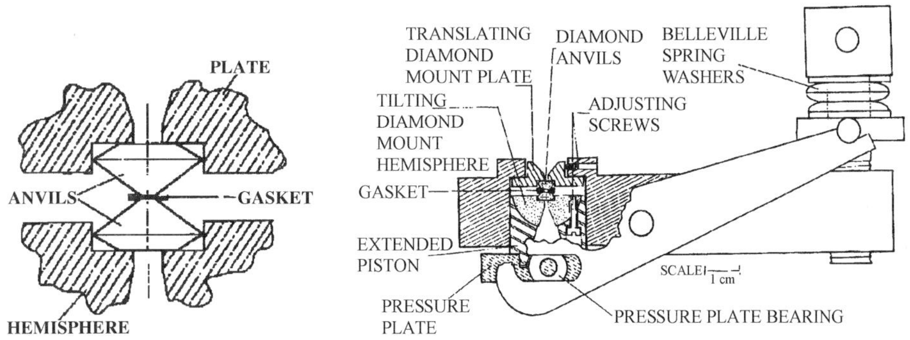

Fig. 1. Typical high-pressure optical diamond cell (DAC) designed for use at room temperature. Details of the anvil assembly are shown in the insert on the left side of the figure. (Source: Piermarini and Block [11].)

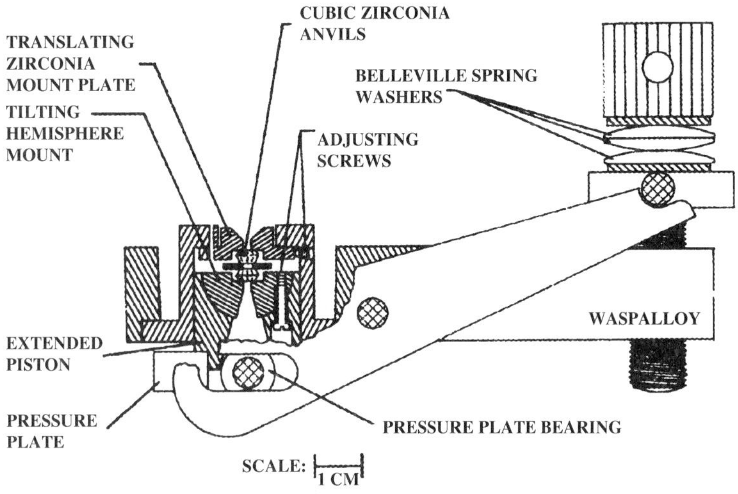

Fig. 2. A schematic diagram of an Inconel 718 diamond anvil cell used for heating a sample to temperatures in the range of 500 °C. The heating is provided by a cylindrical coil furnace located in the recess surrounding the anvil assembly. (Source: Barnett et al. [13].)

2.2. High-Pressure Compaction with the Piston–Cylinder Device

The equipment for the compaction of nanosize powders, originally developed by Chen et al. [8], was designed to perform at least four functions. The first, and most important, was to produce a 3-mm-diameter disk-shaped sample so that accurate measurements of density, hardness, fracture toughness, optical properties, and pore-size distribution could be carried out. Prior DAC samples, because of their miniature size, did not permit many of these measurements to be made easily or with the desired accuracy. The second function was to provide a means of measuring the compaction rheology of powders in the form of the volume (or density) of the compacting sample as a function of the applied pressure, the rate of pressure application, the processing temperature, and the environment. To provide complete information about the compaction process, the authors designed the system so that the previously mentioned parameters were measured continuously and simultaneously. Also, recognizing that nanosize particles are extremely reactive chemically, the third function was to design the compaction chamber to support a vacuum, as well as a positive gas pressure to permit various environments to be used. With such a system, the authors had the capability to experiment with different gases and fluids under selected pressures for protection and lubrication of the surfaces of the nanosize particles prior to compaction. From a theoretical standpoint, it is reasonable to expect that the thickness of the adsorbed coating on the particle surfaces and properties of the coating should depend strongly on two parameters: (1) temperature and (2) gas pressure inside of the environmental chamber. Because of the large amount of data acquired during an experimental run with their device, the authors built in a real-time computer control of all measurements and experimental procedures. To detect and measure accurately phase transitions accompanied by a volume change, they designed the equipment to provide accurate and precise data for the development of pressure–volume (PV) curves. Finally, the fourth function was to design the equipment to withstand temperatures of up to 1000 K for short periods of time. Owing to the importance of this equipment in the high-pressure compaction of nanosize ceramic powders, a detailed description of the compaction apparatus developed by Chen et al. [8] and later used by Gonzalez et al. [14] follows.

3. PISTON–CYLINDER DIE

3.1. Equipment Configuration

A 10,000-kg-capacity screw-driven Instron press with two large parallel platens was used to apply load to the die. The servocontrolled drive system permitted the selection of platen displacement rates ranging from 0.85 to 850 μm/s, a feature that is useful for the study of stress relaxation in a powder during compaction. In this apparatus, a 2100-kg load is required to generate a pressure of 3 GPa on a 3-mm-diameter sample.

The authors designed their equipment to be used under two different modes of operation for the powder compaction experiments: (1) gas lubrication and (2) liquid lubrication. In the first mode, a small amount of lubricating cover gas is introduced into the environmental chamber under controlled pressure. The introduced gas condenses on the surface of the particles and the powder is compacted under dry or semidry conditions. The basic design for this first experimental setup is shown in Figure 3. The environmental chamber (14) is designed for isolating the sample in order to control compaction conditions while the movable piston (13), equipped with two O-ring seals, applies force to the compacting piston (2). Lubricant gas can be introduced from a gas tank into the vacuum chamber (14), and the gas pressure is controlled by a vacuum needle valve (not shown). A thermocouple gauge located outside of the chamber on the supply line measures the gas pressure in the chamber. A single-acting (bottom punch fixed) piston–cylinder-type device (3) is used for the application of variable pressures up to 3 GPa. The piston (2) and the die (3) are made of tungsten carbide and ground to about a 0.1-μm finish on the contact surfaces. The bottom end of the cylindrical hole of the die has a small conical shape for easy removal of the compacted samples. The temperatures can be varied from that of liquid nitrogen, 77 K, to room temperature. A cooling cylinder (4) and a heater (5) are designed with a coil wound on a frame surrounding the die. A cooling loop of LN_2 can pass through a heat-insulating flange into the environmental chamber.

For cryogenic compaction, the applied load is adjusted continually at low-temperatures to maintain the desired pressure. One difficulty is to measure the pressure in nonhydrostatic and low-temperature environments. The frictional force on the wall of the die, the punch, and the sample must be taken into account for pressure calibration. As depicted in Figure 3, two load cells are used for pressure measurement. One, the main load cell, measures the primary force applied by the press and is located on the bottom platen of the press. The other load cell (11) measures the frictional load between the punch and the die. The authors are not aware of any commercially available load cells calibrated for use at extreme temperatures. Therefore, they incorporated into the design for low-temperature compaction a special thermally insulating mechanical structure to maintain the frictional load cell at room temperature. To solve this temperature problem of the load cell, they used

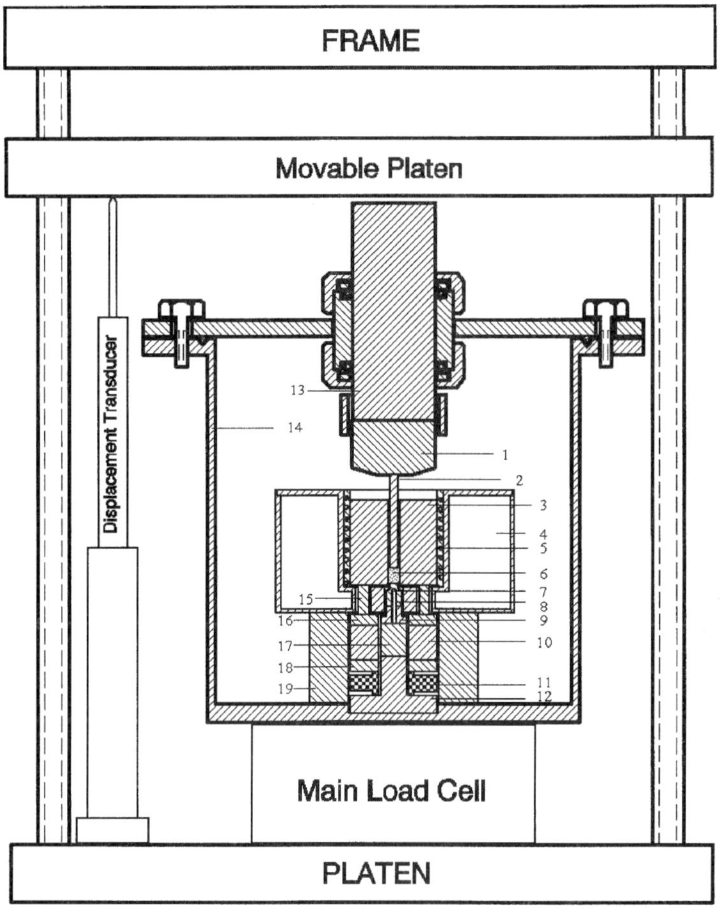

Fig. 3. Configuration of "gas lubrication"–type powder compaction assembly: (1) Bridgman anvil, (2) moving punch, (3) die, (4) cooling cylinder, (5) heater, (6) powder, (7) thermocouple, (8) fixed punch, (9) hardening disk, (10) heat-insulating ring disk, (11) frictional load cell, (12) positioning and heat-conducting disk, (13) pressure transmission piston, (14) vacuum chamber, (15) frictional load transmission pistons, (16) hardening ring disk, (17) heat-insulating disk, (18) hardening ring disk, and (19) plastic alignment ring. (Source: Chen et al. [8].)

frictional force transmission pistons (15) and heat-insulating glass–ceramic (Macor) ring disks (10). The pistons (15) transmit frictional force from the die to the load cell that operates at room temperature. For low-temperature compaction, two ceramic disks (10, 17) prevent heat transfer from the room temperature region to the cold region. The average pressure at the bottom of the sample is calculated from the total and frictional load data.

A K-type thermocouple (7) was placed at the center of the fixed punch as near the sample as possible to provide measurement and control of the temperature. A displacement transducer was positioned between the two platens of the press, as shown in Figure 3, to measure the thickness of the sample versus the force on the sample for the rheological investigations.

An alternative design of their experimental setup to study the effects of lubricants on the compaction of powders is shown in Figure 4. In this design, a cryogenic container (3) holds the die (2) and powder sample (4), and a metal (Inconel 600) foil (6) prevents leakage of LN_2 or other lubricating liquids. The load of the sample (total load minus friction) on the fixed punch (5) is transmitted via the metal foil (6), the hardened-steel load transmission disk (8), the heat-insulating disk (9), and the positioning and heat-conducting disks (13)

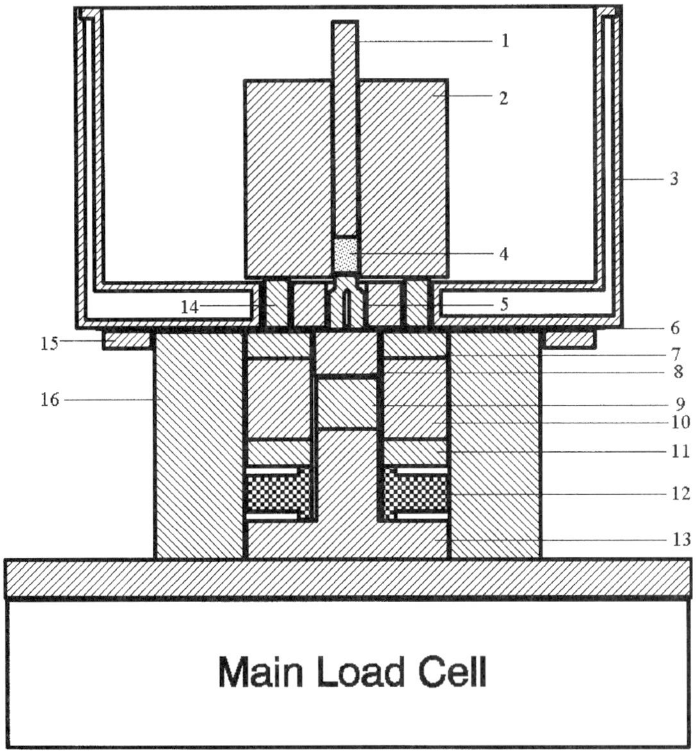

Fig. 4. Configuration of "liquid lubrication"–type powder compaction assembly: (1) moving piston, (2) die, (3) cryogenic container, (4) powder, (5) fixed punch, (6) metal foil, (7) hardening ring disk, (8) hardening disk, (9) heat-insulating disk, (10) heat-insulating ring disk, (11) hardening ring disk, (12) frictional load cell, (13) positioning and heat-conducting disk, (14) frictional load transmission pistons, (15) flange, and (16) plastic alignment ring. (Source: Chen et al. [8].)

to the main load cell as shown in Figure 4. The frictional load on the die (2) is transmitted to the frictional load cell via frictional transmission pistons (14), the metal foil (6), the hardened-steel ring disk (7), the heat-insulating ring disk (10), and the hardened-steel frictional load transmission ring disk.

3.2. Computer Control and Software Development

The authors implemented real-time computer control to measure and/or control (1) the temperature of the chamber, in the case of hot pressing; (2) the total applied and frictional forces; and (3) piston displacement. The applied pressure and rate of pressure application were calculated from measured forces, times, and areas. The configuration of the computer control setup is shown in Figure 5. The computer controls the temperature from remote points with a temperature control unit by means of a digital-to-analog converter card. A digital multimeter combined with a scanner performs the voltage measurements on the load cells, the displacement transducer, and the thermocouples. An IEEE-488 general-purpose interface bus card was used for communicating between the computer and the digital multimeter. The computer controls the sequence of scanning measurements, presets the temperature in the environmental chamber, and conducts autodetermination of steady-state temperature.

The control program uses a built-in clock as the time counter for the measurement of pressure as a function of time, that is, the rate of pressure application. The code of the program for controlling and monitoring the powder processing operation was written in

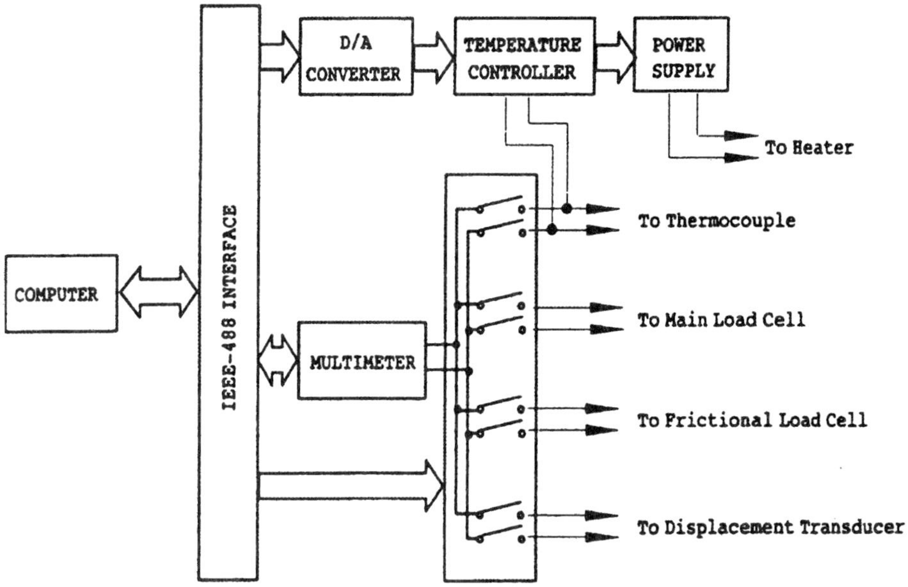

Fig. 5. Block diagram of computer control system. (Source: Chen et al. [8].)

a GFA-BASIC language for MS-Windows on a 486 IBM-compatible personal computer. Eight windows are opened to plot and display experimental data. The following data are displayed graphically in six windows: (1) sample volume vs applied pressure, (2) calculated density vs pressure, (3) calculated density vs time, (4) pressure vs time, (5) temperature vs time, and (6) load fraction (real load on sample or total load minus friction/total load) vs the applied pressure. The numerical data output and picture simulation of the piston position in the die and change of temperature are displayed in windows (7) and (8). During an experimental run, the program estimates the range for each measurement according to the initial scanning measurements and some preset values, and replots all diagrams using an autoscale subroutine to give an appropriate range and boundary for each quantity. The replotting is performed at a preset time interval.

A typical curve of the load fraction (real load/total load) vs pressure for dry compaction of nanosize γ-Al_2O_3 is shown in Figure 6. The load fraction depends on the type of powder, the magnitude of the gap between the punches and the die, the roughness of the contact surfaces, the sample thickness, and the lubrication conditions. The figure shows that the contribution of friction for a real load calculation is larger in the lower-pressure region.

To determine the accuracy of the PV curves measured by this apparatus, two well-studied materials with known pressure–volume characteristics, rubidium bromide and potassium chloride, were selected by the authors. Both alkali halides (RbBr and KCl) at room temperature exhibit face-centered cubic (fcc) structures that undergo a polymorphic transition to a denser simple cubic (sc) form at elevated pressures [15–17]. Both materials were in the form of powders with primary particle size in the range of 1 μm. X-ray powder diffraction analysis indicated that both materials were in the fcc phase in the form of granular crystalline powders and that the degree of hydration was minimal. Hydration is known to cause erroneous results in determining accurate transition pressures for these materials [17]. The results of the compaction of RbBr and KCl at room temperature are presented as pressure vs volume curves in Figures 7 and 8, respectively. These experiments demonstrated the ability and sensitivity of the instrument in measuring volume changes as a function of pressure.

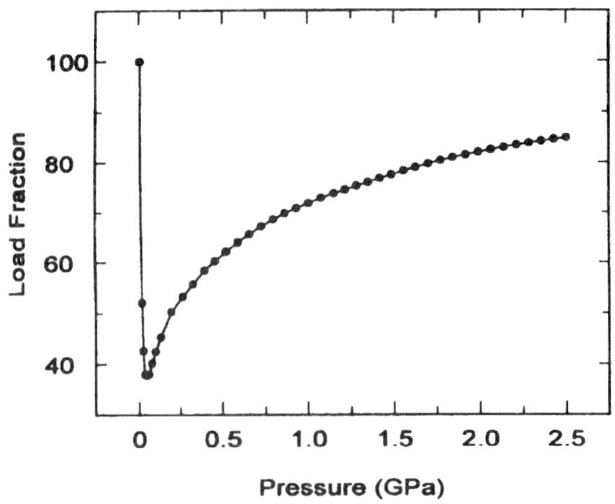

Fig. 6. Fraction of real load (total load minus friction) and total load versus pressure for compaction of nanosize γ-Al_2O_3 powder. (Source: Chen et al. [8].)

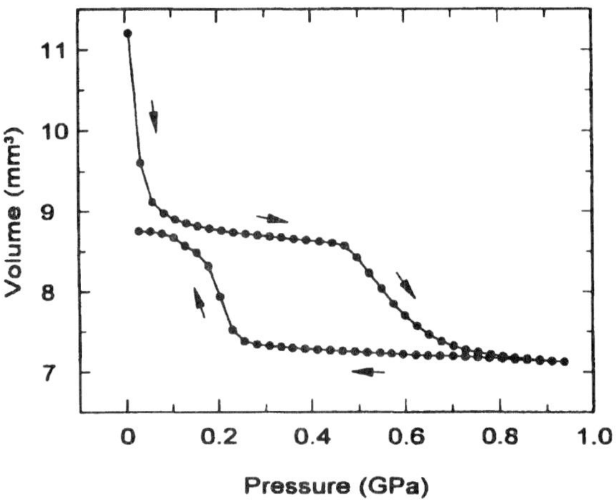

Fig. 7. Hysteresis behavior and volume phase transition of RbBr granular crystalline powder. (Source: Chen et al. [8].)

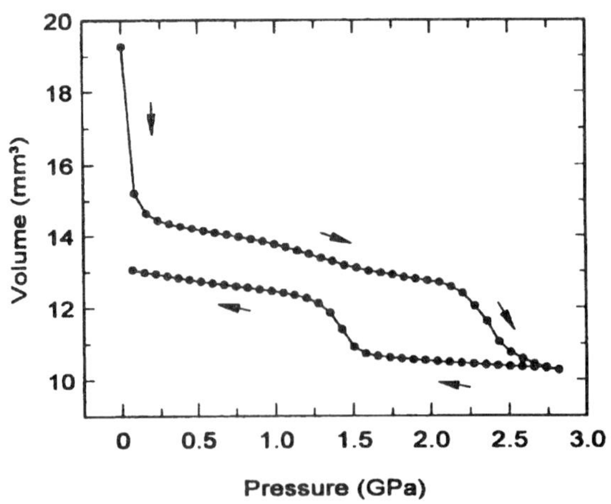

Fig. 8. Hysteresis behavior and volume phase transition of KCl granular crystalline powders. (Source: Chen et al. [8].)

4. COMPACTION AND LUBRICANTS

4.1. Compaction of Si_3N_4 Powder

As previously mentioned, for nanosize particles, the van der Waals attraction forces can prevent the particles from sliding past each other, and, thereby, promote agglomeration. As a result, compacts of nanosize particles usually have low green densities after cold pressing. As demonstrated using several examples in a later section of this chapter, the use of suitable lubricants can improve the packing properties of the nanosize particles, but, in selecting a lubricant, one is severely limited because of the strongly reacting interfaces between particles and the small size of voids in the resulting green microstructure formed by densely packed nanosize particles [10]. Generally speaking, the diameter of the channels that exist in the structure of densely packed particles is approximately 1/10th that of the particle size. Therefore, for a 10-nm-diameter particle, the channel diameter could be approximately 1 nm. Not many molecules exist that can penetrate such small channels. Because the lubricant must be eliminated from the densely packed structure prior to sintering, one must select a lubricant that consists of very small molecules. Moreover, the selected molecule must adsorb on the surface of the nanosize particles strongly enough to provide good sliding properties during cold compaction, but it must also be eliminated easily after achieving a densely packed structure. This could be accomplished by heating in vacuum for an appropriate length of time.

As an example of the effects of lubricants, we will discuss the work done by Pechenik et al. [10] on the compaction of nanosize Si_3N_4. Pechenik et al. used amorphous stoichiometric Si_3N_4 powder produced via a laser-driven gas phase reaction between silane and ammonia. More details about the preparation of this powder can be found elsewhere [18]. The average particle size of this powder was estimated to be between 16 and 17 nm by transmission electron microscopy, and the starting powder had an oxygen mass fraction of approximately 6%.

The initial experiments by Pechenik et al. [10] were conducted in the DAC under three different environmental conditions: (1) RT, (2) under LN_2, and (3) after outgassing at 200 °C. The samples were compacted according to the following procedures. The starting powder was precompacted in a WC/Co die under 0.1 GPa to facilitate the handling and loading of the nanosize powder in the DAC. The precompacted powder achieved a density of approximately 30% of theoretical random close packing. For the room temperature compaction, the powder was loaded in the DAC and pressed to 5 GPa. A molybdenum gasket was used to confine the sample between the two opposed anvils in the DAC and served as the sample container. The maximum size of the compacts fabricated by the DAC in this work was 0.2 mm in diameter by 0.15 mm in thickness. Pressures were measured by the calibrated shift of the ruby fluorescence R_1 line. For this purpose, a small ruby sphere (about 10 μm in diameter) serving as the pressure sensor was included with the sample. In some cases, the samples were hot-pressed for 3 h under a pressure of 5 GPa to further study the effects of compaction.

Cryogenic compaction (LN_2) was performed following a more complex procedure. The gasketed DAC, containing both the Si_3N_4 precompacted sample and the small ruby pressure sensor, was first sealed and isolated from the external environment at RT by applying a nominal load to the gasket. The DAC was then immersed in an LN_2 bath. After about 10 min, it reached equilibrium with the LN_2. When temperature equilibrium was attained, the load on the gasket in the DAC was reduced to zero, thereby breaking the seal between the gasket and the diamond surface and exposing the precompacted Si_3N_4 powder to LN_2. After 1 min, the cell was resealed by applying the load to produce pressures ranging from 1 to 3 GPa on the sample, which now included some LN_2. The pressurized DAC was then removed from the LN_2 bath and warmed to RT. At RT, the sample pressure was measured and defined as the initial compaction pressure. Pressure cycling at RT was then carried out and consisted of reducing the pressure to ambient and then increasing it again to the desired maximum pressure defined as the final compaction pressure (always 5 GPa for these LN_2 experiments). The pressure cycling procedure was carried out usually two or three times until the LN_2 had been eliminated from the sample.

The third group of samples was prepared in a specially designed vacuum chamber that accommodated the DAC and also permitted access to rotation of the load-transmitting screw in the DAC. The loaded cell was heated to 200 °C for 2 h inside the vacuum chamber. The sample was also sealed under vacuum by applying pressures of 0.5–1.0 GPa.

To evaluate the effects of the different compaction conditions, the authors used the hardness of the sample as a measure of compaction efficiency. The authors measured the hardness using a Vickers indenter. Figure 9 summarizes the results for green bodies and sintered samples. It is clear from the plot that the samples processed under LN_2 were always harder. On the contrary, the outgassed samples had the lowest hardness, suggesting that vacuum heating can change the flow properties of the powders during compaction. Pechenik et al. [10] speculated that an adsorbed substance on the surface, either a gas or a liquid, acts as a lubricant during compaction, and that this layer is eliminated during vacuum heat treating. The authors emphasized that samples processed in LN_2 were mostly transparent to visible light and the outgassed samples and samples compacted at RT were typically opaque.

The authors also presented data for compaction at different pressures to show that hardness increases with compaction pressure (Fig. 10). Because these powders are not expected to deform plastically during compaction like metals, the reported increments in hardness would be expected to be a result of a tighter and more efficient packing density. Again, their results indicate that the LN_2-compacted samples were always harder, even after heat treating at high temperatures as indicated in Figure 9.

Chen et al. [8] also studied these powders using the piston–cylinder compaction apparatus. Figure 11 shows a maximum random packing density of approximately 64% for the nanosize amorphous silicon nitride compacted at less than 2.5 GPa pressure under LN_2 using the piston–cylinder apparatus described earlier. At the same pressure, for a dry compaction of the same powder, the green density was about 57% of theoretical. It is important to note that the rate of volume change with pressure is higher for the sample compacted under an LN_2 environment. This suggests that friction between particles has decreased and particle rearrangement is occurring at lower pressures.

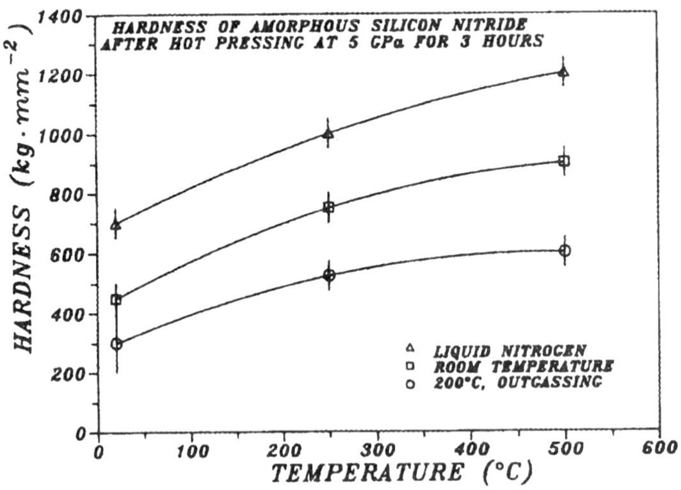

Fig. 9. Comparison of hardness measurements of amorphous Si_3N_4 specimens as a function of hot-pressing temperature. All samples were pressed in the DAC and hot-pressed at 5 GPa for 3 h. Each curve represents a different initial compaction technique; liquid nitrogen, room temperature, and 200 °C outgassing during initial compaction. (Source: Pechenik et al. [10].)

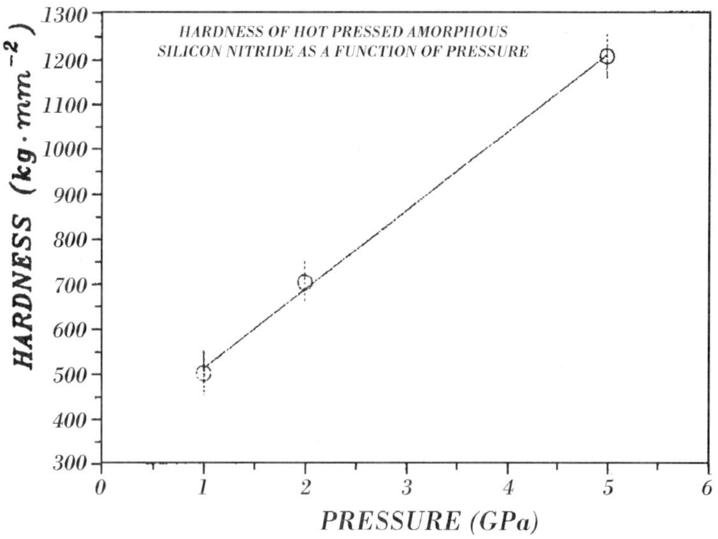

Fig. 10. Hardness of hot-pressed amorphous Si_3N_4 specimens as a function of compaction pressure for samples compacted under liquid nitrogen and hot-pressed in the DAC to the respective pressures at 500 °C for 3 h. (Source: Pechenik et al. [10].)

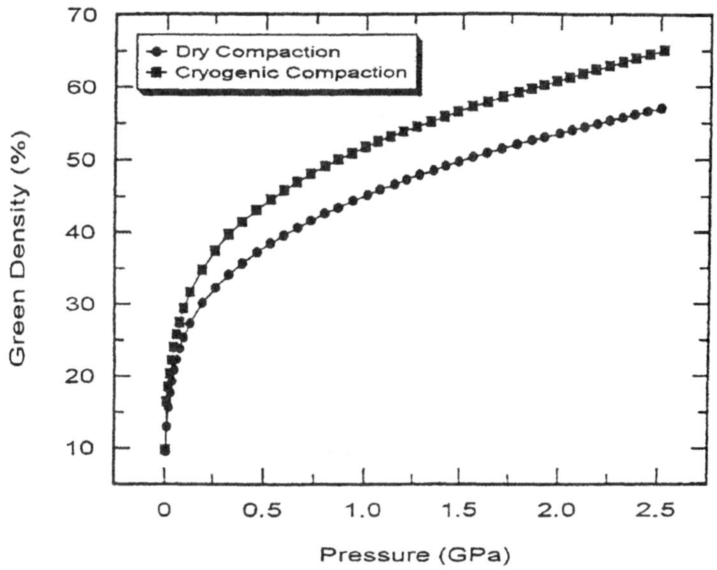

Fig. 11. Variation of density of green compacts of nanosize amorphous Si_3N_4 as a function of compaction pressure. (Source: Chen et al. [8].)

These results clearly indicate that cryogenic compaction is an efficient technique to reduce the compaction pressure necessary to obtain high packing densities and small-scale porosity. The green body produced by LN_2 processing exhibits transparency under visible light, which is an indication of nanoscale, uniform porosity. The actual mechanism of lubrication has not been studied in detail, or at least we are not aware of any report in the literature. Stronger evidence of the effects of lubricants has been reported in the compaction of γ-Al_2O_3 powders.

4.2. Compaction of γ-Al_2O_3 Powder

We have discussed several times throughout this chapter the difficulty of compacting nanosize powders into green bodies that can be easily examined without crumbling. Nanosize γ-Al_2O_3 powder is also very difficult to compact, and high pressures are typically required to obtain structurally sound green bodies. The following discussion presents results reported by Gallas et al. [19] and Gonzalez et al. [14] on a number of compaction attempts of nanosize particles of γ-Al_2O_3. Samples were prepared at RT, and using LN_2 as a lubricant. The main objective of the authors was to produce a compact of γ-Al_2O_3 with uniform and relatively high green density (60%–70%). Without first producing a homogeneous compact of random-closed-packed particles, the green-state body cannot be sintered to high density because coarse porosity is not easily eliminated via surface diffusion, the predominant mechanism for sintering compacts of nanosize particles [20].

As presented in the earlier discussion on the compaction of nanosize particles of silicon nitride, it was shown that dense packing was not achieved by using high pressures alone [10, 21]. It was found that LN_2 provided an excellent lubricant for moving the silicon nitride particles closer together. The hardest compacts were those that were fabricated under LN_2. In light of this previous work, LN_2 also was used to compact nanosize particles of γ-Al_2O_3 and the results were compared with room temperature compactions. Both methods produced transparent green-state compacts, which were then heat treated under vacuum. Hardness was measured before and after heat treatment. In addition, the compacts were analyzed by energy-dispersive X-ray diffraction to identify the crystallographic phases present, and by transmission electron microscopy (TEM) to characterize the microstructure.

The starting nanosize γ-Al_2O_3 powder used was a commercially available material, Aluminum Oxide C, supplied by Degussa AG, Geschäftsbereich Anorganische Chemieprodukt, Frankfurt, Germany. It had an average particle diameter of 20 nm. Angle-dispersive X-ray powder diffraction measurements on the starting powder indicated the presence of primarily γ phase with some δ phase, estimated from diffraction intensities to be less than 10% by volume. The powder was used as received without any prior treatment and was handled in laboratory air.

The experimental procedures used were very similar to the steps followed for the compaction of Si_3N_4 reported earlier. The starting powder was precompacted in a WC/Co die under 0.1 GPa and then subsequently compacted in a DAC under pressures ranging from 1 to 3 GPa. The initial compactions were carried out at two different temperatures: (1) RT and (2) LN_2 temperature. In the case of room temperature compaction, the precompacted starting material was initially pressed to transparency. Typically, transparency was reached at about 1 GPa. However, final sample pressures exceeded this value and ranged from 2 to 4 GPa. The final sample pressure was reached in two ways: (1) by one pressure cycle and (2) by multiple (normally two) pressure cycles. It was found that the hardness of the green-body compact was independent of the number of cycles used to reach the final sample pressure at room temperature. This is an important detail because, as will be discussed in the following paragraph, the procedure for LN_2 compaction requires cycling, and confirms that the increase in hardness is exclusively a lubrication effect.

The compaction of the γ-Al_2O_3 sample under LN_2 (cryogenic compaction) was performed by immersing the DAC in LN_2 as was done with Si_3N_4. Pressure cycling at RT was then carried out and consisted of reducing the pressure to that of the ambient atmosphere and then increasing it again to the desired maximum pressure defined as the final compaction pressure (always 3 GPa for these LN_2 experiments). The pressure cycling procedure was carried out usually two or three times until the LN_2 had been eliminated from the γ-Al_2O_3 sample.

After final compaction and removal from the DAC, the transparent green-body compact of nanosize particles of γ-Al_2O_3 was evaluated by measuring its hardness using a Vickers microhardness indentor at a fixed load of 50 g. Optical clarity was evaluated by microscopy examination. The energy-dispersive X-ray powder diffraction technique was used to determine the polymorphic phases present.

In the final step, the transparent green-body compacts produced by both RT and LN_2 compaction procedures were heat treated in a tube furnace under 100 Pa vacuum at 800 °C for 10 h. Heating and cooling rates of 5 °C/min were used in the heat treatment procedure. The molybdenum gasket was found to be chemically inert to γ-Al_2O_3 at the 800 °C heat treatment condition. After heat treatment, the optical transparency, microhardness, and X-ray characterization were again evaluated. The microstructure of the heat-treated and green-state compacts were also characterized by TEM.

It is important to mention that no phase change occurred during the compaction procedures, and, therefore, the possibility of a different phase of alumina contributing to changes in hardness was ruled out. The authors confirmed this by studying X-ray diffraction patterns collected from both the RT and the LN_2 compacted samples. The authors pointed out that, while optical transparency in the green-state compact is necessary, it is not a sufficient condition for achieving dense packing of nanosize particles [21]. Optical transparency indicates that particles are packed such that the pores are reduced in size to below 100 nm and suggests sample uniformity. Thus, optical clarity alone cannot be used as an indicator for characterizing density or fine-scale homogeneity of particle packing. Similar to the Si_3N_4 study, the shear strength of a compact, characterized by measuring the Vickers hardness, is thought to be a reliable parameter to characterize the density of particle packing because it depends strongly on the coordination number of the particle arrangement. Therefore, the authors used Vickers microhardness numbers as a measure of the quality and efficiency of particle packing.

Figures 12 and 13 summarize hardness measurements done before and after heat treatment (HT) as a function of the compaction pressure applied at RT and at LN_2 temperature, respectively. The HT procedure was the same for all the γ-Al_2O_3 samples (800 °C for 10 h under 100 Pa vacuum) so that relative differences are strictly due to the compaction conditions, for example, RT and LN_2 compaction. The general features of both Figures 12 and 13 are similar. The hardness after heat treatment (H_a) appears to be strongly dependent on the hardness before heat treatment (H_b). In both cases, the ratio H_a/H_b is roughly constant at about 1.5. Furthermore, the hardness before heat treatment appears to be directly dependent on the pressure and temperature of compaction.

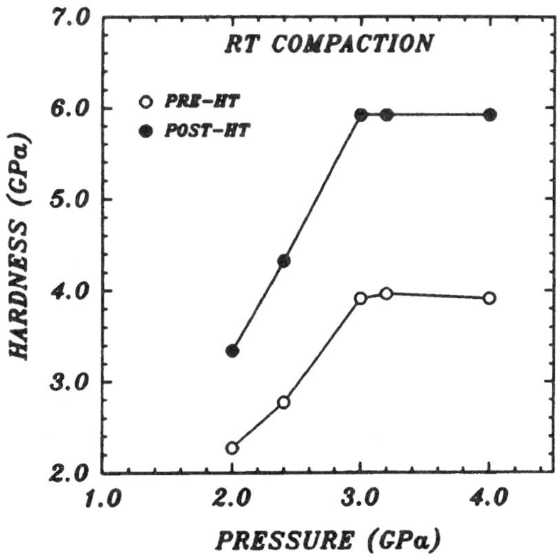

Fig. 12. Hardness of γ-Al_2O_3 compacts as a function of final pressure used to compact the powder at room temperature (RT) before and after a heat treatment of 800 °C for 10 h. (Source: Gallas et al. [19].)

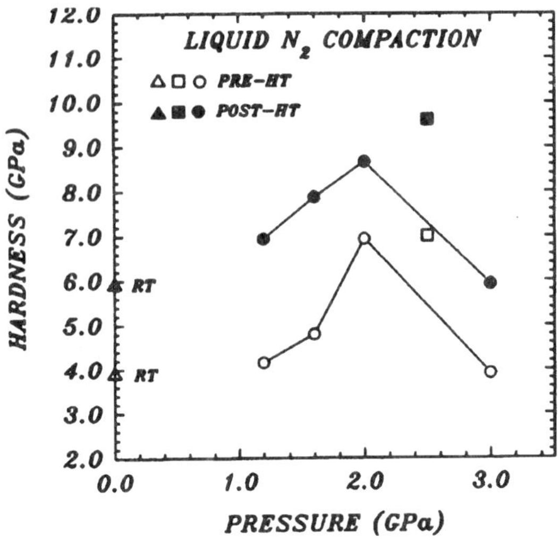

Fig. 13. Hardness of γ-Al_2O_3 compacts as a function of final pressure used to compact the powder at liquid nitrogen temperatures before and after a heat treatment of 800 °C for 10 h. The two triangles in the hardness axis designate the measured limiting hardness for samples pressed at room temperature (Fig. 12). The solid and open square symbols designate measured hardness values for the samples heated at 100 °C to avoid the β phase of N_2 prior to retrieval of the green-state compact. (Source: Gallas et al. [19].)

For samples compacted at RT for pre- and post-HT, the hardnesses were not as high as the values measured in the cryogenic samples, indicating less dense compacts, even though they were also optically transparent (Figs. 12 and 13). For 2 GPa, the hardness of the RT sample is much lower than the hardness measured for the LN_2 sample at this same initial pressure, under both conditions, before and after heat treatment. Another interesting result shown in Figure 12 is that the hardness reaches a limiting value for both the green body and the heat-treated compact and is nearly constant for the compaction pressure between 3 and 4 GPa.

In Figure 13, for the pressure range up to 2 GPa, the hardness increases with the applied initial pressure and then decreases abruptly at 3 GPa. The highest hardness (approximately 9 GPa) was obtained with the initial compaction pressure under LN_2 between 2.0 and 2.5 GPa. When pressures exceeding this range were used, lower hardnesses were obtained. Note that precompaction under LN_2 at 3 GPa followed by ambient compaction at 3 GPa (standard procedure) resulted in a sample hardness comparable to that obtained for RT compaction identified on the hardness axis in Figure 13 as triangles. This result indicates that all beneficial effects of cryogenic compaction are lost when the pressure exceeds roughly 2.5 GPa.

The unexpected decrease in hardness observed for the powder compacted initially at 3 GPa under LN_2 may be explained by referring to the equilibrium pressure/temperature phase diagram for nitrogen (Fig. 14). Of interest here are the successive phase transformations that occur in N_2 with increasing pressure at different temperatures. The temperature of concern here is approximately 77 K, the equilibrium temperature at which the γ-Al_2O_3 powders were initially compacted. According to this pressure/temperature phase diagram for N_2, there are three solid phases, β, γ, and δ, which exist between ambient pressure and 2.5 GPa at 77 K [22–24]. As the temperature increases in the pressure range of concern here, both the γ and the δ phases transform to the β polymorph, which is the form that is in equilibrium with the liquid phase. At RT and 2 GPa, only the liquid phase of nitrogen is stable and is, therefore, present with the powder compact. However, if the powder is pressed initially to 3 GPa at 77 K, then we are on or very near to the phase boundary between the ε and δ polymorph so that either polymorph or a mixture of both solid phases of

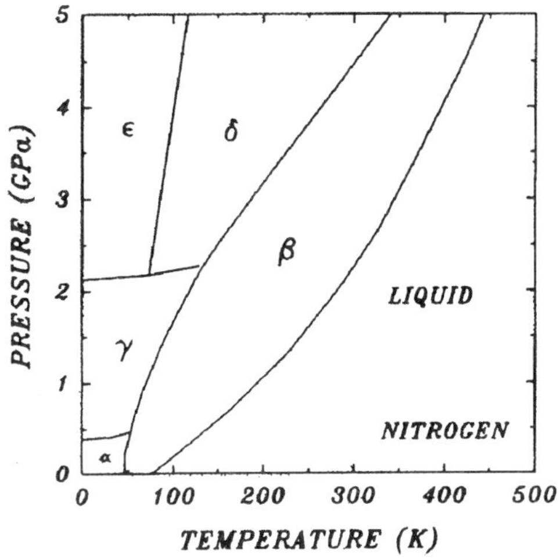

Fig. 14. Equilibrium pressure–temperature phase diagram for nitrogen showing pressure and temperature range pertinent to this work. Reprinted with permission from D. A. Young, "Phase Diagrams of the Elements," pp. 113–117 (© 1991 University of California Press).

nitrogen may exist. When the temperature increases to ambient, the alumina sample may be subjected to the effects of several nitrogen phase transformations, each involving an expansion to a less dense phase. For example, if the ε phase is produced at low temperature by applying a pressure of 3 GPa, then, on warming to RT, the ε phase transforms to the δ phase, which, in turn, transforms to the β phase. At 3 GPa, the stable phase at RT is the β phase rather than the liquid as is the case at lower pressures. Thus, the initial dense packing of the alumina powder achieved by compacting to 3 GPa at low temperatures is significantly altered when the temperature is increased to ambient. This is a consequence of the successive expansions that the solid nitrogen, trapped in the interstices of the powder, undergoes when it transforms from the ε phase, to the δ phase, and, ultimately, to the β phase. However, this may not completely explain the experimental observations. An additional factor must be considered when the solid β phase, trapped in the interstices of the alumina powder, melts.

At RT, LN_2 transforms to solid β-N_2 at 2.44 GPa. The β phase of nitrogen has a molar volume that is about 20% less than the molar volume of the liquid phase [22–24]. When β-N_2 melts on lowering the pressure on the compact at RT, the nitrogen pressure-transmitting medium experiences a 20% expansion. Because most of the solid nitrogen occupies interstices in the powder compact, the resulting expansion upon melting tends to reopen the structure of the compacted powder, thereby decreasing its bulk density. Thus, the mechanism that permits an efficient particle packing under LN_2 must also include the phenomenon of phase transformations that the surrounding nitrogen medium experiences. The solid-to-liquid phase transformation, in particular, involves a significant increase in molar volume of the N_2. However, the manner in which the phase boundary is traversed to produce the melt is also important because it will have a greater or lesser effect on reopening the compressed structure. If the boundary is traversed by lowering the pressure at constant temperature to produce the liquid, as in the case of the 3-GPa run, then there will be a larger net expansion than when the melt is produced by increasing the temperature at constant pressure. In the former case, although the transition volume is the same, to produce the liquid, the pressure has to be lowered and, because of considerable hysteresis in the transition pressure, the final state can be at a substantially lower pressure and, thus, have a larger sample volume than in the latter case. Thus, the particles are not as well packed when the solid/liquid boundary is crossed by lowering the pressure.

To test this theory of the roles that the various phases of nitrogen play in the cryogenic compaction process, the authors fabricated a compact of γ-Al_2O_3 following a procedure that avoided the presence of the solid β-N_2 phase at the final compaction pressure of 3.0 GPa. An initial cryogenic compaction of 2.5 GPa was used. At the final RT compaction pressure of 3.0 GPa, the solid β-N_2 phase was melted by increasing the temperature to 373 K. At this temperature and pressure, only liquid nitrogen exists in the γ-Al_2O_3 sample. On retrieval of the green body from these conditions, the hardness of the transparent compact (H_b) was measured to be 7.0 GPa (open square symbol in Fig. 13). This value is comparable to the hardness found for the green-state sample processed at 2.0 GPa shown also in Figure 13. The green body was then heat treated according to the procedure followed in their earlier experiments. The hardness of the heat-treated transparent compact (H_a) was then measured and found to be 9.6 GPa (solid square symbol in Fig. 13), a value significantly larger than the highest hardness found earlier for the heat-treated sample processed at 2.0 GPa and plotted in Figure 13. It is of interest to compare the H_a/H_b ratio for these two compacts with the other values. The lowest ratio, 1.25, is for the sample with highest hardness, 8.7 GPa, cryogenically compacted at 2.0 GPa. It also has the least difference between the hardness before and after heat treatment. The ratio, 1.37, for the sample of concern here, indicates that increased packing efficiency, and, therefore, higher hardness, can be achieved by avoiding the presence of the solid β phase of nitrogen at the final compaction pressure. Thus, it appears that it is the liquid state of nitrogen that plays the important role in processing nanosize powders of γ-Al_2O_3 by cryogenic compaction.

Comparison of the results plotted in Figures 12 and 13 revealed that cryogenic compaction produces harder samples at much lower pressures. Because all samples were heat treated under the same conditions, the only difference was in the processing before heat treatment, so the microhardness results must be related to the process of compaction. In the earlier work with Si_3N_4 [8], it was clear that LN_2 played an important role as a lubricant in the compaction process. These present results on γ-Al_2O_3 support this earlier conclusion and have also extended the understanding of the mechanism involved in this cryogenic lubrication process. However, it is interesting to note that, in the earlier work with Si_3N_4 [8], the effects of the LN_2 phase transformation on the packing density were not investigated by the authors.

To determine whether the observed differences in hardness with processing conditions can be related to microstructural differences, selected samples were examined by TEM. These samples included one green-body compact, pressed under LN_2 ($H \approx 4$ GPa), and five heat-treated samples. Three of the heat-treated samples were compacted in LN_2 and had hardnesses of approximately 7, 8, and 10 GPa. The other two were compacted at ambient temperature and had hardnesses of approximately 4 and 6 GPa. With the exception of one sample, which was ion milled to electron transparency, TEM sample preparation simply involved fragmenting the compacts and then containing the fragments between two grids with predeposited, holey carbon films. Representative views, illustrating the microstructure of the nanoscale aluminas, are shown in Figures 15 and 16a–c. A typical large-area electron diffraction pattern is included as an insert in Figure 15.

TEM examination of these samples, including the ion-milled sample where there was no fragmentation of the specimen, revealed no significant differences in microstructure. With reference to Figure 16, each sample can be similarly described as a uniform aggregate of nanoscale particles. Selected area electron diffraction gave "spotted" ring patterns, largely consistent with cubic γ-Al_2O_3. The presence of the tetragonal δ-Al_2O_3 phase, established as a minor constituent by angle-dispersive X-ray diffraction, could only be inferred from the excessive widths of individual rings in these patterns.

As illustrated in Figure 16, both green-state (part a) and heat-treated compacts (parts b and c) consist of densely packed particles of irregular, but nearly spherical shape. Particle sizes ranged from approximately 5 to 40 nm, but were most commonly 15 to 20 nm in size, corresponding to the quoted starting powder size. In all cases, the surfaces of individual

Fig. 15. Fragment from compact exhibiting highest hardness (nearly 10 GPa) attained in this study. Note uniform distribution of particles and interconnected porosity. Electron diffraction (insert) confirmed major particle phase is cubic γ-Al_2O_3. (Source: Gallas et al. [19].)

particles appear rough because of fine-scale surface asperities. These asperities are evident not only along the edges of particles, but also along the junctions separating adjacent contacting particles. Although it was not possible by the authors to characterize further the nature of the interparticle junctions, it is important to note that heat treatment did not result in any resolvable changes in these junctions. Finally, all samples, including the ion-milled specimen, appeared to contain a similar, appreciable volume fraction of open or interconnected porosity. Thus, the observed porosity cannot be considered an artifact related to the fracture process used in TEM sample preparation. Distinct differences in porosity could not be reliably established, regardless of differences in processing conditions or measured hardness. Here, however, it should be noted that the void structure can only be defined with reasonable certainty within small regions at the fragment edges, which are about three particle diameters or less in thickness. At best, comparative results from such regions (as in Fig. 16) showed that void dimensions never exceeded the average particle size (15–20 nm) and were typically much smaller. Accordingly, it appears that, for even the lowest compaction pressures, the level of particle packing achieved can be classified as "random dense packing." The significance of this result is that, at such levels of particle packing, the network of interconnected particles becomes load bearing, and stress-induced changes in the structure and density of the compact with increased pressure require relative displacements of the particles, for example, interparticle sliding. Here, structural changes to the particles, resulting from either plastic deformation or fracture of the particles, or from heat treatment (as will be discussed), are not considered relevant based on the observations by the authors.

The absence of discernible differences in porosity in the TEM results complicates discussion of the effect of compaction pressure on the structure and hardness of these compacts. On the one hand, the observed increase in green-state hardness with increased pressure may be due to small, but significant, increases in particle packing density, defined only by correspondingly small decreases in the sizes of distributed voids. On the other hand, it is also possible that the nature of the stresses acting on the compact changes as dense packing is achieved and results in constrained compression rather than pure hydrostatic compression. Under such conditions, deformation of the compact can occur with little change in density. As deformation involves particle motion, realignment occurs and leads to an increased resistance to further deformation because of particle interlocking. This increased resistance to deformation is ultimately reflected in the measured hardness. In any case, it appears that any change in the structure of particles with increased pressure occurs by particle sliding and the associated particle realignment results in more efficient packing. Moreover, because changes in structure occur by interparticle sliding, they will

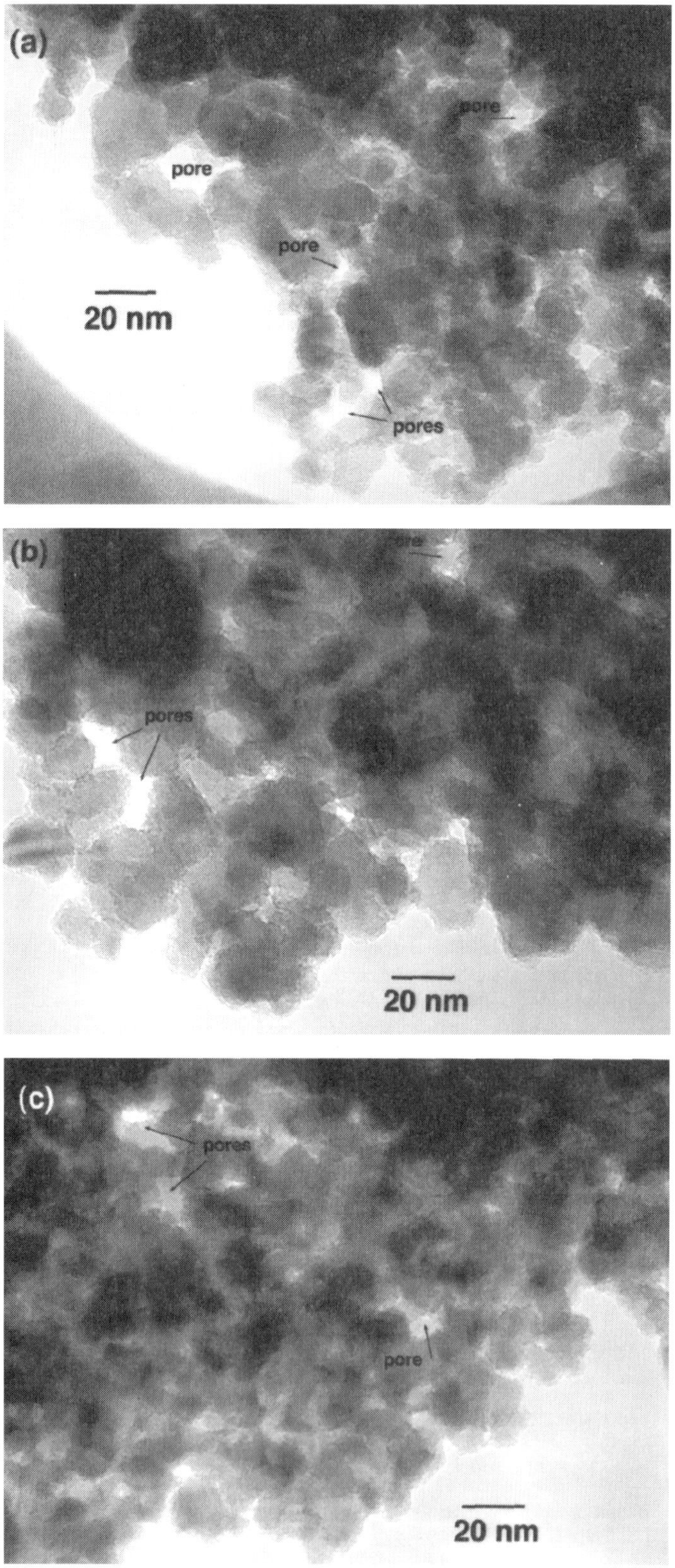

Fig. 16. Comparative views of particle microstructure in compacts produced under different conditions and exhibiting different hardnesses: (a) green body, LN_2 compacted, $H = 4$ GPa; (b) heat treated, ambient temperature compacted, $H = 4$ GPa; and (c) heat treated, LN_2 compacted, $H = 10$ GPa. Note microstructural similarity and dimensions of large pores (P) relative to particle sizes. (Source: Gallas et al. [19].)

be governed largely by the elastic/plastic properties of the particles, the nature of forces binding the particles, and the properties of the interparticle or "void space" phase. On this basis, the pronounced effect of compaction under LN_2 appears directly related to the increase in elastic modulus of the particles at low temperatures and to the presence of a liquid intergranular phase. As has already been suggested, LN_2 appears to act as an effective lubricant, facilitating particle rearrangement to give a densely packed microstructure at low pressures.

The effect of heat treatment on hardness cannot be similarly explained in terms of particle rearrangement or densification. In fact, comparative observations on green-state compacts and heat-treated compacts (e.g., Fig. 16a vs Fig. 16b and c) revealed no apparent changes in particle size, shape, and surface texture or in the appearance of the particle contact junctions. This leads to the conclusion that only extremely limited transport, whether by intrinsic (bulk), surface, or interfacial diffusion, occurred during heat treatment at 800 °C. Thus, the roughly 50% increase in hardness universally realized by this heat treatment cannot be attributed to various microstructural changes that normally occur by diffusion during sintering. Instead, the authors speculated that pressureless heat treatment brings about a change in the nature of the forces acting across the contact junctions of these nanosize particles. Conceivably, heat treatment results in a change from strictly surface force attraction (e.g., van der Waals) to one that includes interatomic bonding (e.g., covalent interactions). Considering the irregular structure of the junctions between particles, interatomic bonding would appear to be limited to the small junctions formed by surface asperity contact. The authors admit that this conclusion is speculative and can only be resolved when complete characterization of the particles, particularly in the surface regions, is carried out.

Finally, we should point out that the authors failed to consider the possibility of any solubility of the γ-Al_2O_3 particles in LN_2 that might promote a solution–reprecipitation mechanism. A solution–reprecipitation process would result in the formation of necks between particles and, therefore, an increase in hardness. In fact, we are aware of claims that water-based solutions can dissolve small amounts of different phases of nanosize aluminas, resulting in the formation of strong agglomerates [25]. However, we must say that TEM observations by the authors do not indicate that this mechanism of sintering is occurring during compaction.

More evidence of the lubricating ability of LN_2 during compaction has been reported by Gonzalez et al. [14]. The authors were studying the transformation of γ-Al_2O_3 to α-Al_2O_3 using the high-pressure DAC to compact the nanosize powder. They observed that, when the γ-Al_2O_3 powder was compacted to high compaction pressures using LN_2 as a lubricating medium and pressureless sintered at high temperatures, the final microstructures were more uniform and the specimens had a smaller average grain size than the same powder compacted under dry conditions. The authors claimed that, when the sample is compacted under LN_2, the load distribution across all particle contact surfaces is more homogeneous because there is less interparticle friction, and, therefore, particle rearrangement and redistribution to a uniform packing occurs more readily. This example is once again clear evidence of the benefits of using LN_2 as a lubricating fluid during the compaction of nanosize γ-Al_2O_3.

It is important to recognize that a lubricating fluid that works for one material does not necessarily work for other powders. For instance, Hahn et al. [26] compacted nanosize TiO_2 using high-pressure and high-vacuum environments to obtain densities greater than 70%. On the other hand, Pechenik et al. [10] established that, in the case of the nanosize amorphous Si_3N_4 powder used in his work, the green density of the compacts was the lowest when the samples were compacted at 200 °C under vacuum, in comparison with RT compaction and LN_2 compaction, with the latter producing the highest green densities. According to Nilsen et al. [27], the Si_3N_4 powder used by Pechenik et al. [10] contained physically absorbed water, nitrogen, and other chemisorbed groups on the surface of the particles. Because these absorbed species can be, in many cases, partially or fully removed by exposing the powder to high temperatures and vacuum, the authors argued that this could explain the different compaction behaviors observed under the three different compaction environments studied by Pechenik et al. [10]. Thus, these results suggest that the presence of these species on the surface of the particles reduces interparticle friction. In fact, it is not hard to visualize these surface modifications having a significant effect on the flow properties of the powder and, therefore, on a powder compaction process. Putting it all together, we have presented clear evidence from the literature that shows that the nature of the particle surface and the ability of the surface to absorb the lubricating fluid, either a liquid, a gas, or the formation and growth of a thermodynamically stable surface layer, is very important. We must say that, even though particle morphology was not discussed in detail by the authors, the morphology of the powder particles will have a strong effect on the flow properties of the powder and will surely have a significant effect on the compaction efficiency when a lubricant is used.

The major emphasis in this chapter up to now has been on how to achieve high random packing densities of nanosize powders in the green body, with the expectation that the resulting sintered material will have a nanoscale dimension microstructure. We now present experimental results on the compaction and sintering of nanosize γ-Al_2O_3 powder that demonstrate that achieving a high-density, random-close-packed green body in the starting material is by no means the ultimate answer to producing a fine-grain nanoscale microstructure in the final sintered material.

4.3. Nanosize γ-Al_2O_3 Powder Processing

Sintering of alumina powders at relatively low temperatures, 1000–1150 °C, has been reported in the literature [28–30]. In some cases, a solid-state phase transformation has been exploited to aid sintering at these temperatures to produce high-density polycrystalline alumina with submicrometer grain size. For example, Kumagai et al. [28] have studied the effects of seeding transitional oxides of alumina (boehmite) with α-Al_2O_3 to lower the sintering temperature and increase the kinetics of the transformation to α-Al_2O_3. Sol–gel techniques were employed in order to obtain homogeneous mixtures of the transitional oxides and the α-Al_2O_3 seeds. Sol–gel methods, however, suffer from severe cracking during the drying or "curing" period. Yeh and Sacks [30] made slurries of fine-grain α-Al_2O_3 by conventional suspension techniques for alumina that were used to slip-cast samples to relatively high bulk densities (69%). The samples, sintered at 1150 °C in air, had relative densities greater than 99.5% and an average grain size of 0.25 μm. In this case, a phase transformation was not used, yet dense samples with submicrometer grain size were obtained. Thus, the phase transition appears to be unnecessary to obtain this result. It also appears that it is important to have a uniform density in the green body to reduce internal stresses that occur during inhomogeneous densification. However, making homogeneous suspensions (mixtures) with submicrometer-sized powders is very difficult, particularly when nanosize particles ($\approx$20 nm) are used. The authors recognized that an alternative approach to achieving high-density green compacts with a nanoscale microstructure is to utilize high-pressure compaction techniques.

Earlier in this chapter, it was demonstrated that, by increasing the compacting pressure on nanosize particles, an increase in green-body bulk density can be achieved. However, it has yet to be demonstrated that the gain in green-body density has a significant effect on the densification process that should take place during sintering. To investigate this problem, Gonzalez et al. [14] studied nanosize γ-Al_2O_3 as a model material. Although the thermodynamically stable phase of alumina, α-Al_2O_3, is obviously the better choice, these authors were unable to obtain nanosize particles of the α phase commercially. This is a common problem with nanosize particles of ceramic materials, because these powders usually exist as thermodynamically unstable amorphous phases (silicon nitride) or polymorphic crystalline phases (γ-Al_2O_3) at ambient conditions of pressure and temperature.

These unstable nanophases readily transform to a stable phase, for instance, by elevating the temperature.

For example, nanosize γ-Al_2O_3 particles transform to the α-Al_2O_3 phase at approximately 1150 °C accompanied by rapid grain growth. The kinetics of the transformation have been studied by Dynys and Halloran [31], who reported that the rate of transition is strongly dependent on the mechanical pretreatment of the powder. They showed that mechanical milling of the powder can reduce the time for complete transformation at 1150 °C. In addition, the authors reported that compaction under different loads changed the transformation kinetics. From their work, it appears that both compaction and milling involve the creation of large contact stresses, which can cause an increase in the internal energy of the material. For example, lattice defects can be created, and these can act as heterogeneous nucleation sites for the transformation. This variability in the transformation kinetics complicates the interpretation and understanding of the compaction and sintering results on γ alumina that will now be presented. Nevertheless, the results on γ-Al_2O_3 reported by Gonzalez et al. [14] give us a closer look at the possible effects that high-pressure compaction and high green-body density have on the sintering of nanosize powders in general.

The preliminary work conducted by the authors consisted of preparing green compacts of γ-Al_2O_3 made using the piston–cylinder device described earlier in this chapter. Two sets of green compacts were produced at room temperature using no lubricant. One set was pressed to 1.0 GPa, the other to 2.5 GPa. Samples from both compaction pressures were sintered in an alumina tube furnace under the rough vacuum of a mechanical pump at 1000, 1100, 1200, or 1300 °C for 5 h, employing heating and cooling rates of 300 °C/h. After the heat treatments, the weight and volume of the samples were remeasured to determine the extent of densification. To further understand the densification process, scanning electron microscopy (SEM) and TEM were used to characterize the samples. In addition to the microscopy work, X-ray diffraction was used to identify phase composition.

4.3.1. Results and Discussion

The average densities for the samples compacted at 1 and 2.5 GPa were 2.00 g/cm^3 and 2.37 g/cm^3, respectively, which correspond to 54% and 65% of the theoretical density (3.67 g/cm^3). The average density at each sintering temperature is summarized in Figure 17. The densities increase monotonically for both compaction pressures and approach each other at a density of approximately 3.3 g/cm^3, or 83% of the theoretical density of α-Al_2O_3. It is also of interest to note that the 2.5-GPa compacted sample sintered at

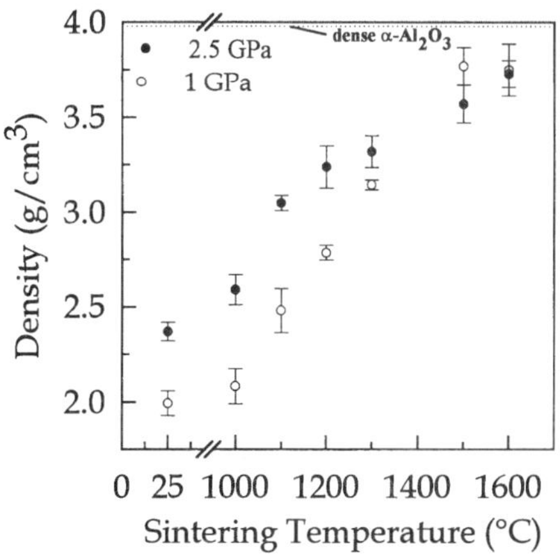

Fig. 17. Density of γ-Al_2O_3 samples compacted to 1 or 2.5 GPa at RT as a function of sintering temperature. (Source: Gonzalez et al. [14].)

1000 °C showed a slight increase in density, which was not as significant in the 1.0-GPa samples. The authors determined the microstructure and phase fraction of the α and γ phases to gain further understanding of the densification process occurring in their specimens.

4.3.2. Microstructure

Figure 18 shows a TEM micrograph of a green body compacted at 2.5 GPa. Both X-ray diffraction and electron diffraction results on these samples confirmed that they are predominantly γ-Al_2O_3. As seen in Figure 18, the γ particles are equiaxed with an average equivalent spherical diameter of approximately 20 nm. More generally, observations on 1.0 and 2.5-GPa compacts indicate a random-dense-packed particle structure with uniform interconnected porosity. For both compaction pressures, the pore dimensions were less than the particle size. As a consequence, the 16% difference in packing density between the samples compacted at 1.0 and 2.5 GPa was not resolved in TEM.

The samples sintered at 1000 °C have a completely different microstructure. Parts a and b of Figure 19 show, in comparison, TEM micrographs of the samples sintered at 1000 °C for the 1.0- and 2.5-GPa compaction pressures. The sample compacted at 1.0 GPa consists of a mixture of equiaxed particles of γ phase and isolated clusters of predominantly α phase. Even at this temperature, necking between γ particles, indicative of γ phase sintering, was not resolved in TEM. The isolated α clusters grow in a wormy or spongy structure with continuous porosity. X-ray diffraction results corroborate that both the γ and the α phases of Al_2O_3 are present in the 1.0-GPa pressed samples. In contrast, the samples pressed at 2.5 GPa are made up of all α phase. The microstructure is spongy or wormy with continuous porosity similar to the isolated α clusters in the 1.0-GPa samples. Again, the authors confirmed by X-ray diffraction that the 2.5-GPa samples contained α-Al_2O_3 exclusively.

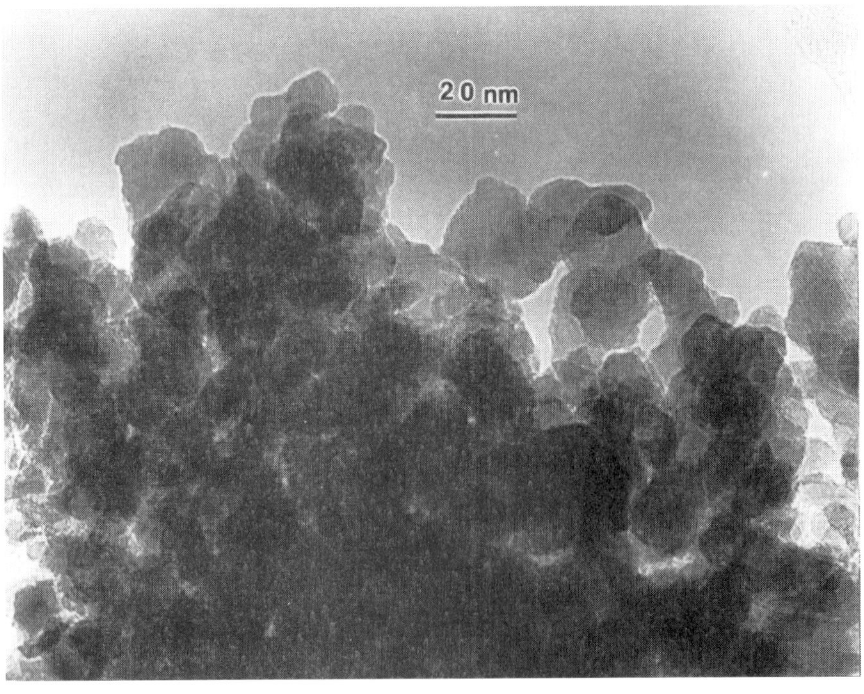

Fig. 18. TEM bright field image shows the microstructure of a green body of γ-Al_2O_3 compacted at 2.5 GPa. (Source: Gonzalez et al. [14].)

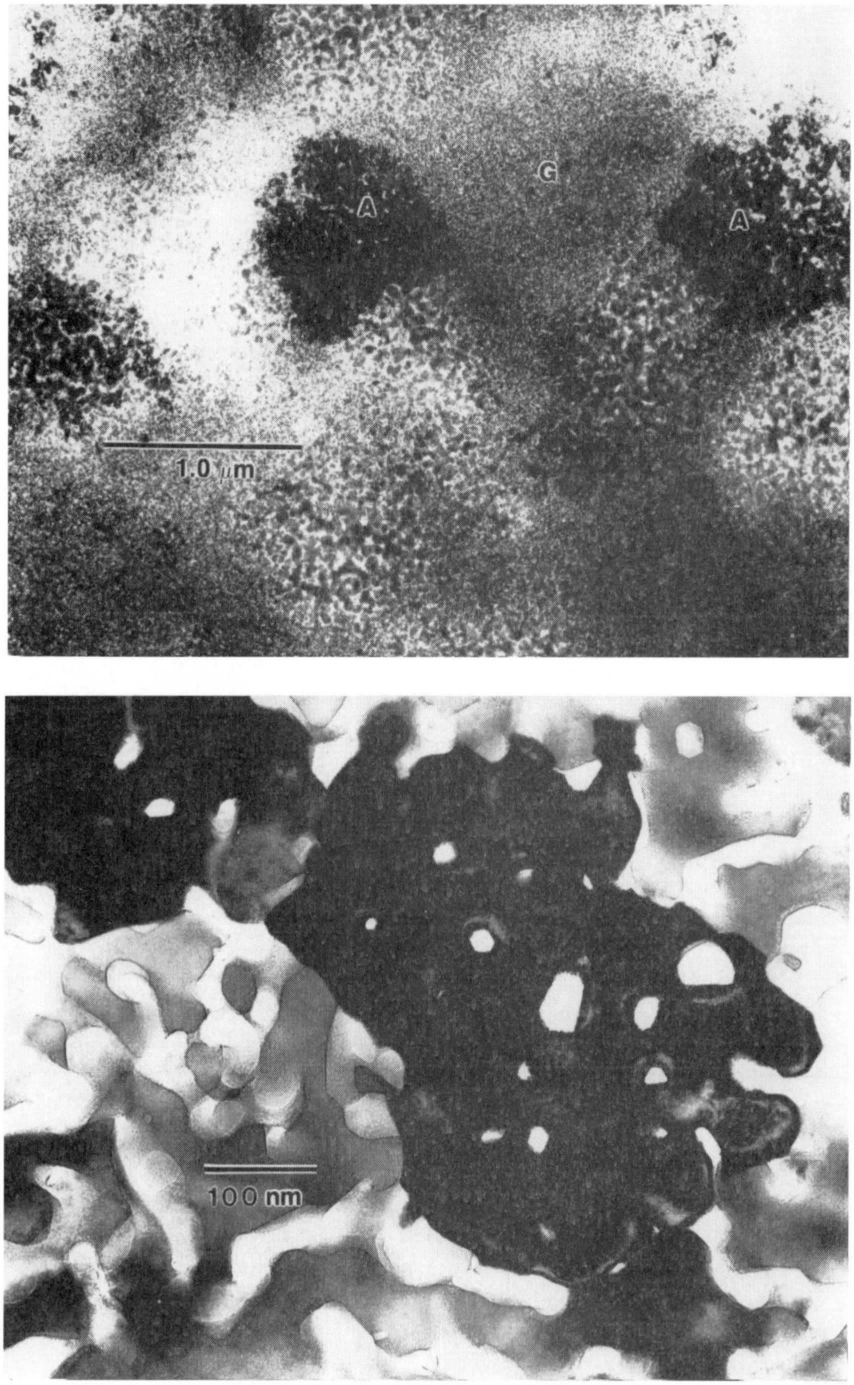

(a)

(b)

Fig. 19. Bright-field image in TEM shows the typical microstructure of samples sintered at 1000 °C for (a) 1 GPa where A and G identify regions of α-Al_2O_3 and γ-Al_2O_3 respectively and (b) 2.5 GPa. (Source: Gonzalez et al. [14].)

It should be noted that the spongy α-phase structure actually consists of interconnected individual grains, ranging in size from about 100 nm to nearly 1 μm. As illustrated in Figure 19b, where regions of dark contrast define areas of common orientation, the grains have a complex morphology, and the boundaries connecting adjacent grains are typically 50–100 nm in length.

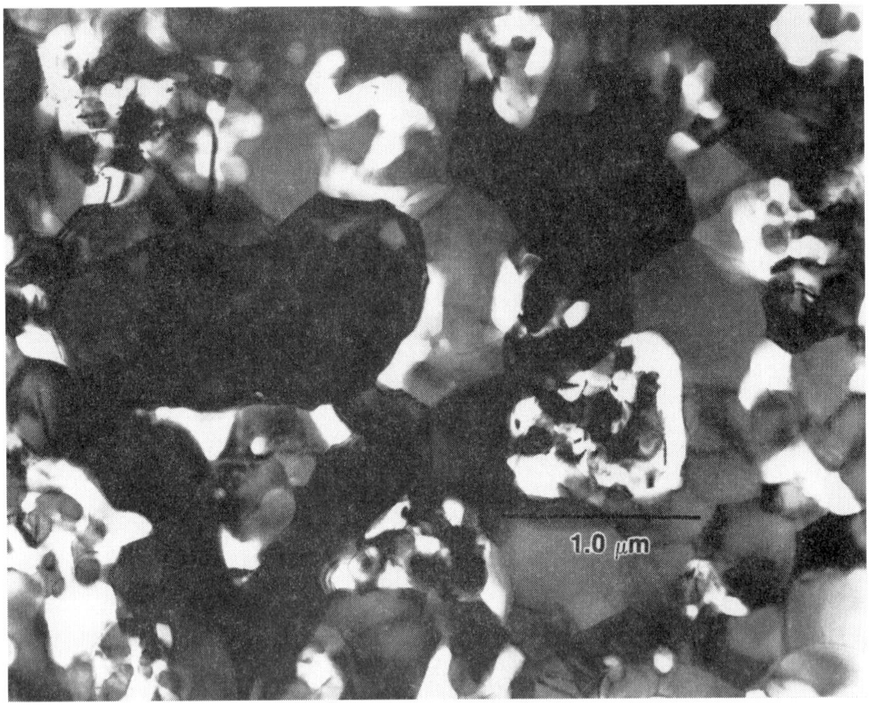

Fig. 20. TEM bright-field image of 1-GPa compacted sample sintered at 1300 °C. (Source: Gonzalez et al. [14].)

The authors also studied the surface area and pore size distributions by the Brunauer–Emmett–Teller (BET) adsorption isotherm on a green, 1000 °C sintered body (2.5-GPa compacted sample). The average pore size for the green body was 5 nm, consistent with a packing coordination number of 4. However, the porosity determined by TEM ranges in size from 100 to 300 nm, which is an order of magnitude larger than the pore size measured in the BET. The average surface area also decreased during sintering, suggesting neck formation and some densification in agreement with the observed microstructure.

As mentioned earlier, the 2.5-GPa samples showed some densification at 1000 °C, whereas the 1.0-GPa samples did not. It is evident that the samples have different grain morphologies, which contribute to the differences in densities. Another reason for this difference in density may be because the theoretical density of α-Al_2O_3 (3.987 g/cm^3) is larger than the density of γ-Al_2O_3 (3.67 g/cm^3). The samples pressed to 2.5 GPa experienced significant shrinkage during the transformation from the γ to the α phase (given the same mass) and, as a result, yielded higher bulk densities. The amount of α phase produced in the 1.0 GPa samples, however, is so small that the change in volume associated with the transition is insignificant and could not be determined by the technique employed.

The microstructural changes in the samples sintered at 1300 °C were also reported. The density of these samples did not exceed 83% of theoretical. X-ray diffraction results indicated that both the 1.0- and the 2.5-GPa compacted samples were 100% α-Al_2O_3. The 1.0-GPa samples are made up of a combination of the wormy grain structure and equiaxed micrometer-sized grains of alumina (Fig. 20). Significant porosity is visible, and necking between grains is evident. In contrast, the 2.5-GPa compact exhibited only an equiaxed grain morphology (Fig. 21). No evidence for the spongy α-phase structure was observed in these samples.

As originally anticipated, the processing of alumina from γ-phase particle compacts is strongly influenced by the γ-to-α transformation. The results reported by Gonzalez

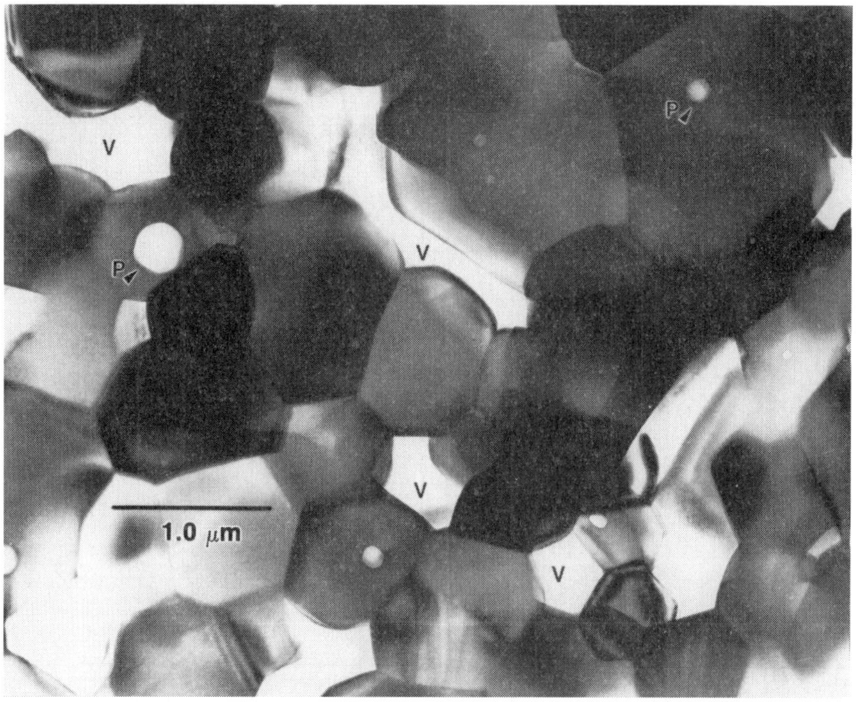

Fig. 21. TEM bright-field image of 2.5-GPa compacted sample sintered at 1300 °C where V and P identify voids and pores respectively. (Source: Gonzalez et al. [14].)

et al. [14] confirm the earlier observations of Dynys and Halloran [31], namely, that agglomerates of nanosize γ particles do not simply transform into agglomerates of similarly sized α particles. Instead, transformation appears to involve heterogeneous nucleation followed by a period of rapid, almost explosive growth. This growth involves the rearrangement of hundreds to many thousands of 20-nm γ particles into single-crystal grains of α phase, ranging in size from 0.1 to 1 mm. With this consolidation of the solid phase, there is a corresponding rearrangement of porosity into pores of micrometer dimensions. This large-scale change in microstructure, brought about by the transformation, clearly complicates the discussion of the effect of initial compaction pressure on the densification that occurs during heat treatment. Nevertheless, the results reported so far do indicate a twofold effect of compaction pressure on the sintering of alumina from nanosize γ particles.

First, increased compaction pressure enhances the γ-to-α transition, reducing the transition temperature or, alternatively, increasing the nucleation rate at temperatures ranging from 1000 to 1150 °C. Evidence for this is based on the observations that, for compacts pressed to 2.5 GPa, complete conversion to the α phase occurs within 5 h at 1000 °C, whereas those pressed to 1.0 GPa showed less than 50% conversion to the α phase. These compacts were completely converted to the α phase at 1100 °C.

Second, increased compaction pressure leads to increased densification upon heat treatment at still higher temperatures. The results show a significant difference in the microstructure, density, and hardness of 1.0- and 2.5-GPa compacts sintered at 1300 °C. For 2.5-GPa compacts, this heat treatment results in equilibrium or faceted α-phase grains of 1 μm or larger dimensions. Adjacent grains tend to be connected by straight grain boundaries of nearly grain size dimensions. Although considerable porosity is retained, it is in the form of distributed voids at multigrained junctions and isolated pores within the grains. By contrast, 1300 °C heat treatment of 1.0-GPa compacts results in limited formation of faceted grains and retention of large (1 μm and larger) porous structures, reminiscent of the

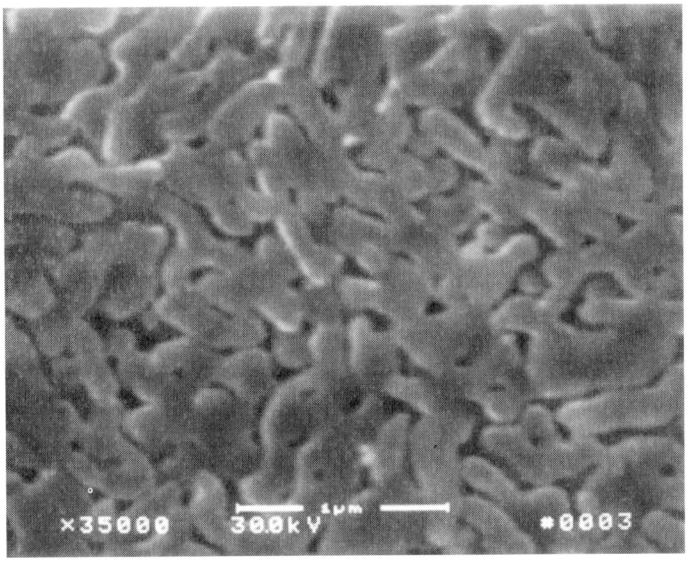

Fig. 22. SEM micrograph of an alumina sample compacted in a DAC to 2.1 GPa and pressureless sintered at 1150 °C for 2 h. (Source: Gonzalez et al. [14].)

initial spongy α structure. The change from what has been termed the spongy α structure to a faceted grain structure must involve diffusive transport as there is a change in both grain shape and size. Accordingly, the observed differences in microstructures for the 1.0- and 2.5-GPa compacts, after heat treatment at 1300 °C, strongly suggests a relative increase in the transport kinetics associated with the increased compaction pressure. Further study, however, will be required to elucidate the underlying reasons for these observed effects.

In the case of γ alumina, the compaction and sintering results are complicated by the large microstructural changes that occur during the γ-to-α phase transformation, which takes place at about 1000–1100 °C, depending on the initial compaction pressure. The authors realized the relationship between the compaction pressure and the transformation. Their results also agreed with other investigators in that the transformation is nucleation limited and that presumably the higher the compaction pressure the more nucleation sites are created. For this reason, they explored even higher compaction pressures. The samples were compacted in the DAC to 3, 4, or 5 GPa. Because an Inconel X750 gasket is used in the DAC technique, the sintering temperatures were limited to 1150 °C for 5 h. In agreement with the piston–cylinder results, the authors observed that samples compacted at 2.1 GPa developed an α-phase vermicular microstructure as shown in Figure 22. In contrast, the formation of this vermicular structure was completely eliminated in samples compacted at 3 GPa and higher. The 3-GPa samples were sintered at 1150 °C for 1 h (Fig. 23). The authors estimated an average grain size of 223 ± 37 nm (one standard deviation) using the line intercept method. The relative density of the 3-GPa samples was 89% and most of the resolved porosity appeared to be isolated at grain interstices. The authors noticed that, as the compaction pressure was increased, the average grain size decreased and the resulting microstructure was more homogeneous after sintering. Evidence of these observations was also seen in the sample precompacted to 4 GPa. The authors estimated an average grain size of 192 ± 22 nm and a relative density of 93% (Fig. 24). In a similar analysis of sintered 5-GPa compacted samples, the average grain size was determined to be 189 ± 26 nm and the relative density was also 93% (Fig. 25). The pores in these samples were exclusively accommodated at grain interstices. To study the effects of time, the authors sintered the 5-GPa samples at 1150 °C for 1 h. The authors explained that the results from these samples support the contention that, once the a phase nucleates, the grains

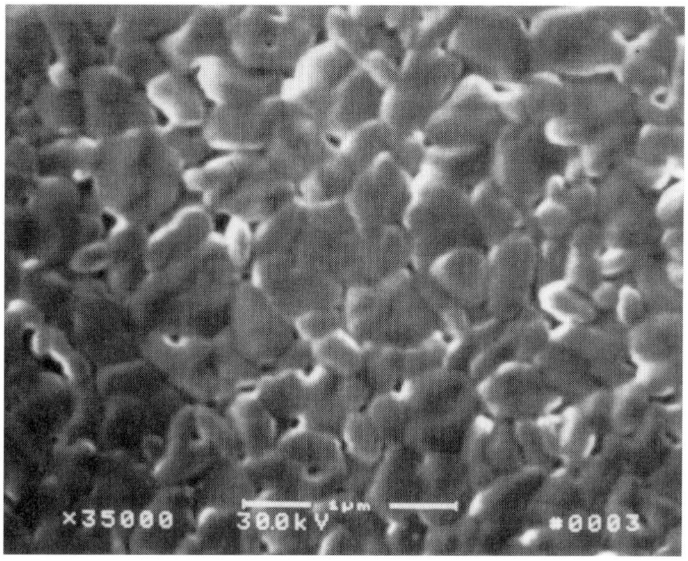

Fig. 23. SEM micrograph of an alumina sample compacted in a DAC to 3.0 GPa and pressureless sintered at 1150 °C for 2 h. (Source: Gonzalez et al. [14].)

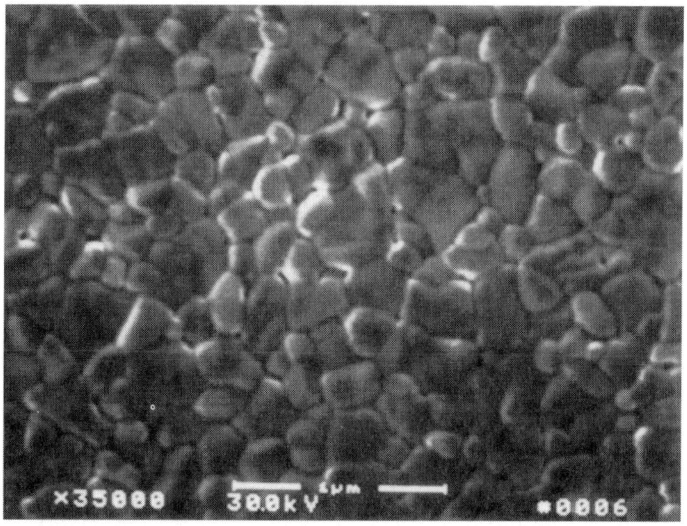

Fig. 24. SEM micrograph of an alumina sample compacted in a DAC to 4.0 GPa and pressureless sintered at 1150 °C for 2 h. (Source: Gonzalez et al. [14].)

grow rapidly. The microstructure, however, was very inhomogeneous. The average grain at the center of the sample was estimated to be 181 ± 21 nm, and 144 ± 16 nm near the edges of the sample. The authors speculate that the nonuniform microstructure is a result of the nonuniform distribution of nucleation sites in the compact that were created by a nonuniform distribution of stress during compaction. Because the authors were aware of the benefits of LN_2, they proceeded to compact specimens to 5 GPa using LN_2 as a lubricant, hoping for a more uniform stress distribution during compaction and a homogeneous distribution of nucleation sites. The microstructure that develops after sintering is shown in Figure 26. In contrast with the samples compacted under dry conditions at room

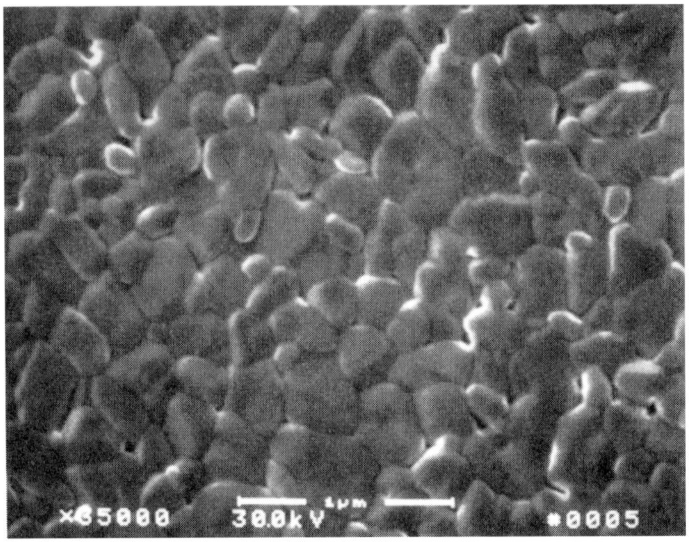

Fig. 25. SEM micrograph of an alumina sample compacted in a DAC to 5.0 GPa and pressureless sintered at 1150 °C for 1 h in air. (Source: Gonzalez et al. [14].)

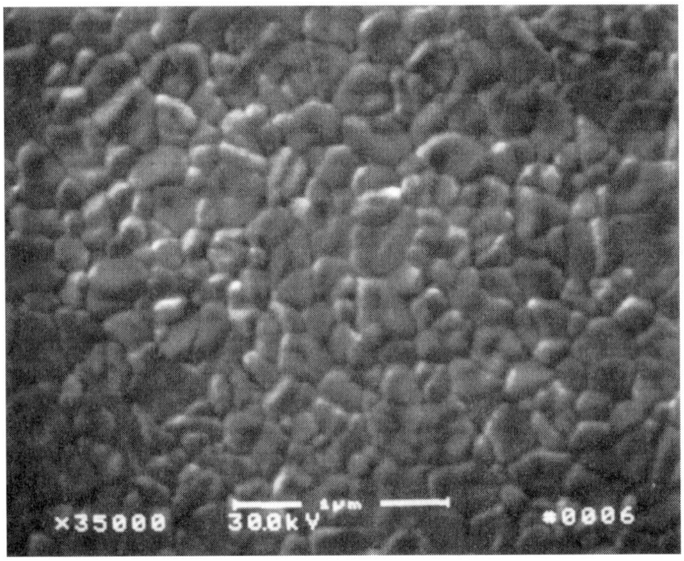

Fig. 26. SEM micrograph of an alumina sample compacted in a DAC to 5.0 GPa and pressureless sintered at 1150 °C for 1 h in liquid nitrogen. (Source: Gonzalez et al. [14].)

temperature, the average grain size (146 ± 15 nm) is uniform throughout the entire cross section of the sample. This result provides additional evidence of the lubricating properties of LN_2 on the compaction of nanosize powders. For a more detailed description of this work, we refer the reader to Gonzalez et al. [14].

4.3.3. Conclusions

Gonzalez et al. [14] were able to list eight conclusions concerning the compaction and sintering of γ alumina nanosize powder relevant to the compaction process. They are as follows:

1. Random dense packing of γ alumina nanosize powder can be achieved at high compaction pressures ($\geqslant$1 GPa).
2. Increased compaction pressure results in a corresponding increase in green-body density, specifically from 54% theoretical at 1.0 GPa to 65% theoretical at 2.5 GPa.
3. Green-body hardness values reflect the final compaction pressure.
4. Increased compaction pressure enhances the γ-to-α phase transformation, effectively increasing the nucleation rate, or, alternatively, lowering the transition temperature.
5. Heterogeneous nucleation of the α phase in nanoscale γ alumina compacts is invariably followed by a period of rapid growth, which results in a coarsening of both the solid phase and the interconnected porosity. Although this microstructural coarsening decreases the internal surface area, reduction of the volume fraction of porosity is not apparent. As a consequence, densification is limited to the solid phase component and corresponds to the relative increase in density, accompanying the γ-to-α phase transition, 3.67 g/cm^3 to 3.99 g/cm^3, respectively.
6. Despite the enhancement of sintering kinetics with increasing green-body density at temperatures below 1300 °C and pressures below 3 GPa, the results, based on density measurement and microstructural evaluation, indicate that the final density is independent of the initial density. The consequence of this is that full densification is not achieved by pressureless sintering. It appears that a similar volume fraction of closed porosity is retained.
7. Increased compaction pressure below 3 GPa (green-body density) appears to have no effect on the final density of the product sintered at temperatures above approximately 1400 °C. Increased compaction density, however, does affect how fast one can achieve 83% of theoretical density when the sintering temperatures are below 1400 °C.
8. Compaction pressures in the DAC over 3 GPa are necessary to avoid the formation of the vermicular structure during sintering.

These are very important conclusions because they have basically changed our views concerning the processing of nanosize particles to form dense nanoscale microstructures. As a consequence, it appears that the notion of achieving full packing density in the green body of a nanosize powder will result in a dense nanoscale grain-sized sintered ceramic is not necessarily a correct one. From what we now know, this is particularly so when a phase transformation occurs during the sintering process, as in the cases of nanosize γ alumina and amorphous silicon nitride powders. In both materials, high-density green bodies did not produce dense nanoscale microstructures when the sintering temperature employed exceeded the phase transformation temperature. Amorphous silicon nitride crystallized to the α phase with a significant increase in density and grain size. As we saw earlier, γ alumina transformed to the α phase also accompanied by an increase in density and grain size. Unfortunately, there are few ceramic powders that are thermodynamically stable as nanosize particles at room temperature and pressure. Thus, until such materials become available and can be tested by these procedures, the notion of sintering fully packed green bodies of single-component nanosize particles to dense nanoscale microstructures remains an open issue.

5. COMPACTION EQUATIONS FOR POWDERS

Considering all the information that has been presented so far in this chapter, the question that now needs to be addressed is how can we explain, in a more detailed or quantitative manner, the effects of lubricants on the compaction efficiency of nanosize powders. Intuition suggests that this should be addressed by studying the process of compaction and the compaction equations that attempt to model these processes. Unfortunately, most mathematical models are exclusively phenomenological and, therefore, offer little insight into the mechanics of compaction. Most analytical problems require information that, in the case of powder compaction, is not available or is difficult to obtain experimentally. For example, many expressions call for information such as (1) the distribution of powder particle-to-particle contact stresses, (2) the flow properties of the powder, (3) the load distribution throughout the compact, and (4) the compressibility and strength of the individual particles [32]. When some of these critical parameters are known and are used in compaction equations, the information gained during compaction is considered to be useful [32]. Chen and Malghan [32] suggest that the compressibility of the material, compaction ratio, and average particle size could be determined from such model equations. Most equations appear to focus on factors such as (1) the different possible stages of the compaction process and (2) the different mechanisms involved, which also can provide useful information. The difficulty lies in putting all these parameters together in one general expression applicable to the general powder system. In the following section, we will briefly discuss the basic ideas behind compaction equations, with special emphasis on the equation derived by Chen and Malghan, which is the most familiar to the authors.

One of the goals in the study of the compaction process in powders is to estimate the pressure necessary for achieving a desired bulk density. The process of compaction of a ceramic powder, however, is very complex, especially for nanosize particulates. Ceramic powders can exhibit a high compressibility during the initial or low-pressure stages of compaction, and a low compressibility at higher pressures because the particles are closer together and the coordination number of individual particles has increased significantly [32]. The former is more pronounced during the compaction of nanosize powders, because the tap densities of these powders can be as low as 30% of theoretical. Because of the absence of a sharp or well-defined transition between these two stages of compaction, the modeling of volume changes over the entire pressure range is difficult. Furthermore, even if an equation describes the compaction behavior of a particular powder, it may not be directly applicable to other powders with different compaction properties.

Chen and Malghan [32] developed a new approach to deduce a compaction equation. The authors conducted a number of compaction experiments with ceramic powders and fitted the experimental data using modified nonlinear and linear least-squares techniques. As they pointed out, even the most basic parameters in compaction equations, for example, initial and final volume, are difficult to obtain. The experimental determination of initial volume is not accurate because factors such as particle size distribution, particle agglomeration (especially for nanosize particles), and flow properties, to name a few, will affect the ability of the powder to pack efficiently without the application of pressure. Similarly, the prediction of the final volume based on theoretical considerations can be incorrect because it is typically measured in the absence of an external load. Chen and Malghan avoided dealing with these errors by choosing these values as fitting parameters in their equations.

Their new compaction equation suggests a classification of voids into two types with the resulting equation containing only five parameters:

$$V = a_1 + a_2 \exp\left(\frac{1}{1 + a_3 P}\right) + \frac{a_4}{1 + a_5 P} \tag{1}$$

where a_i, $i = 1, \ldots, 5$, are parameters, $V_0 = a_1 + a_2 \mathrm{e} + a_4$, $V_\infty = a_1 + a_2$, and P is the applied pressure. V_0 represents the initial volume and V_∞ is the net particle volume under extremely high pressure ($P \to \infty$). The authors compared their equation with that of Cooper and Eaton [33], and found that their equation provided a better fit to the experimental data. It is important to recognize that powder compacts are made of many voids of different sizes and shapes that contribute to the complexity of the modeling. Moreover,

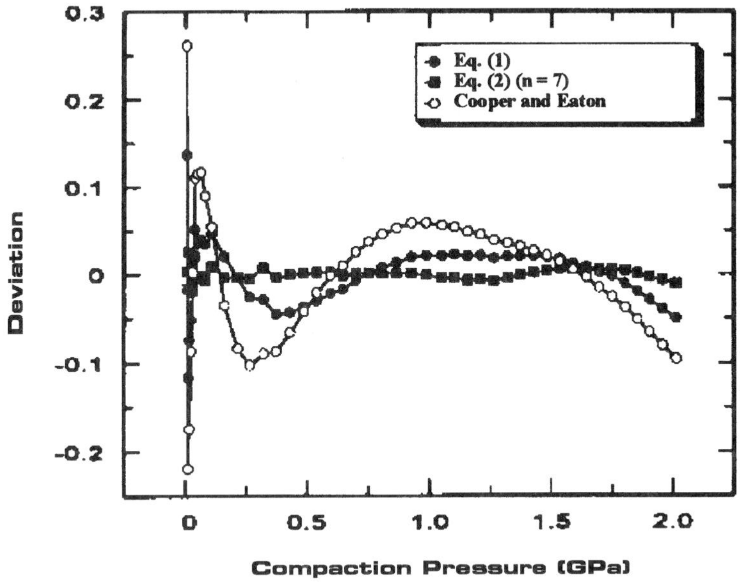

Fig. 27. Deviation of compaction equations and experimental data for α-phase Si_3N_4 powder. (Source: Chen and Malghan [32].)

the size distribution of these voids is changing continuously as pressure increases and particle rearrangement occurs. According to Chen and Malghan, if a compaction equation could consider all possible types of voids, the equation would approach a real compaction process. However, this would require an infinite number of parameters and would make mathematical manipulation impractical.

Chen and Malghan proposed another compaction equation to achieve an even better representation of the different types of voids. The final expression they propose is

$$V = \sum_{i=0}^{n} \frac{a_i}{(1 + u_i \alpha P)} \tag{2}$$

where α is a factor with a dimension of inverse pressure, $a_i = (V_0 - V_\infty)a_i'$, and u_i is the ith zero of the nth-order Laguerre polynomial. The number of terms in the compaction equation is $n + 1$. All parameters a_i in this equation are linear and can be determined by linear least-squares fits, which simplifies the fitting process. For specific details concerning the derivation of this equation, the reader is referred to [32]. Chen and Malghan claim that the advantage of this equation is that the variety of void types is proportional to the number n of parameters, and, therefore, by adjusting n, a better representation of the compaction process is achieved.

Figure 27 shows the divergence of the two equations, one derived by Chen and Malghan [32] and the other by Cooper and Eaton [33], from the experimental data for α-Si_3N_4 powder. It is clear from this figure that the mathematical treatment developed by Chen and Malghan gives a better fit to these data. However, these equations appear to fall into the same trap as all the other equations of powder compaction, because they fail to provide information on transferable physical parameters. For example, there are no individual parameters that provide information on flow properties, lubricating conditions, time dependence, particle size distribution, shape of the powder particles, and, for example, the state of agglomeration of the powder. All these characteristics are basically convoluted in the compaction ratio, defined as

$$\textit{Compaction ratio} = \frac{V_\infty}{V_0} \tag{3}$$

Compaction ratios are useful to some extent because they can provide overall information on compaction characteristics. For example, Chen and Malghan indicated that the compaction ratio of powders with an average particle size between 100 nm and 1.0 mm is approximately 0.4 and approximately 0.1 for a powder with an average particle size below 100 nm. Although compaction ratios are qualitatively useful, they are severely limiting if the intention is to understand fully the compaction process in powders.

In summary, the compaction equations developed by Chen and Malghan are very effective in curve fitting pressure–volume data over the entire pressure range, probably better than other available models. The mathematical manipulation and fitting routines are relatively simple, because the a_i fitting parameters can be determined by linear least squares. It appears that the most important value that can be obtained from these compaction equations is the compaction ratio. Because the compaction ratio is, in essence, a measure of the packing efficiency of the powder, it can provide qualitative information about the compaction process, and is especially relevant to compaction under different lubricating conditions. However, the compaction ratio can only help to create an overall picture of the compaction performance of the powder, because it cannot provide details on the most relevant powder characteristics to the compaction process. Although interpretation of its physical meaning is difficult, it would be useful to study changes in the compaction ratio of powders under different lubrication conditions, including the use of LN_2. Such experiments have not been reported yet.

6. CONCLUSIONS

In recent years, there has been a strong interest in the processing of nanosize ceramic powders because of their promising low-temperature sintering capabilities and because it has been suggested that ceramic pieces made up of nanosize grain structures may exhibit superior mechanical properties. In our understanding, a true nanosize polycrystalline microstructure is defined as a material with an average grain size well below 100 nm. At first glance, to achieve this fine-grain structure might appear trivial, because ceramic powders are readily available with an average particle size of 20 nm or less. However, processing nanosize powders into dense bulk specimens that retain their original nanosize microstructure can be very difficult.

The driving force for densification of nanosize powders is very high because of their small particle size. However, in many situations, their relatively low green-body densities require longer sintering times or higher temperatures than expected to achieve full densification. The longer sintering times and higher temperatures required promote excessive grain growth and/or irregular grain growth, resulting in a microstructure that does not exhibit the desired nanosize grain structure. The homogeneity and density of the green compact is critical in many situations, especially because the cold compaction density of most ceramic nanosize powders is extremely low ($<45\%$ of theoretical). As discussed in this chapter, high-pressure compaction can be a means of increasing the green-body density and structural integrity of the compact. Green densities in excess of 70% were achieved with nanosize amorphous silicon nitride powders. Similar densities were also obtained with γ-Al_2O_3. Even higher densities were achieved when LN_2 was used as a lubricant during compaction. Liquid nitrogen appears to be an excellent lubricant for nanosize particles. As far as we know, a detailed description of the lubrication mechanism of LN_2 for nanosize particles has not been reported yet. However, evidence of its effect on improving the packing density of nanosize particles has been amply demonstrated by several authors and their results are very convincing. Care must be taken when using LN_2 to compact nanosize particles, because it can experience a series of phase transformations, for example, liquid $\rightarrow$ solid, under pressure at low temperatures, which can have undesirable molar volume changes during compaction. These aspects of LN_2 have been discussed in detail here, and are relevant to the researcher interested in using LN_2 as a lubricant.

Finally, as mentioned earlier, obtaining a high green density is not the ultimate accomplishment in the processing of nanosize powders. The reason for this is that, in many cases, ceramic powders exist as metastable phases when they are prepared as very fine powders ($\approx$20 nm, nanosize). These materials are metastable and may transform, typically involving significant volume changes, during high-temperature sintering. Examples are the two materials discussed in this chapter. γ-Al_2O_3 is a metastable phase that transforms to the thermodynamically stable α-Al_2O_3 phase when heated from around 900 to 1100 °C. As far as we know, nanosize Si_3N_4 powder is available only in the amorphous state. It crystallizes upon heating to temperatures in the 1300 °C range. How large an effect such phase transformations have on the microstructure of the sintered product is difficult to predict. Certainly for the two materials presented here, γ-Al_2O_3 and Si_3N_4, the effect is great. Only with further experimental work in this area can we answer this question.

References

1. J. Frenkel, *J. Phys. (USSR)* 8, 386 (1945).
2. C. Herring, *J. Appl. Phys.* 21, 301 (1950).
3. W. H. Rhodes, *J. Am. Ceram. Soc.* 64, 19 (1981).
4. G. Skandan, H. Hahn, and J. C. Parker, *Scr. Metall.* 25, 2389 (1991).
5. T. H. Courtney, "Mechanical Behavior of Materials." McGraw–Hill, New York, 1990.
6. G. Skandan, "Processing of Nanostructured Zirconia Ceramics, Nanostructured Materials," Vol. 5, pp. 111–126. 1995.
7. M. D. Sacks and J. A. Pask, *J. Am. Ceram. Soc.* 65, 70 (1982).
8. W. Chen, A. Pechenik, S. J. Dapkunas, G. J. Piermarini, and S. G. Malghan, *J. Am. Ceram. Soc.* 77, 1005 (1994).
9. P. Duran, M. Villegas, F. Capel, J. F. Fernandez, and C. Moure, *J. Mater. Sci.* 32, 4507 (1997).
10. A. Pechenik, G. J. Piermarini, and S. C. Danforth, *J. Am. Ceram. Soc.* 75, 3283 (1992).
11. G. J. Piermarini and S. Block, *Rev. Sci. Instrum.* 46, 973 (1975).
12. R. A. Forman, G. J. Piermarini, J. D. Barnett, and S. Block, *Science* 176, 284 (1972).
13. J. D. Barnett, S. Block, and G. J. Piermarini, *Rev. Sci. Instrum.* 44, 1 (1973).
14. E. J. Gonzalez, B. Hockey, and G. J. Piermarini, *Mater. Manufacturing Process.* 11, 951 (1996).
15. P. W. Bridgman, *Z. Kristallogr.* 67, 363 (1928).
16. P. W. Bridgman, *Proc. Am. Acad. Arts Sci.* 76, 1 (1945).
17. C. E. Weir and G. J. Piermarini, *J. Res. Natl. Bur. Stand. (U.S.) Phys. Chem.* 68A, 105 (1964).
18. W. Symons and S. C. Danforth, "Ceramic Materials and Components for Engines" (V. J. Tennery, ed.), pp. 67–75. American Ceramic Society, Westerville, OH, 1989.
19. M. R. Gallas, B. Hockey, A. Pechenik, and G. J. Piermarini, *J. Am. Ceram. Soc.* 77, 2107 (1994).
20. J. E. Bonevich and L. D. Marks, *J. Mater. Res.* 7, 1489 (1992).
21. A. Pechenik, G. J. Piermarini, and S. C. Danforth, *Nanostruct. Mater.* 2, 479 (1993).
22. D. A. Young, "Phase Diagrams of the Elements," pp. 113–117. University of California Press, Berkeley, CA, 1991.
23. R. L. Mills, D. H. Liebenberg, and J. C. Bronson, *J. Chem. Phys.* 63, 4026 (1975).
24. W. L. Vos and J. A. Schouten, *J. Chem. Phys.* 91, 6302 (1989).
25. S. Kwon, C. S. Nordahl, and G. L. Messing, *Mater. Manufacturing Process.* 11, 969 (1996).
26. H. Hahn, J. Logas, and R. S. Averback, *J. Mater. Res.* 5, 609 (1990).
27. K. J. Nilsen, R. E. Riman, and S. C. Danforth, *Ceram. Trans.* 1, 469 (1988).
28. M. Kumagai and G. L. Messing, *J. Am. Ceram. Soc.* 68, 500 (1985).
29. G. L. Messing and M. Kumagai, *J. Am. Ceram. Soc.* 72, 40 (1989).
30. T.-S. Yeh and M. D. Sacks, *J. Am. Ceram. Soc.* 71, 841 (1988).
31. F. W. Dynys and J. W. Halloran, *J. Am. Ceram. Soc.* 65, 442 (1982).
32. W. Chen and S. G. Malghan, *Powder Technol.* 81, 75 (1994).
33. A. R. Cooper, Jr., and L. E. Eaton, *J. Am. Ceram. Soc.* 45, 97 (1962).

Chapter 4

SEMICONDUCTOR NANOPARTICLES

Prashant V. Kamat

Notre Dame Radiation Laboratory, Notre Dame, Indiana, USA

Kei Murakoshi, Yuji Wada, Shizo Yanagida

Chemical Process Engineering, Faculty of Engineering, Osaka University, Suita, Osaka, Japan

Contents

Handbook of Nanostructured Materials and Nanotechnology, edited by H.S. Nalwa
Volume 3: Electrical Properties

1. INTRODUCTION

The multidisciplinary research of semiconductor nanoclusters has witnessed exceptional growth during the past decade. A basic understanding of the photophysical and photochemical properties gained from colloidal semiconductor research is beneficial in the design and development of nanostructured semiconductor materials for practical applications. A great deal of attention has been focused on understanding the properties of nanostructured semiconductor systems because of their practical applications in solar energy conversion and photocatalytic degradation of organic contaminants. Some other applied areas, which can benefit from nanostructured semiconductor technology, are chemical sensors, electrooptics, microelectronics, imaging science, and photovoltaics. Possible applications that make use of the unique properties of semiconductor nanoclusters are illustrated in Figure 1.

By making use of the principles of photoelectrochemistry, several semiconductor systems comprised of native and surface-modified semiconducor nanoparticles have been employed in the direct conversion of light energy into chemical or electrical energy [1–5]. The term colloid refers to ultrasmall particles with a particle diameter less than 100 nm with the ability to suspend in an aqueous or nonaqueous medium. These are also referred as *Q* particles, nanoclusters, nanoparticles, or nanophase materials. Under band gap excitation, these semiconductor nanoparticles act as short-circuited microelectrodes and initiate the oxidation and reduction processes of the adsorbed substrate. Alternatively, an adsorbed molecule can inject charge from its excited state into the semiconductor particle and the sensitizer can be regenerated with a redox couple. This process, which is commonly referred to as photosensitization, is extensively used in photoelectrochemistry and imaging science. The application of semiconductor systems in initiating and controlling various photocatalytic processes has been reviewed recently by several researchers [5–18]. Elucidation of the photophysical properties of the semiconductor colloids and the dynamics of the interfacial processes at the semiconductor–electrolyte interface has also been reviewed by many researchers [11, 19–46].

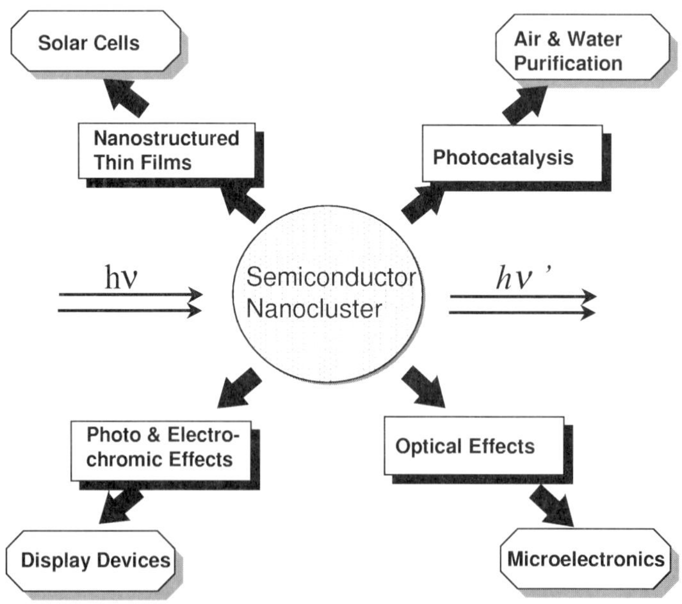

Fig. 1. A diversifying area of semiconductor nanocluster research.

2. PREPARATION AND CHARACTERIZATION

2.1. Size Control

Several approaches have been considered to prepare size-controlled semiconductor nanocrystallites with narrow size distribution. Many of these preparation methods of the size-controlled nanocrystallites control the growth process by using stabilizers such as thiols [47–65], phosphate [22, 66], and phosphine oxide [67] and restricting the reaction space in matrices such as zeolites [68–76], glasses [71, 77–79], polymers [70, 80–84], reverse micelles [85–92], vesicles [93–102], LB films [103–112], multilayer film [113, 114], xerogels [115], β-cyclodextrin [116, 117], and silica [118, 119]. Each step in the growth process, for example, nucleation, growth (propagation), and ripening, can be controlled by changing the preparation conditions, such as the choice of the starting materials and media (solvents), composite ratios of the reactants, and preparation temperature and time. These synthetic methods for the smaller nanocrystallites (clusters) were much more successful in the preparation and characterization of thiol-capped CdS clusters. Highly purified clusters form single crystals of $(Cd_{10}S_4(SC_6H_5)_{16})^{4-}$ [120], $(Cd_{17}S_4(SC_6H_5)_{28})^{2-}$ [121], $Cd_{17}S_4(SCH_2CH_2OH)_{26}$ [122], $Cd_{32}S_{14}(SCH_2CH(OH)CH_3)_{36}{\cdot}4H_2O$ [123], and $Cd_{32}S_{14}(SC_6H_5)_{24}{\cdot}4DMF$ [124], which can be used as model compounds of the smallest CdS nanocrystallites. Nosaka et al. [49] reported size control of colloidal CdS nanocrystallites by changing concentration of the surface stabilizer (C_2H_5SH) in a reaction bath. Henglein and co-workers also reported the dependence of CdS size on concentration of long-chain alkanethiols at the stage of nucleation and growth of the nanocrystallites [52].

Apart from chemical precipitation and hydrolysis of suitable precursors, several other methods have also been considered in the synthesis of semiconductor colloids by chemical methods [11, 125–127]. Electrochemical and photoelectrochemical etching of silicon wafers can yield porous film composed of nanostructured silicon [128, 129]. Some other techniques such as radiolysis [11, 45, 51, 130–136] and sonolysis [137] have also been employed. Sonochemical formation of Q-CdS colloids and dissolution of larger colloidal CdS particles has also been attempted by Grieser and co-workers [138].

Despite these various attempts to control the size and shape of the nanoclusters, two crucial problems still remain as challenging issues. (i) The size distributions of nanocrystallites prepared by the previous methods are not monodisperse (standard deviation more than 10%). A relatively narrow size distribution (standard deviation ~8%) has been achieved in the preparation of CdS nanocrystallites in vesicles of phosphatidylcholine [101]. (ii) The use of stabilizers or confined matrices often yield byproducts, such as metal–thiolate complexes, which makes it difficult to quantitatively assess the microscopic structure of nanocrystallites.

Efforts have been made to narrow the size distribution by size fractionation using exclusion chromatography [139–141] and capillary zone electrophoresis [141–144]. The separation in electrophoresis is based on the fact that the charge-to-size ratios are different for different size particles. Size-selective precipitation [48, 67] or photocorrosion has also been considered to minimize the dispersity in particle size [145, 146]. Yoneyama et al. [145] applied the method of size-selective photocorrosion using monochromatic light irradiation. Narrower size distribution of CdS nanoparticles was achieved by irradiating light whose wavelength was close to the absorption onset of nanocrystallites in air-saturated sodium hexametaphosphate solution [145]. Photocorrosion of CdS nanocrystallites using Ar ion laser irradiation was also useful in preparing smaller nanocrystallites [146]. Recently, Murakoshi et al. [147] reported the size control of CdS nanocrystallites over the range from 1.8 to 4.2 nm while keeping the *hexagonal* crystalline structure by varying the sulfur source (H_2S and Na_2S) and by regulating the preparation temperature. Strongly terminating sulfur species stabilize the nanocrystallites, thus leading to the formation of smaller size particles even in the absence of any stabilizers or matrices. Bawendi and co-workers [67] have recently adopted a novel approach of synthesizing size-controlled CdSe nanocrystallites of highest monodispersity (standard deviation less than 5%). Although these methods are useful for the size control of semiconductor nanocrystallites, further development of a novel and simple preparation method that excludes chemical capping or host matrices would be the topic of future research efforts.

2.2. Crystalline Phase Control

II–VI semiconductor bulk materials, such as CdS and ZnS, are well known to have two crystalline structures, cubic (zincblende) and hexagonal (wurtzite) types. For the bulk materials, CdS has highly stable hexagonal phase from room temperature to the melting point (1750 °C) [148], while bulk ZnS has the cubic structure. Because the difference in the crystalline structure of the bulk semiconductor materials leads to considerable change in the effective masses of e and h in their electronic bands [149], the crystalline structure of semiconductor nanocrystallites should play a dominant role in determining their photochemical, photocatalytic, and photophysical properties.

The crystalline structure of nanocrystallites is largely influenced by the preparation conditions. Although the hexagonal phase is the most stable phase of CdS, for example, only a few reports on the preparation of *hexagonal* CdS nanocrystallites have been published [48, 67, 150–153]. *Hexagonal* CdS nanocrystallites can be prepared by reacting Cd^{2+} and S^{2-} (H_2S or Na_2S) in *N,N*-dimethylformamide (DMF) without using any surface stabilizers [147]. The mean diameter of CdS nanocrystallites can be rigorously controlled over the range from 1.8 to 4.2 nm by varying the sulfur source (H_2S and Na_2S) and by controlling the reacton temperature (Figs. 2 and 3). In these studies, the use of Na_2S as a sulfur source produces smaller CdS nanocrystallites than those prepared using H_2S. The size control and stability of CdS nanocrystallites are discussed in terms of the surface structures that were elucidated using *in situ* Cd K-edge extended X-ray absorption fine structure (EXAFS) analysis. The strong terminal binding of the sulfur atoms to the surface cadmium atoms resulted in the weaker solvation of DMF molecules and the controll of their size and crystalline phase.

Phase transformation of semiconductor nanocrystallites can be used to obtain a desirable crystalline structure [44]. The phase transition of semiconductor nanocrystallites can

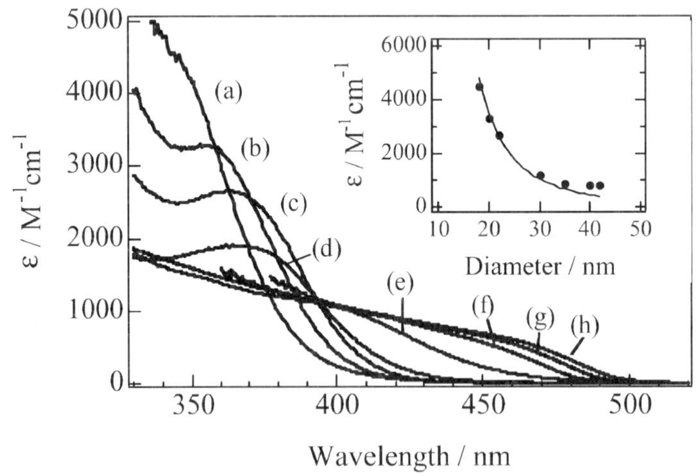

Fig. 2. Room temperature absorption spectra of $CdS–Na_2S$, which were prepared at (a) −40 °C, (b) −20 °C, (c) 0 °C, (d) 30 °C, and (e) 60 °C. Those of $CdS–H_2S$ were prepared at (f) 0 °C, (g) 30 °C, and (h) 60 °C. The inset shows the dependence of the molar extinction coefficient on the size of $CdS–H_2S$ and $CdS–Na_2S$. The experimental data were fitted in proportion to the inverse cube of the Bohr radius divided by the CdS radius $(a_B/R)^3$, $a_B = 26$ Å). (Source: Reprinted with permission from [147].)

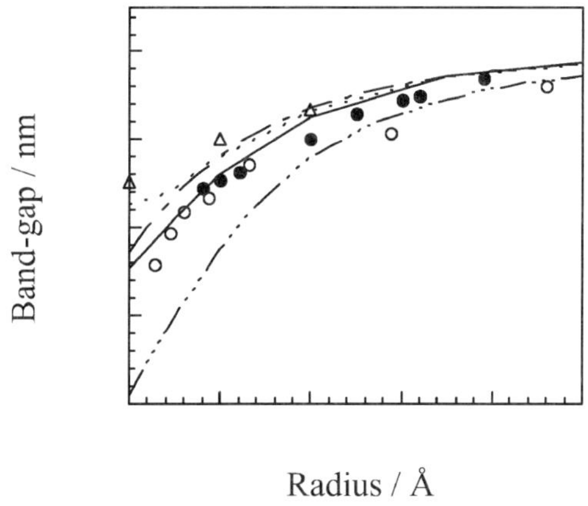

Fig. 3. Band gap dependence on the CdS radius. The experimental data are from this study (•), from Vossmeyer et al. (○), and from Herron et al. (△). The curves are predicted by the effective mass approximations with infinite (–· · ·–) and finite potential well (–·–), and the pseudopotential calculations for hexagonal (—) and cubic CdS lattice (· · ·). (Source: Reprinted with permission from [147].)

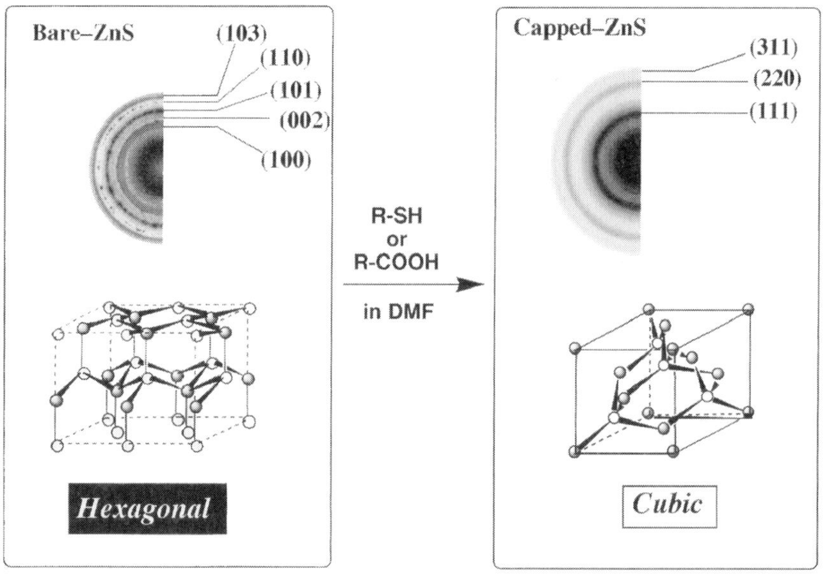

Fig. 4. Electron diffraction patterns of (a) ZnS–DMF and (b) ZnS–DMF combined with C_6F_5SH. (Source: Reprinted with permission from [158].)

be achieved under high temperature [154] and pressure [155–157]. Unfortunately, these extreme conditions often cause unexpected structural deformations, such as the growth of nanocrystallites and change of their chemical composition [154]. Alivisatos and coworkers carried out a four- to six-coordinate phase transformation of CdS [155] and CdSe [156, 157] nanocrystallites without altering their size or their chemical composition. The control of crystalline structures of quantum-confined semiconductor nanocrystallites was achieved at ambient temperature and pressure through the chemical modification of the ZnS nanocrystallite surface with organic molecules (Fig. 4) [158]. Pentafluorothiophenol, thiophenol, 1-decanethiol, 1-hexanethiol, benzoic acid, and cyanoacetic acid, which chemically bind to the surface of ZnS nanocrystallites, are capable of changing the hexagonal phase of ZnS nanocrystallites to the cubic. On the other hand, similar phase transitions could not be observed with unreactive octanoic acid, acetic acid, and phenol, which do not bind to ZnS. The formation of the chemical bonding between surface zinc atoms of ZnS nanocrystallites and the added thiolate or carboxylate anion should lead to a change in the surface energy of ZnS nanocrystallites. This change in the surface energy could induce the phase transition from the metastable phase (hexagonal) to the stable phase (cubic) at ambient temperature and pressure. [The difference in the internal energy of ZnS between the hexagonal and the cubic phase is quite small (3.2 kcal mol^{-1})].

2.3. Size Quantization Effects

A great deal of attention has been focused on the transformation from solid-state electronic properties to molecular properties of various semiconductor nanoparticles [11, 34, 35, 39–41, 74, 149, 159–167]. These crystallites are molecular clusters in which complete electron delocalization has not yet occurred. Quantization in these ultrafine particles arises from the confinements of charge carriers in semiconductors with potential wells of small dimensions (less than the De Broglie wavelength of the electrons and holes). Under these conditions, the energy levels available for the electrons and holes in the conduction and valence bands become discrete. In addition to the very large effects on optical properties, size quantization also leads to major changes in the effective redox potential of the photogenerated carriers. Semiconductor particles that exhibit size-dependent optical and electronic properties are termed quantized particles (or *Q* particles) or semiconductor clusters. The particles with diameters greater than 150 Å typically behave as bulk semiconductors. However, similarly prepared crystallites of approximately 15–50 Å diameter that possess a bulk lattice structure differ considerably in electronic properties.

Size quantization effects have been experimentally demonstrated in various laboratories around the world by controlling the synthetic procedure and monitoring the size-dependent optical effects. A few representative examples include [11, 24, 25, 34–36, 40, 42, 49, 51, 52, 79, 133, 137, 139, 160–162, 164, 165, 168–207]. Optical effects caused by size quantization in layered-type semiconductors have also been reported [131, 170, 199, 200, 208, 209]. The absorption spectra of CdS colloids of various particle size are compared in Figure 5 [173]. The onset of absorption for larger colloids closely matches the absorption edge of bulk CdS. The absorption edge shifts to blue for smaller size particles as a result of the increase in the effective band gap.

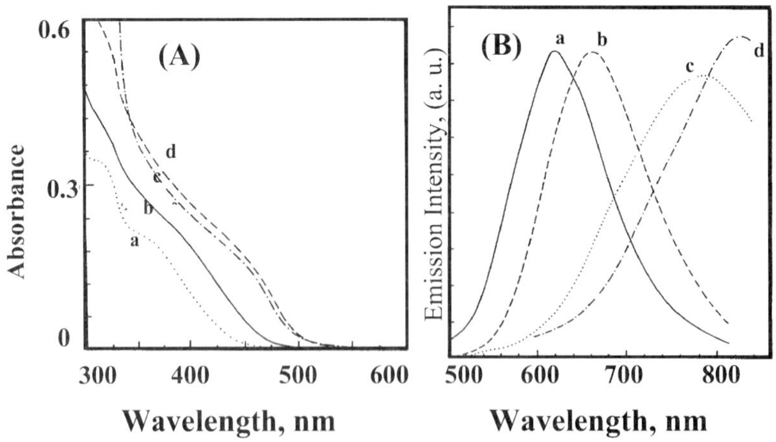

Fig. 5. (A) Absorption and (B) emission spectra of colloidal CdS in acetonitrile. The average particle diameters were, respectively, (a) <25, (b) 30, (c) 37, and (d) >42 Å. (Source: Reprinted with permission from [173]. © 1987 American Chemical Society.)

Brus [210, 211], Henglein [22, 24], Anderson [212], and Friesner [149] and co-workers have modified the quantum mechanical treatment to explain the various experimental results of wavelength threshold as a function of the diameter of CdS particles. A comparison of various models and the experimental results has been established. A plot of the wavelength of the onset of absorption versus the function of particle diameter seems to agree well with the proposed model [22, 24].

Size quantization effects on the optical properties of nanocrystalline silicon have been discussed by Brus and co-workers [36, 188, 213]. A comparison is made to explain the behavior of porous Si that is prepared from electrochemical etching. Nanocrystalline Si and porous Si are indirect-gap-type materials with oscillator strengths that are not markedly increased with respect to bulk Si. The indirect gap in these Si nanocrystallites can be as high as 1.0 eV. An enhanced luminescence is seen in these materials because of the spatial confinement of electrons and holes. Zunger and co-workers [214–217] have carried out theoretical modeling to elucidate the electronic structure of Si quantum dots and wires. The published work related to porous Si is quite vast and readers may refer to a recently published review article [218].

2.4. Nonlinear Optical Properties

Semiconductor nanoparticles as materials for generating nonlinear optical effects have drawn the attention of many researchers [219–227]. Photoinduced blue shifts in the absorption have been reported for the colloids of CdS [228–239], Cd_3P_2 [234], In_2S_3 [240], PbS [84, 171, 197, 241], CdSe [83, 242, 243], In_2Se_3 [244], FeS_2 [233], TiO_2 [177], and ZnO [206, 245, 246]. Such photoinduced blue shifts appear as a bleaching in the transient absorption spectrum. Transient absorption spectroscopy has been employed to probe the transient bleaching in CdS [236, 237, 247] and GaAs [248] colloids. Subpicosecond events in colloidal $Zn_{1-x}Cd_xS$ have been characterized by Ernsting et al. [247]. In the case of CdS/HgS/CdS colloids with a quantum dot–quantum well structure, the transient photobleaching follows spectrally the newly evolving 1s–1s electronic transition of the composite particles [249].

The observed transient bleaching in these nanoclusters was attributed to photophysical processes, such as the dynamic Burstein effect, increase in excitonic energy due to surface-trapped electrons, or Coulombic screening by photogenerated free carriers [230–232, 235–237, 250]. An alternate explanation based on the Stark effect has also been proposed recently. Charge separation occurs when semiconductor colloids are subjected to band gap excitation with a laser pulse. The intense electric field created as a result of photoinduced charge separation is likely to cause a blue shift in the band edge of these colloids. This effect has also been observed by depositing excess charge on CdS colloids or by applying an electric field to CdSe film [251, 252].

Third-harmonic generation in CdS-doped poly(vinyl alcohol) (PVA) film has also been investigated [253]. Wang and co-workers have extensively investigated the third-order optical nonlinearity in quantum-confined CdS clusters [219, 220]. They have demonstrated nonresonant third-order optical nonlinearity using third-harmonic generation from 1.91–0.64 μm. The nonlinearity increased with increasing cluster size. The intrinsic, nonresonant $\chi^{(3)}$ for 30- and 15-Å-diameter CdS clusters were 3.3×10^{-10} and 2.5×10^{-11} esu, respectively. The nonlinearity can be controlled by passivating the CdS surface with ammonia.

2.5. Emission Characteristics

Considerable attention has been given in recent years to the study of the emission behavior of various semiconductor colloids: ZnO [191, 206, 254–257], ZnS [98, 168, 258–260], CdSe [83, 243, 261–265], CdS [22, 57, 66, 71, 96, 98, 150, 173, 207, 229, 231, 238, 253,

257, 259, 260, 266–287], and MoS_2 [288]. Luminescence of porous silicon has also been the theme of a number of recent studies [36, 289–301]. Photoluminescence in semiconductors occurs when a photogenerated minority carrier recombines with a previously existing majority carrier. Hence, emission measurements give valuable information regarding the nature of surface states and the recombination of charge carriers. The analysis of luminescence decay gives important information on the kinetic and mechanistic details of charge carrier recombination in colloidal semiconductors. The photophysical properties of II–VI semiconductor nanoclusters have been extensively investigated in this regard. The emission spectra of CdS colloids recorded under steady-state irradiation excitation are shown in Figure 5. Under steady-state irradiation conditions, red emission appears as a result of sulfur vacancies. The emission maximum is dependent on the particle size and is blue-shifted in smaller size particles. The short-lived excitonic emission can be seen when a high concentration of charge carriers is attained with laser excitations. The excitonic emission has been considered to arise via detrapping of electrons. The traps work as a reservoir for electrons and thus contribute to the delayed emission [143]. With decreasing temperature, an increase in the intensity and a blue shift in the maximum have been observed for the Q-CdS excitonic band. A thermal repopulation phenomenon is also investigated by monitoring band edge luminescence in CdS Q-particles [276, 277, 302]. A similar size-dependent shift in the excitonic emission band at 77 K has been observed for quantized PbI_2 particles embedded in porous silica films [209]. Emission spectroscopy has also been used to resolve the spectra of quantization effects in CdS- [303] and CdTe-doped glasses [304].

A model that involves strong resonance between three-dimensional quantized internal states and the intrinsic surface states because of reconstruction has also been outlined by Brus and co-workers [262, 305]. Time, wavelength, and polarization resolution of the emission indicate a complex kinetics for the band edge luminescence in 32-Å-diameter CdSe particles. A strong resonant mixing between the standing wave molecular orbital delocalized inside the crystallite and the intrinsic surface Se lone pair states has been postulated.

2.6. Trapping of Charge Carriers

Semiconductor nanoclusters synthesized from a common bench top recipe contain a high density of defect sites, usually at the semiconductor surface, and the nature of these defect sites depends strongly on the method of chemical synthesis. Upon optical excitation, the free carriers get trapped at the defect sites and these trapped charge carriers undergo further radiative and radiationless recombination:

$$TiO_2 + h\nu \rightarrow TiO_2\left(e^-_{CB} + h^+_{VB}\right) \tag{1}$$

$$e^-_{CB} \rightarrow e^-_{tr} \tag{2}$$

$$h^+_{VB} \rightarrow h_t \tag{3}$$

where e_{CB} and h_{VB} are the free charge carriers in the conduction and valence bands and e_t and h_t are the trapped electrons and holes, respectively. Such an accumulation of charges in the minute reaction space of a semiconductor colloid is important in improving the efficiency of interfacial redox reactions.

Bard and co-workers [306–308] demonstrated the effect of charge trapping on the photoelectrochemical properties of TiO_2 particles under illumination. They carried out electrophoretic and photocurrent measurements in the presence and absence of electron acceptors. A buildup of negative charge on the TiO_2 particles under illumination was observed when reducible species were excluded. The dynamics of charge carrier trapping, recombination, and photophysical processes have been studied by several researchers [143, 236, 238, 268, 302, 309–334]. These trapped charge carriers exhibit characteristic absorption bands in the infrared (IR) and ultraviolet (UV) regions. These absorption bands have also been independently confirmed from radiolysis and electrochemical experiments [335].

By selectively scavenging the photogenerated holes in a semiconductor colloid, it is possible to monitor the behavior of trapped electrons. Electron paramagnetic resonance (EPR) studies of photoexcited anatase TiO_2 particles have shown that the photogenerated electrons are trapped at Ti^{4+} sites within the bulk [336–338]. Electron trapping in ZnO [206], WO_3 [183], and SnO_2 [339] colloids has been investigated by laser flash photolysis. Similarly, trapped electrons in colloidal metal chalcogenides such as CdS undergo chemical changes by reacting with Cd^{2+} sites. Zhang and co-workers [326, 340] and Bowman and co-workers [332] have concluded that the electron trapping in TiO_2 and CdS colloids occurs in the range of 90 to 180 fs. A mechanism that allows for slower electron transport via surface trap-to-trap hopping has been considered in these studies.

The basic hydroxide groups at the surface of TiO_2 can be considered deep traps for valence band holes. Although the majority of these trapped holes recombine with conduction band and trapped electrons, a small fraction of O^- undergoes dimerization at the surface to form stable titanium peroxide. An indirect method of hole trapping was achieved in this study by reacting TiO_2 colloids with pulse radiolytically generated $OH^\bullet$ radicals. The transient spectrum recorded after irradiation of aqueous colloidal TiO_2 suspension with electron pulse showed an absorption band with onset at 470 nm and a maximum around 350 nm. A slightly red-shifted band ($\lambda_{max} = 420$ nm) has also been observed in UV-irradiated TiO_2 colloids [309, 323]. Electron spin resonance studies have identified these trapped holes as oxygen anion radicals covalently bound to titanium ions, Ti^{IV}–O–Ti^{IV}–$O^\bullet$ [338, 341–344]. Time-resolved microwave conductivity measurements have also been carried out to probe the electronic processes in semiconductor materials [315, 330, 345–351]. Based on the analysis of luminescence decay, Brus has estimated that hole localization on the CdSe surface occurs within 150 fs [305].

The chemical changes associated with trapping of charge carriers in CdS, PbS, and CdSe have also been time resolved [238, 241, 268, 318, 352]. The observed growth in the absorption band at wavelengths greater than 460 nm is attributed to the formation of S^- at the CdS surface, the identity of which has been confirmed earlier by pulse radiolysis measurements [192].

3. INTERFACIAL CHARGE TRANSFER PROCESSES IN COLLOIDAL SEMICONDUCTOR SYSTEMS

The principle of electron and hole transfer at the photoexcited semiconductor particle has already been described in Figure 6. Both free and trapped charge carriers participate in these interfacial redox processes. By decreasing the particle size, it is possible to shift the conduction band to more negative potentials and the valence band to more positive

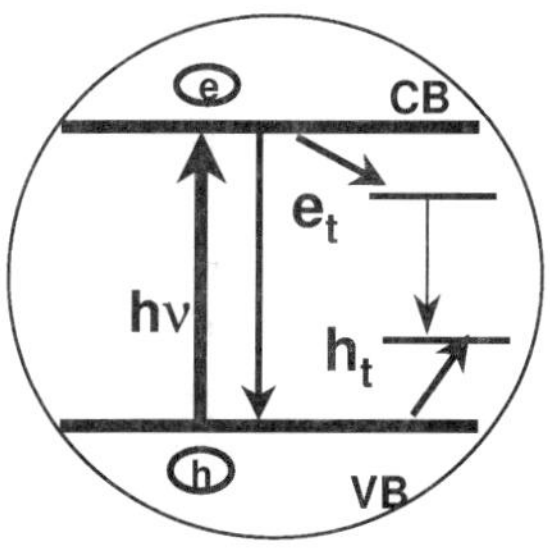

Fig. 6. Photoinduced charge transfer processes in a semiconductor particle. E_{CB} and E_{VB} correspond to the conduction and valence bands, and e_t and h_t correspond to trapped electrons and trapped holes, respectively. These charge carriers are responsible for the interfacial redox processes.

potentials as a result of quantization effects. Hence, redox processes that cannot occur in bulk materials can be facilitated in quantized semiconductor particles.

3.1. Reductive Process

In the early years, semiconductor nanoparticles were extensively employed in water-splitting reactions [7, 353–357]. Although this area still continues to draw the attention of several scientists [358–364], recent studies have focused on the mechanistic and kinetic study of the interfacial charge transfer processes. Laser flash photolysis has been extensively employed to probe the reduction of viologen in colloidal TiO_2 [23, 75, 309, 354, 365–377] and metal chalcogenide [62, 173, 195, 240, 242, 378–386] systems. The reduction of several organic dyes in various colloidal semiconductor systems has also been investigated. Some of these dye reductions include thiazine [387], oxazine [388], phenoxazine [389], squaraine [390], and azo [391, 392] dyes. Band gap excitation of the semiconductor leads to one-electron reduction of the dye to yield a semireduced dye, which then disproportionates to yield a two-electron reduction product (leuco dye). Recently, one-electron reduction of fullerenes (C_{60} and C_{70} [393–396], C_{76} and C_{78} [397], and C_{84} [398]) has been carried out in ZnO and TiO_2 colloidal suspensions. The anion formed as a result of direct reduction of fullerene at the excited semiconductor colloid surface has been monitored with both steady-state and laser flash photolysis experiments. The energetics of the conduction band and the redox couple act as a driving force for heterogeneous electron transfer at the semiconductor colloid.

The reduction of oxygen in a semiconductor suspension plays an important role in photocatalytic applications [399]. The formation of $O_2^{-\bullet}$ has recently been confirmed from chemiluminescence measurements [400]. $O_2^{-\bullet}$ was deactivated mainly by an electron transfer to surface defects or vacancy sites on the TiO_2 surface.

The simple double and triple bonds can be photocatalytically hydrogenated with Pt-loaded TiO_2 using alcohol as an electron donor [401]. TiO_2 can be also used as an electron donor for asymmetric reduction of carbonyl compounds with chiral Rh(I) complexes as an electron mediator [402]. Photoreduction of aldehydes [403], and Shiff bases [404] in TiO_2 suspensions have also been reported. Photolysis of dimethyl maleate in deuterated methanol revealed that CdS nanocrystallites catalyze sequential two-electron-transfer photoreduction, affording deuterated dimethyl succinate [405].

Under visible-light-induced photolysis, the methanolic CdS suspension becomes brown owing to the reduction of lattice Cd^{2+} to Cd^0 (metallic Cd). The effective formation of alcohols from ketones and dihydro compounds from alkenes has been observed as a result of this reduction step. The presence of excess sulfide ions (S^{2-}) in the CdS nanocrystallite system, however, suppresses the formation of lattice Cd^0. The photogenerated lattice Cd^0 can also function as the active site for the two-electron reduction [406]. The relationship between light intensity and product distribution has been investigated in the case of the photocatalytic reduction of nanoscale CdS nanocrystallites. The yield of one-electron reduction products increases with decreasing light intensity. Cd^0 formation is proportional to the square of the relative light intensity. Consequently, the chemoselectivity in the photocatalysis using nanoscale CdS depends on the amount of Cd^0 formation, which, in turn, is controlled by the excitation light intensity [407].

3.2. Oxidative Process

Photogenerated holes can be utilized to oxidize adsorbed substrates at the semiconductor surface. The oxidation of halide ions has been investigated in various colloidal semiconductor suspensions by laser flash photolysis [173, 240, 367, 408, 409]. The energetics of the semiconductor valence band and the oxidation potential of the redox couple influence the photocatalytic oxidation. For example, the enhancement in the efficiency of halide oxidation at TiO_2 follows the sequence $Cl^- < Br^- < I^-$, correlating with the decrease in the oxidation potential [367, 408]. Shifting the valence band of TiO_2 shifts to more positive potentials, by decreasing the pH and thus results in the enhancement of quantum efficiency of halide oxidation. The femtosecond dynamics of the oxidation of SCN^- has been investigated recently in a colloidal TiO_2 system [410]. These studies confirm the previous finding that the surface complexation between SCN^- and TiO_2 plays an important role in the efficient charge transfer at the semiconductor interface.

The oxidation of several other organic compounds such as 2-propanol [388, 411], diethyldithiocarbamate [412], and methyl orange [413] in colloidal semiconductor suspensions has also been investigated. The oxidation of the N_3^- to the $N_3^\bullet$ radical in CdS and ZnO suspensions has been investigated by the ESR spin-trapping the technique using phenyl-tert-butylnitrone (PBN) as a spin trap [414]. Draper and Fox [372, 415] have employed a diffuse reflectance laser flash photolysis technique to elucidate the mechanism of oxidation of several inorganic and organic compounds in TiO_2 particulate systems. Becker and co-workers [416] have discussed the factors that control the photocatalytic oxidations in nonaqueous media.

By simultaneously capturing holes and electrons at the semiconductor surface, it is possible to decrease the charge recombination and thus enhance the quantum yield of the charge transfer process [367, 417]. The importance of back reaction suppression for improving the quantum efficiency of charge transfer has been illustrated by Curran and Lamouche [418]. They have presented a kinetic model to derive the quantum efficiency based on the particle radius, the Faradaic rate constants, and the diffusion length in the semiconductor.

3.3. Kinetics of Interfacial Electron Transfer

The kinetic analysis of heterogeneous electron transfer in colloidal semiconductor systems is often complex and it is difficult to apply a generalized kinetic expression for these photocatalytic reactions. Lewis [419] has used both steady-state (equilibrium) and pulsed (nonequilibrium) experimental data to propose a unified kinetic model for heterogeneous electron transfer at the semiconductor interface. A more detailed discussion on the kinetic aspects of interfacial electron transfer has been presented in a recent review article [167]. Apart from the energetics of the conduction band of the semiconductor and the redox potential of the acceptor, factors such as the surface charges of the colloids, adsorption of the substrates, participation of surface states, and competition with charge recombination influence the rate of heterogeneous charge transfer at the semiconductor interface. This fact is evident from the widely differing rates of experimentally observed charge transfer in various colloidal semiconductor systems. (The time scale reported for charge transfer varies from picoseconds to milliseconds under different experimental conditions in various laboratories.) In general, interfacial charge transfer can be classified into two categories: static and dynamic quenching processes.

The static process dominates when the electron acceptor (or donor) substrate is adsorbed strongly on the semiconductor surface. Electron transfer under these conditions is rapid and directly competes with charge recombination. Both the adsorption equilibrium and the quick removal of the product from the surface is important in controlling the efficiency of charge transfer. Laser flash photolysis has been employed to probe the ultrafast charge transfer between an excited semiconductor colloid and the adsorbed substrate [310, 318, 323, 420–423]. The time-resolved transient absorption spectra recorded in a picosecond laser flash photolysis of a CdS–ZV system have indicated the rise time of $ZV^{\bullet -}$ formation to be less than 20 ps. The rate constant for the interfacial electron transfer is thus expected to be greater than $5 \times 10^{10}\ s^{-1}$. This estimate is in good agreement with the value ($k = 2$–$5 \times 10^{10}\ s^{-1}$) proposed by Rossetti and Brus [424] in their picosecond resonance Raman experiments. Such a large value of the rate constant indicates the interfacial electron transfer from the photoexcited CdS to the adsorbed viologen as rapid. Strong adsorbate–adsorbent interaction is essential to carry out the reduction process.

A major fraction of $ZV^{\bullet -}$ decayed within 500 ps as a result of its reaction with trapped holes (or S^{-}_{surf}). A fraction of $ZV^{\bullet -}$ survives as it diffuses away from the surface. Negatively charged Nafion, which is used to stabilize the colloids, drives away $ZV^{\bullet -}$ from the particle surface by electrostatic repulsion. The role of the ionic charge of the polymeric stabilizer in influencing the rate of heterogeneous electron transfer has been studied recently [385, 425–427]. The influence of electrostatic interaction on the stabilization of the viologen radical in negatively charged TiO_2-coated SiO_2 colloids [426] and functionalized polymeric systems [428] has also been discussed. The conditions for carrying out efficient redox processes on a semiconductor particle have been discussed by Gerischer [429, 430]. A large increase in the reduction efficiency can be achieved by locally contacting the particles with suitable metals, owing to the formation of an accumulation layer at this contact.

Similar static processes can also be observed in hole transfer processes. For example, the formation of $(SCN)_2^{\bullet -}$ was completed within the laser pulse duration as a result of the association of SCN^- with TiO_2 [417]. Scavenging of holes by SCN^- had a marked effect on the reduction of tetranitromethane [367] and oxazine dyes [417] at the TiO_2 colloidal particle. Simultaneous capture of holes and electrons at the TiO_2 surface with suitable scavengers enhanced the efficiency of interfacial charge transfer by a factor of 2. The holes trapped at the TiO_2 surface have been shown to survive for a long duration [309, 310, 343, 344, 394]. Surface complexation of colloidal TiO_2 by monodentate and bidentate benzene derivatives can enhance the interfacial electron transfer to methyl viologen [431]. The oxidation of diethyldithiocarbamate anion ($NCEt_2S_2^-$) in colloidal CdS suspension has been investigated [318, 412] to resolve the kinetic components arising from the reactions of the valence band and trapped holes.

The formation of $Et_2NCS_2^{\bullet}$ following the laser pulse excitation of CdS colloids was monitored from its absorption band at 600 nm. The rate constants for these two processes were estimated to be greater than 5×10^{10} s^{-1} and approximately 10^{10} s^{-1}, respectively. Only a small fraction (~15%) of $NCEt_2S_2^{\bullet}$ decayed during the time period of 1 ns probably by reacting with trapped electrons, and the rest underwent dimerization as it diffused away from the CdS surface ($k = 4.6 \times 10^9$ M^{-1} s^{-1}) [412].

Dynamic electron transfer is observed when the adsorption between the electron acceptor (or donor) molecules and the semiconductor colloid is poor. Under these conditions, the substrate molecules have to diffuse from solution to the semiconductor surface to accept the charge. Because photogenerated charge carriers undergo rapid charge recombination, it is necessary to accumulate one of the charges (electrons or holes) within the semiconductor particle. By employing sacrificial electron donors (e.g., alcohols) or electron acceptors (O_2), it is possible to accumulate electrons or holes within the semiconductor particles as trapped charge carriers. The kinetics of dynamic charge transfer is controlled by factors such as particle size, surface charge, and the thermodynamic driving force for the charge transfer.

One of the initial demonstrations, of dynamic charge transfer involved the reduction of tetranitromethane [367] and methyl viologen [354] in aqueous TiO_2 sols stabilized with polyvinyl alcohol. Recently, it has been possible to carry out a systematic kinetic study of heterogeneous electron transfer to several dyes from excited WO_3 colloids [183]. The efficiency of electron trapping in these colloids was very high ($\phi = 0.5$) as the holes were quickly scavenged by a sacrificial electron donor, oxalic acid. Dyes such as thionine could easily be reduced in excited WO_3 suspension. The bimolecular rate constants were determined by monitoring the pseudo-first-order growth of the dye anion radical, k_d, at various dye concentrations and by the expression

$$k_d = k_0 + k_{et}\,(\text{dye}) \tag{4}$$

The rate constants for the heterogeneous electron transfer, k_{et}, can be obtained from the slope of a straight-line plot of k_d versus dye concentration. The rate constants, k_{et}, which vary from 0.7–2.4 $\times 10^9$ M^{-1} s^{-1}, are close to the value for a diffusion-controlled electron transfer reaction. Because the reduction potential of these dyes is positive compared to the flat band potential of WO_3, we expect ΔG to be sufficiently negative to drive the electron transfer by a diffusion-controlled process.

Albery and Bartlett [432, 433] and Grätzel and Frank [434] have shown that the effective second-order rate constant, k_E (cm^3 mol^{-1} s^{-1}), was related to the electrochemical rate constant, k_{et}, as follows:

$$k_E = 4\pi r^2 N k_{et} \tag{5}$$

where r is the radius and N is Avagadro's number. The larger particles are expected to react faster than the smaller particles because of the dependence of the electron transfer rate constant on the surface area. Thus, the discrete range of particle sizes in a colloidal semiconductor system leads to the multiexponential kinetic behavior of the heterogeneous electron transfer because k_{et} is dependent on the particle size. Albery and co-workers [433, 435–437] have proposed a general model that considers a Gaussian distribution of the logarithm of the rate constant about some mean and the width of the distribution as an additional parameter. In the case of colloidal semiconductors, the dispersion in the kinetic transients is caused by the dispersion in the radii. Good agreement for the radial dispersion between the value found from dynamic light scattering and the value found from kinetic analysis has been demonstrated for colloidal semiconductor systems.

However, Darwent and co-workers [23] and Meisel and co-workers [438] have shown that the potential distribution around colloidal oxides coupled with the charge of the reducing species could significantly affect the rate of electron transfer to and from the colloid. Thus, while investigating pH dependence, it is necessary to consider both the thermodynamic driving force and the electrostatic effects. The observed electrochemical rate constant, k_{et}, can be related to the surface charge by the expression,

$$\log k_{et} = \log k_{et}^0 + \alpha n(\text{pH–PZZP}) \tag{6}$$

where k_{et}^0 is the electrochemical rate constant when the surface charge is zero and PZZP is the point of zero ζ potential. Such a model can explain the deviation of results from monoexponential kinetic behavior in several other studies. Good agreement between the calculated rate constants based on electrophoretic data and the experimentally obtained rate constant for the electron transfer between $ZV^{\bullet -}$ (and $MV^{+\bullet}$) and iron (III) oxide colloids at various pH values has been demonstrated in these studies. The manipulation of the surface potential by control of the pH and ionic strength can improve the efficiency of charge transfer processes.

4. PHOTOCATALYTIC APPLICATIONS

In one of the early articles in 1955, Markham [439] pointed out the photocatalytic properties of zinc oxide, titanium oxide, and antimony trioxide with UV irradiation [439]. In aqueous solutions, the holes are scavenged by surface hydroxyl groups to generate $^\bullet$OH radicals that then promote the oxidation of organics [338, 341–344]. The $^\bullet$OH radical–mediated oxidation has been successfully employed in the mineralization of several hazardous chemical contaminants. Interestingly, the electron scavenging becomes a rate limiting factor in such oxidation processes [399, 440–445]. As described in an earlier section, failure to scavenge electrons with a sacrificial electron acceptor results in the promotion of charge recombination. Alternative methods have also been suggested to promote the charge separation by applying an electrochemical bias to a TiO_2 particulate film electrode [444]. On the other hand, both reductive and oxidative steps can be beneficially used to degrade aromatic compounds such as trinitrophenol [446].

4.1. Organic Synthesis

Early contributions in the area of organic phototransformation came from Kato and Masuo [447, 448], who carried out photocatalyzed oxidation of tetralin in UV-irradiated TiO_2 suspensions. Bard and co-workers [341, 449, 450] demonstrated that photocatalytic processes such as the photo-Kolbe reaction and amino acid synthesis can be readily performed with platinized TiO_2 under band gap excitation. Developments in this area have been extensively reviewed by Fox [8, 448, 451–453].

CdS-mediated cis–trans isomerization of several derivatives of styrene have been reported by de Mayo and co-workers [454, 455]. The conversion of cis to trans isomer proceeded via the olefin radical cation intermediate, which was formed at the photoexcited CdS surface. The mechanism of cis–trans photoisomerization of electron-deficient alkenes on CdS has also been investigated in the presence of electron donors [456]. The reaction accompanies the formation of the dihydro compound as a two-electron reduction product. The photoisomerization does not occur at all in the absence of triethylamine (TEA). Two pathways involving the photoreduction of alkenes have been proposed. One is through the back electron transfer from the radical anion of the alkene toward the radical cation of TEA, both formed by photoexcited conduction band electrons and holes of CdS, respectively. The other is the reoxidation of a radical intermediate, formed by protonation of alkenes by the TEA cation radical.

Recently, Li and co-workers [457] have initiated reduction of nitro compounds for synthesizing specialty chemicals. The role of various semiconductors in initiating and controlling organic reactivity has been explored by several research workers. Such transformations on photoexcited semiconductor surfaces include oxidation of olefins, arenes, alkanes, amines, and alcohols; cycloadditions and retrocycloadditions; and geometric isomerizations. For example, double and triple bonds can be photocatalytically hydrogenated by Pt-loaded TiO_2 in the presence of alcohol as an electron donor [401]. Addition reaction of polychlorocompounds to carbon–carbon bonds catalyzed by TiO_2 under photoirradiation is considered to proceed through photoreduction [458]. TiO_2 can also be used as an electron supplier for asymmetric reduction of carbonyl compounds in the presence of chiral Rh(I) complexes as an electron mediator [402]. Photoreduction of aldehydes [403], azo dyes [391, 392], and Shiff bases [404] was observed by TiO_2. The developments in this area have been extensively reviewed by Fox [8, 451, 452].

Freshly prepared CdS suspensions consisting of quantized particles and their loose aggregation catalyze photoreduction of aromatic ketones and olefins in methanol under visible light irradiation using triethylamine as a sacrificial electron donor, yielding alcohols and dihydro compounds, respectively, which are more selective than photocatalysis of commercially available crystalline CdS [405]. Deuterium incorporation experiments in the photolysis of dimethyl maleate in methanol-O-D revealed that CdS^- nanocrystallites catalyze sequential two-electron-transfer photoreduction, affording deuterated dimethyl succinate.

4.2. Fixation of Carbon Dioxide into Organic Compounds

Photoreduction of carbon dioxide (CO_2) to organic compounds is an important topic for fuel production and resolving the greenhouse effect. The development of artificial photosynthetic systems is one of the ultimate goals in the reduction of CO_2. Photoassisted reductions of CO_2 in semiconductor particulate systems have been studied, since Inoue et al. [459] first reported photoreduction of CO_2 with suspensions of particulate semiconductors such as WO_3, TiO_2, ZnO, CdS, GaP, and SiC. However, most semiconductor-catalyzed photoreduction of CO_2 using water as the electron donor suffers from very low photoconversion efficiency and poor selectivity of reduction products. Until now, semiconductor particulate systems have been used to reduce CO_2 to a small range of compounds such as formic acid, CO, methanol, or methane. Photofixation of CO_2 into large organic substrates will be a challenging topic that needs to be tackled in the future.

Photocatalytic fixation of CO_2 into organic compounds can also be achieved using visible light ($\lambda > 400$ nm) irradiation of CdS nanocrystallites suspended in CO_2-saturated DMF solutions and using triethylamine as a sacrificial electron donor [152, 460]. Benzilic acid (BPCOOH), atrolactic acid (APCOOH), and phenylacetic acid (BZCOOH) were produced from benzophenone (BP), acetophenone (AP), and benzyl halides. The formation of a CO_2 anion radical ($CO_2^{-\bullet}$) was confirmed by EPR measurements using DMPO as a spin-trapping agent. Both the formation of $CO_2^{-\bullet}$ and the concurrent one-electron reduction of the organic substrates were evident in the photofixation process. This suggests that the photofixation proceeds via their (bimolecular) coupling on the surface of CdS nanocrystallites. The photocatalytic fixation of CO_2 of pyruvic acid to malic acid is achieved in TiO_2 or CdS particle suspensions using malic enzyme as the catalyst and methyl viologen as the electron mediator [461].

4.2.1. Photocatalytic Reduction of Carbon Dioxide with Metal Oxides

Among the many metal oxides examined as photocatalysts, TiO_2 is most commonly used as a photocatalyst for the reduction of CO_2 in combination with metal complexes working as photosensitizers [462–464]. Ru(II) complexes along with EDTA-bipyridine or EDTA-o-phenanthroline are useful as photosensitizers as they inject electrons from the excited state into the conduction band of TiO_2. These photoinjected electrons further lead to the reduction of CO_2 into HCOOH and HCHO [465]. Similar photosensitization experiments have also been reported in the photocatalytic reduction of CO_2 using chemically bonded eosin on the TiO_2 surface [464]. In the system consisting of TiO_2 and immobilized Pd, the photogenerated conduction band electrons reduce HCO_3^- to formate [466]. However, the fate of the photoproduced holes in these experiments is unclear. A possibility exists for these holes to react with various organic compounds contained as ligands of the complexes or stabilizers such as cyclodextrin. In a recent work, Anpo and co-workers [467, 468] report a Ti–mesoporous zeolite system in which CO_2 is reduced to methanol under UV irradiation. Photocatalyzed CO_2 reduction over pure oxide or oxides doped with other transition metal oxides have also been reported [362, 469, 470]. The production of CO in the system containing ZrO_2 as a photocatalyst and $NaHCO_3$ as an efficient promoter should be interesting because of the stoichiometric evolution of oxygen with respect to reduction products, namely, H_2 and CO [362].

4.2.2. Photocatalytic Reduction of Carbon Dioxide with Metal Sulfides

Metal chalcogenides, especially metal sulfides, constitute another class of the photocatalysts used for CO_2 reduction [471–474]. Although they are attractive as visible light catalysts, metal sulfides suffer from the problem of photocorrosion. Photogenerated holes if not scavenged efficiently induce anodic corrosion. Efforts have been made to overcome this problem using electron donors and/or organic reaction media [475–480]. The colloidal ZnS suspensions effectively catalyze photoreduction of CO_2 in Na_2S solution (pH 7 buffer) [475, 476]. The evolution of H_2 as well as the formation of formate and a very small quantity of CO were observed. An apparent quantum yield of $\Phi_{1/2}(HCOO^-) = 0.24$ was observed at 313 nm, where $H_2PO_2^-$ was quantitatively photooxidized to HPO_3^{2-}. The high efficiency of this system was attributed to the low density of surface defects on quantized ZnS crystallites.

Another system uses ZnS and CdS nanocrystallites prepared in N,N-dimethylformamide (DMF) at 0 °C [151, 479–481]. CO_2 undergoes effective photoreduction in the presence of triethylamine as a sacrificial electron donor. These systems show high quantum efficiency for the reduction of CO_2 under UV light ($\lambda > 290$ nm) irradiation, giving formate and CO. The formation of formate is observed when ZnS–DMF is prepared stoichiometrically [479, 481]. These colloids show a blue emission at approximately 325 nm arising from the conduction band or the shallow electron trap sites of ZnS–DMF. The yield and the lifetime of the blue emission are enhanced by the addition of CO_2 to the system. This enhancement has been explained on the basis of stabilization of photogenerated electrons in the conduction band or at trap sites via the adsorptive interaction with CO_2 molecules on the ZnS–DMF surface. The addition of a zinc ion to this system changes the product distribution without a significant decrease in the photoconversion efficiency or the emission behavior. The competitive formation of CO with formate and the appearance of red emission at approximately 460 nm under continuous light excitation could also be seen. The selectivity of the two products, formic acid and CO, in the photocatalytic reduction of CO_2 can be controlled by the introduction of In^{3+} cations in the CdS colloidal system [153].

CO_2 is reduced to CO on hexagonal CdS nanocrystallites (mean diameter of 4 nm) in DMF under visible light irradiation. A quantum yield of 0.098 at $\lambda = 405$ nm has been observed using triethylamine as a sacrificial electron donor [151, 479, 480]. Solvation of the surface of CdS nanocrystallites by DMF or related amide molecules should control the crystalline growth and stabilize the favorable morphology with a size quantization effect. Addition of Cd^{2+} or H_2S influences the emission behavior and CO production, thereby suggesting the important role that surface structures of CdS–DMF plays in the photocatalytic reduction of CO_2.

One of the important aspects of understanding photocatalytic properties of semiconductor nanoclusters is to establish the surface active sites. Stable suspension of colloidal CdS (CdS–DMF) [152] and ZnS (ZnS–DMF) [476] nanocrystallites showed a relatively high catalytic activity for the photoreduction of CO_2 [151–153, 460, 476, 481–483]. The stability of the metal chalcogenide nanocrystallites prepared in DMF was found to be higher than those prepared in other solvents such as acetonitrile, methanol, and water. This observation supports the fact that adsorbed DMF molecules stabilize the nanocrystallites via effective binding to the semiconductor surface. In the case of CO_2 reduction by metal complexes and electrodes, the adsorptive interaction between CO_2 molecules and the metal center strongly depends on their orbital parentage, the charge of the metal atoms, and the microscopic structures of neighboring molecules. The adsorptive interaction of CO_2 molecules with the surface of semiconductor nanocrystallites also plays a crucial role in determining their photocatalytic activity.

The effect of the surface sulfur vacancy on photocatalytic activity has been examined thoroughly for the CdS nanocrystallites [460]. A remarkable increase in the photocatalytic activity of CdS was achieved by the addition of excess Cd^{2+} to the colloidal system (Fig. 7). The emission behavior, which depends on the amount of excess Cd^{2+} in the system, suggests the formation of sulfur vacancies on the surface of nanocrystallites because of the adsorption of excess Cd^{2+} to the surface. The formation of the sulfur vacancies on the surface was supported by *in situ* Cd K-edge EXAFS analysis of the nanocrystallites in

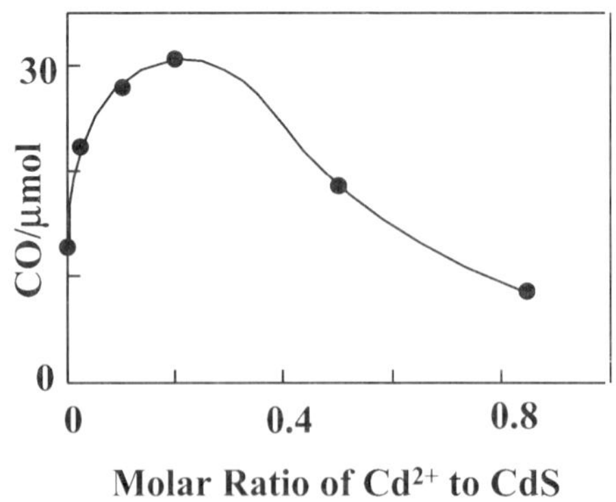

Fig. 7. Effect of the addition of excess Cd^{2+} on the efficiency of CdS–DMF-catalyzed CO formation. (Source: Reprinted with permission from [460]. © 1994 American Chemical Society.)

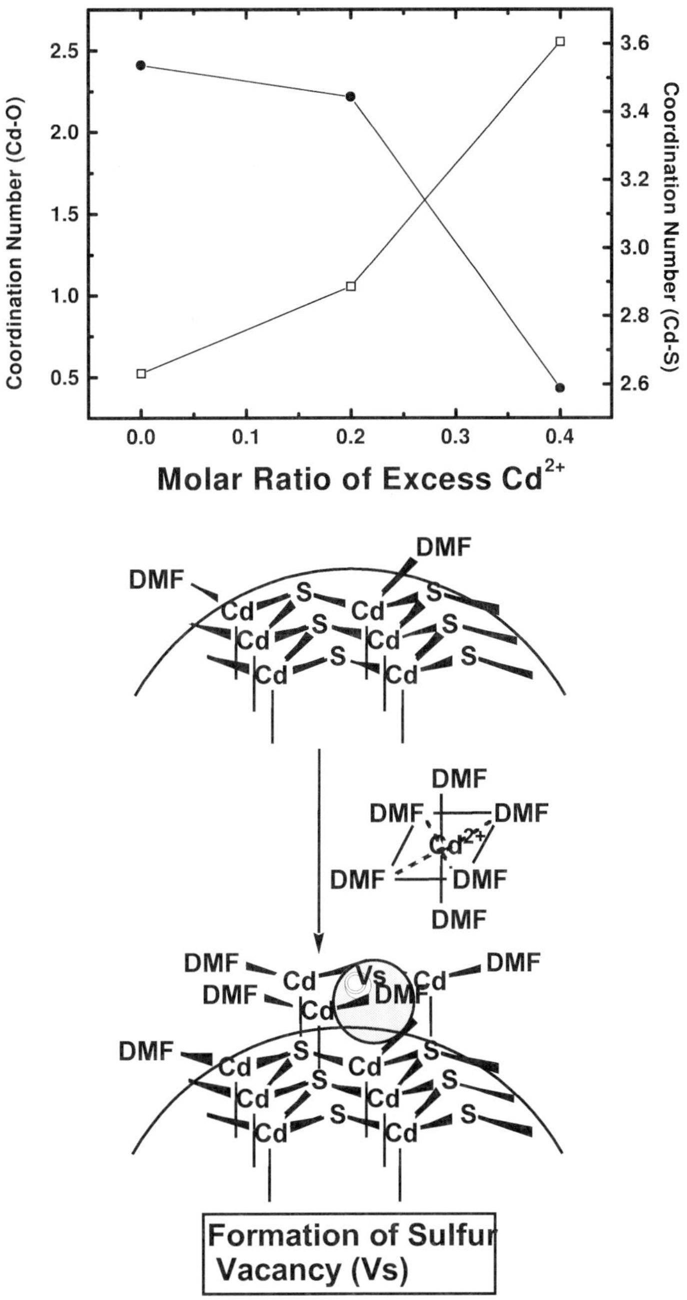

Fig. 8. Effect of the excess Cd^{2+} addition on the coordination number of the Cd–O (•) and Cd–S (○) shell obtained by EXAFS measurement. (Source: Reprinted with permission from [460]. © 1994 American Chemical Society.)

solution as changes in the coordination numbers of cadmium–sulfur and cadmium–oxygen (Fig. 8). Theoretical molecular orbital calculations using a density-functional method support the preferential adsorptive interaction of a CO_2 molecule with a Zn atom in the vicinity of a sulfur vacancy on hexagonal ZnS. Furthermore, the calculations for CdS models suggest the preferential bidentate-type absorption of CO_2 with the Cd atom in the vicinity of the sulfur vacancy.

4.3. Reduction of Nitrogen

Nitrogen fixation with irradiated TiO_2 semiconductor nanoparticles in aqueous solution was confirmed from the production of NH_3 [484]. The hydrous oxides of samarium(III) and europium(III) are found to catalyze the photoreduction of dinitrogen in aqueous suspensions [485]. The chemisorption of nitrogen and the sufficiently negative flat band potential of oxides were found to play an important role in determining the efficiency of photocatalytic reduction for TiO_2 systems. Photosensitized reduction of nitrogen to ammonia in the presence of sacrificial electron donors has been attempted by Taqui-Khan and co-workers [463] using a Ru(II) complex–TiO_2 system. The complex containing nitrogen as a ligand has also been reduced by the photogenerated electrons in Pt/Cd, giving ammonia [486]. This is another approach for nitrogen fixation using semiconductors.

4.4. Decomposition of Nitrogen Oxides and Their Anions

Conversion of nitrogen oxides to less toxic compounds is also important from the viewpoint of global environmental pollution. Nitrogen oxides can be converted to N_2 and other nitrogen compounds by reduction. CdS catalysts can reduce nitrite to ammonia using sodium sulfate and sodium sulfite as sacrificial agents in the photoefficiency of 2.6% [487]. The activity is enhanced by loading noble metals, such as Ru, Pd, or Ir, or by the use of hole-transferring agents, such as RuO_2. TiO_2 prepared by the sol–gel method also reduces nitrate and nitrite ions to ammonia at higher activity than TiO_2 prepared by other conventional methods [488].

Anpo and Yamashita [468, 489, 490] discovered that TiO_2-loaded zeolites and the vanadium silicalite-1 decompose NO under irradiation. Although the vanadium silicalite-1 requires propane as a reducing reagent [490], TiO_2 included in zeolite cavities results in complete decomposition into N_2 and O_2 without any reducing reagent [468]. Both the nitrogen fixation and the decomposition of nitrogen oxides have just appeared on the scene as a new aim for research in the field of semiconductor photocatalysis. Accumulation of much knowledge will promote the application of semiconductors to new reaction systems.

4.5. Photocatalytic Degradation of Organic Contaminants

The photocatalytic properties of anatase TiO_2 particles in degrading undesirable organics from air and water are well documented [15, 17, 18, 491–494]. Organic materials such as hydrocarbons, haloaromatics, phenols, halogenated biphenyls, surfactants, and textile dyes in TiO_2 slurries have been successfully mineralized. For reactor applications, it is convenient to immobilize the semiconductor particles on a suitable substrate. Several studies have been reported with TiO_2 particles immobilized on glass substrates [492, 495–498].

The oxidation reactions are usually initiated by the hydroxyl groups generated at the semiconductor surface. The reaction mechanism leading to mineralization of aromatic molecules is a multistep process and the pathways are often complex to analyze. A detailed discussion on this topic is beyond the scope of the present chapter. The reader can refer to recent review articles for the developments in this area [1, 6, 12, 14–18, 491–493, 495, 499–511].

In a single-crystal semiconductor (n-type)-based photoelectrochemical cell, the problem of charge recombination is easily overcome by applying an anodic bias as was first demonstrated by Fujishima and Honda [512]. Using a single crystal of TiO_2, they were able to carry out the photoelectrolysis of water under the influence of an anodic bias. Thin semiconductor particulate films coated on a conducting surface provide a convenient way of manipulating the photocatalytic reaction by electrochemical methods [444, 492]. We have thus succeeded in achieving better charge separation by applying an anodic bias to the immobilized semiconductor nanocrystallites [444, 513, 514]. In an electrochemically assisted photocatalytic process (ECAP), the externally applied anodic bias greatly improves the efficiency of charge separation by driving the photogenerated electrons via the external circuit to the counterelectrode compartment.

The photocatalytic studies of 4-chlorophenol degradation provide a representative example of the use of ECAP in elucidating the mechanism of degradation [444, 513]. By controlling the applied bias potential, it is possible to control the degradation rate. The photocatalytic degradation occurs at a faster rate when the applied potential is maintained at +0.6 V, while little degradation is seen when the potential is maintained at −0.6 V versus saturated calomel electrode (SCE). Because the charge separation in the TiO_2 particulate film increases when an anodic bias is applied to the optically transparent electrode (OTE)/TiO_2 electrode, one observes a higher efficiency for photocatalytic degradation. At potentials close to the flat band potential (−0.6 V versus SCE), all the electron–hole pairs are lost in the recombination process and, hence, it is not possible to carry out the oxidation of 4-chlorophenol. In a slurry system, the irradiated particles behave as short-circuited microelectrodes and thus the interfacial charge transfer competes with the charge recombination process. This situation closely resembles the experimental conditions in which the OTE/TiO_2 is maintained at 0.0 V. Thus, nanostructured semiconductor films are useful in carrying out electrochemically assisted photocatalysis and overcome the limitation of electron scavenging, which one encounters in the slurry system. A similar technique has also been extended to several other systems [446, 515–517].

The advantage of using the ECAP technique is not just limited to the faster degradation rates. The use of an anodic bias to separate the charge carriers obviates the need for oxygen as an electron scavenger and makes it possible to carry out the photocatalytic reaction in anaerobic conditions. The electrochemical arrangement also provides unique opportunity to separate the anodic and cathodic processes and thereby isolate the various reactions occurring in photocatalytic systems.

5. SURFACE MODIFICATION OF SEMICONDUCTOR COLLOIDS

5.1. Deposition of Metals on Semiconductors

The selectivity and efficiency of a photoelectrochemical reaction can be improved by modifying the surface of the semiconductor particle with a suitable noble metal. For example, an order of magnitude enhancement in the efficiency of photocatalytic hydrogen production from water has been achieved upon platinization of semiconductor particles. The metal deposit on the semiconductor acts as a sink for the photogenerated electron and catalyzes the production of hydrogen. Platinization can be done either by direct photoreduction of $PtCl_6^{2-}$ on the semiconductor or particle [518, 519] or by stirring the suspensions of semiconductor colloids and platinum colloids together [256, 520–524]. The effect of platinum loading on the photocatalytic activity of semiconductor particles has also been investigated [525]. Photoinduced oxidation of bromide to bromine on irradiated platinized TiO_2 powders and platinized TiO_2 particles supported in Nafion films has also been carried out [526].

Other noble metals and metal oxides (e.g., RuO_2) have also been deposited on semiconductor particles to improve the efficiency of the photocatalytic reaction [527–534]. The effect of various transition metal ion doping on the dynamics of photoinduced charge transfer processes in colloidal semiconductor systems has been probed [535–539]. RuO_2 clusters in the native form [540–543] and deposited on semiconductor particles [465, 527, 529, 530, 533, 534, 544–546] improve the efficiency of photocatalytic redox processes such as H_2 production from water or H_2S and fixation of nitrogen and carbon dioxide. Similarly, In_2O_3 particles have also been found to be useful for oxygen generation [547]. Efforts have been made to bind semiconductor nanoclusters to metal surfaces using a self-assembled monolayer approach [58] or to synthesize multilayered metal nanoclusters [548]. Recently, an

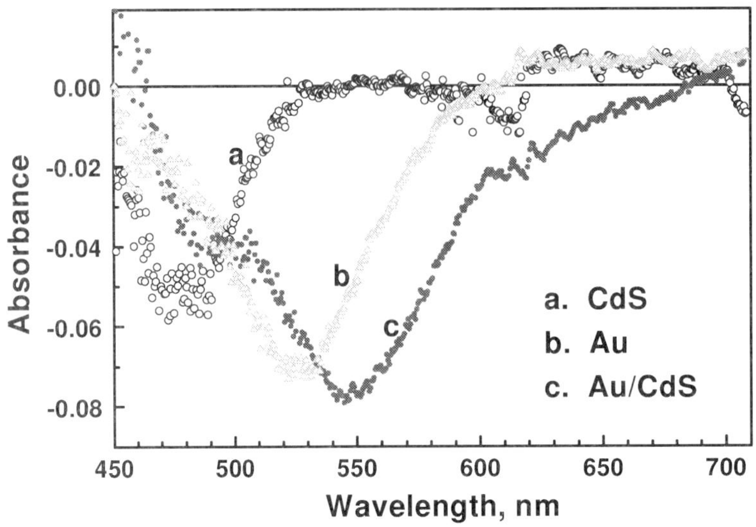

Fig. 9. Plasmon bleach of the CdS/Au system. (Source: Reprinted with permission from [460]. © 1994 American Chemical Society.)

attempt has been made to investigate the optical properties of gold-capped CdS composite nanoclusters [549]. The photophysical study of metal clusters continues to reveal new interesting properties and the reader is refered to some of the recently published articles [11, 45, 222, 550–563].

The kinetics and the mechanism of interparticle electron transfer between semiconductor and metal nanocrystallites were elucidated using picosecond laser flash photolysis. Capping of gold colloids with ultrasmall CdS nanoclusters (particle diameter ~4 nm) significantly alters the picosecond dynamics of the gold core. The bleaching of the surface plasmon absorption of the gold core is achieved by exciting the CdS shell (Fig. 9). The major fraction of the interparticle electron transfer between CdS and Au nanoclusters is completed within the laser pulse duration of 18 ps.

5.2. Capping with Organic and Inorganic Molecules

Redox species that strongly interact with a semiconductor surface are of great interest in improving the performance of photoelectrochemical activity of the semiconductor system [564, 565]. Such interactions are also useful in controlling the size of the colloidal particle as observed for the colloidal CdSe and CdS. Steigerwald et al. [50] have monitored the CdSe nanocrystal growth in the presence of added reagents and passivation by the addition of organoselenides. Similarly, surface-capped CdS colloids have been prepared in the presence of various thiols [62, 63, 133, 142, 566–569]. For example, Herron et al. [54] have shown that CdS clusters in the quantum confinement regime (particle diameter <50 Å) can be prepared by adjusting the ratio of sulfide to thiophenol during synthesis. The competitive reaction chemistry between the growth of the CdS core and the surface capping by thiophenolate leads to a CdS core whose surface has been passivated by covalently attached phenyl groups. Figure 10 shows the absorption spectra of CdS colloids prepared with and without a surface-capping procedure. Fox and co-workers have investigated the optical and electrochemical properties of cadmium benzenethiolate clusters as a function of cluster size [570, 571]. These clusters, which show structured absorption, represent molecular models for metal chalcogenides. Similarly, amorphous polysilsesquioxanes [115], aminocalixarenes [184], charged polymers [83, 207, 229, 425, 427, 572, 573], and surfactants [574] have also been employed to control the particle size in the quantum

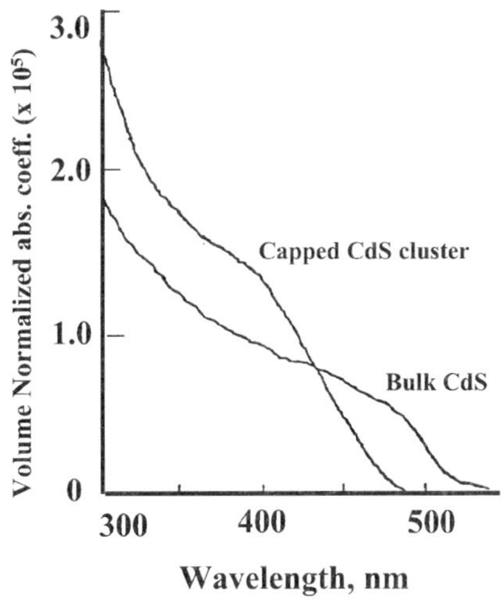

Fig. 10. Absorption coefficients of 30-Å CdS/SPh cluster solid. Also shown is the coefficient of bulk CdS in a medium of dielectric constant 30, calculated with the Lorentz–Mie formula. (Source: Reprinted with permission from [54]. © 1990 American Chemical Society.)

size regime. The preparation of MoS_2 nanoclusters has also been facilitated by capping with a layer of dialkylthiophosphate [575].

Significant efforts have also been made to functionalize CdS colloids with electron donors and acceptors. These include viologen derivatives [62, 385], ferrocene [576], amines [172, 577, 578], and diethyldithiocarbamate anion [412]. A bifunctional mercaptocarboxylic acid has also been employed to bridge CdS and TiO_2 particles [65]. The interaction between the semiconductor surface and the surface modifier often alters their absorption characteristics. These spectral changes include the displacement or broadening of the absorption bands, the appearance of new charge transfer bands, and changes in the extinction coefficient of absorption. For example, surface complexation of a TiO_2 colloid with SCN^- results in the appearance of a new absorption band at 285 nm [417]. Recently, Lauerhaas and Sailor [297] have investigated the effect of adsorbates on the quenching of porous silicon luminescence. The visible luminescence from porous silicon has been found to be highly surface sensitive, and the surface interactions can be tuned by specific chemical transformations.

Enhancement in the emission yield has been achieved by modifying the colloidal CdS surface with triethylamine [172, 578, 579], indole [286], and mercaptoethylamine [567] or by activation of the surface with Ag_2S or Cd^{2+} ions [66, 269]. Similarly, coating the CdS particles with a $Cd(OH)_2$ layer resulted in a strong enhancement of the emission yield and stabilized against photocorrosion. Surface defects, which promote radiationless recombination of the charge carriers, can be blocked upon such modifications.

A decrease in the emission lifetime of semiconductor colloids observed when an electron acceptor or a surface modifier is adsorbed onto the semiconductor surface. For example, the average emission lifetime of CdS colloids in acetonitrile at 640 nm decreased from 156 to 26 ns when the semiconductor surface was modified with the diethyldithiocarbamate anion [412]. It has also been shown that viologen compounds, thiazine dyes, or ethylenediaminetetraacetate [375] can efficiently quench the emission of CdS colloids, and by using Poisson statistics one can determine the number of molecules aggregated in the CdS colloid [150, 173, 196, 580]. A decrease in the emission lifetime observed upon adsorption of Cu^{2+}, I_3^-, and S^{2-} ions on a CdS surface has been explained on the basis of increased surface recombination velocity [312, 581–583]. However, in the case of ZnO colloids, hole scavengers such as I^- and SCN^- quenched the green emission as they interacted with the surface anionic vacancies [206].

5.3. Surface Modification with Sensitizing Dyes

Another important aspect of surface modification of a semiconductor colloid is to extend its response toward low-energy excitations. This is often achieved by attaching sensitizer molecules by direct adsorption or covalent linkage. The interaction between the sensitizer and the semiconductor surface can be probed with the electronic absorption and emission spectra of the sensitizer as the energetics of its ground and excited states are altered.

Extensive efforts have been made in various laboratories to modify the semiconductor surface with various dyes. The process of charge injection from the excited state into the conduction band of the semiconductor has been studied with respect to extending the photoresponse of large-band-gap semiconductor materials and has applications in imaging science and solar energy conversion [39, 584–589]. The energy level diagram, which compares the band energies of semiconductors and the oxidation potentials of excited states, is illustrated in Figure 11. Spectroscopic techniques such as emission, resonance Raman, diffuse reflectance, microwave absorption, and nanosecond and picosecond laser flash photolysis are often employed to probe the interfacial charge transfer between an excited sensitizer and the semiconductor particle.

Dyes such as erythrosin B [590], eosin [464, 591–594], rose bengal [389, 595], rhodamines [596–600], cresyl violet [601–603], thionine [604], chlorophyllin [605, 606], anthracene-9-carboxylate [607], 8-hydroxyquinoline [608], porphyrins [609], phthalocyanines [423, 610], perylenes [611–613], transition metal cyanides [614, 615], $Ru(bpy)_3^{2+}$ and its analogs [38, 339, 377, 465, 588, 616–621], cyanines [587, 622–629], squaraines [352, 630, 631], and phenylfluorone [632], which have high extinction coefficients in the visible, are often employed to extend the photoresponse of the semiconductor in photoelectrochemical systems. Visible light sensitization of platinized TiO_2 photocatalyst by surface-coated polymers derivatized with ruthenium tris(bipyridyl) complex has also been attempted [633, 634]. Because the singlet excited state is short lived, it becomes essential

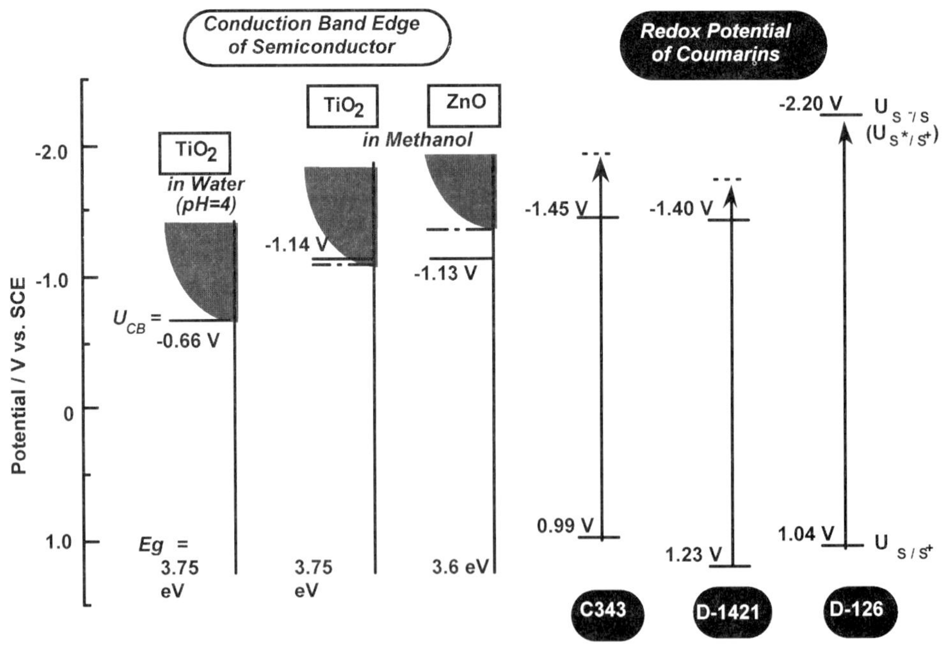

Fig. 11. Energy diagram of TiO_2 and ZnO nanocrystallites in methanol or water. Redox potentials of coumarin dyes, C343, D-1421, and D-126, are also shown. Absorption coefficients of 30-Å CdS/SPh cluster solid. Also shown is the coefficient of bulk CdS in a medium of dielectric constant 30, calculated with the Lorentz–Mie formula. (Source: Reprinted with permission from [612]. © 1996 American Chemical Society.)

to adsorb these dyes on the semiconductor surface with an electrostatic, hydrophobic, or chemical interaction so that electron transfer can occur within the lifetime of the singlet excited state. The sensitization of semiconductor crystallites (e.g., silver halides) by dye aggregates has important applications in imaging science. The charge injection from an excited dye aggregate into the semiconductor crystallite is an important step in latent image formation. Internal reflection spectroscopy has been employed by Natoli et al. [625] to probe the mechanism of sensitization of a ZnO single-crystal electrode with J aggregates of cyanine dyes. Quantum efficiencies up to 0.2 were observed for sensitized photocurrent generation in these systems. Similarly, the interaction between J aggregates of squaraine dyes and semiconductor nanocrystallites has been probed by both spectral and photocurrent measurements [630, 631]. The dependence of quantum efficiency (ϕ_{et}) of photoinduced electron transfer from sensitizing dyes (cyanine dyes) to silver bromide microcrystals on the sensitizer reduction potential has been discussed [627, 635, 636]. The dependence of ϕ_{et} on the reduction potential of the dye has been explained on the basis of Marcus theory with a small rearrangement energy (0.05 eV) and without the inverted region. Spitler [636] has employed a quantitative photoconductive model to explain the data of Tani et al. [627]. The model predicts an energetic threshold for spectral sensitization, correlating a high speed of spectral sensitization of a dye with a highly negative electrochemical reduction potential. Temperature-independent heterogeneous electron transfer reactions in the inverted Marcus region have been reported for the interfacial electron transfer between the conduction band of a TiO_2 nanocrystallite and an oxidized sensitizer [637]. Recently, efforts have also been made to understand the role of H aggregates of cresyl violet [638–643] and rhodamine 6G [644] in sensitizing SnO_2 colloids and thin films. The efficiency of net charge transfer was significantly lower than that observed for corresponding monomers. This was attributed to fast intersystem crossing of the excited singlet state and back electron transfer processes.

Coadsorption of a photosensitizer, such as ruthenium(II) trisbipyridyl complex, $Ru(bpy)_3^{2+}$, and an electron transfer on the colloid surface opens up the possibility of initiating semiconductor-mediated electron transfer. For example, optical excitation of $Ru(bpy)_3^{2+}$ results in the population of a relatively long-lived metal-to-ligand charge transfer (MLCT) excited state ($\tau = 580$ ns in aqueous solution at room temperature), which is simultaneously more reducing and more oxidizing than the precursor ground state [645–651]. On a semiconductor surface, electrons are injected from the excited $Ru(bpy)_3^{2+}$ and subsequently transferred to an electron acceptor such as an H aggregate of cresyl violet [639]. A similar strategy of using a semiconductor nanoparticle as a mediator for transporting electrons between an electron donor and acceptor has been demonstrated in several other systems [607, 652, 653].

Other techniques have also been employed to obtain mechanistic information on the interfacial charge transfer process. The mechanism of cation radical formation in the colloidal TiO_2–eosin system has also been established by time-resolved resonance Raman spectroscopy [592, 594]. Diffuse reflectance laser flash photolysis [389, 654–658] and microwave absorption [351, 595, 605] studies have also been carried out in our laboratory to probe the charge injection process. The net charge transfer efficiency in the case of the eosin–TiO_2 system is approximately 38% [591]. However, much smaller ($<1\%$) efficiencies have been observed when anthracene-9-carboxylic acid [607] or chlorophyllin [605] is chosen as a sensitizer.

Very few examples of a triplet excited state being the electron donor in the charge injection process exist in the literature. These include eosin (EO)–TiO_2 [593], thionine (TH^+)–ZnO [604], chlorophyllin–TiO_2 [606], cresyl violet aggregate–TiO_2 [659], and squaraine–TiO_2 [641] systems. The difference in reactivity of the singlet and triplet excited states in different colloidal semiconductor systems depends on the interaction between the sensitizer and the semiconductor [641, 659, 660]. When the sensitizer is strongly bound to the semiconductor with a strong electronic coupling (e.g., charge transfer complexation), the singlet excited state dominates the charge injection process. And when the interaction between the sensitizer and the semiconductor is very weak, one can expect the triplet excited state to participate in a diffusion-controlled charge injection process.

The sensitizer molecules adsorbed on the TiO_2 surface have a significantly shorter fluorescence lifetime than in the homogeneous solution and this decrease in lifetime has been attributed to the charge injection process [339, 597–599, 601–603, 606, 621, 628, 661–664]. Heterogeneous electron transfer rate constants in the range of 10^7 to 10^{11} have been reported in these studies. Microwave absorption and luminescence decay measurements have been independently carried out to monitor the charge injection from excited $Ru(bpy)_2(dcbpy)^{2+}$ into SnO_2, ZnO, and TiO_2 nanocrystallites [351]. Because microwave conductivity arises as a result of mobile charge carriers within semiconductor particles, it is possible to probe the charge injection process by monitoring the growth in the microwave absorption [595].

5.4. Ultrafast Charge Injection into Semiconductor Nanocrystallites

Excitation of sensitizing dyes adsorbed on semiconductor nanoclusters results in the production of a dye cation radical ($S^{\bullet +}$). Picosecond and femtosecond laser flash photolysis experiments have shown that the formation of the dye cation is an ultrafast process and occurs within a few hundred femtoseconds to several picoseconds. Recently, the dynamics of charge injection from an excited sensitizer into semiconductor nanoclusters has been characterized by femtosecond laser flash photolysis and emission experiments [612, 643, 665–669]. Ford and Rodgers have elucidated the kinetics of nitroxyl radical oxidation by the oxidized sensitizer $Ru(bpy)_3^{3+}$ following photosensitization of antimony-doped tin oxide colloidal particles [670].

Dye sensitization from photoexcited strongly anchored dye molecules on the semiconductor surface often achieve a quantum efficiency close to unity [671]. Theories have been developed to explain the ultrafast rate and the mechanism of the phenomenon at the interfaces [419, 671, 672]. It was proposed that the description for the process at the heterogeneous interface is directly related to Fermi's golden rule when nuclear degrees of freedom are neglected. The rate of electron transfer can be related to the electronic density of states and the electronic coupling of the system. The quasicontinuum electronic states in the semiconductor conduction band accept electrons from excited dye molecules and thus contribute to delocalization of the electronic wave function via electronic dephasing. The electronic dephasing that occurs within a sub-50-fs time scale is significantly faster than the typical relaxation of the nuclear degrees of freedom for most molecular species ($\sim$100 fs) in a homogeneous system. A time scale of 10–50 fs was suggested in the theoretical estimation for electron transfer at a heterogeneous interface [673]. In spite of the advantage of providing a well-defined, flat, and smooth surface, the optical signals of time-correlated single-photon counting from an ordered monolayer of oxazine SnS_2 led to a relatively low signal-to-noise ratio. Direct determination of the electron transfer time as well as characterization of the adsorbed structure seems to be difficult in this system. Recent progress in femtosecond laser spectroscopy has allowed us and others to observe such a fast electron transfer event on the semiconductor surface [673]. McLendon et al. [666] reported a charge injection time of less than 200 fs for a coumarine dye, C343, adsorbed on TiO_2 nanocrystallites. Willig et al. [612] showed that the perylene dye adsorbed on a TiO_2 porous electrode had a time of less than 190 fs in ultrahigh vacuum (UHV). They minimized the dye aggregation on the semiconductor surface by functionalizing the dye with phosphonate groups. Durrant et al. [667] reported a lifetime of less than 150 fs for the Ru complex (*cis*-di(thiocyanato)-*N,N*′-bis(2,2′-bipyridyl-4,4′-dicarboxylicacid)-ruthenium(II)dihydrate, $\{RuL_2(NCS)_2\}\cdot 2H_2O$) on a TiO_2 porous electrode. Fast injection from a natural dye of blackberries exceeding 100 fs was also reported by Zhang and co-workers [674].

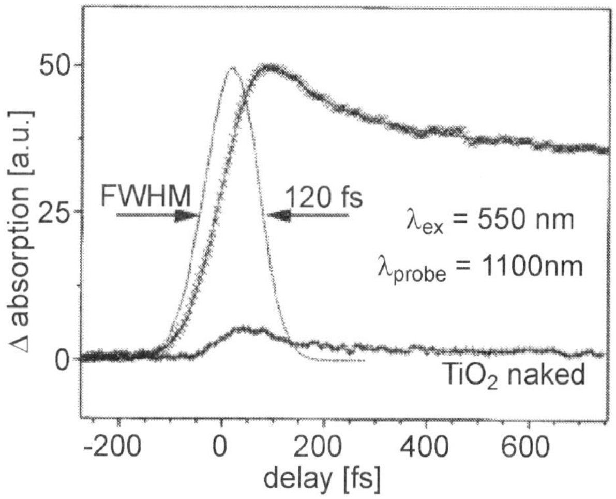

Fig. 12. Transient absorption signal detected at 1100 nm for the Ru-dye–covered colloidal TiO_2 electrode mounted in ultrahigh vacuum. This signal is due to light absorption by the injected electrons. The instrumental function is shown with 120 fs half-width. (Source: Reprinted with permission from [669]. © 1997 American Chemical Society.)

In a recent study, Willig et al. [669] observed a transient absorption signal of the injected hot electron with a rise time less than 25 fs for the system of ruthenium(II) *cis*-di(isothochanato)-bis(2,2′-bipyridyl-4,4′-dicarboxylate)/TiO_2 porous nanocrystalline electrode in ultrahigh vacuum (Fig. 12). They have used the UHV condition to measure the temperature dependence on the rate of the charge injection process. A fast charge injection on the time scale of 10–50 fs was also available for the system involving several coumarine dyes on TiO_2 and ZnO nanocrystallites. Murakoshi et al. [675, 676] also observed a charge carrier injection rate of less than 20 fs for the coumarine 343 adsorbed on 53-Å-diameter TiO_2 nanocrystals that were suspended in water. The dependence of the band energies of semiconductor nanocrystallites on the solvents and redox potential of excited dye molecules indicated that the ultrafast electron transfer injection rate should not only occur because of the strong adsorption, but also because of large overlap between the conduction band levels and the donor states of the excited dye molecules.

Based on the fact that injection at TiO_2 is relatively slower than that at ZnO in methanol, the effect of the density of states of the conduction band as well as the conduction band edge due to proton concentration was considered (see Fig. 11 for an energy level diagram). The density of states for the conduction band of TiO_2, originating from high-density d orbitals, is larger than that of ZnO, originating from relatively simple s orbitals, by more than two orders of magnitude. Such a high density of states of TiO_2 should increase the states overlapping for electron transfer. They concluded that the electronic orbital character of the semiconductor is also an essential factor in determining the efficiency of dye sensitization.

5.5. Designing Multicomponent Semiconductor Systems

The feasibility of coupling a metal oxide nanocrystallite with one or more semiconductors by chemical precipitation or electrochemical deposition opens up a wide array of possibilities in utilizing these materials for microelectronics. For example, in a recent study, we have shown that it is possible to rectify the flow of electrons in nanostructured TiO_2 film by coating a thin CdSe film [677]. Uekawa et al. [678] have also succeeded in developing a p–n junction diode. The advantage of using coupled semiconductor films in photoelectrochemical cells is twofold: (i) to extend the photoresponse by coupling a large-band-gap

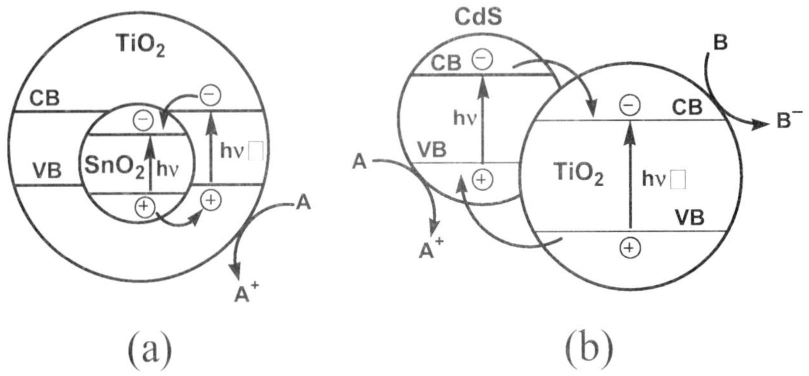

Fig. 13. Photoinduced charge separation in semiconductor heterostructures: (a) coupled and (b) capped semiconductor nanocrystallites. Photogenerated charge carriers move in the opposite directions, thus facilitating charge rectification.

semiconductor with a short-band-gap semiconductor and (ii) to retard the recombination of photogenerated charge carriers by injecting electrons into the lower-lying conduction band of the large-band-gap semiconductor.

Both capped and coupled geometries have been employed to develop semiconductor heterostructures. The principle of charge separation in coupled and capped semiconductor systems is illustrated in parts a and b of Figure 13, respectively. While the mechanism of charge separation in a capped semiconductor system is similar to that in coupled semiconductor systems, the interfacial charge transfer or charge collection at this multicomponent semiconductor system is significantly different. In coupled semiconductor systems, the two particles are in contact with each other and both holes and electrons are accessible for selective oxidation and reduction processes on different particle surfaces. On the other hand, capped semiconductors have a core–shell geometry. If the shell is relatively thick, a charge rectification occurs and only one of the charge carriers is accessible at the surface. This makes selective charge transfer possible at the semiconductor electrolyte interface. The other charge carrier (i.e., the electron in Fig. 13b) gets trapped within the inner semiconductor particle and is not readily accessible.

5.5.1. Coupling of Two Semiconductor Colloids

A variety of interesting properties of such coupled semiconductor systems have recently been reported by several researchers (see, e.g., [679, 680]). Evidently, the rectification properties of these coupled semiconductor systems are attractive enough to explore their feasibility in designing molecular devices and integrated chemical systems for solar energy conversion. Use of coupled semiconductor particles in improving the photocurrent generation in TiO_2–CdS [680–683], ZnO–CdS [257], TiO_2/CdSe [677, 679], and SnO_2/CdSe [684] systems has been demonstrated. Such coupled semiconductor thin films not only extend the photoresponse of large-band-gap semiconductors but also rectify the flow of photogenerated charge carriers [679] and improve the efficiency of dye sensitization [685].

Electron transfer between two different semiconductor colloids can be achieved by coupling them with or without surface modification. For example, in a colloidal CdS–TiO_2 system, a photogenerated electron can be injected from CdS into a TiO_2 particle while the holes remain in the CdS particle (Fig. 13b). The difference in the energy levels of the two semiconductor systems plays an important role in achieving such a charge separation. The feasibility of sensitizing large-band-gap semiconductor ZnO with a deposit of CdS film has been demonstrated by Gerischer and Lübke [681].

One of the early demonstrations of this concept came from Nozik and co-workers [686], who developed a simple photochemical diode by sandwiching p- and n-type semiconductors. Recently, considerable interest has focused on studying coupled semiconductor systems. These include CdS–TiO_2 [274, 681, 687, 688], SnO_2/TiO_2 [684], CdS–PbS [689], CdS–ZnO [257, 688], CdS–AgI [274], Cd_3P_2–TiO_2 and Cd_3P_2–ZnO [255], ZnO–ZnS [254], ZnO–ZnSe [206], AgI–Ag_2S [195], and CdS/HgS [690]. In our earlier studies, we have shown that excitation of CdS colloids coupled with a metal oxide colloid such as TiO_2 [274] or ZnO [257] results in the injection of electrons into the lower-lying conduction band of the metal oxide.

The charge injection process in a coupled semiconductor system was further demonstrated by carrying out methyl viologen reduction at a TiO_2 semiconductor, but with excitation of a CdS semiconductor in the CdS–TiO_2 system [688]. The efficient quenching of CdS emission and high quantum yield for reduction of MV^{2+} in the presence of colloidal TiO_2 indicated an efficient charge injection process with a quantum efficiency of 100%. Similarly, enhanced efficiency of SCN^- oxidation has been observed in a ZnO–ZnSe coupled semiconductor system [206].

The primary photochemical events associated with the charge separation in CdS–TiO_2 and ZnO systems were also probed with transient absorption measurements [257, 274]. The transient absorption spectra recorded immediately after the laser pulse excitation of CdS in a CdS–TiO_2 system are shown in Figure 14. The broad absorption in the region of 500–760 nm corresponds to the charge trapped at the Ti^{4+} site of a TiO_2 colloid. This transient formation was completed within the laser pulse duration of 18 ps. Recent studies have indicated that such an inter-particle electron transfer occurs within 500 fs to 2 ps [320, 691]. This suggests that the rate constant for the charge injection process is greater than 5×10^{11} s^{-1}.

The improved charge separation in the coupled semiconductor system also enhances the process of hole trapping. This process can be followed by monitoring the absorbance of $S^{\bullet -}_{surf}$. The photocatalytic applications of coupled semiconductor systems have yet to be explored fully. But it is evident that the rectification properties of these coupled semiconductor systems are attractive enough to explore their feasibility in designing molecular devices and integrated chemical systems for solar energy conversion. Use of coupled semiconductor particles for improving the photocurrent generation of thin TiO_2–CdS [682, 683]

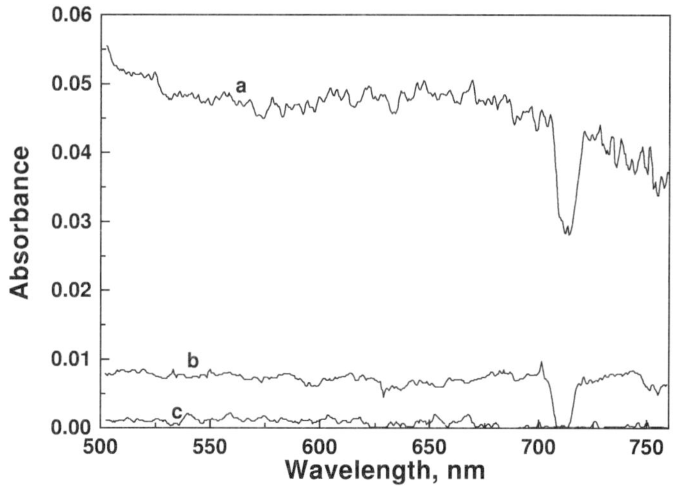

Fig. 14. Transient absorption spectra recorded after 355-nm laser pulse excitation of colloidal suspension ($\Delta t = 0$ ps): (a) 0.75 mM CdS + 1.15 mM TiO_2, (b) 0.75 mM CdS, and (c) 1.5 mM TiO_2. (Source: Reprinted with permission from [274]. © 1990 American Chemical Society.)

and ZnO–CdS [257] films and photocatalytic degradation of an azo dye with TiO_2–SnO_2 film [514] has been demonstrated recently.

5.5.2. *Capped Semiconductor Colloids*

Capped semiconductor particles facilitate an interesting way to change electronic structure by combining the properties of two semiconductor materials. Recently, Brus and co-workers [53] have synthesized and characterized composite semiconductor crystallites involving CdSe grown on a ZnS seed, and vice versa. They have demonstrated nucleation and partial growth of composite particles despite the fact that the two materials have bond lengths differing by 13%. Eychmüller and co-workers [249, 690] have demonstrated interesting fluorescence properties of CdS colloids capped with HgS. Elucidation of the energetic structure and charge carrier dynamics of this system led to the conclusion that its electronic properties are significantly different from the sum of the separate particles. Haus and co-workers [692] have modeled quantum confinement effects in nanometer-size semiconductor particles coated with another semiconductor. By using the band offsets and effective masses, the overlap of electron and hole functions can be altered by changing the core radius of the particles. Development of such semiconductor heterostructures are considered to be interesting because by changing the thickness of the shell and particle radius the overlap of the wave functions and band gap can be altered.

Matijevic and co-workers [693] have prepared spherical ZnO particles coated with uniform layers of TiO_2. By making use of the principle described in Figure 13a, it was possible to enhance the efficiency of SCN^- oxidation in a ZnSe-capped ZnO system [206]. Such a capping process is not only convenient to build tandem semiconductor structures but also provides an economical way to deposit expensive semiconductor material on an inexpensive support.

Charge transfer processes in a TiO_2-capped SnO_2 (TiO_2@SnO_2) and TiO_2-capped SiO_2 (TiO_2@SiO_2) colloids have also been investigated [684]. The effectiveness of SnO_2@TiO_2 colloids as an improved oxidative photocatalyst was demonstrated by oxidizing I^- in a laser flash photolysis experiment. Upon excitation with a 308-nm laser pulse, the photogenerated electrons accumulate at the inner core of SnO_2 and holes at the outer TiO_2 shell. Surface-adsorbed scavengers such as I^- can quickly scavenge away the holes at the interface and forms $I_2^{\bullet -}$ (reaction 7):

$$TiO_2(h_t) + I^- \rightarrow TiO_2 + I^{\bullet}\text{—}(I^-) \rightarrow I_2^- \qquad (7)$$

The absorption bands observed at 380 and 700 nm confirmed the formation of $I_2^{-\bullet}$. The relative quantum efficiency for I^- oxidation at a SnO_2@TiO_2 colloid was two- to threefold higher than SiO_2@TiO_2 or TiO_2 colloid alone (Fig. 15). These results further demonstrate the usefulness of SnO_2@TiO_2 colloids in carrying out photocatalytic oxidation reactions. A simple variation in the synthetic procedure results in significant changes in the photoelectrochemical properties of the multicomponent semiconductor system. If employed in a suitable configuration, multicomponent semiconductor systems can pave the way toward developing materials for microelectronics or photosensitive devices. It should be noted that the capping layer of TiO_2 studied in these experiments is relatively thick and does not exhibit size quantization effects. Such systems differ from other systems [53, 249, 689, 690] in which the capping layer was size quantized (4–5 Å thick).

6. ORDERED NANOSTRUCTURES USING SEMICONDUCTOR NANOCRYSTALLITES AND THEIR FUNCTIONALITY

Considerable effort has been made in recent years to synthesize and characterize the properties of thin semiconductor films prepared from colloidal semiconductor suspensions [694–697]. This method of preparation is relatively simpler and inexpensive compared

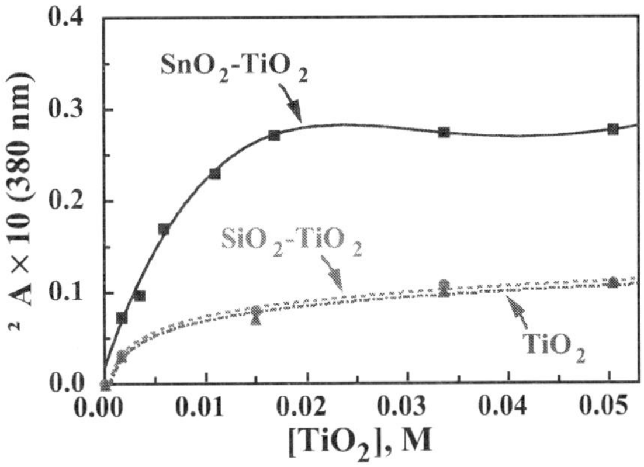

Fig. 15. Dependence of $I_2^{-\bullet}$ yield on the concentration of TiO_2: The maximum absorbance at 380 nm was monitored following the 308-nm excitation of (a) SnO_2@TiO_2, (b) SiO_2@TiO_2, and (c) TiO_2 colloidal suspension containing 0.05 M KI (deaerated with N_2). The concentrations of SnO_2 (29 mM) and SiO_2 (15 mM) were kept constant in the corresponding set of experiments. (Source: Reprinted with permission from [684]. © 1995 American Chemical Society.)

to the other existing methods such as chemical vapor deposition or molecular beam epitaxy. By controlling the preparative conditions of the semiconductor colloid precursors, it is possible to tailor the properties of these semiconductor films. These thin metal oxide films exhibit interesting photochromic, electrochromic, and photocatalytic and photoelectrochemical properties that are inherited from the native colloids.

Preparation of semiconductor nanoclusters in polymer films [80, 83, 84, 203, 227, 229, 532, 572, 698–703] and Langmuir Blodgett films [94, 103, 105, 108, 110, 111, 704–711] have also been considered. The sol-gel technique has been found useful in developing nanostructured semiconductor membranes with either a 2D [712] or a 3D configuration [180, 224, 620, 712–723]. Organic template mediated synthesis has been employed to develop nanoporous tin(IV) sulfide materials [724]. Fabrication and characterization of CdS-nanoparticle mono- and multilayers on a self-assembled monolayer of alkanedithiols on gold has also been reported recently [725].

The nanostructured films are highly porous and can easily be surface modified with sensitizers, redox couples, and other nanostructured semiconductors [143, 257, 339, 513, 629, 630, 638, 677–680, 685, 694, 696, 704, 711, 726–747]. Attempts have been made to sensitize large-band-gap semiconductors such as TiO_2 and SnO_2 with short-band-gap semiconductors [679, 680, 682, 685] and composite semiconductor films with Ru(II) polypyridyl complex [742]. Photoconversion efficiencies in the range of 11% to 15% in diffuse daylight are obtained for the sensitized photocurrent generation using nanostructured TiO_2 films modified with a ruthenium polypyridyl complex [588, 748]. Recent efforts have also focused on utilizing nanostructured semiconductor materials for sensors and microelectronic, photovoltaic, and electrooptic devices. Examples include a p–n junction rectifier using metal oxide films [678], gas sensors using SnO_2 films [749], a Schottky barrier solar cell fabricated from chemically deposited CdSe and Sb_2S_3 films [750], and a light-emitting diode developed from CdSe nanocrystalline film and a semiconducting polymer [751].

6.1. Preparation and Characterization of Nanostructured Semiconductor Films

6.1.1. From Preformed Colloidal Suspensions

The synthetic procedure involves preparation of ultrasmall semiconductor particles (particle diameter 2–10 nm) in aqueous or ethanolic solutions by controlled hydrolysis. A con-

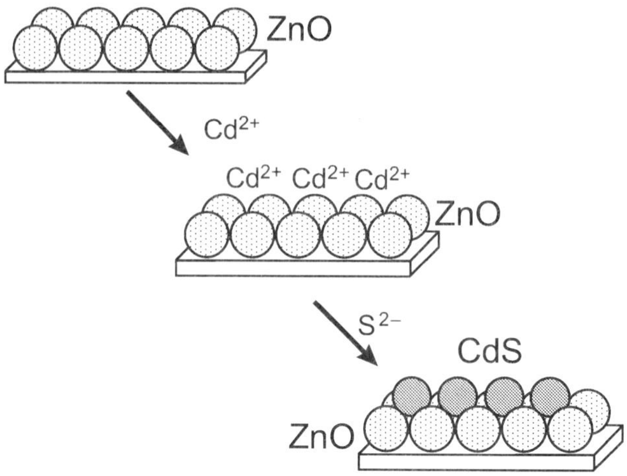

Fig. 16. Designing semiconductor heterostructures with selective adsorption of ions from solution.

centrated TiO_2 is one of the most commonly used semiconductor suspensions for casting thin nanostructured films. A colloidal suspension of SnO_2 (particle diameter 5–15 nm) is also commercially available (Johnson Matthey). The colloidal suspension of the metal oxide semiconductor is coated onto a conducting glass plate [also referred to as an optically transparent electrode (OTE)], and dried on a warm plate. The film is then annealed at 200–400 °C in air for about 1–2 h. This simple approach of coating preformed colloids on a desired surface and annealing produces a thin semiconductor film that is robust with excellent stability in both acidic and alkaline media (pH range 1–13). Usually, one can achieve a thickness in the range of 0.1 to 10 μm. It may be necessary to optimize the concentration of precursor colloid for a particular application because higher colloid concentrations lead to cracking of the film. This procedure can be repeated several times to cast thicker films. (For example, see the references for the methodology of preparing SnO_2 [339], ZnO [257], TiO_2 [444], and WO_3 [752] films.)

Transmission electron micrographs of nanostructured films prepared from colloidal suspensions show a three-dimensional network of SnO_2 nanocrystallites of particle diameter less than 5 nm [339]. All these particles are in direct contact with each other, often resembling a *grape-bunch* type of assembly. No significant particle growth or sintering effects could be seen during the annealing process. X-ray diffraction (XRD) analysis has also confirmed the crystallinity of these nanostructured films.

6.1.2. Chemical Precipitation

Chemical precipitation of the semiconductor directly on the surface of a desired substrate is another convenient approach to growing thin films of II–VI compound semiconductor nanocrystallites (Fig. 16) [257, 682]. For example, by successively dipping the ZnO film in Cd^{2+} and S^{2-} solutions, one can cast a thin film of CdS nanocrystallites. The yellow coloration of the film indicates the formation of CdS on the ZnO surface. Such a chemical precipitation technique is especially useful for designing coupled or multicomponent semiconductor films. The scanning electron microscope picture (Fig. 17) shows the formation of uniform size CdS clusters of diameter approximately 20 nm.

Electrochemical deposition is also a convenient technique for casting thin films of II–VI compound semiconductors, namely, CdS, CdSe, CdTe, and several mixed semiconductor films. Quantum size TiO_2 crystallites have also been deposited on conductive surfaces by anodic oxidative hydrolysis of $TiCl_3$ [753]. The details of the synthesis of electrochemically deposited semiconductor films are discussed elsewhere [679, 694, 726, 754]. Elec-

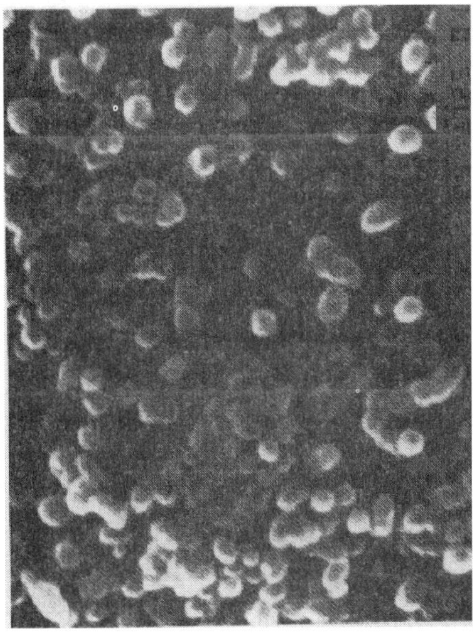

Fig. 17. Scanning electron micrograph of CdS nanocrystallites deposited on a ZnO thin film. The chemical precipitation was carried out by dipping ZnO film in $Cd(ClO_4)_2$ and Na_2S solutions successively. (Source: Reprinted from [685], with permission of Elsevier Science B.V.)

trochemically deposited films are strongly adherent to the substrate and are composed of aggregated nanocrystallites. The nanocrystal size distribution (isolated or aggregated) can be controlled by the deposition current and temperature. Recently, a new hybrid electrochemical/chemical (E/C) method has been employed to synthesize β-CuI nanocrystallites (10–180 Å) on an atomically smooth graphite basal plane surface [755]. The method consists of electrochemical deposition of copper nanocrystallites onto an electrode surface followed by their oxidation to Cu_2O and then displacement of oxygen in Cu_2O by iodide in aqueous solution. A similar method has also been employed to synthesize CdS nanocrystalline thin films [756].

6.1.3. Self-Organized Ordered Structure

Three- or two-dimensional periodic superstructures formed by colloidal crystals can be applied to important technological fields for optical filters [757, 758], switches [118], and photonic band gap materials [759–761]. Recently, methods have been developed for self-assembly of nanocrystal superlattices or nanocrystallite arrays of metal [710, 762–764], semiconductors [110, 765–769], oxides [770, 771], and sulfide clusters [725, 772, 773]. A highly ordered structure of nanometer-sized nanocrystallites is expected to induce collective excitation in ordered superlattice structures, possibly leading to unusual electronic interactions between the adjacent quantum dots [774]. Bawendi et al. [765] reported the self-organization of CdSe nanocrystallites capped with trialkylphosphine chalcogenide into a three-dimensional superlattice. They successfully prepared ordered thin films at solid and liquid interfaces using CdSe quantum dots with diameters of 63, 54, 47, 39, and 35 Å. The distance between the quantum dots was controlled with near atomic precision using various capping reagents. Capping reagents of tributylphosphine oxide (four carbons per chain), trioctylphosphine chalcogenide (eight carbons per chain), and trihexadecylphosphate (16 carbons per chain) gave separation distances of 7, 11, and 17 Å, respectively. Their preparation method onto solid surfaces can be applied to the integration of dot superlattice structures into existing Si-based microelectronic architectures. Nanometer-size templates, such as the liquid-crystalline phase formed from nonionic organic amphiphiles [775] and the bacterial S layer, have also been found useful for the synthesis of superlattices [766]. Use of molecular templates seems to be a promising method to fabricate "tailored" superlattice structures, because the template can be synthesized or modified with a desirable structure using a cell protein monomer via an ordinary organic synthetic method. Fabrication for a scale larger than nanometer size was also achieved using nanotubes of chrysotile asbestos [776], porous aluminum oxide films [777], and commercially available membrane filter [778]. In the near future, novel superlattice structures using colloidal semiconductor nanocrystallites will be synthesized or fabricated for optical/magnetic/electronic devices with ultrasmall ordered structure.

6.1.4. *Surface Modification*

The porous metal oxide semiconductor films prepared from colloidal suspensions have a great affinity for interaction with organic dyes, redox couples, and organometallic complex molecules. This technique is especially convenient for extending the photoresponse of large-band-gap semiconductors with sensitizing dye molecules or for making the nanostructured films electrochemically active [339, 638, 748]. Because of the high porosity of the nanocrystalline film, large amounts of the sensitizing molecules (up to 0.1 M) can be incorporated in a nanocrystalline film of thickness of approximately 1 μm. Significant interaction between the adsorbed molecules leads to the aggregation effects. Similarly, deposition of monomer and aggregate forms of chlorophyll and chlorophyllin on ZnO, TiO_2, and SnO_2 films has been carried out [685, 728, 779]. A monolayer of aromatic amines anchored on nanocrystalline metal oxide films has been used to investigate the lateral electron transport in nanostructured materials [780].

6.2. Electron Storage and Photo- and Electrochromic Effects

Thin films made from metal oxide colloids such as WO_3, SnO_2, and TiO_2 exhibit reversible electrochromic and photochromic effects [183, 752, 781–784]. A blue coloration quickly develops as a result of electron storage at the trap sites when these films are subjected to UV irradiation or to an electrochemical (cathodic) bias. The electrons injected into the conduction band of the semiconductor get quickly trapped at the surface defects. Excess metal ions (e.g., Ti^{4+} in TiO_2) present at these defect sites accept the electrons and the reduced metal ions give rise to the blue coloration.

The changes in the absorption observed in a WO_3 nanostructured film at different applied potentials are shown in Figure 18. The film turns blue as we increase the applied bias. This is indicated by the increase in the absorption at wavelengths greater than 500 nm. The original color of the film can be restored by reversing the applied potential. The onset potential at which the electrochromic effect is observed corresponds to the flat band potential of the corresponding semiconductor and is dependent on the pH of the medium. By monitoring the optical absorbance, Fitzmaurice and co-workers [782, 785–788] have determined the flat band potentials of nanostructured films of TiO_2 and GaAs and the extinction coefficient of the trapped electrons. On the other hand, ZnO [246] and CdS [232] films exhibit bleaching at potentials more negative than the flat band potential. The accumulation of electrons in the conduction band results in the blue shift of the absorption edge (Burstein–Moss shift).

When dry films of WO_3 and TiO_2 were subjected to UV irradiation, they turned blue as a result of electron trapping. Spectrochemical and microwave absorption experiments suggest that trapped electrons are the major species responsible for the blue coloration of the film [729, 752]. Because the trapped electrons have a long lifetime, they provide a convenient method of storing electrons. Investigations of the trapping process by picosecond laser flash photolysis indicate this trapping process occurs on a subnanosecond

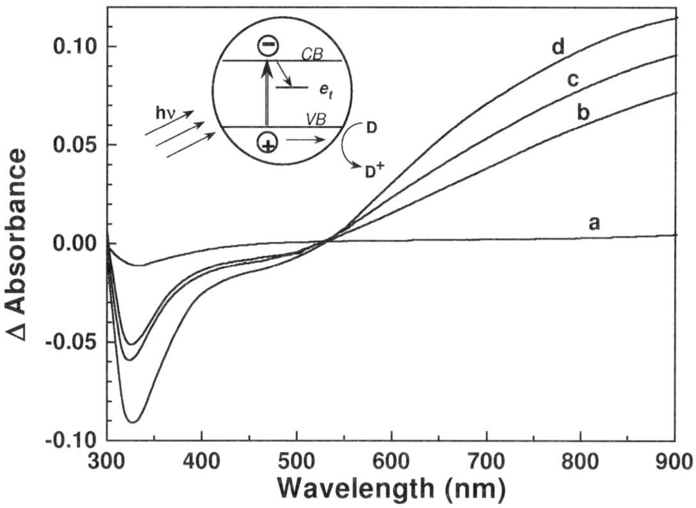

Fig. 18. Photoelectrochromic effect at WO_3 particulate films. (Source: Reprinted with permission from [752]. © 1994 American Chemical Society.)

time scale [183]. Annealing at higher temperature (>473 K) destroys the surface traps that are responsible for trapping photogenerated electrons [718, 752]. The electron storage effects in such semiconductor nanostructures have potential applications not only in electrochromic devices, but also in building electronic devices such as microcapacitors.

6.3. As a Photosensitive Electrode

A nanostructured semiconductor film is capable of generating photocurrent when employed as an electrode in a photoelectrochemical cell. The operation of a photoelectrochemical cell employing nanostructured semiconductor film as a photoanode is shown in Figure 19. The nanostructured semiconductor films exhibit photoelectrochemical characteristics similar to those of a polycrystalline semiconductor material. Under illumination with UV light, these metal oxide films undergo charge separation to form electron–hole pairs. Because the formation of an ideal space charge layer in semiconductor nanocrystallites with particle diameters less than 10 nm is less likely, the differing rates of electron or hole transfer into the electrolyte results in the accumulation of one of the charge carriers within the film [339, 444, 679, 726, 731, 789]. For example, one can accumulate electrons in a nanocrystalline TiO_2 film if photogenerated holes are scavenged by hydroxide ions. The varying degree of electron accumulation changes the pseudo-Fermi level and creates a potential gradient to drive away the electrons toward the collecting surface. The photocurrent generation as represented by IPCE (incident photon-to-current efficiency) in SnO_2, TiO_2, and SnO_2/TiO_2 films is compared in Figure 20. The onset of photoresponse in these films corresponds to the band gap energy of respective semiconductors.

The photoelectrochemical properties of these films are susceptible to surface adsorbed species such as oxygen [444], charge recombination at the grain boundaries [339, 731], and leakage of electrons into the solution instead of generating photocurrent [679]. However, a properly designed semiconductor particulate system and a suitably matched redox couple can greatly improve the efficiency of charge separation and thus the photoelectrochemical performance of the nanostructured semiconductor film.

The generation of photocurrent as a result of charge separation is strongly dependent on the applied potential. The applied anodic bias shifts the pseudo-Fermi level in such a way that it creates a potential gradient within the nanocrystalline film and drives away the photogenerated electrons and holes in opposite directions. With increasing negative

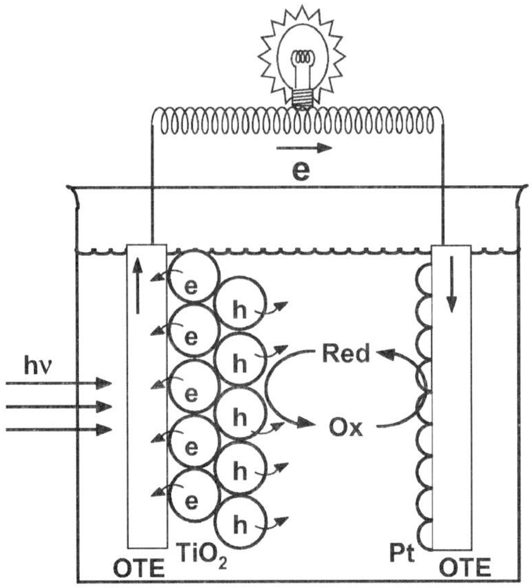

Fig. 19. Mechanism of photocurrent generation in a nanostructured semiconductor film. The electrons accumulated in TiO_2 nanocrystallites are collected at the optically transparent electrode surface and are transferred to the counterelectrode via the external circuit.

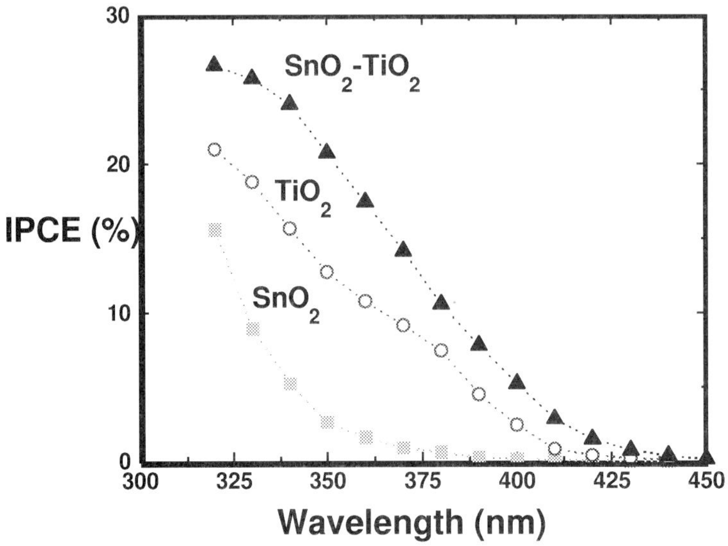

Fig. 20. Photocurrent action spectra of SnO_2, TiO_2, and SnO_2/TiO_2 composite films. (Source: Reprinted with permission from [814]. © 1996 American Chemical Society.)

potential, the pseudo-Fermi level increases, resulting in the accumulation of electrons. At a potential corresponding to the flat band potential, the photocurrent drops to zero as all the photogenerated charge carriers are lost in recombination.

A Schottky barrier model has been employed to analyze photovoltage and photocurrent results [257, 727]. For a photoelectrochemical cell operating on a Schottky barrier principle, V_{oc} can be related to i_{sc} by the following expression [10]:

$$V_{oc} = \frac{nkT}{q} \ln\left\{\frac{i_{sc}}{i_0} + 1\right\} \tag{8}$$

where k, T, and q are, respectively, the Boltzmann constant, absolute temperature, and electric charge, and n and i_0 are the diode quality factor and the reverse saturation current, respectively.

Efforts have also been made to understand the role of surface traps in influencing the photocurrent generation. For example, time-resolved photovoltage measurements carried out with a 337-nm laser pulse excitation indicate fast and slow growth components. The initial rise in photovoltage, which is completed within the time constant of the apparatus (500 ns), is attributed to the photoinduced charge separation between the conduction and valence bands. The slow growth was attributed to the existence of trapped charge carriers that exist in equilibrium with the charge carriers in the conduction and valence bands. Both fluorescence [83, 191, 302, 790] and microwave absorption [173, 315, 345, 347, 349, 605, 791] measurements have shown the importance of surface states in extending the lifetime of charge carriers.

One can observe novel effects from the thin films consisting of semiconductor structures in the form of multiple quantum wells (MQWS) or superlattices (SLS). Nozik and co-workers [792–795] have carried out a photoelectrochemical study on a 250-Å-thick $GaAs/GaAs_{0.5}P_{0.5}$ superlattice film. The photocurrent action spectra of lattice-matched superlattice electrodes in photoelectrochemical cells show structure with stepped waves that correspond to the quantum states in the quantum wells of the superlattice. The question regarding the hot-electron transfer from upper quantum states into solution has also been addressed with both a kinetic model and experimental results [795–797]. The flat band potential and photocurrent spectra of a single quantum well was also investigated in these studies.

6.4. Sensitization of Large-Band-Gap Semiconductors

The nanostructured films of metal oxides respond only in the UV region because they are large-band-gap semiconductors ($E_g > 3.0$ eV). One way to extend their photoresponse is to modify the semiconductor surface with sensitizing dyes that absorb strongly in the visible. The process of utilizing sub-band-gap excitations with dyes is referred to as photosensitization and conveniently employed in color photography and other imaging science applications. This approach of light energy conversion is similar to plant photosynthesis, in which chlorophyll molecules act as light-absorbing antenna molecules. In other words, the dye-modified semiconductor films provide an efficient method to mimic the photosynthetic process. Bignozzi et al. [798] have presented a supramolecular approach for designing photosensitizers. By optimizing the molecular design, it should be possible to suppress the interfacial charge recombination and improve the cross section for light absorption.

The high porosity and strong surface bonding property of the nanostructured semiconductor films facilitate surface modification with organic dyes and organometallic complexes such as bis(2,2′-bipyridine)(2,2′-bipyridine-4,4′dicarboxylic acid)ruthenium(II). The nanostructured TiO_2 films modified with a ruthenium complex exhibit photoconversion efficiencies in the range of 10% to 15% in diffused daylight [748], which is comparable to that of amorphous silicon-based photovoltaic cells.

The photocurrent response evaluated in terms of the photon-to-current efficiency (IPCE) of SnO_2 film with and without surface modification is shown in Figure 21 [339]. The IPCE maximum of the surface-modified SnO_2 film closely matches the absorption maximum of the sensitizer. The SnO_2 film, which is sensitive only to UV excitation prior to surface modification, responds to the visible light (wavelengths greater than 400 nm) as a result of surface modification. This shows that a photosensitization mechanism is operative in extending the photocurrent response of the SnO_2 film. When the electrode is illuminated with visible light, the sensitizer molecules absorb light and inject electrons into the SnO_2 particles. These electrons are then collected at the conducting glass surface to generate anodic photocurrent. The redox couple (e.g., I_3^-/I^-) present in the electrolyte quickly regenerates the sensitizer. By choosing an appropriate sensitizer, it is possible to tune the

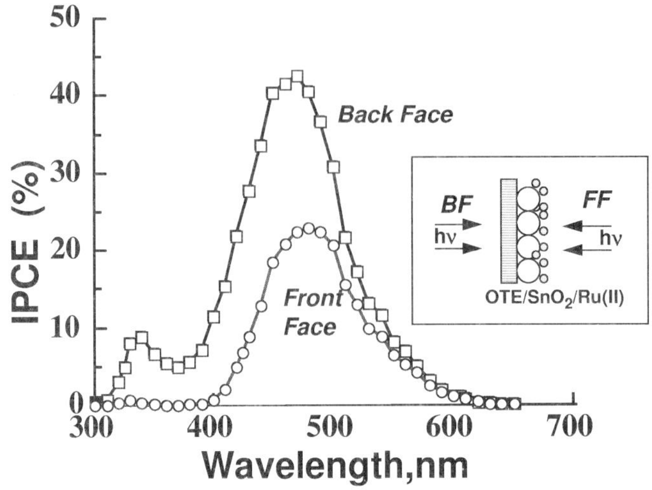

Fig. 21. Sensitized photocurrent generation at a SnO_2 nanocrystalline semiconductor film modified with a ruthenium complex in a photoelectrochemical cell containing 0.04 M I_2 and 0.5 M LiI in acetonitrile as electrolyte. (a) SnO_2 film before modification and (b) after modification with a ruthenium complex. (Source: Reprinted with permission from [339]. © 1994 American Chemical Society.)

photoresponse of these nanostructured semiconductor films. For example, sensitizing dyes such as chlorophyll a and b [685, 728], squaraines [630], and oxazines [638] can extend the photoresponse of SnO_2 films to the red-infrared region. The maximum IPCE in the example discussed in Figure 24 (around 50%) shows that nearly half of the injected charge from the excited sensitizer is lost as a result of recombination with the oxidized sensitizer. By optimizing the operating conditions, it is possible to improve the performance of the IPCE of the sensitizer-based cells. Ru complex–modified TiO_2 nanostructured films exhibit an IPCE of nearly 90% under optimized light-harvesting conditions [748]. Both experimental and theoretical evaluations of these cells have been carried out and the efficiency limiting factors have been identified [799, 800].

The varying degree of electron accumulation within the semiconductor particles alters the energetics of the quasi-Fermi level and creates a potential gradient within the thin film (Fig. 22). The formation of such a potential gradient provides the necessary driving force for the electron transport to the collecting surface of OTE. Although it is difficult to establish the exact nature of this overall potential gradient, the experimental results indicate it to be qualitatively similar to that of the Schottky barrier observed in a single-crystal semiconductor system [339, 629, 728]. Because this potential gradient is not an ideal type of Schottky barrier, significant loss of electrons is encountered during the transit because of recombination at the grain boundaries. This is evident from the relatively high reverse saturation photocurrents observed in these examples.

The excited state lifetime of the sensitizer adsorbed on semiconductor films is significantly lower than that adsorbed on an insulator surface such as alumina. The kinetic evaluation of the multiexponential luminescence decay of the excited Ru complex adsorbed on SnO_2 suggests that multiple injection/adsorption sites exist on the surface of a semiconductor nanocrystallite. The rate constants for heterogeneous electron transfer between excited Ru complex and semiconductor crystallites such as SnO_2, TiO_2, and ZnO are in the range of 10^7 to 10^9 s^{-1} [339, 621, 658]. Independent microwave absorption and luminescence measurements have been carried out to monitor the charge injection from excited $Ru(bpy)_2(dcbpy)^{2+}$ into SnO_2, ZnO, and TiO_2 nanocrystallites [351]. The growth of microwave absorption was delayed from the laser pulse by a process showing a similar rate constant to the fast decay portion of the luminescence. The appearance of microwave

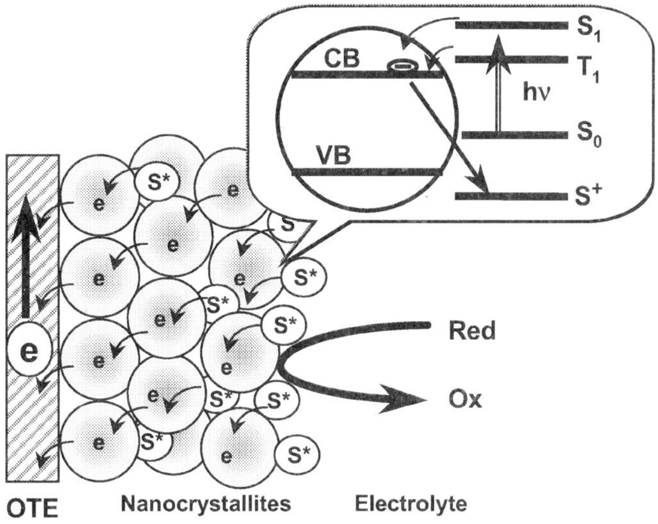

Fig. 22. Mechanism of sensitized photocurrent generation in a nanocrystalline semiconductor film.

conductivity at rates corresponding to the luminescence directly confirms the fast component of the heterogeneous electron rate constant to be in the range of 1 to 3×10^8 s^{-1}.

The charge injection from a singlet excited sensitizer into the conduction band of a large-band-gap semiconductor is usually considered to be an ultrafast process occurring in the picosecond time domain. The charge injection process in the case of organic dyes such as anthracene carboxylate [801], squaraines [802], and cresyl violet [602, 603] has been shown to occur within 20 ps. Similar fast electron transfer has also been noted for $Ru(H_2O)_2^{2-}$ on a TiO_2 surface at very low coverage [603]. On the contrary, relatively smaller charge injection rate constants (10^8–10^9 s^{-1}) have been reported by several research groups investigating the photophysical behavior of ruthenium complexes adsorbed on various semiconductor surfaces [339, 621, 658, 662, 663, 803, 804]. It has been suggested that electron trapping at surface defects may be a contributing factor in controlling the heterogeneous electron transfer at the semiconductor surface [805]. Similarly, the charge injection from the triplet excited dyes into TiO_2 [593] and ZnO [604] colloids has also been shown to occur on a slower time scale. The results presented here suggest that the electron transfer from the excited $Ru(bpy)_2(dcbpy)^{2+}$ occurs with a relatively slower rate than the singlet excited organic dyes but is comparable to the triplet excited dyes. It should be noted that the excited state of $Ru(bpy)_2(dcbpy)^{2+}$ involves metal-to-ligand charge transfer state and the implications are that such an electronic configuration of the excited state plays an important role in controlling the electron injection rates. The possibility of a connection between the multiplicity of the excited sensitizer and the rate constant for charge injection deserves more careful study.

6.5. Single-Electron Tunneling Devices

Controlling single-electron tunneling (SET) phenomena in nanostructured materials is gaining interest, because of its importance in developing extremely small electronic devices with high speed and low power consumption. When the size of the junction becomes small enough to have extremely small capacitance, C, the charging energy resulting from a single electron at the junction, $W = q^2/2C$, becomes larger than the thermal energy, kT, thus preventing additional electron injection to the junction (Fig. 23) [806]. Assembling a semiconductor junction using 1–2-nm nanoparticles results in the construction of single-electron devices that can be operated at room temperature [807, 808]. The charging energy

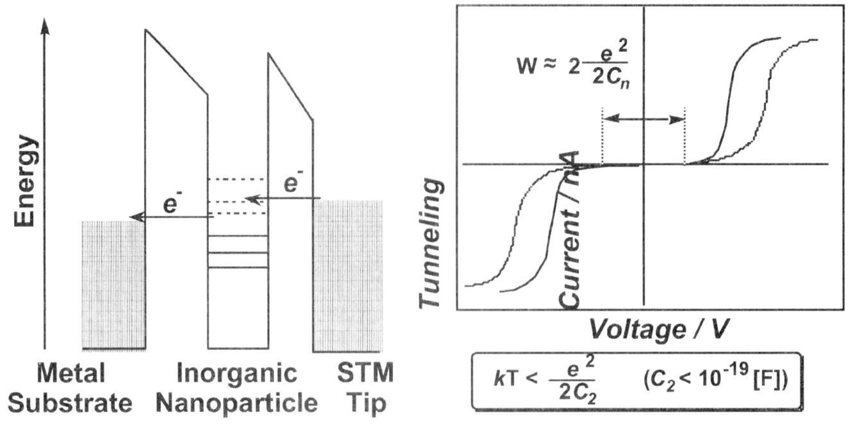

Fig. 23. Energy diagram of single-electron tunneling junction and typical voltage–current characteristics. (Source: Reprinted with permission from [806].)

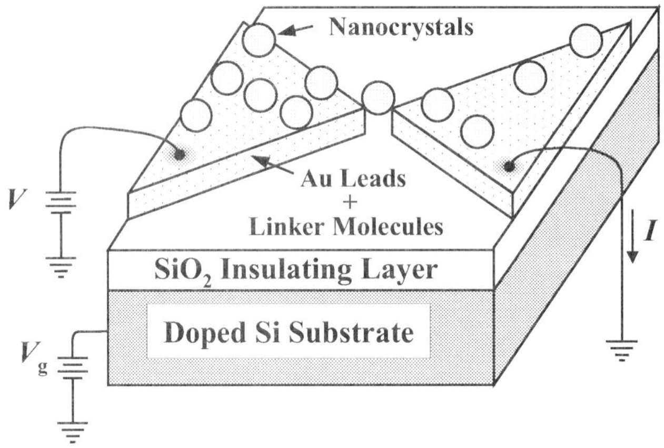

Fig. 24. Diagram of single-electron tunneling device. The electrochemical potential of the nanocrystallites between bridges can be tuned by applying a gate voltage to the substrate. (Source: Reprinted with permission from [813]. © 1997 Macmillan, Magazines Limited.)

of the junction, which consists of metal nanoparticles in the size range of 1 to 2 nm, becomes larger than the thermal energy at room temperature because of the extremely small capacitance of the metal nanoparticle ($C < 10^{-19}$ F). Several reports are available for the observation of the Coulomb blockade phenomenon at room temperature using systems of small gold nanoparticles on self-assembled organic monolayers [807, 808], polysilicon dots [809], small gold nanoparticles sandwiched by lamellar inorganic solids [810], and ultrasmall microelectrodes [811].Efforts have now gradually become to be focused on the development of techniques to construct such ultrasmall junctions with organized structure, because generally their dimensions are much smaller than those using ordinary lithographic techniques at the present. The techniques of colloidal chemistry could be a promising way to construct/synthesize/assemble such ultrasmall junctions. Controlling the size, crystalline structure, and structure anisotropy should be applied to this field. The use of various organic surface-modified reagents should also be helpful in controlling the nature of junction gap characteristics.

Alivisatos et al. [812, 813] reported interesting results showing the future possibility of applying colloidal semiconductors to single-electron transistors. Electrical transport properties of a single-electron transistor using 5.5-nm CdSe nanocrystallines was investigated at 77 K (Fig. 24). They observed clearly the Coulomb oscillation spacing at 30–60 meV using the devices. The observed values correspond to estimated energy values of 50 meV for a 5.5-nm-diameter metallic sphere. The number of charges on the nanocrystallites was tuned precisely by the gate voltage, and changed the blockade energy reflecting their degeneration of the state. They pointed out the possibility of the device application for the measurement of energy level spectra of small electronic systems [813]. Their reports on CdSe nanocrystallites are the first examples of applying semiconductor nanocrystallites for ultrasmall devices using single-electron phenomenon. In the near future, smaller nanocrystallites with well-defined crystalline structure could be applied to the devices and operated at room temperature.

7. CONCLUDING REMARKS

The ability of semiconductor nanoclusters to carry out redox processes with greater efficiency and selectivity than in homogeneous solutions has made them potential candidates for the conversion and storage of solar energy and environmental remediation. The future challenge lies in designing novel microheterogeneous assemblies for artificial photosynthesis. The possibility of casting thin nanostructured semiconductor films on conducting glass plates has brought us new opportunities to explore their properties and novel applications.

Acknowledgments

PVK acknowledges the support of the Office of Basic Energy Sciences of the U.S. Department of Energy. This is contribution 4068 from the Notre Dame Radiation Laboratory. PVK also acknowledges the Japan Society for Promotion of Science for the award of JSPS fellowship, which made the visit to Osaka University possible.

References

1. A. J. Bard, *J. Photochem.* 10, 59 (1979).
2. A. J. Nozik, *Faraday Discuss. Chem. Soc.* 17 (1980).
3. A. J. Bard, *J. Phys. Chem.* 86, 172 (1982).
4. Y. Y. Gurevich and Y. V. Pleskov, *Russ. Chem. Rev.* 52, 563 (1983).
5. P. V. Kamat and N. M. Dimitrijevic, *Sol. Energy* 44, 83 (1990).
6. M. A. Fox, *Acc. Chem. Res.* 16, 314 (1983).
7. K. Kalyanasundaram, M. Grätzel, and E. Pelizzetti, *Coord. Chem. Rev.* 69, 57 (1986).
8. M. A. Fox, *Top. Curr. Chem.* 142, 72 (1987).
9. A. Heller, *New J. Chem.* 11, 187 (1987).
10. R. Memming, *Top. Curr. Chem.* 143, 79 (1988).
11. A. Henglein, *Chem. Rev.* 89, 1861 (1989).
12. M. Anpo, *Res. Chem. Intermed.* 11, 67 (1989).
13. E. N. Savinov, Y. A. Gruzdkov, and V. N. Parmon, *Int. J. Hydrogen Energy* 14, 1 (1989).
14. N. Serpone and E. Pelizzetti, "Photocatalysis Fundamentals and Applications." Wiley, New York, 1989.
15. M. A. Fox, *CHEMTECH* 22, 680 (1992).
16. D. F. Ollis and H. Al Ekábi, "Photocatalytic Purification of Water and Air." Elsevier Science, Amsterdam, 1993.
17. M. R. Hoffmann, S. T. Martin, W. Choi, and D. W. Bahnemann, *Chem. Rev.* 95, 69 (1995).
18. U. Stafford, K. A. Gray, and P. V. Kamat, *Heterogeneous Chem. Rev.* 3, 77 (1996).
19. M. Grätzel, "Energy Resources through Photochemistry and Catalysis." Academic Press, New York, 1983.
20. A. Henglein, *Pure Appl. Chem.* 56, 1215 (1984).
21. G. Hodes and M. Grätzel, *Nouv. J. Chim.* 8, 509 (1984).
22. A. Fojtik, H. Weller, U. Koch, and A. Henglein, *Ber. Bunsen-Ges. Phys. Chem.* 88, 969 (1984).
23. G. T. Brown, J. R. Darwent, and P. D. I. Fletcher, *J. Am. Chem. Soc.* 107, 6446 (1985).
24. H. Weller, H. M. Schmidt, U. Koch, A. Fojtik, S. Baral, A. Henglein, W. Kunath, K. Weiss, and E. Dieman, *Chem. Phys. Lett.* 124, 557 (1986).
25. A. Henglein, *Prog. Colloid Polym. Sci.* 73, 1 (1987).

26. J. J. Ramsden, *Proc. R. Soc. London, Ser. A* 410, 89 (1987).
27. M. Grätzel and K. Kalyanasundaram, "Kinetics and Catalysis in Microheterogeneous Systems." Marcel Dekker, New York, 1991.
28. M. Grätzel, "Heterogeneous Photochemical Electron Transfer." CRC Press, Boca Raton, FL, 1989.
29. N. S. Lewis, *Acc. Chem. Res.* 23, 176 (1990).
30. M. D. E. Forbes and N. S. Lewis, *J. Am. Chem. Soc.* 112, 3682 (1990).
31. L. M. Peter, *Chem. Rev.* 90, 753 (1990).
32. P. V. Kamat, in "Kinetics and Catalysis in Microheterogeneous Systems" (M. Grätzel and K. Kalyanasundaram, eds.), p. 375. Marcel Dekker, New York, 1991.
33. C. A. Koval and J. N. Howard, *Chem. Rev.* 92, 411 (1992).
34. L. Brus, *J. Phys. Chem.* 90, 2555 (1986).
35. M. L. Steigerwald and L. E. Brus, *Acc. Chem. Res.* 23, 183 (1990).
36. L. Brus, *J. Phys. Chem.* 98, 3575 (1994).
37. L. Brus, *Colloid Interface Sci.* 1, 197 (1996).
38. Y. I. Kim, S. J. Atherton, E. S. Brigham, and T. E. Mallouk, *J. Phys. Chem.* 97, 11802 (1993).
39. P. V. Kamat, *Chem. Rev.* 93, 267 (1993).
40. H. Weller, *Adv. Mater.* 5, 88 (1993).
41. H. Weller, *Angew. Chem. Int. Ed. Engl.* 32, 41 (1993).
42. O. I. Micic, T. Rajh, J. M. Nedeljkovic, and M. I. Comor, *Isr. J. Chem.* 33, 59 (1993).
43. G. N. Ryba, C. N. Kenyon, and N. S. Lewis, *J. Phys. Chem.* 97, 13814 (1993).
44. P. Alivisatos, *J. Phys. Chem.* 100, 13226 (1996).
45. J. Belloni, in "Homogeneous Photocatalysis" (M. Chanon, ed.), p. 170. Wiley, New York, 1997.
46. P. V. Kamat and D. Meisel, in "Studies in Surface Science and Catalysis," p. 474. Elsevier Science, Amsterdam, 1997.
47. T. Rajh, O. I. Micic, and A. J. Nozik, *J. Phys. Chem.* 97, 11999 (1993).
48. T. Vossmeyer, L. Katsikas, M. Giersig, I. G. Popovic, K. Diesner, A. Chemseddine, A. Eychmüller, and H. Weller, *J. Phys. Chem.* 98, 7665 (1994).
49. Y. Nosaka, K. Yamaguchi, H. Miyama, and H. Hayashi, *Chem. Lett.* 605 (1988).
50. M. L. Steigerwald, A. P. Alivisatos, J. M. Gibson, T. D. Harris, R. Kortan, A. J. Muller, A. M. Thayer, T. M. Duncan, D. C. Douglass, and L. E. Brus, *J. Am. Chem. Soc.* 110, 3046 (1988).
51. D. Hayes, O. I. Micic, M. T. Nenadovic, V. Swayambunathan, and D. Meisel, *J. Phys. Chem.* 93, 4603 (1989).
52. C. H. Fischer and A. Henglein, *J. Phys. Chem.* 93, 5578 (1989).
53. A. R. Kortan, R. Hull, R. L. Opila, M. G. Bawendi, M. L. Steigerwald, P. J. Carroll, and L. E. Brus, *J. Am. Chem. Soc.* 112, 1327 (1990).
54. N. Herron, Y. Wang, and H. Eckert, *J. Am. Chem. Soc.* 112, 1322 (1990).
55. V. Swayambunathan, D. Hayes, K. H. Schmidt, Y. X. Liao, and D. Meisel, *J. Am. Chem. Soc.* 112, 3831 (1990).
56. S. Yanagida, T. Enokida, A. Shindo, T. Shiragami, T. Ogata, T. Fukumi, T. Sakaguchi, and T. Sakata, *Chem. Lett.* 1773 (1990).
57. U. Resch, A. Eychmueller, M. Haase, and H. Weller, *Langmuir* 8, 2215 (1992).
58. V. L. Colvin, A. N. Goldstein, and A. P. Alivisatos, *J. Am. Chem. Soc.* 114, 5221 (1992).
59. S. Mahamuni, A. A. Khosravi, M. Kundu, A. Kshirsagar, A. Bedekar, D. B. Avasare, P. Singh, and S. K. Kulkarni, *J. Appl. Phys.* 73, 5237 (1993).
60. T. Torimoto, H. Uchida, T. Sakata, H. Mori, and H. Yoneyama, *J. Am. Chem. Soc.* 115, 1874 (1993).
61. T. Torimoto, T. Sakata, H. Mori, and H. Yoneyama, *J. Phys. Chem.* 98, 3036 (1994).
62. T. Torimoto, K. Maeda, J. Maenaka, and H. Yoneyama, *J. Phys. Chem.* 98, 13658 (1994).
63. H. Inoue, N. Ichiroku, T. Torimoto, T. Sakata, H. Mori, and H. Yoneyama, *Langmuir* 10, 4517 (1994).
64. S. Yanagida, H. Kawakami, Y. Midori, H. Kizumoto, C. Pac, and Y. Wada, *Bull. Chem. Soc. Jpn.* 68, 1811 (1995).
65. D. Lawless, S. Kapoor, and D. Meisel, *J. Phys. Chem.* 99, 10329 (1995).
66. L. Spanhel, M. Haase, H. Weller, and A. Henglein, *J. Am. Chem. Soc.* 109, 5649 (1987).
67. C. Murray, D. Norris, and M. Bawendi, *J. Am. Chem. Soc.* 115, 8706 (1993).
68. T. L. Pettit and M. A. Fox, *J. Phys. Chem.* 90, 1353 (1986).
69. Y. Wang and N. Herron, *J. Phys. Chem.* 91, 257 (1987).
70. M. A. Fox, *Res. Chem. Intermed.* 15, 153 (1991).
71. R. D. Stramel, T. Nakamura, and J. K. Thomas, *J. Chem. Soc., Faraday Trans.* 84, 1287 (1988).
72. N. Herron, Y. Wang, M. M. Eddy, G. D. Stucky, D. E. Cox, K. Moller, and T. Bein, *J. Am. Chem. Soc.* 111, 530 (1989).
73. K. Moller, M. M. Eddy, G. D. Stucky, N. Herron, and T. Bein, *J. Am. Chem. Soc.* 111, 2564 (1989).
74. Y. Wang and N. Herron, *Res. Chem. Intermed.* 15, 17 (1991).
75. X. Liu, K. K. Iu, and J. K. Thomas, *J. Chem. Soc., Faraday Trans.* 89, 1861 (1993).
76. A. Jentys, R. W. Grimes, J. D. Gale, and C. R. A. Catlow, *J. Phys. Chem.* 97, 13535 (1993).
77. A. I. Ekimov, A. L. Efros, and A. A. Onushchenko, *Solid State Commun.* 56, 921 (1985).

78. J. Kuczynski and J. K. Thomas, *J. Phys. Chem.* 89, 2720 (1985).
79. T. Rajh, M. I. Vucemilovic, N. M. Dimitrijevic, O. I. Micic, and A. J. Nozik, *Chem. Phys. Lett.* 143, 305 (1988).
80. M. Krishnan, J. R. White, M. A. Fox, and A. J. Bard, *J. Am. Chem. Soc.* 105, 7002 (1983).
81. N. Kakuta, J. M. White, A. Campion, A. J. Bard, M. A. Fox, and S. E. Webber, *J. Phys. Chem.* 89, 48 (1985).
82. Y. Wang, A. Suna, W. Mahler, and R. Kasowski, *J. Chem. Phys.* 87, 7315 (1987).
83. K. R. Gopidas and P. V. Kamat, *Mater. Lett.* 9, 372 (1990).
84. R. Tassoni and R. R. Schrock, *Chem. Mater.* 6, 744 (1994).
85. P. Lianos and J. K. Thomas, *Chem. Phys. Lett.* 125, 299 (1986).
86. C. Petit and M. P. Pileni, *J. Phys. Chem.* 92, 2282 (1988).
87. C. Petit, P. Lixon, and M. P. Pileni, *J. Phys. Chem.* 94, 1598 (1990).
88. L. Motte, F. Billoudet, and M. P. Pileni, *J. Phys. Chem.* 99, 16425 (1995).
89. J. Cizeron and M. P. Pileni, *J. Phys. Chem.* 99, 17410 (1995).
90. L. Levy, J. F. Hochepied, and M. P. Pileni, *J. Phys. Chem.* 100, 18322 (1996).
91. S. Modes and P. Lianos, *J. Phys. Chem.* 93, 5854 (1989).
92. E. Stathatos, P. Lianos, F. Del Monte, D. Levy, and D. Tsiourvas, *Langmuir* 13, 4295 (1997).
93. Y. M. Tricot and J. H. Fendler, *J. Am. Chem. Soc.* 106, 7359 (1984).
94. R. Rafaeloff, Y. M. Tricot, F. Nome, and J. H. Fendler, *J. Phys. Chem.* 89, 533 (1985).
95. Y. M. Tricot, A. Emeren, and J. H. Fendler, *J. Phys. Chem.* 89, 4721 (1985).
96. Y. M. Tricot and J. H. Fendler, *J. Phys. Chem.* 90, 3369 (1986).
97. H. J. Watzke and J. H. Fendler, *J. Phys. Chem.* 91, 854 (1987).
98. H. C. Youn, S. Baral, and J. H. Fendler, *J. Phys. Chem.* 92, 6320 (1988).
99. A. C. Chang and J. H. Fendler, *J. Phys. Chem.* 93, 2538 (1989).
100. M. I. Khramov and V. N. Parmon, *J. Photochem. Photobiol., A* 71, 279 (1993).
101. B. A. Korgel and H. G. Monbouquette, *J. Phys. Chem.* 100, 346 (1996).
102. B. A. Korgel and H. G. Monbouquette, *J. Phys. Chem.* 100, 346 (1996).
103. X. Zhang and A. J. Bard, *J. Phys. Chem.* 92, 5566 (1988).
104. R. Zhu, G. Min, Y. Wei, and H. J. Schmitt, *J. Phys. Chem.* 96, 8210 (1992).
105. K. C. Yi and J. H. Fendler, *Langmuir* 6, 1519 (1990).
106. J. H. Fendler, in "Electrochemistry in Colloids and Dispersions" (R. A. Mackay and J. Texter, eds.), p. 475. VCH Publishers, New York, 1992.
107. J. H. Fendler, *Isr. J. Chem.* 33, 41 (1993).
108. N. J. Geddes, R. S. Urquhart, D. N. Furlong, C. R. Lawrence, K. Tanaka, and Y. Okahata, *J. Phys. Chem.* 97, 13767 (1993).
109. X. Peng, R. Lu, Y. Y. Zhao, L. H. Qu, H. Y. Chen, and T. J. Li, *J. Phys. Chem.* 98, 7052 (1994).
110. J. H. Fendler, *Chem. Mater.* 8, 1616 (1996).
111. J. H. Fendler, in "Semiconductor Nanoclusters—Physical, Chemical and Catalytic Aspects" (P. V. Kamat and D. Meisel, eds.), p. 261. Elsevier Science, Amsterdam, 1997.
112. J. Yang, F. C. Meldrum, and J. H. Fendler, *J. Phys. Chem.* 99, 5500 (1995).
113. I. Ichinose, N. Kimizuka, and T. Kunitake, *J. Phys. Chem.* 99, 3736 (1995).
114. I. Ichinose, T. Asai, S. Yoshimura, N. Kimizuka, and T. Kunitake, *Chem. Lett.* 1837 (1994).
115. K. M. Choi and K. J. Shea, *J. Phys. Chem.* 98, 3207 (1994).
116. I. Willner and Y. Eichen, *J. Am. Chem. Soc.* 109, 6862 (1987).
117. I. Willner, Y. Eichen, and A. J. Frank, *J. Am. Chem. Soc.* 111, 1884 (1989).
118. S. Y. Chang, L. Liu, and S. A. Asher, *J. Am. Chem. Soc.* 116, 6745 (1994).
119. S. Y. Chang, L. Liu, and S. A. Asher, *J. Am. Chem. Soc.* 116, 6739 (1994).
120. I. G. Dance, A. Choy, and M. L. Scudder, *J. Am. Chem. Soc.* 106, 6285 (1984).
121. G. S. Lee, D. C. Craig, I. Ma, M. L. Scudder, T. D. Bailey, and I. G. Dance, *J. Am. Chem. Soc.* 110, 4863 (1988).
122. T. Vossmeyer, G. Reck, L. Katsikas, E. T. K. Haupt, B. Schulz, and H. Weller, *Science* 267, 1476 (1995).
123. T. Vossmeyer, G. Reck, B. Schulz, L. Katsikas, and H. Weller, *J. Am. Chem. Soc.* 117, 12881 (1995).
124. N. Herron, J. C. Calabrese, W. E. Farneth, and Y. Wang, *Science* 259, 1426 (1993).
125. D. M. Wilhelmy and E. Matijevic, *J. Chem. Soc., Faraday Trans. 1* 80, 563 (1984).
126. E. Matijevic, *Langmuir* 2, 12 (1986).
127. C. Hayashi, *Phys. Today* 44 (1987).
128. C. Levy-Clement, A. Lagoubi, D. Ballutaud, F. Ozanam, and J.-N. Chazalviel, *Appl. Surf. Sci.* 65/66, 408 (1993).
129. C. Levy-Clement, A. Lagoubi, and M. Tomkiewicz, *J. Electrochem. Soc.* 141, 958 (1994).
130. T. Rajh, O. I. Micic, and A. J. Nozik, *Langmuir* 2, 477 (1986).
131. O. I. Micic, L. Zongguan, G. Mills, J. C. Sullivan, and D. Meisel, *J. Phys. Chem.* 91, 6221 (1987).
132. Y. Haruvy, M. Grätzel, and L. A. Rajbenbach, *Radiat. Phys. Chem.* 31, 843 (1988).
133. D. Hayes, D. Meisel, and O. I. Micic, *Colloids Surf.* 55, 121 (1991).
134. D. Hayes, K. H. Schmidt, and D. Meisel, *J. Phys. Chem.* 93, 6100 (1989).

135. D. Meisel, in "Radiation Research, A Twentieth Century Perspective," Proceedings of the 9th International Congress Radiation Research, Vol. 2, p. 121. Academic Press, New York, 1992.
136. J. Khatouri, M. Mostafavi, and J. Belloni, in "Photochemistry and Radiation Chemistry. Complementary Methods for Studying Electron Transfer" (J. Wishart and D. Nocera, eds.), p. XX. American Chemical Society, Washington, DC, 1997.
137. R. A. Hobson, P. Mulvaney, and F. Grieser, *J. Chem. Soc., Chem. Commun.* 823 (1994).
138. J. Z. Sostaric, R. A. Caruso-Hobson, P. Mulvaney, and F. Grieser, *J. Chem. Soc., Faraday Trans.* 93, 1791 (1997).
139. C. H. Fischer, H. Weller, A. Fojtik, P. C. Lume, E. Janata, and A. Henglein, *Ber. Bunsen-Ges. Phys. Chem.* 90, 46 (1986).
140. C. H. Fischer, J. Lilie, H. Weller, L. Katsikas, and A. Henglein, *Ber. Bunsen-Ges. Phys. Chem.* 93, 61 (1989).
141. C. H. Fischer, H. Weller, L. Katsikas, and A. Henglein, *Langmuir* 5, 429 (1989).
142. Y. Wang, M. Harmer, and N. Herron, *Isr. J. Chem.* 33, 31 (1993).
143. H. Weller, A. Eychmuller, R. Vogel, L. Katsikas, A. Hasselbarth, and M. Giersig, *Isr. J. Chem.* 33, 107 (1993).
144. H. Matsumoto, H. Uchida, T. Matsunaga, K. Tanaka, and T. Sakata, *J. Phys. Chem.* 98, 11549 (1994).
145. H. Matsumoto, T. Sakata, H. Mori, and H. Yoneyama, *J. Phys. Chem.* 100, 13781 (1996).
146. T. Torimoto, H. Nishiyama, T. Sakata, H. Mori, and H. Yoneyama, *J. Electrochem. Soc.* 145, 1964 (1998).
147. K. Murakoshi, H. Hosokawa, M. Saito, Y. Wada, T. Sakata, H. Mori, and S. Yanagida, *J. Chem. Soc., Faraday Trans.* 4, 579 (1998).
148. D. R. Lide, in "CRC Handbook of Chemistry and Physics," 75th ed., p. 4. CRC Press, Boca Raton, FL, 1995.
149. M. V. Ramakrishna and R. A. Friesner, *J. Chem. Phys.* 95, 8309 (1991).
150. J. J. Ramsden, S. E. Webber, and M. Grätzel, *J. Phys. Chem.* 89, 2740 (1985).
151. M. Kanemoto, K. Ishihara, Y. Wada, T. Sakata, H. Mori, and S. Yanagida, *Chem. Lett.* 835 (1992).
152. M. Kanemoto, H. Ankyu, Y. Wada, and S. Yanagida, *Chem. Lett.* 2113 (1992).
153. M. Kanemoto, M. Nomura, Y. Wada, T. Akano, and S. Yanagida, *Chem. Lett.* 1687 (1993).
154. R. J. Bandaranayake, G. W. Wen, J. Y. Lin, H. X. Jiang, and C. M. Sorensen, *Appl. Phys. Lett.* 67, 831 (1995).
155. M. Haase and A. P. Alivisatos, *J. Phys. Chem.* 96, 6756 (1992).
156. S. H. Tolbert and A. P. Alivisatos, *Science* 265, 373 (1994).
157. S. H. Tolbert and A. P. Alivisatos, *J. Chem. Phys.* 102, 4642 (1995).
158. K. Murakoshi, H. Hosokawa, N. Tanaka, M. Saito, Y. Wada, T. Sakata, H. Mori, and S. Yanagida, *Chem. Commun.* 3, 321 (1998).
159. L. Brus, in "Photoinduced Electron Transfer. Part B: Experimental Techniques and Medium Effects" (M. A. Fox and M. Chanon, eds.), p. 314. Elsevier, Amsterdam, 1988.
160. L. Brus, *New J. Chem.* 11, 123 (1987).
161. A. Henglein, *Top. Curr. Chem.* 143, 113 (1988).
162. A. Henglein and H. Weller, in "Photochemical Energy Conversion" (J. R. Norris, Jr. and D. Meisel, eds.). Elsevier, New York, 1989.
163. A. I. Ekimov, A. L. Efros, T. V. Shubina, and A. P. Skvortsov, *J. Lumin.* 46, 97 (1990).
164. Y. Wang and N. Herron, *J. Phys. Chem.* 95, 525 (1991).
165. Y. Wang, in "Photochemical Conversion and Storage of Solar Energy" (E. Pelizzetti and M. Schiavello, eds.). Kluwer Academic, Dordrecht, The Netherlands, 1991.
166. L. Banyai and S. W. Koch, "Semiconductor Quantum Dots." World Scientific, River Edge, NJ, 1993.
167. P. V. Kamat, *Prog. React. Kinet.* 19, 277 (1994).
168. H. Weller, U. Koch, M. Gutierrez, and A. Henglein, *Ber. Bunsen-Ges. Phys. Chem.* 88, 649 (1984).
169. R. Rossetti, R. Hull, J. M. Gibson, and L. E. Brus, *J. Chem. Phys.* 83, 1406 (1985).
170. A. J. Nozik, F. Williams, M. T. Nenadovic, T. Rajh, and O. I. Micic, *J. Phys. Chem.* 89, 397 (1985).
171. O. I. Micic, M. T. Nenadovic, T. Rajh, N. M. Dimitrijevic, and A. J. Nozik, *NATO ASI Ser. C* 26 (1986).
172. T. Dannhauser, M. O'Neil, K. Johansson, D. Whitten, and G. McLendon, *J. Phys. Chem.* 90, 6074 (1986).
173. P. V. Kamat, N. M. Dimitrijevic, and R. W. Fessenden, *J. Phys. Chem.* 91, 396 (1987).
174. T. Rajh, M. W. Peterson, J. A. Turner, and A. J. Nozik, *J. Electroanal. Chem. Interfacial Electrochem.* 228, 55 (1987).
175. A. Fojtik and J. Jirkovsky, *Chem. Phys. Lett.* 137, 226 (1987).
176. H. Weller, M. Haase, L. Spanhel, and A. Henglein, *Prog. Colloid Polym. Sci.* 76, 24 (1988).
177. C. Kormann, D. W. Bahnemann, and M. R. Hoffmann, *J. Phys. Chem.* 92, 5196 (1988).
178. D. Meisel and W. A. Mulac, *Colloids Surf.* 35, 179 (1989).
179. H. Yoneyama, S. Haga, and S. Yamanaka, *J. Phys. Chem.* 93, 4833 (1989).
180. L. Spanhel and M. A. Anderson, *J. Am. Chem.* Soc. 112, 2278 (1990).
181. H. Weller, *Ber. Bunsen-Ges. Phys. Chem.* 95, 1361 (1991).
182. M. P. Pileni, L. Motte, and C. Petit, *Chem. Mater.* 4, 338 (1992).
183. I. Bedja, S. Hotchandani, and P. V. Kamat, *J. Phys. Chem.* 97, 11064 (1993).

184. J. L. Coffer, R. R. Chandler, C. D. Gutsche, I. Alam, R. F. Pinizzotto, and H. Yang, *J. Phys. Chem.* 97, 696 (1993).
185. A. Mews, A. Eychmuller, M. Giersig, D. Schooss, and H. Weller, *J. Phys. Chem.* 98, 934 (1994).
186. R. Rossetti, J. L. Ellison, J. M. Gibson, and L. E. Brus, *J. Chem. Phys.* 80, 4464 (1984).
187. D. W. Banemann, *Isr. J. Chem.* 33, 115 (1993).
188. K. A. Littau, P. J. Szajowski, A. J. Muller, A. R. Kortan, and L. E. Brus, *J. Phys. Chem.* 97, 1224 (1993).
189. H. Weller, A. Fojtik, and A. Henglein, *Chem. Phys. Lett.* 117, 485 (1985).
190. A. Fojtik, H. Weller, and A. Henglein, *Chem. Phys. Lett.* 120, 552 (1985).
191. U. Koch, A. Fojtik, H. Weller, and A. Henglein, *Chem. Phys. Lett.* 122, 507 (1985).
192. S. Baral, A. Fojtik, H. Weller, and A. Henglein, *J. Am. Chem. Soc.* 108, 375 (1986).
193. A. Henglein, *J. Chim. Phys. Phys.-Chim. Biol.* 84, 1043 (1987).
194. A. Henglein, A. Fojtik, and H. Weller, *Ber. Bunsen-Ges. Phys. Chem.* 91, 441 (1987).
195. A. Henglein, M. Gutierrez, H. Weller, A. Fojtik, and J. Jirkovsky, *Ber. Bunsen-Ges. Phys. Chem.* 93, 593 (1989).
196. P. Lianos and J. K. Thomas, *J. Colloid Interface Sci.* 117, 505 (1987).
197. J. M. Nedeljkovic, M. T. Nenadovic, O. I. Micic, and A. J. Nozik, *J. Phys. Chem.* 90, 12 (1986).
198. O. I. Micic, M. T. Nenadovic, M. W. Peterson, and A. J. Nozik, *J. Phys. Chem.* 91, 1295 (1987).
199. M. W. Peterson, O. I. Micic, and A. J. Nozik, *J. Phys. Chem.* 92, 4160 (1988).
200. M. W. Peterson, M. T. Nenadovic, T. Rajh, R. Herak, O. I. Micic, J. P. Goral, and A. J. Nozik, *J. Phys. Chem.* 92, 1400 (1988).
201. D. W. Bahnemann, C. Kormann, and M. R. Hoffmann, *J. Phys. Chem.* 91, 3789 (1987).
202. E. S. Smotkin, C. Lee, A. J. Bard, A. Campion, M. A. Fox, T. E. Mallouk, S. E. Webber, and J. M. White, *Chem. Phys. Lett.* 152, 265 (1988).
203. Y. Nosaka, K. Yamaguchi, H. Yokoyama, and H. Miyama, in "Photoresponsive Materials," MRS International Meeting on Advanced Materials, Vol. 12, p. 155. Materials Research Society, 1989.
204. Y. Nosaka, N. Ohta, and H. Miyama, *J. Phys. Chem.* 94, 3752 (1990).
205. M. Grätzel, in "Electrochemistry in Colloids and Dispersions" (R. A. Mackay and J. Texter, eds.), p. 333. VCH Publishers, New York, 1992.
206. P. V. Kamat and B. Patrick, *J. Phys. Chem.* 96, 6829 (1992).
207. K. R. Gopidas and P. V. Kamat, *Proc. Ind. Acad. Sci. (Chem. Sci.)* 105, 505 (1993).
208. Y. Nosaka, T. Fukuyama, M. Horiuchi, and N. Fujii, *Isr. J. Chem.* 155, 71 (1993).
209. E. Lifshitz, M. Yassen, L. Bykov, and I. Dag, *J. Phys. Chem.* 99, 1245 (1995).
210. L. E. Brus, *J. Chem. Phys.* 80, 4403 (1984).
211. L. E. Brus, *J. Chem. Phys.* 79, 5566 (1983).
212. V. Mohan and J. B. Anderson, *Chem. Phys. Lett.* 156, 520 (1989).
213. W. L. Wilson, P. F. Szajowski, and L. E. Brus, *Science* 262, 1242 (1993).
214. L.-W. Wang and A. Zunger, *J. Chem. Phys.* 100, 2394 (1994).
215. L.-W. Wang and A. Zunger, *Phys. Rev. Lett.* 73, 1039 (1994).
216. S. B. Zhang, C.-Y. Yeh, and A. Zunger, *Phys. Rev. B* 48, 11204 (1993).
217. C.-Y. Yeh, S. B. Zhang, and A. Zunger, *Appl. Phys. Lett.* 64, 3545 (1994).
218. M. J. Sailor, J. L. Heinrich, and J. M. Lauerhaas, in "Semiconductor Nanoclusters—Physical, Chemical and Catalytic Aspects" (P. V. Kamat and D. Meisel, eds.), p. 209. Elsevier Science, Amsterdam, 1997.
219. L.-T. Cheng, N. Herron, and Y. Wang, *J. Appl. Phys.* 66, 3417 (1989).
220. Y. Wang, N. Herron, W. Mahler, and A. Suna, *J. Opt. Soc. Am. B* 6, 808 (1989).
221. L. Brus, *Appl. Phys. A* 53, 465 (1991).
222. O. A. Aktsipetrov, P. V. Elyutin, A. A. Fedyanin, A. A. Nikulin, and A. N. Rubtsov, *Surf. Sci.* 325, 343 (1995).
223. Y. Wang, *Adv. Photochem.* 19, 179 (1995).
224. Y. Nosaka, N. Tohriiwa, T. Kobayashi, and N. Fujii, *Chem. Mater.* 5, 930 (1993).
225. E. J. Heilweil and R. M. Hochstrasser, *J. Chem. Phys.* 82, 179 (1985).
226. L. L. Beecroft and C. K. Ober, *Chem. Mater.* 9, 1302 (1997).
227. Y. Wang, in "Semiconductor Nanoclusters—Physical, Chemical and Catalytic Aspects" (P. V. Kamat and D. Meisel, eds.), p. 277. Elsevier Science, Amsterdam, 1997.
228. W. J. Albery, G. T. Brown, J. R. Darwent, and E. Saivar-Iranizad, *J. Chem. Soc., Faraday Trans. 1* 81, 1999 (1985).
229. J. P. Kuczynski, B. H. Milosavljevic, and J. K. Thomas, *J. Phys. Chem.* 88, 980 (1984).
230. A. Henglein, A. Kumar, E. Janata, and H. Weller, *Chem. Phys. Lett.* 132, 133 (1986).
231. M. Haase, H. Weller, and A. Henglein, *J. Phys. Chem.* 92, 4706 (1988).
232. C. Liu and A. J. Bard, *J. Phys. Chem.* 93, 3232 (1989).
233. C. Liu and A. J. Bard, *J. Phys. Chem.* 93, 7047 (1989).
234. M. Haase, H. Weller, and A. Henglein, *Ber. Bunsen-Ges. Phys. Chem.* 92, 1103 (1988).
235. P. V. Kamat, N. M. Dimitrijevic, and A. J. Nozik, *J. Phys. Chem.* 93, 2873 (1989).
236. E. F. Hilinski, P. A. Lucas, and Y. Wang, *J. Chem. Phys.* 89, 3435 (1988).

237. Y. Wang, A. Suna, J. McHugh, E. F. Hilinski, P. A. Lucas, and R. D. Johnson, *J. Chem. Phys.* 92, 6927 (1990).
238. T. Rajh, O. I. Micic, D. Lawless, and N. Serpone, *J. Phys. Chem.* 96, 4633 (1992).
239. A. V. Barzykin and M. A. Fox, *Isr. J. Chem.* 33, 21 (1993).
240. P. V. Kamat, N. M. Dimitrijevic, and R. W. Fessenden, *J. Phys. Chem.* 92, 2324 (1988).
241. M. T. Nenadovic, M. I. Comor, V. Vasic, and O. I. Micic, *J. Phys. Chem.* 94, 6390 (1990).
242. N. M. Dimitrijevic and P. V. Kamat, *J. Phys. Chem.* 91, 2096 (1987).
243. N. M. Dimitrijevic, *J. Chem. Soc., Faraday Trans. 1* 83, 1193 (1987).
244. N. M. Dimitrijevic and P. V. Kamat, *Langmuir* 3, 1004 (1987).
245. M. Haase, H. Weller, and A. Henglein, *J. Phys. Chem.* 92, 482 (1988).
246. G. Redmond, A. O'Keeffe, C. Burgess, C. MacHale, and D. Fitzmaurice, *J. Phys. Chem.* 97, 11081 (1993).
247. N. P. Ernsting, M. Kaschke, H. Weller, and L. Katsikas, *J. Opt. Soc. Am. B* 7, 1630 (1990).
248. H. Uchida, C. J. Curtis, P. V. Kamat, K. M. Jones, and A. J. Nozik, *J. Phys. Chem.* 96, 1156 (1992).
249. A. Eychmüller, T. Vobmeyer, A. Mews, and H. Weller, *J. Lumin.* 58, 223 (1994).
250. L. Banyai and S. W. Koch, *Phys. Rev. Lett.* 57, 2722 (1986).
251. C. Luangdilok, D. Lawless, and D. Meisel, in "Fine Particles Science and Technology" (E. Pelizzatti, ed.), p. 457. Kluwer Academic, Boston, 1996.
252. V. L. Colvin and A. P. Alivisatos, *J. Chem. Phys.* 97, 730 (1992).
253. Y. Nosaka, K. Tanaka, and N. Fujii, *J. Appl. Polym. Sci.* 47, 1773 (1993).
254. J. Rabani, *J. Phys. Chem.* 93, 7707 (1989).
255. L. Spanhel, A. Henglein, and H. Weller, *Ber. Bunsen-Ges. Phys. Chem.* 91, 1359 (1987).
256. M. Anpo, K. Chiba, M. Tomonari, S. Coluccia, M. Che, and M. A. Fox, *Bull. Chem. Soc. Jpn.* 64, 543 (1991).
257. S. Hotchandani and P. V. Kamat, *J. Phys. Chem.* 96, 6834 (1992).
258. W. G. Becker and A. J. Bard, *J. Phys. Chem.* 87, 4888 (1983).
259. M. F. Finlayson, K. H. Park, N. Kakuta, A. J. Bard, A. Campion, M. A. Fox, S. E. Webber, and J. M. White, *J. Lumin.* 39, 205 (1988).
260. Z. Alfassi, D. Bahnemann, and A. Henglein, *J. Phys. Chem.* 86, 4656 (1982).
261. S. A. Empedocles, D. K. Norris, and M. G. Bawendi, *Phys. Rev. Lett.* 77, 3873 (1996).
262. M. G. Bawendi, P. J. Carroll, W. L. Wilson, and L. E. Brus, *J. Chem. Phys.* 96, 946 (1992).
263. B. O. Dabbousi, M. G. Bawendi, O. Onitsuka, and M. F. Rubner, *Appl. Phys. Lett.* 66, 1316 (1995).
264. J. S. Massa, G. S. Buller, A. C. Walker, G. Horsburgh, J. T. Mullins, K. A. Prior, and B. C. Cavenett, *Appl. Phys. Lett.* 66, 1346 (1995).
265. M. A. Hines and P. Guyot-Sionnest, *J. Phys. Chem.* 100, 468 (1996).
266. R. Rossetti and L. Brus, *J. Phys. Chem.* 86, 4470 (1982).
267. A. Cornet, J. Collet, T. Amand, M. Pugnet, B. S. Razbirin, and G. Michailov, *J. Phys. Chem. Solids* 44, 53 (1983).
268. W. J. Albery, G. T. Brown, J. R. Darwent, and I. E. Saievar, *J. Chem. Soc., Faraday Trans. 1* 81, 1999 (1985).
269. L. Spanhel, H. Weller, A. Fojtik, and A. Henglein, *Ber. Bunsen-Ges. Phys. Chem.* 91, 88 (1987).
270. M. Hiramoto, K. Hashimoto, and T. Sakata, *Chem. Phys. Lett.* 133, 440 (1987).
271. Y. Wang and N. Herron, *J. Phys. Chem.* 92, 4988 (1988).
272. M. I. Vucemilovic, N. Vukelic, and T. Rajh, *J. Photochem. Photobiol., A* 42, 157 (1988).
273. Y. Y. Aaviksoo, Y. E. Lippmaa, A. M. Freiberg, and S. F. Savikhin, *Sov. Phys. Solid State* 31, 462 (1989).
274. K. R. Gopidas, M. Bohorquez, and P. V. Kamat, *J. Phys. Chem.* 94, 6435 (1990).
275. A. I. Ekimov, I. A. Kudryavtsev, M. G. Ivanov, and A. L. Efros, *J. Lumin.* 46, 83 (1990).
276. A. Eychmueller, A. Haesselbarth, L. Katsikas, and H. Weller, *Ber. Bunsen-Ges. Phys. Chem.* 95, 79 (1991).
277. A. Eychmueller, A. Haesselbarth, L. Katsikas, and H. Weller, *J. Lumin.* 49, 745 (1991).
278. M. Hiramoto, K. Hashimoto, and T. Sakata, *Chem. Phys. Lett.* 182, 139 (1991).
279. K. Misawa, H. Yao, T. Hayashi, and T. Kobayashi, *Chem. Phys. Lett.* 183, 113 (1991).
280. A. Kumar and S. Kumar, *J. Photochem. Photobiol., A* 69, 91 (1992).
281. A. Eychmueller, A. Haesselbarth, and H. Weller, *J. Lumin.* 53, 113 (1992).
282. S. R. Bigham and J. L. Coffer, *J. Phys. Chem.* 96, 10581 (1992).
283. O. Horvath and J. H. Fendler, *J. Phys. Chem.* 96, 9591 (1992).
284. T. Miyoshi, *Jpn. J. Appl. Phys.* 31, 375 (1992).
285. A. Haesselbarth, A. Eychmueller, and H. Weller, *Chem. Phys. Lett.* 203, 271 (1993).
286. A. Kumar and S. Kumar, *J. Photochem. Photobiol., A* 83, 251 (1994).
287. H. S. Zhou, H. Sasahara, I. Honma, H. Komiyama, and J. W. Haus, *Chem. Mater.* 6, 1534 (1994).
288. F. Parsapour, D. F. Kelley, S. Craft, and J. P. Wilcoxon, *J. Chem. Phys.* 104, 4978 (1997).
289. C. Delerue, M. Lannoo, and G. Allan, *J. Lumin.* 57, 249 (1993).
290. P. M. Fauchet, *J. Lumin.* 70, 294 (1996).
291. M. Ben-Chorin, A. Kux, and I. Schechter, *Appl. Phys. Lett.* 64, 481 (1994).
292. L. Brus, *Jpn. J. Appl. Phys.* 2, 34, Suppl. 34-1 (1994).
293. L. Brus, *Phys. Rev. B* 53, 4649 (1996).

294. A. Bsiesy, F. Muller, M. Ligeon, F. Gaspard, R. Herino, R. Romestain, and J. Vial, *Phys. Rev. Lett.* 71, 637 (1993).
295. A. Bsiesy, F. Muller, M. Ligeon, F. Gaspard, R. Herino, R. Romestain, and J. Vial, *Appl. Phys. Lett.* 65, 3371 (1994).
296. E. Kooij, R. Despo, and J. Kelly, *Appl. Phys. Lett.* 66, 2552 (1995).
297. J. M. Lauerhaas and M. J. Sailor, *Science* 261, 1567 (1993).
298. K. Li, C. Tsai, J. Sarathy, and J. Campbell, *Appl. Phys. Lett.* 62, 3192 (1993).
299. H. Noguchi, T. Kondo, and K. Uosaski, *J. Phys. Chem. B* 101, 4978 (1997).
300. K. Uosaki, H. Noguchi, K. Murakoshi, and T. Kondo, *Chem. Lett.* 667 (1995).
301. K. Uosaki, H. Noguchi, K. Murakoshi, and T. Kondo, *J. Phys. Chem.* 100, 4564 (1996).
302. M. O'Neil, J. Marohn, and G. McLendon, *J. Phys. Chem.* 94, 4356 (1990).
303. J. Warnock and D. D. Awschalom, *Appl. Phys. Lett.* 48, 425 (1986).
304. C. R. M. de Oliveira, A. M. de Paula, F. O. P. Filho, J. A. Medeiros Neto, L. C. Barbosa, O. L. Alves, E. A. Menezes, J. M. M. Rios, H. L. Fragnito, and C. H. Brito Cruz, *Appl. Phys. Lett.* 66, 439 (1995).
305. L. Brus, *Isr. J. Chem.* 33, 9 (1993).
306. W. W. Dunn, Y. Aikawa, and A. J. Bard, *J. Am. Chem. Soc.* 103, 3456 (1981).
307. M. D. Ward and A. J. Bard, *J. Phys. Chem.* 86, 3599 (1982).
308. M. D. Ward, J. R. White, and A. J. Bard, *J. Am. Chem. Soc.* 105, 27 (1983).
309. D. Bahnemann, A. Henglein, J. Lilie, and L. Spanhel, *J. Phys. Chem.* 88, 709 (1984).
310. G. Rothenberger, J. Moser, M. Grätzel, N. Serpone, and D. K. Sharma, *J. Am. Chem. Soc.* 107, 8054 (1985).
311. C. Arbour, D. K. Sharma, and C. H. Langford, in "Photochemistry and Photophysics of Coordination Compounds." Springer-Verlag, Berlin, 1987.
312. D. Benjamin and D. Huppert, *J. Phys. Chem.* 92, 4676 (1988).
313. J. J. Kasinski, J. L. A. Gomez, L. Min, Q. Bao, and R. J. D. Miller, *J. Lumin.* 41, 555 (1988).
314. K. Bitterling, F. Willig, and F. Decker, *J. Electroanal. Chem. Interfacial Electrochem.* 228, 29 (1987).
315. J. M. Warman, M. P. de Haas, S. W. F. M. van Hovell tot Westerflier, J. J. M. Binsma, and Z. I. Kolar, *J. Phys. Chem.* 93, 5895 (1989).
316. J. R. Morgan and L. V. Natarajan, *J. Phys. Chem.* 93, 5 (1989).
317. G. J. Meyer, L. K. Leung, J. C. Yu, G. C. Lisensky, and A. B. Ellis, *J. Am. Chem. Soc.* 111, 5146 (1989).
318. P. V. Kamat, T. W. Ebbesen, N. M. Dimitrijevic, and A. J. Nozik, *Chem. Phys. Lett.* 157, 384 (1989).
319. M. O'Neil, J. Marohn, and G. McLendon, *Chem. Phys. Lett.* 168, 208 (1990).
320. R. Kietzmann, F. Willig, H. Weller, R. Vogel, D. N. Nath, R. Eichberger, P. Liska, and J. Lehnert, *Mol. Cryst. Liq. Cryst.* 194, 169 (1991).
321. Y. Rosenwaks, B. R. Thacker, A. J. Nozik, Y. Shapira, and D. Huppert, *J. Phys. Chem.* 97, 10421 (1993).
322. Y. Rosenwaks, B. R. Thacker, R. K. Ahrenkiel, and A. J. Nozik, *J. Phys. Chem.* 96, 10096 (1992).
323. G. P. Lepore, C. H. Langford, J. Vichova, and A. J. Vlcek, *J. Photochem. Photobiol., A* 75, 67 (1993).
324. C. Liu, H. Pan, M. A. Fox, and A. J. Bard, *Science* 261, 897 (1993).
325. M. P. Cantao, J. I. Cisneros, and R. M. Torresi, *J. Phys. Chem.* 98, 4865 (1994).
326. J. G. Zhang, R. H. O'Neil, and T. W. Roberti, *J. Phys. Chem.* 98, 3859 (1994).
327. X. Marguerettaz and D. Fitzmaurice, *J. Am. Chem. Soc.* 116, 5017 (1994).
328. X. Liu, G. Zhang, and J. K. Thomas, *J. Phys. Chem.* 99, 10024 (1995).
329. N. Serpone, D. Lawless, and R. Khairutdinov, *J. Phys. Chem.* 99, 16646 (1995).
330. H. Herrmann, S. Martin, and M. Hoffmann, *J. Phys. Chem.* 99, 16641 (1995).
331. N. Serpone, D. Lawless, R. Khairutdinov, and E. Pelizzetti, *J. Phys. Chem.* 99, 16655 (1995).
332. D. E. Skinner, D. P. J. Colombo, J. J. Cavaleri, and R. M. Bowman, *J. Phys. Chem.* 99, 7853 (1995).
333. G. K. Boschloo and A. Goossens, *J. Phys. Chem.* 100, 19489 (1996).
334. P. V. Kamat and K. Vinodgopal, in "Fine Particles Science and Technology: From Micro to Nanoparticles" (E. Pelizzetti, ed.), p. 303. Kluwer Academic, Dordrecht, The Netherlands, 1996.
335. M. T. Nenadovic, T. Rajh, O. I. Micic, and A. J. Nozik, *J. Phys. Chem.* 88, 5827 (1984).
336. E. Serwicka, M. W. Schlierkamp, and R. N. Schindler, *Z. Naturforsch.* 32 (1981).
337. R. F. Howe and M. Grätzel, *J. Phys. Chem.* 91, 3906 (1987).
338. R. F. Howe and M. Grätzel, *J. Phys. Chem.* 89, 4495 (1985).
339. I. Bedja, S. Hotchandani, and P. V. Kamat, *J. Phys. Chem.* 98, 4133 (1994).
340. J. Z. Zhang, R. H. O'Neil, T. W. Roberti, J. L. McGowen, and J. E. Evans, *Chem. Phys. Lett.* 218, 479 (1994).
341. B. Kraeutler, C. D. Jaeger, and A. J. Bard, *J. Am. Chem. Soc.* 100, 4903 (1978).
342. M. Anpo, T. Shima, and Y. Kubokawa, *Chem. Lett.* 1799 (1985).
343. O. I. Micic, Y. Zhang, K. R. Cromack, A. D. Trifunac, and M. C. Thurnauer, *J. Phys. Chem.* 97, 13284 (1993).
344. O. I. Micic, Y. Zhang, K. R. Cromack, A. D. Trifunac, and M. C. Thurnauer, *J. Phys. Chem.* 97, 7277 (1993).
345. J. M. Warman, M. P. de Haas, M. Grätzel, and P. P. Infelta, *Nature* 310, 306 (1984).
346. J. M. Warman, M. P. de Haas, and H. M. Wentinck, *Radiat. Phys. Chem.* 34, 581 (1989).

347. J. M. Warman, M. P. de Haas, P. Pichat, T. P. M. Koster, A. E. A. van der Zouwen, A. Mackor, and R. Cooper, *Radiat. Phys. Chem.* 37, 433 (1991).
348. J. M. Warman, M. P. de Haas, P. Pichat, and N. Serpone, *J. Phys. Chem.* 95, 8858 (1991).
349. S. T. Martin, H. Herrmann, W. Choi, and M. R. Hoffmann, *Faraday Trans.* 90, 3315 (1994).
350. S. T. Martin, H. Herrmann, and M. R. Hoffmann, *Faraday Trans.* 90, 3323 (1994).
351. R. W. Fessenden and P. V. Kamat, *J. Phys. Chem.* 99, 12902 (1995).
352. P. V. Kamat, K. R. Gopidas, and N. M. Dimitrijevic, *Mol. Cryst. Liq. Cryst.* 183, 439 (1990).
353. A. Henglein, *J. Phys. Chem.* 86, 2291 (1982).
354. D. Duonghong, J. Ramsden, and M. Grätzel, *J. Am. Chem. Soc.* 104, 2977 (1982).
355. T. Watanabe, A. Fujishima, and K. Honda, in "Energy Resour. Photochem. Catal." (M. Grätzel, ed.). Academic Press, New York, 1983.
356. N. Serpone, E. Pelizzetti, and M. Gratzel, *Coord. Chem. Rev.* 64, 225 (1985).
357. M. Grätzel, *NATO ASI Ser. C* 174, 91 (1986).
358. V. N. Parmon, *Adv. Hydrogen Energy* 8, 801 (1990).
359. Z.-C. Bi, Y.-Y. Qian, J. Huang, Z. Xiao, and J.-Y. Yu, *J. Photochem. Photobiol., A* 77, 37 (1994).
360. K. Sayama and H. Arakawa, *J. Photochem. Photobiol., A* 77, 243 (1994).
361. K. Sayama and H. Arakawa, *Chem. Lett.* 6 (1992).
362. K. Sayama and H. Arakawa, *J. Phys. Chem.* 97, 531 (1993).
363. E. Amouyal, *Sol. Energy Mater. Sol. Cells* 38, 249 (1995).
364. A. J. Bard and M. A. Fox, *Acc. Chem. Res.* 28, 141 (1995).
365. M. A. Fox, B. Lindig, and C. C. Chen, *J. Am. Chem. Soc.* 104, 5828 (1982).
366. A. Harriman, G. Porter, and P. Walters, *J. Photochem.* 19, 183 (1982).
367. D. Bahnemann, A. Henglein, and L. Spanhel, *Faraday Discuss. Chem. Soc.* 78, 151 (1984).
368. K. Chandrasekaran and J. K. Thomas, *J. Colloid Interface Sci.* 106, 532 (1985).
369. J. R. Darwent and A. Lepre, *J. Chem. Soc., Faraday Trans. 1* 82, 2323 (1986).
370. N. Serpone, D. K. Sharma, J. Moser, and M. Grätzel, *Chem. Phys. Lett.* 136, 47 (1987).
371. N. Getoff, *Adv. Hydrogen Energy* 93, 383 (1990).
372. R. B. Draper and M. A. Fox, *Langmuir* 6, 1396 (1990).
373. O. Enea, P. Crouigneau, J. Moser, M. Grätzel, and S. Huenig, *New J. Chem.* 15, 267 (1991).
374. J. A. Navio, F. J. Marchena, M. Roncel, and M. A. De la Rosa, *J. Photochem. Photobiol., A* 55, 319 (1991).
375. T. Uchihara, T. Nakamura, and A. Kinjo, *J. Photochem. Photobiol., A* 66, 379 (1992).
376. T. Uchihara, Y. Asato, and A. Kinjo, *J. Photochem. Photobiol., A* 67, 101 (1992).
377. I. Willner, Y. Eichen, A. J. Frank, and M. A. Fox, *J. Phys. Chem.* 97, 7264 (1993).
378. J. R. Darwent, *J. Chem. Soc., Faraday Trans. 1* 80, 183 (1984).
379. J. J. Ramsden and M. Grätzel, *Chem. Phys. Lett.* 132, 269 (1986).
380. R. D. Stramel and J. K. Thomas, *J. Colloid Interface Sci.* 110, 121 (1986).
381. P. V. Kamat and N. M. Dimitrijevic, *Proc. Electrochem. Soc.* 6, 90 (1988).
382. S. Baral and J. H. Fendler, *J. Am. Chem. Soc.* 111, 1604 (1989).
383. A. Mills, P. Douglas, and G. Williams, *J. Photochem. Photobiol., A* 48, 397 (1989).
384. A. Green and A. Mills, *J. Photochem. Photobiol., A* 64, 211 (1992).
385. Y. Nosaka, K. Yamaguchi, A. Kuwabara, H. Miyama, R. Baba, and A. Fujishima, *J. Photochem. Photobiol., A* 64, 375 (1992).
386. H. Matsumoto, H. Uchida, T. Matsunaga, K. Tanaka, T. Sakata, H. Mori, and H. Yoneyama, *J. Phys. Chem.* 98, 11549 (1994).
387. P. V. Kamat, *J. Photochem.* 28, 513 (1985).
388. P. V. Kamat, *J. Chem. Soc., Faraday Trans. 1* 81, 509 (1985).
389. K. R. Gopidas and P. V. Kamat, *J. Phys. Chem.* 93, 6428 (1989).
390. B. Patrick, M. V. George, P. V. Kamat, S. Das, and K. G. Thomas, *J. Chem. Soc., Faraday Trans.* 88, 671 (1992).
391. K. Vinodgopal, I. Bedja, S. Hotchandani, and P. V. Kamat, *Langmuir* 10, 1767 (1994).
392. C. Nasr, K. Vinodgopal, S. Hotchandani, A. K. Chattopadhyay, and P. V. Kamat, *Res. Chem. Intermed.* 23, 219 (1997).
393. P. V. Kamat, *J. Am. Chem. Soc.* 113, 9705 (1991).
394. P. V. Kamat, I. Bedja, and S. Hotchandani, *J. Phys. Chem.* 98, 9137 (1994).
395. P. V. Kamat, I. Bedja, and S. Hotchandani, in "Recent Advances in the Chemistry and Physics of Fullerenes and Related Materials" (K. M. Kadish and R. S. Ruoff, eds.), Vol. 94-24, p. 964. Electrochemical Society, Pennington, NJ, 1994.
396. A. Stasko, V. Brezova, S. Biskupic, K.-P. Dinse, P. Schweitzer, and M. Baumgarten, *J. Phys. Chem.* 99, 8782 (1995).
397. D. Guldi, D. Liu, and P. V. Kamat, *J. Phys. Chem. A* 101, 6195 (1997).
398. G. Sauve, P. V. Kamat, and R. S. Ruoff, *J. Phys. Chem.* 99, 2162 (1995).
399. J. M. Kesselman, G. A. Shreve, M. R. Hoffmann, and N. S. Lewis, *J. Phys. Chem.* 98, 13385 (1994).
400. K. Ishibashi, Y. Nosaka, K. Hashimoto, and A. Fujishima, *J. Phys. Chem. B* 102, 2117 (1998).
401. H. Yamataka, N. Seto, J. Ihichara, T. Hanafusa, and S. Teratani, *J. Chem. Soc., Chem. Commun.* 788 (1985).

402. H. Wang, T. Sakata, M. Azumo, T. Ohta, and H. Takaya, *Chem. Lett.* 1331 (1990).
403. C. Joyce-Pruden, J. K. Pross, and Y. Li, *J. Org. Chem.* 57, 5087 (1992).
404. B. Ohtani, T. Goto, S. Nishimoto, and T. Inui, *J. Chem. Soc., Faraday Trans.* 92, 4291 (1996).
405. T. Shiragami, C. Pac, and S. Yanagida, *J. Phys. Chem.* 94, 504 (1990).
406. T. Shiragami, H. Ankyu, S. Fukami, C. Pac, S. Yanagida, H. Mori, and H. Fujita, *J. Chem. Soc., Faraday Trans.* 88, 1055 (1992).
407. T. Shiragami, S. Fukami, Y. Wada, and S. Yanagida, *J. Phys. Chem.* 97, 12882 (1993).
408. J. Moser and M. Grätzel, *Helv. Chim. Acta* 65, 1436 (1982).
409. D. J. Fitzmaurice, M. Eschle, H. Frei, and J. E. Moser, *J. Phys. Chem.* 97, 3806 (1993).
410. D. P. J. Colombo and R. M. Bowman, *J. Phys. Chem.* 100, 18445 (1996).
411. A. Henglein, *Ber. Bunsen-Ges. Phys. Chem.* 86, 241 (1982).
412. P. V. Kamat and N. M. Dimitrijevic, *J. Phys. Chem.* 93, 4259 (1989).
413. G. T. Brown and J. R. Darwent, *J. Phys. Chem.* 88, 4955 (1984).
414. R. Amadelli, A. Maldotti, C. Bartocci, and V. Carassiti, *J. Phys. Chem.* 93, 6448 (1989).
415. R. B. Draper and M. A. Fox, *J. Phys. Chem.* 94, 4628 (1990).
416. W. G. Becker, M. M. Truong, C. C. Ai, and N. N. Hamel, *J. Phys. Chem.* 93, 4882 (1989).
417. P. V. Kamat, *Langmuir* 1, 608 (1985).
418. J. S. Curran and D. Lamouche, *J. Phys. Chem.* 87, 5405 (1983).
419. N. Lewis, *Ann. Rev. Phys. Chem.* 42, 543 (1991).
420. N. Serpone, D. K. Sharma, M. A. Jamieson, M. Grätzel, and J. J. Ramsden, *Chem. Phys. Lett.* 115, 473 (1985).
421. Y. Nosaka, H. Miyama, M. Terauchi, and T. Kobayashi, *J. Phys. Chem.* 92, 255 (1988).
422. Y. Nosaka and M. A. Fox, *J. Phys. Chem.* 92, 1893 (1988).
423. C. Arbour, D. K. Sharma, and C. H. Langford, *J. Phys. Chem.* 94, 331 (1990).
424. R. Rossetti and L. E. Brus, *J. Phys. Chem.* 90, 558 (1986).
425. Y. Nosaka and M. A. Fox, *Langmuir* 3, 1147 (1987).
426. A. J. Frank, I. Willner, Z. Goren, and Y. Degani, *J. Am. Chem. Soc.* 109, 3568 (1987).
427. T. Rajh and J. Rabani, *Langmuir* 7, 2054 (1991).
428. T. Nakahira and M. Grätzel, *J. Phys. Chem.* 88, 4006 (1984).
429. H. Gerischer, *J. Phys. Chem.* 88, 6096 (1984).
430. H. Gerischer, in "Photocatalytic Purification and Treatment of Water and Air" (D. F. Ollis and H. Al-Ekabi, eds.), p. 1. Elsevier Science, Amsterdam, 1993.
431. J. Moser, S. Punchihewa, P. P. Infelta, and M. Grätzel, *Langmuir* 7, 3012 (1991).
432. W. J. Albery and P. N. Bartlett, *J. Electroanal. Chem. Interfacial Electrochem.* 131, 137 (1982).
433. W. J. Albery and P. N. Bartlett, *J. Electrochem. Soc.* 131, 315 (1984).
434. M. Grätzel and A. J. Frank, *J. Phys. Chem.* 86, 2964 (1982).
435. P. Lemasson, J. Gautron, W. J. Albery, P. N. Bartlett, E. M. P. Dare, and A. Hammett, *J. Electrochem. Soc.* 129, 1269 (1982).
436. W. J. Albery, *J. Electrochem. Soc.* 131, 2896 (1984).
437. W. J. Albery, P. N. Bartlett, C. P. Wilde, and J. R. Darwent, *J. Am. Chem. Soc.* 107, 1854 (1985).
438. P. Mulvaney, V. Swayambunathan, F. Grieser, and D. Meisel, *J. Phys. Chem.* 92, 6732 (1988).
439. S. M. C. Markham, *J. Chem. Ed.* 32, 540 (1955).
440. H. Gerischer and A. Heller, *J. Phys. Chem.* 95, 5261 (1991).
441. C. M. Wang, A. Heller, and H. Gerischer, *J. Am. Chem. Soc.* 114, 5230 (1992).
442. H. Gerischer and A. Heller, *J. Electrochem. Soc.* 139, 113 (1992).
443. M. W. Peterson, J. A. Turner, and A. J. Nozik, *J. Phys. Chem.* 95, 221 (1991).
444. K. Vinodgopal, S. Hotchandani, and P. V. Kamat, *J. Phys. Chem.* 97, 9040 (1993).
445. J. M. Kesselman, A. Kumar, and N. S. Lewis, in "Photocatalytic Purification and Treatment of Water and Air" (D. F. Ollis and H. Al-Ekabi, eds.), p. 19. Elsevier Science, Amsterdam, 1993.
446. D. Schmelling, K. A. Gray, and P. V. Kamat, *Environ. Sci. Technol.* 30, 2547 (1996).
447. S. Kato and F. Masuo, *Kogyo Kagaku Zasshi* (*Ind. Chem. J.*) 67, 1136 (1964).
448. K. Tokumaru, H. Sakuragi, T. Kanno, T. Oguchi, H. Misawa, Y. Shimamura, and Y. Kuriyama, *ACS Symp. Ser.* 55, 43 (1985).
449. B. Kraeutler and A. J. Bard, *J. Am. Chem. Soc.* 100, 2239 (1978).
450. B. Kraeutler and A. J. Bard, *Nouv. J. Chim.* 3, 31 (1979).
451. M. A. Fox, in "Topics in Organic Electrochemistry," p. 177. Plenum Press, New York, 1986.
452. M. A. Fox, in "Advances in Electron Transfer Chemistry" (P. S. Mariano, ed.), Vol. 1, p. 1. Jai Press, Greenwich, CT, 1991.
453. Y. Li and L. Wang, in "Nanocrystalline Semiconductor Materials" (P. V. Kamat and D. Meisel, eds.), p. 391. Elsevier Science, Amsterdam, 1997.
454. H. Al-Ekabi and P. de Mayo, *J. Phys. Chem.* 89, 5815 (1985).
455. T. Hasegawa and P. de Mayo, *Langmuir* 2, 362 (1986).
456. T. Shiragami, S. Fukami, C. Pac, Y. Wada, and S. Yanagida, *Bull. Chem. Soc. Jpn.* 66, 2461 (1993).
457. F. Madhavi, T. C. Burton, and Y. Li, *J. Org. Chem.* 58, 744 (1993).

458. M. Mitani, T. Kiriyama, and T. Kuratate, *J. Org. Chem.* 59, 1279 (1994).
459. T. Inoue, A. Fujishima, S. Konishi, and K. Honda, *Nature* 277, 637 (1979).
460. H. Fujiwara, H. Hosokawa, K. Murakoshi, Y. Wada, S. Yanagida, T. Okada, and H. Kobayashi, *J. Phys. Chem. B* 101, 8270 (1997).
461. H. Inoue, M. Yamachika, and H. Yoneyama, *J. Chem. Soc., Faraday Trans.* 88, 2215 (1992).
462. M. W. Rophael and L. B. Khalil, *Vacuum* 41, 143 (1990).
463. M. M. Taqui Khan, D. Chatterjee, M. Krishnaratnam, and M. Bala, *J. Mol. Catal.* 72, 13 (1992).
464. V. Heleg and I. Willner, *J. Chem. Soc., Chem. Commun.* 2113 (1994).
465. M. M. Taqui Khan, D. Chatterjee, A. Hussain, and M. A. Moiz, *J. Photochem. Photobiol., A* 76, 97 (1993).
466. Z. Goren, I. Willner, A. J. Nelson, and A. J. Frank, *J. Phys. Chem.* 94, 3784 (1990).
467. S. G. Zhang, H. Yamashita, K. Koano, T. Tatsumi, and M. Anpo, *Chem. Lett.* 659 (1997).
468. M. Anpo, H. Yamashita, Y. Ichihashi, Y. Fujii, and M. Honda, *J. Phys. Chem. B* 101, 2632 (1997).
469. K. M. Vijayakumar and N. N. Lichtin, *J. Catal.* 90, 173 (1984).
470. Y. Kohno, T. Tanaka, T. Funabiki, and S. Yoshida, *Chem. Lett.* 993 (1997).
471. S. M. Aliwi and J. K. F. Al, *Sol. Energy Mater.* 18, 223 (1989).
472. H. Kisch and G. Twadzik, *Chem. Ber.* 124, 1161 (1991).
473. T. S. Dzhabiev, B. B. Tarasov, and A. M. Uskov, *Catal. Today* 13, 695 (1992).
474. H. Inoue, H. Moriwaki, K. Maeda, and H. Yoneyama, *J. Photochem. Photobiol., A* 86, 191 (1995).
475. M. Kanemoto, T. Shiragami, C. Pac, and Y. Yanagida, *Chem. Lett.* 931 (1990).
476. M. Kanemoto, T. Shiragami, C. Pac, and Y. Yanagida, *J. Phys. Chem.* 96, 3521 (1992).
477. S. Yanagida, M. Kanemoto, T. Shiragami, and C. Pac, in "Proceedings of the IS&T 44th Annual Conference," 1991.
478. M. Kanemoto, K. Ishihara, Y. Wada, T. Sakata, H. Mori, and S. Yanagida, *Chem. Lett.* 6 (1992).
479. S. Yanagida, M. Kanemoto, H. Hosokawa, Y. Wada, and M. Ishikawa, *Trans. Mater. Res. Soc. Jpn., B* 16, 1575 (1994).
480. S. Yanagida, M. Kanemoto, K. Ishihara, Y. Wada, T. Sakata, and H. Mori, *Bull. Chem. Soc. Jpn.* 70, 2063 (1997).
481. M. Kanemoto, H. Hosokawa, Y. Wada, K. Murakoshi, S. Yanagida, T. Sakata, H. Mori, M. Ishikawa, and H. Kobayashi, *J. Chem. Soc., Faraday Trans.* 92, 2401 (1996).
482. H. Fujiwara, H. Hosokawa, K. Murakoshi, Y. Wada, S. Yanagida, T. Okada, and H. Kobayashi, *J. Phys. Chem. B* 101, 8270 (1997).
483. H. Fujiwara, M. Kanemoto, H. Ankyuu, M. Murakoshi, Y. Wada, and S. Yanagida, *J. Chem. Soc., Perkin Trans.* 2, 317 (1997).
484. G. N. Schrauzer and T. D. Guth, *J. Am. Chem. Soc.* 99, 7189 (1977).
485. K. Tennakone, O. A. Ileperuma, J. M. S. Bandara, and C. T. K. Thaminimulla, *Sol. Energy Mater.* 22, 319 (1991).
486. N. N. Rao, *J. Mol. Catal.* 93, 23 (1994).
487. K. T. Ranjit, R. Krishnamoorthy, T. K. Varadarajan, and B. Viswanathan, *J. Photochem. Photobiol., A* 86, 185 (1995).
488. K. T. Ranjit, B. Viswanathan, and T. K. Varadarajan, *J. Mater. Sci. Lett.* 15, 874 (1996).
489. H. Yamashita, Y. Ichihashi, M. Anpo, M. Hashimoto, C. Louis, and M. Che, *J. Phys. Chem.* 100, 16041 (1996).
490. S. Higashimoto, S. G. Zhang, H. Yamashita, Y. Matsumura, S. Souma, and M. Anpo, *Chem. Lett.* 1127 (1997).
491. D. F. Ollis, E. Pelizzetti, and N. Serpone, *Environ. Sci. Technol.* 25, 1523 (1991).
492. W. A. Zeltner, C. G. Hill, Jr., and M. A. Anderson, *CHEMTECH* 21 (1994).
493. K. Rajeshwar, *J. Appl. Electrochem.* 25, 1067 (1995).
494. P. V. Kamat and K. Vinodgopal, in "Organic and Inorganic Photochemistry" (V. Ramamurthy and K. Schanze, eds.), Vol. 2, p. 307. Marcel Dekker, New York, 1998.
495. R. W. Matthews, *Sol. Energy* 38, 405 (1987).
496. H. Al-Ekabi and N. Serpone, *J. Phys. Chem.* 92, 5726 (1988).
497. S. Yamagata, R. Baba, and A. Fujishima, *Bull. Chem. Soc. Jpn.* 62, 1004 (1989).
498. Y. Ohko, D. A. Tryk, K. Hashimoto, and A. Fujishima, *J. Phys. Chem. B* 102, 2699 (1998).
499. D. F. Ollis, *Environ. Sci. Technol.* 19, 480 (1985).
500. P. Pichat, *NATO ASI Ser. C* 174, 533 (1986).
501. D. F. Ollis, E. Pelizzetti, and N. Serpone, in "Photocatalysis. Fundamentals and Applications" (N. Serpone and E. Pelizzetti, eds.), p. 603. Wiley, New York, 1989.
502. M. Grätzel, in "Photocatalysis. Fundamentals and Applications" (N. Serpone and E. Pelizzetti, eds.). Wiley, New York, 1989.
503. P. Pichat, in "Photochemical Conversion and Storage of Solar Energy" (E. Pelizzetti and M. Schiavello, eds.). Kluwer Academic, Dordrecht, The Netherlands, 1991.
504. N. Serpone, D. Lawless, R. Terzian, C. Minero, and E. Pelizzetti, in "Photochemical Conversion and Storage of Solar Energy" (E. Pelizzetti and M. Schiavello, eds.). Kluwer Academic, Dordrecht, The Netherlands, 1991.

505. N. Serpone, D. Lawless, R. Terzian, and D. Meisel, in "Electrochemistry in Colloids and Dispersions" (R. A. Mackay and J. Texter, eds.), p. 399. VCH Publishers, New York, 1992.
506. J. M. Herrmann, C. Guillard, and P. Pichat, *Catal. Today* 17, 7 (1993).
507. M. A. Fox and M. T. Dulay, *Chem. Rev.* 93, 341 (1993).
508. O. Legrini, E. Oliveros, and A. M. Braun, *Chem. Rev.* 93, 671 (1993).
509. A. Wold, *Chem. Mater.* 5, 280 (1993).
510. D. Bahnemann, J. Cunningham, M. A. Fox, E. Pelizzetti, P. Pichat, and N. Serpone, in "Aquatic and Surface Chemistry" (G. R. Helz, R. G. Zepp, and D. G. Crosby, eds.), p. 261. Lewis Publishers, Boca Raton, FL, 1994.
511. N. Serpone, *J. Photochem. Photobiol., A* 104, 1 (1997).
512. A. Fujishima and K. Honda, *Nature* 238 (1972).
513. K. Vinodgopal, U. Stafford, K. A. Gray, and P. V. Kamat, *J. Phys. Chem.* 98, 6797 (1994).
514. K. Vinodgopal and P. V. Kamat, *Environ. Sci. Technol.* 29, 841 (1995).
515. D. H. Kim and M. A. Anderson, *Environ. Sci. Technol.* 28, 479 (1994).
516. N. Serpone, E. Borgarello, and E. Pelizzetti, *J. Electrochem. Soc.* 135, 2760 (1988).
517. I. U. Haque and J. F. Rusling, *Chemosphere* 26, 1301 (1997).
518. B. Kraeutler and A. J. Bard, *J. Am. Chem. Soc.* 100, 4317 (1978).
519. R. Palmans and A. J. Frank, *J. Phys. Chem.* 95, 9438 (1991).
520. D. Duonghong, E. Borgarello, and M. Grätzel, *J. Am. Chem. Soc.* 103, 4685 (1981).
521. A. Mills and G. Porter, *J. Chem. Soc., Faraday Trans. 1* 78, 3659 (1982).
522. T. Sakata, T. Kawai, and K. Hashimoto, *Chem. Phys. Lett.* 88, 50 (1982).
523. C. P. Lafrance, S. Kaliaguine, P. C. Roberge, and P. Pichat, *Stud. Surf. Sci. Catal.* 17, 309 (1984).
524. R. Baba, S. Nakabayashi, A. Fujishima, and K. Honda, *J. Phys. Chem.* 89, 1902 (1985).
525. T. Uchihara, M. Matsumura, A. Yamamoto, and H. Tsubomura, *J. Phys. Chem.* 93, 5870 (1989).
526. R. Dabestani, X. Wang, A. J. Bard, A. Campion, M. A. Fox, S. E. Webber, and J. M. White, *J. Phys. Chem.* 90, 2729 (1986).
527. D. H. M. W. Thewissen, R. M. Eeuwhorst, K. Timmer, A. H. A. Tinnemans, and A. Mackor, *Recl.: J. R. Neth. Chem. Soc.* 101, 79 (1982).
528. M. M. Taqui Khan, R. C. Bhardwaj, and C. M. Jadhav, *J. Chem. Soc., Chem. Commun.* 1690 (1985).
529. C. Minero, E. Lorenzi, E. Pramauro, and E. Pelizzetti, *Inorg. Chim. Acta* 91, 301 (1984).
530. K. Kalyanasundaram, E. Borgarello, and M. Grätzel, *Helv. Chim. Acta* 64, 362 (1981).
531. W. Gissler, A. J. McEvoy, and M. Grätzel, *J. Electrochem. Soc.* 129, 1733 (1982).
532. D. Gningue, G. Horowitz, J. Roncali, and F. Garnier, *J. Electroanal. Chem. Interfacial Electrochem.* 269, 337 (1989).
533. G. Blondeel, A. Harriman, G. Porter, D. Urwin, and J. Kiwi, *J. Phys. Chem.* 87, 2629 (1983).
534. E. Borgarello, K. Kalyanasundaram, M. Grätzel, and E. Pelizzetti, *Helv. Chim. Acta* 65, 243 (1982).
535. E. Serwicka and R. N. Schindler, *Z. Phys. Chem. (Wiesbaden)* 127, 79 (1981).
536. J. S. Thorp and H. S. Eggleston, *J. Mater. Sci.* 20, 2369 (1985).
537. Y. F. Lee, M. Olshavsky, and J. Chrysochoos, *J. Less-Common Met.* 148, 259 (1989).
538. S. C. Martin, C. L. Morrison, and M. R. Hoffmann, *J. Phys. Chem.* 98, 13695 (1994).
539. W. Choi, A. Termin, and M. R. Hoffmann, *J. Phys. Chem.* 98, 13669 (1994).
540. D. Duonghong, W. Erbs, L. Shuben, and M. Grätzel, *Chem. Phys. Lett.* 95, 266 (1983).
541. U. Resch and M. A. Fox, *J. Phys. Chem.* 95, 6316 (1991).
542. U. Resch and M. A. Fox, *J. Phys. Chem.* 95, 6169 (1991).
543. U. Resch and M. A. Fox, in "Electrochemistry in Colloids and Dispersions" (R. A. Mackay and J. Texter, eds.), p. 417. VCH Publishers, New York, 1992.
544. E. Borgarello, N. Serpone, P. Liska, W. Erbs, M. Grätzel, and E. Pelizzetti, *Gazz. Chim. Ital.* 115, 599 (1985).
545. K. Rajeshwar and M. Kaneko, *J. Phys. Chem.* 89, 3587 (1985).
546. M. M. Taqui Khan, R. C. Bhardwaj, and C. Bhardwaj, *Indian J. Chem., Sect. A* 25, 1 (1986).
547. W. Erbs, J. Kiwi, and M. Grätzel, *Chem. Phys. Lett.* 110, 648 (1984).
548. P. Mulvaney, M. Giersig, and A. Henglein, *J. Phys. Chem.* 97, 7061 (1993).
549. B. Shanghavi and P. V. Kamat, *J. Phys. Chem. B* 101, 7675 (1997).
550. A. Henglein, *J. Phys. Chem.* 97, 5457 (1993).
551. A. Henglein, *Ber. Bunsen-Ges. Phys. Chem.* 99, 903 (1995).
552. M. Mostafavi, N. Keghouche, M.-O. Delcourt, and J. Belloni, *Chem. Phys. Lett.* 167, 193 (1990).
553. J. Khatouri, M. Mostafavi, J. Amblard, and J. Belloni, *Chem. Phys. Lett.* 191, 351 (1992).
554. O. Platzer, J. Amblard, J. L. Marignier, and J. Belloni, *J. Phys. Chem.* 96, 2334 (1992).
555. T. S. Ahmadi, S. L. Logunov, and M. A. El-Sayed, *J. Phys. Chem.* 100, 8053 (1996).
556. S. L. Logunov, T. S. Ahmadi, M. A. El-Sayed, J. T. Khoury, and R. L. Whetten, *J. Phys. Chem. B* 101, 3713 (1997).
557. P. Mulvaney, in "Semiconductor Nanoclusters—Physical, Chemical and Catalytic Aspects" (P. V. Kamat and D. Meisel, eds.), p. 99. Elsevier Science, Amsterdam, 1997.
558. P. Mulvaney, *Langmuir* 12, 788 (1996).

559. U. Kreibig and M. Vollmer, "Optical Properties of Metal Clusters." Springer-Verlag, Berlin, 1995.
560. U. Kreibig, M. Gartz, A. Hilger, and H. Hovel, in "Fine Particles Science and Technology" (E. Pelizzatti, ed.), p. 499. Kluwer Academic, Boston, 1996.
561. I. Lisiecki, F. Billoudet, and M. P. Pileni, *J. Phys. Chem.* 100, 4160 (1996).
562. P. V. Kamat, M. Flumiani, and G. Hartland, *J. Phys. Chem. B* 101, 3123 (1997).
563. M. M. Alvarez, J. T. Khoury, T. G. Schaaff, M. N. Shafigullin, I. Vezmar, and R. L. Whetten, *J. Phys. Chem. B* 101, 3706 (1997).
564. J. W. Thackeray, M. J. Natan, P. Ng, and M. S. Wrighton, *J. Am. Chem. Soc.* 108, 3570 (1986).
565. M. J. Natan, J. W. Thackeray, and M. S. Wrighton, *J. Phys. Chem.* 90, 4089 (1986).
566. A. Fojtik, P. Mulvaney, T. Linnert, M. Giersig, and A. Henglein, *Ber. Bunsen-Ges. Phys. Chem.* 95, 770 (1991).
567. P. V. Kamat, M. de Lind, and S. Hotchandani, *Isr. J. Chem.* 33, 47 (1993).
568. D. Meisel, *Isr. J. Chem.* 33, 53 (1993).
569. R. R. Chandler and J. L. Coffer, *J. Phys. Chem.* 97, 9767 (1993).
570. T. Tuerk, U. Resch, M. A. Fox, and A. Vogler, *J. Phys. Chem.* 96, 3818 (1992).
571. R. Tuerk, U. Resch, M. A. Fox, and A. Vogler, *Inorg. Chem.* 31, 1854 (1992).
572. F. R. F. Fan, H. Y. Liu, and A. J. Bard, *J. Phys. Chem.* 89, 4418 (1985).
573. M. D. Butterworth, S. P. Armes, and A. W. Simpson, *J. Chem. Soc., Chem. Commun.* 2129 (1994).
574. J. Kuczynski and J. K. Thomas, *Langmuir* 1, 158 (1985).
575. Z. J. Zhang, J. Zhang, and Q. J. Xue, *J. Phys. Chem.* 98, 12973 (1994).
576. R. R. Chandler, J. L. Coffer, S. J. Atherton, and P. T. Snowden, *J. Phys. Chem.* 96, 2713 (1992).
577. G. C. Lisensky, R. L. Penn, C. J. Murphy, and A. B. Ellis, *Science* 248, 840 (1990).
578. C. J. R. Cowdery, D. G. Whitten, and G. L. McLendon, *Chem. Phys.* 176, 377 (1993).
579. K. Johansson, R. Cowdery, M. O'Neil, J. Rehm, G. McLendon, A. Marchetti, and D. G. Whitten, *Isr. J. Chem.* 33, 67 (1993).
580. J. J. Ramsden and M. Grätzel, *J. Chem. Soc., Faraday Trans.* 80, 919 (1984).
581. M. Evenor, S. Gottesfeld, Z. Harzion, D. Huppert, and S. W. Feldberg, *J. Phys. Chem.* 88, 6213 (1984).
582. Y. Rosenwaks, L. Burstein, Y. Shapira, and D. Huppert, *J. Phys. Chem.* 94, 6842 (1990).
583. P. P. Bessler, D. Huppert, Y. Rosenwaks, and Y. Shapira, *J. Phys. Chem.* 95, 4370 (1991).
584. H. Meier, *J. Photochem. Photobiol.* 16, 219 (1972).
585. H. Gerischer and F. Willig, *Top. Curr. Chem.* 61, 31 (1976).
586. M. Spitler and B. A. Parkinson, *Langmuir* 2, 549 (1986).
587. M. A. Ryan and M. T. Spitler, *J. Imaging Sci.* 33, 46 (1989).
588. B. O'Regan and M. Grätzel, *Nature (London)* 353, 737 (1991).
589. M. Grätzel, *Coord. Chem. Rev.* 111, 167 (1991).
590. P. V. Kamat and M. A. Fox, *Chem. Phys. Lett.* 102, 379 (1983).
591. J. Moser and M. Grätzel, *J. Am. Chem. Soc.* 106, 6557 (1984).
592. R. Rossetti and L. E. Brus, *J. Am. Chem. Soc.* 106, 4336 (1984).
593. M. A. Ryan, E. C. Fitzgerald, and M. T. Spitler, *J. Phys. Chem.* 93, 6150 (1989).
594. S. Umapathy, A. M. Cartner, A. W. Parker, and R. E. Hester, *J. Phys. Chem.* 94, 8880 (1990).
595. R. W. Fessenden and P. V. Kamat, *Chem. Phys. Lett.* 123, 233 (1986).
596. T. Takizawa, T. Watanabe, and K. Honda, *J. Phys. Chem.* 82, 1391 (1978).
597. K. Hashimoto, M. Hiramoto, and T. Sakata, *Chem. Phys. Lett.* 148, 215 (1988).
598. K. Hashimoto, M. Hiramoto, and T. Sakata, *J. Phys. Chem.* 92, 4272 (1988).
599. Y. Liang, P. F. Moy, J. A. Poole, and A. M. Ponte Goncalves, *J. Phys. Chem.* 88, 2451 (1984).
600. K. Bitterling and F. Willig, *J. Electroanal. Chem. Interfacial Electrochem.* 204, 211 (1986).
601. R. L. Crackel and W. S. Struve, *Chem. Phys. Lett.* 120, 473 (1985).
602. F. Willig, R. Eichberger, N. S. Sundaresan, and B. A. Parkinson, *J. Am. Chem. Soc.* 112, 2702 (1990).
603. R. Eichberger and F. Willig, *Chem. Phys. Lett.* 141, 159 (1990).
604. B. Patrick and P. V. Kamat, *J. Phys. Chem.* 96, 1423 (1992).
605. P. V. Kamat, J. P. Chauvet, and R. W. Fessenden, *J. Phys. Chem.* 90, 1389 (1986).
606. A. Kay, R. Humphry-Baker, and M. Grätzel, *J. Phys. Chem.* 98, 952 (1994).
607. P. V. Kamat, *J. Phys. Chem.* 93, 859 (1989).
608. V. H. Houlding and M. Grätzel, *J. Am. Chem. Soc.* 105, 5695 (1983).
609. K. Kalyanasundaram, N. Vlachopoulos, V. Krishnan, A. Monnier, and M. Grätzel, *J. Phys. Chem.* 91, 2342 (1987).
610. F. R. F. Fan and A. J. Bard, *J. Am. Chem. Soc.* 101, 6139 (1979).
611. J. D. Luttmer and A. J. Bard, *J. Electrochem. Soc.* 126, 414 (1979).
612. B. Burfeindt, T. Hannappel, W. Storck, and F. Willig, *J. Phys. Chem.* 100, 16463 (1996).
613. S. Ferrere, A. Zaban, and B. A. Gregg, *J. Phys. Chem. B* 101, 4490 (1997).
614. E. Vrachnou, N. Vlachopoulos, and M. Grätzel, *J. Chem. Soc., Chem. Commun.* 868 (1987).
615. E. Vrachnou, M. Grätzel, and A. J. McEvoy, *J. Electroanal. Chem. Interfacial Electrochem.* 258, 193 (1989).
616. P. K. Ghosh and T. G. Spiro, *J. Am. Chem. Soc.* 102, 5543 (1980).

617. D. A. Gulino and H. G. Drickamer, *J. Phys. Chem.* 88, 1173 (1984).
618. R. Dabestani, A. J. Bard, A. Campion, M. A. Fox, T. E. Mallouk, S. E. Webber, and J. M. White, *J. Phys. Chem.* 92, 1872 (1988).
619. M. K. Nazeeruddin, P. Liska, J. Moser, N. Vlachopoulos, and M. Grätzel, *Helv. Chim. Acta* 73, 1788 (1990).
620. B. O'Regan, J. Moser, M. Anderson, and M. Grätzel, *J. Phys. Chem.* 94, 8720 (1990).
621. W. E. Ford and M. A. J. Rodgers, *J. Phys. Chem.* 98, 3822 (1994).
622. T. Iwasaki, S. Oda, H. Kamada, and K. Honda, *J. Phys. Chem.* 84, 1060 (1980).
623. H. Hada, Y. Yonezawa, and H. Inaba, *Ber. Bunse-Ges. Phys. Chem.* 85, 425 (1981).
624. C. Kavassalls and M. T. Spitler, *J. Phys. Chem.* 87, 3166 (1983).
625. L. M. Natoli, M. A. Ryan, and M. T. Spitler, *J. Phys. Chem.* 89, 1448 (1985).
626. Y. Hayashi, S. Ogawa, M. Sanada, and R. Hirohashi, *J. Imaging Sci.* 33, 124 (1989).
627. T. Tani, T. Suzumoto, and K. Ohzeki, *J. Phys. Chem.* 94, 1298 (1990).
628. R. Kietzmann, A. Ehret, M. Spitler, and F. Willig, *J. Am. Chem. Soc.* 115, 1930 (1993).
629. C. Nasr, S. Hotchandani, P. V. Kamat, S. Das, K. George Thomas, and M. V. George, *Langmuir* 11, 1777 (1995).
630. S. Hotchandani, S. Das, K. G. Thomas, M. V. George, and P. V. Kamat, *Res. Chem. Intermed.* 20, 927 (1994).
631. Y.-S. Kim, K. Liang, K.-Y. Law, and D. G. Whitten, *J. Phys. Chem.* 98, 984 (1994).
632. H. Frei, D. J. Fitzmaurice, and M. Grätzel, *Langmuir* 6, 198 (1990).
633. T. Nakahira and M. Grätzel, *Makromol. Chem., Rapid Commun.* 6, 341 (1985).
634. T. Nakahira, Y. Inoue, K. Iwasaki, H. Tanigawa, Y. Kouda, S. Iwabuchi, K. Kojima, and M. Grätzel, *Makromol. Chem., Rapid Commun.* 9, 13 (1988).
635. T. Tani, T. Suzumoto, K. Kemnitz, and K. Yoshihara, *J. Phys. Chem.* 96, 2778 (1992).
636. M. T. Spitler, *J. Imaging Sci.* 35, 351 (1991).
637. J. E. Moser and M. Grätzel, *Chem. Phys.* 176, 493 (1993).
638. D. Liu and P. V. Kamat, *J. Electrochem. Soc.* 142, 835 (1995).
639. D. Liu, G. L. Hug, and P. V. Kamat, *J. Phys. Chem.* 99, 16768 (1995).
640. D. Liu and P. V. Kamat, *Langmuir* 12, 2190 (1996).
641. D. Liu and P. V. Kamat, *J. Chem. Phys.* 105, 965 (1996).
642. D. Liu, R. W. Fessenden, G. L. Hug, and P. V. Kamat, *J. Phys. Chem. B* 101, 2583 (1997).
643. I. Martini, G. Hartland, and P. V. Kamat, *J. Phys. Chem. B* 101, 4826 (1997).
644. C. Nasr, D. Liu, S. Hotchandani, and P. V. Kamat, *J. Phys. Chem.* 100, 11054 (1996).
645. D. M. Klassen and G. A. Crosby, *J. Chem. Phys.* 48, 1853 (1968).
646. C. Creutz and N. Sutin, *J. Am. Chem. Soc.* 98, 6384 (1976).
647. J. Van Houten and R. J. Watts, *J. Am. Chem. Soc.* 98, 4853 (1976).
648. D. Meisel, M. S. Matheson, W. A. Mulac, and J. Rabani, *J. Phys. Chem.* 81, 1449 (1977).
649. J. C. Casper and T. J. Meyer, *J. Am. Chem. Soc.* 105, 5583 (1983).
650. Y. Komada, S. Yamauchi, and N. Hirota, *J. Phys. Chem.* 92, 6511 (1988).
651. E. Krausz, *Comments Inorg. Chem.* 7, 139 (1988).
652. D. J. Fitzmaurice and H. Frei, *Langmuir* 7, 1129 (1991).
653. V. Kurshev and L. Kevan, *Langmuir* 13, 225 (1997).
654. P. V. Kamat, K. R. Gopidas, and D. Weir, *Chem. Phys. Lett.* 149, 491 (1988).
655. K. R. Gopidas, P. V. Kamat, and M. V. George, *Mol. Cryst. Liq. Cryst.* 183, 403 (1990).
656. K. Vinodgopal and P. V. Kamat, *J. Phys. Chem.* 96, 5053 (1992).
657. M. Gevaert and P. V. Kamat, *J. Chem. Soc., Chem. Commun.* 1470 (1992).
658. K. Vinodgopal, X. Hua, R. L. Dahlgren, A. G. Lappin, L. K. Patterson, and P. V. Kamat, *J. Phys. Chem.* 99, 10883 (1995).
659. D. Liu, P. V. Kamat, K. George Thomas, K. J. Thomas, S. Das, and M. V. George, *J. Chem. Phys.* 106, 6404 (1997).
660. P. V. Kamat and B. Patrick, in "Electrochemistry in Colloids and Dispersions" (R. A. Mackay and J. Texter, eds.), p. 447. VCH Publishers, New York, 1992.
661. K. Itoh, Y. Chiyokawa, M. Nakao, and K. Honda, *J. Am. Chem. Soc.* 106, 1620 (1984).
662. K. Hashimoto, M. Hiramoto, T. Kajiwara, and T. Sakata, *J. Phys. Chem.* 92, 4636 (1988).
663. K. Hashimoto, M. Hiramoto, A. B. P. Lever, and T. Sakata, *J. Phys. Chem.* 92, 1016 (1988).
664. K. Kemnitz, N. Nakashima, K. Yoshihara, and H. Matsunami, *J. Phys. Chem.* 93, 6704 (1989).
665. K. Misawa, K. Minoshima, and T. Kobayashi, *J. Raman Spectrosc.* 26, 553 (1995).
666. J. M. Rehm, G. L. McLendon, Y. Nagasawa, K. Yoshihara, J. Moser, and M. Grätzel, *J. Phys. Chem.* 100, 9577 (1996).
667. Y. Tachibana, J. E. Moser, M. Grätzel, D. R. Klug, and J. R. Durrant, *J. Phys. Chem.* 100, 20056 (1996).
668. I. Martini, J. Hodak, G. Hartland, and P. V. Kamat, *J. Chem. Phys.* 107, 8064 (1997).
669. T. Hannappel, B. Burfeindet, W. Storck, and F. Willig, *J. Phys. Chem. B* 101, 6799 (1997).
670. W. E. Ford and M. A. J. Rodgers, *J. Phys. Chem.* 101, 930 (1997).
671. F. Willig, in "Surface Electron Transfer Process" (R. J. Dwayne Miller, G. L. McLendon, A. J. Nozik, W. Scmickler, and F. Willig, eds.), p. 167. VCH Publishers, New York, 1995.

672. R. J. D. Miller, in "Surface Electron Transfer Process" (R. J. D. Miller, G. L. McLendon, A. J. Nozik, W. Scmickler, and F. Willig, eds.), p. 95. VCH Publishers, New York, 1995.
673. J. M. Lanzafame, R. J. D. Miller, A. A. Muenter, and B. A. Parkinson, *J. Phys. Chem.* 96, 2820 (1992).
674. N. J. Cherepy, G. P. Smestad, M. Grätzel, and J. Z. Zhang, *J. Phys. Chem. B* 101, 9342 (1997).
675. K. Murakoshi, S. Yanagida, M. Capel, and E. W. Castner, in "Nanostructured Materials: Clusters, Composites, and Thin Films" (M. Moskovits, ed.), p. 221. American Chemical Society, 1997.
676. J. Randy, R. J. Ellingson, J. B. Asbury, S. Ferrere, H. N. Ghosh, J. R. Sprague, T. Lian, and A. J. Nozik, *J. Phys. Chem. B* 102, 6455 (1998).
677. D. Liu and P. V. Kamat, *J. Electroanal. Chem. Interfacial Electrochem.* 347, 451 (1993).
678. N. Uekawa, T. Suzuki, S. Ozeki, and K. Kaneko, *Langmuir* 8, 1 (1992).
679. D. Liu and P. V. Kamat, *J. Phys. Chem.* 97, 10769 (1993).
680. R. Vogel, P. Hoyer, and H. Weller, *J. Phys. Chem.* 98, 3183 (1994).
681. H. Gerischer and M. Lübke, *J. Electroanal. Chem. Interfacial Electrochem.* 204, 225 (1986).
682. R. Vogel, K. Pohl, and H. Weller, *Chem. Phys. Lett.* 174, 241 (1990).
683. S. Kohtani, A. Kudo, and T. Sakata, *Chem. Phys. Lett.* 206, 166 (1993).
684. I. Bedja and P. V. Kamat, *J. Phys. Chem.* 99, 9182 (1995).
685. S. Hotchandani and P. V. Kamat, *Chem. Phys. Lett.* 191, 320 (1992).
686. A. J. Nozik, *Appl. Phys. Lett.* 30, 567 (1977).
687. N. Serpone, E. Borgarello, and M. Grätzel, *J. Chem. Soc., Chem. Commun.* 342 (1984).
688. L. Spanhel, H. Weller, and A. Henglein, *J. Am. Chem. Soc.* 109, 6632 (1987).
689. H. S. Zhou, I. Honma, and H. Komiyama, *J. Phys. Chem.* 97, 895 (1993).
690. A. Haesselbarth, A. Eychmüller, R. Eichberger, M. Giersig, A. Mews, and H. Weller, *J. Phys. Chem.* 97, 5333 (1993).
691. J. E. Evans, K. W. Springer, and J. Z. Zhang, *J. Chem. Phys.* 101, 6222 (1994).
692. J. W. Haus, H. S. Zhou, I. Honma, and H. Komiyama, *Phys. Rev.* 47, 1359 (1993).
693. M. Ocana, W. P. Hsu, and E. Matijevic, *Langmuir* 7, 2911 (1991).
694. G. Hodes, *Isr. J. Chem.* 33, 95 (1993).
695. G. J. Meyer and P. C. Searson, *Interface* 23 (1993).
696. P. V. Kamat, *CHEMTECH* 22 (1995).
697. A. Hagfeldt and M. Grätzel, *Chem. Rev.* 95, 49 (1995).
698. A. W. H. Mau, C. B. Huang, N. Kakuta, A. J. Bard, A. Campion, M. A. Fox, J. M. White, and S. E. Webber, *J. Am. Chem. Soc.* 106, 6537 (1984).
699. K. Honda, A. Kuwano, K. Chiba, A. Ishikawa, and H. Miyama, *Chem. Lett.* 195 (1988).
700. M. Kaneko, T. Okada, S. Teratani, and K. Taya, *Electrochim. Acta* 32, 1405 (1987).
701. O. V. Salata, P. J. Dobson, P. J. Hull, and J. L. Hutchison, *Thin Solid Films* 251, 81 (1994).
702. M. Gao, Y. Yang, B. Yang, F. Bian, and J. Shen, *J. Chem. Soc., Chem. Commun.* 2779 (1994).
703. Y. Wang and N. Herron, *J. Lumin.* 70, 48 (1996).
704. J. H. Fendler, *Chem. Rev.* 87, 877 (1987).
705. W. Kautek and H. Gerischer, *Electrochim. Acta* 27, 355 (1982).
706. D. V. Paranjape, M. Sastry, and P. Ganguly, *Appl. Phys. Lett.* 63, 18 (1993).
707. R. S. Urquhart, C. L. Hoffmann, D. N. Furlong, N. J. Geddes, J. F. Raboldt, and F. Grieser, *J. Phys. Chem.* 99, 15987 (1995).
708. L. Sereno, J. Silber, L. Otero, M. Bohorquez, A. L. Moore, T. A. Moore, and D. Gust, *J. Phys. Chem.* 100, 814 (1996).
709. F. Grieser, D. N. Furlong, R. S. Urquhart, and D. J. Elliot, in "Fine Particles Science and Technology" (E. Pelizzatti, ed.), p. 733. Kluwer Academic, Boston, 1996.
710. L. Motte, F. Billoudet, E. Lacaze, J. Douin, and M. P. Pileni, *J. Phys. Chem.* 101, 138 (1997).
711. S. G. Yan and J. T. Hupp, *J. Phys. Chem. B* 101, 1493 (1997).
712. I. Moriguchi, H. Maeda, Y. Teraoka, and S. Kagawa, *J. Am. Chem. Soc.* 1139 (1995).
713. Q. Xu and M. A. Anderson, *J. Mater. Res.* 6, 1073 (1991).
714. S. Tunesi and M. Anderson, *J. Phys. Chem.* 95, 3399 (1991).
715. J. Sabate, M. A. Anderson, M. A. Aguado, J. Gimenez, M. S. Cervera, and C. G. J. Hill, *J. Mol. Catal.* 71, 57 (1992).
716. G. Dagan and M. Tomkiewicz, *J. Phys. Chem.* 97, 12651 (1993).
717. M. N. Kamalasanan, N. D. Kumar, and S. Chandra, *J. Appl. Phys.* 74, 679 (1993).
718. S. Sakohara, L. D. Tickanen, and M. A. Anderson, *J. Phys. Chem.* 96, 11086 (1992).
719. D. Papoutsi, P. Lianos, P. Yianoulis, and P. Koutsoukos, *Langmuir* 10, 1684 (1994).
720. B. O'Regan and D. T. Schwartz, *Chem. Mater.* 7, 1349 (1995).
721. D. M. Antonelli and J. Y. Ying, *Angew. Chem., Int. Ed. Engl.* 34, 2014 (1995).
722. N. Negishi, T. Iyoda, K. Hashimoto, and A. Fujishima, *Chem. Lett.* 841 (1995).
723. C. Anderson and A. J. Bard, *J. Phys. Chem.* 99, 9882 (1995).
724. T. Jiang, A. L. Lough, G. A. Ozin, and D. Young, *Chem. Mater.* 7, 245 (1995).
725. T. Nakanishi, B. Ohtani, and K. Uosaki, *J. Phys. Chem. B* 102, 1571 (1998).
726. G. Hodes, I. D. J. Howell, and L. M. Peter, *J. Electrochem. Soc.* 139, 3136 (1992).

727. S. Hotchandani and P. V. Kamat, *J. Electrochem. Soc.* 139, 1630 (1992).
728. I. Bedja, S. Hotchandani, R. Carpentier, R. W. Fessenden, and P. V. Kamat, *J. Appl. Phys.* 75, 5444 (1994).
729. I. Bedja, S. Hotchandani, R. Carpentier, K. Vinodgopal, and P. V. Kamat, *Thin Solid Films* 247, 195 (1994).
730. U. Bjorksten, J. Moser, and M. Grätzel, *Chem. Mater.* 6, 858 (1994).
731. A. Hagfeldt, S. E. Lindquist, and M. Grätzel, *Sol. Energy Mater. Sol. Cells* 32, 245 (1994).
732. C. Nasr, S. Hotchandani, and P. V. Kamat, *Proc. Ind. Acad. Sci.* 107, 699 (1995).
733. P. V. Kamat, in "Nanoparticles and Nanostructural Films" (J. Fendler, ed.), p. 207. Wiley–VCH, New York, 1998.
734. I. Bedja, S. Hotchandani, and P. V. Kamat, *J. Electroanal. Chem.* 401, 237 (1996).
735. I. Bedja, P. V. Kamat, and S. Hotchandani, *J. Appl. Phys.* 80, 4637 (1996).
736. K. Vinodgopal, D. Wynkoop, and P. V. Kamat, *Environ. Sci. Technol.* 30, 1660 (1996).
737. H. Lindstrom, H. Rensmo, S. Sodergren, A. Solbrand, and S. Lindquist, *J. Phys. Chem.* 3084 (1996).
738. N. Papageorgiou, Y. Athanassov, M. Armand, P. Bonhote, H. Petterson, A. Azam, and M. Grätzel, *J. Electrochem. Soc.* 143, 3099 (1996).
739. S. G. Yan and J. T. Hupp, *J. Phys. Chem.* 100, 6867 (1996).
740. L. Kavan, M. Grätzel, J. Rathousky, and A. Zukal, *J. Electrochem. Soc.* 143, 394 (1996).
741. A. Zaban, S. Ferrere, J. Sprague, and B. A. Gregg, *J. Phys. Chem. B* 101, 55 (1997).
742. C. Nasr, S. Hotchandani, W. Y. Kim, R. H. Schmehl, and P. V. Kamat, *J. Phys. Chem. B* 101, 7480 (1997).
743. P. V. Kamat, in "Progress in Inorganic Chemistry Series" (J. Meyer, ed.), Vol. 44, p. 273. Wiley, New York, 1997.
744. J. J. Fang, J. X. Wu, X. Lu, S. Y., and Z. Lu, *Chem. Phys. Lett.* 270, 145 (1997).
745. C. Nasr, S. Hotchandani, and P. V. Kamat, in "Photoelectrochemistry" (K. Rajeshwar, ed.). Electrochemical Society, Pennington, NJ, 1997.
746. I. Bedja, P. V. Kamat, X. Hua, A. G. Lappin, and S. Hotchandani, *Langmuir* 13, 2398 (1998).
747. C. Nasr, S. Hotchandani, and P. V. Kamat, *J. Phys. Chem. B* 102, to appear.
748. M. K. Nazeeruddin, A. Kay, I. Rodicio, B. R. Humphry, E. Mueller, P. Liska, N. Vlachopoulos, and M. Grätzel, *J. Am. Chem. Soc.* 115, 6382 (1993).
749. R. Ionescu, C. Moise, and A. Vancu, *Appl. Surf. Sci.* 84, 291 (1995).
750. O. Savadogo and K. C. Mandal, *J. Electrochem. Soc.* 141, 2871 (1994).
751. V. L. Colvin, M. C. Schlamp, and A. P. Alivisatos, *Nature* 370, 354 (1994).
752. S. Hotchandani, I. Bedja, R. W. Fessenden, and P. V. Kamat, *Langmuir* 10, 17 (1994).
753. L. Kavan, T. Stoto, M. Grätzel, D. J. Fitzmaurice, and V. Shklover, *J. Phys. Chem.* 97, 9493 (1993).
754. K. Rajeshwar, R. O. Lezna, and N. R. deTacconi, *Anal. Chem.* 64, 429 (1992).
755. G. S. Hsiao, M. G. Anderson, S. Gorer, D. Harris, and R. M. Penner, *J. Am. Chem. Soc.* 119, 1439 (1997).
756. M. A. Anderson, S. Gorer, and R. M. Penner, *J. Phys. Chem. B* 101, 5895 (1997).
757. E. A. Kamenetzky, L. G. Mangliocco, and H. P. Panzer, *Science* 263, 207 (1994).
758. H. B. Sunkara, J. M. Jethmalani, and W. T. Ford, *Chem. Mater.* 6, 362 (1994).
759. W. L. Vos, R. Sprik, A. van Blaaderen, A. Imhof, A. Lagendijk, and G. H. Wegdam, *Phys. Rev. B* 53, 16231 (1996).
760. I. I. Tarhan and G. H. Watson, *Phys. Rev. Lett.* 76, 315 (1996).
761. C. R. Kagan, C. B. Murray, and M. G. Bawendi, *Phys. Rev. B* 54, 8633 (1996).
762. J. R. Heath, C. M. Knobler, and D. V. Leff, *J. Phys. Chem. B* 101, 189 (1997).
763. G. Chmanov, K. Sokolov, and T. M. Cotton, *J. Phys. Chem.* 100, 5166 (1996).
764. C. A. Mirkin, R. L. Letsinger, R. C. Mucic, and J. J. Storhoff, *Nature* 382, 607 (1996).
765. C. Murray, C. Kagan, and M. Bawendi, *Science* 270, 1335 (1995).
766. W. Shenton, D. Pum, U. B. Sleytr, and S. Mann, *Nature* 389, 585 (1997).
767. L. Cusack, X. Marguerettaz, S. N. Rao, J. Wenger, and D. Fitzmaurice, *Chem. Mater.* 9, 1765 (1997).
768. L. Cusack, R. Rizza, A. Gorelov, and D. Fitzmaurice, *Angew. Chem., Int. Ed. Engl.* 36, 848 (1997).
769. T. Moritz, J. Reiss, K. Diesner, D. Su, and A. Chemseddine, *J. Phys. Chem. B* 101, 8052 (1997).
770. J. S. Yin and Z. L. Wang, *J. Phys. Chem. B* 101, 8979 (1997).
771. J. S. Yin and Z. L. Wang, *Phys. Rev. Lett.* 79, 2570 (1997).
772. L. Motte, F. Billoudet, E. Lacaze, and M.-P. Pileni, *Adv. Mater.* 8 (1996).
773. S. Ogawa, K. Hu, F.-R. F. Fan, and A. J. Bard, *J. Phys. Chem. B* 101, 5707 (1997).
774. T. Takagahara, *Surf. Sci.* 267, 310 (1992).
775. P. V. Braun, P. Osenar, and S. I. Stupp, *Nature* 380, 325 (1996).
776. S. G. Romanv, C. M. Sotomayor Torres, H. M. Yates, M. E. Pemble, V. Butko, and V. Tretijakov, *J. Appl. Phys.* 82, 380 (1997).
777. R. Routkevitch, T. Bigioni, M. Martin, and J. M. Xu, *J. Phys. Chem.* 100, 14037 (1996).
778. J. D. Klein, R. D. Herrick, II, D. Palmer, and M. J. Sailor, *Chem. Mater.* 5, 902 (1993).
779. A. Kay and M. Grätzel, *J. Phys. Chem.* 97, 6272 (1993).
780. P. Bonhote, E. Gogniat, S. Tingry, C. Barbe, N. Vlachopoulos, F. Lenzmann, P. Comte, and M. Grätzel, *J. Phys. Chem. B* 102, 1498 (1998).
781. C. M. Lampert, *Solar Energy Mater.* 11, 1 (1984).
782. B. O'Regan, M. Grätzel, and D. Fitzmaurice, *Chem. Phys. Lett.* 183, 89 (1991).

783. A. Hagfeldt, N. Vlachopoulos, and M. Grätzel, *J. Electrochem. Soc.* 141, L82 (1994).
784. C. Bechinger, S. Ferrere, A. Zaban, J. Sprague, and B. A. Gregg, *Nature* 383, 608 (1996).
785. B. O'Regan, M. Grätzel, and D. Fitzmaurice, *J. Phys. Chem.* 95, 10525 (1991).
786. G. Redmond, D. Fitzmaurice, and M. Grätzel, *J. Phys. Chem.* 97, 6951 (1993).
787. G. Redmond and D. Fitzmaurice, *J. Phys. Chem.* 97, 1426 (1993).
788. L. Butler, G. Redmond, and D. Fitzmaurice, *J. Phys. Chem.* 97, 10750 (1993).
789. S. Sodergren, A. Hagfeldt, J. Olsson, and S.-E. Lindquist, *J. Phys. Chem.* 98, 5552 (1994).
790. N. Chestnoy, T. D. Harris, R. Hull, and L. E. Brus, *J. Phys. Chem.* 90, 3393 (1986).
791. H. Tributsch, G. Schlichthoerl, and L. Elstner, *Electrochim. Acta* 38, 141 (1993).
792. A. J. Nozik, B. R. Thacker, J. A. Turner, and J. M. Olson, *J. Am. Chem. Soc.* 107, 7805 (1985).
793. A. J. Nozik, B. R. Thacker, and J. M. Olson, *Nature* 316, 51 (1985).
794. A. J. Nozik, B. R. Thacker, J. A. Turner, J. Klem, and H. Morkoc, *Appl. Phys. Lett.* 50, 34 (1987).
795. A. J. Nozik, B. R. Thacker, J. A. Turner, and M. W. Peterson, *J. Am. Chem. Soc.* 110, 7630 (1988).
796. A. J. Nozik, J. A. Turner, and M. W. Peterson, *J. Phys. Chem.* 92, 2493 (1988).
797. C. A. Parsons, M. W. Peterson, B. R. Thacker, J. A. Turner, and A. J. Nozik, *J. Phys. Chem.* 94, 3381 (1990).
798. C. A. Bignozzi, R. Argazzi, T. Indelli, and F. Scandola, *Solar Energy Mater.* 32, 229 (1994).
799. G. Smestad, C. Bignozzi, and R. A. Argazzi, *Solar Energy Mater.* 32, 259 (1994).
800. G. Smestad, *Solar Energy Mater.* 32, 273 (1994).
801. P. V. Kamat, *Langmuir* 6, 512 (1990).
802. P. V. Kamat, S. Das, K. G. Thomas, and M. V. George, *Chem. Phys. Lett.* 178, 75 (1991).
803. H. Takemura, T. Saji, M. Fujihira, S. Aoyagui, K. Hashimoto, and T. Sakata, *Chem. Phys. Lett.* 122, 496 (1985).
804. K. Hashimoto, M. Hiramoto, T. Sakata, H. Muraki, H. Takemura, and M. Fujihira, *J. Phys. Chem.* 91, 6198 (1987).
805. K. Schwarzburg and F. Willig, *Appl. Phys. Lett.* 58, 2520 (1991).
806. H. Grabert and M. H. Devoret, "Single Charge Tunneling: Coulomb Blockade Phenomena in Nanostructures." Plenum Press, London, 1992.
807. R. P. Andres, T. Bein, M. Dorogi, S. Feng, J. I. Henderson, C. P. Kubiak, W. Mahoney, R. G. Osifchin, and R. Reifenberger, *Science* 272, 323 (1996).
808. R. P. Andres, S. Datta, M. Dorogi, J. Gomez, J. I. Henderson, D. B. Jones, V. R. Kolagunta, C. P. Kubiak, W. Mahoney, and R. G. Osifchin, *J. Vac. Sci. Technol., A* 14, 1178 (1996).
809. L. Guo, E. Leobandung, and S. Y. Chou, *Science* 275, 649 (1997).
810. D. L. Fedheim, K. C. Grabar, M. J. Natan, and T. E. Mallouk, *J. Am. Chem. Soc.* 118, 7640 (1996).
811. F. F. Fan and A. J. Bard, *Science* 277, 1791 (1997).
812. D. L. Klein, P. L. McEuen, J. E. B. Katari, R. Roth, and P. Alivisatos, *Appl. Phys. Lett.* 68, 2574 (1996).
813. D. L. Klein, R. Roth, A. K. L. Lim, P. Alivisatos, and P. L. McEuen, *Nature* 389, 699 (1997).
814. K. Vinodgopal, I. Bedja, and P. V. Kamat, *Chem. Mater.* 8, 2180 (1996).

Chapter 5

COLLOIDAL QUANTUM DOTS OF III–V SEMICONDUCTORS*

Olga I. Mićić, Arthur J. Nozik

Center for Basic Sciences, National Renewable Energy Laboratory, Golden, Colorado, USA

Contents

1. INTRODUCTION

Quantization effects in semiconductor structures were first demonstrated in the early 1970s in III–V quantum wells [1, 2]; these structures consisted of a thin epitaxial film of a smaller band gap (E_g) semiconductor (e.g., GaAs, $E_g = 1.42$ eV) sandwiched between two epitaxial films of a larger band gap semiconductor (e.g., $Al_{0.3}Ga_{0.7}As$, $E_g = 2.0$ eV). The conduction and valence band offsets of the two semiconductor materials produce potential barriers for electrons and holes, respectively; the smaller band gap semiconductor constitutes the quantum well region and the larger band gap material the potential barrier region. If the film of the smaller band gap material is sufficiently thin (thickness less than the De Broglie wavelength of the charge carriers, which typically requires thicknesses less than about 300 Å for III–V semiconductors), then the charge carriers are confined in one dimension by the potential barriers and quantization of the energy levels for both electrons and holes can occur (Fig. 1).

*Parts of this manuscript have been excerpted with permission from [17] and [24]. © 1995 and 1997 American Chemical Society.

Handbook of Nanostructured Materials and Nanotechnology, edited by H.S. Nalwa
Volume 3: Electrical Properties

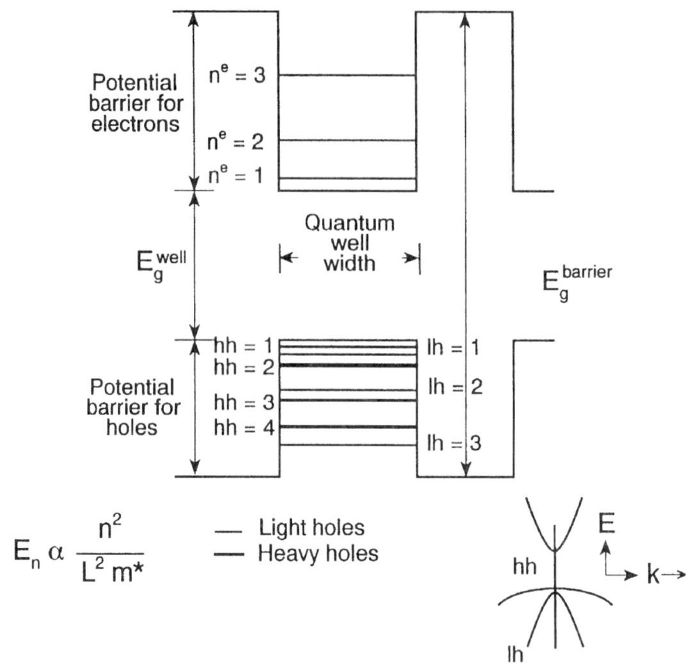

Fig. 1. Quantization in quantum wells (quantum films).

In general, three regimes of quantization can be defined depending on whether the charge carriers are confined in one, two, or three dimensions. Confinement in one dimension, described previously, creates structures that have been termed quantum wells; this is because the first such structures [1] could be described by the elementary quantum mechanics of a particle in a one-dimensional box; these structures have also been labeled as "quantum films." Carrier confinement in two dimensions produces "quantum wires," and confinement in three dimensions produces "quantum dots" (also called nanocrystals or quantum boxes). The dimensionality of confinement affects many aspects of quantization; Figure 2 shows the effect on the density of states. Many reviews of quantum wells or films [3–5], quantum wires [6–8], and quantum dots [6, 9–15] are available.

Quantum wells are produced by the growth of high-quality epitaxial layers of semiconductors by molecular beam epitaxy (MBE) or metallorganic chemical vapor deposition (MOCVD). Quantum dots (QDs) are usually produced by one of two approaches: (1) epitaxial growth and/or nanoscale patterning and (2) formation of nanosized semiconductor particles through colloidal chemistry. This chapter will deal only with the preparation, characterization, spectroscopy, and applications of QDs of III–V semiconductors produced via colloidal chemistry. Some important characteristics of the colloidal route to nanocrystalline QDs are: (1) very small sized QDs can be readily formed (20–100 Å); (2) colloidal QDs can be easily formed into closed-packed QD arrays to create either amorphous QD solids or crystalline QD solids (three-dimensional superlattices and opal-like structures); (3) QD size distributions can be made relatively narrow (5%–10% variation in average QD size); (4) high-quality QD nanocrystallites can be formed; and (5) colloidal QDs can be studied either in solution, as solid powders, or as solid QD arrays in films. These characteristics have been realized in II–VI [9–15] and III–V QDs [16–24]. Other semiconductors that have been prepared as colloidal QDs include groups I–VII, IV, IV–VI, and layered chalcogenides [6, 25].

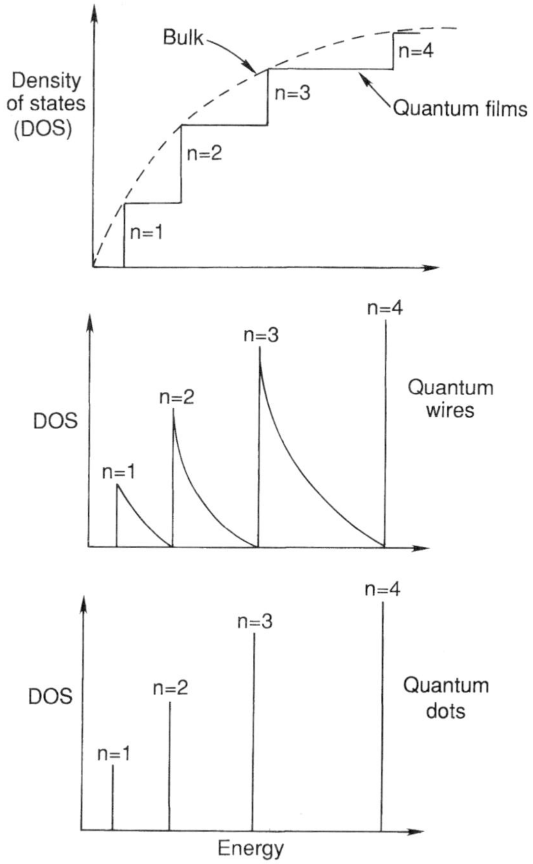

Fig. 2. Effects of dimensionality of quantization on the electronic density of states.

2. SYNTHESIS OF COLLOIDAL QUANTUM DOTS

The most common approach to the synthesis of colloidal QDs is the controlled nucleation and growth of particles in a solution of chemical precursors containing the metal and the anion sources (controlled arrested precipitation) [26–28]. The technique of forming monodisperse colloids is very old, and can be traced back to the synthesis of gold colloids by Michael Faraday in 1857. A common method for II–VI colloidal QD formation is to rapidly inject a solution of chemical reagents containing the group II and group VI species into a hot and vigorously stirred solvent containing molecules that can coordinate with the surface of the precipitated QD particles [26, 28, 29]. Consequently, a large number of nucleation centers are initially formed, and the coordinating ligands in the hot solvent prevent or limit particle growth via Ostwald ripening (the growth of larger particles at the expense of smaller particles to minimize the higher surface free energy associated with smaller particles). Further improvement of the resulting size distribution of the QD particles can be achieved through selective precipitation [28, 29], whereby the slow addition of a nonsolvent to the colloidal solution of particles causes precipitation of the larger-sized particles (the solubility of molecules with the same type of chemical structure decreases with increasing size). This process can be repeated several times to narrow the size distribution of II–VI colloidal QDs to a few percent of the mean diameter [28, 29].

The synthesis of colloidal III–V QDs is more difficult than that of II–VI QDs. The reason is that III–V semiconductors are more covalent compounds and high temperatures are required for their synthesis. In progressive, more covalent series of semiconductors, the synthesis of QDs by colloidal techniques becomes increasingly difficult. To use the colloidal chemical method for the synthesis of QDs, it is important that the stabilizer and solvent do not decompose during the reaction time in order to ensure good solubility of QDs after synthesis and to avoid formation of different trapping states on the surface. The preparation method that we used is a compromise between these two requirements. One difference with the synthesis described previously for II–VI materials is that more than 1 day of heating at reaction temperature is required to form crystalline III–V QDs, whereas II–VI QDs form immediately upon injection of the reactants into the hot coordinating solution. The synthesis must be conducted in rigorously air-free and water-free atmospheres; it generally requires higher reaction temperatures. The best results to date have been obtained for InP QDs [17, 18, 20, 22–24].

2.1. Synthesis of Colloidal InP Quantum Dots

In this synthesis, an indium salt [$In(C_2O_4)_3$, InF_3, or $InCl_3$] is reacted with trimethylsilylphosphine [$P(Si(CH_3)_3)_3$] in a solution of trioctylphosphine oxide (TOPO) and trioctylphosphine (TOP) to form a soluble InP organometallic precursor species that contains In and P in a 1:1 ratio [17, 22]. The following procedure can be used: First, indium salts (oxalate, 1.15 g), tris(trimethylsilyl)phosphine (0.75 g), and a colloidal stabilizer (10 g of a mixture of trioctylphosphine oxide and trioctylphosphine in a TOPO:TOP ratio of 0.1:1) are mixed together at room temperature to form a transparent solution of an InP precursor; then, the precursor solution is heated at 250–290 °C for 1–6 days, depending on the desired QD properties. Use of TOPO/TOP as a colloidal stabilizer is based on the work of Murray et al. [29], who showed the remarkable ability of TOPO/TOP to stabilize semiconductor CdSe QDs at high temperature. Different particle sizes of InP QDs can be obtained by changing the temperature at which the solution is heated. The duration of heating only slightly affects the particle size, but does improve the QD crystallinity. The precursor has a high decomposition temperature (>200 °C); this is advantageous for the formation of InP quantum dots because the rate of QD formation is controlled by the rate of decomposition of the precursor. This slow process leads to InP QDs with a narrow size distribution. p-doped (InP:Zn), n-doped (InP:S), and undoped QDs of InP were synthesized as well-crystallized zincblende nanocrystals. For the synthesis of p-doped InP QDs, indium salt is first mixed with zinc acetate in a ratio such that the resulting QDs contain one Zn^{2+} ion per particle. For example, for 40-Å QDs, the ratio of In:Zn is 10^3:1. For n-doped InP QDs, $P(SiMe_3)_3$ is mixed with bis(trimethylsilyl)sulfide and then added to an indium salt solution. After heating, the clear reaction mixture contains InP QDs, byproducts of the synthesis, products resulting from TOPO/TOP thermal decomposition, and untreated TOP and TOPO. Anhydrous methanol is then added to the reaction mixture to flocculate the InP nanocrystals. The flocculate is separated and completely redispersed in the mixture of hexane (0.9) and 1-butanol (0.1) containing 1% TOPO to produce an optically clear colloidal solution. The process of dispersion in the mixture of hexane and 1-butanol and flocculation with anhydrous methanol is repeated several times to purify and isolate a pure powder of InP nanocrystals that are capped with TOPO. Repetitive selective flocculation by methanol gradually strips away the TOPO capping group; thus, TOPO (1%) is always included in the solvent when the QDs are redissolved in order to maintain the TOPO cap on the QDs.

Fractionation of the QD particles into different sizes was obtained by selective precipitation methods [29]. This technique will narrow the size distribution of the initial colloid preparation; the narrower the latter, the narrower will be the final size distribution.

The resulting InP QDs contain a capping layer of TOPO, which can be readily exchanged for several other types of capping agents, such as thiols, furan, pyridines, amines, fatty acids, sulfonic acids, and polymers. The size distribution of the InP QDs can be further narrowed to less than 10% through selective precipitation techniques. Finally, they can be studied in the form of colloidal solutions, powders, or dispersed in transparent polymers or organic (for low-temperature studies) glasses; capped InP QDs recovered as powders can also be redissolved to form transparent colloidal solutions.

In addition to InP QDs, similar methods have been used to produce QDs of GaP [16, 17], $GaInP_2$ [16, 17], GaAs [16, 17, 21, 30, 31], and InAs [19]; higher temperatures (400 °C) are required for GaP and $GaInP_2$ QDs. However, the quality (especially narrow size distributions) of these other III–V QDs is not yet as high as that obtained for InP QDs; this is primarily because less effort has been spent thus far on these other QD preparations.

2.2. Etching of Colloidal InP Quantum Dots with HF

Intense band edge emission from InP QDs can be achieved after etching the particles with a dilute alcoholic solution of HF [23]. The method described by Waldeck et al. [32] was adapted for etching InP QD colloidal solutions; in the former work, HF was used to remove the surface native oxide from single-crystal InP to obtain a clean surface. For colloidal solutions of InP QDs with a diameter of 20–40 Å, about 1–10 μL of methanolic solution was used; it contained 5% HF and 10% H_2O, which was added to 4 mL of a mixture of 2,2,4-trimethylpentane and acetonitrile (1:1) that also contains InP QDs and 2–5% oleyamine or dodecanethiol. Two liquid phases are formed and the dots are dispersed in the upper nonpolar phases of 2,2,4-trimethylpentane, while HF, methanol, and H_2O are in the acetonitrile phase. The mixture is shaken and left overnight and then the 2,2,4-trimethylpentane phase with the colloids is separated and used. For fresh colloidal InP QDs solutions, lower amounts of HF are required than for solutions that were exposed in air or stored for a long period of time. Freshly prepared solutions give the best quantum yields, and adsorption spectra only slightly shift to the ultraviolet (UV). It was also noticed that etching of smaller particles with a diameter of 25–35 Å is much faster than etching of particles larger than 50 Å. After etching, solutions were stored in an air-free atmosphere. The solution slowly loses emission intensity with time. Solid samples are more stable in air. Solid samples are prepared after drying etched InP QD solutions in vacuum and dissolving the residue in toluene and 5% polyvinyl butyryl under vacuum. Alternatively, after etching the QDs can be treated with octanethiol, and the QDs (now capped with octanethiol) can be redissolved in tetrahydrofuran to produce a colloidal QD solution that is stable in air and that shows intense band edge emission after a month in contact with air.

2.3. Synthesis of Colloidal GaP Quantum Dots

Quantum dots of GaP were synthesized by mixing $GaCl_3$ and $P(SiMe_3)_3$ in a molar ratio of Ga:P of 1:1 in toluene at room temperature to form a GaP precursor species and then heating this precursor in a high-boiling solvent. Wells et al. [33–35] first synthesized and characterized the yellow GaP precursor, $[Cl_2GaP(SiMe_3)_2]_2$, that is formed from $GaCl_3$ and $P(SiMe_3)_3$. The GaP precursor solution is then added to a toluene solution of TOPO, stirred for 1 h, and then transferred to a high-boiling solvent; toluene was removed using vacuum and the resulting transparent viscous solution was slowly heated. The colloidal 30 Å GaP QDs reported here were prepared by first mixing 0.18 g of $GaCl_3$ with 0.25 g of $P(SiMe_3)_3$ in 0.1 g of TOPO and 2 g of TOP and heating to 270–320 °C. At this stage, the GaP is amorphous. To prepare crystallized QDs, the amorphous material is mixed with 10 g of TOPO (or 5 g of triphenylphosphine) and heated at 360 °C for 3 days. The GaP QDs were precipitated with CH_3OH, separated, washed with CH_3OH, and redissolved in toluene; when triphenylphosphine or tris(2-diphenylphosphinoethyl) phosphine were used instead of TOPO, the QD precipitate only partially dissolved in toluene.

2.4. Synthesis of Colloidal $GaInP_2$ Quantum Dots

Quantum dots of $GaInP_2$ were synthesized by mixing chlorogallium oxalate and chloroindium oxalate complexes and $P(SiMe_3)_3$ in a molar ratio of Ga:In:P of 1:1:2.6 in toluene at room temperature, followed by heating. The specific conditions for the 25-Å QDs were to heat 0.19 g of chlorogallium oxalate and 0.23 g of chloroindium oxalate with 0.52 g of $P(SiMe_3)_3$ in 0.1 g of TOPO and 0.4 g of diphenyltetraphosphine at 400 °C for 3 days. The $GaInP_2$ QDs were separated, purified with CH_3OH, and partially redissolved in a 0.1% solution of TOPO in toluene. The larger, more crystalline 65-Å sample was obtained by heating a flask containing the isolated material in a yellow flame and redissolving in a similar manner.

3. PROPERTIES OF III–V QUANTUM DOTS

3.1. InP Quantum Dots

3.1.1. X-Ray Diffraction and Transmission Electron Microscopy

X-ray diffraction patterns of 26-, 35-, and 46-Å InP QD particles formed into a film by drying the colloids are shown in Figure 3. Powder X-ray diffraction spectra were collected on a Rigaku 300 Rotaflex diffractometer operating in the Bragg configuration using Cu Kα radiation. Samples for X-ray diffraction were prepared from approximately 50–100 mg of thoroughly washed and dried nanocrystalline powder. Diffraction peaks from the ⟨111⟩, ⟨220⟩, and ⟨311⟩ planes of crystalline zincblende InP are seen at 2θ of $26.2 \pm 0.2°$, $46.3 \pm 0.2°$, and $51.7 \pm 0.2°$, respectively. The mean particle diameter can be estimated from the broadening of the diffraction peaks using the Debye–Scherer formula. These diameters are in agreement with the values obtained from transmission electron microscopy (TEM)

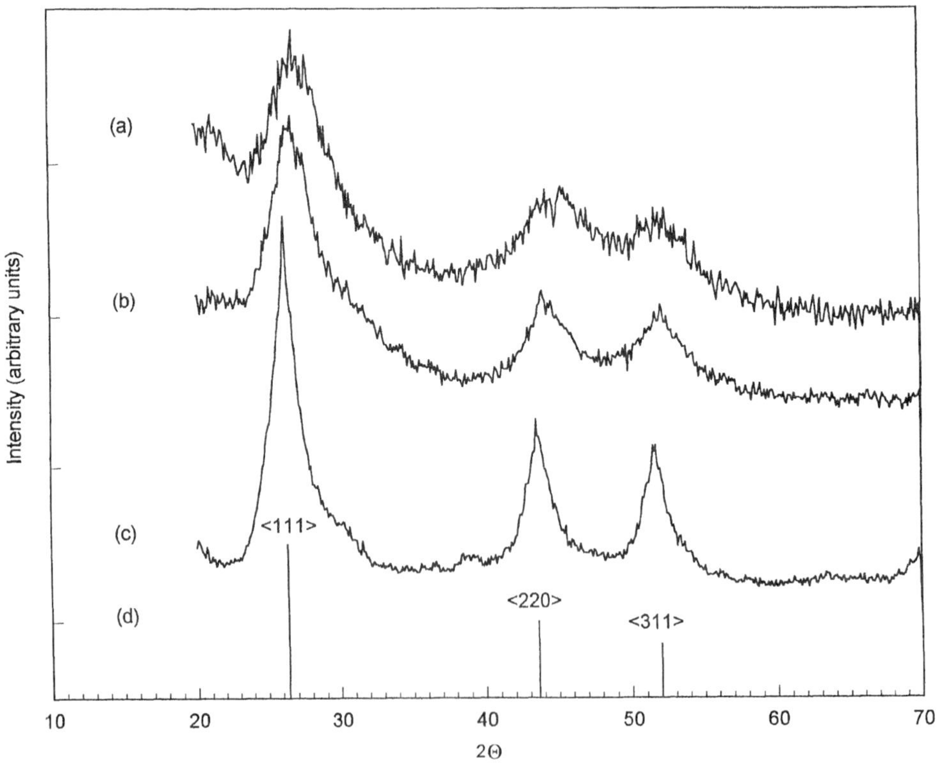

Fig. 3. X-ray diffraction of dried colloidal InP QDs with diameters (a) 25 Å, (b) 35 Å, and (c) 45 Å compared with bulk zincblende InP. (Source: Reprinted with permission from [17]. © 1995 American Chemical Society.)

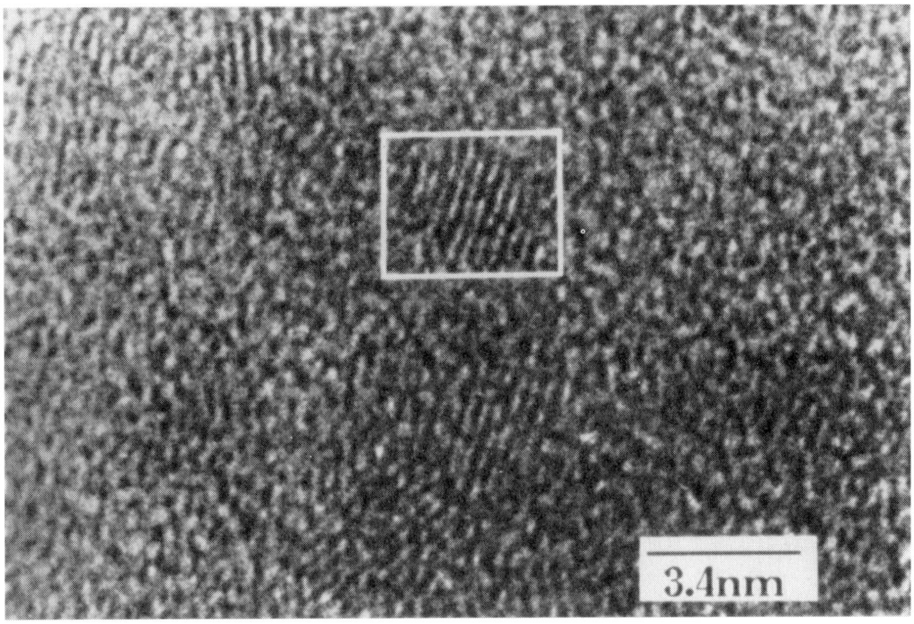

Fig. 4. TEM of InP QD. (Source: Reprinted with permission from [22]. © 1994 American Chemical Society.)

and from small-angle X-ray scattering (SAXS) data [17, 22]. In the absence of the TOPO stabilizer, the particles grow large and the sharp peaks of bulk InP are obtained.

The shape and size distribution of the InP QDs can be determined by TEM [22]. A Philips CM-30 electron microscope operating at 200 kV was used for TEM. Imaging was carried out in bright field with an objective aperture selected to permit lattice imaging of the ⟨111⟩ zincblende plane. TEM pictures of InP preparations with TOPO that were only heated to 220 °C for 3 days do not show the formation of either amorphous or crystalline InP. Upon heating to 240 °C for three days, the formation of zincblende nanocrystallites becomes evident, but the product is primarily amorphous. However, when the preparation is heated at 270 °C for 2 days, electron diffraction patterns show the ⟨111⟩, ⟨220⟩, and ⟨311⟩ planes of zincblende InP [17, 22]. The InP QDs generally show an elliptical shape (see Fig. 4).

3.1.2. Absorption and Luminescence Spectroscopy

3.1.2.1. Absorption and Global Photoluminescence

The absorption and emission spectra of initially prepared InP QDs with a mean diameter of 32 Å are shown in Figure 5. Optical absorption spectra were collected at room temperature using a Cary 5E UV-VIS-NIR spectrophotometer with 0.2-cm quartz cuvettes. Samples were prepared by dispersing washed QDs in toluene. Photoluminescence spectra were obtained at room temperature using a SPEX Fluorolog-2 spectrometer.

The absorption spectrum shows an excitonic peak at about 590 nm. The photoluminescence (PL) spectrum (excitation at 500 nm) shows two emission bands: a weaker one near the band edge with a peak at 655 nm and a second, stronger, broader band that peaks above 850 nm. The PL band with deep red-shifted subgap emission peaking above 850 nm is attributed to radiative surface states on the QDs produced by phosphorous vacancies [23]. These surface defects and the accompanying red-shifted emission can be completely removed by etching the InP QDs with HF. Intense band edge emission from our InP QDs can be achieved after etching the particles with a diluted methanolic solution of HF [23].

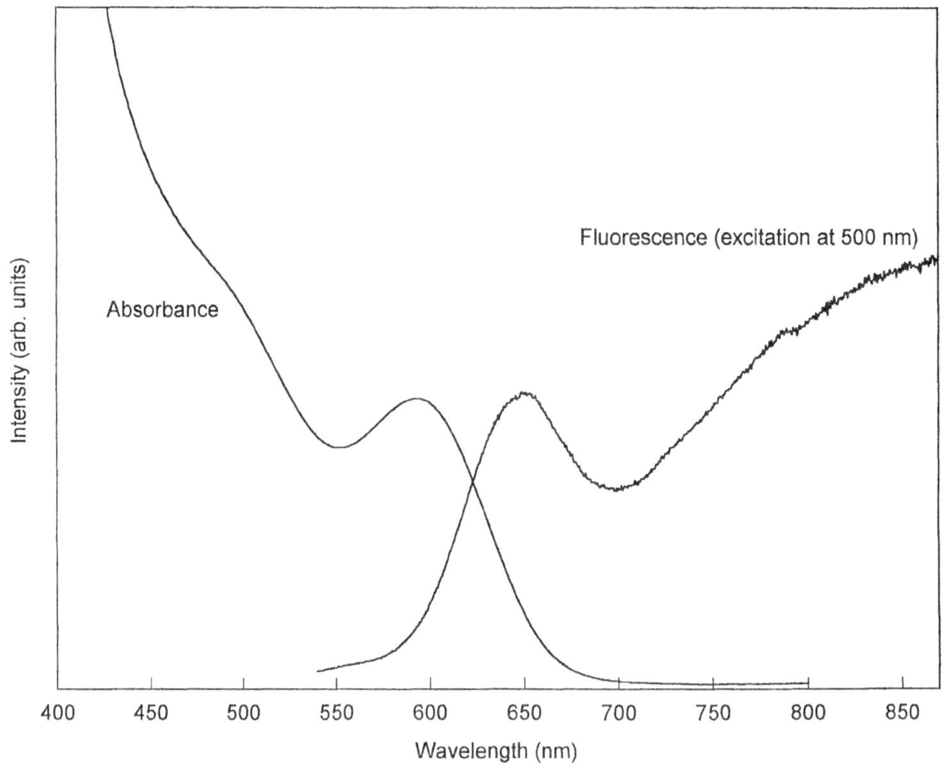

Fig. 5. Absorption and emission spectra of InP QDs as initially prepared. (Source: Reprinted with permission from [23].

After the etching of fresh solution, the PL intensity increases by a factor of 10, and the near-infrared emission produced by deep surface traps is completely removed [23]. The near–band edge quantum yield increases from 30% to 60% as the temperature decreases from 300 to 10 K [23].

The room temperature absorption spectra as a function of QD size ranging from 25 to 54 Å (measured by TEM) are shown in Figure 6. The absorption spectra show one or more broad excitonic peaks that reflect substantial inhomogeneous line broadening arising from the QD size distribution; as expected, the spectra shift to higher energy as the QD size decreases [24]. The color of the InP QD samples changes from deep red (1.7 eV) to green (2.4 eV) as the diameter decreases from 60 to 26 Å. Bulk InP is black with a room temperature band gap of 1.35 eV and an absorption onset at 918 nm. Higher energy transitions above the first excitonic peak in the absorption spectra can also be easily seen in QD samples with mean diameters equal to or greater than 30 Å. The spread in QD diameters is generally about 10% and is somewhat narrower in samples with larger mean diameters; this is why higher energy transitions can be resolved for the larger-sized QD ensembles. All of the prepared QD nanocrystallites are in the strong confinement regime because the Bohr radius of bulk InP is about 100 Å. Figure 7 shows typical room temperature absorption and global emission spectra of the InP colloids with mean particle diameters ranging from 26 to 60 Å as measured by TEM; the excitation energy for all QD sample ensembles in Figure 7 was 2.48 eV, well above their absorption onset in each case.

The global emission peaks in Figure 7 show an increasing red shift from the first excitonic absorption peak as the mean QD size decreases from 60 to 26 Å; the results are summarized in Figure 7. The global ("nonresonant") red shift is as large as 300 meV for samples with the smallest mean diameter [24]. We define global PL as that observed when

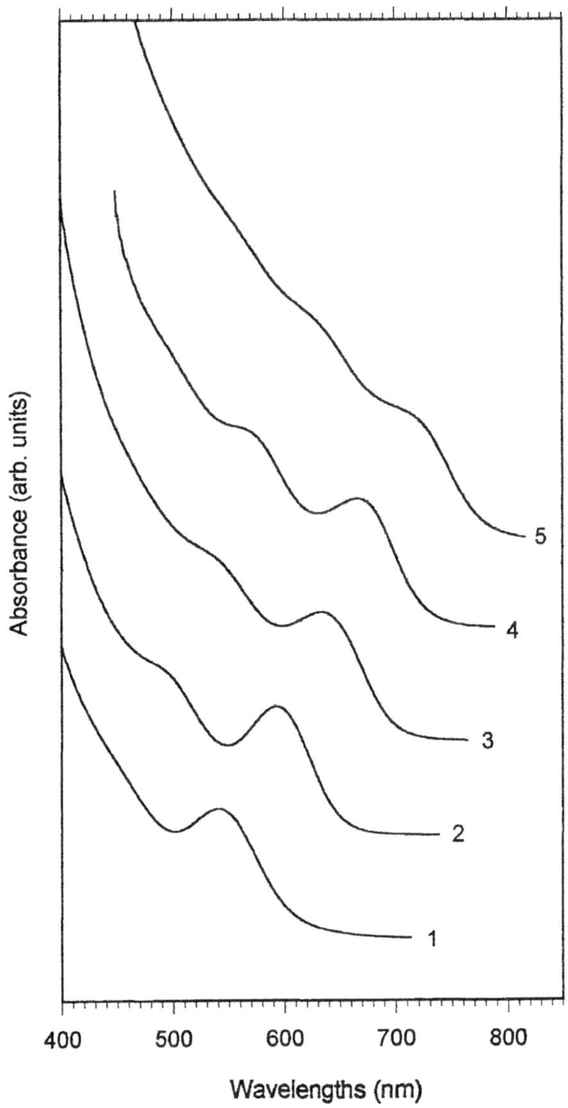

Fig. 6. Absorption spectra for InP QD colloids with different diameters (1) 25 Å, (2) 30 Å, (3) 35 Å, (4) 45 Å, and (5) 55 Å.

the excitation energy is much higher than the energy of the absorption threshold exhibited in the absorption spectrum produced by the ensemble of QDs in the sample. That is, the excitation wavelength is well to the blue of the first absorption peak for the QD ensemble, and, therefore, a large fraction of all the QDs in the sample are excited. The particle diameters that are excited range from the largest in the ensemble to the smallest, which has a diameter that produces a blue-shifted band gap equal to the energy of the exciting photons. The resulting global PL shows a very broad peak (line width of 175–225 meV) that is red-shifted by 100–300 meV from the first absorption peak (Figs. 7 and 8). The broad PL line width is caused by inhomogeneous line broadening arising from the approximately 10% size distribution sampled in the global PL experiment. The large global red shift is caused by the volume dominance of the larger particles in the size distribution; the larger QDs will absorb most of the incident photons and will also show large red shifts because the PL excitation energy is well above their lowest transition energy.

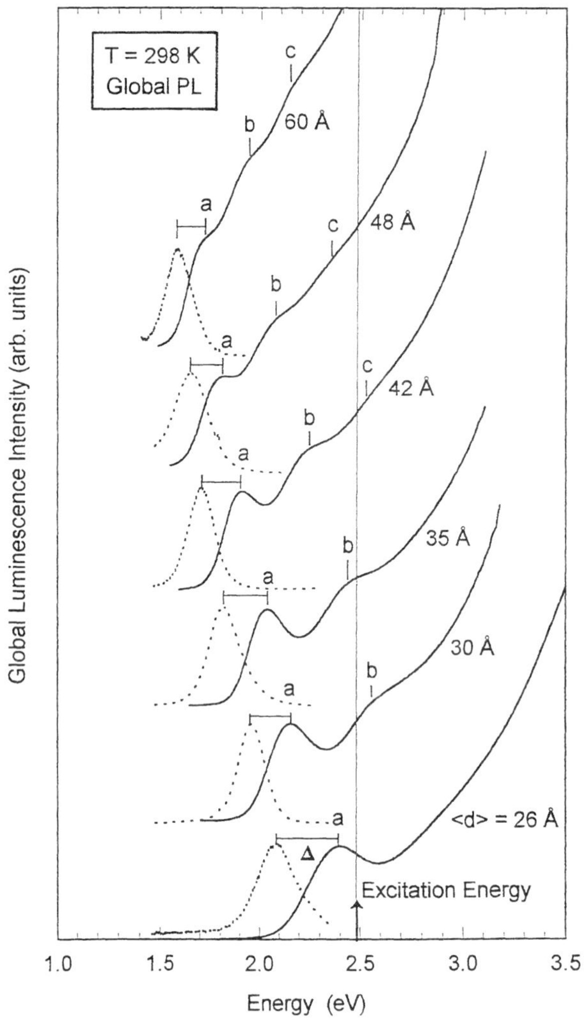

Fig. 7. Absorption and emission spectra of InP QDs as a function of QD size after treatment with HF. (Source: Reprinted with permission from [24]. © 1997 American Chemical Society.)

3.1.2.2. Size-Selected Photoluminescence

If the excitation energy is restricted to the onset region of the absorption spectrum of the QD ensemble, then a much narrower range of QD sizes is excited; these QDs will have the larger particle sizes in the ensemble. Consequently, the PL spectra from this type of excitation show narrower line widths and smaller red shifts with respect to the excitation energy. This technique is termed fluorescence line narrowing (FLN)—the resulting PL spectra being considerably narrowed.

FLN spectra at 10 K are shown in parts a–e of Figure 9 for InP QDs with a mean diameter of 32 Å. FLN/PL spectra are shown for a series of excitation energies (1.895–2.07 eV) spanning the absorption tail near the onset of absorption for this sample [24]. Also shown is the global PL spectrum produced when the excitation energy (2.41 eV) is deep into the high energy region of the absorption spectrum (Fig. 9f).

FLN spectra can be combined with photoluminescence excitation (PLE) spectra to determine the resonant red shift [24]. The experiment is done as follows: (a) first, a photon energy is selected in the onset region of the absorption spectrum of the QD ensemble spectrum, and this energy is set as the detected photon energy in the PLE; (b) the PLE spectrum

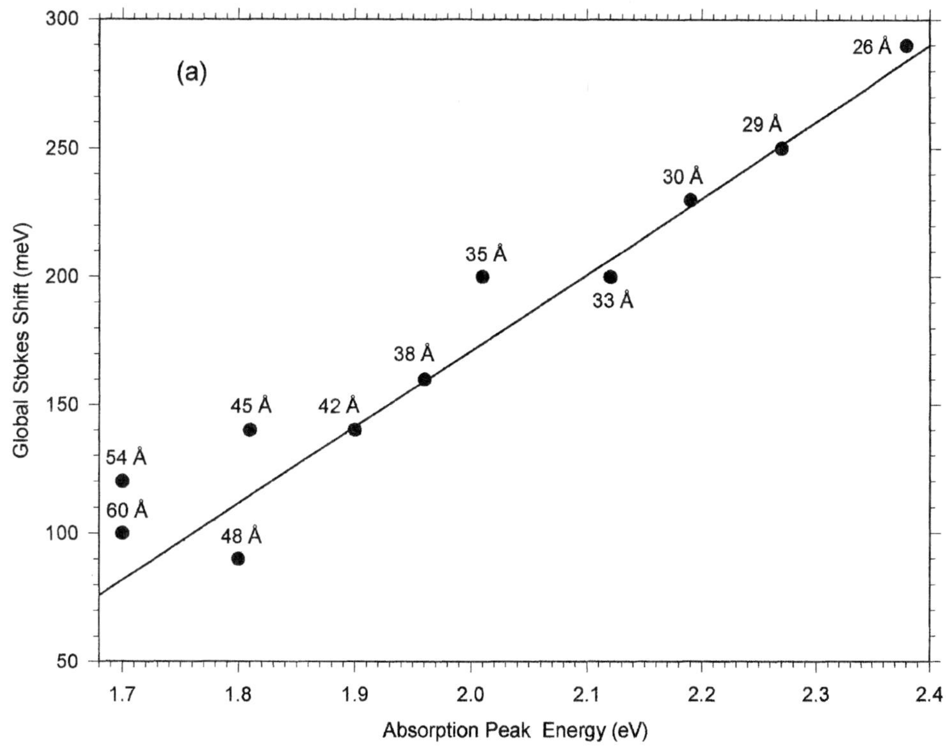

Fig. 8. The global red shift for different colloidal ensembles of InP QDs as a function of the position of the first peak in their absorption spectra. (Source: Reprinted with permission from [24]. © 1997 American Chemical Society.)

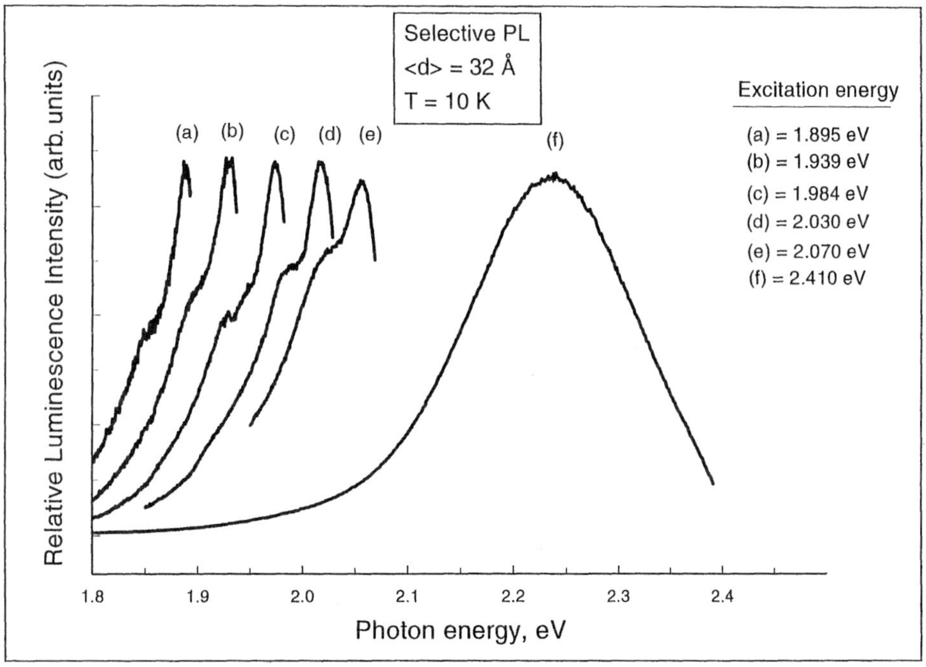

Fig. 9. PL spectra at 10 K for an ensemble of InP QDs with a mean diameter of 32 Å for different excitation energies. PL curves (a)–(e) result from excitation (1.895–2.070 eV) in the red tail of the onset region of the absorption spectrum; curve (f) is a global PL spectrum. (Source: Reprinted with permission from [24]. © 1997 American Chemical Society.)

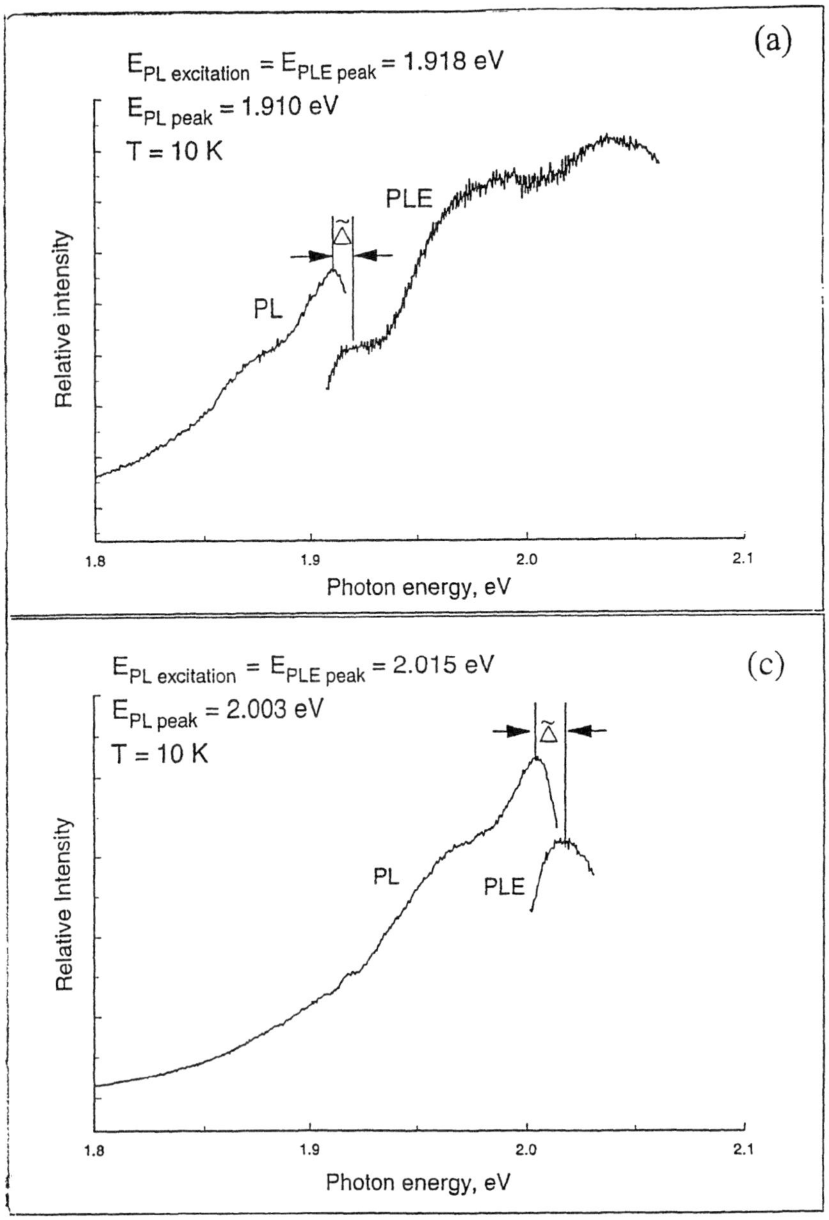

Fig. 10. Representative pairs of PL (FLN) and PLE spectra for the InP ensemble with a mean diameter of 32 Å. (Source: Reprinted with permission from [24]. © 1997 American Chemical Society.)

is then obtained, and the first peak of the PLE spectrum is taken to be the lowest energy excitonic transition for the QDs capable of emitting photons at the selected energy; and (c) an FLN spectrum is then obtained with excitation at the first peak of the PLE spectrum. The energy difference between the first FLN peak and the first PLE peak is then defined as the resonant red shift for the ensemble of QDs represented by the selected PL excitation energy. This process is repeated across the red tail of the absorption onset region of the absorption spectrum to generate the resonant red shift as a function of QD size. Typical FLN and PLE spectra are shown in Figure 10 for InP QDs with a mean diameter of 32 Å. The resultant $T = 11$ K resonant red shift as a function of PL excitation energy is presented in Figure 11 and ranges from 6 to 16 meV.

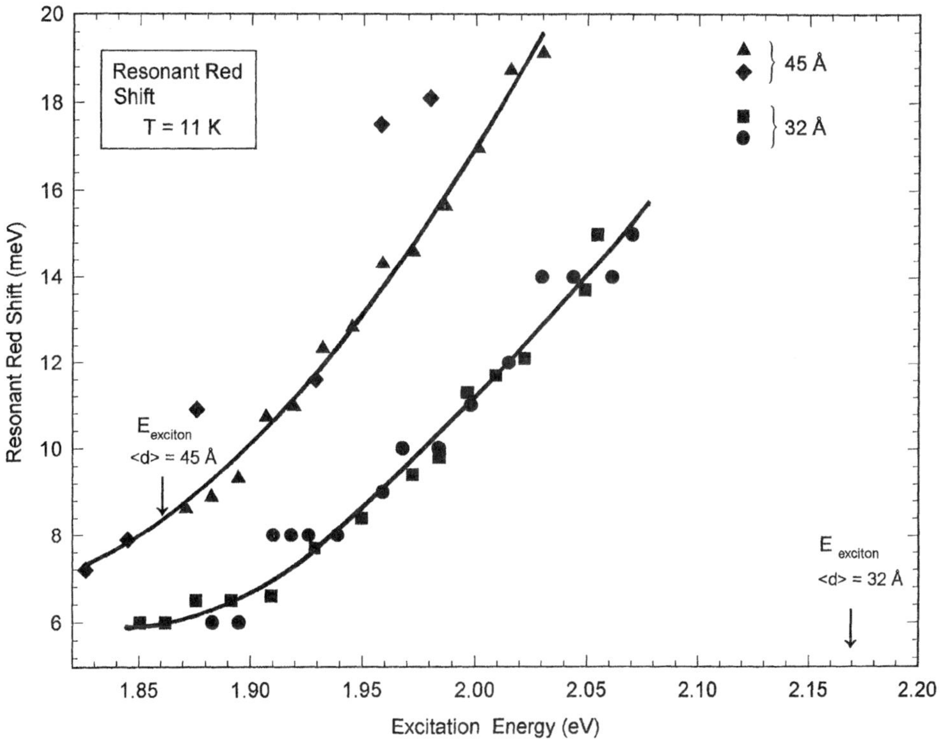

Fig. 11. The red shift for InP QD ensembles at 10 K for 32 Å and 45 Å QDs. (Source: Reprinted with permission from [24]. © 1997 American Chemical Society.)

As shown in Figures 9 and 10, the PL obtained by exciting into the red tail of the QD absorption spectrum, and thereby selectively exciting only the largest particles in the distribution, shows much smaller line widths (15–30 meV) and smaller red shifts compared to the global PL (compare Figs. 8 and 11). However, as our analysis shows, even the size-selected PL/FLN spectra contain effects caused by a residual size distribution. We have extracted the resonant red shifts for QDs of a single size from the experimental PL/FLN spectra [24]. The results of this analysis [24] show that the effective single-dot resonant red shift at 10 K exhibited by InP quantum dots that have been etched in HF to passivate surface states ranges from 4 meV for an excitation energy of 1.85 eV (corresponding to a QD size of 53 Å) to 9.7 meV for an excitation energy of 2.06 eV (corresponding to a QD size of 34 Å).

Our results show that for InP QDs formed via colloidal chemical processes, very large effects caused by a residual QD size distribution remain manifested in the photoluminescence spectroscopy of QD ensembles even after the colloidal samples have been subjected to size-selective precipitation techniques and size-selective photoexcitation. This behavior is similar to that reported for colloidal CdSe QD ensembles [36–46].

3.1.2.3. Photoluminescence Lifetime

The PL lifetimes were measured in a sample of HF-treated InP QDs immobilized in a PVB film at 298 and 13 K; the mean QD diameter was 30 Å. The emission lifetimes were determined by time-correlated single-photon counting. A cavity-dumped synchronously pumped dye laser (Spectra-Physics 3500) operating at 585 nm provided pump pulses of 10 ps. A Hamamatsu microchannel plate detector provided a typical instrument response function of 70 ps. The PL decay as a function of time for this sample is shown in Figure 12;

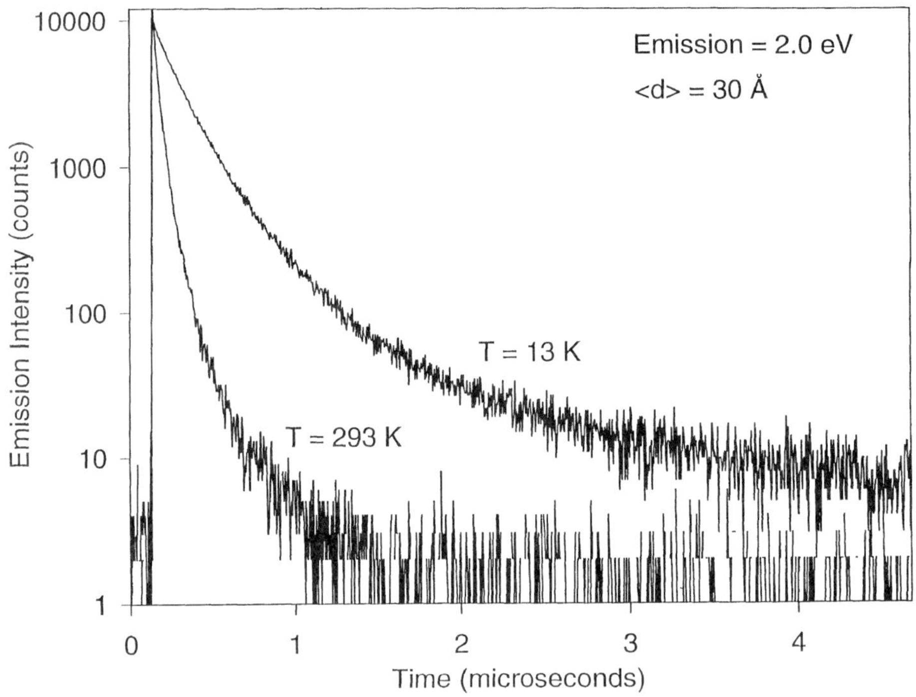

Fig. 12. PL decay for HF-treated InP QDs at 298 and 13 K. (Source: Reprinted with permission from [24]. © 1997 American Chemical Society.)

the excitation was at 585 nm and the emission was monitored at 620 nm. The decay is multiexponential and the data were fit to three exponentials. At 298 K, most of the decay (91%) can be described by two time constants of 28 and 73 ns; at 13 K, most of the decay (98%) can be described by two time constants of 173 and 590 ns.

3.1.2.4. Origin of Resonant Red Shift

The origin of the resonant red shift in InP has been recently analyzed theoretically [47, 48]. The methodology used was to treat a passivated quantum dot as a "giant molecule" in its own right, rather than an object drawn from an infinite crystal surrounded by an infinite potential barrier [36, 37, 39–43]. To this end, infinite-barrier k·p approaches [36, 37, 39–43] were avoided in favor of a pseudopotential supercell approach [47, 48], in which a dot of any selected shape and size is modeled explicitly, and passivating atoms decorate all surface sites. To examine possible surface effects, cation passivants and anion passivants were selectively removed, and the electron structure was recalculated. Four possible models have been examined as to their ability to explain the resonant red shift: (1) emission from an intrinsic, spin-forbidden state, split from its singlet counterpart because of screened electron–hole exchange; (2) emission from an intrinsic, orbitally forbidden conduction band state, for example, X_{1c} (rather than Γ_{1c}); (3) emission to an intrinsic, orbitally forbidden valence band state (e.g., p-like); and (4) emission from extrinsic surface defects (e.g., surface vacancies). The experimental results reported here are quantitatively consistent with model 1 when the exchange interactions are screened by a distance-dependent dielectric function [24]. In model 1, an enhanced (relative to bulk) electron–hole exchange interaction splits the exciton state into a lower-energy spin-forbidden state (triplet) and a higher-energy spin-allowed singlet. Absorption occurs into the upper state, followed by relaxation to and emission from the lower state; the difference between these two states is the resonant red shift [30, 31, 36–42, 47–49]. The value of the single-dot resonant red shift (as a function of QD size) derived from the experimental data is in excellent agreement with the theoretical predictions [48].

Model 2 was rejected [48] because it was shown that the conduction band minimum in InP dots is not derived from an indirect X_{1c}-like state as in small GaAs particles [48], and model 3 was rejected [48] because the symmetry of the envelope function of the valence band maximum was found to be 1s-like and not 1p-like as expected from simple k·p models. Model 4 shows that *fully* passivated QDs have no surface states, despite the large surface-to-volume ratio. However, in the event that some of the surface atoms were not capped by a passivant (because of, e.g., steric hindrance by large passivating molecules), model 4 shows that surface *defect states* (caused by surface-uncapped In or surface-uncapped P) could appear inside the QD band gap. These surface defects lead to large red shifts extending from 100 to a few hundred meV depending on the surface conditions; such large red shifts are observed in unetched InP QDs but are removed upon HF etching [23]. The magnitude of the observed resonant red shifts reported here after etching is not consistent with the surface defects present in the initial QD synthesis; these are removed or passivated and do not affect the PL. Although surface defect states do not explain the less than 10 to 7 meV *resonant* red shift, direct theoretical modeling of such states [48] shows that they (i) affect the quantum efficiency (through nonradiative recombination) and (ii) lead to a significant hybridization with the ordinary, corelike band edge states, thus affecting the radiative emission rate from these states. Furthermore, because these hybridized states reflect the properties of the uncapped site (i.e., P or In "dangling bond") rather than those of the passivating molecules around this site, it was predicted [48] that surface defect states are mostly independent of the passivant and have size-dependent lifetimes. The degree of mixing of surface defect wave functions with the ordinary corelike band edge states remains unknown at this time.

The relatively long lifetimes of InP QDs with a mean diameter of 30 Å (28–73 ns at 298 K and 173–590 ns at 13 K) are also consistent with model 1 because the spin-forbidden lowest excitonic state has a small probability for radiative transitions to the ground state.

Additional experiments that can be done for InP QDs to check the validity of model 1 include measuring the PL lifetime as a function of magnetic field [36, 37, 40, 41], measuring the intensity ratio of the zero-phonon PL line to its replica as a function of magnetic field [37], and measuring the degree of linear polarization of the PL [40, 41]. In model 1, the PL lifetime should decrease with increasing magnetic field, the zero-phonon PL line intensity should increase relative to the one-phonon replica with magnetic field, and the degree of linear polarization should be negative. These experiments will be done in the future.

The preceding technique does not directly produce the *single-dot* resonant red shift as a function of QD size because the data still reflect the effects of a finite size distribution. The line widths of our FLN spectra are typically 15–30 meV; although these line widths are significantly narrower than the 175–225-meV line widths typically obtained from global (nonresonant) PL excitation, they are still much broader than the line widths reported from PL measurements on single dots. For a variety of II–VI and III–V QDs, single-dot PL line widths have been reported to range from 40 to 1000 μeV [35, 50–54]. The broader line widths in our FLN spectra are attributed to the significant QD size and shape distribution that still remains in the FLN experiment; this is evident in Figure 9 from the decreasing FLN line widths obtained as the excitation energy is moved to lower energies and, hence, to narrower QD size distributions (and also to larger QD diameters). Also, from Figure 6 it can be seen that the absorption peak moves about 35–45 meV for every 1-Å change in QD diameter. Thus, the PL line width of 15–30 meV for our FLN spectra reflects a QD diameter variation of less than 1 Å! It is apparent that the PL line width is extremely sensitive to the spread in QD diameters and that true line widths require PL data obtained from single dots. The experimental PL line widths observed for single QDs are smaller than a line widths of we used by a factor of about 2 to 50. However, this does not affect the extracted red shifts we report here for single dots because the additional PL line broadening represented by a line width of about 2 meV reflects a variation in the single-dot diameter of less than 0.1 Å.

The FLN spectra show a shoulder that is displaced from the highest-energy PL peak by 30–35 meV; this is attributed to replicate PL lines caused by phonon emission. However, the bulk LO phonon energy for InP is 43 meV; further work is required to understand this difference.

3.1.3. *InP Quantum Dots Arrays*

Arrays of close-packed InP QDs can be formed by slowly evaporating colloidal solutions to form solid films [55]. It is known that monodispersed colloidal solutions have a natural tendency to self-organize to form colloidal crystal. Recently, Bawendi and co-workers [28, 43] fabricated rather perfect three-dimensional (3D) superlattices of CdSe QDs that are size selected to 3% to 4%. This is a novel quantum dot configuration where QDs form a crystal lattice similar to atoms in a solid. III–V QD preparations have a size distribution of about 10%, and, with such a size distribution, it is possible to prepare short-range two-dimensional (2D) ordered arrays [55].

Close-packed glass solid films of InP QDs can also be also fabricated that are completely optically transparent. In the films, the QDs are randomly ordered.

3.1.4. *Applications in Solar Cells*

Photosensitization of nanocrystalline TiO_2 semiconductor electrodes by adsorbed dyes has been extensively studied recently because of potential application as a new type of solar cell [44]. The key feature of this system is the use of nanocrystalline TiO_2 films that have extremely large surface-to-volume ratios because they are formed from nanocrystallites. This allows for greatly increased dye coverage in the TiO_2 film and produces very high quantum yields for photon-to-electronic current flow (above 80%) and solar conversion efficiencies above 10% [44, 45].

The band edge offset between InP QDs and TiO_2 allows for efficient photoinduced electron transfer from InP to TiO_2 because the conduction band of InP QDs can be 0.1–0.5 eV above that for TiO_2, depending on the degree of quantization. It has been found that InP QDs adsorb strongly on nanocrystalline TiO_2 electrode films and that InP QD particles can be utilized for photosensitization of nanocrystalline TiO_2 electrodes in solar cells [46]. The potential attraction of InP QDs for this application is that they can have a high absorption coefficient over a larger region of the solar spectrum, and they are photochemically stable.

3.2. GaP Quantum Dots

Quantum dots of GaP can be synthesized by mixing $GaCl_3$ (or the chlorogallium oxalate complex) and $P(Si(CH_3)_3)_3$ in a molar ratio of Ga:P of 1:1 in toluene at room temperature to form the GaP precursor species and then heating this precursor in TOPO at 400 °C for 3 days [17]. The mean particle diameters of our GaP QD preparations were estimated from the line broadening of their X-ray diffraction patterns and from TEM. Figure 13 shows the X-ray diffraction patterns for GaP QD samples that were heated at different temperatures. For diffraction pattern d, the temperature was 400 °C and the size was estimated to be about 30 Å; for diffraction pattern c, the temperature was 360 °C and the size was estimated to be about 20 Å. The diffraction patterns for the lower-temperature preparations (a and b) were too broad and ill defined to permit an estimate of the particle size.

The absorption spectrum in Figure 14 of the 30-Å-diameter GaP QD colloid (heated at 400 °C) exhibits a shoulder at 420 nm (2.95 eV) and a shallow tail that extends out to about 650 nm (1.91 eV) [17]. For 20-Å-diameter GaP QDs (heated at 370 °C), the shoulder is at 390 nm (3.17 eV) and the tail extends to about 550 nm.

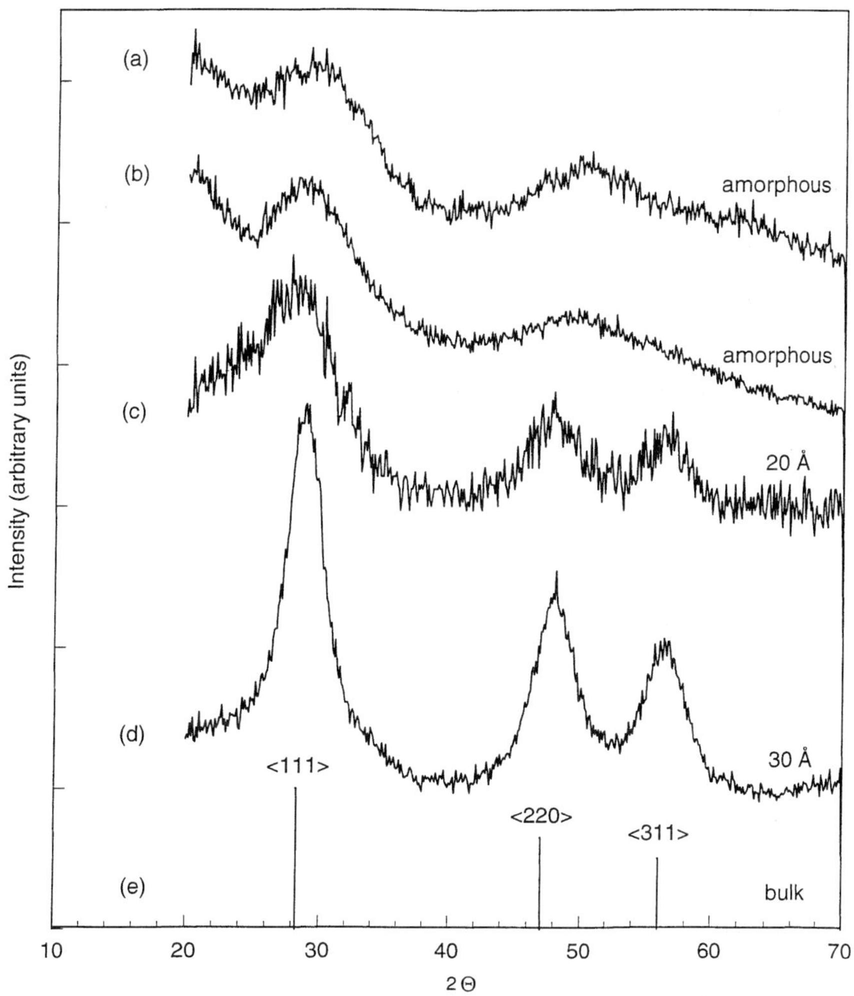

Fig. 13. X-ray diffraction of GaP QDs prepared at different temperatures: (a) 270 °C, (b) 320 °C, (c) 360 °C, and (d) 400 °C. (Source: Reprinted with permission from [17]. © 1995 American Chemical Society.)

Bulk GaP is an indirect semiconductor with an indirect band gap of 2.22 eV (559 nm) and a direct band gap of 2.78 eV (446 nm). Theoretical calculations [56] on GaP QDs show that the increase of the indirect band gap with decreasing QD size is much less pronounced than that for the direct gap; for 30-Å-diameter GaP QDs, the direct and indirect band gaps are predicted to be 3.35 eV and 2.4 eV, respectively. Below 30 Å, the direct band gap is predicted to *decrease* with decreasing size while the indirect band gap continues to increase. As a result, GaP is expected to undergo a transition from an indirect semiconductor to a direct semiconductor below about 20 Å.

The steep rise in absorption and the shoulder at 420 nm in the absorption spectrum of GaP QDs [17] is attributed to a direct transition in the GaP QDs; the shallow-tail region above 500 nm is attributed to the indirect transition. Also, the absorption tail extends below the indirect band gap of bulk GaP [17]. The origin of this subgap absorption is not understood at the present time; it could be caused either by a high density of subgap states in the GaP QDs, by impurities created by the high decomposition temperature, or by Urbach-type band tailing produced by unintentional doping in the QDs [57]. We note that such subgap absorption below the band gap was also observed in GaP nanocrystals that were prepared in zeolite cavities by the gas phase reaction of trimethylgallium and phosphine

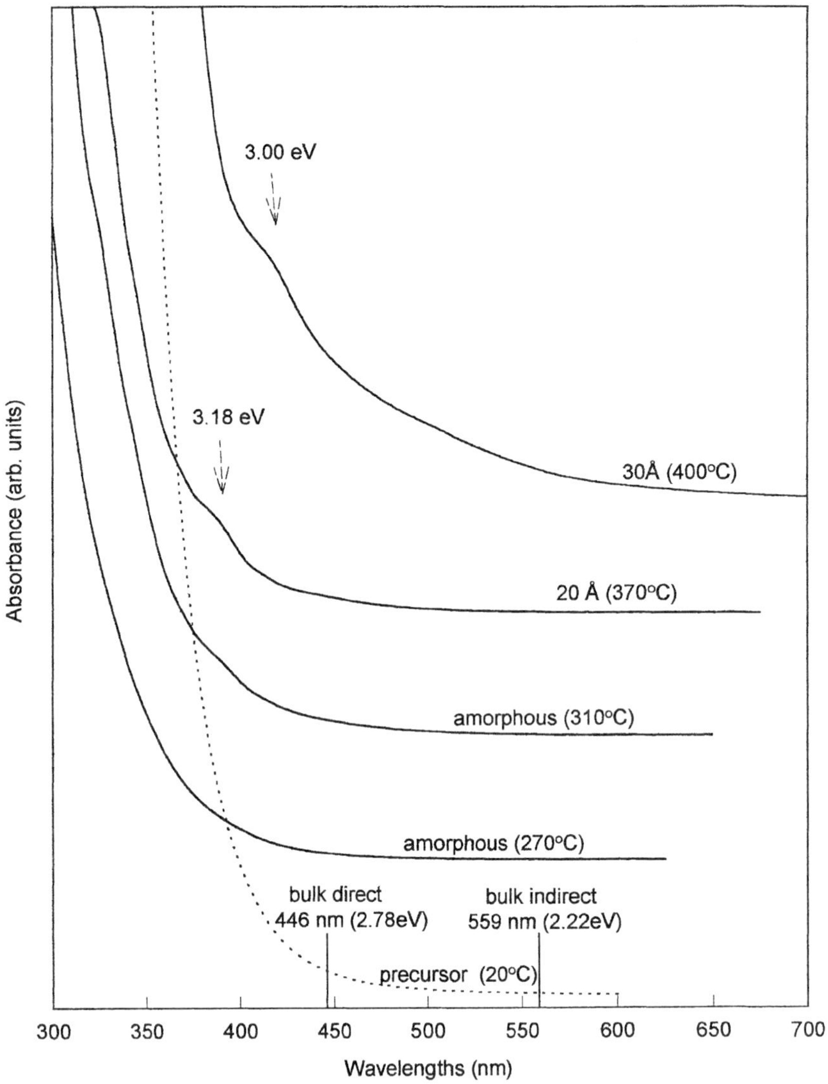

Fig. 14. Optical absorption spectra at room temperature of GaP QDS heated at different temperatures and dissolved in toluene; also shown (dotted line) is the absorption spectrum of the GaP precursor $\{[Cl_2GaP(SiMe_3)_2]_2\}$. The direct and indirect band gaps of bulk GaP are indicated by the vertical line markers on the abscissa. (Source: Reprinted with permission from [17]. © 1995 American Chemical Society.)

at temperatures above 225 °C [58]. This latter result implies that the subgap absorption in GaP QDs is related to their intrinsic behavior and is not caused by synthetic byproducts or impurities.

3.3. $GaInP_2$ Quantum Dots

Quantum dots of $GaInP_2$ can be synthesized by mixing chlorogallium oxalate and chloroindium oxalate complexes and $P(Si(CH_3)_3)_3$ in the molar ratio of Ga:In:P of 1:1:2.6 in toluene at room temperature, followed by heating in TOPO [17]. Heating at 400 °C for 3 days was required to form 25-Å QDs. An X-ray diffraction pattern for a 65-Å sample is shown in Figure 15 along with that of the GaP and InP QDs. As expected for $GaInP_2$, the lattice spacings of these QDs are approximately the average of that of GaP and InP.

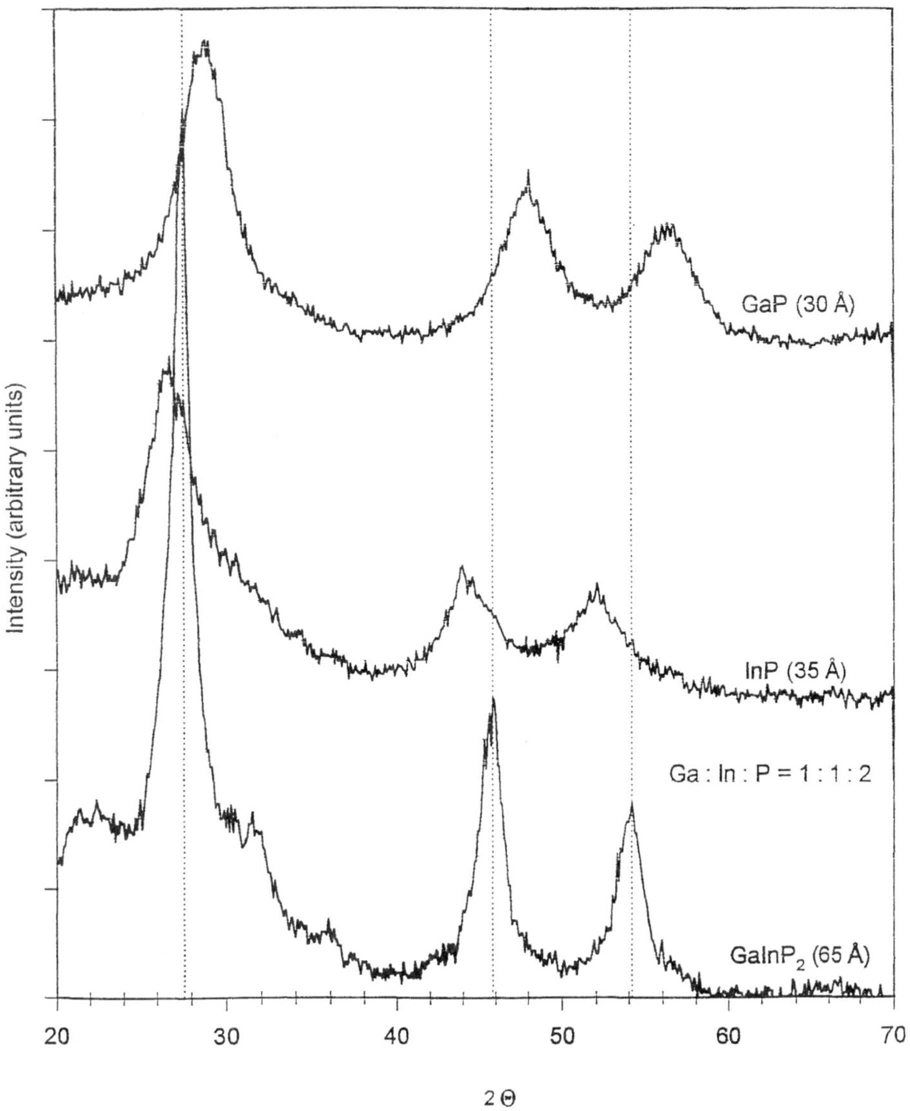

Fig. 15. X-ray diffraction of $GaInP_2$ QDs compared to InP and GaP QDs. (Source: Reprinted with permission from [17]. © 1995 American Chemical Society.)

The ternary Ga–In–P system forms solid solutions that can exhibit direct band gaps ranging from 1.7 to 2.2 eV, depending on the composition and growth temperature [59–65]. At the composition $Ga_{0.5}In_{0.5}P$, the structure can be either atomically ordered or disordered (random alloy) [59–65]; the band gap is direct, but it can range from about 1.8 to 2.0 eV, depending on the degree of atomic ordering. Although one issue of interest is how size quantization will affect the dependence of band gap on atomic ordering, at the present time it is not possible to specify the degree of atomic order in the $GaInP_2$ QDs.

The absorption spectrum of 25-Å $GaInP_2$ QDs does not show any excitonic structure (Fig. 16) [17]; this is caused by a wide size distribution that masks the excitonic peaks. An estimate of the direct band gap of the $GaInP_2$ QDs from a plot of the square of the absorbance times photon energy versus photon energy indicates a value of about 2.7 eV; this value is blue-shifted from the bulk value of 1.8–2.0 eV.

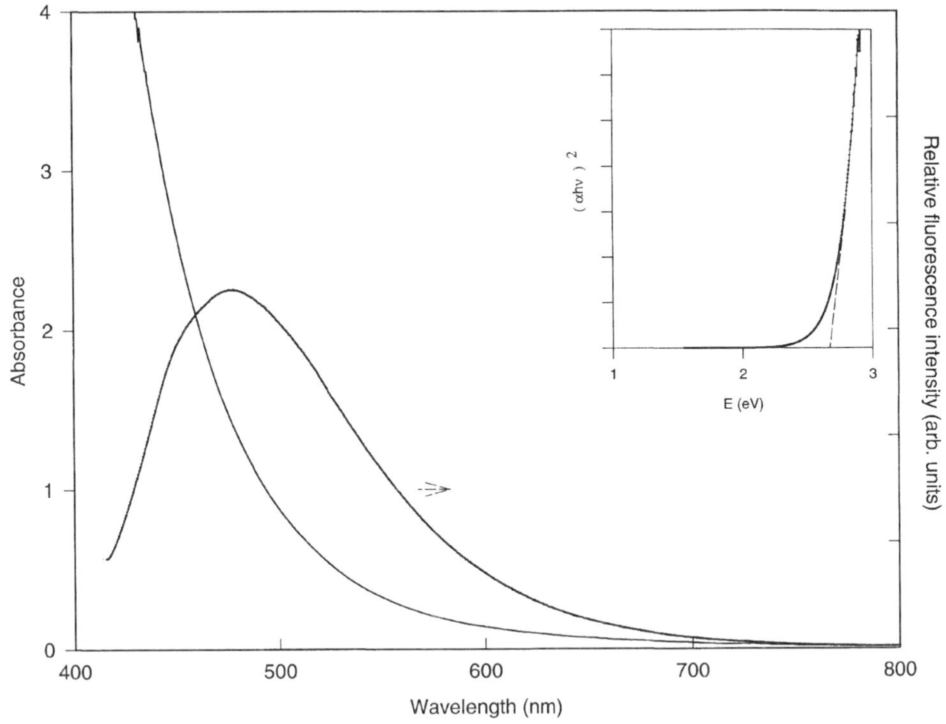

Fig. 16. Absorption spectra for 25-Å $GaInP_2$ colloids; insert shows that the colloid exhibits a direct band gap of 2.7 eV. (Source: Reprinted with permission from [17]. © 1995 American Chemical Society.)

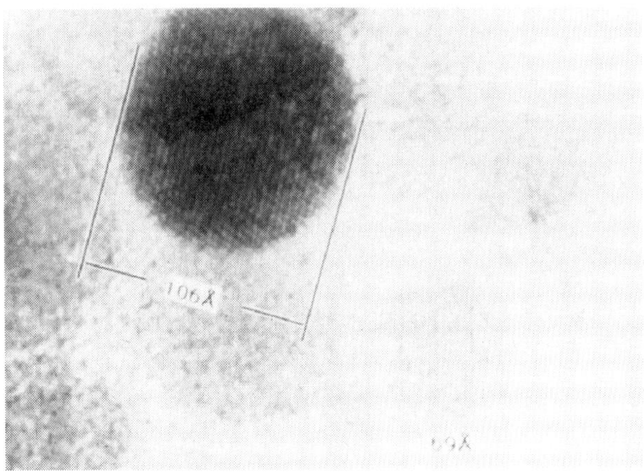

Fig. 17. TEM of GaAs QDs. (Source: Reprinted with permission from [30]. © 1992 American Chemical Society.)

3.4. GaAs Quantum Dots

GaAs quantum dots can be formed by first reacting Ga(III) acetylacetonate and $As(Si(CH_3)_3)_3$ at reflux (216 °C) in triethylene glycol dimethyl ether (triglyme) [16, 21, 30, 31]. This produces an orange-brown turbid slurry that can be filtered through a 700-Å ultrafilter to produce the GaAs colloidal QD solution. A TEM image of GaAs QDs prepared in this way, except that quinoline was used instead of triglyme, is shown in Figure 17. Electron diffraction data show clear *hkl* zincblende GaAs patterns of (111), (220), (311), and (422).

The observed lattice plane spacing of 3.2 Å in TEM images corresponds to the d(111) of GaAs.

A series of ultrafiltrations through 700-, 100-, and 15-Å pore-size filters can be performed to separate particles with different sizes. The optical absorption spectra of the GaAs colloid passed through the 700- and 100-Å pore-size filters exhibit similar absorption spectra: The onset of absorption occurs at about 600 nm; a shallow rise with decreasing wavelength steepens at about 470 nm and peaks at about 440 nm [16, 21, 30, 31]. The absorbance of the 100-Å filtrate was slightly less than that of the 700-Å filtrate. The spectrum for the 15-Å filtrate was similar to that of the blank sample consisting of Ga acetylacetonate refluxed in triglyme. The absorbance begins at about 525 nm and rises very weakly without any peaks.

The particle size distribution of the GaAs QDs was not sufficient to observe excitonic transitions in the absorption or emission spectra [16, 21, 30, 31]. Therefore, more detailed studies of these GaAs QDs were not conducted.

4. SUMMARY

Quantum dots of the III–V semiconductors InP, GaP, $GaInP_2$, InAs, and GaAs can be prepared by nonaqueous colloidal chemical synthesis at temperatures ranging from about 250 to 400 °C, depending on the specific compound and the desired QD size. The QD size for the synthesis can range from about 20 to 100 Å, and the size distribution around the mean diameter can be typically from about 5% to 10%. The synthesis requires capping the quantum dots with a stabilizing molecular species, such as, for example, trioctylphosphine oxide, to control the QD size and the size distribution. The capping species can be subsequently replaced with different molecules of varying length and type, or eliminated completely. As a result, the QDs can be formed into powders or suspended in various liquids or solids (such as polymers or organic and inorganic glasses); they can be studied at low or high temperatures.

The best colloidal III–V QDs synthesized to date are InP QDs; treatment with HF results in the elimination of deep surface traps and produces band edge emission with very high quantum yields (60% at 10 K). The color of these dots ranges from deep red (band gap = 1.7 eV) to green (band gap = 2.4 eV), corresponding to a size range of 60 to 26 Å, respectively. Structure in the absorption and emission spectra reflects the existence of quantized energy levels in the QDs. Size-selective spectroscopy (fluorescence line narrowing and photoluminescence excitation spectroscopy) has been done to help determine the energy level structure in the QDs. Solid arrays of close-packed InP QDs have been prepared; ordered hexagonal packing occurs in some regions.

Acknowledgment

This work was supported by the U.S. Department of Energy, Office of Basic Energy Sciences, Division of Chemical Sciences.

References

1. R. Dingle, W. Wiegmann, and C. H. Henry, *Phys. Rev. Lett.* 33, 827 (1974).
2. R. Dingle, *Feskörperprobleme* 15, 21 (1975).
3. C. Weisbuch and B. Vinter, "Quantum Semiconductor Structures." Academic Press, New York, 1991.
4. G. Bastard, "Wave Mechanics Applied to Semiconductor Heterostructures." Halsted Press, New York, 1988.
5. M. Jaros, "Physics and Applications of Semiconductor Microstructures." Oxford University Press, Oxford, U.K., 1989.
6. A. D. Yoffe, *Adv. Phys.* 42, 173 (1993).
7. K. Kash, *J. Lumin.* 46, 69 (1990).

8. S. P. Beaumont and C. M. Sotomayor Torres, *NATO ASI Ser. B* 1 (1990).
9. P. V. Kamat and D. Meisel, in "Studies in Surface Science and Catalysis," Vol. 103. Elsevier, Amsterdam, 1997.
10. A. P. Alivisatos, *Science* 271, 933 (1996).
11. A. P. Alivisatos, *J. Phys. Chem.* 100, 13226 (1996).
12. M. G. Bawendi, *NATO ASI Ser. B* 339 (1995).
13. L. Brus, *Appl. Phys. A* 53, 465 (1991).
14. Y. Wang, in "Advances in Photochemistry" (D. C. Neckers, D. H. Volman, and V. Bünau, eds.), Vol. 19, p. 179. Wiley, New York, 1995.
15. L. E. Brus, A. L. Efros, and T. Itoh, *J. Lumin.* 70, 1 (1996).
16. O. I. Mićić and A. J. Nozik, *J. Lumin.* 70, 95 (1996).
17. O. I. Mićić, J. R. Sprague, C. J. Curtis, K. M. Jones, J. L. Machol, and A. J. Nozik, *J. Phys. Chem.* 99, 7754 (1995).
18. U. Banin, G. Cerullo, A. A. Guzelian, C. J. Bardeen, A. P. Alivisatos, and C. V. Shank, *Phys. Rev. B* 55, 7059 (1997).
19. A. A. Guzelian, U. Banin, A. V. Kadavanich, X. Peng, and A. P. Alivisatos, *Appl. Phys. Lett.* 69, 1432 (1996).
20. A. A. Guzelian, J. E. B. Katari, A. V. Kadavanich, U. Banin, K. Hamad, E. Juban, A. P. Alivisatos, R. H. Wolters, C. C. Arnold, and J. R. Heath, *J. Phys. Chem.* 100, 7212 (1996).
21. M. A. Olshavsky, A. N. Goldstein, and A. P. Alivisatos, *J. Am. Chem. Soc.* 112, 9438 (1990).
22. O. I. Mićić, C. J. Curtis, K. M. Jones, J. R. Sprague, and A. J. Nozik, *J. Phys. Chem.* 98, 4966 (1994).
23. O. I. Mićić, J. R. Sprague, Z. Lu, and A. J. Nozik, *Appl. Phys. Lett.* 68, 3150 (1996).
24. O. I. Mićić, H. M. Cheong, H. Fu, A. Zunger, J. R. Sprague, A. Mascarenhas, and A. J. Nozik, *J. Phys. Chem. B.* 101, 4904 (1997).
25. M. Peterson and A. J. Nozik, in "Photoelectrochemistry and Photovoltaics of Layered Semiconductors" (A. Aruchamy, ed.), p. 297. Kluwer Academic, Boston, 1992.
26. H. Weller and A. Eychmüller, in "Semiconductor Nanoclusters" (P. V. Kamat and D. Meisel, eds.), Vol. 103, p. 5. Elsevier Science B.V., Amsterdam, 1996.
27. J. T. G. Overbeek, *Adv. Colloid Interface Sci.* 15, 251 (1982).
28. C. B. Murray, Synthesis and Characterization of II–VI Quantum Dots and Their Assembly into 3D Quantum Dot Superlattices, Ph.D. Thesis, Massachusetts Institute of Technology, 1995.
29. C. B. Murray, D. J. Norris, and M. G. Bawendi, *J. Am. Chem. Soc.* 115, 8706 (1993).
30. H. Uchida, C. J. Curtis, P. V. Kamat, K. M. Jones, and A. J. Nozik, *J. Phys. Chem.* 96, 1156 (1992).
31. A. J. Nozik, H. Uchida, P. V. Kamat, and C. Curtis, *Isr. J. Chem.* 33, 15 (1993).
32. Y. Gu, Z. Lin, V. Smetkowski, R. Butera, and D. H. Waldeck, *Langmuir* 11, 1849 (1995).
33. S. R. Aubuchon, A. T. McPhail, R. L. Wells, J. A. Giambra, and J. R. Browser, *Chem. Mater.* 6, 82 (1994).
34. R. L. Wells, M. F. Self, A. T. McPhail, S. R. Aubuchon, R. C. Wandenberg, and J. P. Jasinski, *Organometallics* 12, 2832 (1993).
35. S. A. Empedocles, D. J. Norris, and M. G. Bawendi, *Am. Phys. Soc.* 77, 3873 (1996).
36. M. Nirmal, D. J. Norris, M. Kuno, M. G. Bawendi, A. L. Efros, and M. Rosen, *Phys. Rev. Lett.* 75, 3728 (1995).
37. A. L. Efros, M. Rosen, M. Kuno, M. Nirmal, D. J. Norris, and M. Bawendi, *Phys. Rev. B* 54, 4843 (1996).
38. D. J. Norris and M. G. Bawendi, *Phys. Rev. B* 53, 16338 (1996).
39. D. J. Norris, A. L. Efros, M. Rosen, and M. G. Bawendi, *Phys. Rev. B* 53, 16347 (1996).
40. M. Chamarro, C. Gourdon, P. Lavallard, O. Lublinskaya, and A. I. Ekimov, *Phys. Rev. B* 53, 1336 (1996).
41. M. Chamarro, C. Gourdon, and P. Lavallard, *J. Lumin.* 70, 222 (1996).
42. U. Woggon, F. Gindele, O. Wind, and C. Klingshirn, *Phys. Rev. B* 54, 1506 (1996).
43. C. B. Murray, C. R. Kagan, and M. G. Bawendi, *Science* 270, 1335 (1995).
44. B. O'Regan and M. Grätzel, *Nature* 353, 737 (1991).
45. M. K. Nazeeruddin, A. Kay, I. Rodicio, R. Humphry-Baker, E. Müller, P. Liska, N. Vlachopoulos, and M. Grätzel, *J. Am. Chem. Soc.* 115, 6382 (1993).
46. A. Zaban, O. I. Mićić, B. A. Gregg, and A. J. Nozik, *Langmuir* 14, 3153 (1998).
47. H. Fu and A. Zunger, *Phys. Rev. B* 55, 1642 (1997).
48. H. Fu and A. Zunger, *Phys. Rev. B* 56, 1496 (1997).
49. A. Resta, *Phys. Rev. B* 16, 2717 (1977).
50. A. Forchel, R. Steffen, T. Koch, M. Michel, M. Albrecht, and T. L. Reinecke, *Semicond. Sci. Technol.* 11, 1529 (1996).
51. D. Gammon, E. S. Snow, and D. S. Katzer, *Appl. Phys. Lett.* 67, 2391 (1995).
52. M. Grundmann, J. Christen, N. N. Ledentsov, J. Bohrer, D. Bimberg, S. S. Ruvimov, P. Werner, U. Richter, U. Gosele, J. Heydenreich, V. M. Ustinov, A. Y. Egorov, A. E. Zhukov, P. S. Kop'ev, and Z. I. Alferov, *Phys. Rev. Lett.* 74, 4043 (1995).
53. L. Samuelson, N. Carlsson, P. Castrillo, A. Gustafsson, D. Hessman, J. Lindahl, L. Montelius, A. Petersson, M.-E. Pistol, and W. Seifert, *Jpn. J. Appl. Phys.* 34, 4392 (1995).
54. Y. Nagamune, H. Watabe, M. Nishioka, and Y. Arakawa, *Appl. Phys. Lett.* 67, 3257 (1995).
55. O. I. Mićić, K. M. Jones, A. Cahill, and A. J. Nozik, *J. Phys. Chem. B* 102, 9791 (1998).

56. M. V. Rama Krishna and R. A. Friesner, *J. Chem. Phys.* 95, 525 (1991).
57. J. I. Pankove, "Optical Processes in Semiconductors." Dover, New York, 1971.
58. J. E. MacDougall, H. Eckert, G. D. Stucky, N. Herron, Y. Wang, K. Moller, T. Bein, and D. Cox, *J. Am. Chem. Soc.* 111, 8006 (1989).
59. M. C. DeLong, W. D. Ohlsen, I. Viohl, P. C. Taylor, and J. M. Olson, *J. Appl. Phys.* 70, 2780 (1991).
60. S.-H. Wei and A. Zunger, *Phys. B* 39, 3279 (1989).
61. S.-H. Wei, L. G. Ferreira, and A. Zunger, *Phys. Rev. B* 41, 8240 (1990).
62. S. Froyen and A. Zunger, *Phys. Rev. Lett.* 66, 3132 (1991).
63. A. Mascarenhas and J. M. Olson, *Phys. Rev. B* 41, 9947 (1990).
64. G. S. Horner, A. Mascarenhas, S. Froyen, R. G. Alonso, K. A. Bertness, and J. M. Olson, *Phys. Rev. B* 47, 4041 (1993).
65. J. M. Olson, S. R. Kurtz, A. E. Kibbler, and P. Faine, *Appl. Phys. Lett.* 56, 623 (1990).

Chapter 6

STRAINED-LAYER HETEROEPITAXY TO FABRICATE SELF-ASSEMBLED SEMICONDUCTOR ISLANDS

W. H. Weinberg, C. M. Reaves, B. Z. Nosho, R. I. Pelzel, S. P. DenBaars

Departments of Chemical Engineering and Materials, University of California, Santa Barbara, California, USA

Contents

Handbook of Nanostructured Materials and Nanotechnology, edited by H.S. Nalwa
Volume 1: Synthesis and Processing

1. INTRODUCTION

1.1. Trends in Semiconductor Nanostructures: Smaller in All Dimensions

In the study of physical properties, nanostructures often provide the best or the only testing ground for phenomena in fields such as quantum mechanics and condensed-matter physics. In electronic devices, there is a trend to use smaller numbers of electrons to get a task done. With devices that emit light such as laser diodes, the emission wavelengths must be controlled. To improve both types of devices, structures that exploit quantum mechanical behavior are an option. This is achieved by reducing the size of the structure. If only one dimension is made small, the electron will only be partially confined; it will still behave as a free electron in the remaining two large dimensions. Quantum structures are, therefore, classified by how many dimensions provide confinement or, inversely, how many dimensions allow free-electron behavior. If a structure provides confinement in one dimension, it is called a quantum well. If a structure provides confinement in two dimensions, it is called a quantum wire. If a structure provides confinement in three dimensions, it is called a quantum box or quantum dot. Although the quantum mechanics are well established, the creation of quantum structures, in particular, quantum dots, is difficult.

This difficult work is pursued because quantum dots emulate a single atom. One atomic property is that an atom has discrete energy levels. If a nanostructure that confines electrons can be fabricated small enough, then discrete energy levels can be observed. This has been done with thin semiconductor structures for several decades [1]. However, thin structures only provide quantum confinement, and discrete energy states, in one dimension. The goal with quantum dots is to achieve quantum confinement in all dimensions. One motivation for this arises from the concept of density of states. The total energy of an electron has kinetic energy components resulting from motion (momentum) in three Cartesian directions. To account for a particular amount of energy, there is usually a number of combinations of momentum components that can be considered. Even in a quantum well or a quantum wire, the discrete energy levels only partially define the energy; momentum in the unconfined dimensions can lead to a range of allowed states. A quantum dot provides confinement in all dimensions. The allowed energy states are completely defined by the quantum confinement, and the resulting density of states is, therefore, a delta function [2–4].

Why is this well-defined density of states so desirable? One reason is the increased accuracy in the energy. With the widely used quantum well, the energy of an electron can be narrowed down to a *minimum* of the lowest energy state. There are several instances where this uncertainty can be a problem. Quantum structures are often used in physical measurements to understand better quantum mechanics and the properties of materials. What if an external field (e.g., magnetic, electric, or stress) is applied to a sample and the experimentalist is looking for a shift in the quantized energy level? A shift in the measured energy of the electron could possibly be due to redistribution among allowed energy states, not just a shift in the level. Quantum structures are often used in electronic and optoelectronic devices. In the case of lasers, a partially continuous density of states may lead to the transformation of electrical energy into light at an unwanted energy. For many applications, the distribution of electronic energy resulting from thermal considerations can be a problem, yet would be nearly eliminated if the density of states were a delta function [5].

Other chapters in this book discuss the fabrication and use of many types of nanostructures, many of which use semiconductor epitaxy. One approach to nanostructure fabrication is to take advantage of the unique aspects of epitaxy to form self-assembled nanostructures. Such structures have several advantages. One major advantage is that they are formed during the epitaxial growth with no processing.

1.2. Processing: The Good and the Bad

Regardless of the technique, there are two key requirements in fabricating nanostructures: (1) achieving the desired size, shape, density, and spatial distribution and (2) maintaining high material quality. These requirements can be difficult to achieve. Producing structures that are nanometer sized in one dimension is relatively simple. Epitaxy grows material on an atomic level of a control. Thin layers such as quantum wells have been readily fabricated for about two decades [1]. They can be found in a number of commercial microelectronic devices and have been used in a range of physical studies.

Thin semiconductor layers can be formed simply by epitaxy. What about nanostructures that are wires and boxes? These structures require control not only in the epitaxial growth direction but also laterally. There are several methods to achieve this. There are many successful approaches that involve common processing steps such as lithography and etching. Some of these techniques are illustrated in Figure 1. Other methods involve processing after growth. A semiconductor sample, often starting with a thin layer, is patterned by placing a patterned etch mask (often a photoresist or a dielectric layer) on the sample, as shown in Figure 1a. The sample is then etched with either solution-based wet techniques or reactive-ion dry techniques such that the thin layer is laterally defined. This technique has been used to create a quantum box laser [6].

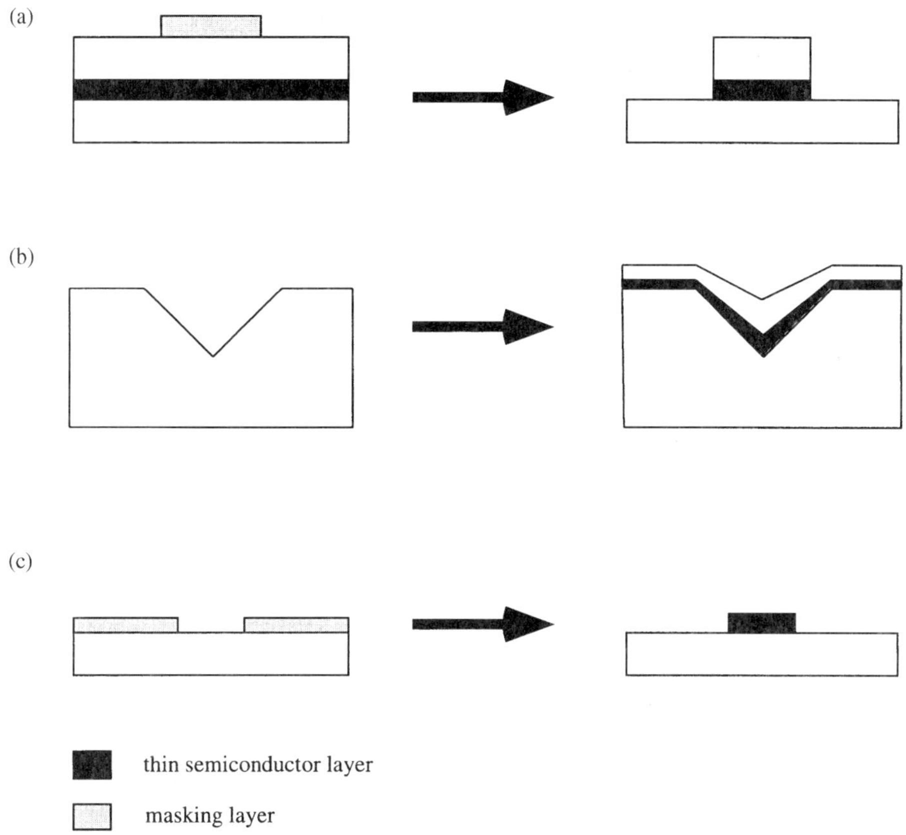

Fig. 1. Typical methods of semiconductor nanostructure fabrication via processing. A thin layer sandwiched between other layers is etched (a) to reduce the lateral dimensions. Epitaxial growth can occur over a patterned surface (b) where variations in thickness will occur, leading to a wirelike region at the bottom of the groove. Epitaxial growth can also occur over a masked surface (c) such that the new material is only deposited in the exposed region.

A processed surface can also be used for semiconductor overgrowth to obtain laterally small nanostructures. For example, a V-shaped groove, shown schematically in Figure 1b, can be etched into a semiconductor by taking advantage of the etch selectivity of different crystallographic planes. These crystallographic planes will often exhibit different behavior during growth. Some work with this approach has formed material that is thicker at the bottom of the groove than on the side wall. The electronic behavior of such a structure has been used to make quantum wires [7]. Another approach that involves processing before growth consists of placing a patterned mask, often a dielectric, over the surface, as shown in Figure 1c. The growth can be done such that the new semiconductor material only deposits in the open area. The dielectric can later be removed to leave a laterally small semiconductor nanostructure.

There are many advantages of processing approaches to achieve semiconductor nanostructures. They can be very good in determining the shape, density, and spatial distributions of the structures. In some cases, these are major concerns. The size of the structure can also be controlled down to a lower limit. Lithographic techniques are evolving to smaller and smaller sizes and the 100–1000-Å range is readily achievable. In some cases, however, smaller nanostructures may be desired.

One disadvantage with processing techniques is that the commonly used etching processes often cause damage to the remaining material [2]. This damage may lower the material quality of the nanostructure. Another problem is exposing the sample to air between steps. For example, growth is usually done in one chamber, the sample removed, taken to another chamber for dielectric deposition, then removed, taken to a lithography system, patterned, then etched either in a beaker or in a chamber, removed, and so on. These multiple steps are often done in different environments. Changes in environments can introduce oxidation and contamination, also compromising material quality. One approach is to connect a processing chamber to a deposition chamber such that oxidation and contamination are reduced [8].

Many of the problems with processing routes to nanostructures are being addressed with ongoing research. A complementary approach is to find methods to fabricate semiconductor nanostructures without processing.

1.3. An Alternative: Self-Assembled Structures

Self-assembling approaches to quantum structures have the advantage that the structures are formed in the growth environment and no processing is needed either before or after the growth. There is no processing-related damage or contamination and the nanostructures can be smaller than lithographic dimensions, yet there is little direct control over the size, shape, density, and spatial distribution.

There have been several different types of self-assembled quantum structures. One class that can be fabricated *in situ* during growth are lateral superlattices. Two approaches have been demonstrated. One approach relies on the fact that some alloys undergo atomic ordering, leading to low band gap and high band gap regions [9]. This has been observed in the case of GaInP, which separates into gallium phosphide–rich and indium phosphide–rich regions that extend in one dimension within the sample. If the layer is thin, the resulting structures are quantum wires [10].

The other approach is to deposit fractional monolayers of one material alternately on a vicinal surface. During the step-flow growth mode [11, 12], new adatoms that adsorb on a terrace will attach to the up-step edge. Hence, if a half monolayer of material A is deposited on a vicinal surface followed by a half monolayer of material B and this process is repeated, a lateral superlattice can be formed consisting of vertical regions of different materials. This technique has been used to form tilted superlattices [13] and serpentine superlattices [14], which exhibited quantum wire behavior and have been used in laser structures [15].

There has also been considerable work in forming self-assembled quantum dots not using traditional epitaxy, but solution chemistry and other techniques to form small clusters [16, 17]. Although there have been a number of successes with this work, including precise control of cluster size by selection of template molecules, there are several disadvantages with producing quantum dots with these techniques. Passivation steps are vital to prevent a significant fraction of the cluster from oxidizing, and these clusters would also be difficult to integrate with traditional semiconductor structures.

The types of self-assembled quantum structures discussed in this chapter involve the formation of defect-free three-dimensional islands during strained-layer epitaxy. The previously mentioned techniques involving heteroepitaxy were demonstrated in lattice-matched materials systems.

1.4. Outline of the Chapter

Because the formation of these island nanostructures is highly dependent on the growth process, the basics of heteroepitaxy will be reviewed next (Section 2). Then comments will be made on the common experimental techniques used to fabricate and study these structures (Section 3). During the growth, there is an abrupt transition between two-dimensional growth and three-dimensional growth. One way to classify the self-assembling islands is to divide them into those that form before the transition (Section 4) and those that form after the transition (Section 5). Before summarizing, a brief discussion of the properties and applications of these islands will be given (Section 6).

2. BASICS OF HETEROEPITAXY

Heteroepitaxy is the process of depositing one crystalline material on a different material with an interface that is nearly perfect. The process is widely used, not only for research, but for manufacturing semiconductor devices such as lasers, light-emitting diodes, and transistors. With its well-established position in semiconductor research and manufacturing, using heteroepitaxy to fabricate nanostructures is a natural extension. Although epitaxy has been studied for many years, it is still not fully understood [18]. There have been several books, chapters, and reviews on the topic, a subset of which is listed here [19–24], and these can be consulted for more in-depth information. In this section, we review some basic concepts of heteroepitaxy that are important in understanding how self-assembled islands can be made. The basic surface processes will be reviewed and the most common growth modes will be introduced.

2.1. Fundamental Processes during Epitaxy

The key surface processes that occur during epitaxy are shown schematically in Figure 2 [22]. Regardless of the growth technique, atoms (and molecules) are delivered to the substrate surface, and a large fraction of these species adsorb on the surface. Once adsorbed, there are three things that can happen to the adatom. It can either form a strong chemical bond to the surface where it is trapped, diffuse on the surface to find an energetically preferred location prior to strong chemical bonding, or desorb. Once adsorbed chemically, the adatoms can also diffuse on the surface, and this diffusion can be highly anisotropic, depending on the symmetry and nature of the surface. These chemisorbed adatoms diffuse until they either (1) desorb from the surface, (2) find another adatom and nucleate into an island, (3) attach, or aggregate, into an existing island, (4) diffuse *into* the surface, or (5) react at defect sites. The last two effects are often considered relatively minor occurrences in epitaxy but are mentioned here for completeness. Diffusion into the surface, or interdiffusion, can be significant at times. The extent of interdiffusion can be thought of

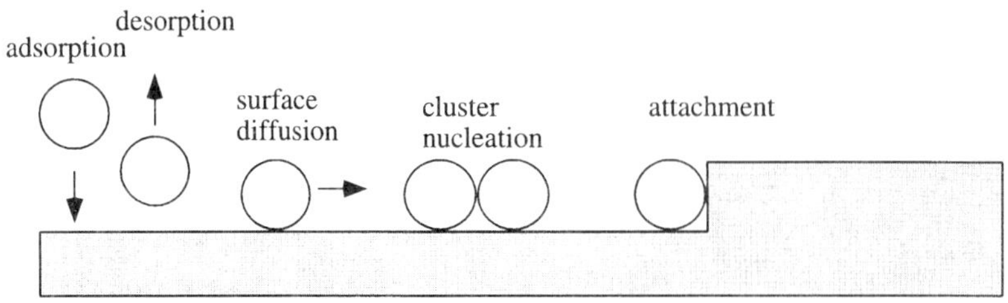

Fig. 2. Basic processes during epitaxy.

as the solubility of one material into the other and clearly has a strong dependence on the material system. The segregation of atoms into other layers can be seen, for example, in the case of In segregation into surrounding barrier materials [25] and also in atomic diffusion from delta-doped layers [26]. The reactions at defect sites are often important. For example, reactions at step edges (a defect with respect to a perfect surface) are the foundations of step-flow growth.

The formation of clusters and the attachment of atoms to existing structures and clusters are important in the formation of self-assembled islands. When diffusing adatoms find each other, they can nucleate and form an island. Island growth continues either when other diffusing adatoms attach themselves or by direct impingement of gas phase atoms onto existing islands. Adatoms that directly impinge on an island can either incorporate into the island or lead to the next-layer growth, depending on the surface potential. Although the diffusion of adatoms attached to islands can be significantly reduced because of the local surface potential, it is still possible for adatoms to detach from the islands. Thus, islands have a "critical size" associated with them, at which they become "stable" with respect to "evaporation." Here, stable means that the islands are sufficiently large that the rate of attachment to the islands is the same or greater than the rate of detachment from the islands [27–31]. As the islands continue to grow further, and possibly migrate, they can find other islands and coalesce into one large island. The evolution of island formation can, therefore, be visualized as a progression through three different growth regimes. Initially, there is a high concentration of adatoms or monomers diffusing on the surface, resulting in a high probability of island nucleation. This is the nucleation regime, where the density of islands on the surface increases with coverage. The density continues to increase until the probability of a diffusing adatom finding an island is much higher than the probability of finding another adatom. The number of nucleation events is substantially reduced as the adatom diffusion length becomes large relative to the average island spacing, and, thus, the majority of events occurring are adatoms attaching to the existing islands, hence defining the aggregation regime. As further growth continues in the aggregation regime, the island density remains relatively constant while the islands continue to grow in size. Eventually, the islands will begin to merge with one another and enter into the coalescence regime, which is signified by a decrease in the island density with increasing coverage.

There has been a considerable amount of work in the literature in trying to describe analytically the atomic processes involved in thin-film growth through the use of kinetic rate equations. In these equations, each of the atomic processes can be represented by writing expressions for the time-dependent changes in the densities of single adatoms, clusters of a given size, and stable clusters. The specifics of these equations will be not be discussed in detail here as comprehensive reviews on this topic are widely available in the literature [22, 23, 27, 32, 33]. To illustrate the point, however, a possible equation for the time rate of change of the single adatom density is given by

$$\frac{dn_1}{dt} = U_{\text{dep}} + U_{\text{dis}} - U_{\text{evap}} - 2U_1 - U_{\text{cap}} \tag{1}$$

where similar expressions could be written for the rate of change of islands of a given size. The single adatom density can increase by the rate of deposition and dissociation (or detachment) from larger islands as embodied by U_{dep} and U_{dis}, respectively, or decrease by the rate of evaporation, the nucleation of two single adatoms, or the capture of a single adatom by a larger island as embodied by U_{evap}, $2U_1$, and U_{cap}, respectively. The factor of 2 in front of U_1 is to account for the two adatoms that nucleation requires. Clearly, additional terms could be added to represent diffusion into the surface, or reaction at defect sites, or any other surface process one could imagine.

To continue with this description, explicit expressions for the various elementary rates must be determined. The terms describing the deposition rate and the evaporation rate are fairly straightforward. The other terms involving the nucleation and aggregation of islands are functions of the diffusion coefficient, the densities of the single adatoms, the densities of islands of any given size, and a "capture number," which is a variable that takes into account the local distribution of adatoms around an island. To help discard some of the terms, certain regimes of the growth are studied to find terms that are minimal in that regime and, thus, reduce the rate equations. For example, in considering the aggregation regime of growth, an assumption could be made that the density of single adatoms on the surface is much smaller than the total density of islands, and the rate equations can be modified accordingly. By simplifying the rate equations sufficiently, the variables can be separated and then integrated to give general expressions for the densities. With the appropriate approximations, the equations describing the densities of the single adatoms and islands can be expressed as simple functions of the coverage and the ratio of flux to diffusion. This is the basis for the scaling relations derived for thin-film growth, and they have been used extensively in attempting to model epitaxial growth [28–30, 34–36].

2.2. Heteroepitaxial Growth Models

There are three general ways in which one material, say B, can grow epitaxially on a dissimilar material, say A [19]. These growth modes can be described by the equilibrium morphology, as determined from the surface free energies [18, 19, 21]. Following the notation used by Tsao [21], the three surface free energies considered are the energies associated with the substrate–vacuum, substrate–epilayer, and epilayer–vacuum interfaces and are denoted by $\gamma_{\text{sub–vac}}$, $\gamma_{\text{sub–epi}}$, and $\gamma_{\text{epi–vac}}$, respectively. The $\gamma_{\text{sub–vac}}$ term can be thought of as the initial energy term before the epilayer formation, and the remaining two terms, $\gamma_{\text{epi–vac}}$ and $\gamma_{\text{sub–epi}}$ are associated with the epilayer formation. Based on work by Bruinsma and Zangwill [37], a "spreading pressure" can be defined as

$$S = \gamma_{\text{sub–vac}} - \gamma_{\text{sub–epi}} - \gamma_{\text{epi–vac}} = \gamma_{\text{sub–vac}} - (\gamma_{\text{sub–epi}} + \gamma_{\text{epi–vac}})$$

It is evident that the relative contributions from these terms change as the epilayer evolves, resulting in a competition to determine the lowest-energy surface.

We will start the discussion with the two extremes. If $\gamma_{\text{sub–epi}} + \gamma_{\text{epi–vac}} < \gamma_{\text{sub–vac}}$, then material B will grow in a layer-by-layer fashion. This means that the B atoms will try to cover completely the surface of A because this growth minimizes the surface free energy. During the growth of the first layer of B, new atoms landing on the bare A surface will diffuse on the surface until they attach to existing clusters of B atoms. If a new B atom lands on top of an existing cluster of B atoms, it will diffuse to the edge of the structure and jump down, attaching to the edge of the cluster. Growth occurs in a two-dimensional fashion. Thus, each complete layer is thermodynamically stable as a two-dimensional layer. This is known as the Frank–van der Merwe growth mode and is illustrated in Figure 3a.

On the other hand, when $\gamma_{\text{sub–epi}} + \gamma_{\text{epi–vac}} > \gamma_{\text{sub–vac}}$, it is thermodynamically unfavorable for the epilayer to be flat and the B atoms will cluster and form islands to try to minimize the interface between A and B. Water beading on a waxed car is a result of a high interfacial energy. New atoms that land on top of an existing cluster of B atoms will

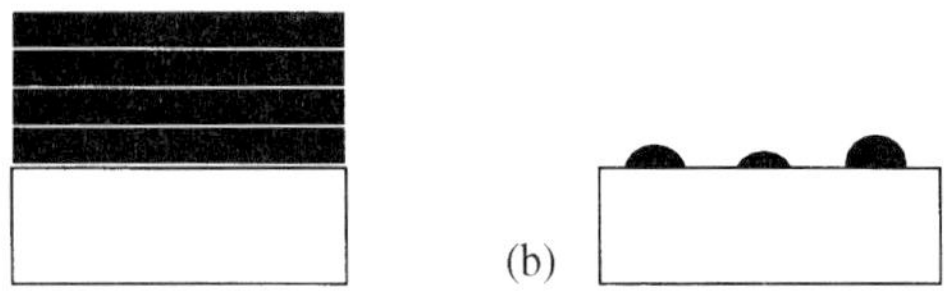

Fig. 3. Comparison of Volmer–Weber (a) and Frank–Van der Merwe (b) growth modes.

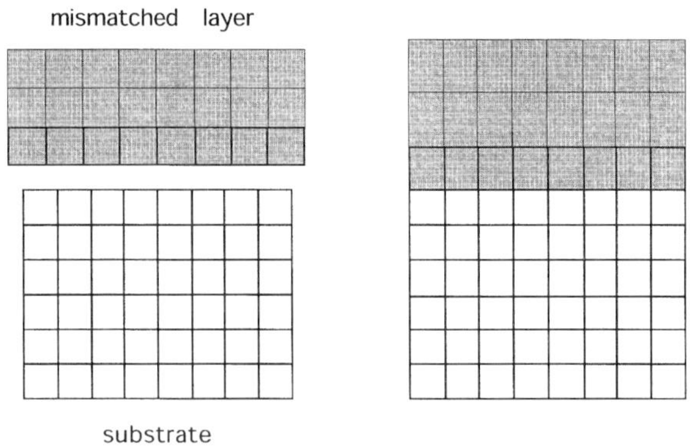

Fig. 4. Heteroepitaxy of lattice-mismatched materials. Note the (tetragonal) deformation of the epitaxial layer.

remain on top instead of jumping down. Growth occurs in a three-dimensional fashion; new material will add to the height of the existing islands more than the lateral size. This is known as the Volmer–Weber growth mode and is illustrated in Figure 3b.

Before describing the third growth mode, we should discuss one important point that has been excluded from the previous discussion, namely, the concept of lattice mismatch. If the lattice structure or lattice constant of A and B is dissimilar, then elastic strain must be considered. The material in the layer being deposited must stretch or compress, as shown in Figure 4, to match the lattice of the underlying material. The fraction of lattice mismatch, f, is given by

$$f = \frac{a_f - a_s}{a_s}$$

where a_f is the lattice constant of the film and a_s is the lattice constant of the substrate. This definition is widely used [38] although other similar definitions do exist [39, 40].

There are two paths that lattice-mismatched films can take. For small lattice mismatches, approximately 2% or less, the growth will occur in a layer-by-layer fashion for many layers. At some point, the strain energy will build up and bonds within the sample, often at the heterointerface, will break. These patterns of broken bonds are known as dislocations. Dislocations in semiconductors and other materials are a widely studied field [40, 41]. Materials that contain dislocations and other crystal defects are often referred to as incoherent, in contrast to a coherent material with no defects.

For moderate to large lattice mismatches, approximately 3% and larger, the growth also initially occurs in a layer-by-layer fashion. In some cases, the growth of the first layer is heavily impacted by strain, a topic that is discussed further in Section 4. Returning to our discussion based on thermodynamics, consider a material system where the initial stages of growth resemble the Frank–van der Merwe growth mode. Recall that in this scenario $\gamma_{\text{sub–epi}} + \gamma_{\text{epi–vac}} < \gamma_{\text{sub–vac}}$. After the first layer is grown, we should now replace "substrate" terms with wetting layer terms, such that $\gamma_{\text{WL–epi}}$, $\gamma_{\text{epi–vac}}$, and $\gamma_{\text{WL–vac}}$ are the relevant free energies resulting from the wetting layer–epilayer, epilayer–vacuum, and wetting layer–vacuum interfaces, respectively. These new terms take into account the strained wetting layer. As the film thickness increases and strain builds up, the contribution from the wetting layer–epilayer interface will begin to dominate such that $\gamma_{\text{WL–epi}} + \gamma_{\text{epi–vac}} > \gamma_{\text{WL–vac}}$. As in the Volmer–Weber growth mode, the surface will form three-dimensional islands to minimize the free energy and accommodate the strain. This is known as the Stranski–Krastanov growth mode [19], and is the most often observed growth mode in the lattice-mismatched heteroepitaxy of semiconductors.

The phenomenon of three-dimensional island formation during epitaxy has been documented for many decades [19]. However, there are two reasons why the islands so formed were not readily explored as potential quantum structures. First, there is a bias in semiconductor epitaxy toward flat (smooth) surfaces and interfaces. One reason behind this bias is that, with thin layers such as quantum wells and electron tunneling barriers, the thickness of the layer is critical to the performance of the device, such as the emission wavelength of a laser. If the interfaces are rough, leading to thickness variations of the layer, the emission energy will vary. Traditionally in semiconductor epitaxy, there has been an emphasis on developing and using flat surfaces and interfaces [1]. In more recent efforts, the formation of islands during strained-layer growth was seen as a problem, namely, a rough surface. In point of fact, considerable work has been done to suppress island formation by varying growth conditions [42] and by using surface treatments (surfactants) during growth [43, 44]. Second, the islands that were observed, often in metal epitaxy, contained dislocations and other defects. It was assumed that all such islands would be dislocated and, hence, not be suitable for quantum structures. However, many lattice-mismatched heteroepitaxial systems have been found to grow in a Stranski–Krastanov growth mode, where three-dimensional islands evolve after the formation of the two-dimensional wetting layer. Often, these systems possess a coverage range, or "window," in which defect formation is suppressed and three-dimensional coherent islands form. These coherent islands are the topic of Section 5. Eventually, the islands will dislocate; however, the coverage regime for coherent structures is easily obtained with current epitaxial growth techniques.

3. COMMON EXPERIMENTAL TECHNIQUES

Before discussing the details of the self-assembled islands, a few comments will be made on the growth and characterization techniques commonly used. These techniques will be codified briefly with appropriate references. The techniques to be discussed will be molecular beam epitaxy (MBE) and chemical vapor deposition (CVD) for synthesis, and reflection high-energy electron diffraction (RHEED), transmission electron microscopy (TEM), scanning tunneling microscopy (STM), and atomic force microscopy (AFM) for characterization.

3.1. Synthesis Techniques

3.1.1. Molecular Beam Epitaxy [21, 45–47]

Molecular beam epitaxy refers to the growth of a crystalline material in ultrahigh vacuum (UHV) using collimated gas phase reactants. The UHV environment facilitates the growth of extremely pure materials. The sources for growth can range from solids to gases and can be either elemental or compound. If the precursors are solid or liquid, they are heated in crucibles and their vapor is used to generate a molecular beam, or gaseous sources can be used directly. The geometry of the MBE system is such that there is a line of sight between the source and dopant beams and a temperature-controlled rotating substrate. Because growth occurs in UHV, the mean free path of the molecules is rather large, ensuring that the source molecules impinge onto the substrate directly. Growth is controlled by varying such parameters as substrate temperature, source flux, the sequence and duration of source beam(s) [i.e., alternating beam epitaxy and migration-enhanced epitaxy (MEE)], and group V overpressure (for III–V growth). Fluxes are controlled by modulating the molecular beams, usually through the use of high-speed mechanical shutters. These shutters provide the control necessary to deposit the desired quantity of material with better than 0.05 monolayer (ML) accuracy.

3.1.2. Chemical Vapor Deposition [48–50]

In chemical vapor deposition, growth occurs at a much higher pressure (1–760 torr) than in MBE. Often, the sources for growth are organometallic compounds (e.g., trimethyl gallium and arsine or tertiarybutylarsine for GaAs growth). These CVD techniques using organometallic sources go by a variety of names, two of the most common of which are organometallic vapor phase epitaxy (OMVPE) and metalorganic chemical vapor deposition (MOCVD) [48]. During growth, precursor compounds are flowed (using an inert carrier gas) over a substrate located on a heated susceptor. Flow rates are usually such that transport is governed by mass transport within a boundary layer that is present near the substrate surface. Growth usually occurs at relatively high temperatures (600–1000 °C) such that the metalorganic precursors are cracked in the boundary layer, facilitating the diffusion of the alkyl fragments through the boundary layer (away from the surface) into the free-stream flow of the carrier gas. Growth is controlled by varying such things as substrate temperature and reagent flow rates. Flow rates are controlled by metering the gaseous sources through flow controllers and fast switching valves.

3.2. Characterization Techniques

3.2.1. Reflection High-Energy Electron Diffraction [51–54]

One distinct advantage of MBE growth in comparison to CVD techniques is the ability to monitor MBE growth (*in situ*) using reflection high-energy electron diffraction. RHEED can be used to infer information about surface cleanliness, surface order and smoothness, and the growth rate. For RHEED, monoenergetic electrons (3–15 keV) are diffracted from the substrate (angle of incidence $<1°$) onto a fluorescent screen. The small angle of incidence used in RHEED corresponds to a relatively small penetration depth (a few monolayers), making RHEED a very surface-sensitive technique. In essence, RHEED is similar to X-ray diffraction for a surface. (The theoretical analysis for RHEED is identical to the formalism used to explain X-ray diffraction.) Thus, RHEED offers insight about the periodicity present at the substrate surface. Furthermore, by monitoring the intensity of the oscillations of a single RHEED spot, one is able to determine the growth rate because a single oscillation corresponds to the deposition of one monolayer.

3.2.2. Transmission Electron Microscopy [55–57]

The foundation of TEM is the wave behavior exhibited by electrons. In an experiment, periodic atomic planes within a thin crystalline sample ($\sim$1000 Å) diffract a monochromatic electron beam, typically accelerated by a 60–200-kV potential. Because TEM is sensitive to variations in the spacing of atomic planes [58], it is a useful technique for the study of islands formed during strained-layer epitaxy. Strain effects can be clearly seen as well as the presence of crystal defects [59, 60].

3.2.3. Scanning Tunneling Microscopy [61–66]

As the name suggests, STM relies on the quantum mechanical phenomenon of electron tunneling. In an STM experiment, a sharp metallic tip is brought sufficiently close to a conducting surface that electron tunneling between the sample and the tip will occur. A bias voltage (either positive or negative) is applied to the tip, and a tunneling current flows from the surface to the tip (for negative sample bias) or from the tip to the surface (for positive sample bias). In the constant-current mode, the tip is rastered in the plane of the surface and the tip–sample separation is altered, using a feedback circuit, such that the tunneling current is maintained at a constant value. The tip position in this rastering procedure follows a constant density of electronic states topograph of the surface. It is important to realize that an STM image is not a map of atomic position. Rather, it represents a constant density of electronic states contour of the surface for a given bias voltage. Even so, by varying the bias voltage, the electronic states of the surface involved in tunneling are changed. Furthermore, both negative and positive sample biases can be used to probe the filled and empty states. Thus, bias-dependent imaging can be used to infer details about the atomic structure and electronic nature of the surface. In practice, STM has been an invaluable technique used in the study of surface reconstructions and atomic scale features, providing information at resolutions unavailable by almost any other technique.

3.2.4. Atomic Force Microscopy [64, 67–69]

In an AFM experiment, a sharp tip mounted on the end of a flexible cantilever is brought into sufficiently close proximity with the sample that a detectable force is generated. Detection of the tip–sample force has been achieved in different ways. One common method involves the deflection of a laser beam reflected from the back of the cantilever. The tip is rastered above the surface and a feedback loop is used to keep the separation between the tip and the sample at a constant value through the actuation of a piezoelectric translator that moves the sample (in the z direction perpendicular to the sample).

As mentioned, the AFM relies on force detection. Depending on the tip–sample separation, the type and magnitude of the tip–sample force will vary. In the so-called contact mode, the separation distance between the tip and the sample is small, and the detected force is a result of core–core repulsion (the Pauli principle). Although contact mode AFM can provide atomic resolution, it requires a very rigid substrate. In the noncontact mode, the tip–sample separation is larger than for contact mode AFM, and the gradient of the van der Waals potential is the relevant force. Generally, the spatial resolution of noncontact mode AFM is inferior to the resolution achievable with contact mode AFM. Yet, noncontact mode AFM is less susceptible to imaging artifacts resulting from deformation of the sample by the tip. To image a surface nonintrusively and still achieve a high level of spatial resolution, tapping mode AFM was developed. In the tapping mode, the cantilever of the AFM is forced to oscillate at a certain distance from the sample while the probe is scanned laterally. When the tip encounters a surface feature, the resonant frequency of the cantilever changes. A change in the resonant frequency of the cantilever will result in a change in the amplitude of oscillation, which is detected by the laser reflecting from the back of the cantilever. When this happens, the feedback loop lowers the sample so that the original amplitude of cantilever oscillation is restored. An on-line computer records the voltages applied to the z-direction piezoelectric actuator for feedback control and converts them into a topographic map of the surface.

4. TWO-DIMENSIONAL GROWTH AND ISLAND FORMATION BEFORE TRANSITION TO THREE-DIMENSIONAL GROWTH

There has been much recent interest in creating reduced-dimensional structures with carrier confinement in two and three dimensions by self-assembling mechanisms in lattice-mismatched heteroepitaxy. It is hoped that the *in situ* formation of these low-dimensional structures can be utilized as a suitable alternative to damage-inducing *ex situ* processes such as etching and lithography [70–74]. Many lattice-mismatched heteroepitaxial systems follow the Stranski–Krastanov growth mode, where at least one monolayer grows in a layer-by-layer mode, forming a two-dimensional layer, sometimes known as the two-dimensional layer. The strain from the lattice mismatch is accommodated in elastic deformation and, thus, determines the critical thickness of the two-dimensional layer. In the layer-by-layer regime of growth, two-dimensional islands (islands that are one monolayer in height) can form as a mechanism to help minimize surface energy. As the film thickness increases beyond the critical thickness, three-dimensional islands form on the two-dimensional layer. Considering that high-quality materials preclude dislocation formation, a given heterostructure is limited by the amount of coherent material that can be deposited before dislocation formation becomes energetically favorable. Hence, it is of great importance to determine the mechanisms of strain relief for various technologically important semiconductor systems because a better understanding of the initial stages of strained-layer heteroepitaxy is crucial for the further development of this technology.

The importance of two-dimensional islanding should also be extended to the understanding of interfaces. Interfaces can play a critical role in the device performance of heteroepitaxial systems. As relevant length scales decrease, the importance of morphological issues at these interfaces becomes even more crucial. For example, resonant tunneling devices (RTDs) are currently of considerable interest because of their potential use in high-speed terahertz devices. RTDs are double-barrier heterostructures, where the barriers are typically on the order of 5 to 10 monolayers (ML) thick. A theoretical study examined the effects of islands at the interface on the performance of RTDs, as predicted by the peak-to-valley ratios determined from calculated I–V curves as a function of terrace sizes [75]. A modest amount of lattice mismatch can be tolerated because the barrier thickness can be substantially smaller than the critical thickness to three-dimensional growth. However, new issues arise when two compounds with different group-V components are grown together. As an example, consider the case of InAs and AlSb, which are used in RTDs. AlSb has a lattice constant slightly larger than InAs (1% mismatch) and also a significant band offset, making it an ideal candidate as a barrier material to InAs. The interface bonds for this system must be either In–Sb or Al–As. Most MBE growths occur under group-V overpressures, and the traditional approach is to "soak" the surface under the new group-V flux when changing materials at an interface. In either case, the InSb-like or AlAs-like material on the InAs is significantly lattice mismatched to the InAs substrate, thus making strain issues relevant.

In this section, we will review the issues relating to the initial stages of Stranski–Krastanov growth. In particular, we will first discuss the beginning stages of lattice-mismatched heteroepitaxial growth as the two-dimensional layer forms up through the transition from two-dimensional to three-dimensional growth. Furthermore, the effects of reconstruction and surface orientation on the formation and morphology of two-dimensional islands will be discussed. Clearly, the literature is far too rich to attempt to review every material system that has been studied throughout the years. Rather, we will discuss specific examples from the literature that we believe are representative of the concepts we bring forth. What we wish to emphasize are concepts, because the ideas are general for lattice-mismatched heteroepitaxy. We will focus this discussion on elemental or compound semiconductor systems with the diamond or zincblende structure.

4.1. Initial Stages of the Two-Dimensional Layer Formation

Insight into the mechanisms involved in the formation of the two-dimensional layer is needed to understand some of the fundamental aspects of lattice-mismatched heteroepitaxial growth. Figure 5 shows STM images of the initial stages of InAs growth on GaAs(001)–(2×4) at various submonolayer coverages. Parts a and b of Figure 5 show a coverage of 0.15 and 0.29 ML, respectively. The InAs forms two-dimensional islands, which are one monolayer in height above the GaAs surface. Higher-resolution STM images show that these islands also exhibit a $c(4 \times 4)$ reconstruction, which further distinguishes the islands

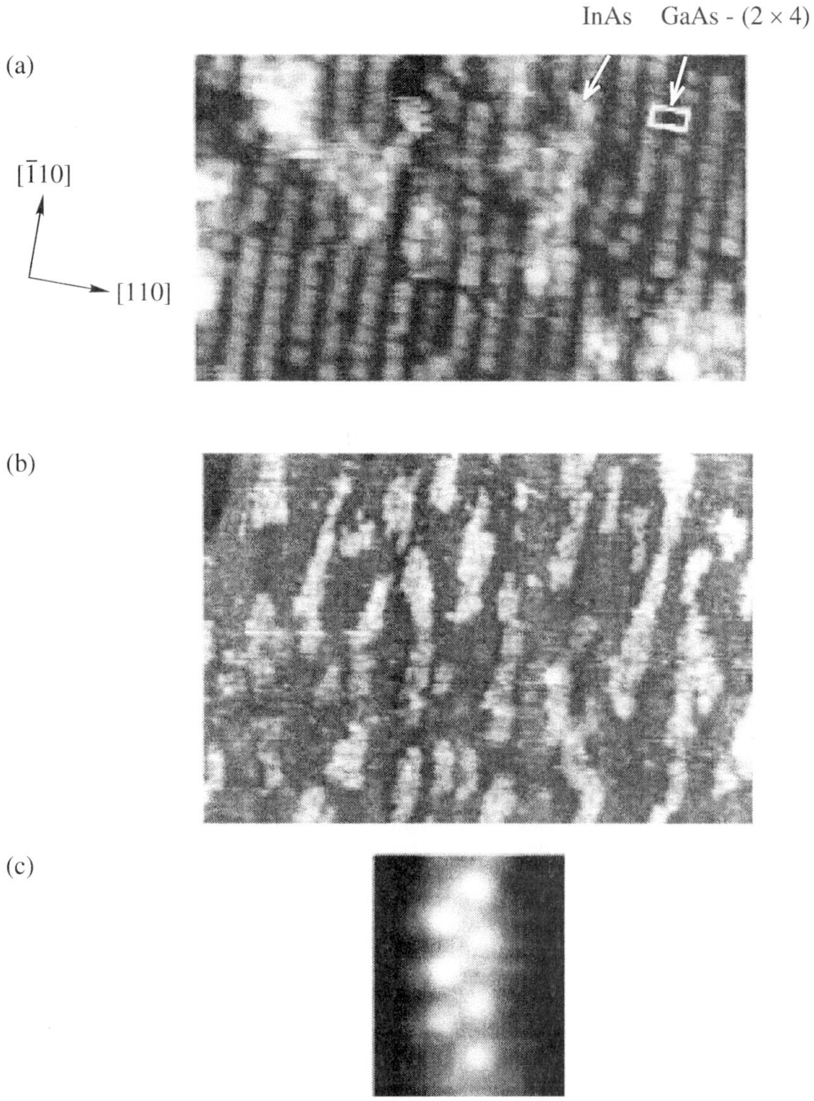

Fig. 5. Filled-state STM images of InAs islands on GaAs(001)–(2 × 4): (a) 30-nm × 15-nm image of 0.15 ML of InAs deposition and (b) 100-nm × 70-nm image of 0.29 ML of InAs deposition. The islands in both images are elongated in the [$\bar{1}$10] direction and exhibit a c(4 × 4) reconstruction and easily distinguishable from the (2 × 4) reconstruction still observed on the bare substrate. Reprinted with permission from V. Bressler-Hill et al., *Phys. Rev. B* 50, 8479 (© 1994 American Physical Society) and from J. Tersoff and R. M. Tromp, *Phys. Rev. Lett.* 70, 2782 (© 1993 American Physical Society).

from the initial (2 × 4) reconstructed surface (cf. Fig. 5c). In addition, the islands are observed to be anisotropic and elongated in the [$\bar{1}$10] direction at all submonolayer coverages studied. Qualitatively, there appears to be a difference in the energetics of the two types of island edges. The edges that run along the [$\bar{1}$10] direction are relatively straight compared to the edges that are parallel to the [110] direction, which are seen to have more kinks (cf. Fig. 5b). This indicates that rearrangement along the steps in the [$\bar{1}$10] direction is relatively rapid. A statistical analysis on the island dimensions shows that the InAs islands maintain a most probable width of approximately 4 nm in the [110] direction at all coverages examined, whereas the islands appear to grow freely in the [$\bar{1}$10] direction with increasing coverage. From this, it follows that the aspect ratio of these islands will significantly increase with coverage until the onset of coalescence. It should be mentioned that anisotropic island shapes are common in semiconductor systems, and are even seen in homoepitaxial systems. Several factors could lead to this observed anisotropy, including an anisotropy in the step energies, a preferential growth direction resulting from different step edge reactivities, anisotropic barriers to surface diffusion, or some combination of them all [76–78]. Strain, in addition to the other possible anisotropies mentioned, is clearly expected to influence the surface morphology for the case of lattice-mismatched heteroepitaxial growth.

One other noteworthy result seen in the aforementioned InAs/GaAs study was found when growing InAs islands on a 1° vicinal, B-type (anion-terminated steps running parallel to the [110] direction) GaAs(001)–(2×4) surface [79]. On this substrate, the steps limited the In diffusion in the $[\bar{1}10]$ direction and, hence, the extent of growth possible in the $[\bar{1}10]$ direction. Moreover, the B-type steps were found to be more reactive than the A-type steps (cation-terminated steps running parallel to the $[\bar{1}10]$ direction) and to act as adatom sinks. Even with these growth constraints, the islands still showed a preferred width of around 4 nm in the [110] direction, indicating that the strain is likely being accommodated in that crystallographic direction. The observation of a preferred two-dimensional island size is consistent with work done by Massies and Grandjean [80], who considered the effects of nontetragonal elastic distortion at the edges of two-dimensional islands. A simplified version of the valence force field model was used in which the shape of a one-dimensional island on top of a completely rigid substrate could be calculated. For a surface coverage of 0.5 ML of $In_{0.49}Ga_{0.51}As$ islands on GaAs (lattice mismatch of 4%), the model predicted a preferred island size of 11 unit cells, or 4.4 nm. This preferred size would be expected to be smaller for the increased lattice mismatch of InAs on GaAs(001).

The observation of islands with a preferred size is also consistent with the theoretical predictions of Tersoff and Tromp [81], who proposed a shape transition for coherently strained islands. From their model, small islands initially evolve in a compact shape, but at a critical size, the islands take on a rectangular shape and become progressively elongated, where the width of these islands approaches a constant. The islands can then minimize their energy by keeping the optimal size in one direction, while continuing to grow in the other direction. The question then arises as to what determines which direction will elongate. On a surface terminated with group-V atoms, dimers form with the dimer bonds aligned in the $[\bar{1}10]$ direction. Deposition of group-III atoms will form bonds between the group-III and the group-V atoms that will also be aligned in the $[\bar{1}10]$ direction. If the atoms wish to move to help accommodate strain, they could modestly move perpendicular to the directions of their back bonds to the surface; that is, they could move in the [110] direction. Thus, strain energy could be partially relieved by the same nontetragonal deformation model of Massies and Grandjean [80] in the [110] direction.

4.2. Transition from the Two-Dimensional Layer to Three-Dimensional Islands

Following the Stranski–Krastanov growth mechanism, the two-dimensional layer will eventually give way to the formation of three-dimensional islands. The exact mechanism of this transition is presently not well understood and is a subject of much study [82–93]. It is reasonable to believe that the size and spatial distribution of the two-dimensional islands formed in the two-dimensional layer of these systems is somehow related to the distribution of three-dimensional islands formed after further material deposition. In other words, the two-dimensional islands, or platelets, are precursors to three-dimensional islands and, ultimately, grow into the larger structures. Priester and Lannoo [90] have proposed a model that follows this line of reasoning in an attempt to describe the formation of three-dimensional InAs islands on GaAs(001). In their model, a complete monolayer of InAs would first cover the surface forming the two-dimensional layer. At a total coverage of 1.4 ML, the model predicts the formation of large two-dimensional islands that are randomly distributed across the surface. These two-dimensional islands are uniform in size and are a single monolayer in height. This prediction of uniformly sized two-dimensional islands is reported by Chen and Washburn [84], who believe that the increase in strain at island edges affects the surface potential near the island edges and, subsequently, makes it increasingly difficult for adatoms to aggregate further into existing islands. Similarly, in kinetically driven models, such as Monte Carlo simulations, the detachment probabilities of adatoms from islands as a function of island size have been studied [30, 36]. As more material is deposited, the islands continue to grow and eventually evolve into uniformly sized three-dimensional islands.

In an attempt to study the two-dimensional-to-three-dimensional morphology transition, Heitz et al. [85] have examined this same InAs/GaAs system at InAs coverages ranging from 0.87 to 1.61 ML on the GaAs(001)–c(4 × 4) surface. Images from their study are shown in Figure 6. There is no evidence of the large platelets that were predicted by Priester and Lannoo, but rather a surprising result regarding the two-dimensional-to-three-dimensional transition. At a coverage of approximately 1.15 ML, features 2–4 ML high begin to appear on the surface. The density of these features increases until an approximate coverage of 1.35 ML, whereupon these three-dimensional features disappear. Upon further deposition to 1.45 ML, the three-dimensional islands reappear, but at a much higher density. This type of appearance, disappearance, and subsequent reappearance of three-dimensional islands at coverages below the critical film thickness has not been previously reported for systems evolving in the Stranski–Krastanov growth mode. To study this phenomenon further, samples were grown for photoluminescence (PL) measurements with InAs deposition ranging from 1 to 2 ML. Peaks attributed to three-dimensional islands are observed at 1.15 and 1.25 ML, where the peak at 1.25 ML is shifted toward lower energies because of the larger island sizes. However, at 1.45 ML, this peak disappears and does not reappear until about 1.55 ML of deposition, whereupon it evolves into the peak typically observed at the critical point.

Other examples of the transition from two-dimensional to three-dimensional growth can be found in the growth of antimony-based materials on GaAs. Thibado et al. [91] studied GaSb on GaAs(001)–c(4 × 4), and found that, after 1 ML of GaSb deposition, the surface was covered with two-dimensional islands, or platelets, which were approximately 10 nm in diameter, although slightly anisotropic in the $[\bar{1}10]$ direction. An image in a related study of the same material system is shown in Figure 7, where the islanded two-dimensional layer is clearly observed. Furthermore, they found that adding a second monolayer of material primarily added onto the existing platelets, making them 2 ML high while approximately maintaining the same diameter. After 3 ML of deposition, they observed the formation of three-dimensional islands. Imaging the areas between the three-dimensional islands showed that the two-dimensional layer composed of the network of two-dimensional islands (2 ML in height) was still intact. Furthermore, a rough calculation based on the apparent island dimensions (neglecting convolution effects from the tip) showed that approximately 0.6 ML of material was incorporated into the three-dimensional islands, which is consistent with the two-dimensional layer remaining intact and not necessarily incorporating itself into the three-dimensional islands. Voigtländer and Kästner [92] have also tried to address the issues of three-dimensional island formation from two-dimensional growth by studying Ge growth on Si using *in vivo* STM during growth. They concluded that there did not appear to be any type of special morphology (e.g., step edges, large or high islands in the two-dimensional layer, domain boundaries, etc.) where the three-dimensional islands would nucleate and evolve, but rather it simply appeared to occur at random locations on the surface.

Based on these experimental observations, it appears that the large two-dimensional platelets are not the precursors to three-dimensional islands as predicted by equilibrium-based theoretical models [90]. In an earlier and related work to that of Heitz et al. [85], Ramachandran et al. [86] considered mass transfer and kinetics in the formation of

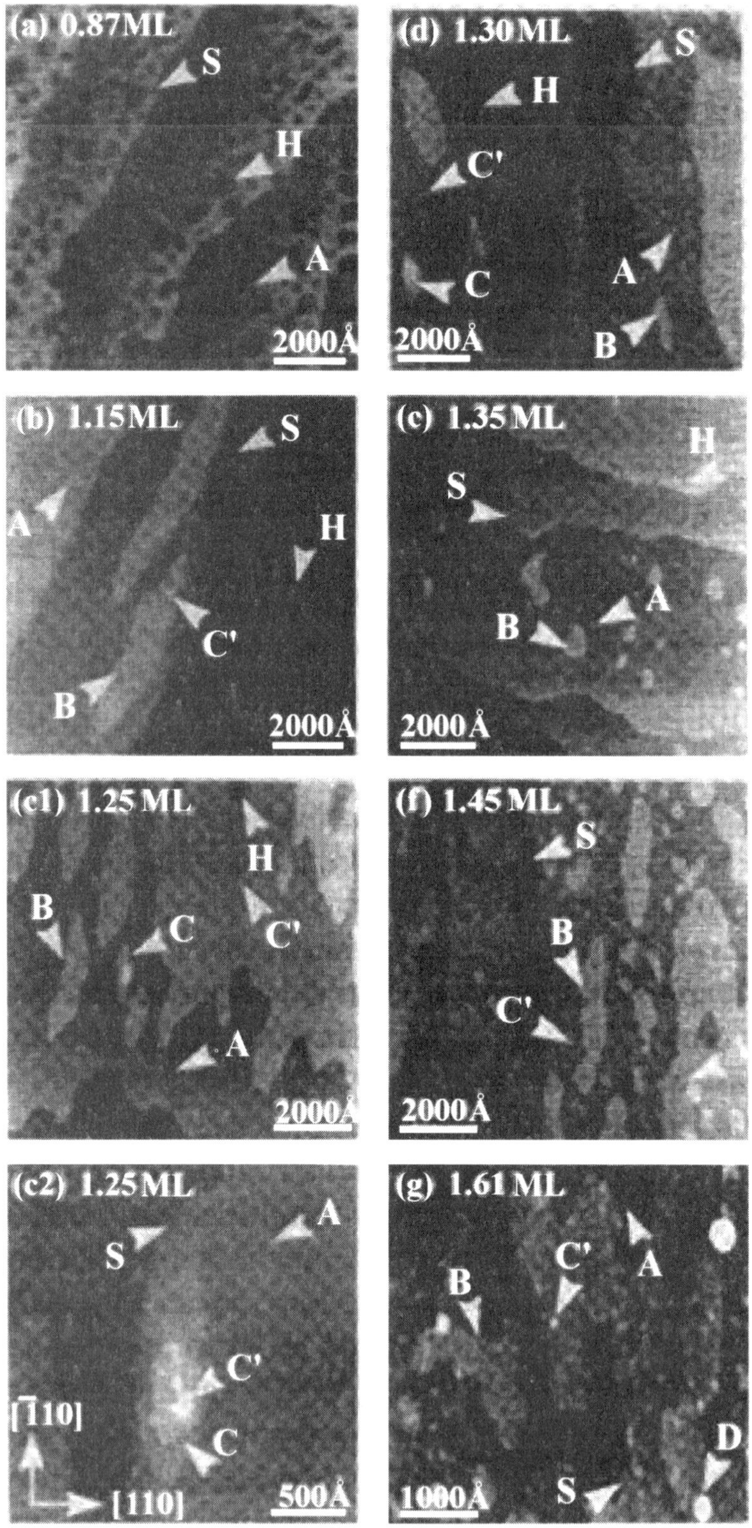

Fig. 6. Filled-state STM images showing the surface evolution for InAs deposition on GaAs(001)–c(4 × 4) at coverages of (a) 0.87, (b) 1.15, (c1, c2) 1.25, (d) 1.30, (e) 1.35, (f) 1.45, and (g) 1.61 ML. The labels represent small two-dimensional islands less than 20 nm in width (A), large two-dimensional islands greater than or equal to 50 nm (B), small multilayer (2–4 ML) clusters up to 20 nm wide (C′), larger multilayer clusters up to 50 nm wide (C), three-dimensional islands (D), atomic steps (S), and 1-ML-deep holes (H). Reprinted with permission from R. Heitz et al., *Phys. Rev. Lett.* 78, 4071 (© 1997 American Physical Society).

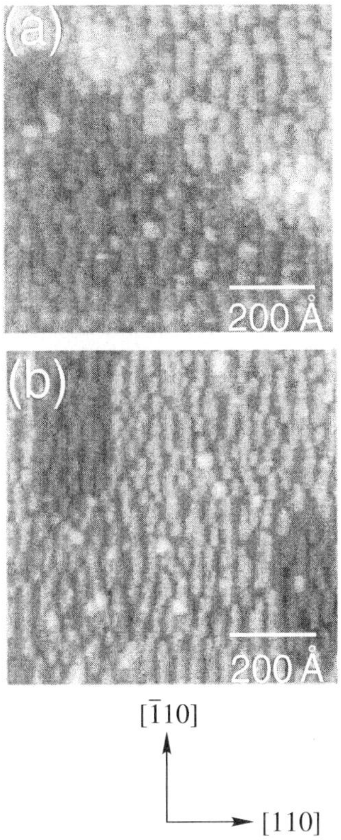

Fig. 7. Filled-state STM images of GaSb on GaAs(001)–c(4×4) at coverages of (a) 1.0 and (b) 3.5 ML. The image in (b) shows a region between three-dimensional islands that has already formed by that coverage. Reprinted from *J. Cryst. Growth*, 175/176, 888, B. R. Bennett et al. (© 1997), with kind permission of Elsevier Science–NL, Sara Borgerhartstraat 25, 1055 KV Amsterdam, The Netherlands.

three-dimensional InAs islands on GaAs. Again, no direct relationship between the large two-dimensional islands and the three-dimensional islands was observed. The major observation was that the highest observed density of the coherent three-dimensional islands was approximately an order of magnitude greater than the highest density of the two-dimensional islands. Some hints of a possible precursor to the three-dimensional islands were observed and denoted "quasi"-three-dimensional islands, or islands 2–4 ML in height (cf. Fig. 6, the features labeled C). These features are relatively high in density, nearly twice that of the coherent three-dimensional islands. These "quasi"-three-dimensional islands appear about 0.2 ML before the formation of the true three-dimensional islands, where upon they quickly disappear within about 0.2 ML prior to additional deposition. Thus, the changing densities of these different types of islands over very small changes in the total coverage indicate a mass transfer mechanism from the two-dimensional to the three-dimensional islands, which leads to a significant reorganization of mass on the surface. The authors conclude that to study and understand the formation and evolution of three-dimensional islands, the strain-dependent kinetics of the system must be considered [86]. Clearly, this is plausible as epitaxy by nature is a kinetically driven process. Furthermore, there may be island–island interactions that affect the surface morphology and that result from strain-related phenomena [86, 94].

4.3. Effects of Surface Reconstruction

Semiconductors often have a different symmetry at the surface compared to the bulk solid. This change of symmetry at the surface is a result of the displacement of atoms with respect to their bulk positions, thus creating a new unit cell on the surface. This rearrangement of surface atoms is referred to as a surface reconstruction, and significant effort has been devoted to understanding this phenomenon [95]. The motivation for the restructure itself is to minimize its free energy by forming bonds at the surface and maintaining charge neutrality. It should be noted that the reconstruction that is observed is not necessarily the lowest-energy surface, but rather the lowest-energy surface that is obtainable under the given set of kinetic conditions. Because the surface reconstruction represents a surface energy and symmetry (geometry), it is reasonable to believe that it will influence subsequent deposition of materials on the surface.

To illustrate the effects that surface reconstruction can have on the formation of islands, consider the study by Belk et al. [96] of InAs deposition at various coverages up to 1 ML on the GaAs(001)–c(4 × 4) surface at a substrate temperature of 420 °C. A number of STM images showing the progression of island formation from 0.1 to 1 ML are shown in Figure 8. At 0.1 ML of InAs, new domains of (1 × 3) reconstruction are observed at the step edges, and it appears that the (1 × 3) areas are on the same layer as the c(4 × 4) ones. As the coverage is increased to 0.3 ML (cf. Fig. 8b), two-dimensional islands be-

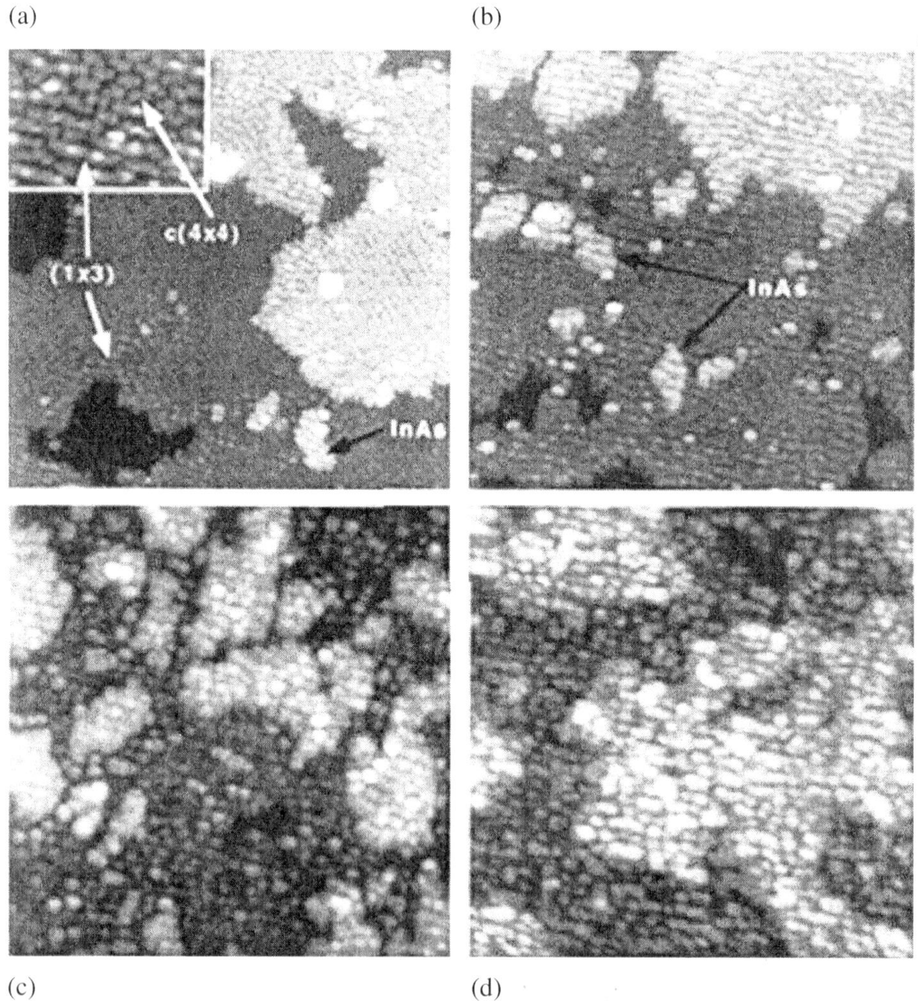

Fig. 8. InAs islands grown on GaAs(001)–c(4 × 4) shown at fractional coverages of (a) 0.1, (b) 0.3, (c) 0.6, and (d) 1.0 ML. The image sizes are (a) 50 nm × 50 nm and (b)–(d) 40 nm × 40 nm. Reprinted from *Surf. Sci.*, 365, 735, J. G. Belk et al. (© 1996), with kind permission of Elsevier Science-NL, Sara Burgerhartstraat 25, 1055 KV Amsterdam, The Netherlands.

gin to appear on the surface, which, at this point, is a mixture of (1×3) and $c(4 \times 4)$ domains. These islands are 1 ML in height and show the same (1×3) reconstruction. In contrast to the highly anisotropic, two-dimensional InAs islands formed on the (2×4) surface and shown in Figure 5, the islands formed on the $c(4 \times 4)$ surface have a different reconstruction and appear to be much more isotropic. At 0.6 ML (cf. Fig. 8c), the surface is entirely covered by the (1×3) structure. The RHEED pattern in the $[\bar{1}10]$ direction at 0.3 ML showed a mixture of 1/2- and 1/3-order spots, corresponding to the $c(4 \times 4)$ and the (1×3) reconstructions, respectively. By 0.6 ML, the 1/2-order spots were gone and the RHEED showed a clear (1×3). At 1.0 ML (cf. Fig. 8d), the surface is composed of large islands and terraces as some of the smaller islands have coalesced and the growth has proceeded in a reasonable, layer-by-layer fashion. Further study of the system revealed that the amount of InAs required to fill the surface completely with the (1×3) reconstruction varied with surface temperature. On this basis, the authors postulated that the (1×3) domains were actually composed of an (In, Ga)As alloy. The exact mechanism for this reconstruction/transformation remains unknown.

Bennett et al. [82, 83] have studied the deposition of InSb on different reconstructions of GaAs(001). In particular, they deposited 1.5 ML of InSb on both a (2×4) and a $c(4 \times 4)$ reconstructed surface. Because of the significant lattice mismatch between InSb and GaAs (14.6%), this deposition is beyond the critical thickness and three-dimensional islands are present. The differences resulting from deposition on the different initial surface reconstructions can be seen in Figure 9. Figure 9a shows 1.5 ML of deposition on GaAs(001)–$c(4 \times 4)$, whereas Figure 9b shows the equivalent deposition on GaAs(001)–(2×4). The islands were observed to be anisotropic in both cases, with the elongation in the $[\bar{1}10]$ direction. The island separations, as determined from an autocorrelation analysis, were found to be 50 Å for growth on the $c(4 \times 4)$ surface and 40 Å for growth on the (2×4) surface.

The islands formed in the two-dimensional layers of these systems appear to be somewhat dependent on the reconstruction of the initial growth surface. Islands that are formed on the (2×4) surface are highly anisotropic (cf. Figs. 5 and 9b), whereas islands that are formed on the $c(4 \times 4)$ surface appear to be slightly more isotropic (cf. Figs. 8a–d and 9a). This dependence on the initial surface reconstruction could be due to several different factors. To consider how the reconstruction of the initial surface could play a part in subsequent island formation, it would be sensible to consider first differences in the atomic structure of the reconstructions. Although there have been several different proposals for the (2×4) reconstruction [97, 98], the commonly accepted structures of the (2×4) surface are shown in Figure 10a [95, 99]. The surface is composed of two As dimers and two missing dimers on the top level, with the exposed As atoms in the third layer forming dimers. There is a missing row of gallium atoms in the second layer, leaving 0.75 ML of Ga atoms in that layer. A proposed model for the $c(4 \times 4)$ reconstructed surface is shown in Figure 10b [100]. There is no missing row of gallium atoms in this case, but rather a complete layer of cations with 1.75 ML of arsenic on top of it. The top layer, composed of 0.75 ML of arsenic, forms the characteristic "brickwork" pattern of the $c(4 \times 4)$ reconstruction. Clearly, there are significant differences in the concentrations of group-III and group-V elements present at the surface for the different reconstructions.

There are a number of possibilities concerning how these differences could affect subsequent island morphology. Because these structures are typically grown in a group-V overpressure, it would seem that the concentration of group-III material could be the key variable. As previously noted, the (2×4) reconstruction has an incomplete layer of cations at the surface. Thus, significant variations in the energy barriers for diffusion on the surface may well result from the different atomic structures. These variations could lead to changes in the rate of diffusion of cations on the surface and also along step edges. Furthermore, the variations in the surface potential could easily produce sites of varying reactivity. This can be seen clearly in a study by Köhler et al. [88] in which Ge was deposited on the Si(111)–(7×7). When the deposition was at room temperature, Ge clusters with various

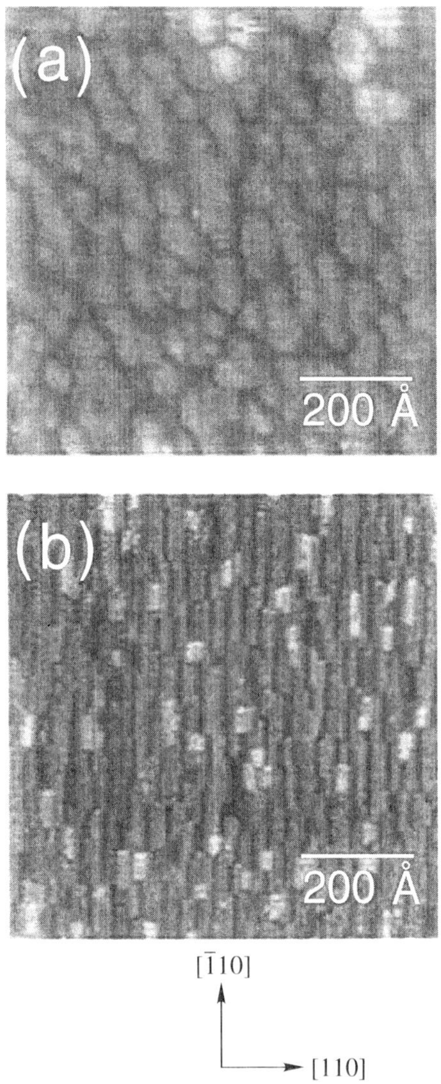

Fig. 9. Filled-state STM images of 1.5 ML of InSb deposition on (a) GaAs(001)–c(4 × 4) and (b) GaAs (001)–(2 × 4). The islands in (b) are slightly more elongated in the [$\bar{1}$10] direction than in (a). Monolayer-high steps are also observed in both images. Reprinted from *J. Cryst. Growth*, 175/176, 888, B. R. Bennett et al. (© 1997), with kind permission of Elsevier Science-NL, Sara Burgerhartstraat 25, 1055 KV Amsterdam, The Netherlands.

sizes were observed to be randomly distributed on the surface. However, the clusters rarely occupied a corner hole in the (7 × 7) reconstruction, indicating a different reactivity at that site. Along this same line of reasoning, there may be differences in the structures that the surface will try to form to minimize the surface energy; that is, the equilibrium structures of the surface may be different.

4.4. Effects of Surface Orientation

Lattice-mismatched heteroepitaxy on a number of different (001) surfaces has been shown in many systems to follow the Stranski–Krastanov growth mode and eventually form coherently strained three-dimensional islands. Because the growth mode is dependent on the surface energy and reconstruction, another variable that can be used to affect the strained-layer growth of heteroepitaxial systems is the surface orientation. Obviously, the electronic structure and symmetry of the surface can lead to different reactivities and island shapes.

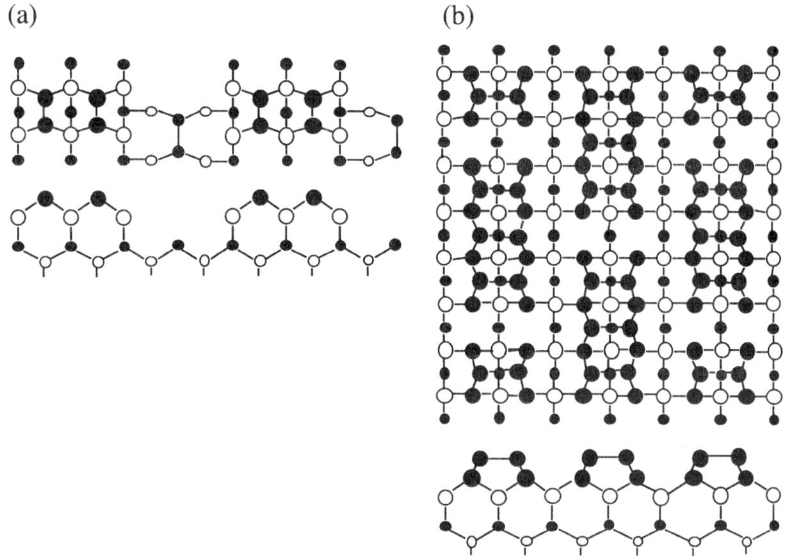

Fig. 10. Ball-and-stick models of (a) (2×4) and (b) $c(4 \times 4)$ reconstructions.

For example, the Si(111) surface has threefold symmetry in the surface plane. Both Si homoepitaxy [88, 92] and Ge deposition [101] on the Si(111) surface result in triangularly shaped islands, reflecting the symmetry of the surface. On the other hand, both the (001) and the (110) surfaces are rectangular, with in-plane nearest neighbors in the $[\bar{1}10]$ and [110] directions. In GaAs homoepitaxy on the (001) surface, anisotropic islands elongated in the $[\bar{1}10]$ direction are observed. These islands are random in shape apart from their anisotropy [76, 102]. However, GaAs homoepitaxy on a (110) surface results in triangularly shaped islands pointing in the [110] direction with elongated sides running in the $\langle 113 \rangle$ and $\langle 115 \rangle$ directions. This island shape indicates a preference for adatoms to attach at the base of existing islands (edges that run in the $[\bar{1}10]$ direction) and illustrates the difficulty in predicting the shapes that islands will form on a given surface from surface symmetry alone [103].

In zincblende structures, certain crystal planes can actually have two types of surfaces because of the different ways in which the surface can be terminated. This is simply a result of the bilayer nature of zincblende structures. Diamond structures, such as silicon, do not exhibit this type of dependence because of their higher symmetry. To illustrate these differences, consider the {111} surfaces of a diamond and a zincblende structure. Figure 11a shows the unit cell of a zincblende structure. For the sake of discussion, let the white dots represent the anions and let the black dots represent the cations. The diamond structure, on the other hand, would have the same atomic positions, but, obviously, there would be no distinction between the types of atoms. The models for the (111)A and (111)B surfaces shown in parts b and c of Figure 11, respectively. For the unreconstructed compound surface, it can be seen that a surface will be either cation rich for the (111)A surface or anion rich for the (111)B surface. Clearly, different reactivities can be expected between the A- and B-type surfaces. For the diamond structure, the (111) surface would appear identical in either case because of its symmetry.

In some heteroepitaxial systems, growth on different surface orientations can dramatically change the growth mode. For example, in the case of InAs deposition on GaAs(001), which was discussed earlier, the growth was shown to proceed in a layer-by-layer manner, to the formation of coherent three-dimensional island and then, finally, to dislocated three-dimensional islands. When InAs is deposited on GaAs(110) and GaAs(111)A, however,

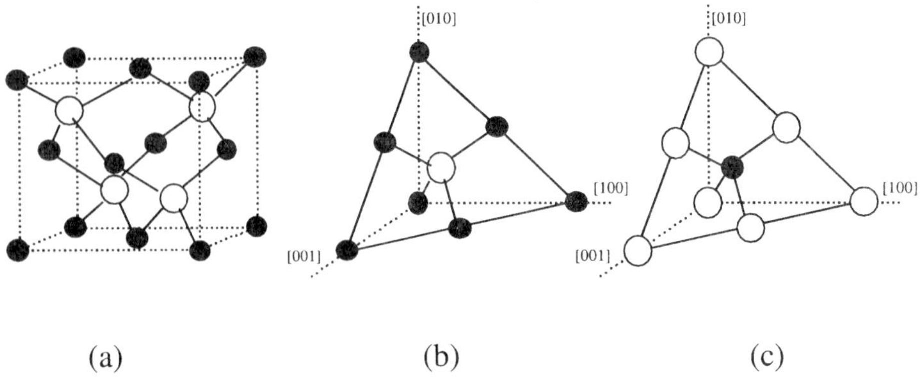

Fig. 11. Ball-and-stick models of the (a) zincblende unit cell, (b) (111)A surface, and (c) (111)B surface. The white dots represent the cations and the black dots represent the anions. Two different (111) planes are possible because of the zincblende structure, with the difference being either cation [cf. (b)] or anion [cf. (c)] termination of the plane.

no coherent three-dimensional islands are observed [104–106]. On both surfaces, two-dimensional islands are observed in a layer-by-layer growth mode, but the formation of three-dimensional islands is replaced by the introduction of misfit dislocations as a strain-relieving mechanism. Interestingly, on the GaAs(111)A surface, Gonzalez-Borrero et al. [107] report quantum dot formation for InGaAs deposition. Altogether, these authors grew $In_{0.5}Ga_{0.5}As$ quantum dots on $(n11)$A/B surfaces, where n was 1, 2, 3, 5, and 7, and compared the samples by PL measurements. In their work, they demonstrated that the $(n11)$B surfaces had a higher integrated PL intensity compared to the $(n11)$A surfaces, with the (311)B surface showing the most uniformity in island sizes as judged from PL. Nishi et al. [108] studied the $In_{0.5}Ga_{0.5}As$ quantum dot formation on the (311)B surface and found the dots to have an approximate diameter of 25 ± 2 nm and a height of 13.7 ± 2.2 nm, as determined from AFM. Luminescence peaks resulting from dots grown on the (311)B surface were shown to be much narrower than those measured from dots on a (100) surface.

Germanium deposition on different silicon surface orientations has been an area of recent interest because of its potential uses in optoelectronic devices. High-index silicon surfaces have been of particular interest because of their unusual density of states, which results in promising optical [109, 110] and transport properties [111, 112]. One drawback of high-index surfaces is that they tend to be high-energy surfaces and, consequently, form low-index facets upon annealing [113, 114]. One high-index surface that has been of particular interest is the (113) surface. One study by Gibson et al. [115] showed that annealing Si(110) produced various low-index planes, as well as (113) facets. Also, the Ge "hut clusters" that have been observed are faceted structures composed primarily of {113} planes [88, 89]. Thus, it appears that the Si(113) surface is a relatively low-energy surface and may be suitable as an MBE growth surface. One example of Ge growth on Si(113) was reported by Knall and Pethica [87]. In this work, a number of STM images were presented at various coverages ranging from clean Si(113) up through 5 ML, by which point three-dimensional islands had begun to evolve on the surface. Furthermore, they performed similar growths on Si(100) to contrast the differences resulting from the initial growth surface. The Ge growth on the (113) surface was predominantly two dimensional up to coverages of approximately 3 ML, whereas on the (100) surface second-layer growth of Ge islands began at submonolayer coverages. By 2.5 ML of deposition, at least four distinct levels could be observed on a given terrace. The differences in the growths were attributed to the preferred nucleation of islands at antiphase domain boundaries in the (100) surface reconstruction. Other studies have also shown that the nucleation of islands at antiphase domain boundaries can be a dominant growth mechanism [116, 117].

5. THREE-DIMENSIONAL ISLANDS

In the previous sections, the Stranski–Krastanov mode was introduced and the transition from layer-by-layer growth to three-dimensional island growth was discussed. In recent years, these islands have been utilized as nanostructures. These self-assembled islands will be the focus of this section.

5.1. Early Work

Although the Stranski–Krastanov growth mode has been known for many decades, it was not considered until relatively recently as a method to fabricate useful nanostructures. As previously mentioned, there is an emphasis in epitaxy on the creation of smooth surfaces and interfaces. Moreover, much of the early work in studying growth modes was done with metal-on-metal epitaxy (e.g., Ag/Mo) or metal-on-semiconductor epitaxy (e.g., Ag/Si) [19, 24]. These structures were not considered to be useful nanostructures because defects readily formed in these islands. With more work in strained-semiconductor epitaxy, evidence evolved that these islands were not defected.

One early report that showed evidence of island formation was presented by Goldstein et al. in 1985 [118]. They were studying InAs/GaAs superlattices and some samples exhibited three-dimensional nucleation. During growth, the RHEED pattern changed from a diffuse pattern to a spotty pattern, indicating three-dimensional growth for an InAs thickness greater than 2 ML. In TEM images, they observed localized strain fields that, in retrospect, were caused by the InAs islands that were present. Because the islands were surrounded by other material, direct observation of the islands was not possible; yet there was no evidence of dislocations.

In 1990, two important reports were published that peaked interest in the field of self-assembled semiconductor islands. In the first report, Eaglesham and Cerullo [59] examined Ge deposition onto Si(100). Their goal was to disprove the assumption that islands formed during Stranski–Krastanov growth are always dislocated. Using TEM, which is sensitive to dislocations and structural defects, they explored the islands. In plan-view images, they observed a strain contrast feature (cf. Fig. 12), consistent with a coherent particle within a lattice-mismatched matrix [58]. From their work, they developed the concept of the *coherent* Stranski–Krastanov growth mode. Similar results were reported later in this material system by Krishnamurthy and co-workers [119].

The second paper, by Guha and co-workers [60], presented coherent $Ga_{0.5}In_{0.5}As$ islands formed on GaAs, examined by cross-sectional TEM. Many of the islands, as shown

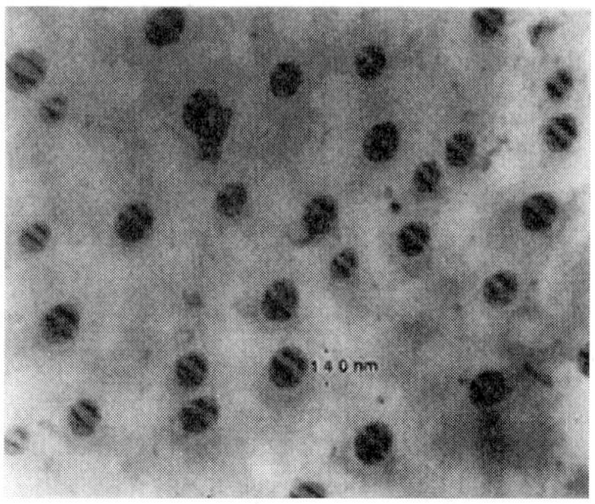

Fig. 12. Plan-view TEM image showing Ge islands surrounded by Si. The strain contrast features indicate that the islands are defect free. Reprinted with permission from D. J. Eaglesham and M. Cerullo, *Phys. Rev. Lett.* 64, 1943 (© 1990 American Physical Society).

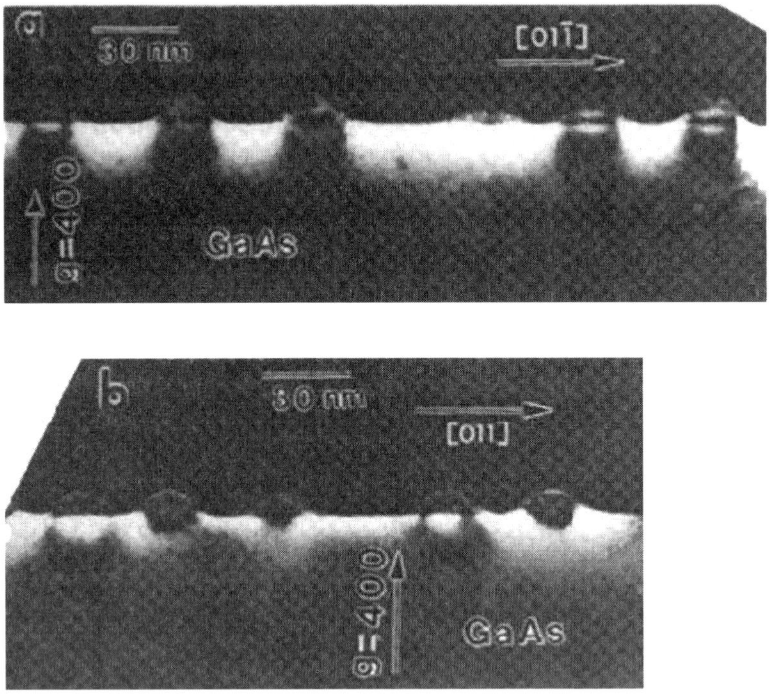

Fig. 13. Cross-sectional TEM image showing InAs on a GaAs surface. No defects are observed in these islands. Reprinted with permission from S. Guha et al., *Appl. Phys. Lett.* 57, 2110 (© 1990 American Institute of Physics).

in Figure 13, were found to be defect free. Those that were defected exhibited dislocations and stacking faults that were injected from the edge of the islands. These results not only support those seen for Ge on Si, but extend the phenomenology to results for compound semiconductor systems.

Semiconductor nanostructures are often considered candidates for quantum structures such as quantum dots. Such structures are discussed in detail elsewhere in this book as well as in the literature [16, 70]. Leonard et al. [120] first demonstrated the application of coherent Stranski–Krastanov islands as quantum dots in 1993. In this work, $Ga_{0.5}In_{0.5}As$ was deposited on GaAs. As in the case of the previous work, the GaInAs was deposited by MBE until the RHEED pattern indicated a three-dimensional growth. The quantized properties of the islands were determined from photoluminescence. This work was the precursor to additional studies of these self-assembled islands as quantum dots, which is ongoing today, because it is extremely promising for many applications. Additional comments concerning applications of such islands are discussed in the following section.

Many additional reports have more recently disclosed self-assembled semiconductor islands during multilayer strained heteroepitaxy. The remainder of this section will address some topics in this field, including how the islands provide strain relief, different types of three-dimensional islands, the impact of deposition conditions, and the formation of islands on different surface orientations. This section will finish with a few unique approaches to arranging the islands.

5.2. Strain Relief from the Islands

Islands form as a way for the system to relieve strain. The complete details of how this occurs are not known, but a few observations and theories exist. In the Stranski–Krastanov growth mode, the material will begin to grow in a layer-by-layer fashion. At some point, the strain energy can be initially relieved by surface roughening, as shown in Figure 14 [121].

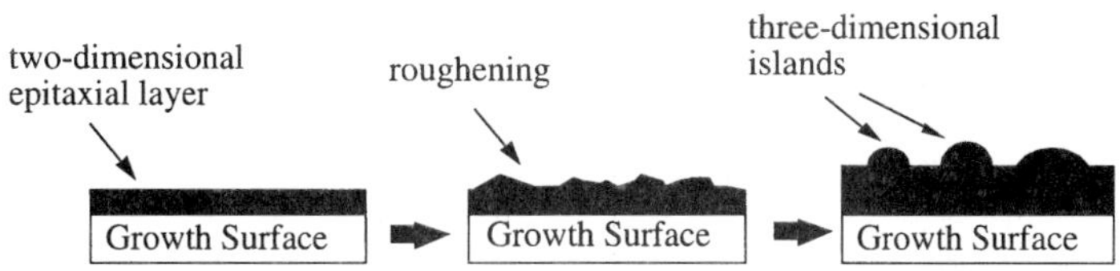

Fig. 14. Stranski–Krastanov growth mode. The growth is initially two dimensional. As the strain energy increases, the surface starts to roughen. With additional deposition, three-dimensional islands form to provide additional surface area to relieve strain. In many cases, the roughening cannot be clearly observed before island formation.

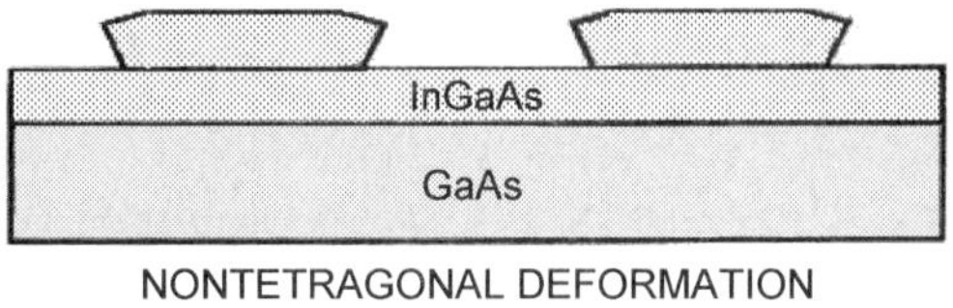

Fig. 15. Schematic diagram showing nontetragonal deformation of three-dimensional islands. The layer relaxes toward its natural lattice constant with distance from the heterointerface. Reprinted from *J. Cryst. Growth* 134, 51, N. Grandjean and J. Massies (© 1993), with kind permission of Elsevier Science–NL, Sara Burgerhartstraat 25, 1055 KV Amsterdam, The Netherlands.

A rough surface has a larger surface area than a smooth surface; free surfaces can provide strain relief [122]. This roughening has been observed to occur in some material systems. In other material systems, the growth goes directly to the formation of three-dimensional islands at a low coverage without a clear roughening stage.

Regardless of whether a layer undergoes roughening, the islands form to relieve strain. This relief results most likely from a combination of two effects. First, three-dimensional islands have a larger surface-to-volume ratio than a flat layer. Although this increased area has a related increase in the total surface energy, there is more freedom for the lattice within the island to relax. This relaxation is often referred to as nontetragonal deformation [80, 121]. In a strained two-dimensional layer, there is tetragonal deformation, as shown in Figure 4. The lateral or in-plane lattice constant is constant in each layer. In an island, however, the in-plane lattice constant may be different on the top of a unit cell than on the bottom; that is, the cell is trying to return to its natural lattice size, as shown in Figure 15. In a study of GaInAs islands on GaAs, Guha et al. [60] measured the lattice spacing within the islands and found that it increases with the distance from the base or interface.

Second, the material surrounding the islands can accommodate some of the strain [59]. The lattice constant of the material surrounding the island deviates from its natural value; the strain energy is distributed over a larger area. A schematic representation of this is shown in Figure 16, where the substrate lattice is locally deformed. This has been observed in several transmission electron microscopy studies [59, 123].

5.3. Different Types of Islands

Several stages of islanding have been observed in heteroepitaxial growth. These stages are often overlapping with different types or sizes of islands coexisting. The three types of islands are the following: (1) small islands, also known as precursor [90] or quasi-two-dimensional [86] islands; (2) medium-sized three-dimensional islands, which are still lattice-matched to the substrate (i.e., which are still strained); and (3) large, defected islands, which are no longer lattice matched to the substrate (i.e., which have relieved their strain by defect formation).

One example of these different types of islands was observed in the work by Reaves et al. [123] for InP islands formed on GaInP/GaAs(100) surfaces (cf. Fig. 17). The small

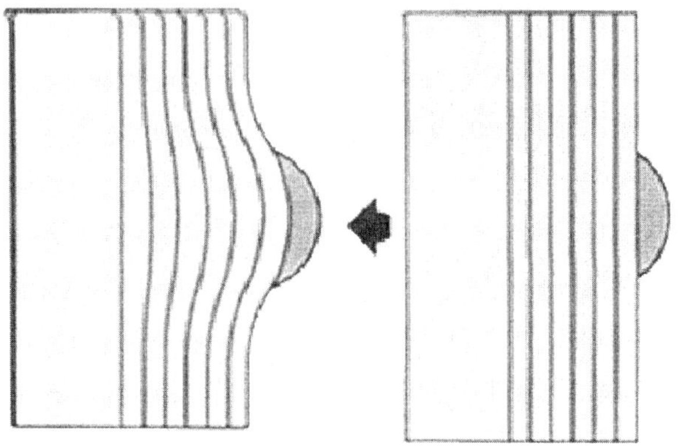

Fig. 16. Schematic diagram showing deformation of the substrate resulting from concentration of the strain by a three-dimensional coherent island. Reprinted with permission from D. J. Eaglesham and M. Cerullo, *Phys. Rev. Lett.* 64, 1943 (© 1990 American Physical Society).

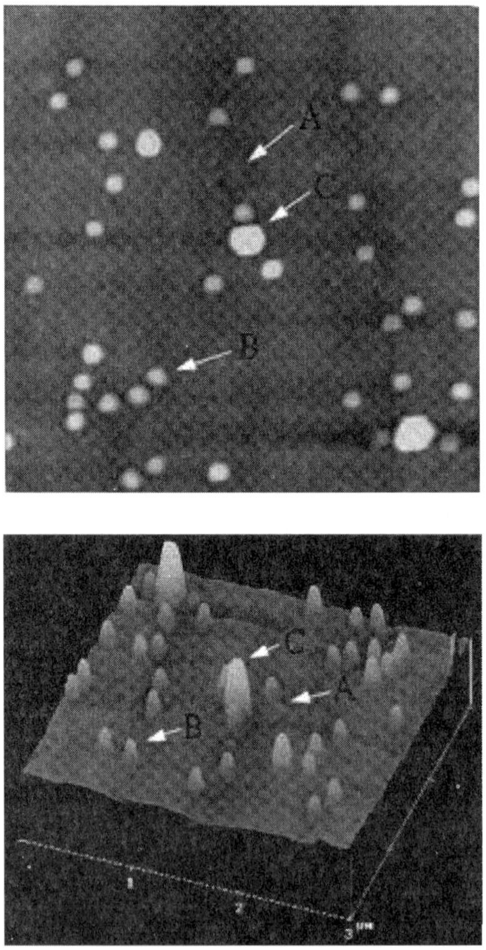

(a)

(b)

Fig. 17. Example of the three different types of islands. In this atomic force micrograph of InP growth on GaInP/GaAs(100), small islands are labeled A, medium-sized islands are labeled B, and large defected islands are labeled C. Reprinted from *Surf. Sci.*, 326, 209, C. M. Reaves et al. (© 1995), with kind permission of Elsevier Science–NL, Sara Burgerhartstraat 25, 1055 KV Amsterdam, The Netherlands.

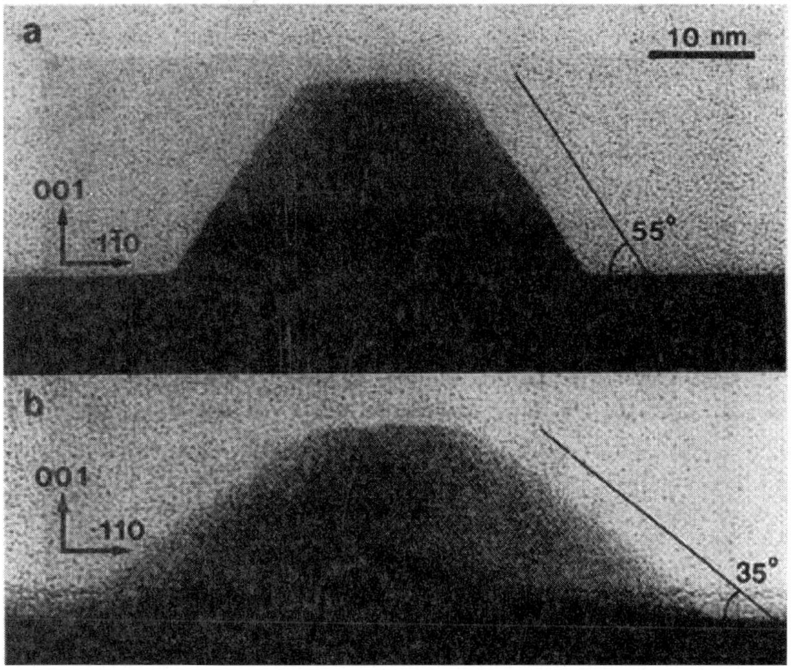

Fig. 18. Deduced shape of coherent InP islands. Reprinted with permission from K. Georgsson et al., *Appl. Phys. Lett.* 67, 2981 (© 1995 American Institute of Physics).

islands, an example of which is labeled A, are about 20 Å high and have a base width of 1200 Å. The medium-sized islands, an example of which is labeled B, are about 240 Å high and have a base width of 1200 Å. The defected islands, an example of which is labeled C, can vary in size, depending on the extent of deposition. The defected islands continue to grow with deposition, with each new dislocation in an island allowing a significant jump in the island size [124]. These islands become more obvious after they have grown to be much larger than the other islands that are present. Note that, for a given type, the islands appear to be similar in size. Size distribution studies for this and other material systems illustrate this [120, 125].

The exact shape of self-assembled islands is difficult to measure. One study discussing the shape of these islands was presented by Georgsson et al. [126] for InP islands. Many reports of islands observed by AFM or low-resolution electron microscopy conclude that the islands are cap-shaped and featureless. As shown in Figure 18, however, the islands are polyhedral. The tops are flat and the side walls consist of {110} and {111} planes. The shape varies according to the different crystallographic orientations that are present.

5.4. Impact of Deposition Conditions

In fabricating nanostructures for a particular application, the size is often important. What determines the size of self-assembled islands? Because they form as a way for a heteroepitaxial layer to relieve strain, one thought would be that only the lattice mismatch and the elastic properties of the layers should matter. This is not the case, however, because the island size can vary for the same material system deposited under different conditions. This implies that the formation of the islands is kinetically controlled under technologically relevant growth conditions. A few observations on these variations in island size and density will be discussed next.

One interesting observation is that the size of the islands for a given material system and set of deposition conditions will lock into a defined size. Only defected islands increase in

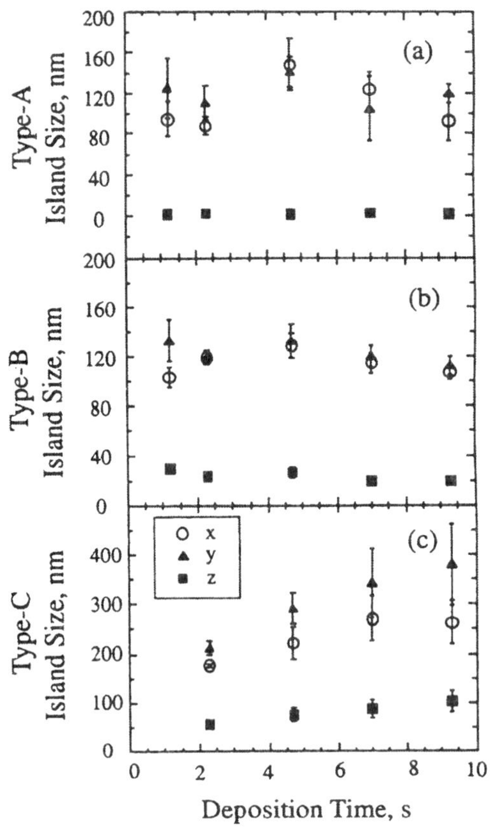

Fig. 19. Island size vs deposition time for different types of InP islands. Note that the size of the small (type A) and medium-sized (type B) islands does not change with additional deposition. Reprinted from *Surf. Sci.*, 326, 209, C. M. Reaves et al. (© 1995), with kind permission of Elsevier Science–NL, Sara Burgerhartstraat 25, 1055 KV Amsterdam, The Netherlands.

size as additional material is deposited. This was quantified for the InP/GaInP/GaAs(100) system [123]. The height and base width of the small and medium-sized coherent islands are constant with additional deposition, as shown in Figure 19. Such observations indicate that coherent islands have a preferred size and remain at that size until they develop dislocations. As indicated by the error bars in Figure 19, the size distributions for the coherent islands, particularly for the medium-sized islands, are narrow.

If the islands are not getting bigger, then where is the newly deposited material going? Although some of this material goes into the growth of defected islands, the primary result is that the density of the islands increases. Reports in the literature show that the onset of three-dimensional islands, that is, the medium-sized islands, is abrupt. For the case of InAs/GaAs(100), the critical coverage for this transition is around 1.5 ML [86, 127]. Within 0.1 ML of additional coverage, the density of the medium-sized islands is already 10^9 cm^{-2}. This density continues to increase throughout the deposition [123, 128, 129].

Hence, the formation of these islands is governed by kinetics. With this in mind, the substrate temperature is expected to be a major influence on the size and density of the islands. Indeed, this has been observed to be the case. At low temperatures, island formation can be suppressed; diffusion is insufficiently rapid for islands to nucleate and grow [129]. The island size, in particular, the height, increases with increasing substrate temperature [125, 128, 130]. One explanation for this observation is that adatoms have a larger diffusion range at higher temperatures. They are more likely to find an existing island and add to that island rather than nucleate a new island. As a result, a lower density of larger islands is expected at higher temperatures. This is observed to be the case in reports where both island size and density are tracked as a function of substrate temperature [125, 128, 130].

Another deposition condition that can influence island size is the deposition rate or, equivalently, the arrival flux of the growth species. If the deposition rate is relatively high (relative to the rate of surface diffusion), the number of unattached adatoms on the surface is high and, therefore, the probability of nucleation of a new island is high. This would lead to a higher density of smaller islands. Surprisingly, there have been few studies of the impact of deposition rate with few consistent trends having been observed to date [128, 131].

5.5. Impact of Surface Orientation

As discussed in the previous section, the surface orientation will impact many of the processes occurring during epitaxy. A large fraction of semiconductor epitaxy is carried out on the (100) surface. The dominance of this surface extends into self-assembled islands formed during strained-layer epitaxy. Studies of growth on other surface orientations have, in general, shown some interesting results such as unique reconstructions [132] and different step-flow growth modes [133, 134]. Interesting results have also been observed in the case of self-assembled islands.

One of the more interesting results for island growth on higher-index surfaces has been demonstrated by Nötzel, Temmyo, and their co-workers [135–139]. For example, on a GaAs(n11)B substrate, with $n = 1$–5, GaInAs has been deposited on AlGaAs layers. What happens is that the GaInAs moves below the AlGaAs surface and forms coherent disks. Such structures, as shown in Figure 20, are uniform in size and often well aligned. The self-assembled nanostructures formed on (311)B surfaces have exhibited the best alignment. Similar disks have also been formed on InP(311)B surfaces. Several reasons have been suggested for these phenomena, based on the high surface energies of these surfaces [140]. These structures have been used for several optical studies as well as the active region of injection lasers [141].

Self-assembled islands, produced by the coherent Stranski–Krastanov growth mode, also form on high-index planes. The (311) surface, for example, has been used to form coherent islands in several systems [108, 142]. One of the more obvious differences with

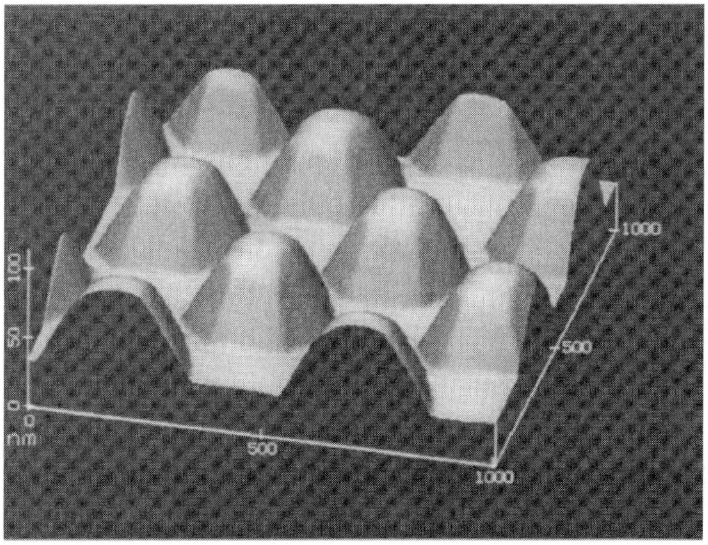

Fig. 20. Self-organized quantum disks formed by interlayer mixing of $Al_{0.5}Ga_{0.5}As$ on top of $In_{0.2}Ga_{0.8}As$ on a GaAs (311)B surface. Reprinted with permission from R. Nötzel, *Semicond. Sci. Technol.* 11, 1365 (© 1996 IOP Publishing).

respect to the (100) surface is that the islands are smaller and denser on the (311) surface. For example, with InP islands on GaInP/GaAs, the density of islands on the (311)A surface is 10^{10} cm^{-2}, whereas, on the (100) surface, it is 10^9 cm^{-2} [108, 142]. The height of these islands decreases from 240 Å on (100) to 60 Å on (311)A. Island growth has also been observed for InAs on GaAs(111)A. These islands appear, however, to be incoherent from the initial stages of growth [106].

As self-assembled islands are used more frequently for physical studies and device applications, the ability to adjust their size and density will become increasingly important. One of the disadvantages of self-assembled approaches with respect to lithographic avenues is the lack of direct control on sample morphology. The situation with self-assembled islands is not hopeless, however. There is clear tunability of island morphology, not just with deposition parameters, but also with surface orientation.

5.6. Controlling the Location of Self-Assembled Islands

There have been several interesting phenomena observed with respect to where self-assembled coherent islands are located on the growth surface. These phenomena can readily be traced back to basic epitaxial processes. One case is the vertical alignment of islands grown in sequential layers. If an array of islands is formed and then overgrown with buffer materials, the next layer of islands that is formed will be positioned directly above the first array of islands. This will continue to occur with additional layers of islands, as shown in Figure 21. This phenomenon has been observed by several groups [143–148]. Because

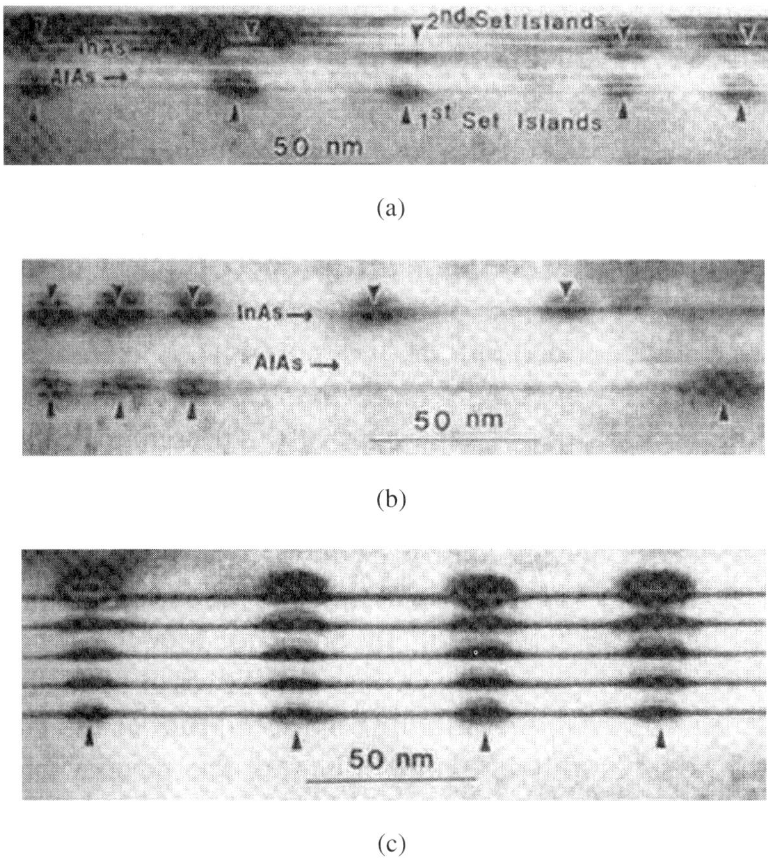

Fig. 21. Vertical alignment of InAs islands formed during sequential layers of island growth. Reprinted with permission from Q. Xie et al., *Phys. Rev. Lett.* 75, 2542 (© 1995 American Physical Society).

the strain fields from the islands extend into the surrounding materials, the thickness and modulus of the spacer layer are important. For the case of InAs islands separated by GaAs layers, spacers thinner than 150 Å lead to good vertical alignment [144]; for Ge islands separated by Si, spacers thinner than 1000 Å are needed [147]. This alignment arises from the impact of strain on surface diffusion [144]. The strain fields from an island can extend into the capping layer. One simple picture is that for a compressively strained island, the strain field in the spacer layer will make the local lattice constant larger. Adatoms for the next layer of islands will preferentially locate in these regions above previous islands. One result of this strain propagation is that the islands may grow larger because of less local lattice constant mismatch. This can be seen clearly in Figure 21.

Surface morphology will also impact where islands form. Vicinal surfaces with monolayer steps are common as epitaxial growth substrates. Two reports for InAs islands on GaAs(100) have presented evidence that these steps are impacting island formation. Ikoma and Ohkouchi [149] found islands aligning along $[1\bar{1}0]$ steps on a surface that had a 1° vicinal miscut towards [110]. A similar preferential nucleation of islands at step edges was observed by Leonard et al. [127] for a nominally flat surface. Moison et al. [150], however, have observed that the majority of islands nucleate on the terraces away from the step edges. It is known that surface steps influence surface diffusion and attachment [151, 152]. These influences may provide local variations in adatom concentrations that increase the probability of island formation. Such variations in the basic epitaxial growth process would also arise from surface features larger than monolayer steps.

Under certain growth conditions, various surface features will form. These are commonly identified as bunches of monolayer-high steps [133, 134, 153]. When these step bunches are present, they will act as nucleation sites for self-assembled islands. This has been observed for InAs/GaAs(100) [127, 154], for InGaAs/GaAs(100) [155], and for InP/GaInP/GaAs(100) [156]. In these reports, strings of several islands were found to be aligned along these surface features.

Large surface features can also be formed by patterning a substrate. Several groups have demonstrated various results. Mui et al. [157] performed InAs deposition on a GaAs(100) surface that had etched ridges. For $[01\bar{1}]$ ridges, the islands only formed on the (100) planes at the top of and in the valleys between the ridges. The islands were observed to form along the side wall of [011] ridges with no islands formed on the (100) top and valley planes of the ridges. Saitoh et al. [158] etched tetrahedral pits into a GaAs(111)B surface and then formed InAs islands. Depending on the deposition temperature, the islands were observed to form either on the side wall of the pit or only in the bottom vertex.

6. PHYSICAL PROPERTIES AND APPLICATIONS OF SELF-ASSEMBLED ISLANDS

In this chapter, there has been a focus on the formation and structural properties of self-assembled semiconductor islands formed during heteroepitaxy. One motivation for work on this topic is the unique physical properties of the islands. These properties are discussed in detail elsewhere in this book and in other reviews [159, 160]. In this section, a few properties and uses of self-assembled islands will be discussed.

6.1. Physical Properties: Some Examples

As previously mentioned, semiconductor nanostructures are of interest for their often unique physical properties. These structures can be engineered such that they can confine electrons into small regions, leading to potentially useful effects. A few studies and applications of these effects will be discussed here.

Luminescence is the optical emission resulting from an electronic relaxation [161, 162]. The electrons can be initially excited with higher-energy photons as in photoluminescence or with higher-energy electrons as in cathodoluminescence. Such techniques are commonly used to study the properties of solids and engineered structures. They also provide some insight into how materials will behave in some applications such as lasers and light-emitting diodes. There has been some effort to study the luminescence from a single quantum dot. As previously discussed for self-assembled islands, the sizes have a narrow, yet finite, distribution. The quantization of energy levels in these nanostructures is a function of size. Even small variations in size will lead to a range of electron energy levels. The resulting luminescence from an ensemble of islands is, therefore, broadened. This convolution can hide the true properties of the nanostructures.

There were two early reports in the literature regarding luminescence from small ensembles of self-assembled islands. Marzin et al. [163] have studied self-assembled InAs quantum dots with photoluminescence. Using electron beam lithography, the authors were able to pattern a sample containing the nanostructures into separated mesas. Because each mesa had a limited number of islands, the luminescence spectrum, shown in Figure 22, exhibits isolated peaks. Grundmann et al. [164] have made similar studies with cathodoluminescence in which the size of the excitation electron beam could be decreased to a 50 nm diameter. Their results from a self-assembled InAs island sample, shown in Figure 23, also exhibit isolated peaks. Both studies conclude that the narrow peaks (<0.15 meV) arise from single quantum dots. The spacing between peaks has been linked to an equivalent island size difference of 16 InAs molecules [164].

Some interesting studies have also been done that consider the charge storage of self-assembled quantum dots. A traditional capacitor relies on the storage and discharge of electrons. By using capacitance spectroscopy to charge and discharge the dots in a controlled way, the energetics of electrons in InAs quantum dots formed on GaAs(100) substrates were studied [165]. The difference between the first energy level and the second energy level in these 200 Å islands was measured to be 41 meV. It was also observed that the Coulombic repulsion between the first and second electrons in an energy level (with opposite spins) was small. By using both n-type and p-type structures, electron and hole

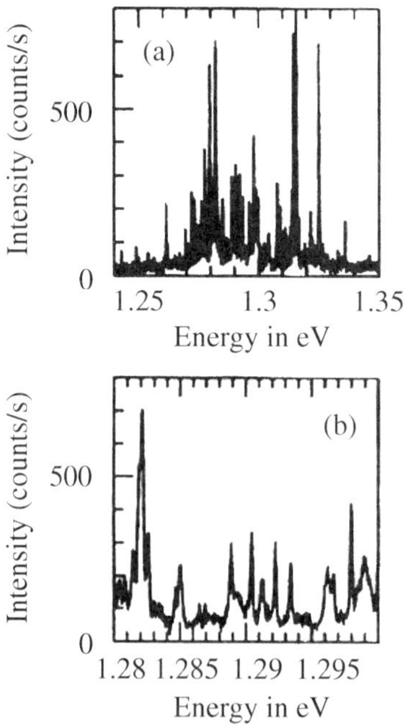

Fig. 22. Photoluminescence spectrum from self-assembled InAs quantum dots. The narrow lines are believed to arise from single dots. Reprinted with permission from J.-Y. Marzin et al., *Phys. Rev. Lett.* 73, 716 (© 1994 American Physical Society).

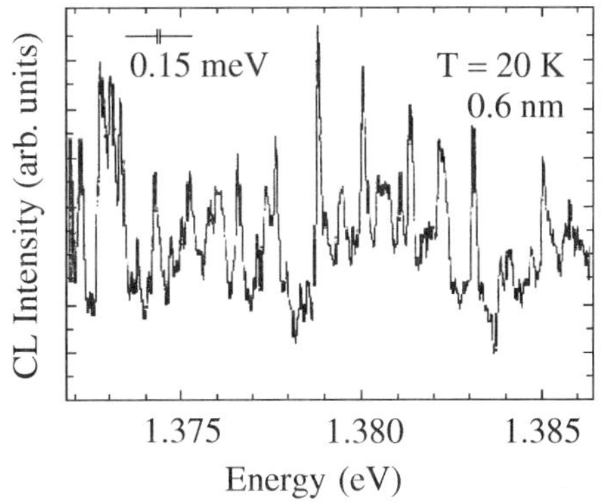

Fig. 23. Cathodoluminescence spectrum from self-assembled InAs quantum dots. The narrow lines are believed to arise from single dots. Reprinted with permission from M. Grundmann et al., *Phys. Rev. Lett.* 74, 4043 (© 1995 American Physical Society).

energy levels can be separately explored [166]. For example, InAs self-assembled islands may have two electron states, but only one hole state.

6.2. Self-Assembled Islands in Devices

One of the more exciting applications for quantum dots is the fabrication of an injection laser. Although lasers have been made from quantum wells and quantum wires, making one from a quantum dot has been difficult [167]. One problem with conventional lasers is that, as they heat up, because of either operation or ambient temperature, the energy distribution for carriers spreads. This spreading populates unwanted energy levels. The recombination from these levels does not add to the primary recombination, lowering efficiency. One way in which this is characterized is that the laser is tested at different temperatures, and the threshold current density for the start of stimulated emission is determined as a function of temperature. This behavior can be fitted by an exponential function of the form $\exp(T/T_0)$, where T_0 is a characteristic temperature [5, 168]. The higher this value, the less temperature sensitive is the laser. Because of the delta-function density of states in a quantum dot, thermal spreading of the carrier energy distribution is not allowed. Only well-defined energy transitions are allowed. The result is that the ideal quantum dot laser is insensitive to temperature variations, effectively leading to a T_0 of infinity. Along with the improved temperature performance, quantum dot lasers will have better gain properties; yet, one critical challenge is to obtain a very high density of dots to provide the necessary material [169]. Multiple layers of islands may be one approach to achieve this.

Some of the work using self-assembled islands to create lasers has exhibited large values of T_0 in an InAs-based device [170, 171]. Other work with lasers and optical structures based on islands has focused on devices that emit at 1.3 μm, a wavelength important for fiberoptic telecommunications. In these systems, either InAs or InGaAs alloys are used to form the islands [172]. Islands may also be used for integrating compound semiconductor devices with silicon. One report by Gérard et al. [173] used InAs islands formed on silicon substrates to obtain strong photoluminescence. When III–V semiconductor layers are grown on silicon, there is a large number of defects, and the resulting structures are often of low quality. This work found that, for InGaAs quantum wells grown on silicon, the photoluminescence was weak, whereas self-assembled quantum dot structures exhibited luminescence as strong as control samples deposited on GaAs. It is believed that self-assembled islands localize electrons and prevent them from finding defects where they may undergo nonradiative recombination.

Another potential application of quantum dots is in data storage. Most memory devices involve the storage of a charge: the larger that charge is, the more time that is needed to read and write to the memory. It has been observed that some self-assembled quantum dots only store one electron. If such devices can be used in the heart of a memory unit, then faster and higher-capacity memories can be created. Phenomena such as spin–spin interactions can be used for certain logic features currently not exploited [174]. For data storage, will each dot be solely responsible for a single bit of data? This would be the more economical way, but this also requires high fault tolerance for each quantum dot. This may be impractical with each dot storing such a small charge. Another approach is to have each memory cell based on hundreds or thousands of quantum dots [175]. If some dots fail, there is replication. This use of multiple dots will be limited when device density increases to the extreme point where feature sizes are 1–10 nm and each dot must be an individual, reliable device [176].

6.3. Use of Islands to Make Other Nanostructures

Although self-assembled islands are themselves high-quality nanostructures that can be used directly as quantum dots, they can also be used for other purposes. Some of these applications include making other nanostructures. One such application takes advantage of the fact that strain alone can modify the electronic band structure of a material [177]. This effect is realized by using the resulting strain field from a self-assembled island to modulate the band structure of the underlying material. In some cases, this is a quantum well located below the islands. The resulting potential in the quantum well resembles a parabolic quantum dot and exhibits good luminescence properties [178].

It has also been observed that, when material A is deposited on an islanded surface consisting of material B, and if these two materials are naturally lattice matched, then material A will initially grow exclusively on top of the islands of material B. The specific example for which this was demonstrated is that islands of InP were formed on a GaAs substrate. The alloy $Ga_{0.47}In_{0.53}As$ (which has the same lattice constant as bulk InP) was deposited on this surface and the new material was observed to grow only on top of the InP islands. This technique has been used to demonstrate quantum structures [179].

These two methods of using self-assembled islands arise from the different paths by which islands provide strain relief. The strain modulation is an example of substrate deformation, and the selective growth on the islands is likely a result of nontetragonal relaxation of the islands. As more is learned about self-assembled islands and strain relief, these and other applications will be further developed.

As mentioned earlier, one of the problems with etching to achieve quantum structures is the limitation of lithography. One alternate approach is to use self-assembled islands as an etch mask. This was done to create small GaAs structures by etching through a surface covered with InAs islands using chlorine gas [180]. These structures can then be overgrown and used as quantum dots.

7. SUMMARY

Although there are many methods for fabricating nanostructures discussed elsewhere in the literature and in this book, they have their limitations. New approaches are desired to complement these established methods. Self-assembled islands formed during strained-layer epitaxy constitute a unique and promising approach to the fabrication of semiconductor nanostructures. This chapter has both reviewed the basic concepts of strained-layer epitaxy needed to understand the formation of these self-assembled islands and described the techniques commonly used to fabricate and characterize these nanostructures.

During growth, the deposited material grows in a two-dimensional fashion until a transformation occurs and three-dimensional growth begins. The two-dimensional islands that grow during the early stages are often anisotropic and depend on the nature of the deposition surface. The transition is abrupt and the resulting islands are uniform in size. These islands have special properties and are being used more frequently for physical studies and device fabrication.

Regardless of whether the desired islands are being formed before or after the transition to three-dimensional growth, the fundamental aspects of epitaxy must be considered. Thermodynamic issues such as elastic strain and kinetic issues such as surface diffusion are part of the complex process of self-assembled island formation. This topic is still in its early stages of study. With time, a better understanding will evolve and an improved control of these nanostructures will almost certainly be possible.

Acknowledgment

The financial support of the NSF (Grant DMR-9504400) and QUEST, an NSF Science and Technology Center (Grant DMR-9120007), is very much appreciated.

References

1. C. Weisbuch and B. Vinter, "Quantum Semiconductor Structures: Fundamentals and Applications," Chap. 1. Academic Press, San Diego, 1991.
2. C. Weisbuch and B. Vinter, "Quantum Semiconductor Structures: Fundamentals and Applications," Chap. 6. Academic Press, San Diego, 1991.
3. Y. Arakawa and A. Yariv, *IEEE J. Quantum Electron.* QE-22, 1887 (1986).
4. G. Bastard, "Wave Mechanics Applied to Semiconductor Heterostructures." Halsted Press, New York, 1988.
5. Y. Arakawa and H. Sakaki, *Appl. Phys. Lett.* 40, 939 (1982).
6. H. Hirayama, K. Matsunaga, M. Asada, and Y. Suematsu, *Electron. Lett.* 30, 142 (1994).
7. E. Kapon, *Proc. IEEE* 80, 398 (1992).
8. T. A. Strand, B. J. Thibeault, D. S. L. Mui, L. A. Coldren, P. M. Petroff, and E. L. Hu, *Appl. Phys. Lett.* 66, 1966 (1995).
9. K. Y. Cheng, K. C. Hsieh, and J. N. Baillargeon, *Appl. Phys. Lett.* 60, 2892 (1992).
10. A. C. Chen, A. M. Moy, P. J. Pearah, K. C. Hsieh, and K. Y. Cheng, *Appl. Phys. Lett.* 62, 1359 (1993).
11. W. K. Burton, N. Cabrera, and F. C. Frank, *Philos. Trans.* 243, 299 (1951).
12. W. W. Mullins and J. P. Hirth, *J. Phys. Chem. Solids* 24, 1391 (1963).
13. M. Tsuchiya, P. M. Petroff, and L. A. Coldren, *Appl. Phys. Lett.* 54, 1690 (1989).
14. M. S. Miller, H. Weman, C. E. Pryor, M. Krishnamurthy, P. M. Petroff, H. Kroemer, and J. L. Merz, *Phys. Rev. Lett.* 68, 3464 (1992).
15. S. Y. Hu, J. C. Yi, M. S. Miller, D. Leonard, D. B. Young, A. C. Gossard, N. Dagli, P. M. Petroff, and L. A. Coldren, *IEEE J. Quantum Electron.* QE-31, 1380 (1995).
16. A. P. Alivisatos, *Science* 271, 933 (1996).
17. J. R. Heath, *Science* 270, 1315 (1995).
18. E. G. Bauer, B. W. Dodson, D. J. Ehrlich, L. C. Feldman, C. P. Flynn, M. W. Geis, J. P. Harbison, R. J. Matyi, P. S. Peercy, P. M. Petroff, J. M. Phillips, G. B. Stringfellow, and A. Zangwill, *J. Mater. Res.* 5, 852 (1990).
19. E. Bauer and H. Poppa, *Thin Solid Films* 12, 167 (1972).
20. W. A. Tiller, "The Science of Crystallization: Microscopic Interfacial Phenomena." Cambridge University Press, New York, 1991.
21. J. Y. Tsao, in "Materials Fundamentals of Molecular Beam Epitaxy." Academic Press, San Diego, 1993.
22. J. A. Venables and G. L. Price, in "Epitaxial Growth, Part B" (J. W. Matthews, ed.), pp. 381–436. Academic Press, San Francisco, 1975.
23. J. A. Venables, G. D. T. Spiller, and M. Hanbucken, *Rep. Prog. Phys.* 47, 399 (1984).
24. R. Kern, G. LeLay, and J. J. Metois, in "Current Topics in Materials Science" (E. Kaldis, ed.), Vol. 3, pp. 131–419. North-Holland, New York, 1979.
25. M. Pfister, M. B. Johnson, S. F. Alvarado, H. W. M. Salemink, U. Marti, D. Martin, F. Morier-Genoud, and F. K. Reinhart, *Appl. Surf. Sci.* 104/105, 516 (1996).
26. P. M. Koenraad, M. B. Johnson, H. W. M. Salemink, W. C. van der Vleuten, and others, *Mater. Sci. Eng., B* 35, 485 (1995).
27. J. A. Venables, *Philos. Mag.* 27, 697 (1973).
28. J. G. Amar and F. Family, *Phys. Rev. Lett.* 74, 2066 (1995).
29. M. Schroeder and D. E. Wolf, *Phys. Rev. Lett.* 74, 2062 (1995).
30. C. Ratsch, A. Zangwill, and P. Smilauer, *Surf. Sci.* 314, L937 (1994).
31. J. Tersoff, A. W. Denier van der Gon, and R. M. Tromp, *Phys. Rev. Lett.* 72, 266 (1994).
32. J. A. Venables, G. G. Hembree, J. Lui, and J. S. Drucker, in "Thirteenth International Congress on X-Ray Optics and Microanalysis," 1993, p. 415.

33. J. A. Venables, *Surf. Sci.* 299/300, 798 (1994).
34. G. S. Bales and D. C. Chrzan, *Phys. Rev. B* 50, 6057 (1994).
35. J. W. Evans and M. C. Bartelt, *Surf. Sci. Lett.* 284, L437 (1993).
36. C. Ratsch, P. Smilauer, A. Zangwill, and D. D. Vvedensky, *Surf. Sci.* 329, L599 (1995).
37. R. Bruinsma and A. Zangwill, *Europhys. Lett.* 4, 729 (1987).
38. V. Swaminathan and A. T. Macrander, "Materials Aspects of GaAs and InP Based Structures," Chap. 3. Prentice–Hall, Englewood Cliffs, NJ, 1991.
39. L. B. Freund, *Mater. Res. Soc. Bull.* 52 (1992).
40. J. W. Matthews, in "Epitaxial Growth: Part B" (J. W. Matthews, ed.), Chap. 8. Academic Press, San Francisco, 1975.
41. J. P. Hirth and J. Lothe, "Theory of Dislocations." Wiley, New York, 1982.
42. C. W. Snyder, B. G. Orr, and H. Munekata, *Appl. Phys. Lett.* 62, 46 (1993).
43. B. Voigtländer and A. Zinner, *Surf. Sci.* 351, L233 (1996).
44. C. W. Oh, E. Kim, and Y. H. Lee, *Phys. Rev. Lett.* 76, 776 (1996).
45. R. F. C. Farrow, in "Molecular Beam Epitaxy." Noyes Publications, Park Ridge, NJ, 1995.
46. M. A. Herman and H. Sitter, "Molecular Beam Epitaxy: Fundamentals and Current Status." Springer-Verlag, New York, 1989.
47. M. B. Panish and H. Temkin, "Gas Source Molecular Beam Epitaxy." Springer-Verlag, Berlin/Heidelberg, 1993.
48. G. B. Stringfellow, "Organometallic Vapor-Phase Epitaxy: Theory and Practice." Academic Press, San Diego, 1989.
49. H. O. Pierson, "Handbook of Chemical Vapor Deposition." Noyes Publications, Park Ridge, NJ, 1992.
50. T. P. Pearsall, "GaInAsP Alloys." Wiley, New York, 1982.
51. B. A. Joyce, J. H. Neave, P. J. Dobson, and P. K. Larsen, *Phys. Rev. B* 29, 814 (1984).
52. P. J. Dobson, B. A. Joyce, J. H. Neave, and J. Zhang, *J. Cryst. Growth* 81, 1 (1987).
53. T. Sakamoto, T. Kawamura, S. Nago, G. Hashiguchi, G. Sakamoto, and K. Kuniyoshi, *J. Cryst. Growth* 81, 59 (1987).
54. M. Prutton, "Surface Physics." Clarendon Press, Oxford, UK, 1983.
55. J. W. Edington, "Practical Electron Microscopy in Materials Science." TechBooks, Herndon, VA, 1976.
56. P. R. Busek, J. M. Cowley, and L. Eyring, "High-Resolution Transmission-Electron Microscopy." Oxford University, New York, 1988.
57. L. Reimer, "Transmission Electron Microscopy." Springer-Verlag, Berlin/Heidelberg, 1984.
58. P. Hirsch, A. Howie, R. Nicholson, D. W. Pashley, and M. J. Whelan, "Electron Microscopy of Thin Crystals." Krieger Publishing, Malabar, FL, 1977.
59. D. J. Eaglesham and M. Cerullo, *Phys. Rev. Lett.* 64, 1943 (1990).
60. S. Guha, A. Madhukar, and K. C. Rajkumar, *Appl. Phys. Lett.* 57, 2110 (1990).
61. G. Binnig and H. Rohrer, *Helv. Phys. Acta* 55, 726 (1982).
62. G. Binnig, H. Rohrer, C. Gerber, and E. Weibel, *Appl. Phys. Lett.* 40, 178 (1982).
63. J. A. Kubby and J. J. Boland, *Surf. Sci. Rep.* 26, 63 (1996).
64. S. N. Magonov, *Appl. Spectrosc. Rev.* 28, 1 (1993).
65. J. Tersoff and D. R. Hamann, *Phys. Rev. Lett.* 50, 1998 (1983).
66. J. Tersoff and D. R. Hamann, *Phys. Rev. Lett. B* 31, 805 (1985).
67. G. Binnig, C. F. Quate, and C. Gerber, *Phys. Rev. Lett.* 56, 930 (1986).
68. E. T. Yu, *Mater. Sci. Eng., R* 17, 147 (1996).
69. Q. Zhong, D. Inniss, K. Kjoller, and V. B. Elings, *Surf. Sci.* 290, L688 (1993).
70. C. Weisbuch and B. Vinter, "Quantum Semiconductor Structures: Fundamentals and Applications." Academic Press, San Diego, 1991.
71. U. A. Griesinger, S. Kronmuller, M. Geiger, D. Ottenwalder, F. Scholz, and H. Scheizer, *J. Vac. Sci. Technol., B* 14, 4058 (1996).
72. T. Koster, B. Hadam, J. Gondermann, B. Spangenberg, H. G. Roskos, H. Kurz, J. Brunner, and G. Abstreiter, *Microelectron. Eng.* 30, 341 (1996).
73. M. Illing, G. Bacher, T. Kummell, A. Forchel, D. Hommel, B. Jobst, and G. Landwehr, *J. Vac. Sci. Technol., B* 13, 2792 (1995).
74. M. Lopez, N. Tanaka, I. Matsuyama, and T. Ishikawa, *Solid-State Electon.* 40, 627 (1996).
75. P. Roblin, R. C. Potter, and A. Fathimulla, *J. Appl. Phys.* 1996, 2502 (1996).
76. E. J. Heller and M. G. Lagally, *Appl. Phys. Lett.* 60, 2675 (1992).
77. M. G. Lagally, Y.-W. Mo, R. Kariotis, B. S. Swartzentruber, and M. B. Webb, in "Kinetics of Ordering and Growth at Surfaces" (M. G. Lagally, ed.), pp. 145–168. Plenum, New York, 1990.
78. Y.-W. Mo, B. S. Swartzentruber, R. Kariotis, M. B. Webb, and M. G. Lagally, *Phys. Rev. Lett.* 63, 2393 (1989).
79. V. Bressler-Hill, A. Lorke, S. Varma, P. M. Petroff, K. Pond, and W. H. Weinberg, *Phys. Rev. B* 50, 8479 (1994).
80. J. Massies and N. Grandjean, *Phys. Rev. Lett.* 71, 1411 (1993).
81. J. Tersoff and R. M. Tromp, *Phys. Rev. Lett.* 70, 2782 (1993).

82. B. R. Bennett, P. M. Thibado, M. E. Twigg, E. R. Glaser, R. Magno, B. V. Shanabrook, and L. J. Whitman, *J. Vac. Sci. Technol., B* 14, 2195 (1996).
83. B. R. Bennett, B. V. Shanabrook, P. M. Thibado, L. J. Whitman, and R. Magno, *J. Cryst. Growth* 175/176, 888 (1997).
84. Y. Chen and J. Washburn, *Phys. Rev. Lett.* 77, 4046 (1996).
85. R. Heitz, T. R. Ramachandran, A. Kalburge, Q. Xie, I. Mukhametzhanov, P. Chen, and A. Madhukar, *Phys. Rev. Lett.* 78, 4071 (1997).
86. T. R. Ramachandran, R. Heitz, P. Chen, and A. Madhukar, *Appl. Phys. Lett.* 640, 640 (1997).
87. J. Knall and J. B. Pethica, *Surf. Sci.* 265, 156 (1992).
88. U. Köhler, O. Jusko, G. Pietsch, B. Müller, and M. Henzler, *Surf. Sci.* 248, 321 (1991).
89. Y.-W. Mo, D. E. Savage, B. S. Swartzentruber, and M. G. Lagally, *Phys. Rev. Lett.* 65, 1020 (1990).
90. C. Priester and M. Lannoo, *Phys. Rev. Lett.* 75, 93 (1995).
91. P. M. Thibado, B. R. Bennett, M. E. Twigg, B. V. Shanabrook, and L. J. Whitman, *J. Vac. Sci. Technol., B* 14, 885 (1996).
92. B. Voigtländer and M. Kästner, *Appl. Phys. A* 63, 577 (1996).
93. M. Berti, A. V. Drigo, G. Rossetto, and G. Torzo, *J. Vac. Sci. Technol., B* 15, 1794 (1997).
94. N. P. Kobayashi, T. R. Ramachandran, P. Chen, and A. Madhukar, *Appl. Phys. Lett.* 68, 3299 (1996).
95. C. B. Duke, *Chem. Rev.* 96, 1237 (1996).
96. J. G. Belk, J. L. Sudijono, D. M. Holmes, C. F. McConville, T. S. Jones, and B. A. Joyce, *Surf. Sci.* 365, 735 (1996).
97. H. H. Farrell and C. J. Palmstrøm, *J. Vac. Sci. Technol., B* 8, 903 (1990).
98. D. J. Chadi, *J. Vac. Sci. Technol., A* 5, 834 (1987).
99. A. R. Avery, D. M. Holmes, J. Sudijono, T. S. Jones, and B. A. Joyce, *Surf. Sci.* 323, 91 (1995).
100. D. K. Biegelsen, R. D. Bringans, J. E. Northrup, and L.-E. Swartz, *Phys. Rev. B* 41, 5701 (1990).
101. B. Voigtländer and A. Zinner, *Appl. Phys. Lett.* 63, 3055 (1993).
102. J. L. Sudijono, M. D. Johnson, M. B. Elowitz, C. W. Snyder, and B. G. Orr, *Surf. Sci.* 280, 247 (1993).
103. D. M. Holmes, J. G. Belk, J. L. Sudijono, J. H. Neave, T. S. Jones, and B. A. Joyce, *Surf. Sci.* 341, 133 (1995).
104. J. G. Belk, J. L. Sudijono, H. Yamaguchi, X. M. Zhang, D. W. Pashley, C. F. McConville, T. S. Jones, and B. A. Joyce, *J. Vac. Sci. Technol., A* 15, 915 (1997).
105. J. G. Belk, J. L. Sudijono, X. M. Zhang, J. H. Neave, T. S. Jones, and B. A. Joyce, *Phys. Rev. Lett.* 78, 475 (1997).
106. H. Yamaguchi, M. R. Fahy, and B. A. Joyce, *Appl. Phys. Lett.* 69, 776 (1996).
107. P. P. Gonzalez-Borrero, D. I. Lubyshev, E. Marega, Jr., E. Petitprez, and P. Basmaji, *J. Cryst. Growth* 169, 424 (1996).
108. K. Nishi, R. Mirin, D. Leonard, G. Medeiros-Ribeiro, P. M. Petroff, and A. C. Gossard, *J. Appl. Phys.* 80, 3466 (1996).
109. L. Sham, S. J. Allen, A. Kamgar, and D. C. Tsui, *Phys. Rev. Lett.* 40, 472 (1978).
110. D. Tsui and E. Gornik, *Appl. Phys. Lett.* 32, 365 (1978).
111. T. Cole, A. Lakhani, and P. J. Stiles, *Phys. Rev. Lett.* 38, 722 (1977).
112. T. J. Thornton, J. M. Fernandez, S. Kaya, P. W. Green, and K. Fobelets, *Appl. Phys. Lett.* 70, 1278 (1997).
113. B. Z. Olshanetsky and V. I. Mashanov, *Surf. Sci.* 111, 414 (1981).
114. T. Berghaus, A. Brodde, H. Neddermeyer, and S. Tosch, *J. Vac. Sci. Technol., A* 6, 478 (1988).
115. J. M. Gibson, M. L. McDonald, and F. C. Unterwald, *Phys. Rev. Lett.* 55, 1765 (1985).
116. A. Oral and R. Ellialtioglu, *Surf. Sci.* 323, 295 (1995).
117. M. J. Bronikowski, Y. Wang, and R. J. Hamers, *Phys. Rev. B* 48, 12361 (1993).
118. L. Goldstein, G. Glas, J. Y. Marzin, M. N. Charasse, and G. L. Rous, *Appl. Phys. Lett.* 47, 1099 (1985).
119. M. Krishnamurthy, J. S. Drucker, and J. A. Venables, *J. Appl. Phys.* 69, 6461 (1991).
120. D. Leonard, M. Krishnamurthy, C. M. Reaves, S. P. DenBaars, and P. M. Petroff, *Appl. Phys. Lett.* 63, 3203 (1993).
121. N. Grandjean and J. Massies, *J. Cryst. Growth* 134, 51 (1993).
122. D. J. Srolovitz, *Acta Metall.* 37, 621 (1989).
123. C. M. Reaves, V. Bressler-Hill, S. Varma, W. H. Weinberg, and S. P. DenBaars, *Surf. Sci.* 326, 209 (1995).
124. F. K. LeGoues, M. C. Reuter, J. Tersoff, M. Hammar, and R. M. Tromp, *Phys. Rev. Lett.* 73, 300 (1994).
125. V. Bressler-Hill, C. M. Reaves, S. Varma, S. P. DenBaars, and W. H. Weinberg, *Surf. Sci.* 341, 29 (1995).
126. K. Georgsson, N. Carlsson, L. Samuelson, W. Seifert, and L. R. Wallenberg, *Appl. Phys. Lett.* 67, 2981 (1995).
127. D. Leonard, K. Pond, and P. M. Petroff, *Phys. Rev. B* 50, 11687 (1994).
128. M. Sopanen, H. Lipsanen, and J. Ahopelto, *Appl. Phys. Lett.* 67, 3768 (1995).
129. M. Taskinen, M. Sopanen, H. Lipsanen, J. Tulkki, T. Tuomi, and J. Ahopelto, *Surf. Sci.* 376, 60 (1997).
130. D. Leonard, M. Krishnamurthy, S. Fafard, J. L. Merz, and P. M. Petroff, *J. Vac. Sci. Technol., B* 12, 1063 (1994).
131. C. M. Reaves, V. Bressler-Hill, W. H. Weinberg, and S. P. DenBaars, *J. Electron. Mater.* 24, 1603 (1995).
132. A. Zangwill, "Physics at Surfaces." Cambridge University Press, New York, 1988.

133. M. Krishnamurthy, A. Lorke, M. Wassermeier, D. R. M. Williams, and P. M. Petroff, *J. Vac. Sci. Technol.*, *B* 11, 1384 (1993).
134. K. Pond, A. Lorke, J. Ibbetson, V. Bressler-Hill, R. Maboudian, W. H. Weinberg, A. C. Gossard, and P. M. Petroff, *J. Vac. Sci. Technol.*, *B* 12, 2689 (1994).
135. R. Nötzel, J. Temmyo, and T. Tamamura, *J. Cryst. Growth* 145, 990 (1994).
136. R. Nötzel, J. Temmyo, and T. Tamamura, *Jpn. J. Appl. Phys.* 33, L275 (1994).
137. R. Nötzel, T. Fukui, H. Hasegawa, J. Temmyo, and T. Tamamura, *Appl. Phys. Lett.* 65, 2854 (1994).
138. J. Temmyo, A. Kozen, T. Tamamura, R. Nötzel, T. Fukui, and H. Hasegawa, in "Proceedings of the Indium Phosphide and Related Materials Conference," 1995, p. 766.
139. J. Temmyo, A. Kozen, T. Tamamura, R. Nötzel, T. Fukui, and H. Hasegawa, *J. Electron. Mater.* 25, 431 (1996).
140. R. Nötzel, *Semicond. Sci. Technol.* 11, 1365 (1996).
141. J. Temmyo, E. Kuramochi, M. Sugo, T. Nishiya, R. Nötzel, and T. Tamamura, *IEICE Trans. Electron.* E79, 1495 (1996).
142. C. M. Reaves, R. I. Pelzel, G. C. Hsueh, W. H. Weinberg, and S. P. DenBaars, *Appl. Phys. Lett.* 69, 3878 (1996).
143. G. S. Solomon, M. C. Larson, and J. J. S. Harris, *Appl. Phys. Lett.* 69, 11897 (1996).
144. Q. Xie, A. Madhukar, P. Chen, and N. P. Kobayashi, *Phys. Rev. Lett.* 75, 2542 (1995).
145. Q. Xie, N. P. Kobayashi, T. R. Ramachandran, A. Kalburge, P. Chen, and A. Madhukar, *J. Vac. Sci. Technol.*, *B* 14, 2203 (1996).
146. G. S. Solomon, J. A. Trezza, A. F. Marshall, and J. J. S. Harris, *J. Vac. Sci. Technol.*, *B* 14, 2208 (1996).
147. B. Rahmati, W. Jäger, H. Trinkaus, R. Loo, L. Vescan, and H. Lüth, *Appl. Phys. A* 62, 575 (1996).
148. N. N. Ledenstov, V. A. Shchukin, M. Grundmann, N. Kristaedter, J. Böhrer, O. Schmidt, D. Bimber, V. M. Ustinov, A. Y. Egorov, A. E. Zhukov, P. S. Kop'ev, S. V. Zaitsev, N. Y. Gorbeev, Z. I. Alferov, A. I. Borovkov, A. O. Kosogov, S. S. Ruvimov, P. Werner, U. Gösele, and J. Heydenreich, *Phys. Rev. B* 54, 8743 (1996).
149. N. Ikoma and S. Ohkouchi, *Jpn. J. Appl. Phys.* 34, L724 (1995).
150. J. M. Moison, L. Leprince, F. Barthe, F. Houzay, N. Lebouchè, J. M. Gèrard, and J. Y. Marzin, *Appl. Surf. Sci.* 92, 526 (1996).
151. R. L. Schwoebel and E. J. Shipsey, *J. Appl. Phys.* 37, 3682 (1966).
152. J. Ishizaki, S. Goto, M. Kishida, T. Fukui, and H. Hasegawa, *Jpn. J. Appl. Phys.* 33, 721 (1994).
153. M. Kitamura, M. Nishioka, J. Oshinowo, and Y. Arakawa, *Appl. Phys. Lett.* 66, 3663 (1995).
154. S. Jeppesen, M. S. Miller, D. Hessman, B. Kowalski, I. Maximov, and L. Samuelson, *Appl. Phys. Lett.* 68, 2228 (1996).
155. M. Kitamura, M. Nishioka, J. Oshinowo, and Y. Arakawa, *Appl. Phys. Lett.* 66, 3663 (1995).
156. S. Varma, C. M. Reaves, V. Bressler-Hill, S. P. DenBaars, and W. H. Weinberg, *Surf. Sci.* 393, 24 (1997).
157. D. S. L. Mui, D. Leonard, L. A. Coldren, and P. M. Petroff, *Appl. Phys. Lett.* 66, 1620 (1995).
158. T. Saitoh, A. Tanimura, and K. Yoh, *Jpn. J. Appl. Phys.* 35, 1370 (1996).
159. A. C. Gossard and S. Fafard, *Solid State Commun.* 92, 63 (1994).
160. P. M. Petroff and S. P. DenBaars, *Superlattices Microstruct.* 15, 15 (1994).
161. V. Swaminathan and A. T. Macrander, "Materials Aspects of GaAs and InP Based Structures," Chap. 5. Prentice–Hall, Englewood Cliffs, NJ, 1991.
162. J. I. Pankove, "Optical Processes in Semiconductors," Chap. 6. Dover, New York, 1971.
163. J.-Y. Marzin, J.-M. Gérard, A. Izraël, D. Barrier, and G. Bastard, *Phys. Rev. Lett.* 73, 716 (1994).
164. M. Grundmann, J. Christen, N. N. Ledentsov, J. Böhrer, D. Bimberg, S. S. Ruvimov, P. Werner, U. Richter, U. Gösele, J. Heyenreich, V. M. Ustinov, A. Y. Egorov, A. E. Zhukov, P. S. Kop'ev, and Z. I. Alferov, *Phys. Rev. Lett.* 74, 4043 (1995).
165. H. Drexler, D. Leonard, W. Hansen, J. P. Kotthaus, and P. M. Petroff, *Phys. Rev. Lett.* 73, 2252 (1994).
166. G. Medeiros-Ribeiro, D. Leonard, and P. M. Petroff, *Appl. Phys. Lett.* 66, 1767 (1995).
167. R. F. Service, *Science* 271, 920 (1996).
168. L. A. Coldren and S. W. Corzine, "Diode Lasers and Photonic Integrated Circuits," Chap. 2. Wiley, New York, 1995.
169. K. J. Vahala, *IEEE J. Quantum Electron.* QE-24, 523 (1988).
170. N. Kirstaedter, N. N. Ledentsov, M. Grundmann, D. Bimberg, V. M. Ustinov, S. S. Ruvimov, M. V. Maximov, P. S. Kop'ev, Z. I. Alferov, U. Richter, P. Werner, U. Gosele, and J. Heydenreich, *Electron. Lett.* 30, 1416 (1994).
171. N. Kirstaedter, O. G. Schmidt, N. N. Ledentsov, D. Bimberg, V. M. Ustinov, A. Y. Egorov, A. E. Zhukov, M. V. Maximov, P. S. Kop'ev, and Z. I. Alferov, *Appl. Phys. Lett.* 69, 1226 (1996).
172. A. Tackeuchi, Y. Nakata, S. Muto, Y. Sugiyama, T. Inata, and N. Yokoyama, *Jpn. J. Appl. Phys.* 34, L405 (1995).
173. J. M. Gérard, O. Cabrol, and B. Sermage, *Appl. Phys. Lett.* 68, 3123 (1996).
174. S. Bandyopadhyay and V. P. Roychowdhury, in "1995 International Conference on Solid State Devices and Materials," Osaka, Japan, p. 180.
175. R. S. Williams, personal communication.

176. N. Yokoyama, S. Muto, K. Imamura, M. Takatsu, T. Mori, Y. Sugiyama, Y. Sakuma, H. Nakao, and T. Adachihara, *Solid-State Electron.* 40, 505 (1996).
177. V. Swaminathan and A. T. Macrander, "Materials Aspects of GaAs and InP Based Structures." Prentice–Hall, Englewood Cliffs, NJ, 1991.
178. M. Sopanen, H. Lipsanen, and J. Ahopelto, *Appl. Phys. Lett.* 66, 2364 (1995).
179. H. Lipsanen, J. Ahopelto, T. Koljonen, and M. Sopanen, *J. Cryst. Growth* 145, 988 (1994).
180. G. Yusa, H. Noge, Y. Kadoya, T. Someya, T. Suga, P. Petroff, and H. Sakaki, *Jpn. J. Appl. Phys.* 34, L1198 (1995).

Chapter 7

HYBRID MAGNETIC–SEMICONDUCTOR NANOSTRUCTURES

François M. Peeters

Departement Natuurkunde, Universiteit Antwerpen, Antwerpen, Belgium

Jo De Boeck

IMEC, Leuven, Belgium

Contents

Handbook of Nanostructured Materials and Nanotechnology, edited by H.S. Nalwa
Volume 3: Electrical Properties

1. INTRODUCTION

During the last half century, there has been a rapid development in the field of solid-state electronics, which started with the discovery of the transistor and subsequently was fueled by the integration of many circuit elements onto one semiconductor chip. Magnetic materials, on the other hand, have been developed separately, independently from the semiconductor systems. They have been very important for information storage (tape, disk, magnetooptic disk) and for magnetic circuits.

Recently, one started to incorporate magnetic materials into planar integrated electronic circuitry. In those integrated devices, the semiconductor properties are combined with and enhanced by the presence of magnetic elements. In a first level of integration, one fabricates on the same substrate a hybrid system consisting of separated magnetic and electronic components. A much higher level of integration is realized when the magnetic elements modify and enhance the behavior of the electronic devices by becoming part of them. This may ultimately lead to completely new classes and concepts of devices.

The outline of the present chapter follows the preceding division. In the first part (Sections 2–8), we consider systems consisting of a semiconductor [typically containing a two-dimensional electron gas (2DEG) as found, for example, in heterojunctions] with a patterned ferromagnetic material on top of it. The electrons move in local nonhomogeneous magnetic fields that alter the orbital motion of the electrons through the Lorentz force, that is, $\mathbf{F} = q\mathbf{v} \times \mathbf{B}$. Such a magnetic field will also induce a shift in the energy of the electrons. Because in typical III–V semiconductors the effective g factor is small, the spin will be of secondary importance in this case.

In the second part (Sections 9–12), structures are considered in which the electron (or hole) current passes through the ferromagnetic/semiconductor (FM/SC) interface. Here, one relies on spin-dependent scattering and the Zeeman spin splitting as the fundamental mechanisms on which the action of the device is based. Spin-up and spin-down electrons now become different types of carriers. Several device concepts have been proposed, including an all-metal spin transistor based on the accumulation of a spin-up or spin-down population, an analog of the electrooptic modulator based on a precession of the electron spin, and systems where one uses the FM/SC analog of the giant magnetoresistance.

As in semiconductor research and technology, in present-day magnetic thin-film research the technology to fabricate submicrometer structures is very important. Electron beam (e-beam) lithography and other techniques are used for fabricating submicrometer features. The characterization of magnetic nanowires, magnetic dot arrays, and so forth will lead to an increased understanding of the physics of micromagnetism and magnetoelectronics. Small feature sizes will also be essential in applications such as magnetic random-access memory (MRAM) and intriguing device applications such as spin transport devices.

In the first part of this overview, we focus on the motion of two-dimensional (2D) electrons in nonhomogeneous magnetic fields. In Section 3, we show how such nonhomogeneous magnetic field profiles can be created experimentally and we calculate the perpendicular magnetic field component that is felt by the 2DEG. The ballistic motion of 2D electrons through different magnetic field profiles are studied in Section 4, while the diffusive regime is discussed in Section 5. Periodic magnetic structures are considered in Section 6 [one-dimensional (1D) magnetic modulation] and Section 7 (2D periodic and random modulation). Different devices based on the Hall effect in combination with nonhomogeneous magnetic fields are discussed in Section 8.

In the second part of this overview, we will focus on ferromagnetic/semiconductor (FM/SC) combinations that try to take advantage of the combination of the electronic transport of both materials for new device structures. The emphasis will be on technological issues for the fabrication of structures in which the electronic current is passing through at least one interface between the ferromagnetic and the semiconductor. In Section 10, a device structure exploiting spin-dependent scattering is briefly discussed. The next section deals with spin-polarized current in an FM/SC structure. The other sections are devoted to materials issues in trying to realize optimum FM/SC combinations for future spin-dependent transport devices. These issues include the control of the FM/SC interface and the fabrication of magnetic nanostructures.

2. ELECTRONS IN MICROSCOPICALLY INHOMOGENEOUS MAGNETIC FIELDS

The behavior of electrons in *macroscopically* homogeneous magnetic fields has been used extensively to obtain experimental information on the properties of charge carriers [1] such as, for example, their density and the Fermi surface [through the Shubnikov–de Haas (SdH) effect] and their mass (e.g., using cyclotron resonance). The scattering of electrons on magnetic impurities form the other limit in which electrons feel locally (on an angstrom scale) strong magnetic fields (i.e., *microscopically* inhomogeneous) that may act as scattering centers in, for example, diluted semimagnetic materials [2].

Between these limits lie inhomogeneous magnetic fields on the *nanometer* scale. They have been realized with the creation of magnetic dots, the integration of ferromagnetic materials with semiconductors where the patterning of such films was recently demonstrated experimentally. This new technology will add a new functional dimension to the present semiconductor technology and will open new avenues for new physics and possible applications such as switches based on the Lorentz force and nonvolatile memories based on the Hall voltage generated by a local magnetic field. A different route to create inhomogeneous magnetic fields is through the integration of superconducting materials with semiconductors. This was realized experimentally using type II superconductors, which were deposited on a metal-oxide-semiconductor field-effect transistor (Si-MOSFET) or a GaAs/AlGaAs heterojunction. Magnetic flux lines penetrate the two-dimensional electron gas (2DEG) that acts as nanoscale scattering centers for the electrons, offering the possibility of studying the weak localization and the dynamics of vortices. Using lithographic techniques, these superconducting films can be patterned into any desired form. The geometry of the patterning determines the geometry of the inhomogeneous magnetic field.

In general, the shape anisotropy of the magnetic film (or the stripes) will force the magnetization in the plane of the film. Other mechanisms can be active that can lead to a magnetization vector perpendicular to the film, which is the situation we are mostly interested in. Out-of-plane magnetization has been realized in ultrathin layers of Fe on Ag or Cu; compounds such as MnAlGa, Co/Ni multilayers, and ultrathin MnGa films; and the metastable τ-MnAl phase, which can be grown epitaxially on GaAs/AlAs heterostructures using molecular beam epitaxy (MBE).

3. MAGNETIC FIELD PROFILES

3.1. One-Dimensional Profiles

As schematically illustrated in Figure 1, a one-dimensional magnetic barrier can be created by the deposition, on top of a heterostructure, of a ferromagnetic strip with magnetization (a) perpendicular and (b) parallel to the 2DEG, (c) of a conducting strip with a current driven through it, and (d) of a type I superconducting plate interrupted by a strip. In all cases, the 2DEG is situated at a distance z_0 below the strip whose thickness and height are d and h, respectively. The resulting magnetic field profile is obtained from the following Maxwell equation:

$$\begin{aligned} \operatorname{div} \mathbf{B} &= -4\pi \operatorname{div} \mathbf{M} = 4\pi \rho_M(\mathbf{r}) \\ \mathbf{B} &= -\operatorname{grad} \Phi_M \end{aligned} \tag{1}$$

which can be integrated and results in the magnetic potential

$$\Phi_M(\mathbf{r}) = \int d^3r' \, \frac{\rho_M(\mathbf{r}')}{|\mathbf{r} - \mathbf{r}'|} \tag{2}$$

For illustrative purposes, we consider perpendicular magnetization (Fig. 1a) in which the width of the magnetic strip d is small such that we can replace it by a dipole line with magnetic charge density $\rho_M(\mathbf{r}) = -M_0\delta(x)(d/dz)\delta(z)$. Integrating (2) results in the magnetic potential $\Phi_M(\mathbf{r}) = 2M_0 z/(x^2 + z^2)$, which leads to the magnetic field distribution

$$B(x) = 2M_0 \frac{z_0^2 - x^2}{(z_0^2 + x^2)^2} \tag{3}$$

and the vector potential $A(x) = 2M_0 x/(z_0^2 + x^2)$. If we have strips of width d instead of wires, we have to integrate Eq. (2) numerically in the region $-d/2 \leqslant x' \leqslant d/2$. It turns out

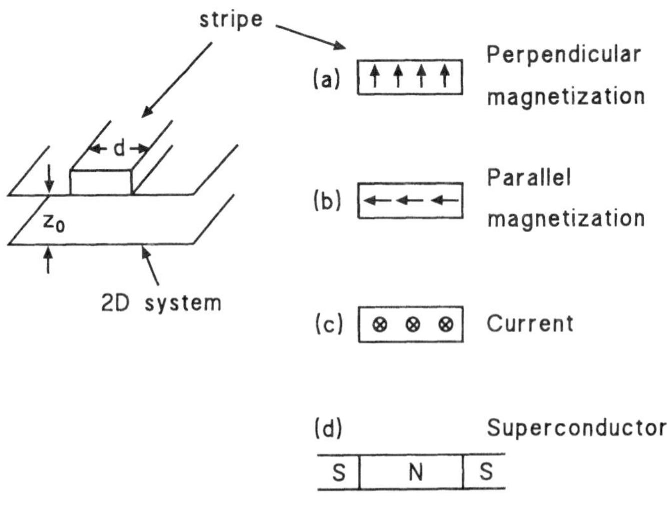

Fig. 1. Layout of the system for the creation of a one-dimensional magnetic field profile in a 2DEG.

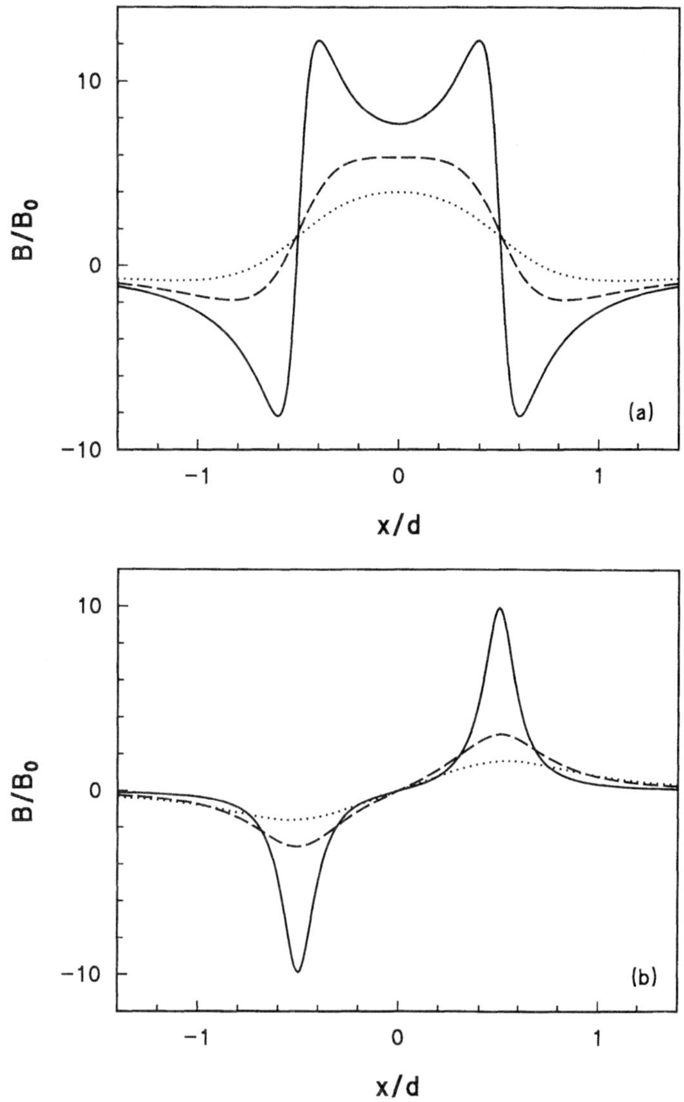

Fig. 2. Magnetic field under the strip corresponding to the four different configurations shown in Figure 1 for three different distances from the magnetic strip: $z_0 = 0.1$ (solid curve), $z_0 = 0.3$ (dashed curve), and $z_0 = 0.5$ (dotted curve).

that in the limit z_0, $h \ll d$, the magnetic field distribution takes the simple form (h is the height of the magnetic strip)

$$B(x, z_0) = B_0\big(K(x + d/2, z_0) - K(x - d/2, z_0)\big) \tag{4}$$

with $B_0 = Mh/d$ and $K(x, z) = 2xd/(x^2 + z^2)$, which is depicted in Figure 2a for three values of z_0. Similarly for parallel magnetization, we found $K(x, z) = -zd/(x^2 + z^2)$, which is depicted in Figure 2b for three values of z_0.

For completeness, we also calculate the magnetic field resulting from a finite size magnetic film of finite thickness that is located at $a < x < b$, $c < y < d$, $e < z < f$. The magnetization of such a ferromagnetic film is

$$\begin{aligned}\mathbf{M}(x, y, z) &= \theta(x)\big(1 - \theta_b(x)\big)\theta(y)\big(1 - \theta_c(y)\big)\theta(z)\big(1 - \theta_d(y)\big)\\ &\quad \times (m_z\mathbf{e}_x + m_y\mathbf{e}_y + m_z\mathbf{e}_z)\end{aligned} \tag{5}$$

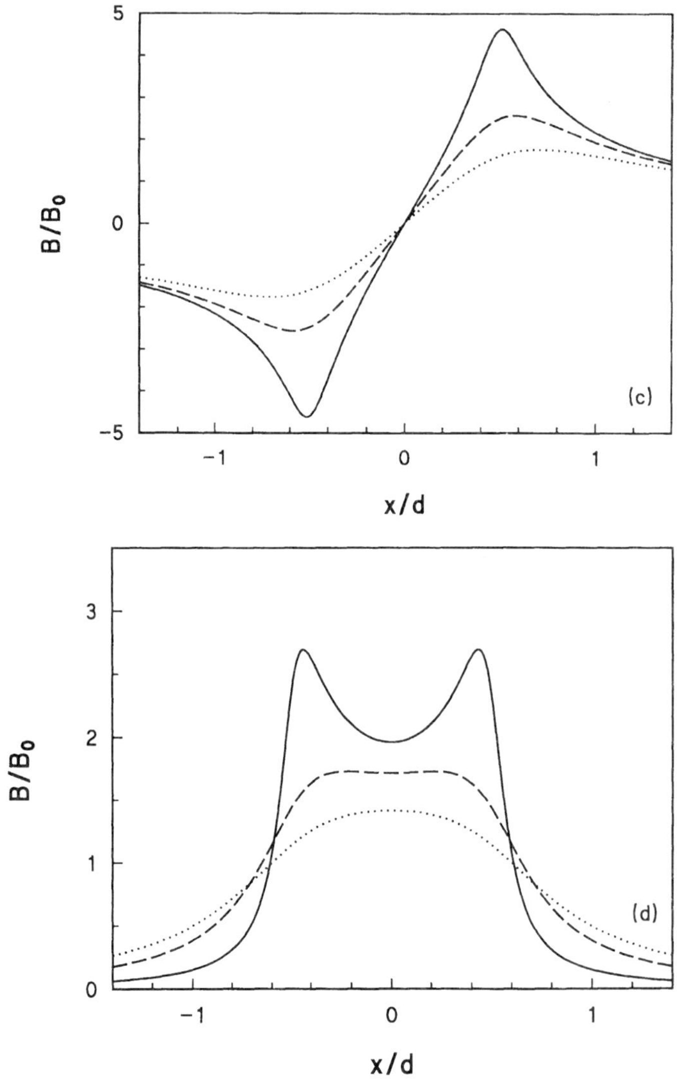

Fig. 2. (*Continued.*)

where $\theta_a(x) = \theta(a - x)$ and $\theta(x)$ is the Heaviside step function. We insert this expression into Eq. (2) and take as an example the magnetization along the x direction, that is, $m_y = m_z = 0$. After performing the differentiation and the integration, we obtain the following exact solution for the z component of the magnetic field [3]:

$$\begin{aligned} B_z(x, y, z) = m_x\big[&-\big(f_{adf}(x, y, z) - f_{acf}(x, y, z)\big) + \big(f_{ade}(x, y, z) - f_{ace}(x, y, z)\big) \\ &+ \big(f_{bdf}(z, y, z) - f_{bcf}(x, y, x)\big) - \big(f_{bde}(x, y, z) - f_{bcd}(x, y, x)\big)\big] \end{aligned} \tag{6}$$

with $f_{\alpha\beta\gamma}(x, y, z) = \operatorname{arcsinh}((\beta - y)/\sqrt{(x-\alpha)^2 + (z-\gamma)^2})$.

For a wire with a current I going through it (Fig. 1c), the magnetic field is determined by the Maxwell equation $\operatorname{rot}\mathbf{B} = 4\pi\mathbf{j}/c$, which results in the angular magnetic field component $B_\phi = 2I/cr$ at a radial distance r from the wire. In the plane of the 2DEG a distance z_0 from the wire, this leads to the magnetic field profile

$$B(x) = B_z(x, z_0) = B_0 \frac{x}{z_0^2 + x^2} \tag{7}$$

with $B_0 = 2I/c$, which is depicted in Figure 2c. The vector potential is obtained by integrating the equation $\mathbf{B} = \mathrm{rot}\,\mathbf{A}$, which gives $A(x) = A_y(x, z_0) = \frac{1}{2}B_0 \ln((z_0^2 + x^2)/(z_0^2 + R^2))$, where R is some distance away from the wire where the opposite current is flowing. For a strip of finite width d such that $z_0, h \ll d$, we can cast the result in the form of Eq. (4) with $K(x, z) = \ln[(x^2 + z^2)/d^2]$. This magnetic field profile is shown in Figure 2c.

For a superconducting strip in a magnetic field, we have to solve the Maxwell equations in the superconductor: $\mathbf{j} = -(ne^2/mc)\mathbf{A}$ and $\nabla^2\mathbf{H} = (1/L_0^2)\mathbf{H}$, where $L_0^2 = mc^2/4\pi ne^2$ and we made use of $\mathrm{rot}\,\mathbf{H} = -(4\pi ne^2/mc^2)\mathbf{A}$. Outside the superconductor, we have the equation $\nabla^2\Phi_\mathrm{M} = 0$, with the boundary condition on the superconductor $B_n = -\partial\Phi_\mathrm{M}/\partial n = 0$. This 2D potential problem can be solved by the conformal mapping $x + iz = \sin(\pi w)$ where we introduce the complex potential $W = \Phi_\mathrm{m} + i\Psi$. The solution of the problem [4] is $W = iw$ and the magnetic potential is given by $\Phi_\mathrm{M} = \mathrm{Re}\,W = -\mathrm{Im}(\arcsin(x + iz))$, which results in the magnetic field

$$B = B_0\,\mathrm{Re}\left(\frac{1}{\sqrt{1-(x+iz)^2}}\right) \tag{8}$$

which is depicted in Figure 1d.

3.2. Periodic Structures

As a last example, we consider a periodic structure and calculate the magnetic field resulting from a ferromagnetic thin film in which the magnetic domains are in the form of a periodic array of parallel strips (see Fig. 3a). The magnetization of each domain strip can be either in the plane of the film or perpendicular to it.

First, we consider the case depicted in Figure 3a but where we consider only a single magnetic domain strip with magnetization perpendicular to the (x, y) plane with boundaries $x \in (-a, a)$, $y \in (-\infty, \infty)$, $z \in (-d, d)$, which implies

$$\rho_\mathrm{m} = -\nabla \cdot \mathbf{M} = M\big[\delta(z-d) - \delta(z+d)\big]\big[\theta(x+a) - \theta(x-a)\big] \tag{9}$$

and the magnetic field at the point $\mathbf{r}$ is obtained from Eqs. (1) and (2) as [5]:

$$\mathbf{B}(\mathbf{r}) = \frac{\mu_0 M}{2\pi}\big[\mathbf{B}_*(x, y, z+d) - \mathbf{B}_*(x, y, z-d)\big] \tag{10}$$

where

$$\mathbf{B}_*(\mathbf{r}) = \frac{1}{2}\ln\left(\frac{(x+a)^2 + z^2}{(x-a)^2 + z^2}\right)\mathbf{e}_x + \left[\arctan\left(\frac{x+a}{z}\right) - \arctan\left(\frac{x-a}{z}\right)\right]\mathbf{e}_z \tag{11}$$

For in-plane magnetization, the solution can be obtained by simply rotating the magnetic strip by 90° about the y axis with the corresponding interchange of the x and z coordinates in the previous expression for the magnetic field. That is, Eqs. (10) and (11) become, respectively,

$$\mathbf{B}(\mathbf{r}) = \frac{\mu_0 M}{2\pi}\big[\mathbf{B}_*(x+a, y, z) - \mathbf{B}_*(x-a, y, z)\big] \tag{12}$$

where

$$\mathbf{B}_*(\mathbf{r}) = \frac{1}{2}\ln\left(\frac{x^2 + (z+d)^2}{x^2 + (z-d)^2}\right)\mathbf{e}_z + \left[\arctan\left(\frac{z+d}{x}\right) - \arctan\left(\frac{z-d}{x}\right)\right]\mathbf{e}_x \tag{13}$$

The magnetic field profile resulting from a periodic array of ferromagnetic domain strips is the superposition of the contributions of each of the individual domain strips at that point. In particular, for the situation of Figure 3a, we have

$$\mathbf{B}_\mathrm{total}(\mathbf{r}) = \sum_{n=-\infty}^{\infty} (-1)^n \mathbf{B}(x - nl/2, y, z) \tag{14}$$

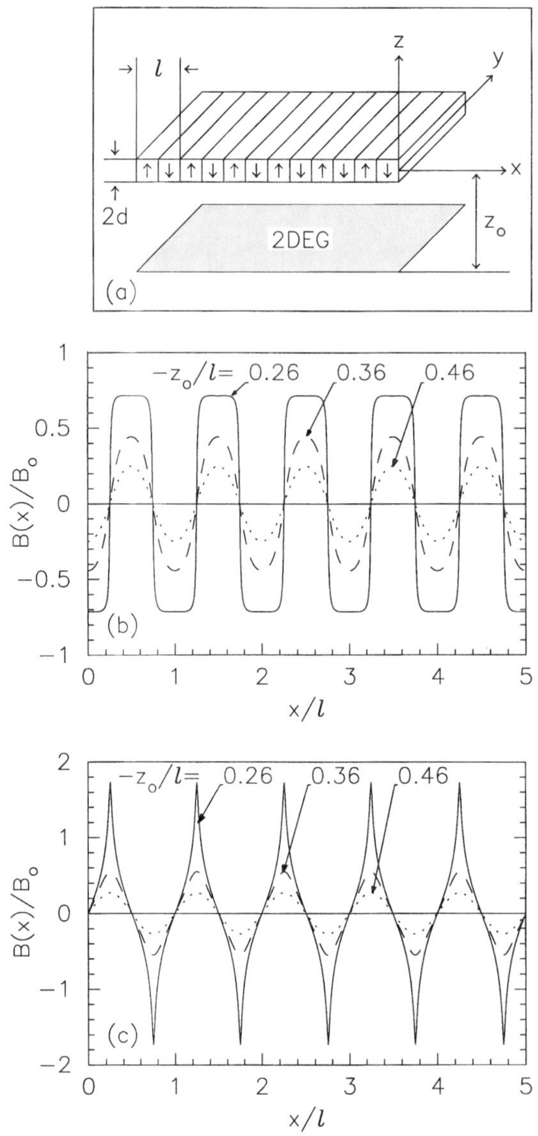

Fig. 3. (a) Superlattice of magnetic strips with alternating sign of magnetization above a 2DEG. (b) Resulting magnetic field in the plane of the 2DEG for three different values of the setback distance. (c) Magnetic field profile when the magnetization of the strips is in plane along the x direction with alternating sign.

where $\mathbf{B}(\mathbf{r})$ is given by Eq. (10) and the period $l = 4a$. Using Poisson's method for the summation of infinite series, we can write the z component of this expression also as

$$\mathbf{B}_z(\mathbf{r}) = \sum_{n=0}^{\infty} \mathbf{B}_n(kz)\cos(kx) \tag{15}$$

with

$$\mathbf{B}_n(kz) = \frac{(-1)^n}{(2n+1)}\left(e^{-kd} - e^{kd}\right)e^{-kz}\mathbf{e}_z \tag{16}$$

where $k = (2n+1)2\pi/l$. This result agrees with the solution obtained directly from Poisson's equation using the Fourier series expansion for ρ_{m} and the method of separation of variables. The resulting magnetic field profile is shown in Figure 3b for different values of the distance z_0 from the area of the magnetic strips.

In the case the magnetic domains are aligned in the plane, but with alternating direction, we have

$$\mathbf{B}_{\text{total}}(\mathbf{r}) = \sum_{n=-\infty}^{\infty} (-1)^n \mathbf{B}(x - nl/2, y, z) \tag{17}$$

but where $\mathbf{B}(\mathbf{r})$ is now given by Eq. (12). Using Poisson's summation method, we find the z component to be

$$\mathbf{B}_z(\mathbf{r}) = \sum_{n=0}^{\infty} \mathbf{B}_n(kz)\sin(kx) \tag{18}$$

with k and $\mathbf{B}_n(kz)$ identical as given previously. The resulting magnetic field profile is shown in Figure 3c for different values of the distance z_0 from the area of the magnetic strips.

Similarly, one can study periodic two-dimensional magnetic strip structures. This was done by Van Roy et al. [6] in which only a numerical solution for the magnetic field profile can be obtained.

4. QUANTUM MOTION IN NONHOMOGENEOUS MAGNETIC FIELDS

First, we will show some remarkable aspects of the energy spectrum and the motion of electrons in nonhomogeneous magnetic fields. We consider a 2DEG moving in the (x, y) plane with a magnetic field $\mathbf{B}$ along the z direction. In the single-particle approximation, such a system is described by the Hamiltonian

$$H = \frac{1}{2m}\left(\mathbf{p} + \frac{e}{c}\mathbf{A}\right)^2 \tag{19}$$

For 1D nonhomogeneous magnetic field profiles, we take the vector potential in the Landau gauge $\mathbf{A} = (0, A, 0)$ and the magnetic field is uniform along the y direction but modulated along the x direction and thus $B_z = B(x) = dA(x)/dx$.

Let us introduce the following characteristic parameters [7]: (i) the frequency $\omega_c = eB_0/mc$ with B_0 some typical magnetic field, and (ii) the length $l_B = \sqrt{\hbar c/eB_0}$. For GaAs and an estimated $B_0 = 0.1$ T, we have $l_B = 813$ Å, $\hbar\omega_c = 0.17$ meV, and $l_B\omega_c = 1.4$ m/s. From now on, we will express all quantities in dimensionless units: (1) the magnetic field $B(x) \to B_0 B(x)$, (2) the vector potential $A(x) \to B_0 l_B A(x)$, (3) the time $t \to t/\omega_c$, (4) the coordinate $\mathbf{r} \to l_B\mathbf{r}$, (5) the velocity $\mathbf{v} \to l_B\omega_c\mathbf{v}$, and (6) the energy $E \to \hbar\omega_c E$.

In these dimensionless units, the two-dimensional (2D) Schrödinger equation becomes

$$\left\{\frac{\partial^2}{\partial x^2} + \left(\frac{\partial}{\partial y} + iA(x)\right)^2 + 2E\right\}\Psi(x, y) = 0 \tag{20}$$

Because of the special form of the gauge, the system is translational invariant along the y direction and, as a consequence, we can choose the following form for the wave function:

$$\Psi(x, y) = e^{-iqy}\psi(x) \tag{21}$$

where $-q = k_y$ is the wavevector of the electron in the y direction, which is a conserved quantity. This does *not* imply that v_y is conserved. The wave function $\psi(x)$ actually satisfies the following 1D Schrödinger equation:

$$\left\{\frac{d^2}{dx^2} - \left(A(x) - q\right)^2 + 2E\right\}\psi(x) = 0 \tag{22}$$

where the function

$$V(x) = \frac{1}{2}\big(A(x) - q\big)^2 \tag{23}$$

can be interpreted as a q-dependent electrical potential. Note that in the case of one-dimensional (1D) magnetic field modulation, there is an analogy between the magnetic field and the potential given by the following relation $B(x) = (1/\sqrt{2V(x)})(dV(x)/dx)$. A jump in the magnetic field will result in a discontinuity in the derivative of the potential $V(x)$.

4.1. Magnetic Step

First, let us consider the most simple shape for a nonhomogeneous magnetic field: the magnetic step. We will show that this leads to a *new type of bound state*. In this situation, the magnetic field fills the half space $x > 0$, which is described by $B(x) = \theta(x)$, with the corresponding vector potential $A(x) = x\theta(x)$, where $\theta(x) = 1$ $(x \geqslant 0)$, 0 $(x < 0)$ is the step function. A continuous version of this magnetic step profile was studied by Calvo [8]. There are two different cases that we have to consider.

In Case 1 ($q > 0$), the potential $V(x) = \frac{1}{2}(x\theta(x) - q)^2$ has the form of an asymmetric quantum well that deepens with increasing q. It is well known that such a well can have a bound state if the well is sufficiently deep. Thus, we have to consider separately: (a) $E < q^2/2$ where bound eigenstates are expected to appear in the region $x \sim q$, and (b) $E > q^2/2$, which corresponds to scattered states, describing the electron reflection by the magnetic step. The appearance of bound states makes this system essentially different from the usual potential step problem where only scattered states exist.

In Case 2 ($q < 0$), the equivalent potential is a constant $V(x) = q^2/2$ for $x < 0$, and a barrier $V(x) = \frac{1}{2}(x\theta(x) + |q|)^2$ in the region $x > 0$, which is unbounded for $x \to \infty$. In this case, there are only scattered states that correspond to electron reflection by the magnetic barrier.

For $x > 0$, the Schrödinger equation takes the form

$$\left\{\frac{d^2}{dz^2} - \frac{z^2}{4} + p + \frac{1}{2}\right\}\psi(z) = 0 \tag{24}$$

with $z = \sqrt{2}(x - q)$ and $p = E - 0.5$. The solutions are the Weber functions $D_p(z)$, which have the following asymptotic behavior: $D_p(z)|_{z\to+\infty} \to 0$. In the region $x > 0$, the wave function becomes $\psi(x) \sim D_{E-0.5}(\sqrt{2}(x - q))$, up to a normalization constant, while, for $x < 0$, there is no magnetic field and the wave function is proportional to $\psi(x) \sim \exp(x\sqrt{q^2 - 2E})$ when $E < q^2/2$. Matching the wave function and its first derivative at $x = 0$, we obtain the following equation:

$$\sqrt{q^2 - 2E} = \frac{d}{dx} \ln D_{E-0.5}\big(\sqrt{2}(x - q)\big)\Big|_{x=0} \tag{25}$$

whose solutions lead to the electron eigenvalues $E = E_n(q)$ with the corresponding wave function $\psi_{nq}(x)$.

Once we know the eigenvalues and the corresponding wave function, we can obtain the other characteristics of the bound states. The average electron velocity of the bound state along the magnetic step (y direction)

$$-v_n(q) \equiv \int_{-\infty}^{\infty} dx\, \psi_{nq}^2(x)\big(q - A(x)\big) = \frac{d}{dq}E_n(q) \tag{26}$$

where the minus sign results from the definition $q = -k_y$. The mean electron position along the x axis $X_n(q)$ has the simple form

$$X_n(q) = \int_{-\infty}^{\infty} dx\, x\psi_{nq}^2(x) = q + \left(1 + \frac{1}{2q\sqrt{q^2 - 2E_n(q)}}\right) v_n(q) \tag{27}$$

The numerical results of the solution of Eq. (25) are depicted in Figure 4a by the solid curves, for the lowest three eigenvalues. These curves start at a certain q value (denoted by the solid dot in Fig. 4a), which is a function of n. The corresponding results for the average electron velocity $v_n(q)$ (solid curves) and mean electron position $X_n(q)$ (dashed curves) are shown in Figure 4b. Notice that the eigenvalues asymptotically, that is, $q \to \infty$, reach the values $(n + \frac{1}{2})$ for Landau levels in a homogeneous magnetic field as it should be. In this asymptotic limit, the mean electron position approaches $X_n(q) \sim q$ and the average

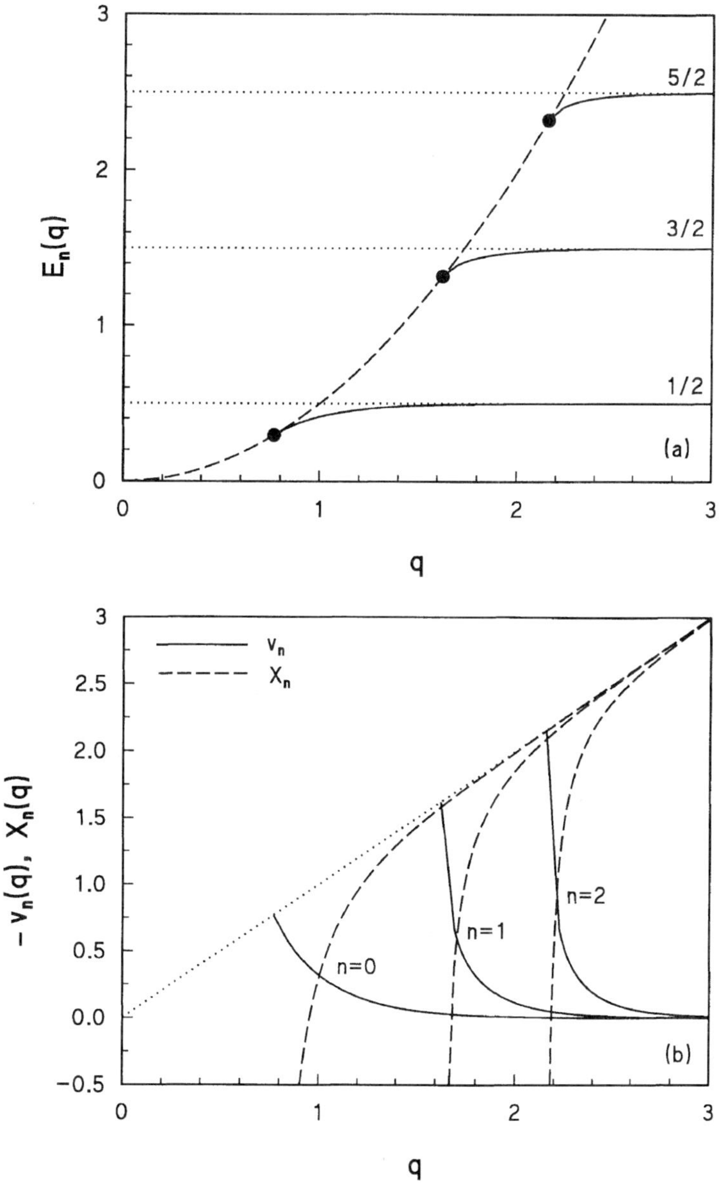

Fig. 4. (a) Energy spectrum for the bound states (solid curves) and (b) corresponding average velocity along the magnetic step $v_n(q)$ (solid curves) and the electron average position $X_n(q)$ along the x axis (dashed curves) for the magnetic step problem.

electron velocity tends to zero. In this limit, the electron is situated far from the magnetic step and is not influenced by the $x < 0$ region. With decreasing q, the electron wave function starts to feel the magnetic step: (1) Its energy decreases, because part of the wave function will be situated in a region with zero magnetic field where the electron will have a smaller kinetic energy; (2) its average position is less than q, because the wave function is sucked into the $x < 0$ region; and (3) its velocity increases and the electron runs along the step. From parts a and b of Figure 4, we notice that the width of the transition region, that is, the q region where $E_n < (n + \frac{1}{2})$, is narrower with increasing n. The preceding properties of these bound states force us to make the analogy with edge states [9]. Nevertheless, there are a number of differences: (1) The available q space for edge states increases with increasing Landau level number n, which is opposite to the behavior of the present bound states; (2) the direction of the velocity is opposite as compared to that of the usual edge states; and (3) the magnitude of the velocity satisfies $|v_n(q)| \leqslant q$, which is different from edge states that do not have an upper bound on their velocity.

From Figure 4, we notice that there exist critical values q_n^* such that for $q < q_n^*$ no bound states are found. These points are indicated by the dots in Figure 4a and are situated on the free-electron spectrum curve $E = q^2/2$ (dashed curve in Fig. 4a). For the plotted curves, we found the critical values $q_0^* = 0.768$, $q_1^* = 1.623$, $q_2^* = 2.155$ at which the eigenenergy curve $E_n(q)$ is tangent to the $E = q^2/2$ curve. At these points, the electron velocity equals the free-electron value $-v_n = q$ and $X_n(q) \to -\infty$.

Notice that in the present magnetic step case, the transmission coefficient is always zero. Independent of the strength of the magnetic field and the magnitude of the electron energy, an electron impeding on the magnetic barrier will always be reflected, which is a consequence of the Lorentz force acting on the electron. In this respect, this system is different from the textbook potential step problem in which the reflection coefficient becomes different from zero when the electron energy is larger than the potential barrier height.

4.2. Magnetic Barrier

The magnetic step can be used as a building block, from which more complicated structures can be built [4, 7, 10]. As a first example, we consider the magnetic barrier in which the magnetic field is different from zero in a strip of width d. In this case, the magnetic field has the following form in dimensionless units: $B(x) = \theta(d^2/4 - x^2)$, and we choose the vector potential as follows: $A(x) = -d/2$ $(x < -d/2)$, x $(|x| \leqslant d/2)$, $d/2$ $(x > d/2)$. The analogous potential $V(x)$ of (23) depends on the value of the wavevector q: When $|q| < d/2$, the potential consists of an asymmetric well of finite height; when $|q| > d/2$, it is a gradual step. The problem is symmetric under the substitution $q \to -q$ (and $x \to -x$) and, consequently, we may limit ourselves to the case $q \geqslant 0$.

There are three different energy regions important to us: (1) $0 \leqslant E \leqslant (d/2 - q)^2/2$, where bound eigenstates can exist; (2) $(d/2 - q)^2/2 < E \leqslant (d/2 + q)^2/2$, which is the reflection region; and (3) $(d/2 + q)^2/2 < E$, where the electron is transmitted through the magnetic barrier.

First, let us concentrate on the situation in which we have bounded electron states. In this case, the electron wave function in the barrier region, that is, $|x| < d/2$, is a linear combination of Weber functions

$$\psi(x) = a D_{E-0.5}\big(\sqrt{2}(x - q)\big) + b D_{E-0.5}\big(\sqrt{2}(q - x)\big) \tag{28}$$

which we must match (and its first derivative) to the free-electron wave functions at the points $x = \pm d/2$. This matching results in a nonlinear equation for the electron energy that was solved numerically. The results for a wide magnetic barrier ($d = 5$) are shown in Figure 5a by the solid curves that end at the solid dots. The latter are situated on the $E = (q - d/2)^2/2$ curve (dashed curve). Notice that the spectrum resembles the one of the magnetic step case (see Fig. 4a) with the distinction that the latter has an infinite number of branches, while the one for a magnetic barrier has a finite number of bound states for each q. For $d = 5$, there are only three branches in the energy spectrum. The number of energy branches decreases with decreasing barrier width d. Irrespective of the value of d, there is always *at least one* discrete energy value for $q = 0$. This is a consequence of the fact that for $q = 0$ the potential $V(x)$ is one dimensional and symmetric. Such a potential is known [11] to have at least one discrete eigenvalue irrespective of the size of the potential well. The value of the lowest branch in the spectrum is plotted in Figure 5b for $q = 0$ as function of the barrier width d. Notice that when $d < 1$ (i.e., when

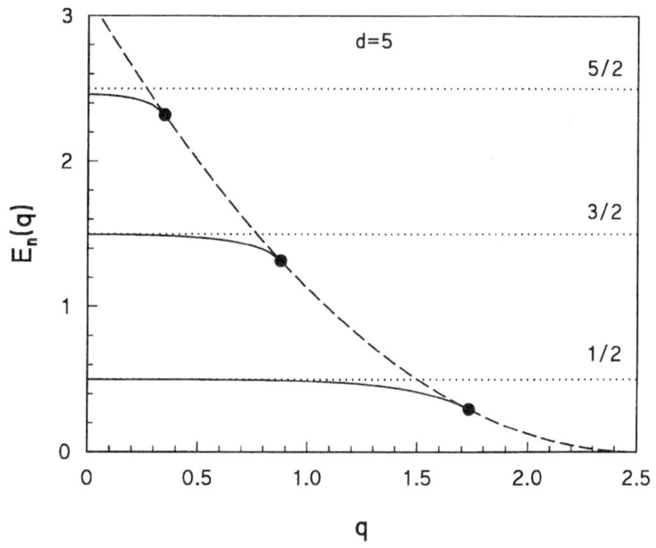

(a)

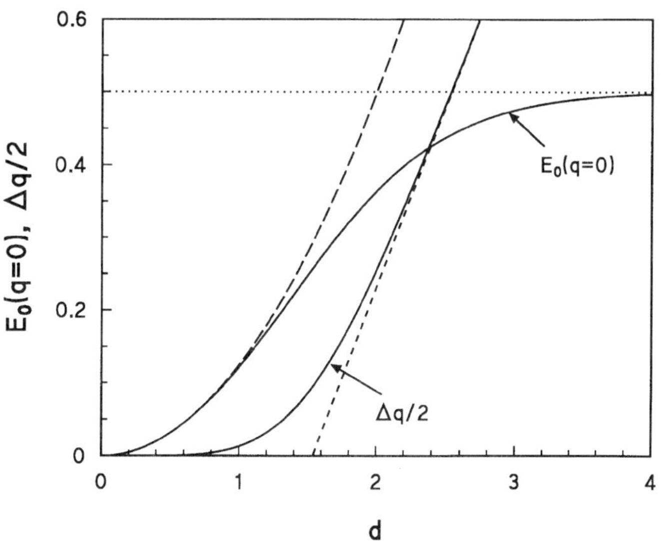

(b)

Fig. 5. (a) Energy spectrum for the bound states (solid curves) in a magnetic barrier. The dashed curve indicates the free-electron spectrum. (b) Lowest eigenvalue of the bound state in a magnetic barrier as function of the barrier width for $q = 0$ and the width Δq of the lowest energy branch in q space.

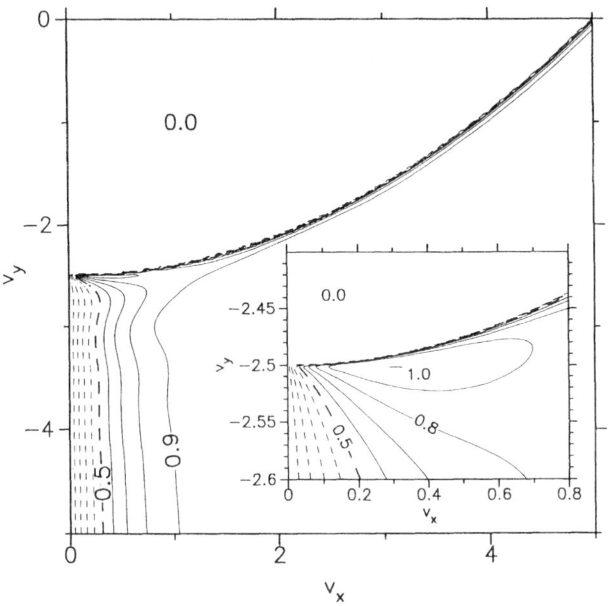

Fig. 6. Contour plot of the transmission coefficient through a magnetic barrier in the incident electron velocity space (v_x, v_y).

the magnetic barrier width is less than the magnetic length l_B) the eigenvalue approaches $E_0(q=0) \approx (d/2)^2/2$, which is shown by the long-dashed curve in Figure 5b. Although the electron is bound to the barrier, in the case of small d values the electron wave function is situated mainly outside the barrier and, consequently, its energy approaches the height of the potential well $V(d/2)$. The width in q space (Δq) of the lowest energy branch is also given in Figure 5b. It is seen that this width decreases rapidly to zero when $d < 1$ and in the opposite case (when $d \to \infty$) it asymptotically reaches the line $\Delta q/2 = d/2 - q_0^*$ (short-dashed line), where $q_0^* = 0.768$ is the value as obtained from the magnetic step spectrum. Another distinction as compared to the magnetic barrier spectrum (see Fig. 4a) is that the energy eigenvalues are smaller in magnitude than those in the magnetic step case.

For the unbounded states, we have calculated the transmission coefficient, which now depends not only on the electron energy but also on the electron wavevector q in the y direction. In the present case, tunneling is a two-dimensional process in which the total electron wavevector and the electron energy is conserved but the direction of the wavevector is altered. A contour plot of the transmission coefficient $T(q, E)$ versus the initial electron velocity components (v_x, v_y) is shown in Figure 6 for a magnetic barrier of width $d = 5$. The quantum transition coefficient is zero above the line $v_y = (v_x^2 - d^2)/2d$, which is the result one would obtain from classical mechanics and which defines a semiinfinite transmission window. Below this line, we have classically $T = 1$, but quantum mechanically $T(q, E)$ gradually increases with increasing electron energy. For rather thick barriers (as in the case of $d = 5$), there is some additional structure at low energy, which is enlarged in the inset of Figure 6. There is an additional peak around $(v_x, v_y) = (0.3, -2.5)$, which is a consequence of the presence of a resonant energy level above the quantum well $V_q(x)$.

The continuous magnetic barrier problem with magnetic field profile $\mathbf{B} = (0, 0, B/\cosh^2((x - x_0)/\delta))$ was discussed by Hudák [12], who was able to solve the single-particle problem for this case exactly. The same discussion was given later on by Calvo [8]. The situation of a magnetic barrier in a ballistic quantum wire was discussed in [13, 14], where the conductance was calculated and found to be quantized and to oscillate with changing magnetic field. Ramaglia and co-workers [15, 16] studied the related problem of quantum tunneling of a 3DEG through a potential barrier in a transverse magnetic field where the magnetic field has the same extension as the barrier. They found that for sufficiently wide barriers the tunneling current and the differential conductivity exhibits steps and resonances as a function of the applied voltage. A model consisting of a continuous potential well together with a continuous magnetic barrier was discussed by Calvo [17, 18], who obtained exact expressions for the energy spectrum.

4.3. Magnetic Quantum Well

The inverse situation of the previous problem is the magnetic well case, which we will discuss now. Because of the essential 2D character of the electron motion in a magnetic field, we should rather speak of a magnetic wire. In dimensionless units, the magnetic field is given by $B(x) = 0$ $(|x| \leqslant d/2)$, 1 $(|x| > d/2)$, and the corresponding vector potential is $A(x) = x - d/2$ $(x > d/2)$, 0 $(|x| \leqslant d/2)$, $x + d/2$ $(x < -d/2)$. The value of the vector potential is now unbounded, that is, $A(x)|_{x\to\pm\infty} \to \pm\infty$, and, as a consequence, the potential satisfies $V(x)|_{x\to\pm\infty} \to \infty$, which implies that the electron motion is confined in the x direction and *all the states are bound* at least in this direction.

The corresponding wave functions are constructed by matching the quasi-free-electron wave function in the region $|x| < d/2$ with the Weber functions $\psi(x) = D_{E-0.5}(\pm\sqrt{2}(x \mp d/2 - q))$, which are valid in the regions $|x| > d/2$. This matching of the wave function and its first derivative leads to a nonlinear algebraic equation for the eigenvalues that is solved numerically.

The results of the numerical solution of this equation are presented in Figure 7 for a wide well (i.e., $d = 5$). There are clearly two distinct regions that are separated by the free-electron energy $E = q^2/2$ curve (dashed curve in Fig. 7). For $E \ll q^2/2$, the energy spectrum consists of Landau levels. The electron is mainly located in the barrier where there exists a uniform magnetic field. For small q values, that is, $E \gg q^2/2$, the spectrum consists of bands with free-electron-like motion in the y direction. This is similar to the case of the well-known quantum wire with electrical potential barriers. When we decrease the width of the well, the two regions are less distinct. For example, for $d = 1$ the well is narrower than the width of the electron wave function and, consequently, there is always an appreciable overlap of the wave function with the magnetic barrier region and the energy

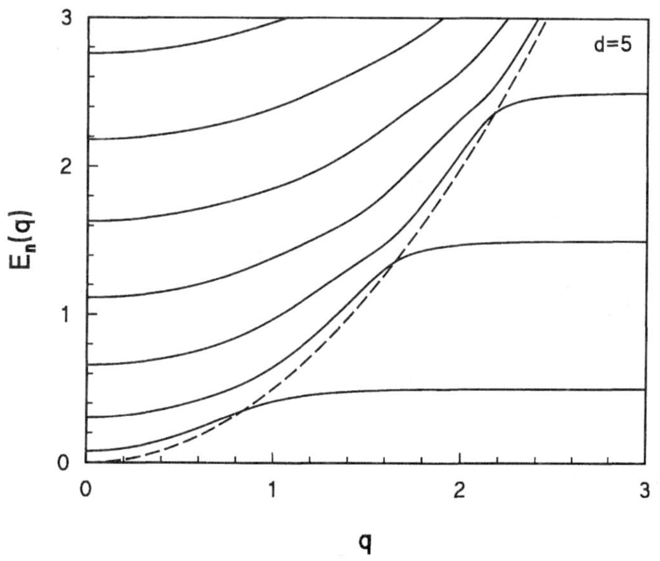

Fig. 7. Energy spectrum of a magnetic well of width $d = 5$.

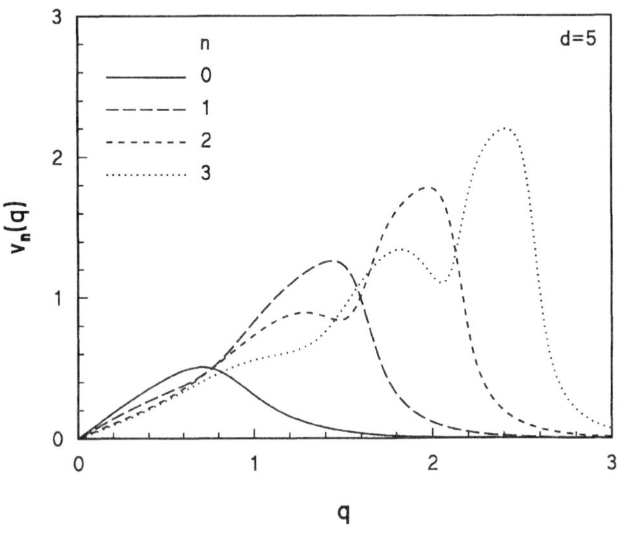

Fig. 8. Electron average velocity corresponding to the energy spectrum of Figure 7 for the different branches.

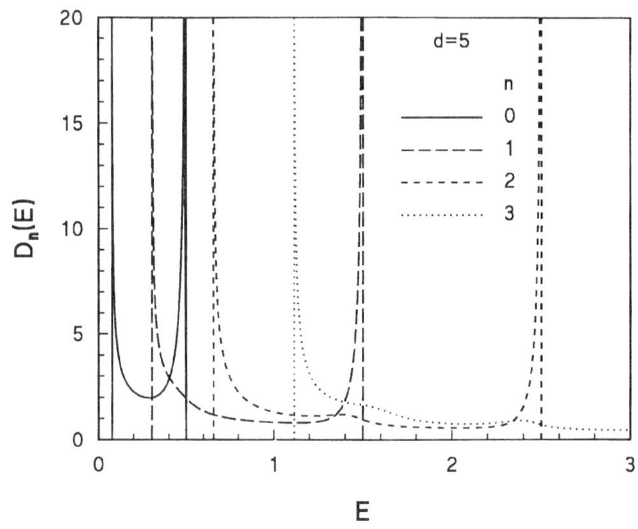

Fig. 9. Density of states of the one-electron states in the magnetic well corresponding to Figure 7.

levels will have almost no dispersion (see, e.g., [7]). The different behavior between the two cases is also illustrated in Figure 8 where the electron velocity is shown for the different states. Notice that the velocity exhibits a maximum near $E = q^2/2$ and it diminishes fast for $q \gg \sqrt{2E}$, which is the region where the electron is mainly located inside the magnetic barrier. For wide wells, the velocity curve $v_n(q)$ can have several local maxima, which is a consequence of the repulsion of the different energy levels as shown in Figure 7. In the case of the usual quantum wire constructed from walls consisting of potential barriers, the electron velocity is $v_n = \hbar k_y = -q$ and is independent of the energy level index n and is a uniform increasing function of the electron wavevector. The behavior of $v_n(q)$ as depicted in Figure 8 is also different from that of the edge states in which $v_n(q)$ is a uniform increasing function of q.

The density of states (DOS) is depicted in Figure 9. Notice that, as in the usual quantum wire case, the DOS exhibits singularities at the onset of each energy level. There is a difference, however, in that the width in energy space of each level is finite and bounded by a singularity in the DOS. Suppose we have a system in which we are able to increase the Fermi energy gradually. Starting from zero, we first populate the quantum wire states; the electrons are mainly situated in the well region. Further increasing the Fermi energy, we see that for $d = 5$ we first start to populate the next energy level which consists initially of states located inside the well. For a narrower well, for example, $d = 1$, on the other hand, we start to populate states that are situated in the magnetic barrier region and that are nothing else then 2D Landau states. Thus, by changing the Fermi level, we are able to have 1D states or 2D states at the Fermi level, which will have considerable influence on the electrical properties of the system. The 1D states are quasi-free, while the 2D states are localized on Landau orbits and can only move if scattering is involved.

4.4. Resonant Tunneling Structures

In previous sections, we have made a detailed study of the nature of the electron states in different magnetic barrier and well structures. In this section, we will consider different tunneling structures where we will focus on the tunneling current going through them [10, 19].

For simplicity, we now consider electron tunneling through a magnetic barrier of constant height B_0 and width $d = x_+ - x_-$ surrounded by regions of zero magnetic field. The free-electron wave function on the left-hand side of the barrier ($x < x_-$) is $\psi_-(x) = A\exp[ik_-(x - x_-)] + B\exp[-k_-(x - x_-)]$ and on the right-hand side of it ($x > x_+$) $\psi_+(x) = \exp[ik_+(x - x_+)]$, where $k_\pm = \sqrt{2[E - V(\pm\infty)]}$ is the x component of the electron wavevector on the corresponding side of the barrier. Under the barrier, there are two solutions for $\psi(x)$ that can be written as a linear combination of the Weber function $D_p(x)$ and its derivative $D'_p(x)$. Next, we construct the transition matrix

$$T(x, x_0) = \begin{pmatrix} u(x) & v(x) \\ u'(x) & v'(x) \end{pmatrix} \tag{29}$$

where we defined the functions $u(x) = c\{D'_p(\sqrt{2q})D_p(z) + D'_p(-\sqrt{2q})D_p(-z)\}$ and $v(x) = c\{D_p(\sqrt{2q})D_p(z) - D_p(-\sqrt{2q})D_p(-z)\}$, with $p = E - \frac{1}{2}$ and $z = \sqrt{2}(x - q)$, which satisfies the boundary conditions $u(x_0) = 1$, $u'(x_0) = 0$, $v(x_0) = 0$, and $v'(x_0) = 1$. Matching the wave function at the edges of the barrier, $x_\pm$, by means of the preceding matrix, we obtain

$$A = T_{11}^{-1} + \frac{k_+}{k_-}T_{22}^{-1} + i\left(\frac{1}{k_-}T_{21}^{-1} - k_+T_{12}^{-1}\right) \tag{30}$$

The electron transmission through the barrier $t(E, q)$ is given by

$$t(E, q) = \frac{k_+}{k|A|^2} \tag{31}$$

where T^{-1} stands for the inverse of the matrix $T \equiv T(x_+, x_-)$. For complex structures involving several barriers of constant height, the total T matrix is a product of the T matrices that correspond to the separate barriers and the one describing the free-electron propagation between the barriers. As for the electron current through such a structure, it can be calculated, in the ballistic regime, by introducing the conductance G as the electron flow averaged over half the Fermi surface [20]:

$$G = G_0\int_{-\pi/2}^{\pi/2} t\big(E_F, \sqrt{2E_F}\sin\phi\big)\cos\phi\, d\phi \tag{32}$$

where ϕ is the angle of incidence relative to the x direction. Furthermore, $G_0 = e^2mv_Fl/\hbar^2$, where l is the length of the structure in the y direction and v_F is the Fermi velocity.

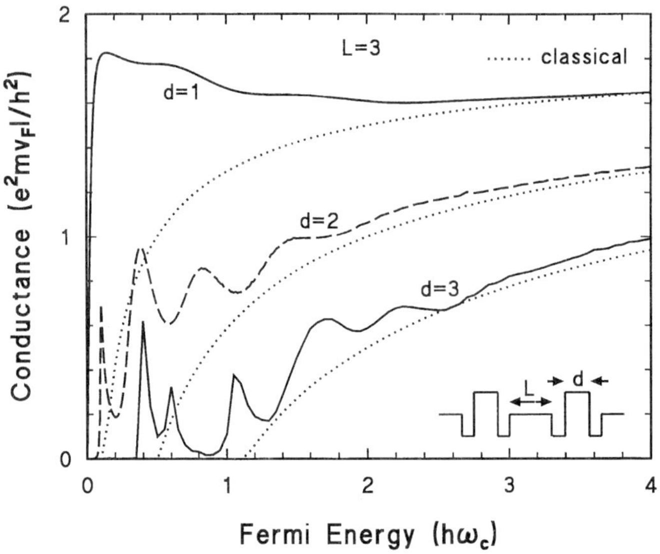

Fig. 10. Conductance through the barrier structure shown in the inset for different values of the barrier parameter. The dotted curves are the results from a classical calculation.

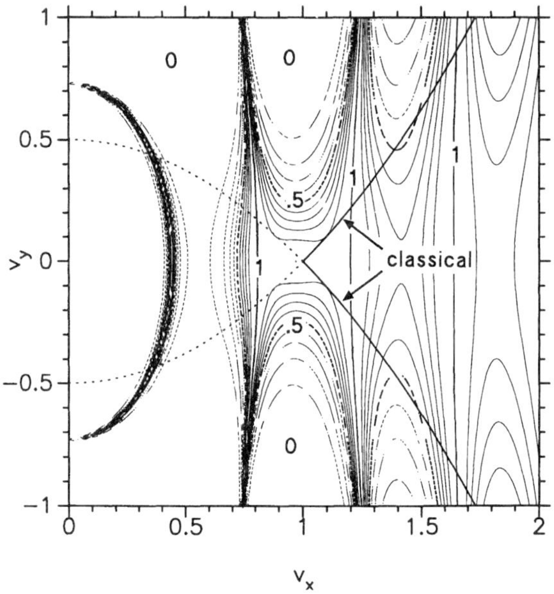

Fig. 11. Contour plot of the electron transmission probability in the (v_x, v_y) plane for the structure of Figure 10 with $d = 2$ and $L = 3$ together with the corresponding classical result.

To reveal the main qualitative features of tunneling through these barriers, we restrict ourselves to complex structures composed of rectangular magnetic barriers that are used as a building block to make a double-barrier-like structure. The contour plot of the transmission through a complex structure, shown in the inset of Figure 10, is presented in Figure 11 for $L = 3$ and $d = 1$. Notice that the quantum calculation and the classical calculation give drastically different results. In the classical calculation, the electrons satisfy Newton's equation, which is solved in the presence of the magnetic barriers, that is, for a spatially dependent magnetic field. Notice the sharp resonances in the velocity contour plot and the strong transmission anisotropy. As can be seen, the system exhibits *wavevector filtering* properties.

Having seen the transmission results, one may wonder to what extent their structure is reflected in measurable quantities that involve some kind of averaging. In Figure 10, we show the conductance, as given by Eq. (32), for the previous tunneling structure shown in the inset of the figure, together with the corresponding classical result (dotted curves). Despite the averaging of $t(E, q)$ over half the Fermi surface, we have again a strong resonant structure. This structure will become sharper if one can select the wavevectors that give the sharpest resonance in the transmission. As for the classical result, we see again that they are determined only by the first barrier in each structure.

Although our consideration of electron tunneling through rectangular magnetic barrier structures gives only a qualitative picture, nevertheless these resonant tunneling spikes should be present in the more realistic cases with barriers of smooth shape. This was studied in [21]. The effect of spin, which was neglected here, on quantum tunneling was studied in [22] and the influence of the asymmetry of the magnetic barriers was investigated in [23]. Indeed, these resonant spikes do not depend on the actual shape of the magnetic barrier but only on the presence of barriers in the potential $V(x)$.

Other magnetic field profiles were studied by several groups. Vil'ms and Éntin [24] presented a theoretical analysis of the energy spectrum of 2D electrons near domain walls and near narrow parallel magnetic strips. The latter system is the continuous version of the single and multiple magnetic barrier system discussed here and in [4, 7]. Linear varying magnetic fields in one direction were studied in [25–27]. The energy spectrum and the transport properties of a parabolic varying magnetic field in a quantum wire and in a Hall cross were studied in [28]. A circular symmetric magnetic field profile was considered by Grosse et al. [29], who proved some general properties of the energy spectrum. A problem related to the motion of electrons in nonhomogeneous magnetic fields is the one of electrons moving on different topological surfaces in the presence of a homogeneous magnetic field. Electrons moving on a sphere in an axial magnetic field were studied in [30, 31]. Foden et al. [32] proved that a curved 2DEG in a uniform magnetic field induces quantum magnetic confinement.

4.5. Magnetic Dot

Microfabrication techniques have made it possible to further confine a 2DEG in a quantum dot through built-in electrostatic potentials. Such quantum dots have a discrete spectrum and because of their analogy with atoms are called artificial atoms. Using nonhomogeneous magnetic fields, it is also possible to confine electrons in a dot structure. One way to realize this is by considering a system consisting of a heterostructure with a superconducting disk on top of it. This new quantum dot system is fundamentally different from the usual quantum dot system because now (1) the electrons are confined magnetically, (2) the confinement potential is inherently nonparabolic, and (3) the filling of the dot with electrons is a discrete function of the strength of the confinement.

First, we consider the magnetic field profile around a superconducting disk with radius a. We solve Maxwell equations outside the disk: $\operatorname{div}\mathbf{B} = 0$ and $\operatorname{rot}\mathbf{B} = 0$. In analogy with the electric field case, we can introduce a magnetic potential $\Phi(x, y, z)$, which is related to the magnetic field through $\mathbf{B} = -\operatorname{div}\Phi$. This reduces the problem to the solution of the Poisson-like equation $\Delta\Phi = 0$ with the boundary condition $B_z = \partial\Phi/\partial z = 0$ taken at the surface of the disk. The resulting magnetic field profile is shown in Figure 12 for different values of the distance between the 2DEG and the superconducting disk. Notice that the magnetic field under the disk is very small because of the Meissner effect, while far from the disk it becomes equal to the external magnetic field strength B_0. At the edge of the disk, there is an overshoot of the magnetic field strength, which becomes larger with decreasing value of z/a. When the 2DEG is farther away from the superconducting disk, almost no overshoot appears and there is a gradual increase of B when we go away from the center.

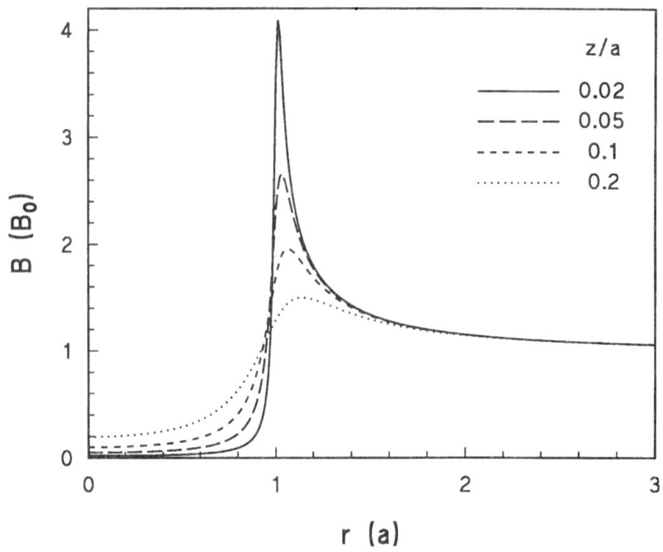

Fig. 12. Magnetic field profile at a distance z under a superconducting disk of radius a.

In principle, we can use the previous magnetic field profile, $\mathbf{B}(\rho = \sqrt{x^2 + y^2}, z)$, to determine the vector potential, $\mathbf{A}(\rho, z)$, which we have to insert into the Schrödinger equation $[-i(\hbar/m^*)\nabla - (e/c)\mathbf{A}(\mathbf{r})]^2\Psi = E\Psi$. The problem can only be solved numerically. Furthermore, we have an additional parameter in our problem: z/a, the distance between the superconducting disk and the 2DEG. Therefore, we found it more convenient to consider two-model systems, which correspond to two extreme situations but which contain the essential physics of the problem. Because of the cylindrical symmetry of the system, we introduce cylindrical coordinates (ρ, φ). The vector potential is given by $\mathbf{A} = \mathbf{B} \times \mathbf{r}/2$. In the cylindrical coordinate system, we have $\mathbf{A} = (A_r, A_\varphi)$ with $A_r = 0$ in case the magnetic field is directed along the z axis with $B_z = (\partial(\rho A_\varphi)/\partial\rho)/\rho$. The two models we consider [33] are defined by the following profiles: (1) $B_z = B(\rho) = B_0\theta(\rho - a)$, which is a limiting case for the situation that the distance between the 2DEG and the superconducting disk is large. It leads to the vector potential $A_\varphi = B_0a^2/2\rho$ $(\rho < a)$, $B_0\rho/2$ $(\rho > a)$. The spectrum of this system was discussed in [33] and also in [34, 35] where the analogy with classical paths was pointed out; (2) $A_\varphi = (B_0\rho/2)\theta(\rho - a)$, which results in a magnetic field profile with a delta function overshoot: $B(\rho) = 0$ $(\rho < a)$, $(B_0\rho/2)\delta(\rho - a)$, B_0 $(\rho > a)$. In [33–35], one refers to the present system as the magnetic dot system, stressing the magnetic confinement in this quantum dot. From the magnetic field point of view, one has a circular hole inside a uniform magnetic field profile and, therefore, one can also call the present system a magnetic antidot.

The wave function for the single-electron states can be written as $\Psi(\rho, \varphi) = e^{im\varphi}\psi(\rho)$ where the radial part is determined by the Schrödinger equation

$$\left(\frac{1}{\rho}\frac{d}{d\rho}\rho\frac{d}{d\rho} - V(m, \rho) + 2E\right)\psi(\rho) = 0 \tag{33}$$

with the angular momentum–dependent effective potential $V(m, \rho) = (A_\varphi(\rho) - m/\rho)^2$. The numerical results for the energy spectrum are shown in Figure 13a for the case with overshoot and in Figure 13b for the model without overshoot. Although the limiting behavior of the spectrum for $B_0 \to 0$ and $B_0 \to \infty$ is very similar for both models, the intermediate field behavior is very different. This distinct behavior is made more visible by plotting the energy in units of $\hbar\omega_c$ as a function of $a/l_B \sim \sqrt{B_0}$. This different behavior between the two models can be understood as follows. For $m > 0$, the electron wave func-

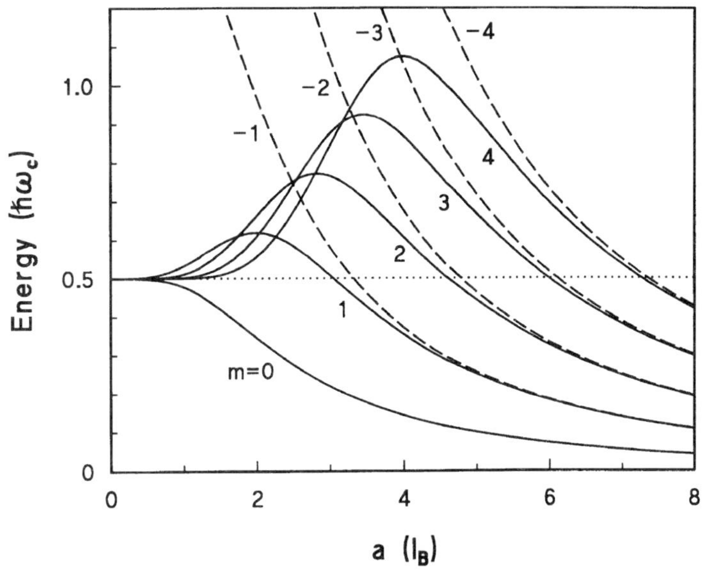

(a)

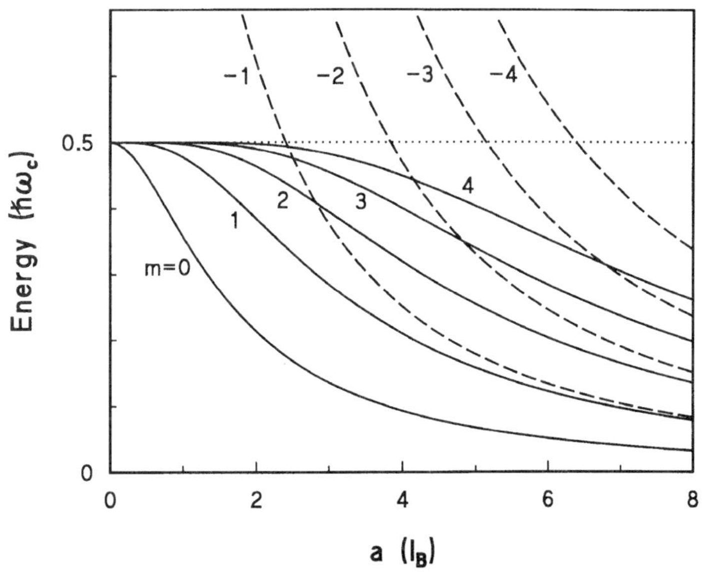

(b)

Fig. 13. One-electron energy levels in the magnetic quantum dot: (a) with magnetic field overshoot near the edge of the dot and (b) without overshoot.

tion exhibits a maximum at $\rho = \rho^* > 0$. With increasing a/l_B, this maximum shifts toward the center of the dot. In case there is an overshoot in the magnetic field profile, when the maximum of the electron wave function is situated near $\rho = a$, the electron energy will be increased, which results in a local maximum as shown in Figure 13a. From this interpretation, it is easy to understand that the maximum in $E_{n,m}/\hbar\omega_c$ shifts to larger a/l_B values with increasing angular momentum m. For large values of a/l_B, most of the wave function will be situated in the dot where there is no magnetic field and, consequently, the energy decreases.

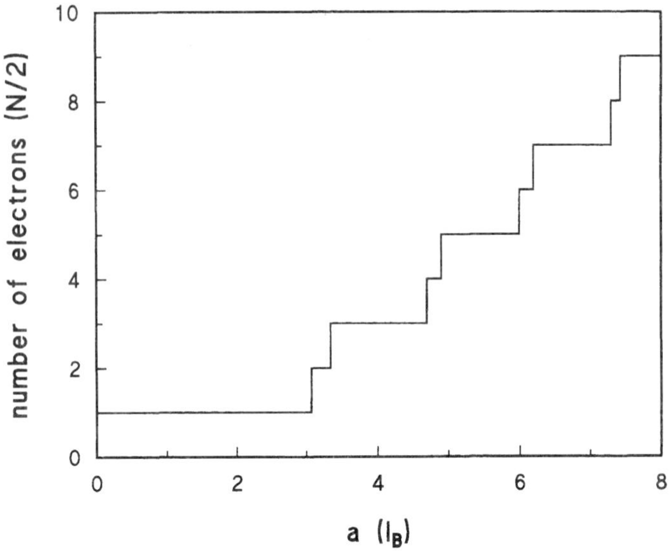

Fig. 14. Filling of the magnetic dot in case of magnetic field overshoot.

This behavior of the electron energy has important consequences for the filling of the dot. Outside the quantum dot, the magnetic field is B_0 and the electron lowest energy state is $\hbar\omega_c/2$. An electron will only be situated in the dot when its energy is lower than in the region outside the dot. For the moment, we will neglect the electron–electron interaction. From Figure 13b, we note that for a dot without magnetic field overshoot at its edges there are an infinite number of states, that is, the states with $m \geqslant 0$ for $n = 0$, which have an energy less than $\hbar\omega_c/2$ and, consequently, the electrons will be attracted toward the dot. For the system with magnetic overshoot, the situation is totally different. The $|0, 0\rangle$ state has energy below $\hbar\omega_c/2$ and two electrons (two because of spin) will be able to occupy the dot. When we add more electrons to the system, we see that for small a/l_B values, the electrons are repelled and are forced outside the dot. Only when a/l_B is sufficiently large will quantum dot states become available with energy less than $\hbar\omega_c$. Consequently, as a function of a/l_B, we observe a discrete filling of the dots. This is depicted in Figure 14 where we show the number of electrons in the dot as a function of $a/l_B \sim \sqrt{B_0}$.

Including the real magnetic profile as shown in Figure 12 will not alter our conclusions qualitatively. For example, the discrete filling of the dot as shown in Figure 14 will still be present. The exact position at which the number of electrons jump to higher values will be a function of the exact magnetic field profile, and, in particular, it will strongly depend on the sharpness of the magnetic overshoot.

5. DIFFUSIVE TRANSPORT OF ELECTRONS THROUGH MAGNETIC BARRIERS

In previous sections, it was assumed that the electron transport through the magnetic barriers was ballistic, and, consequently, the electrons exhibit quantum tunneling. This is valid when the width of the barriers (W) is much smaller than the mean free path (l_e) of the electrons and the cyclotron radius $R_c > W$. For wide magnetic field barriers, the charge transport is better described by a diffusive theory, which is also valid at higher temperature where the mobility is low and the electron motion is diffusive.

Until now, no experiments were available of quantum transport through single or multiple magnetic barriers. Recently, Leadbeater et al. [36] reported an alternative technique to

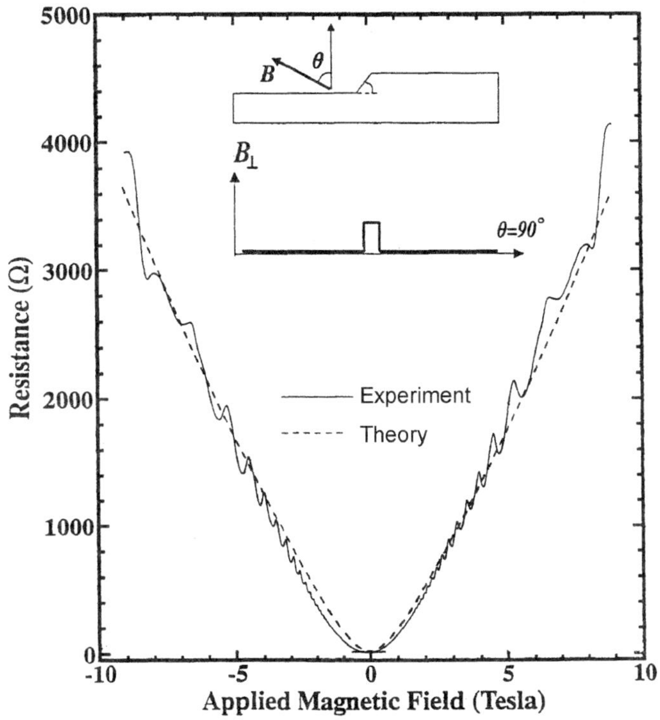

Fig. 15. Experimental (solid curve) and theoretical (dashed curve) results for the magnetoresistance as a function of the applied magnetic field. The magnetic barrier (inset) was created by faceting the 2DEG plane where the facet makes an angle of 20° to the substrate and the applied magnetic field is in the plane of the substrate (i.e., $\theta = 90°$).

produce effective spatially varying magnetic fields of much larger strength and gradients than could be obtained by lithographic patterned superconducting or ferromagnetic films. They constructed a nonplanar two-dimensional electron gas (2DEG) that was fabricated by growth of a GaAs/(AlGa)As heterojunction on a wafer prepatterned with facets at 20° to the substrate. Applying a uniform magnetic field (B) produces a spatially nonuniform field component perpendicular to the 2DEG (see inset of Fig. 15). With the field in the plane of the substrate, an effective magnetic barrier is created located at the facet. The resistance measured across such an etched facet showed oscillations that are periodic in $1/B$ and that are on top of a positive magnetoresistance background, which increases quadratically with the magnetic field for small B and quasi-linearly in B for large B (see full curve in Fig. 15).

5.1. Theoretical Formalism

To explain, quantitatively, the main features of the experimental measurements of [36–38], namely, the smooth background of the magnetic field dependence of the resistance, we will rely on a classical model for magnetotransport through such a wide (typically on the order of a micrometer) magnetic barrier. This theory is also valid for 2D diffusive transport in case magnetic barriers are created through the structuring of ferromagnetic material on top of a heterostructure. The 2DEG situated in the (x, y) plane is bounded by the edges of the Hall bar where a small part of the 2DEG is subjected to a perpendicular magnetic field in the z direction. The $B_z \neq 0$ region corresponds to the facet region in the experimental system, that is, $B_z = B\sin(\theta)$, $\theta = 20°$ is the facet angle, and B is the externally applied magnetic field in the plane of the substrate. To calculate the spatial distribution of the electrostatic potential, the electric field, and the current density, we start from the stationary continuity equation, which expresses charge conservation

$$\nabla \cdot \mathbf{J} = 0 \tag{34}$$

which is supplemented by Ohm's law

$$\mathbf{J} = \sigma \mathbf{E} \tag{35}$$

In the steady state we have $\nabla \times \mathbf{E} = 0$ and the electric field can be written as the gradient of a potential, that is, $\mathbf{E} = -\nabla\phi$. The previous system of equations reduces to the following 2D elliptic partial differential equation for the electrical potential ϕ:

$$\nabla \cdot \left[\sigma(x, y)\nabla\phi(x, y)\right] = 0 \tag{36}$$

where $\sigma(x, y)$ is a spatially dependent conductivity tensor. For a homogeneous system and in the absence of a magnetic field $\sigma(x, y) = \text{constant}$, we recover the Laplace equation. In our case, the conductivity tensor is no longer constant owing to the presence of the finite magnetic barrier:

$$\sigma(x, y) = \begin{pmatrix} \sigma_{xx} & \sigma_{xy}(x, y) \\ \sigma_{yx}(x, y) & \sigma_{yy} \end{pmatrix} = \frac{\sigma_0}{1 + (\mu B_z(x, y))^2} \begin{pmatrix} 1 & \mu B_z(x, y) \\ -\mu B_z(x, y) & 1 \end{pmatrix} \tag{37}$$

where $\sigma_0 = n_s e\mu$ is the Drude conductivity and $B_z = 0$ outside the facet. The 2D partial differential equation is cast into a finite-difference form and solved numerically [39] using the accelerated Gauss–Seidel iteration scheme with the boundary conditions $\phi(x, 0) = 0$ and $\phi(x, L) = V_0$ (L is the length of the sample and V_0 is the applied voltage) and the condition that no current can flow out the sides of the sample, that is, $j_x = 0$ for $x = 0$, and $x = W$. The distances are normalized by the width of the sample (typically $W = 40$ μm taken along the x axis) and the voltages are normalized by the total voltage drop between the current probes. The magnetoresistance between any two points a and b along one side of the sample is given by $R = V_{ab}/I_{cd}$, with the voltage drop $V_{ab} = \phi(x, b) - \phi(x, a)$, and the total current flowing normal to the facet is obtained through $I_{cd} = \int_c^d j_y(x, y)\,dx$, where c and d are any two points on the opposite sides of the sample. The Hall resistance at a distance y along the sample length is given by $R_H = V_H/I_{cd}$, where $V_H = \phi(W, y) - \phi(0, y)$.

There exist alternative approaches to solve the differential equation (36). Badalian et al. [40] used the conformal mapping method in order to find an analytic expression for the magnetic potential in an infinite long wire in the presence of a magnetic barrier. Jou and Kriman [41] developed a spectral expansion approach in which the magnetic potential was expanded into an orthogonal basis, which consisted of solutions of the Laplace equation in the different regions that satisfy the boundary conditions. Continuity of the magnetic potential at the boundary between the separate regions determined the expansion coefficients. In [41], this approach was applied to a single magnetic barrier and to a periodic array of them.

5.2. Single Magnetic Barrier

In Figure 15, we show both the experimental (solid curve) and the theoretical (dashed curve) traces for the magnetoresistance across a single facet of width 3 μm where the voltage probes are situated 10 μm apart across the facet and the current probes are more than 900 μm apart for a sample with width $W = 40$ μm. Notice that apart from the Shubnikov–de Haas (SdH) oscillations, which result from the quantizing effect of the magnetic field

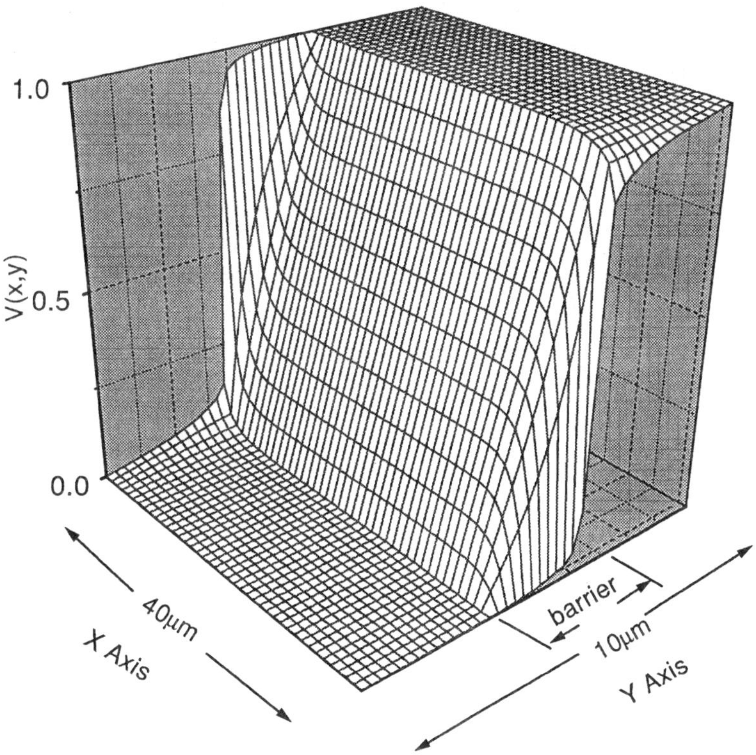

Fig. 16. Potential distribution in the sample for $B = -2$ T.

at low temperature, the theoretical curve accounts nicely for the overall behavior of the magnetoresistance. The experimental curve is slightly asymmetric around $B = 0$, which is due to the fact that the voltage probes are not exactly equidistant from the facet. The classical origin of the positive magnetoresistance was confirmed experimentally where it was found that it persists even for temperatures above 100 K. Note that the experimental configuration is effectively a two-terminal measurement where the measured resistance is determined by the Hall resistance as well as the magnetoresistance. For small B fields, the Hall resistance is small and thus the resistance is determined by the magnetoresistance and is, consequently, quadratic in B. For larger magnetic fields, a quasi-linear behavior of the resistance as a function of B is found, which is due to the fact that now the Hall resistance mainly limits the current.

The theoretical electric potential distribution in the sample is shown in Figure 16 for an applied magnetic field of $B = -2$ T, which gives $B_z = -2\sin(20°) = -0.6840$ T. Notice that almost all the potential drop takes place across the magnetic barrier. In the barrier region and just outside it, there is a voltage difference between the edges of the sample (i.e., across the x axis), which is nothing else than a spatially dependent Hall voltage. This is in accord with the concept that the ($B_z = 0$ regions) can be thought of as extended high mobility contacts to a short and wide Hall bar (the facet region) that tends to short out most of the voltage immediately outside the facet region. Particularly interesting is the development of the Hall voltage between the opposite edges of the facet. This becomes very small but nonzero outside the facet region and gives a steep increase of the Hall potential profile at the edges of the $B_z \neq 0$ region, which is reminiscent of the potential profile investigated experimentally and theoretically in [42, 43] in a conventional Hall bar under the conditions of the quantum Hall regime and in the middle of a plateau in the Hall resistance.

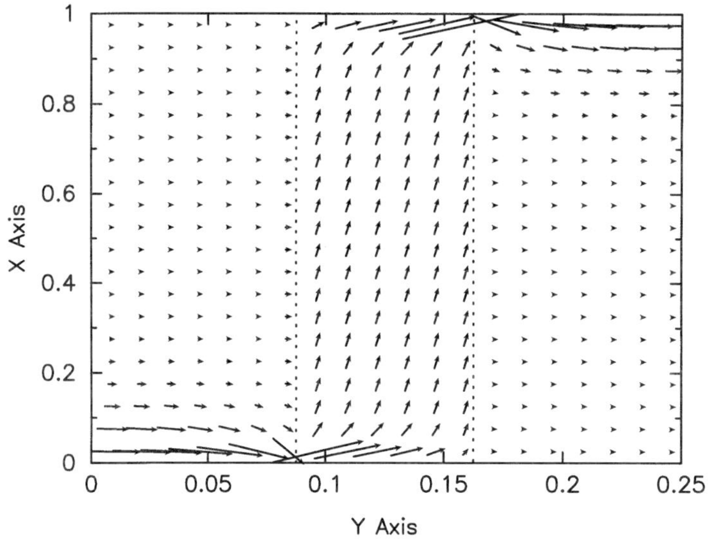

Fig. 17. Current flow in the sample for $B = -2$ T corresponding to the situation of Figure 16. The magnetic barrier region is delimited by the two dashed lines.

The spatial distribution of the components of the electric field are obtained as follows: $E_x = -\partial\phi(x, y)/\partial x$ and $E_y = -\partial\phi(x, y)/\partial y$. From Figure 16, it is clear that both components are very small outside the barrier region. Inside the magnetic barrier, E_x becomes very large close to the edges especially at the diagonally opposite corners and vanishingly small in the middle where E_y is finite and more uniform, singular at the diagonally opposite corners, and very small at the other two corners. Accordingly, the largest part of the current will enter the magnetic barrier region from the corner where both electric field components are large and exit the barrier from the diagonally opposite corner (see Fig. 17). Once inside the barrier region, the guiding center of the electron cyclotron orbits will drift along the equipotential lines (see Fig. 16) according to the $\mathbf{E} \times \mathbf{B}$ drift with velocity $\mathbf{v}_{\text{drift}} = -(\nabla\phi \times \mathbf{B})/B^2$. Electrons entering or exiting the small regions of the corners of the barrier will have large velocities, which are proportional to the electric field at these locations, to account for current conservation. There are larger number of electrons drifting with slow and uniform velocities in the middle of the barrier where the electric field is smaller and more uniform. This picture is graphically represented in Figure 17 where we show the calculated results for the current distribution $\mathbf{J}(x, y) = -\sigma(x, y)\nabla\phi(x, y)$ corresponding to the experimental situation of Figure 15. Notice that even well outside the barrier the current distribution is already modified by the presence of the magnetic field barrier in the facet region and it is concentrated closer to the edges of the sample. At the diagonally opposite corners, it is strongly peaked. These results for the field and current distribution are consistent with those of [44], which were calculated for a conventional Hall geometry in the case of very low-aspect-ratio. This problem of low-aspect-ratio Hall devices in a homogeneous magnetic field was studied earlier in the context of applications for magnetic sensors (see, e.g., [45] and the references therein).

5.3. Magnetic Barriers in Series

In [37], a ridge geometry (see top figure of the inset of Fig. 18) was fabricated resulting in two magnetic barriers in series (bottom figure of the inset of Fig. 18), each having the same B_z but with opposite sign. In Figure 18, both the experimental (solid) trace and the theoretical (dash-dotted and dashed) curves are shown. The experimental trace was claimed to be for $2S = 1$ μm base length of the ridge (in [37], it is noted that the etch depth for a

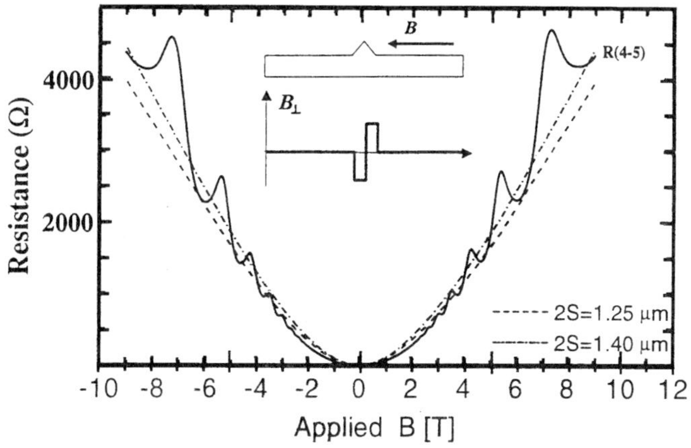

Fig. 18. Experimental (solid curve) and theoretical (dashed and dash-dotted curves) results for the magnetoresistance of the magnetic barrier created by the ridge (see inset).

ridge is less than the depth of the regrown material, which may produce some planarization during regrowth), while for the theoretical curves we found that a larger base length gives closer agreement with experiment. Notice that with this renormalization of the width of the magnetic barrier we obtain rather good agreement for the positive magnetoresistance part of the experimental curves. The oscillatory part in the experimental curves are again due to quantum effects.

The more general case of two magnetic barriers in series was studied in [39]. This can be realized experimentally by producing a ridge with a flat top, which is equivalent to introducing a zero magnetic field region between the two facets. It was shown that the relative sign of the magnetic field of the two magnetic barriers has important consequences for the magnetoresistance. For the case of two separate barriers with the same sign of B_z, the magnetoresistance at high magnetic fields is almost twice that of a single barrier and, consequently, classically the magnetoresistance of multiple barriers is additive. For not too large magnetic fields (the $R \sim B^2$ region), the current path spreads across the magnetic–nonmagnetic interface and, consequently, the current path is shorter, leading to a smaller magnetoresistance, and the simple rule of addition of resistances in series is no longer valid [38]. For barriers in series with opposite direction of the B_z field, the situation is different. The removal point of current from the first barrier is at the same side of the sample as the injection point of the current into the second barrier and, consequently, the resistance is not sensitive to the distance of separation between the two barriers.

6. ONE-DIMENSIONAL MAGNETIC MODULATION

The magnetoresistance oscillations of the two-dimensional electron gas (2DEG) subject to periodic *electric* (or potential) weak modulations, along one or two directions, also called *Weiss* oscillations, are now well established both experimentally [46–53] and theoretically [54–59]. The situation is mostly clear in the case of one-dimensional (1D) modulations where the oscillations reflect the commensurability between two length scales: the cyclotron diameter at the Fermi level, $2R_c = 2\sqrt{2\pi n_e} l^2$ (where n_e is the electron density and $l = \sqrt{\hbar/eB}$ is the magnetic length), and the period a of the modulation.

In this section, we consider electrical magnetotransport of a 2DEG in the presence of a weak 1D periodic modulation (of strength B_0) of the magnetic field and the extreme case of a strong magnetic modulation in which the average magnetic field is zero. We will show that both situations are essentially different from each other.

Different methods of establishing a periodic magnetic field on a micrometer or nanometer scale have been used: (i) using periodically arranged flux tubes in a type II superconductor (Abrikosov lattice) that penetrate the underlying 2DEG [60–63], (ii) using patterned superconducting gates that partially shield the external magnetic field [64–66], (iii) microfabricated ferromagnetic structures whose magnetic polarization adds to the external field [65–78], or (iv) nonplanar two-dimensional electron systems [36–38, 79, 80]. The first method is limited to low magnetic fields and, furthermore, a periodic flux lattice is, in practice, not achievable in an evaporated superconducting film because of the presence of pinning centra, which leads to a rather random array of flux lines. The last method of nonplanar 2D systems can result in large magnetic field modulations, the strength of which is determined by the external magnetic field. Also sign-alternating magnetic modulations are attainable. For the second method, different superconducting periodic arrays have been made using lead [64], and niobium [66]. Different ferromagnetic materials have been patterned on top of the 2DEG, for example, nickel [65, 66, 69–72, 74, 76, 81–83], dysprosium [67, 75, 77, 82], and cobalt [73, 83]. Alternatively, metal organic chemical vapor deposition with a tunneling microscopic tip [84] was used to deposit small magnetic particles.

The magnetic metallic strips also induce an electric modulation in the 2DEG and often also a strain-induced electric modulation at the 2DEG occurs because of the differential thermal contraction of the deposited magnetic material and the semiconductor. Initially [81], these built-in potential modulations dominated any effects caused by the magnetic modulation. By depositing a thin metallic film between the 2DEG and the magnetic strips, the electric modulation from the metallic strips can be shielded. For example, Weiss et al. [82] used a 10-nm thin NiCr film between the 2DEG and the magnetic strips in order to define an equipotential plane. The strain-induced modulation was removed [65, 72] by orienting the strips normal to the [100] direction of GaAs, which is nonpiezoelectric.

The magnetic metallic strips act as a periodic gate, which causes both scalar and vector potential modulation. By applying a bias to the ferromagnetic gates, the electric field modulation can be altered. In this way, one can drive the system from a predominantly electric-dominated modulation to a magnetic field–dominated modulation.

6.1. Weak Magnetic Modulation

There already exists a number of theoretical studies on the effect of a periodic magnetic modulation on the electrical transport properties of a 2DEG. Vasilopoulos and Peeters [85] predicted phase-shifted Weiss oscillations, which were later reobtained in [86]. This study was later extended [87] to the case of combined electric and magnetic 1D modulations (see also [88]). Li et al. [89] considered more general periodic magnetic field profiles by including several Fourier components in the expression for $\mathbf{B}(x)$. The corresponding quasi-classical band conductivity problem was studied by Gerhardts [90]. In [81], a numerical calculation of the energy spectrum for a 1D magnetic modulation of arbitrary strength was presented and the corresponding transport properties were discussed in [91]. Collective excitations of a 2DEG in a unidirectional magnetic field modulation were discussed by Wu and Ulloa [92–94].

Consider a 2DEG in the (x, y) plane, subject to the magnetic field $\mathbf{B} = (B + B_0 \cos Kx)\mathbf{e}_z$, where $K = 2\pi/a$, and a is the modulation period. Only the first Fourier component of the periodic magnetic field is retained, which is sufficiently accurate in most situations. Here we consider only the case of a weak modulation, that is, $B_0 \ll B$. In the Landau gauge, the vector potential $\mathbf{A}$ that describes this periodic magnetic field is $\mathbf{A} = (0, Bx + (B_0/K)\sin(Kx), 0)$ with $B_0 \ll B$. In the effective-mass (m^*) approximation, the corresponding one-electron Hamiltonian is given by $H = (\mathbf{p} + e\mathbf{A})/2m^*$ and can be written as

$$H = \frac{1}{2m^*}\left[p_x^2 + (p_y + eBx)^2\right] + \frac{\omega_0}{K}(p_y + eBx)\sin Kx + \frac{m^*\omega_0^2}{4K^2}(1 - \cos 2Kx) \quad (38)$$

where p_μ $(\mu = x, y)$ is the momentum operator, $\omega_0 = eB_0/m^* \ll \omega_c = eB/m^*$, and ω_c is the cyclotron frequency. The first term on the right-hand side of Eq. (38) is the Hamiltonian of a free electron in a magnetic field. The corresponding normalized eigenvectors are $\psi_{nk_y} = \exp(ik_y y)\phi_n(x + x_0)/\sqrt{L_y}$, where $x_0 = -l^2k_y = -\hbar k_y/eB$ is the center coordinate of the cyclotron orbit, L_y is the width of the sample in the y direction, and n is the Landau level index. $\phi_n(x)$ is the well-known harmonic oscillator wave function and the corresponding eigenvalues are $E_{nk_y} = (n + 1/2)\hbar\omega_c$ and are degenerate with respect to the wavevector k_y along the y direction.

Because of the inequality $B_0 \ll B$, we can consider the last two terms on the right-hand side of Eq. (38) as perturbation and evaluate the energy correction to E_{nk_y} given previously by first-order perturbation theory using the unperturbed eigenstates ψ_{nk_y}. The result is

$$E_{nk_y} = \left(n + \frac{1}{2}\right)\hbar\omega_c + \hbar\omega_0 G_n(u)\cos Kx_0 + \left(\frac{m^*\omega_0^2}{4K^2}\right)\left[1 - e^{-2u}L_n(4u)\cos 2Kx_0\right] \quad (39)$$

where $u = K^2l^2/2$, $G_n(u) = e^{-u/2}[L_n(u)/2 + L_{n-1}^1(u)]$, and $L_n(z)$ is a Laguerre polynomial.

Because $x_0 = -l^2k_y$, the k_y degeneracy of the energy levels is lifted and the energy levels broaden into bands. In the following, we will retain only terms of order ω_0 in the energy spectrum and neglect terms of higher order. Because a potential modulation can be present, we also consider the electric modulation problem. The correspondence with the *electric* case [54] for a modulation strength V_0 is as follows: $G_n(u) = -\partial F_n(u)/\partial u \to F_n(u) = \exp(-u/2)L_n(u)$. In a typical experiment, the magnetic field B is weak and many Landau levels are occupied. It then turns out that it is a good approximation to take the large n limit of the Laguerre polynomials that appear in $G_n(u)$ and $F_n(u)$. For the *electric* modulation, the width of the Landau level at the Fermi energy is given by [54]:

$$2V_0\left|F_n(u)\right| \approx 2V_0\left(\frac{2}{\pi KR_c}\right)^{1/2}\left|\cos\left(\frac{KR_c - \pi}{4}\right)\right| \quad (40)$$

whereas now V_0 is replaced by $\hbar\omega_0$ and $F_n(u)$ by $G_n(u)$. The resulting bandwidth for large n is

$$2\hbar\omega_0\left|G_n(u)\right| \approx 2\hbar\omega_0\left(\frac{ak_F}{2\pi}\right)\left(\frac{2}{\pi KR_c}\right)^{1/2}\left|\sin\left(\frac{KR_c - \pi}{4}\right)\right| \quad (41)$$

where $R_c = (n_F + \frac{1}{2})^{1/2}l$ is the cyclotron orbit at the Fermi energy and n_F is defined as the largest integer contained in $(E_F - \frac{1}{2})/\hbar\omega_c$. From Eq. (41), we obtain the flat-band condition as $2R_c/a = i + \frac{1}{4}$; the maximum bandwidth occurs for $2R_c/a = i + \frac{3}{4}$, with $i = 0, 1, 2, \ldots$.

Equations (40) and (41) were obtained using the asymptotic expressions for the Laguerre polynomials and are valid in the low-magnetic-field limit, typically $B \leqslant 1$ T, in which case $n \gg 1$. Comparing Eq. (40) with Eq. (41), we see that (1) the bandwidths are *out of phase*, and (2) the amplitude in the *magnetic* modulation case is larger by a factor of $ak_F/2\pi = \sqrt{n_e a^2/2\pi}$ ($= 8.6$ for $n_e = 3 \times 10^{11}$ cm^{-2} and $a = 3820$ Å) than the corresponding amplitude in the *electric* case for equal modulation strengths. This is illustrated in Figure 19 for the width of the Landau level at the Fermi energy. For the *magnetic* modulation, we show the result $2|G_n(u)|$ (solid curve) as well as the asymptotic expression Eq. (41) (dotted curve). The jumps in the solid curve are due to the fact that the integer n_F changes discontinuously with B. The bandwidth for the *electric* modulation, when $V_0 = \hbar\omega_0$, is shown as the short-dashed curve.

It is instructive to give a classical derivation of the energy correction resulting from the modulation because the magnetic field is weak and the quantum number n large. In classical mechanics, the electron motion along the x and y directions is described by $x(t) = x_0 + R_c\sin(\omega_c t + \varphi)$ and $y(t) = y_0 + R_c\cos(\omega_c t + \varphi)$, respectively, where x_0, y_0 are the center coordinates and φ is a phase factor. Without loss of generality,

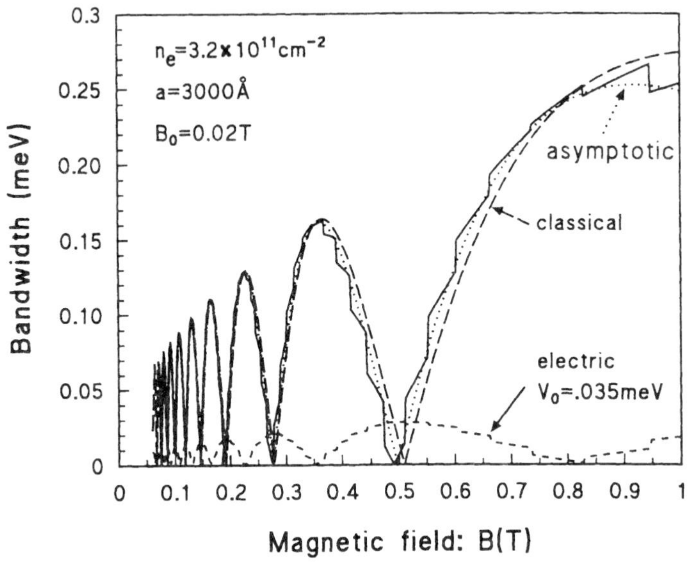

Fig. 19. Width of the Landau level at the Fermi energy as a function of the applied magnetic field. The solid curve is the exact numerical result for a constant magnetic field modulation of strength $B_0 = 0.02$ T. The dotted and long-dashed curves are the corresponding asymptotic and classical results. The short-dashed curve is the result for an electric modulation of equal strength, $V_0 = \hbar\omega_0$.

φ can be taken equal to zero. The contribution to the potential energy U caused by the modulation of the magnetic field is $U = -e\mathbf{A}_0 \cdot \mathbf{v}$, where $\mathbf{v}$ is the electron velocity and $\mathbf{A}_0 = (eB_0/K)\sin(Kx)\mathbf{e}_y$. The mean value $\langle U \rangle$ over the period t_0 of the cyclotron motion is

$$\begin{aligned}\langle U \rangle = \frac{1}{t_0}\int_0^{t_0} U(t)\,dt &= eB_0\omega_c\left(\frac{R_c}{K}\right)\cos Kx_0\,\frac{1}{t_0}\int_0^{t_0}\sin\omega_c t\,\sin(KR_c\sin\omega_c t)\,dt \\ &= \hbar\omega_0\left(\frac{R_c}{Kl^2}\right)J_1(KR_c)\cos Kx_0 \end{aligned} \tag{42}$$

where $J_1(z)$ is the Bessel function of order one. In the limit $KR_c \gg 1$, that is, for weak magnetic fields, $\langle U \rangle$ becomes equal to the second term of Eq. (39), thus resulting in the bandwidth given by Eq. (41). The bandwidth resulting from Eq. (42), prior to taking the limit $KR_c \gg 1$, is shown in Figure 19 by the long-dashed curve and is indicated as "classical."

In the absence of modulation, we have $v_x = v_y = 0$ for all eigenstates. Because the modulation lifts the k_y degeneracy, we find, to order $\hbar\omega_0$,

$$v_y = \frac{1}{\hbar}\frac{\partial E_{n,k_y}}{\partial k_y} = -\frac{2\omega_0}{K}uG_n(u)\sin Kx_0 \tag{43}$$

while $v_x = 0$. The fact that v_y is no longer zero has important consequences for transport which will be detailed later.

For weak electrical fields, that is, for linear responses, and weak scattering, the conductivity tensor $\sigma_{\mu\nu}$ has been evaluated, for the case of *electric* modulations, in [54, 55]. In general, we have $\sigma^{\mathrm{I}}_{\mu\nu}(\omega) + \sigma^{\mathrm{II}}_{\mu\nu}(\omega)$, where ω is the frequency. The relevant quantity $\sigma^{\mathrm{I}}_{\mu\nu}(0) \equiv \sigma_{\mu\nu}$ for direct-current (dc) transport is, in general, the sum of a diffusive (dif)

$$\sigma^{\mathrm{dif}}_{\mu\nu} = \frac{\beta e^2}{\Omega}\sum_{\zeta} f_\zeta(1-f_\zeta)\tau(E_\zeta)v^\zeta_\mu v^\zeta_\nu \tag{44}$$

and a collisional (col) contribution. The collisional contribution is the only contribution in the case of transport in a uniform 2DEG because $\mathbf{v}^\zeta = 0$ and leads to the well-known Shubnikov–de Haas (SdH) oscillations. Because in the modulated case $v_y \neq 0$, a new conduction channel opens up. In Eq. (44), we used the unperturbed one-electron states $\mid \zeta\rangle \equiv \psi_{nk_y}$ with the corresponding eigenvalue given by Eq. (39), $\tau(E_\zeta)$ is the relaxation time, f_ζ is the Fermi–Dirac function, and the velocity in the state $\mid \zeta\rangle$: $v_\mu^\zeta = \langle \zeta \mid v_\mu \mid \zeta\rangle$.

The dc component $\sigma_{\mu\nu}^{\mathrm{II}}$ is given by [54, 55]:

$$\sigma_{\mu\nu}^{\mathrm{II}} = i\hbar\frac{e^2}{\Omega}\sum_{\zeta\neq\zeta'} f_\zeta(1-f_{\zeta'})\langle\zeta \mid v_\mu \mid \zeta'\rangle\langle\zeta' \mid v_\nu \mid \zeta\rangle \frac{1-\exp[\beta(E_\zeta - E_{\zeta'})]}{(E_\zeta - E_{\zeta'})^2} \tag{45}$$

and is independent of scattering in the limits taken.

A detailed evaluation of formulas (44) and (45) has been carried out in [95] for scattering by randomly distributed impurities of density N_{I} and of interaction strength, in Fourier space, $U_q = (2\pi e^2/\epsilon)/\sqrt{q^2+k_{\mathrm{s}}^2} \approx (2\pi e^2/\epsilon)/k_{\mathrm{s}} = U_0$ in the limit $q \ll k_{\mathrm{s}}$; here k_{s} is the inverse screening length and ϵ the dielectric constant, which gives approximately $\tau(E_\zeta) \approx (\pi l^2\hbar^2/N_{\mathrm{I}}U_0^2)^{1/2}$.

In the absence of modulation, we have $v_x = v_y = 0$ and the diffusion contribution vanishes. In its presence, we have $v_x = 0$ but $v_y \neq 0$; cf. Eq. (43). Thus, $\sigma_{xx}^{\mathrm{dif}}(0) = 0$, whereas $\sigma_{yy}^{\mathrm{dif}}(0) \equiv \sigma_{yy}^{\mathrm{dif}} \neq 0$. To order B_0^2, we find

$$\sigma_{yy}^{\mathrm{dif}} = \frac{e^2}{h}\frac{2\pi^2\tau}{\hbar}\frac{l^2}{a^2}\sum_n \left[\hbar\omega_0 G_n(u)\right]^2\left(-\frac{\partial f(E)}{\partial E}\right)_{E=E_n} \tag{46}$$

where $E_n = (n + 1/2)\hbar\omega_{\mathrm{c}}$. Comparing this result with that for *electric* modulation of strength V_0, Eq. (13) of [55], we see that it has the same structure: $V_0F_n(u)$ is replaced by $\hbar\omega_0 G_n(u)$. In either case, this contribution is proportional to the square of the bandwidth.

The resistivity tensor $\rho = \sigma^{-1}$ is evaluated from the conductivity tensor using the standard expressions $\rho_{xx} = \sigma_{yy}/S$, $\rho_{yy} = \sigma_{xx}/S$, and $\rho_{xy} = -\sigma_{yx}/S$ with $S = \sigma_{yy}\sigma_{xx} - \sigma_{xy}\sigma_{yx}$.

For $n \gg 1$, that is, when many Landau levels are occupied, the results of (46) can be cast into a form that exhibits explicitly both the Weiss (at low B) and the Shubnikov–de Haas (SdH) (at higher B) oscillations. The procedure has been detailed in [55, 87] and consists of using the asymptotic expressions for the Laguerre polynomials, for $n \gg 1$, and of the density of states $D(E) = D_0[1 - 2\exp(-\pi/\omega_{\mathrm{c}}\tau_{\mathrm{f}})\cos(2\pi E/\hbar\omega_{\mathrm{c}})]$ with the prescription $\sum_n \to 2\pi l^2 \int D(E)\,dE$. The parameter τ_{f} is the electron quantum lifetime. Following verbatim this procedure and retaining only the leading terms, we obtain that Eq. (46) takes the form

$$\frac{\sigma_{yy}^{\mathrm{dif}}}{\sigma_0} = \frac{ak_{\mathrm{F}}}{2\pi^2}\frac{\hbar\omega_0}{\hbar\omega_{\mathrm{c}}}\frac{\hbar\omega_0}{E_{\mathrm{F}}}\left[G - 2\exp\left(\frac{-\pi}{\omega_{\mathrm{c}}\tau_{\mathrm{f}}}\right)A\left(\frac{T}{T_{\mathrm{c}}}\right)\cos\left(\frac{2\pi E_{\mathrm{F}}}{\hbar\omega_{\mathrm{c}}}\right)\sin^2\left(\frac{2\pi R_{\mathrm{c}}}{a} - \frac{\pi}{4}\right)\right] \tag{47}$$

where

$$G = \left[1 - A\left(\frac{T}{T_{\mathrm{a}}}\right)\right]\Big/2 + A\left(\frac{T}{T_{\mathrm{a}}}\right)\sin^2\left(\frac{2\pi R_{\mathrm{c}}}{a} - \frac{\pi}{4}\right) \tag{48}$$

Here $\sigma_0 = n_{\mathrm{e}}e^2\tau/m^*$ is the conductivity at zero magnetic field B and $A(x) = x/\sinh(x)$. The characteristic temperatures T_{a}, for the Weiss oscillations, and T_{c}, for the SdH oscillations, are defined by $k_{\mathrm{B}}T_{\mathrm{a}} = (\hbar\omega_{\mathrm{c}}/4\pi^2)ak_{\mathrm{F}}$ and $k_{\mathrm{B}}T_{\mathrm{c}} = \hbar\omega_{\mathrm{c}}/2\pi^2$, respectively, with k_{F} being the Fermi wavevector. In typical experiments, we have $T_{\mathrm{a}}/T_{\mathrm{c}} = ak_{\mathrm{F}}/2 \gg 1$; for example, for $n_{\mathrm{e}} = 3 \times 10^{11}$ cm^{-2} and $a = 3500$ Å, we have $ak_{\mathrm{F}} \approx 24$. As a result, the SdH oscillations disappear much faster with temperature than the Weiss oscillations. We further

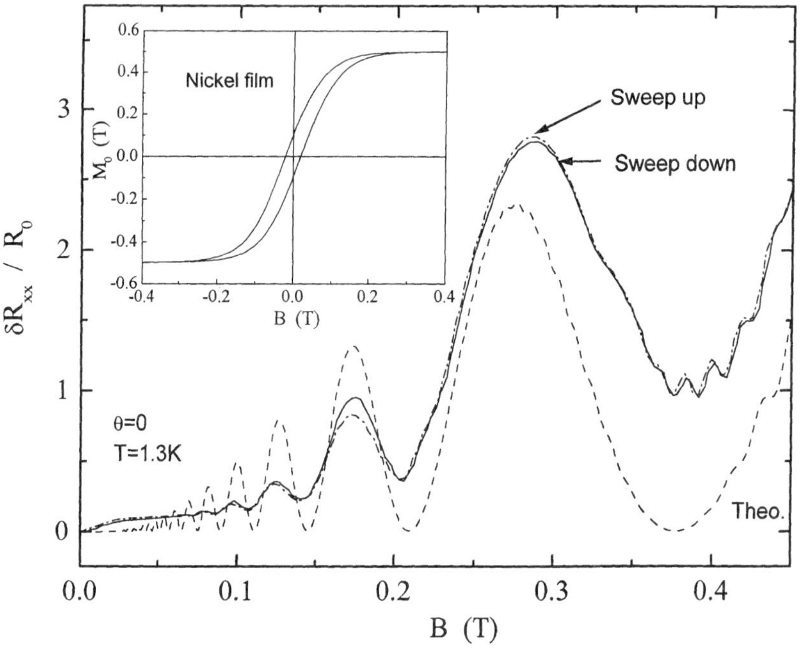

Fig. 20. Longitudinal magnetoresistance at 1.3 K for a 1D magnetic modulation of a 2DEG for "up" (dot-dashed curve) and "down" (solid curve) field sweeps as compared to the calculated behavior (dashed curve). The inset shows the magnetization of the Ni strips that is responsible for the magnetic modulation. (Source: Reprinted with permission from [72]. © 1997 American Physical Society.)

remark that in comparison with the *electric* case, cf. Eq. (34) of [55], the present result differs essentially by a term $(ak_F/2\pi)^2 \approx 60$ in the prefactor. Correspondingly, the amplitude will be larger in the *magnetic* case by this factor.

In Figure 20, the commensurability oscillations in the longitudinal magnetoresistance, $\delta R_{xx} = R_{xx}(B) - R_0$, of a 2DEG with mobility $\mu = 7 \times 10^5$ cm^2/V s (corresponding to a mean free path of $l_e = 7.7$ μm) that is subjected to a 2D magnetic modulation are shown. The 2D magnetic modulation was produced [72] by a periodic array ($a = 500$ nm) of Ni strips of width $d = 200$ nm and height $h = 100$ nm that are situated a distance of 35 nm from a 22-nm-wide GaAs/AlGaAs quantum well. These magnetoresistance oscillations result from the preceding commensurability effects between the diameter of the cyclotron orbit at the Fermi level $2R_c$ and the period of the magnetic modulation. Notice that these magnetoresistance oscillations are clearly visible and their position agrees very well with the previous theoretical result (dashed curve). For $B > 0.3$ T, small SdH oscillations are superimposed on top of the Weiss oscillations. Note that the experimental result exhibits a small hysteretic behavior that is associated with the hysteresis of the magnetization of the Ni strips (see inset of figure). In the calculation (dashed curve), the strength of the magnetic modulation (B_0) was taken from the magnetization (M_0) given by the model hysteresis curve shown in the inset of the figure. Note that there is not only a good agreement with the position of the oscillations, confirming that the oscillations originate from a predominantly magnetic modulation, but also the magnitude of the last commensurability peak at approximately 0.27 T is well reproduced. Above about 0.2 T, the theoretical prediction involves no free parameters so this agreement confirms the accuracy of the calculated saturation modulation amplitude. Below 0.2 T, the decrease in amplitude will depend on the exact form of the $M_0(B_{ext})$ curve and damping caused by scattering, which is not included in the theoretical curve, will be important.

6.2. Electric and Magnetic Modulations

Magnetic modulations are often accompanied by an electric modulation. Therefore, we have to study how the results of a pure magnetic modulation are modified when an electric modulation with the same period is present. Two cases are of interest: that where the two modulations are *in phase* and that where they are *out of phase*.

6.2.1. In-Phase Modulations

If a weak *electric* modulation described by the periodic potential $V_0 \cos(Kx)$, which is *in phase* with the *magnetic* one, is present, Eq. (39) will have an additional term $V_0 F_n(u) \cos(Kx_0)$. As a result, v_y, given by Eq. (43), will have an additional term $-(2V_0/\hbar K) u F_n(u) \sin(Kx_0)$. The bandwidth will be the sum of the two bandwidths. At the Fermi energy, the latter is given by the sum of those of Eqs. (40) and (41) and is equal to

$$2V_0 \sqrt{\frac{2}{\pi K R_c}} \sqrt{1+\delta^{-2}} \left| \sin\left(\frac{2\pi R_c}{a} - \frac{\pi}{4} + \phi\right) \right| \tag{49}$$

where $\delta = 2\pi V_0 / a k_F \hbar\omega_0 = \tan(\phi)$. Notice that the flat-band condition now reads $2R_c/a = i + 1/4 - \phi/\pi$ and depends on the relative strength of the two modulations.

The changes mentioned previously will be reflected in the transport coefficients as well. As an example, the diffusion contribution to the conductivity, Eq. (46), now takes the form

$$\sigma_{yy}^{\text{dif}} \approx \frac{e^2}{h} \frac{2\pi^2 \tau}{\hbar} \frac{l^2}{a^2} \sum_n \left[\hbar\omega_0 G_n(u) + V_0 F_n(u)\right]^2 \left(-\frac{\partial f(E)}{\partial E}\right)_{E=E_n} \tag{50}$$

and its asymptotic expression, obtained in the manner described earlier, reads

$$\begin{aligned} \frac{\sigma_{yy}^{\text{dif}}}{\sigma_0} \approx{}& \frac{a k_F}{2\pi^2} \frac{\hbar\omega_0}{\hbar\omega_c} \frac{\hbar\omega_0}{E_F} (1+\delta^2) \\ &\times \left\{ 1 - A\left(\frac{T}{T_a}\right) + \left[A\left(\frac{T}{T_a}\right) - 2\exp\left(\frac{-\pi}{\omega_c \tau_f}\right) A\left(\frac{T}{T_c}\right) \cos\left(\frac{2\pi E_F}{\hbar\omega_c}\right) \right] \right. \\ &\left. \times \sin^2\left(\frac{2\pi R_c}{a} - \frac{\pi}{4} + \phi\right) \right\} \end{aligned} \tag{51}$$

Notice that the influence of the combined effect of the two modulations is to introduce the phase factor ϕ in the Weiss oscillations.

In Figure 21, we plot $\Delta\rho$, that is, the change in resistance caused by the modulations, as a function of B (solid curve) for $B_0 = 0.02$ T and $V_0 = 0.2$ meV for a 2DEG with $n_e = 3 \times 10^{11}$ cm^{-2} at $T = 4.2$ K with a modulation period of $a = 3000$ Å. For comparison, we show the results when only the *magnetic* (dotted curve) or the *electric* (dashed curve) modulation is present. In line with Eq. (51) that was used for the evaluation of $\Delta\rho$, the solid curve is shifted from the other curves because of the phase factor ϕ. We also note the $\pi/2$ phase difference between the dashed and dotted curves, which reflects that of the corresponding bandwidths shown in Figure 19.

The dependence of $\Delta\rho$ on B and the phase factor ϕ is shown in Figure 22. B_0 is again kept constant ($B_0 = 0.02$ T) but V_0 is varied as indicated between positive and negative values; δ and ϕ change accordingly. We notice that (1) the position of the peaks depends on the specific value of δ, and (2) there is a π phase difference between large positive and large negative values of δ.

Such in-phase modulation of the electric and magnetic period was studied by Iye et al. [69]. In their experiment, a GaAs/AlGaAs heterojunction containing a 2DEG, with density $n_e = 4 \times 10^{11}$ cm^{-2} and mobility $\mu = 6 \times 10^5$ cm^2/V s at $T = 4.2$ K, was placed 75 nm below a periodic array ($a = 0.5$ μm) of Ni strips with height 150 nm. By changing the voltage on the gate consisting of the Ni strips, the strength of the electric field modulation is

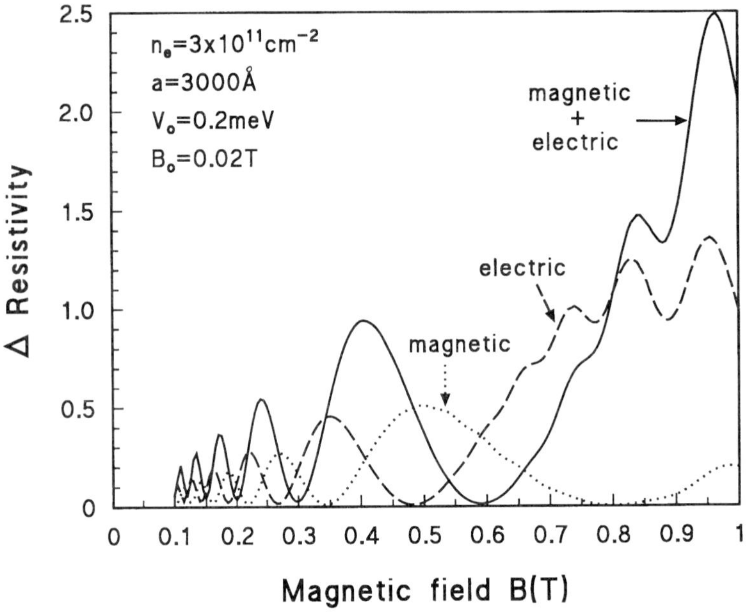

Fig. 21. Increase of the magnetoresistance (solid curve) resulting from magnetic and in-phase electric modulations with $B_0 = 0.02$ T and $V_0 = 0.2$ meV. The dotted and dashed curves represent the increase caused by the separate magnetic and electric modulation, respectively.

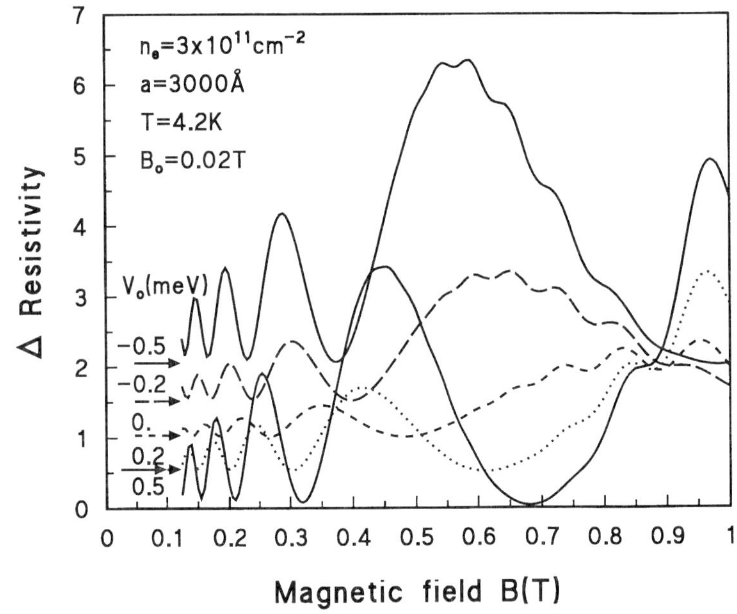

Fig. 22. Change of the magnetoresistance when magnetic and electric in-phase modulations are present for $B_0 = 0.02$ T and for different strengths of the electric modulation. The different curves are shifted upwards for clarity.

altered as shown in Figure 23. The positions of the two distinct minima shift to larger values of B/k_F (applying a gate voltage to the 2DEG changes the electron density, which is the reason why the result is plotted as a function of B/k_F). At $V_g = -250$ mV, the minima are almost at the positions with $\phi = -\frac{1}{4}$, and, consequently, the potential modulation is

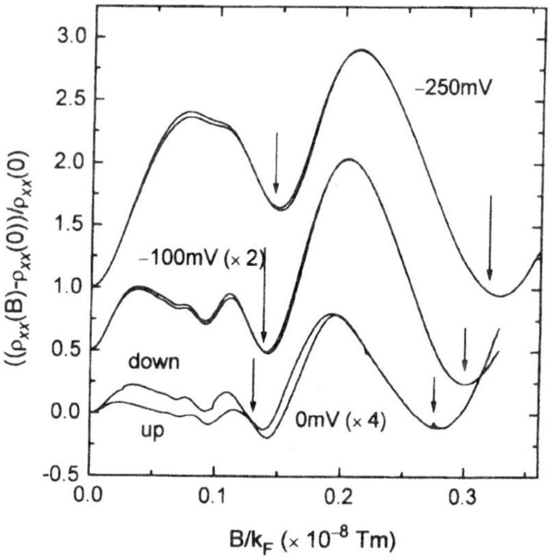

Fig. 23. Magnetoresistance for a 1D magnetic modulation for different values of the gate bias V_g. The arrows indicate the calculated positions of the minima. The curves are offset for clarity. (Source: Reprinted from [69], with permission from Elsevier Science.)

almost purely electric. Increasing the gate voltage shifts the position of the minima to a lower field because of the increase of the phase factor (ϕ) toward $\frac{1}{2}$, which is the value for the pure magnetic modulated case. Iye et al. [69] estimated that the magnetic modulation amplitude is $B_0 \approx 13$ mT at $B = 0.254$ T on the down-sweep. Similarly, they estimated that the electric potential modulation is 0.31, 0.85, and 1.8 meV for $V_g = 0, -100$, and -250 mV, respectively.

6.2.2. Out-of-Phase Modulations

If light is shined on top of a magnetically modulated heterostructure, the light pulses will ionize DX centers in the AlGaAs layer between the gates (e.g., strips of magnetic materials) where $B_0 = 0$. This will create an *electric* modulation $\pi/2$ *out of phase* with the *magnetic* one. Assuming that the *electric* modulation of the gates is much smaller than that due to the light, we are led to consider transport in the presence of the two modulations that are $\pi/2$ *out of phase* [87].

We take the same *electric* modulation $V_0 \cos(Kx)$ and assume a *magnetic* one given by $B(x) = B + B_0 \sin(Kx)$. To first order in V_0 and B_0, the eigenvalue E_{nk_y} and velocity v_y are given by

$$E_{nk_y} = \left(n + \frac{1}{2}\right)\hbar\omega_c + \hbar\omega_0 G_n(u)\sin(Kx_0) + V_0 F_n(u)\cos(Kx_0) \tag{52}$$

and

$$v_y = -\left(\frac{2u}{\hbar K}\right)\left[\hbar\omega_0 G_n(u)\cos(Kx_0) + V_0 F_n(u)\sin(Kx_0)\right] \tag{53}$$

respectively. These apparently smooth changes can have important consequences for transport. For instance, the combined bandwidth at the Fermi energy is now equal to

$$2V_0\sqrt{\frac{2}{\pi K R_c}}\sqrt{1 + \left(\delta^{-2} - 1\right)\sin^2\left(\frac{2\pi R_c}{a} - \frac{\pi}{4}\right)} \tag{54}$$

If $\delta = \pm 1$, the bandwidth no longer oscillates as a function of the magnetic field; that is, the Weiss oscillations are washed out.

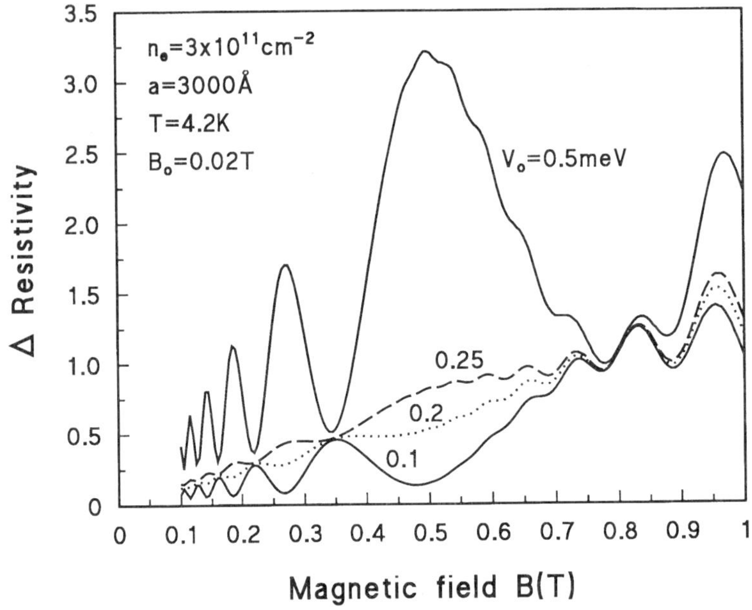

Fig. 24. Increase of the magnetoresistance when magnetic and electric out-of-phase modulations are present for different values of the strength of the electric modulation.

The result for the diffusion contribution reads

$$\sigma_{yy}^{\rm dif} = \frac{e^2}{h}\frac{2\pi^2\tau}{\hbar}\frac{l^2}{a^2}\sum_n \left\{\left[V_0 F_n(u)\right]^2 + \left[\hbar\omega_0 G_n(u)\right]^2\right\}\left(-\frac{\partial f(E)}{\partial E}\right)_{E=E_n} \tag{55}$$

There is no cross term involving the product $\hbar\omega_0 V_0$ in Eq. (55), as expected from Eqs. (44) and (53), because the integral over k_y vanishes for this term if we neglect the very weak k_y dependence of the argument of the Fermi functions.

The asymptotic expression for $\sigma_{yy}^{\rm dif}$ is

$$\frac{\sigma_{yy}^{\rm dif}}{\sigma_0} = \frac{\hbar\omega_0}{E_{\rm F}}\frac{\hbar\omega_0}{\hbar\omega_{\rm c}}\frac{ak_{\rm F}}{2\pi^2}\left\{G + \delta^2 F - 2\exp\left(\frac{-\pi}{\omega_{\rm c}\tau_{\rm f}}\right)A\left(\frac{T}{T_{\rm c}}\right)\cos\left(\frac{2\pi E_{\rm F}}{\hbar\omega_{\rm c}}\right)D\right\} \tag{56}$$

where

$$D = \delta^2 + (\delta^2 - 1)\sin^2\left(\frac{2\pi R_{\rm c}}{a} - \frac{\pi}{4}\right) \tag{57}$$

G is given by Eq. (48) and F by Eq. (48) after replacing $\sin^2(\ )$ by $\cos^2(\ )$. Comparing Eq. (56) with Eq. (51), we see that changing δ does not change the position of the extrema of $\Delta\rho$ as a function of B because the phase factor ϕ is absent from Eq. (56). This is illustrated in Figure 24 where $\Delta\rho$ is plotted as a function of B for $B_0 = 0.02$ T with variable V_0. For the upper two curves we have $\delta > 1$, whereas, for the lower two curves, δ is smaller than 1. This results in an *antiphase* between the two groups of curves and reflects the corresponding behavior of the bandwidth as given by Eq. (54).

6.3. Magnetic Minibands

The electronic band structure in a periodic magnetic field with zero average was investigated by several authors. Ibrahim and Peeters [96] calculated the energy spectrum of a periodic array of a delta function magnetic field oscillating in sign, which was generalized in [97] to an array with a more general unit cell. The energy spectrum and the conductivity through sinusoidal [5, 98] and different periodic magnetic fields profiles was studied in [5]. Ballistic quantum transport through a periodic alternating step magnetic field profile and through a sinusoidal magnetic field structure was investigated in [99] and through a quasi-periodic magnetic superlattice in [100]. The surface states of a lateral magnetic superlattice were calculated in [101].

6.3.1. Periodic Sinusoidal Magnetic Fields

Directing the magnetic field more and more in plane reduces the background component (B) and enhances the modulation component (B_0). In the limit of $B = 0$, we have a pure periodic magnetic field with zero average. As an example of such a case, consider the physical system depicted in Figure 3. The magnetic field profile in the plane of the 2DEG is sinusoidal when the distance between the 2DEG and the ferromagnetic thin film is sufficiently large (see dotted curve in Fig. 3b). In such a case, only the first term of the Fourier series is important. In this subsection, we study the magnetic field profile [5, 98, 102]:

$$\frac{B(x)}{B_0} = \sin\left(\frac{2\pi x}{l}\right) \tag{58}$$

which has the corresponding vector potential (Fig. 25)

$$\frac{A(x)}{A_0} = -\frac{l}{2\pi}\cos\left(\frac{2\pi x}{l}\right) \tag{59}$$

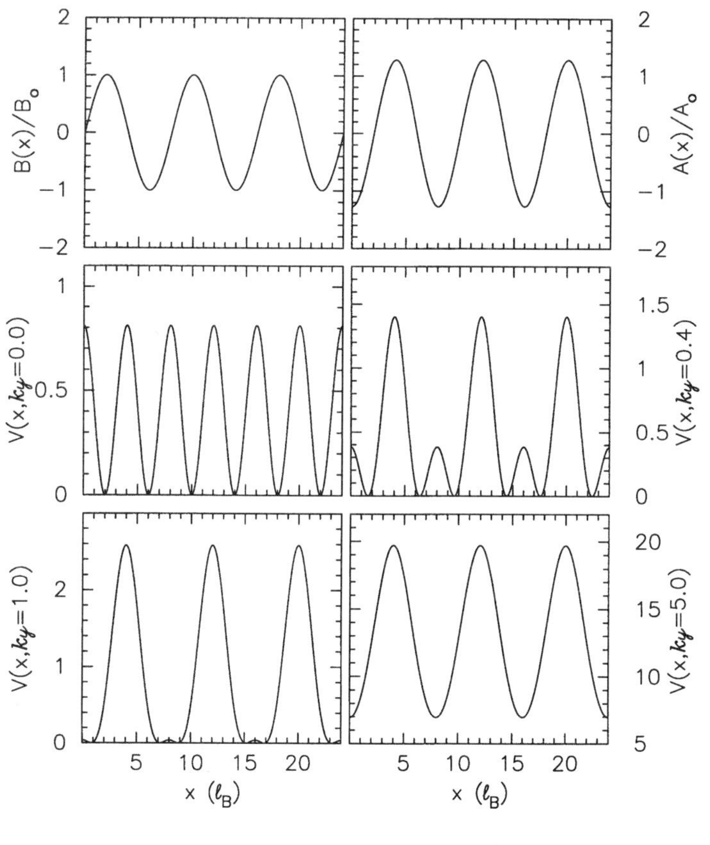

Fig. 25. Profiles for the magnetic field, the vector potential, and the effective potential for different values of k_y in the case of a sinusoidal magnetic field.

and which leads to the following Schrödinger equation:

$$\left\{\frac{d^2}{dx^2} + 2E - \left[k_y - \frac{l}{2\pi}\cos\left(\frac{2\pi x}{l}\right)\right]^2\right\}\psi(x) = 0 \tag{60}$$

The profiles for $B(x)$ and $A(x)$ are depicted in Figure 25 together with the effective potential for different k_y values. When solving the Schrödinger equation, we can restrict ourselves to one period and impose periodic boundary conditions. The potential profile $V(x, k_y)$ is shown in Figure 25 and satisfies the symmetry relation $V(x, k_y) = V(x + l/2, -k_y)$.

After making the substitutions $\theta = \pi x/l$, $\omega = (l/\pi)^2$, $\rho = lk_y/\pi$, and $\eta = (l/\pi)^2(2E - k_y^2 - 1/4)$, the preceding equation is cast into the following form:

$$\left\{\frac{d^2}{d\theta^2} + \left(\eta + \left(\frac{\omega^2}{8}\right) + \omega\rho\cos(2\theta) - \left(\frac{\omega^2}{8}\right)\cos(4\theta)\right)\right\}\psi(\theta) = 0 \tag{61}$$

which is the Wittaker–Hill equation. It is interesting to note that, in the case $k_y = 0$, Eq. (61) reduces to Mathieu's equation with period $l/2$.

The numerical solution of Eq. (61) for E versus k_y is shown in Figure 26 for $l = 8$ and in the inset for a larger period $l = 16$. To understand the energy spectrum, we show in Figure 9 the potential $V(x, k_y)$ for different k_y values. First, note that $V(x, k_y) = V(x + l/2, -k_y)$ results in spatially separated motions for the $+k_y$ and $-k_y$ electrons moving with $E < V_{max}$. For $k_y = 0$, the profile for $V(x, k_y)$ (see Fig. 25) is a periodic array of harmonic-like oscillators with finite depth and period $l/2$. In this case, the first Brillouin zone (FBZ) edge is at $k = 2\pi/l$ and gaps in the spectrum appear only at $k = 2n\pi/l$ where n is the band index. When k_y takes values different from zero, the periodicity of $V(x, k_y)$ becomes twice as large and equals l and, consequently, the FBZ edge becomes $k = \pi/l$ and the

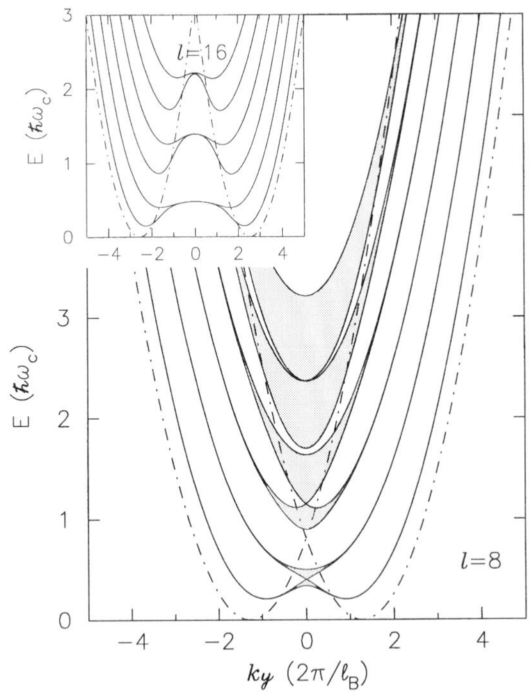

Fig. 26. Energy versus k_y dispersion relation for a single 2D electron in the presence of a sinusoidal magnetic field modulation with period $l = 8$ and $l = 16$ (inset). The shaded regions are the lowest six energy bands. The dash-dotted curves are the maximum and minimum values of the effective potential $V(x, k_y)$.

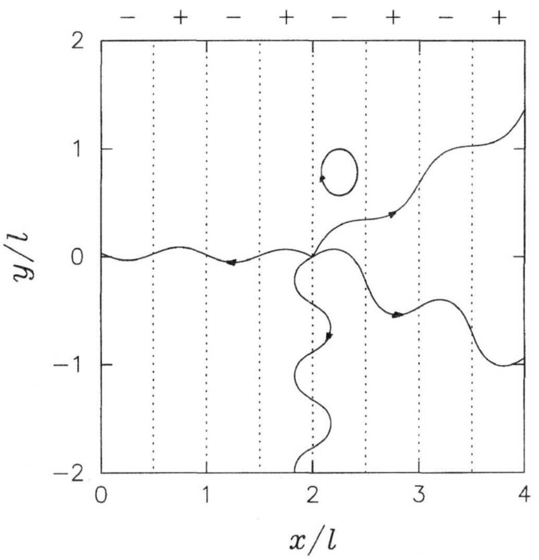

Fig. 27. Possible classical trajectories with the same initial position but different initial velocities for the magnetic field model of Figure 25.

number of energy bands and gaps doubles. The bands with energy below the barrier height (the dash-dotted curves in Fig. 26 indicate the bottom and top of the potential profile), that is, $E < V_{\max}$, correspond to open orbits traveling along the y direction and oscillating around the boundary separating the two magnetic strips, which are also called *snake orbits*. Electrons at adjacent boundaries move in the opposite y direction. The 1D states below $V_{\max}$ are similar to the so-called magnetoelectric states. Here, also, the electron motion along the $\pm y$ directions are spatially separated by half a period $l/2$.

In Figure 27, examples of the three kinds of possible classical orbits are given: (1) 0D cyclotron motion, which occurs when the electron energy is sufficiently small such that its cyclotron diameter is less than the width of the magnetic strip (because of the spatial variation of the magnetic field, the orbit in this case is not a circle); (2) 1D drift parallel to the magnetic strips for states with the center of their cyclotron orbit near the interface between the $+B_0$ and $-B_0$ regions, called snake orbits; and (3) 2D motion in the plane when the electron energy is larger than the magnetic field barriers, which results in a cyclotron diameter larger than the width of the magnetic strips.

The zero-temperature density of states (DOS) for the system considered is calculated by numerically integrating the inverse of the magnitude of the energy gradient in momentum space along the constant energy contours at the Fermi energy and within the first Brillouin zone

$$\frac{D(E)}{D_0} = \frac{1}{2\pi} \sum_n \int_{S_n} \frac{dS_n}{|\nabla E_n(k)|} \tag{62}$$

where $D_0 = m^*/\pi\hbar^2$ is the DOS of the free 2DEG, n is the band index, and S_n is the constant energy contour. Spin splitting of the energy levels is neglected. The spin degeneracy is taken into account and no scattering is assumed other than the interaction with the magnetic field.

The diagonal components of the electric conductivity tensor are calculated concurrently with the DOS and along the same constant energy contours where we used the expression

$$\frac{\sigma_{ii}}{\sigma_0} = \sum_n \int_{S_n} v_{n,i} v_{n,i} \, dS_n \tag{63}$$

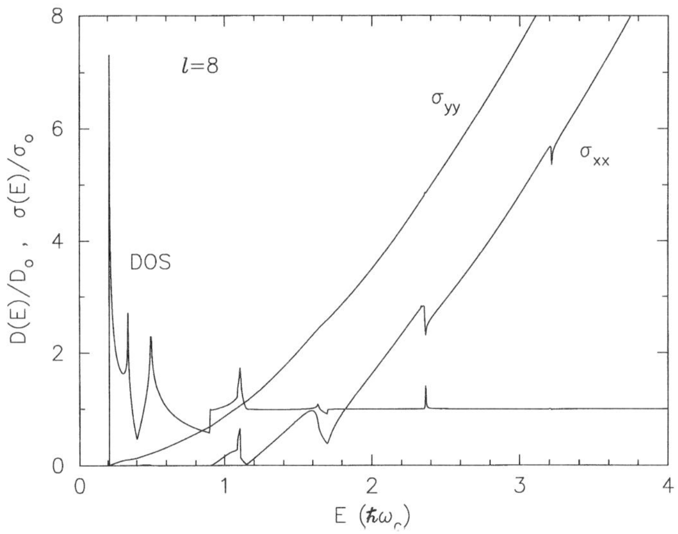

Fig. 28. Density of states (DOS) and the diagonal components of the electric conductivity tensor versus Fermi energy for 2D electrons moving in a sinusoidal magnetic field.

where $v_{n,i} = \partial E_n/\partial k_i$ is the drift velocity and σ_0 is the Drude conductivity of the free 2DEG. This formula is valid for a diffusive type of transport at zero temperature. Notice that because there is no net magnetic field, that is, $\langle B\rangle = 0$, we do not expect any Hall resistance and, consequently, $\sigma_{xy} = 0$.

The density of states (DOS) (Fig. 28) is on the average the one of a free 2DEG except for singular points (1) resulting from the edges of the minibands, which occur for $k_x = (2n+1)\pi/l$, $n = 0, 1, 2, 3, \ldots$; and (2) resulting from the local minima in the energy spectrum (see Fig. 26), which occur for small k_y values where $v_y = 0$. At these points in momentum space, the conditions for Bragg reflection are met for motion in the x direction and the average velocity in the y direction is zero. Electrons with these energies and wavevectors form standing waves with zero average velocity. This effect results in dips in the σ_{xx} conductivity (Fig. 28). Note that no such clear structure is apparent in the σ_{yy} conductivity because motion in the y direction does not exhibit this Bragg reflection. The σ_{yy} conductivity is obviously the least affected by the presence of the magnetic field modulations along the x direction, because there are always states available for motion along the y direction regardless of the value of the Fermi energy. Furthermore, the number of these states increases with increasing energy. In contrast, the miniband structure has a pronounced effect on the σ_{xx} component owing to the existence of energy gaps for motion along the x direction, where σ_{xx} shows downward dips corresponding to the edges of the Brillouin zone where $v_x \approx 0$. The magnitude of σ_{xx} does not become appreciable until the Fermi energy is near or above the magnetic potential barrier at $k_y = 0$. The σ_{xx} is always less than σ_{yy} even for high energies. The reason is that the effect of Bragg reflections, which does not influence σ_{yy}, persists up to high energies—a purely quantum mechanical effect.

6.3.2. Snakelike Orbits

The effects of these snake orbits on the magnetotransport were investigated experimentally in [72]. They measured the magnetoresistance for different angles θ between the external magnetic field and the 2DEG plane, which is shown in Figure 29. The results are plotted as a function of the z component of the applied field B_z, which is the component that affects the motion of the electrons in the 2DEG. The most striking feature is the appearance of

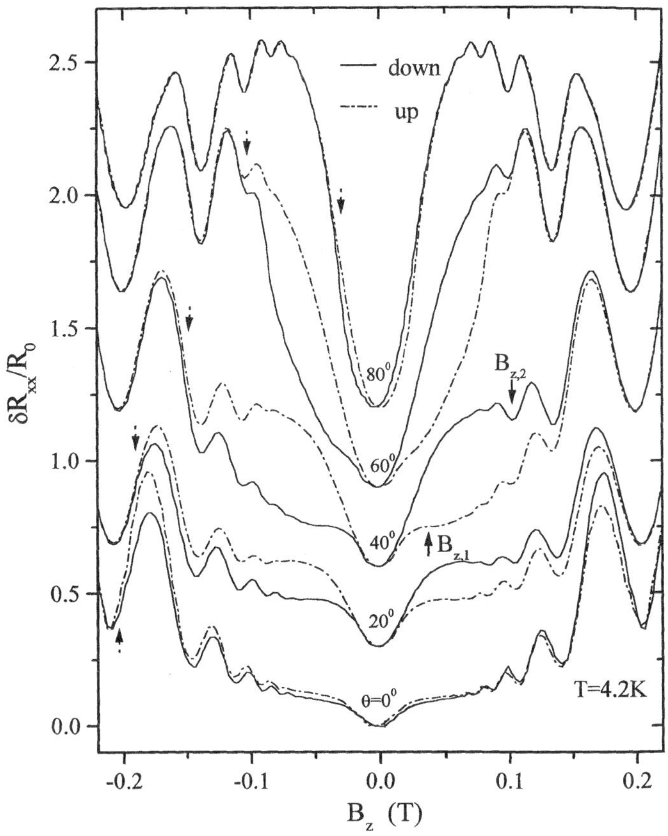

Fig. 29. Low-field magnetoresistance at 4 K as a function of the magnetic field component perpendicular to the 2DEG for different tilt angles θ of the magnetic field. (Source: Reprinted with permission from [72]. © 1997 American Physical Society.)

a positive low-field magnetoresistance that becomes much stronger and extends to larger B_z as θ increases. R_{xx} increases by a factor of up to approximately 2 on application of a B_z of only 50 mT. This can be understood from Figure 28 where increasing the magnetic field implies reducing $(E = E_F)/\hbar\omega_c$, which results in a decrease of the conductivity and, consequently, an increase of the resistance. The strong positive magnetoresistance is still observable well above 200 K, which suggests its semiclassical origin.

The observed low-field magnetoresistance can be understood in terms of channeling of electrons along lines of zero magnetic field within the sample. These are the so-called snake orbits discussed previously. In this situation, two kinds of electron states coexist. Electrons that have a sufficiently large initial velocity component perpendicular to the magnetic strips will propagate across the sample in the same direction they would have in the absence of the modulation. Electrons with smaller initial velocity components cannot pass through the magnetic barriers and will be channeled in the y direction along snakelike orbits centered on the lines of zero B.

The guiding center drift correction to the D_{yy} diffusion coefficient [58] leads to a magnetoresistance

$$\frac{\Delta R_{xx}}{R_0} = 2(\omega_c\tau)^2\frac{\langle v_d^2\rangle}{v_F^2} \tag{64}$$

where τ is the elastic scattering time and $\langle v_d^2\rangle$ is the appropriate average of the square drift velocity. This average was calculated [72] by numerically integrating the classical equation

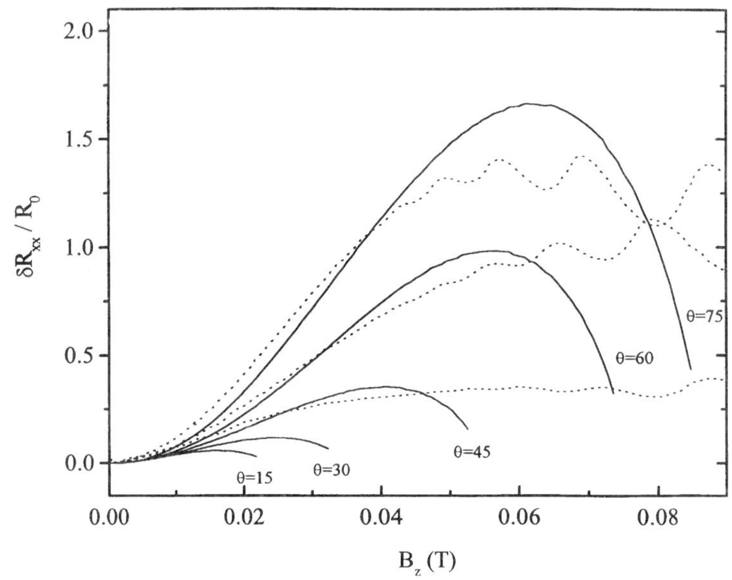

Fig. 30. Measured positive magnetoresistance (dotted curves) for different tilt angles of the magnetic field (see Fig. 29) as compared to the calculated contribution (solid curves) from the different classical orbits. (Source: Reprinted with permission from [72]. © 1997 American Physical Society.)

of motion for electrons traveling through the magnetic profiles. The numerical results are compared with the experimental results in Figure 30, and the dependence on the tilt angle is in good agreement with the experimental results, proving the existence of 1D electron states propagating along lines of zero magnetic field, that is, the snakelike orbits.

7. TWO-DIMENSIONAL MAGNETIC MODULATION

7.1. Periodic Two-Dimensional Modulation

The energy spectrum of electrons moving in a 2D lattice in the presence of a perpendicular homogeneous magnetic field is determined by fascinating commensurability effects. When the flux through a unit cell of the lattice $\Phi = Ba^2$ is a rational multiple of the flux quantum $\Phi_0 = ch/e$, that is, $\Phi/\Phi_0 = p/q$, then the resulting energy spectrum consists of broadened Landau levels, each having the same internal structure, which is called the Hofstadter butterfly [103] when plotted versus the inverse flux ratio. The width of the Landau bands oscillates as a function of the Landau quantum number and the flux ratio Φ/Φ_0. So far, this gap structure has not been observed directly in experiments. This band structure was investigated theoretically by several groups [104–108]. Recently, Gerhardts et al. [109] investigated this Hofstadter butterfly for the mixed 2D periodic electrostatic and magnetic modulation problem where they included higher harmonics and they found that the overall gap structure is similar to the one found in [103]. The classical dynamics was investigated in [110, 111]. The magnetoresitance of electrons in a periodic 2D magnetic field for different shapes of the magnetic field profile was investigated in [88, 90].

The experimental investigation of this system is still in its infancy. A 2D array of magnetic dysprosium dots on top of a GaAs/AlGaAs heterostructure was investigated in [74, 75]. The Dy micromagnets had a typical diameter of approximately 200 nm and a height of approximately 200 nm with a lattice period of $a = 500$ nm. No effects associated with the Hofstadter energy spectrum could be identified. The magnetoresistance was found [74] to exhibit commensurability oscillations very similar to those found in the case of 1D modulation (see the previous section) with minima in the magnetoresistance occurring near $2R_c/a = n + \frac{1}{2}$ with n an integer. Also effects resulting from a strain-induced electrostatic periodic potential were observed [75]. By rotating the magnetization of the Dy dots, it was possible to induce a phase change of the magnetic stray field pattern with respect to the strain-induced electrostatic potential [75, 82].

7.2. A Random Array of Identical Magnetic Disks

Here we consider a new system in which electrons are scattered by a random array of identical circular magnetic field profiles that are created by magnetic disks located a certain distance from a 2DEG. This problem is different from the magnetic flux tube problem that is encountered when a type II superconducting film is deposited on top of a heterostructure [112]. The essential difference is that in the present problem the local inhomogeneous magnetic fields have zero average magnetic field strength. Here the circular magnetic field profiles have an inner core with a magnetic field that is opposite to the outer part in such a way that the average magnetic field strength is zero.

Recently, Van Roy et al. [6, 113] successfully fabricated a grating structure of magnetic material with a pattern of magnetic dots and antidots with period of about 600 nm. A similar structure with Dy micromagnets on top of a GaAs/AlGaAs heterostructure was realized by Ye et al. [82]. An alternative approach was demonstrated by McCord and Awschalom [114], who directly deposited nanometer-scale magnetic dots using a scanning tunneling microscope. They fabricated magnetic dots of diameter ranging from 10 to 30 nm and heights from 30 to 100 nm.

The perpendicular magnetic field component produced by such a ferromagnetic disk of radius R with magnetic moment along the z direction is shown in Figure 31a for different distances between the 2DEG and the ferromagnetic disk. The corresponding vector potential is inserted into the Schrödinger equation. The radial equation contains an effective potential $V(m, \rho)$ [see Eq. (33)], which is depicted in Figure 31b for the case of $z/R = 0.02$. It is clear that such a potential can only have scattered states. The scattering properties are described in terms of a phase shift δ_m, which is determined by solving the Schrödinger equation numerically. The zero-temperature magnetoresistance and Hall resistance are determined by (see, e.g., [115])

$$\frac{R_{xx}}{R_0} = 2\sum_m \sin^2(\delta_m - \delta_{m+1}) \tag{65}$$

and

$$\frac{R_{xy}}{R_0} = \sum_m \sin\bigl(2(\delta_m - \delta_{m+1})\bigr) \tag{66}$$

where $R_0 = (\hbar/e^2)(n_0/n)$ with n_0 the density of ferromagnetic disks and n is the electron density. The numerical results are shown in Figure 32 for $b = (e/\hbar c)Ma\pi R = 2$, where M is the magnetization of the ferromagnetic disk and $z/R = 0.1$.

Notice that for small Fermi energies R_{xy} increases with k_F as is also the case when a constant magnetic field is present. This implies that the electrons are only able to probe the outer part of the magnetic field profile of the magnetic disk. For large k_F values, the electrons have sufficient energy such that they are able to penetrate the inner part of the magnetic field profile where the magnetic field is in the opposite direction. As a consequence, the average magnetic field felt by the electrons diminishes, which results in a reduction of R_{xy}. The oscillatory structure is a consequence of quantum mechanical resonances, which are a consequence of the quasi-bound states in the potential $V(m, \rho)$. Note also that at resonance R_{xy} is reduced while R_{xx} exhibits a local maximum. The reason for the different behavior is that R_{xx} measures the actual resistance while R_{xy} is a measure of the Lorentz force.

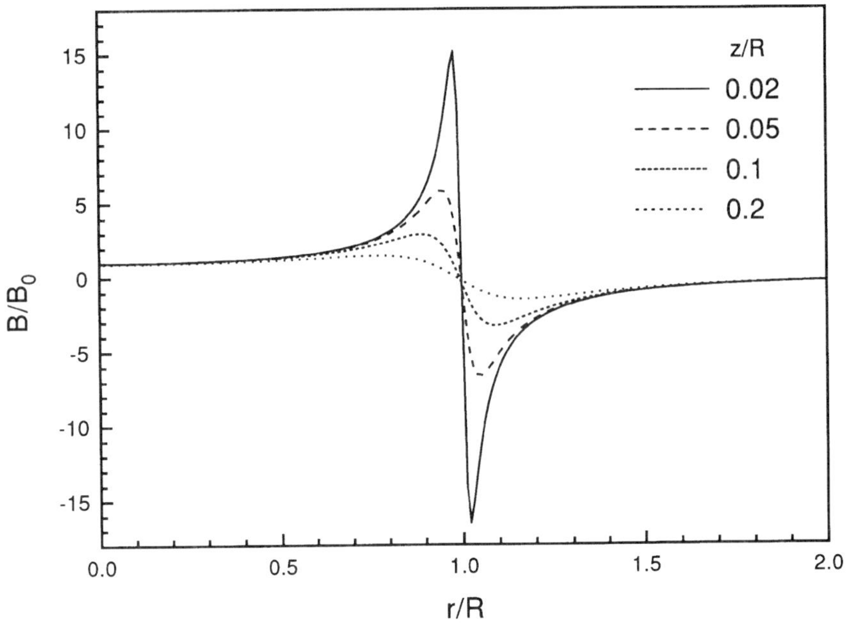

(a)

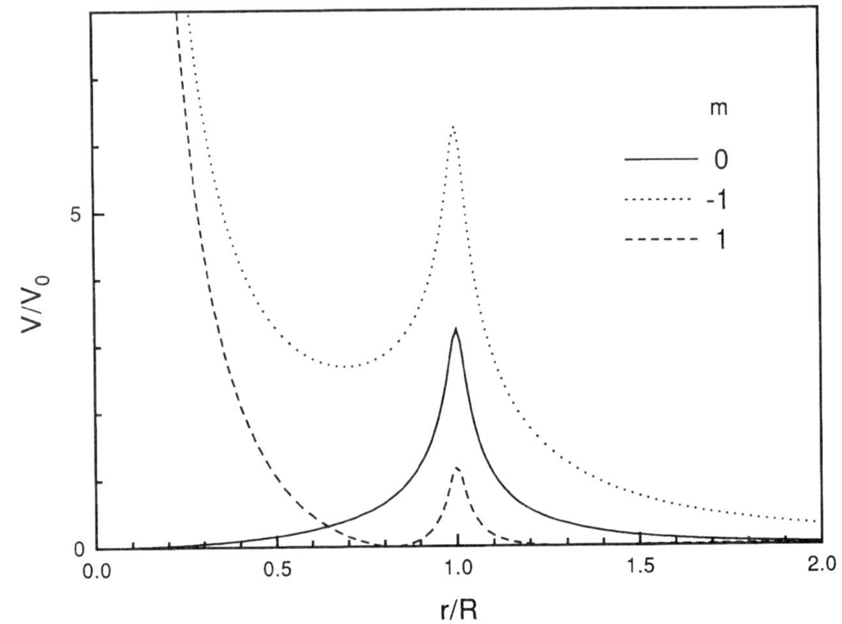

(b)

Fig. 31. (a) Magnetic field profile at a distance z under a ferromagnetic disk of radius R and (b) effective radial potential (in units of $V_0 = M/m_0\hbar^2$) for $z/R = 0.02$ and different values of the angular momentum m.

7.3. Random Magnetic Fields

The problem of electrons moving in a 2D static random magnetic field has generated great experimental [60–63, 79, 80, 112, 116–120] and theoretical [121–123] interest. It is now understood to be distinctly different from transport in systems with ordinary potential dis-

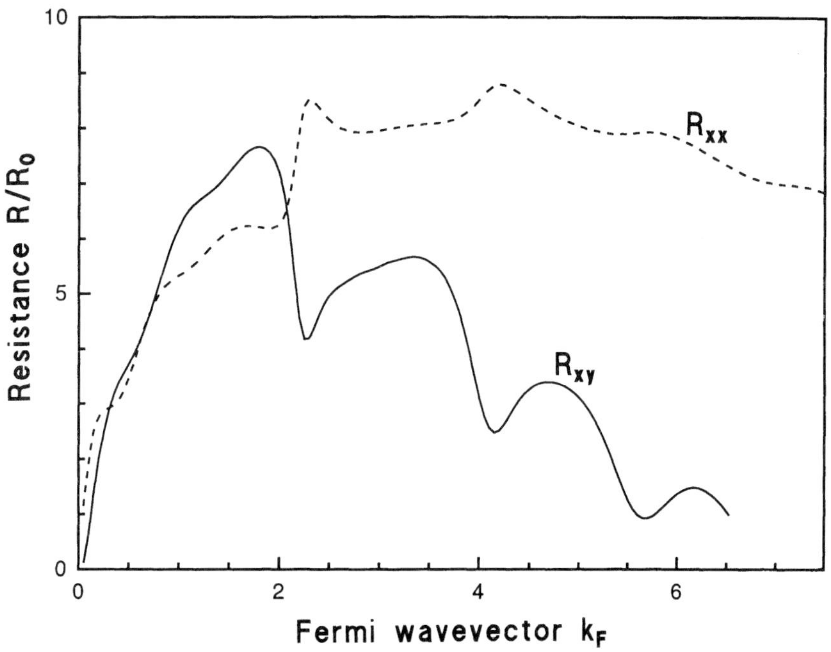

Fig. 32. Magnetoresistance and Hall resistance of a 2DEG under the influence of a random area of magnetic field profiles produced by ferromagnetic disks.

order. This problem is closely related to the 2D quantum Hall system near half integer filling factor [124–127] and the gauge field description of the high T_c superconductivity problem [128–131]. Both problems can be mapped into a system of charged particles moving in the background of a random magnetic field. The essential issue in the theoretical studies is to determine how the random magnetic field affects the transport properties. The localization of electrons in random magnetic fields is still a controversial issue. In principle, all states of a 2D system are exponentially localized according to the scaling theory of localization [132]. But the presence of a magnetic field that destroys time-reversal symmetry makes a stand against localization. Perturbation renormalization group calculations show that all states remain localized [133, 134] but numerical simulations have not led to a definite conclusion (see, e.g., [135, 136]). It is generally accepted that, except for weak localization effects [137] yielding small quantum corrections to the magnetoresistance at weak fields and at low temperatures, the semiclassical approach is an adequate framework when the typical length over which the random magnetic field varies is much larger than the Fermi wavelength of the electrons. In [138, 139], the effect of the random magnetic field was included in the collision term of the Boltzmann equation. A different approach in which the random magnetic field was included in the driving force of the transport equation was followed in [140] (see also [123]) using perturbation theory. Khveshchenko [141] analyzed the electron motion in a random magnetic field with long correlation length in the absence of impurities and derived a nonperturbative result based on an eikonal approximation. Calvo [122] used an ergodic hypothesis that allowed him to study the random magnetic field problem for arbitrary correlation length and in both the diffusive and ballistic regime.

Random magnetic fields in a 2DEG have been realized by using stochastically distributed vortices in a type II superconducting film [60–63, 112, 142–145], superconducting grains [118, 119], 2DEG grown on substrates with prepatterned submicrometer dimples [79, 80], rough macroscopic magnets [116, 117], and a random distribution of mesoscopic micromagnets [75]. The latter was achieved by depositing dysprosium dots

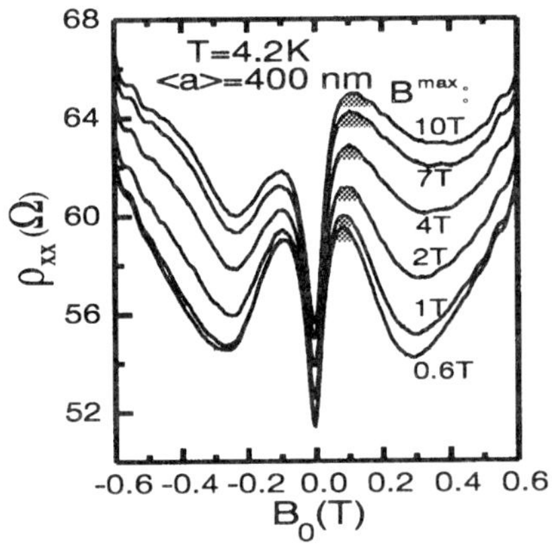

Fig. 33. Magnetoresistance of a 2DEG resulting from a random area of Dy disks with average separation $\langle a \rangle = 400$ nm as a function of the external magnetic field B_0 for different strengths of the magnetization of the disks that are determined by the conditioning field $B_{\max}$. The asymmetry is due to the hysteresis loops of the micromagnets. (Source: Reprinted with permission from [75].)

with submicrometer diameters randomly on top of a high-mobility GaAs/AlGaAs heterojunction. This system is essentially different from the flux tube system because now the local inhomogeneous magnetic fields have zero average field strength. The circular magnetic field profiles of the dot have an inner core with a magnetic field that is opposite to the outer part in such a way that the average magnetic field strength is zero.

As a typical experiment, we consider the results of Ye et al. [75]. They started from a high-mobility GaAs/AlGaAs heterostructure ($\mu = 1.4 \times 10^6$ cm^2/V s, $n_e = 2.2 \times 10^{11}$ cm^{-2} at $T = 4.2$ K) in which the 2DEG was located approximately 100 nm underneath the sample surface. A 10-nm-thick NiCr film was evaporated on top of the device which defines an equipotential plane to the 2DEG. A random distribution of Dy dots with height 200 nm and with different density was defined by electron beam lithography on top of the NiCr gate. In Figure 33, the magnetoresistance of the 2DEG at 4.2 K is displayed for a random distribution of Dy dots with an average interparticle distance of $\langle a \rangle = 400$ nm. The different traces correspond to different magnitudes of the magnetization of the micromagnets. The random magnetic field is probably accompanied by a strain-induced electrostatic random potential as discussed before for the periodic modulated magnetic field case. These random magnetic fields give rise to (1) an increase of the zero magnetic field resistance and (2) a pronounced positive magnetoresistance, both of which increase with increasing strength of the magnetization of the micromagnets. Similar behavior has been found by others in other 2DEGs with random magnetic fields.

8. HALL EFFECT DEVICES

When a magnetic field is applied to a metallic system that carries a current, a Hall voltage develops in the direction perpendicular to the magnetic field and the current because of the Lorentz force. This phenomenon is called the Hall effect [146] and has been used very successfully to obtain information on the properties of the charge carriers, for example, the sign and the density of those carriers. In most previous investigations of the Hall effect, the applied magnetic field is uniform and its strength is known. In [112, 147–150], the Hall effect of a known 2DEG was used as a probe of the magnetic field. In such a case, the experimental system consists of a microfabricated Hall bar in which an inhomogeneous magnetic field is piercing through the Hall cross. This inhomogeneous magnetic field may result, for example, from magnetic or superconducting clusters [112, 149] that are placed on top of the Hall bar.

Such Hall probes have become increasingly popular as local magnetometers for superconducting and ferromagnetic materials because of their (1) noninvasive nature; (2) high magnetic field sensitivity; (3) very small active region, that is, the cross region of the Hall bar, which, using modern microfabrication techniques, can be of submicrometer dimensions; and (4) broad temperature and magnetic field region over which they can be used.

In [151–153], a quantitative theory was presented that relates the measured Hall voltage to the local flux through the Hall probe. This is needed in order to interpret the experimental data obtained from the Hall magnetometer and to understand the mechanism behind the functioning of Hall devices. Such a theory was given in the ballistic regime [152], which is applicable at low temperatures (up to liquid nitrogen temperatures) and for small Hall probes, and in the diffusive regime [148, 153], which is more applicable for larger Hall crosses and/or for higher temperatures. We find that in the ballistic regime the relation between the measured Hall voltage and the probed inhomogeneous profile is able to provide much more detailed information than the equivalent system in the diffusive regime.

In this section, we investigate the effect of micrometer and submicrometer spatial variations of the magnetic field on a 2DEG. Knowing the response of a 2DEG to these magnetic field inhomogeneities is an important fundamental research problem in its own right but, in addition, has direct consequences on the development of microelectronic devices and applications that, in recent years, have witnessed a growing interest in exploiting the versatility and added functionality of the Lorentz force on a submicrometer scale. For instance, (i) a magnetoelectronic device in the shape of a Hall cross was recently shown to have the potential of operating as a bistable memory cell or logic gate component, (ii) with Hall probe microscopy it is possible to observe the dynamics of a single flux vortex and domain walls in superconducting films and magnetic materials, and (iii) Hall components are used as magnetic field sensors for distance and speed detection in brushless engines such as used in floppy and CD-ROM disc drives.

8.1. Ballistic Hall Magnetometry

In the inset of Figure 34a, we show the system under study, which consists of a Hall bar with four identical leads and a circular magnetic field profile situated in the middle of the cross junction. As an example, we consider two kinds of circular magnetic field profiles. The first profile is a superconducting disk in an external magnetic field. Because of the Meissner effect, the magnetic field will be expelled from the superconducting disk. We model the field distribution underneath the disk as zero inside a region of radius d, and constant outside it. In this way, we neglect the overshoot of the magnetic field at the edge of the disk, which is sensitive to several system parameters such as the thickness of the superconducting disk and the distance of the disk from the 2DEG. The second profile is a ferromagnetic disk that we model by a magnetic dipole placed a distance z_0 above the cross junction in the absence of an external magnetic field.

The Hall resistance is calculated numerically using the semiclassical formalism. The current in lead i is denoted by I_i, which can be expressed according to the Landauer–Büttiker formula as [20]:

$$I_i = \frac{2e}{h}\left[(N_i - R_i)\mu_i - \sum_{j\neq i} T_{ij}\mu_j\right] \tag{67}$$

where T_{ij} is the transmission probability for an electron from lead j to lead i and R_i is the reflection probability for returning back into the same lead i. In practice, these probabilities are calculated at the Fermi energy ϵ_F, and satisfy $\sum_{j\neq i} T_{ij} + R_i = N_i$, according to the condition of current conservation, where N_i is the number of propagating modes in lead i.

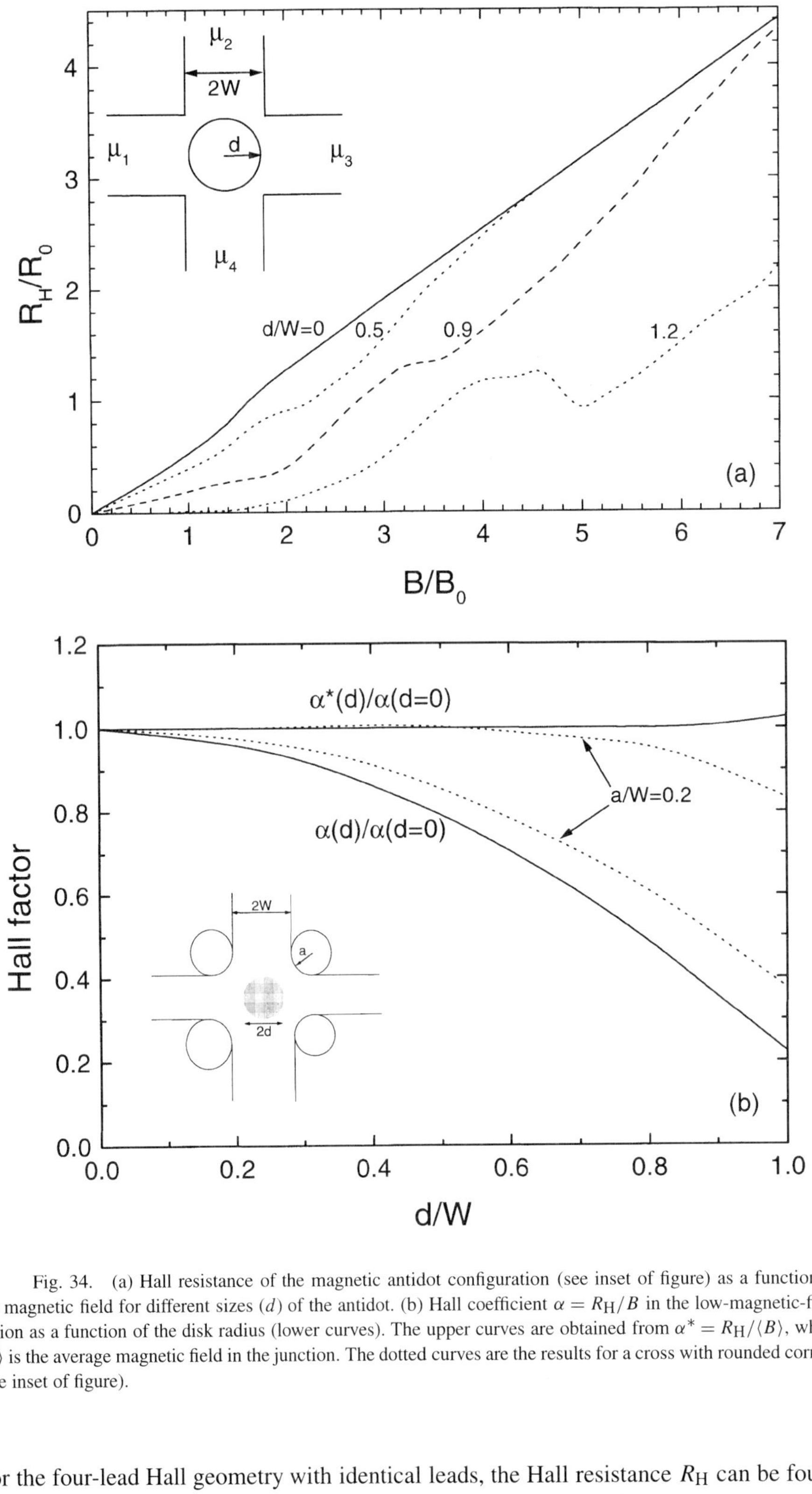

Fig. 34. (a) Hall resistance of the magnetic antidot configuration (see inset of figure) as a function of the magnetic field for different sizes (d) of the antidot. (b) Hall coefficient $\alpha = R_H/B$ in the low-magnetic-field region as a function of the disk radius (lower curves). The upper curves are obtained from $\alpha^* = R_H/\langle B \rangle$, where $\langle B \rangle$ is the average magnetic field in the junction. The dotted curves are the results for a cross with rounded corners (see inset of figure).

For the four-lead Hall geometry with identical leads, the Hall resistance R_H can be found from Eq. (67) by setting $I_1 = -I_3 = I$ and $I_2 = I_4 = 0$ and results in

$$R_H = \frac{(\mu_2 - \mu_4)/e}{I} = \frac{h}{2e^2} \frac{T_{21}^2 - T_{41}^2}{Z} \tag{68}$$

where $Z = [T_{21}^2 + T_{41}^2 + 2T_{31}(T_{31} + T_{21} + T_{41})](T_{21} + T_{41})$. For the asymmetric Hall system with either unidentical leads or asymmetric magnetic field in the cross junction, the simple formula (68) breaks down. In this case, the Hall and bend resistances should be solved from Eq. (67) by setting the same boundary conditions for the currents as used in the derivation of Eq. (68).

To obtain the probabilities T_{ij} and R_i, we follow the semiclassical approach developed by Beenakker and van Houten [154]. Thus, we neglect quantum interference effects, which are expected not to be important when $\lambda_F \ll d, W$, where λ_F is the Fermi wavelength, d the size of the magnetic field inhomogeneity, and W the width of the Hall probes. In our numerical simulation, we inject a large number of electrons ($N_e \geqslant 10^5$) toward the junction through lead 1, and follow their classical trajectories to determine the probabilities $T_{j1} = N_j/N_e$, where N_j is the number of electrons collected in lead j. Note that for the case of unidentical leads or asymmetric magnetic field profile, similar procedures should be followed for each of the four leads. The electrons are injected uniformly over lead 1, with a Fermi velocity $v_F = \sqrt{2\epsilon_F/m}$ and an angular distribution $P(\theta) = \frac{1}{2}\cos\theta$, where $\theta \in (-\pi/2, \pi/2)$ is the injecting angle with respect to the channel axis.

In the following, we will express the magnetic field in units of $B_0 = mv_F/2eW$ and the resistance in $R_0 = (h/2e^2)\pi/2k_F W$, where W is the half-width of the lead, m the mass of the electron, $k_F = (2m\epsilon_F/\hbar^2)^{1/2}$ the Fermi wavevector, and $v_F = \hbar k_F/m$ the Fermi velocity. For electrons moving in GaAs ($m = 0.067m_e$) and for a typical channel width of $2W = 1$ μm and a Fermi energy of $\epsilon_F = 10$ meV ($n_e = 2.8 \times 10^{11}$ cm^{-2}), we obtain $B_0 = 0.087$ T and $R_0 = 0.308$ kΩ.

First, we discuss the magnetic field profile that models the situation with a superconducting disk above the junction, which may be called a *magnetic antidot*. In Figure 34a, we show the Hall resistance as a function of the external applied magnetic field for different disk sizes. Notice that there exists a critical magnetic field $B_c = B_c(d)$, such that for $B > B_c$ the Hall resistance in the presence of the disk coincides with that for the homogeneous magnetic field case (i.e., $d = 0$). When $B < B_c$, the Hall resistance is sensitive to the presence of the disk. At this critical magnetic field B_c, the diameter of the cyclotron orbit, $2R_c = 2v_F/\omega_c$, equals the distance between the edge of the dot and the corner of the cross junction: $2R_c = \sqrt{2}W - d$. For $B > B_c$, the motion of the electron is determined by skipping orbits, located along the edge of the device, which do not feel the presence of the $B = 0$ region in the cross junction and the Hall resistance equals the classical 2D value: $R_H/R_0 = 2B/\pi$. We have $B_c/B_0 = 4.4, 7.8, 18.7$ for $d/W = 0.5, 0.9, 1.2$, respectively. Notice that in the absence of the magnetic antidot, that is, $d = 0$, we find $B_c/B_0 = \sqrt{2} = 1.41$, which is the field where the 2D Hall and bend resistance is recovered. For $B/B_0 \ll 1.4$, the Hall resistance is determined by the geometry of the cross [155] and for a cross with sharp corners is given by [156] $R_H/R_0 = B/2B_0$. The Hall resistance has a slope that is a factor $4/\pi = 1.27$ smaller than the classical 2D result. For $B < B_c$, the electron trajectories are only weakly bent and, consequently, they sample the $B = 0$ region.

Even in the presence of the magnetic antidot, the Hall resistance is linear for small magnetic fields (i.e., $B \ll B_c$), with a slope that decreases with the disk size, that is, with the amount of expelled magnetic flux. We analyzed the Hall factor $\alpha = R_H/B$, which is shown in Figure 34b, as a function of the disk radius (lower solid curve). α decreases with d because the average magnetic field in the cross junction decreases. Therefore, we calculate the average field in the cross junction $\langle B \rangle = [1 - \pi(d/2W)^2]B$, from which we define an effective Hall coefficient $\alpha^* = R_H/\langle B \rangle$, which is shown by the upper solid curve in Figure 34b. Notice that α^* is independent of d for $d/W < 1.0$, which indicates that for low magnetic fields the Hall resistance is completely determined by the average field in the cross region and is independent of the detailed distribution of the field. This conclusion is still true [152] (i) for a more continuous magnetic field profile such as, for example, a Gaussian; (ii) the case of asymmetric leads, that is, when the current and voltage leads have different widths; and (iii) for an off-center location of the magnetic antidot. The effect

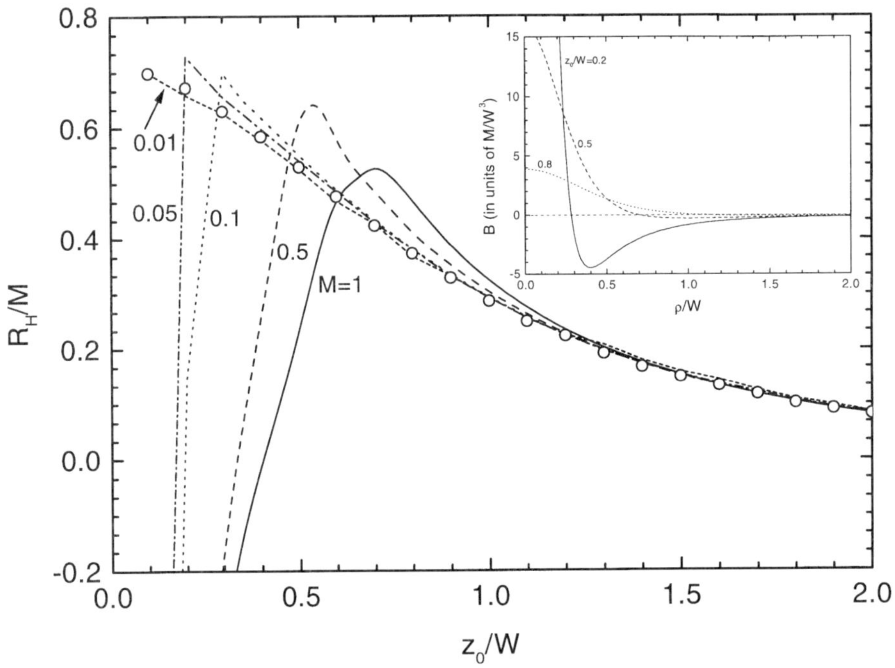

Fig. 35. Hall resistance resulting from a magnetic dipole in the center of the junction as a function of the distance (z_0) between the magnetic dipole and the 2DEG in the cross junction for different values of the magnetic moment. The dots are calculated using $R_H = \langle B \rangle/2$ (see Fig. 34). The inset shows the magnetic field profile for the dipole for different values of z_0.

of rounding the corners of the cross junction on the Hall factor is illustrated by the dashed curves in Figure 34b. We considered circular rounded corners with radius $a/W = 0.2$. It is well known [155, 156] that rounding the corners has a very important effect on the low-magnetic-field Hall resistance but we found that even for a/W up to 0.2 the low-field Hall resistance is still determined by the average magnetic field for $d/W < 0.7$.

Next, we investigate the effect of the exact magnetic field profile on the Hall measurement. Therefore, we consider a magnetic dipole for which the component of the magnetic field perpendicular to the 2D plane is given by $B(\rho) = M(3\cos^2\theta - 1)/R^3$, where $R^2 = \rho^2 + z_0^2 = x^2 + y^2 + z_0^2$, $\cos^2\theta = z_0^2/R^2$, z_0 is the distance between the dipole and the 2D plane, and M is the dipole moment. In Figure 35, we show the scaled Hall resistance R_H/M as a function of the distance z_0 for different values of M, the strength of the dipole. Notice that, for large z_0, the Hall resistance R_H scales perfectly with M. This feature can be understood by the fact that the Hall resistance is proportional to the average magnetic field over the junction region. The result $R_H/R_0 = \langle B \rangle/2B_0$ is shown by the open dots in Figure 35 and reproduces our numerical results. However, with decreasing z_0, the Hall resistance deviates from the preceding simple formula, because of the strongly nonuniform distribution of the magnetic field (see inset of Fig. 35). The validity range of the simple formula for R_H is determined by the maximum magnetic field in the cross junction, that is, $B(\rho = 0)$, such that the corresponding cyclotron diameter equals the width of the lead, that is, $R_c = W$, which leads to $z_0 = (2M)^{1/3}$. For $M = 0.01, 0.05, 0.1, 0.5, 1$, this gives, respectively, $z_0/W = 0.3, 0.5, 0.6, 1, 1.3$, which agrees with Figure 35. For small z_0, $B(\rho)$ has a large positive peak at the origin (see inset of Fig. 35) that is not sampled by the electrons and the Hall resistance is determined only by the negative magnetic field region, which results in a negative Hall resistance. In this regime, the Hall resistance is no longer determined completely by the average field in the cross junction.

8.2. Hall Magnetometry in the Diffusive Regime

In the diffusive regime, the approach outlined in Section 5 has to be used. Thus, the continuity equation is solved numerically using finite-difference methods subjected to the mixed boundary conditions $\phi(x, 0) = 0$ and $\phi(x, L) = V_0$ at the current contacts and the condition that the normal component of the current is zero (i.e., $\mathbf{J}_n = 0$) at all boundaries of the system other than the current contacts. In our numerical analysis, we took the length of the voltage leads to be the same as the length of the current leads of the Hall cross and each with aspect ratio one. We checked that our results are practically not altered by increasing the length of the voltage leads [153].

As in the previous subsection, we consider first a magnetic antidot with a magnetic field profile: $B(r) = B_0\theta(r - r_0)$, where r_0 is the radius of the antidot with origin at the geometrical center of the cross junction. In Figure 36, the steady-state equipotential contours in the cross-junction system for two magnetic field strengths are shown: (a) $\mu B_0 = 1$ and (b) $\mu B_0 = 10$ and for a disk radius of $r_0/W = 0.3$. The current is injected at $y/W = 0$ and leaves the system at $y/W = 3$. The zero magnetic field region (i.e., the antidot) is the white disk in the center of the Hall cross. Note that for $\mu B_0 = 10$ these contours are nearly the same as the current flow lines followed by the center coordinate of the cyclotron orbits while drifting according to the $\mathbf{E} \times \mathbf{B}$ force with drift velocity $v_d = E/B$. The Hall angle in the nonzero field region is $\theta_H = 84.3°$ owing to the finite mobility. There is a Hall voltage buildup between the voltage contacts at $x/W = 0$ and $x/W = 3$ and the potential inside the antidot region is flat; that is, the electric field inside the antidot is almost zero and, hence, there is no current inside the $B = 0$ antidot region. This flat potential region occurs because the zero field conductivity inside the antidot is larger than any of the components of the conductivity tensor outside the antidot. Consequently, the charges inside the antidot are more mobile and shorten out the potential difference inside the disk, resulting in nearly zero electric field in the antidot and a dipolar excess charge around the periphery of the antidot that screens the inside of the disk from the electric field. In Figure 36a, the Hall angle is 45° and there is still significant potential drop across the antidot, making charge flow possible through the antidot. There is an alternative physical explanation for the behavior of the equipotential contours in parts a and b of Figure 36. Because of the Lorentz force, the current flows along the edges of the 2DEG. This is more so the larger the magnetic field. At the cross junction, the current will prefer to flow along the edge of the voltage leads and in this way will avoid the central cross region and, consequently, the $B = 0$ region.

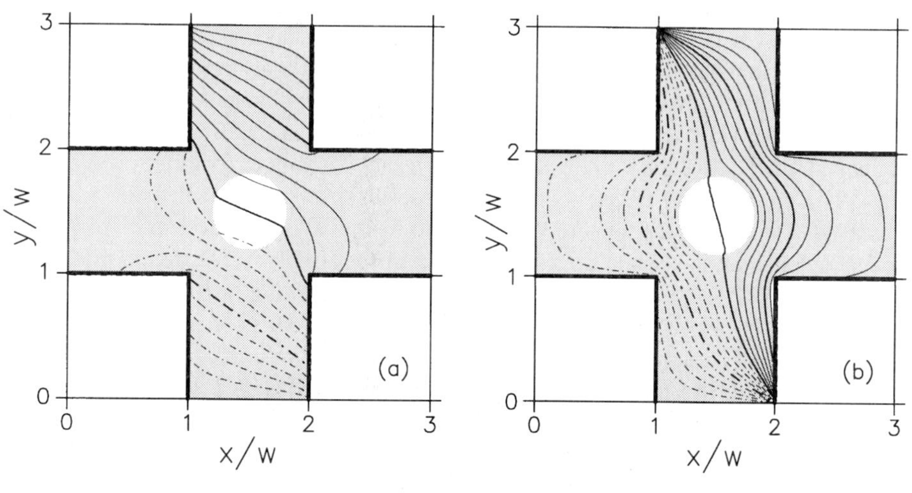

Fig. 36. Steady-state equipotential contours in the Hall junction with a magnetic antidot (white disk) in the center of the cross junction. The gray shaded region is the region of nonzero magnetic field. The radius is $r_0/W = 0.3$ and the magnetic field strength: (a) $\mu B_0 = 1$ and (b) $\mu B_0 = 10$.

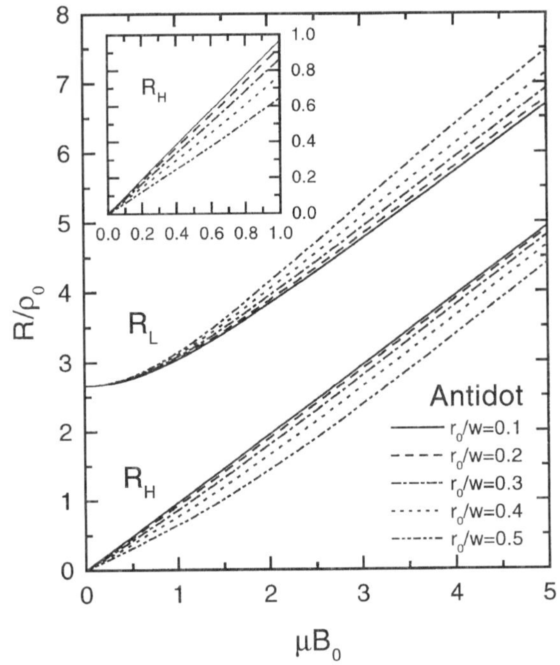

Fig. 37. Hall resistance and the longitudinal resistance versus μB_0 for the magnetic antidot. Inset shows the Hall resistance for small μB_0 values.

The Hall resistance and the longitudinal resistance are shown in Figure 36 as a function of μB_0 for different values of r_0/W. Notice that when $\mu B_0 < 1$ (see inset of Fig. 37) the Hall resistance is linear in μB_0 with a decreasing slope with increasing r_0. For $r_0 = 0$, we recover the well-known result $R_H = B/n_s e = \rho_0 \mu B$. The quenching of the Hall resistance increases as r_0 increases because the average magnetic field in the active area of the junction decreases as the magnetic flux is removed in proportion to the area of the antidot (i.e., πr_0^2). In contrast to the Hall resistance, the longitudinal resistance increases with increasing radius of the antidot, which can be explained by the fact that the contribution to the dissipative transport increases with the zero magnetic field area and the amount of transport current going through it, which is inversely proportional to the field strength. Another contribution to the longitudinal resistance is due to the fact that the current paths followed by the charge carriers become longer at high fields as the antidot radius is increased. To investigate how the Hall resistance changes with r_0, we plot the Hall factor $\alpha_H = R_H/(\rho_0 \mu B)$ for different values of B_0 as a function of r_0 in Figure 38. In general, the Hall factor is defined for $B_0 \to 0$, but we notice from Figure 38 that even for $\mu B_0 = 0.5$ the Hall factor is still very close to its $B_0 \approx 0$ result. Following [152], we calculate $\alpha_H^\circ = R_H/\langle B \rangle$ where the magnetic field is averaged over the cross junction (the cross junction has a unit area if expressed in units of W^2). We found that, in general, $\alpha_H^\circ \langle B \rangle / B \geqslant \alpha_H \geqslant \frac{1}{2}\alpha_H^\circ \langle B \rangle / B$, which is in contrast to the ballistic regime where it was found that $\alpha_H \approx \alpha_H^\circ \langle B \rangle / B$ for r_0/W up to almost 0.4–0.5. Therefore, in Figure 38, we also show another effective Hall factor defined as $\alpha_H^* = R_H/\langle B \rangle_2$ where now the magnetic field $\langle B \rangle_2$ is averaged over the region $W/2 < x < 5W/2$, $W < y < 2W$, which is twice the junction area of the cross. For the antidot system with $r_0/W \leqslant 0.5$, we have $\langle B \rangle = [1 - \pi(r_0/W)^2]B_0$, while $\langle B \rangle_2 = [1 - (\pi/2)(r_0/W)^2]B_0 \geqslant \langle B \rangle$. Notice that α_H^* is independent of the size of the disk for $r_0/W \leqslant 0.5$. Therefore, given the previous effective Hall factor, a measurement of the Hall resistance allows the experimentalist to obtain information on the size of the antidot.

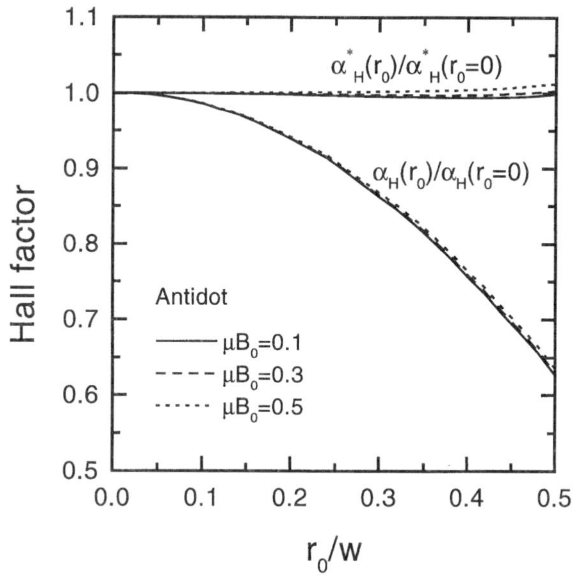

Fig. 38. Hall coefficient α_H and the effective Hall coefficient α_H^* for the magnetic antidot as a function of the antidot radius R_0 for different μB_0 values.

As compared to the ballistic regime, in the diffusive regime the Hall factor for the magnetic antidot system is larger, implying that in the diffusive regime the active area of the Hall cross is larger than in the ballistic regime. This is due to the spread of part of the transport current into the voltage probes, leading to an increased active area. We found numerically that the effective area in the diffusive regime is about two times larger than the cross area. A similar factor of 2 was recently found in [147, 148] for the case of a magnetic flux vortex penetrating a Hall junction.

The longitudinal resistance is found to be quadratic in B in the small B region and linear in B for large B values. Notice that at $B_0 = 0$ we have $R_L = 2.666\rho_0$. For a macroscopic sample, we have $R_L = (L/W)\rho_0$ with L the length and W the width of the 2DEG. Here $L = 3W$ and, consequently, one would expect $R_L = 3\rho_0$ at $B_0 = 0$. The presence of the voltage probes makes the 2DEG locally wider and, consequently, the effective width of the 2DEG is larger than W, resulting in a smaller longitudinal resistance. Furthermore, R_L increases with increasing disk radius r_0, which is opposite to the behavior of R_H. Notice that for large B values the slopes of R_L and R_H become equal because the longitudinal resistance includes contributions from both the magnetoresistance and the Hall resistance. At small magnetic fields, the magnetoresistance gives the dominant contribution and, consequently, R_L is quadratic in B. When B is large, the contribution from the magnetoresistance is less important and the Hall resistance becomes the dominant part of the longitudinal resistance and, consequently, R_L is linear in B. Notice that in the present units we have $dR_H/dB = dR_L/dB = 1$ for large magnetic fields, which implies that R_L and R_H are linear in B with coefficient $1/n_s e$. Notice also that $B/n_s e$ is nothing else than the classical 2D Hall resistance in the absence of the antidot. Both R_H and R_L are, for large B, shifted by a constant value that depends on r_0. The dependence of the Hall resistance on the position of the antidot was investigated in [148, 153].

As a last example, we studied the magnetic field profile resulting from a magnetic dipole placed above the junction and aligned perpendicular to the junction plane, that is, $B(r) = M(3\cos^2\theta - 1)/R^3$ where $R^2 = r^2 + z_0^2 = x^2 + y^2 + z_0^2$, $\cos^2\theta = z_0^2/R^2$, z_0 is the distance between the dipole and the plane of the junction, and M is the dipole moment. This system models the magnetic profile of a circular single-domain ferromagnetic cluster

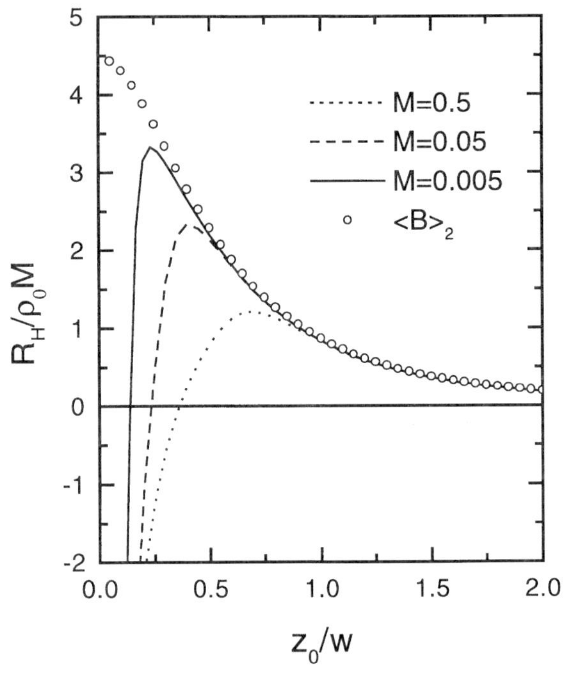

Fig. 39. Hall resistance versus the distance of the magnetic dipole from the Hall bar z_0 for different values of the strength of the dipole. The circles are obtained from $R_H = \langle B \rangle_2$.

that is placed above the Hall cross. Figure 39 shows the Hall resistance as a function of the perpendicular distance of the dipole above the junction z_0, where the Hall resistance is normalized by the magnetization M. For large z_0 values, the Hall resistance scales perfectly with the magnetization and is, to a good approximation, given by the average field in the effective area of the junction. This situation is valid up to smaller values of z_0 when the magnetization is decreased. Notice that for small z_0 values R_H is no longer related to the average field in the effective area of the junction and, in fact, it has even the opposite sign, which is the sign of the dipole field in the current and voltage probes. This implies that, as in the ballistic regime [152], the electrons do not sample the whole magnetic field profile in the junction. The reason is that they cannot penetrate into the strong magnetic field peak region in the center of the junction, which gives the dominant contribution to the average magnetic field in the active area. Consequently, $\langle B \rangle_2$ leads even to the wrong sign for R_H and, therefore, the simple expression $R_H = \langle B \rangle_2$ is valid only when z_0 is large or the magnetic dipole strength is weak.

Next, we discuss the effect of changing the relative ratio of the widths of the Hall voltage probes (W_y) to the current probes ($W_x = W$) on the Hall response of the cross-junction device. As an example of a magnetic field profile, we take the one of a magnetic dot, that is, $B(r) = B_0\theta(r_0 - r)$. In Figure 40, the Hall resistance and the effective Hall factor of the magnetic dot are shown as a function of the voltage probe width for $r_0/W = 0.25$ and $\mu B_0 = 0.3$. Notice that the effective Hall factor decreases with increasing deviation of the Hall probes from the symmetric case. For $W_y/W \leqslant 1.0$, this can be understood as being due to the decreasing averaging magnetic field in the Hall cross, that is, $\langle B \rangle = B_0\pi(r_0/W)^2/(W_y/W)$, with increasing W_y/W. Indeed, for $0.5 \leqslant W_y/W \leqslant 1.5$, we find $\alpha^\circ = R_H/\langle B \rangle \approx 0.35$–$0.5$ practically independent of W_y for $r_0/W < 0.5$. For Hall probes that are narrower than the current probes, the Hall resistance decreases faster with W_y/W than expected from $\langle B \rangle$. This implies that in such a case the effective area is smaller than twice the Hall cross region. The Hall factor is nearly half of the one given by the average

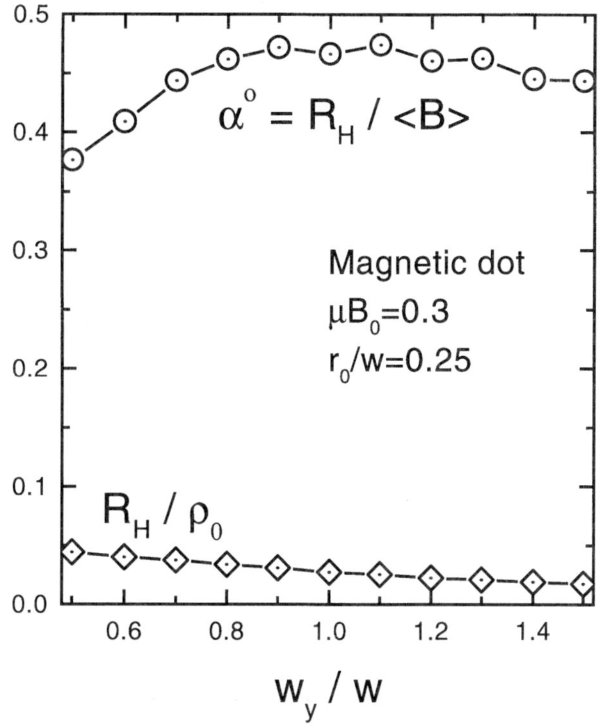

Fig. 40. Hall factor $\alpha_{\mathrm{H}}^{\circ}$ and the Hall resistance R_{H} as a function of $W_y/(W_x = W)$ for a magnetic dot with a radius of $r_0/W = 0.25$ and a magnetic field strength of $\mu B_0 = 0.3$.

magnetic field $\langle B \rangle$ when $W_y/W = 1$, but decreases slightly when the ratio of the width of the leads deviates from unity. This is different from the ballistic regime [152] where it was found that the small magnetic field Hall resistance is determined by the average magnetic field in the cross junction independent of the relative size of the Hall probes.

In the diffusive regime, the Hall resistance depends on the relative average velocity between the magnetic flux and the electric charges through the Lorentz expression of the induced electric field $\mathbf{E} = -\mathbf{v} \times \mathbf{B}(r)$, where the average velocity in this expression is influenced by the width of the Hall voltage probe (see also [153]). By changing the width of the voltage probe, we are also changing the average magnetic field in the junction area of the cross (i.e., the active area), and, therefore, the Hall resistance depends on both the average magnetic field and the average velocity in the active area.

8.3. Hybrid Hall Effect Device

As an example of the application of the Hall effect in novel magnetoelectronic devices, we consider the system recently proposed by Johnson et al. [157] and Monzon et al. [158]. They fabricated a new hybrid Hall device, incorporating a microstructured ferromagnetic film and a micrometer-scale Hall cross, which has potential applications as a magnetic field sensor, nonvolatile storage cell, or logic gate. Their experimental measurements showed this new device to have excellent scaling properties and output characteristics at room temperature.

The top view of the device configuration is shown in Figure 41. In [157], a thin metal ferromagnetic film F with thickness $d_{\mathrm{f}} = 0.15\ \mu$m was deposited a distance $h = 0.076\ \mu$m above a Hall cross made from a high-mobility InAs film [157] (or a GaAs/AlGaAs heterostructure [158]) containing a 2DEG. F and the 2DEG are electrically isolated from each other. The ferromagnetic film has an in-plane magnetization in the x direction (taken

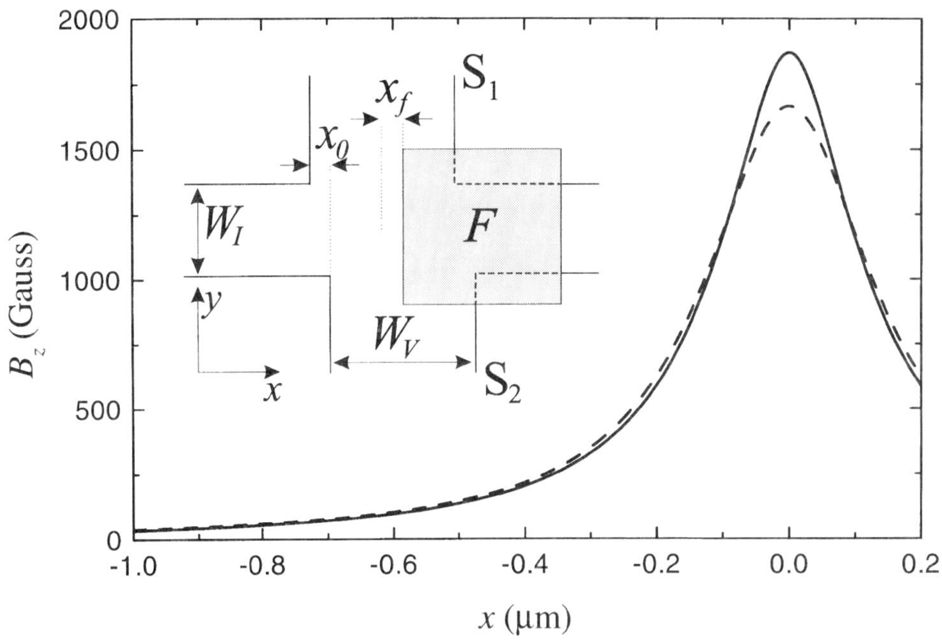

Fig. 41. Comparison between the analytical calculated magnetic field profile B_z in the 2DEG along the center of the current probe (solid curve) and the one based on the line charge model (dashed curve). The inset shows the top view of the configuration.

along the current direction), which will result in an inhomogeneous B_z profile in the 2DEG. When a voltage is applied in the x direction, a current will flow, and this will generate a measurable Hall voltage V_H over the voltage probes S_1 and S_2. An external applied magnetic field in the x direction sweeps the magnetization, resulting in changes in V_H. The current probes and the voltage probes of the Hall cross have, respectively, widths W_I and W_V, which are in the micrometer region. S_1 and S_2 may be shifted a distance x_0 from each other, for reasons described later. F is square shaped with typical dimension $W_I + 2\ \mu$m and is located with one edge at distance x_f from the center of the Hall cross. In the y direction, the configuration is symmetric.

The experiments of [157] were done at room temperature and, consequently, our earlier approach [39, 153], which relies on a classical diffusive model to describe the electrical transport, is applicable. In the following, we restrict the calculations to a $(3W_V, 3W_I)$ region centered around $(0, 0)$, with the boundary conditions $\phi(-3W_V/2, y) = 0$ and $\phi(3W_V/2, y) = 1$, and the condition that no current can flow out of the sides of the sample, that is, $j_y(x, -3W_I/2) = j_y(x, 3W_I/2)$. We checked that our results were practically not altered by increasing the length of the voltage leads. The Hall resistance is given by $R_H = V_H/I$, where $V_H = \phi(x, -3W_I/2) - \phi(x, 3W_I/2)$, and $I = \int j_x(x, y)\, dy$ is the total current flowing in the direction of the applied field. In our numerical analysis, we used $\sigma_0 = 173.611\ \Omega$, which corresponds to the mobility $\mu = 20{,}000$ cm^2/V s and the electron density $n_s = 1.8 \times 10^{12}$ of the experimental system of [157] at $T = 300$ K.

The magnetic field profile in the 2DEG, resulting from the ferromagnetic film, was calculated (see Section 3) and is shown in Figure 41. We consider the situation of full saturation of the magnetization, which for the Permalloy material used in [157] yielded $m_x = 860 \pm 20$ emu/cm^3 at room temperature. This result is compared with the line charge model used in [157, 158], which gives $B_z(x) = 2\lambda_m R/((x-a)^2 + R^2)$, where $\lambda_m = m_x d_f$ is the magnetic charge density, $x - a$ is the lateral distance to the edge of F, and $R = h + d_f/2$ is the distance of the approximated line of magnetic poles to the 2DEG. Note that

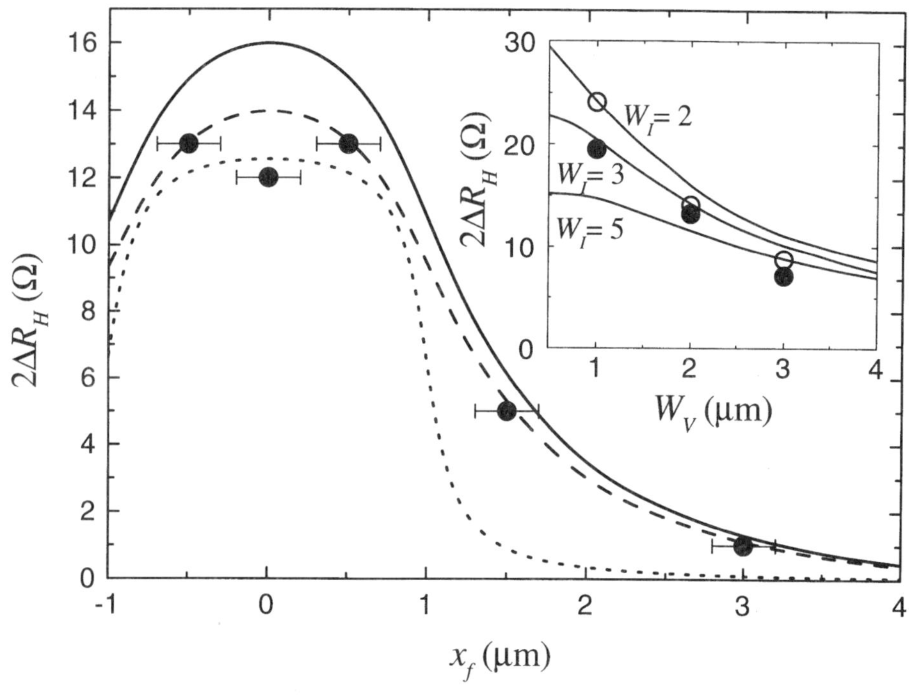

Fig. 42. Maximum change of the Hall resistance $2\Delta R_{\mathrm{H}}$ as a function of x_{f} for the $(W_V, W_I) = (2, 3)$-μm configuration at $T = 300$ K. The symbols are the experimental results and the theoretical results are given by the solid curve and the dashed curve when we alter the magnetic film thickness to $d_{\mathrm{f}} = 0.13$ μm. The dotted curve (which is reduced by a factor of 2) is based on the simple formula $2\Delta R_{\mathrm{H}} = 2\langle B_z\rangle/n_{\mathrm{s}}e$. The inset shows $2\Delta R_{\mathrm{H}}$ for various sizes of the current Hall probes W_I as a function of W_V (solid curves). The experimental values (solid dots) are compared with the calculated values (hollow dots).

there is good agreement between both curves (on a 10%–15% level), and using the line charge model was, at least in this case, justified.

Inserting this calculated magnetic profile in (37), we solve the partial differential equation (36) numerically. We only consider the case when the magnetization is saturated. First, we will study the symmetric configuration, where $x_0 = 0$. In Figure 42, the peak magnitude $2\Delta R_{\mathrm{H}}$ is plotted as a function of x_{f} using the parameters as given by Johnson et al. Our results (solid curve) slightly overestimate the experimental results (solid circles). Nevertheless, these calculations are in much better agreement than those based on the approximation that the Hall resistance is given by the average magnetic field in the cross region (dotted curve, which is reduced by a factor of 2), $2\Delta R_{\mathrm{H}} \approx 2\langle B_z\rangle/ne$. Note that the latter approach only works well in the ballistic regime [152]. Possible reasons for the discrepancy between experiment and theory are (1) it is possible that the magnetization of the film is not yet totally saturated, so $m_x < 860$ emu/cm^2; (2) the distance h between the magnetic film and the 2DEG or (3) the thickness d_{f} of the magnetic film can be different from that given in [157]. To find agreement, we have to increase h from 0.076 μm to 250 μm, which is unrealistically large, or we have to decrease the thickness of the magnetic film d_{f}. Decreasing this thickness to $d_{\mathrm{f}} = 0.13$ μm gives (see dashed curve in Fig. 42) excellent agreement with experiment. This also could be the reason why in the inset of Figure 42, where the experimental peak magnitudes $2\Delta R_{\mathrm{H}}$ (solid circles) for systems with different dimensions (1 μm, 2 μm), (2 μm, 3 μm), and (3 μm, 5 μm) are plotted, are theoretically (open circles) overestimated.

When the Hall cross is fabricated symmetrically, that is, $x_0 = 0$, the hybrid Hall device is characterized by a bipolar output $\pm\Delta R_{\mathrm{H}}$ as B is swept from positive to negative.

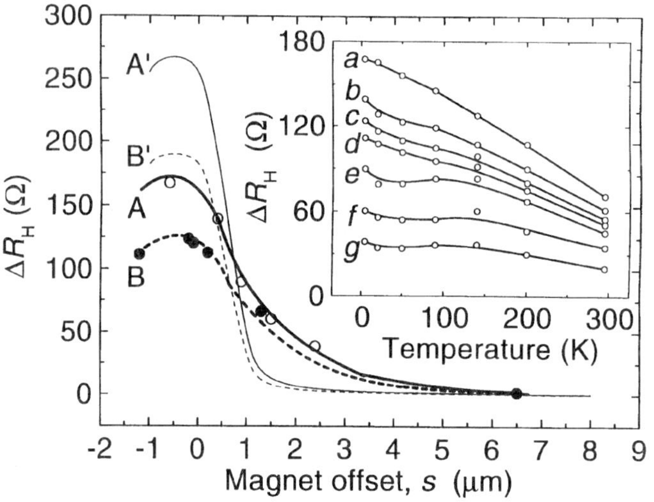

Fig. 43. Same as Figure 44 but now for $T = 4.2$ K. The experimental curves (A and B) are for two different thicknesses of the magnetic film. The theoretical results (A′ and B′) are obtained by averaging the magnetic field in the cross junction. The inset shows the temperature dependence of $\Delta R_{\rm H}$ for devices with different values of the magnetic offset $s = x_{\rm f}$: (a) $s = -0.6$, (b) 0.4, (c) -1.2, (d) -0.2, (e) 0.9, (f) 1.5, and (g) 2.4 μm. (Source: Reprinted with permission from [158]. © 1997 American Institute of Physics.)

Introducing a small asymmetry, a shift x_0 of voltage probes S_1 and S_2, results in a small offset resistance R_0. As shown in [157], this makes it possible to shift the output values and to obtain output characteristics varying, for example, between 0 Ω and $2\Delta R_{\rm H}$.

The room temperature experiments of [157] were extended in [158] to liquid helium temperatures and the full temperature range of the output signal was investigated. They found that the magnitude of $\Delta R_{\rm H}$ decreases with increasing temperature (see inset of Fig. 43). This can be attributed to (i) an increase of the electron density ($R_{\rm H} \sim 1/n_{\rm e}$) with increasing temperature, and (ii) the transition from the ballistic regime at low temperatures to the diffusive regime at high temperatures, which decreases the Hall resistance as discussed in the previous two subsections because of the larger effective area over which the magnetic field has to be averaged. The dependence of $\Delta R_{\rm H}$ on the relative magnetic film position s (cf. Fig. 42 for $T = 300$ K) as obtained at $T = 4.2$ K ($n_{\rm e} = 1.5 \times 10^{11}$ cm^{-2}, $\mu = 1.1 \times 10^6$ cm^2/V s) from ten devices on two separate chips (A and B, with five devices each) is shown in Figure 43. The overall behavior is similar to that seen at room temperature. The line charge model was used to obtain the magnetic field distribution inside the 2DEG and perpendicular to the 2DEG (see [158]). For chip A (B), they used for the saturation magnetization $m_x \approx 838$ emu/cm^3 (838), $d = 150$ nm (105) and for the position of the line charge above the 2DEG 152 nm (130). The curves labeled A′ and B′ are the theoretical results for $\Delta R_{\rm H}$ obtained by averaging the magnetic field over the Hall cross region using the magnetic field distribution obtained from the line charge model in which the magnetic charge was distributed over a distance $\delta = 1$ μm. For $s < 1$ μm, the Hall resistance is overestimated by about a factor 1.5 for both cases. For $s > 1$ μm, the theoretical model predicts a substantially smaller output then found experimentally. The latter is more similar to the output found at room temperature [157] (see Fig. 42). This is surprising in view of the theoretical response of the ballistic Hall magnetometer discussed in [152] and in Section 8.1 where the response was found to be due to the average magnetic field in the cross. It is possible that the magnetic field strength is too large such that the electrons do not sample the complete magnetic field profile as discussed in Section 8.1 for the magnetic dipole profile. The discrepancy may also suggest that the local magnetic field profile is more complicated than the one given by the distributed line charge model.

9. NONPOLARIZED CURRENT INJECTION FROM SEMICONDUCTOR INTO FERROMAGNETS

Since the discovery of the so-called giant magnetoresistance (GMR) effect in metallic magnetic multilayers [159], a huge amount of research has been done on the properties of spin-dependent scattering in magnetic thin films and multilayers [160]. Device demonstrations where semiconductor and ferromagnetic layers are both in the current-carrying parts of solid-state devices are still very few.

One recently demonstrated device [161] is, in essence, a metal base hot-electron transistor and relies on the spin-dependent scattering of hot electrons in a magnetic metal multilayer [162, 163]. In this respect, the device structure is different from the standard all-metal GMR structures where conduction takes place at the Fermi level. The vertical spin valve transistor is a semiconductor-based device comprising a magnetic multilayer base, with a Si emitter and collector (Fig. 44). It was shown that this type of device structure can be fabricated successfully using low-temperature wafer bonding technology in a UHV sputter deposition chamber [164].

Let us take a closer look at the operation of this component. Hot electrons are launched by the Schottky barrier of the emitter into the base. In the base, the carriers are subject to spin-dependent scattering that determines their mean free path. The degree of this scattering will determine the collected carriers in the collector and can be controlled by controlling the magnetic state of the base. The combination of the perpendicular trajectory of the electrons and the exponential dependence of the collector current on the mean free path of the electrons in the multilayer is responsible for a very large change in collector current upon change of the magnetic state of the base. The metal base is typically a spin valve multilayer that is known to give relative resistance changes of 1% to 10% upon magnetization change. The spin valve transistor has been shown to give a collector current change of over 100%. Although the control of the magnetic base layer thickness is nanoscale, the lateral dimensions are currently technology limited to some tens of micrometers. In a new generation of such magnetic hot-electron devices, a metal–insulator tunnel emitter [165, 166] may replace the emitter semiconductor Schottky barrier, leading to a planar technology for the realization of this component, improving the application potential for integrated circuits (e.g., for MRAM).

The electrons arriving in the semiconductor collector of the structure may have a certain degree of spin polarization. The device operation, however, does not depend on the spin

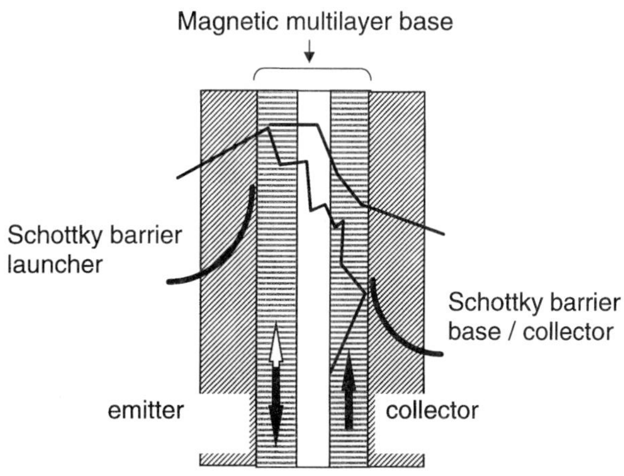

Fig. 44. Schematic cross section of a hot-electron spin valve transistor. Semiconductor emitter and collector are separated by a magnetic multilayer base.

polarization of this current. In the following sections, we will discuss device structures where spin-polarized electron transport across FM/SC interfaces is essential to their operation.

10. SPIN INJECTION FERROMAGNETIC/SEMICONDUCTOR STRUCTURES

10.1. Spin-Polarized Electronic Current from Ferromagnets

Because of the imbalance in spin population of a ferromagnetic metal, the electronic current emanating from a ferromagnetic electrode is spin polarized. This has been experimentally demonstrated in the pioneering work of Messervey et al., who injected current from a ferromagnetic into a superconductor (Al) across an Al_2O_3 barrier [167]. The degree of spin polarization was found to be 10% to 50% for Ni–Fe alloys in such a tunneling experiment with a superconductor/tunnel oxide/ferromagnetic metal trilayer [168]. The process of injecting an electric current with a spin imbalance is commonly referred to as spin injection.

A straightforward way to inject spin-polarized electrons in a nonmagnetic metal is to use a ferromagnetic contact and drive current through the interface [169]. For an ideal interface, this current would carry magnetization (spin information from the ferromagnetic): $\mathrm{Im} = \eta\beta I_e/e$. This formula simply states that with each electron crossing the interface (I_e/e), one Bohr magneton β is associated and that there will be an imbalance in the spin population of the current to a degree of η ($|\eta| \leqslant 1$). The (sign of the) coefficient η depends, in general, on the contributions from both spin bands, the hybridization of bands and interface effects. Associated with this magnetically polarized electric current is a spin accumulation and an induced nonequilibrium magnetization in the normal metal.

In principle, placing a second ferromagnetic contact to the nonmagnetic metal will enable one to probe (in an open-circuit voltage measurement) the polarization of the carriers at the position of that second ferromagnet [170]. The analysis of the coupling of the electron charge and the magnetization and the experimental demonstration of the injection/detection effect were reported by Johnson and Silsbee [171–173]. In this ferromagnetic/normal metal/ferromagnetic geometry, in which the current traverses the device's interfaces perpendicularly [so-called, current perpendicular to the plane (CPP) geometry], there are two approaches to theoretically explain the interface resistance associated with the spin-dependent signature of the transport measurements, based on thermodynamic arguments [174] and based on the theory and experiments of a spin scattering and relaxation in giant magnetoresistance multilayers [175]. The conclusion, however that the spin-polarized current can be used in electronic devices including ferromagnetic injectors is correct and exciting.

10.2. Optical Detection of Spin-Polarized Tunnel Current

The preceding experiments introducing spin injection are performed using either all-metal or metal/superconductor structures, but spin injection will also work in a semiconductor.

In fact, it was shown by optical means that a current from a magnetic (nickel) needle across a vacuum barrier into GaAs is spin polarized [176]. The geometry in this experiment, which makes use of scanning tunneling microscopy (STM), is as follows. The injector needle's magnetization is along its length axis, pointing perpendicular to the semiconductor surface. Carriers were injected from the tip over a vacuum barrier into the surface depletion zone of a GaAs wafer. The spin orientation direction of the injected electrons will also be perpendicular to the semiconductor surface. The ensemble of electrons injected in the semiconductor will have a net polarization, provided that the spin information is not lost while traversing the vacuum/semiconductor interface. Because of the spin polarization of the injected electrons, their radiative recombination in the depletion region results in circularly polarized light emission from the semiconductor. The degree of polarization of the luminescence is a measure of the injected spin imbalance, taking into account the spin rotation processes and spin relaxation in the depletion region.

Note that in this experiment the sign of the polarization was found to be negative ($\eta = -0.31$). In another tunneling experiment involving Ni as a spin injector [177] into superconducting Al, the spin polarization was found to be positive. The latter result agrees with theoretical considerations [178–180] that in tunneling from a ferromagnetic transition metal alloy, the current is dominated by itinerant majority-spin carriers from the hybridized s–d band although around the Fermi level the minority spins have a higher density of available states. This contradicts the experimental results of Alvarado and Renaud [176], who showed that conduction from the 3d states at the Fermi level is dominant.

10.3. Spin-Polarized Electronic (Tunnel) Current from Optically Pumped Semiconductors

Also, the inverse configuration of the preceding experiment has received considerable attention: the emission of spin-polarized electrons from an optically pumped GaAs structure. Because of the spin–orbit splitting energies and the selection rules in GaAs, circularly polarized light induces electronic spin imbalance in GaAs.

The spin polarization of electrons photoemitted from negative-electron-affinity (NEA) thin GaAs photocathodes is well known. Highly doped p-type gallium arsenide with a thin surface coating of cesium with oxygen is used to create emitters with a negative electron affinity. Theoretical studies predict the intensity and spin polarization of photoelectrons excited by circularly polarized radiation from the GaAs(110) surface [181].

It has been demonstrated that these NEA (Al)GaAs sources [182] have a limited ($\approx$40% to 50%) but rather unstable polarization. To prevent the degradation of the electrode surface, oxide barrier films were suggested to protect the emitting surface [183]. The vacuum level and the surface potential barrier play a decisive role in the degree of spin polarization of the photoemission [184].

An increased polarization can be achieved by introducing strain in the semiconductor, affecting the energy band picture. The residual lattice misfit strain splits the degeneracy of the electron states in the valence band. The dependence of the amount of splitting on the strain was confirmed by the high spin polarization of photoelectrons [185]. Strained GaAs layers have been grown on GaAsP or InGaAsP [186] buffer layers, or vice versa [187]. Pseudomorphically strained epitaxial layers reveal a higher degree of polarization [188]. In the case of GaAs epilayers, a polarization of 85% could be obtained (at 866 nm wavelength) with a quantum efficiency of about 1.0%. Also, (strained) AlGaAs/GaAs superlattices reveal comparable numbers of polarization and quantum efficiency [189]. Further, structures have been proposed to increase the quantum efficiency at a predetermined wavelength, incorporating a distributed Bragg reflector [190] in the semiconductor structure. Strained GaAsP has been used to study the inverse photoemission from Fe(100) [191] and as a polarized electron source in microtron facilities [192].

Also, optically generated spin imbalance in an ultraclean cleaved GaAs tip leads to emission of spin-polarized electrons from the semiconductor tip [193]. Studies on the relation between the tunnel current, applied voltage, incident optical power, and tip–sample distance have been reported [194]. The experimental results are well described by the model that includes carrier capture in the semiconductor surface states.

Such a tip is considered very useful in spin-polarized scanning tunneling microscopy (SP-STM) [195] because an optically activated semiconductor spin source does not have difficulty controling the magnetic domain structure and would not alter the magnetic state of the sample under study through fringing fields. The construction of an SP-STM is being attempted and first results have been reported [196].

10.4. Spin-Polarized Current from Magnetic Contacts to Semiconductors

10.4.1. Spin Field Effect Transistor Concept

The next challenge is to inject spin-polarized electrons from a direct ferromagnetic contact into a semiconductor channel. It is clear that the semiconductor component would have a higher resistance than its all-metal counterparts. Also, the carrier density and mobility can be engineered in a semiconductor heterostructure. The combination of ferromagnetic and semiconductor structures in a single device will lead to possibilities of signal amplification and more exotic properties involving, for example, magnetooptic effects.

An FM/SC/FM device that resembles the structure of a field effect transistor (FET) has been proposed by Datta and Das [197]. Their spin FET is schematically represented in Figure 45. It relies on spin injection from a ferromagnetic source contact and detection at a ferromagnetic drain contact. A gate is present to impose a surface electric field that will cause a precession of the injected spins. The analysis of this mechanism is as follows.

10.4.2. Rashba Mechanism

An effective magnetic field is felt by a moving electron in a two-dimensional electron gas (2DEG, by convention in the x–y plane) when a perpendicular electric field E_z is present. The magnetic field is perpendicular to the electron motion and in the plane of the 2DEG. Assuming that the electrons are confined in one subband, the spin degeneracy of the subband can be lifted by a spin-splitting energy caused by the interaction of the electron with the effective magnetic field.

In InAs or InGaAs heterostructures [198–200], the main origin of the spin-splitting energy is the surface electric field in the heterostructure, referred to as the Rashba mechanism [201, 202]. This electric field can be modulated by an externally applied (gate) voltage, which can change the band-bending and carrier density in the heterostructure.

The spin–orbit interaction parameter α is proportional to the expectation value of the electric field $\langle E_z \rangle$ perpendicular to the 2DEG: $\alpha = b\langle E_z \rangle$. The energy associated with the Rashba mechanism is a linear function of this spin–orbit interaction parameter α and the Fermi wave number k_F:

$$\Delta R = 2\alpha k_F \tag{69}$$

The precession angle $\Delta\theta$ of the electron spin in semiconductor heterostructures is dependent on α and is a function of the length of the channel between source (injector) and drain (detector), L [197]:

$$\Delta\theta = \frac{2m^*\alpha L}{\hbar 2} \tag{70}$$

Experimentally, the spin–orbit interaction parameter for a 2DEG in a semiconductor heterostructure with sufficient mobility can be obtained from transverse magnetoresistance measurements. The experimental method relies on the occurrence of beat patterns in the Shubnikov–de Haas oscillations at moderate magnetic fields. In moderate fields, the Zeeman energy contribution is smaller than the effect of the Rashba mechanism.

10.4.3. Experimental Study to Support the Rashba Mechanism

In experiments by Heida et al. [203], it was found that the splitting energy δ_R in an AlSb/InAs heterostructure can be changed by an applied voltage. But the origin of the change was attributed to a change in k_F rather than a modulation of α by the voltage. In $In_{0.53}Ga_{0.47}As/In_{0.52}Al_{0.48}As$ heterostructures investigated by Nitta et al. [204], however, the parameter α was found to be modulated by the gate voltage V_g. A factor of 1.5 difference in α at $V_g = -1.0$ V and at $V_g = +1.54$ V was found. This would lead to a spin

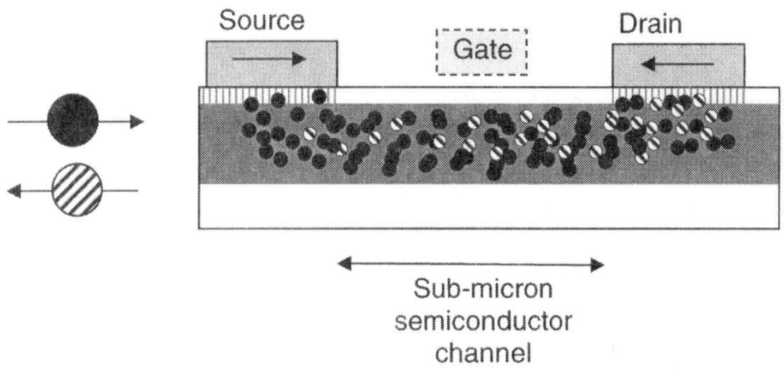

Fig. 45. Schematic representation of a lateral spin injection field effect transistor–like component with two ferromagnetic contacts and a gate (spin FET). (Source: Adapted from [197].)

precession change from π at the source to 1.5π at the drain for a gate length of 0.4 μm in a lateral spin transistor.

From this analysis, it is clear that the scattering events that change the direction of the electron motion will change the effective magnetic field for this electron. Hence, even if spin flip does not occur, scattering would disturb the coherent spin precession. Therefore, a ballistic device geometry would be ideal for the gated spin FET. In this respect, submicrometer structures fabricated from high-mobility one- or two-dimensional electron gas systems are desirable. The inverted $In_{0.53}Ga_{0.47}As/In_{0.52}Al_{0.48}As$ heterostructure and high-mobility near-surface InAs two-dimensional electron gas structures [205] are promising candidates for spin channels in the FET.

Associated with the gate electrode is the requirement of a stable insulator to separate it from the channel. In such metal–insulator–semiconductor (MIS) structures, the insulator is often a semiconductor layer with a larger band gap, while in some cases an SiO_x has been applied successfully [204, 206].

10.4.4. Spin Switch Concept

Making abstraction of the gate voltage–induced spin precession, a spin-dependent signature in the transport measurements is expected when the relative magnetization orientation of the source and drain is switched from parallel to antiparallel. Such a geometry is the equivalent of the all-metal spin switch discussed previously.

To make this spin switch work, the channel mobility requirement is somewhat more relaxed, as long as the spin diffusion length in the channel is sufficiently large. This is the case for appropriately designed semiconductor (hetero)structures (spin diffusion lengths of 1–10 μm). Nevertheless, devices with submicrometer channel length are desired to minimize the loss of spin coherence during channel transit. This spin disordering can be of particular importance for semiconductors with a large spin–orbit coupling, associated with the crystal asymmetry and in channels with a large number of magnetic scatterers where the number of spin-flip events is large.

The first data showing a "spin switch effect" have been reported [207] for a two-terminal device consisting of two sets of interdigitated Ni contacts on an n-type bulk Si wafer (three-dimensional channel). In this case, the change in relative orientation of the magnetization in the two sets of Ni electrodes is realized by making one set of the submicrometer Ni finger contacts narrower than the other set. As expected, the coercivity (or magnetic switching field) for the narrower one is larger. During the sweep of the magnetic field, an antiparallel magnetic situation will now appear when the softer contacts follow the external field, whereas the switching field of the harder contacts has not been attained.

Optical experiments are very useful to clarify the process of spin injection from a semiconductor into a ferromagnetic contact, for example, through an oxide tunnel barrier [208], or in the case of a direct contact interface [209]. In the latter experiment, the photoyield through a thin Fe overlayer on GaAs was found to be 1.7 times larger when the orientation of the Fe is parallel to the spin orientation of the optically generated polarized electrons from the GaAs, compared to the antiparallel orientation.

10.4.5. Technological Aspects of Ferromagnetic/Semiconductor Spin Injection Structures

As can be inferred from Figure 45, a respectable set of "show stoppers" for the realization of a spin FET or spin switch can be pictured. It has become clear from the previous description that the magnitude and the sign of the polarization of the injected current depend strongly on the detailed properties of the band structure and relative contributions to the conduction of carriers of the s, p, d (hybridized) bands [179], the ferromagnet's crystal orientation versus the current direction, and the spin transparency of the ferromagnetic/semiconductor interface. Therefore, the ideal ferromagnetic current injector is a single crystal with controllable crystal orientation, having a well-defined (single) magnetic domain structure. Clearly, these observations push the ideal contact size to the nanoscale regime, putting the same requirements on the lithographic tools as in other fields of microelectronics. Another reason for the ultrasmall contact sizes is the spin coherence of the injected current. For smaller contacts, however, it will become harder to switch the magnetization direction. The geometry and size difference in the source and drain contacts can actually be used to realize different coercive fields (see the preceding discussion on the Ni/Si injector geometry).

Moreover, instead of working with different size contacts, two ferromagnetic materials with different H_c may be a more desirable choice for realizing antiparallel alignment of the contact's magnetization. Such a requirement may be a tough challenge for lateral components in the nanoscale regime.

Magnetic contacts cause fringing fields either from edge states or from domain walls. Although it is unlikely that the ferromagnetic contacts will be in close enough proximity to suffer from magnetostatic coupling effects, the fringing fields close to the contacts may result in spin disorder of the injected current, through the Hanle effect.

As a final point, one has to consider the FM/SC interface. Given the importance of the control of the magnetic properties of the magnetic layers serving as contacts and the need for an extremely well controlled interface (structural, magnetic, and electrical), we will treat aspects of the FM/SC interface in a separate section later.

11. FERROMAGNETIC/SEMICONDUCTOR EXPERIMENTAL STRUCTURES

11.1. The Need for Epitaxy

Most metals used in semiconductor devices are polycrystalline. When the device dimensions shrink, these dimensions approach the grain size of the metals. In this case, the uniformity of the device characteristics is compromised. Inferior reliability and a variation in the physical properties of the device may also result. As an example of the latter, it was already mentioned that magnetic properties are strongly dependent on the crystal orientation (anisotropy, degree of spin polarization). Further, controlled deposition techniques are required to optimize the FM/SC interface properties.

Uniformity at nanoscale, single-crystalline structures and extremely well controlled interfaces are tough requirements to meet for most standard metal deposition methods. With the advent of epitaxial growth techniques such as molecular beam epitaxy, the preceding requirements seem to be within reach. Epitaxy of metals on semiconductors has received considerable attention over the past decades. First, we will describe some general aspects of metal epitaxy on semiconductors. We continue with a review of the achievements of ferromagnetic/semiconductor epitaxial layered structures. In this case, control over the structural and magnetic properties (magnetization and magnetic anisotropy) is the focus.

11.2. General Metal Epitaxy Criteria

Many successful attempts have been made to deposit stable, single-crystal metallic thin films on semiconductor surfaces. An excellent review of this research for metal/III–V semiconductor combinations has been published by Sands et al. [210]. Because most of this chapter is related to III–V semiconductors, we will take this technology as an example to discuss the problems.

The items of concern are schematically shown in Figure 46. We finally want to arrive at a pinhole-free, monocrystalline metal. The interface should remain well defined after possible heat treatment during device fabrication or operation. Eventually, one may attempt to fabricate a metal/semiconductor/metal stack multilayer, in which case the epitaxy criterion is of importance to maintain good-quality semiconductor overgrowth.

The result after annealing under the conditions of case A of Figure 46 is the typical case for a metal contact to a semiconductor structure, in, for example, MESFET technology. After annealing, many phases may form and the metal layer is polycrystalline in nature. In the as-grown condition, there is no tendency to form a continuous layer because of failure to meet the epitaxy criterion.

In case B (Fig. 46), there exists some epitaxial relation to the semiconductor, which helps to stabilize the metal layer. After heat treatment, phase separation occurs because of thermodynamical instability. When solving both the epitaxy and the thermodynamical criteria, as in case C, the annealing will still cause problems as a result of insufficient morphological stability. The alloy phase that was deposited epitaxially as grown will ball up into a highly textured polycrystalline film. Notice that metal–III or metal–V alloys

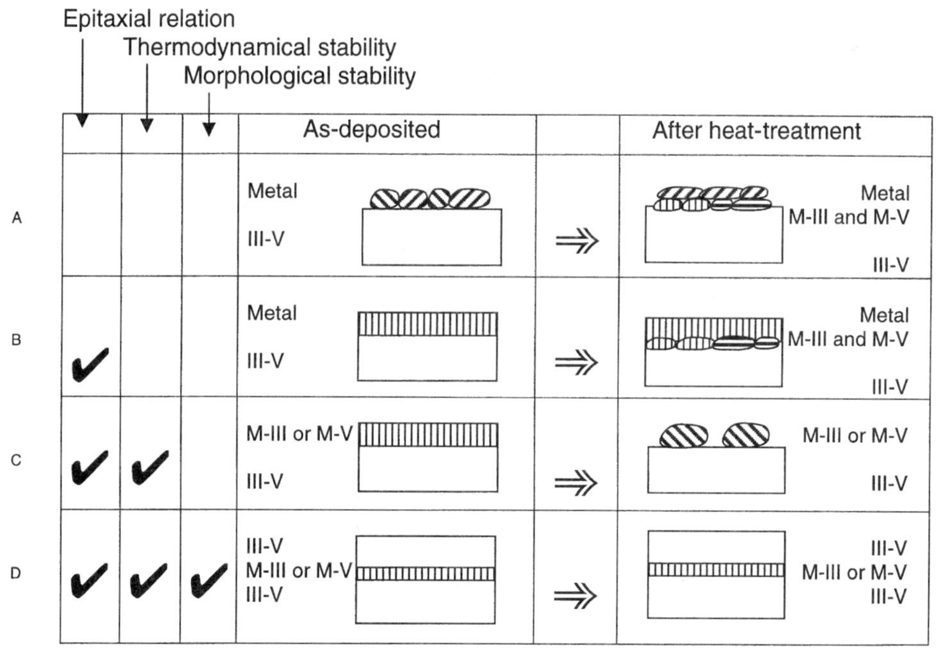

Fig. 46. Overview of the stability criteria of metal/semiconductor heterostructures. (Source: Adapted from [210].)

often have a higher chance of remaining thermodynamically stable. In the final case (D), all three criteria are fulfilled and a stable metal/III–V heterostructure results. In this case, semiconductor overgrowth can be attempted with a reasonable chance of success. Examples of the latter include GaAs/ErAs/GaAs resonant tunneling structures [211, 212] and GaAs/AlAs/NiAl/AlAs/GaAs [213]. In the case of the GaAs/ErAs/GaAs heterostructures [214, 215], a negative differential resistance was observed at room temperature for ErAs thicknesses in the range of 2.6 to 5 nm [216]. The success was attributed to the use of a Mn template monolayer deposited prior to ErAs deposition.

11.3. Elemental Ferromagnetic Metal Epitaxy on Semiconductors

The report on the epitaxy of Co on GaAs by Prinz [217] can be considered as the seed for research activity on magnetic/semiconductor epitaxy [218]. A large amount of research on metal/ferromagnet epitaxial superlattices on semiconductor substrates has been performed to clarify aspects of superlattice magnetism and giant magnetoresistance (GMR). In this section, we will limit the discussion to the recent studies of epitaxy of elemental Fe and Co on semiconductor substrates, in most cases GaAs. For spin injection devices, this material combination is an important test structure to gain understanding of the structural, magnetic, and electrical properties at a ferromagnetic/semiconductor interface. Figure 47 shows a set of examples of metallic and ferromagnetic structures that have been grown epitaxially on GaAs. In the case of Co, epitaxy on GaAs can lead to the body-centered cubic (bcc) phase; some aspects of this widely studied structure will be briefly discussed. At the end of this section, the Mn-based magnetic layers will be briefly discussed.

11.3.1. Epitaxial Layers of Iron and Cobalt on Gallium Arsenide

A few research groups have been very active in the study of the epitaxy of elemental magnetic thin films of Co and Fe on GaAs (for a review, see also [160]). Such studies have addressed in detail the evolution of the magnetic anisotropy of thin (a few monolayers) Fe [219] and Co films on GaAs. The substrate conditions prior to metal deposition are crucial in the control of this anisotropy because contributions to the net magnetic anisotropy arise from the formation of the interface and early stages of metal film growth. The determining factors are the initial metal adsorption sites and subsequent bond or site filling,

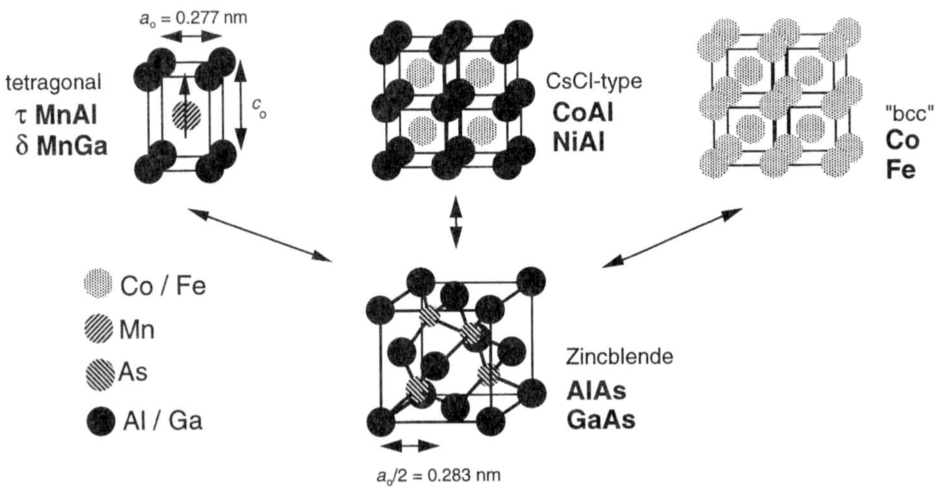

Fig. 47. Overview of some magnetic and compound metal structures suitable for growing epitaxially on GaAs (AlGaAs) semiconductors. The structural and magnetic properties of such epitaxial structures are determined by the epitaxial relationship and by the structure of the FM/SC interface.

and are strongly dependent on the semiconductor surface reconstruction [220, 221]. Also, anisotropies can be influenced by surface topography, as in the case of epitaxy on cross-hatched InGaAs epilayers [222], and by the early stages of the growth (e.g., step bunching and island formation). The domain structure and magnetization reversal processes, dictated by the induced magnetic anisotropy, have been studied *in situ* and *ex situ*. Such experiments are typically performed using (a combination of) the following characterization tools:

- Magnetoresistance measurements using anisotropic magnetoresistance (AMR) [223, 224]
- Kerr microscopy [225–228]
- Brillouin light scattering [229–231]
- Ferromagnetic resonance (FMR); sometimes photothermally modulated FMR to achieve lateral resolution [232, 233]
- Magnetic force microscopy (MFM) [234]
- Lorentz microscopy [235]

Almost all epitaxy experiments are performed under UHV deposition conditions using molecular beam epitaxy. However, sputter beam methods are occasionally reported. Examples are Fe(001) single-crystal films on GaAs(001) [236] and Fe(111) on GaAs(111) [237]. Also, electrodeposition has gained much interest in magnetic multilayer research [238] and the formation of FM/GaAs Schottky barriers [239] and single-crystal magnetic (NiCu) layers have been reported [240].

For device structures, it will be important to know the magnetic properties after patterning the films for contact formation into (sub-)micrometer or even nanoscale dimensions. Magnetic properties of arrays of submicrometer ferromagnetic lines have been studied using magnetoresistance measurements [241]. For lateral dimensions down to about 500 nm, far-field Kerr microscopy has been used to determine the magnetization reversal in arrays of magnetic wires [242].

11.3.2. Body-Centered Cubic Cobalt

In the case of cobalt epitaxy on GaAs, a forced phase [243] of Co, the bcc phase [217], can be realized because of the epitaxial relationship. This bcc Co phase has received a lot of attention over the past decade. A review of the bcc Co related work until 1991 was given by Prinz [244]. More recently, the study of bcc Co as seed layers for epitaxial metal structures on GaAs has been reported, as well as further studies on the structural [226] and magnetic properties [245] of bcc Co and CoFe alloys [246]. The appearance of bcc Co in superlattices of Co/Fe [247–249], Co/Pd(111) [250], and Co/τ-MnAl [251] has been demonstrated. In the case of Co/Fe superlattices, nuclear magnetic resonance (NMR) measurements on the short-range chemical order in combination with hyperfine field experiments have illustrated the limits of stability. In superlattices with Co thicknesses above 2 nm, alloy formation with Fe at the interfaces is responsible for the observation of the bcc phase. From reflection high-energy electron diffraction (RHEED) during the growth of the Co, it was concluded that the bcc phase is not sustained beyond about 1 nm. Oscillations of the Fe and Co magnetic moments at the (110) Fe/Co interfaces have recently been studied [252].

11.4. Magnetic and Electrical Properties of Ferromagnets at the Ferromagnetic/Semiconductor Interfaces

For spin injection experiments, it is essential to know the details of the interface, such as the formation of extra phases, the magnetic character of the first metal layers, the stability in time [253], and the electrical properties.

Sulfur passivation of GaAs surfaces has been employed to prevent As interdiffusion into Fe overlayers [254, 255]. It has been suggested that Au be used as an epitaxial barrier for Fe on GaAs(110) [256]. The Au layer is lattice matched and does not suppress magnetism at the interface. It will also work as a surfactant to support layer-by-layer growth. Recent work by Filipe et al. [257] showed the existence of a nearly half-magnetized phase at the Fe/GaAs interface, which was suggested to be the Fe_3GaAs compound. For Co, an amorphous phase tends to occur at the Co/GaAs interface because of a solid phase regrowth [258]. The use of O_3 treatment of the GaAs surface prior to growth inhibits the As interdiffusion and results in bcc Fe overlayers that are ferromagnetic when thicker than 4 monolayers [259]. Another study reports the onset of ferromagnetism to be at a coverage of 4 to 8 monolayers of Fe on a GaAs(001)–2 × 4 reconstructed surface [260]. In these experiments, it was shown that the initial nucleation proceeds through 2D island formation along the [110] direction. The easy axis of such films turned out to be perpendicular to this direction, demonstrating the limited importance of shape anisotropy induced by the initial nuclei.

The interface properties will change the electrical behavior of the metal/semiconductor heterostructure. These electrical characteristics are normally designed by changing the electronic properties of the semiconductor using doping and band gap engineering such that a tunnel barrier, an ohmic contact, or a Schottky barrier is formed. An uncontrolled interface resulting from phase separation or severe intermixing leads to unpredictable changes in the device performance. The structural and magnetic information on the ferromagnetic/semiconductor interface is quite substantial, but detailed electrical characterization is rather limited. Recently, enhanced carrier lifetimes at room temperature and suppression of midgap states in GaAs were found at an Fe/GaAs interface [261]. This observation leads to the very promising conclusion that the epitaxial Fe film is a good ferromagnetic contact and an excellent choice as a surface passivation layer that does not pin the Fermi level. This was explained in terms of an atomic model of the Fe/GaAs interface in which the Fe is bonded to an As monolayer after dissolution of the As dimers of the GaAs(001)–2 × 4 surface [262].

11.5. Properties of Manganese-Based Epitaxial Magnetic Layers on III–V Semiconductors

When a metal alloy contains either a group III or a group V element, improved interface characteristics can be expected. MnAs is such an alloy that has been grown epitaxially on GaAs [263, 264], for which it was shown that the preparation of the semiconductor surface before growth changes the orientation of the epilayer with respect to the crystal orientation of the substrate. Another example of this class is MnSb epitaxy, which has been demonstrated recently [265, 266]. As in the case of elemental metals, the magnetization of these epitaxially grown alloys is in the plane of the film. Another interesting class of epitaxial magnetic films is that of Mn–III alloys, in which the magnetization is out of the plane. Examples of this class are Mn-based magnetic thin films such as τ-MnAl [267, 268] and δ-MnGa [269, 270] (see Fig. 47). In these cases, the epitaxial growth on a III–V semiconductor substrate (such as GaAs) is the key to obtaining the desired perpendicular magnetization. The metastable phase τ-MnAl can be stabilized by coherence to the semiconductor substrate, and because of the epitaxial relationship, the c axis of the tetragonal lattice is normal to the growth surface. As a consequence, the easy axis of the magnetization of the thin film points out of the plane. The flexibility in engineering new epitaxial multilayers and new magnetic properties using this class of ferromagnetic films has been demonstrated: MnGa/NiGa [269] with multistepped hysteresis loops and ferrimagnetic [271] τ-MnAl/Co superlattices on GaAs [272], in which both magnetic sublayers of the superlattice have a perpendicular magnetization, which are aligned antiparallel at zero field. Very large antiferromagnetic interface coupling strength is observed, together with a strong perpendicular magnetic anisotropy for both the τ-MnAl and the Co layers [271]. Kerr hysteresis loops of

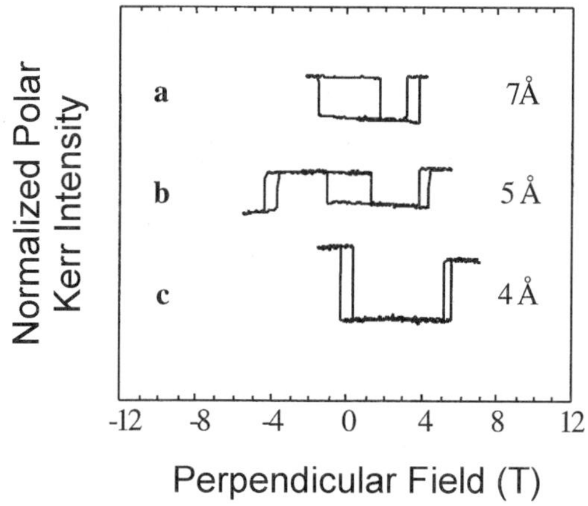

Fig. 48. Polar magneto-optic Kerr effect (MOKE) measurements with the field applied perpendicular to the sample surface of a wedged $6 \times \tau$-MnAl(20 Å)/Co(x Å) lattice grown on GaAs. For the curves in the figure, $x = 7$ (a), $x = 5$ (b), and $x = 4$ (c). (Source: Adapted from [272].)

τ-MnAl/Co superlattices showing the abrupt transitions from antialigned to aligned states are given in Figure 48. These Co layers are found to be partially in the bcc phase.

In alloys such as MnAl and MnGa, a large imbalance in the density of states for majority and minority spins exists about 1 eV above the Fermi level [273, 274]. This property makes them interesting for ballistic magnetotransport studies in ferromagnet/semiconductor hybrid devices. A drawback is the large coercive field in these magnetic films, which may limit their applicability in circuit applications such as magnetic random-access memory (MRAM) [275]. For a more complete review, see [276].

11.6. Semiconductor/Ferromagnetic/Semiconductor Multilayers

As the technology of building artificial structures improves, even more challenging structures may become possible such as epitaxial multilayers of magnetic metal and semiconductors (case D in Fig. 46). Van Roy et al. [277, 278] reported magnetoresistance and coupling effects as a function of the (Ga, Mn, As) thickness in MnGa/(Ga, Mn, As)/MnGa trilayers. When such structures are further optimized, control of the magnetic properties with external modulation sources, such as optical power, may become reality. Announcements of light- (or heat-) induced ferromagnetism [279] and antiferromagnetic coupling [280] in Fe/FeSi have been made recently. This FeSi/Si structure is still under extensive investigation; see, for example, [281].

Butler et al. [282] have performed calculations of the electronic structure of a Fe/GaAs/Fe structure. They conclude that the simple models relying on potential barriers or density of states from the bulk may not apply to such FM/SC/FM structures.

The magnetooptical properties of Co(100)/Ge/Co trilayers have been reported [238]. The coercivity and the anisotropy of these structures oscillate with the Ge thickness. CoPt/(Si, Ge)/Co trilayer and multilayer structures have been fabricated and analyzed using CPP magnetotransport [284]. In these structures, a small steep CPP resistance change has been found that was ascribed to the change in the relative spin orientation of the Co (magnetic soft) and the CoP (magnetic hard) layers.

Although these structures show promising results, often stability at higher temperatures (even room temperature) is a critical issue. In the following section, we discuss an alternative approach to fabricate intimate ferromagnetic/semiconductor hybrid structures, using nanoscale magnetic clusters.

12. NANOSCALE MAGNETS

12.1. Introduction

In all of the previously described structures, the nanoscale dimensions are either controlled in the vertical dimension by deposition or advanced lithography tools are required to realize the nanoscale in the lateral dimension. One alternative that deserves special attention is that of self-organization or self-assembly to realize nanoscale structured materials.

Also popular in semiconductor epitaxial technology is self-organization of nanostructures by growth on structured substrates, step-flow techniques, three-dimensional nucleation, and phase separation. Sort-like experiments in micromagnetics have been reported, nanowires of Co on stepped Cu surfaces [285], nanometer-scale needle-shaped magnetic multilayer wires [286, 287] fabricated using electrodeposition, and a variety of granular structures. Alternatives to incorporating nanoscale magnets in a III–V semiconductor matrix are described in the following discussion. The degrees of freedom in the design of structures incorporating such nanoscale magnets are discussed.

12.2. Self-Organized Magnetic Nanostructures in Semiconductor Thin Films

Self-assembly of nanoscale magnetic clusters in semiconductor materials has gained much interest recently as a route to fabricate magnetic/semiconductor structures that show spin-dependent transport.

Mn or Fe ion implantation in GaAs and subsequent annealing leads to phase separation and formation of magnetic precipitates: MnGa [288–290], MnAs by implantation in As-rich GaAs [291], and Fe_3GaAs [292, 293]. Typical annealing conditions are 600 to 920 °C, for a few seconds to 20 min for Mn-based structures, and 650 °C, 30 s for Fe-implanted layers.

In MBE-grown (Fe, Mn)(In, Ga)As structures, precipitation of FeAs [294, 295], MnAs [296–298], and ErAs [299] has been demonstrated.

Comparing these two techniques leads to the following advantages of ion implantation over MBE growth:

- Lateral dot, wire, or grating structures can be fabricated by "writing" a pattern with the ion beam or by deposition through a lithographically defined mask [289].
- There is no need for an epitaxial layer (unless a specially designed buffer layer is needed, e.g., low-temperature GaAs with an As-rich condition [291].
- Various cluster materials have been demonstrated, see MnGa, MnAs, and Fe_3GaAs mentioned previously.

On the other hand, an MBE process has the following interesting features:

- Buried layers (ferromagnetic/semiconductor) are easier because the epitaxy of the Mn-rich structures is compatible with the growth of III–V semiconductor structures.
- Extreme thickness control of the magnetic layers, down to thicknesses of 5 nm, is possible, and vertical magnet/semiconductor superlattices can be grown.
- The anealing temperatures are compatible with MBE growth (500–650 °C) and annealing can be performed *ex situ* or *in situ*.
- In both fabrication methods, the cluster sizes can be tailored, but, typically, larger near-surface clusters are formed when using ion implantation. Next, we will discuss some experimental data illustrating the flexibility of the MBE procedure.

12.3. Experimental Conditions for Thin Films with Nanoclusters by Molecular Beam Epitaxy + Annealing

12.3.1. Nanoscale Clusters by Molecular Beam Epitaxy: Fabrication and Magnetic Properties

In brief, the fabrication procedure to obtain MnAs precipitates in GaAs is as follows. It is well known that metallic arsenic precipitates form upon annealing of low-temperature grown GaAs (LT-GaAs) using MBE [300]. After the growth of an LT-GaAs buffer layer of typically 200 nm, a Mn flux is supplied to grow LT-$Ga_{1-x}Mn_xAs$. Annealing experiments are performed either *ex situ* or *in situ* while growing a III–V layer or held under As pressure. Under the proper conditions, magnetic particles are formed that are found to be MnAs. The α-MnAs [301] phase is a ferromagnetic phase with NiAs-type hexagonal structure ($a = 0.3724$ nm, $c = 0.5706$ nm). In the bulk, a reversible α-MnAs to β-MnAs (paramagnetic, orthorombic) phase transition takes place at 40 °C.

Magnetization measurements are performed for various annealing treatments. Figure 49 shows a compilation of alternating gradient force magnetometry data for a 0.72-μm-thick $Ga_{0.94}Mn_{0.06}As$ layer [276]. The evolution of the magnetization is described later, where the measured moment is divided by the total epilayer volume to obtain the magnetization value. The values of the saturation magnetization M_{sat} are measured at 796 kA/m (10^4 Oe).

The as-grown films (curve a in Fig. 49) show very little saturation magnetization, M_{sat} of 3 kA/m (3 emu/cc). Upon annealing at 625 °C for 10 s (curve b of Fig. 49), the M_{sat} rises to 22 kA/m, while the remanent magnetization M_r is only 0.8 kA/m and the coercive field $H_c = 0.96$ kA/m (12 Oe). This curve suggests a superparamagnetic behavior of ultrasmall magnetic particles dispersed in the semiconductor. Annealing at 650 °C for 30 s (curve c in Fig. 49) does not change M_{sat}, but M_r increases to 6 kA/m and H_c increases to 10.4 kA/m (131 Oe). The particle size has increased and the film shows remanence and a clear hysteretic behavior. The decrease of saturation magnetization with subsequent higher annealing is ascribed to the change in composition or phase in the coarsened metallic clusters.

Figure 50 illustrates the phase separation, comparing an as-grown $Ga_{0.94}Mn_{0.06}As$ (Fig. 50a) with the same layer after anneal (600 °C, 30 s, Figure 50b). For $Ga_{1-x}Mn_xAs$

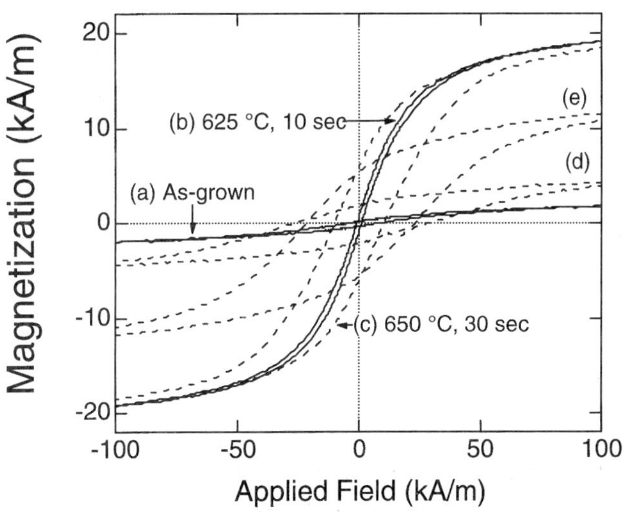

Fig. 49. Magnetization curves of a 0.72-μm-thick $Ga_{0.94}Mn_{0.06}As$ layer as grown and after annealing. The conditions for (a)–(c) are given in the figure. Subsequent annealing at higher tempertures leads to inferior magnetic properties (d) and (e). (Source: Adapted from [276].)

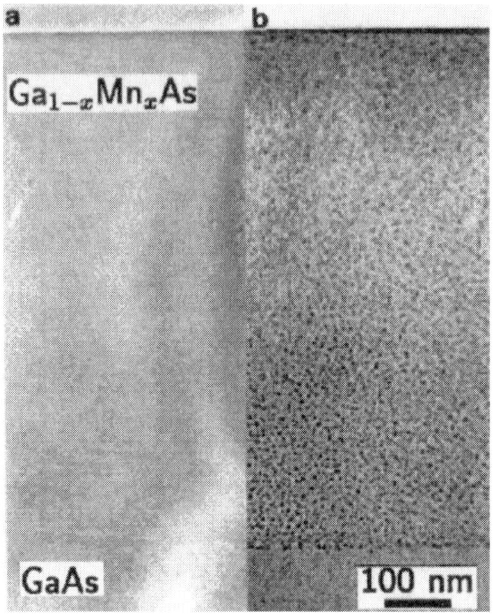

Fig. 50. TEM cross section, illustrating the phase separation in a $Ga_{0.94}Mn_{0.06}As$ layer. No particles are observed in the layer as grown (a) but after anneal (600 °C, 30 s) MnAs precipitates are present (b). (Source: Adapted from [276].)

($x = 0.06$) samples without high-temperature treatment, no particles can be observed using tunneling electron microscopy (TEM). After annealing, the size of the clusters varies with Mn concentration and annealing conditions. For $x = 0.06$, we observe diameters from 3 to 20 nm for heat treatments in the range of 600 to 625 °C (10–30 s, Fig. 50) to 30 nm for treatments exceeding 650 °C (30 s). The surrounding GaAs semiconductor crystal is found to be free of structural defects.

12.3.2. Good-Quality III–V Quantum Structures on Buried Magnetic Buffer Layers

Good-quality semiconductor heterostructures can be fabricated on top of the magnetic MnAs-rich buffers. This is demonstrated by a structure consisting of a 750-nm-thick $Ga_{0.94}Mn_{0.06}As$ buffer layer followed by a $10 \times \{GaAs(5\ nm)/AlAs(5\ nm)\}$ multiquantum well (MQW). This sample was terminated by another LT-grown $Ga_{0.96}Mn_{0.04}As$ layer (300 nm thick). The substrate temperature for the epitaxy of the homogeneous $Ga_{0.96}Mn_{0.04}As$ buffer layer was 230 °C. The substrate temperature was then increased to 600 °C for the initiation of the MQW growth. During this growth, the LT-$Ga_{0.96}Mn_{0.04}As$ buffer experienced an *in situ* annealing of 90 min total at 600 °C and MnAs cluster formation occurred. Magnetization measurements on these samples show clear magnetic behavior.

Figure 51 shows the TEM cross-sectional images of this structure, as-grown and after annealing (650 °C, 30 s). The defects shown in Figure 51a are introduced because of improper growth parameters at the initiation of the LT-$Ga_{0.96}Mn_{0.04}As$ layer. They are mostly annealed out after the heat treatment (Fig. 51b). Figure 51c is a TEM closeup of the MQW after anneal Fig. 51(b). The MQW is of excellent structural quality and the magnetic precipitates are well confined in the original Mn-rich epilayers. TEM images reveal as-grown particle sizes of 25–30 nm diameter. Photoluminescence characterization demonstrated good optical quality of the MQW, illustrating the potential of vertically combining magnetic and semiconductor heterostructures using this technique.

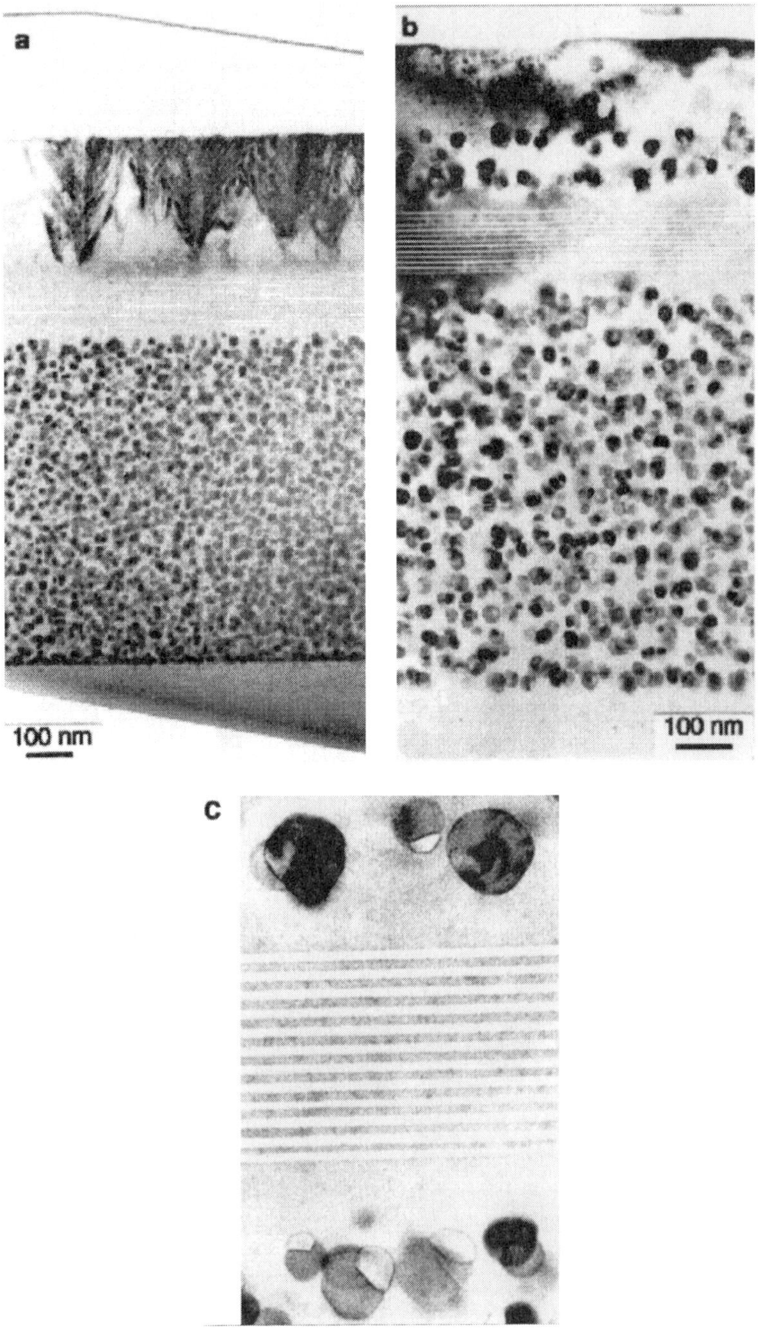

Fig. 51. TEM cross section of sample A (see text for description) (a) as grown and (b) after anneal (650 °C, 30 s). A high-resolution image of the MQW sandwiched by magnetic MnAs-rich layers (c). (Source: Adapted from [276].)

12.3.3. Magnetotransport in Thin Heterogeneous Films

ErAs clusters in GaAs grown by MBE have been recently reported to show large current in the plane (CIP) magnetoresistance (MR) at low temperature [299]. The magnetoresistance properties at low temperatures of clustered InGaAs/FeGaAs [302] and GaAs:MnAs [303] have been recently reported, for the current in the plane of the thin films (CIP geometry). In the case of FeGaAs, the spin transport signal can only be obtained after subtraction of the MR of a reference InGaAs sample. In the GaAs:MnAs case, up to 1.5 negative MR was found at 30 K in a magnetic field of 1 T, which was attributed to spin-dependent transport properties caused by the presence of the MnAs precipitates [303]. The flexibility of the MBE procedure allows the fabrication of more complex structures such as magnet/semiconductor superlattices in which stronger MR effects may be demonstrated. Next, we will discuss the potential for fabricating superlattices with MnAs-rich heterogeneous layers.

13. SUPERLATTICES OF NANOSCALE MAGNET LAYERS AND SEMICONDUCTORS

13.1. Engineering Aspects of Superlattices of Nanoscale Magnet Layers and Semiconductors

For the realization of spin transport structures involving magnetic clusters, magnet/semiconductor superlattices are a possible candidate. Recently, the fabrication and characterization of magnetic {GaAs:MnAs}/{InGaAs} superlattices was described [304]. In such structures, magnetotransport phenomena can be studied with the current applied perpendicular to the plane (so-called CPP geometry). Such a structure may have many advantages: a vertical structure that is promising for future device applications, control over the vertical distance of the clusters, and the possibility of having a semiconductor spacer of different composition between the clusters, to act as a barrier or a conducting layer (using, e.g., delta-doping). This may help to overcome one of the prime problems in achieving interesting spin-dependent transport, namely, the high resistance of such layers owing to the presence of a Schottky barrier around the metallic precipitates.

The right choice of layer structure may enable modulation of the magnetoelectronic properties of the superlattice by electrically or optically varying the semiconductor properties. These degrees of freedom may be applied in novel device concepts. For clarity, in the text the term magnetic layer is used for the heterogeneous layers with the magnetic clusters {GaAs:MnAs}.

13.2. Structural and Magnetic Properties of the Superlattices

As an example of such superlattices, Figure 52 shows a TEM cross section of a superlattice consisting of 40 periods of {5 nm $In_{0.03}Ga_{0.97}As$/5 nm GaAs:MnAs} after a 500 °C anneal. The MnAs clusters have formed and they are nicely confined within the 5-nm-thick original GaMnAs layers. The clusters are arranged in a 2D sheet and they are spherical with a narrow diameter distribution around 5 nm. As grown, the sample shows negligible magnetization and, after annealing to 500 °C, the room temperature magnetization increases a little. Also, after a 550 °C anneal, superparamagnetic behavior occurs with an increased total moment. Annealing at 600 °C resulted in a ferromagnetic behavior at room temperature, a consequence of large cluster formation (15–30 nm in diameter).

13.3. Current Perpendicular to the Plane Magnetotransport

These superlattice samples are very well suited to test CPP magnetotransport in clustered structures. In a simple model, one can visualize the current in such CPP transport as in Figure 53. As grown (Fig. 53A), the superlattice consists of LT-grown InGaAs and the dilute magnetic semiconductor (DMS) GaMnAs, comparable to a recently reported structure by Shen et al. [305]. In this case, the MR will be dominated by the DMS behavior, which shows a ferromagnetic transition between 40 and 90 K depending on the Mn concentration. Below this transition temperature, hopping conductivity and large negative MR are observed [306].

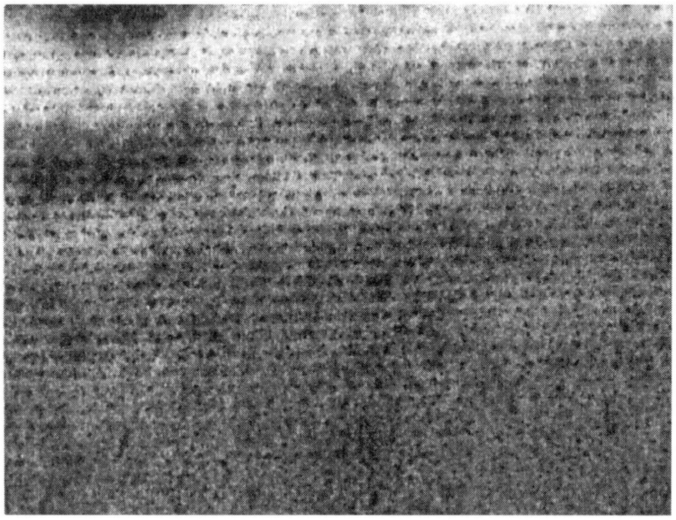

Fig. 52. High-resolution TEM image of an InGaAs(5 nm)/GaAs:MnAs(5 nm) superlattice. (Source: Adapted from [304].)

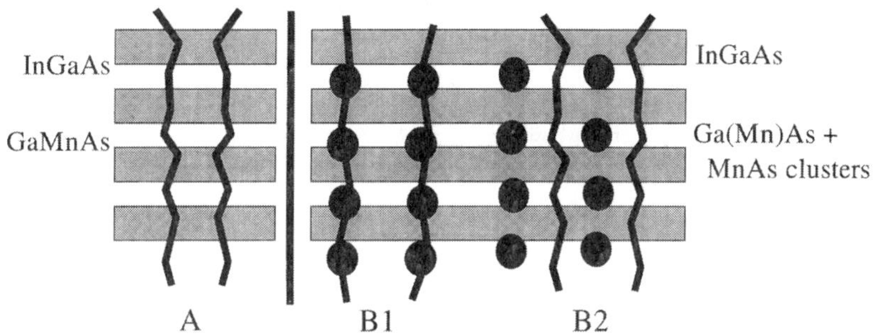

Fig. 53. CPP magnetotransport devices can be fabricated from magnet/semiconductor superlattices. The interpretation of the magnetotransport data is complicated because of the various possible contributions and current paths in the device structure (as grown, A; after anneal, B). (Source: Adapted from [304].)

When clusters are present (Fig. 53B), the idealized model shows two possible current paths in the CPP orientation: the path of interest in this study is B1 where the current samples the clusters and crosses the magnet/semiconductor interfaces, hence becoming dependent on the magnetic state of the clusters. In path B2, the situation is comparable to A, except for the much lower Mn concentration in the Ga(Mn)As layers of B2.

13.4. Conclusions

It is clear that self-assembled hybrid nanoscale magnetic/semiconductor heterostructures can be designed to show exciting new physical, electronic, and magnetooptical [307–312] properties. When the magnetic/semiconductor structures are reaching the nanoscale dimensions, effects such as Coulomb blockade may start to appear. First attempts are being made today in tunneling into magnetic clusters to observe a magnetotransport phenomenon in structures working in the Coulomb blockade regime [313]. It is to be expected that nanomagnet structures will become important in the demonstration of spin-dependent transport phenomena for device applications.

Acknowledgments

JDB would like to thank G. Borghs for his encouragement and many fruitful discussions and for a critical reading of the manuscript. FMP is supported as a Research Director with the Flemish Science Foundation. The following co-workers are kindly acknowledged for their important contribution to parts of the work described: H. Akinaga, K. Attenborough, M. Behet, H. Boeve, G. Borghs, C. Bruynseraede, K. Dessein, B. Dutta, I. S. Ibrahim, A. Matulis, J. Reijniers, V. Schweigert, A. Van Esch, W. Van Roy, P. Vasilopoulos, T. Sands, and M. Tanaka. This work was supported by the Interuniversity Micro-Electronics Center (IMEC), the Flemish Science Foundation (FWO-Vl), and IUAP-IV. JDB acknowledges the financial support of the European Commission through funding of the ESPRIT long-term research program SPIDER, Spin-dependent nanoelectronics (No. 23.307), in which ferromagnetic/semiconductor combinations are investigated.

References

1. D. Shoenberg, "Magnetic Oscillations in Metals." Cambridge University Press, Cambridge, U.K., 1984.
2. For a review, see, for example, J. K. Furdyna, *J. Appl. Phys.* 64, R29 (1988).
3. J. Reijniers and F. M. Peeters, *Appl. Phys. Lett.* 73, 357 (1998).
4. F. M. Peeters and A. Matulis, *Braz. J. Phys.* 24, 283 (1994).
5. I. S. Ibrahim and F. M. Peeters, *Phys. Rev. B* 52, 17321 (1995).
6. W. Van Roy, J. De Boeck, and G. Borghs, *Appl. Phys. Lett.* 61, 3056 (1992).
7. F. M. Peeters and A. Matulis, *Phys. Rev. B* 48, 15166 (1993).
8. M. Calvo, *Phys. Rev. B* 48, 2365 (1993).
9. J. Hajdu, in "Physics of the Two-Dimensional Electron Gas" (J. T. Devreese and F. M. Peeters, eds.), p. 27. Plenum Press, New York, 1987.
10. A. Matulis, F. M. Peeters, and P. Vasilopoulos, *Phys. Rev. Lett.* 72, 1518 (1994).
11. L. D. Landau and E. Lifshitz, "Quantum Mechanics," p. 66. Pergamon Press, Oxford, U.K., 1977.
12. O. Hudák, *Z. Phys.* 88, 233 (1992).
13. H. Yoshioka and Y. Nagaoka, *J. Phys. Soc. Jpn.* 59, 2884 (1990).
14. B. Y. Gu and X. D. Fan, *Z. Phys. B* 101, 527 (1996).
15. V. M. Ramaglia and F. Ventriglia, *J. Phys.: Condens. Matter* 3, 4881 (1991).
16. V. M. Ramaglia, A. Tagliacozzo, F. Ventriglia, and G. P. Zucchelli, *Phys. Rev. B* 43, 2201 (1991).
17. M. Calvo, *J. Phys.: Condens. Matter* 6, 3329 (1994).
18. M. Calvo, *Phys. Rev. B* 51, 2268 (1995).
19. A. Matulis, F. M. Peeters, and P. Vasilopoulos, *Surf. Sci.* 305, 434 (1994).
20. M. Büttiker, *Phys. Rev. Lett.* 57, 1761 (1986).
21. J. Q. You, L. Zhang, and P. K. Ghosh, *Phys. Rev. B* 52, 17243 (1995).
22. A. Majumdar, *Phys. Rev. B* 54, 11911 (1996).
23. Y. Guo, B. L. Gu, W. H. Duan, and Y. Zhang, *Phys. Rev. B* 55, 9314 (1997).
24. P. P. Vil'ms and M. V. Éntin, *Sov. Phys. Semicond.* 22, 1209 (1988) [*Fiz. Tekh. Poluporovodn.* 22, 1905 (1988)].
25. J. M. Repko, W. W. Repko, and A. Saaf, *Am. J. Phys.* 59, 652 (1991).
26. J. E. Müller, *Phys. Rev. Lett.* 68, 385 (1992).
27. E. Hofstetter, J. M. C. Taylor, and A. MacKinnon, *Phys. Rev. B* 53, 4676 (1996).
28. Y. Takagaki and K. Ploog, *Phys. Rev. B* 53, 3885 (1996).
29. H. Grosse, A. Martin, and J. Stubbe, *Phys. Lett. A* 181, 7 (1993).
30. D. Huang and G. Gumbs, *Phys. Lett. A* 167, 396 (1992).
31. C. L. Foden, M. L. Leadbeater, and M. Pepper, *Phys. Rev. B* 52, 8646 (1995).
32. C. L. Foden, M. L. Leadbeater, J. H. Burroughes, and M. Pepper, *J. Phys.: Condens. Matter* 6, L127 (1994).
33. (a) F. M. Peeters, A. Matulis, and I. S. Ibrahim, *Physica B* 227, 131 (1996); (b) J. Reijniers, F. M. Peeters, and A. Matulis, *Phys. Rev. B* 59, 2817 (1999).
34. L. Solimany and B. Kramer, *Solid State Commun.* 96, 471 (1995).
35. H. S. Sim, K. H. Ahn, K. J. Chang, G. Ihm, N. Kim, and S. J. Lee, *Phys. Rev. Lett.* 80, 1501 (1998).
36. M. L. Leadbeater, C. L. Foden, T. M. Burke, J. H. Burroughes, M. P. Grimshaw, D. A. Ritchie, L. L. Wang, and M. Pepper, *J. Phys.: Condens. Matter* 7, L307 (1995).
37. M. L. Leadbeater, C. L. Foden, J. H. Burroughes, M. Pepper, T. M. Burke, L. L. Wang, M. P. Grimshaw, and D. A. Ritchie, *Phys. Rev. B* 52, R8629 (1995).
38. M. L. Leadbeater, C. L. Foden, J. H. Burroughes, T. M. Burke, D. A. Ritchie, and M. Pepper, in "Proceedings of the 23rd International Conference on the Physics of Semiconductors" (M. Scheffler and R. Zimmermann, eds.), p. 2219. World Scientific, Singapore, 1996.

39. I. S. Ibrahim, V. A. Schweigert, and F. M. Peeters, *Phys. Rev. B* 56, 7508 (1997).
40. S. M. Badalian, I. S. Ibrahim, and F. M. Peeters, in "Proceedings of the 12th International Conference on High Magnetic Fields in the Physics of Semiconductors II" (G. Landwehr and W. Ossau, eds.), p. 327. World Scientific, Singapore, 1997.
41. H. Jou and A. M. Kriman, *Superlattices Microstruct.* 19, 203 (1996).
42. P. F. Fontein, J. A. Kleinen, P. Hendriks, F. A. P. Blom, J. H. Wolter, H. G. M. Lochs, F. A. J. M. Driessen, L. J. Giling, and C. W. J. Beenakker, *Phys. Rev. B* 43, 12090 (1991).
43. A. H. MacDonald, T. M. Rice, and W. F. Brinkman, *Phys. Rev. B* 28, 3648 (1983).
44. R. W. Rendell and S. M. Girvin, *Phys. Rev. B* 23, 6610 (1981).
45. J. Heremans, *J. Phys. D: Appl. Phys.* 26, 1149 (1993).
46. D. Weiss, K. von Klitzing, K. Ploog, and G. Weiman, *Europhys. Lett.* 8, 179 (1989).
47. R. R. Gerhardts, D. Weiss, and K. von Klitzing, *Phys. Rev. Lett.* 62, 1173 (1989).
48. R. W. Winkler, J. P. Kotthaus, and K. Ploog, *Phys. Rev. Lett.* 62, 1177 (1989).
49. E. S. Alves, P. H. Beton, M. Henini, L. Eaves, P. C. Main, O. H. Hughes, G. A. Toombs, S. P. Beaumont, and C. D. W. Wilkinson, *J. Phys.: Condens. Matter* 1, 8257 (1989).
50. P. H. Beton, E. S. Alves, P. C. Main, L. Eaves, M. W. Dellow, M. Henini, O. H. Hughes, S. P. Beaumont, and C. D. W. Wilkinson, *Phys. Rev. B* 42, 9229 (1990).
51. P. H. Beton, P. C. Main, M. Davison, M. Dellow, R. P. Taylor, E. S. Alves, L. Eaves, S. P. Beaumont, and C. D. W. Wilkinson, *Phys. Rev. B* 42, 9689 (1990).
52. H. Fang and P. J. Stiles, *Phys. Rev. B* 41, 10171 (1990).
53. A. Toriumi, K. Ismail, M. Burkhardt, D. A. Antoniadis, and H. I. Smith, *Phys. Rev. B* 41, 12346 (1990).
54. P. Vasilopoulos and F. M. Peeters, *Phys. Rev. Lett.* 63, 2120 (1989).
55. F. M. Peeters and P. Vasilopoulos, *Phys. Rev.* B 46, 4667 (1992).
56. C. T. Cui, V. Fessatidis, and N. J. M. Horing, *Phys. Rev. Lett.* 63, 2598 (1989).
57. P. Středa and A. H. MacDonald, *Phys. Rev. B* 41, 11892 (1990).
58. C. W. J. Beenakker, *Phys. Rev. Lett.* 62, 2020 (1989).
59. C. Zhang and R. R. Gerhardts, *Phys. Rev. B* 41, 12850 (1990).
60. G. H. Kruithof and T. M. Klapwijk, in "Proceedings of the 19th International Conference on the Physics of Semiconductors" (W. Zawadzki, ed.), p. 499. Polish Academy of Science, Wroclaw, Poland, 1988.
61. G. H. Kruithof, P. C. van Son, and T. M. Klapwijk, *Phys. Rev. Lett.* 67, 2725 (1991).
62. S. J. Bending, K. von Klitzing, and K. Ploog, *Phys. Rev. Lett.* 65, 1060 (1990).
63. A. K. Geim, S. J. Bending, I. V. Grigorieva, and M. G. Blamire, *Phys. Rev. B* 49, 5749 (1994).
64. H. A. Carmona, A. K. Geim, A. Nogaret, P. C. Main, T. J. Foster, M. Henini, S. P. Beaumont, and M. G. Blamire, *Phys. Rev. Lett.* 74, 3009 (1995).
65. S. Carlton, A. Nogaret, H. A. Carmona, B. L. Gallagher, P. C. Main, and S. P. Beaumont, in "Proceedings of the 23rd International Conference on the Physics of Semiconductors" (M. Scheffler and R. Zimmermann, eds.), p. 1537. World Scientific, Singapore, 1996.
66. P. D. Ye, D. Weiss, R. R. Gerhardts, K. von Klitzing, K. Eberl, H. Nickel, and C. T. Foxon, *Semicond. Sci. Technol.* 10, 715 (1995).
67. P. D. Ye, D. Weiss, R. R. Gerhardts, M. Seeger, K. von Klitzing, K. Eberl, and H. Nickel, *Phys. Rev. Lett.* 74, 3013 (1995).
68. H. A. Carmona, A. Nogaret, A. K. Geim, P. C. Main, T. J. Foster, M. Henini, S. P. Beaumont, H. McLelland, and M. G. Blamire, *Solid-State Electron.* 40, 217 (1996).
69. Y. Iye, A. Endo, S. Izawa, M. Kato, and S. Katsumoto, *Physica B* 227, 122 (1996).
70. S. I. Izawa, S. Katsumoto, A. Endo, and Y. Iye, *J. Phys. Soc. Jpn.* 64, 706 (1995).
71. S. I. Izawa, S. Katsumoto, A. Endo, and Y. Iye, *Jpn. J. Appl. Phys.* 34, 4306 (1995).
72. A. Nogaret, S. Carlton, B. L. Gallagher, P. C. Main, M. Henini, R. Wirtz, R. Newbury, M. A. Howson, and S. P. Beaumont, *Phys. Rev. B* 55, R16037 (1997).
73. N. Overend, A. Nogaret, B. L. Gallagher, P. C. Main, M. Henini, R. Wirtz, R. Newbury, M. A. Howson, and S. P. Beaumont, in "ICM'97."
74. P. D. Ye, D. Weiss, K. von Klitzing, and K. Eberl, *Appl. Phys. Lett.* 67, 1441 (1995).
75. P. D. Ye, D. Weiss, G. Lütjering, R. R. Gerhardts, K. von Klitzing, and K. Eberl, in "Proceedings of the 23rd International Conference on the Physics of Semiconductors" (M. Scheffler and R. Zimmermann, eds.), p. 1529. World Scientific, Singapore, 1996.
76. P. D. Ye, D. Weiss, R. R. Gerhardts, K. von Klitzing, K. Eberl, and H. Nickel, *Surf. Sci.* 361/362, 337 (1996).
77. P. D. Ye, D. Weiss, R. R. Gerhardts, G. Lütjering, K. von Klitzing, and H. Nickel, *Semicond. Sci. Technol.* 11, 1613 (1996).
78. P. D. Ye, D. Weiss, R. R. Gerhardts, K. von Klitzing, and S. Tarucha, in "Proceedings of the Electronic Properties of Two-Dimensional Systems," *Physica B* (1998).
79. G. M. Gusev, U. Gennser, X. Kleber, D. K. Maude, J. C. Portal, D. I. Lubyshev, P. Basmaji, M. de P. A. Silva, J. C. Rossi, and Yu. V. Nastaushev, *Superlattices Microstruct.* 18, 67 (1995).
80. G. M. Gusev, V. Gennser, X. Kleber, D. K. Maude, J. C. Portal, D. I. Lubyshev, P. Basmaji, M. de P. A. Silva, J. C. Rossi, and Yu. V. Nastaushev, *Phys. Rev. B* 53, 13641 (1996).

81. R. Yagi and Y. Iye, *J. Phys. Soc. Jpn.* 62, 1279 (1993).
82. D. Weiss, P. D. Ye, and R. R. Gerhardts, in "Proceedings of the 23rd International Conference on the Physics of Semiconductors" (M. Scheffler and R. Zimmermann, eds.), p. 2367. World Scientific, Singapore, 1996.
83. N. Overend, A. Nogaret, B. L. Gallagher, P. C. Main, M. Henini, C. H. Marrows, M. A. Howson, and S. P. Beaumont, *Appl. Phys. Lett.* 72, 1724 (1998).
84. D. D. Awschalom, M. A. McCord, and G. Grinstein, *Phys. Rev. Lett.* 65, 783 (1990).
85. P. Vasilopoulos and F. M. Peeters, *Superlattices Microstruct.* 7, 393 (1990).
86. D. P. Xue and G. Xiao, *Phys. Rev. B* 45, 5986 (1992).
87. F. M. Peeters and P. Vasilopoulos, *Phys. Rev. B* 47, 1466 (1993).
88. R. R. Gerhardts, in "Proceedings of the 12th International Conference on High Magnetic Fields in the Physics of Semiconductors II" (G. Landwehr and W. Ossau, eds.), p. 289. World Scientific, Singapore, 1997.
89. T. Z. Li, S. W. Gu, X. H. Wang, and J. P. Peng, *J. Phys.: Condens. Matter* 8, 313 (1996).
90. R. R. Gerhardts, *Phys. Rev. B* 53, 11064 (1996).
91. R. Menne and R. R. Gerhardts, *Phys. Rev. B* 57, 1707 (1998).
92. X. G. Wu and S. E. Ulloa, *Solid State Commun.* 82, 945 (1992).
93. X. G. Wu and S. E. Ulloa, *Phys. Rev. B* 47, 7182 (1993).
94. X. G. Wu and S. E. Ulloa, *Phys. Rev. B* 47, 10028 (1993).
95. F. M. Peeters and P. Vasilopoulos, *Phys. Rev. B* 42, 5899 (1990).
96. I. S. Ibrahim and F. M. Peeters, *Am. J. Phys.* 63, 171 (1995).
97. A. Krakovsky, *Phys. Rev. B* 53, 8469 (1996).
98. R. B. S. Oakeshott and A. MacKinnon, *J. Phys.: Condens. Matter* 5, 9355 (1993).
99. J. Q. You and L. Zhang, *Phys. Rev. B* 54, 1526 (1996).
100. Y. Guo, B. L. Gu, Z. Q. Li, J. L. Zhu, and Y. Kawazoe, *J. Phys.: Condens. Matter* 10, 1549 (1998).
101. A. Akjouj and B. Djafari-Rouhani, *Solid State Commun.* 103, 161 (1997).
102. I. S. Ibrahim and F. M. Peeters, *Surf. Sci.* 361/362, 341 (1996).
103. R. D. Hofstadter, *Phys. Rev. B* 14, 2239 (1976).
104. B. A. Dubrovin and S. P. Novikov, *Sov. Phys. JETP* 52, 511 (1980) [*Zh. Eksp. Teor. Fiz.* 79, 1006 (1980)].
105. D. Yoshioka and Y. Iye, *J. Phys. Soc. Jpn.* 56, 448 (1987).
106. A. Barelli, J. Bellissard, and R. Rammal, *J. Phys. France* 51, 2167 (1990).
107. P. Hedegård and A. Hoff, *Physica B* 194, 1075 (1994).
108. M. C. Chang and Q. Niu, *Phys. Rev. B* 50, 10843 (1994).
109. R. R. Gerhardts, D. Pfannkuche, and V. Gudmundsson, *Phys. Rev. B* 53, 9591 (1996).
110. G. J. O. Schmidt, *Phys. Rev. B* 47, 13007 (1993).
111. P. Schmelcher and D. L. Shepelyansky, *Phys. Rev. B* 49, 7418 (1994).
112. A. K. Geim, S. J. Bending, and I. V. Grigorieva, *Phys. Rev. Lett.* 69, 2252 (1992).
113. W. Van Roy, E. L. Carpi, M. Van Hove, A. Van Esch, R. Bogaerts, J. De Boeck, and G. Borghs, *J. Magn. Magn. Mater.* 121, 197 (1993).
114. M. A. McCord and D. D. Awschalom, *Appl. Phys. Lett.* 57, 2153 (1990).
115. M. Nielsen and P. Hedegård, *Phys. Rev. B* 51, 7679 (1995).
116. F. B. Mancoff, R. M. Clarke, C. M. Marcus, S. C. Zhang, K. Campman, and A. C. Gossard, *Phys. Rev. B* 51, 13269 (1995).
117. F. B. Mancoff, L. J. Zielinski, C. M. Marcus, K. Campman, and A. C. Gossard, *Phys. Rev. B* 53, R7599 (1995).
118. P. Omling, H. Linke, K. Deppert, L. Samuelson, L. T. Hansen, and P. E. Lindelof, *Jpn. J. Appl. Phys.* 34, 4575 (1995).
119. A. Smith, R. Taboryski, L. T. Hansen, C. B. Sørensen, P. Hedegård, and P. E. Lindelof, *Phys. Rev. B* 50, 14726 (1994).
120. D. F. Yue, G. Banerjee, A. E. Miller, and S. Bandyopadhyay, *Superlattices Microstruct.* 19, 191 (1996).
121. A. G. Aronov, E. Altshuler, A. D. Mirlin, and P. Wölfle, *Europhys. Lett.* 29, 239 (1995).
122. M. Calvo, *Phys. Rev. B* 57, R4241 (1998).
123. L. Zielinski, K. Chaltikian, K. Birnbaum, C. M. Marcus, K. Campman, and A. C. Gossard, *Europhys. Lett.* 42, 73 (1998).
124. B. I. Halperin, P. A. Lee, and N. Read, *Phys. Rev. B* 47, 7312 (1993).
125. V. Kalmeyer and S. C. Zhang, *Phys. Rev. B* 46, 9889 (1992).
126. W. Kang, H. L. Stormer, L. N. Pfeiffer, K. J. Boldwin, and K. W. West, *Phys. Rev. Lett.* 71, 3850 (1993).
127. R. L. Willet, R. R. Ruel, K. W. West, and L. N. Pfeiffer, *Phys. Rev. Lett.* 71, 3846 (1993).
128. P. W. Anderson, *Science* 235, 1196 (1987).
129. L. B. Ioffe and A. I. Larking, *Phys. Rev. B* 39, 8988 (1989).
130. N. Nagaosa and P. A. Lee, *Phys. Rev. Lett.* 64, 2450 (1990).
131. N. Nagaosa and P. A. Lee, *Phys. Rev. B* 46, 5621 (1992).
132. E. Abrahams, P. W. Anderson, D. C. Liciardello, and V. Ramakrishnan, *Phys. Rev. Lett.* 42, 673 (1979).
133. S. Hikami, *Prog. Theor. Phys. Suppl.* 107, 213 (1992).
134. A. G. Aronov, A. D. Mirlin, and P. Wölfle, *Phys. Rev. B* 49, 16609 (1994).

135. T. Sugiyama and N. Nagaosa, *Phys. Rev. Lett.* 70, 1980 (1993).
136. D. N. Sheng and Z. Y. Weng, *Phys. Rev. Lett.* 75, 2388 (1995).
137. J. Rammer and A. L. Shelankov, *Phys. Rev. B* 36, 3135 (1987).
138. L. Brey and H. A. Fertig, *Phys. Rev. B* 47, 15961 (1993).
139. A. Khaetskii, *J. Phys.: Condens. Matter* 3, 5115 (1991).
140. P. Hedegård and A. Smith, *Phys. Rev. B* 51, 10869 (1995).
141. D. Khveshchenko, *Phys. Rev. Lett.* 77, 1817 (1996).
142. A. K. Geim, *JETP Lett.* 50, 389 (1989) [*Pis'ma Zh. Eksp. Teor. Fiz.* 50, 359 (1989)].
143. A. K. Geim, S. V. Dubonos, and A. V. Khaetskii, *JETP Lett.* 51, 121 (1990) [*Pis'ma Zh. Eksp. Teor. Fiz.* 51, 107 (1990)].
144. A. K. Geim, *JETP Lett.* 50, 389 (1989) [*Pis'ma Zh. Eksp. Teor. Fiz.* 50, 359 (1989)].
145. A. K. Geim, V. I. Falko, S. V. Dubonos, and I. V. Grigorieva, *Solid State Commun.* 82, 831 (1992).
146. E. Hall, *Am. J. Math. II* 287 (1879).
147. A. Oral, S. J. Bending, and M. Henini, *Appl. Phys. Lett.* 69, 1324 (1996).
148. S. J. Bending and A. Oral, *J. Appl. Phys.* 81, 3721 (1997).
149. A. K. Geim, S. V. Dubonos, J. G. S. Lok, I. V. Grigorieva, J. C. Maan, L. T. Hansen, and P. E. Lindelof, *Appl. Phys. Lett.* 71, 2379 (1997).
150. A. K. Geim, I. V. Grigorieva, S. V. Dubonos, J. G. S. Lok, J. C. Maan, A. E. Filippov, and F. M. Peeters, *Nature (London)* 390, 259 (1997).
151. F. M. Peeters, X. Q. Li, and A. Matulis, in "Proceedings of the International Conference on Quantum Devices and Circuits" (K. Ismail, S. Bandyopadhyay, and J. P. Leburton, eds.), p. 253. World Scientific, Singapore, 1997.
152. F. M. Peeters and X. Q. Li, *Appl. Phys. Lett.* 72, 572 (1998).
153. I. S. Ibrahim, V. A. Schweigert, and F. M. Peeters, *Phys. Rev. B* 57, 15416 (1998).
154. C. W. J. Beenakker and H. van Houten, *Phys. Rev. Lett.* 63, 1857 (1989).
155. M. L. Roukes, A. Scherer, S. J. Allen, Jr., H. G. Craighead, R. M. Ruthen, E. D. Beebe, and J. Harbison, *Phys. Rev. Lett.* 59, 3011 (1987).
156. H. U. Baranger, D. P. DiVincenzo, R. A. Jalabert, and A. D. Stone, *Phys. Rev. B* 44, 10637 (1991).
157. M. Johnson, B. R. Bennett, M. J. Yang, M. M. Miller, and B. V. Shanabrook, *Appl. Phys. Lett.* 71, 974 (1997).
158. F. G. Monzon, M. Johnson, and M. L. Roukes, *Appl. Phys. Lett.* 71, 3087 (1997).
159. M. N. Baibich, J. M. Broto, A. Fert, F. Ngyuen van Dau, F. Petroff, P. Etienne, G. Creuzet, A. Friederich, and J. Chazelas, *Phys. Rev. Lett.* 61, 2472 (1988).
160. B. Heinrich and J. A. C. Bland, eds., "Ultrathin Magnetic Films I and II." Springer-Verlag, Berlin, 1994, and the many references therein.
161. D. J. Monsma, J. C. Lodder, Th. J. A. Popma, and B. Bieny, *Phys. Rev. Lett.* 74, 5260 (1995).
162. R. J. Celotta, J. Unguris, and D. T. Pierce, *J. Appl. Phys.* 75, 6452 (1994).
163. H. J. Drouhin, A. J. van der Sluijs, Y. Lassaly, and G. Lampel, *J. Appl. Phys.* 79, 4734 (1996).
164. D. J. Monsma, R. Vlutters, T. Shimatsu, E. G. Keim, R. H. Mollema, and J. C. Lodder, *IEEE Trans. Magn.* 33, 3495 (1997).
165. K. Mizuyama, T. Kinno, T. Yamauchi, and K. Tanaka, *IEEE Trans Magn.* 33, 3500 (1997).
166. T. Kinno, K. Tanaka, and K. Mizushima, *Phys. Rev. B* 56, R4391 (1997).
167. M. Messervey, P. M. Tedrow, and P. Fulde, *Phys. Rev. Lett.* 25, 1270 (1970).
168. R. Messervey, D. Paraskevopoulos, and P. M. Tedrow, *Phys. Rev. Lett.* 37, 858 (1976).
169. A. G. Aronov, *JETP Lett.* 24, 32 (1976) [*Pis'ma Zh. Eksp. Teor. Fiz.* 24, 37 (1976)].
170. R. H. Silsbee, *Bull. Magn. Reson.* 2, 284 (1980).
171. M. Johnson and R. H. Silsbee, *Phys. Rev. Lett.* 55, 17 (1985).
172. M. Johnson and R. H. Silsbee, *Phys. Rev. B* 37, 5312 (1988).
173. M. Johnson and R. H. Silsbee, *Phys. Rev. B* 37, 5326 (1988).
174. M. Johnson, *Phys. Rev. Lett.* 67, 3594 (1991), and references therein.
175. A. Fert and S. F. Lee, *J. Magn. Magn. Mater.* 165, 115 (1997), and references therein.
176. S. Alvarado and P. Renaud, *Phys. Rev. Lett.* 68, 1387 (1992).
177. P. M. Thedrow and R. Messervey, *Phys. Rev. Lett.* 26, 192 (1971).
178. J. A. Hertz and K. Aoi, *Phys. Rev. B* 8, 3252 (1973).
179. M. B. Stearns, *J. Magn. Magn. Mater.* 5, 167 (1977).
180. R. Messervey, *J. Appl. Phys.* 61, 3709 (1987).
181. J. Braun and G. Borstel, *Surf. Sci.* 307–309, 1118 (1994).
182. T. Maruyama, E. L. Garwin, R. A. Mair, R. Prepost, J. S. Smith, and J. D. Walker, *J. Appl. Phys.* 73, 5189 (1993).
183. G. K. Zyryanov, I. A. Pchelkin, and M. S. Lepestkina, *Techn. Phys. Lett.* 20, 122 (1994).
184. M. Tawada, T. Baba, Y. Kurihara, M. Mizuta, T. Nakanishi, K. Nishitani, S. Okumi, T. Omori, C. Suzuki, Y. Takeuchi, K. Togawa, and M. Yoshioka, *Jpn. J. Appl. Phys., Part 1* 36, 2863 (1997).
185. T. Saka, T. Kato, T. Nakanishi, H. Aoyagi, T. Kosugoh, S. Nakamura, M. Tawada, M. Tsubata, H. Horinaka, and Y. Kamiya, *J. Cryst. Growth* 124, 346 (1992).

186. W. G. Bi and C. W. Tu, *J. Vac. Sci. Techol., B* 14, 2282 (1996).
187. Yu. B. Bolkhovityanov, A. M. Gilinskii, N. V. Nomerotskii, E. M. Trukhanov, and A. S. Yaroshevich, *Techn. Phys. Lett.* 19, 399 (1993).
188. E. Pelucchi, S. De-Rossi, and F. Ciccacci, *J. Electron Spectrosc. Relat. Phenom.* 76, 505 (1995).
189. T. Nakanishi, H. Aoyagi, S. Nakamura, S. Okumi, C. Suzuki, C. Takahashi, Y. Tanimoto, M. Tawada, K. Togawa, M. Tsubata, Y. Kurihara, T. Omori, Y. Takeuchi, M. Yoshioka, T. Kato, T. Saka, T. Baba, and M. Mizuta, *AIP Conf. Proc.* 343, 622 (1995).
190. T. Saka, T. Kato, T. Nakanishi, M. Tsubata, K. Kishino, H. Horinaka, Y. Kamiya, S. Okumi, C. Takahashi, Y. Tanimoto, M. Tawada, K. Togawa, H. Aoyagi, S. Nakamura, *Jpn. J. Appl. Phys.* 32, L1837 (1993).
191. F. Ciccacci, S. De Rossi, E. Pelucchi, and A. Tagliaferri, *Rev. Sci. Instrum.* 68, 1841 (1997).
192. P. Drescher, H. G. Andresen, K. Aulenbacher, J. Bermuth, T. Dombo, H. Fischer, H. Euteneuer, N. N. Faleev, M. S. Galaktionov, D. von-Harrach, P. Hartmann, J. Hoffmann, P. Jennewein, K. H. Kaiser, S. Koebis, O. V. Kovalenkov, H. J. Kreidel, J. Langbein, Y. A. Mamaev, C. Nachtigall, M. Petri, S. Pluetzer, E. Reichert, M. Schemies, H. J. Schoepe, K. H. Steffens, M. Steigerwald, A. V. Subashiev, H. Trautner, D. A. Vinokurov, Y. P. Yashin, and B. S. Yavich, *Appl. Phys. A* 63, 203 (1996).
193. M. W. J. Prins, R. Jansen, and H. van Kempen, *Phys. Rev. B* 53, 8105 (1996).
194. M. W. J. Prins, R. Jansen, R. H. M. Groeneveld, A. P. van Gelder, and H. van Kempen, *Phys. Rev. B* 53, 8090 (1996).
195. R. Wiesendanger, H.-J. Güntherodt, G. Güntherodt, R. J. Gambina, and R. Ruf, *Phys. Rev. Lett.* 65, 247 (1990), and references therein on spin-polarized analysis.
196. Y. Suzuki, W. Habhan, and K. Tanaka, *Appl. Phys. Lett.* 71, 3153 (1997).
197. S. Datta and B. Das, *Appl. Phys. Lett.* 56, 665 (1990).
198. G. Lommer, F. Malcher, and U. Rössler, *Phys. Rev. Lett.* 60, 728 (1988).
199. J. Luo, H. Munekata, F. F. Fang, and P. J. Stiles, *Phys. Rev. B* 38, 10142 (1988).
200. J. Luo, H. Munekata, F. F. Fang, and P. J. Stiles, *Phys. Rev. B* 41, 7685 (1990).
201. E. I. Rashba, *Sov. Phys. Solid State* 2, 1109 (1960) [*Fiz. Tverd. Tela (Leningrad)* 2, 1224 (1960)].
202. Y. A. Bychov and E. I. Rashba, *J. Phys. C* 17, 6039 (1984).
203. J. P. Heida, B. J. van Wees, T. M. Klapwijk, and G. Borghs, in "Proceedings of the 23rd International Conference on the Physics of Semiconductors," p. 2467. World Scientific, Singapore, 1996.
204. J. Nitta, T. Akazaki, H. Takayanagi, and T. Enoki, *Phys. Rev. Lett.* 78, 1335 (1997).
205. M. Behet, S. Nemeth, J. De Boeck, G. Borghs, J. Tummler, J. Woitok, and J. Geurts, *Semicond. Sci. Technol.* 13, 428 (1998).
206. F. Rahman, B. L. Gallagher, M. Behet, and J. De Boeck, *Appl. Phys. Lett.* 72 (1998).
207. Y. Q. Jia, R. C. Shi, and S. Y. Chou, *IEEE Trans. Magn.* 32, 4707 (1996).
208. M. W. J. Prins, H. van Kempen, H. van Leuken, R. A. de Groot, W. Van Roy, and J. De Boeck, *J. Phys.: Condens. Matter* 7, 9447 (1995).
209. J. C. Grobli, D. Guarisco, S. Frank, and F. Meier, *Phys. Rev. B* 51, 2945 (1995).
210. T. Sands, C. J. Palmstrøm, J. P. Harbison, V. G. Keramidas, N. Tabatabaie, T. L. Cheeks, R. Ramesh, and Y. Silberberg, *Mater. Sci. Rep.* 5, 99 (1990), and references therein.
211. J. G. Zhu, C. B. Carter, C. J. Palmstrøm, and K. C. Garrison, *Appl. Phys. Lett.* 55, 39 (1990).
212. D. E. Brehmer, K. Zhang, Ch. J. Schwarz, S. P. Chau, S. J. Allen, J. P. Ibbetson, J. P. Zhang, C. J. Palmstrom, and B. Wilkens, *Appl. Phys. Lett.* 67, 1268 (1995).
213. T. Sands, J. P. Harbison, N. Tabatabaie, W. K Chan, H. L. Gilchrist, T. L. Cheeks, L. T. Florez, and V. G. Keramidas, *Surf. Sci.* 228, 1 (1990).
214. S. J. Allen, Jr., N. Tabatabaie, C. J. Palmstrøm, G. W. Hull, T. Sands, F. DeRosa, H. L. Gilchrist, and K. C. Garrison, *Phys. Rev. Lett.* 62, 2309 (1989).
215. J. G. Zhu, C. B. Carter, C. J. Palmstrom, and S. Mounier, *Appl. Phys. Lett.* 56, 1323 (1990).
216. M. Tanaka, M. Tsuda, T. Nishinaga, and C. J. Palmstrøm, *Appl. Phys. Lett.* 68, 84 (1996).
217. G. A. Prinz, *Phys. Rev. Lett.* 54, 1051 (1985).
218. G. A. Prinz, *Science* 250, 1092 (1990).
219. M. Gester, C. Daboo, R. J. Hicken, Sj. Gray, A. Ercole, and J. A. C. Bland, *J. Appl. Phys.* 80, 347 (1996).
220. B. T. Jonker, H. Abad, and J. J. Krebs, *J. Appl. Phys.* 76, 6294 (1994).
221. E. Kneedler, P. M. Thibado, B. T. Jonker, B. R. Bennett, B. V. Shanabrook, R. J. Wagner, and L. J. Whitman, *J. Vac. Sci. Technol.* 14, 3193 (1996).
222. J. M. Florczak, P. J. Ryan, J. N. Kuznia, A. M. Wowchak, P. I. Cohen, R. M. White, G. A. Prinz, and E. D. Dahlberg, *Mater. Res. Soc. Symp. Proc.* 300, 213 (1989).
223. D. K. Lottis, G. A. Prinz, and E. D. Dahlberg, *Mater. Res. Soc. Symp. Proc.* 300, 219 (1989).
224. K. T. Riggs, E. D. Dahlberg, and G. A. Prinz, *Phys. Rev. B* 41, 7088 (1990).
225. J. M. Florczak and E. D. Dahlberg, *J. Magn. Magn. Mater.* 104, 399 (1992).
226. S. J. Blundell, M. Gester, J. A. C. Bland, C. Daboo, E. Gu, M. J. Baird, and A. J. R. Ives, *J. Appl. Phys.* 73, 5948 (1993).
227. C. Daboo, R. J. Hicken, D. E. P. Fley, M. Gester, S. J. Gray, A. J. R. Ives, and J. A. C. Bland, *J. Appl. Phys.* 75, 5586 (1994).
228. U. Ebels, M. Gester, C. Daboo, and J. A. C. Bland, *Thin Solid Films* 275, 172 (1996).

229. A. Moghadam, J. G. Booth, D. G. Lord, J. Boyle, and A. D. Boardman, *IEEE Trans. Magn.* 30, 769 (1994).
230. E. Gu, M. Gester, R. J. Hicken, C. Daboo, M. Tselepi, S. J. Gray, J. A. C. Bland, L. M. Brown, T. Thomson, and P. C. Riedi, *Phys. Rev. B* 52, 14704 (1995).
231. R. J. Hicken, D. E. P. Eley, M. Gester, S. J. Gray, C. Daboo, A. J. R. Ives, and J. A. C. Bland, *J. Magn. Magn. Mater.* 145, 278 (1995).
232. R. Meckenstock, O. von Geisau, F. Schreiber, E. C. Da Silva, and J. Pelzl, *J. Phys. IV* 4, 663 (1994).
233. R. Meckenstock, O. von Geisau, J. Pelzl, and J. A. Wolf, *J. Appl. Phys.* 77, 6439 (1995).
234. M. Schneider, S. Muller-Pfeiffer, and W. Zin, *J. Appl. Phys.* 79, 8578 (1996).
235. E. Gu, J. A. C. Bland, C. Daboo, M. Gester, L. M. Brown, R. Ploessl, and J. N. Chapman, *Phys. Rev. B* 51, 3596 (1995).
236. S. Okamoto, O. Kitakami, and Y. Shimada, *Jpn. J. Appl. Phys.* 33, 6164 (1994).
237. Y. Endo, S. Okamoto, O. Kitakami, and Y. Shimada, *J. Appl. Phys.* 81, 344 (1997).
238. M. Alper, K. Attenborough, R. Hart, S. J. Lane, D. S. Lashmore, C. Younes, and W. Schwarzacher, *Appl. Phys. Lett.* 63, 2144 (1993).
239. P. Allongue and E. Souteyrand, *J. Vac. Sci. Technol., B* 5, 1644 (1987).
240. R. Hart, P. A. Midgley, A. Wilkinson, and W. Schwarzacher, *Appl. Phys. Lett.* 67, 1316 (1995).
241. C. Shearwood, S. J. Blundell, M. J. Baird, J. A. C. Bland, M. Gester, H. Ahmed, and H. P. Hughes, *J. Appl. Phys.* 75, 5249 (1994).
242. U. Ebels, O. Adeyeye, M. Gester, R. P. Cowburn, C. Daboo, and J. A. C. Bland, *Phys. Rev. B* 56, 5443 (1997).
243. A. Y. Liu and D. J. Singh, *Appl. Phys. Lett.* 73, 6189 (1993).
244. G. A. Prinz, *J. Magn. Magn. Mater.* 100, 469 (1991).
245. J. A. C. Bland, R. D. Bateson, P. C. Riedi, R. G. Graham, H. J. Lauter, J. Penfold, and C. Shackleton, *J. Appl. Phys.* 69, 4989 (1991).
246. C. J. Gutierrez, J. J. Krebs, and G. A. Prinz, *Appl. Phys. Lett.* 61, 2476 (1992).
247. C. J. Gutierrez, G. A. Prinz, J. J. Krebs, M. E. Filipkowski, V. G. Harris, and W. T. Elam, *J. Magn. Magn. Mater.* 126, 232 (1993).
248. J. Dekoster, E. Jedryka, C. Meny, and G. Langouche, *J. Magn. Magn. Mater.* 121, 69 (1993).
249. J. P. Jay, E. Jedryka, M. Wojcik, J. Dekoster, G. Langouche, and P. Panissod, *Z. Phys. B* 101, 329 (1996).
250. C. D. England, B. N. Engel, and C. M. Falco, *J. Appl. Phys.* 69, 5310 (1991).
251. J. De Boeck, W. Van Roy, C. Bruynseraede, A. Van Esch, H. Bender, and G. Borghs, *Microelectron. J.* 27, 383 (1996).
252. B. Swinnen, J. Dekoster, J. Meersschaut, S. Demuynck, S. Cottenier, G. Langouche, and M. Rots, *J. Appl. Phys.* 81, 4343 (1997).
253. J. Weissman, E. Zolotoyabko, L. Levin, and A. Ginzburg, *Mater. Chem. Phys.* 37, 175 (1994).
254. G. W. Anderson, R. C. Hanf, and P. R. Norton, *Phys. Rev. Lett.* 74, 2764 (1995).
255. G. W. Anderson, M. C. Hanf, X. R. Qin, P. R. Norton, K. Myrtle, and B. Heinrich, *Surf. Sci.* 346, 145 (1996).
256. F. J. Himpsel, W. Drube, A. B. McLean, and A. Santoni, *Appl. Surf. Sci.* 56–58, 160 (1992).
257. A. Filipe, A. Schuhl, and P. Galtier, *Appl. Phys. Lett.* 70, 129 (1997).
258. F. Y. Shiau, S. L. Chen, M. Loomans, and Y. A. Chang, *J. Mater. Res.* 6, 1532 (1991).
259. A. Filipe and A. Schuhl, *J. Appl. Phys.* 81, 4359 (1997).
260. E. M. Kneedler and B. T. Jonker, *J. Appl. Phys.* 81, 4463 (1997).
261. B. T. Jonker, O. J. Glembocki, R. T. Holm, and R. J. Wagner, *Phys. Rev. Lett.* 79, 4886 (1997).
262. B. T. Jonker, E. M. Kneedler, P. Thibado, O. J. Glembocki, L. J. Whitman, and B. R. Bennett, *J. Appl. Phys.* 81, 4362 (1997).
263. M. Tanaka, J. P. Harbison, M. C. Park, Y. S. Park, T. Shin, and G. M. Rothberg, *J. Appl. Phys.* 76, 6278 (1994).
264. M. Tanaka, J. P. Harbison, and G. M. Rothberg, *J. Cryst. Growth* 150, 1132 (1995).
265. H. Akinaga, Y. Suzuki, K. Tanaka, K. Ando, and T. Katayama, *Appl. Phys. Lett.* 67, 141 (1995).
266. H. Akinaga, S. Miyanishi, W. Van Roy, and L. H. Kuo, *Appl. Phys. Lett.* 70, 2472 (1997).
267. T. Sands, J. P. Harbison, M. L. Leadbeater, S. J. Allen, Jr., G. W. Hull, R. Ramesh, and V. G. Keramidas, *Appl. Phys. Lett.* 57, 2609 (1990).
268. W. Van Roy, J. De Boeck, H. Bender, C. Bruynseraede, A. Van Esch, and G. Borghs, *J. Appl. Phys.* 78, 398 (1995).
269. M. Tanaka, J. P. Harbison, J. De Boeck, T. Sands, B. Philips, T. L. Cheeks, and V. G. Keramidas, *Appl. Phys. Lett.* 62, 1565 (1993).
270. K. M. Krishnan, *Appl. Phys. Lett.* 61, 2365 (1992).
271. G. Lauhoff, C. Bruynseraede, J. De Boeck, W. Van Roy, J. A. C. Bland, and G. Borghs, *Phys. Rev. Lett.* 79, 5290 (1997).
272. J. De Boeck, C. Bruynseraede, H. Bender, A. Van Esch, W. Van Roy, and G. Borghs, *J. Cryst. Growth* 150, 1139 (1995).
273. H. van Leuken, A. Lodder, M. T. Czyzyk, F. Springelkamp, and R. A. de Groot, *Phys. Rev. B* 41, 5613 (1990).

274. A. Sakuma, *J. Phys. Soc. Jpn.* 63, 1422 (1994).
275. J. De Boeck, T. Sands, J. P. Harbison, A. Scherer, H. Gilchrist, T. L. Cheeks, M. Tanaka, and V. G. Keramidas, *Electron. Lett.* 29, 421 (1993).
276. J. De Boeck, W. Van Roy, C. Bruynseraede, A. Van Esch, H. Bender, C. Van Hoof, and G. Borghs, *Phys. Scr., T* 66, 183 (1996).
277. W. Van Roy, H. Akinaga, S. Miyanishi, K. Tanaka, and L. H. Kuo, *Appl. Phys. Lett.* 69, 711 (1996).
278. W. Van Roy, H. Akinaga, S. Miyanishi, K. Tanaka, and L. H. Kuo, *J. Magn. Magn. Mater.* 165, 149 (1997).
279. B. Briner and M. Landolt, *Z. Phys. B* 92, 137 (1993).
280. J. E. Mattson, S. Kumar, E. E. Fullerton, S. R. Lee, C. H. Sowers, M. Grimsditch, S. D. Bader, and F. T. Parker, *Phys. Rev. Lett.* 71, 185 (1993).
281. J. Dekoster, H. Bemelmans, S. Degroote, R. Moons, J. Verheyden, A. Vantomme, and G. Langouche, *J. Appl. Phys.* 81, 5349 (1997).
282. W. H. Butler, X. G. Zhang, X. Wang, J. van Eck, and J. M. MacLaren, *J. Appl. Phys.* 81, 5518 (1997).
283. S. S. Kang, G. J. Jin, X. N. Xu, A. Hu, H. R. Zhai, S. S. Jiang, Y. Chen, W. R. Zhu, G. S. Dong, S. M. Zhou, X. Jin, and J. W. Feng, *Appl. Phys. Lett.* 69, 3923 (1996).
284. K. Matsuyama, H. Asada, T. Saeki, Y. Sawamoto, and K. Taniguchi, *J. Appl. Phys.* 81, 5449 (1997).
285. J. de la Figuera, M. A. Huerta-Garnica, J. E. Prieto, C. Ocal, and R. Miranda, *Appl. Phys. Lett.* 66, 1006 (1995).
286. M. Blondel, J. P. Meyer, B. Doudin, and J. Ph. Ansermet, *Appl. Phys. Lett.* 65, 3019 (1994).
287. L. Piraux, J. M. George, J. F. Despres, C. Leroy, E. Ferain, and R. Legras, *Appl. Phys. Lett.* 65, 2484 (1994).
288. J. Shi, J. Kikkawa, R. Proksch, T. Schffer, D. Awschalom, G. Medeiros-Ribeiro, and P. Petroff, *Nature (London)* 377, 707 (1995).
289. J. Shi, D. D. Awschalom, P. M. Petroff, and K. Babcock, *J. Appl. Phys.* 81, 4331 (1997).
290. J. P. Zhang, A. K. Cheetham, K. Sun, J. S. Wu, K. H. Kuo, J. Shi, and D. D. Awschalom, *Appl. Phys. Lett.* 71, 143 (1997).
291. P. J. Wellmann, J. M. Garcia, J. L. Feng, and P. M. Petroff, *Appl. Phys. Lett.* 71, 2532 (1997).
292. J. C. P. Chang, N. Otsuka, E. S. Harmon, M. R. Melloch, and J. M. Woodall, *Appl. Phys. Lett.* 65, 2801 (1994).
293. T. M. Pekarek, B. C. Crooker, D. D. Nolte, J. Deak, M. McElfresh, J. C. P. Chang, E. S. Harmon, M. R. Melloch, and J. M. Woodall, *J. Magn. Magn. Mater.* 169, 261 (1997).
294. M. W. Bench, C. B. Carter, W. Feng, and P. I. Cohen, *Appl. Phys. Lett.* 66, 2400 (1995).
295. J. Katcki, M. Shiojiri, T. Isshiki, K. Nishio, Y. Yabuuchi, and N. Y. Jin-Phillipp, *Electron Technol.* 29, 393 (1996).
296. S. Guha and H. Munekata, *J. Appl. Phys.* 74, 2974 (1993).
297. J. De Boeck, R. Oesterholt, A. Van Esch, H. Bender, C. Bruynseraede, C. Van Hoof, and G. Borghs, *Appl. Phys. Lett.* 68, 2744 (1996).
298. J. De Boeck, R. Oesterholt, H. Bender, A. Van Esch, C. Bruynseraede, C. Van Hoof, and G. Borghs, *J. Magn. Magn. Mater.* 156, 148 (1996).
299. D. R. Schmidt, J. P. Ibbetson, D. E. Brehmer, C. J. Palmstrøm, S. J. Allen, Jr., *Mater. Res. Soc. Conf. Proc.* 475, 251 (1997).
300. See, for example, *Mater. Res. Soc. Proc.* 214 (1992) and *J. Electron. Mater.* 22 (1993).
301. H. Okamoto, *Bull. Alloy Phase Diagrams* 10, 549 (1989).
302. T. M. Pekarek, B. C. Crooker, S. Li, M. McElfresh, J. C. P. Chang, D. Mcinturff, E. S. Harmon, M. R. Melloch, and J. M. Woodall, *J. Appl. Phys.* 81, 4869 (1997).
303. H. Akinaga, J. De Boeck, G. Borghs, S. Miyanishi, A. Asamitsu, W. Van Roy, Y. Tomioka, and L. H. Kuo, *Appl. Phys. Lett.* 72 (1998).
304. J. De Boeck, H. Akinaga, C. Bruynseraede, H. Bender, and G. Borghs, *IEEE Trans. Magn.* (1998).
305. A. Shen, H. Ohno, F. Matsukura, Y. Sugawara, Y. Ohno, N. Akiba, and T. Kuroiwa, *Jpn. J. Appl. Phys.* 36, L73 (1997).
306. A. Van Esch, L. Van Bockstal, J. De Boeck, G. Verbanck, A. S. Van Steenbergen, P. J. Wellman, B. Grietens, R. Bogaerts, F. Herlach, and G. Borghs, *Phys. Rev. B* 56, 20 (1997).
307. A. Berkowitz, J. R. Mitchell, M. J. Carey. A. P. Young, S. Zhang, F. E. Spada, F. T. Parker, A. Hutten, and G. Thomas, *Phys. Rev. Lett.* 68, 3745 (1992).
308. J. Q. Xiao, J. S. Jiang, and C. L. Chien, *Phys. Rev. Lett.* 68, 3749 (1992); *Phys. Rev. B* 46, 9266 (1992).
309. F. W. Smith, H. W. Le, V. Diadiuk, M. A. Hollis, A. R. Calawa, S. Gupta, M. Frankel, D. R. Dykaar, G. A. Mourou, and T. Y. Hsiang, *Appl. Phys. Lett.* 54, 890 (1991).
310. A. C. Warren, N. Katzenellenbougen, D. Grischkowsky, J. M. Woodall, M. R. Melloch, and N. Otsuka, *Appl. Phys. Lett.* 58, 1512 (1991).
311. D. D. Nolte, M. R. Melloch, J. M. Woodall, and S. E. Ralph, *Appl. Phys. Lett.* 62, 1356 (1991).
312. D. D. Nolte, *J. Appl. Phys.* 76, 3740 (1991).
313. R. Desmicht, G. Faini, V. Cros, A. Fert, F. Petroff, and A. Vaurs, *Appl. Phys. Lett.* 72, 386 (1998).

Chapter 8

CARBON NANOTUBES

P. M. Ajayan

Department of Materials Science and Engineering, Rensselaer Polytechnic Institute, Troy, New York, USA

Contents

1. INTRODUCTION

The discovery of fullerenes [1] provided exciting new insights into carbon nanostructures and how architectures built from sp^2 carbon units based on simple geometrical principles can result in new symmetries and structures that have fascinating and useful properties. Carbon nanotubes represent the most striking example. Less than a decade after their discovery [2], the new knowledge available in this field indicates that nanotubes may be used in several practical applications. There have been great improvements in synthesis techniques, which can now produce reasonably pure nanotubes in gram quantities. Studies of structure–property correlation in nanotubes have been supported and, in some cases, preceded by theoretical modeling that provided the insights for experimentalists to seek new directions. The latter assisted the rapid expansion of this field.

Quasi-one-dimensional carbon whiskers or "nanotubes" are perfectly straight tubules (Fig. 1) with diameters in nanometers and properties close to those of an *ideal* graphite fiber [3–6]. Nanotubes were discovered quite accidentally by Sumio Iijima while studying the surfaces of graphite electrodes used in an electric arc discharge. His observation

Handbook of Nanostructured Materials and Nanotechnology, edited by H.S. Nalwa
Volume 5: Organics, Polymers, and Biological Materials

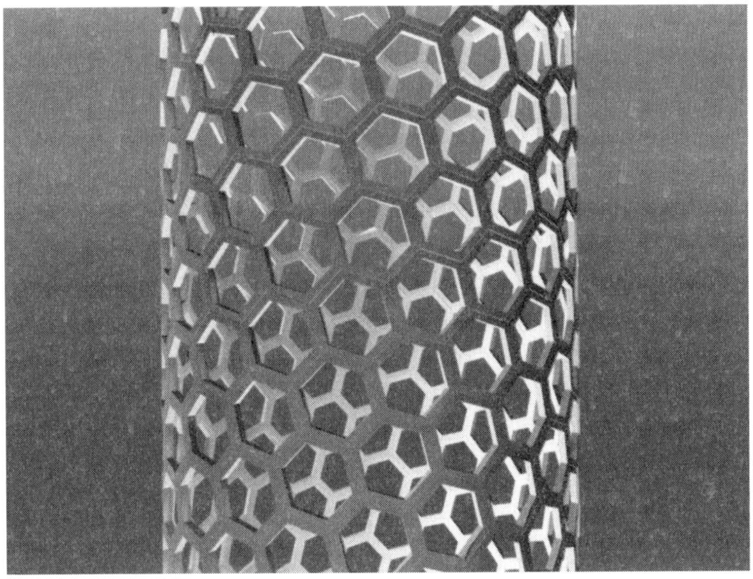

Fig. 1. Schematic of a single layer of carbon nanotube. The cylindrical structure is built from hexagonal honeycomb lattice of sp^2 bonded carbon with no dangling bonds. Imagine taking one layer of graphite and folding it to match the edges that contain the dangling bonds; the cylinder so formed is the basic unit of the nanotube. In multiwalled nanotubes, cylinders of various diameters are arranged concentric to each other with a constant spacing of 0.34 nm between them. In single-walled nanotube ropes, tens of cylinders of near uniform diameter (~1–2 nm) arrange in a triangular lattice. (Source: Reprinted from [6], with permission from IOP Publishing.)

and elucidation of nanotube structure started a new direction in carbon research that complemented the excitement and activities that covered the fullerene research front. These tiny carbon whiskers with incredible strength and fascinating electronic properties appear to be ready to overtake fullerenes in the race toward the technological marketplace. It is the structure, topology, and dimensions of nanotubes that makes their properties exciting compared to the parent, planar graphite.

The uniqueness of the nanotube structure arises from what is termed *helicity* in the arrangement of carbon hexagons on their surface layer honeycomb lattice [2, 7]. The helicity defined by symmetry and the tube diameter (both of which determine the size of the repeating structural unit) introduce significant changes in the electronic density of states and, hence, provide unique electronic character for the nanotubes. The other factor is topology, or the closed geometry of individual layers in each tube [8, 9], which has a profound effect on the physical properties. The combination of size, structure, and topology endows nanotubes with important and unique mechanical (stability, strength, stiffness, and elastic deformability), transport (coherent electron transport), and surface properties [10]. Soon after the discovery of nanotubes in 1991, a number of theoretical papers suggested this strong structure–property correlation, establishing a new level of excitement in the study of this novel material.

Among these fascinating carbon nanotubular structures, two variations exist that contrast in general appearance, structure, and graphitization. There are graphitic multiwalled carbon nanotubes (MWNTs), which were the first type that were found in 1991, and single-walled carbon nanotubes (SWNTs), which were found later in 1993 [11, 12]. The former may be considered to be single-crystal nanosized ideal graphite fibers; the latter are true elongated fullerene tubes.

Since their discovery in 1991, several demonstrations have been carried out that suggest potential applications of nanotubes. These include the use of MWNTs for field emitting devices [13, 14], as nanoprobes when attached to the ends of atomic force microscope (AFM) tips [15], and as efficient supports in heterogeneous catalysis [16] and microelectrodes in electrocatalytic reactions [17, 18], and the use of SWNTs as electronic devices [19], as good media for hydrogen storage [20], and as individual quantum wires [21]. Some of these uses could become real marketable applications in the near or distant future. The lack of availability of bulk amounts of well defined samples and the lack of knowledge in organizing and manipulating objects such as nanotubes have hindered progress in this field to some extent. However, optimism has prevailed in the last two years, which have seen important breakthroughs, resulting in the availability of near uniform samples at high yield. Noticing the way this field has progressed in the last two years, it seems that a set of fully functional devices/structures based on nanotubes is beginning to evolve.

2. STRUCTURE

SWNT consist of singular graphene cylindrical walls with diameters ranging between 1 and 2 nm (Fig. 2a). MWNTs have thicker walls, consisting of several coaxial graphene cylinders separated by a spacing (0.34 nm) that is close to the interlayer distance in graphite. The outer diameters of MWNTs range between 2- and 25 nm (Fig. 2b) and the inner hollows range from $\sim$1 to 8 nm. In MWNTs there is no three-dimensional ordering (as in bulk graphite) between the individual graphite layers, suggesting that the interlayer structure is turbostratic (rotationally disordered). This results from the severe geometrical constraint of having to match the open edges when forming continuous cylindrical geometry and at the same time maintaining a constant interlayer spacing. The aspect ratios of nanotubes vary with diameter, but the average length can be several micrometers. Individual SWNTs have uniform diameter, although when formed they also show a strong tendency to pack together in larger bundles [22, 23]. In fact, the packing occurs in a regular triangular lattice structure with an intertube spacing in each of the bundles that corresponds to about 0.315 nm and a lattice parameter of 1.7 nm. These data have been confirmed by X-ray diffraction [22] studies and they correspond to early theoretical predictions [24]. Raman scattering experiments are also very useful in studying the structure and diameter of SWNTs present in the samples. Experimentally detected Raman lines can be compared to those computed using ab initio calculations [25]. The position and presence of the Raman lines sensitively depends on the nanotube helicity. A low frequency breathing mode in the Raman spectra provides a unique indexing scheme for different nanotubes that are structural isomers. In the mapping of a graphene plane into a cylinder, the boundary conditions around the cylinder can be satisfied only if one of the Bravais lattice vectors (defined in terms of the two primitive lattice vectors and a pair of integer indices (n_1, n_2) of the graphene sheet maps to a whole circumference of the cylinder (Fig. 3). This scheme is very important in characterizing properties of individual nanotubes because it provides the essential symmetry to the nanotube structure. Nanotubes made from lattice translational indices of the form $(n, 0)$ or (n, n) will possess one plane of reflection and hence will have only two helical symmetry operations. All other sets of nanotubes will have all three equivalent helical operations. $(n, 0)$ type of nanotubes are, in general, called *zigzag nanotubes*, whereas the (n, n) types are called *armchair nanotubes*.

The helicity in the structure of carbon nanotubes was discovered by Iijima and revealed in his first landmark paper. The electron diffraction patterns obtained from MWNTs suggested that the top and bottom parts of individual nanotube lattices were sheared with respect to the axis, and as a result the hexagonal arrays in the honeycomb structure had helicity around the circumference of the tubes. There were several sets of the hexagonal $(hk0)$ spots in the diffraction patterns of individual MWNT that strongly suggested this arrangement. Later this was confirmed in the diffraction patterns obtained from individual SWNTs [11]. The top and bottom cylindrical portions of the tubes were indeed rotated with respect to the tube axis due to this helicity and produced, in the diffraction patterns, just two sets of $(hk0)$ spots.

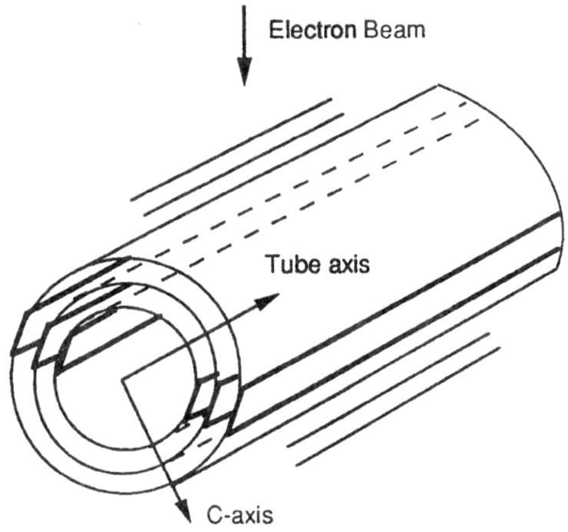

(a)

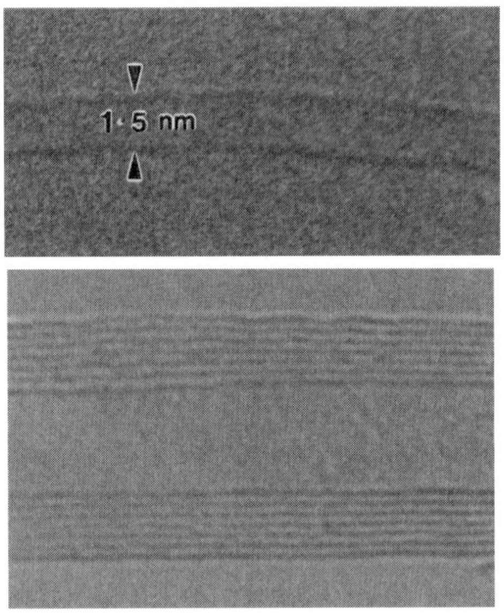

(b)

Fig. 2. High resolution transmission electron microscopy (HRTEM) images of typical SWNTs and MWNTs (b). The images look like cross sections of the tubes because they show the projections of the edges of the cylinders that comprise the nanotubes. The horizontal lines in the image correspond to edges of cylinders (thick lines in the schematic), and the inner space corresponds to the diameter of the inner hollow of the tubes. The separation between the closely spaced fringes in the MWNTs is 0.34 nm, close to the spacing between graphite planes. The schematic shows the imaging geometry of a tube lying with its axis normal to the electron beam.

If nanotubes are made simply by wrapping planar graphite sheets into cylinders, this should result in open ended tubes. However, the nanotubes prepared experimentally are observed to be closed at both ends. This involves the introduction of pentagonal topological defects, a minimum of six, on each end of each cylinder in the tube. Thus, the tubes are essentially made of cylinders attached to halves of large fullerene-like structures at the ends. The morphology of nanotubes observed under electron microscopy provides compelling evidence for the presence of isolated topological defects. Pentagons, heptagons, and 5-7 ring defect pairs have been seen on the body of nanotubes [8, 9], which alter the shape and the dimensions of nanotubes without straining the lattice through lattice

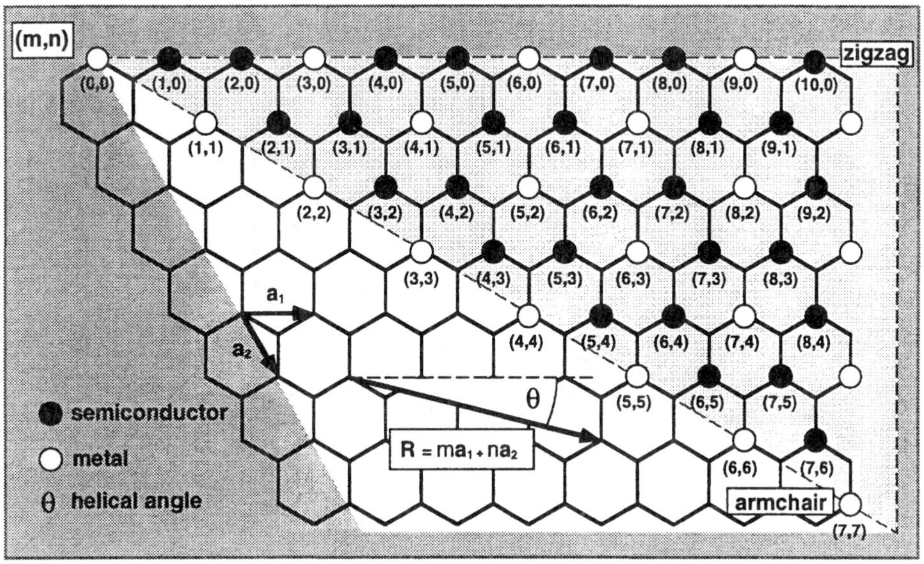

Fig. 3. Indexing scheme that shows the folding procedure for creating nanotube cylinders from planar graphene sheets. When any index shown in the lattice is mapped onto an origin, a nanotubes of that particular index is produced. For example, all the (n, n) type nanotubes form armchair tubes (indicated) and all $(n, 0)$ tubes are zigzag (indicated). The position and length of the lattice vector that connects the origin to the lattice point that defines the nanotube index determine the helicity and diameter of the tube. a_1 and a_2 are the primitive lattice vectors of the hexagonal lattice. (Source: Reprinted from [6], with permission from IOP Publishing.)

distortions (Fig. 4). The presence of closely spaced defect pairs introduces local changes in structural symmetry (a 5-7 defect pair in a hexagonal lattice is akin to a dislocation core). So the helicity of nanotubes can be altered by the insertion of these defect junctions, which could lead to the construction of joined nanotubes of different electronic structure.

It is noted that the in-plane (axial) structure of nanotubes is close to an ideal graphene sheet. This contrasts with traditional carbon fibers made using a variety of growth techniques starting from extrusion of polymer precursor slurries to catalytic chemical vapor deposition. The orientation of graphite layers in all these traditional carbon fibers show poor alignment due to a variety of in-plane defects and poor graphitization. The perfection in the structure and, more importantly, the subtlety in the structure (helicity), possible only due to its inherent defect-free structure, has a strong influence on the physical properties of nanotubes. This ideal structure and the small dimensions of nanotubes have enabled a big effort in theoretical atomistic modeling, which would be impossible for traditional carbon fibers.

3. GROWTH

3.1. Synthesis of Nanotubes

Carbon nanotubes were first noticed at the ends of graphite electrodes used in an electric arc discharge employed in fullerene synthesis. The needlelike structures of MWNTs were found in what looked like graphitic soot on the cathode surface. Later, conditions in the dc arc were optimized to build long cigar-shaped deposits that contained gram quantities of MWNTs [26]. Typically the deposit builds up on the cathode surface from evaporated anode during an arc struck between pure graphite electrodes in 500-torr He pressure at 20 V and 50–100 A. The deposit consists of a hard shell made of pyrolitic graphite and an interior made of a soft black powder containing nanotubes and closed graphite nanoparticles (Fig. 5). The yield of nanotubes depends on the stability of the arc, although the

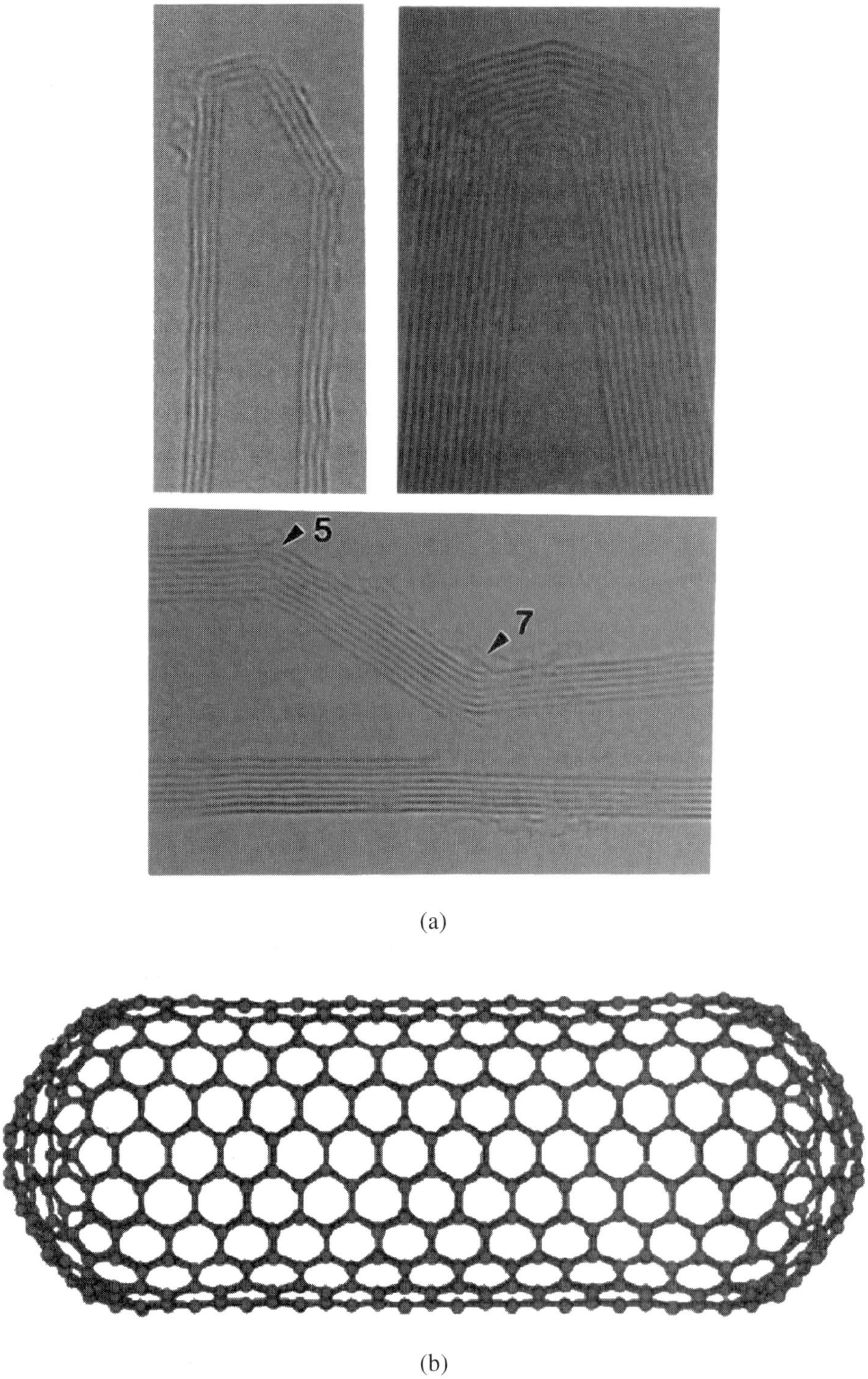

Fig. 4. (a) HRTEM images of MWNT tube tips showing the closed geometry. Also shown is the architecture produced from the positioning of 5-7 defects in the nanotube lattice (positions shown by arrows). Note that the positions of the defects in all the layers are perfectly aligned. The distance between the parallel fringes in the image corresponds to 0.34 nm. SWNTs show similar geometry, but the structures are small and not as vivid as in MWNTs. (Source: Reprinted from [6], with permission from IOP Publishing.) (b) The schematic of a closed (10, 10) SWNT is shown to emphasize the relationship between fullerenes and nanotube ends. (Courtesy of Dr. J.-C. Charlier.)

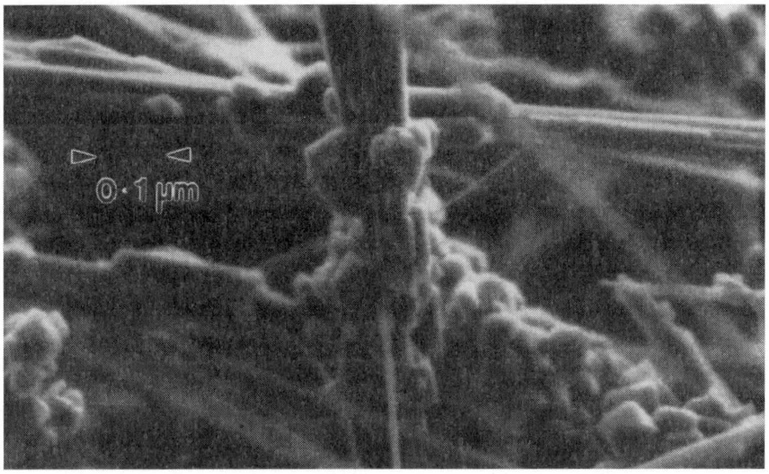

Fig. 5. SEM image of multiwalled nanotube samples grown in an electric arc. Both nanotubes and polyhedral nanoparticles can be seen in the sample. The particles have a range of cross-sectional shapes.

arc conditions still are not clearly understood. One of the problems in this method is that nanoparticles form at least a third of the fraction of the nanotube samples. The arc method used in the production of MWNTs is very similar to a process developed by Bacon [27] in 1960 to produce micrometer sized graphite whiskers, although the conditions used in the latter were vastly different (pressure of 92 atm was used for the efficient production of the fibers).

Decomposition of hydrocarbon gases by transition metal catalyst particles (Co, Fe, Ni) has been used to produce carbon nanofibers that are similar in dimensions to the nanotubes, but far in structural perfection [28]. Several morphological variations to the straight nanotubes have been made using this technique; for example, by the decomposition of acetylene gas on the surface of cobalt particles [29]. The temperatures at which such reactions are carried out fall in the 800–1500 °C range and hence the fibers are not well graphitized. These fibers typically have twisted and complicated shapes, reflecting the large number of structural defects in them, such as low angle tilt boundaries. The catalytic method, however, is a continuous process and helps to produce large quantities (kilograms) of nanofibers. The fibers also can be grown into organized predefined patterns because the catalyst particles can be patterned prior to carbon deposition. Recently it was reported [30, 18] that well-aligned and seemingly well-graphitized (stiff) multiwalled nanofibers can be grown by chemical vapor deposition catalyzed by iron nanoparticles embedded in mesoporous silica. The growth direction of these tubes is controlled by the pores in the substrate from which the fibers grow. Similarly, laser-etched arrays of cobalt nanoparticles on silicon substrate have been used to grow aligned bundles of multiwalled nanofibers, using the pyrolysis of triazine precursors [31]. Large bundles of aligned fibers with fairly uniform diameters (30–50-nm diameter) have been grown to lengths up to 50 μm (Fig. 6). The catalyst technique also has been shown to be capable of producing SWNTs (of larger size, $\sim$5 nm), using a Mo nanoparticle catalyst in a CO atmosphere [32].

SWNTs were first made by the electric arc through the introduction of catalyst species (Fe, Co) into the carbon plasma [11, 12]. Here, a hole is drilled in the center of the anode and packed with mixtures of metal catalysts and graphite powder, the metal being 1–10% by weight. Several catalysts have been tried, but the best yield of nanotubes has been obtained for Co, Ni, and bimetallic systems such as Co–Ni, Co–Pt, and Ni–Y. During the arc, weblike or collarlike structures are formed around the cathode and the outer (cooler) regions of the reaction vessel. These areas contain networks of SWNT ropes that contain 5–100 individual SWNTs codeposited along with amorphous carbon and nanoparticles

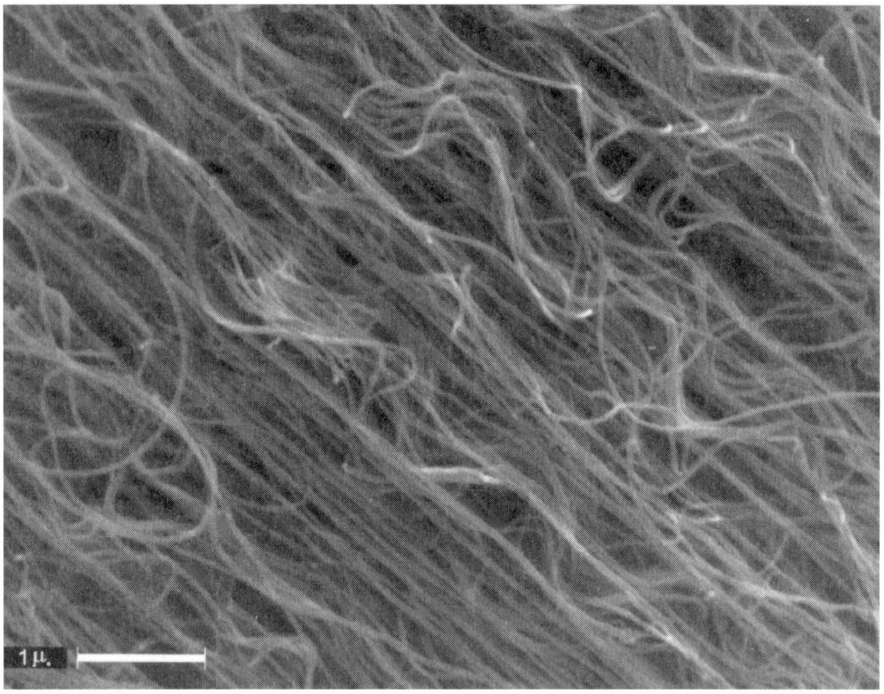

Fig. 6. TEM image showing aligned multiwalled carbon nanofibers made by pyrolysis of triazine precursors on laser-etched Co catalyst particle arrays. Notice the uniformity in diameter of the fibers. (Courtesy of M. Terrones.) (Source: Reprinted with permission from [31]. © 1997 Macmillan Magazines Limited.)

of the catalyst material (Fig. 7). Originally, the amount of SWNTs that could be produced was small, less than milligram quantities. Very recently, the use of a Ni–Y catalyst (in a 4:1 atomic ratio) in the anode provided very good yield (>75%) of SWNT material [23]. Grams of SWNT material (containing >50% nanotubes by weight) can be produced nowadays in few hours time.

Another effective way to produce SWNTs is by using laser evaporation [33, 22]. This method, reported three years ago, had a strong impact on nanotube research because it provided, for the first time, reasonable quantities of pure (<10% impurities) SWNT material. It was also suggested that nearly all the nanotubes produced by this method were armchair tubes. Although this was later proved to be not very accurate, the majority of the tubes are still considered to be armchair type. For the first time, it was shown that nanotubes could be obtained with good uniformity in size and structure (helicity).

Many of the experiments that demonstrate the potential of electronic device applications of nanotubes have been a direct result of the success of the laser ablation method. In this technique, direct laser vaporization of transition metal– (e.g., Co–Ni, 1%) graphite composite electrode targets is done in a helium atmosphere in heated ovens (1200 °C). In the most efficient version of the experiment, the amount of carbon deposited as soot is minimized by the use of two successive laser pulses: the first to ablate the carbon–metal mixture and the second to break up the larger ablated particles and feed them into the growing nanotube structures. This method has been scaled up by using continuous ablation of rotating targets and is capable of making several grams of raw material daily (the yield in the continuous process drops drastically). It has also been shown that, depending on the temperature of the oven in which nanotubes are grown, the tube diameter varies between 1 and 5-nm diameter; the higher the temperature (range between 800 and 1200 °C), the larger the nanotube diameter [34].

Alternative strategies for the synthesis of nanotubes are being developed. As mentioned before, the only commercial means to make nanosized fibers of carbon is through catalytic

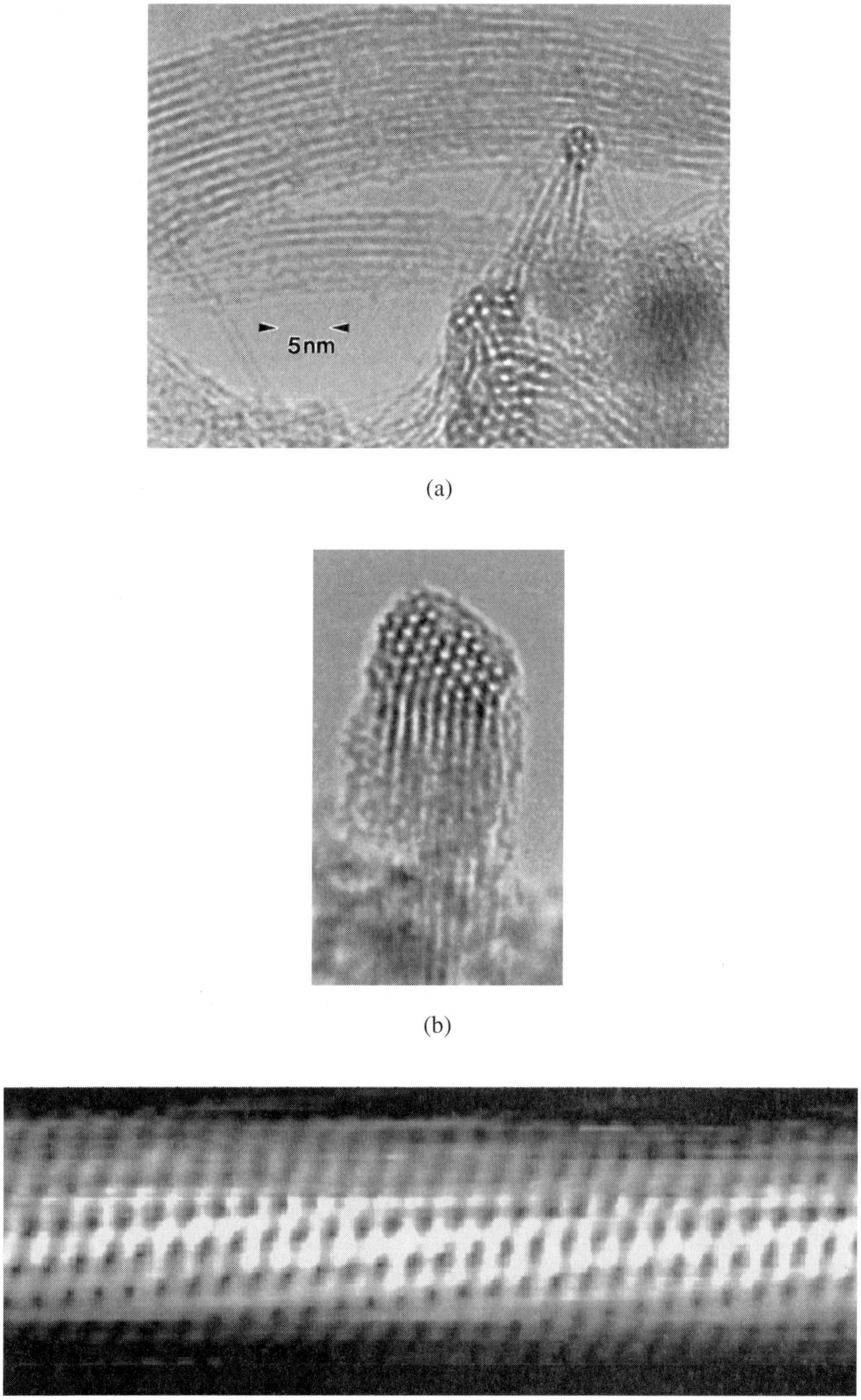

(a)

(b)

(c)

Fig. 7. (a) TEM image of bundles of SWNTs and (b) of an individual bundle that shows the lattice structure of the organized tubes inside. Also shown is (c) a high resolution scanning tunneling microscopy (STM) image showing the lattice structure of a SWNT that is close to an armchair tube. The helical lattice is clearly revealed. The STM image was provided by Dr. C. Dekker. (Source: Reprinted with permission from [68]. © 1996 Macmillan Magazines Limited.)

chemical vapor deposition (CVD) [28]; catalyst particles have been utilized to promote the growth fibers similar to nanotubes by different methods [31, 32]. Well defined pores of porous inorganic membranes (such as alumina) have been used to deposit a disordered form of carbon by CVD and graphitized further at higher temperatures to produce nanotubes [18]. The advantage of such template-based methods is that the size of the particles and the pores, which determine the size of the nanotubes, can be controlled prior to the deposition of carbon. The length of the nanotubes formed can be controlled by adjusting the amount of carbon vapor feedstock supplied and the thickness of the membranes. Recent reports [18] suggest that by removing these templates after nanotube growth, free-standing arrays of highly graphitized nanotubes can be formed.

3.2. Purification of Nanotubes

For bulk measurements as well as applications, the nanotube samples prepared by the arc method and laser ablation need to be purified. MWNTs contain small amounts of graphite impurities and large amounts of polyhedral graphite nanoparticles (with aspect ratio of 1). These substances need to be removed to retain only pure tubular structures. The first method devised to purify MWNTs relied on the oxidation behavior of carbon nanotubes [35, 36]. It was observed that oxidation preferentially occurs at the tube ends and on nanoparticles that have a higher concentration of topological defects. The tube tips and nanoparticles can thus be burnt away (at >700 °C in air or oxygen) and purified MWNT samples can be made (Fig. 8). This, as the method suggests, is a destructive technique and the yield is extremely low (<5%). Oxidation in solvents (like concentrated acids) has proved to be more effective [37], but still has not proved to be a good method for large scale separation. Recently multistep physiochemical methods have been designed to purify nanotubes from the smaller particles. These methods are based on the dispersion of nanotubes in polar solvents assisted by surfactants (e.g., sodium dodecyl sulfate), followed by ultracentrifugation, microfiltration, and size exclusion chromatography [38]. Preliminary results show that size-selected separation of MWNTs may be possible by such methods.

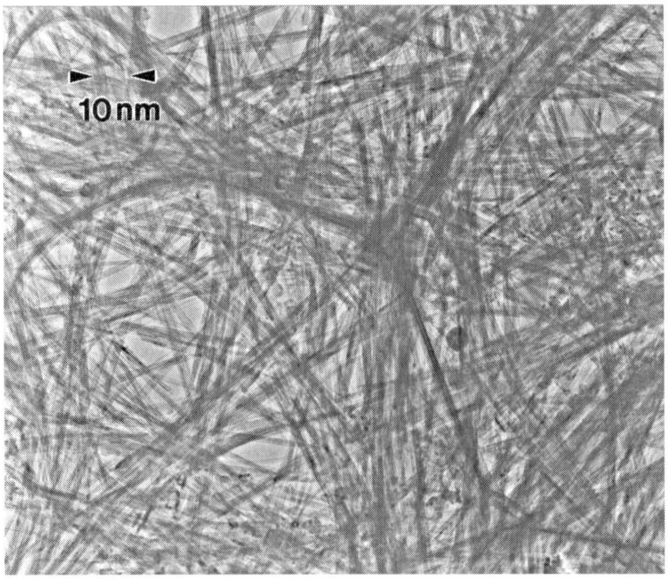

Fig. 8. TEM image of purified MWNT sample. Most of the polyhedral nanoparticles originally present in the samples have been burnt away by oxidation in air at 750 °C for about half an hour.

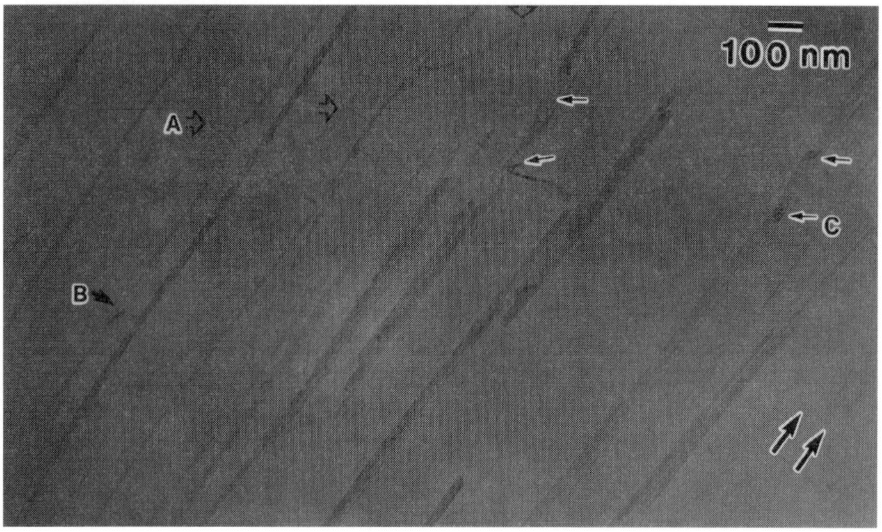

Fig. 9. TEM image of MWNTs aligned in an epoxy polymer matrix. The alignment was accomplished by first making a composite of epoxy and purified nanotubes, and then making thin (a few hundred nanometer) sections of the polymer nanotube composite. Nanotubes align along the cutting direction. Notice the structural defects in the nanotubes (buckling, breaks, twisting, etc., denoted by A, B and C) due to deformation after the cutting was done.

The impurities present in the SWNT samples are amorphous carbon soot, fullerenes, and catalyst particles. Once again a multistep process based primarily on the dissolution of catalyst particles in strong acids and dissolution of fullerenes in organic solvents (CS_2, for example), followed by microfiltration, have resulted in high purity SWNT nanotubes [39, 40]. Very recently, a further development was reported in which the purified SWNTs could be broken down into smaller segments of a few hundred nanometers length through sonication in warm acid mixtures [41]. This development provides, for the first time, the accessibility to manipulate nanotube ends through functionalization and innovative chemistry for creating nanotube-based arrays and organized nanotube-based chemical complexes.

There have been some attempts to align nanotubes after they are produced. The properties of graphite and, hence, nanotubes are highly anisotropic; the in-plane properties are governed by the strong covalent bonding, which is very different from the c axis (normal to the plane) properties governed by weak van der Waals bonding. MWNTs have been aligned by cutting [42] (Fig. 9) and stretching [43] polymer–nanotube composites; the shear stresses produced during the mechanical loading tend to align the tubes. This may suggest that the bonding between nanotubes and the matrix is not very strong.

Orientation and purification of nanotubes have been demonstrated using ac electrophoresis in organic solvents (isopropyl alcohol) [44]. The degree of orientation and separation from other impurity particles depends on the frequency of the field applied. Although these novel techniques only apply to small amounts of the sample, they are definite steps toward development of new techniques to manipulate nanotubes. Oriented growth of arrays of nanotubes over patterned catalyst substrates via CVD is another way of making aligned, organized nanotubes [30, 31].

3.3. Growth Mechanisms

Years of study of catalyst-grown multiwalled carbon fibers indicate that growth occurs via precipitation of dissolved carbon from a moving catalytic particle surface [45]. Growth terminates when the catalyst particle is deactivated somehow or when a stable metal carbide is formed through reaction with metal and carbon. Carbon fibers form in tubular morphology because it is energetically favorable for the newly formed surface of the growing fiber to precipitate as low energy basal planes of graphite rather than as high energy prismatic/edge planes. However, curving of the graphite layers introduces an extra elastic term into the free energy equation of nucleation and growth. The energy consideration puts a lower limit ($\sim$10 nm) on the diameters of carbon fibers that can form from curved graphite layers [46]. This implies that to explain the growth of carbon nanotubes, new mechanisms have to be considered. There are some visible differences in the growth of catalyst-grown fibers and the nanotubes. The ends of carbon fibers always carry metal particles, which is clear evidence for the catalytic role of the particles. In MWNTs, no catalyst is involved during growth and in SWNTs, catalyst particles are never observed at the tube ends. Moreover, the ends of both SWNTs and MWNTs are always closed.

It seems likely that two entirely different mechanisms operate during the growth of MWNTs and SWNTs because the presence of a catalyst species is absolutely necessary for the growth of the latter. Open MWNTs can occasionally be spotted among arc-grown nanotube samples. The simplest scenario could be that by some mechanism, all the growing layers of a tube remain open during growth [47]. Closure of the layer is caused by the nucleation of pentagonal rings due to local perturbations in growth conditions or due to the energetics (stability) between structures that contain hexagons and pentagons [48, 49]. Successive outer layers grow on inner tube templates and the large dimensional anisotropy results from the vastly different rates of growth at the high energy open ends compared to the unreactive basal planes. Several arguments have been invoked to see why the ends of the tubes should remain open during growth, because the dangling bonds would create a large increase in the energy of the system. Present understanding suggests that during growth, the dangling bond energy is minimized through interaction between adjacent layers (lip–lip interaction) [50]. The bonds formed between the layers are dynamic, and incorporation of carbon species during growth easily occurs through continuous breaking and reforming of this lip around the periphery of an open-ended tube.

Since the presence of a catalyst is necessary for SWNT growth, any mechanism that accounts for growth should incorporate the role of the catalyst in the growth. One of the suggestions, after observing the substantial boost in the growth of uniform diameter armchair tubes in the laser ablation, was that atomic species of the catalyst decorate the dangling bonds of the open end of a tube and achieve growth by a scooter mechanism. The catalyst atom scoots around the rim of the tube and absorbs incoming carbon atoms [22]. However, theoretical calculations indicate that tubes as small as the SWNTs will have a strong tendency to close by forming pentagons and ejecting any metal atoms that saturate the dangling bonds [51]. There is no consensus as yet on how the SWNTs grow, and more so why the yield of nanotubes increase drastically when a second element (Y) is added; yttrium do not show any catalytic activity when used alone.

There is yet another idea, where the role of the metal catalyst is seen as retarding the growth of outer layers of the nanotube tube through etching [52]. All the more interesting is the fact that changes in the temperature of the zone in which SWNTs form affect the diameter of the tubes.

4. NANOTUBE PROPERTIES

4.1. Electronic Properties

The most exciting of nanotube properties relates to its electronic band structure. Early calculations revealed that nanotubes could be metallic or semiconducting, depending on their helicity and diameter [53, 54] (Fig. 10). The armchair tubes are always metallic, whereas the zigzag and chiral tubes can be either metallic or semiconducting. These predictions have been major driving forces behind the rapid evolution of this field. It is only recently that some of these predictions have been experimentally verified. From the physicists point

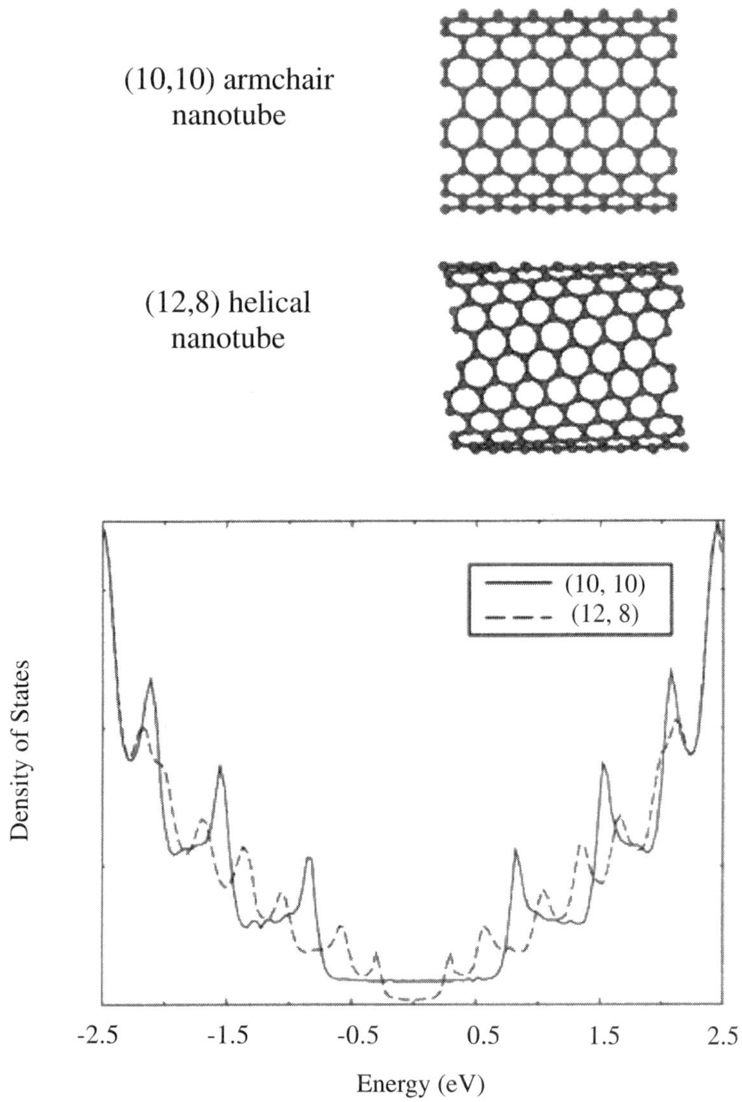

Fig. 10. Calculated electronic density of states of two SWNTs: one with a helical lattice (12, 8) and the other with a nonhelical lattice (10, 10). Note the zero band gap for the armchair tube, whereas a real gap is seen for the helical tube. (Courtesy of Dr. J.-C. Charlier.)

of view, a robust molecular system has become available to test many fundamental quantum transport properties in mesoscopic systems.

The electronic conduction process in nanotubes is quantum confined, because, in the radial direction, the electrons are confined in the singular plane of the graphene sheet. The conduction occurs in armchair (metallic) tubes through gapless modes because the valence and conduction bands always cross each other at Fermi energy for a certain special wave vector (corresponding to the K point in the Brillouin zone) [7]. In most of the chiral tubes, where the unit cell contains a large number of atoms, the one-dimensional band structure shows an opening of the gap at Fermi energy and, hence, has semiconducting properties. In certain cases, however, zigzag and chiral tubes also become conducting when one of the subbands still crosses the K point. When the diameter of the tubes increases, the band gap (which varies inversely with the tube diameter) tends to zero, yielding a zero-gap semiconductor that is essentially equivalent to the planar graphene sheet. Hence, in

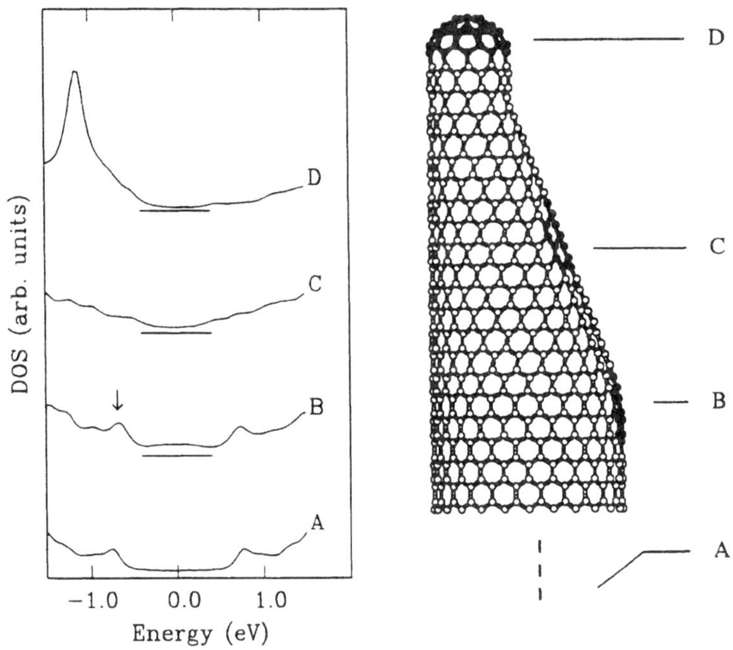

Fig. 11. Calculated DOS, showing the effect of defects in the nanotube structure. The DOS is shown from the tip of a nanotube where topological defects are present. A, B, C, and D denote the positions on the nanotubes where the DOS were calculated. The small arrow corresponds to a weak acceptor state. Note that the large peak that appears at the tip (D) changes in energy depending on the constitution of the tip. (Courtesy of Dr. J.-C. Charlier.) A detailed report can be found in [67]. (Source: Reprinted with permission from [67]. © 1997 American Physical Society.)

a MWNT, the electronic structure of the smallest inner tubes is superimposed by several outer, larger planar graphenelike tubes. Thus it can be assumed that any measurement on the electronic properties of MWNTs is going to result in semimetallic behavior like the parent graphite structure.

Calculations have shown that the topological defects present at the tips can induce sharp resonant peaks in the local density of states near the Fermi energy, thereby closing the gap in semiconducting tubules and making them metallic (Fig 11a). Hence, the ends of all the tubes should show metallic features. Similar metallization of the tubes also can occur through doping the lattice of nanotubes with impurities such as boron and nitrogen (Fig. 11b).

Theoretical predictions need to be validated by experimental testing. Measurements on these nanosized objects are challenging and not always unambiguous because connecting nanotubes to measuring devices using micrometer sized probes can lead to misleading results. However, there are spectroscopic techniques that can probe local electronic structure, and several such techniques have been used to understand carbon nanotubes. Transport measurements also have been carried out using very sophisticated nanolithographic techniques, and all the measurements confirm the quantized conduction behavior predicted by theory. We shall now look at the progress made in these areas that points to the possible use of nanotubes as electronic devices.

4.1.1. Spectroscopic Measurements of Nanotube Electronic Structure

Experimentally, one way to probe the local electronic structure of carbon nanotubes is by electron energy loss spectroscopy (EELS) because it uses a fine electron probe that is compatible with the dimensions of the nanotubes [55–58]. The low energy loss (plasmon loss) spectrum of nanotubes shows peaks that are typical of graphite (7-eV peak corresponding to π electron oscillation and 25–27-eV peaks due to the π-σ plasmon; these peaks shifts to lower energies as the size of the tube decreases), but in addition shows surface plasmons at energies in the range 10–16 eV [59]. In corroboration with experiments, calculations have shown that plasmons in nanotubes exhibit a crossover from one- to three-dimensional character as the number of shells and the tube size increase and become graphitelike in larger MWNTs [60]. The core carbon K-edge energy loss near edge structure (ELNES), which can be related to the atomic environment and local density of electronic states (LDOS), in nanotube spectra show similarities and differences with graphite. A sharp peak at 285.5 eV due to the transitions from the 1s core level to the π^* band (antibonding states) and an edge at 291 eV due to the p-orbital projected part of a broad σ^* band are observed. The difference in the LDOS of SWNTs and MWNTs has been compared using EELS data obtained from nanotubes and LDOS calculated by ab initio methods on corrugated graphite layers [57]. The spectrum from MWNTs looks identical to that from graphite. In the SWNT spectrum, the σ^* transitions look markedly different compared to features in the graphite spectrum. Calculations done to match the EELS data reveal that the removal of interplanar interactions is the primary reason for deviation from graphitelike features in EELS. The curvature of the orbitals in a ~1-nm diameter tube causes broadening of the peaks in the σ^* band and shifts them to lower energies. The π^* peak width is the same as found in graphite. Tight binding calculations [57] indicate that the calculated π^* peak width in nanotubes is different from graphite by 0.3 eV, which is below the experimental resolution and hence is not observed. The results suggest that, the bonding in nanotubes can be taken as covalent and planar to a first approximation. In small diameter nanotubes ($<$0.7 nm), however, a strong σ^*-π^* hybridization has been observed in ab initio local density approximation (LDA) calculations [61]. The effect introduces modified low-lying conduction states into the band gap of nanotubes, increasing the metallicity of the tubes. However, nanotubes that are made experimentally are larger than the size needed to see this effect.

Raman spectroscopy is another powerful tool to characterize carbon materials; it has been used successfully to characterize the diameter and helicity of SWNTs [34]. The high crystallinity and graphitelike features were observed in MWNT samples in early experiments using Raman spectroscopy [62]. It was recently shown that the Raman features from purified SWNTs can be correlated to quantum electronic effects [25]. Several Raman active modes over a wide range of wave numbers from 116 to 1609 cm^{-1} were detected in these experiments and were compared to theoretically calculated values for nanotubes of different size and structure (Fig. 12). The Raman peaks were characterized as strongly diameter dependent (breathing vibrational modes) and diameter independent (in-plane vibrational modes; for example, the strong E_{2g} mode as in graphite). Interestingly, it was also observed that resonant Raman scattering from tubes of different diameters dominates the Raman spectra when the laser energy matches the separation between the one-dimensional bands in the electronic density of states of any particular nanotube; the well separated electronic bands in small diameter SWNT provide initial and final states necessary for the resonance seen in Raman scattering. This diameter selective resonant scattering is evidence for the one-dimensional quantum confinement of electrons in nanotubes.

Electron spin resonance (ESR) studies of carbon nanotubes have been used to detect conduction electrons, to tell whether metallic or very narrow band-gap semiconducting tubes are present in the samples [63, 64]. Some MWNT samples reportedly show no conduction electrons and are thought to be mostly semiconducting, whereas other samples show indications of the presence of metallic nanotubes. Signals that have Curie-like behavior disappear with high temperature annealing and purification of MWNT, and it may be concluded that this is due to defects in raw nanotubes. The line shape of the conduction electrons in nanotubes is quite distinct from graphite because the conduction electrons cannot diffuse out of the skin depth (approximately micrometer); the average MWNT diameter

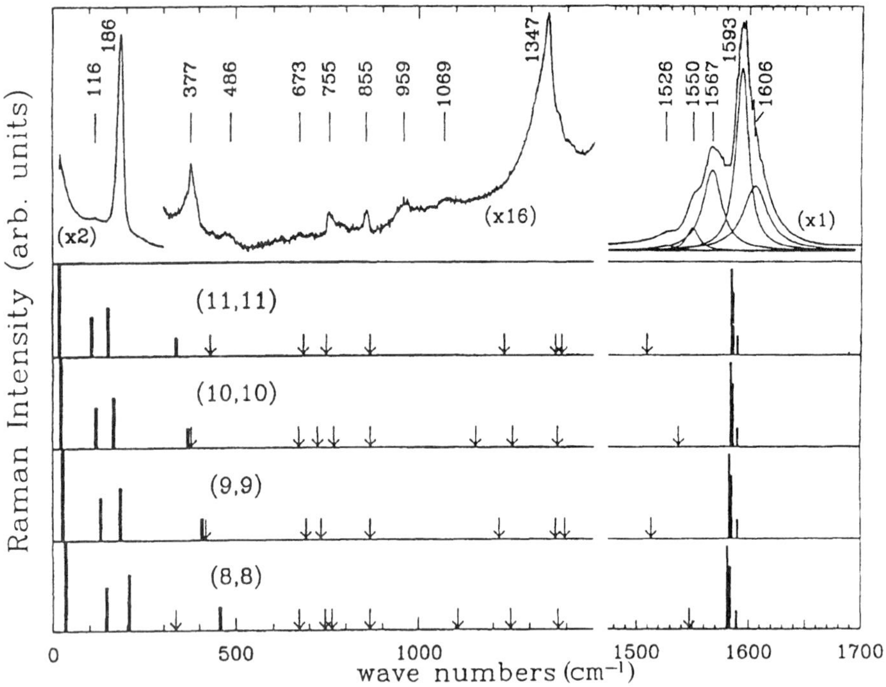

Fig. 12. The top box shows experimental Raman spectra from nearly pure SWNT samples. Several peaks, some of which are structure sensitive and some structure insensitive (see text) are evident. The bottom boxes show calculated Raman spectra for different (n, n) nanotubes. Arrows correspond to positions where weak Raman lines are expected. (Source: Reprinted with permission from [25]. © 1997 American Association for the Advancement of Science.)

is only few tens of nanometers at maximum. Perhaps only in the longest nanotubes that happen to be aligned with the probing field can the conduction electrons diffuse out of the skin depth. For SWNTs, the information is scant, but a room temperature signal is reported to be Dysonian due to the great length of these tubes and their assembly into larger ropes [22].

One of the most important contribution to the understanding of the electronic structure of nanotubes has come from the use of scanning tunneling spectroscopy [65–69]. Scanning tunneling microscopy (STM) and spectroscopy (STS) can be used locate and measure the band gap of individual nanotubes. Earlier studies using STS suggested a range of values for band gaps (200 meV to 1.2 eV) for MWNTs [4]. In general, the band gaps of nanotubes vary inversely with the diameter. Spatially resolved STS has been used to characterize the change in the electronic structure of individual nanotubes as a function of position [67]. Such studies have shown the change in the electronic structure near the tube ends, which behave like metal due to the extra states near the Fermi energy, brought about by the position, distribution, and interaction of odd member defects at the tube ends (Fig. 13a). The assumption has been that in the case of MWNTs, at low voltages, tunneling takes place only between the STM tip and the outermost layer in the nanotube. The localized peaks found in the LDOS of the tube tips could influence nanotube properties, for example, field emission from the tips.

The most striking results are the recent STM/STS studies on SWNTs. For the first time, simultaneous determination of structure (atomically resolved nanotube surface revealing the helicity) and LDOS could be achieved using low temperature STM/STS (Fig. 13b) [68, 69]. Several of the band structure calculations that predated this experimental feat and that linked the electronic structure to helicity and size were finally proved true

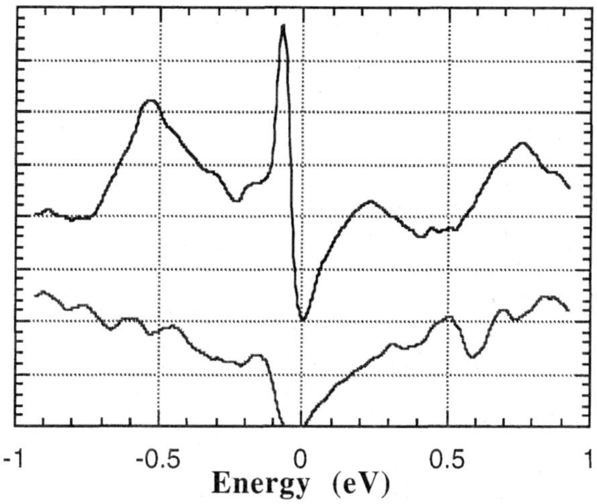

(a)

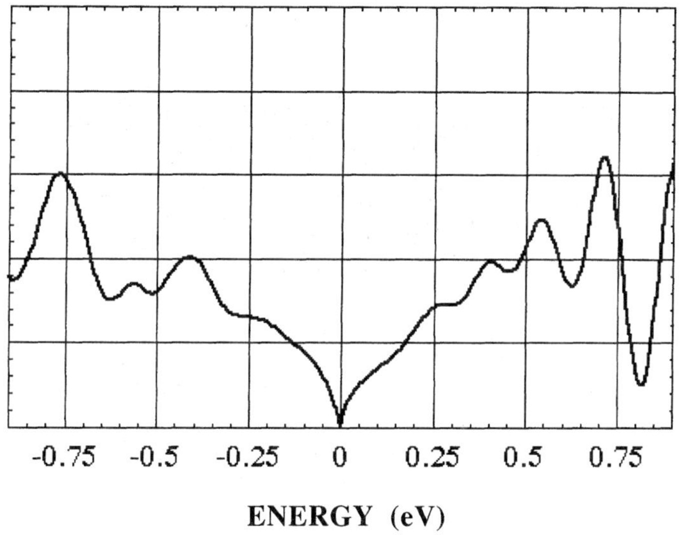

(b)

Fig. 13. (a) Scanning tunneling spectroscopy data on MWNTs, showing variation in the LDOS (plotted on the *y* axis in arbitrary units) on the cylindrical surface (upper curve) and the tip of a nanotube (lower curve). At the tip sharp resonant peaks are seen that close the gap at the Fermi level of semiconducting tubes, making the tips more metallic. (b) STS data of a SWNT in a bundle of tubes. The large peaks in the LDOS are presumably Van Hove-like singularities. (Courtesy of Dr. D. L. Carroll.)

unambiguously. Apart from this important verification, several other interesting features were evident from these studies. The one-dimensional nature of conduction electrons in nanotubes was also confirmed from the observation of van Hove singularities at the onset of the one-dimensional energy bands. Spatially resolved STS studies of nanotube bundles have also confirmed that the band structure changes spatially, and occasionally this can result in a rectifying effects (conducting to nonconducting states) along the tube bundles. It has been observed that contacts between tubes in a bundle could result in breaking of the local symmetry in the conduction paths of electrons [70]. In MWNTs, additional electronic states due to interlayer correlation have been noticed by STS [71].

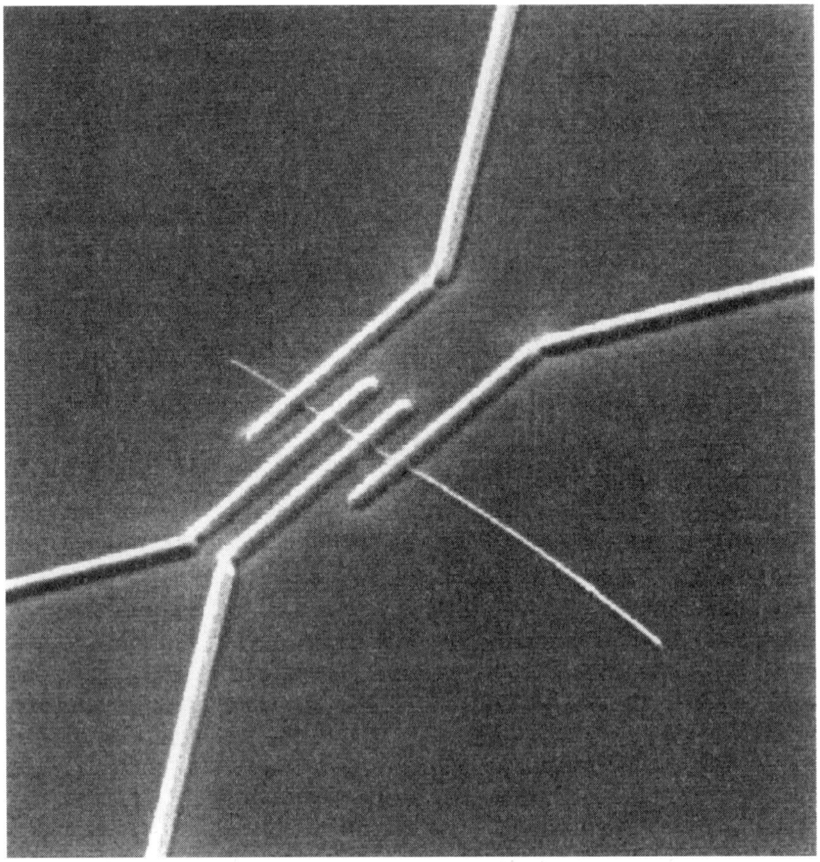

Fig. 14. A four probe measurement geometry for individual nanotubes. Here four 80-nm tungsten leads are fabricated over a MWNT placed on an oxidized Si substrate using automated focused ion beam-induced deposition and patterning. (Source: Reprinted with permission from [73]. © 1996 Macmillan Magazines Limited.)

4.1.2. Transport Measurements in Nanotubes

Given the large aspect ratio and the small dimensions of nanotubes, it is clearly a big challenge to attach at least four electrodes (only four or more probe measurements will be meaningful because contact resistance can be significant) to the nanotubes and measure the current–voltage characteristics. This has been achieved using sophisticated nanolithography techniques and microelectronics technology (Fig. 14). The first measurements performed on single MWNT [72] revealed that the transport properties were consistent with quantum transport in a weakly disordered low-dimensional conductor. The inverse logarithmic dependence of resistance with temperature was explained by two-dimensional weak localization, which appears when the probability for inelastic scattering by lattice defects is much larger than the inelastic carrier–carrier or carrier–phonon scattering. Later, systematic four probe measurements of several individual MWNTs showed a range of conducting behavior (metallic, semiconducting, and semimetallic) in nanotubes [73]. A weak magnetoresistance in these tubes also indicated a very short mean-free path for the conduction electrons, most likely due to scattering by defects. There is always ambiguity in MWNTs that corresponds to whether the measurements made are pertinent only to the outer wall (to which the contact is made) or to the whole tubular assembly. The possibility of having several topological surface defects (e.g., 5-7 defect pairs) and cylinders of different helicity complicates any simple interpretation in MWNT, based on theoretical predictions.

SWNTs comprise a well defined system in terms of electronic properties. The samples available through the arc process or laser ablation, under optimum conditions, consist of uniform diameter nanotubes, a majority of which have a narrow range of helicity (around armchair configuration). Electron transport measurements are hence unambiguous, and great progress has been made (mainly by the experimental mesoscopic group at Delft University in Netherlands) in such measurements of individual SWNTs randomly dropped on prefabricated electrode assemblies. Measurements carried out on individual SWNT in the millikelvin temperature range show important features [21]. Observed Coulomb charging of the nanotube can be suppressed by the application of an appropriate gate voltage. Conduction occurs through well separated discrete electronic states that are coherent over the distance between the probes (a few hundred nanometers). In other words, quantum electron transport occurs in nanotubes over large lengths and nanotubes can be treated as genuine quantum wires. Calculations show that, unlike normal metal wires, conduction electrons in (armchair) nanotubes experience an effective disorder averaged over the tube's circumference, leading to electron mean free paths that increase with nanotube diameter [74]. This increase should result in exceptional ballistic transport properties and localization lengths of several micrometers.

SWNTs form a network of bundles held together by van der Waals forces. These tube bundles also have been studied [75, 76] for resistance and temperature dependencies. At higher temperatures, true metallic behavior is observed ($d\rho/dT > 0$). However, there is a minima above which resistance increases again at low temperature ($d\rho/dT < 0$). The exact cause of this behavior is not clear.

4.2. Mechanical Properties

Apart from unique electronic properties, the mechanical behavior of nanotubes also has provided excitement because nanotubes are seen as the ultimate carbon fiber, which can be used as reinforcements in advanced composite technology [28]. Early theoretical work [77–79] and recent experiments [80–82] on individual nanotubes (mostly MWNTs) have confirmed that nanotubes are one of the stiffest material ever made. Whereas carbon–carbon covalent bonds are one of the strongest in nature, a structure based on a perfect arrangement of these bonds oriented along the axis of nanotubes would produce an exceedingly strong material. Traditional carbon fibers show high strength and stiffness, but fall far short of the theoretical in-plane strength of graphite layers (an order of magnitude lower). Nanotubes come close to being the best fiber that can be made from graphite structure.

Theoretical studies have suggested that SWNTs have Young's modulus as high as 1–5 TPa [77], whereas other researchers have predicted a softening with increasing tube radius [83]. This threoretical estimate can be compared with the in-plane graphite value of 1 TPa (and a tensile stiffness of 0.8 TPa). For MWNTs, the actual strength in practical situations would be affected by the sliding of individual graphene cylinders with respect to each other [78]. There are no easy mechanical testing experiments that can be done on individual nanotubes. However, clever experiments that indirectly provide a glimpse of the mechanical parameters, as well as scanning probe techniques that can manipulate individual nanotubes, have provided some basic answers to the mechanical behavior of nanotubes. Nanotubes projecting out onto holes in a transmission electron microscope (TEM) specimen grid were assumed to be equivalent to clamped homogeneous cantilevers. The horizontal vibrational amplitudes at the tube ends were measured from the blurring of the images of nanotube tips (using TEM) and related to the Young's modulus and vibration energy [80]. The analysis performed on several MWNTs gave average Young's modulus values of 1.8 TPa. Recent experiments using atomic force microscopy in bending nanotubes attached to substrates also indicate [81, 82] modulus and strength values close to what was obtained in the vibration experiments. The high strength and rigidity (stiffness) of nanotubes also were seen in deformation experiments on nanotube polymer composites [84].

The fracture and deformation modes of nanotubes are intriguing [79, 85]. Fracture can occur in a nanotube via collapse of the hollow [86, 87], providing extra absorption of energy and increased toughness in a composite. The high Young's modulus suggests that nanotubes have high bending moments. This is seen from images of nanotubes that always appear straight. Deformation experiments, however, suggest that bending in nanotubes depends on various parameters, such as the size of the hollow interior, nanotube wall thickness, and tube size. Simulations have suggested the most interesting deformation behavior in nanotubes [79] (Fig. 15). Highly deformed nanotubes were seen to reversibly switch into different morphological patterns with abrupt release of energy. The patterns are interesting in all forms of deformation; for example, in torsion, the nanotubes first flatten and then get twisted into a loop. Nanotubes can sustain extreme strains (40%) in tension without showing signs of brittle behavior, plastic deformation, or bond rupture. The reversibility of deformations such as buckling has been recorded under TEM [3, 85] and supports the idea that the tubes can recover from severe structural distortions [82]. This flexibility is related to the in-plane flexibility of planar graphene sheet and the ability for the carbon atoms to rehybridize, where the degree of sp^2–sp^3 rehybridization depends on the strain [88]. Such flexibility of nanotubes under mechanical loading is important for their potential application as nanoprobes in scanning probe microscopes [15].

Very curious plastic behavior has been predicted [89] in nanotubes. It has been suggested that a 5-7 ring pair defect, called a Stone–Wales defect in sp^2 carbon systems [90], on the nanotube lattice (similar to a dislocation core) can initiate deformation and become mobile under the influence of stress. This leads to a stepwise size reduction (localized necking) of the nanotube. This process introduces a change in helicity in the deformed nanotube structure. This extraordinary behavior may lead to another nanotube application: a new type of probe that responds to mechanical stress by changing its electronic character. Such a property derives from the phenomenal constitution of the closed structure and the interesting effects of topological defects on the structure of nanotubes. Fracture of individual nanotubes under tensile loading, evidenced in molecular dynamics simulations, is also unique [10]. Elastic stretching elongates the hexagons on the tube surface until, at high strain, some bonds are broken. This local defect is redistributed easily over the entire surface due to the mobility of defect. A new form of necking slowly sets in until the tube is locally reduced to a linear chain of carbenes (carbon atoms linked by double bonds into a chain). Such shrinking and reconstruction of the tube lattice has been evidenced in electron irradiation studies supported by molecular dynamics simulations, where the tube shrinks continuously during atom loss (through irradiation) and breaks at localized neck regions [91].

Not many successful experiments have been reported on fabricating and testing polymer composites with nanotubes. Very little is known about the interfacial properties of nanotubes and the matrix, especially how nanotubes that are atomically smooth and have dimensions nearly the same as the lineal dimensions of individual polymer chains interact with the matrix. Evidence for possible single nanotube fragmentation was reported [92] when a nanotube composite was tested in tension, indicating a strong interface and an effective load transfer from the matrix to the tubes. Our preliminary results indicate that the response (modulus) of MWNTs as fillers in epoxy-based composites is different in tension and compression [93]. This difference possibly arises from the fact that in tension only the outer layer of the tube is affected during load transfer, whereas in compression all the layers are involved. The possibility that nanotube walls slide with respect to each other in MWNTs [94] or that individual SWNTs slip in tube bundles [78] cannot be neglected and, hence, the actual strength attainable from aggregates of nanotubes could be much lower than what is expected from the ideal structure of isolated tubes.

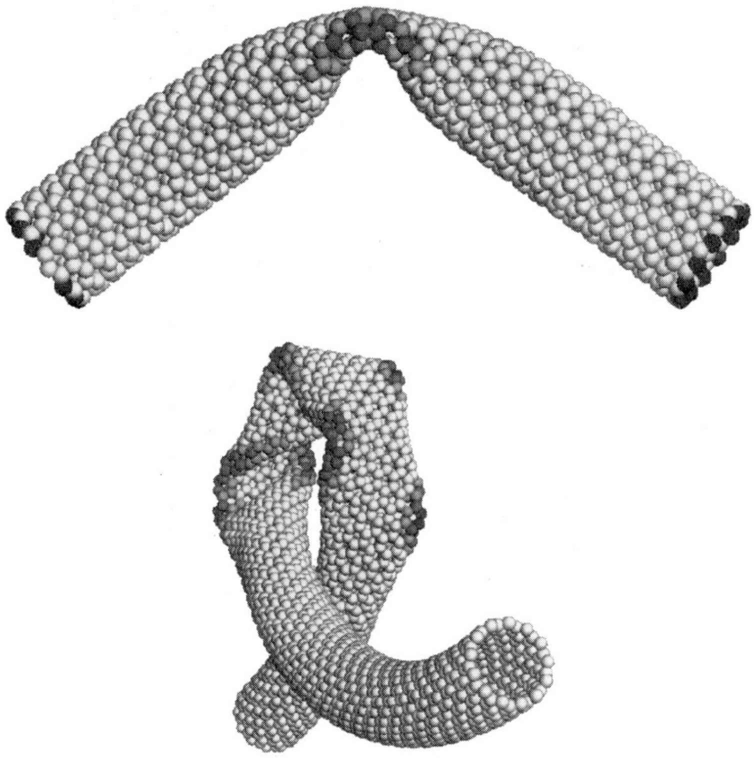

(a)

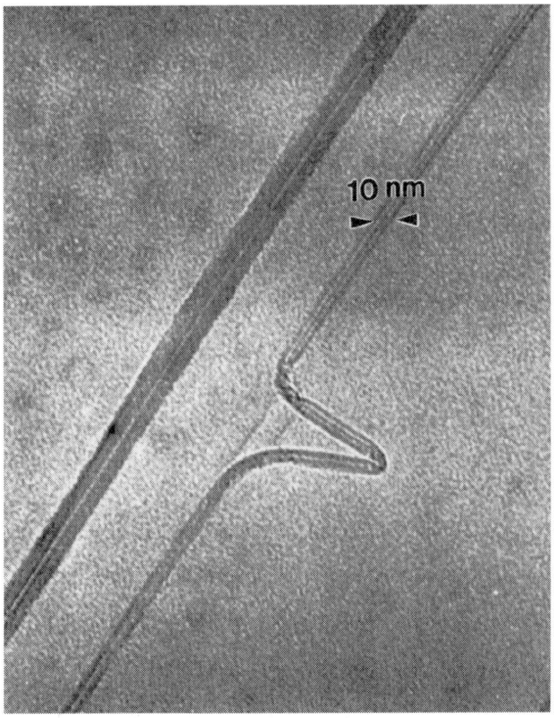

(b)

Fig. 15. (a) Simulated buckling behavior in nanotubes, showing two possible conformations that the nanotubes assume when loaded. See [79] for details of the deformation behavior. (Courtesy of Dr. B. I. Yakabson.) (b) TEM image of a MWNT dispersed in a polymer film showing buckling of the nanotube due to compressive stress produced during cutting of the film. Notice that a thicker tube lying adjacent to the buckled tube has not been deformed.

4.3. Other Properties

The electronic and mechanical properties of nanotubes have received the most attention as yet, but there are other characteristics that make nanotubes a material of interest. The hollow structure of nanotubes makes them very light (density varies from $\sim$0.8 g/cm^3 for SWNTs up to 1.8 g/cm^3 for MWNTs, compared to 2.26 g/cm^3 for graphite) and this is very useful for a variety of lightweight applications from composites to fuel cells. Specific strength (strength/density) is important in the design of structural materials; nanotubes have this value at least 2 orders of magnitude greater than steel. Traditional carbon fibers have specific strength 40 times that of steel. Whereas nanotubes are made of graphitic carbon, they have good resistance to chemical attack and have high thermal stability. Oxidation studies have shown that the onset of oxidation shifts by about 100 °C to higher temperatures in nanotubes compared to high modulus graphite fibers. The oxidation in nanotubes begins at the tube tips and this leads to the possibility of opening nanotubes by oxidation [35, 36, 95]. In vacuum or reducing atmospheres, nanotube structures will be stable to any practical service temperature.

As described previously, electron transport in nanotubes is unique and the tubes are highly conducting in the axial direction. Similarly, the thermal conductivity of nanotubes also should be high in the axial direction and should be close to the in-plane value of graphite (one of the highest among materials) [4]. No experiment to date has tested the thermal conductivity of nanotube material or nanotube composites. In the case of composites, although the high aspect ratio of nanotubes will aid in improving conductivity, the interface between the nanotubes and the matrix could have a deleterious effect. This is emphasized by the large surface area available for interface formation. The surface area of nanotube material has been estimated using Brunauer–Emmett–Teller (BET) techniques and corresponds to 10–20 m^2/g (could be increased by opening the tubes), which is greater than that of graphite, but small compared to activated mesoporous carbon (in the range of a few thousand meters squared per gram) used as catalytic supports.

The catalytic nature of nanotubes surfaces has been studied and the indications are that nanotubes are catalytically active. It has been demonstrated that MWNTs decorated with metals can show high selectivity in heterogeneous catalysis (e.g., hydrogenation reaction of cinnamaldehyde in liquid phase using Ru on nanotubes) compared to the same metals attached on other carbon substrates [16]. Microelectrodes made from nanotubes have been used to carry out bioelectrochemical reactions [17]. Electrodes made from nanotubes show superior properties (reversibility and efficient electron transfer) during in vitro oxidation studies of biomolecules (dopamine) compared to electrodes made from graphite fiber and graphite paste. MWNT microelectrodes were used to study the oxygen reduction reaction in aqueous acid and neutral media [96]. It was observed that electron transfer occurred at a much faster rate (determined from the exchange current density measured at the electrodes) on nanotubes compared to graphite. Ab initio calculations on the electron transfer rates of nanotubes indicate that curvature do not enhance the rates significantly, although the presence of topological defects on the surface can cause a significant improvement in the electron transfer rates, especially at the pentagon sites, which are electrophilic in nature compared to the hexagonal rings. The electrodic efficiency can be increased by electrode activation (treatment of nanotubes in concentrated HNO_3 for a short period) and through the deposition of small amounts of metal catalysts (Pt, Pd, Ag) [18].

4.4. Nanotube Templates

Whereas nanotubes have relatively straight and narrow channels in the center, it was speculated from the beginning that it might be possible to stack atoms of foreign materials in these cavities and fabricate one-dimensional nanowires. Early calculations suggested that strong capillary forces exist in nanotubes, strong enough to hold gases and fluids inside these cavities [97]. This was experimentally demonstrated by filling and solidifying molten

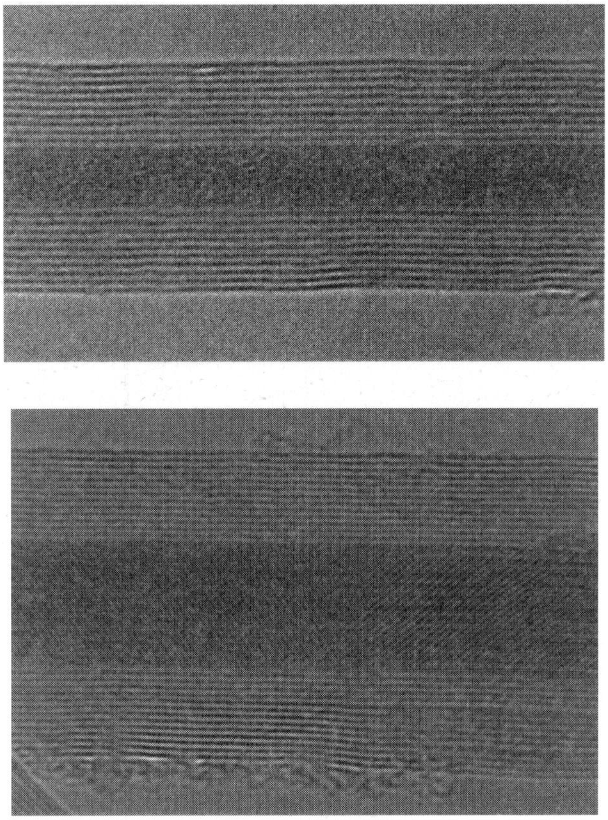

Fig. 16. HRTEM images of MWNTs filled with Pb-based compound. The structure of the filled phase is disordered in smaller hollows of the nanotubes (top) and crystalline in larger ones (bottom). The distance between parallel horizontal fringes in the nanotube walls is 0.34 nm.

lead compound (Pb–C–O) inside the channels of MWNTs [98]. Wires as small as 1.2 nm in diameter were formed inside nanotubes during this experiment (Fig. 16). Various low melting oxides, for example, Bi_2O_3 and V_2O_5, also could be filled later by the same technique [95, 99]. The possibility of making filled nanotubes with different materials led to the exciting use of nanotubes as molds to fabricate nanowires and one-dimensional composites with interesting electrical and mechanical properties.

The critical issue in filling nanotubes is the wetting characteristics of nanotubes [100], which seem quite different from those of planar graphite. Wetting of low melting alloys and solvents occurs quite readily in the internal pores of MWNTs. The main determining parameter for this to happen is the surface tension of the wetted substance; a cutoff value of ~200 mN/m, above which wetting no longer occurs, was found. Liquids such as organic solvents wet nanotubes easily, and it has been proposed that interesting chemical reactions could be performed inside nanotubes [52]. A simple chemical method based on opening and filling nanotubes in solution was developed after the initial experiments on molten material filling were reported [101]. Here an acid is used to open the nanotube tip (through oxidation) and act as a low surface tension carrier for solutes (metal containing salts) to fill the inner hollow. Calcination of solvent-treated nanotubes leaves deposits of oxide material (e.g., NiO) inside the nanotube cavities. The oxides can then be reduced to metals by annealing in a reducing atmosphere. Observation of solidification inside the one-dimensional channels of nanotubes provides a fascinating study of phase stabilization under geometrical constraints [102]. It was found experimentally that when the channel size gets smaller than some critical diameter, solidification results in the disordered phase

Fig. 17. SEM image of a nanocomposite (top) made by annealing MWNT and vanadium pentoxide. Notice the sea-urchin-like microstructure that results from the flow of molten oxide along the fibrous nanotubes. A HRTEM of an individual nanotube covered with thin layers of the oxide (arrows) is shown at the bottom. The distance between nanotube layers is 0.34 nm.

(e.g., V_2O_5) [99]. Crystalline bulk phases are formed in larger cavities. Numerous modeling studies are under way to understand the solidification behavior of materials inside nanotubes and the physical properties of these unique filled nanocomposites [103, 104].

Filled nanotubes also can be synthesized by using composite anodes. During the arc formation of carbon species, encapsulated nanotubular structures are created in abundance. In general, carbide phases are formed by this technique (e.g., transition metal carbides), but in a few cases pure metals also are deposited (weak carbide-forming metals, such as Co, Mn, and Cu, and semiconducting elements such as Sb and Ge [105–111]).

The great potential for nanotubes to be used as templates and in the fabrication of nanocomposite structures (Fig. 17) has been demonstrated. These unique composites are expected to have interesting mechanical and electrical properties due to a combination of dimensional effects and interface properties [103]. Similar to filled nanotubes, finely coated nanotubes with monolayers of certain layered oxides have been made and characterized [99]; for example, vanadium pentoxide. Films of ~1 nm and often one monolayer thick have been successfully deposited on MWNTs. The interface formed between nanotubes and the layered oxide is atomically flat due to the absence of covalent bonds across the interface. Growth phenomena of epitaxial films on nanotubes are fundamentally interesting from two aspects: (1) the epitaxial growth does not depend on lattice mismatch and is probably governed by the electronic structure of the two lattices, and (2) growth proceeds on a highly curved substrate. More interestingly, it has been demonstrated that after coating, the nanotubes can be removed by oxidation, leaving behind freely standing vanadium oxide tubes with fine wall thickness. These novel ceramic tubules made using nanotubes as templates could have applications in catalysis.

There are other ways in which pristine nanotubes can be modified into composite structures. Chemical functionalization has been used to build macromolecular structures from fullerenes. Attachment of organic functional groups on the surface of nanotubes have been achieved [112], and with the recent success in breaking up SWNTs into shorter fragments [41], the possibility of functionalizing and building structures through chemistry has become real. Decoration of nanotubes with metal particles has been achieved for different purposes, most importantly the use of metal-coated nanotubes for heterogeneous catalysis [16]. SWNT bundles have been doped with K, Br, and I_2, resulting in an order of magnitude increase in electrical conductivity [113]. Evidence of charge transfer between the host nanotubes and the intercalates has been seen in these doped nanotube bundles in Raman spectroscopy studies [114]. In some cases, dopants have been observed to form a linear chain and sit in the one-dimensional interstitial channels of the bundles. MWNT systems have also been intercalated with alkali metals [115] and $FeCl_3$ [116]. Similarly, in nanotubes made inside nanoporous ceramic templates, Li intercalation has been carried out successfully and this result could have an impact in battery applications [18]. All the intercalation and doping studies suggest that nanotube systems provide the host lattice for the creation of a whole new set of carbon-based synthetic metallic structures.

Mineralization of nanotubes is another route to create composites using nanotubes as the backbone. When volatile gases such as halogenated compounds or SiO_x are reacted with nanotubes, the tubes are converted into carbide nanorods of similar dimensions [117]. The reactions can be controlled such that the outer nanotube layers can be converted to carbides while the inner graphite layer structure remains intact. The carbide rods so produced (e.g., SiC, NbC) have wide ranging electrical and magnetic properties and could be ideal mechanical reinforcements in composites.

5. APPLICATIONS OF NANOTUBES

Demonstrations in the literature point to possible uses of nanotubes. Most applications are based on the unique electronic structure, mechanical strength, flexibility, and dimensions of nanotubes. Whereas the electronic applications are based on SWNTs, no distinction has been made between SWNTs and MWNTs in other fields of applications.

The application of nanotubes as quantum wires and tiny electronic devices has received the most coveted attention. The Delft group, which pioneered the measurements of electron transport in individual SWNTs [21], has built the first single molecule field-effect transistor based on a semiconducting SWNT [19]. The device, which operates at room temperature, comprises a nanotube bridging two metal electrodes. The band structure suggested for this device is similar to traditional semiconductor devices (two Schottky-type diodes connected back to back), and the performance of this device is comparable to existing devices in terms of switching speeds. Although the demonstration of such a device is exciting, the next stage of integrating devices into circuits will be crucial. None of the procedures developed so far for nanotube fabrication enables the construction of complex architectures that the semiconductor industry needs today. New ideas based on the self-assembly of carbon structures into integrated nanotube assemblies have to be realized before nanotube electronics become practical reality.

The possibility of connecting nanotubes of different helicity (and hence different electronic character) through the incorporation of 5-7 defect pairs could lead to the fabrication of heterojunction devices [7]. Although this concept has never been shown experimentally on an individual nanotube structure, STS studies of nanotube ropes and MWNTs have indicated spatially varying changes in electronic properties along the length of the rope. Similar junction devices can be designed from two nanotube segments, one of which is semiconducting and the other made metallic by doping with impurity elements such as boron [118]. A whole range of nanoscale physics based on nanotube structures is beginning to unfold.

The foundations for this advance have been laid already through theoretical models and predictions.

Another application that has caught the attention of the scientific and engineering communities is the use of nanotubes as electron emitters [13, 14]. Field emission has been observed from arrays of partially aligned MWNTs that have been aligned by pulling a slurry of nanotube dispersion through a ceramic filter. The film of aligned tubes is then transferred onto a substrate and a voltage is applied across the supporting film and a collector. Such nanotube films act like field emission sources with turn-on voltages of a few tens of volts and electron emission at current densities of a few hundred milliamperes per centimeter squared. The nanotube electron source remains stable over several hours of field emission and is air stable. One of the practical issues that still needs to be resolved is obtaining uniformly aligned tubules, which would guarantee uniform emission.

Due to high aspect ratio, mechanical strength, and elasticity, nanotubes could be used as nanoprobes; for example, as tips of scanning probe microscopes. This idea has been demonstrated successfully and a nanotube tip on an atomic force microscope was used to image the topography of TiN-coated aluminum film [15]. A bundle of MWNTs is first attached to a Si cantilever through adhesive bonding and then the bundle is sheared to expose one tube at the end of the bundle that performs as the tip. Due to the flexibility of nanotubes, the nanotube tips do not suffer the common problem of tip crashes. Also, such a slender structure as the nanotube is ideally suited to image deep features like surface cracks. Due to the conducting nature of the tubes, they also can be used as STM tips. Images of charge density waves on TaS_2 has been obtained at high resolution using nanotube tips. Nanotube tips also have been used to image biological systems [119]. The proposal to use nanotube tips is enticing, but the vibration of individual free-standing tips can spoil some of the advantages (resolution) of the small tube dimensions, especially for high resolution imaging.

Filler-based applications of nanotubes for polymer composites is another area being hotly researched. One of the biggest applications of traditional carbon fibers is in reinforcing polymers in high strength, high toughness lightweight structural composites. Epoxy-based MWNT composites have been made and tested, but the results are not very conclusive. Substantial increase in modulus has been reported together with high strain to failure, but the strength of the composite is less than expected. The success of nanotube-reinforced composites depends on how strong the interface (between tubes and the matrix) can be made. The atomically smooth surfaces of nanotubes do not guarantee a strong interface. Molecular interlocking of the nanotubes and polymer chains could happen, but it is unclear how such interactions affect strength. Poor dispersion of the samples can create weak regions in the composite where cracks can originate. So far the failure mode observed in the composites is highly brittle, similar to pure epoxy [93]. Raman experiments show a shift in the first and second order peaks (main peak at 1582 cm^{-1}) in MWNTs due to the strain in the C–C bonds as a result of loading. The shifts show marked differences when the loading occurs in tension and compression, as described in Section 4.2. This could be due to the inherent structure of MWNTs where the inner cylinders are coupled only weakly to the outermost layer, which transfers the load.

High conductivity composites (electrical and thermal) using nanotube-filled polymers could be useful, but the problems is getting well distributed nanotubes in the matrix; heavy settling of nanotubes is seen when larger nanotube epoxy composites are made [120], probably due to the lack of interaction with the tubes and the matrix. One advantage, however, is the negligible breakdown of nanotubes during processing of the composites. This is a big problem in carbon fiber composites because the fibers are extremely brittle.

Other than structural composites, some unique properties are being pursued by physically doping (filling) polymers with nanotubes. Such a scheme was demonstrated in a conjugated luminescent polymer, poly(*m*-phenylenevinylene-co-2,5-dioctoxy-*p*-phenylenevinylene) (PPV), filled with MWNTs and SWNTs [121]. Compared to the pristine polymer, nanotube/PPV composites have shown a large increase in electrical conductivity of nearly 8 orders of magnitude, with little loss in photoluminescence/electroluminescence yield. In addition, the composite is far more robust than the pure polymer with regard to mechanical strength and photobleaching (breakdown of the polymer structure due to thermal buildup). Preliminary studies indicate that the host interacts very little with the embedded nanotubes, but that the nanotubes act like nanometric heat sinks, which prevent the buildup of large thermal effects that usually break down a conjugated polymer. Other new directions for the use of nanotubes in polymer matrices are being discovered for nonlinear optical properties, membrane technologies, and implant materials for biological applications.

The potential of nanotubes for electrode applications is being explored, especially because carbon-based electrodes have been used for decades in important electrode applications such as fuel cells and batteries. The unique surface constitution of nanotubes permits high selectivity for reactions. Early studies of bioelectrochemical reactions (in vitro oxidation of dopamine) using MWNT electrodes showed high reversibility and catalytic activity (shifting of oxidation potential) at the nanotube electrodes [17]. More recent studies indicate that nanotubes can catalyze oxygen reduction reactions, showing electron transfer rates on nanotubes much higher compared to on other carbon electrodes [18, 96]. The catalytic activity of metal-deposited nanotubes (Pt, Pd, Ag) is also superior to metals on traditional carbon (graphite and glassy carbon) supports. Oxygen reduction is an important fuel cell reaction, and the experiments show the potential of nanotube catalysts in energy production and storage.

Finally, another interesting and exciting area of application has been demonstrated in SWNTs: the possibility of hydrogen storage inside the well defined SWNT pores [20]. Temperature-programmed desorption spectroscopy has shown that hydrogen will condense inside SWNTs under conditions that do not induce adsorption in traditional porous carbon material. The hydrogen uptake is high and can be compared to the best presently available material (metal hydrides) for hydrogen storage. If an optimum nanotube diameter can be established for the best intake and release of hydrogen, high energy storage efficiency can be obtained and the process could operate at ambient temperature. Advances in controlling the nanotube size during production can have an impact in this field. Whereas fuel cells are increasingly becoming part of future technology, the role of nanotubes as energy storage material is significant.

6. NANOTUBES MADE FROM NONCARBON MATERIALS

Soon after the discovery of carbon nanotubes, it was shown that nanotubular structures also can be made from other layered structures. For a long time, geologists have known that certain forms of clays (complex layered silicates) are found in nature in tubular form (chrisotile, imogolite) [122]. These tubular structures show structural characteristics typical of carbon nanotubes, such as helicity and rotational disorder. Generally, tubules can be formed from most materials that have a layered structure in their bulk form, although dimensions, morphology, and structural character would depend on the complexity of the pristine structure.

The first noncarbon nanotubes that were made after the discovery of carbon nanotubes were based on the dichalcogenide family. Multiwalled tubular structures of WS_2 and MoS_2 (a few to a few tens of nanometers in diameter and up to a micrometer in length) were successfully synthesized by passing reactive sulfur containing gaseous species (H_2S) over films of W and Mo [123]. These structures consist of alternating layers of metal and sulfur. Their superior lubrication properties and potential industrial application have been demonstrated [124]. These small closed-shell structures lubricate by rolling on the substrate, a more efficient way compared to the conventional sliding lubrication in planar layered structures.

As mentioned in the previous section, carbon nanotubes can be used to make ceramic tubular structures by templating action. V_2O_5 is a good example. Other oxides that have a pseudolayered structure such as MoO_3 also have been made into nanotubular form [125].

Nanotubes made by substituting sp^2 carbon with isoelectronic structures (hexagonal boron nitride, *h*-BN) are another possibility. There exists a whole range of composition of B–C–N compounds that form layered structures, and it has been experimentally demonstrated that nanotubes of $B_xC_yN_z$ can be synthesized [126–128]. In the case of pure BN, both SWNTs and MWNTs have been made [129], where as in BCN, only MWNT varieties have been found [126]. In the latter, a strong atomic segregation to maximize the C–C and B–N bonds in the structure occurs. This results in the formation of separated BN and C cylinders in a MWNT. In the case of dilute boron doping, this tendency of atomic segregation results in the formation of isolated islands of BC_3 (also a layered structure) in a carbon nanotube lattice [118]. These structures were synthesized mainly by using modified versions of the carbon arc technique, with boron and nitrogen in the anode or the atmosphere used in the arc [129, 130]. Note that samples have been made only in small quantities, and high yield for the growth of any particular composite BCN structure has not been achieved so far. However, boron doping enhances graphitization in nanotubes (interlayer spacing decreases to the single crystal graphite value) and increases the average length of the tubes (tubes longer than 50 μm are readily formed). Theory predicts that boron acts like a surfactant, staying the tip of the growing tube and preventing tube closure through pentagon formation. It is expected that the mechanical and electronic properties of these structures are expected to be very different from pure carbon nanotubes [131].

7. CONCLUSIONS

In a short period of time, from discovery in 1991 to present day, carbon nanotubes have caught the fancy of chemists, physicists, and material scientists. Interest in this material has overshadowed that of fullerenes in recent years, although nanotubes still are not as readily available as fullerenes. The market price of nanotubes (about $400 per gram for good quality MWNTs and probably twice as much for SWNTs) is still too high, and new synthesis methods based on a continuous process need to appear for this scenario to change. Theoretical modeling has had a strong impact on the expansion of the nanotube research front. But in pointing to growth mechanisms, so that new fabrication methodologies can be pursued by experimentalists, theory has been rather disappointing. However, it should be noted that the laser method for producing SWNTs reported in 1996 [22] provided a boost by making adequate quantities of pure SWNTs available. Many theoretical predictions were based on modeling SWNTs, and this availability of good quality SWNTs carried the possibility, for the first time, that some of these predictions could be tested. Indeed, most of the models have been confirmed, which is quite remarkable.

It is hard to tell where the future of nanotubes lies. The most promising and fascinating developments have taken place in exploring the potential of nanotube electronics. Is it realistic to imagine that a few years from now, carbon nanotubes will become an integral part of microelectronic circuitry? Important concepts based on nanotube molecular devices have been demonstrated (as mentioned in the sections preceding), but the biggest challenge still remains in building nanotube-based architectures to suit existing or future electronic fabrication technology. Manipulating individual nanotubes and placing them in desired locations and configurations have been reported [132], but this is clearly not the approach that needs to be taken if highly complex architectures are to be built from individual elements.

It must be remembered that carbon fibers have existed for more than three decades. They were never really considered for electronic applications due to a high density of structural defects. However, carbon fibers have found important applications in composite technology and as electrodes for energy conversion. Millions of dollars have been spent to solve some of the engineering problems in this area. Nanotubes can be considered to be the ultimate carbon fiber, and it will be surprising if applications for nanotube are not developed in areas where traditional carbon fibers are abundantly used. One of the problems for nanotube-based composite is the lack of understanding of how the mechanics work around a nanosized inclusion or filler. It will take a lot of dedicated and tedious work before some of the fundamental questions concerning this are answered. One has to differentiate between carbon fibers and nanotubes; the latter is close to a molecular structure and the properties are governed more from interactions at the atomic level. Also, new areas of composite applications (other than structural) are being investigated. As pointed out earlier, properties of polymers such as photoluminescence can be tailored by physically doping with nanotubes. Many new functionalities are emerging from a judicious use of nanotubes in composites.

With extremely small dimensions and mechanical strength, as well as elasticity, one area where nanotubes ultimately may become indispensable is their use as nanoprobes. One could think of such probes being used in a variety of applications, such as high resolution imaging, nanolithography, nanoelectrodes, drug delivery, sensors, and field emitters.

On another note, the number of researchers and groups working in the nanotube field has doubled in the last year or so. Many people have crossed over from the fullerene field bringing their knowledge of carbon-based chemistry. More than a thousand papers have already been published on nanotubes. From the progress made so far, it is highly conceivable that nanotubes may one day become an integral part of our lives through the high technology that it promises.

Acknowledgments

I wish to acknowledge Dr. Sumio Iijima (NEC, Japan), Dr. Thomas W. Ebbesen (NEC, Princeton), Professor Christian Colliex (Laboratoire de Physique des Solides, Orsay), and Professor Manfred Rühle (Max-Planck-Institute für Metallforschung, Stuttgart) for collaboration on several aspects of nanotube research since 1991. Present support from the DMR division of the National Science Foundation (CAREER) to continue work on nanotubes is acknowledged. This chapter benefited greatly from my previous review, in collaboration with Dr. Thomas Ebbesen, that appeared in the *Reports of the Progress of Physics*. I also wish to thank Dr. C. Dekker, Dr. J.-C. Charlier, Dr. B. I. Yakabson, Dr. T. W. Ebbesen, Dr. P. Eklund, and Dr. M. Terrones for supplying some of the images/figures used herein.

References

1. H. W. Kroto, J. R. Heath, S. C. O'Brien, R. F. Curl, and R. E. Smalley, *Nature* 318, 162 (1985).
2. S. Iijima, *Nature* 354, 56 (1991).
3. T. W. Ebbesen (ed.), "Carbon Nanotubes: Preparation and Properties." CRC Press, Boca Raton, FL, 1997.
4. M. S. Dresselhaus, G. Dresselhaus, and P. C. Eklund, "Science of Fullerenes and Carbon Nanotubes." Academic Press, New York, 1996.
5. T. W. Ebbesen, *Phys. Today* June, 26 (1996).
6. P. M. Ajayan and T. W. Ebbesen, *Rep. Prog. Phys.* 60, 1025 (1997).
7. M. S. Dresselhaus, G. Dresselhaus, P. C. Eklund, and R. Saito, *Phys. World* January, 33 (1998).
8. S. Iijima, T. Ichihashi, and Y. Ando, *Nature* 356, 776 (1992).
9. P. M. Ajayan, T. Ichihashi, and S. Ijiima, *Chem. Phys. Lett.* 202, 384 (1993).
10. B. I. Yakabson and R. E. Smalley, *Am. Sci.* July–August, 324 (1997).
11. S. Iijima and T. Ichihashi, *Nature* 363, 603 (1993).
12. D. S. Bethune, C. H. Kiang, M. S. de Vires, G. Gorman, R. Savoy, J. Vazquez, and R. Beyers, *Nature* 363, 605 (1993).
13. W. A. de Heer, A. Chatelain, and D. Ugarte, *Science* 270, 1179 (1995).
14. P. G. Collins and A. Zettl, *Appl. Phys. Lett.* 69, 1969 (1996).
15. H. J. Dai, J. H. Hafner, A. G. Rinzler, D. T. Colbert, and R. E. Smalley, *Nature* 384, 147 (1996).

16. J. M. Planeix, N. Coustel, B. Coq, V. Brotons, P. S. Kumbhar, R. Dutartre, P. Geneste, P. Bernier, and P. M. Ajayan, *J. Am. Chem. Soc.* 116, 7935 (1994).
17. P. J. Britto, K. S. V. Santhanam, and P. M. Ajayan, *Bioelectrochem. Bioenerg.* 41, 121 (1996).
18. G. Che, B. B. Lakshmi, E. R. Fisher, and C. R. Martin, *Nature* 393, 346 (1998).
19. S. J. Tans, A. R. M. Verschueren, and C. Dekker, *Nature* 393, 49 (1998).
20. A. C. Dillon, K. M. Jones, T. A. Bekkedahl, C. H. Kiang, D. S. Bethune, and M. J. Heben, *Nature* 386, 377 (1997).
21. S. J. Tans, M. H. Devoret, H. Dai, A. Thess, R. E. Smalley, L. J. Geerlings, and C. Dekker, *Nature* 386, 474 (1997).
22. A. Thess, R. Lee, P. Nikolaev, H. Dai, P. Petit, J. Robert, C. Xu, Y. H. Lee, S. G. Kim, A. G. Rinzler, D. T. Colbert, G. E. Scuseria, D. Tomanek, J. E. Fischer, and R. E. Smalley, *Science* 273, 483 (1996).
23. C. Journet, W. K. Maser, P. Bernier, A. Loiseau, M. Lamy de la Chappelle, S. Lefrant, P. Deniard, R. Lee, and J. E. Fischer, *Nature* 388, 756 (1997).
24. J.-C. Charlie, X. Gonze, and J.-P. Michenaud, *Europhys. Lett.* 29, 43 (1995).
25. A. M. Rao, E. Richter, S. Bandow, B. Chase, P. C. Eklund, K. A. Williams, S. Fang, K. R. Subbaswamy, M. Menon, A. Thess, R. E. Smalley, G. Dresselhaus, and M. S. Dresselhaus, *Science* 275, 187 (1997).
26. T. W. Ebbesen and P. M. Ajayan, *Nature* 358, 220 (1992).
27. R. Bacon, *J. Appl. Phys.* 31, 283 (1960).
28. M. S. Dresselhaus, G. Dresselhaus, K. Sugihara, I. L. Spain, and H. A. Goldberg, "Graphite Fibers and Filaments" (U. Gonser, A. Mooradian, K. A. Muller, M. B. Panish, and H. Sakaki, eds.). Springer-Verlag, New York, 1988.
29. S. Amelinckx, X. B. Zhang, D. Bernaerts, X. F. Zhang, V. Ivanov, and J. B. Nagy, *Science* 265, 635 (1994).
30. W. Z. Li, S. S. Xie, L. X. Qian, B. H. Chang, B. S. Zou, W. Y. Zhou, R. A. Zhao, and G. Wang, *Science* 274, 1701 (1996).
31. M. Terrones, N. Grobert, J. Olivares, J. P. Zhang, H. Terrones, K. Kordatos, W. K. Hsu, J. P. Hare, P. D. Townshend, K. Prassides, A. K. Cheetham, and H. W. Kroto, *Nature* 388, 52 (1997).
32. H. Dai, A. G. Rinzler, P. Nikolaev, A. Thess, D. T. Colbert, and R. E. Smalley, *Chem. Phys. Lett.* 260, 471 (1996).
33. T. Guo, P. Nikolaev, A. Thess, D. T. Colbert, and R. E. Smalley, *Chem. Phys. Lett.* 243, 49 (1995).
34. S. Bandow, S. Asaka, Y. Saito, A. M. Rao, L. Grigorian, E. Richter, and P. C. Eklund, *Phys. Rev. Lett.* 80, 3779 (1998).
35. T. W. Ebbesen, P. M. Ajayan, H. Hiura, and K. Tanigaki, *Nature* 367, 519 (1993).
36. S. C. Tsang, P. J. F. Harris, and M. L. H. Green, *Nature* 362, 520 (1993).
37. H. Hiura, T. W. Ebbesen, and K. Tanigaki, *Adv. Mater.* 7, 275 (1995).
38. G. S. Duesberg, M. Burghard, J. Muster, G. Phillip, and S. Roth, *Chem. Commun.* 435 (1998).
39. K. Tohji, H. Takahashi, Y. Shinoda, N. Shimizu, B. Jeyadevan, I. Matsuoka, Y. Saito, A. Kasuya, S. Ito, and Y. Nishina, *J. Phys. Chem. B* 101, 1974 (1997).
40. S. Bandow, K. A. Williams, A. Thess, R. E. Smalley, and P. C. Eklund, *J. Phys. Chem. B* 101, 8839 (1997).
41. J. Liu, A. G. Rinzler, H. Dai, J. H. Hafner, R. K. Bradley, P. J. Boul, A. Lu, T. Iverson, K. Shelimov, C. B. Huffman, F. Rodriguez-Macias, Y. Shon, T. R. Lee, D. T. Colbert, and R. E. Smalley, *Science* 280, 1253 (1998).
42. P. M. Ajayan, O. Stephan, C. Colliex, and D. Trauth, *Science* 265, 1212 (1994).
43. O. Zhou, private communication.
44. K. Yamamoto, S. Akita, and Y. Nakayama, *J. Phys. D: Appl. Phys.* 31, L34 (1998).
45. A. Oberlin, M. Endo, and T. Koyama, *J. Cryst. Growth* 32, 335 (1976).
46. G. G. Tibbetts, *J. Cryst. Growth* 66, 632 (1984).
47. S. Iijima, P. M. Ajayan, and T. Ichihashi, *Phys. Rev. Lett.* 69, 3100 (1992).
48. A. Maiti, C. J. Brabec, C. M. Roland, and J. Bernholc, *Phys. Rev. Lett.* 73, 2468 (1994).
49. E. G. Gamaly and T. W. Ebbesen, *Phys. Rev. B* 52, 2083 (1995).
50. Y. K. Kwon, Y. H. Lee, S. G. Kim, P. Jund, D. Tomanek, and R. E. Smalley, *Phys. Rev. Lett.* 79, 2065 (1997).
51. J.-C. Charlier, A. DeVita, X. Blase, and R. Car, *Science* 275, 646 (1997).
52. D. Ugarte, A. Chatelain, and W. A. de Heer, *Science* 274, 1897 (1996).
53. J. W. Mintmire, B. I. Dunlap, and C. T. Carter, *Phys. Rev. Lett.* 68, 631 (1992).
54. N. Hamada, S. Sawada, and A. Oshiyama, *Phys. Rev. Lett.* 68, 1579 (1992).
55. P. M. Ajayan, S. Iijima, and T. Ichihashi, *Phys. Rev. B* 47, 6859 (1993).
56. V. P. Dravid, X. Lin, Y. Wang, X. K. Wang, A. Yee, J. B. Ketterson, and R. P. H. Chang, *Science* 259, 1601 (1993).
57. O. Stephan, P. M. Ajayan, C. Colliex, F. Cyrot-Lackaman, and E. Sandre, *Phys. Rev. B* 53, 13824 (1996).
58. T. Pichler, M. Knupfer, M. S. Golden, J. Fink, A. Rinzler, and R. E. Smalley, *Phys. Rev. Lett.* 80, 4729 (1998).
59. L. A. Bursill, P. A. Stadelmann, J. L. Peng, and S. Prawer, *Phys. Rev. B* 49, 2882 (1994).
60. C. Yannouleas, E. N. Bogachek, and U. Landman, *Phys. Rev. B* 50, 7977 (1994).
61. X. Blase, L. X. Benedict, E. L. Shirley, and S. G. Louie, *Phys. Rev. Lett.* 72, 1878 (1994).

62. H. Hiura, T. W. Ebbesen, K. Tanigaki, and H. Takahashi, *Chem. Phys. Lett.* 202, 509 (1993).
63. M. Kosaka, T. W. Ebbesen, H. Hiura, and K. Tanigaki, *Chem. Phys. Lett.* 225, 161 (1994).
64. M. Kosaka, T. W. Ebbesen, H. Hiura, and K. Tanigaki, *Chem. Phys. Lett.* 233, 47 (1995).
65. M. Ge and K. Sattler, *Science* 260, 515 (1993).
66. C. H. Olk and J. P. Heremans, *J. Mater. Res.* 9, 259 (1994).
67. D. L. Carroll, P. Redlich, P. M. Ajayan, J.-C. Charlier, X. Blase, A. De Vita, and R. Car, *Phys. Rev. Lett.* 78, 2811 (1997).
68. J. W. G. Wildoer, L. C. Venema, A. G. Rinzler, R. E. Smalley, and C. Dekker, *Nature* 391, 59 (1998).
69. T. Odom, J. Huang, P. Kim, and C. Lieber, *Nature* 391, 62 (1998).
70. P. Delaney, H. J. Choi, J. Ihm, S. G. Louie, and M. L. Cohen, *Nature* 391, 466 (1998).
71. D. L. Carroll, P. M. Ajayan, and S. Curran, *J. Mater. Res.* 13, 2389 (1998).
72. L. Langer, V. Bayot, E. Grivei, J.-P. Issi, J. P. Heremans, C. H. Olk, L. Stockman, C. Van Haesendonck, and Y. Bruynseraede, *Phys. Rev. Lett.* 76, 479 (1996).
73. T. W. Ebbesen, H. Lezec, H. Hiura, J. W. Bennett, H. F. Ghaemi, and T. Thio, *Nature* 382, 54 (1996).
74. C. T. White and T. N. Todorov, *Nature* 393, 240 (1998).
75. M. Bockrath, D. H. Cobden, P. L. McEuen, N. G. Chopra, A. Zettl, A. Thess, and R. E. Smalley, *Science* 275, 1922 (1997).
76. J. E. Fischer, H. Dai, A. Thess, R. Lee, N. M. Hanjani, D. DeHaas, and R. E. Smalley, *Phys. Rev. B* 55, 4921 (1997).
77. G. Overney, W. Zhong, and D. Tomanek, *Z. Phys. D* 27, 93 (1993).
78. R. S. Ruoff and D. C. Lorents, *Carbon* 33, 925 (1995).
79. B. I. Yakabson, C. J. Brabec, and J. Bernholc, *Phys. Rev. Lett.* 76, 2511 (1996).
80. M. M. J. Treacy, T. W. Ebbesen, and J. M. Gibson, *Nature* 381, 678 (1996).
81. E. W. Wong, P. E. Sheehan, and C. M. Lieber, *Science* 277, 1971 (1997).
82. M. R. Falvo, C. J. Clary, R. M. Taylor, V. Chi, F. P. Brooks, S. Washburn, and R. Superfine, *Nature* 389, 582 (1997).
83. D. H. Robertson, D. W. Brenner, and J. W. Mintmire, *Phys. Rev. B* 45, 12592 (1992).
84. P. M. Ajayan, O. Stephan, C. Colliex, and D. Trauth, *Science* 265, 1212 (1994).
85. S. Iijima, C. J. Brabec, A. Maiti, and J. Bernholc, *J. Chem. Phys.* 104, 2089 (1996).
86. R. S. Ruoff, J. Tersoff, D. C. Lorents, S. Subramoney, and B. Chan, *Nature* 364, 514 (1993).
87. N. Chopra, L. Benedict, V. Crespi, M. Cohen, S. Luoie, and A. Zettl, *Nature* 377, 135 (1995).
88. H. Hiura, T. W. Ebbesen, J. Fujita, K. Tanigaki, and T. Takada, *Nature* 367, 148 (1994).
89. B. I. Yakabson, *Appl. Phys. Lett.* 72, 918 (1998).
90. A. J. Stone and D. J. Wales, *Chem. Phys. Lett.* 128, 501 (1986).
91. P. M. Ajayan, V. Ravikumar, and J.-C. Charlier, *Phys. Rev. Lett.* 81, 1437 (1998).
92. H. D. Wagner, O. Lourie, Y. Feldman, and R. Tenne, *Appl. Phys. Lett.* 72, 188 (1998).
93. L. S. Schadler, S. C. Giannaris, and P. M. Ajayan, *Appl. Phys. Lett.* 73, 26 (1998).
94. J.-C. Charlier and J. P. Michenaud, *Phys. Rev. Lett.* 70, 1858 (1993).
95. P. M. Ajayan, T. W. Ebbesen, T. Ichihashi, S. Iijima, K. Tanigaki, and H. Hiura, *Nature* 362, 522 (1993).
96. P. J. Britto, K. S. V. Santhanam, V. Alonso, A. Rubio, and P. M. Ajayan, *Adv. Mater.* 11, 154 (1999).
97. M. R. Pederson and J. Q. Broughton, *Phys. Rev. Lett.* 69, 2689 (1992).
98. P. M. Ajayan and S. Iijima, *Nature* 361, 333 (1993).
99. P. M. Ajayan, O. Stephan, Ph. Redlich, and C. Colliex, *Nature* 375, 564 (1995).
100. E. Dujardin, T. W. Ebbesen, H. Hiura, and K. Tanigaki, *Science* 265, 1850 (1994).
101. S. C. Tsang, Y. K. Chen, P. J. F. Harris, and M. L. H. Green, *Nature* 372, 159 (1994).
102. P. Prasad and S. Lele, *Philos. Mag. Lett.* 70, 357 (1994).
103. P. M. Ajayan, Ph. Redlich, and M. Rühle, *J. Microscopy* 185, 275 (1997).
104. A. Rubio, *Condensed Matter News* 6, 6 (1997).
105. R. S. Ruoff, D. C. Lorents, B. Chan, R. Malhotra, and S. Subramoney, *Science* 259, 346 (1992).
106. S. Seraphin, D. Zhou, J. Jiao, J. C. Withers, and R. Loufty, *Nature* 362, 503 (1993).
107. P. M. Ajayan, C. Colliex, J. M. Lambert, P. Bernier, L. Barbedette, M. Tence, and O. Stephan, *Phys. Rev. Lett.* 72, 1722 (1994).
108. C. Guerret-Plecourt, Y. Le Bouar, A. Loiseau, and H. Pascard, *Nature* 372, 761 (1994).
109. Y. Saito, T. Yoshikawa, O. Mitsumasa, F. Naoya, K. Sumiyama, K. Suzuki, A. Kasuya, and Y. Nishina, *J. Phys. Chem. Solids* 54, 1849 (1993).
110. M. Liu and J. W. Cowley, *Carbon* 33, 749 (1995).
111. A. A. Setlur, J. M. Lauerhaas, J. Y. Dai, and R. P. H. Chang, *Appl. Phys. Lett.* 69, 345 (1996).
112. B. C. Satishkumar, A. Govindaraj, J. Mofokeng, G. N. Subbanna, and C. N. R. Rao, *J. Phys. B: At. Mol. Opt. Phys.* 29, 4925 (1996).
113. R. S. Lee, H. J. Kim, J. E. Fischer, A. Thess, and R. E. Smalley, *Nature* 388, 255 (1997).
114. A. M. Rao, P. C. Eklund, S. Bandow, A. Thess, and R. E. Smalley, *Nature* 388, 257 (1998).
115. O. Zhou, R. M. Fleming, D. W. Murphy, C. H. Chen, R. C. Haddon, A. P. Ramirez, and S. H. Glarum, *Science* 263, 1744 (1994).
116. V. Z. Mardkovich, M. Bazendale, R. P. H. Chang, and S. Yoshimura, *Synth. Met.* 86, 2049 (1997).

117. H. J. Dai, E. W. Wong, Y. Z. Lu, S. S. Fan, and C. M. Lieber, *Nature* 375, 769 (1995).

118. D. L. Carroll, X. Blase, J.-C. Charlier, S. Curran, Ph. Redlich, P. M. Ajayan, S. Roth, and M. Rühle, *Phys. Rev. Lett.* 81, 2332 (1998).

119. S. S. Wong, J. D. Harper, P. T. Lansbery, and C. M. Lieber, *J. Am. Chem. Soc.* 120, 603 (1998).

120. M. Goodman, S. Sternstein, L. S. Schadler, and P. M. Ajayan, unpublished.

121. S. Curran, P. Ajayan, W. Blau, D. Carroll, J. Coleman, A. Dalton, A. P. Davey, B. McCarthy, and A. Strevens, *Adv. Mater.* 10, 1091 (1998).

122. W. J. McHardy, "The Electron-Optical Investigation of Clays" (J. A. Gard, ed.), Monograph 3. Mineralogical Society, London, 1971.

123. R. Tenne, L. Margulis, M. Genut, and G. Hodes, *Nature* 360, 444 (1992).

124. L. Rapoport, Yu. Bilik, Y. Feldman, M. Homoyonfer, S. R. Cohen, and R. Tenne, *Nature* 387, 791 (1997).

125. B. C. Satishkumar, A. Govindaraj, E. M. Vogl, L. Basumallik, and C. N. R. Rao, *J. Mater. Res.* 12, 604 (1997).

126. O. Stephan, P. M. Ajayan, C. Colliex, P. Redlich, J. M. Lambert, P. Bernier, and P. Lefin, *Science* 266, 1683 (1994).

127. Z. Weng-Sieh, K. Cherrey, N. G. Chopra, X. Blase, Y. Miyamoto, A. Rubio, M. L. Cohen, S. G. Louie, A. Zettl, and R. Gronsky, *Phys. Rev. B* 51, 11229 (1995).

128. K. Suenaga, C. Colliex, N. Demoncy, A. Loiseau, H. Pascard, and F. Willaime, *Science* 278, 653 (1997).

129. A. Loiseau, F. Willaime, N. Demoncy, G. Hug, and H. Pascard, *Phys. Rev. Lett.* 76, 4737 (1996).

130. Ph. Redlich, J. Loeffler, P. M. Ajayan, J. Bill, F. Aldinger, and M. Ruhle, *Chem. Phys. Lett.* 260, 465 (1996).

131. A. Zettl, *Adv. Mater.* 8, 443 (1996).

132. S. Curran, D. L. Carroll, P. M. Ajayan, P. Redlich, S. Roth, M. Rühle, and W. Blau, *Adv. Mater.* 10, 311 (1998).

Chapter 9

ENCAPSULATION AND CRYSTALLIZATION BEHAVIOR OF MATERIALS INSIDE CARBON NANOTUBES

J. Sloan

Inorganic Chemistry Laboratory and Department of Materials, University of Oxford, Oxford, UK

M. L. H. Green

Inorganic Chemistry Laboratory, University of Oxford, Oxford, UK

Contents

Handbook of Nanostructured Materials and Nanotechnology, edited by H.S. Nalwa
Volume 5: Organics, Polymers, and Biological Materials

1. INTRODUCTION

Studies on the filling of multiple-walled carbon nanotubes (MWTs) and single-walled carbon nanotubes (SWTs) raise the possibility of the creation of modified hybrid materials with technologically important physical properties on a nanometric scale [1–6]. Actual and foreseen applications for such materials include their use in field emission devices [7], SPM tips [8], nanowires [3, 4], and composites [9], as new types of mesoporous catalysts [10], and in biotechnology [11]. Since the discovery of MWTs in 1991 [12], a number of researchers have successfully inserted both crystalline and noncrystalline materials into the capillaries of MWT and SWT cavities using both physical and chemical methods [1–5]. These experiments have recently been extended to SWTs, with their successful intercalation with orientated Ru crystallites [6]. Materials insertion into MWTs also has been achieved by arc evaporation of composite carbon electrodes, resulting in the formation of "nanowires" containing metal carbides formed along the internal capillaries of the MWTs. At present, this method works effectively only for a limited range of transition, main group, and lanthanide metals [3]. Partially filled and defect MWTs also can be obtained by gas–solid deposition of a carbon-containing species, typically a hydrocarbon, onto catalytic metal particles [13–15].

Although MWT and SWT encapsulation has been directed toward the enhancement of the physical properties of both the encapsulated and encapsulating materials, other benefits arising from this research have included the ability to observe in situ the chemistry of the encapsulates [16] and also their restricted crystallization behavior on an approaching atomic scale [17–19]. In this chapter, it is the latter upon which we ultimately wish to focus. Although initially we will concentrate on methodologies pertinent to the synthesis and purification of materials intercalated inside both MWTs and SWTs (and also with respect to the MWTs and SWTs themselves), we wish at the same time to describe and elucidate the relationships between method of insertion and the crystallization behavior obtained in each case. It is hoped that these phenomena will contribute to a knowledge of how insertion of materials inside both MWTs and SWTs can be optimized and also how materials formation can be manipulated on the most intimate scale.

2. METHODS OF OPENING, FILLING, AND PURIFYING MULTIPLE- AND SINGLE-WALLED CARBON NANOTUBES

2.1. Preparation of Multiple-Walled Carbon Nanotubes and Removal of Extraneous Carbon Material

MWTs can be prepared in the bulk by a modified Krätschmer–Huffmann procedure. This technique involves the arc evaporation of graphitic rods under dynamic pumping conditions with a partial pressure of helium according to the general experimental arrangement shown in Figure 1. Typically, MWTs are prepared from 1-cm-diameter graphite rods under ~0.17 atm of helium and from an arc generated by a dc voltage of 30 V and a current of ~180 Å [20]. The MWTs are formed in the carbon distilled on the cathode as a soft deposit covered by a hard cylindrical outer layer. The MWTs can then be removed from the soft inner core of the deposit and the outer shell is discarded. The nature of the products in the distillate obtained by this method can be radically altered by the introduction of additives to the source carbon electrode [21, 22]. However, the most useful application of this modification is in the synthesis of SWTs, as described subsequently.

The MWTs prepared by the modified Krätschmer–Huffmann method are invariably accompanied with 30–70% of other forms of carbon, including nanoparticles (similar to carbon "onions") and microporous carbons. The presence of these extraneous carbons impairs

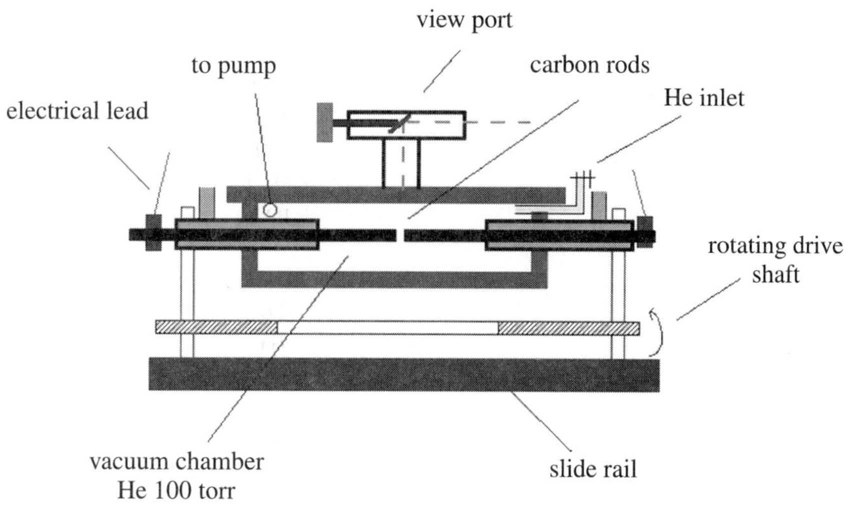

Fig. 1. Krätschmer–Huffman apparatus used in the synthesis of MWTs and SWTs. (Source: Reprinted with permission from [20].)

the characterization of MWTs and also any properties of the composite materials derived from them. Therefore, the separation and extraction of purified MWTs is an essential step for their development as useful practical materials.

Physical methods for MWT purification make use of filtration, chromatography, centrifugation, and sonication, sometimes in combination [23, 24]. Although these techniques have been effective in preparing enriched MWT samples, none of them has achieved complete purification. More recently developed purification methods make use of the fact that the extraneous carbons contain a far higher density of "defects" (i.e., fullerenic moieties or non-six-membered rings) than the MWTs, which contain a relatively lower proportion of defects because of their high aspect ratios, making them less prone to chemical attack [25]. Ebbesen and co-workers [23] have reported a low yield (~1–2%) technique that produces complete purification of MWTs by selective oxidation in air at 750 °C. Hwang [26] has described a liquid-phase oxidation that uses a mixture of $KMnO_4$ and H_2SO_4. However, although the residual carbon sample produced by this method contained pure MWTs, both their walls and caps were found to be severely damaged. Ikazaki et al. [27] have shown that the small graphitic nanoparticles can be removed by first intercalating them with copper chloride, followed by heating in air at 550 °C, and resulting in the preferential destruction of the intercalated carbon. The removal of the residual copper by acid washing then gives purified MWTs.

In this laboratory, we have explored the use of liquid bromine as a reagent for the selective intercalation and thereby disruption of the nanoparticles and their subsequent elimination by selective oxidation [28]. Typically, the cathodic soot containing nanotubes and nanoparticles were treated with bromine for 4 days. High resolution transmission electron microscope (HRTEM) studies performed on the specimen at this stage (see Fig. 2a) showed the presence of significant amounts of large black agglomerates that were subsequently demonstrated by energy dispersive X-ray microanalysis (EDX) to be enriched in bromine. The brominated sample was then heated to 530 °C under a flow of helium containing 4% oxygen for 5 days, after which the sample was found to be composed solely of pure nanotubes (see Fig. 2b). HRTEM showed that the nanotubes surfaces were only slightly damaged and the majority of the tubes are opened at both ends (see succeeding text). The yields depended crucially on the flow rate of oxidant, the scale of the initial experiment, the manner of packing of the carbon, and the quality of the cathodic soot. The yields of MWTs produced by this method were found to vary from between 10 and 20 wt%.

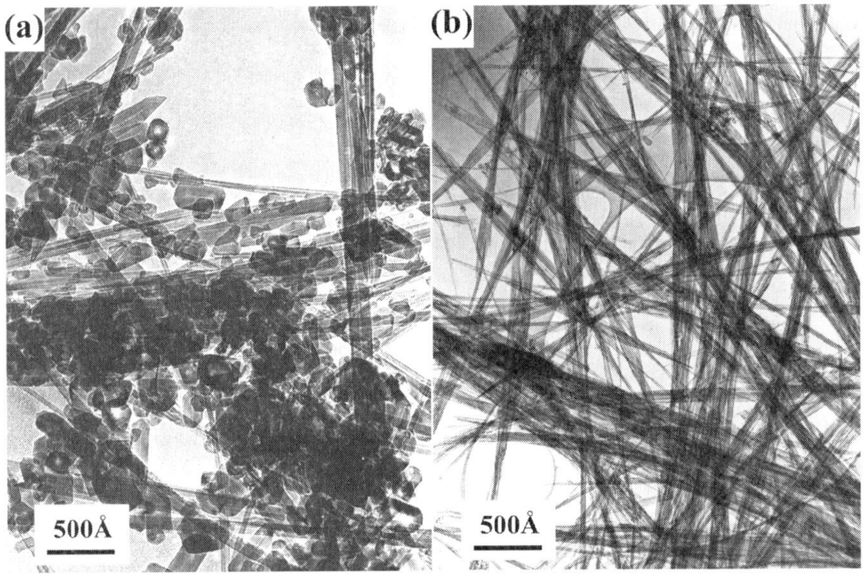

Fig. 2. Micrographs showing (a) as-prepared MWTs after bromination but prior to oxidation and (b) purified MWTs. (Source: Reprinted with permission from [28].)

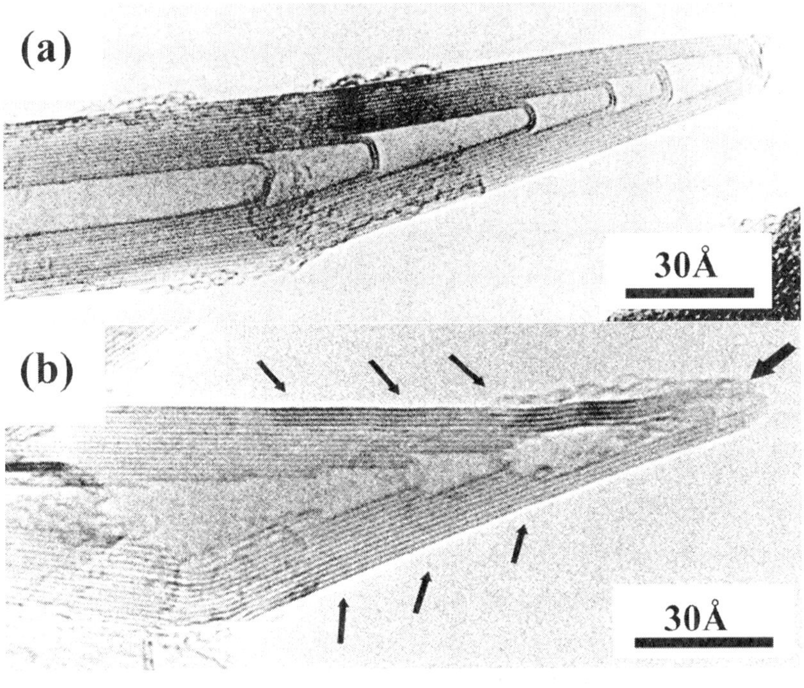

Fig. 3. Micrographs showing (a) a closed MWT with internal caps and (b) a MWT selectively opened at the tip and at internal cap regions (arrowed). (Source: Reprinted with permission from [2]. © 1994 Macmillan Magazines Ltd.)

2.2. Opening and Decarboxylation of Multiple-Walled Carbon Nanotubes

MWTs prepared by arc evaporation are invariably closed at both ends (Fig. 3a). However, the presence of the fullerenic moeties in the external caps at the end of each tube renders

Table I. Surface Acid Groups Determined by TPD [30]

Material	Acidity (acid sites/g)
Closed nanotubes	$< 2.0 \times 10^{18}$
Graphite/HNO_3	5.3×10^{19}
Opened tubes	$1.1–1.7 \times 10^{20}$

them susceptible to selective opening at these sites (Fig. 3b), therefore making the MWT internal capillaries accessible for filling. The first example of opened and filled MWTs was reported by Ajayan and Iijima, who showed that their treatment of closed MWTs with molten lead in the presence of air led to capillary filling with continuous PbO_x [1]. MWT caps also can be opened in a poorly selective oxidation process using either carbon dioxide [29] or oxygen [23]. Both methods give low yields of opened MWTs and cause structural damage to the tube walls. Subsequently, it was shown that the treatment of closed MWTs with refluxing nitric acid gives very high yields of opened tubes without substantial damage to their walls [2]. Subsequently, many materials have been introduced into the capillaries of opened MWTs by a variety of methods, as outlined in the following text.

When MWTs are opened in the presence of acid, the processes of oxidation results in the formation of oxygen-containing functional groups at the sites of acid attack. The types of obtained functionalities appear to be carboxylic or phenolic in nature, and their presence and surface density is revealed by titration with a base (see Table I) [30]. The acidic functionalities may well interfere with attempts to fill the MWT capillaries, and it is often deemed preferable to remove them before filling experiments are attempted. The oxygen-containing groups may be removed most efficiently by heating the oxidized MWTs in flowing argon at temperatures of up to 600 °C with the result that carbon oxides are evolved and the MWTs are decarboxylated without any detrimental effect on the opened nanotube specimen.

3. TECHNIQUES FOR FILLING MULTIPLE-WALLED CARBON NANOTUBES AND SOME REACTIONS OF THE INCLUDED MATERIALS

3.1. Chemical Methods for Filling Multiple-Walled Carbon Nanotubes

MWTs can be filled by a one-step method, which incorporates the opening step described in the previous section, by treating the closed tubes with refluxing nitric acid containing a soluble metal salt, typically a metal nitrate, for 4–24 h. The resulting insoluble black carbon materials are dried in air at 60 °C overnight and then calcined, typically at ~600 °C, under argon for several hours. At this temperature, the metal nitrates are decomposed to the oxides and, in many cases, it is possible to unambiguously identify the encapsulated oxide materials by direct observation of the characteristic lattice fringes of single crystals (see Table II) [2, 5]. Figure 4a–c shows HRTEM images of encapsulated crystallites of Ni (reduced from the oxide; see following text), Sm_2O_3, and Nd_2O_3, all of which were encapsulated via the one-step method. MWTs containing oxides of the metals Eu, La, Ce, Y, and Cd have also been prepared using this method [5]. Mixed metal oxides may be prepared in a similar manner. Thus, refluxing closed MWTs in a nitric acid solution

Table II. HRTEM Observed *d*-Spacings of MWT Encapsulated Metal Oxides, Sulfides, and Chlorides [5]

Encapsulated materials	Starting materials	Lattice spacing	Correlating planes $\{hkl\}$	Literature spacing
Eu_2O_3	$Eu(NO_3)_3{\cdot}6H_2O$	3.12	{222}	3.14
La_2O_3	$La(NO_3)_3{\cdot}6H_2O$	2.8, 3.5	{−303}, {202}	2.799, 3.59
Pr_2O_3	$Pr(NO_3)_3{\cdot}6H_2O$	2.8, 3.1, 3.5, 4.1	{012}, {302}, {400}, {102}	2.875, 3.15, 3.55, 4.17
CeO_2	$Ce(NO_3)_3{\cdot}6H_2O$	3.08	{111}	3.123
Y_2O_3	$Y(NO_3)_3{\cdot}5H_2O$	2.65	{400}	2.652
Nd_2O_3	$Nd(NO_3)_3{\cdot}xH_2O$	3.01	{002}	2.998
Sm_2O_3	$Sm(NO_3)_3{\cdot}6H_2O$	3.21	{222)	3.155
$FeBiO_3$	$Fe(NO_3)_3{\cdot}9H_2O$/ $Bi(NO_3)_3{\cdot}5H_2O$	2.8, 3.9	{110} or {−110}, {100}	2.811 or 2.783, 3.95
UO_{2-x}	$UO_2(NO_3)_2{\cdot}6H_2O$	3.15	{111}	Stoichiometry varies
NiO	$Ni(NO_3)_2{\cdot}6H_2O$	2.40	{111}	2.41
MoO_3	MoO_3	3.85	{110}	3.81
MoO_2	MoO_3	3.44, 2.47	{−111}, {111}	3.42, 2.43
ZrO_2	$ZrCl_4$	3.16, 2.89	{−111}, {1 − 11}	3.165, 2.84
ZrO_2	$ZrO(NO_3)_2{\cdot}xH_2O$	2.98	{111}	2.96
SnO_2	$Sn(SO_4)$	3.36, 2.65	{110}, {101}	3.351, 2.644
CdO	$Cd(NO_3)_2{\cdot}4H_2O$	1.9, 2.6	{220}, {200}	1.877, 2.655
CdS	$CdO + H_2S$	2.5, 1.8, 1.7	{102}, {200}, {004}	2.450, 1.791, 1.679
Ni metal	NiO	2.05 ± 0.05	{111}	2.034
Re metal	$KReO_4$	2.30	{002}	2.226
Pd metal	$Pd(NO_3)_2$	2.26	{111}	2.245
Ag metal	$Ag(NO_3)$	2.40	{111}	2.359
Au metal	$AuCl_3$	2.325	{111}	2.355
AuCl	$AuCl_3$	5.32	{101}	5.33

containing equimolar amounts of $Fe(NO_3)_3{\cdot}9H_2O$ and $Bi(NO_3)_3{\cdot}5H_2O$ gives crystals of the mixed oxide, $FeBiO_3$ (Fig. 4d) [5].

In the two-step method, the closed nanotubes are first opened with nitric acid and dried at 160 °C. These tubes contain acidic functionalities, which are then removed by the decarboxylation procedure, as described previously. The empty, opened, and decarboxylated MWTs are then filled by stirring with a concentrated solution of the chosen substrate (e.g., a metal salt). For example, stirring nanotubes with a concentrated solution of $H_4SiW_{12}O_{40}$ in deionized water for 16 h gives tubes filled with the same material [5]. When this method is applied to solutions of $AuCl_3$ or $AgNO_3$, a high percentage (~70%) of opened nanotubes are found to be filled with the gold chloride or silver nitrate, respectively [5, 31]. Calcination of the gold samples gives spherical gold crystals of diameter 10–50 Å (Fig. 4d). Calcination of the silver specimen gives crystallites of Ag ranging between 35 and 85 Å in

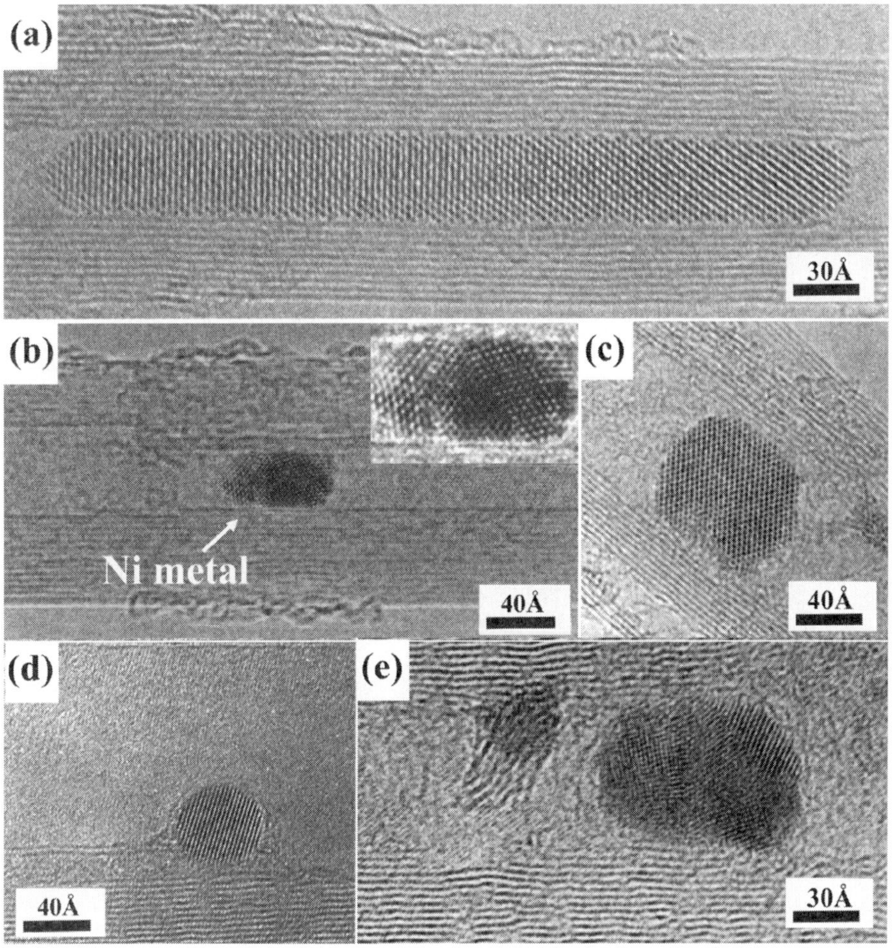

Fig. 4. (a) Elongated Sm_2O_3 crystal encapsulated inside and MWT capillary. (Source: Reprinted with permission from [5].) (b) A Ni metal particle (inset shows faceting) prepared by in situ reduction of NiO. (Source: Reprinted with permission from [5].) (c) A Nd_2O_3 crystal inside MWT capillary. (Source: Reprinted with permission from [5].) (d) A Au particle inside large MWT capillary. (Source: Reprinted with permission from [31]. © 1996 American Chemical Society.) (e) A faceted Ag crystallite inside MWT capillary. (Source: Reprinted with permission from [31]. © 1996 American Chemical Society.)

diameter (Fig. 4e). Visible in the micrograph is a region of amorphous filling, which was demonstrated by EDX microanalysis to contain AuCl. Some crystalline AuCl also was observed inside carbon nanotubes. Both the crystalline and amorphous AuCl are completely converted to Au metal after treatment with H_2 at 300 °C.

In a modification to the two-step procedure [17], polycrystalline SnO has been encapsulated by mixing opened MWTs with $SnCl_2{\cdot}2H_2O$ in hot concentrated HCl, followed by dropwise addition of an aqueous Na_2CO_3 solution. At a pH of 10.2, precipitation occurs and, following refluxing (3 h) and drying (160 °C for 12 h), the MWTs are filled with polycrystalline SnO. HRTEM and EDX studies indicate that 80% of the nanotubes contained polycrystalline SnO. The packing of this material inside the MWT capillaries has proven to be a good model system for the investigation of crystallization inside nanotubes (vide infra).

3.2. Filling Multiple-Walled Carbon Nanotubes with Molten Media

Opened nanotubes can be filled via capillary action using molten media [1, 9]. Typically, the tubes are stirred in either a crucible or a sealed silica quartz tube, depending on the air

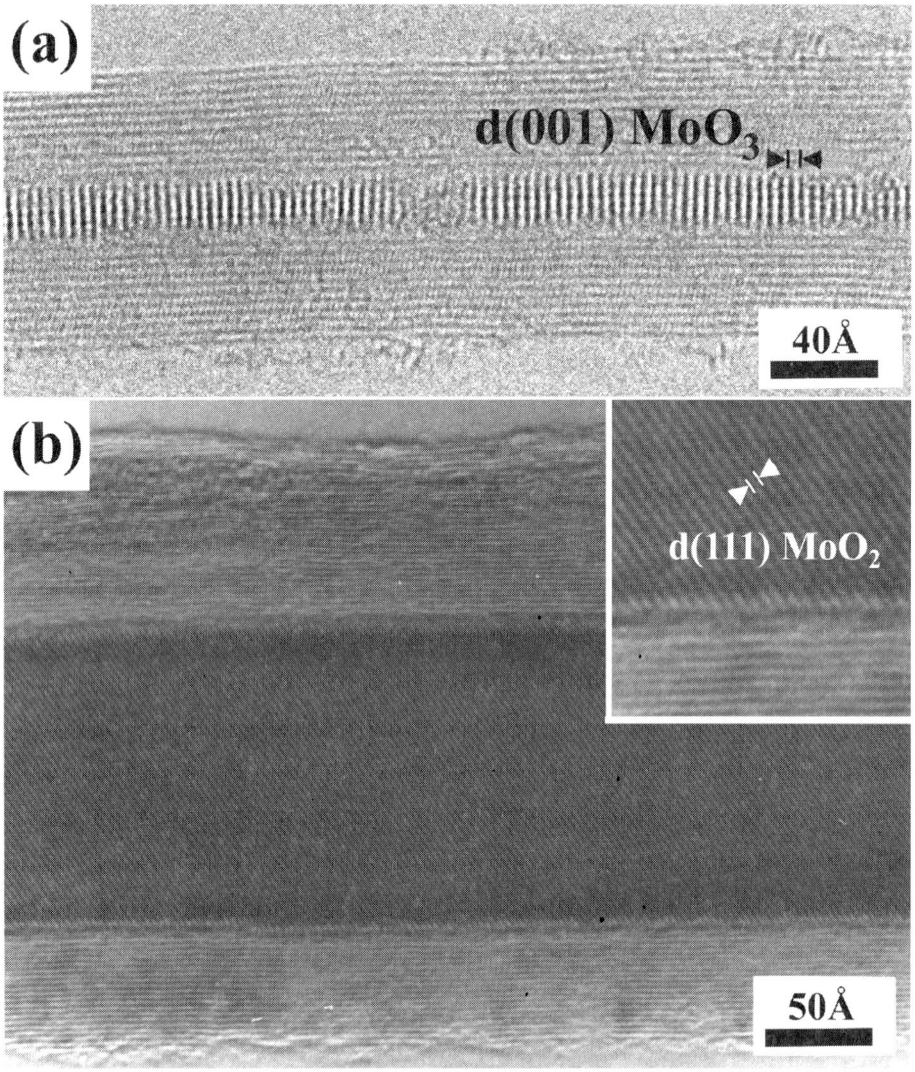

Fig. 5. Micrographs showing (a) continuous capillary filling with MoO_3 and (b) MoO_2 product formed by H_2 reduction of MoO_3. The inset shows MoO_2 (111) planes oriented at ~60° to the MWT wall. (Source: Reprinted with permission from [4].)

sensitivity of the filling medium, containing a 1:1 mixture of MWTs with either the molten metal oxide or a metal halide. The surface tension of the melt should be less than ~100–200 mN/m [9, 32]. This filling method generally gives long, continuous crystals that appear to occupy the entire internal volume of the nanotubes, as has been observed for the oxide phase of lead [1] and also V_2O_5 [9]. In this laboratory, we have studied this filling method using molten molybdenum trioxide and various mixtures of KCl and UCl_4 (see subsequent text) [4, 18]. In a typical procedure, MoO_3 was mixed with opened tubes and heated to 800 °C for 3 h. Approximately 50% of the nanotubes were observed to be continuously filled with single crystals of MoO_3, which are typically several hundred nanometers long, as shown by the example in Figure 5a. Subsequent reduction of the MoO_3 by H_2 leads to the formation of MoO_2 (Fig. 5b). Attempts to fill empty tubes with molten metals directly were unsuccessful. It has been suggested that the surface tensions of most liquid metals are too high to wet and fill the tubes [30]. A further problem is that the melting point of many metals is simply too high to attempt filling by this method without causing catastrophic

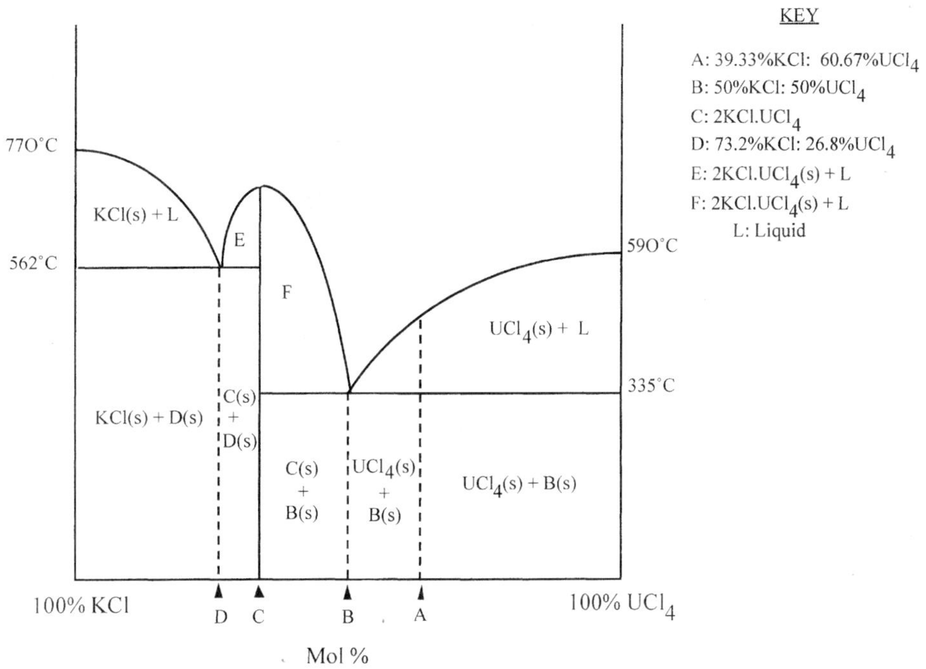

Fig. 6. Pseudobinary phase diagram constructed for the KCl–UCl_4 system. (Source: Reprinted from [18], based on data from [33].)

damage to the MWTs. However, elemental metals can be encapsulated inside MWTs by other routes, as subsequently discussed.

Another example of the molten media method is the use of low-melting mixtures of phases in a eutectic system that contains at least one melting temperature such that the resulting composition (i) has an overall surface tension lower than the threshold value (100–200 mN/m), (ii) has a sufficiently low melting temperature (<900 °C) such that the carbon MWTs are not damaged thermally, and (iii) does not attack the MWTs chemically. We have described a study of molten media filling of MWTs achieved using various eutectic and noneutectic mixtures of KCl with UCl_4 [18]. The technique makes direct use of the phase relationships of the KCl–UCl_4 system, represented here by the pseudobinary phase diagram depicted in Figure 6 [33]. This study also made use of surface tension/composition data recorded by Janz et al. for the KCl–UCl_4 system (the same authors also have published data on similar low melting systems [34]), collectively defined in terms of

$$s = a - bT \tag{3.1}$$

where s is the surface tension in millinewtons per meter, T is the temperature (in kelvin), and a and b are coefficients determined for a range of compositions and temperatures (Table III). For example, the surface tension of a noneutectic 39.33-mol% KCl:60.67-mol% UCl_4 mixture (composition A, Fig. 6) is 46.94 mN/m at 900 K, which is below the threshold value for which filling of MWTs with molten media is predicted and is also too low to cause them significant thermal damage. In addition, the surface tension of pure UCl_4 at 900 K is 38.00 mN/m which is also below the threshold value for which filling of MWTs is predicted. Filling experiments were attempted by heating each of the mixtures indicated in Figure 6 to ~150 K above the liquidus of the respective composition together with a 1:1 mass ratio of opened, decarboxylated MWTs. The mixtures were then allowed to cool in the furnace to room temperature.

The modified molten media method was successful in forming a variety of fillings inside MWT capillaries that individually depended on the KCl:UCl_4 molar ratio employed

Table III. Compositions, a and b Coefficients, and Temperature Ranges for the KCl–UCl_4 Surface Temperature (ST) Relationship (Eq. (3.1)) [34]

KCl–UCl_4 (mol%)	a	b	Temperature range (K)
0–100	204.95	0.185	880–960
39.33–60.67	102.83	0.0621	890–980
56.04–43.96	93.04	0.04808	880–990
64.31–35.69	59.73	0.00865	930–1010
73.68–26.32	77.59	0.01349	870–1050
96.00–4.00	136.65	0.06732	1020–1150
97.33–2.67	164.82	0.08871	1050–1090
100–0	182.51	0.0782	1080–1170

Table IV. Compositions Used to Fill MWTs (see also Fig. 6)

Compositions (see Fig. 1)	KCl:UCl_4 (mol%)	Liquidus temperature (K)	Calculated ST (mN/m)
UCl_4	204.5	0.185	38.00 at 900 K
A	102.83	0.0621	46.94 at 900 K
B	93.04	0.04808	56.98[a] at 750 K
C	59.73	0.00865	51.51[b] at 950 K
D	77.59	0.01349	64.37[c] at 980 K
KCl	182.51	0.0782	94.93 at 1120 K

[a] Estimated from 56.04% KCl:43.96% UCl_4 composition; probably lower at eutectic composition.

[b] Estimated from 64.31% KCl:35.69% UCl_4 composition; may be higher due to formation of 2KCl·UCl_4 (K_2UCl_6).

[c] Estimated from 73.68% KCl:26.32% UCl_4 composition.

for a given experiment. An important factor also, however, was the melting properties of the particular composition used to fill the MWT capillaries, as summarized in Table VI. When either pure UCl_4 or KCl was used, no significant crystalline filling was obtained. However, when UCl_4-enriched composition A was used (Fig. 6), for example, preferential filling with polycrystalline UCl_4 was obtained (Fig. 7a and b). When filling was attempted with either one of the eutectic compositions (compositions B and D, Fig. 6), continuous capillary filling with a homogeneous product corresponding to the starting composition was obtained (Fig. 7c). The exception to this was when filling was obtained using a thin capillary. In this instance, the filling was observed as a (presumably) metastable crystalline product (Fig. 7d). The crystallization processes of these various compositions inside MWT capillaries is discussed in more detail in the following text.

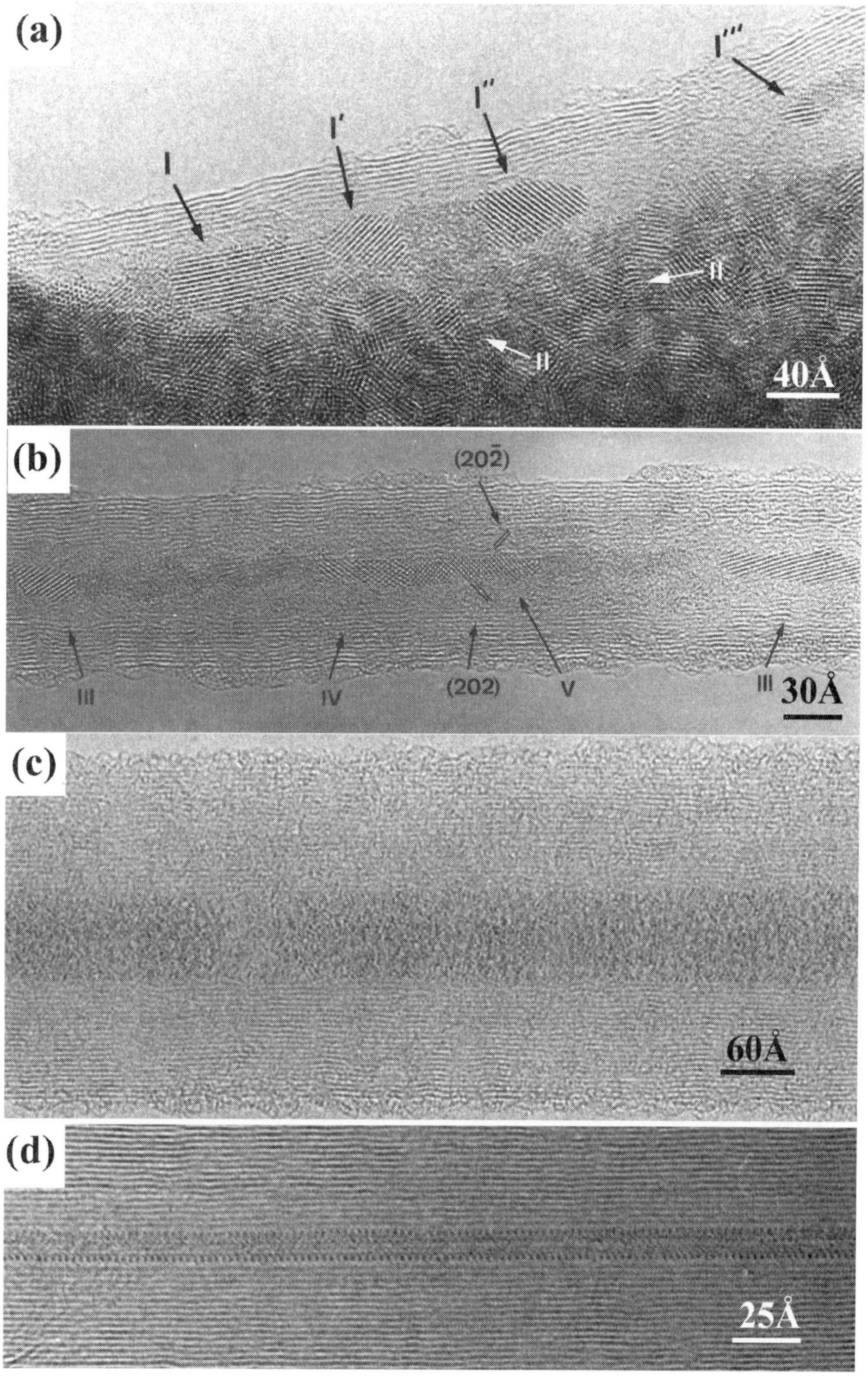

Fig. 7. Micrographs showing (a) MWT filled with polycrystalline UCl_4/KCl by composition A (see Fig. 6) and excess material on the outside, (b) continuous polycrystalline UCl_4 filling inside an MWT capillary, (c) thick capillary MWTs filled with 1:1 KCl–UCl_4 eutectic mixture (composition B, Fig. 6), and (d) thin capillary MWT filled with composition B from the same specimen as in (c). Note the continuous crystalline nature of product. (Source: Reprinted with permission from [18].)

3.3. Arc and Catalytic Methods for Filling Multiple-Walled Carbon Nanotubes

Closed nanotubes and the related closed carbon nanoparticles filled with metal carbides can been formed by in situ arc evaporation of composite carbon electrodes made by blending graphite powder with the appropriate metal or metal oxide [3, 35–40]. The resulting products normally consist of nanoparticles with closed carbon shells with the encapsulated materials normally corresponding to the metal carbides. The experimental technique consists of packing the bore of a hollowed-out graphite anode with a mixture of the appropriate element and powdered graphite, and then proceeding with the Krätschmer–Huffman evaporation experiment as previously described. During the arcing process, the tip of the anode and its contents are rendered into the vapor phase and the carbon shell then grows catalytically from the condensing species. The metals Cr, Fe, Co, Ni, Pd, Gd, Dy, and Yb all form nanowires inside MWT-like nanotubes [3]. All other metals tend to form single crystals of either the metals or the respective metal carbide inside the carbon nanoparticles. A mixture of both types of filling is observed in the case of the arc-encapsulated metals or carbides of Y [35, 40] and Mn [40].

Defect MWTs filled with either metal carbides or catalytic metal particles also can be produced by direct gas–solid carbon deposition onto catalytic metal species [13–15]. The carbon carrier can vary from hydrocarbon gases, such as methane [15] and acetylene [41], to aromatics, including benzene [42], to polymers, such as polyethylene [43], and even more complex organic species [44, 45]. Certain metals efficiently promote carbon growth from the gas-phase hydrocarbons and these include Co, Fe, Ni, Pd, Pt, Ti, W, Cu, or Ge [13–15, 41, 45–47]. In nearly all cases, the goal has been to produce either pure MWTs, SWTs, or defect nanotubes, rather than the encapsulated species, which, in these cases, are considered to be an unwanted by-product. However, Dai et al. [45] have deliberately prepared encapsulated nanowires containing copper and germanium by pyrolysis of polycyclic aromatic hydrocarbons over finely divided copper and germanium catalytic particles, respectively.

3.4. Chemical Reactions inside Multiple-Walled Carbon Nanotubes

In favorable circumstancse, chemical reactions can be identified inside MWTs. At the current time this may be achieved most efficiently either by direct HRTEM imaging, when the chemical transformations are gross enough, or by electron probe techniques, such as energy dispersive X-ray microanalysis (EDX) [48]. The transformation of $AuCl_3$ to its base metal, as previously described, is an example of the observation of reduction, because the latter is clearly distinguishable from the former due to its relatively higher contrast in the HRTEM. Such a reaction also can be followed by EDX, provided that a sufficiently small and dense electron probe is available [48]. The observation of the reduction of NiO to nickel metal in real time also has been performed by controlled environment electron microscopy (CETEM) [16]. The reducing medium was 5-mbar H_2 under differential pumping conditions, and the direct observation of the formation of faceted nickel crystals from the oxide (Fig. 4b) was seen inside the MWT capillaries. The prominent lattice fringes of 2.05 Å in the faceted crystallite correspond to the spacing between the (111) planes of Ni metal.

Reductions of continuous MWT fillings also have been followed by HRTEM observation of continuous MoO_3 capillary filling both prior to and following reduction [4]. Treatment of the nanotubes filled with continuous crystals of MoO_3 (Fig. 5a) with hydrogen at 450–500 °C causes reduction to pure elongated MoO_2 filling (see Fig. 5b). The MoO_2 crystals show a strong preferred orientation with the [111] direction aligning parallel to the MWT axis and with the (100) planes also lying at an angle of 60° to the same axis. It is perhaps surprising that the reduction of long MoO_3 crystals to similarly long MoO_2 crystals proceeds without apparent degradation of the crystallinity or morphology of the latter [4]. Hydrogen must penetrate along the entire length of the MWT capillary and water vapor subsequently must be expelled. The loss of the water must then be accompanied by shrinkage of the original dimensions of the MoO_3 crystal with a longitudinal rather than a cross-sectional contraction. The properties of MoO_2-filled MWTS are potentially interesting because single crystals of MoO_2 have a resistivity of $\sim 2.7 \times 10^{-4}$ W cm^{-1} at 400 K and a higher conductivity at all temperatures than many of the metallic elements [49].

As stated earlier, it has not been possible to insert metals directly into MWTs via the capillary method. However, suitable metal compounds can be reduced to their respective metals by reduction with hydrogen gas at elevated temperatures as shown for the case of Ni formed from NiO. Similarly, encapsulated crystals of pure Pd, Ag, and Au [30, 31] have been prepared inside MWTs. The d-spacing data obtained by the direct imaging of lattice planes are also summarized in Table II. Hydrogen reduction of tubes filled with $KReO_4$ gives crystallites of rhenium metal with a measured lattice d-spacings of 2.30 Å, corresponding to the (002) plane of Re metal [5]. In this specimen, residual crystalline potassium oxide was not observed and is presumed to be present as amorphous material observed inside the MWT capillary.

Apart from hydrogen reductions, other kinds of reactions also can be carried out inside nanotubes. An example is the formation of cadmium sulfide crystals by the in situ treatment of cadmium oxide crystals with hydrogen sulfide at 400 °C. Electron diffraction patterns obtained from these crystals correspond to the thermodynamically stable form of hexagonal CdS [5]. The formation of Au_2S_3 from $AuCl_3$ by a similar reaction has been reported also [5]. In the case of UCl_4 selectively deposited inside MWTs, as described previously, oxidation can be detected in the specimen after it has been exposed to the atmosphere for a period of ~2 weeks. An unidentified oxidized crystalline product with large fringe d-spacings (5.12 Å) and the composition $U(Cl, O)_x$ was observed inside the MWT capillaries [18].

3.5. Purification of Multiple-Walled Carbon Nanotubes from External Material Following Encapsulation

The various described methods of filling MWTs all can provide samples of the desired filled tubes. However, always associated with these specimens is a considerable amount of material that is external to the nanotubes (e.g., see Fig. 7a). Attempts to remove such material by washing with normal solvents almost invariably results in the simultaneous removal of the material on the inside of the nanotubes. We have demonstrated a method for the removal of material external to the filled nanotubes that uses reversed micelles in nonpolar solvents [50]. Opened carbon nanotubes were stirred with a concentrated, aqueous solution of $RhCl_3$. The resulting solid material was dried overnight at 60 °C and then reduced to Rh metal by heating at 550 °C under H_2. HRTEM studies revealed that approximately 70% of the opened nanotubes were filled with discrete crystals of Rh. No chlorine was present, either inside the nanotubes or in the bulk specimen, as was demonstrated by EDX. Figure 8a shows a HRTEM micrograph of this specimen following reduction. Both small spherical and longer ellipsoidal crystallites can be observed encapsulated inside the nanotubes, as well as a large number of spherical metal particles deposited outside the carbon nanotubes. Attempts to remove the external rhodium material, prior to hydrogen reduction, by washing with water resulted in removal of rhodium material from both inside and outside the nanotubes. However, by washing the nanotubes encapsulated with $RhCl_3$ prior to reduction with the reversed micelle medium formed from dodecylammonium propionate (DAP) solubilized water in benzene, the extraneous chloride could be selectively removed from the outside of the nanotubes. The optimum conditions were found to be a 3–7 $\times 10^{-3}$ M concentration of DAP with stirring for 3 days at 30 °C. Under these conditions, and following filtration and calcination at 500 °C under H_2 for 5 h, cleaned and filled nanotubes containing rhodium metal were produced. HRTEM studies of the resulting nanotubes showed rhodium metal crystals to be present solely inside the nanotubes and with

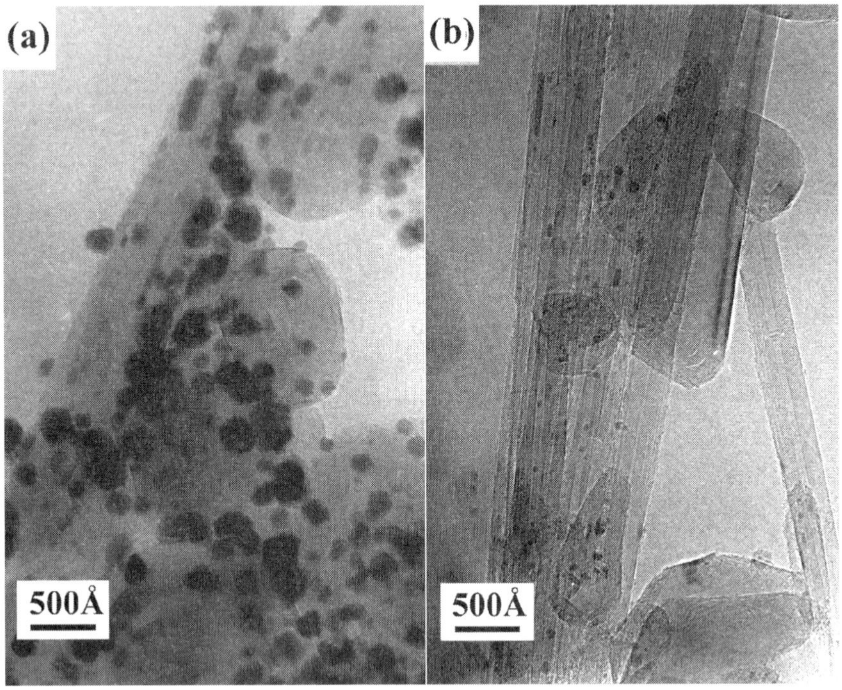

Fig. 8. (a) HRTEM micrograph showing Rh crystallites on the inside and outside of MWTs. (b) Micrograph showing Rh filled MWTs following removal of extraneous material. (Source: Reprinted with permission from [50].)

no crystals of rhodium exterior to the nanotubes (see Fig. 8b). Elemental analysis results, obtained after treatment with the reverse micelle solution, showed a rhodium content of ~0.25%, as compared to ~2% for the unwashed samples.

4. SYNTHESIS, PURIFICATION AND FILLING OF SINGLE-WALLED CARBON NANOTUBES

4.1. Methods for Preparing Purified Samples of Single-Walled Carbon Nanotubes

SWTs can be synthesized by coevaporating composite carbon rods mixed with certain metals [46, 51–54], as in a modified Krätschmer–Huffmann-type experiment, and by laser vaporization of metal-doped graphite targets [55, 56]. SWTs also have been produced during thinning of MWTs using CO_2 [29]. In terms of bulk synthesis, the most efficient methods are laser vaporisation [55, 56] and arc evaporation employing mixed metal combinations, such as Fe/Ni and Co/Ni [54]. Currently, these methods can produce yields in the range of 50–70%. No technique currently is capable of producing a completely pure product in a single step. As with the synthesis of MWTs, SWT products are accompanied by extraneous materials that include carbon-encapsulated metal particles and microporous carbons. Various approaches have been developed with respect to their purification.

Most techniques of SWT purification involve a method of isolating the undesired extraneous materials followed by some process of selective filtration. Bandow et al. [57] have separated SWTs from the other materials by use of a cationic surfactant in aqueous solution, which renders the latter into suspension, while the SWTs are trapped on a porous membrane. Amorphous and crystalline carbon impurities also can be removed from SWT samples by ultrasonically assisted microfiltration [58]. This method also "cuts up" the normally long, high aspect ratio SWTs into smaller lengths and, furthermore, encourages their propensity to form aligned bundles. Another reported purification method involves the hydrothermal treatment of the impure SWT starting materials, followed by extraction of the SWTs and selective oxidation and dissolution of the impurity metal particles [59].

4.2. Filling of Single-Walled Carbon Nanotubes with Ruthenium Metal

SWTs display a higher degree of uniformity with respect to their physical dimensions compared to MWTs because they consist only of a single cylindrical graphitic layer and exhibit both a smaller range of diameters and far fewer defects than their multiwalled counterparts. Similarly, therefore, a greater degree of uniformity may be anticipated from their physical properties and also those of composite materials resulting from their encapsulation. We have been able to fill SWTs successfully with Ru metal by wet chemical techniques [6]. A milder methodology was necessary than the oxidation methodology employed for opening MWTs (refluxing concentrated HNO_3), which resulted in the destruction of the SWTs.

The SWTs used in the filling experiments were produced by the evaporation of cobalt doped in a modified arc-discharge chamber (also employed to make MWTs) according to the general method described by Bethune et al. [51]. The yield in our apparatus was improved from ~10 to ~30% by confining the external environment of the electrodes by a 10-cm-i.d. steel cylinder 20 cm in length. For the filling experiments, samples of SWTs were suspended in concentrated HCl and stirred for 8 h. After centrifuging for 5 min, a light blue solution containing $CoCl_2$ was removed and the specimen was dried at 70 °C for 8 h and then washed with deionized water. A saturated solution of $RuCl_3$ (~5 mL) was then added to the SWT-containing specimen and the mixture was stirred for 5 h at 35 °C, after which the black residue was dried at 60 °C overnight. The sample was then heated in a stream of H_2 at a rate of 5 °C/min to 45 °C and kept at this temperature for 3 h. HRTEM micrographs and EDX microanalysis revealed that a small proportion (~5%) of the opened SWTs contained Ru metal. Other particles also were observed on the exterior of SWTs, and these were shown by EDX to be either "loose" Ru particles or carbon-encapsulated Co particles, which were presumably protected from acid attack by their carbon shells. Encapsulated and elongated crystallites also were observed inside the SWT capillaries, and these were again identified by EDX to be Ru metal. Figure 9a shows an elongated Ru

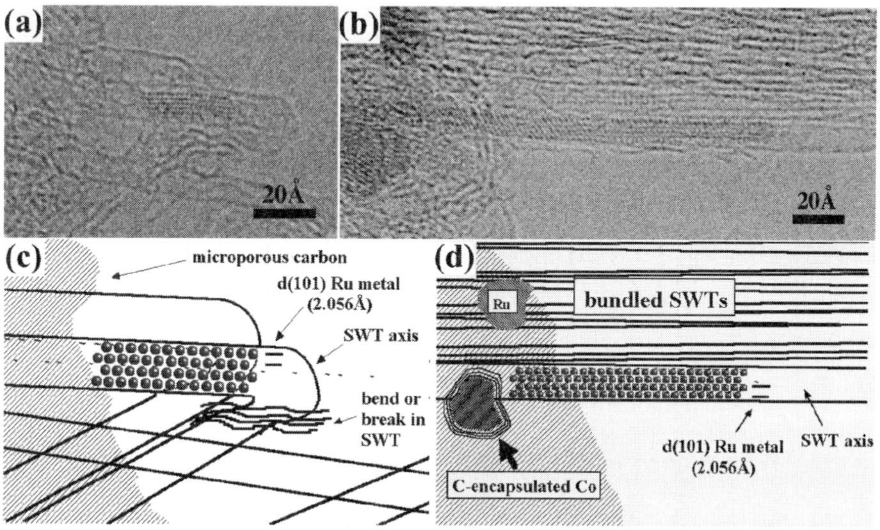

Fig. 9. Micrographs showing (a) an elongated Ru crystal encapsulated near the elbow of a bent or damaged SWT capillary and (b) an elongated Ru crystal encapsulated inside a single SWT on the periphery of a bundle of SWTs. (c) and (d) Schematic representations of (a) and (b), respectively. (Source: Reprinted with permission from [6].)

crystallite encapsulated near the elbow of a bent or damaged SWT. Figure 9b shows an elongated Ru crystallite growing along the bore of one SWT at the edge of a "bundle" of SWTs. Both encapsulated crystallites exhibit preferred orientations, with their (101) lattice fringes aligning parallel with the SWT axes as shown schematically in Figures 9c and 9d.

5. CRYSTALLIZATION BEHAVIOR INSIDE MULTIPLE- AND SINGLE-WALLED CARBON NANOTUBES

5.1. Control over Crystallite Morphology and Orientation in Multiple- and Single-Walled Carbon Nanotubes

The behavior of crystalline materials formed by precipitation inside MWTs allows for the study of the morphological and orientational control of their capillaries over crystallite formation. Initially, crystallization will proceed according to conventional nucleation and growth mechanisms, but once the size of the encapsulated crystallite approaches that of the internal diameter of the nanotube cavity, then the MWT capillaries must start to exert control over any future growth. The precise nature of the interaction between MWT capillaries over crystal growth at this juncture raises some interesting questions because the whole process is taking place in confined volumes of approaching atomic scale dimensions. On this scale, two factors are likely to be of importance in terms of influencing the crystal growth of the encapsulated crystallites: first of all, the morphological influence of the MWT capillaries and, second, the influence of the atomic structure of the capillary walls themselves. In this and the next three sections, we outline the various kinds of crystallization behavior observed in MWTs and SWTs.

Polycrystalline SnO forms spherical or ellipsoidal encapsulates with diameters ranging from 20 to 60 Å [17] inside MWT capillaries with internal diameters that also range between 20 and 90 Å. The packing of SnO crystallites inside MWTs thus provides a useful model system for the investigation of capillary control over crystallization. Figure 10a shows a HRTEM micrograph of an agglomeration of randomly oriented SnO crystallites inside a nanotube with a large internal diameter (~90 Å). In this instance, the nanotube exerts morphological control over the agglomeration, but none over individual crystallites. In Figure 10b, a second nanotube with a smaller diameter (~35 Å) can be seen in which two well resolved and slightly elongated crystallites I and II reside in the central cavity. A third, less well resolved crystallite (III) is also visible. Whereas crystallite I has its growth axis (arrowed) aligned parallel with the nanotube axis, crystallite II is tilted at angle of ~25° to the tube axis. Regardless of where the nucleation of crystallite II started, crystal growth in either of the directions indicated in Figure 10b is clearly constrained by the MWT capillary. By contrast, crystallite I can potentially continue growth along its current axis until either it meets an obstruction or the crystallization process is terminated. In Figure 10c a micrograph showing a MWT closely packed with SnO crystallites is shown in which the smaller crystallites (denoted IV) observe apparently random orientations, while the larger ones (denoted V) have their (101) lattice fringes oriented at 90° to the nanotube axis. The nanotubes capillaries shown in Figure 10b and c apparently exert influence over both the morphology and the orientation of their encapsulated SnO crystallites.

Other types of polycrystalline filling show types of orientational control similar to SnO. Figure 11a and b shows examples of filling with polycrystalline SnO_2 and ZrO_2, respectively [5]. Whereas polycrystalline SnO_2 crystallites form small crystallites with diameters approximately equal to the internal diameter of the MWTs and which apparently randomly orient with respect to the capillary (Fig. 11a), polycrystalline ZrO_2 crystallites are elongated and several crystallites in Figure 11b can be seen to align with their lattice fringes arranged parallel to the nanotube axis.

When materials are inserted into nanotubes by capillary action, oriented crystal growth is almost invariably obtained as Ajayan et al. [9] have noted for V_2O_5 and Chen et al. [4]

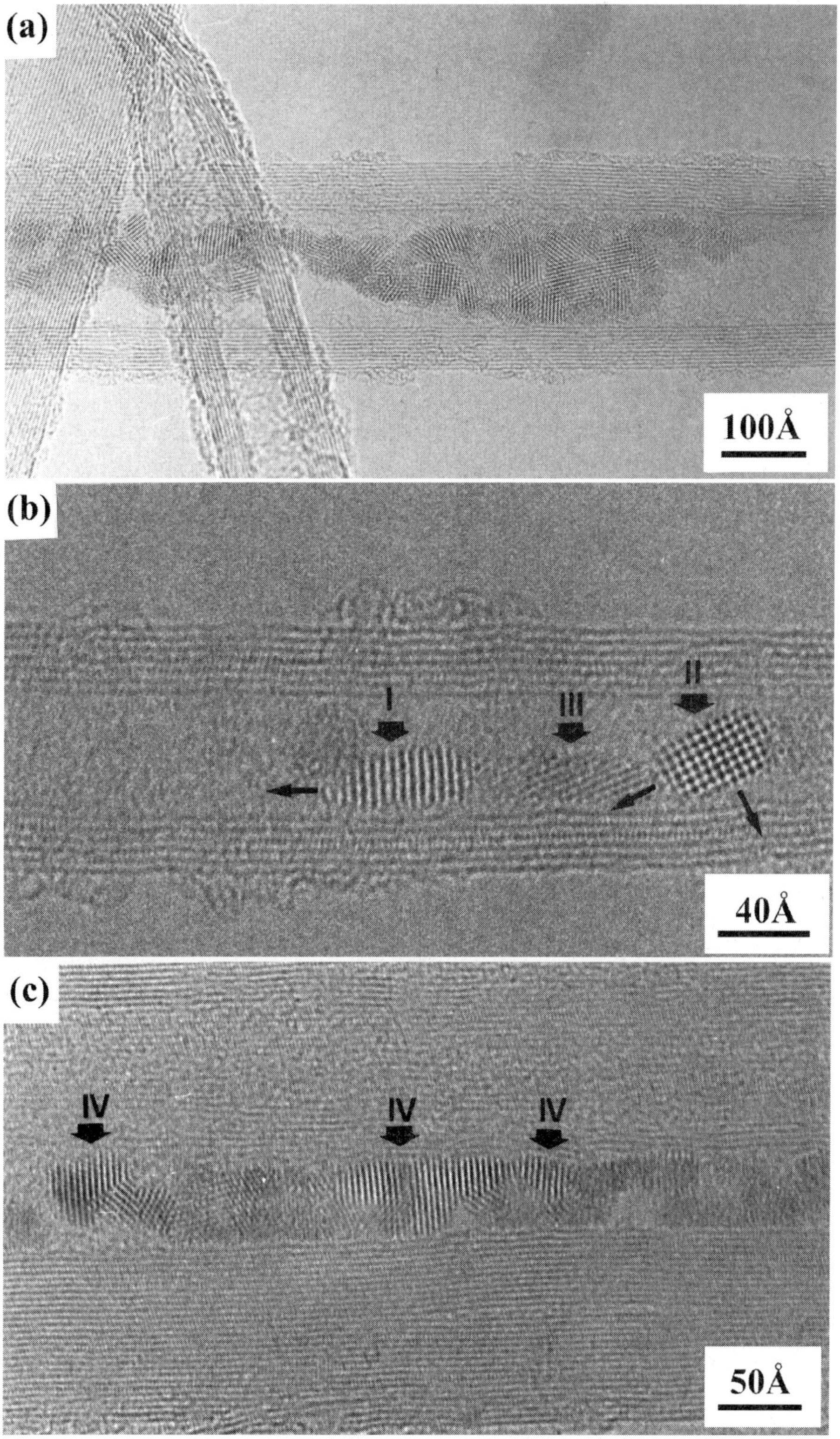

Fig. 10. Micrographs showing (a) packing of an agglomeration of SnO crystallites inside a wide capillary MWT, (b) MWT capillary control over SnO crystallite orientation (see text for explanation). (Source: (a) and (b) Reprinted with permission from [19].) (c) Preferentially aligned SnO crystallites (denoted V) inside an MWT capillary. (Source: (c) Reprinted with permission from [17].)

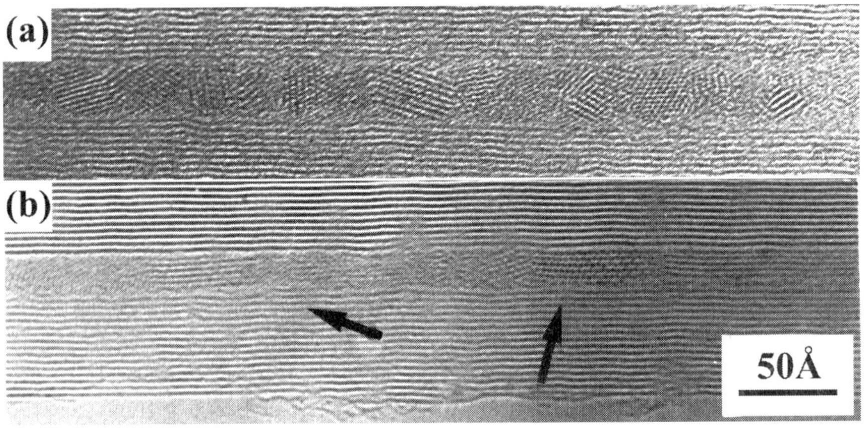

Fig. 11. (a) Randomly oriented SnO_2 crystallites and (b) preferentially oriented ZrO_2 crystallites (arrowed), both observed inside a MWT capillary. (Source: Reprinted with permission from [19].)

have noted for MoO_3 (e.g., see Fig. 5a). The exception to this is when the filling is achieved using a noneutectic mixture in a two component system, as described for UCl_4, when the filling is polycrystalline [18]. Figure 7a shows a region of bulk polycrystalline UCl_4 with a partially filled MWT on the surface. All of the encapsulated crystallites denoted I have lattice fringes conforming to the (211) planes of UCl_4. Several crystallites denoted II in the material on the exterior of the nanotube exhibit the same d-spacing. Two of the encapsulated crystallites denoted I can be seen with their (211) lattice fringes oriented parallel to the MWT axis. Two further (211) crystallites, denoted I′ and I″, can be distinguished in which the respective orientations are 32° and 44°. A crystallite with a smaller diameter than the nanotube capillary, I‴, can be seen with an orientation of 10° to the MWT axis. Figure 7b shows MWT containing continuous polycrystalline UCl_4 filling. The crystallites denoted III and IV can be assigned to the (202) and (200) planes of UCl_4, respectively. For the crystallite denoted V, the (202) and $(20\bar{2})$ planes are both resolved at 90° to each other, indicating that this crystallite is in [010] projection. The elongated crystallites denoted III have their (202) lattice fringes oriented at $\sim$15° with respect to the nanotube wall, whereas the elongated crystallites denoted IV (200) and V (202) have their respective lattice fringes oriented at 47° and 32° with respect to the nanotube wall. These results and those for SnO, SnO_2, and ZrO_2 all seem to indicate that larger elongated crystallites exhibit a definite, though not exclusive, tendency to align preferentially with the MWT capillary axis. This trend is not observed for smaller crystallites (with longitudinal dimensions smaller than the MWT internal diameters) which appear to be randomly oriented within MWT capillaries.

In the case of SWTs, only one successful filling with Ru metal has been reported, but that result appears to be consistent with what has already been observed in the case of MWTs. As shown in Figure 9a and b, the (101) planes of both the encapsulated crystallites align parallel with the SWT axis, as shown in schematic representation in Figure 9c and d, respectively. Clearly these crystallites are exhibiting preferred orientations, but it remains to be seen whether this same type of behavior will be exhibited by any other materials that can be encapsulated within SWTs.

5.2. Spiraling Crystal Growth inside Multiple-Walled Carbon Nanotubes

Polycrystalline materials encapsulated inside MWTs, whether inserted from solution or by capillary action, occasionally show a tendency to form spirals or "zigzags" inside their capillaries. Figure 12a and b shows a micrograph and schematic representation of a crystal

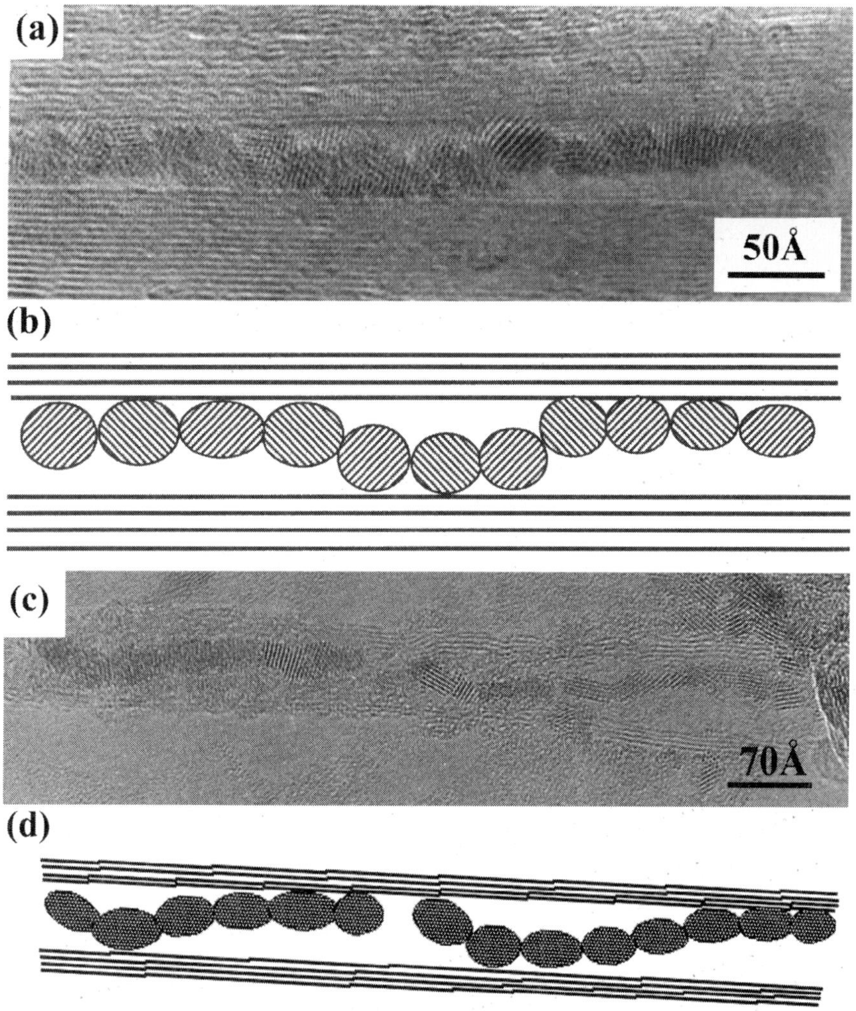

Fig. 12. (a) Micrograph and (b) schematic representation of spiraling crystal growth observed for SnO inside an MWT capillary. (Source: (a) and (b) Reprinted from [17], with permission from Elsevier Science.) (c) Micrograph and (d) schematic representation of spiraling crystal growth observed for UCl_4 inside an MWT capillary. (Source: (c) and (d) Reprinted with permission from [18].)

spiral formed from a chain of single SnO crystallites of approximately equal size observed inside a MWT capillary [17]. Figure 12c and d shows a micrograph and schematic representation of the same phenomenon observe for UCl_4, which was selectively deposited inside MWTs using composition A in Figure 6 [18].

The crystallization of polycrystalline SnO, which occurred by pH-controlled solution deposition, can be thought of as occurring according to conventional nucleation and growth mechanisms that begin on inner surface of the MWT capillary wall. A single step of this crystallization process can be seen "frozen in time" in Figure 10b. Presumably, to produce a chain of crystallites that forms a spiral, the nucleation of crystallites from solution along a given MWT capillary must occur simultaneously along a row of evenly spaced sites on the capillary wall, thus producing a chain of crystallites of approximately equal size. Then spiraling can occur according to one of the two possible mechanisms discussed in the next paragraph. In the instance of the spirals formed from UCl_4 (i.e., from composition A in Fig. 6), the growth mechanism is probably similar to that observed from aqueous solution except that the UCl_4 solid component is crystallizing preferentially from a melt also

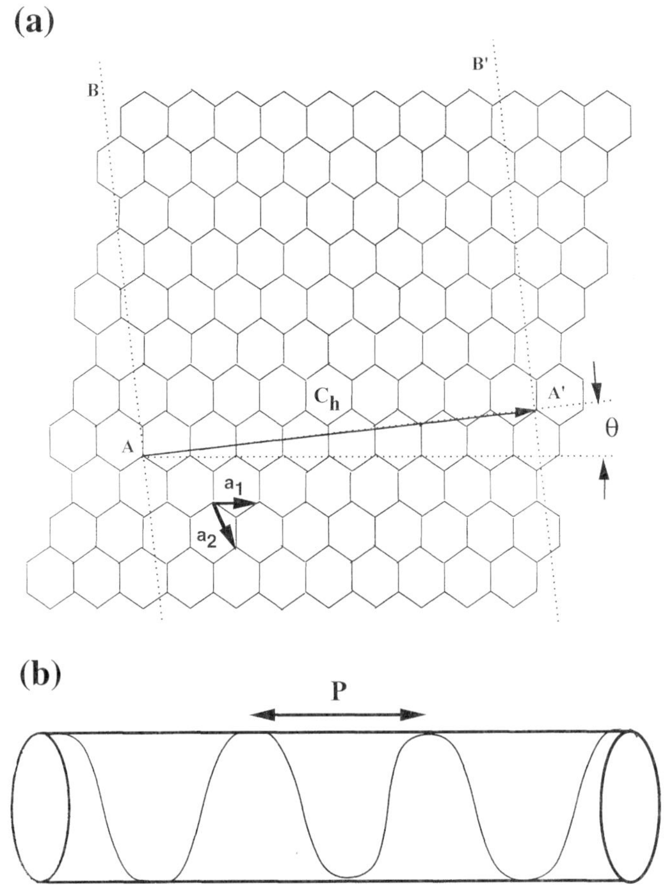

Fig. 13. (a) Schematic showing the origin of the "roll-up" vector C_h. A helical nanotube is created by connecting the two points A and A' to form a cylinder. The fundament lattice vector **T** is defined along AB. (Source: Reprinted with permission from [60]. © 1992 American Physical Society.) (b) Schematic showing the pitch **P** of a helical nanotube.

also contains the low-melting liquid component (which does not intercalate into the MWT capillaries) in the temperature range between the solidus and liquidus in Figure 6.

Two mechanisms can be put forward to explain why chains of SnO and UCl_4 crystallites form spirals inside MWT capillaries. First, the chains can spiral as a result of a Van der Waals-type interaction between the helically arranged MWT walls and the crystallite chains. If we assume a "Russian doll"-type structure for the encapsulating MWT, the diameter and helicity of the innermost graphene cylinder are uniquely characterized by the so-called roll-up vector, given by

$$C_h = n\mathbf{a}_1 + m\mathbf{a}_2 \tag{5.1}$$

that connects crystallographically equivalent sites on a two-dimensional graphene sheet, where $\mathbf{a}_1$ and $\mathbf{a}_2$ are the graphene lattice vectors, and n and m are integers (Fig. 13a) [60]. In the (n, m) notation, the vectors $(n, 0)$ denote "zigzag" tubules and the vectors (n, n) denote "armchair" tubules, whereas all other vectors (n, m) correspond to chiral or helical tubules [61]. In terms of determining the relationship between the helicity of the nanotube and the spiral arrangement of the encapsulated crystallites, the most important quantity will be given by the fundamental lattice vector, **T**, which is normal to C_h and is defined as

$$\mathbf{T} = \sqrt{3}C_h/d_R \tag{5.2}$$

where

$$d_R = \begin{cases} d & \text{if } n - m \neq 3dr \\ 3d & \text{if } n - m = 3dr \end{cases} \quad (5.3)$$

where d is the greatest common devisor of n and m [62]. For a helical nanotube, **T** must be directly related to the "pitch" of the helix of the nanotube, illustrated schematically as **P** in Figure 13b. If it can be demonstrated that **P** is related either fundamentally or by some harmonic ratio with regard to the corresponding pitch of an encapsulated spiral, then the relationship between helicity and crystal spiraling can be demonstrated. However, there are practical difficulties with establishing this relationship, not the least of which is the determination of the helicity of the innermost (interfacial) MWT cylinders, which are the ones most likely to influence the growth behavior of the encapsulated chains. This problem is further complicated by the fact that the multiple concentric cylinders within a Russian doll MWT exhibit multiple helicities [12] and, as a result, it is difficult to determine unambiguously the helicity of the innermost cylinders. The best chance of ascertaining the relationship between helicity and crystal spiraling exists for SWTs, where the helicity can be unambiguously determined. However, spiraling crystal growth has yet to be established for SWT encapsulates.

There is an alternative explanation for the spiraling behavior, which is at least as likely as the first. According to this second mechanism, chains of crystals can form spirals or zigzags as a result of localized compressive forces caused by the crystallites growing and eventually pressing together along the nanotube capillary during the crystallization process. This effect can be modeled by observing the packing of hard spheres (e.g., ball bearings) of diameter x inside a smooth vertical glass capillary tube of diameter $>0.5x$, but less than $2x$ under the influence of gravity. Under such conditions, the chain of ball bearings will always form a spiral or zigzag as the forces due to gravitation on each of the spheres is minimized. To demonstrate this relationship unequivocally, however, one must also determine the helicity of the innermost encapsulating graphene cylinder and thereby eliminate any causal relationship between MWT/SWT helicity and crystal spiraling. Other methods include the theoretical determination of the forces likely to exist during crystallization inside a nanotube capillary. Any such calculations should take into account the internal energy surfaces of the nanotube capillaries because they still may have a profound influence upon crystallization.

5.3. Crystallization Observed in Catalytically Formed Multiple-Walled Carbon Nanotubes

There is still much uncertainty concerning the mechanism of formation of materials encapsulated by the arc method. Giuerret-Piécort et al. [3] claimed that the propensity for the formation of long metallic "nanowires" rather than small encapsulates inside nanocapsules is correlated with the existence of an incomplete shell in the most stable ionic state of the element. Saito et al. [63] have indicated that there is an additional correlation for rare earths between their volatility and their ability to form encapsulates. However, Seraphin et al. [64] have indicated that neither of these models is wholly without exceptions and have advanced their own model defined in terms of the interfacial compatibility of the carbide with the encapsulating graphitic network. The phenomenon of arc encapsulation is further complicated by the fact that, in some cases, mixed and often metastable products are obtained. In the case of encapsulated manganese carbides, for example, Liu and Cowley [40] observed no fewer than four different encapsulated carbides, Mn_3C, Mn_5C_2, Mn_7C_3, and $Mn_{23}C_6$, some of which are incommensurate and presumably metastable. Giuerret-Piécort et al. [3] also observed the formation of both microcrystalline ytterbium and "spiral" dysprosium products in their nanowires, both of which indicate that their respective formation process is complex. In the case of the latter, the authors suggested that the growths of the spiral dysprosium product and helical encapsulating MWT are coupled, which is interesting in view of the preceding discussion. However, it seems unlikely that the phenomena of spiralling crystal growth observed to occur as a result of crystallization from solution (or a from a melt) and by condensation from an electric arc are directly related, given the vast differences between the respective conditions of formation.

In view of the shear complexity of products formed in the case of arc-formed encapsulates, it seems unlikely that one model or explanation alone will suffice to account for all of the crystallization behavior observed. A particular practical difficulty is that there is, at present, no effective way to observe the encapsulation process in situ. Perhaps a better approach is to consider each case individually, also taking into consideration the complex kinetic and thermodynamic mechanisms that can be obtained within a particular system.

The nature of the encapsulated products obtained in the case of gas-phase deposition onto catalytic particles is relatively simpler to interpret than those formed by arc codeposition. In general, encapsulation has been proposed to proceed via a solution-precipitation mechanism [14, 65], which involves absorption of carbon onto the surface of the catalytic particle, resulting in the formation of a small amount of interfacial carbide, thus leaving the remainder of the encapsulate in its native elemental state.

5.4. Relationship between Graphene Wall Periodicity and Crystallization inside Multiple- and Single-Walled Carbon Nanotubes

As mentioned before, Ajayan et al. have investigated the crystallization behavior of V_2O_5 inside MWTs [9]. They found that V_2O_5 continuously oriented with the [010] zone axis, or *b* repeat (equivalent to 3.564 Å), of the orthorhombic unit cell aligning parallel to the axis of the MWT capillary. The behavior of MoO_3 is very similar (Fig. 5a) with continuously oriented filling in which the 3.7-Å *c* lattice repeat is aligned with the axis of the MWT capillary [4]. These phenomena beg the question: what is the precise nature of the influence of the nanotube wall upon crystallization? Obviously considerations such as the morphological effect of the MWT capillaries, as discussed previously, are significant, but can we confirm or rule out any influence of the atomic periodicity of the MWT walls themselves?

To answer the foregoing question, we need to look again at the structure of the MWT walls, but, for simplicity, we will consider only the innermost MWT cylinders. For SWTs, only one cylinder need be considered. A zigzag or $(n, 0)$ graphene cylinder consists of a repeating graphitic lattice with a longitudinal periodicity (i.e., with respect to the nanotube axis) of 2.13 Å, corresponding to the (100) [or (010)] plane of graphite (Fig. 14a). For an armchair, or (n, n) graphene cylinder, the basic periodicity is given by the $(\bar{2}10)$ plane, that is, 1.228 Å, also indicated in Figure 14a. Between these two extreme conformations, there also will be a range of graphene cylinders whose essential structure is given by Eq. (5.1) and whose gross translational periodicity will be defined by Eq. (5.2). The latter quantity, however, will extend over several tens of graphitic units and will be unlikely to influence directly the localized periodicity within the encapsulate. The most important consideration for such nanotubes is that, about the circumference of a helical graphene cylinder, the position of the carbon network relative to an encapsulate will vary incrementally in accordance with the helicity. In other words, to the encapsulate, the inner surface of the graphene cylinder will appear like a screw thread rather than like a uniformly periodic cylinder.

We now look at how the encapsulated materials themselves interact at the MWT/encapsulate interface. In this respect, it is instructive to look at well resolved HRTEM images of continuous crystalline filling. Figure 4a shows an elongated Sm_2O_3 crystallite that completely fills the internal volume of a nanotube for a distance of ~600 Å. In Figure 14b and c, we see two enlargements taken from regions at an interval of about 200 Å along the capillary. In the enlargements, the image contrast reveals precise arrangement of the Sm^{3+} cations (which image as dark dots) along the MWT wall at the nanotube/encapsulate interface. As can be seen, the cations at the interface are arranged in a triangular motif that

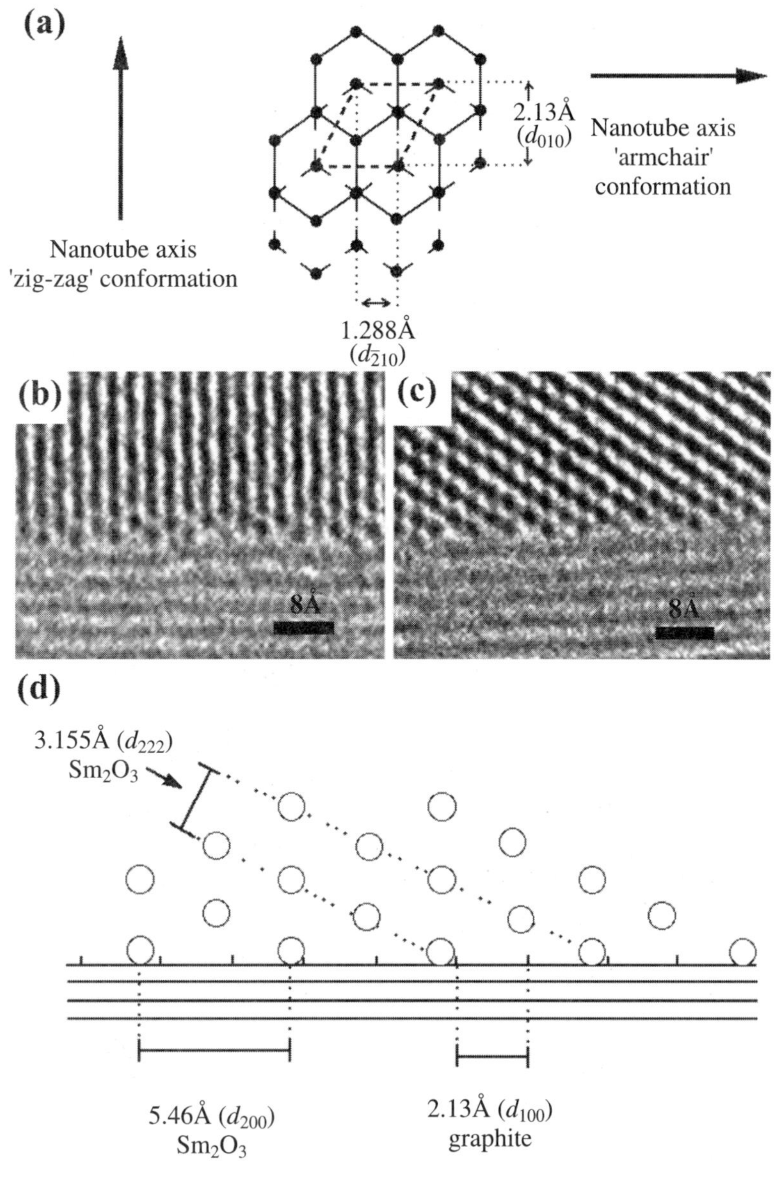

Fig. 14. (a) Schematic showing graphitic periodicity in zigzag and armchair graphene tubules. (b) and (c) Enlargements of the Sm_2O_3 MWT wall interface showing periodic stacking of Sm^{2+} cations. (Source: Reprinted with permission from [19].) (d) Schematic representation of the termination of Sm_2O_3 (222) lattice planes along the MWT wall. (Source: Reprinted with permission from [19].) The periodicity of the zigzag graphitic network indicated in (a) is shown for comparison.

extends along the wall of the carbon nanotube and is visible in both enlargements. The motif represents the point at which parallel Sm_2O_3 (222) lattice planes, arranged at an angle of ~30° to the MWT wall, terminate at precise intervals of 5.46 Å along the capillary. This distance also corresponds to d_{200} of Sm_2O_3. The atomic arrangement (cations only) is depicted schematically in Figure 14d. If we look again at the two-dimensional projection of the graphitic network corresponding to the MWT wall (Fig. 14a), we see that, for the zigzag and armchair conformations, the graphitic periodicity will repeat at 2.13 Å (shown in Fig. 14d) or 1.228 Å (not shown in Fig. 14d) along the axis of the carbon nanotube. Both are incompatible (or incommensurate) with the periodicity of the (222) lattice plane terminations as viewed in Figure 14b. Furthermore, we see that the (200) lattice planes (visible as vertical lines in Fig. 4a and 14b) are arranged at 90° with respect to the MWT wall. This precise arrangement also will be incompatible with the continuously shifting lattice about the "screw thread" along a nanotube axis for a helical graphene cylinder of any (n, m) conformation. By inspection, we see that the same principle holds for V_2O_5 and MoO_3, which observe regular periodicities inside MWT capillaries of 3.564 and 3.7 Å, respectively (see preceding text). In the case of SWTs, we have only the example of encapsulated Ru metal in which the (101) layers are apparently staggered within the SWT capillaries (Fig. 9a and b). Such an arrangement possibly would be favored inside helical SWT capillaries. However, more experimental evidence is needed before any firm conclusions can be drawn about the relationship between this type of filling and the atomic periodicity of SWT capillary walls.

In light of the foregoing observations, can we definitely rule out the direct influence of the periodicity of nanotube capillary walls upon crystallization? Perhaps a caveat should be introduced at this stage. In Figure 7c we see the homogeneous product obtained when a eutectic 1:1 KCl–UCl_4 mixture is inserted inside a MWT capillary. This is the type of product that one would expect if the same mixture had been observed crystallized outside the MWT capillary. If we now look at the same mixture (obtained from the same sample) inside a thin capillary, we see that the mixture has both crystallized and formed a continuous filling along the length of the capillary. Although this phenomenon possibly could be explained in terms of phase separation, it is important to note that the same behavior was noted for the second eutectic composition (composition C, Fig. 6) in the KCl–UCl_4 system [18] and invariably occurred in MWTs of less than 10 Å in diameter. Although this phenomenon still cannot be linked directly to wall periodicity (or helicity), it appears that the severe constriction in smaller nanotube capillaries is causing the eutectic mixtures in both cases to form metastable crystalline products.

6. CONCLUDING REMARKS

The study of crystallization inside MWTs and SWT capillaries gives some intriguing insights into the phenomenon of constricted crystal growth occurring at near atomic scale dimensions. Although the practical application of such observations may not be immediately apparent, it is quite feasible that they will be important in terms of the development of nanoscale materials in the near future. For example, the study of the morphological control of nanotube capillaries over crystallite formation contributes to an understanding of how encapsulates such as nanowires can form inside nanotubes, including both SWTs and MWTs. Furthermore, it is evident that such processes need not be restricted to nanoscale materials based solely on carbon. The study of these crystallization processes may also aid the intercalation of other fullerene-related materials, including the B–N and B–C–N nanotubes, and also other mesoporous-type materials, including zeolites, MCM-41, and related materials. Given the huge level of interest in carbon nanotubes and other nanoscale materials as a whole, this is an area of investigation that is likely to remain active for years to come.

Acknowledgments

The authors are indebted to Dr. P. J. F. Harris and Dr. S. C. Tsang of the Catalysis Research Centre at the University of Reading for the provision of Figure 3a and b and other referenced figures that appear elsewhere in this chapter.

References

1. P. M. Ajayan and S. Iijima, *Nature* 361, 6410 (1993).
2. S. C. Tsang, Y. K. Chen, P. J. F. Harris, and M. L. H. Green, *Nature* 372, 159 (1994).

3. C. Guerret-Piècourt, Y. Le Bouar, A. Loiseau, and H. Pascard, *Nature* 372, 761 (1994).
4. Y. K. Chen, M. L. H. Green, and S. C. Tsang, *J. Chem. Soc., Chem. Commun.* 2489 (1996).
5. Y. K. Chen, A. Chu, J. Cook, M. L. H. Green, P. J. F. Harris, R. Heesom, M. Humphries, J. Sloan, S. C. Tsang, and J. C. F. Turner, *J. Mater. Chem.* 7, 545 (1997).
6. J. Sloan, J. Hammer, M. Zweifka-Sibley, and M. L. H. Green, *J. Chem. Soc., Chem. Commun.* 348 (1998).
7. W. A. Deheer, A. Chatelain, and D. Ugarte, *Science* 270, 1179 (1995).
8. H. J. Dai, J. H. Hafner, A. G. Rinzler, D. T. Colbert, and R. E. Smalley, *Nature* 384, 147 (1996).
9. P. M. Ajayan, O. Stephan, P. Redlich, and C. Colliex, *Nature* 375, 564 (1995).
10. M. Freemantle, *Chem. Eng. News* 73, 6 (1995).
11. S. C. Tsang, J. J. Davis, M. L. H. Green, H. A. O. Hill, Y. C. Yeung, and P. J. Sadler, *J. Chem. Soc., Chem. Commun.* 2579 (1995).
12. S. Iijima, *Nature* 354, 56 (1991).
13. J. Rostrup-Nielsen, *Catal. Sci. Tech.* 5, 1 (1984).
14. R. Lamber, N. Jaeger, and G. Schulz-Ekloff, *Surf. Sci.* 197, 402 (1988).
15. S. C. Tsang, J. B. Claridge, and M. L. H. Green, *Catal. Today* 23, 3 (1995).
16. E. G. Bithell, A. Rawcliffe, S. C. Tsang, M. J. Goringe, and M. L. H. Green, *Inst. Phys. Conf. Ser.* 147, 361 (1995).
17. J. Sloan, J. Cook, J. R. Heesom, M. L. H. Green, and J. L. Hutchison, *J. Cryst. Growth* 173, 81 (1997).
18. J. Sloan, J. Cook, A. Chu, M. Zweifka-Sibley, M. L. H. Green, and J. L. Hutchison, *J. Solid State Chem.* 140, 83 (1998).
19. J. Sloan, J. Cook, M. L. H. Green, J. L. Hutchison, and R. Tenne, *J. Mater. Chem.* 7, 1089 (1997).
20. Y. Ando and S. Iijima, *Jpn. J. Appl. Phys.* 32, L107 (1998).
21. S. Seraphin, D. Zhou, J. Jiao, M. A. Minke, S. Wang, and J. C. Withers, *Chem. Phys. Lett.* 217, 191 (1994).
22. N. M. Rodriguez, A. Chambers, and R. T. K. Baker, *Langmuir* 11, 3963 (1995).
23. T. W. Ebbesen, P. M. Ajayan, H. Hiura, and K. Tanigaki, *Nature* 367, 510 (1994).
24. K. Hernadi, A. Fonesca, J. B. Nagy, D. Bernaerts, J. Riga, and A. Lucas, *Synth. Met.* 77, 31 (1996).
25. P. M. Ajayan, T. W. Ebbesen, T. Ichihashi, S. Iijima, K. Tanigaki, and H. Hiura, *Nature* 362, 522 (1993).
26. K. C. Hwang, *J. Chem. Soc., Chem. Comm.* 173 (1995).
27. F. Ikazaki, S. Ohshima, Y. Kuriki, H. Hayakawa, Y. Yumura, K. Takahashi, and K. Tojima, *Carbon* 32, 2539 (1994).
28. Y. K. Chen, M. L. H. Green, J. L. Griffin, J. Hammer, R. M. Lago, and S. C. Tsang, *Adv. Mater.* 8, 1012 (1996).
29. S. C. Tsang, P. J. F. Harris, and M. L. H. Green, *Nature* 362, 520 (1993).
30. R. M. Lago, S. C. Tsang, K. L. Lu, Y. K. Chen, and M. L. H. Green, *J. Chem. Soc., Chem. Commun.* 1355 (1995).
31. A. Chu, J. Cook, R. Heesom, J. L. Hutchison, M. L. H. Green, and J. Sloan, *Chem. Mater.* 8, 2751 (1996).
32. T. W. Ebbesen, *J. Phys. Chem. Solids* 57, 951 (1996).
33. V. N. Desyatnik and S. P. Raspopin, *Russ. J. Inorg. Chem.* 20, 780 (1975).
34. G. J. Janz, *J. Phys. Chem. Ref. Data* 17, 109 (1988).
35. S. Serpahin, D. Zhou, J. Jiao, J. C. Withers, and R. Loufty, *Nature* 362, 503 (1993).
36. Y. Saito, T. Yoshikawa, M. Okuda, K. Sumiyama, K. Suzuki, A. Kayasuya, and Y. Nishina, *J. Phys. Chem. Solids* 54, 1849 (1993).
37. D. Ugarte, *Chem. Phys. Lett.* 209, 99 (1993).
38. P. M. Ajayan, C. Colliex, J. M. Lambert, P. Bernier, L. Barbedette, M. Tencé, and O. Stephan, *Phys. Rev. Lett.* 72, 1722 (1994).
39. S. Subramoney, R. S. Ruoff, D. C. Lorentz, B. Chan, R. Malhorta, M. J. Dyer, and K. Parvin, *Carbon* 32, 507 (1994).
40. M. Liu and J. W. Cowley, *Carbon* 33, 749 (1995).
41. S. Motojima, S. Asakura, and M. Hirata, *Mater. Sci. Eng., B* 34, L9 (1995).
42. M. Endo, K. Takeuchi, K. Kobori, K. Takahashi, H. W. Kroto, and A. Sarkar, *Carbon* 33, 873 (1995)
43. N. A. Kiselev, J. Sloan, D. N. Zakharov, E. F. Kukovitski, J. L. Hutchison, J. Hammer, and A. S. Kotosonov, *Carbon* 36, 1149 (1998).
44. M. Yudasaka, R. Kikuchi, T. Matsui, Y. Ohki, S. Yoshimura, and E. Ota, *Appl. Phys. Lett.* 67, 2477 (1995).
45. J. Y. Dai, J. M. Lauerhaas, A. A. Setlur, and R. P. H. Chang, *Chem. Phys. Lett.* 258, 547 (1996).
46. C. H. Kiang, W. A. Goddard, R. Beyers, J. R. Salem, and D. S. Bethune, *J. Phys. Chem.* 98, 6612 (1994).
47. D. Bernaerts, X. B. Zhang, X. F. Zhang, S. Amelinkx, G. Vantendeloo, J. Vanlanduyt, V. Ivanov, and J. B. Nagy, *Philos. Mag. A* 71, 605 (1995).
48. J. Cook, J. Sloan, A. Chu, R. Heesom, M. L. H. Green, J. L. Hutchison, and M. Kawasaki, *JEOL News* 32E, 2 (1996).
49. R. C. Weast (ed.), "Handbook of Chemistry and Physics," 66th ed., p. F-120. CRC Press, Boca Raton, FL, 1985.
50. J. Cook, J. Sloan, R. Heesom, J. Hammer, and M. L. H. Green, *J. Chem. Soc., Chem. Commun.* 2673 (1996).
51. D. S. Bethune, C. H. Kiang, M. S. Devries, G. Gorman, R. Savoy, J. Vasquez, and R. Beyers, *Nature* 363, 605 (1993).

52. P. M. Ajayan, J. M. Lambert, P. Bernier, L. Barbedette, C. Colliex, and J. M. Planeix, *Chem. Phys. Lett.* 215, 509 (1993).
53. S. Iijima and T. Ichihashi, *Nature* 363, 603 (1993).
54. S. Seraphin and D. Zhou, *Appl. Phys. Lett.* 64, 2087 (1994).
55. T. Guo, P. Nikolaev, A. Thess, D. T. Colbert, and R. E. Smalley, *Chem. Phys. Lett.* 243, 49 (1995).
56. A. Thess, R. Lee, P. Nikolaev, H. Dai, P. Petit, J. Robert, C. Xu, Y. H. Lee, S. G. Kim, A. G. Rinzler, D. T. Colbert, G. E. Scuseria, D. Tomanek, J. E. Fischer, and R. E. Smalley, *Science* 273, 483 (1996).
57. S. Bandow, A. M. Rao, K. A. Williams, A. Thess, R. E. Smalley, and P. C. Eklund, *J. Phys. Chem. B: At. Mol. Opt. Phys.* 101, 8839 (1997).
58. K. B. Shelimov, R. O. Esenaliev, A. G. Rinzler, C. B. Huffman, and R. E. Smalley, *Chem. Phys. Lett.* 282, 429 (1998).
59. K. Tohji, H. Takahashi, Y. Shinoda, N. Shimizu, B. Jeyadevan, I. Matsuoka, Y. Saito, A. Kasuya, S. Ito, and Y. Nishina, *J. Phys. Chem. B: At. Mol. Opt. Phys.* 11, 1974 (1997).
60. M. S. Dresselhaus, G. Dresselhaus, and R. Saito, *Phys. Rev. B* 45, 6234 (1992).
61. M. S. Dresselhaus, G. Dresselhaus, and R. Saito, *Carbon* 60, 2204 (1992).
62. P. C. Eklund, J. M. Holden, and R. A. Jishi, *Carbon* 33, 959 (1995).
63. Y. Saito, M. Okuda, T. Yoshikawa, A. Kasuya, and Y. Nishina, *J. Phys. Chem.* 98, 6696 (1994).
64. S. Seraphin, D. Zhou, and J. Jiao, *J. Appl. Phys.* 381, 772 (1996).
65. P. E. Nolan, D. C. Lynch, and A. H. Cutler, *Carbon* 34, 817 (1996).

Chapter 10

SILICON-BASED NANOSTRUCTURES

Tamim P. Sidiki, Clivia M. Sotomayor Torres

Institute of Materials Science and Department of Electrical Engineering, University of Wuppertal, 42097 Wuppertal, Germany

Contents

1. INTRODUCTION

Well-defined nanostructures are fabricated in a variety of materials to modify material properties. The laws of quantum mechanics determine that reducing the dimensions of an electronic system to something comparable to the De Broglie wavelength, that is, the nanometer scale, will modify the electronic states because of confinement and will thus lead to different properties compared to those of the bulk material, such as optical absorption, electrical conductivity, or mechanical properties. In the last 15 years, progress in scanning probe microscopy techniques opened the way to visualizing single atoms and manipulating them. As a result, the concept of atomic engineering of mate-

Handbook of Nanostructured Materials and Nanotechnology, edited by H.S. Nalwa
Volume 3: Electrical Properties

rials has become a real technological method that has attracted much scientific and engineering effort. In parallel, several different methods were developed to fabricate and characterize structures with nanometer scales and clusters with a well-defined number of atoms.

The search for suitable device and systems applications of nanostructures is an ongoing process. In this respect, the driving force to fabricate nanostructure devices can be split into three different aspects: speed, versatility, and increased integration. An important point in fabricating nanometer active areas is the opportunity to fabricate electronic devices operating at frequencies in the 100 GHz up to the terahertz range. In optoelectronics, the realization of structures on the submicrometer scale, where quantum effects become more interesting, enables new material properties to be investigated making use of exciton confinement in a three-dimensional (3D) quantum box with distinct energy levels and their associated optical transitions yielding distinct emission lines. Furthermore, nanometer-sized active regions in electronic components could allow an increase of integration in circuits up to 10^{10} elements/cm^2.

The vast knowledge about silicon together with existing production lines makes silicon the most favored material for many potential applications of nanostructures. The efforts in nanoelectronics are to be understood in the context of future possible needs of the silicon industry. These are well documented in a number of journals, conference proceedings, and patents. In the optoelectronics area, research progress is driven by, for example, the demand to address the electrical wire bottleneck and the exploration of using Si for optoelectronic devices integrated into a Si platform. In this chapter, we focus mainly, but not exclusively, on research results aimed at fabricating Si-based nanostructures with improved optical properties.

The need for faster and more complex microelectronic devices leads unavoidably to an increase in miniaturization of devices and, consequently, to higher packing densities in integrated circuits (ICs). Current feature sizes are about 250 nm and are expected to reach 70 nm by the year 2007. However, this trend brings with itself a design problem because the density of aluminum wires connecting the various ports of the microchips increases dramatically, too, leading to what is called an "electrical wire bottleneck" in the microelectronics industry. One strategy is to replace electrical interconnects with optical interconnects between chips and between circuit boards for ultrafast connection with the expectation of overcoming the bottleneck problem. Many research teams worldwide are looking at the hybrid gallium arsenide on silicon technology. An alternative is to investigate optical interconnects based on silicon technology. Several optical components based on silicon and silicon oxide already exist, such as photodetectors and waveguides. However, for a realistic approach, it is necessary to address two basic devices: (1) a silicon-based light emitter and (2) a silicon-based detector.

Despite their indirect band gap, silicon-based structures continue to be intensively investigated for direct integration of a light-emitting diode onto chips. There are two main geometries considered for potential silicon-based optoelectronic devices: vertical structures, such as top- or bottom-emitting devices, suitable for interplane connections, and edge-emitting devices potentially suitable for in-plane optical connections. Some device ideas are based on laterally structured quantum wires and quantum dots. However, the fabrication of one- or quasi-zero-dimensional devices for device research, as opposed to physics research, remains a significant challenge mainly because of the strong device specifications expected.

Since 1990, porous silicon has evoked great interest because it shows a bright luminescence in the visible. Porous silicon is fabricated by electrochemical etching of crystalline silicon, which leads to the creation of quantum wirelike structures with modified electronic states. This has stimulated many different approaches for realizing nanostructures. A review of silicon-based optoelectronics has been published in the April 1998 issue of the *Materials Research Society Bulletin*.

This chapter is organized as follows: A brief overview of the relevant optical properties of Si and Si-based materials is given in Section 2 to lay down some of the terminology, concepts, and phenomena to be discussed later. In Section 3, we focus on structures exhibiting quantum confinement. Surfaces are intrinsically two-dimensional systems and thus constitute a good starting point. However, given that there are some excellent reviews about two-dimensional Si-based structures, we focus mainly on one- and zero-dimensional nanostructures. Much of the work referred to here aims at realizing a light-emitting device. We discuss hybrid integration of direct-gap alloys with silicon for optical as well as electronic devices. The electronic properties of nanostructures are treated in other chapters of this book; thus, our review of them will be a brief mention of the single-electron transistor in the context of Si nanostructures in Section 4. This chapter would not be complete without considering the progress made in microtechnologies; therefore, in Section 5 we discuss as a special example atomic force microscopy and field emission tips. Although not yet on the nanometer scale, it is instructive to realize how micromechanical devices are fabricated with high throughput.

2. OPTICAL PROPERTIES OF SILICON AND RELATED MATERIALS

2.1. General Remarks

Lasers and light-emitting diodes are used, for example, in optoelectronics, display devices, printers, and image processing and information storage on compact discs. Except for various kinds of silicon-based detectors, most optoelectronic devices are made of III–V compounds.

The development of modern optoelectronics and optical communications has been made possible by the outstanding light-emitting properties of III–V semiconductor compounds combined with efficient waveguiding by optical fibers. On the other hand, current microelectronic technology relies almost entirely on the use of silicon, which requires a different processing technology making the integration of optoelectronics (III–V) and microelectronic silicon devices at best cumbersome and at worst very expensive. However, bulk silicon being an indirect-band-gap semiconductor, is known to be a poor light emitter. The advent of a suitable Si-based light emitter could change this, particularly if the fabrication method of such a light emitter is compatible with silicon industry production lines. A comprehensive summary of optical and other properties of silicon can be found in the EMIS data review series volume on Si and in [1].

During the last few years, enormous progress has been made in the development of SiGe-based devices, such as double heterostructure bipolar transistors (HBTs), optical detectors, and many other devices. Good control in the preparation of this lattice-mismatched material by sophisticated growth techniques, such as molecular beam epitaxy (MBE), metallorganic chemical vapor deposition (MOCVD), and chemical vapor deposition (CVD), enables specific engineering of the desired semiconductor properties. One example is band gap engineering, realized by controlling the layer thickness and composition, resulting in a well-controlled strained or strain-symmetrized heterostructure. Although the use of SiGe strained layers has increased the electrical speed of p and n logics by means of the SiGe heterobipolar transistor, there is still no useful optical device based on Si because its indirect band gap.

Before summarizing the different approaches to realizing nanostructures, a brief overview of the optical properties of silicon is essential to understanding the effort put into silicon-based light emitters. Silicon crystallizes in the diamond structure, which consists of two face-centered cubic lattices (fcc) shifted by a quarter of the body diagonal of the unit cell. By contrast, GaAs crystallizes in the zincblende structure and, therefore, it differs from Si because Ga and As atoms each occupy a separate sublattice. This leads

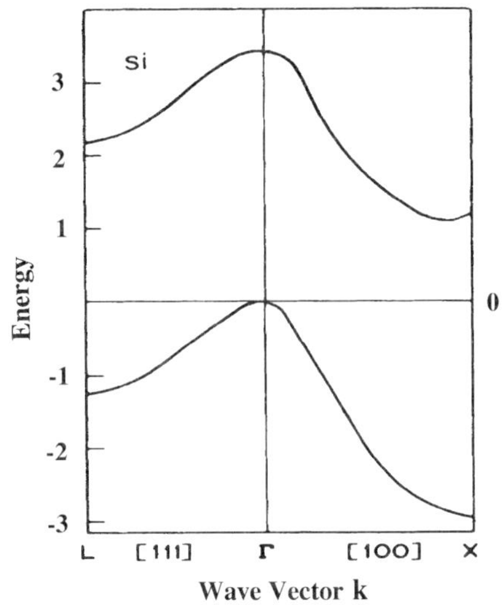

Fig. 1. Schematic band structure of Si. The top of the valence band lies at $k = 0$ and the minimum of the conduction band is at approximately $0.85X$ along the (001) direction. Thus, a phonon is involved in the recombination process.

to a break in the inversion symmetry of the GaAs lattice, which is not the case in Si [2]. The different electronic and optical properties of Si and GaAs can be derived from this structural difference, which is reflected in their respective band structure. The band structure is determined by the crystal structure, the bond type between the atoms, their bond lengths, the chemical species, the electronegativity, the stiffness, and the elasticity [3]. The band structure is the key to the dispersion relation of the charge carriers. Figure 1 shows a schematic dispersion relation for electrons and holes in Si within the first Brillouin zone. For a detailed calculation, see [4].

Note that positive and negative energies refer to electrons and holes, respectively. The valence band structure exhibits a maximum at $k = 0$ known as the Γ-symmetry point. Optical transitions are divided into direct and indirect ones, depending on whether conduction band minimum and valence band maximum occur at the same k value. Restricting the reciprocal space to the first Brillouin zone, Figure 1 shows that the conduction band minimum does not occur at $k = 0$ but at $k \approx 0.85X$ along the (001) direction. In a direct-band gap semiconductor, exciting an electron from the top of the valence band to the bottom of the conduction band, leaving a hole in the valence band, will lead to the emission of a photon upon recombination of these two carriers. The energy of this photon is the same as that of the band gap. In the case of indirect semiconductors, for example, Si, owing to translational symmetry such direct recombination is not possible because the bulk selection rule $\Delta k = 0$ has to be fulfilled. The transition from the highest occupied level of the valence band to the lowest unoccupied level of the conduction band is forbidden unless a momentum-conserving phonon participates. Such a three-body interaction is less likely than a two-body interaction. Thus, the recombination rate in pure Si is much slower compared to GaAs [5]. Therefore, the main drawback concerning Si as a light emitter is its extremely low efficiency of only 1 photon emitted per 10^7 generated electron–hole pairs.

2.2. Group IV Heterostructures: Electronic Zone Folding

One method to overcome the poor luminescence of bulk silicon was proposed in 1974 by Gnutzmann and Clausecker [6]. By building an artificial order perpendicular to the surface, a new potential term is introduced in the Hamiltonian for electron motion in the semiconductor material. The idea was to introduce this extra periodicity by growing superlattices

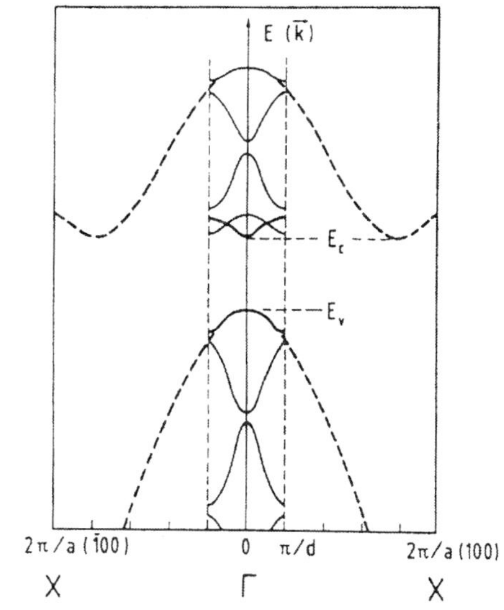

Fig. 2. Schematic illustration of the effect of Brillouin zone folding of the Si band dispersion induced by a superlattice period of about 10 monolayers of Si. The conduction band minimum near the X point is folded back onto the Γ point. (Source: Reprinted with permission from [7].)

of Si and Ge, the period d of which is given by the sum of the thickness of each layer. If the period of the superlattice is smaller than the electron mean free path, "minizones," not observed in the bulk crystal, are formed in momentum space as shown in Figure 2.

The idea of zone folding is based on an increase of the coupling of the conduction band minimum with the valence band maximum at the Γ point, resulting from the superperiodicity of the superlattice. Choosing the superlattice period as $d \approx 5/2a$, where a is the crystal lattice, one can fold the conduction band minimum onto the valence band maximum, resulting in a quasi-direct band structure (see Fig. 2). In a rigorous treatment, the presence of interfaces and the effect of strain as well as band offsets would have to be considered.

The hydrostatic pressure within the strained layers leads to a splitting of the sixfold conduction band X into two degenerate longitudinal states $X_c^{\perp}$ and four degenerated states $X_c^{\parallel}$, which show different folding behavior. The strain-induced tetragonal distortion permits the folding of the $X_c^{\perp}$ states, whereas the $X_c^{\parallel}$ states do not fold onto the Γ point [8]. For Si_nGe_n grown on Si(001), $X_c^{\perp}$ is shifted above the nonfolding $X_c^{\parallel}$ band and thus the alignment is indirect. However, Froyen et al. [8] have calculated that, using a substrate with a larger lattice constant than Si, such as $Si_{0.5}Ge_{0.5}$, will produce a nearly direct-band-gap superlattice. In this case, the Ge layers are less strained and the tetragonal distortion can be neglected. As the system Si/Ge is of type II, the transition will remain indirect in real space, even if a direct band alignment is achieved via zone folding in reciprocal space. Because increasing the superlattice period would lead to a decrease of the optical matrix element, zone folding tends to be restricted to ultrashort-period superlattices.

The first observation of superlattice-induced interband transitions was reported by Pearsall et al. [9] using electroreflectance spectroscopy of Si_4Ge_4 superlattices grown on Si(001). This observation stimulated a great deal of interest based on theoretical band structure calculations [8, 10]. Menczigar et al. [11, 12] have grown Si_mGe_n superlattices on step-graded 650-nm-thick alloy buffers followed by a 500-nm $Si_{0.6}Ge_{0.4}$ alloy film. Figure 3 shows photoluminescence spectra of Si_9Ge_6, Si_6Ge_4, Si_3Ge_2, and the $Si_{0.6}Ge_{0.4}$ alloy. The inset shows the effective mass calculations of the energy positions of the no-phonon (NP) peak. The two lines observed from the superlattices are attributed to the NP line and the transverse optical (TO) phonon-assisted interband transitions of excitons

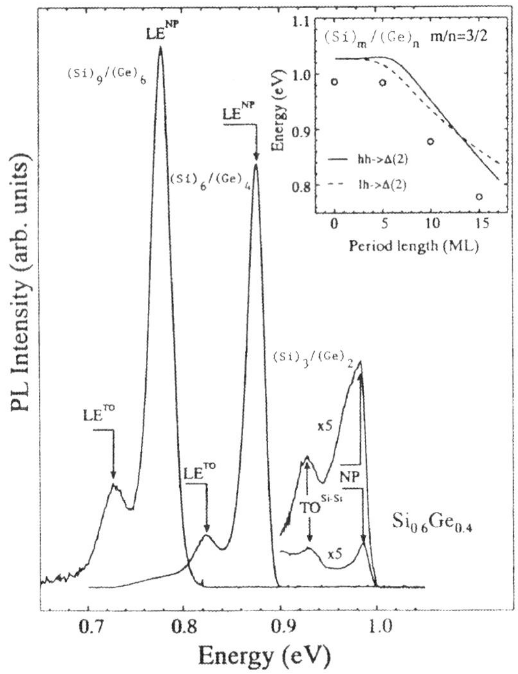

Fig. 3. Photoluminescence spectra of Si_9/Ge_6, Si_6/Ge_4, Si_3/Ge_2 superlattices and a $Si_{0.6}Ge_{0.4}$ alloy. The inset shows the energy positions of the NP lines compared to effective mass calculations. (Source: Reprinted with permission from [12]. © 1993 American Physical Society.)

localized at alloy fluctuations. The systematic blue shift of the peak positions with increasing superlattice period length fits well with the effective mass calculations, indicating the superlattice modifications of the luminescence transitions observed by Pearsall et al. [9].

Although the high luminescence intensity and the promising emission wavelength range suggested the use of Si/Ge superlattices as light emitters, the extremely low luminescence quenching temperature of about 30 to 50 K hindered applications. Afterwards, these authors also reported a reduction in luminescence intensity as well as an increased superlattice band gap after thermal annealing, which led to interdiffusion of the Si and Ge layers. Engvall et al. [13] attributed their TO Ge luminescence to Si–Ge modes originating from the interface. Their annealed samples luminesced up to about 200 K. This provided a hint that further structural improvements could eventually lead to room temperature luminescence. Two recent reviews of the properties of SiGe have appeared [7, 14] and the reader is encouraged to consult them.

2.3. The Direct-Gap Material $FeSi_2$ as a Silicon-Based Light Emitter

Whereas alloying Si with Ge or more recently also with C [15, 16] has led to modified optical and electrical properties, the fundamental band gap remained indirect. The situation changes significantly if Si or Ge is alloyed with Sn [3, 17] or if Si is alloyed with Fe to form $FeSi_2$. Both materials are predicted to exhibit a direct band alignment.

Although $NiSi_2$ and $CoSi_2$ are metallic in their stable fluorite structure, $FeSi_2$ has a 0.85-eV direct gap at room temperature in its stable structure. Hence, there is significant technological interest evoked by this feature [18, 19] as the observed band gap matches quite well with the absorption minimum of optical fibers at 1.5 μm. Because of a local Jahn–Teller distortion, determined by electron–lattice coupling, the cubic fluorite structure is modified to acquire the stable orthorhombic configuration [20]. This deformation, which is responsible for the direct gap, breaks the symmetry and, unfortunately, the

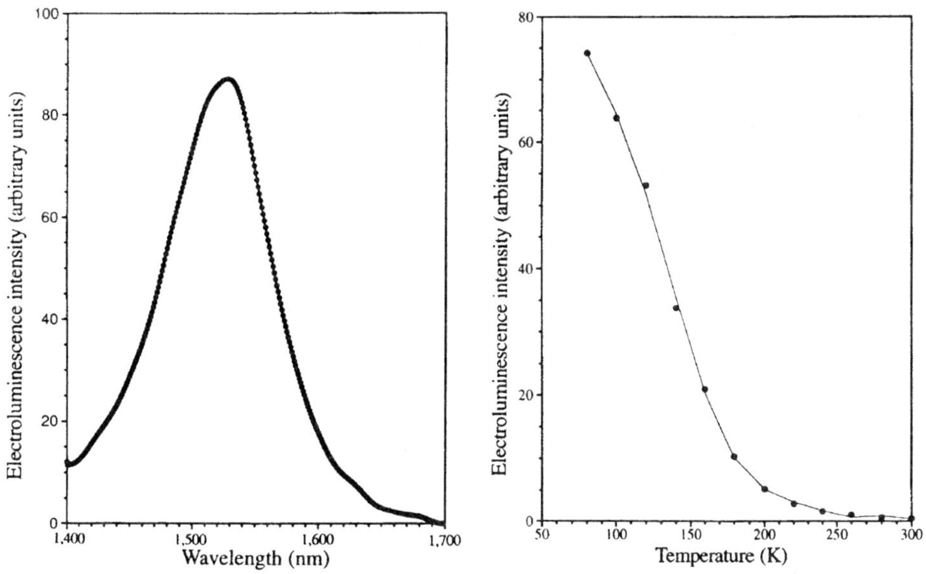

Fig. 4. Electroluminescence spectrum of a β-iron disilicide–based LED recorded at $T = 80$ K under 15 mA of forward current (left). Temperature dependence of the electroluminescence intensity, still detectable at room temperature (right). (Source: Reprinted by permission from [22]. © 1997 Macmillan Magazines Ltd.)

lattice matching to silicon (001) substrates. Thus, the crystalline quality of experimentally achieved $FeSi_2$ films is so far rather low. A possible solution would be a suitable tailoring of the Si(001) surface in analogy to graded SiGe buffer layers leading to coherent epitaxial growth.

Leong et al. [21, 22] have demonstrated a β-iron disilicide-based light-emitting diode (LED) operating at 1.54 μm that incorporates β-$FeSi_2$ into a conventional silicon bipolar junction. The silicon p–n junction was grown by molecular beam epitaxy and consisted of a 1.0-μm-thick B-doped p-type region (1×10^{17} cm^{-3}) grown on a 0.4-μm-thick Sb-doped n^+-type region (5×10^{18} cm^{-3}). The substrate was n^+-Si(001) with a resistivity of 0.008–0.02 Ω cm. The direct-gap iron disilicide is incorporated into the recombination region at the p side of the depletion region by ion beam synthesis with an implant dose of 2×10^{16} cm^{-2} and an ion energy of 960 keV. The implantation produces isolated single-crystalline precipitates after annealing at 900 °C for 18 h with typical sizes of several tens of nanometers. Because of the large size of the crystallites, there are no quantum confinement effects involved in this process. The onset of room temperature electroluminescence was determined to be around 0.8 V.

Figure 4 (left) shows the electroluminescence spectrum of a β-iron disilicide-based LED obtained at $T = 80$ K under a forward current $I = 15$ mA. The luminescence peaks at 1.54 μm with a full width at half-maximum (FWHM) of 50 meV. Figure 4 (right) shows the temperature quenching of the electroluminescence. Besides the decrease by nearly a factor of 70 from 80 to 300 K, there is still some luminescence observable at room temperature. A red shift of the electroluminescence with temperature (50 meV between 80 and 300 K) indicates a band edge–related emission and follows the band gap shift determined from optical absorption studies [23]. The external quantum efficiency of the LED was estimated to be approximately 0.1%. The LED shows no deterioration after several hundred hours of operation under continuous-wave mode at various temperatures.

The possibility of increasing the quantum efficiency of the device, the compatibility with conventional microelectronics processing technologies, and the direct-gap nature of β-$FeSi_2$ make this material system very promising for the integration of a light-emitting diode or an injection laser with an integrated circuit.

2.4. Erbium-Doped Silicon Light Emitters

Because of its potential application in Si-based optoelectronics, Er doping of amorphous and crystalline Si has recently attracted much interest. An intra-4f shell transition between the first excited state ($^4I_{13/2}$) and the ground state ($^4I_{15/2}$) of Er in its trivalent charge state (Er^{3+}) arises from a host electron–hole-mediated excitation when Er is inserted in a Si matrix [24]. The free Er ion intra-4f transitions are forbidden because of the parity selection rules. However, when Er is incorporated into a host matrix, the Er states are mixed, leading to a relaxation of the selection rule. The deexcitation process, with its sharp atomlike light emission at 1.54 μm, is of special interest as it corresponds to the minimum absorption in silica optical fibers. Ennen et al. [25] reported electroluminescence from Er-doped Si diodes at 77 K. However, because of the strong temperature quenching and the absence of luminescence at room temperature, the Er:Si system was not seriously considered for optoelectronics until very recently. Additionally, Xie et al. [40] reported a very low modulation speed of about 1 kHz resulting from the long lifetime of Er in Si ($\sim$1 ms). Owing to this long lifetime, nonradiative competing processes, that is, energy back transfer to the Si host and Auger deexcitation, strongly reduce the luminescence intensity and the quenching temperature.

Several groups have shown sharp photoluminescence [26–28] and electroluminescence [29, 30] up to room temperature. The use of Er:Si in optoelectronics was limited by the strong temperature quenching of three orders of magnitude between 77 K and room temperature [31]. Polman et al. [31] reported a very weak low-temperature luminescence of Er-implanted float-zone Si, indicating that a very small fraction of Er is in the optical active 3+ state or that only a small fraction can be pumped through electron–hole recombination. Codoping Er with C, F, O, and N modifies the Er surrounding, enhancing the Er luminescence [32] and increasing the quenching temperature [27]. Priolo et al. [33] reported a detailed analysis on the erbium–impurity interaction and its effects on the 1.54-μm luminescence in crystalline silicon and concluded that O or F codoping modifies the chemical environment of Er in a way that increases the temperature quenching and enhances the luminescence intensity owing to the enhanced donor behavior of Er.

Franzò et al. [30] fabricated p^+–n^+ room temperature silicon LEDs by erbium ion implantation. An Er concentration of 1×10^{19} cm^{-3} was implanted into the n^+ active region at 0.5–5 MeV. 1×10^{20} cm^{-3} O or F were then implanted at 0.15–1.4 MeV to modify the chemical environment of the Er ions. To recrystallize the approximately 2.3-μm-thick amorphous layer after ion implantation, the samples were annealed at 620 °C for 3 h in a N_2 atmosphere. To activate the dopants and to passivate the Si surface by an approximately 50-nm oxide, the samples were then annealed for 30 min at 900 °C in N_2. The electroluminescence intensity under reverse bias in the breakdown regime was reported to be about 20 times higher than under forward bias. Measuring the electroluminescence intensity as a function of current density reveals a linear increase without any saturation for both forward and reverse bias, the slope of the reverse-biased curve being larger. Increasing the temperature from 100 K to room temperature decreases the electroluminescence intensity in forward bias and the photoluminescence intensity for both O and F doping by a factor of 30 to 50. Recalling the three-orders-of-magnitude decrease reported by Michel et al. [26] and Polman et al. [31] codoping Er has drastically increased the quenching temperature, an effect that was also reported by Ni et al. [34].

Coffa et al. [27] suggested that the electron–hole-mediated recombination process for photoluminescence, which excites the Er^{3+} ion, is due to the recombination of excitons bound to the neutral donor level of the Er ions. Ren et al. [35] alternatively suggested that Auger recombination of a hole in the valence band with an electron in an Er-related trap level transfers energy to the intra-4f Er^{3+} shell. However, only those Er^{3+} ions, which introduce deep levels in the silicon band gap, at which electron–hole (e–h) recombination can transfer energy to the Er ion, can contribute to efficient luminescence. Owing to the similarity in intensity and temperature quenching of photo- and electroluminescence under forward bias, these authors conclude that a similar process must be responsible for the electroluminescence.

Under reverse bias, the temperature is quenched only by a factor of 7 between 77 K and room temperature. The authors attribute the luminescence process under reverse bias to the impact excitation of Er^{3+} ions from the $^4I_{15/2}$ to the $^4I_{13/2}$ level. Here band-to-band tunneling produces carriers that can be accelerated in the electric field over the depletion region to gain the required energy of approximately 0.8 eV, necessary for impact excitation. Coffa et al. [36] reported an internal electroluminescence quantum efficiency of 1.5×10^{-5} under forward bias and 5×10^{-4} under reverse bias at room temperature. Owing to its impact excitation origin, the quantum efficiency is higher under reverse bias, as Auger processes are strongly reduced because of a lower free-carrier concentration in the depletion regime. In the case of ion implantation, a subsequent annealing step is necessary to remove induced lattice damage and to activate optically the Er atoms. Enhancement of the Er incorporation in high-quality crystalline Si is one of the main targets concerning this issue. Stimmer et al. [37] and Sterna et al. [38] have reported low-temperature MBE growth of Er-doped Si layers in an oxygen ambient. Matsuoka and Tohno [39] achieved an Er concentration of 10^{19} cm^{-3} by ion beam epitaxy.

Ni et al. [34] have investigated the luminescence properties of Er/O and Er/F in MBE-grown silicon Schottky diodes. An advantage of Schottky-type LEDs is that electro- and photoluminescence can both be obtained from the same sample, enabling a direct comparison and a better understanding of the excitation and deexcitation mechanisms of Er^{3+} ions. In this work, Er was directly incorporated together with oxygen and fluorine during MBE growth by evaporating Er_2O_3 and ErF_3. Structural properties were characterized *in situ* by reflection high-energy electron diffraction, low-energy electron diffraction, and Auger electron spectroscopy; the *ex situ* characterization was performed by secondary ion mass spectroscopy, photoluminescence, electroluminescence, and cross-sectional transmission electron microscopy. The epitaxial growth was performed at a growth rate of 0.5–1 Å s^{-1} with substrate temperatures varying between 350 and 600 °C. The authors could increase the concentration of Er incorporation to 5×10^{19} cm^{-3} without Er-related precipitates. This is already more than one order of magnitude higher than the solid-state solubility limit for Er in Si reported by Xie et al. [40]. A strong Er luminescence at 1.54 μm was detected for Er/O- and Er/F-doped layers.

Figure 5 shows a comparison of the low-temperature photoluminescence with the room temperature electroluminescence spectra recorded on the same sample. At elevated temperatures, the photoluminescence is seen to decrease strongly. In contrast to the relative broad photoluminescence peak (FWHM $\approx$ 4.5 meV) from Er/O-doped samples, the Er/F samples show a number of very fine lines in the spectrum. A few of these fine lines remain with higher intensity especially after annealing. From X-ray diffraction studies, the authors concluded that Er/O doping destroys the lattice more efficiently than Er/F.

The authors have demonstrated that the very thin emission lines from Er/F-doped samples disappear both in the photo- and electroluminescence spectra of samples with high F concentration ($0.5–2 \times 10^{20}$ cm^{-3}). Furthermore, the authors found, by electrical characterization, that F-related complexes act as acceptors in Si, leading to p-type material [42]. In this kind of diode, driven in reverse bias, holes would be accelerated as hot carriers. Because of the four times lower mobility of holes in contrast to electrons, the kinetic energy that the holes could gain crossing the depletion region, is lower, leading to a more difficult excitation of the Er atoms and thus to a strong-intensity quenching already at temperatures less than 200 K. Thus, for impact excitation, Er/O doping is usually favored. Franzò et al. [30] found relatively strong electroluminescence from reverse-biased Schottky diodes at room temperature. As in the case of p^+–n^+ diodes [30], the origin of the luminescence in Schottky diodes is also ascribed to electron impact excitation. The electroluminescence line width is comparable to that from a reverse-biased Er-doped Si p^+–n^+

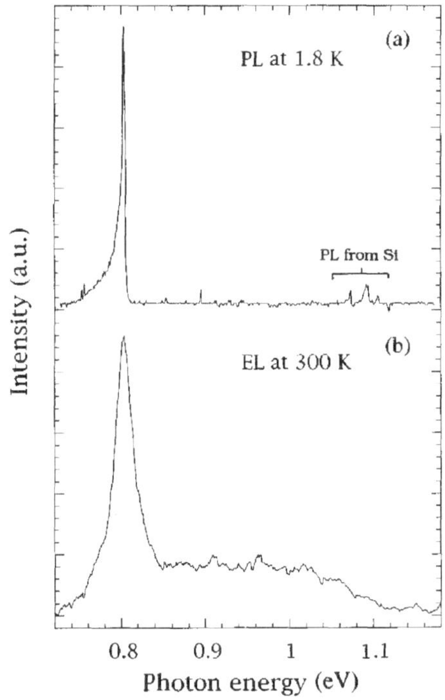

Fig. 5. (a) Low-temperature photoluminescence spectrum of an Er/O-doped sample grown at 420 °C showing intense Er luminescence compared to silicon-related emission. (b) Room temperature electroluminescence of the same sample excited by electron impact. (Source: Reprinted with permission from [34]. © 1997 American Institute of Physics.)

structure (24 meV) [37]. The broad bump at higher energies (see Fig. 5) is related to hot carrier transitions [41].

Time-dependent electroluminescence experiments on forward-biased Er-doped Si p–i–n LEDs at 30 K have revealed a slow component ($\sim$700 μs) and a fast component ($\sim$4 μs). By increasing the driving current, the fraction of the fast decay can be increased. The long decay can be attributed to the spontaneous Er emission. Ni et al. [42] fabricated a novel Er-doped edge-emitting LED with a SiGe waveguide. A B-doped p-SiGe layer acts as a waveguide for this edge-emitting device. The Er and the O concentrations were both 1×10^{20} cm^{-3}. Additionally, approximately 1% C was incorporated, which resulted in the narrowing of the band gap. Although these are recent developments, the external quantum efficiency could be increased to about 5×10^{-5}. The efficiency remains rather low for applications as a light emitter. However, work is in progress to fabricate a lasing structure, which would increase the intensity by several orders of magnitude and would open the field of real applicable optical interchip connects for Er-doped emitters.

3. QUANTUM CONFINEMENT

3.1. Two-, One-, and Zero-Dimensional Confinement

The particle–wave duality allows us to regard electrons and holes as waves. This situation is extended to other excitations in the crystal such as phonons and excitons. The wave description allows us to associate a De Broglie wavelength to charge carriers and these excitations. Quantum confinement (see Fig. 6) of charge carriers or phonons or excitons occurs when the sizes of the volume where these waves are found is comparable to the De Broglie wavelength.

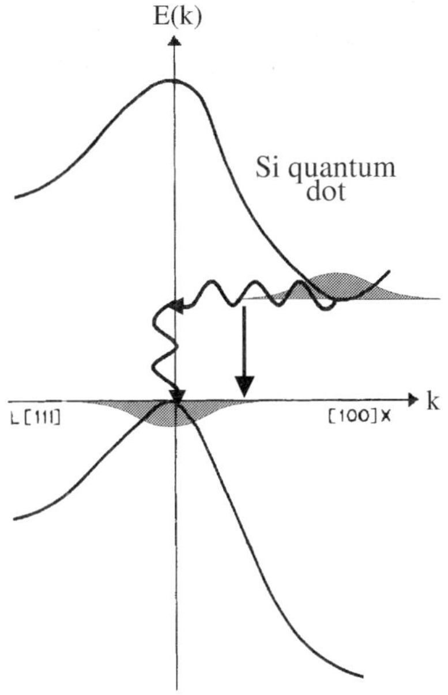

Fig. 6. Schematic illustration of the effect of spatially confining electrons and holes leading to an enhanced overlap of the carriers wave functions as discussed in the text.

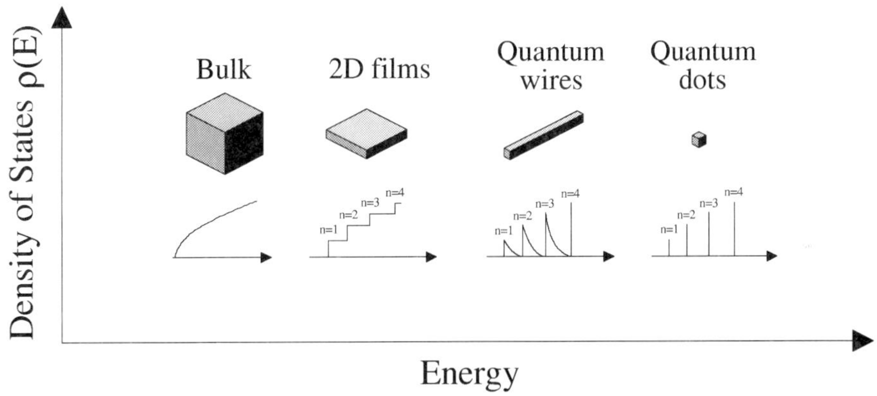

Fig. 7. Change in the idealized density of states $\rho(E)$ while reducing the system dimensionality from a three-dimensional bulk crystal to a two-dimensional layer and, finally, via a one-dimensional wire to a zero-dimensional quantum box.

The physics of low-dimensional semiconductors has been treated by several authors and the reader is pointed to some of these works [43, 44]. A key aspect of semiconductor nanostructures is the modification of the energy levels and the density of states owing to the confinement of charge carriers (see Fig. 7). The carriers behave as the well-known quantum mechanical particle in a box. Thus, only discrete energy levels are allowed, leading to a higher density of states and, theoretically, in the context of this chapter, to well-defined emission lines.

The overlap of the electron and hole wave functions plays an essential role in the radiative recombination efficiency. Because confining these quasiparticles leads to an increased overlap of wave functions in real space, there is also a concentration of the oscillator strength into the few optical transitions allowed in a quantum-confined electronic system. In addition, the recombination probability via nonradiative channels decreases

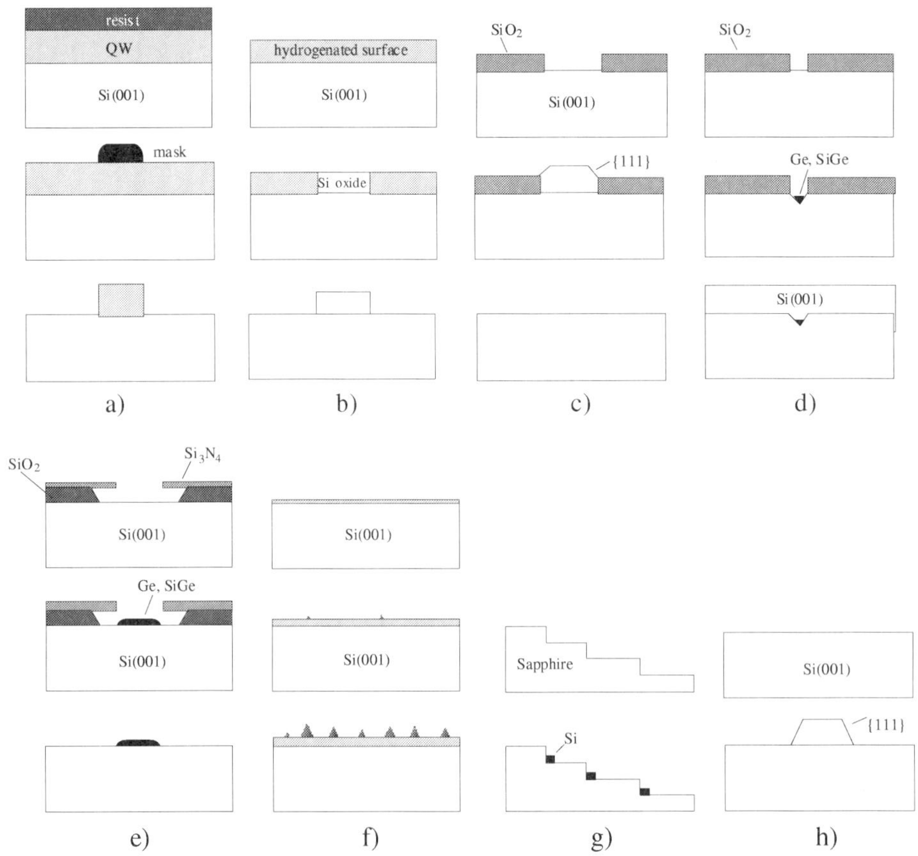

Fig. 8. Sketch of different techniques to fabricate local nanostructures: (a) electron beam or other lithography and reactive ion etching; (b) scanning probe lithography by, for example, using a scanning tunneling microscope; (c) selective epitaxy; (d) growth in V grooves; (e) growth through shadow masks; (f) self-organized growth; (g) growth on structured surfaces; and (h) anisotropic wet etching.

simply because of the lower probability of the carriers reaching nonradiative centers. However, as confinement of many particles leads to Auger nonradiative recombination processes, ideally one wants only one carrier per quantum dot. One practical approach is to use low-excitation-power densities in the case of photoluminescence experiments or lower applied current densities in the case of current injection to create fewer electrons and holes per dot and thus keep the Auger recombination probability small.

Figure 8 shows a sketch of the different fabrication processes for nanostructures. Local growth techniques are of special interest because of the possibility of achieving defect-free strain relaxation and avoiding the introduction of defects through reactive ion etching (RIE). Later in this section, we discuss these techniques and present some experimental results.

3.2. Si–SiGe Quantum Wells

Before turning to one-dimensional (1D) and zero-dimensional (0D) structures, some aspects of two-dimensional (2D) systems have to be discussed. The five group IV elements C, Si, Ge, Sn, and Pb are characterized by their large lattice mismatch resulting from the large difference in their covalent radii. This lattice mismatch leads to strain fields in the heterostructures, which, in turn, determines many characteristics such as the effective mass, the dielectric constant, and the band gap. The lattice mismatch restricts the realization of

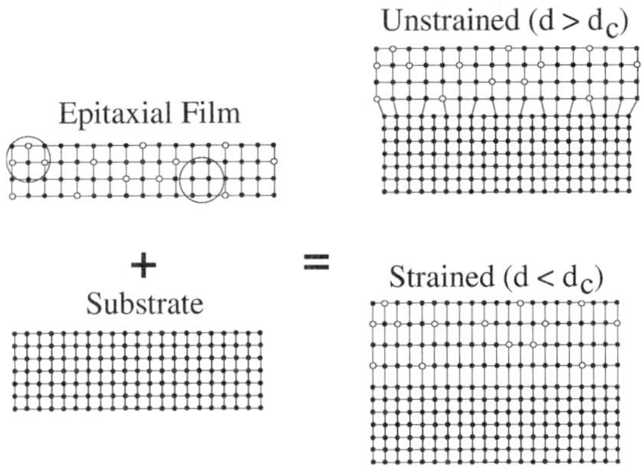

Fig. 9. Schematics of two different natural ways to accommodate lattice mismatch. The full circles denote Si atoms; the open circles represent Ge atoms. The translational symmetry is broken because of compositional disorder (e.g., in the circled regions).

high-quality heterostructures to short-period Si/Ge or to Si/SiGe superlattices. For a detailed review on the Si/SiGe system, we refer to the extensive literature [7, 14, 45]. Some promising progress has been reported on band gap engineering of SiC/Si and SiCGe/SiGe heterostructures [16].

Despite the 4.17% lattice mismatch between Si and Ge, the growth of $Si/Si_{1-x}Ge_x$ heterostructures is possible with a very high epitaxial quality. The in-plane stress is compensated by a tetragonal distortion of the lattice. There is a crossover from this elastic mismatch compensation to a plastic accommodation, if the layer thickness exceeds the critical thickness. In this case, misfit dislocations are created at the interface or threading dislocations can pass through the whole layer, leading to a relaxation of the stored elastic energy. Figure 9 illustrates these two fundamental mismatch accommodation modes.

The rapid development of Si/SiGe heterostructures, relying mainly on improved growth techniques such as MBE or CVD, enabled precise band gap engineering in epitaxially grown wafers. This feature has been used effectively in electronic applications to realize new high-speed devices such as modulation-doped field effect transistors (MODFETs) or heterojunction bipolar transistors (HBTs). Top values for f_{max} reported by IBM and Daimler Benz [46] are well in the range of 160 GHz and are comparable to state-of-the-art GaAs transistors. As many of the following Si-based nanostructures discussed in this chapter are characterized optically, we briefly discuss the major parts of the photoluminescence (PL) spectrum of Si and SiGe. Photoluminescence is very sensitive to nonradiative defects and thus it is used, in general, to probe the quality of undoped or low-doped semiconductors. The use of photoluminescence to characterize 3D and 2D semiconductors has been reviewed by Perkowitz [47]. There are a number of delicate issues such as exciton localization, exciton confinement, and their roles in the emission process. Such themes are the subject of many international conferences (e.g., the conference series "Optics of Excitons in Confined Systems") and will not be treated here.

Figure 10 shows a typical low-temperature band edge–resolved PL spectrum of a $Si/Si_{0.7}Ge_{0.3}/Si$ multiple quantum well recorded under an excitation power $P = 60$ mW. At $\lambda = 1089.9$ and 1107.7 nm, Si substrate emission lines are observed originating from the no-phonon (NP) bound-exciton (BE) recombination (Si_{BE}^{NP}) and the transverse acoustical (TA) phonon–assisted exciton emission (Si_{BE}^{TA}). The strong emission at 1128.7 nm is due to the transverse optical (TO) phonon–assisted bound-exciton recombination (Si_{BE}^{TO}), which is commonly observed in Si crystals [1]. At 1147.7 nm, the Si electron–hole droplet (EHD), emission (Si^{EHD}) is observed under higher excitation power densities. Intense luminescence lines labeled $SiGe^{NP}$ (1219.9 nm) and $SiGe^{TO}$ (1290.3 nm) are the no-phonon

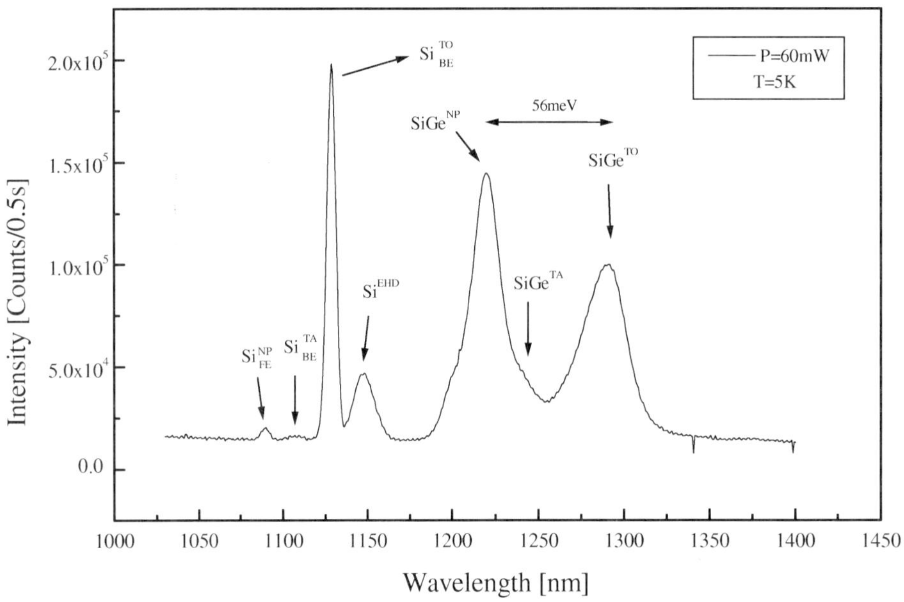

Fig. 10. 5-K PL spectrum of a 15-period Si(3nm)/$Si_{0.7}Ge_{0.3}$(3nm)/Si(3nm) multiple quantum well structure recorded under an excitation power of 60 mW.

and the transverse optical phonon–assisted emission, respectively, originating from SiGe layers, as shown by several authors. Note that the energy difference of 56 meV between the NP and TO lines closely matches the Si–Si TO phonon energy as expected. The Si–Ge and Ge–Ge TO lines are not resolved in this spectrum. Transverse acoustical phonons are also observed in the same spectrum between the NP and TO emission lines.

The strong no-phonon emission is a manifestation of the broken translational symmetry within the SiGe layers, as has already been reported by Weber and Alonso [48]. This occurs via scattering because of compositional disordering (Fig. 9) in the SiGe well layer, leading to a momentum-conserving transition. However, the emission intensity from SiGe quantum wells is still too low for use in LEDs.

Nanofabrication of these two-dimensional systems is under investigation to confine the excitons further and in more dimensions in an attempt to increase the emission intensity.

3.3. Porous Silicon

Current interest in porous silicon was initiated by the demonstration that a simple electrochemical etching process leads to intense luminescence from silicon in the visible region of the spectrum even at room temperature [49]. As the fabrication by an anodic dissolution in a HF-based electrolyte is very simple, porous silicon has been intensely investigated. The structural characteristics of porous silicon exhibit extremely fine tubular structures, viewed as a mesh of wires or dots, which are small enough to exhibit some quantum confinement effects.

Transmission electron microscopy (TEM) studies [50] have demonstrated that under suitable conditions this etching process can generate isolated columns of width less than 5 nm (see Fig. 11). With such small sizes, quantum size effects are expected, which Canham [49] has suggested are responsible for the higher luminescence intensity and emission energy of porous Si in contrast to bulk silicon. The small cross section of the wires also means that the number of nonradiative centers per unit length of wire is very small, irrespective of these centers being distributed over the surface or throughout the bulk [3]. Migration of excitons to these nonradiative defects is suppressed by exciton localization [49].

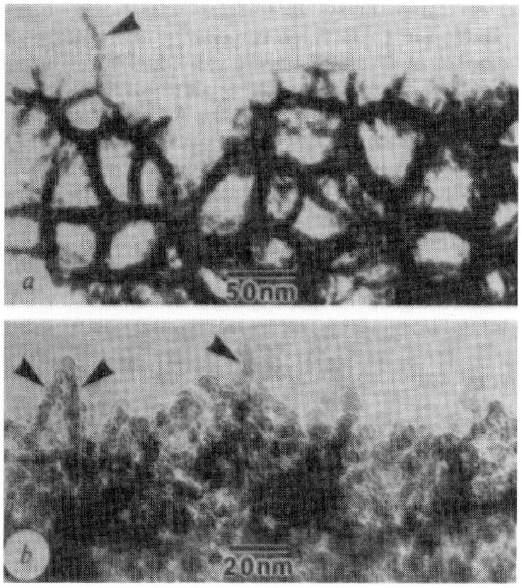

Fig. 11. Transmission electron micrographs of thin Si layers with 80% porosity. The narrow Si columns within the irregular matrix are marked with arrows. (a) Sample with weak luminescence and (b) sample with strong luminescence. (Source: Reprinted by permission from [50]. © 1991 Macmillan Magazines Ltd.)

The assignment of the visible photoluminescence of porous silicon to quantum-confined carriers in crystalline silicon was controversial for a long time but is now generally accepted. Alternative suggestions were, for example, that the photoluminescence was due to siloxene [51], a silicon-based polymer ($Si_6O_3H_6$), or due to polysilanes or hydrates at the surface [52]. The problem with these models was the experimental result that the material exhibited intense red photoluminescence even after high oxidation temperatures, which modified the silicon surface. Other groups agree with the quantum confinement model, but propose quantum dot structures [53] instead of Canham's quantum wires. Detailed characterization by several researchers has confirmed that the filaments are single-crystallites. Bassous et al. [54] have demonstrated a clear quantum confinement effect as there is a blue shift of the photoluminescence signal with reduction of the porous silicon features sizes caused by subsequent HF etching.

The width of the rather broad luminescence peaks ($\approx$0.4–0.6 eV) results from quantum size fluctuations. It is quite possible that the luminescence is at least partly surface defect related, as hydrogen passivation leads to an enhancement of the luminescence. This point is backed up by the observation that high annealing temperatures cause a strong quenching of the luminescence as hydrogen is desorbed. To reduce the surface recombination velocity and thus to keep the radiative lifetime long in comparison with the lifetime of nonradiative channels, it is necessary to terminate the dangling bonds at the silicon surface with hydrogen [49]. The temperature dependence of the photoluminescence intensity hints at some interesting possibilities to explain the luminescence process in more detail. The maximum in the photoluminescence intensity at $T = 200\,°\mathrm{C}$ [55] and the fact that the luminescence is completely quenched at low temperatures indicate the possibility of a defect level quenching or an interplay of radiative and nonradiative channels, where the latter is dominant at low temperatures.

However, it is more likely that the still indirect character of the radiative transition involves additional phonons for momentum conservation reasons and that these phonons are depopulated at low temperatures. Using PL spectroscopy under resonant photoexcitation, Calcott et al. [56] showed that phonons are involved in the luminescence process and that porous silicon thus remains an indirect band gap material. From the experimental point of view, the origin of the luminescence can be traced to the existence of quantum size crystalline structures. Detailed theoretical studies have been carried out to identify the confinement character of the luminescence.

Based on experimental results, Buda et al. [57] have performed first-principle calculations of the electronic properties of (001)-oriented silicon quantum wires with diameters $d = 0.76$, 1.14, and 1.56 nm using the density functional theory within the local-density approximation for the exchange correlation energy. As in the experimental case, they have assumed the silicon dangling bonds to be passivated at the surface by hydrogen atoms. In contrast to bulk silicon with its indirect gap of 1.17 eV, here a direct gap of 1.84 eV appears at $k = 0$. For the external field polarized in the direction of the axis of the wires, calculations of the imaginary part of the dielectric function $\varepsilon_2(\omega)$ exhibit a peak in the range of visible light. This feature cannot be identified with any bulklike electronic state of crystalline silicon. The dependence of this peak on the wire size was correlated with the behavior of the luminescence peak in porous silicon. Moreover, the extrapolation of the luminescence energy as a function of the wire thickness covers the energy range in which porous silicon is seen to luminesce.

Although the breakdown of the momentum selection rule enhances the interband optical matrix element, this is not the main reason for the radiative efficiency of porous silicon. Read et al. [58] have calculated the radiative lifetime of localized excitons in quantum wires to be in the range of 380 μs, which is much longer than the lifetime of 1 ns for bound excitons in GaAs [58, 59]. Here the enhancement of the exciton binding and localization energies by confinement and the low rate of nonradiative recombination are considerable. However, the luminescence behavior of porous silicon does depend strongly on the exact sample treatment during sample preparation. In the case of an array of quantum wires in porous silicon, samples etched for a shorter time should have a distribution of larger wire diameters. However, Pearsall et al. [60] did not identify any red shift of the luminescence peak with decreasing etch time. Instead, they only measured a decrease in the luminescence intensity as the etch time decreases. The results of this group suggested that a film of strained [61] luminescent material such as a fractal phase of silicon or a silicon complex [62] is created or redeposited on the surface during etching.

Quantum confinement effects in porous silicon are no longer disputed. The simple etching of silicon does not work if using n-doped silicon as the required holes for the etching process are no longer available. To fabricate porous silicon out of an n-doped wafer, one has to illuminate the sample with light to create electron–hole pairs. Among others, Eisebitt et al. [63] have prepared porous silicon under different wavelengths of illumination. Using soft X-ray emission, absorption spectroscopy, and X-ray fluorescence measurements, they could selectively probe the silicon core of oxidized nanostructures by suitable choice of the excitation wavelength. The changes in the valence and conduction bands could be explained by the changes in the electronic structure associated with quantum confinement in silicon nanostructures.

The interpretation of the luminescence from porous Si is made more difficult by the presence of defects in the silicon oxide that act as radiative centers. To analyze this, Ge nanoparticles were fabricated by Nozaki et al. [64] using the cluster beam evaporation technique. These nanoparticles have the same diamond structure and they also have an indirect band gap. In contrast to Si, however, Ge shows a less natural oxidation. Saito [65] has reported a mixture of diamond and tetragonal structure in Ge nanoparticles when their size is smaller than 20 nm. This phase transition occurs due to a minimization of the surface energy, as the diamond structure is less closely packed [64, 65]. Nozaki et al. [64] attributed the blue luminescence ($E = 2.9$ eV) of the Ge nanoparticles to defect centers in the Ge oxide formed by photooxidation. In addition, quantum confinement could be excluded as there was no energy shift with adjacent photooxidation.

Although some progress has been made concerning the potential integration of porous silicon–based light emitters with microelectronics, two fundamental problems affecting

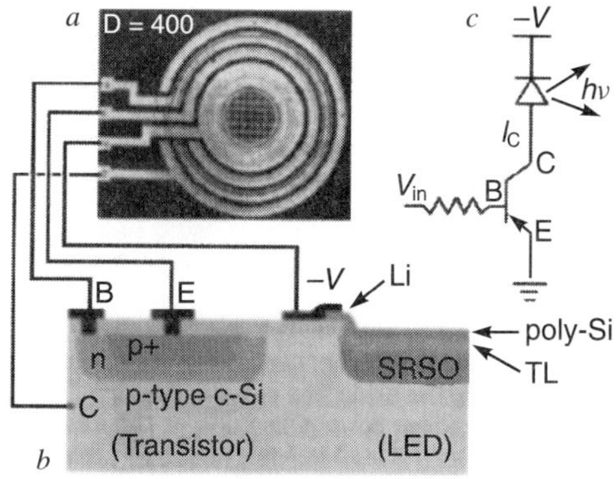

Fig. 12. Sketch of the integrated porous silicon–based LED with the driving bipolar transistor. (a) Top view of the structure with a 400-μm active light-emitting area. (b) Cross section from the center of the LED at the right side to the transistor at the left side. (c) Electronic circuit. (Source: Reprinted by permission from [68]. © 1996 Macmillan Magazines Ltd.)

porous silicon are its extreme fragility, which leads to a degradation of the more than 5% PL quantum efficiency in about 1 h under ambient conditions [66], and the relative low modulation speed. Moreover, the reported values of electroluminescence (EL) efficiency were more than 100 times smaller than that of PL, with even faster degradation times. The reason behind this is inefficient carrier transport through the irregularly built structure of porous silicon. The lack of robustness during high-temperature cycles may seriously limit the potential use of this material. However, Loni et al. [67] enhanced the room temperature EL efficiency to more than 10^{-3}.

First Integration of a Porous Silicon–Based Light-Emitting Diode with a Conventional Bipolar Transistor. Despite its fragility and low-temperature resistivity, the integration of porous silicon with conventional silicon technology was recently demonstrated. Hirschman et al. [68] demonstrated the integration of a porous silicon–based LED with a conventional bipolar transistor as a driving transistor. They enhanced the thermal (~1000 °C) and chemical stability of porous silicon by partial oxidation [69]. With this silicon-rich silicon oxide (SRSO) process, the light emission and electrical characteristics still retain the desirable device stability, which is higher than that of porous silicon [68]. Figure 12 shows the optoelectronic circuit that integrates the SRSO-based LED with a surrounding bipolar driver transistor. In an emitter configuration, a base input signal is amplified by the transistor and controls the current flow through the diode. Thus, the diode can be directly modulated.

These authors integrated several diodes with active areas ranging from 0.005 to 2 mm^2. Typical of porous silicon is the rather broad electroluminescence ranging from 1.4 to 2.3 eV. The luminescence onset lies at 2 V. The highest noted external power efficiency is 0.1%. Although the device stability has been tested over several weeks without any degradation, the modulation bandwidth of 1 MHz still remains extremely low.

3.4. Postgrowth Nanofabrication by Lithography and Etching

The feature size limit set by current lithography is well within the quantum size regime using state-of-the-art pattern transfer sizes down to a few 10 nm. At present, nanostructure devices are fabricated by a pattern definition technique, for example, electron beam lithography, followed by a pattern transfer using an etching technique such as reactive ion etching. An outline of this process is shown in Figure 13. Considering a positive resist such as polymethylmethacrylate (PMMA), the fabrication process basically consists of (a) spinning the resist on the as-grown 2D substrate, for example, a multiple quantum well. A pattern consisting of wires and or dots can be written on the resist with a focused

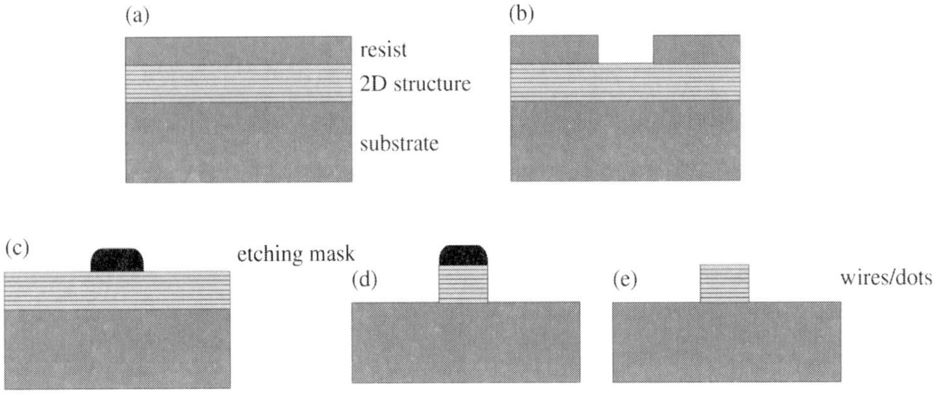

Fig. 13. Schematic view of the nanofabrication process used for fabricating quantum wires or quantum dots in the range of approximately 50 nm (see text).

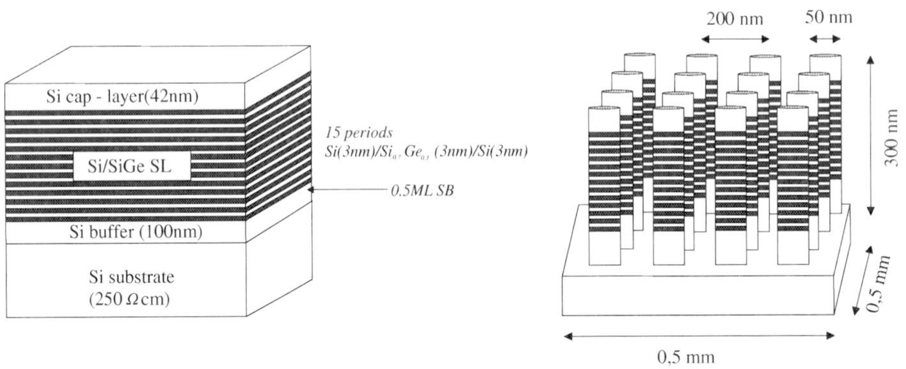

Fig. 14. Sketches of the as-grown $Si_{1-x}Ge_x$ multiple quantum wells and the dot array.

electron beam and then undergo development in a solvent (b). After depositing an etching mask, the resist is lifted off (c). A highly anistropic reactive ion etch process is then used to transfer the pattern onto the as-grown wafer (d). In the final step, the etch mask is removed if necessary. The advantage of this technique is the good size control of the nanostructures and flexibility in the nanostructure pattern design. However, the nonnegligible damage introduced by the etching procedure could alter dramatically the electrical and optical characteristics of the material. Even with an optimized low-damage RIE recipe, damage, such as an amorphous layer on the surface of the nanostructures, will persist along with possible point defects inside the semiconductor.

Tang et al. [70, 71] reported an increase in the electroluminescence properties of reactive ion-etched $Si/Si_{1-x}Ge_x$ quantum dots by a factor of 100. For this work, the structures shown in Figures 14 to 18 were used. Figure 14 illustrates the as-grown wafer and nanostructured dots. The highest luminescence intensity was observed for wafers consisting of a 15-period $Si/Si_{0.7}$(3 nm)$Ge_{0.3}$(3 nm) multiple quantum well structure. The pitch size of the dots was about 200 nm and each dot had a diameter of 50–70 nm and a height of 300 nm. Thus, very dense dot arrays could be fabricated. The single-array areas with a high dot density were about $500 \times 500\ \mu m^2$.

Figure 15 shows a scanning electron microscopy (SEM) micrograph of Si/SiGe dots fabricated by electron beam lithography and RIE in $SiCl_4$. At this stage, photoluminescence experiments can be performed to characterize the dot array, provided the patterned sample is nominally undoped. In several samples grown by MBE on Si(001) and on strain

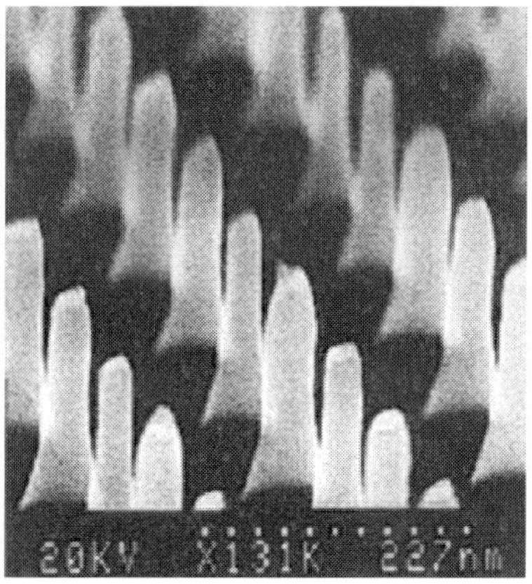

Fig. 15. SEM micrograph of a $Si/Si_{1-x}Ge_x$ quantum dot array.

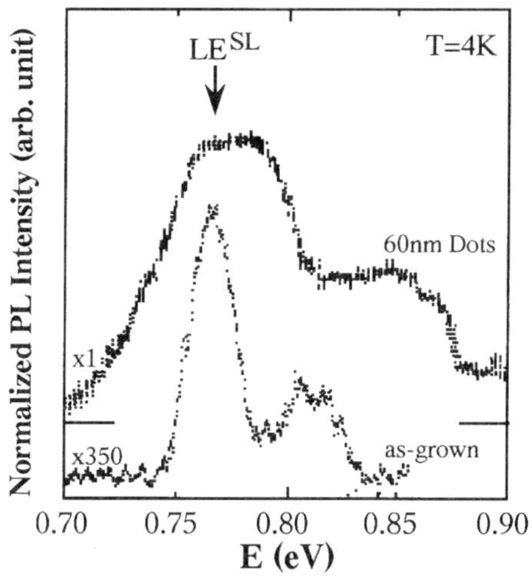

Fig. 16. Comparison of 4-K photoluminescence spectra from an as-grown Si_9/Ge_6 superlattice and a 60-nm dot array from the same sample.

symmetrized substrates of Si with a SiGe superlattice, a similar observation was made: The 4-K photoluminescence intensity of samples patterned into dot arrays is 100–200 times stronger than that of the as-grown samples. One example is shown in Figure 16, which shows a photoluminescence increase of more than two orders of magnitude of photoluminescence from 60-nm dots compared to the as-grown sample. The lower curve is the reference spectrum. The intense peak at 0.7654 eV is attributed to the localized intrinsic exciton from the superlattice (LE^{SL}) as already reported by Menczigar et al. [12]. The broadening of this peak after nanofabrication is explained as an effect of process-induced strain.

For electroluminescence, electrical contacts on the top and at the bottom are needed. Prior to depositing the top contact, the dots have to be planarized with a suitable low-stress insulator. Tang et al. [72] have reported the crucial role of strain on the observed enhanced light emission. A 3-nm-thick $Si_{0.78}Ge_{0.22}$ quantum well was grown on float-zone (FZ) Si and capped by 120-nm Si. After nanofabrication into 50-nm dots, different dot arrays were coated with amorphous SiN_x by PECVD at 300 °C. The built-in stress of the SiN_x coating was varied by changing the radio frequency (rf) power, and measured by laser interferometry on a simultaneously codeposited SiN_x film on a glass substrate.

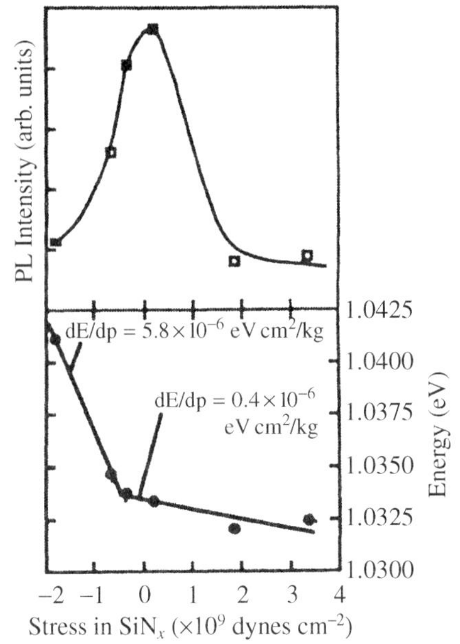

Fig. 17. Photoluminescence intensity and peak energy of the SiN_x-coated dots as a function of the stress in the planarizing layer.

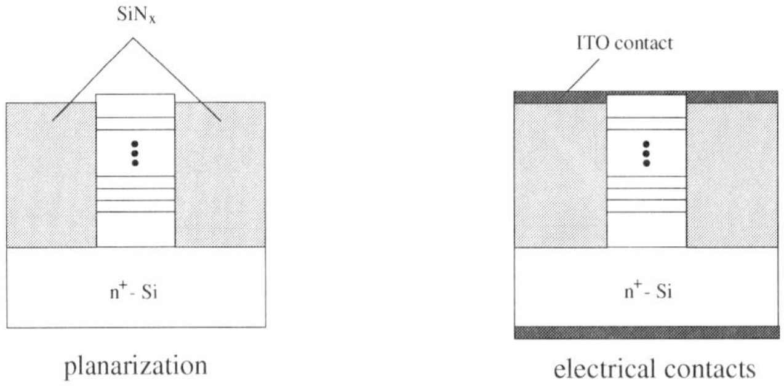

Fig. 18. Planarization of the dot array with SiN_x (left) and electrical contacts with ITO (right).

Figure 17 shows the photoluminescence intensity of the dot arrays recorded at 6 K together with the peak energy as a function of the corresponding stress of the SiN_x film. When the stress in the SiN_x is nearly zero, the spectrum shows the maximum photoluminescence efficiency, close to that of uncoated dots. External induced tensile as well as compressive strain drastically decreases the luminescence intensity. Thus, there is an optimum regime for the SiN_x deposition with respect to maximum luminescence yield. These experiments on the stress of the SiN_x coating were very important, as in previous tests polyimide had led to unstable and shifted current–voltage (I–V) characteristics because of a leakage current under bias. Using high-stress PECVD SiN_x removed these problems, but decreased the emission. On the other hand, stress-free SiN_x reported in [72] did not alter the light emission strength and exhibited acceptable I–V characteristics. Figure 18 (left) shows a schematic of the planarized structure.

Finally, a 70-nm NiCr layer was deposited as electrical top contact. Experiments are in progress to replace NiCr by an optically transparent and conducting indium tin oxide (ITO), to be used as a mask and ohmic contact in future tests. In Figure 18, NiCr has already been replaced by ITO. Figure 19 shows the electroluminescence intensity of the

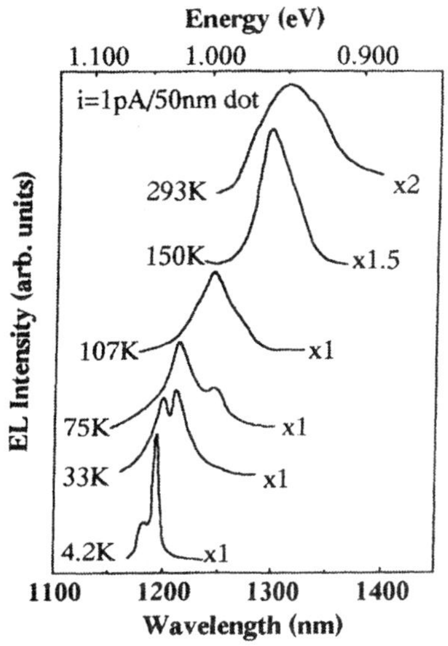

Fig. 19. Electroluminescence spectrum of the $Si/Si_{1-x}Ge_x$ dot array at different temperatures. This array had polyimide as the insulating material and a 70-nm NiCr top contact. For reference, the spectrum of a large mesa structure is also shown.

dots at different temperatures, compared to a low-temperature spectrum of the as-grown wafer. These spectra were recorded at an injection current of 1 pA/dot under reverse bias, 0.5 V at 293 K and 2.0 V at 4.2 K. At room temperature, an external quantum efficiency of 0.14% was estimated [70].

The results of the luminescence data have shown the crucial role of strain during the fabrication of low-dimensional structures. Tang et al. [73] found that the smaller the nanostructures, the bigger is the strain relaxation. This observation was deduced from the Raman shift of the Ge–Ge, Si–Ge, and Si–Si modes of 50-nm dots with different Ge concentration, which were accounted for by a 50% strain relaxation.

Figure 20 shows the Ge content–dependent Raman shifts of the Si–Si, Si–Ge, and Ge–Ge modes for different dot arrays of similar size. The solid line is the calculation of a corresponding unstrained pseudomorphic structure; the dashed line is the calculation with a 50% strain relaxation [74]. The solid circles represent the experimental data. In addition to the strain relaxation resulting from material removal through etching, a thin SiGe layer, redeposited on the walls of the dots, may force the atoms to rearrange and induce a lattice distortion inside the dots [75]. This hypothesis is supported by Raman spectra from etched regions away from the dot arrays pointing to the existence of a thin SiGe alloy being redeposited all over the remaining silicon substrate.

To investigate the effect of size reduction on strain relaxation inside the Si/SiGe nanostructures, a series of 2.5-μm-long and 10-to-500-nm-wide wires were fabricated of a modulation-doped p^+-$Si/Si_{1-x}Ge_x$ heterojunction structure [76]. Figure 21 shows room temperature photoreflectance spectra for wires with different widths compared to a mesa control sample. In this spectrum, $11h$ represents Γ-like 2D interband transitions between the first electron subband and the first heavy-hole subband. On reducing the wire width down to about 100 nm, the $11h$ transition is red-shifted, indicating a strain relaxation of the original comprehensive strain in the SiGe layer of the heterostructure during wire fabrication. Further reduction of the wire width to 40 and 15 nm leads to a blue shift because of lateral confinement effects, indicating a clear 2D–1D transition [77]. Similar phenomena were also observed in wires cut from Si–$Si_{1-x}Ge_x$ multiple quantum wells [78]. However, a quantitative analysis of the strain and confinement shifts was not carried out.

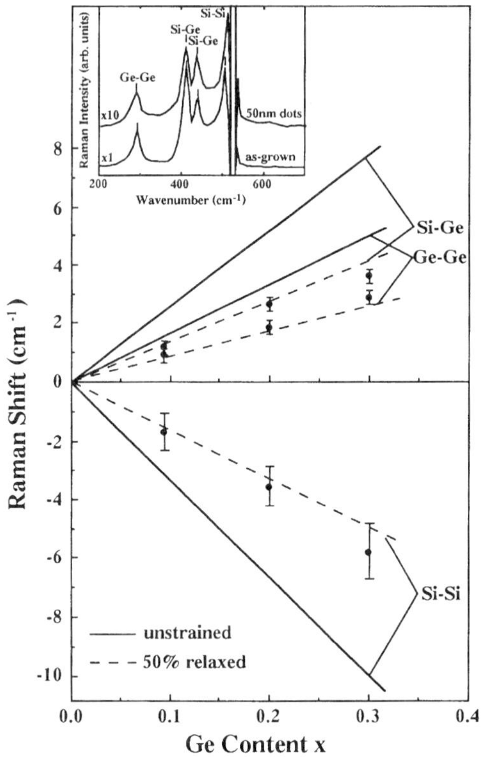

Fig. 20. Ge content dependence of the Raman peak shift of the dot arrays compared to the as-grown superlattice. The inset on the top left shows the Raman spectra of a dot array and the corresponding as-grown samples cut from a Si–$Si_{0.7}Ge_{0.3}$ superlattice. The sharp features at around 521 cm^{-1} are due to the removal of the Si TO phonon peak from the substrate.

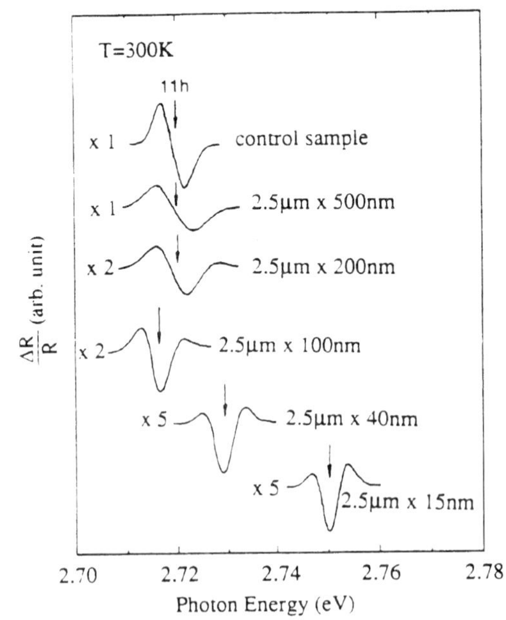

Fig. 21. Photoreflectance spectra of a control sample and etched wires with different widths. The top spectrum corresponds to the control sample. Wire widths range from 500 nm (second scan from top) to 15 nm (bottom spectrum).

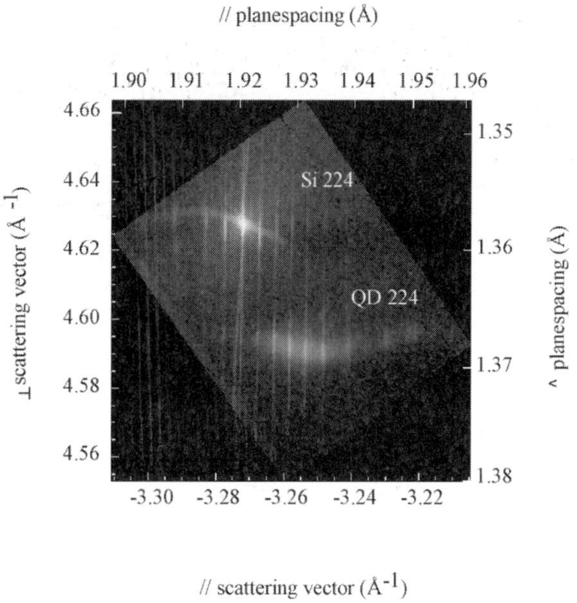

Fig. 22. Intensity distribution of a 2D RSM obtained around the (224) reflection from an array consisting of 50-nm dots. (Source: Courtesy of W.-X. Ni and J. Birch.)

Generally, the smaller the nanostructure, the more the strain relaxes in strained-layer heterostructures.

Nondestructive X-ray diffraction is widely used to map the two-dimensional reciprocal space to characterize ordering, crystallinity, or layer sequence in superlattices. However, the relatively low diffraction intensity from low-dimensional structures makes necessary the use of synchrotron radiation for structural analysis. In comparison to reciprocal space mapping (RSM) of two-dimensional superlattices, which shows satellites resulting from the periodicity of the superlattice in the q_z direction, in the case of quantum dot arrays additional satellites are found in the $q_{\|}$ direction, which originate from the periodicity of the dots within the array. Figure 22 shows the 2D reciprocal space maps around the Si(224) reflection recorded at the European Synchrotron Radiation Facility. The spacing of these satellites is, as usual, related to the inverse of the dot distance in real space. Ni et al. [79] have found a reduction of the mean lattice mismatch of the dots to the substrate in the growth direction from 0.0113 for a fully strained 30-period Si(3 nm)/$Si_{0.7}Ge_{0.3}$(3 nm) 2D superlattice to 0.0070 caused by strain relaxation after structuring the sample into 50-nm dots with a height of 300 nm and a pitch size of 130 nm. Such a lattice relaxation was only detected within dot structures compared to an approximately 0.1×1 mm^2 reference 2D Si/SiGe area. From 2D RSM around an asymmetric reflection, the authors also determined a mean in-plane lattice mismatch of 0.0059 to the substrate.

A partial strain relaxation consistent with the previously mentioned Raman experiments led to the difference between these two values. The clear observation of higher-order satellites of the superlattice within the dots indicates that the superlattice layers are still well ordered after the partial strain relaxation. However, the existence of some weak diffuse scattering in between the superlattice reflections raises the possibility of the presece of amorphous layers surrounding the dots. This may be the result of crystal damage caused by the rather crucial reactive ion etching or from a redeposited SiGe alloy surrounding the dots.

Using bulk elastic theory for a homogeneous deformation, the authors found a partial relaxation in the SiGe layers, whereas the Si layers are biaxially expanded. This strain

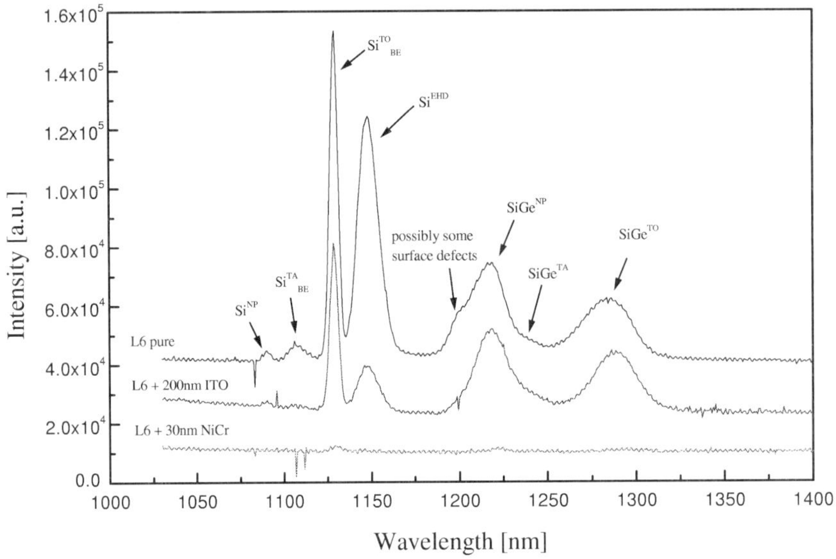

Fig. 23. Photoluminescence spectra of a $Si/Si_{1-x}Ge_x$ mesa structure (top curve) and of the same structure coated with 200-nm ITO (middle curve) and 30 nm NiCr (bottom curve).

symmetrization affects the band offsets at the heterointerfaces in becoming type II. The strain induces a splitting of the sixfold degenerated conduction band. Edge electrons in a $\Delta(2)$ band located near $k_{\|} \approx 0$ can recombine with holes located at $k_{\|} \approx 0$ across the interface without additionally involving a phonon for the required momentum conservation. This is probably a reason for the reported enhanced luminescence properties of Si/SiGe quantum dots.

Considering the almost 94% absorption of infrared photons in the top NiCr layer, the luminescence intensity could potentially be increased by a factor of 10 by replacing NiCr with ITO. Figure 23 shows a 5-K PL spectrum of a $Si/Si_{1-x}Ge_x$ mesa structure and of the same structure coated with ITO and NiCr. The low absorption of about 5% with a still good sheet resistance of 2000 $\mu\Omega$/square is noteworthy. An additional advantage of this layer is also pointed out in Figure 23. Higher excitation powers revealed surface defects, which are visible in the PL spectrum as a small shoulder at the smaller wavelength side of the $SiGe^{NP}$ line. The shoulder, which is only observed at higher excitation powers, is not detectable when the ITO layer is deposited. This is possibly due to oxygen absorption out of the ITO layer, which passivates the Si surface.

Lithographic techniques combined with reactive ion etching provide well-defined spatial structures. However, lateral dimensions at which quantum confinement effects are observed at room temperature (~5–10 nm) cannot directly be generated. An additional thinning of the structures by dry oxidation at elevated temperatures leads to much smaller structures.

Nassiopoulos et al. [80] have used deep ultraviolet (UV) lithography and reactive ion etching in a mixture of SF_6 and CHF_3 to fabricate Si nanopillars (see Fig. 24). Their intention was to relate the red photoluminescence observed from porous silicon to the quantum confinement model. PMMA and Cr were both used as a mask layer. In both cases, a remarkable aspect ratio of height to diameter greater than 40:1 was achieved. The diameter of the pillars was in the range of 20 to 40 nm and could be reduced to below 10 nm by thermal oxidation at 900–950 °C followed by a dip in HF to remove the oxide layer.

From these vertical nanowires, photoluminescence in the visible was obtained at room temperature. The peak position was between 580 and 650 nm, depending on the chemical

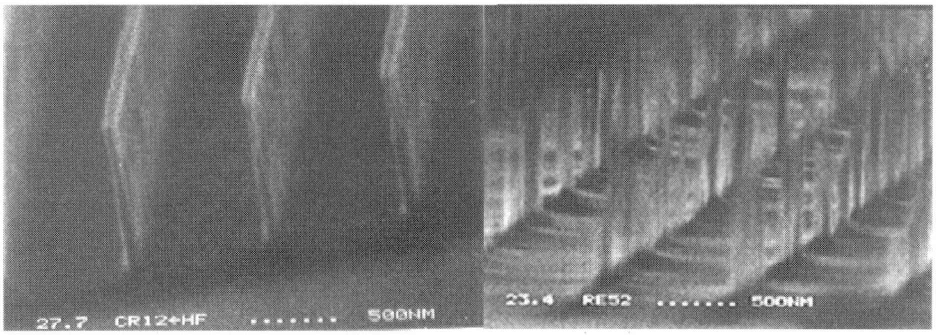

Fig. 24. SEM micrographs of Si pillars (right) and ridges (left) fabricated by high-resolution deep-UV lithography and highly anisotropic silicon etching based on fluorine chemistry. (Source: Reprinted with permission from [81].)

preparation and the exposure to ambient conditions [81]. In a next step, the pillars were planarized and etched back by an oxygen plasma in order to set free the pillar tops. A thin (150 Å) Au layer was deposited as the top ohmic contact. The electroluminescence spectrum under forward bias peaks at approximately 650 nm and is visible to the naked eye. The onset bias voltage for electroluminescence was determined to be 12–14 eV. The luminescence is attributed to electrons injected from the n-type substrate into the wires. During the drift in the electric field to the metal electrode, some of these electrons can recombine radiatively.

Chen and Ahmed [82] have used electron beam (e-beam) lithography and reactive ion etching in $SiCl_4$ and CF_4 to fabricate Si nanocolumns with an aspect ratio of height to diameter of about 7:1. The tip diameters of the approximately 40-nm-high dots were between 5 and 7 nm, the bottom diameters were 9 to 11 nm. However, no luminescence of the dots was reported. Liu et al. [83] have fabricated 50-nm-wide and 0.6-μm-high Si columns by e-beam lithography and NH_3 reactive ion etching. To remove the reactive ion etch damage, the samples were oxidized in dry O_2 at 800 °C and then dipped in HF. The oxidation process was monitored by transmission electron microscopy over a temperature range from 800 to 1200 °C. At oxidation temperatures below 950 °C, a self-limiting oxidation process was found. Thus, a silicon core remained even after 10 h of oxidation. The vertical Si nanowire depicted in Figure 25 was oxidized for 10 h at 875 °C. The inner Si core is reduced to its limiting dimension $2r_c = 2$ nm.

With increasing oxidation time, the Si core diameter decreased, whereas the oxide thickness increased until both reached saturation after a few hours. This saturation, which was absent for planar oxides, depends on the starting column diameter. The limiting Si core diameter had a stronger dependence on the starting column diameter at lower temperatures. The authors attributed this saturation effect to a diffusion-limited oxidation. During the oxidation process, the oxygen diffuses to the Si/SiO_2 interface and new oxide is formed. The previously formed oxide is thus pushed to the surface, creating stress. The counterplay of diffusion and stress can lead to an equilibrium state where diffusion stops. At high temperatures, the viscous flow of the oxide diminishes all stress. Therefore, at temperatures higher than 950 °C, there is no self-limiting oxidation observed.

As vertical nanopillars require additional, challenging technology for contacting, Kedzierski et al. [84] have fabricated planar silicon nanowires down to 10 nm in diameter using self-limited oxidation. The 30–40-nm-wide wires were defined on silicon-on-insulator (SOI) with electron beam lithography followed by a nichrome liftoff process and a silicon plasma etch. These lines were then oxidized at a low temperature (875 °C for 5–7 h) to thin the silicon core. As reported by Liu et al. [83], the stress created at the Si/SiO_2 interface slows down and eventually stops further oxidation. In this way, wires

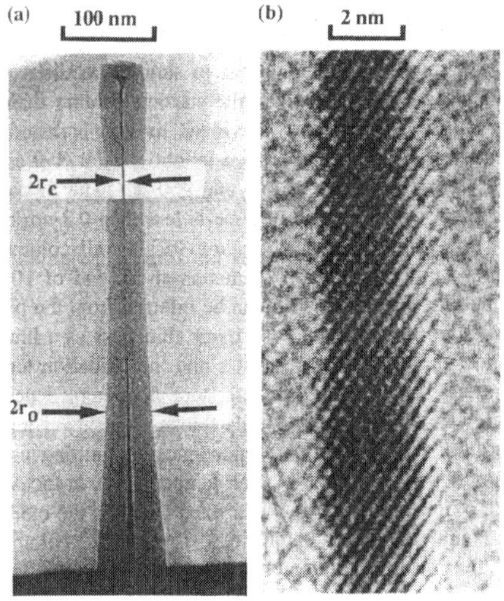

Fig. 25. (a) Transmission electron micrograph of an oxidized Si nanocolumn. The inner silicon core has reached its limiting dimensions. (b) High-resolution image of the inner 2-nm-thick Si core. (Source: Reprinted from [83]. © 1994 American Institute of Physics.)

with 25-nm-high and 5-nm-wide Si cores were fabricated along the $\langle 110 \rangle$ direction and annealed afterwards to improve the interface quality between the Si and the oxide. The shape of the oxidized structures strongly depends on the crystal orientation. The oxide growth rate for (111) planes is greater than for (110) planes. (100) planes have the lowest growth rate. From TEM cross sections of $\langle 110 \rangle$ oriented wires, the authors reported a factor of 2 higher oxidation rate for the (110) planes as compared to the (100) planes. They concluded a stronger (100) linear oxidation rate dependence on stress.

Reactive ion etching suffers from contamination effects and crystal damage. Therefore, it is worthwhile investigating appropriate direct epitaxial growth techniques for the realization of silicon-based nanostructures, which are discussed next.

3.5. Self-Organized Growth

Considerable efforts are in progress to produce quantum wires and dots from crystalline Si and crystalline $Si_{1-x}Ge_x$, as has been done for GaAs-based systems, but, to date, the available lithographic and etching techniques have not been able to produce nanostructures small enough to observe substantial confinement effects [85]. In c-Si, dimensions much smaller than the free-exciton Bohr radius of approximately 5 nm are required for such effects [86].

Another approach to obtain confinement is by appropriate epitaxial growth, such as quantum wire formation along wafer edge steps or by self-organized dot growth [87]. Most of the reported studies of Ge and SiGe islands on Si use physically determined deposition methods such as solid source MBE [88, 89]. Few reports describe the growth by chemical vapor deposition at pressures lower than about 0.1 torr such as gas source MBE [87, 90], ulrahigh vacuum (UHV) CVD [91], and low-pressure CVD [92]. Other reports describe the formation of islands by liquid phase epitaxy [93, 94]. Yanagiya et al. [95] have grown self-organized Si(110) wires on ultrasmooth sapphire $(01\bar{1}2)$ substrates by disilane (Si_2H_6) gas source MBE. The original sapphire surface has been imaged by AFM image and is shown in Figure 26a. After annealing the substrate at 1400 °C for 1 h and cleaning the surface in order to remove carbon contamination, atomic steps were obtained as depicted in Figure 26b. The distance between the flat terraces was 350 nm and the step height was 0.5 nm. The Si growth was performed at 700 °C at a Si_2H_6 pressure of 7×10^{-3} Pa.

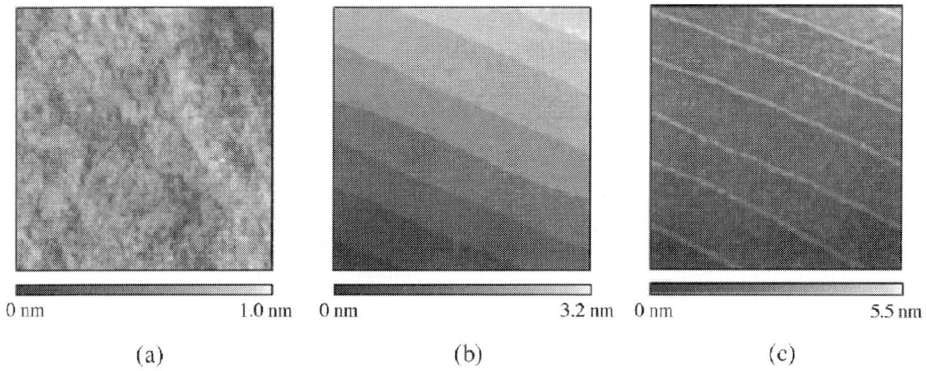

Fig. 26. AFM images (2 μm × 2 μm) of (a) as-supplied sapphire substrate, (b) substrate after 1 h annealing at 1400 °C showing approximately 350-nm-wide and 0.5-nm-high terraces, and (c) 50-nm-wide and 1-nm-high Si wires grown at 700 °C for 2 min. (Source: Reprinted from [95]. © 1997 American Institute of Physics.)

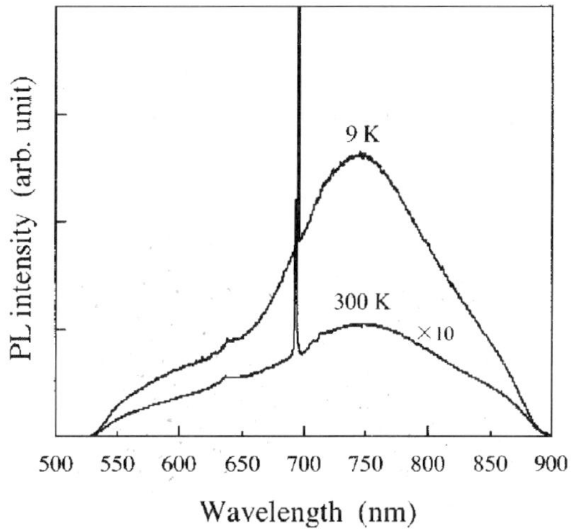

Fig. 27. Photoluminescence spectra at 9 and 300 K of self-organized silicon nanowires on the ultra smooth annealed sapphire substrate. (Source: Reprinted from [95]. © 1997 American Institute of Physics.)

Figure 26c shows the obtained Si wires aligned along the steps and oriented along the (110) direction. The width of the uniformly shaped wires was 50 nm and the height was 1 nm.

From X-ray diffraction, it could be concluded that the wires were monocrystalline. The sapphire surface could be modified by altering the annealing conditions [96]. Thus, in combination with the growth temperature, the size of the Si wires could be controlled. The wires grew according to the step-flow mode. Figure 27 shows photoluminescence signals at 9 and 300 K. At room temperature, the luminescence is quenched by more than a factor of 20. Both spectra peak at 750 nm with an FWHM of about 100 nm. The luminescence is quite similar to that of porous Si. The authors explain the broad luminescence band by the three-region model proposed by Kanemitsu et al. [97] that explains the luminescence of Si nanocrystallites in the system Si/SiO_2. According to this model, the electron–hole pairs are excited in the Si wires. The photons are emitted through radiative recombination centers that exist in the disordered localized state in the Si/Al_2O_3 interface region. As the distribution of the localized states is broad, this also leads to a broad luminescence signal. The sharp feature at 695 nm does not originate from the wires and is attributed to a chromium ion in ruby, which is also seen in the spectrum of porous Si formed on sapphire [98].

Besides the fabrication of quantum dots by lithographic tools and the growth on patterned substrates, *in situ* self-organized growth of islands has recently attracted much interest. Here the Stranski–Krastanow growth mode in the heteroepitaxy was used. We will only briefly discuss this technique. A detailed review is given in the chapter by Weinberg in this Handbook. The main advantage of self-organization is the process simplicity, the possibility of growing defect-free structures with dot sizes well below 100 nm. A drawback of this approach is making the dots conducting. A further problem is the rather large size and distance fluctuation between the dots compared to those fabricated by lithographic means. Therefore, the major application envisaged at the moment is restricted mainly to optical devices. We will see later on that, under certain conditions, quantum dot superlattices will point to some possible solutions to overcome these problems and to use the strain created by embedded islands to enhance the self-organization and to obtain more homogeneous islands with respect to size and spacing.

The driving force for nucleation of self-assembled islands is the strain resulting from the lattice mismatch between the substrate and the epilayer. According to the Stranski–Krastanow mode, the growth first starts in the layer-by-layer mode. If the thickness of this wetting layer exceeds a certain critical value (typically 4–6 monolayers), the strain will relax either through misfit dislocations or, under appropriate growth conditions, by a 2D $\rightarrow$ 3D transition. Because of to the relatively high misfit, the systems Ge or SiGe on Si and InAs on GaAs are the most investigated ones showing this effect. The strain depends on both the thickness of the layer and the alloy composition. Schittenhelm et al. [88, 99] have grown self-assembled $Si_{1-x}Ge_x$ dots on Si(001) by molecular beam epitaxy. The size, areal density, and homogeneity of the dots were investigated by varying several parameters including the growth temperature, the growth rate, the composition of the dots, and the thickness of the deposited layer. The authors found an increase of the dot height (8–33 nm) with increasing Ge concentration x while the areal density and the diameter remained nearly unchanged. They attributed this effect to the reduced strain with reduced Ge concentration. As the total amount of material is the same, the dots will automatically become smaller if the wetting layer becomes thicker because of the smaller strain.

With increasing coverage, the authors reported an increased formation of dots until the areal density starts to saturate for layers with more than 7 monolayers (ML). After this stage, the dots start to grow in height and diameter. Increasing the growth temperature also increases the surface diffusion length of the incoming atoms. As already existing islands represent a region of lower strain [100], the atoms nucleate preferentially at these islands, which leads to fewer but larger islands at higher growth temperatures. Darhuber et al. [101] determined the 2D $\rightarrow$ 3D transition for Ge on Si(001) to be between 5.33 and 5.67 ML by performing X-ray reflectivity experiments, which are in good agreement with photoluminescence experiments on the same samples [88, 89].

For electronic applications, uniformly sized quantum dots are needed. Therefore, there has been a growing interest in self-organized three-dimensional stacking of self-assembled dots. For technological applications, besides the size homogeneity, a high spatial density as well as spatial homogeneity is also required. Several groups [102–106] have grown quantum dot superlattices of InAs dots embedded in GaAs. They reported a spatial correlation in the following layers. The strain field at the surface arising from the underlying embedded dot makes the new dot nucleate directly above the embedded island. Yao et al. [104] suggested that such an alignment of the islands in the growth direction results if the spacer layers are not thick enough, and they associated the alignment with a lattice strain relaxation. Thus, one can conclude from the alignment that the island is a region of lower strain. Xie at al. [106] considered the effect of strain on the surface diffusion of incoming atoms to explain the island distribution. Tersoff and Le Goues [107] considered the exponential dependence of the nucleation rate on the nucleation barrier, which, in turn, depends on the

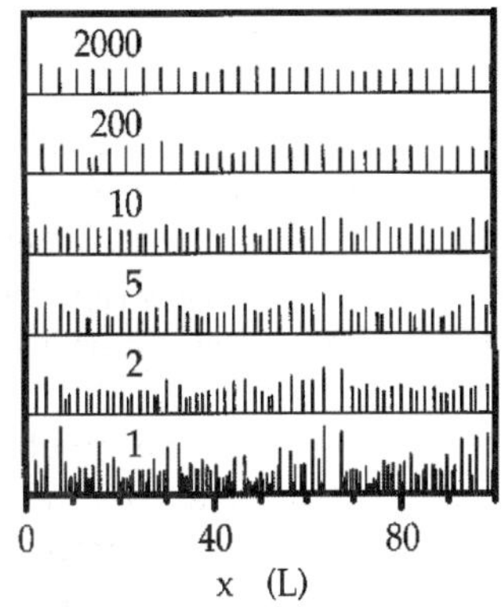

Fig. 28. Calculated island position and size for different layers of a dot stack. The island distance and size becomes more homogeneous in subsequent layers. (Source: Reprinted with permission from [108]. © 1996 American Physical Society.)

strain. Therefore, an island nucleates most favorably at those positions where the strain at the surface reduces the mismatch between surface and island to a local minimum.

However, a detailed analysis is rather complex and many parameters such as flux, temperature, and strain field must be known. Tersoff et al. [108], therefore, developed a generic model based on the continuum elastic theory considering Ge dots embedded in Si, which is also valid for SiGe, GaAs, or InAs islands. The buried dots are treated as spherical inclusions in an isotropic elastic matrix. Their size is assumed to be small relative to their lateral and vertical separation. The nucleation only occurs at each local misfit minimum. Following experimental results [103, 105], they assumed a nearly flat Si layer covering the Ge islands. Only some local strain resulting from the buried islands is present. Figure 28 shows the calculated position and the height of the Ge islands. At the beginning (layer 1), the islands are randomly distributed on the Si, substrate. After covering the islands by Si, new Ge islands are assumed to form at local strain minima. The development of a vertical correlation is clearly visible. By changing the spacer layer thickness L and the starting distribution of the islands, one can tune the resulting height and spatial homogeneity of the islands. After several layers are grown, the final lateral island spacing is about $3.5L$. Thus, the spacer thickness allows direct control of the final island spacing. If the initial island spacing is larger than $3.5L$, the new islands will also nucleate in the space between the prior islands. If the initial spacing is smaller than $3.5L$, the strain minimum will not be distinct and several islands will be replaced by a single new island. This explains the better homogeneity of the following layers. It must be mentioned that for real systems the model may only be applicable in certain growth regimes as parameters such as the surface diffusion length are not taken into account. Furthermore, for three-dimensional real systems, the strain minima at the surface may be too weak or may vanish. However, the surface diffusion length can be manipulated by varying the substrate temperature and the molecular flux during growth.

Figure 29 shows a $Si_{0.25}Ge_{0.75}$ dot layer deposited on (001)Si (part a) and the surface of a 20-period superlattice (part b). It is obvious that the islands at the top are more uniform in size and space. Qualitatively, this rather simple model can describe the experimental case even if many parameters are not yet taken into account.

Following prior experimental work on the vertical stacking of SiGe dots [105] and the theoretical modeling of Tersoff et al. [108], Schittenhelm et al. [99] have grown a 10-period Si(30 nm)/5–6-ML Ge multiple-dot layer at 670 °C at a growth rate of 0.075 Å s^{-1}. These parameters were chosen according to results from single layers to get an optimum size and spatial homogeneity already in the first layer. Atomic force microscopy analysis of the height distribution of the dots at the topmost uncapped layer revealed a variation of only

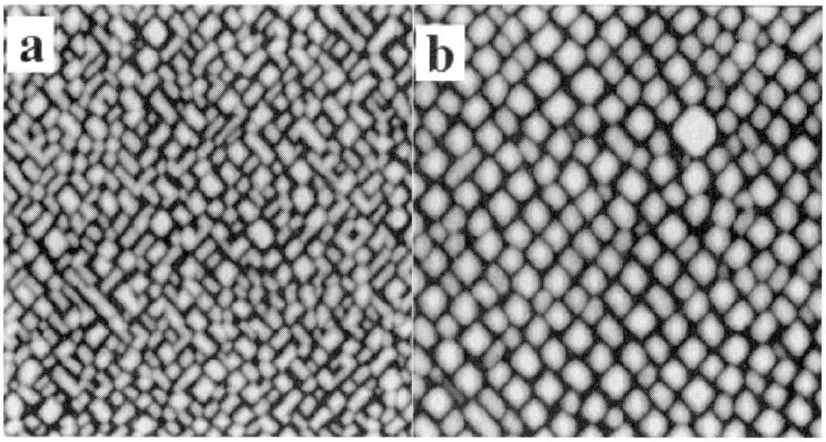

Fig. 29. AFM images of SiGe dot layers: (a) single layer and (b) topmost dot layer of a 20-period dot stack. (Source: Reprinted with permission from [108]. © 1996 American Physical Society.)

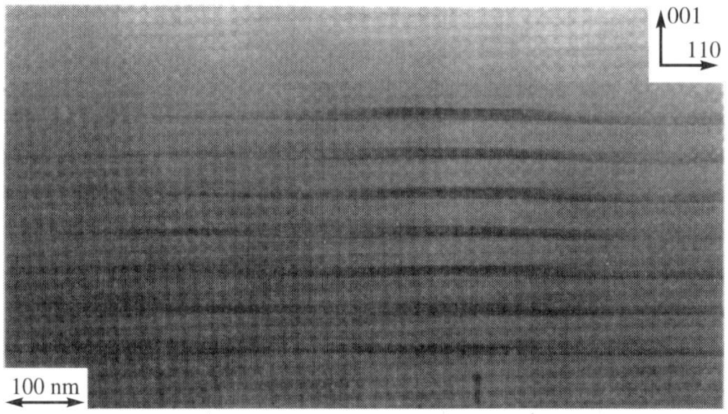

Fig. 30. Cross-sectional transmission electron micrograph of a seven-period 35-nm Si/12-ML Ge stack. The dots are seen to exhibit vertical order. (Source: Reprinted from [99], with permission from Elsevier Science.)

6% for the 5-ML sample and 8% for the 6-ML sample. Figure 30 shows a cross-sectional transmission electron micrograph of a Si(35 nm)/12-ML Ge multiple dot layer. The strong vertical correlation of the dots as simulated by Tersoff et al. [108] is clearly visible.

3.6. Selective Epitaxial Growth

Vescan et al. [105, 109–111] have grown strained SiGe/Si quantum well dots and wires by the selective epitaxial growth technique using low pressure chemical vapor deposition (LPCVD). They grew an approximately 500-Å-thick thermal oxide on 2000-Ω cm p-type Si(001) wafers and patterned them into arrays of rectangular holes parallel to the ⟨100⟩ or ⟨110⟩ direction by optical lithography. Figure 31 shows the process schematically. The 5-mm-long wires had a width between 0.4 and 300 μm; the square holes had a similar sizerange. The growth was performed at 700 °C using $SiCl_2H_2$ and GeH_4 as the source gases, and the carrier gas was H_2. The schematic structure is shown in Figure 31.

Here the basic idea is to form a facet of the selective epitaxial growth (SEG) area as narrow as possible. During SEG, facets usually form at the edges of the growth area. For the surface energies of the different Si crystallographic planes, one obtains $E_{001} > E_{110} > E_{311} > E_{111}$ [112, 113]. In the case of SiGe, one gets the same relations [114]. Therefore, the {111}, {110}, and {311} facets are favored during growth. Figure 32 shows a TEM

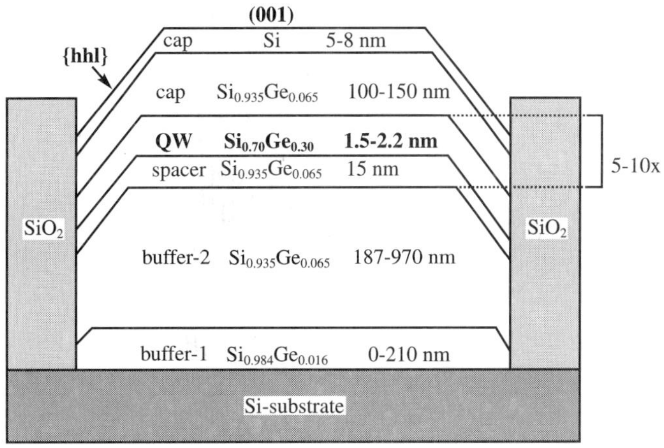

Fig. 31. Schematics of the QW structure within the oxide window. (Source: Reprinted from [110], with permission from Elsevier Science.)

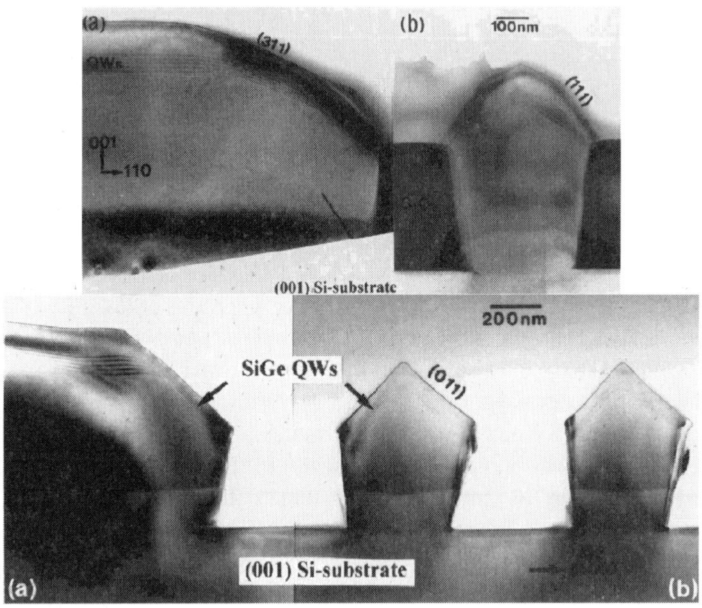

Fig. 32. Transmission electron micrograph cross section of wires with ⟨110⟩ side-wall orientation: (a) edge of a 10-μm wire and (b) 400-nm wire. (Source: Reprinted from [110], with permission from Elsevier Science.)

micrograph of the cross section of 10-μm (part a) and 400-nm (part b) wires with side-wall orientations of ⟨110⟩. The (001) quantum wells are planar and the {311} and (111) facets are also flat. In contrast to the wires, the dots all show the facets during growth, because the holes in the SiO_2 do not have perfect square shapes. Facet formation in this growth technique is needed to reduce the lateral dimensions of the quantum wells. However, to realize these facets, structures much thicker than the critical thickness are needed.

Stoica and Vescan [115] attributed this higher critical thickness to an eventually hindered defect multiplication in the local structures. For $x = 0.065$, the size reduction leads to an increase in the critical thickness of approximately 700 nm for 30-μm-wide wires and to even more than 1 μm for 300-μm dots. The characterization of the grown structures

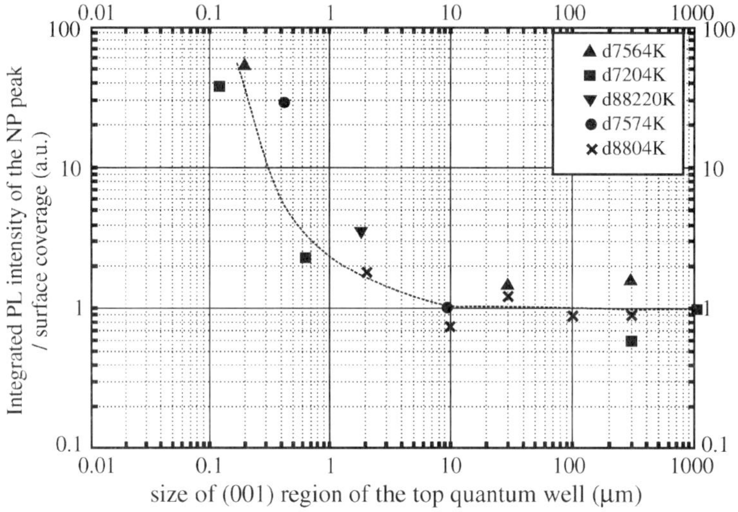

Fig. 33. Integrated intensity of the $SiGe^{NP}$ line, normalized to the surface of the dots, for different dot samples against the size of the (001) region of the top quantum well in the dots. (Source: Reprinted from [110], with permission from Elsevier Science.)

was performed optically by photoluminescence. From the presence of dislocation-related emission lines, the authors concluded that wires remain strained for sizes less than or equal to 30 μm, whereas dots are strained for diameters less than or equal to 300 μm. The photoluminescence intensity normalized to the patterned area increased by a factor of 50 for the dots less than or equal to 400 nm (see Fig. 33). This is attributed to the reduced role of nonradiative recombination channels by exciton localization. This suggests a further increase of intensity through fabricating smaller dots when the probability of the excitons reaching nonradiative center is decreased.

Hobart et al. [116] have used solid-source MBE to grow Si on nonplanar Si substrates. The prestructured Si substrate surface consisted of truncated pyramids with {111} sides and (100) tops that were formed by anisotropic wet etching in KOH or EDA solutions. The height of the pyramids was about 1 μm; the width of the tops ranged from 0.4 to 1 μm. The pitch size was about 10 μm. Because of undercutting of the SiO_2 layer used as a mask for the anisotropic etching, the mesa structure typically also had {331} facets. Si was deposited on such templates at various temperatures and growth rates. The authors found no structural change for temperatures between 470 and 550 °C. In a temperature range between 650 and 700 °C, {113} facets formed that laterally reduced the (100) tops of the pyramids to less than 20 nm. At higher temperatures around 800 °C, only {113} and {100} facets were formed. {113} facets have also been observed by Booker and Joyce [117] and Yang and Williams [118] during CVD deposition of Si, whereas the Si surface was carbon contaminated. Hirayama et al. [119] have observed {113} facets during selective epitaxial growth through a patterned SiO_2 mask. The occurrence of {113} facets was explained by pinning of the step flow at defect sites [120] or at the edges of the SiO_2 mask used for the selective growth [119]. Hirayama et al. [119] and Oshiyama [121] attributed the formation of {113} facets on Si(001) to bunching of double height steps. Thus, during chemical vapor deposition, surface orientation–dependent growth rates occur because of different silane and disilane deposition rates, whereas for MBE growth dynamical processes determine the facet growth.

Hobart et al. [116] have used the previously described Si templates to fabricate $Si_{1-x}Ge_x$ quantum dots. Thus, a 20-period Si(20 nm)/$Si_{0.8}Ge_{0.2}$(4 nm) multiple quantum well was grown at 550 °C on top of a mesa with a Si(100) surface reduced by the {113} facets to 10 nm. Although no optical or electrical characterization was performed to confirm low dimensionality, transmission electron micrographs indicated that quantum boxes may be realized on the top as the SiGe growth rate on the {113} facets was 85% of that on the (100) facets for geometrical reasons.

3.7. V-Groove Growth

Kapon et al. [122] have demonstrated the growth of quantum wire arrays on V-groove-patterned GaAs substrates by MOCVD. Performing cathodoluminescence (CL), they showed the existence of quantized states resulting from the quantum wires at the bottom of the grooves. Tsukamoto et al. [123] fabricated GaAs "arrowhead-shaped" quantum wires by selective growth using MOCVD.

Usami et al. [124] were the first to demonstrate SiGe/Si quantum wires on V-groove-patterned Si substrates. They have grown Si/SiGe/Si double-heterostructure quantum wires on V-groove-patterned p-Si(001) wafers (resistivity 5–10 Ω) by gas source MBE at 740 °C using silane and germane. Using thermal oxidation, a SiO_2 film was first grown as a mask layer for the chemical etching and for the subsequent selective epitaxial growth. The fabrication process is depicted in Figure 34.

With electron beam lithography and wet chemical etching in a N_2H_4-based solution, line patterns along the ⟨110⟩ direction with a period of 0.6 μm were fabricated. Because of the directional selectivity of wet etching, the V grooves have {111} side facets (Fig. 34a). The SiO_2 mask layer was then removed by a HF dip (Fig. 34c) and subsequently the whole structure was overgrown by a thick Si overlayer (Fig 34d). Figure 35 shows a TEM micrograph of a 15-nm-high SiGe/Si crescent-shaped quantum wire revealing no misfit dislocations. The shape and the vertical height, which is higher than the nominal one, are due to anisotropy in the different orientations during gas source Si MBE. As the surface migration of adatoms on (111) surfaces is higher than on (100) surfaces, one gets an enhanced mass transport from Si(111) toward the bottom of the V grooves.

The SiGe transverse optical phonon–assisted transition and the no-phonon line are clearly visible in low-temperature photoluminescence spectra, indicating a high crystalline structure in the wires. The broadening of these two peaks in the case in the wires in comparison with spectra of a two-dimensional reference structure results from the SiGe width and Ge compositional variations. The SiGe luminescence quenching temperature was determined to be 70 K. The electroluminescence spectrum is comparable to the photoluminescence spectrum, indicating that the recombination process is identical in both cases.

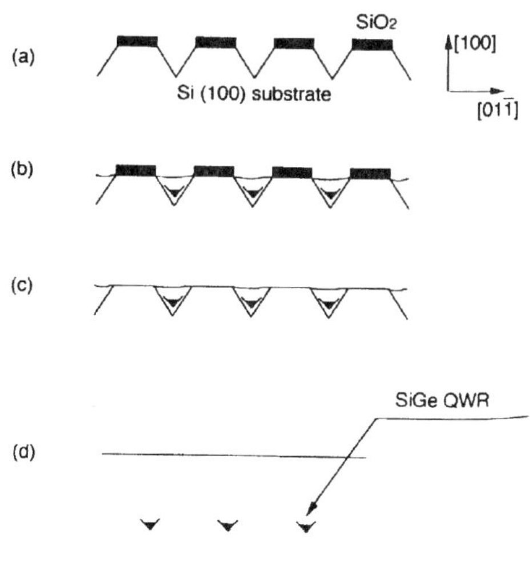

Fig. 34. Schematics of the processing steps of SiGe quantum wires by selective epitaxial growth. (Source: Reprinted from [125]. © 1994 American Institute of Physics.)

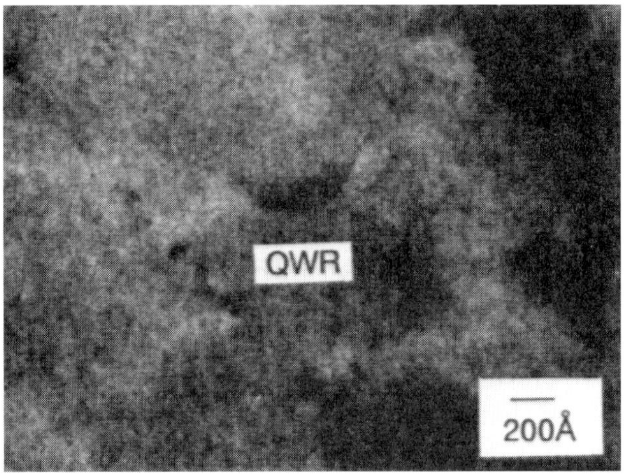

Fig. 35. Transmission electron micrograph of a Si/SiGe/Si wire grown on a V-groove patterned substrate. (Source: Reprinted from [124]. © 1993 American Institute of Physics.)

Polarized electroluminescence was obtained in the direction of the cross section of the wires and in the growth direction. Although the wires do not show any polarization in the cross section, the single quantum well shows a strong in-plane polarization as expected from the selection rule of the $hh1$–$e1$ transition. In the direction perpendicular to the sample surface, the situation changes to the opposite. Here the quantum well is optically isotropic and the wires show strong anisotropy. The polarization experiments clearly indicate successfully realized luminescent SiGe wires of high crystalline quality. From TEM micrographs, the lateral and vertical wire widths were determined to be 30 and 8 nm. However, the peak positions do not fit these dimensions as the distribution of strain and the Ge atomic fluctuations lead to a shift of these peaks.

A main disadvantage of the V-groove growth technique is its limit for further reduction to quantum dots. From a scientific point of view, this method seems to be an elegant way to produce quantum wires 30–40 nm in width with high crystallinity.

3.8. Local Growth of Dots and Wires through Shadow Masks

The method of local epitaxial growth through shadow masks for the homoepitaxy of Si epilayers was reported by Hammerl et al. in 1991 [126]. Brunner et al. [127] used this technique to grow directly wires and dots with SiGe quantum wells through a patterned SiO_2/Si_3N_4 shadow mask. The dots and wires were physically detached from the surrounding material and after overgrowth they were completely embedded in Si. The dots and wires were grown on (100) Si wafers with a resistivity of 1200 Ω cm. Figure 36 shows the schematics of this growth technique.

After fabrication of the LPCVD-deposited Si_3N_4 100-nm-thick shadow mask [126], the opened window is cleaned and the wafer transferred to the MBE reactor. The thickness of the SiO_2 spacer layer is 1 μm. The nitride is underetched up to 5 μm. The shape of the wires and dots is determined by the growth parameters and by the angle of the incident beam with respect to the sample surface as shown in parts b and c of Figure 36. Finally, the SiO_2 layer is lifted off by selective wet etching (part d).

The structures mentioned here were grown at 720 °C under an incident angle of 20°, consisting of a 140-nm Si buffer layer, a Ge concentration of 25% in the quantum well, and a 160-nm-thick Si cap layer. The minimum lateral dimension was 200 nm [128]. An SEM cross-sectional micrograph of a dot array is shown in Figure 37. The mask layer is not yet removed and it can be clearly seen that there is no direct contact of the structures with

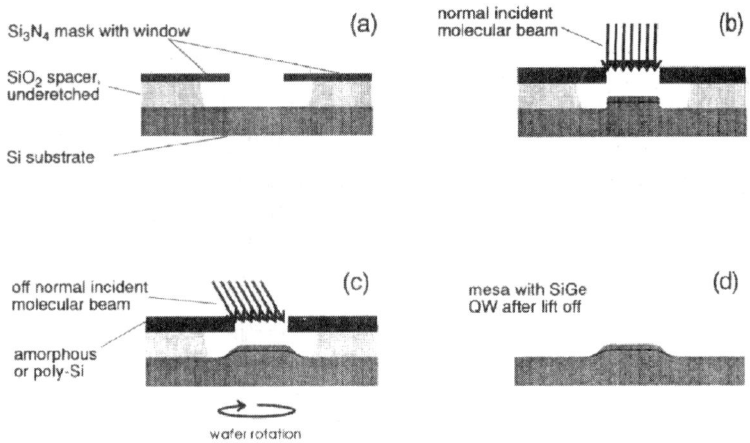

Fig. 36. Schematics of the local MBE growth through micro shadow masks: (a) underetched SiO_2 with the patterned silicon nitride, (b, c) local epitaxy with normal and off-normal incident molecular beams, and (d) mesa with smooth side walls after liftoff. (Source: Reprinted with permission from [127]. © 1994 American Institute of Physics.)

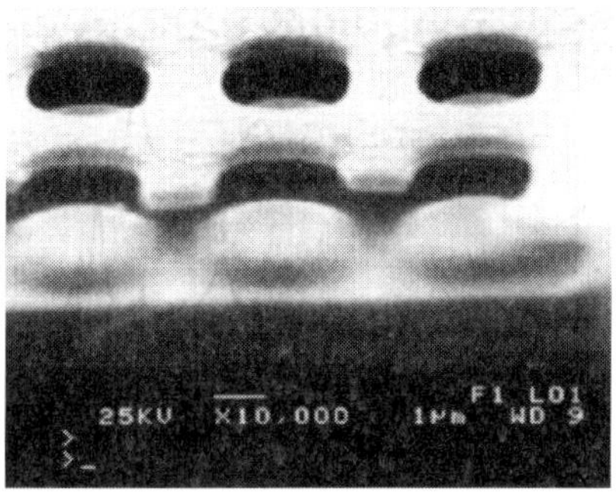

Fig. 37. Scanning electron micrograph of locally grown Si/SiGe/Si 2-μm × 5-μm mesas. Note that there is no pillar support of the nitride mask over the whole dot group. (Source: Reprinted with permission from [127]. © 1994 American Institute of Physics.)

the walls; thus, structural imperfections can be avoided during growth. Figure 38 shows a TEM micrograph of the edge of a wire with the embedded SiGe quantum well [129]. As the quantum well is completely buried in silicon, the potential problem of surface states at the side walls is eliminated.

The thickness of the quantum well (QW) decreases toward the edge of the mesa; thus, the energy gap increases toward the edges confining carriers in the middle part of the mesa structure. This results in an effective passivation of the side walls without mediation of surface states. Photoluminescence experiments excited with 457 nm reveal the typical SiGe TO phonon–assisted and no-phonon lines and reflect the high structural quality of these dots and wires. The exact shape of the side walls can be controlled by the growth conditions to be nearly vertical or smoothly ascending [130]. Rotating the substrate during growth and positioning the e-beam evaporator off axis leads to smoothly ascending side walls as shown in Figure 36c.

Figure 39 shows the peak intensities of the SiGe TO phonon replica as a function of wire width normalized to the fractional coverage of the mesa islands [131]. For comparison, the intensity of a homogeneously grown quantum well reference sample is also included. The increasing luminescence intensity with decreasing wire width from 1 μm to 200 nm has been attributed to an effective passivation of the side walls and to confinement of excitons

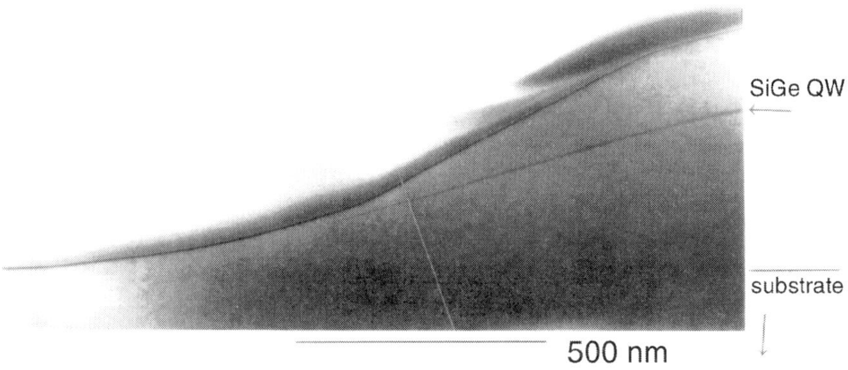

Fig. 38. Cross-sectional transmission electron micrograph of a SiGe QW completely embedded in silicon. (Source: Reprinted from [129], with permission from Elsevier Science.)

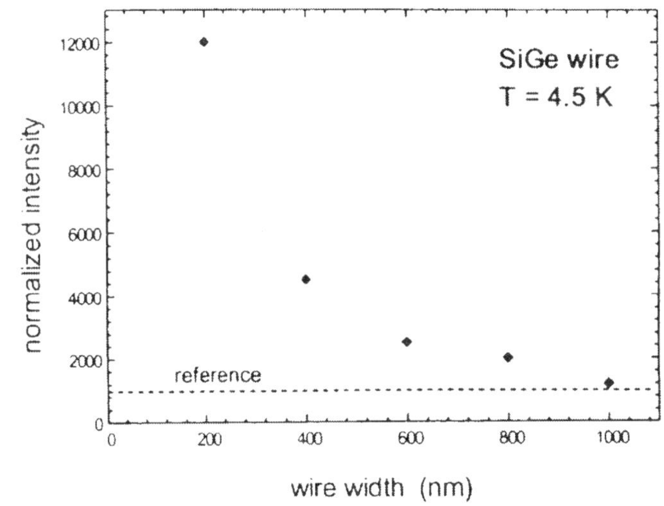

Fig. 39. Normalized TO phonon–assisted luminescence transition intensity of locally grown SiGe wires as a function of wire width. The reference line is the intensity of a homogeneously grown quantum well. (Source: Reprinted from [131], with permission from Elsevier Science.)

within the mesa structure and thus to a smaller probability of reaching nonradiative channels. This trend is similar to observations made in GaAs–GaAlAs wires [132].

However, as in the case of all normalized spectra of this type, one has to keep in mind that excitons in silicon have a diffusion length of about 400 μm. Thus, once generated, they can diffuse to the SiGe wires and recombine there. This effect will always contribute to an increase in intensity unless the SiGe layer is excited with light below the Si absorption edge. Some simulations to investigate this effect may be helpful to estimate the general relevance. Spatially resolved photoluminescence was used to investigate the luminescence intensity distribution and thus the exciton density distribution along a 1-μm-long wire. From fitting the data to an exponential decay, a decay length of about 10 μm could be estimated.

If growth is carried out with the incident beam parallel to the mask normal direction, one obtains faceted side walls [129]. In the case of (100) substrates, the slightly lower (111) surface energy leads to $\{111\}$ facet formation. Eisele et al. [129] have grown wires and pyramidal dots through submicrometer masks with the edges oriented in the $\langle 110 \rangle$ direction. Figure 40 shows SEM micrographs of this facet formation for a cleaved wire

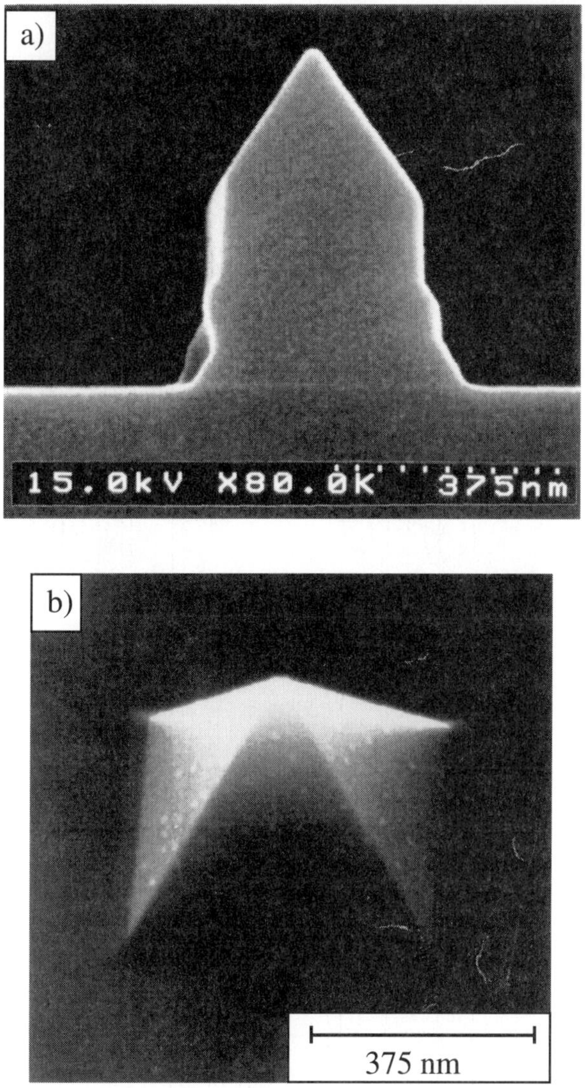

Fig. 40. Scanning electron micrographs of {111} facet formation in (a) a wire structure grown at 500 °C through a 310-nm mask and (b) a pyramidal dot structure grown at 500°C through a 350-nm square mask. (Source: Reprinted from [129], with permission from Elsevier Science.)

(part a) and a dot structure (part b). Both structures were grown at 500 °C with a rate of 0.1 nm/s. In the case of the wire, a 310-nm mask was used, and a 350-nm square mask was used for the dot. The radius of the dot tip was estimated to be less than 5 nm. The maximum length of the {111} facets is diffusion limited. When impinging atoms have reached their diffusion limit underneath the mask, vertical (011) side walls grow as can be seen in Figure 40a. Neither thermal oxidation at 800 °C nor thermal annealing at 1000 °C under nitrogen atmosphere led to any facet degradation. From capacitance–voltage characteristics of metal-on-silicon (MOS) capacitors, the presence of a dense oxide with negligible leakage could be deduced.

Using the same technique, Gossner et al. [133] grew silicon nanostructures through micro shadow masks. Figure 41 shows the patterned nitride film and the Si mesa structures grown through this mask. At a growth temperature of 500 °C, {111} facets formed on the side walls. Wedge-like mesas with a top radius of about 5 nm with perfect {111}

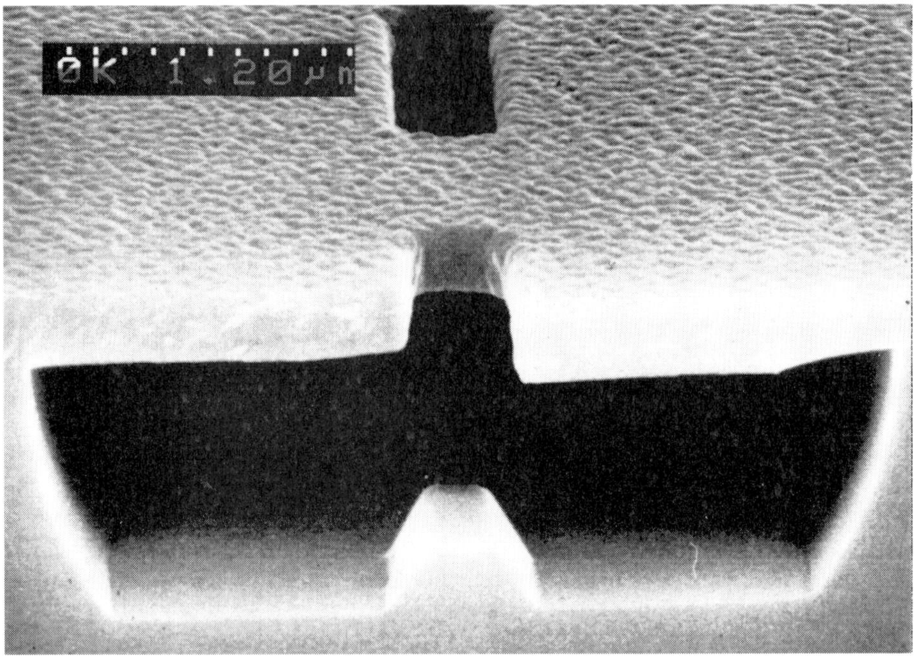

Fig. 41. Scanning electron micrograph of a Si mesa grown through a nitride micro shadow mask at 500 °C. (Source: Reprinted from [133], with permission from Elsevier Science.)

facets were grown through ⟨110⟩-oriented line-shaped masks. The formation of a perfect pyramidical shape was independent of the orientation and shape of the mask aperture.

The technique of local growth through a mask allows a variety of nanostructures to be grown, as well as the realization of laboratory-scale devices as, for example, a field effect transistor with built-in one- and zero-dimensional electrical confinement (see Section 4), ultrathin tips for scanning probe microscopy and direct nanostructuring of surfaces.

3.9. Silicon Nanocrystallites

3.9.1. General Remarks

Another way of realizing nanostructures is forming nanocrystallites in, for example, a matrix. Nanocrystallites are made up of one or more clusters. A cluster consists of agglomerations of up to 10^5 atoms, which, in conventional terms, can be regarded as neither a bulk material nor as atoms or molecules. Clusters in the size range of 1 to 1000 nm exhibit new physical and chemical properties, which can be directly modified by controlling the cluster size. In the limit of very small clusters, the addition of a single atom can drastically modify electronic, magnetic, and other physical properties. To investigate cluster properties as a function of cluster size, it is necessary to separate the cluster by its mass. Electronic and optical properties can be probed by photoelectron, absorption, fluorescence, or photoluminescence spectroscopy, among other techniques. Chemical reactions of clusters with surfaces can be studied by depositing the cluster onto different substrates.

Clusters can be classified according to their chemical bonding. For example, in a van der Waals cluster, the induced dipole–dipole interaction is the force keeping the cluster together. The stability and structure of the cluster is driven by geometric packing effects leading to closed packages with 12 nearest neighbors. Group IV semiconductor clusters, especially C, Si, and Ge, are extensively investigated. Their bonding is covalent and the clusters are not closely packed. Instead, the coordination number is given by the number of valence electrons or by the degree of hybridization. A prediction of the cluster structure is very difficult because of the spatially oriented bondings. Different hybridization of the valence orbitals leads to single, double, or triple bonding, forming carbon chains, rings, or nanotubes [134]. In semiconductors, II–VI and III–V compounds have been widely investigated [135, 136] and silver halides have been widely used in the photography industry [137].

Silicon nanoparticles are found to luminesce in the red and blue region of the visible spectrum when illuminated by ultraviolet radiation or when a current is passed through them [138, 139]. The origin of the red luminescence is generally accepted to be due to a quantum confinement shift of the bulk silicon band gap, whereas the blue emission is not yet fully understood.

Since the discovery of porous Si [49] and the interpretation of its luminescence originating from quantum confinement in nanostructures, different methods have been used to synthesize nanostructures. In the reduction of $SiCl_4$, $HSiCl_3$, or octyl $SiCl_3$ by Na [140] in nonpolar organic solvents, the synthesis is performed at temperatures around 385 °C at relatively high pressures of around 100 atm for several days. After cooling and washing out the Na, NaCl, and hydrocarbon residue, the product is dried in vacuum. Analysis mainly reveals single crystals of 5–3000 nm with a hydrogen-terminated surface. Using octyl $SiCl_3$, one can achieve a size distribution of only 2–9 nm. Other groups have separated nanocrystals from a porous Si sample [141–144]. Here the porous silicon layer on top of the single crystalline wafer is removed mechanically. After ultrasonic agitation within a solvent for several days, one gets a colloid with particles ranging from 1 to 10 nm and some agglomerates up to 50 nm. The disadvantage of this method is the lack of size control. Thermal evaporation [145] and decomposition of silanes have also been used to fabricate silicon nanoclusters [146–148]. Well-controlled silicon nanopowder with particle sizes varying between less than 10 and 200 nm can be synthesized from gas phase reactions in a radio frequency discharge in silane. The size of the particles can be controlled by the gas flow, gas composition, and plasma characteristics, as reported by Kobayashi et al. [149] and Yasuda [150]. Dutta et al. [151] have published a review article on this technique. In the following, we describe some of the recently reported techniques with a high potential for crystallite size control.

3.9.2. Silicon Nanocrystals Produced by a Laser Vaporization Technique

Li et al. [152] have produced weblike aggregates of coalesced Si nanoparticles on glass through laser vaporization and subsequent controlled condensation. A pulsed Nd:YAG laser with 15–30 mJ/pulse and 10^{-8} s pulses evaporates Si in a chamber consisting mainly of two metal plates facing each other at different temperatures. The chamber is filled with different carrier gases such as Ar. The temperature gradient between the two metal plates leads to a strong convection. Nucleation of the Si vapor leads to the growth of Si nanoparticles, which are driven to the cold plate by convection. The temperature gradient between the two plates controls the convection and the supersaturation of the Si vapor. High supersaturation leads to smaller nuclei, which are required for the condensation of nanoparticles. The convection drives away the small nanoparticles before they can grow further.

Although SEM analysis revealed spherical particles of about 10 nm in diameter connected by a weblike structure with a wall thickness of 10–20 nm, the Raman shift suggested an average particle size of approximately 4 nm. X-ray powder diffraction of these samples revealed distinct Si(111), (220), and (311) reflections, but, as the powder sample was not rotated during the scan, a possible texturing of the nanoparticles cannot be ruled out. Fourier transform infrared (FTIR) spectroscopy indicated an oxidized surface layer after exposing the nanoparticles to air consisting of SiO_x with $x < 2$. The thickness of this SiO_x was estimated to be 1–2 nm. Annealing the particles at 300 °C in air leads to $x = 2$. The freshly deposited particles do not show any luminescence until they are exposed to

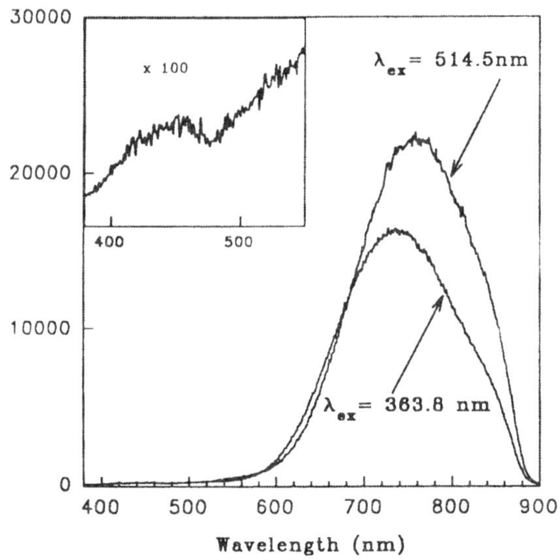

Fig. 42. Photoluminescence spectrum of Si nanocrystals excited with 363.8- and 415.5-nm radiation. The inset shows the weak blue luminescence excited with a 363.8-nm wavelength similar to the luminescence of SiO_2 nanoparticles. (Source: Reprinted with permission from [152]. © 1997 American Chemical Society.)

air and their surface is passivated by SiO_x. The luminescence, which shows a peak in the red and a smaller one in the blue, was excited with 363.8- and the 514.5-nm laser lines, respectively. The quantum efficiency of the red line is about 1.3%. Based on a quantum confinement model [153], the authors concluded that the longer wavelength excitation selects the larger particle resulting in a stronger emission to the red. The inset of Figure 42 shows a weak blue emission at 450 nm using an excitation wavelength of 363.8 nm, which is quite similar to the emission of SiO_2 nanoparticles and is related to the oxidized surface layer of Si nanoparticles [154].

Based on the results of the FTIR experiments and bearing in mind the approximately 8-eV band gap of SiO_2, this surface layer was said to be nonstoichiometric SiO_x with x between 1.4 and 1.6, resulting in a band gap of around 3–4 eV, corresponding well to the blue emission [155]. Defects in SiO_2 cannot be ruled out because they luminesce in the red and blue spectral regions [156]. In contrast to this, pure SiO_2 does not show red luminescence. Its absorption spectra in the range of 200 to 800 nm suggest that the radiative transitions are indirect. From fitting the data with the relation $(\alpha h\nu)^{1/2} = B(h\nu - E_g)$, the authors obtained a band gap of 1.78 eV, corresponding well to the 700-nm red luminescence. The blue shift compared to the 1.1 eV of bulk Si is attributed to quantum confinement. The great potential of this technique lies in the good size control through a simple technique. Applications such as Si powders seem reasonable. However, for any device application, the integration into the Si circuit remains an open question.

3.9.3. Si/SiO$_2$/Si Multilayers

Abeles and Tiedje [157] demonstrated the first amorphous semiconductor/insulator superlattices. In contrast to porous silicon, these regular periodic structures do not show a size distribution and suffer minimal contamination effects. Growing Si nanostructures by MBE enables a detailed study of the correlation of luminescence properties to crystallinity, interface chemistry, and nanostructure size. The commonly used insulators are SiO_2, CaF_2, and Al_2O_3.

Lockwood et al. [86, 158, 159] fabricated a number of Si/SiO_2 superlattices with a silicon layer thickness $d_{Si} = 1$–30 nm. The samples were grown by solid-source MBE and

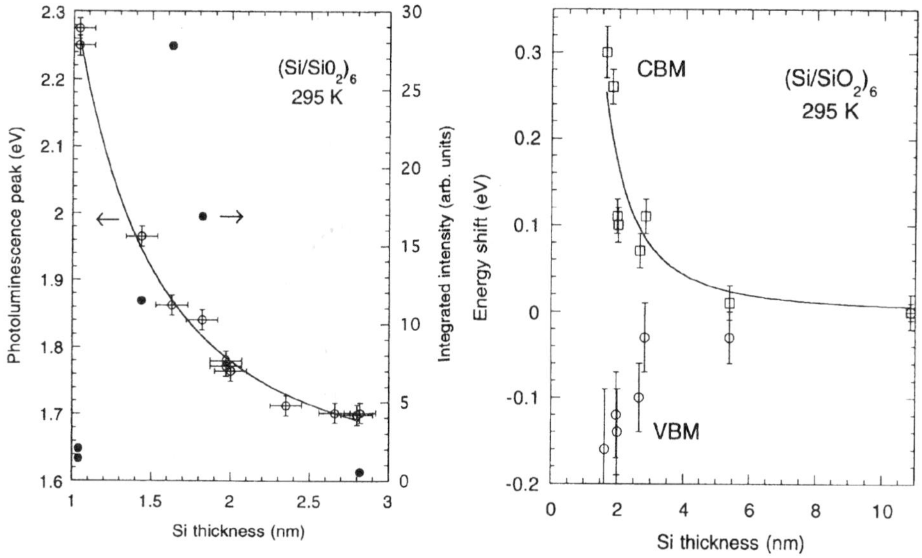

Fig. 43. Photoluminescence energy (open circles) and integrated intensity (full circles) for different Si layer thicknesses (left) and shift of the conduction and valence band of Si/SiO_2 superlattices at room temperature for different Si layer thicknesses. The solid line in both pictures is the best fit using effective mass theory. (Source: Reprinted with permission from [159]. © 1996 American Physical Society.)

characterized by Auger electron spectroscopy (AES), X-ray photoelectron spectroscopy (XPS), photoluminescence (PL), Raman scattering, soft X-ray Si–$L_{2,3}$ near-edge absorption spectroscopy (XANES), TEM, and X-ray reflectivity. The density ratio of the amorphous Si layers to c-Si was found to be 0.97, indicating little film porosity and low contamination level. Visible-to-near-infrared PL was observed at 295 K in all superlattices with $d_{\text{Si}} < 3$ nm, with the maximum of the emission peak at 665 nm (1.86 eV) with a full width at half-maximum (FWHM) of 0.38 eV.

The dependence of the PL peak energy and intensity on d_{Si} is shown on the left-hand side of Figure 43, indicating a clear blue shift with decreasing Si layer thickness. The PL signal is attributed to the Si layers as the SiO_2 layer thickness remains constant. The blue shift and the increasing intensity with decreasing d_{Si} indicates quantum confinement effects [57, 58]. Using effective mass theory together with the assumption of infinite potential barriers, this energy gap dependency on layer thickness is fitted to

$$E = E_g + \frac{\pi^2 h^2}{2d^2}\left(\frac{1}{m_e^*} + \frac{1}{m_h^*}\right) \tag{1}$$

where E_g is the bulk material energy gap and m_e^* and m_h^* are the electron and hole effective masses. The band edge E_g obtained from fitting is in good agreement with the bulk value for amorphous Si (1.5–1.6 eV at 295 K). To obtain the shift of the conduction and valence band, XANES and XPS measurements were performed. Figure 43 (right) shows the conduction band minimum upshifted by approximately 0.3 eV and the valence band maximum downshifted by approximately 0.1 eV when the Si layer thickness is reduced to $d_{\text{Si}} = 1.5$ nm. These results are consistent with the band gap shift obtained from PL experiments (Fig. 43, left) and for the first time demonstrate unambiguously quantum confinement and direct band-to-band transitions.

Figure 44 shows a photo of the National Research Council logo made of a Si/SiO_2 superlattice. To obtain the PL image, the mesa structure was excited from the back by a laser. The red luminescence is visible to the naked eye. The complete integrability with conventional silicon-based technology and the good size controllability makes this technique a strong candidate for the realization of a silicon-based light-emitting diode. Quantum

Fig. 44. Visible photoluminescence from a Si/SiO_2 superlattice mesa structure excited from the back. (Source: Reprinted from [160]. © 1996 American Institute of Physics.)

efficiency values, not published up to now, would be very desirable to gauge the technological potential of Si/SiO_2 superlattices.

3.9.4. Nanocrystalline Si/CaF_2 Multilayers

D'Avitaya et al. [161, 162] have grown nanocrystalline Si/CaF_2 multilayers with periods ranging from 50 to 100 by solid-source MBE at room temperature on CaF_2 and Si(111) substrates. CaF_2 was chosen as the barrier layers because of its large band gap of about 12 eV and its small lattice mismatch of 0.6% to Si. For a review of Si/CaF_2 superlattices, see [163]. As the nanocrystallinity of the whole multilayer is driven by the ionic character of CaF_2 and the low growth temperature, the authors expect no effect of the substrate. However, to date, there are no published results on the growth of nanocrystalline Si/CaF_2 on technologically relevant Si(100) surfaces. Whereas the CaF_2 layer thickness is kept constant at 3 nm, the Si layer thickness is varied between 0.3 and 5 nm. From extended X-ray absorption fine structure (EXAFS) experiments on Si(2 nm)/CaF_2(3 nm) multilayers, the Si grain size of the multilayers could be determined to be smaller than 1.5 nm. Schuppler et al. [164] have reported the same upper limit for the grain size in porous Si also determined by EXAFS experiments.

After allowing hydrogen and oxygen to passivate the dangling bonds of the Si nanostructures, intense PL was observed at room temperature, with a peak maximum at approximately 1.65 eV and an FWHM of approximately 0.4 eV, very similar to the observed spectra of porous Si. The luminescence could be observed only if the Si layer thickness was less than 2.5 nm, similar to the 3-nm-thick Si layer reported by Lockwood et al. [159] for Si/SiO_2 multilayers. Thus, in combination with the observed blue shift as the Si layer thickness decreased, the authors attributed the luminescence to a quantum confinement effect. Optical absorption measurements made on the same structures gave a corresponding gap value of 2.4 eV. This energy difference between the absorption and emission was ascribed to the Stokes shift. Time-resolved photoluminescence experiments revealed a slow ($t > 1$ μs) and a fast ($t < 1$ μs) component [165]. These two emission bands, their lifetime, and their spectral positions are very similar to porous silicon, although the monotonic temperature dependence of PL intensity differs from that of porous silicon. Moreover, Vervoort et al. [165] have reported a significant increase of the PL intensity upon aging of the multiple quantum wells of Si/CaF_2 in air, again an effect also observed in porous silicon. This effect was attributed to a possible oxygen saturation of Si dangling bonds migrating along the grain boundaries, which most likely led to the passivation of nonradiative centers.

A mesa light-emitting diode was fabricated with a transparent indium tin oxide (ITO) ohmic contact. Under forward bias (7–8 V), room temperature EL was observed. It was also reported that ITO deposition has a strong effect on the luminescence. First, the PL peak position is shifted to lower energies, then the line width decreases to 100 nm, and, finally, the luminescence intensity is higher by a factor of 3. These effects are attributed to the absorption of oxygen out of the ITO layer, which modifies the surface passivation and also affects the size of the Si crystallites. This could correspond to the reported aging effect. In the case of EL, the authors attribute the excitation to tunneling of carriers through the dielectric under the high applied electric field.

Bassani et al. [166] have reported a more pronounced blue shift of the optical absorption edge with decreasing Si layer thickness, adding support to the interpretation of the emission being determined by quantum confinement of carriers within Si nanocrystallites. The CaF_2 thickness for all these samples is kept constant. Nevertheless, the nanocrystalline Si layers are assumed to have an indirect band gap. From Tauc's relationship $(\alpha h\nu)^{1/2} = B(h\nu - E_g)$, the optical pseudogap could be obtained by extrapolating the experimental data to $\alpha = 0$. The dependence of the optical band gap on the Si layer thickness is consistent within the effective mass theory. The huge Stokes shift of 1 eV observed for the thinnest Si layer (1.2 nm) was attributed to relaxation of the Si atoms near the surface forming a deep luminescent center. Numerical simulations by Allan et al. [167] and Baierle et al. [168] have predicted a Stokes shift of approximately 1 eV for crystalline diameters of 1.5 nm, which agrees well with that reported by Bassani et al. [166]. The smaller blue shift of photoluminescence compared to the absorption spectrum was related to near-surface localized states, which are less influenced by the silicon grain sizes [166]. However, quantum confinement remains a necessary condition for luminescence. Borisenko et al. [169] have used an improved self-consistent linear combination of atomic orbitals (LCAO) method with the extended Huckel Hamiltonian to calculate the electronic band structure of nanometer-thick (111) monocrystalline and nanocrystalline freestanding silicon films.

These structures are very similar to the Si/CaF_2 system. The Si dangling bonds are passivated by hydrogen. Figure 45 shows the evolution from a monocrystalline film to a "pseudofilm" and finally to a "pseudowire" and a "pseudodot" structure. On the right-hand side of Figure 45, the corresponding electronic band structure is depicted, showing the transition from an indirect to direct band gap resulting from silicon grains being formed. The calculated band gap of 2.07 eV of the "pseudodot" grain can be considered as a lower limit. For an isolated grain, the calculated band gap is 4.61 eV, which can be considered as an upper limit. Thus, the value of 2.4 eV for the gap in Si/CaF_2 multiple quantum wells with grain size in the range of 0.5 to 1.5 nm would fall within the likely band gap energies in this material.

3.10. Si/III–V Light-Emitting Nanotips

So far, the route favored by industry to solve temporarily the electrical wire bottleneck is to combine III–V direct band gap materials with indirect band gap silicon complementary metal on silicon (CMOS) platforms. This approach explores the excellent optical properties of III–V compounds and combines them with the existing technology of the silicon microelectronics industry. There are essentially two different ways to realize a hybrid Si/III–V device. The first is to grow directly GaAs or GaAs-based structures on Si. The second is to fabricate first the III–V-based optoelectronic circuit and to mount it by wafer bonding, fusion, epitaxial liftoff, or flip-chip bonding on the Si wafer. The first approach faces three different fundamental challenges [3]:

1. The high lattice mismatch of 3.88% between GaAs and Si leads to a high density of misfit dislocations and to relaxed GaAs layers.
2. There is also a polar–nonpolar surface mismatch. Whereas III–V compounds are polar, different sides of the nonpolar silicon crystal cannot be distinguished. This

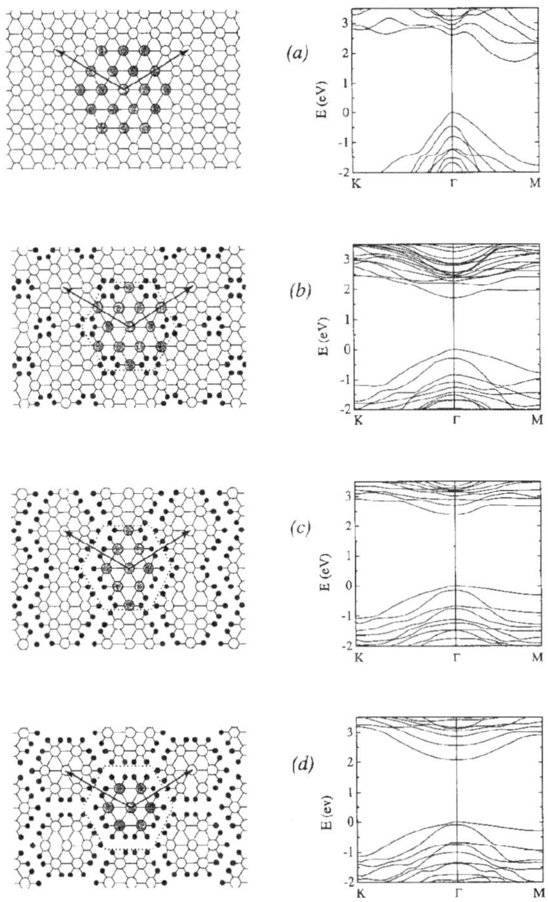

Fig. 45. The evolution from a monocrystalline film (a) to a pseudodot structure (d) is shown on the left-hand side. The corresponding electronic band structure is shown on the right-hand side. The small circles represent H atoms; the large circles represent Si atoms. The unit cell is denoted by a dotted line. (Source: Reprinted from [169]. © 1997 American Institute of Physics.)

mismatch leads to the growth of antiphase boundaries, as there are always atomic steps on the silicon wafer. Thus, during growth, a phase difference occurs in the stacking of GaAs between domains on opposite sites of the silicon steps, leading to Ga–Ga and As–As bonds at the so-called antiphase boundary in contrast to the Ga–As bonds.

3. Finally, there is the problem of charging of the interface. An interface that is abruptly terminated between a polar and a nonpolar semiconductor results in a heavily charged plane modifying strongly the electronic characteristics of a device.

There are several groups exploring these two approaches and the reader is referred to [170–174] for recent progress in this area. One novel approach, though not specific for optical interconnects, has been proposed by Shealy et al. [175], who selectively grew a III–V alloy (GaInP) by organometallic vapor phase epitaxy on sharp silicon tips. As the thin tip behaves as a compliant substrate, the creation of misfit dislocations can be avoided as the authors showed by photoluminescence studies. The basic idea was to fabricate field-emitting devices for high-resolution-color flat panels without the need of phosphorous

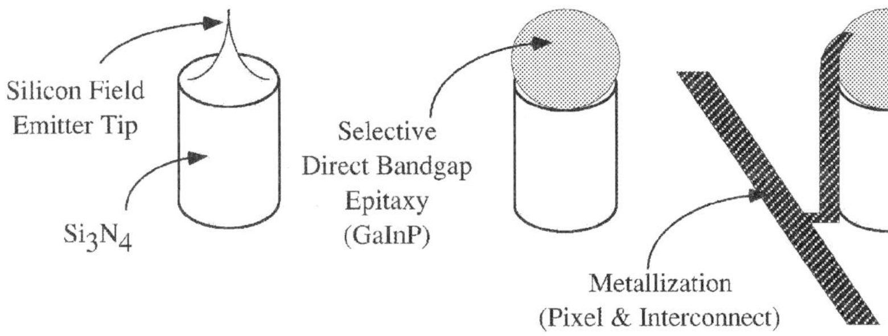

Fig. 46. Schematics of the fabrication process of a single pixel of Si–GaInP field-emitting tip proposed for flat panel display. (Source: Reprinted with permission from [175]. © 1997 American Institute of Physics.)

Fig. 47. SEM micrographs of submicrometer silicon tips coated by GaInP after (a) 1 min, (b) 3 min, and (c) 5 min. At the beginning, growth takes place only on top of the tip, but, at longer deposition times, material is also deposited in between the tips. (Source: Reprinted with permission from [175]. © 1997 American Institute of Physics.)

screens and vacuum conditions. The silicon tip can thus directly field emit electrons into the conduction band of the direct band gap material. Figure 46 shows schematically the pixel fabrication process. More details on the fabrication of the silicon tips with an onset voltage of 30 V for field emission into a vacuum are given elsewhere [176, 177].

It is important to cover all the bare silicon surrounding the sharp tips with Si_3N_4 to avoid nucleation of the III–V material. After growth of the direct gap material, the metallization is performed to connect each single pixel of the array. Fabrication of silicon field emission arrays proceeded by selective lateral thermal oxidation, which is described in more detail in [177]. A tip-to-tip distance of 3 μm for the tips with a radius of curvature of 10–20 nm was thus achieved. Figure 47 shows an SEM micrograph of a silicon tip after different GaInP coating times.

Photoluminescence in the visible consisted of single-peak spectra and a peak shift to higher wavelengths as compared to a GaInP reference layer grown on GaAs, indicating at least different alloy composition. Other effects may also contribute to this red shift. Both photoluminescence and cathodoluminescence images showed luminescence from the area between the tips for longer deposition times, probably coming from III–V material deposited between the tips as shown in the SEM micrographs. Thus, short deposition times, of around 1 min, have to be chosen to realize spatially well-defined tip luminescence. As the widths of the cathodoluminescence image have nearly the same dimensions as the SEM features, the authors excluded surface recombination processes.

4. SINGLE-ELECTRON ELECTRONICS

An electrically isolated nanostructure, such as a quantum dot, can be charged by electrons tunneling into the dot, which then become localized if the tunnel resistance is larger than the resistance quantum h/e^2 and if the energy level separation $e^2/2C$ within the dot is bigger than the thermal energy kT [178]. The second condition requires nanostructures less than or equal to 5 nm to reduce the total capacitance C of the island if Coulomb blockade effects are to be observed at room temperature. The charging energy $e^2/2C$ blocks further electrons from tunneling into the dot until the applied voltage exceeds the critical value $e/2C$ and the energy is higher than the next energy level in the dot. Thus, between $-e/2C$ and $+e/2C$, the energy level separation of $e^2/2C$ within the dot blocks any carrier transport and steplike I–V characteristics are obtained. Extending the device structure to a single-electron transistor (SET) by applying a gate voltage V to the dot via a capacitance C to modulate the field leads to oscillations of the current through the dot as a function of the applied gate voltage. Thus, if $CV/e = n$ is an integer value, the current is blocked. For $CV/e \pm \frac{1}{2} = n$, the current can flow through the dot [178].

Silicon-based SETs are regarded as particularly important because of their potential role in future electronic circuits. Single-electron electronics based on silicon-on-insulator (SOI) SETs have been discussed first by Ali and Ahmed [179] and Takahashi et al. [180]. Kurihara et al. [181] reduced the size of the islands to obtain Coulomb oscillations at nearly room temperature. Yano et al. [182] used a thin polysilicon film with small islands and demonstrated a room temperature operating single-electron memory device. Carrier transport through double-barrier structures has been demonstrated with SiO_2 barriers and well layers of amorphous Si [183], single-crystalline Si [184], microcrystalline Si [185, 186], and polycrystalline Si [187]. However, evidence of room temperature Coulomb blockade was not given in these tunneling studies.

Matsumoto et al. [188] used scanning tunneling microscopy (STM) lithography to fabricate a room temperature operating SET with a Coulomb staircase of approximately 150 mV. The width of the 30–50-nm-long TiO_x lines was 15–25 nm. The island size was $30\text{–}50 \times 30\text{–}50$ nm^2. The relevant tunnel junction area was only 2–3 nm (the thickness of the Ti film) $\times$ 30–50 nm (the length of the TiO_x line). For such small sizes, the tunnel capacitance is about 10^{-19} F, which enables room temperature operation of SETs with well-defined size.

Possible applications of SETs, such as very accurate current standards, sensitive electrometers [189], and single-electron memories [190, 191], have been demonstrated. For a detailed review of single-electron devices, we refer to the chapter by Chou on quantum dot transistors and memories in this book.

Eisele et al. [129] realized the use of Si wires and dots grown through micro shadow masks (see Section 3.8) for field-effect-controlled one- and zero-dimensional electronic confinement.

Figure 48 shows the schematics of a MOS device on a nonplanar silicon surface. No dry etching is used in this process. The (111) faceted silicon side walls are formed by a self-assembling process. In this structure, the nonplanar surface regions exhibit different threshold voltages for inversion, where the surface potential is larger than the difference between the intrinsic semiconductor Fermi level E_i and the gate material Fermi level. By solving Poisson's equation numerically for p-type silicon, the authors showed that the surface potential is lowered at convex edges, thus forming a quasi-one-dimensional wire before the side walls start to invert. Eisele et al. [1129] fabricated vertical MOS transistors with 85-nm channel lengths but different planar geometries as shown in the inset of Figure 49. Thus, they realized two or four edges within the width of the transistor. Figure 49 shows the conductance against applied gate voltage for both transistors. Although the wide transistor ($W = 80$ μm) shows a steplike behavior, it is not possible to resolve the structures because of the relatively large subthreshold current. Decreasing the width to 9 μm results in smaller

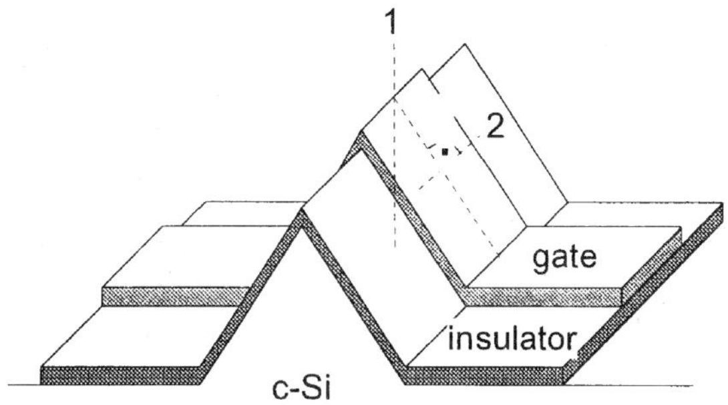

Fig. 48. Schematical cross section of the self-assembled growth mechanism on a nonplanar surface for 1D and 0D silicon devices. (Source: Reprinted from [129], with permission from Elsevier Science.)

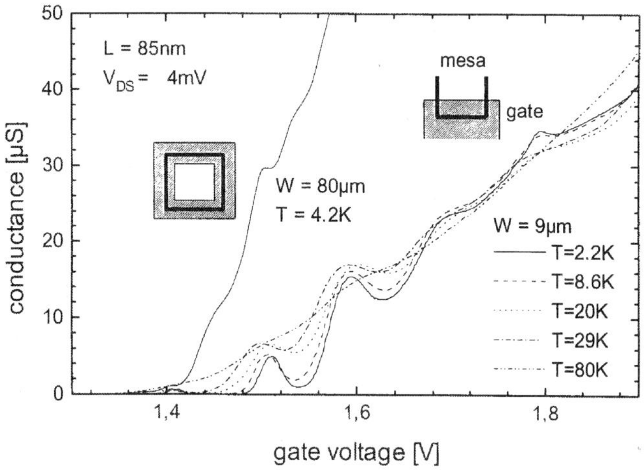

Fig. 49. Conductance versus varying gate voltage for two different gate configurations ($W = 80\ \mu$m and $W = 9\ \mu$m). The temperature dependence is related to the narrow device. The applied source–drain voltage was 4 meV. (Source: Reprinted from [129], with permission from Elsevier Science.)

parasitic 2D current, allowing resolution of these conductance oscillations, which are not observed for transistors without edges. The authors explained the observed steplike conductance as arising from the 1D subband crossing of the Fermi level with increasing gate voltage. The mobility near the subband edges is reduced because of carrier intersubband scattering into the next subband. From the presence of the oscillations up to about 80 K, the authors deduced an energy level separation corresponding to kT of 6 meV.

Coulomb staircases in the I–V characteristics [192] and Coulomb blockade effects [193, 194] were observed at room temperature owing to the strong confinement in silicon quantum dots. Fukuda et al. [193] fabricated self-assembled Si quantum dots by LPCVD on native and thermally grown SiO_2 that had been etched back to a thickness of 1 nm. The height and diameter of the dots varied between 1.5 to 8 nm and 10 to 20 nm, respectively. A top barrier layer of 1 nm thickness was achieved by the same means. The I–V characteristics were measured on a single dot using AFM with a conductive cantilever. The dots had a height of 2.7 and 2.8 nm. The I–V characteristics were also simulated using the well-known transfer matrix approach [102, 195]. Some clear discrepancies were attributed to a possible leakage current through the SiO_2 barrier and to Si/SiO_2 interface states and traps, which smear out the resonance current.

As a quantum dot is a three-dimensional confined electronic system, its discrete energy spectrum potentially permits the storage of single electrons. Nakajima et al. [194, 196]

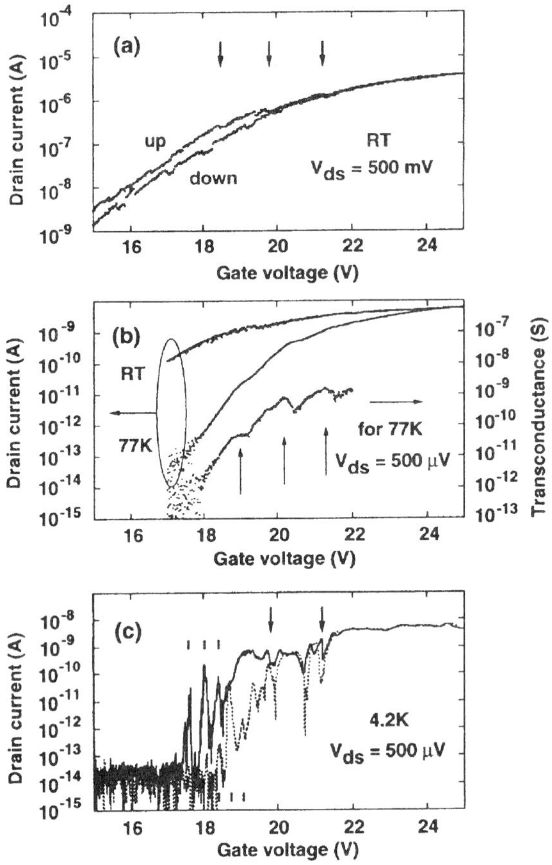

Fig. 50. Drain current versus gate voltage of a single-electron tunneling transistor at (a) room temperature ($V_{DS} = 500$ mV), (b) room temperature and 77 K with transconductance versus gate voltage ($V_{DS} = 500\ \mu$V), and (c) 4.2 K with down sweeps represented by the dotted line. The Coulomb oscillation peaks are denoted by the arrows. (Source: Reprinted from [194]. © 1997 American Institute of Physics.)

realized a Si single-electron tunneling transistor with a nanoscale floating dot stacked on a Coulomb island by a self-aligned process as a possible electron memory device. The control gate consisted of polysilicon. The Coulomb island had a width of about 40–50 nm and a length of 50–60 nm. The floating dot on top of the Coulomb island had a side length of about 20–30 nm.

Figure 50 shows the drain current versus the control gate voltage at 4.2 K, 77 K, and room temperature. The drain current exhibits oscillations, denoted by the arrows, because of the Coulomb blockade effect. Sweeping the voltage up and down in parts a and c of Figure 50, denoted by the dotted line for down sweeping, leads to a hysteresis effect. The oscillations in the drain current do not occur at the same voltages. Nakajima et al. [194, 196] demonstrated the Coulomb blockade effect via tunneling of a single electron from the channel to the floating dot gate. If one can control the device design such that the drain current change from a valley to a peak is influenced by only a single electron, this device could be used to detect single electrons and, because of the hysteresis effect in the I–V characteristics, as an ultralow-power-consumption memory.

Guo et al. [197–199] fabricated a silicon SET memory that operates at room temperature and is suitable for ultralarge-scale integration (ULSI). A conventional floating-gate MOS memory was scaled down in order to realize a single-electron MOS memory (SEMM).

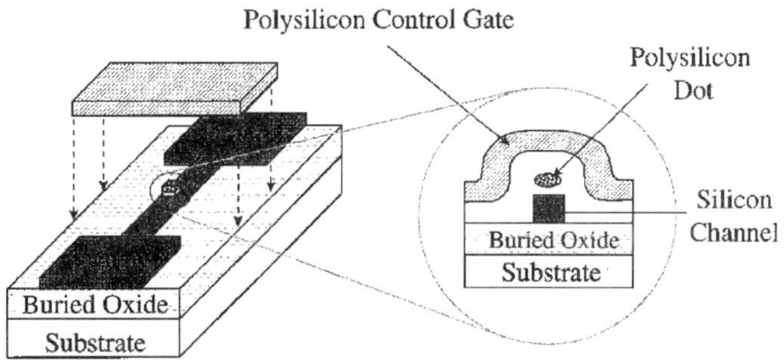

Fig. 51. Schematic drawing of a single-electron MOS memory using single-electron charging of a polysilicon nanodot to switch the gate threshold. (Source: Reprinted from [197]. © 1997 American Institute of Physics.)

However, the channel width was narrower than the Debye screening length of a single electron and the floating gate was a nanoscale square dot. Figure 51 shows a schematic view of this SEMM. As a floating gate, Guo et al. deposited an 11-nm-thick polysilicon film on top of the 35-nm-thick crystalline silicon layer of the SOI wafer. In between the crystalline silicon and the polysilicon, there was an approximately 1-nm-thick native oxide layer. No additional tunnel oxide was formed to enable fast charging of the dot through the channel. Furthermore, the number of electrons stored on the dot was regulated by the Coulomb blockade mechanism, as the potential difference between the channel and the dot is minimized for a given charging voltage. In a two-level step using electron beam lithography and reactive ion etching, the widths of the floating gate and the silicon channel were patterned first; then patterning in the perpendicular direction was carried out to form the squarelike polysilicon dot.

For the smallest device, the width of the channel was roughly estimated at approximately 10 nm. The approximately 7-nm × 7-nm square dot had a height of about 2 nm. To determine the electrical properties of the smallest device, a source–drain voltage of 50 meV was applied. Despite a continuous charging voltage, the threshold voltage of the device made discrete shifts of 55 meV, corresponding to a charging voltage interval of approximately 4 V. Because of the very thin native oxide in between the channel and the floating dot, the charge could be stored for about 5 s after setting the gate potential to ground. This shifted the threshold voltage to its original value. The calculated capacitance for an approximately 7-nm × 7-nm square dot is about 4.4×10^{-20} F, giving a single-electron charging voltage of 3.6 V, which corresponds well to the experimental value of about 4 V. The shift of the threshold voltage was estimated from the single-electron Debye screen length (~70 nm) and the channel thickness (~26 nm) to be approximately 64 meV, also corresponding well with the experimental value of 55 meV.

For single-electron computing, many interesting circuits have been proposed setting up Boolean logic circuits using Coulomb interactions between bistable charge polarization states [200, 201]. The main drawback of most of the proposed schemes is the missing isolation between the input and output terminals so that the input can thus drive the output but not vice versa. As the Coulomb interaction between two identical charge polarizations is reciprocal, the output can also drive the input, violating the necessity of unidirectionality and thus leading to the so-called "catastrophic failure" [202, 203].

To realize von Neumann computing architectures, proposals for dissipationless computing [204] have been made. The drawback of this scheme is the fault tolerance. Bandyopadhyay and Roychowdhury [203, 205, 206] have proposed a nanoelectronic architecture to overcome these drawbacks. However, it is still too early to comment on the chances of success of any given approach for logic operations.

5. TIPS FOR ATOMIC FORCE MICROSCOPY AND FIELD EMISSION

The displacement of single atoms on a surface is already possible using scanning probe techniques allowing, for example, new molecular arrangements to be realized for a variety of purposes. Some of these include integrated microscopic gears, pumps, motors, membranes, and nanotweezers. Whereas silicon-based microelectronic devices switch electron fluxes through thin metallic wires, micromechanical devices could establish the contact between microelectronics and the physical world. Because of the predominance of silicon-based microelectronics, this material is favored in micromechanics research and development. For a comprehensive overview, we refer the reader to *Scientific American, Special Dossier: Microsystem Technology 1996*.

Silicon is also the favored material for cold field emitter devices for nanolithography and for flat panel devices. Recent studies [207–210] have shown that miniaturization of the electron beam column leads to a well-focused and bright electron beam source. Such microcolumns can be used for scanning electron microscopy and electron beam lithography with high resolution [210, 211]. The main disadvantage of electron beam lithography having a low throughput could potentially be overcome, using arrays of electron beam microcolumns. The tip sharpness, gate diameter, and gate-to-tip distance influence the turn-on voltage and the emission current of field emission devices. Thus, very sharp tips with a small gate diameter and a small gate–tip distance are desired. Rakhshandehroo and Pang [212] have fabricated 100 single Si tip arrays with a spacing of 1 μm. Using dry etching and low-temperature plasma oxidation led to a sharpening of the tips with radius from 67 down to 8 nm, leading, in turn, to a lower turn-on voltage and to a higher emission current because of the higher electric field at the tip apex.

To reduce further the voltage applied to the anode, Jung et al. [213] fabricated nanosize Si tips with a tip diameter smaller than 20 nm. After reactive ion etching, a tip radius of about 300 nm was determined. To sharpen the tip to 20 nm, a low-temperature oxidation at 900 °C with a subsequent dip in BHF was carried out. Low-temperature oxidation leads to an enhanced oxide growth in the regions with high curvature. The diameter of the tip was reduced in this fashion to less than 20 nm. A 30-nm amorphous hydrogen free diamond-like carbon (DLC) film was deposited on top of this structure by PECVD. The film is composed of a diamond-like sp^3 and a graphite-like sp^2 phase.

The emission current of the DLC film is twice as high as the pure Si tip, and the onset of the field emission is shifted to smaller voltages from 700 to 600 V. At higher voltages, the emission current of the pure Si tip was zero because of deterioration, whereas the DLC-coated tip exhibited an emission current of approximately 100 μA at 1.5 kV.

Figure 52 shows a silicon tip with a silicon cantilever used in atomic force microscopy (AFM) fabricated by Dr. O. Wolter GmbH in Germany. The pyramidal tip points to the $\langle 100 \rangle$ direction. The tips have a radius less than or equal to 15 nm and are 10–15 μm high. The thickness of the cantilever with a trapezoidal cross section ranges from 0.3 to 8, 20, 60, 125, and 450 μm, depending on the specific applications. A sharp tip of 2 nm radius has been achieved (see the right-hand side of Fig. 52). The half cone angle is better than 10° at the last 200 nm of the tip and its mechanical robustness permits fast scan rates. The ability to control single atoms could potentially enable the storage of 10^{12} GB/cm^2 on a silicon chip. Such a data storage system would beat all conventional systems because of its increased capacity and lack of movable parts such as a rotating disk.

MacDonald et al. [214] have fabricated an STM tip with a diameter of 20 nm mounted on a floating frame with four motors, each 400 μm in diameter. These motors can pull or push the tip vertically and horizontally up to one million times per second and even tilting of the tip is possible. With a matrix of several hundred tips, a direct surface structuring in a parallel and thus a very fast process could be realized.

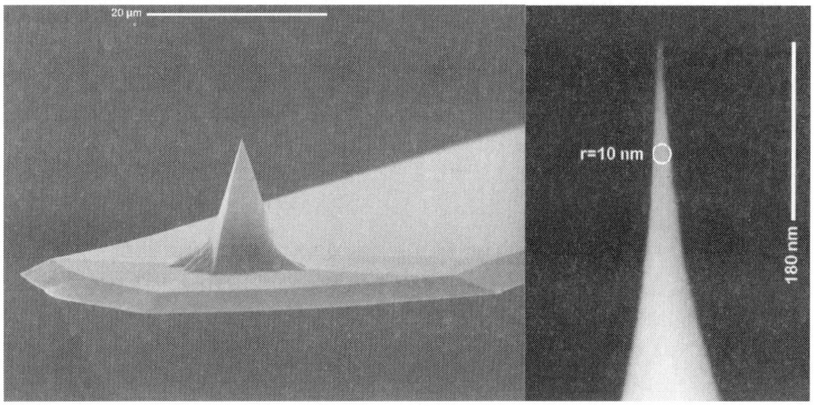

Fig. 52. Silicon tip together with a monolithic cantilever used in atomic force microscopy (left). Supersharp silicon tip for high-resolution imaging of nanostructures and microroughness. The tip radius is only 2 nm (right). (Source: Reprinted with permission from NANOSENSORS.)

The use of Si tips in scanning probe–based lithography is increasing fast although formally still an the laboratory scale. Madsen et al. [215] wrote patterns with feature sizes down to 50 nm with a scanning near-field optical microscope (SNOM) using Al-coated fiber probes. Nagahara et al. [216] used the tip to expose a resist to extremely low energy electrons. Minne et al. [217] used a mix-and-match lithography to fabricate a metal oxide semiconductor field effect transistor (MOSFET). The electrodes at the gate were defined by atomic force microscopy (AFM), whereas the remaining levels were defined by optical lithography. Line widths between 60 and 700 nm were achieved. Abeln et al. [218] used STM lithography to fabricate structures smaller than 10 nm. Using STM, Matsumoto et al. [188] fabricated an SET with line widths of 15–25 nm.

Field-enhanced oxidation as a lithographic tool was first introduced by Dagata et al. [219]. They used a 5–10-V negatively charged STM tip under ambient air conditions. The huge electric field beneath the tip selectively desorbs passivating hydrogen atoms and molecules from the silicon surface, allowing bare silicon to oxidize under the influence of ambient air in the presence of huge fields generated at the tip. Lyding et al. [220] wrote 1.5-nm line widths in an UHV system pushing the resolution of this approach. During pattern transfer, by wet or dry etching, a high etch selectivity is essential as the oxide thickness is less than 5 nm.

In many cases, AFM tips consist of Si_3N_4 coated with a thin layer such as Ti to lengthen its lifetime. Birkelung et al. [221] have used Ni tips coated with a 5-nm Ti adhesion layer and a 100-nm Au cap layer as a plating base between the Ni and Si. They achieved 80-nm line widths writing over 1000 lines before the tip degraded. The partial Au deposition during the first few scans onto the surface had to be removed by etching.

Snow et al. [222] used AFM for local oxidation and subsequent RIE to fabricate approximately 10-nm-wide lines with a height of 30 nm. They used an n-type Si wafer hydrogen passivated in 10% aqueous solution after ozone treatment. Using a Si_3N_4 AFM tip coated with 30-nm Ti, they oxidized locally the Si surface by applying a negative voltage between 4 and 7 V under a scanning speed of 1–10 μm/s. To obtain fine patterns, they worked with low voltages and used slow scanning speeds. The ambient humidity was kept constant to 50% as this is a key process parameter.

The addition of oxygen has been found to increase the etch selectivity to more than 20. Figure 53 shows an AFM image of several lines of varying width after dry etching to a depth of 30 nm. The two lines marked by arrows in Figure 53 have a width between 25 to 30 nm (right) and 35 and 40 nm (left), respectively. Subtracting the convolution with the 20-nm tip diameter, a real line width for the smallest line of less than 10 nm is obtained.

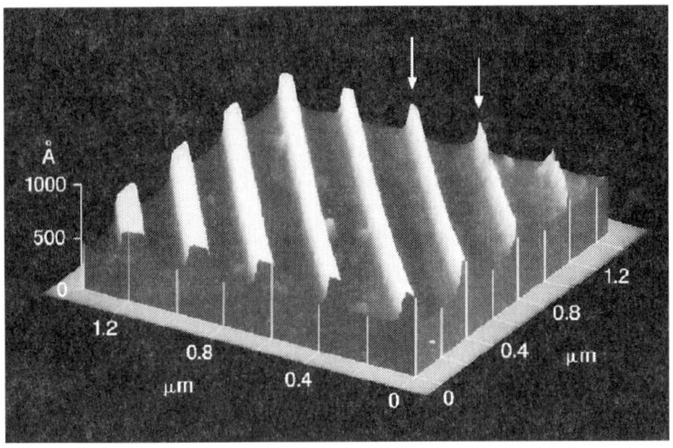

Fig. 53. AFM image of dry-etched Si nanowires of varying width etched to a depth of 30 nm. The arrows indicate two lines with widths between 25–30 nm and 35–40 nm. The real line widths are about 20 nm less because of a convolution effect with the tip diameter of the AFM. (Source: Reprinted from [222]. © 1995 American Institute of Physics.)

Snow et al. [223] demonstrated the use of silicon as a resist material for patterning GaAs using scanning probe lithography. Thus, combining this technique with an UHV system, true nanostructures may be realized in Si as well as in III–V wafers.

6. CONCLUSIONS

We have attempted to provide an overview concerning recent developments involving silicon-based nanostructures. It is almost certain we have not done justice to the many scientists and engineers who have contributed to the field. This is mainly a representive list of examples determined by the usual constraints of space, time, and energy.

In the quest to realize a variety of nanostructures using silicon and silicon-related semiconductors, the trend to miniaturization in electronics plays an overwhelming role. To some degree, this places an enormous burden on other developments that could potentially make use of nanostructures for other applications. The range of other applications comes under the umbrella of nanotechnology as demonstrated by the work on nanotips. It is clear that extension of the use of Si nanostructures into the biological and environmental areas has begun; however, this has not been touched upon here. The scientific and technological challenges are still formidable and there are no clear winners yet in, for example, a silicon-based light emitter suitable for optical interconnects. Our understanding is still too incomplete and we continue to learn about the impact of the local environment, in the nanometer scale, of optically active centers in silicon. On the other hand, the wealth of technological expertise developed collectively in silicon electronics ensures that any relevant scientific development in this area, which survives the acid test of industrial suitability and device specifications, could be made available to potential users because the connections to the outside world are provided by existing input and output Si-based circuitry and technology. Whereas this may well be the case in sensors and optical devices, there are significant challenges remaining in the area of quantum nanoscale electronic circuits, which is bound to keep the Si nanoelectronics community very active for years to come.

Acknowledgments

We are indebted to the many authors and colleagues who kindly provided us with photos and figures of their work as well as authors who gave permission to reproduce their published figures. Of course, any misinterpretation is our sole responsibility. The dry-etched

Si–SiGe dot work reviewed here was mostly carried while one of the authors (CMST) was at the University of Glasgow jointly with Y.-S. Tang and C. D. W. Wilkinson and G. V. Hansson and W.-X. Ni of Linköping University. Samples for the work at Glasgow University were also kindly provided by E. H. C. Parker of Warwick University and H. Presting of Daimler-Benz Research Laboratories in Ulm. The more recent work on deep-etched Si–SiGe dots has been supported by the European Commission ESPRIT Project 22644 SIBLE. We thank T. Maka, T. Köpke and our colleagues for their help with the preparation of this chapter. This work was partially supported by the University of Wuppertal.

References

1. G. Davies, *Phys. Rep.* 176, 83 (1989).
2. D. J. Lockwood, in "Semiconductors and Semimetals" (R. K. Williardson and E. R. Weber, eds.), Vol. 49, p. 1. Academic Press, San Diego, 1997.
3. S. S. Iyer and Y.-H. Xie, *Science* 260, 40 (1993).
4. J. R. Chelikowsky and M. L. Cohen, *Phys. Rev. B* 14, 556 (1976).
5. L. C. Kimberling, K. D. Kolenbrander, J. Michel, and J. Palm, *Solid State Phys.* 50, 333 (1997).
6. U. Gnutzmann and K. Clausecker, *Appl. Phys.* 3, 9 (1974).
7. E. Kasper and F. Schäffler, in "Semiconductors and Semimetals" (T. P. Pearsall, ed.), Vol. 33, p. 223. Academic Press, Boston, 1991.
8. S. Froyen, D. M. Wood, and A. Zunger, *Phys. Rev. B* 36, 4547 (1987).
9. T. P. Pearsall, J. Bevk, L. C. Feldman, J. M. Bonar, J. P. Mannaerts, and A. Ourmazd, *Phys. Rev. Lett.* 58, 729 (1987).
10. R. J. Turton and M. Jaros, *Mater. Sci. Eng., B* 7, 37 (1990).
11. U. Menczigar, J. Brunner, E. Friess, M. Gail, G. Abstreiter, H. Kibbel, H. Presting, and E. Kasper, *Thin Solid Films* 222, 227 (1992).
12. U. Menczigar, G. Abstreiter, J. Olajos, H. Grimmeiss, H. Kibbel, H. Presting, and E. Kasper, *Phys. Rev. B* 47, 4099 (1993).
13. J. Engvall, J. Olajos, G. Grimmeiss, H. Kibbel, and H. Presting, *Phys. Rev. B* 51, 2001 (1995).
14. S. C. Jain, in "Germanium–Silicon Strained Layers and Heterostructures." Academic Press, Boston, 1994.
15. K. Brunner, K. Eberl, and W. Winter, *Phys. Rev. Lett.* 76, 303 (1996).
16. K. Brunner, W. Winter, and K. Eberl, *Appl. Phys. Lett.* 69, 1279 (1996).
17. S. Oguz, W. Paut, T. F. Deutsch, B.-Y. Tsaur, and D. V. Murphy, *Appl. Phys. Lett.* 43, 848 (1983).
18. J. Derrien, J. Chevrier, V. Le Thanh, T. E. Crumbaker, J. Y. Natoli, and I. Berbezier, *Appl. Surf. Sci.* 70–71, 546 (1993).
19. M. C. Bost and J. E. Mahan, *J. Appl. Phys.* 58, 2696 (1996).
20. L. Miglio, F. Tarazza, and G. Malegori, *Appl. Phys. Lett.* 67, 1 (1995).
21. D. N. Leong, M. A. Harry, K. J. Reeson, and K. P. Homewood, *Appl. Phys. Lett.* 68, 1649 (1996).
22. D. Leong, M. Harry, K. J. Reeson, and K. P. Homewood, *Nature* 387, 686 (1997).
23. Z. Yang, K. P. Homewood, M. S. Finney, M. Harry, and K. J. Reeson, *J. Appl. Phys.* 78, 1958 (1995).
24. H. Ennen, J. Schneider, G. Pomrenke, and A. Axmann, *Appl. Phys. Lett.* 43, 943 (1983).
25. H. Ennen, G. Pomrenke, A. Axmann, K. Eisele, W. Haydl, and J. Schneider, *Appl. Phys. Lett.* 46, 381 (1985).
26. J. Michel, J. L. Benton, R. F. Ferrante, D. C. Jacobson, D. J. Eaglesham, E. A. Fitzgerald, Y.-H. Xie, J. M. Poate, and L. C. Kimerling, *J. Appl. Phys.* 70, 2672 (1991).
27. S. Coffa, G. Franzò, F. Priolo, A. Polman, and A. Serna, *Phys. Rev. B* 49, 16313 (1994).
28. G. N. van den Hoven, J. H. Shin, A. Polman, S. Lombardo, and S. U. Campisano, *J. Appl. Phys.* 78, 2642 (1995).
29. B. Zheng, J. Michel, F. Y. G. Ren, L. C. Kimerling, D. C. Jacobson, and J. Poate, *Appl. Phys. Lett.* 64, 2842 (1994).
30. G. Franzò, F. Priolo, S. Coffa, A. Polman, and A. Carnera, *Nucl. Instrum. Methods Phys. Res., Sect. B* 96, 374 (1995).
31. A. Polman, J. S. Custer, E. Snoeks, and G. N. van den Hoven, *Nucl. Instrum. Methods Phys. Res., Sect. B* 80/81, 653 (1993).
32. D. C. Adler, D. C. Jacobson, D. J. Eaglesham, M. A. Marcus, J. L. Benton, J. M. Poate, and P. H. Citrin, *Appl. Phys. Lett.* 61, 2181 (1992).
33. F. Priolo, G. Francò, S. Coffa, A. Polman, S. Libertino, R. Barklie, and D. Carey, *J. Appl. Phys.* 78, 3874 (1995).
34. W.-X. Ni, K. B. Joelsson, C.-X. Du, I. A. Buyanova, G. Pozina, W. M. Chen, G. V. Hansson, B. Monemar, J. Cardenas, and B. G. Svensson, *Appl. Phys. Lett.* 70, 3383 (1997).

35. F. Y. G. Ren, J. Michel, Q. Sun-Paduano, B. Zheng, H. Kitagawa, D. C. Jacobson, J. M. Poate, and L. C. Kimerling, *Mater. Res. Soc. Symp. Proc.* 301, 87 (1993).
36. S. Coffa, G. Franzò, and F. Priolo, *MRS Bull.* 23, 25 (1998).
37. J. Stimmer, A. Reittinger, J. F. Nützel, G. Abstreiter, H. Holzbrecher, and C. Buchal, *Appl. Phys. Lett.* 68, 3290 (1996).
38. R. Sterna, J. H. Shin, M. Lohmeier, E. Vlieg, A. Polman, and P. F. A. Alkemade, *J. Appl. Phys.* 79, 2658 (1996).
39. M. Matsuoka and S.-I. Tohno, *Appl. Phys. Lett.* 66, 2751 (1995).
40. Y.-H. Xie, E. A. Fitzgerald, and Y. J. Mii, *J. Appl. Phys. Lett.* 70, 3223 (1990).
41. S. Coffa, F. Priolo, G. Franzò, A. Polman, S. Libertino, M. Saggio, and A. Carnera, *Nucl. Instrum. Methods Phys. Res., Sect. B* 106, 386 (1995).
42. C.-X. Du, W.-X. Ni, K. B. Joelsson, F. Duteil, and G. V. Hansson, Proc. of E-MRS'98 Spring Symposium "Light Emission from Silicon: Progress towards Si-Based Optoelectronics" (in press).
43. C. Weisbuch and B. Vinter, in "Quantum Semiconductor Structures." Academic Press, Boston, 1991.
44. J. H. Davies and A. R. Long, in "Proceedings of the 38th Scottish Universities Summer School in Physics" (P. Osborne, ed.), Vol. 38. SUSSP, 1991.
45. E. Kasper, in "EMIS Datareviews," Vol. 12. INSPEC, London, 1988.
46. A. Schuppen, U. Erben, A. Gruhle, H. Kibbel, H. Schumacher, and U. Konig, *Tech. Digest Int. Electron Device Meeting* 743 (1995).
47. S. Perkowitz, in "Optical Characterization of Semiconductors: Infrared, Raman and Photoluminescence Spectroscopy" (N. H. March, ed.). Academic Press, London, 1993.
48. J. Weber and I. Alonso, *Phys. Rev. B* 40, 5683 (1989).
49. L. T. Canham, *Appl. Phys. Lett.* 57, 1046 (1990).
50. A. G. Cullis and L. T. Canham, *Nature* 353, 335 (1991).
51. M. S. Brandt, H. D. Fuchs, M. Stutzmann, J. Weber, and M. Cardona, *Solid State Commun.* 81, 307 (1992).
52. S. M. Prokes, O. J. Glembocki, V. M. Bermudez, R. Kaplan, L. E. Friedersdorf, and P. C. Perason, *Phys. Rev. B* 45, 13788 (1992).
53. Y.-H. Xie, W. L. Wilson, F. M. Ross, J. A. Mucha, E. A. Fitzgerald, and J. M. Macaulay, *J. Appl. Phys.* 71, 2403 (1992).
54. E. Bassous, M. Freeman, J.-M. Halbout, S. S. Iyer, V. P. Kesan, P. Munguia, S. F. Pesarcik, and B. L. Williams, *Mater. Res. Soc. Symp. Proc.* 256, 23 (1992).
55. V. P. Kesan, E. Bassous, P. Munguia, S. F. Pesarcik, M. Freeman, and S. S. Iyer, *J. Vac. Sci. Technol., A* 11, 1736 (1993).
56. P. D. G. Calcott, K. J. Nash, L. T. Canham, M. J. Kane, and D. Brumhead, *J. Phys.: Condens. Matter* 5, 91 (1993).
57. F. Buda, J. Kohanoff, and M. Parrinello, *Phys. Rev. Lett.* 69, 1272 (1992).
58. A. J. Read, R. J. Needs, K. J. Nash, L. T. Canham, P. D. J. Calcott, and A. Qteish, *Phys. Rev. Lett.* 69, 1232 (1992).
59. C. J. Hwang, *Phys. Rev. B* 8, 646 (1973).
60. T. P. Pearsall, J. C. Adams, J. N. Kidder, Jr., P. S. Williams, S. A. Chambers, J. Lach, D. T. Schwartz, and B. Z. Nosho, *Thin Solid Films* 222, 200 (1992).
61. G. Bai, K. H. Kim, and M.-A. Nicolet, *Appl. Phys. Lett.* 57, 2247 (1990).
62. T. P. Pearsall, J. C. Adams, J. E. Wu, B. Z. Nosho, C. Aw, and J. C. Patton, *J. Appl. Phys.* 71, 4470 (1992).
63. S. Eisebitt, J. Luning, J.-E. Rubensson, T. Van Buuren, S. N. Patitsas, T. Tiedje, M. Berger, R. Arens-Fischer, S. Frohnhoff, and W. Eberhardt, *Solid State Commun.* 97, 549 (1996).
64. S. Nozaki, S. Sato, H. Ono, H. Morisaki, and M. Iwase, *Nucl. Instrum. Methods Phys. Res., Sect. B* 121, 455 (1997).
65. Y. Saito, *J. Cryst. Growth* 47, 61 (1979).
66. M. A. Tischler, R. T. Collins, J. H. Stathis, and J. C. Tsang, *Appl. Phys. Lett.* 60, 639 (1992).
67. A. Loni, A. J. Simons, T. I. Cox, P. D. J. Calcott, and L. T. Canham, *Electron. Lett.* 31, 1288 (1995).
68. K. D. Hirschman, L. Tsybeskov, S. P. Duttagupta, and P. M. Fauchet, *Nature* 384, 338 (1996).
69. L. Tsybeskov, S. P. Duttagupta, K. D. Hirschman, and P. M. Fauchet, *Appl. Phys. Lett.* 68, 2058 (1996).
70. Y.-S. Tang, W.-X. Ni, C. M. Sotomayor Torres, and G. V. Hansson, *Electron. Lett.* 31, 1385 (1995).
71. Y. S. Tang, C. M. Sotomayor Torres, W. X. Ni, and G. V. Hansson, *Superlattices Microstruct.* 20, 505 (1996).
72. Y.-S. Tang, S. E. Hicks, C. M. Sotomayor Torres, C. D. W. Wilkinson, W.-X. Ni, and G. V. Hansson, in "23rd International Conference on the Physics of Semiconductors, Berlin, 1996," p. 1628. World Scientific, Singapore, 1996.
73. Y. S. Tang, C. M. Sotomayor Torres, S. Nilsson, B. Dietrich, W. Kissinger, T. E. Whall, E. H. C. Parker, H. Presting, and H. Kibbel, *Solid State Electron.* 40, 383 (1996).
74. Y.-S. Tang, C. M. Sotomayor Torres, B. Dietrich, W. Kissinger, T. E. Whall, and E. H. C. Parker, *Solid State Commun.* 94, 369 (1995).
75. Y.-S. Tang, C. D. W. Wilkinson, D. W. Smith, T. E. Whall, and E. H. C. Parker, *J. Appl. Phys.* 33, 2348 (1994).

76. Y.-S. Tang, C. D. W. Wilkinson, C. M. Sotomayor Torres, D. W. Smith, T. E. Whall, and E. H. C. Parker, *Solid State Commun.* 85, 199 (1993).
77. Y.-S. Tang, *J. Appl. Phys.* 69, 8289 (1991).
78. Y.-S. Tang, C. D. W. Wilkinson, C. M. Sotomayor Torres, D. W. Smith, T. E. Whall, and E. H. C. Parker, *Appl. Phys. Lett.* 63, 497 (1993).
79. W.-X. Ni, J. Birch, Y.-S. Tang, K. B. Joelsson, C. M. Sotomayor Torres, Å. Kvick, and G. V. Hansson, *Thin Solid Films* 294, 300 (1997).
80. A. G. Nassiopoulos, S. Grigoropoulos, and D. Papadimitriou, *Appl. Phys. Lett.* 69, 2267 (1996).
81. A. G. Nassiopoulos, S. Grigoropoulos, D. Papadimitriou, and E. Gogolides, *Phys. Status Solidi B* 190, 91 (1995).
82. W. Chen and H. Ahmed, *Appl. Phys. Lett.* 63, 1116 (1993).
83. H. I. Liu, D. K. Biegelsen, F. A. Ponce, N. M. Johnson, and R. F. W. Pease, *Appl. Phys. Lett.* 64, 1383 (1994).
84. J. Kedzierski, J. Bokor, and C. Kisielowski, *J. Vac. Sci. Technol., B* 15, 2825 (1997).
85. A. S. Chu, S. H. Zaidi, and S. R. J. Brueck, *Appl. Phys. Lett.* 63, 905 (1993).
86. D. J. Lockwood, *Solid State Commun.* 92, 101 (1994).
87. H. Sunamura, N. Usami, Y. Shiraki, and S. Fukatsu, *Appl. Phys. Lett.* 66, 3024 (1995).
88. P. Schittenhelm, M. Gail, and G. Abstreiter, *J. Cryst. Growth* 157, 260 (1995).
89. P. Schittenhelm, M. Gail, J. Brunner, J. F. Nützel, and G. Abstreiter, *Appl. Phys. Lett.* 67, 1292 (1995).
90. H. Sunamura, S. Fukatsu, N. Usami, and Y. Shiraki, *J. Cryst. Growth* 157, 265 (1995).
91. B. Cunningham, J. O. Chu, and S. Akbar, *Appl. Phys. Lett.* 59, 3574 (1991).
92. A. Hartmann, L. Vescan, C. Dieker, and H. Lüth, *Mater. Sci. Technol.* 11, 410 (1995).
93. P. O. Hansson, M. Albrecht, W. Dorsch, H. P. Strunk, and E. Bauser, *Phys. Rev. Lett.* 73, 444 (1994).
94. M. Albrecht, S. Christiansen, J. Michler, W. Dorsch, and H. P. Strunk, *Appl. Phys. Lett.* 67, 1232 (1995).
95. S.-I. Yanagiya, S. Kamimura, M. Fujii, M. Ishida, Y. Moriyasu, and M. Matsui, *Appl. Phys. Lett.* 71, 1409 (1997).
96. M. Yoshimoto, T. Maeda, T. Ohnishi, H. Koinuma, O. Ishiyama, M. Shinohara, M. Kubo, R. Minra, and A. Miyamoto, *Appl. Phys. Lett.* 67, 2615 (1995).
97. Y. Kanemitsu, *Phys. Rev. B* 49, 16845 (1994).
98. W. B. Dubbelday, D. M. Szaflarski, R. L. Shimabukuro, S. D. Russel, and M. J. Sailor, *Appl. Phys. Lett.* 62, 1694 (1993).
99. P. Schittenhelm, G. Abstreiter, A. Darhuber, G. Bauer, P. Werner, and A. Kosogov, *Thin Solid Films* 294, 291 (1997).
100. J. Eaglesham and M. Cerullo, *Phys. Rev. Lett.* 64, 1943 (1990).
101. A. A. Darhuber, J. Stangl, G. Bauer, P. Schittenhelm, and G. Abstreiter, *Thin Solid Films* 294, 296 (1997).
102. T. M. Kalotas and A. R. Lee, *Am. J. Phys.* 59, 48 (1991).
103. T. S. Kuan and S. S. Iyer, *Appl. Phys. Lett.* 59, 2242 (1991).
104. J. Y. Yao, T. G. Andersson, and G. L. Dunlop, *J. Appl. Phys.* 69, 2224 (1991).
105. L. Vescan, W. Jäger, C. Dieker, K. Schmidt, A. Hartmann, and H. Lüth, *Mater. Res. Soc. Symp. Proc.* 263, 23 (1992).
106. Q. Xie, A. Madhukar, P. Chen, and N. P. Kobayashi, *Phys. Rev. Lett.* 75, 2542 (1992).
107. J. Tersoff and F. K. Le Goues, *Phys. Rev. Lett.* 72, 3570 (1994).
108. J. Tersoff, C. Teichert, and M. G. Lagally, *Phys. Rev. Lett.* 76, 1675 (1996).
109. L. Vescan, C. Dieker, A. Hartmann, and A. van der Hart, *Semicond. Sci. Technol.* 9, 387 (1994).
110. L. Vescan, *Thin Solid Films* 294, 284 (1997).
111. L. Vescan, C. Dieker, A. Souifi, and T. Stoica, *J. Appl. Phys.* 81, 6709 (1997).
112. A. Aoyama, T. Ikarashi, K. Miyanaga, and T. Tatsumi, *J. Cryst. Growth* 136, 349 (1994).
113. P. Rai-Choudhury and D. K. Schroeder, *J. Electrochem. Soc.* 120, 664 (1973).
114. L. Vescan, W. Jäger, C. Dieker, K. Schmidt, A. Hartmann, and H. Lüth, *Mater. Res. Soc. Symp. Proc.* 263, 23 (1992).
115. T. Stoica and L. Vescan, *J. Cryst. Growth* 131, 32 (1993).
116. K. D. Hobart, F. J. Kub, H. F. Gray, M. E. Twigg, D. Park, and P. E. Thompson, *J. Cryst. Growth* 157, 338 (1995).
117. G. R. Booker and B. A. Joyce, *Philos. Mag.* 14, 301 (1969).
118. Y.-N. Yang and E. D. Williams, *J. Vac. Sci. Technol., A* 8, 2481 (1990).
119. H. Hirayama, M. Hiroi, and T. Ide, *Phys. Rev. B* 48, 17331 (1993).
120. Y.-N. Yang and E. D. Williams, *J. Vac. Sci. Technol., A* 8, 2481 (1990).
121. A. Oshiyama, *Phys. Rev. Lett.* 74, 130 (1995).
122. E. Kapon, K. Kash, E. M. Clausen, Jr., D. M. Hwang, and E. Colas, *Appl. Phys. Lett.* 60, 477 (1992).
123. S. Tsukamoto, Y. Nagamune, M. Nishioka, and Y. Arakawa, *Appl. Phys. Lett.* 62, 49 (1993).
124. N. Usami, T. Mine, S. Fukatsu, and Y. Shiraki, *Appl. Phys. Lett.* 63, 2789 (1993).
125. N. Usami, T. Mine, S. Fukatsu, and Y. Shiraki, *Appl. Phys. Lett.* 64, 1126 (1994).
126. E. Hammerl, F. Wittmann, J. Messarosch, I. Eisele, V. Huber, and H. Oppolzer, *Mater. Res. Soc. Symp. Proc.* 220, 27 (1991).
127. J. Brunner, T. S. Rupp, H. Gossner, R. Ritter, I. Eisele, and G. Abstreiter, *Appl. Phys. Lett.* 64, 994 (1994).

128. J. Brunner, W. Jung, P. Schittenhelm, M. Gail, G. Abstreiter, J. Gonderman, B. Hadam, T. Koester, B. Spangenberg, H. G. Roskos, H. Kurz, H. Gossner, and I. Eisele, *J. Cryst. Growth* 157, 270 (1995).
129. I. Eisele, H. Baumgärtner, and W. Hansch, *J. Cryst. Growth* 157, 248 (1995).
130. H. Gossner, G. Fehlauer, W. Kiunke, I. Eisele, M. Stolz, M. Hintermaier, and E. Knapek, *Mater. Res. Soc. Symp. Proc.* 351, 393 (1994).
131. J. Brunner, P. Schittenhelm, J. Gonderman, B. Spangenberg, B. Hadam, T. Köster, H. G. Roskos, H. Kurz, H. Gossner, I. Eisele, and G. Abstreiter, *J. Cryst. Growth* 150, 1060 (1995).
132. W. E. Leitch, C. M. Sotomayor Torres, D. Lootens, S. Thoms, P. van Daele, C. R. Stanley, P. Demeester, and S. P. Beaumont, *Surf. Sci.* 263, 622 (1992).
133. H. Gossner, T. Rupp, and I. Eisele, *J. Cryst. Growth* 157, 308 (1995).
134. R. O. Jones and G. Seifert, *Phys. Rev. Lett.* 79, 443 (1997).
135. M. L. Steigerwald and L. B. Brus, *Ann. Rev. Mater. Sci.* 19, 471 (1989).
136. A. Henglein, *Chem. Rev.* 89, 1861 (1989).
137. T. Takagaharar and K. Takeda, *Phys. Rev. B* 46, 15578 (1992).
138. L. Brus, *J. Phys. Chem.* 98, 3575 (1994).
139. Ö. Dag, A. Kuperman, and G. A. Ozin, *Adv. Mater.* 7, 72 (1995).
140. J. R. Heath, *Science* 258, 1131 (1992).
141. J. L. Heinrich, C. L. Curtis, G. M. Credo, K. L. Kavanagh, and M. J. Sailor, *Science* 255, 66 (1992).
142. S. Berhane, S. M. Kauzlarich, K. Nishimura, R. L. Smith, J. E. Davis, H. W. H. Lee, M. L. S. Olson, and L. L. Chase, *Mater. Res. Soc. Symp. Proc.* 298, 99 (1993).
143. R. A. Bley, S. M. Kauzlarich, H. W. H. Lee, and J. E. Davis, *Mater. Res. Soc. Symp. Proc.* 351, 275 (1994).
144. R. A. Bley, S. M. Kauzlarich, J. E. Davis, and H. W. H. Lee, *Chem. Mater.* 8, 1881 (1996).
145. A. Matsunawa, S. Katayama, and Y. J. Arata, *J. High Temp. Soc.* 13, 69 (1987).
146. D. Zhang, R. M. Kolbas, P. Mehta, A. K. Singh, D. J. Lichtenwalner, K. Y. Hsieh, and A. I. Kingon, *Mater. Res. Soc. Symp. Proc.* 256, 35 (1992).
147. K. A. Littau, P. J. Szajowshki, A. J. Muller, A. R. Kortan, and L. E. Brus, *J. Phys. Chem.* 97, 1224 (1993).
148. A. Fojtik and A. Henglein, *Chem. Phys. Lett.* 221, 363 (1994).
149. H. Kobayashi, A. T. Bell, and M. Shen, *Macromolecules* 7, 277 (1974).
150. H. K. Yasuda, in "Plasma Polymerization." Academic Press, Orlando, 1985.
151. J. Dutta, H. Hofmann, C. Hollenstein, and H. Hofmeister, in "Nanoparticles and Nanostructured Films" (J. H. Fendler, ed.). Wiley–VCH, Weinheim, Germany, 1998.
152. S. Li, S. J. Silvers, and M. S. El-Shall, *J. Phys. Chem. B* 101, 1794 (1997).
153. L. J. Brus, P. F. Szajowski, W. L. Wilson, T. D. Harris, S. Schuppler, and P. H. Cirtin, *J. Am. Chem. Soc.* 117, 2915 (1995).
154. M. S. El-Shall, S. Li, T. Turkki, D. Graiver, U. C. Pernisz, and M. E. Baraton, *J. Phys. Chem.* 99, 17805 (1995).
155. R. A. Bley and S. M. Kauzlarich, in "Synthesis of Silicon Nanoclusters" (J. H. Fendler, ed.). Wiley–VCH, Weinheim, Germany, 1998.
156. L. N. Skuja, A. N. Streletsky, and A. B. Pakovich, *Solid. State Commun.* 150, 1069 (1994).
157. B. Abeles and T. Tiedje, *Phys. Rev. Lett.* 51, 2003 (1983).
158. D. J. Lockwood, G. C. Aers, L. B. Allard, B. Bryskiewicz, S. Charbonneau, D. C. Houghton, J. P. McCaffrey, and A. Wang, *Can. J. Phys.* 70, 1184 (1992).
159. D. J. Lockwood, Z. H. Lu, and J.-M. Baribeau, *Phys. Rev. Lett.* 76, 539 (1996).
160. B. T. Sullivan, D. J. Lockwood, H. J. Labbé, and Z.-H. Lu, *Appl. Phys. Lett.* 69, 3149 (1996).
161. F. A. d'Avitaya, L. Vervoort, F. Bassani, S. Ossicini, A. Fasolino, and F. Bernadini, *Europhys. Lett.* 31, 25 (1995).
162. F. A. d'Avitaya, F. Bassani, I. Mihalcesu, and A. G. Nassiopoulos, in "Physics, Chemistry and Application of Nanostructures," p. 3. World Scientific, Singapore, 1997.
163. S. Sinharoy, *Thin Solid Films* 187, 231 (1990).
164. S. Schuppler, S. L. Friedman, M. A. Marcus, D. L. Adier, Y.-H. Xie, F. M. Ross, Y. J. Chabal, T. D. Harris, W. L. Brown, E. E. Chaban, P. F. Szajowski, S. B. Christman, and P. H. Citrin, *Phys. Rev. B* 52, 4910 (1995).
165. L. Vervoort, F. Bassani, I. Mihalcescu, J. C. Vial, and F. A. d'Avitaya, *Phys. Status Solidi B* 190, 123 (1995).
166. F. Bassani, S. Ménard, and F. A. d'Avitaya, *Phys. Status Solidi A* 165, 49 (1998).
167. G. Allan, C. Delerue, and M. Lanoo, *Phys. Rev. Lett.* 76, 2961 (1996).
168. R. J. Baierle, M. J. Caldas, E. Molinari, and S. Ossicini, *Solid State Commun.* 102, 545 (1997).
169. A. B. Filonov, A. N. Kholod, V. A. Novikov, V. E. Borisenko, L. Vervoort, F. Bassani, A. Saul, and F. A. d'Avitaya, *Appl. Phys. Lett.* 70, 744 (1997).
170. K. Maehashi, H. Nakashima, F. Bertram, P. Veit, and J. Christen, *Jpn. J. Appl. Phys.* 37, 39 (1998).
171. M. Tachikawa, T. Yamada, T. Sasaki, and H. Mori, in "Proceedings of the 24th State-of-the-Art Program on Compound Semiconductors," Los Angeles, 1996, p. 115.
172. J. M. Kuo, Y. C. Wang, K. W. Goossen, L. M. F. Chirovsky, S. P. Hui, B. T. Tseng, J. Walker, A. L. Lentine, R. E. Leibenguth, G. Livescu, W. Y. Jan, J. E. Cunningham, L. A. D'Asaro, A. Ron, D. Dahringer, D. Kossives, D. D. Bacon, R. L. Morrison, R. A. Novotny, D. B. Buchholz, and W. E. Mayo, *J. Cryst. Growth* 175–176, 971 (1997).
173. M. J. Jurkovic, J. Alperin, Q. Du, W. I. Wang, and M. F. Chang, *Electron. Lett.* 33, 1658 (1997).

174. J. Liang, P. Li, Y. Gao, and J. Zhao, *J. Mater. Sci.* 32, 4377 (1997).
175. J. R. Shealy, N. C. Macdonald, Y. Xu, K. L. Whittingham, D. T. Emerson, and B. L. Pitts, *Appl. Phys. Lett.* 70, 3458 (1997).
176. Z. L. Zhang and N. C. Macdonald, *J. Vac. Sci. Technol., B* 11, 437 (1993).
177. J. P. Spallas, J. H. Das, and N. C. Macdonald, *J. Vac. Sci. Technol., B* 11, 2538 (1993).
178. H. Ahmed, *J. Vac. Sci. Technol., B* 15, 2101 (1997).
179. D. Ali and H. Ahmed, *Appl. Phys. Lett.* 64, 2119 (1994).
180. Y. Takahashi, M. Nagase, H. Namatsu, K. Kurihara, K. Iwdate, Y. Nakajima, S. Horiguchi, K. Murase, and M. Tabe, *Electron. Lett.* 31, 136 (1995).
181. K. Kurihara, H. Namatsu, M. Nagase, and T. Takino, *Microelectron. Eng.* 35, 261 (1997).
182. K. Yano, T. Ishii, T. Hashimoto, T. Kobayashi, F. Murai, and K. Seki, *Techn. Digest Int. Electron Device Meeting* 541 (1993).
183. A. C. Seabaugh, C.-C. Cho, R. M. Steinhoff, T. S. Moise, S. Tang, R. M. Wallace, E. A. Beam, and Y.-C. Kao, in "2nd International Workshop on Quantum Functional Devices," Tokyo, 1995, p. 32.
184. K. Yuki, Y. Hirai, K. Morimoto, K. Inouse, M. Niwa, and J. Yasui, *Jpn. J. Appl. Phys.* 34, 860 (1995).
185. S. Y. Chou and A. E. Gordon, *Appl. Phys. Lett.* 60, 1827 (1992).
186. Q. Ye, R. Tsu, and E. H. Nicollian, *Phys. Rev. B* 44, 1806 (1991).
187. M. Hirose, M. Morita, and Y. Osaka, *Jpn. J. Appl. Phys.* 16, 561 (1977).
188. K. Matsumoto, M. Ishii, K. Segawa, Y. Oka, B. J. Vartanian, and J. S. Harris, *Appl. Phys. Lett.* 68, 34 (1996).
189. L. J. Geerlings, V. F. Anderegg, P. A. M. Holweg, J. E. Mooij, H. Pothier, D. Esteve, C. Urbina, and M. H. Devoret, *Phys. Rev. Lett.* 64, 2691 (1990).
190. K. Nakazato, R. J. Blaikie, J. R. A. Cleaver, and H. Ahmed, *Electron. Lett.* 29, 384 (1993).
191. K. Yano, T. Ishii, T. Hashimoto, T. Kobayashi, F. Murai, and K. Seki, *IEEE Trans. Electron Devices* ED4, 1628 (1994).
192. A. Dutta, M. Kimura, Y. Honda, M. Otobe, A. Itoh, and S. Oda, *Jpn. J. Appl. Phys.* 36, 4038 (1997).
193. M. Fukuda, K. Nakagawa, S. Miyazaki, and M. Hirose, *Appl. Phys. Lett.* 70, 2291 (1997).
194. A. Nakajima, T. Futatsugi, K. Kosemura, T. Fukano, and N. Yokoyama, *Appl. Phys. Lett.* 71, 353 (1997).
195. R. Tsu and L. Esaki, *Appl. Phys. Lett.* 22, 562 (1973).
196. A. Nakajima, T. Futatsugi, K. Kosemura, T. Fukano, and N. Yokoyama, *Appl. Phys. Lett.* 70, 1742 (1997).
197. L. Guo, E. Leobandung and S. Y. Chou, *Appl. Phys. Lett.* 70, 850 (1997).
198. L. Guo, E. Leobandung, L. Zhuang, and S. Y. Chou, *J. Vac. Sci. Technol., B* 15, 2840 (1997).
199. L. Guo, E. Leobandung, and S. Y. Chou, *Science* 275, 649 (1997).
200. C. S. Lent, P. D. Tougaw, W. Porod, and G. H. Bernstein, *Nanotechnology* 4, 49 (1993).
201. T. Tanamoto, R. Katoh, and Y. Naruse, *Jpn. J. Appl. Phys.* 33, 1502 (1994).
202. A. N. Korotkov, *Appl. Phys. Lett.* 67, 2412 (1995).
203. S. Bandyopadhyay, B. Das, and A. E. Miller, *Nanotechnology* 5, 113 (1994).
204. S. Lloyd, *Science* 261, 1569 (1993).
205. S. Bandyopadhyay and V. Roychowdhury, *Phys. Low Dimension. Syst.* 8/9, 29 (1995).
206. S. Bandyopadhyay and V. Roychowdhury, *Jpn. J. Appl. Phys.* 35, 3350 (1996).
207. T. H. P. Chang, L. P. Muray, U. Staufer, and D. P. Kern, *Jpn. J. Appl. Phys.* 31, 4232 (1992).
208. T. H. P. Chang, D. P. Kern, and L. P. Muray, *J. Vac. Sci. Technol., B* 6, 2743 (1992).
209. J. Y. Park, H. J. Choi, Y. Lee, S. Kang, K. Chun, S. W. Park, and Y. Kuk, *J. Phys. IV* C5, 258 (1996).
210. J. Y. Park, H. J. Choi, Y. Lee, S. Kang, K. Chun, S. W. Park, and Y. Kuk, *J. Vac. Sci. Technol., A* 15, 1499 (1997).
211. T. H. P. Chang, M. G. R. Thomson, E. Kratschmer, H. S. Kim, M. L. Yu, K. Y. Lee, S. A. Rishton, and B. W. Hussey, *J. Vac. Sci. Technol., B* 14, 3774 (1996).
212. M. R. Rakhshandehroo and S. W. Pang, *J. Vac. Sci. Technol., B* 15, 2777 (1997).
213. M.-Y. Jung, D. W. Kim, S. S. Choi, Y.-S. Kim, Y. Kuk, K. C. Park, and J. Jang, *Thin Solid Films* 294, 157 (1997).
214. G. Stix, *Scientific American* 77 (February 1995).
215. S. Madsen, S. I. Bozhevolvyi, K. Birkelund, M. Müllenborn, J. M. Hvam, and F. Grey, *J. Appl. Phys.* 82, 493 (1997).
216. L. A. Nagahara, P. I. Oden, A. Majumdar, J. P. Carrejo, J. Graham, and K. Alexander, *Proc. SPIE* 1639, 171 (1992).
217. S. C. Minne, H. T. Soh, P. Flueckiger, and C. F. Quate, *Appl. Phys. Lett.* 66, 703 (1995).
218. G. C. Abeln, T. C. Shen, J. R. Tucker, and J. W. Lyding, *Microelectron. Eng.* 27, 23 (1995).
219. J. A. Dagata, J. Scheir, H. H. Harary, C. J. Evans, M. T. Postek, and J. Bennett, *Appl. Phys. Lett.* 56, 2001 (1990).
220. J. W. Lyding, T.-C. Shen, J. S. Hubacek, T. R. Tucker, and G. C. Abeln, *Appl. Phys. Lett.* 64, 2010 (1994).
221. K. Birkelung, E. V. Thomsen, J. P. Rasmussen, O. Hansen, P. T. Tang, P. Møller, and F. Grey, *J. Vac. Sci. Technol., B* 15, 2912 (1997).
222. E. S. Snow, W. H. Juan, S. W. Pang, and P. M. Campbell, *Appl. Phys. Lett.* 66, 1729 (1995).
223. E. S. Snow, P. M. Campbell, and B. V. Shanabrook, *Appl. Phys. Lett.* 63, 3488 (1993).

Chapter 11

ELECTRONIC TRANSPORT PROPERTIES OF QUANTUM DOTS

M. A. Reed, J. W. Sleight*, M. R. Deshpande†

Departments of Physics, Applied Physics, and Electrical Engineering, Yale University, New Haven, Connecticut, USA

Contents

1. INTRODUCTION

1.1. Fabricated Quantum Dots: Vertical and Horizontal Systems

Tunneling in low-dimensional semiconductor structures has been a very active research field, both experimentally [1–10] and theoretically [11–19]. Traditionally, there have been two different approaches to achieving a three-dimensionally confined system (Fig. 1).

In one approach, horizontal dots [7–10], electrons are first confined to a two-dimensional (2D) electron layer formed at the interface of GaAs/$Al_xGa_{1-x}As$. Band bending

*Present address: IBM Semiconductor Research & Development Center, New York.
†Current affiliation: Motorola Corporate Research Laboratories, 2100 East Elliot Road, Tempe, AZ 85284.

Handbook of Nanostructured Materials and Nanotechnology, edited by H.S. Nalwa
Volume 3: Electrical Properties

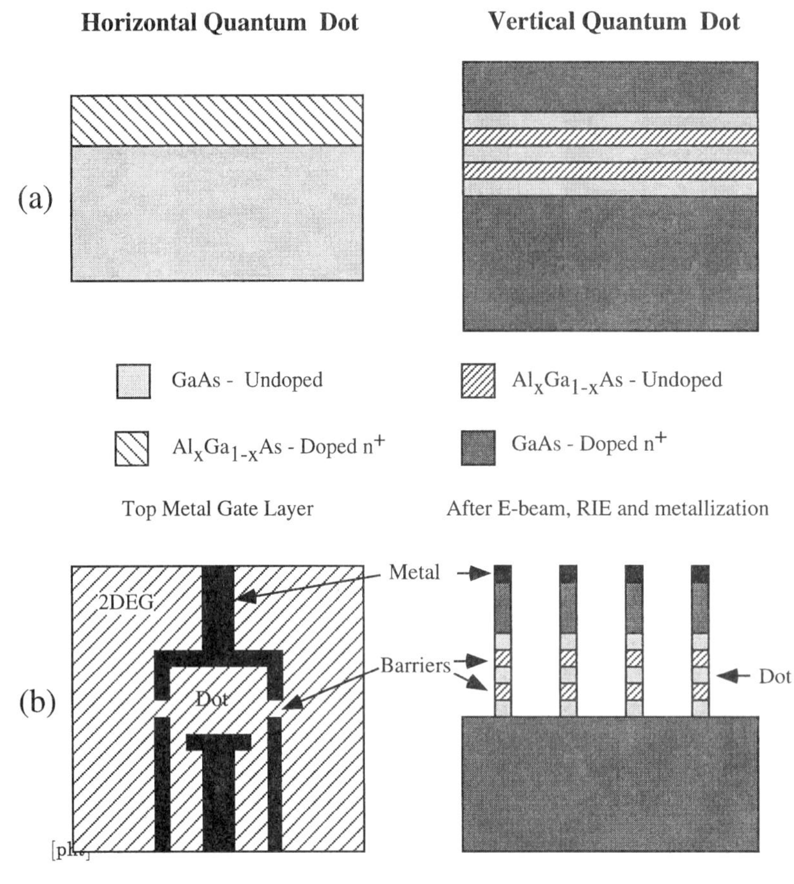

Fig. 1. (a) Comparison of the starting epitaxially grown material for a horizontal quantum dot structure (left) to a vertical structure (right). (b) Same structures after the fabrication that forms the quantum dot showing the top view of the horizontal dot and the side view of the vertical dot.

at the interface results in the formation of a triangular potential well in the GaAs region. Electrons are confined in this well and their motion in the direction perpendicular to the interface (z direction) is restricted. The number of electrons in the well can be controlled by suitably doping the barrier $Al_xGa_{1-x}As$ material sufficiently far from the interface. The electrons released by these donors accumulate in the well formed in the GaAs. These electrons are free to move in the lateral 2D plane and, therefore, this system is called a 2D electron gas. One can also control the density of the electrons by putting a gate electrode on top of the $Al_xGa_{1-x}As$ barrier and applying a suitable bias to it. A negative voltage applied to the gate repels electrons from underneath it. A suitable pattern of gate electrodes will electrostatically confine the electrons in a small area of the 2D layer. A full three-dimensional confinement is achieved as a result of the interface band bending and electrostatics. The electronic states of this confined dot can be probed by the tunneling of electrons into and out of the dot from the 2D electron gas surrounding the dot. The barrier between the dot and the 2D electron gas is due to the electrostatic potential of the fringing electric field of the patterned gate electrodes. The transport in this system is thus within the interface plane and, hence, this system is known as a horizontal quantum dot.

Another approach, vertical dots [1–6], starts with a double-barrier resonant tunneling structure consisting of heterostructure layers such as, GaAs/$Al_xGa_{1-x}As$/GaAs/$Al_xGa_{1-x}As$/GaAs. In this system, electrons are confined along one direction, the z direction, in the quantum well region (GaAs) in between the two barriers ($Al_xGa_{1-x}As$).

Reactive ion etching is used to physically reduce the lateral size of the device (Fig. 1), thus confining electrons along the remaining two (x and y) dimensions. The electronic states of this confined dot can be probed by tunneling of electrons into and out of the dot from the GaAs electrodes on either side of the device. The barrier between the dot and the GaAs electrodes is composed of the $Al_xGa_{1-x}As$ layers. The transport in this system is thus perpendicular to the interface and, therefore, this system is known as the vertical quantum dot.

In both techniques, the confinement potential spatially localizes electrons in a region, and quantizes the allowed energy levels in this region. However, there are major differences between the two systems. The barriers in the vertical dot system are well defined, atomically precise, and rigidly formed during growth of the layers. In contrast, the barriers in the horizontal dot are electrostatically defined and are soft barriers as they are ill defined, smooth, and flexible. The vertical dot is essentially a two-terminal system, whereas in the horizontal dot system it is easily possible to incorporate a gate electrode that makes it a three-terminal system. Usually, the total number of electrons residing in a vertical dot is very small (from 0 up to approximately 10), while that number is typically much larger in the horizontal dot system. In the case of horizontal dots, the single-electron charging energy, U_C, is usually much greater than δE, the spacing between single-particle states in the dot. In the case of vertical dots, δE and U_C are of the same order [2], or U_C can be less than δE. In the regime where they are on the same order, it is difficult to distinguish between the transport phenomena caused by each of these effects [2, 17] and proper modeling [15] is required to appropriately assign the observed structure to either spatial quantization or single-electron charging.

In this chapter, we restrict ourselves to the discussion of vertical quantum dots only. The vertical quantum dots described previously were the first structures where the effects of fully three-dimensionally confined electron states were observed [3]. These devices are small in size and, hence, have potential for high-density device applications. It is thus important to fully understand and characterize this and similar systems.

1.2. Impurity Dot System: Coulomb Potential Confinement

There is another way in which one can create a three-dimensionally confined experimental system consisting of discrete, single-electron states. This is by having a small number of donor impurities in the quantum well regions of large-area resonant tunneling diodes [20–23, 38]. The Coulomb potential of the ionized donor atoms gives rise to shallow, hydrogenic bound states (Fig. 2).

These localized states are physically similar to discrete quantum dot states. The $I(V)$ characteristics of this system are similar to those of a quantum dot system. The impurity system has certain advantages over the quantum dot system. It is a truly 3D–0D–3D

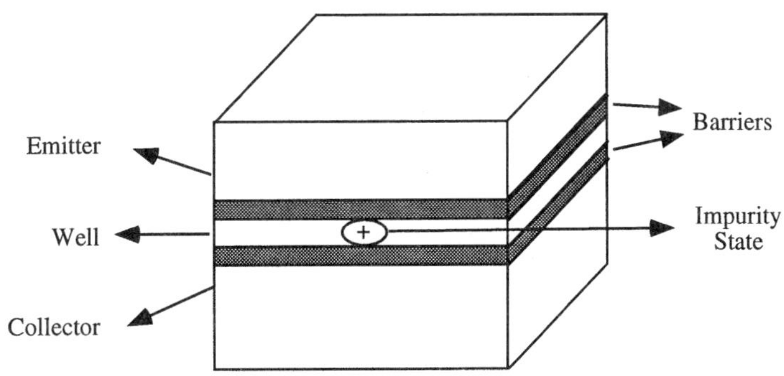

Fig. 2. Schematic of an impurity system showing an impurity state in the quantum well.

tunneling system unlike a 1D–0D–1D system of the fabricated quantum dot. There is no fabrication-imposed unknown potential in the emitter and collector regions. The Coulomb potential experienced by the electron resulting from the impurity is a known potential and the impurity states and energies are better characterized. It is possible to get devices with only a few isolated impurities and thus investigate the basic properties of a single, discrete state without having to worry about other states and their occupancy. The physical extent of the impurity state in a GaAs quantum well is on the order of 10 nm, which is smaller than the lateral extents of fabricated quantum dot eigenstates. An understanding of the basic physics of this impurity system is thus valuable for the understanding of the fabricated quantum dot system.

In Section 2, we first discuss the theory of the quantum dot eigenstates, the charging energy, and the effect of a magnetic field. We then discuss the eigenstates of the impurity potential. In Section 3, we discuss growth and fabrication details and, in Section 4, we present experimental results and, finally, conclude in Section 5.

2. THEORY

2.1. Energy States of a Fabricated Quantum Dot

2.1.1. Quantum Size Confinement Effects

In a fabricated quantum dot as described previously, the potential an electron experiences along the z direction is due to the semiconductor heterostructure band alignments and is modeled as a finite square well potential as in a quantum well. The potential the electron experiences along the lateral (x–y) directions depends critically on the highly anisotropic reactive ion etching process. This confinement potential, which is enhanced because of Fermi-level pinning on the side walls [2, 3], may be modeled in a few different ways.

If the lateral shape of the dot is asymmetric, then each eigenstate of the system is, in general, nondegenerate (except for the spin degeneracy). If the dot has cylindrical symmetry, then the various eigenstates reflect that symmetry. An additional effect to be considered is the Fermi-level pinning of the exposed side walls of the device. The Fermi level in the exposed semiconductor material gets pinned at a fixed value near the midgap. This induces bending of the bands and depletion of carriers from the region near the surface. The effective lateral size of the electrically active device is thus smaller than its physical dimensions. The potential an electron experiences along the lateral directions because of band bending and depletion is often modeled as a cylindrically symmetric parabolic potential [2, 3] as

$$\Phi(r) = \Phi_{\rm T}\left[\frac{1-(R-r)}{W}\right]^2 \tag{1}$$

where $\Phi_{\rm T}$ is the Fermi-level pinning energy at the side walls, R is the lateral physical dimension, r is the radial coordinate, and W is the depletion depth.

Schrödinger's equation for such a structure is best expressed in cylindrical coordinates. There is no θ dependence, and, because the potential is separable along the radial and the z directions, $\Phi(r, z) = \Phi(r) + \Phi(z)$, separation of variables is possible. The single-electron eigenenergies are given by $E_N = E_z + E_{n,l}$. E_z comes from the vertical quantization, and this will almost always be E_0, the quantum well ground state in the vertical direction. $E_{n,l}$ are the eigenenergies that result from the parabolic potential. $E_{n,l} = (2n + |l| + 1)\hbar\omega_0$ where the radial quantum numbers $n = 0, 1, 2, \ldots$ and the azimuthal quantum numbers $l = 0, \pm 1, \pm 2, \ldots$. If we assume that we are in the first vertical subband, E_0, and that the lateral dimension is reduced to $2W$ or less, then the energy spacing between states is

$$\Delta E = \hbar\omega_0 = \left(\frac{2\Phi_{\rm T}}{m^*}\right)^{1/2}\frac{\hbar}{R} \tag{2}$$

For a GaAs well, with AlGaAs barriers, typical values for energy state separation are on the order of 25 meV, for $\Phi_T = 0.8$ eV and $R \approx 500$ Å.

2.1.2. *Magnetic Field Effects*

The dependence of electronic transport on magnetic field is a valuable tool in understanding the nature of localized states in semiconductor nanostructures. If a magnetic field oriented parallel to the current direction (perpendicular to the heterointerfaces) is applied, according to first-order perturbation theory, the shift of the energy levels is given by

$$\delta E_{B\|I} = \left(\frac{e\hbar}{2m^*}\right)\ell B + \frac{e^2 B^2 \langle r_0^2 \rangle}{8m^*} \tag{3}$$

Note that spin is neglected in Eq. (3), as the spin splitting energy, $ge\hbar B/2m$, is only 0.25 meV at 10 T. The shift for a localized level with $\ell = 0$ is *entirely* due to the diamagnetic term, $\delta E_{\text{diamagnetic}} = e^2 B^2 \langle r_0^2 \rangle / 8m^*$, where $\langle r_0^2 \rangle$ is the spatial extent of the localized wave function. Therefore, the observation of an experimental diamagnetic shift is a *direct measure* of $\langle r_0^2 \rangle$.

For a magnetic field oriented perpendicular to the current, the first-order perturbation term is more difficult to evaluate, and numerical methods are required [24, 25]. Numerical results indicate an increase in dot energy states by a diamagnetic shift term, which is predicted to be less than 0.5 meV at 10 T for a GaAs/AlGaAs quantum dot system with the same well width (50 Å) as in this study. The small value for the state energy shift in this field orientation is due to the interaction of a relatively weak magnetic potential with the much stronger confinement potential in the epitaxial confinement (z) direction.

2.1.3. *Charging Effects*

Charging effects become important as the lateral device area is scaled down. Charging can be introduced into the usual single-electron model by modeling the Coulomb charging energy of the quantum dot as arising from a single effective capacitance C [1]. This approach uses the semiclassical geometric capacitance (the sum of an emitter capacitance C_e and a collector capacitance C_c):

$$C = C_e + C_c \approx \frac{\varepsilon\varepsilon_0 \pi a^2}{4}\left(d_e^{-1} + d_c^{-1}\right) \tag{4}$$

where d_e and d_c are the thicknesses of the emitter and collector barriers, respectively, and a is the dot effective electrical diameter. For the structures in this paper, $C \approx 2.6 \times 10^{-16}$ F using $\varepsilon_{\text{AlGaAs}} \approx 11.6$, and $a \approx 800$ Å. The value a is the electrical diameter, which is the extrapolated value determined from current density measurements on large-area samples. In this case, the estimated charging energy, $E_C = e^2/2C$, is approximately 0.31 meV.[1] More generally, $E_{C,N} = e^2/C(N - \frac{1}{2})$ is the Coulomb charging energy for N electrons residing on the dot [1, 26].

2.2. Energy States of the Impurity Dot

2.2.1. *Energy of a Donor Impurity in Bulk GaAs*

In GaAs, a group IV atom (such as Si) at the Ga site acts as an electron donor. Upon ionization, the impurity site has an unbalanced positive charge, which introduces a Coulomb potential that is screened by the available free carriers in the semiconductor. The attractive Coulomb potential gives rise to hydrogen-like energy states that can bind free carriers.

[1] In [1], the expression for capacitance incorrectly includes a factor of 4π in the denominator. E_C is given as e^2/C, whereas it should be $e^2/2C$.

To determine the energy states of such a semiconductor hydrogenic atom, we follow the treatment of Bastard [28].

The wave function of a donor state in a semiconductor can be expressed in the form (from Kohn [29])

$$\psi(\mathbf{r}) = \sum_{i=1}^{N} \alpha_i F_i(\mathbf{r})\phi_i(\mathbf{r}) \tag{5}$$

where N is the number of equivalent conduction band minima, $\phi_i(\mathbf{r})$ are the Bloch wave functions, and $F_i(\mathbf{r})$ are the envelope functions. In GaAs for the conduction band that is nondegenerate and isotropic with parabolic dispersion relations ($N = 1$ and single effective mass m^*), the envelope functions, $F(r)$, of the impurity states fulfill the equation

$$\left[\frac{p^2}{2m^*} - \frac{e^2}{4\pi\varepsilon_0 kr}\right] F(\mathbf{r}) = EF(\mathbf{r}) \tag{6}$$

where p is the momentum, m^* is the carrier effective mass, k is the dielectric constant of the semiconductor, e is the electronic charge, ε_0 is the permittivity of space, and E is the energy of the impurity state. The ground bound state of this system is the 1S hydrogenic wave function:

$$F_{1S}(\mathbf{r}) = \frac{1}{(\pi a_b^3)^{1/2}} \exp\left(-\frac{r}{a_b}\right) \tag{7}$$

where a_b is the effective three-dimensional Bohr radius of the semiconductor hydrogenic impurity:

$$a_b = \frac{4\pi\varepsilon_0 k\hbar^2}{m^* e^2} = 0.53 \times k \times \frac{m_0}{m^*}\ \text{Å} \tag{8}$$

The binding energy of this 1S state, R_b, is given by

$$R_b = \frac{m^* e^4}{32\pi^2\varepsilon_0^2 k^2\hbar^2} = \frac{1}{k^2} \times \frac{m^*}{m_0} \times 13.6\ \text{eV} \tag{9}$$

In GaAs, $k = 12.85$ and $m^*/m_0 = 0.067$ giving us $a_b \approx 101$ Å and $R_b \approx 5.5$ meV.

We can see that the binding energy of hydrogenic donors in bulk GaAs has a very small energy compared to its band gap of 1.5 eV. Therefore, this state is called a shallow donor state. An electron occupying this state has an energy of -5.5 meV as measured from the bottom of the conduction band.

2.2.2. Energy of a Donor Impurity in a Quantum Well

In contrast to the bulk material, the binding energy of an impurity in a quantum well depends on the properties of the well, in particular, its width L and its barrier height V_0. The impurity binding energy increases as the well width decreases as long as the penetration of the quantum well wave function $[\chi(z)]$ in the barriers remains small. This seems surprising at first because we intuitively associate higher kinetic energy (and, hence, lower binding energy) with the localization of a particle in a finite region of space. This is true of the energy value of the ground state of the quantum well, $E_1(L)$, and so also of the ground state of the impurity, $\varepsilon(L)$, when measured with respect to some fixed reference such as the bottom of the well. But the binding energy of the impurity $E_b(L) = E_1(L) - \varepsilon(L)$ actually increases as L decreases because for the impurity state the confinement causes the electron to stay near the attractive center, thus experiencing a higher potential energy. In the limiting case of an infinite barrier ($V_0 = \infty$) and zero well width ($L = 0$), we reach the two-dimensional limit when the ground state binding energy of a hydrogenic impurity is $4R_b = 22$ meV.

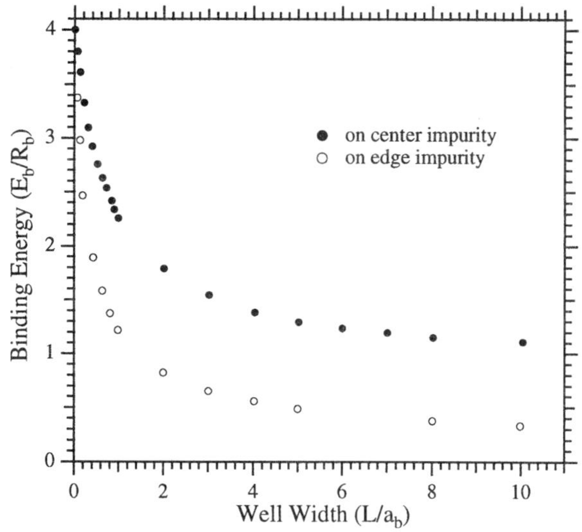

Fig. 3. Calculated dependences of the on-center and on-edge hydrogenic donor binding energies versus the well thickness L in a quantum well with an infinite barrier height ($V_0 = \infty$). R_b and a_b are the bulk effective Rydberg and Bohr radius, respectively. (Source: Adapted from [30].)

Another important feature of impurities in quantum wells is that the impurity binding energy explicitly depends on the precise location of the impurity along the growth axis (z direction) as there is no translational invariance along that axis. This energy depends on whether the impurity is at the center of the well or at the edge of the well or within the barrier. The wave function for an impurity at the center of the well ($z = 0$) approaches the 1S wave function of the bulk. But the wave function for an impurity at the edge of the well ($z = L/2$) approaches a truncated $2p_z$ wave function as the barrier potential forces the impurity wave function to almost vanish at the interface. The edge impurity has a lower binding energy than the center impurity. If the impurity is located within the barrier, it is still able to bind an electron in the well. The electron is physically separated from the impurity charge and such a system has lower binding energy than even an impurity at the edge.

The impurity binding energy in a quantum well can be obtained numerically as has been done by Bastard [28]. The results of such a calculation for an infinite barrier quantum well ($V_0 = \infty$) are shown in Figure 3 for an impurity located at the center and for an edge impurity. The binding energy is in the units of the bulk impurity Rydberg ($R_b = 5.5$ meV for GaAs) and the well width is in the units of the bulk impurity Bohr radius ($a_b = 101$ Å for GaAs).

For our experimental system where the well width is 44 Å, this figure gives us a binding energy for center impurity of approximately 16 meV and a binding energy for edge impurity of approximately 10 meV.

Because the impurities are randomly distributed in the well, it is statistically possible to have two impurities close to each other particularly in large-area samples. If the separation between the two impurities is on the order of the Bohr radius of a single impurity (100 Å), then the energy state of the electron is substantially modified. This is analogous to a hydrogen molecular ion problem. These impurity pairs give rise to higher binding energy states with the energy depending on the impurity separation. In the limiting case when the two impurities overlap, the situation is similar to a He^+ system with the resulting binding energy of the electron being four times as much as the single impurity binding energy.

So far, we have assumed that the barriers are infinite. For real systems, this is not true and the binding energy of an impurity does depend on the barrier height V_0. For finite barriers, as the well width decreases beyond a certain limit ($\approx 0.2a_b$), the impurity energy becomes comparable to the barrier height and the impurity wave function gets more and more delocalized. The effect of the Coulomb potential gets smaller and, hence, the binding energy decreases for L less than approximately $0.2a_b$ (unlike the trend observed in Fig. 3). For the samples under study where $L = 44$ Å, an infinite barrier approximation is shown to be a reasonable approximation [31].

2.2.3. *Coulomb Charging Energy in the Impurity System*

In the impurity system, the lateral size of the device is large (a few micrometers) and the charging energy of one electron is negligible. If the impurity concentration in the device is low and the impurities are separated by a large distance (in the micrometer range), then we can assume that each impurity channel is independent of the others as the presence or absence of an electron in one would not have any effect on the others. A given hydrogenic impurity has many eigenstates. In particular, the ground state of the impurity is spin degenerate in zero magnetic field. Although we have two states with the same energy, only one of them can be occupied by an electron at a given time because of charging effects. This has important consequences on the $I(V)$ characteristics as will be discussed later.

2.3. Current–Voltage Characteristics of Vertical Dot: Fabricated and Impurity Systems

The electronic structure of the dots can be probed by studying the two-terminal current–voltage [$I(V)$] characteristics as in a quantum well resonant tunneling diode. If the heterostructure barriers along the z direction have finite thickness, then the dot eigenstates (localized states) couple to the electronic states in the emitter and the collector regions on either side of the dot along the z direction.

Let us first consider the simple case when the Coulomb charging energy of the dot is small compared to the size quantization energy. Any electron incident on such a structure at an energy equal to one of the localized state energies will see an increased transmission probability as it is able to couple to the eigenstate and tunnel through the structure. As bias is applied across the device, the localized states are pulled down in energy toward the emitter Fermi level. As a level crosses the Fermi energy, electrons in the emitter having the same energy as the localized state experience an enhanced transmission probability and thus there is a sharp increase in current. This current is due to single electrons tunneling one at a time through the quantum dot eigenstate and it can be written as

$$\Delta I \approx \frac{e}{\tau} \tag{10}$$

where e is the electronic charge and τ is the lifetime of the dot state. The lifetime (τ) depends on the heterostructure barrier thicknesses, the bias, and the density of available, occupied electronic states in the emitter. As the bias increases slightly, to first order, the current through a given localized state remains the same and, therefore, causes a current plateau or step in the $I(V)$ characteristics. A further increase in bias brings another localized state below the Fermi level in the emitter introducing another channel for electrons to tunnel through. The $I(V)$ characteristics of the device thus resemble a staircase structure. In the case of the fabricated dot, the bias locations of the steps correspond to the energy spectrum of the quantum dot. In the case of the impurity dot, the bias locations of the steps are random as each step is attributed to a separate impurity and the impurities are randomly distributed. This simple picture gets modified when the Coulomb charging energy becomes comparable to the size quantization energy. An additional bias, corresponding to the charging energy of a single electron, is necessary for the threshold of conduction. The energy spectrum of the dot also depends on the number of electrons occupying the dot states. The staircase structure of the $I(V)$ characteristics is modified and the bias locations of the steps no longer just correspond to the energy spectrum of the uncharged dot but also depend on the occupancy of the dot.

3. SAMPLE GROWTH AND FABRICATION

Samples are grown using molecular beam epitaxy on a Si-doped (100) GaAs substrate [32]. For the resonant tunneling diodes of the impurity system, the epitaxial layers consist of a 1.8×10^{18}-cm^{-3} Si-doped GaAs contact, a 150-Å undoped GaAs spacer layer, an undoped $Al_{0.27}Ga_{0.73}As$ bottom barrier of width 85 Å, a 44-Å undoped GaAs quantum well, an undoped $Al_{0.27}Ga_{0.73}As$ top barrier of the same width 85 Å, a 150-Å undoped GaAs spacer layer, and a 1.8×10^{18}-cm^{-3} Si-doped GaAs top contact. Square mesas with lateral dimensions ranging from 2 to 64 μm are fabricated using standard photolithography techniques.

In the case of the fabricated quantum dots, the active region consists of a 50-Å $In_{0.1}Ga_{0.9}As$ quantum well, enclosed by a pair of 40-Å-thick $Al_{0.25}Ga_{0.75}As$ barriers. This region, along with 100-Å spacer layers of GaAs that contact the barriers, is undoped. The spacer layers are contacted by GaAs doped with Si at a density of 3×10^{18}-cm^{-3}. Small ($\sim$100 nm) AuGe/Ni/Au ohmic top-contact dots are defined by electron beam lithography on the surface of the grown resonant tunneling structure. A bilayer polymethylmethacrylate (PMMA) resist and lift-off method are used. The metal dot ohmic contact serves as a self-aligned etch mask for highly anisotropic reactive ion etching (RIE) using BCl_3 as an etch gas. The resonant tunneling structure is etched through to the bottom n^+ GaAs contact. Contact to the top of the structures is achieved through a planarizing/etch-back process employing polyimide and an O_2 RIE [3]. A gold contact pad is then evaporated over the columns. Bottom contact is achieved through the conductive substrate.

Note that in the fabricated quantum dot samples $In_{0.1}Ga_{0.9}As$ is used as the well material rather than the standard GaAs, although the contact electrodes are still GaAs. $In_{0.1}Ga_{0.9}As$ has a lower band gap than GaAs. The quantum eigenstates in the $In_{0.1}Ga_{0.9}As$ well thus have lower energies (measured from the bottom of the contact electrodes) as opposed to a conventional sample, which would have GaAs well material with the same well width. By using an InGaAs well, the equilibrium dot quantum states are brought down closer to, or below, the Fermi level. Lower quantum levels in the well imply longer lifetimes (i.e., less intrinsic energy broadening). Additionally, because this reduces the resonant bias voltage far less distortion of the emitter/dot/collector potential occurs while examining the density of states of the dot as compared to a conventional GaAs vertical quantum dot. Owing to the lower applied bias required, there is far less power dissipation and local electron heating in the dot region.

Two-terminal direct-current (dc) $I(V)$ characteristics are measured in a dilution refrigerator with a mixing chamber temperature (T_{mix}) of 35 mK.

4. EXPERIMENTAL RESULTS

4.1. Current–Voltage Characteristics

4.1.1. Fabricated Quantum Dots

Several different types of transport measurements were performed to investigate the electronic properties of the vertical InGaAs quantum dot structures.

Large-area resonant tunneling diodes (RTDs) are fabricated to examine the peak positions and current densities for the epitaxial material. Figure 4a shows a current versus voltage trace for a 1024-μm^2 sample measured at 4 K. A symmetric response is observed that is consistent with the symmetric epitaxial structure. The negative differential resistance (NDR) occurs at $\pm$23 mV with a peak current density of 56 A/cm^2. The low bias conductance is 25 mS for the 1024-μm^2 device, and this value is observed to scale linearly with device area, as expected, providing a means of extrapolating the upper limits for the electrical sizes of the smaller devices.

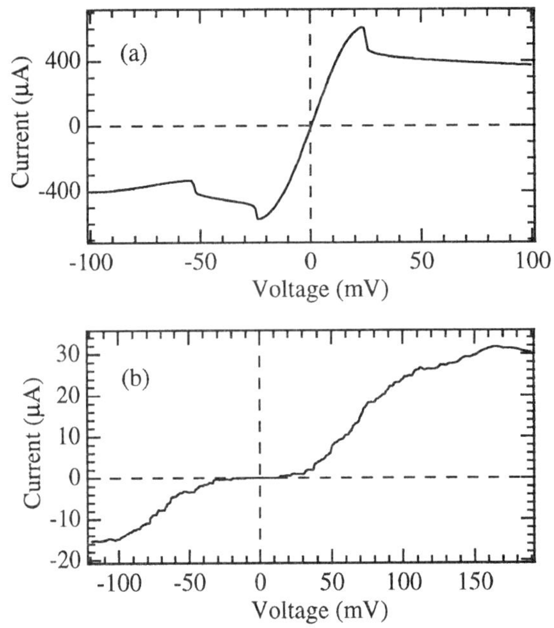

Fig. 4. (a) Current versus voltage for a large-area (1024 μm^2) sample fabricated from the same epitaxial material used for the small-area quantum dots ($T = 4.2$ K). (b) Current versus voltage for an $In_{0.1}Ga_{0.9}As$ quantum dot. Note the steplike structure in current near zero bias ($T = 50$ mK, mixing chamber).

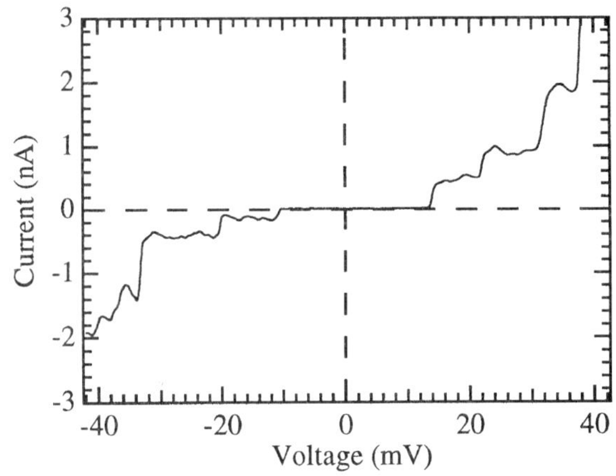

Fig. 5. Expansion of the low-bias regime of Figure 4b.

The small-area quantum dot samples are measured at millikelvin temperatures in a dilution refrigerator. Figure 4b shows a current versus voltage [$I(V)$] curve for a sample ($R \approx 100$ nm) under zero magnetic field, at a mixing chamber temperature of 50 mK. Figure 5 shows an expansion of the zero-bias region. The conductance for this sample ($\sim$0.4 μS) yields a value of approximately 800 Å for the electrical diameter, a. A staircase structure in current is observed in Figures 4b and 5 in both bias directions, especially at low biases.

A similar structure in the low-bias regime for double-barrier resonant tunneling structures (DBRTS) has been previously reported by other groups [2, 27]. The major difference in our work is the use of an InGaAs dot, as well as the use of a symmetric epitaxial

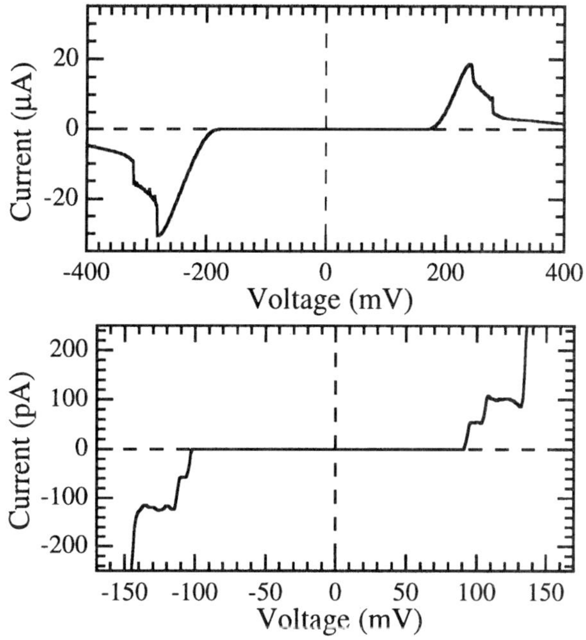

Fig. 6. $I(V)$ characteristics (zero magnetic field) at 1.4 K of the quantum well device showing the main resonance peaks (top). The magnified prethreshold region shows two steplike structures resulting from two isolated impurities (bottom).

structure. The current steps are relatively flat at low bias, where the transmission coefficient is not a strong function of applied bias.

We also note that there is an oscillatory structure on the current steps. This structure appears random but is very reproducible even after thermal cycling of the device. We will refer to this structure as the "fine structure" and discuss it in detail later.

4.1.2. Impurity Quantum Dot

Figure 6 shows the $I(V)$ characteristics for a typical resonant tunneling diode of the impurity system, (64 μm^2, 1.4 K), showing the main quantum well resonance peaks (top). Magnification of the current in the prethreshold region (bottom) shows two sharp current steps for both forward- and reverse-bias directions. This step structure is observed to be sample specific, but for a given sample it is exactly reproducible from one voltage sweep to another and independent of the voltage sweep direction. The steps are reproduced even after thermal cycling of the sample, except for slight threshold voltage shifts (<1 mV). Similar steps are observed in other devices with different barrier thickness. These steps look similar to the steps in Figure 5 and are attributed to tunneling through single impurity states. Note that these steps also show an oscillatory fine structure on the current plateaus.

To analyze the electron spectroscopy of the dot systems in greater detail, we need to determine the bias-to-energy conversion factor, α. The value of α is a measure of the amount of voltage that is actually dropped between the emitter and the quantum dot state. As will be shown, α can be determined from the temperature dependence of the plateau edges.

4.2. Variable-Temperature Measurements

As expressed in Eq. (10), the magnitude of the current step is given by $\Delta I = e/\tau$, where τ is the lifetime of the 0D state. This τ depends, among other factors, on the available

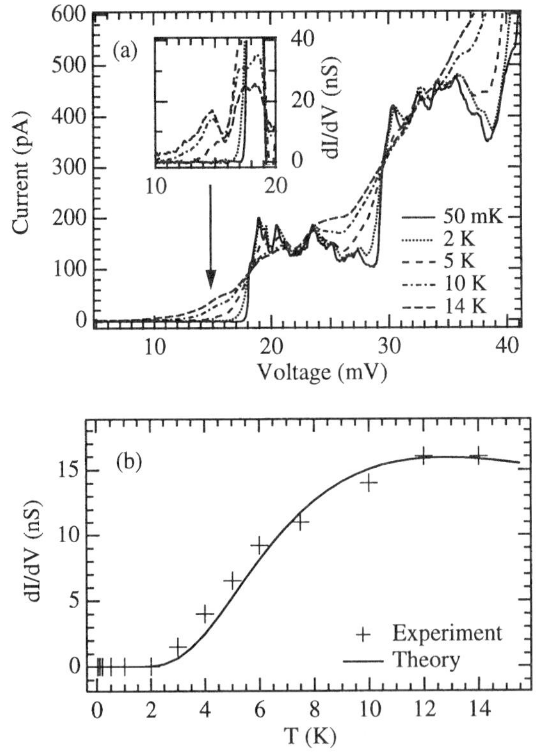

Fig. 7. (a) Reverse bias $I(V)$ characteristics of the fabricated quantum dot at different temperatures. The inset shows the subthreshold peak (in conductance) that arises as temperature is increased. (b) Experimental conductance (crosses) versus temperature [for the peak in the inset of (a)] and the theoretical fit (solid line).

and occupied density of states in the emitter. The occupancy of the electronic states in the emitter is determined by the Fermi distribution function and thus we expect the current ΔI to be proportional to the Fermi function. The sharpness of the current plateau edges is expected to broaden as the temperature increases, because of the broadening of the emitter Fermi distribution function. The $I(V)$ characteristics as a function of temperature for the fabricated quantum dot in reverse bias are shown in Figure 7a.

The voltage-to-energy conversion factor, α, is calculated by fitting the Fermi function to the first current plateau (I_0 is the value of the current on the plateau) in the variable-temperature data, that is,

$$I(V,T) = I_0 f(\alpha V) = \frac{I_0}{1 + \exp[-e\alpha(V - V_{\text{th}})/(kT)]} \tag{11}$$

Here, V_{th} is the threshold voltage for the current plateau and can be accurately determined from the intersection point of the $I(V,T)$ curves in Figure 7a, because the current given by Eq. (11) does not depend on the temperature for $V = V_{\text{th}}$. The value of α is determined to be 0.37 in the forward-bias direction and 0.50 in the reverse-bias direction for the fabricated quantum dot. The different α values imply an asymmetry in the emitter and collector contacts of the quantum dot device (either in doping or in the barrier or spacer layer thickness) in contrast to the symmetric response seen in the large-area device (Fig. 4a). This is not surprising as we are now probing a very localized region instead of averaging over nonuniformities in barriers or series doping, as in the large-area device.

The current plateau edges of the impurity system also exhibit characteristic Fermi-level thermal broadening (Fig. 8). From a Fermi function fit for this device, we get $\alpha = 0.48 \pm 0.02$ for forward bias and 0.42 ± 0.02 for reverse bias. Once again, the fits are done for the

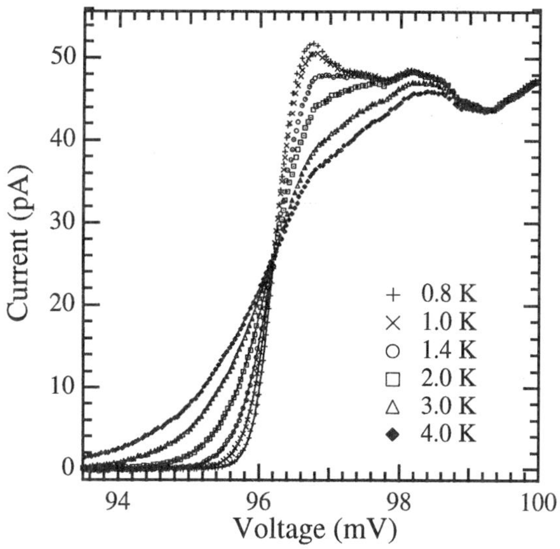

Fig. 8. *I*(*V*) characteristics of the first impurity current step edge of the impurity device at different temperatures showing the Fermi-level broadening and the Fermi fit to these *I*(*V*) traces.

region $V \leqslant V_{\text{th}}$ when the current is small and not affected by the occupancy of the impurity state.

We note the presence of an oscillatory, reproducible fine structure on the current plateaus in both the impurity and the fabricated quantum dot systems. Recently, it has also been investigated in detail by Schmidt et al. [38]. This structure (Figs. 7a and 8) exhibits significantly less dependence on temperature, especially for those regions located far away from the plateau edges. This indicates that the structure is not due to other states in the quantum dot or the quantum well itself, as current associated with tunneling through those states would exhibit Fermi broadening. The temperature insensitivity indicates that the current under consideration is from tunneling of the electrons from below the emitter Fermi level, where the emitter state occupation is not a function of temperature. Therefore, the fine structure is attributed to emitter states below the Fermi level, which pass into and out of resonance with the narrow 0D dot levels, as the applied bias is varied. At very low temperatures, only states in the emitter at or below the Fermi level, E_{F}, are occupied. These are the only emitter states available for tunneling. Electrons in the states nearest to the Fermi level will tunnel into the dot state first. States below the Fermi level then contribute to the tunneling as the bias is increased. If this model is correct, we expect that, as the sample temperature is increased, states above the Fermi level in the emitter will be populated. Figure 9 schematically shows how the occupation of discrete emitter levels (depicted as black lines in parts a and b of Fig. 9) is affected by increasing temperature.

At finite temperature, occupation of the emitter states above the Fermi level is possible and, therefore, thermally activated resonances *below* the first plateau should be observed. The inset of Figure 7a shows such a resonance effect, resulting in a subthreshold thermally activated conductance peak. Because $I(V) = I_0 f(\alpha V)$, where I_0 is a constant prefactor dependent on the transmission coefficient and $f(\alpha V)$ is the Fermi function, the conductance, $dI/dV = I_0\, df(\alpha V)/dV$. A fit of this to the subthreshold conductance peak strength (inset Fig. 7a), assuming an emitter state above the Fermi level becomes thermally activated, is shown in Figure 7b. The energy difference between the thermally activated emitter state and Fermi level (1.7 meV) is known from measuring the difference between the plateau threshold voltage at low temperatures (which determines the emitter Fermi-level position) and the thermally activated conductance peak position at higher temperatures. The *only* fitting parameter is I_0, which is within 5% of the value used for the first

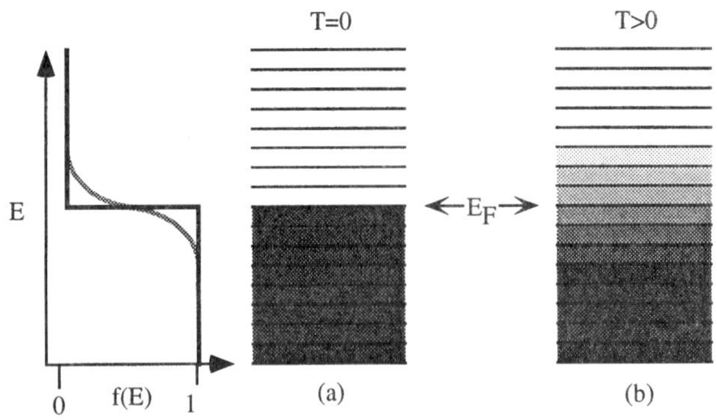

Fig. 9. (a) At $T = 0$, the Fermi function, $f(E)$, is a step function (black line) that results in a very sharp transition from occupied to unoccupied discrete emitter states. (b) At finite temperature, $f(E)$ becomes rounded near the Fermi energy, E_F, causing a gradual transition from states that are mostly occupied to states that are mostly unoccupied.

reverse-bias current plateau (at $V = V_{th}$). A similar thermally activated subthreshold peak is observed in the forward-bias direction as well. Similar thermally activated, prethreshold peaks are not observed in the impurity system. We will discuss the fine structure in more detail after investigating its magnetic field properties.

4.3. Magnetotunneling Measurements: Diamagnetic Shifts and Current Suppression

In addition to variable-temperature measurements, magnetic field is a valuable spectroscopy tool to examine bound states. Figures 10 and 11 show the forward-bias $I(V)$ characteristics in magnetic field parallel to current ranging from 0 to 9 T for the fabricated quantum dot and the impurity system, respectively. The traces are offset by a constant current value for clarity.

Note that there is a diamagnetic movement of all steps to higher bias with the magnetic field parallel to the current. This movement can be better observed in the fan diagrams shown in Figures 19 and 21. Plots of the plateau energy shift versus magnetic field squared (not shown) yield straight lines for all plateaus. At 9 T, the diamagnetic shift is approximately 1.4 to 2.6 meV (2–4 mV), depending on the current step/dot level for the fabricated quantum dot system, while it is about 0.4 meV (0.9 mV) for the current steps in the impurity system, implying that the impurity states are more localized. In the case of the fabricated quantum dot, different current steps are due to different energy eigenstates of the quantum dot. We expect these different eigenstates to show different shifts in a magnetic field and, hence, for the rest of this section we would concentrate on the diamagnetic shift of the fabricated dot only.

Table I lists the values for the diamagnetic shifts and current plateau widths for the first six steps in the reverse- and forward-bias directions of the fabricated quantum dot. From Eq. (3) ($\ell = 0$) and slope of the experimentally measured diamagnetic energy shift versus B^2 (accounting for α), the radial wave function extent is determined to be approximately 100 Å for the ground state in both bias directions. The implications of the radial wavefunction extent on the dot electron spectroscopy will be elaborated on after discussing the results for the magnetic field oriented perpendicular to the current.

It should be noted here that all resonances shift to *higher* bias with *only* a diamagnetic trend. This is in contrast to the single-electron theories for two-dimensionally and three-dimensionally confined, nearly cylindrical quantum dots in a magnetic field [36]. These theories show that some states increase in energy with magnetic field, while other states decrease in energy. Starting from zero field, the ground state *always* shifts upward

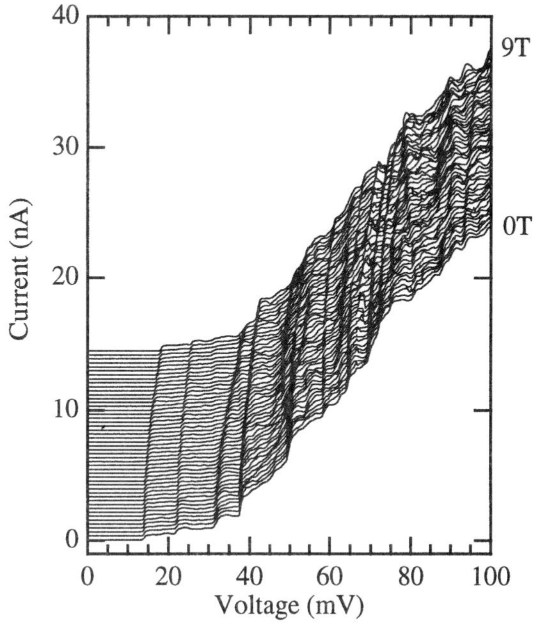

Fig. 10. Current–voltage characteristics in a dilution fridge with a mixing chamber temperature of 35 mK, in a magnetic field (0–9 T in 0.1875-T increments) parallel to the current for the forward-bias direction of the fabricated quantum dot. Traces are offset by a constant current value for clarity.

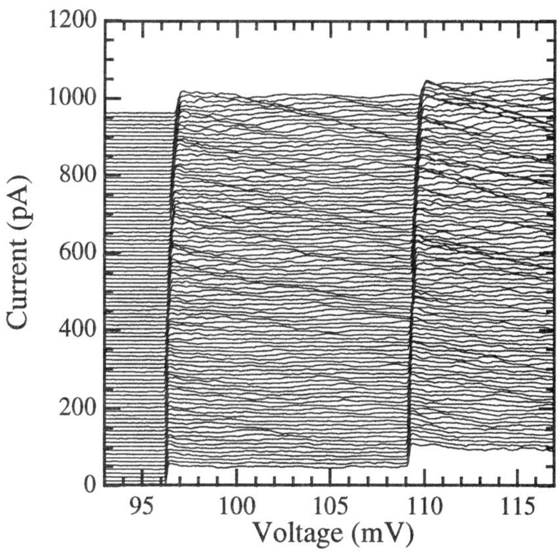

Fig. 11. Current–voltage characteristics in a dilution fridge with a mixing chamber temperature of 35 mK, in a magnetic field (0–9 T in 0.094-T increments) parallel to the current for the forward-bias direction of the impurity system. Traces are offset by a constant current value for clarity.

in energy, while the second state *always* shifts downward (ignoring spin). This is due to the Zeeman term (the term linear in B) in Eq. (3). The absence of Zeeman effects in the experimental data (i.e., peaks splitting and shifting in energy with a linear term) implies that either we cannot probe $\ell > 0$ states, possibly because of orthogonality between the emitter wave functions and $\ell > 0$ dot levels, or this simply tells us that ℓ is not a good quantum number for this system. This implies that the assumed cylindrical symmetry has

Table I. Experimentally Observed Diamagnetic Shifts and Current Plateau Widths for the Reverse- and Forward-Bias Directions for the Fabricated Quantum Dot Structure

Plateau index (N)	$\delta E_{\text{diamagnetic}}$ (meV, $B = 9$ T) (reverse, forward bias)	Plateau width (meV) (reverse, forward bias)
1	1.4, 1.6	5.0, 3.3
2	1.5, 1.6	6.5, 3.3
3	1.9, 2.6	6.5, 2.6
4	2.0, 1.6	9.0, 4.4
5	2.1, 1.6	1.1, 3.0
6	2.1, 1.0	3.5, 2.2

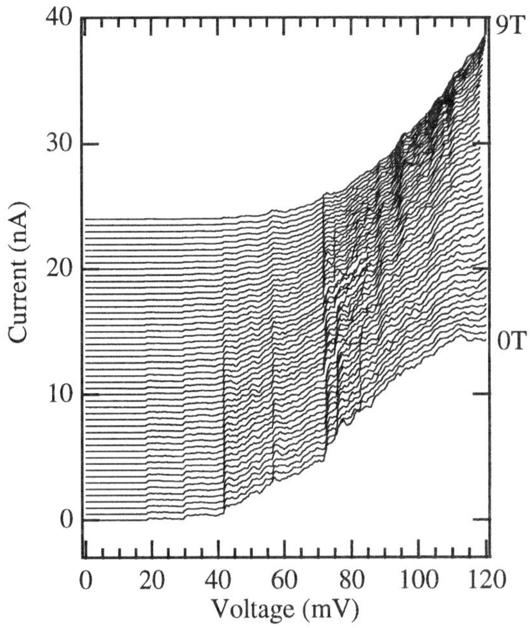

Fig. 12. Current–voltage characteristics in a magnetic field (0–9 T in 0.1825-T steps) perpendicular to the current for the fabricated quantum dot in the reverse-bias direction.

been distorted significantly. It should be noted that other groups [1, 2, 37] investigating similar systems also observe a diamagnetic dependence only.

Figures 12 and 13 show the $I(V)$ characteristics in the perpendicular magnetic field, for the fabricated quantum dot (reverse bias) and the impurity system (forward bias) at a mixing chamber temperature of 35 mK. As expected, the bias locations of all the current steps attributed to 0D dot states are not greatly affected by the perpendicular field, in contrast to the diamagnetic movement observed in the parallel field. This is confirmation that the current steps are the result of tunneling through the laterally localized dot states that reside in the strong confinement of the quantum well.

In addition to the difference in the response of the dot energy levels (i.e., the current plateau edges) to parallel and perpendicular magnetic fields, suppression of the plateau

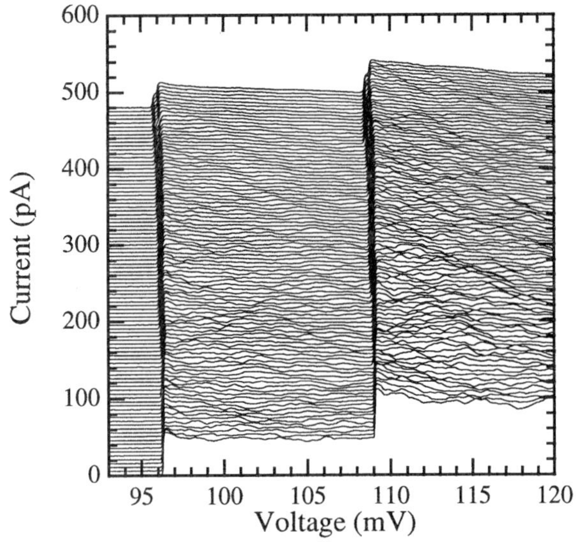

Fig. 13. Current–voltage characteristics in a magnetic field (0–9 T in 0.094-T steps) perpendicular to the current for the impurity system in the forward-bias direction.

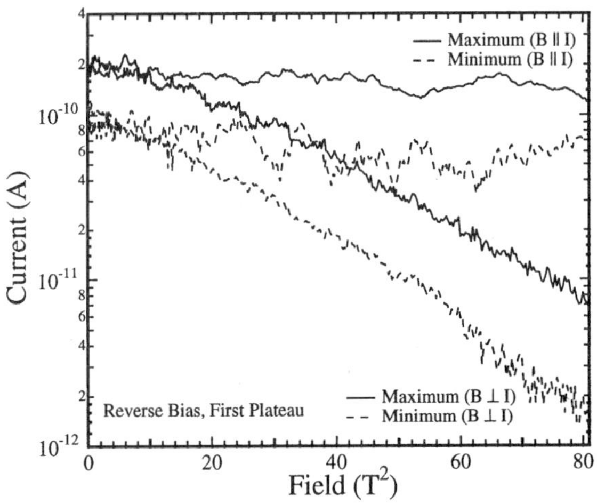

Fig. 14. Maximum (solid) and minimum (dashed) current values for the first current plateau of the fabricated quantum dot in reverse bias as a function of perpendicular and parallel magnetic fields.

current is observed for both bias directions in magnetic field oriented perpendicular to the current for both dot systems. Especially noteworthy are the lower plateaus of the fabricated dot system (Fig. 12), which appear to vanish because of the current scale. Figure 14 shows how dramatic these effects are. Current versus magnetic field squared is shown for the maximum and minimum current observed on the first plateau in the reverse-bias direction, in both parallel and perpendicular field orientations. In parallel field, the difference between the current minimum and maximum remains constant on a logarithmic scale over the field range in both bias directions, and only slight suppression is observed at 9 T. The current suppression for perpendicular field at 9 T is very large for the reverse bias (a factor of 100 for the plateau current minimum and 25 for the current maximum). In forward bias, a factor of approximately 10 is observed for both the current plateau minimum and the current plateau maximum.

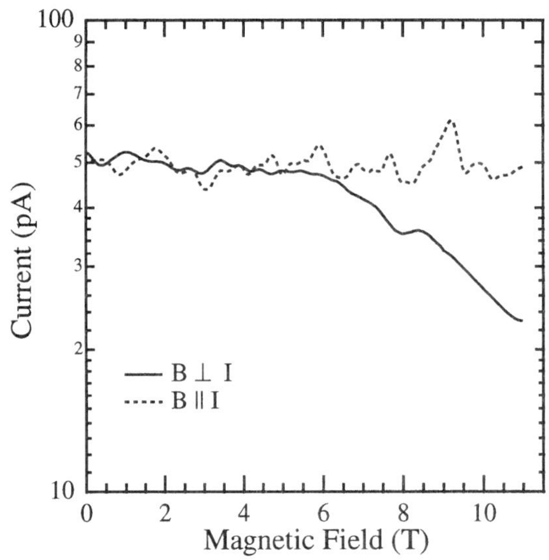

Fig. 15. Maximum current values for the first current step of the impurity system in forward bias as a function of perpendicular and parallel magnetic fields.

A similar current suppression is also observed in the case of the impurity system, although the effect is less dramatic (Fig. 15). In forward bias, the current gets suppressed by a factor slightly larger than 2 and, in reverse bias, by a factor slightly smaller than 2 when the field is perpendicular to the current. In parallel field, the current is approximately of the same field range except for some fluctuations. These fluctuations are attributed to the fine structure on the current plateaus and its movement in a magnetic field.

The current suppression has been observed previously in a coupled-dot system [24], and was attributed to a decreasing 0D–0D transition probability between the coupled dots. In our case, for the fabricated quantum dot, because there is one quantum well, the perpendicular field can *only* be decreasing the 1D emitter to a 0D dot transition probability. If the emitter and dot lateral wave functions are modeled as Gaussian functions, the observed linear Log(I) versus B^2 trend can be accounted for. The current is proportional to the overlap between the emitter and the dot wave functions

$$I \propto \int \psi_{\text{Emitter}} \psi_{\text{Dot}} \, dy = \lambda \sqrt{\frac{\pi}{2}} \tag{12}$$

where both the emitter and the dot wave functions are assumed to have a Gaussian wave function, $\psi \propto \exp(-y^2/\lambda^2)$. Under the application of a perpendicular magnetic field that shifts the emitter wave function with respect to the dot wave function by a displacement y_0,

$$\psi_{\text{emitter}} \propto \exp\left[-\frac{(y-y_0)^2}{\lambda^2}\right] \tag{13}$$

In this case,

$$\int \psi_{\text{Emitter}} \psi_{\text{Dot}} \, dy = \left(\lambda \sqrt{\frac{\pi}{2}}\right) \exp\left(-\frac{y_0^2}{2\lambda^2}\right) \tag{14}$$

As current is proportional to this overlap, Log$(I) \propto y_0^2$. If $y_0 \propto B$, we have the Log$(I) \propto B^2$ behavior observed experimentally.

To consider whether this trend is reasonable, note that coupling to the strong epitaxial dot potential in the z direction prevents a substantial shift of the dot wave function center.[2] The shift of the emitter wave function, y_0, may be estimated using a parabolic lateral potential approximation in magnetic field and assuming that in magnetic field this potential is offset by y_0, that is,

$$V = \frac{1}{2}\omega_c\left(y^2 - y_0^2\right) \tag{15}$$

As a result of introducing the potential into the Hamiltonian and applying Schrödinger's equation in Cartesian coordinates, the displacement y_0 is given by

$$y_0 = \frac{p_z\omega_c}{m^*(\omega_c^2 + \omega_0^2)} \tag{16}$$

For $\omega_0 > \omega_c$, this term varies as B, as required.

4.3.1. Spectroscopic Analysis of Fabricated Dot States

The wide-bias magnetotransport studies point out some clear inconsistencies with the expected quantum dot system. The magnitude of the diamagnetic shift implies a dot wave function spatial extent of approximately 200 Å in diameter. For this dimension, a charging energy of 5 meV is expected,[3] and a quantization energy, E_Q, of 35 to 45 meV between the ground state and the first excited state is expected (the lower limit is the parabolic result; the upper limit is the hard-wall result). While the charging seems reasonable to account for the observed plateau edge–to–plateau edge separations (see Table I), the quantization energy is *far* too large to account for the spacing from the second to the third plateau. As the ground state is only doubly degenerate because of spin, we expect to observe only two plateaus, separated by 5 meV/α. Subsequent plateaus in the $I(V)$ should then not be observed until at least 70 mV higher in bias. Clearly, a model beyond simple charging and quantization is required to explain the experimental data.

To explain similar diamagnetic shifts in previous work, a model of conducting filaments through the dot has been proposed by Tewordt and co-workers [1]. The filaments are assumed to be caused by the potential created by dopant atoms near or in the dot, which enhances the lateral quantization already present in the dot. Because, on average, all filament states are laterally localized by about the same amount, all of the peaks should exhibit similar diamagnetic shifts, which is observed. This also yields a more complicated density of states for the dot. However, transport through the system is still proceeding via tunneling through highly localized dot states. Note that this model also provides for the distortion of the cylindrical symmetry that is needed to prevent observation of $\ell > 0$ levels. Therefore, this model is consistent with our experimental data.

Applying this to our case, we expect to have a series of Coulomb-split plateau pairs with similar current step heights (ΔI_n), current suppression (for $B \perp I$), and diamagnetic shifts. Restated, the transport model is that conduction is proceeding through N dots in parallel with the emitter. For this specific sample, $N \approx 2$–3. In examining Table I, note that for reverse bias the diamagnetic shift is the same for $N = (1, 2)$ and for $N = (3, 4)$, and that the energy spacing between all of these is close to what is expected for the charging energy. For forward bias, a similar trend is observed for the $N = (1, 2)$ and $N = (4, 5)$ pairs. Figure 16 shows the conceptual energy ladder diagram for this transport process.

[2] If the wave function in the dot shifted significantly, the confinement energy would also change, which is not seen experimentally (i.e., no plateau shift for magnetic field perpendicular to current).

[3] In estimating the charging energy, E_C, the crucial parameter is the junction area. The geometrical capacitance argument assumes a metallic density of states, which is a questionable assumption in the emitter and collector electrodes, and certainly suspect in the dot. To provide a better estimate, a charging radius R_C is assumed. R_C is estimated from the wave function extent determined from the diamagnetic shift (100 Å), which results in $E_C \approx 5.0$ meV.

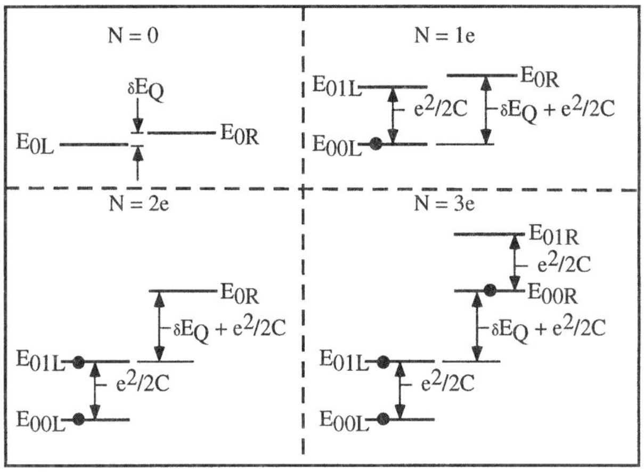

Fig. 16. Energy ladder diagram for the two-dot model. For no electrons in the dot ($N = 0$), there is a ground state spin-degenerate level for the left and right dot (E_{0L} and E_{0R}), separated by a small offset, δE_Q. When the dot is occupied by one electron ($N = 1$), it will occupy the lowest state available, E_{0L}, and the spin degeneracy of that state will be split by the Coulomb energy, $e^2/2C$, into E_{00L} and E_{01L}. The right state is also assumed to be lifted by the charging energy. The second electron ($N = 2$) will enter the next lowest energy state (E_{01L}). The third electron ($N = 3$) will then enter the ground state of the right dot.

Table II. Experimentally Observed Current Step Values in Zero Field and Also for $B = 9$ T ($B \perp I$). Suppression Values are the Ratio of $\Delta I_n(B = 0)/\Delta I_n(B = 9\text{ T})$

Plateau index (N)	ΔI_n (pA) 0 T ($-, +$ bias)	ΔI_n (pA) $B = 9$ T ($-, +$ bias)	Current suppression ($-, +$ bias)
1	200, 600	8, 70	25, 8.6
2	200, 500	11, 60	18, 8.3
3	800, 1100	80, 240	10, 4.6
4	2300, 1600	500, 570	4.6, 2.8
5	2500, 2500	1400, 580	1.8, 4.3

The current suppression and current step height (ΔI_n) for these pairs should also track together. Table II lists the approximate ΔI_n at 0 and 9 T ($B \perp I$) and suppression values for reverse and forward bias for the first five plateaus. There is some degree of uncertainty in estimating ΔI_n, as the assumption is that each current plateau is independent of the others and that the background line shape (caused by the preceding plateaus) is flat. This approximation begins to break down around the fourth plateau in both bias directions. For the $N = (1, 2)$ pairs in both bias directions, there is good agreement between both the ΔI_n values and the suppression. For the $N = (3, 4)$ pair in reverse bias and the $N = (4, 5)$ pair in forward bias, the agreement is not as good, although, as previously mentioned, there is great uncertainty in estimating ΔI_n in this regime. However, it is interesting to note that the $N = 3$ plateau in forward bias (which exhibits a large diamagnetic shift compared to the other steps) exhibits a suppression value different from the other pairs. This could indicate that it is the result of a third dot; however, its absence in reverse bias cannot be explained. While the N-dot model cannot be proven conclusively with these data, it is consistent with many of the features observed (i.e., state separation, magnetic field dependence, current suppression, and the lack of Zeeman shifts).

4.4. Magnetotunneling Measurements: Fine Structure

We mentioned earlier that we observe an oscillating, temperature-independent fine structure on the current plateaus of the $I(V)$ characteristics in both the fabricated dot and the impurity systems. A comparison between the observations of the two systems gives insight into the origin of the fine structure. To better characterize this, we investigate the effect of magnetic field on this structure. Figure 17 shows the evolution of the $I(V)$ characteristics of the fabricated quantum dot in forward bias. The magnetic field is parallel to the current and ranges from 0 T (bottom) to 2.25 T (top). The $I(V)$ traces have been vertically offset by a constant current value, so the movement of the fine structure in magnetic field is apparent. Similar fine-structure behavior is observed at higher magnetic fields (up to 9 T) and in a magnetic field perpendicular to the current. The fine-structure characteristics are reproducible, even after thermally cycling the sample to 300 K. It is apparent that the field dependence of the fine structure is different from that of the dot energy states (plateau edge behavior). Figure 18 shows the evolution of the $I(V)$ characteristics in a parallel magnetic field ranging from 6 to 9 T. Other groups [1, 2, 40, 41] have seen similar fine structure; however, their field steps were on the order of teslas, rather than 100 Gauss, so the dependence of this fine structure upon field appeared random.

To better view the data, fan diagrams are generated that plot the plateau edge and fine-structure peak voltage positions versus magnetic field. Figures 19 and 20 show fan diagrams for the fabricated quantum dot for the reverse bias in a parallel field ($B \parallel I$) and a perpendicular field ($B \perp I$), respectively. In a parallel field, three lines resulting from the current plateau edges at 10, 20, and 32 mV ($B = 0$ T) are clearly visible. Between plateau edges, the fine-structure motion clearly correlates with varying magnetic field. The fine-structure peaks form a complex lattice of crossing lines. From measuring the slope of these lines, we find that the fine structure typically moves at a rate of 2.5 mV/T (1.25 meV/T) in the reverse-bias direction and 5.0 mV/T (1.85 meV/T) in the forward bias direction. Both

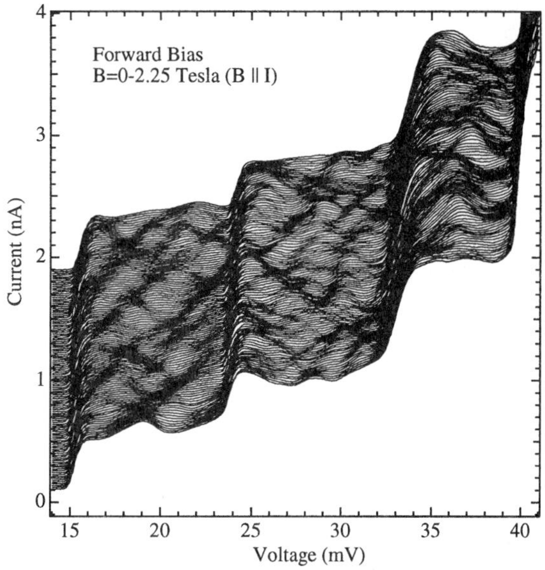

Fig. 17. $I(V)$ data for the fabricated quantum dot in parallel field, ranging from 0 T (bottom) to 2.25 T (top). Curves are vertically offset by a constant current value for clarity.

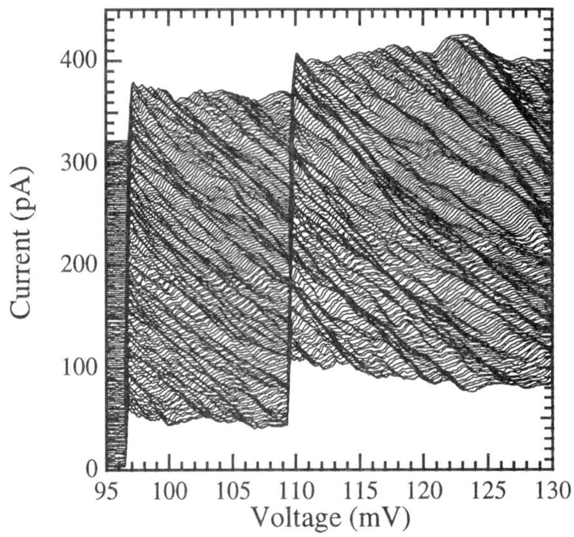

Fig. 18. $I(V)$ data for the impurity dot in parallel field, ranging from 6 T (bottom) to 9 T (top). Curves are vertically offset by a constant current value for clarity.

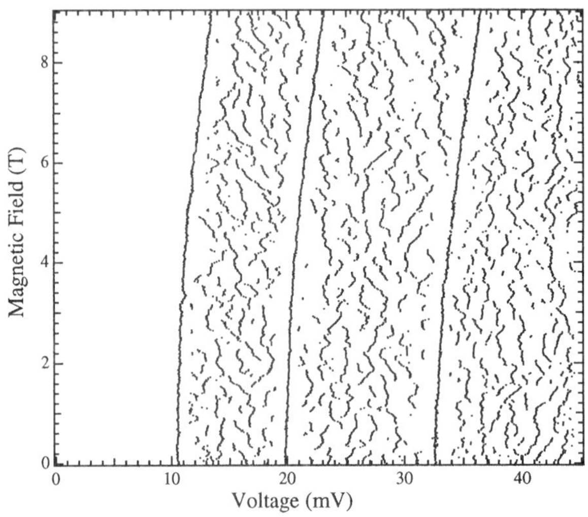

Fig. 19. Fan diagram showing peak voltage location versus magnetic field parallel to the current for the reverse-bias direction for the fabricated quantum dot.

of these values are on the order of the value for eB/m^* (1.8 meV/T in $In_{0.1}Ga_{0.9}As$ and 1.7 meV/T in GaAs). Similar field dependence of the fine structure is seen in the perpendicular orientation.

Figures 21 and 22 show fan diagrams for the impurity dot for the forward bias in a parallel field ($B \parallel I$) and a perpendicular field ($B \perp I$), respectively. Two lines resulting from the two current plateau edges are clearly visible and show only small movement with magnetic field. Between plateau edges, the fine-structure peaks show distinct movement to lower voltage as the field is increased. This movement is most prominent at fields greater than 2 T. The rate of this movement is approximately 5 meV/T for $B \parallel I$ and 3.5 meV/T for $B \perp I$. A similar shift is observed in reverse bias.

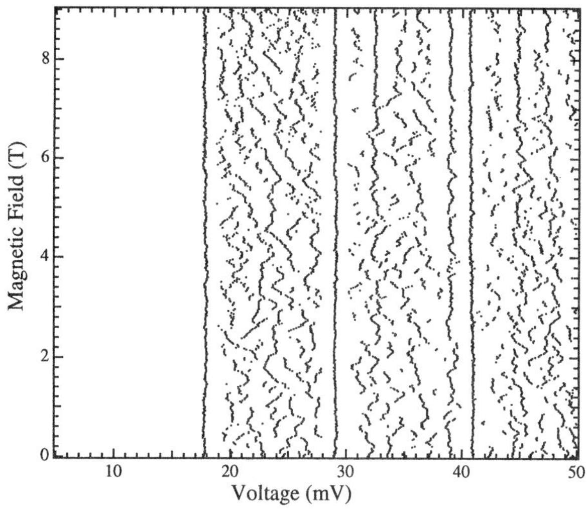

Fig. 20. Fan diagram showing peak voltage location versus magnetic field perpendicular to the current for the reverse-bias direction for the fabricated quantum dot.

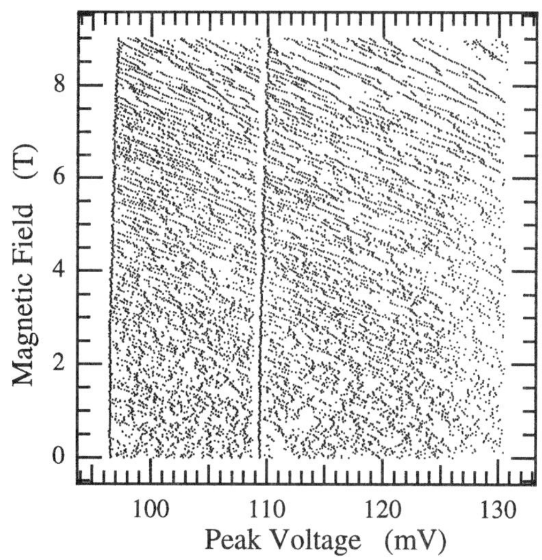

Fig. 21. Fan diagram showing peak voltage location versus magnetic field parallel to the current for the forward-bias direction for the impurity dot system.

While the fine structure, the number of plateaus observed, and the plateau widths are consistent through device thermal cycling, the spectra are sometimes offset in voltage. We believe this is due to the different donor charge distributions in the emitter and collector upon cool down, which may change from run to run. In contrast, dot state current plateau edges only shift in magnetic field parallel to current. This confirms that the source of the fine structure does not reside in the dot and implies the presence of emitter states.

The average energy separation between the fine-structure peaks on the first plateau in both bias and field directions has been calculated by performing a numerical Fourier transform upon the fan diagrams. Table III shows the values for the mean fine-structure separation, $\bar{x}$, and the standard deviation, σ. As expected, no dramatic change in the peak separation is observed when the field orientation is changed.

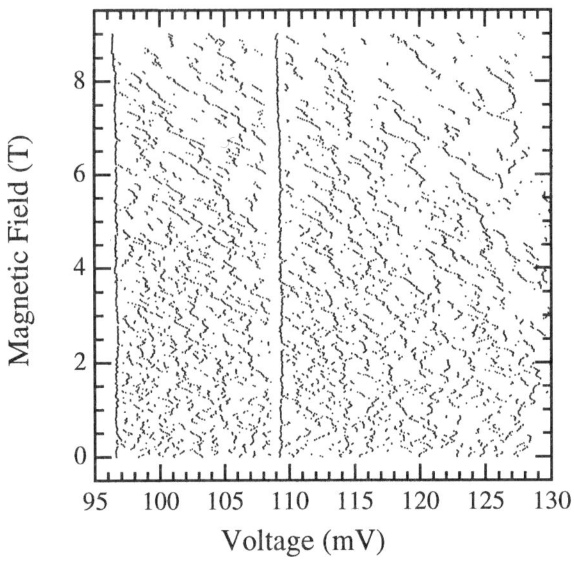

Fig. 22. Fan diagram showing peak voltage location versus magnetic field perpendicular to the current for the forward-bias direction for the impurity dot system.

Table III. Average Fine-Structure Peak-to-Peak Separation and Standard Deviation for the Fabricated Quantum Dot

Plateau (bias, field)	Fine-structure separation $\bar{x} \pm \sigma$ (meV)
Reverse, $B \perp I$	0.91 ± 0.4
Reverse, $B \| I$	0.83 ± 0.3
Forward, $B \perp I$	0.58 ± 0.2
Forward, $B \| I$	0.81 ± 0.3

In previous work by Su et al. [2, 40], the fine structure was observed to quench in magnetic field. This was viewed as evidence against the fine structure being attributed to a discrete density of states in the emitter. This is not the case observed for the samples studied in our work. Even at fields of up to 11 T the fine structure is still strong and clearly present. In a strong perpendicular field, the fine structure is observed to quench at the *same* rate as the current steps. Both are suppressed because of a decrease in the emitter–dot transition probability in a perpendicular field, and this is the expected result if the fine structure is due to the density of states in the emitter.

Several factors indicate that the fine structure results from the emitter contact. The weak temperature dependence shows that the fine structure is not due to states in the dot and is due to tunneling processes from states below the Fermi level in the emitter. The observation of thermally activated preresonant fine-structure peaks is consistent with this model. Also, other groups have shown that the fine structure is due to emitter states by studying transport in asymmetric quantum dots [37, 38]. The motion of the fine structure in magnetic field, which is on the order of $\hbar\omega_c$, is consistent with what one expects for states in GaAs when the confinement energy is weak. The lack of dependence upon magnetic field orientation excludes states localized in the dot, as states in the dot exhibit very different response in parallel ($B\|I$) versus perpendicular ($B \perp I$) magnetic field orientations owing to the strong quantum well confinement. The observed fine-structure peak energy separations of 0.6–0.9 meV are too small to arise from 1D subbands in the leads. Other possible models that account for this close energy separation, including interference effects [42], universal conductance fluctuations (UCF) [43], or effects resulting from chaotic transport [44], can be individually excluded.

The small energy separation between the fine-structure peaks in conductance indicates that they may be due to the effect that individual donors have on the 1D density of states in the emitter. For the dopant levels used in this structure, there are approximately two impurities in a 100-Å cube in the emitter (the value of 100 Å is used as it is the mean free path in the doped lead at low temperature). This places about 100 dopant atoms with electrons contributing to the degenerate doping of the conduction band in this volume. For a calculated Fermi level of 80 meV, this results in level spacings that are approximately 0.8 meV in energy (2.2 mV in forward bias, 1.6 mV in reverse bias). This in good agreement with the observed 0.6–0.9-meV spacing.

4.5. Magnetotunneling Measurements: Spin Splitting and g^* Factor

So far in this chapter, we ignored the spin of the electron. We expect the ground states of the fabricated or the impurity dot systems to be spin degenerate. In a magnetic field, this degeneracy is lifted. The energy separation between these split states is given by $\Delta E = g^* \mu_b B$, where B is the magnetic field strength, μ_b is the Bohr magneton, and g^* is the effective spin (Landé g) factor of electrons in the semiconductor material. This g factor of electrons in a material depends on the band structure and is different from the free-electron value of 2. In bulk GaAs, it is -0.44. In a confined system, the g^* factor changes further as the confinement affects the band structure. Thus, an accurate experimental determination of the spin g^* factor of electrons in a confined region serves as an important test of the band theory of solids. In what follows, we present the observation of spin splitting in the impurity dot system only. A similar effect is not observed in the fabricated dot systems for the several following reasons. The spin splitting energy is very small ($\approx$150 μeV at 10 T). First, in order to observe this small splitting, we have to have sharp current steps and low temperatures. The observed spin splitting in the impurity dot system could not be resolved at a temperature above 300 mK (Fig. 28). It is possible that the electrons in the fabricated dot samples did not get as cold as those in the impurity dot samples. Second, the nature of the emitter in the two systems is different. In the impurity system, we have a 3D emitter with a continuum of density of states. Therefore, electrons of both spin orientations are available for tunneling at any energy even in a magnetic field. We do not have to worry about the spin splitting in the emitter if the emitter is 3D. On the other hand, the fabricated dot samples have a 1D emitter with a discrete density of states. The splitting of these 1D subbands would have to be taken into account as well and the net splitting would be a difference between the emitter subband splitting and the quantum dot state splitting. We will now discuss the observation of spin splitting in the impurity dot system and how it allows us to determine the spin g^* factor and the electron tunneling rates.

Figure 23 shows an expanded view of one such current step with and without a magnetic field when the sample is in a dilution fridge with the base temperature of the fridge (T_{mix}) at 35 mK.

At zero field, the ground state of the impurity is spin degenerate causing a single current step. Upon lifting of the degeneracy at finite field, a splitting of the current step is observed. Note that the splitting is not the same for the two different field orientations. This splitting is observed in all devices and for all the current steps in a given device investigated at low temperatures. Figure 24 depicts the $I(V)$ characteristics for the same device at higher bias showing six different current steps. Each of those steps are observed to split in a magnetic field in both forward and reverse bias.

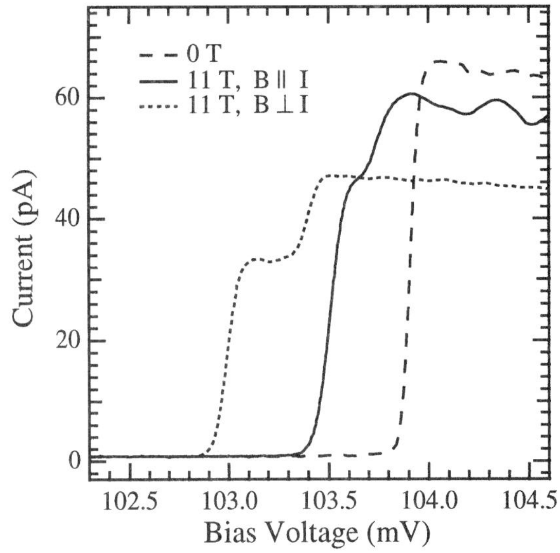

Fig. 23. $I(V)$ characteristics of the first current step edge ($T_{\text{mix}} = 35$ mK) in reverse bias at 0 T and at 11 T under both magnetic field orientations (field parallel to current and field perpendicular to current). The three curves are arbitrarily offset along the x axis for clarity.

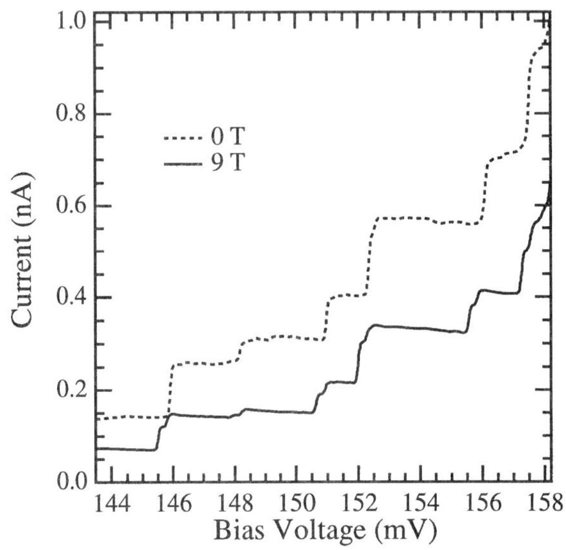

Fig. 24. $I(V)$ characteristics at $T_{\text{mix}} = 35$ mK showing six different current steps in zero field (dashed line) and with a field of 9 T oriented perpendicular to current (solid line).

Figure 25 shows the evolution of the $I(V)$ characteristics upon increasing magnetic field. Note that the spin splitting increases as the field is swept from 0 T (bottom) to 9 T (top).

Figure 26 shows the voltage separation between the corresponding two spin-split conductance peaks, which increases linearly with the magnetic field strength as expected. Because of the finite widths of the conductance peaks, it is not possible to resolve the splitting for magnetic fields less than 5.5 T. The best line fit to the data (Fig. 26) closely intersects $\Delta V = 0$ at $B = 0$ and has a slope of $g^* \mu_b / \alpha$, where μ_b is the Bohr magneton, g^* is the effective spin g factor of the impurity, and α is the voltage-to-energy conversion factor [2, 6, 34, 35]. The observed spin splitting is different for the two field orientations and we denote

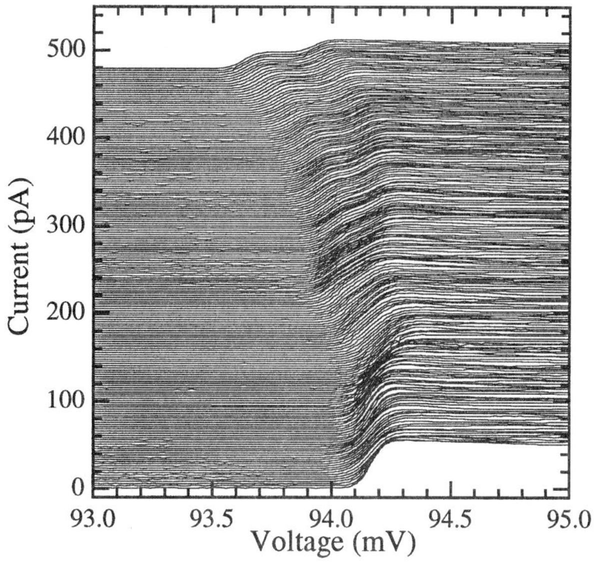

Fig. 25. $I(V)$ characteristics ($T_{\text{mix}} = 35$ mK) of the first current step edge in forward bias in increasing magnetic fields (field perpendicular to current) from 0 T (bottom) to 9 T (top). The successive field curves are offset a constant amount along the y axis for clarity.

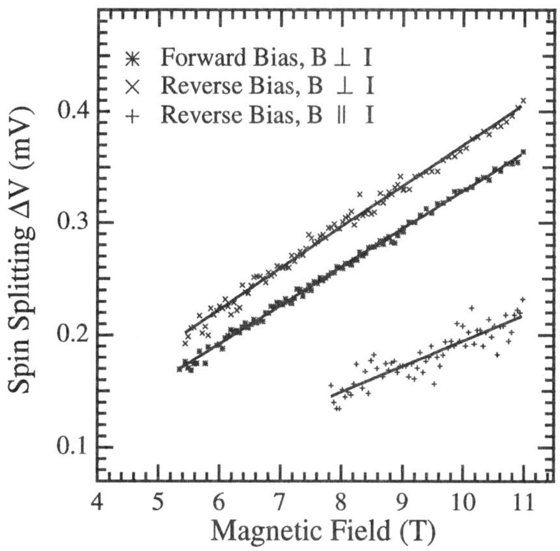

Fig. 26. Experimental spin splitting versus magnetic field for the first current step for the different bias and magnetic field orientations at $T_{\text{mix}} = 35$ mK. The solid lines are linear fits to the data.

$g_{\perp}^{*}$ and $g_{\parallel}^{*}$ to be the effective g factors for the field being perpendicular and parallel to the current flow direction, respectively.

To determine the spin g^* factor, it is necessary to know the voltage-to-energy conversion factor, α, which is accurately determined by studying the thermal broadening of the current steps as discussed earlier in Section 4.2.

From these values of α and the measured slopes of the fits in Figure 26, the absolute magnitude of the spin $g_{\perp}^{*}$ and $g_{\parallel}^{*}$ values of the impurities can be calculated. Table IV lists the measured g^* values for the various devices and current steps. For a given device, we

Table IV. α and g^* Values as Calculated for the Various Devices and Current Steps

	α		$g^*_\perp$		$g^*_\parallel$	
Current step	+bias	−bias	+bias	−bias	+bias	−bias
First step	0.48 ± 0.02	0.42 ± 0.02	0.28 ± 0.02	0.28 ± 0.02	0.14 ± 0.04	0.13 ± 0.04
Second step			0.28 ± 0.02	0.27 ± 0.02		

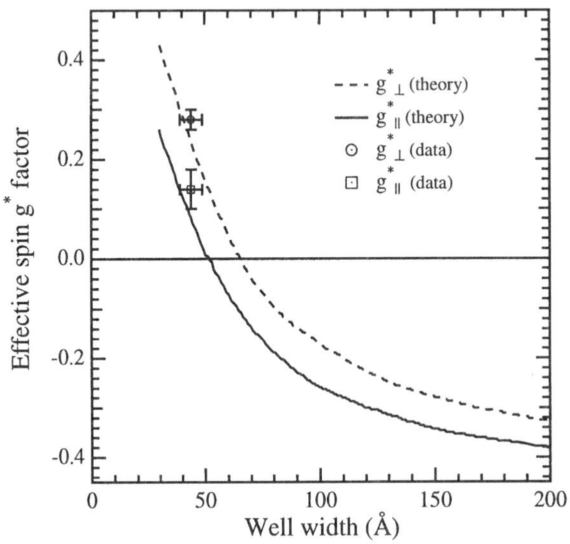

Fig. 27. Comparison of the measured g^* factors with the theoretical predictions of Ivchenko and Kiselev [45]. We assume that the sign of the measured quantities is positive.

determine the g^* factors in both the forward- and the reverse-bias directions. Note that the slopes of the two lines in Figure 26 for the first step in forward bias and reverse bias are different, but so are the corresponding α values, giving us the same g^* values. The α value in this device is not determined for the second current step independently. We assume that the α value for the second step is the same as that of the first step. This is a good assumption, as the α values are not very sensitive to small bias changes.

As discussed earlier, g^* is theoretically predicted to be a strong function of the quantum well width and is calculated to change from the bulk value of -0.44 to greater than $+0.4$ for well widths less than 30 Å and it also becomes asymmetric [45]. In Figure 27, we show our data with respect to the preceding theoretical predictions of Ivchenko and Kiselev [45].

Our measurements are consistent with this prediction and with recent experimental results [46, 48]. We cannot determine the sign of the g^* factors and we have assumed here that it is positive. Note that the predictions are made for quantum wells with $Al_xGa_{1-x}As$ barriers with $x = 0.35$, while our measured data are on quantum wells with $Al_xGa_{1-x}As$ barriers with $x = 0.27$. We expect the effect of this difference in Al concentration to be small. Also note that, strictly speaking, the measured g^* factor is that of an electron bound to a Coulomb potential in the quantum well and not that of free electrons in the quantum well. We expect this difference to be small, too, as the impurity state is a shallow impurity state.

So far, we have neglected the spin splitting of the electrons in the emitter. Naively, one may expect that observed splitting of the current steps is due to the difference in the spin

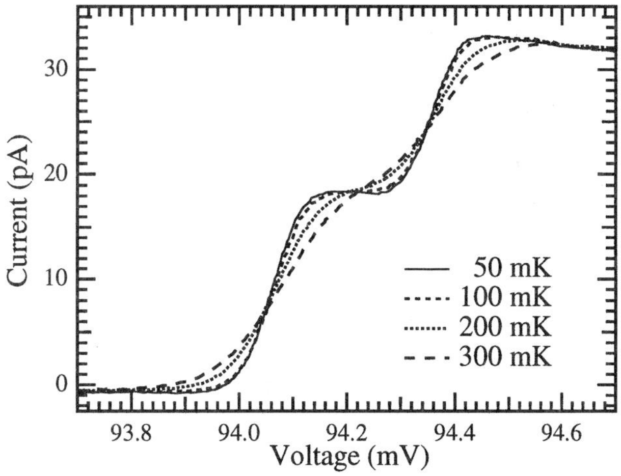

Fig. 28. $I(V)$ characteristics at 9 T showing the spin-split first current step edge in forward bias at different temperatures. Both fragments of the split edge exhibit Fermi broadening with the temperature.

splitting of the emitter electrons and the impurity state. This is not the case as the localized impurity state is an individual, isolated state, while the emitter is a 3D sea of electrons with a continuum of occupied electron states. All electron levels in the emitter split in an applied magnetic field but, owing to the continuum, at any given energy electrons of both spin orientations are available for tunneling. Because the spin splitting energy ($\approx$0.15 meV at 10 T) is much smaller than the Fermi energy in the emitter ($\approx$40 meV), we do not expect to observe any spin polarization effects in the emitter. Figure 28 shows thermal broadening of the first current step edge split at 9 T. Separate Fermi broadening of the two fragments of the split step indicates that in both cases electrons involved in tunneling are the Fermi electrons in the emitter. The observed experimental voltage difference is entirely due to the spin splitting of the impurity state only, and is not affected by the g^* factor of the electrons in the emitter.

If the Fermi energy in the emitter is small and weakly localized states are formed in the emitter, the g^* factor of these weakly localized states would have to be taken into account as reported by Sakai et al. [22, 49].

We also note that electron–electron interactions will not alter the relative energy of the two spin states and thus the measured g^* factor in this experiment. There are two possible mechanisms for electron–electron interactions. One is between the electrons in the well and the other is between the tunneling electron and the electrons in the emitter. The first mechanism can be ignored because, at the bias of interest, there is very little electron density in the well. It has been observed in transport studies of 2D electron gases that there is an enhancement in the g^* factor of electrons, which is attributed to electron–electron interactions; however, such an enhancement is not observed when single-electron energy levels of the 2D electron gas are probed (as shown by Dobers et al. [47]). In our experiment, we do not have a 2D electron gas in the quantum well, and we are probing single-electron states. The second mechanism, the Coulomb interaction between the tunneling electron and the electrons in the emitter, causes a rise in the step current near the threshold at low temperatures, which is observed (as first reported by Geim et al. [34]), but it should not affect the energy separation of the two spin states.

4.6. Magnetotunneling Measurements: Electron Tunneling Rates

Analysis of the tunneling current through a system yields information about the tunneling rates through the system. We now have a unique opportunity to investigate a two-state tun-

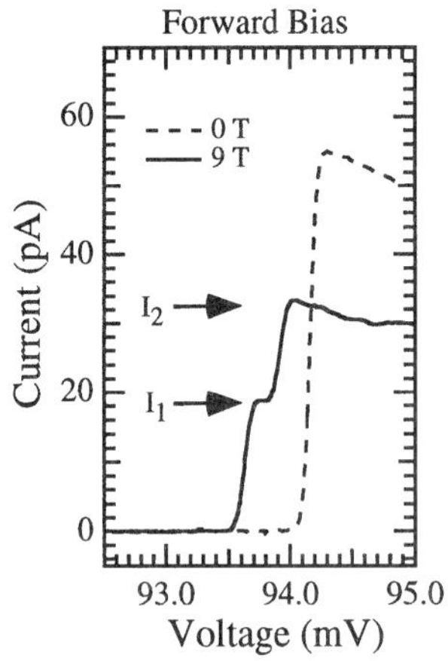

Fig. 29. $I(V)$ characteristics at $T_{\mathrm{mix}} = 35$ mK of the first current step edge of the 85-Å-thick barrier device in forward bias (left) and reverse bias (right) at 0 T (dashed line) and 9 T (solid line). The magnetic field is oriented perpendicular to the current direction. I_1 and I_2 denote the current values at 9 T as shown. I_1 gives the current of the first fragment, while I_2 is the net current of both fragments of the split step edge.

neling system, which can give more information than just a single-state tunneling system. Figure 29 shows that the two fragments of the spin-split step at 9 T are observed *not* to have the same current magnitude, that is, $I_2 - I_1 \neq I_1$, where I_1 and I_2 denote the current values as shown. I_1 gives the current of the first fragment, while I_2 is the net current of both fragments of the split step edge. This difference is more prominent in reverse bias. An understanding of this asymmetry gives us the electron tunneling rates through each of the two barriers of the double-barrier heterostructure individually.

To analyze this, we define T_{b} and T_{t} to be the tunneling rates for an electron to tunnel through the bottom and the top (referred to as the growth direction) potential barriers, respectively, of the double-barrier heterostructure. We also define T_{cl} and T_{em} to be the electron tunneling rates through the collector (downstream) and the emitter (upstream) (referred to as the electron flow direction) potential barriers. In forward bias, electron injection is through the top barrier into the quantum well ($T_{\mathrm{em}} \equiv T_{\mathrm{t}}$, $T_{\mathrm{cl}} \equiv T_{\mathrm{b}}$), while, in reverse bias, electron injection is through the bottom barrier into the quantum well ($T_{\mathrm{em}} \equiv T_{\mathrm{b}}$, $T_{\mathrm{cl}} \equiv T_{\mathrm{t}}$). We also define p to be the occupation probability for an electron in a localized state in the quantum well. In the sequential tunneling picture, $p = f(E)T_{\mathrm{em}}/(T_{\mathrm{em}} + T_{\mathrm{cl}}) = f(E)p_0$. Because the spin splitting energy ($\Delta E \leqslant 150$ μeV) is much smaller than the barrier potential energy ($\approx$300 meV), we assume that the tunneling rates are the same for electrons tunneling through the spin-up or the spin-down states irrespective of the slight energy difference. We also assume that the emitter electrons are not spin polarized, which is a good assumption because the Fermi energy ($\sim$40 meV) is much larger than the spin splitting energy even at 10 T.

In high magnetic fields and at low temperatures, when the Fermi level is sharp ($p = p_0$, Fig. 28), we can adjust the bias near a given impurity to obtain the following two conditions as noted schematically in Figure 30. At a lower bias, V_1, only the lower energy spin state channel is active for conduction, and the current is given by

$$I_1 = peT_{\mathrm{cl}} \tag{17}$$

At a higher bias, V_2, the higher energy spin state channel is also active for conduction, and the current is given by

$$I_2 = p'eT_{\mathrm{cl}} = \left(2p - p^2\right)eT_{\mathrm{cl}} \tag{18}$$

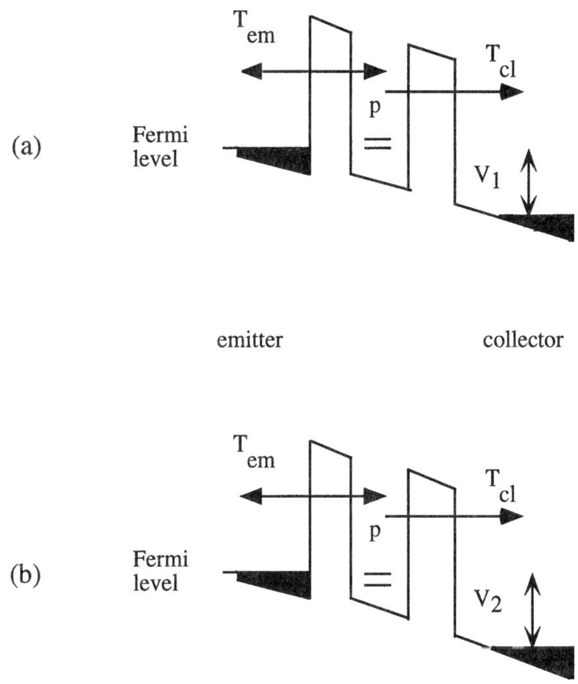

Fig. 30. Schematic of electron tunneling through the two-state system.

where $p' = (2p - p^2)$ is the probability of occupying either the lower or the higher state, but not both of them. Both states cannot be occupied at the same time because of the large Coulomb energy required for another electron to simultaneously occupy the second state. For this system, the single-electron Coulomb charging energy ($U_C = e^2/2C$, where C is the effective capacitance of the double-barrier device) is much larger than the spin splitting energy ΔE. We can estimate the capacitance of the system to be $C = (\varepsilon_0 k \pi r_0^2)(d_t^{-1} + d_b^{-1})$, where k is the dielectric constant, d_t and d_b are the top and bottom barrier thicknesses, and r_0 is the radius of the lateral region affected by the localized impurity state in the quantum well. We take r_0 as the Bohr radius of hydrogenic impurities in GaAs ($\approx$100 Å) and obtain $U_C \approx 9$ meV. Although taking r_0 as the Bohr radius is only an approximation, we obtain $U_C \geqslant \Delta E$ even for r_0 seven times larger.

In Eqs. (17) and (18), we have assumed that the tunneling rates are the same at the two different biases V_1 and V_2, which is a good assumption because the bias difference $(V_1 - V_2)$ is much smaller than the barrier potential. In the extreme limits, these equations indicate that, for $T_{cl} \gg T_{em}$, $p \approx 0$ and $I_2 \approx 2I_1$, while, for $T_{cl} \ll T_{em}$, $p \approx 1$ and $I_2 \approx I_1$. This qualitatively explains the behavior observed in Figure 29. To get a quantitative understanding, we solve Eqs. (17) and (18), using the experimentally measured I_1 and I_2, to determine p.

4.6.1. Effect of Spin Degeneracy at Zero Magnetic Field

The preceding analysis of the tunneling current through the two-state system is valid even at zero magnetic field when the two spin states are degenerate. We, hence, expect to observe some effect of the finite occupancy, p, of the impurity state. This effect is very nicely observed by investigating the thermal broadening of the current steps. In Section 4.2, we mentioned that current steps broaden as the temperature increases and this broadening was modeled by the Fermi equation [Eq. (11)]. A fit of this equation to the data is shown in Figure 8. This fit was done for a bias region $V \leqslant V_{th}$. If we extrapolate these fits to the

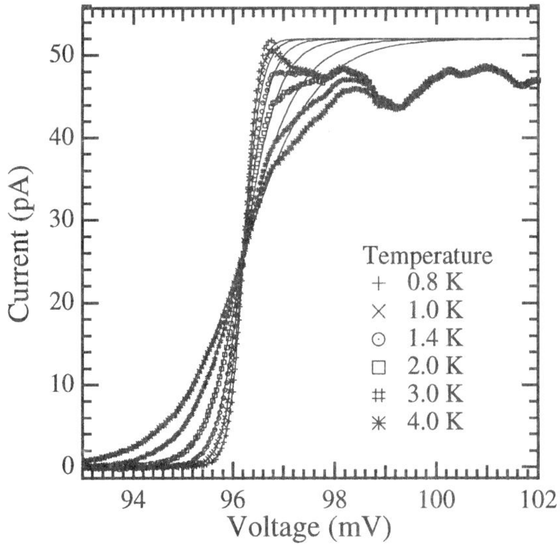

Fig. 31. $I(V)$ characteristics of the first current step edge in forward bias of the impurity system with 85-Å-thick barrier at different temperatures and the extrapolation of the Fermi fits [Eq. (11)] to these $I(V)$ traces as obtained in Figure 8 to voltages greater than the threshold ($V \geqslant V_{th}$). The experimentally measured current can be seen to be less than the theoretically expected current on the plateau.

region above the threshold (Fig. 31), we observe that the experimentally measured current is lower than the predicted value. We attribute this deviation to the Fermi behavior of the spin degeneracy of the localized state and Coulomb charging effects.

This fit is strictly valid for the case of tunneling through a single state only. The current in the case of tunneling through two degenerate states can be expressed as in Eq. (18) [assuming $p = p_0 f(E)$]:

$$\Delta I = A p_0 f(E)\big(2 - p_0 f(E)\big) \tag{19}$$

where A is a constant and $p_0 = T_{em}/(T_{em} + T_{cl})$ is the occupancy of the electron in either one of the two localized degenerate states when $f(E) = 1$. p_0 depends on the relative tunneling rates of the two potential barriers. Equation (19) can be fit to the data. Once again, the fits are done only for bias voltages $V \leqslant V_{th}$ because of the presence of the "fine structure" on the current plateaus for $V \geqslant V_{th}$. However, in Figure 32, we directly show the extrapolation of the corrected fits to all bias voltages. V_{th} is obtained from the common intersection point of the curves at different temperatures. A and p_0 are obtained from the measured current values of the plateau current ($I_{plateau}$) and the threshold current I_{th} in the following way. If $V \gg V_{th}$, then $f(E) = 1$ and

$$\Delta I = I_{plateau} = A p_0 (2 - p_0) \tag{20}$$

If $V = V_{th}$, then $f(E) = \frac{1}{2}$ and

$$\Delta I = I_{th} = A\left(\frac{p_0}{2}\right)\left(\frac{2 - p_0}{2}\right) \tag{21}$$

Solving these two equations simultaneously from the measured values of $I_{plateau}$ and I_{th}, we get A and p_0. We get for the data shown in Figure 32, $A = 81.4$ pA and $p_0 = 0.35$. Thus, once again, the only free parameter in the fit is α, which then can be accurately determined. The α values obtained from this fit for this step is 0.5, which is within the experimental error of that obtained from the previous fit (0.48). This is consistent as both the fits are done at $V \leqslant V_{th}$ [$f(E) < \frac{1}{2}$], when the occupancy [$p = p_0 f(E)$] is small, and Eq. (19)

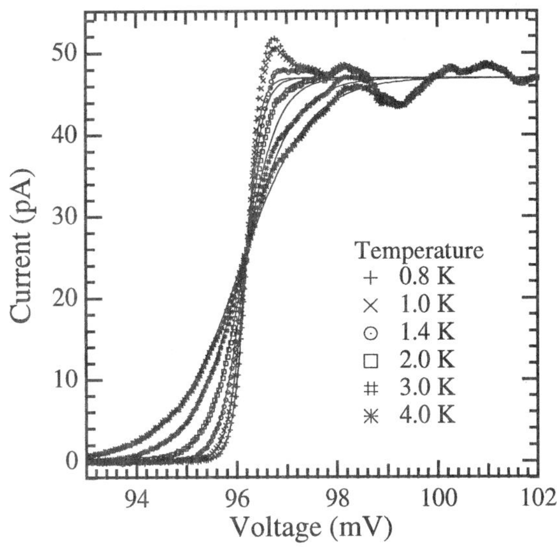

Fig. 32. $I(V)$ characteristics of the first current step edge in forward bias of the impurity system with 85-Å-thick barrier thickness at different temperatures and the extrapolation of the corrected fits [Eq. (19)] to these $I(V)$ traces to voltages greater than the threshold ($V \geqslant V_{\text{th}}$).

reduces to that of Eq. (11). These fits and their extrapolation to bias voltages $V \geqslant V_{\text{th}}$ now accurately determine the current plateau values as can be seen from Figure 32.

It is also possible to determine the occupancy for the localized states in reverse bias. For the same device in reverse bias, we obtain $p_0 = 0.57$ and $\alpha = 0.42$. This difference in forward and reverse bias is attributed to a slight asymmetry in the growth of the device and will be discussed in detail later.

The importance of this corrected fit becomes even more apparent while investigating the thermal broadening of current steps on which the fine structure is not very prominent. In such cases, the fit can be carried out over the entire range of the data unlike the earlier case where it was restricted to $V \leqslant V_{\text{th}}$. This is shown in Figure 33 for the 65-Å-thick barrier device in forward bias. From the fit, we obtain $p_0 = 0.57$, $A = 0.56$ nA, and $\alpha = 0.42$ for this particular step.

Note that doing a Fermi fit only [Eq. (11)] to the data of Figure 33 would be very inaccurate. The ratio of the current at the threshold, the common intersection point of all the curves, and the current on the plateau is $I_{\text{th}}/I_{\text{plateau}} = 0.275$ nA/0.458 nA $= 0.6$. This is much larger than 0.5 as one would have expected if one had a perfect Fermi function behavior.

From the experimental values of A and p_0, one can determine the electron tunneling rates T_{em} and T_{cl}. $p_0 = T_{\text{em}}/(T_{\text{em}} + T_{\text{cl}})$ and $A = eT_{\text{cl}}$. Thus, for the first current step in forward bias of the 85-Å-thick barrier device (Fig. 32), $T_{\text{cl}} = 509$ MHz and $T_{\text{em}} = 274$ MHz. Similarly, for the first current step in forward bias of the 65-Å-thick barrier device (Fig. 33), $T_{\text{cl}} = 3.6$ GHz and $T_{\text{em}} = 4.7$ GHz. Similar measurements of the tunneling rates can also be obtained from the magnetic field studies. Those measurements, however, require really low temperatures (less than 300 mK) and high magnetic fields. The calculation described here requires relatively moderate temperatures (greater than 1 K) and, hence, is important.

4.6.2. Tunneling Rates in a Magnetic Field

For the remainder of this chapter, we confine ourselves to a discussion of the results of only one device (the device with the 85-Å barrier and data from the first current step only).

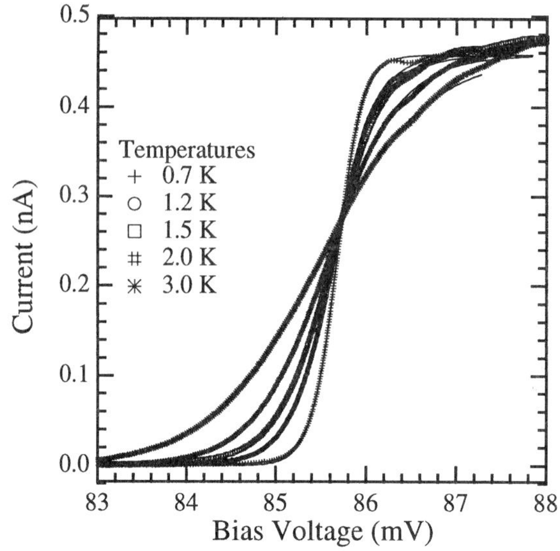

Fig. 33. $I(V)$ characteristics of the first current step edge in forward bias of the impurity system with 65-Å-thick barrier at different temperatures and the corrected fits [Eq. (19)] to these $I(V)$ traces. The fits are done over the entire bias range spanning the step.

The results of the other device (65-Å barrier) show similar trends. From the data shown in Figure 29 at 9 T (field perpendicular to current), we get $p = 0.21$ for forward bias and $p = 0.62$ for reverse bias. A high p value indicates that the electron tunneling rate through the collector (downstream) barrier (T_{cl}) is lower than that through the emitter (upstream) barrier (T_{em}), causing an accumulation in the well. A higher p value for reverse bias (as compared to forward bias) suggests an asymmetry in the heterostructure growth with one barrier being slightly thicker than the other barrier. In forward bias, the top barrier is the emitter barrier, while, in reverse bias, the top barrier is the collector barrier. This implies that the top barrier is slightly thicker than the bottom barrier. This is consistent with the difference in the measured α values, observed asymmetry in the $I(V)$ characteristics for this sample (the main resonance peak voltage and peak current values are higher for reverse bias as compared to forward bias), and is in agreement with previous characterization [32, 33].

We also obtain the absolute magnitude of the electron tunneling rates through the two potential barriers and study their dependence on the magnetic field. Figures 34 and 35 show the tunneling rates and the probability of occupation as a function of the magnetic field perpendicular to the current in the forward-bias and reverse-bias orientations, respectively. Note that T_{em} is smaller than T_{cl} in the forward-bias orientation, while T_{cl} is smaller than T_{em} in the reverse-bias orientation. This is because p is smaller than 0.5 in forward bias, while it is larger than 0.5 in reverse bias. As mentioned before, this asymmetry suggests that the top barrier of the heterostructure is slightly thicker than the bottom barrier.

Despite the asymmetry, we observe that, in either orientation, T_{em} (and p) decreases with field strength, while T_{cl} is approximately constant. T_{em} roughly decreases by a factor of 2 as the field increases from 6 to 11 T in both bias orientations. The probability of occupation p also decreases as the field strength increases. The oscillations in the tunneling rate are probably due to the fine structure observed on the current plateaus and its movement in magnetic field [38, 50]. The suppression of T_{em} results in the observed plateau current suppression with magnetic field.

Figures 34 and 35 and the previous discussion indicate that the magnetic field affects the emitter-to-well tunneling process (T_{em}) more than it affects the well-to-collector tunneling process (T_{cl}). We, therefore, expect more current suppression when the current is emitter

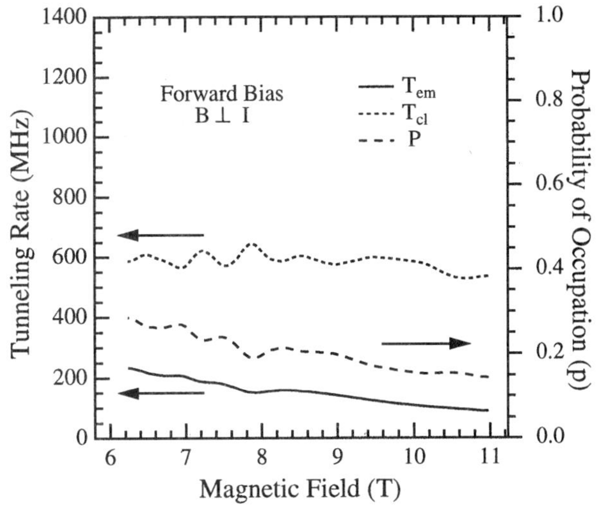

Fig. 34. Tunneling rates T_{em} and T_{cl} and the probability of occupation, p, as a function of the magnetic field strength perpendicular to current for the 85-Å-thick barrier device in forward-bias orientation.

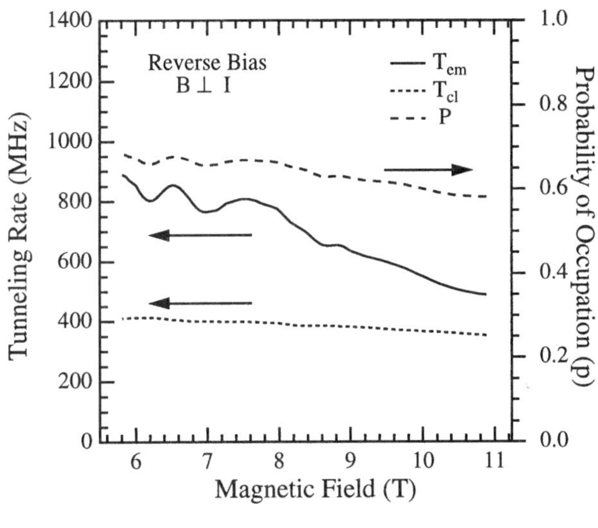

Fig. 35. Tunneling rates T_{em} and T_{cl} and the probability of occupation, p, as a function of the magnetic field strength perpendicular to current for the 85-Å-thick barrier device in reverse-bias orientation.

barrier limited than when it is collector barrier limited, which is observed (Fig. 29). For forward bias (thicker emitter), the current at 9 T is suppressed by a factor of 41% (as compared to 0 T), while the suppression is only 20% for reverse bias (thicker collector).

We can also study the effect on the tunneling rates of magnetic field parallel to current. Figures 36 and 37 show the tunneling rates and the probability of occupation, p, as a function of the magnetic field parallel to the current in forward bias. The oscillations observed in the tunneling rate and, hence, in p (Fig. 36) are due to the "fine structure" on the current plateaus and its movement in the magnetic field [38, 39, 50]. Nevertheless, it is possible to extrapolate the measured value of p to zero field and compare it with those obtained from the thermal broadening measurements.

For the data shown in this chapter for forward bias of the 85-Å-thick barrier device, we estimate the zero-field occupancy $p \sim 0.3$ from magnetic field studies. This is in good

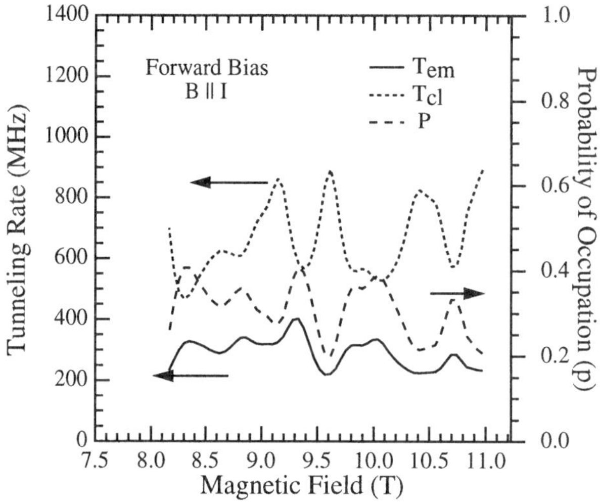

Fig. 36. Tunneling rates T_{em} and T_{cl} and the probability of occupation, p, as a function of the magnetic field strength parallel to current for the 85-Å-thick barrier device in forward-bias orientation.

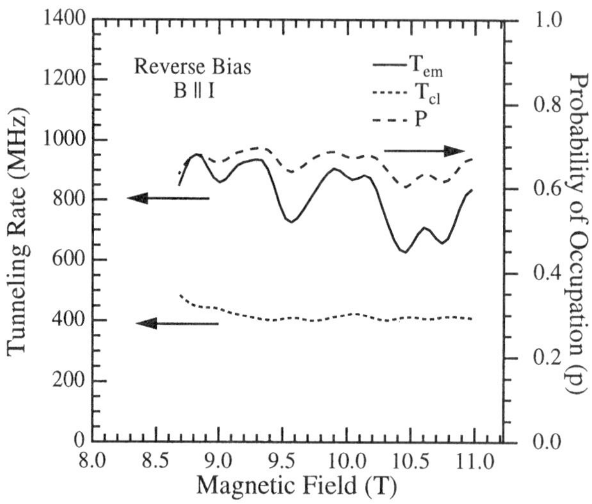

Fig. 37. Tunneling rates T_{em} and T_{cl} and the probability of occupation, p, as a function of the magnetic field strength parallel to current for the 85-Å-thick barrier device in reverse-bias orientation.

agreement with the measured value of $p = 0.35$ from the temperature studies. Similarly, in reverse bias, we get $p \sim 0.65$ from magnetic field measurements and $p = 0.57$ from temperature studies. A single parameter p, the occupancy of the impurity state, thus explains the non-Fermi thermal broadening and also the asymmetry in the magnetic field behavior. These two independent measurements thus give an added proof that the theory is correct.

5. CONCLUSIONS

In this chapter, we have presented both the basic theory and the experimental data for the electron spectroscopy as a function of magnetic field and temperature for two vertical quantum dot systems, the fabricated quantum dot and the impurity quantum dot. Current steps in the $I(V)$ characteristics are observed in both systems. In the fabricated quantum dot, these steps are due to the discrete nature of the 0D dot density of states. In the impurity system, these steps are attributed to the discrete 0D states resulting from the Coulomb potential of a single or a pair of impurities in the quantum well region. In both systems, in magnetic fields parallel to the current, the step edges exhibit a diamagnetic shift as expected, while a suppression of the current plateaus is observed in magnetic fields perpendicular to the current, which is attributed to the decreasing overlap between emitter and dot wave functions. In the impurity system, we also investigate the Zeeman splitting of these current steps and determine the effective spin g^* factor of electrons in the quantum well region, which is observed to be asymmetric. An analysis of the tunneling current through the two-state spin system allows us to determine the tunneling rate across each of the two barriers of the device individually. In the case of the fabricated quantum dot, the inconsistencies in the step spacings, the diamagnetic shifts, and the lack of Zeeman splittings indicate that transport is proceeding through more than one channel in parallel (i.e., two dots or more). A model incorporating this type of transport is proposed and found to be consistent with the first few plateaus.

Using the highly localized dot states as spectroscopic probes, the electronic structure of the emitter electrode is measured through the oscillatory fine structure observed on top of the normal current plateaus of both systems. The fine structure is observed to be reproducible, both upon repeated traces and upon thermal cycling. Variable-temperature results indicate that this fine structure originates from below the Fermi level in the emitter. In both parallel and perpendicular magnetic field orientations, the fine structure peaks are observed to shift to lower bias, which is, again, consistent with emitter states. With increasing temperature, states *above* the Fermi level are populated. In the case of the fabricated dot, we observe a preresonant thermally activated peak as temperature increases, which is in excellent agreement with the emitter discrete-state model.

Acknowledgments

We would like to thank Professor R. G. Wheeler and Ellen S. Hornbeck for experimental assistance and many useful discussions; W. R. Frensley, C. L. Fernando, and R. C. Bowen for help with the modeling of the devices; and J. N. Randall and R. J. Matyi for growth and fabrication. This work was supported by NSF DMR-9112497 and 9216121 and the Office of Naval Research.

References

1. M. Tewordt, L. Martin-Moreno, M. Pepper, J. E. F. Frost, V. J. Law, M. J. Kelly, R. Newbury, D. A. Ritchie, and G. A. C. Jones, *Phys. Rev. B* 46, 3948 (1992).
2. B. Su, V. Goldman, and J. Cunningham, *Phys. Rev. B* 46, 7644 (1992).
3. M. Reed, J. N. Randall, R. J. Aggarwal, R. J. Matyi, T. M. Moore, and A. E. Wetsel, *Phys. Rev. Lett.* 60, 535 (1988); M. Reed, J. N. Randall, J. H. Luscombe, W. R. Frensley, R. J. Aggarwal, R. J. Matyi, T. M. Moore, and A. E. Wetsel, *Festköperprobleme* 29, 267 (1989).
4. M. Dellow, P. H. Beton, M. Henini, P. C. Main, and L. Eaves, *Electron. Lett.* 27, 134 (1991).
5. S. Tarucha, Y. Hirayama, T. Saku, and T. Kimura, *Phys. Rev. B* 41, 5459 (1990); S. Tarucha and Y. Hirayama, *Phys. Rev. B* 43, 9373 (1991).
6. J. W. Sleight, E. S. Hornbeck, M. R. Deshpande, R. G. Wheeler, M. A. Reed, R. C. Bowen, W. R. Frensley, J. N. Randall, and R. J. Matyi, *Phys. Rev. B* 53, 15727 (1996).
7. U. Meirav, M. Kastner, and S. Wind, *Phys. Rev. Lett.* 65, 771 (1990).
8. P. McEuen, E. B. Foxman, U. Meirav, M. A. Kastner, Y. Meir, and N. S. Wingreen, *Phys. Rev. Lett.* 66, 1926 (1991).
9. L. P. Kouwenhoven, N. C. van der Vaart, A. T. Johnson, W. Kool, C. J. P. M. Harmans, J. G. Williamson, A. A. M. Staring, *Z. Phys. B* 85, 367 (1991); L. P. Kouwenhoven, A. T. Johnson, N. C. van der Vaart, A. van der Enden, C. J. P. M. Harmans, and C. T. Foxon, *Z. Phys. B* 85, 381 (1991).
10. A. Ramdane, G. Faini, and H. Launois, *Z. Phys. B* 85, 389 (1991).
11. J. H. Luscombe, *Nanotechnology* 4, 1 (1993).
12. H. Liu and G. Ayers, *J. Appl. Phys.* 65, 4908 (1989).
13. G. Bryant, *Phys. Rev. B* 39, 3145 (1989).
14. H. van Houten and C. Beenakker, *Phys. Rev. B* 63, 1893 (1989).

15. D. Averin, A. Korotkov, and K. Likharev, *Phys. Rev. B.* 44, 6199 (1991).
16. Y. Meir, N. Wingreen, and P. Lee, *Phys. Rev. Lett.* 66, 3048 (1991).
17. A. Groshev, T. Ivanov, and V. Valtchinov, *Phys. Rev. Lett.* 66, 1082 (1991).
18. L. Chen and C. Ting, *Phys. Rev. B* 44, 5916 (1991).
19. N. Johnson and M. Payne, *Phys. Rev. Lett.* 67, 9 (1991).
20. M. W. Dellow, P. H. Beton, C. J. G. M. Langerak, T. J. Foster, P. C. Main, L. Eaves, M. Henini, S. P. Beaumont, and C. D. W. Wilkinson, *Phys. Rev. Lett.* 68, 1754 (1992).
21. M. R. Deshpande, N. H. Dekker, J. W. Sleight, J. L. Huber, E. S. Hornbeck, M. A. Reed, R. J. Matyi, Y. C. Kao, C. J. L. Fernando, and W. R. Frensley, in "Proceedings of the IEEE/Cornell Conference on Advanced Concepts in High Speed Semiconductor Devices and Circuits," 1993, p. 177.
22. J. W. Sakai, P. C. Main, P. H. Beton, N. LaScala Jr., A. K. Geim, L. Eaves, and M. Henini, *Appl. Phys. Lett.* 64, 2563 (1994).
23. M. R. Deshpande, E. S. Hornbeck, P. Kozodoy, N. H. Dekker, J. W. Sleight, M. A. Reed, C. L. Fernando, and W. R. Frensley, *Semicond. Sci. Technol.* 9, 1919 (1994).
24. M. Tewordt, R. J. F. Hughes, L. Martin-Moreno, J. T. Nicholls, H. Asghi, M. J. Kelly, V. J. Law, D. A. Ritchie, J. E. F. Frost, M. Pepper, and G. A. C. Jones, *Phys. Rev. B* 49, 8071 (1994).
25. G. M. G. Oliveira, V. M. S. Gomes, A. S. Chaves, and J. R. Leite, *Phys. Rev. B* 35, 2896 (1987).
26. V. Goldman, D. Tsui, and J. Cunningham, *Phys. Rev. B* 35, 9387 (1987).
27. M. Tewordt, V. J. Law, R. T. Syme, M. J. Kelly, R. Newbury, M. Pepper, J. E. F. Frost, D. A. Ritchie, and G. A. C. Jones, in "Proceedings of the 20th International Conference on the Physics of Semiconductors," (J. Joannopoulos, Ed.), p. 2455. World Scientific, Singapore, 1991.
28. G. Bastard, "Wave Mechanics Applied to Semiconductor Heterostructures." Halsted Press, New York, 1988.
29. W. Kohn, in "Solid State Physics" (F. Seitz and D. Turnbull, Eds.), Vol. 5, p. 258. Academic Press, New York, 1957.
30. G. Bastard, *Phys. Rev. B* 24, 4714 (1981).
31. C. Priester, G. Bastard, G. Allan, and M. Lannoo, *Phys. Rev. B* 30, 6029 (1984).
32. M. A. Reed, W. R. Frensley, R. J. Matyi, J. N. Randall, and A. C. Seabaugh, *Appl. Phys. Lett.* 54, 1256 (1989).
33. M. A. Reed, W. R. Frensley, W. M. Duncan, R. J. Matyi, A. C. Seabaugh, and H.-L. Tsai, in "Proceedings of the International Symposium on GaAs and Related Compounds," Atlanta, GA, 1988, p. 587.
34. A. K. Geim, P. C. Main, N. LaScala Jr., L. Eaves, T. J. Foster, P. H. Beton, J. W. Sakai, and F. W. Sheard, *Phys. Rev. Lett.* 72, 2061 (1994).
35. M. R. Deshpande, J. W. Sleight, M. A. Reed, R. G. Wheeler, and R. J. Matyi, *Phys. Rev. Lett.* 76, 1328 (1996).
36. V. Fock, *Z. Phys.* 47, 446 (1928); C. Silorski and U. Merkt, *Phys. Rev. Lett.* 62, 2164 (1989); W. Hansen, T. P. Smith III, K. Y. Lee, J. A. Brum, C. M. Knoedler, J. M. Hong, and D. P. Kern, *Phys. Rev. Lett.* 62, 2168 (1989).
37. T. Schmidt, M. Tewordt, R. H. Blick, R. J. Haug, D. Pfannkuche, and K. von Klitzing, *Phys. Rev. B* 51, 5570 (1995).
38. T. Schmidt, R. J. Haug, V. I. Fal'ko, K. von Klitzing, A. Forster, and H. Luth, *Europhys. Lett.* 36, 61 (1996).
39. T. Schmidt, R. J. Haug, V. I. Fal'ko, K. von Klitzing, A. Forster, and H. Luth, *Phys. Rev. Lett.* 78, 1540 (1997).
40. V. Goldman, B. Su, and J. Cunningham, in "Nanostructures and Mesoscopic Systems," pp. 173–182. Academic Press, San Diego, 1992.
41. M. Tewordt, V. J. Law, T. J. Nicholls, L. Martin-Moreno, D. A. Ritchie, M. J. Kelly, M. Pepper, J. E. F. Frost, R. Newbury, and G. A. C. Jones, *Solid-State Electron.* 37, 793 (1994).
42. G. Bryant, *Phys. Rev. B* 44, 12837 (1991); A. Szafer and A. D. Stone, *Phys. Rev. Lett.* 62, 300 (1989).
43. A. D. Stone, in "Physics and Technology of Submicron Structures," p. 108. Springer-Verlag, Berlin, 1988.
44. R. A. Jalabert, H. Baranger, and A. D. Stone, *Phys. Rev. Lett.* 65, 2442 (1990); T. M. Fromhold, M. L. Leadbeater, L. Eaves, T. J. Foster, P. C. Main, and F. W. Sheard, *Surf. Sci.* 305, 511 (1994); T. Fromhold, L. Eaves, F. W. Sheard, M. L. Leadbeater, T. J. Foster, and P. C. Main, *Phys. Rev. Lett.* 72, 2608 (1994).
45. E. L. Ivchenko and A. A. Kiselev, *Sov. Phys. Semicond.* 26, 827 (1992).
46. M. J. Snelling, G. P. Flinn, A. S. Plaut, R. T. Harley, A. C. Tropper, R. Eccleston, and C. C. Phillips, *Phys. Rev. B* 44, 11345 (1991).
47. M. Dobers, K. von Klitzing, and G. Weimann, *Phys. Rev. B* 38, 5453 (1988).
48. R. M. Hannak, M. Oestreich, A. P. Heberle, W. W. Rühle, and K. Köhler, *Solid State Commun.* 93, 313 (1995).
49. J. W. Sakai, T. M. Fromhold, P. H. Beton, L. Eaves, M. Henini, P. C. Main, and F. W. Sheard, *Phys. Rev. B* 48, 5664 (1993).
50. M. R. Deshpande, J. L. Huber, N. H. Dekker, P. Kozodoy, J. W. Sleight, M. A. Reed, C. L. Fernando, W. R. Frensley, R. J. Matyi, and Y. C. Kao, in "Proceedings of the 2nd International Conference on the Physics of Semiconductors" (D. J. Lockwood, Ed.), p. 1899. World Scientific, Singapore, 1995.

Chapter 12

PHOTOREFRACTIVE SEMICONDUCTOR NANOSTRUCTURES

D. D. Nolte*, M. R. Melloch†, Y. Ding*, M. Dinu*, K. M. Kwolek*, I. Lahiri*

**Department of Physics, and †School of Electrical and Computer Engineering, Purdue University, 1396 Physics Building, West Lafayette, Indiana, USA*

Contents

Photorefractive semiconductor nanostructures are nonlinear optical devices that alter their optical properties in response to time- and space-varying light intensity patterns. They are used to perform dynamic holography under ultralow-light intensities much smaller than those used for traditional nonlinear optical materials. These devices contain nanometer-scale features that enhance their optical properties and alter their electronic transport relative to bulk behavior. They also contain high densities of deep level defects that make the devices semi-insulating. These defects trap and store photogenerated carriers that accumulate into space-charge densities that match the intensity patterns. The trapped charge

Handbook of Nanostructured Materials and Nanotechnology, edited by H.S. Nalwa
Volume 4: Optical Properties

produces electric fields that modify the optical properties of the nanostructure. The translation of temporally and spatially varying light intensity patterns into physical changes in the optical properties of the material, matching the intensity patterns, is called the photorefractive process. Photorefractive nonlinear optics works at low intensities because the light information is stored in the form of the trapped charge. The charge builds up over time, allowing the photogenerating intensity to be low.

1. OVERVIEW

Semiconductors have several length scales that are intrinsic to the material and are independent of thermal or disorder effects. One of these is the interatomic spacing. Another is the Bohr radius of the free exciton. Both of these length scales are determined by quantum mechanics, and both participate directly in the physical properties of semiconductor nanostructures. The interatomic spacing plays the central role in the physical assembly of nanostructures by epitaxial growth techniques in which atoms are added in layers to construct the crystal. The exciton Bohr radius plays the central role in the optical properties of semiconductor nanostructures. When the size of a nanostructure component (layer, box, dot, etc.) approaches the size of the Bohr radius, the excitonic optical transitions become sensitive to the physical size of the structure. The spectral shape of the optical absorption changes, and the integrated absorption can increase. The spectral properties can also become less sensitive to temperature, allowing discrete energy features to persist from low to room temperature. These discrete spectral features are the origin of the enhanced optical properties that make such semiconductor nanostructures attractive for the study of many optical phenomena and applications.

One of the primary goals of nonlinear optics is to allow light to control light, much in the same way that electrons control electrons in electronic devices. The difficulty encountered is that Maxwell's equations, that govern the propagation of light, are essentially linear in free space—precluding optical interaction to occur. Only when light propagates through materials, and the material properties are a function of the light intensity, can nonlinear optical effects arise. Even in these cases, the optical interactions are usually weak, requiring light intensities in excess of kW/cm^2 to generate appreciable interaction in most materials [1].

Contributing to the weak interaction of light with light is the dependence of the instantaneous material polarizability on the instantaneous intensity of the light. Much higher sensitivity would be expected if light intensity can be integrated over time. This would allow a weak optical beam to gradually accumulate its influence over a time that is characteristic of the material. This accumulation process occurs for resonant optical nonlinearities, in which real material excitations are generated that have an average lifetime. The effect of the light is therefore integrated over the lifetime of the excitation. Continuing this line of reasoning, much longer lifetimes would allow much weaker light beams to integrate their effects on the optical properties of the material.

The photorefractive effect [2] may be viewed as a limiting case of resonant nonlinear optics. In the photorefractive effect, charge carriers are generated in bright fringes by photoionization. These transport to dark fringes, where they are trapped at defect sites. When enough charge accumulates, a steady-state space-charge electric field is generated. This field alters the optical properties of the material through an electro-optic effect. This process has several steps, which include the storage of optical information as trapped space charge. The lifetime of the trapped space charge is set by the dielectric relaxation time of the material. This relaxation time varies over many orders of magnitude depending on the photorefractive material, and is dependent on intensity, but is typically between 1 ms and 1 μs for semiconductors for intensities of mW/cm^2 for bulk semiconductor crystals.

The sensitivity of the photorefractive effect is set by the smallest light intensity that is necessary for the photoconductivity to exceed the dark conductivity of the material. Therefore, materials with the highest sensitivity have the highest dark resistivity, the highest absorption, and the longest lifetime over which the optical signal is integrated. The saturation intensity for the formation of a fully developed space-charge field in a photorefractive crystal is

$$I_{\text{sat}} = \frac{n_d}{\alpha \tau} \tag{1.1}$$

where n_d is the carrier density in the dark, α is the absorption coefficient, and τ is the carrier lifetime. The dark carrier density in semiconductors is given by

$$n_d = n_i \exp \frac{-E_g}{2k_B T} \tag{1.2}$$

where n_i is the intrinsic carrier density and E_g is the semiconductor bandgap. The exponential dependence on the semiconductor bandgap E_g produces a rapidly increasing photorefractive sensitivity with increasing gap.

Photorefractive semiconductor nanostructures, or photorefractive quantum wells, most fully take advantage of the photorefractive properties of semiconductors [3–5]. There are several advantages to the performance of photorefractive nanostructures. The first specific advantage of photorefractive nanostructures is the large absorption α of the quantum-confined excitons, leading to high sensitivity by Eq. (1.1). Absorption coefficients in AlGaAs/GaAs quantum wells are typically $\alpha \approx 1.0 \times 10^4$–$1.5 \times 10^4$ cm^{-1} for wavelengths at or near the excitonic transition. Because device thicknesses are limited to $L \approx 1/\alpha$, the thickness of the nanostructures are typically on the order of a few microns.

The second specific advantage of photorefractive nanostructures is in the electroabsorption properties of the quantum-confined excitons. Electric fields alter the absorption of the excitonic transition, producing a field-dependent $\Delta\alpha$ that varies quadratically with the electric field. By the requirements of causality, the absorption change is accompanied by a change in the index of refraction Δn. The efficiency by which combined absorption and index gratings in a thin-film diffract light is given by

$$\eta = \left[\left(\frac{\pi \Delta n L}{\lambda}\right)^2 + \left(\frac{\Delta\alpha L}{4\lambda}\right)^2\right] \exp(-\alpha L) \tag{1.3}$$

where η is the diffraction efficiency, and λ is the wavelength of the diffracted light. Semiconductor nanostructures have enhanced electroabsorption properties, and therefore have enhanced diffraction properties. These occur, however, at the expense of high absorption, given by the exponential term. Part of the physics and device development of photorefractive quantum-well structures is therefore the interplay between absorption and electroabsorption.

The third and last specific advantage of photorefractive nanostructures is in the incorporation of the defects that trap and store the space charge. Defects can be incorporated during the highly nonequilibrium process of epitaxial growth. They can be chosen and placed virtually at will as a function of distance along the growth axis. In addition, the small sizes of the nanostructures make it possible to introduce defects through ion implantation. Therefore, defect engineering can be performed with a high degree of flexibility in photorefractive nanostructures.

These three specific advantages of photorefractive nanostructures combine to produce highly sensitive, fast, and compact semiconductor devices that have considerable design flexibility. The end products are general-purpose dynamic holographic thin films that can be used in a wide array of applications. This article will outline the basic structures of photorefractive quantum wells, and will describe the process of photorefractive grating formation. The optical properties and diffraction physics are key aspects of the device performance, and are explained in detail. There are several advanced properties of photorefractive nanostructures that are unique to these devices. These include hot-electron transport effects, complex spatiotemporal dynamics, and microcavity effects and enhancements. In addition to these, the nanostructure presents an opportunity to explicitly consider quantum mechanical effects in the optical properties, making it possible to engineer optical density of states. Finally, one of many current applications of photorefractive quantum wells is in femtosecond laser pulse processing. Some of these applications are described at the end of the article.

2. PHOTOREFRACTIVE QUANTUM-WELL STRUCTURES

2.1. Molecular Beam Epitaxy Growth of Epilayers, Heterostructures, and Quantum Wells

Molecular beam epitaxy (MBE) is a film growth process where the sources are thermally evaporated onto a heated substrate. The film growth takes place in a vacuum so that surface analysis by reflection high-energy electron diffraction can take place during film growth. Shutters are used to abruptly turn sources on and off. The MBE of GaAs is schematically depicted in Figure 1 with the shutters to the gallium and arsenic furnaces open, and to the aluminum furnace closed. By opening the shutter in front of the aluminum furnace, which adds a flux of aluminum atoms, the alloy $Al_xGa_{1-x}As$ can be grown, resulting in an abrupt GaAs/AlGaAs heterojunction. For the MBE of GaAs and AlGaAs at a rate of 1 μm per hour, the gallium furnace temperature is around 900 °C and the aluminum

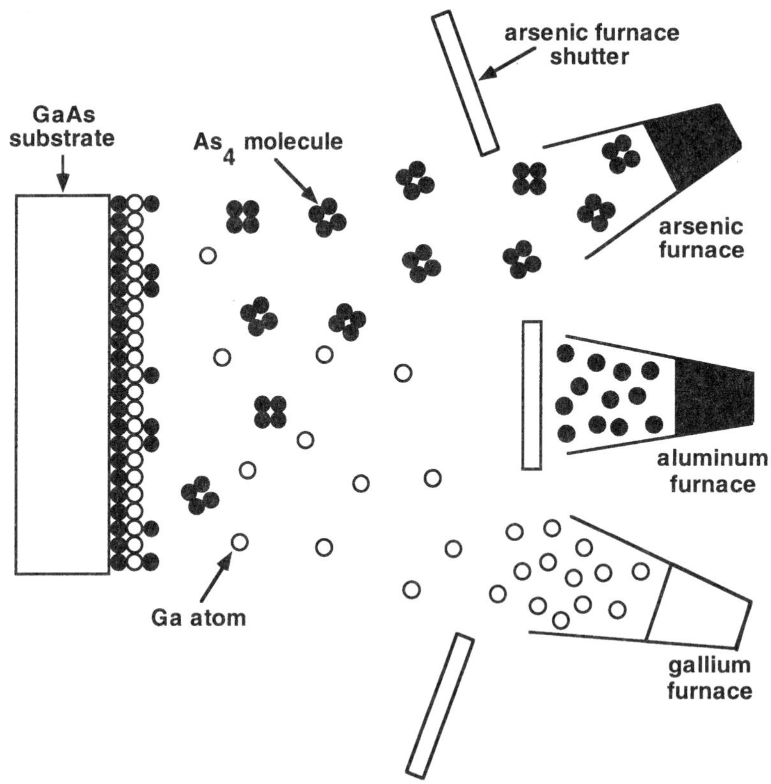

Fig. 1. Schematic of molecular beam epitaxy (MBE).

furnace temperature is around 1000 °C. For growth of stoichiometric GaAs/AlGaAs the typical substrate temperature is 600 °C, allowing every Ga or Al atom that arrives at the substrate to remain. Arsenic sublimes as a tetramer molecule. The typical As_4 to Ga beam-equivalent-pressure (as measured with an ion gauge in the growth position) used during MBE is 10–20:1 and requires an arsenic furnace temperature around 250 °C. An arsenic atom remains on the surface only if there is a Ga or an Al atom with which to bond. Therefore the Ga and Al fluxes determine the growth rate and an arsenic overpressure ensures growth of stoichiometric material.

2.2. Defect Engineering

The photorefractive effect requires the presence of relatively high densities of deep level defects that pin the Fermi level in the middle of the bandgap. The defects have two functions. First, by pinning the Fermi level midgap, the defects render the material semi-insulating giving it high resistivity and long dielectric relaxation times. The long dielectric relaxation times allow long times for the optical flux to accumulate. The second function of the defects is to provide sites to trap and store space charge. The trapped space charge represents the stored optical information, and produces space-charge electric fields that modify the optical properties of the semiconductor. For these reasons, defect engineering plays a central role in the technology of photorefractive semiconductor nanostructures.

The most common techniques to introduce high concentrations of deep level defects are ion implantation, doping with deep impurities during standard growth, and nonstoichiometric MBE growth.

2.2.1. Proton Implantation

The first photorefractive quantum-well devices were standard AlGaAs/GaAs multiple quantum wells that were made semi-insulating using proton implantation [3]. The protons themselves play no role in the defect structure, but the high implant energies produce radiation damage that consists predominantly of dangling bonds. These dangling bonds can trap either electrons or holes, and can provide the sites for the space charge that constitutes the hologram. Protons are used predominantly because they have deep penetration depths because of their light mass. For instance, an acceleration energy of 200 keV will give the protons a range of nearly 1.5 μm in GaAs.

For photorefractive effects, the important property of the implant is the density of generated traps. Although the calculation of the average number of knock-out atoms is straightforward [6], many of the knock-out atoms recombine. Furthermore, many of the defects self-compensate by creating pairs of donors and acceptors. Only a residual number of defects are formed permanently at room temperature. It was found empirically that each proton removed approximately three electrons [6, 7] and one valence hole [8]. Therefore, for an implant energy of 150 keV, with an average range of 1.2 μm, a dose of $[H^+] = 10^{12}$ cm^{-2} produces approximately 3×10^{16} cm^{-3} deep donors and 8×10^{15} cm^{-3} deep acceptors. This implant dose also produces electron free-carrier lifetimes of approximately 100 ps [9, 10].

2.2.2. Low-Temperature Growth

By reducing the substrate temperature during MBE, one can start to incorporate excess arsenic into the crystal. As much as 2% excess arsenic can be incorporated if the substrate temperature is reduced to around 200 °C, yet the epilayer is still of high crystal quality as determined by X-ray diffraction [11]. This excess arsenic is in the form of point defects—arsenic antisites [12] and gallium vacancies [13]. If these low-temperature-grown (LTG)

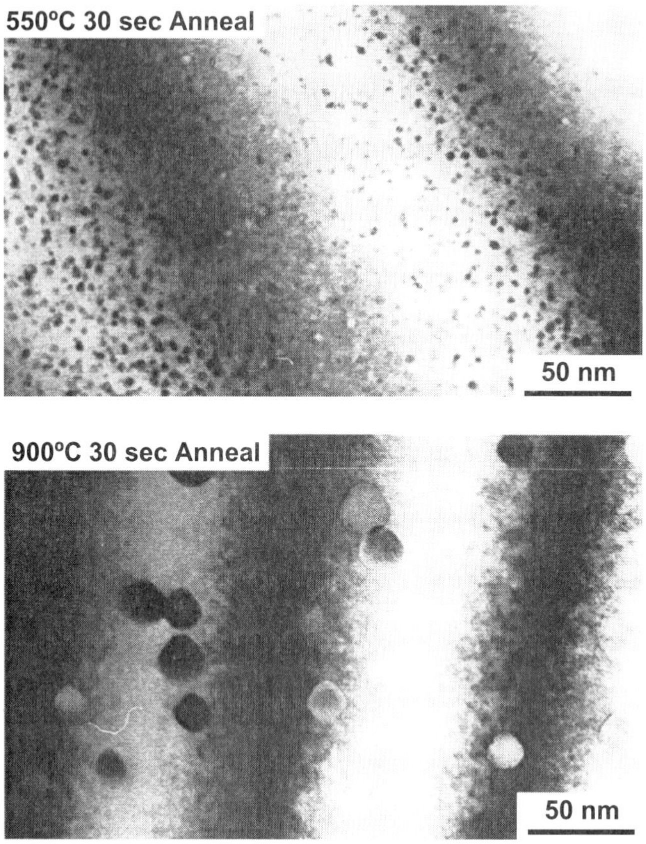

Fig. 2. TEM micrographs of arsenic precipitates in low-temperature-grown (LTG) GaAs.

epilayers are annealed at temperatures of 500 °C or higher, the excess arsenic precipitates [14]. Shown in Figure 2 is an example of a LTG-GaAs epilayer that was annealed at 550 °C for 30 s and 900 °C for 30 s. The arsenic precipitates are clearly seen in the transmission electron micrograph (TEM) with a high density of small precipitates at the lower anneal temperature and a much lower density of much larger precipitates at the higher anneal temperature. (The broad bands seen in Figure 2 are diffraction fringes caused by the wedge shape of the TEM sample.) After formation of the arsenic precipitates, they coarsen with further anneal [15, 16]. The change in the Gibbs function with anneal for this two-phase system of arsenic precipitates in a GaAs matrix is described by

$$dG = \gamma_G \, d\sigma_G - S \, dT \tag{2.1}$$

where dG is the change in the Gibbs function, γ_G is the interfacial energy per unit area, $d\sigma_G$ is the change in the surface area of the precipitates, and S is the entropy. The coarsening is a result of the two-phase system of the arsenic precipitates in the GaAs matrix lowering its Gibbs function by reducing the overall arsenic precipitate-to-GaAs matrix surface area and hence surface energy. This coarsening is illustrated in Figure 3 where the average density and diameter of the arsenic clusters are shown as a function of temperature for a 30 s anneal for four different epilayers. Three of these epilayers are GaAs, one with 0.9% excess arsenic, one with 0.7% excess arsenic, and one with 0.3% excess arsenic. The fourth epilayer is $Al_{0.25}Ga_{0.75}As$ containing 0.3% excess arsenic. The amount of excess arsenic in these epilayers was fixed by the substrate temperature during MBE [17]. Figure 3 illustrates the control one has of the final composite material by controlling the amount of

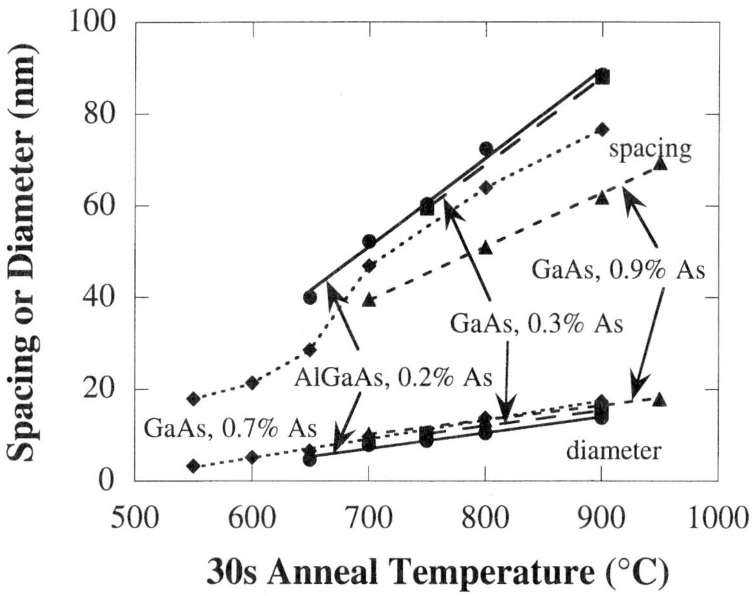

Fig. 3. The dependence of precipitate diameter and spacing on anneal temperature for 30 s anneals. The data represent four different growths at different substrate temperatures, leading to different degrees of nonstoichiometry.

excess arsenic in the epilayer with substrate temperature during MBE and the amount of precipitate coarsening with subsequent anneal.

The arsenic antisite is a donor near the middle of the bandgap in GaAs and the gallium vacancy is an acceptor about 0.4 eV above the valence band edge. Look et al. [18] reported that in as-grown LTG-GaAs the concentration of the deep donor is about 10 × higher than the concentration of the deep acceptor. For LTG-GaAs with excess arsenic of the order of 1%, there is considerable hopping conduction at room temperature because of the high concentration of midgap states [19]. As the excess arsenic precipitates with anneal, the midgap states become too far apart for hopping conduction and the Fermi level goes to midgap. The material therefore becomes semi-insulating with resistivities of about 10^6 Ω-cm [19]. Two mechanisms ensure that the Fermi level is at midgap in annealed LTG-GaAs, the larger concentration of the remaining arsenic antisites compared to the concentration of the remaining gallium vacancies [20], and the arsenic precipitates that form [21]. The arsenic precipitates act as internal Schottky barriers with a barrier height of 0.7 eV, which depletes the surrounding GaAs matrix material of carriers [21].

An interesting effect occurs when heterojunctions are present during arsenic precipitate coarsening. Shown in Figure 4 is a TEM of an LTG-GaAs quantum well after a 700 °C, 5 min anneal and after a 700 °C, 1.5 h anneal. The 100 nm AlGaAs barriers and the 10 nm GaAs wells were grown to contain 1% excess arsenic. With increasing anneal, the arsenic forms precipitates that preferentially coarsen to the GaAs well regions [15, 16, 22]. The change in the Gibbs function with anneal for this LTG quantum-well sample is described by

$$dG = \gamma_G \, d\sigma_G + \gamma_A \, d\sigma_A - S \, dT \tag{2.2}$$

where the symbols are the same as in Eq. (2.1) and the subscript G denotes GaAs and the subscript A denotes AlGaAs. The arsenic precipitates preferentially coarsen to the GaAs regions because of the lower interfacial energy of an arsenic precipitate in a GaAs region than in an AlGaAs region, which lowers the Gibbs function. It should be noted that a point could be reached where a lowering of the Gibbs function occurs by a reduction of the

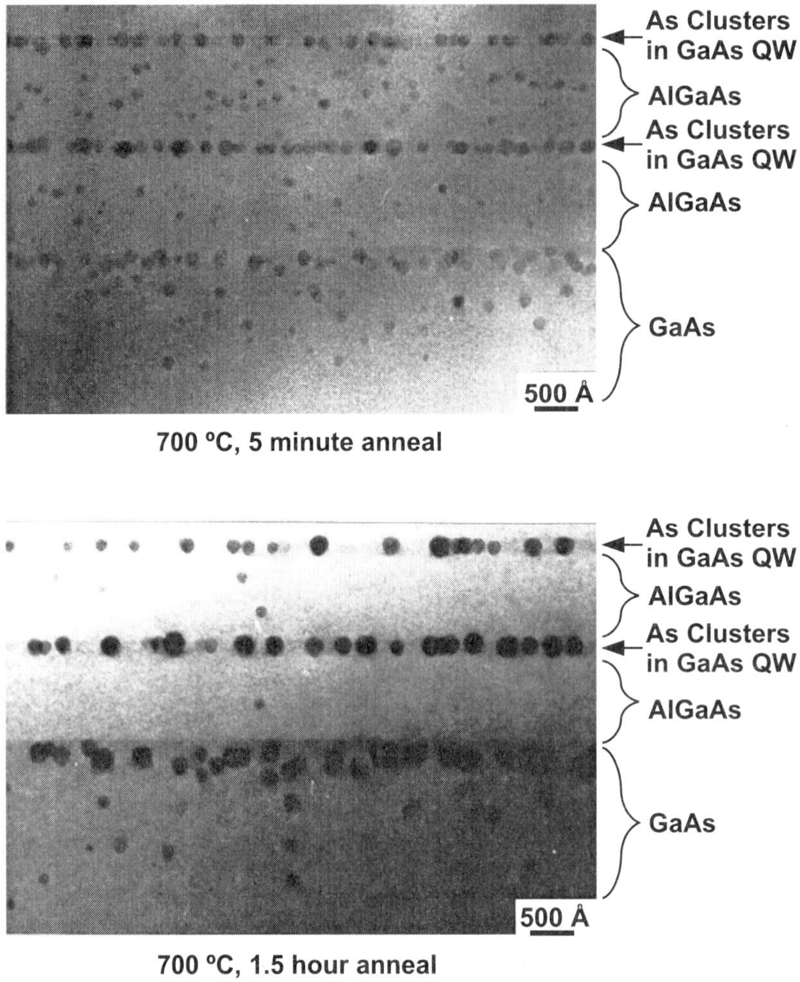

Fig. 4. Micrographs of precipitates that form preferentially in GaAs layers sandwiched between AlGaAs layers.

density of the arsenic precipitates, resulting in the arsenic precipitates that are in the GaAs well regions extending into the AlGaAs barrier regions. Complete reviews of the MBE and material science of LTG materials can be found in Refs. [23] and [24].

Annealed LTG-GaAs is semi-insulating and can have carrier lifetimes on the order of 1 ps. These are properties that make LTG-GaAs attractive for use as a high-speed photoconductive switch. Many uses of LTG-GaAs as a photoconductive switch have been reported. Chen et al. [25] demonstrated an interdigitated photoconductive detector on LTG-GaAs with a 375 GHz bandwidth and a responsivity of 0.1 A/W. Motet et al. [26] utilized the high resistivity and breakdown field of LTG-GaAs to demonstrate generation of 825 V 4 ps full-width-at-half-maximum (FWHM) pulses on a transmission line by photoconductive switching. Warren et al. [27] generated electromagnetic pulses of 0.71 ps FWHM using antennas fabricated on LTG-GaAs to launch and detect electromagnetic pulses. Rahman et al. [28] demonstrated transient scattering measurements of cavity resonances from targets by photoconductively switching coplanar-strip horn antennas on LTG-GaAs. Groeneveld et al. [29, 30] developed a scanning tunneling microscope (STM) tip from LTG-GaAs that can be photoconductively switched for detection of picosecond electrical transients. High-speed photodetectors based on LTG-GaAs have also become commercially available from MellWood Laboratories, Inc. and Newport Research Corporation.

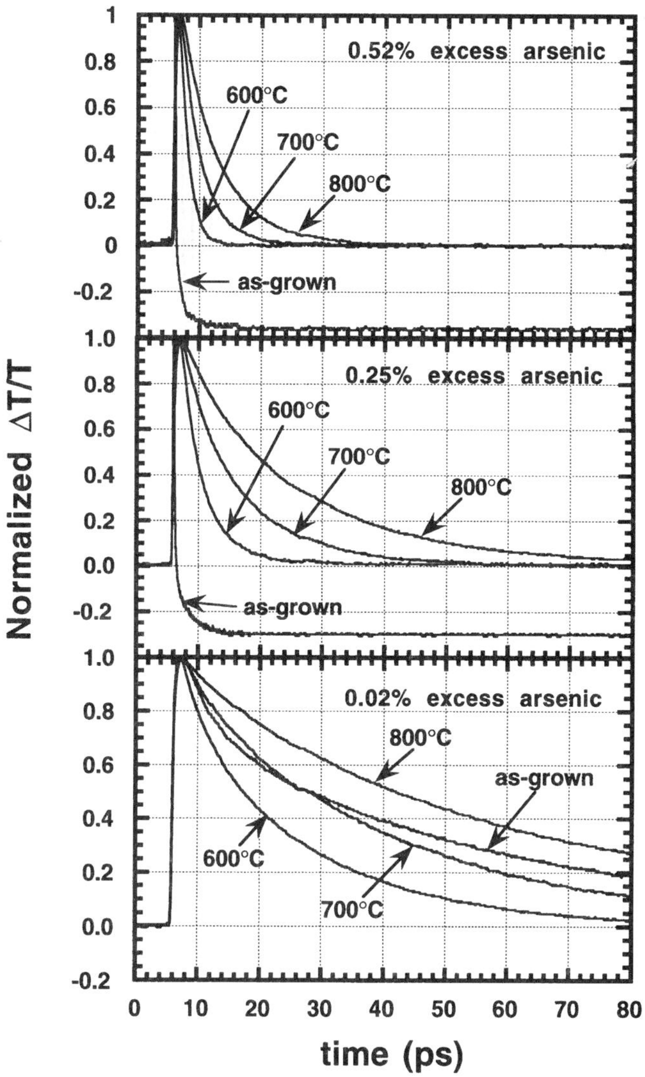

Fig. 5. Differential transmission for pump-probe experiments on LTG GaAs. Results are shown for samples with different amounts of excess arsenic, and annealed at several different temperatures for 30 s.

The removal of photogenerated carriers from the conduction band of LTG-GaAs can be measured with a pump–probe differential transmission technique [31]. The substrate is first removed to allow measurement of optical transmission through the epilayer. A pump pulse with photon energy near the GaAs bandgap creates electron-hole pairs in the epilayer. This increases the transmission through the epilayer because of filling of the conduction band states with electrons. As the electrons are removed from the conduction band the transmission decreases. This change in transmission is monitored with a lower intensity probe pulse. The transmission of this probe pulse is a function of the electron concentration in the conduction band. Normalized pump–probe differential transmission transients are shown in Figure 5 for as-grown samples, plus samples after anneals of 600, 700, and 800 °C for 30 s, from three LTG-GaAs epilayers containing different amounts of excess arsenic, 0.52, 0.25, and 0.02% [32]. The general trend observed in Figure 5 is a slower transient with increasing anneal temperature and a faster transient with increasing excess arsenic in the epilayer. The natural logarithm of the normalized differential transmission transients for

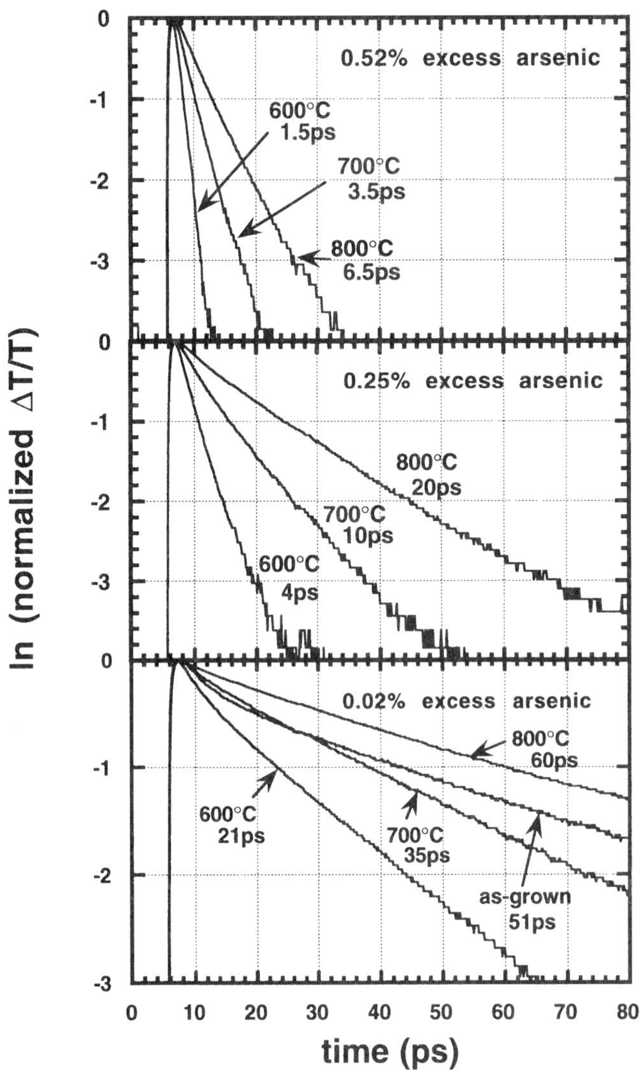

Fig. 6. Logarithmic plots of the differential transmission showing single exponential decay with associated decay times.

the annealed samples are shown in Figure 6 indicating single exponential decays. The lifetimes determined from Figure 6 transients are indicated in the figure and show a spectrum of lifetimes from 1.5 to 60 ps. The range of composite structures in Figure 6 is a small subset of LTG-material sample space, indicating the large variation of lifetimes that can be engineered into these materials.

For the as-grown 0.25 and 0.52% excess arsenic samples, there is an initial fast transient followed by a much slower transient. The initial fast transient is likely due to trapping of the photogenerated electrons and holes on separate point defects, but may not represent an actual recombination. The photogenerated electrons are likely being trapped on the arsenic antisites and the holes on the gallium vacancies, resulting in the long second transient observed. As was seen in Figure 6, single exponential transients are observed after the anneals. Most of the excess arsenic is precipitating with these anneals, indicating the arsenic precipitates may be the source of carrier removal from the conduction band in annealed LTG-GaAs. The role arsenic precipitates play in carrier removal from the conduction band in annealed LTG-GaAs is further illustrated by the transients for the epilayer with 0.02%

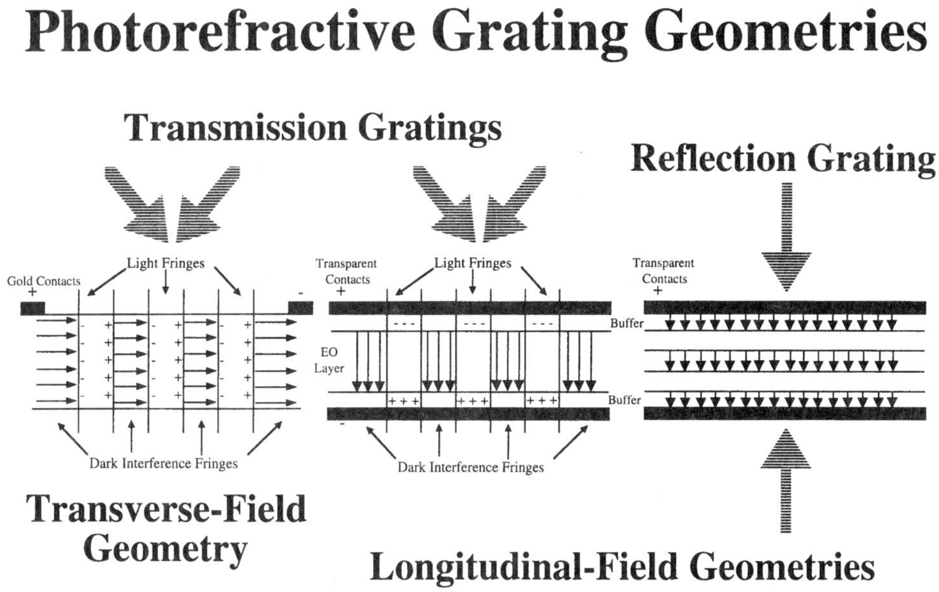

Fig. 7. Photorefractive quantum-well geometries.

excess arsenic. The transient for the 0.02% excess arsenic as-grown sample is much slower than after the 600 or 700 °C anneals. With these anneals there is a significant reduction in the point defect concentration in the epilayer, yet the removal of electrons from the conduction band is becoming faster. Arsenic precipitates are forming with anneal, again strongly suggesting they are the source of this increased rate of removal of electrons from the conduction band.

2.3. Photorefractive Quantum-Well Geometries

There are several ways that electric fields can be applied to quantum-well structures, and several ways that optical beams can be incident on the devices to write holograms. The various configurations can be summarized by three basic field and grating geometries. These are shown in Figure 7. Two of the structures are transmission structures with the optical beams incident on the same face, and one is a reflection geometry with the beams incident from opposite faces. Two of the structures have electric fields applied in the longitudinal direction parallel to the growth axis, and one has the electric field applied in the plane of the wells. The electric field direction and grating direction combine to give each of these geometries unique properties.

2.3.1. Transverse-Field Transmission Geometry

The transverse-field transmission geometry photorefractive multiple quantum-well (MQW) devices utilize charge drift created by the application of an external electric field parallel to the direction of the quantum-well planes. The layered structure of a representative device is shown in Figure 8. The multiple quantum-well region consists of ~100 periods of 75 Å GaAs wells and 100 Å $Al_xGa_{1-x}As$ barriers where x is typically 0, 0.1, or 0.3. This MQW region is sandwiched between two low Al-fraction AlGaAs spacer layers used to control the overall thin-film thickness. The other layers depicted in Figure 8 are used in the device processing procedure.

For use as a transmission device, the absorbing and conducting GaAs substrate must be removed, and must leave a smooth surface. One way to remove the substrate is with

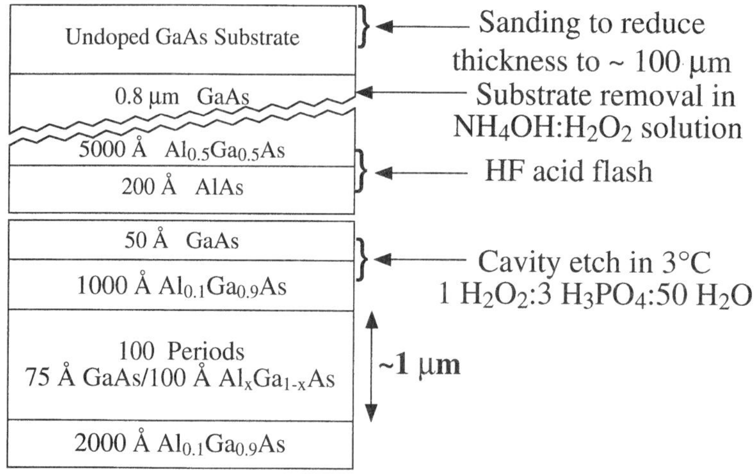

Fig. 8. General transverse-field geometry MQW device structure.

epitaxial lift-off [33] that uses HF acid to etch through the 200 Å AlAs layer to float off the MQW–spacer region. Another method of substrate removal utilizes chemical etching techniques. Chemical fabrication begins with the adhesion of the proton-implanted MQW structure to a suitable transparent glass or sapphire substrate that provides support for the final 2–4 μm thick device. To reduce adhesive contact time with chemical etches and to facilitate electrical contacting, approximately 300 μm of substrate is lapped off with 3 μm alumina grit. The remaining substrate is removed in a 1 NH_4OH:19 H_2O_2 etch that preferentially etches GaAs over $Al_{0.5}Ga_{0.5}As$. In order to level the nonuniform thickness of material left after the substrate etch, the device is placed in concentrated HF acid to remove any remaining $Al_{0.5}Ga_{0.5}As$ stop-etch and AlAs lift-off layer material. The selectivity of the HF between high (>40%) and low (<40%) Al-fraction AlGaAs has been measured to be $>10^8$ [34]. As a result, the uneven and slanted surface left after the substrate etch is removed by the HF, leaving a smooth, flat surface of GaAs or low Al-fraction AlGaAs. Cavity etching of the low Al-fraction AlGaAs layer to further tune the thickness for operation as an asymmetric Fabry–Pérot (ASFP), described in Section 6.3, is achieved with a 1 H_2O_2:3 H_3PO_4:50 H_2O solution at $\sim$3 °C.

Most imaging applications of the MQW devices require perfect surface quality over large areas. In particular, surface scatter can be a problem in imaging systems, such as for biomedical applications [35, 36], while nonuniform film thickness can compromise the ASFP enhancement effects described in Section 6.3. A pressurized chemical jet has been successfully used in other applications to achieve better surface quality [37]. For use in studying ASFP effects, this chemical etching technique was modified for use in an ultrasonic agitator. For improved surface quality over large areas the substrate etch may be modified for use in an 18 °C ultrasonic agitator bath with a concentration ratio of 1 NH_4OH: 22 H_2O_2. Combined with ultrasonic agitation, this solution yields significantly improved surface quality over the nonagitated substrate etch described previously by increasing oxide removal, and thereby dramatically reducing surface scatter [36], with the additional bonus of preferentially etching the device edges. For low-temperature-grown (LTG) devices this helps to remove any remaining noninsulating etch-stop material that might be left behind after standard processing.

The final fabrication step is the deposition of metal pads for use as electrical contacts. In the transverse-field or Franz–Keldysh geometry MQW devices the $\sim$10 kV/cm electric field is applied along the planes of the quantum wells. For this configuration, pads of Au or

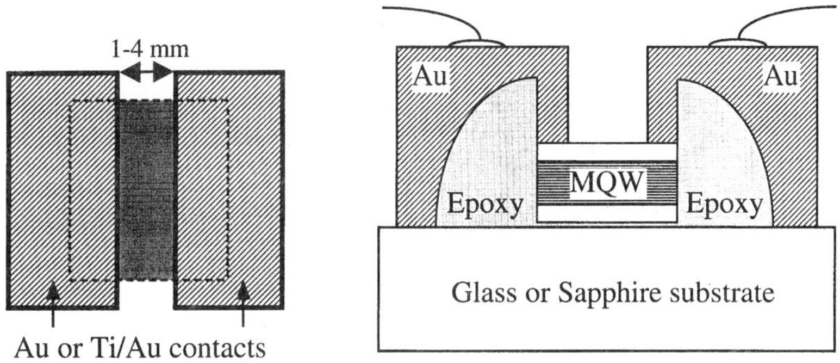

Fig. 9. Top (left) and side views of a fabricated transverse-field MQW device.

Ti/Au are evaporated onto the exposed surface of the device with an intercontact distance of 1–4 mm, shown in Figure 9.

2.3.2. Longitudinal Transmission Geometry

The longitudinal-field Stark geometry operates with photocarrier transport perpendicular to the grating vector, and requires insulating cladding layers to isolate the electro-optic quantum-well layer from the electrical contacts. These devices are grown as p-i-n photorefractive diodes to give the device low leakage currents when the device is reverse-biased during operation. The samples are grown by molecular beam epitaxy using a flux of As_4 on n^+ substrates. The doped n-type and p-type contacts are used [38, 39] to replace semitransparent contacts used in previous designs [40, 41] and permit the operation of all-semiconductor devices without any artificial means (such as SiO_2 and SiN layers) to reduce leakage currents. The active photoconductive electro-optic layer as well as the cladding layers are composed of low-temperature-grown AlGaAs [42]. The low-temperature-growth (LTG) materials have ultrafast lifetimes that extend down to picoseconds [31, 43], which produce spatial resolutions down to 1 μm in the transverse-field geometry [44], and approximately 10 μm in the longitudinal field geometry [45–47]. The low-temperature growth (LTG) introduces large defect concentrations into the cladding layers [48], allowing the layers to perform as active buffer layers that trap and store charge. In all previous device designs, buffer layers with large defect concentrations were used [41, 47, 49, 50], based on the assumption that traps were required in these layers to efficiently trap and store space charge to screen the applied fields. However, large defect densities in the buffer layers prevented the electric field from penetrating uniformly through the device. Much of the voltage dropped across the buffer rather than dropping across the electro-optic layer. However, defects within the quantum-well electro-optic layer can perform the function of charge trapping and storage to screen applied fields for device operation without the need for defects in the buffer layers. This allows greater voltage drops across the active electro-optic layer and hence device operation at lower voltages [47].

A typical device structure grown by molecular beam epitaxy is shown in Figure 10. The photorefractive diode is grown on an n^+ GaAs substrate. A layer of n-GaAs ([Si] = 1×10^{18} cm^{-3}) approximately 5000 Å thick is grown at 600 °C to planarize the wafer to ensure good epitaxy. Next a layer of n-$Al_{0.50}Ga_{0.50}As$ ([Si] = 1×10^{18} cm^{-3}) is grown for 5000 Å, which acts as an etch-stop layer during wet chemical etching used in device fabrication. An AlAs layer is grown which permits the use of another technique of device fabrication known as epitaxial lift-off [33, 34]. Another layer of n-GaAs ([Si] = 1×10^{18} cm^{-3}) is grown which performs a dual role. It acts as an etch-stop layer during an hydrofluoric acid etch and as a spacer layer which determines the final thickness of the device. The

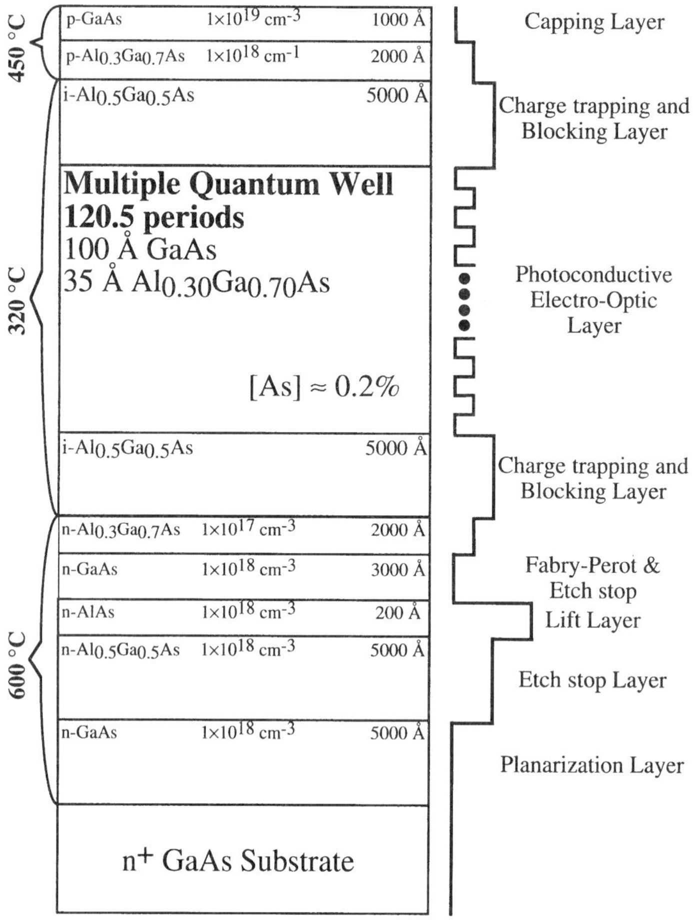

Fig. 10. MBE structure for a representative longitudinal-field transmission photorefractive quantum well.

final thickness of the devices are carefully tuned to meet exact Fabry–Pérot conditions to enhance the performance of the photorefractive diode [51, 52]. A spacer layer of n-$Al_{0.30}Ga_{0.70}As$ ([Si] $= 1 \times 10^{17}$ cm^{-3}) is grown for 3000 Å to prevent depletion of the doped contact. The active photoconductive electro-optic layer consists of a 120.5 period multiple quantum well consisting of 100 Å GaAs wells and 35 Å $Al_{0.30}Ga_{0.70}As$ barriers grown at 320 °C. The low growth temperature results in $\approx$0.2% excess arsenic in the multiple quantum-well region. Cladding layers of 5000 Å of $Al_{0.50}Ga_{0.50}As$ grown at 320 °C sandwich the electro-optic layer and perform the function of charge trapping and also prevent carrier sweepout to the doped contacts. A 2000 Å p-$Al_{0.30}Ga_{0.70}As$ ([Be] $= 1 \times 10^{18}$ cm^{-3}) layer followed by a 1000 Å top p-GaAs ([Be] $= 1 \times 10^{19}$ cm^{-3}) layer were grown at 450 °C. The 450 °C growth temperature for the p-doped layers acts as a weak *in situ* anneal of the previously grown LTG layers and results in the formation of As precipitates in the active electro-optic layer [14, 16]. The arsenic clusters deplete free carriers from the surrounding material, rendering it high resistivity [53]. Thus semi-insulating quantum wells are achieved which retain excellent electro-optic properties [43] and require no postgrowth processing.

The devices are fabricated as transmission devices using the following steps. The wafer is cleaved into 8 $\times$ 5 mm samples. In order to reduce surface leakage currents a mesa is etched into the quantum-well region using a phosphoric acid etch described in Section 2.3.1. Gold is evaporated on a corner of the device before the samples are epoxied

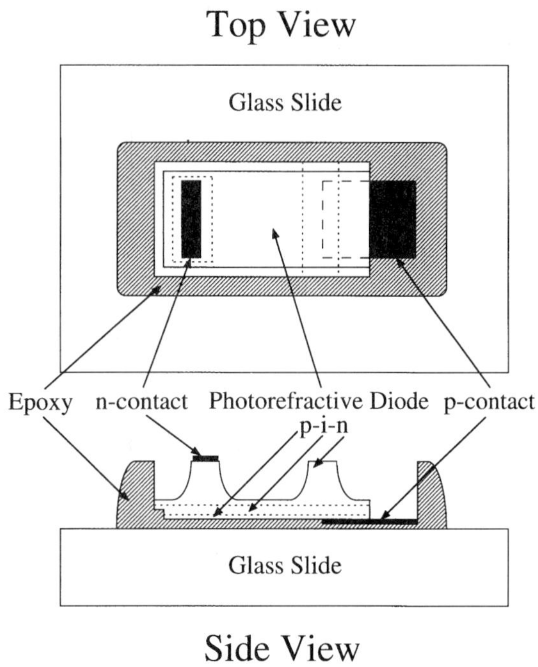

Fig. 11. Longitudinal-transmission device construction.

to a glass slide and are lapped using a fine alumina grit to a thickness of $\approx 100\ \mu m$. A small area ($\approx 0.5 \times 2$ mm) is covered with black wax to fabricate a mesa for the n-contact. The samples are then subject to a nonagitated ammonium hydroxide etch (19 parts H_2O_2:1 part NH_4OH described earlier) at room temperature. This etch selectively removes the remaining GaAs substrate and the GaAs planarization layer and stops somewhere in the $Al_{0.50}Ga_{0.50}As$ stop-etch layer. After the etch the black wax is removed and gold is evaporated on the resulting mesa to form the n-contact. An ammonium hydroxide etch is used to etch to the buried p-contact. A top and side view of the final device is shown in Figure 11. Electrical leads are connected to the contacts using silver paint.

2.3.3. Longitudinal Reflection Geometry

The third photorefractive quantum-well geometry uses an electric field applied across the quantum wells along the growth axis, and has the hologram-writing beams incident from opposite faces. This geometry has the special property that the fringe spacing is extremely short and the holograms are volume holograms. The counterpropagating beams produce a fringe spacing given by $\Lambda = \lambda/2n$, where n is the refractive index of the semiconductor material. The large index $n = 3.5$ produces fringe spacings of only 120 nm for $\lambda = 840$ nm. In the 1–2 μm thickness of the devices, this produces only 10–15 fringes. Despite this small number of fringes, these fringes constitute volume holograms, in contrast to the two transmission geometries that produce thin gratings.

The extremely short fringe spacing places severe constraints on the design of this geometry. First, extremely high defect densities are needed to trap sufficient space charge in a fringe to screen the applied field of 10^5 V/cm. Second, vertical transport must be suppressed to prevent overwriting the grating. These two requirements are both met by a LTG AlAs/GaAs multiple quantum-well structure [43]. The low-temperature growth produces high defect concentrations, and the AlAs barriers produce small vertical mobilities.

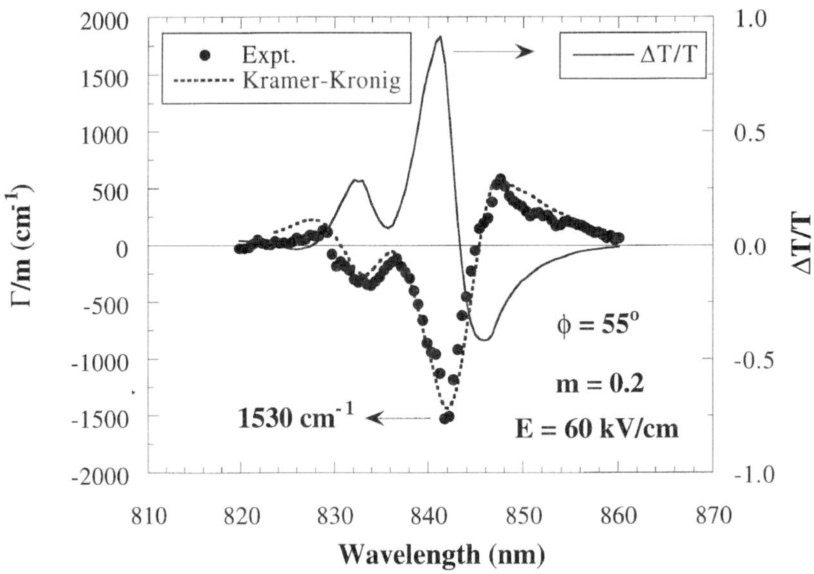

Fig. 12. Two-beam coupling spectrum for a longitudinal reflection geometry photorefractive quantum well.

Two-wave mixing was performed in this longitudinal reflection device [54], and the results are shown in Figure 12 for an applied field of 60 kV/cm. The two-beam coupling is compared against the differential transmission of the sample. The two-beam coupling has a similar spectrum as the differential spectrum, but has an opposite sign. Fitting of the coupling yielded a relatively large internal photorefractive phase shift of $\phi = 55°$ that was insensitive to large applied fields. This large photorefractive phase shift is likely due to trap limitation, in which insufficient traps are present to fully screen the applied electric field, despite the low growth temperature. The trap limitation and low photorefractive gains in this device, combined with the difficulty of working with such short fringe spacing, place interesting challenges on the further development of this device structure.

3. ELECTRONIC TRANSPORT AND GRATING FORMATION

3.1. Dielectric Relaxation Time

The response speed of photorefractive materials depends on the rate at which charge is generated, undergoes transport from the illuminated regions of the sample, and is screened and/or trapped. The ultimate limit to the speed of the photorefractive effect is the rate at which charge is photogenerated [55], which is directly dependent on the incident light intensity. In practice, the speed of grating formation is determined mainly by the screening of charge in the material. Once an image is recorded in the holographic medium (the photorefractive semiconductor), it can only be updated after the response time of the material, which is proportional to the dielectric relaxation time.

Dielectric relaxation describes the evolution of a charge distribution in a material where charge recombination or trapping occurs on a slower time scale than charge transport; this feature makes possible the nonlocal response of photorefractive materials to nonuniform illumination, which is the origin of photorefractive gain [56]. The temporal evolution of a charge density distribution ρ can be modeled using Gauss's equation for the electric

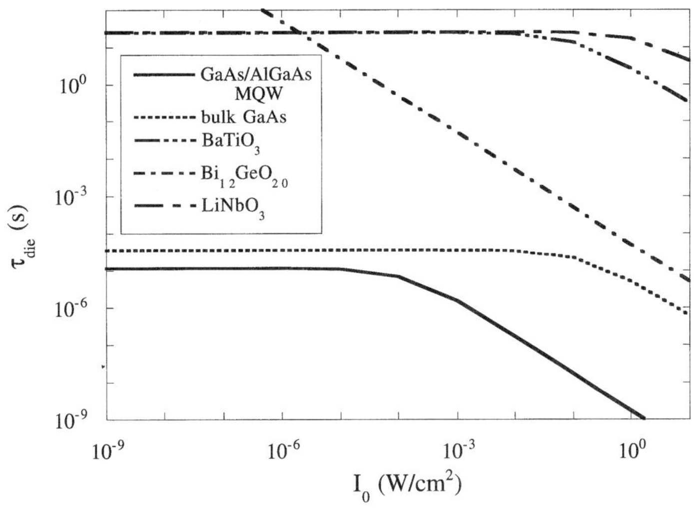

Fig. 13. Dielectric relaxation time versus laser intensity for photorefractive multiple quantum wells in comparison with several bulk photorefractive materials. Parameters from Ref. [5] (GaAs/AlGaAs MQW), Ref. [147] (GaAs, $BaTiO_3$, $Bi_{12}GeO_{20}$, $LiNbO_3$).

field, together with the continuity equation and a constitutive relation for the current density $\mathbf{j}$:

$$\begin{aligned} \nabla\mathbf{j} &= \frac{\partial\rho}{\partial t} \\ \nabla\mathbf{E} &= \frac{1}{\varepsilon\varepsilon_0}\rho \\ \mathbf{j} &= \sigma\mathbf{E} \end{aligned} \tag{3.1}$$

where ε is the dielectric permittivity, and σ is the conductivity. The charge density follows the equation,

$$\frac{\partial\rho}{\partial t} - \frac{\sigma}{\varepsilon\varepsilon_0}\rho = 0 \tag{3.2}$$

Therefore, the charge density decays exponentially as $\rho = \rho_0 \exp(-t/\tau_{\text{die}})$, where the dielectric relaxation time constant is $\tau_{\text{die}} = \varepsilon\varepsilon_0/\sigma$. In photoconductive materials such as photorefractive media, the conductivity consists of two terms,

$$\sigma = \sigma_{\text{dark}} + \sigma_{\text{photo}} = \sigma_{\text{dark}} + \frac{\alpha e}{h\nu}\mu\tau I \tag{3.3}$$

where α is the absorption coefficient, $h\nu$ is the photon energy, $\mu\tau$ is the mobility-lifetime product of the dominant carrier, and I is the incident intensity. The dielectric time constant is plotted in Figure 13 versus incident light intensity for a few representative photorefractive materials. For real-time holographic applications, materials with small dielectric relaxation times are desirable. The large absorption coefficients for interband transitions and the moderately high dark conductivities of semiconductor quantum wells enable the attainment of small relaxation time constants for low incident intensities.

3.2. The Two-Band One-Defect Model

The photorefractive effect consists in the generation of a spatially varying refractive index pattern (photorefractive grating) because of nonuniform illumination of a material. The process can be thought to take place in several stages. Free charge carriers (electrons or holes) are generated in the illuminated areas of the crystal. The photogeneration can take place from deep-level defects within the forbidden gap, or, in the case of the resonant photorefractive effect in semiconductor quantum wells, predominantly across the gap at or above the exciton transition energy. The photogenerated carriers undergo transport from the illuminated areas of the sample to the dark regions, where they are trapped at other deep-level defect sites; the defect levels that the photocarriers are generated from and those that the carriers are trapped at may be different defect species or may be different ionization states of the same defect species. Through the trapping of the free photogenerated charge at fixed defect sites, the nonuniform illumination creates a spatially varying space-charge pattern, which in general may be spatially shifted with respect to the illumination.

The immobile ionized traps produce a space-charge field which gives rise to a pattern of refractive index variation. In photorefractive materials operating far from the interband resonance, the mechanism involved is the linear electro-optic (Pockels) effect. In the case of semiconductor quantum-well devices operating at wavelengths near the gap, the space-charge field modifies the exciton resonance in a spatially varying manner, either through the Franz–Keldysh effect or through the Stark effect, depending on the photorefractive geometry. The Franz–Keldysh effect or Stark shift translate into a change in absorption at the exciton energy, which by virtue of causality is related to a change in refractive index through the Kramers–Kronig transform. Therefore, in the case of semiconductor quantum wells operating at the exciton resonance, absorption and refractive index gratings are simultaneously present.

Models of the photorefractive effect in semiconductor quantum wells have to take into account the generation with equal rates of electrons and holes across the bandgap. The simplest such model is the one-defect two-band model.

In the one-defect two-band model of the photorefractive effect, a single species of deep-level defect interacts with both the conduction and valence bands. A schematic of the carrier generation and recombination processes involved is shown in Figure 14. For example, if the defect is a deep-level donor, electrons are photo- and thermally generated from the neutral defects with concentration N_D^0, and holes are photo- and thermally generated from ionized donors (N_D^+). However, in quantum wells the dominant carrier generation mechanism is the photogeneration of electron-hole pairs across the bandgap, with the corresponding absorption coefficient α.

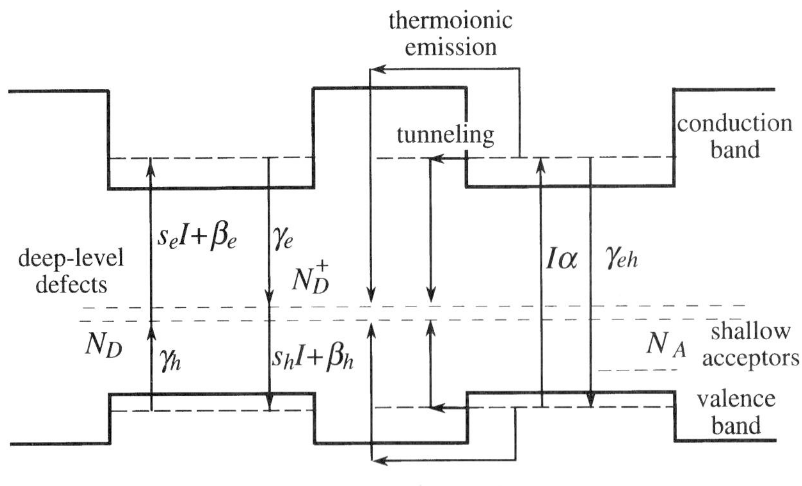

Fig. 14. Schematic of the carrier generation and transition processes in the one-defect two-band model for photorefractive quantum wells, showing the relevant transition rates.

The equations that describe the photogeneration, transport, and charge trapping processes were used for the first time for the modeling of the photorefractive effect by Kukhtarev et al. [57]. In the transverse-field (Franz–Keldysh) photorefractive geometry, when two types of carriers are generated across gap, these equations are

$$\frac{\partial n}{\partial t} = (s_e I + \beta_e) N_D^0 + I\alpha - \gamma_e n N_D^+ - \gamma_{eh} np + \frac{1}{e}\nabla \mathbf{j}_e \tag{3.4}$$

$$\frac{\partial p}{\partial t} = (s_h I + \beta_h) N_D^+ + I\alpha - \gamma_h p N_D^0 - \gamma_{eh} np - \frac{1}{e}\nabla \mathbf{j}_h \tag{3.5}$$

$$\mathbf{j}_e = en\mu_e \mathbf{E} + eD_e \nabla n \tag{3.6}$$

$$\mathbf{j}_h = ep\mu_h \mathbf{E} - eD_h \nabla p \tag{3.7}$$

$$\frac{\partial}{\partial t}\left(p - n + N_D^+ - N_A\right) = \frac{1}{e}\nabla(\mathbf{j}_e + \mathbf{j}_h) \tag{3.8}$$

$$N_D = N_D^0 + N_D^+ \tag{3.9}$$

$$\nabla \mathbf{E} = \frac{e}{\varepsilon\varepsilon_0}\left(p - n + N_D^+ - N_A\right) \tag{3.10}$$

Here n and p are the concentrations of electrons and holes, respectively, N_D is the total concentration of donors, and N_A is the concentration of compensated shallow acceptors. The electrons (holes) are photo- and thermally generated from the defect levels with cross-sections $s_{e(h)}$, respectively, $\beta_{e(h)}$, and recombine with the defect levels according to the probabilities $\gamma_{e(h)}$. These coefficients are considered to take into account real space transport and generation effects such as tunneling into the quantum-well barriers and thermoionic generation across the barrier followed by capture at the defect levels. The electrons and holes are generated across the gap with the rate $I\alpha$, where I is the photon flux and α is the absorption coefficient, and recombine across the gap with probability γ_{eh}. The electron (hole) mobilities are $\mu_{e(h)}$; in the GaAs/AlGaAs system, the electron mobilities are a function of the electric field for high applied electric fields, leading to transport nonlinearities which provide a distinct mechanism for the photorefractive phase shift [58]. However, in this section the carrier mobilities will be considered independent of field. The diffusion coefficients are related to the mobilities through Einstein's equation $D_{e(h)} = k_B T \mu_{e(h)}/e$.

3.2.1. *Linearized Solution*

The previous set of coupled nonlinear equations cannot be solved analytically in the general case, although numerical solutions can be found. Analytical solutions exist under several simplifying conditions, which can nevertheless be relevant for practical experimental conditions. The case of sinusoidal illumination $I(x) = I_0 + I_1 \mathrm{Re}\{\exp(iKx)\}$, where the modulation depth $m = I_1/I_0$ is small, is one such example. Here $K = 2\pi/\Lambda$ is the grating vector, where Λ is the interference fringe spacing. The assumption of small modulation depths implies that all relevant quantities can also be considered to have spatially sinusoidal variation, and higher order terms can be neglected. Although in practice the modulation index m can approach unity, the small m approximation is still relevant for the understanding of the photorefractive grating formation.

The zero-order solutions for the carrier concentrations are found assuming a constant illumination I_0, and are

$$\begin{aligned} n_0 &= \frac{(s_e I_0 + \beta_e)(N_D - N_A - n_0 + p_0) + I_0\alpha}{\gamma_e(N_A + n_0 - p_0) + \gamma_{eh} p_0} \\ p_0 &= \frac{(s_h I_0 + \beta_h)(N_A + n_0 - p_0) + I_0\alpha}{\gamma_h(N_D - N_A - n_0 + p_0) + \gamma_{eh} n_0} \end{aligned} \tag{3.11}$$

The interband absorption term in the numerator usually predominates, and the electron and hole concentrations, for moderate laser intensities (mW/cm^2), are typically much smaller than the defect concentrations N_D, N_A,

$$\begin{aligned} n_0 &= \frac{I_0\alpha}{\gamma_e N_A + \gamma_{eh} p_0} \\ p_0 &= \frac{I_0\alpha}{\gamma_h(N_D - N_A) + \gamma_{eh} n_0} \\ N_{D0}^+ &= N_A + n_0 - p_0 \end{aligned} \tag{3.12}$$

The first-order solutions in the steady-state case are obtained by linearizing the photorefractive equations, expressing the free-carrier and ionized donor concentrations as $n(x) = n_0 + n_1 \,\mathrm{Re}\{\exp(iKx)\}$, and so forth, and the electric field as $E(x) = E_0 + E_1 \,\mathrm{Re}\{\exp(iKx)\}$. The implicit assumption is that the charge transport is one-dimensional, which is not strictly true in the case of photorefractive quantum wells; due to carrier tunneling into the barriers and to thermoionic ionization over the barriers, charge transport is not completely confined within the wells in the Franz–Keldysh geometry.

The first Fourier component of the net space-charge concentration is

$$N_{sc} = N_{D1}^+ - n_1 + p_1 = \frac{\varepsilon\varepsilon_0}{e} iKE_1 \tag{3.13}$$

Eqs. (3.4)–(3.10) can be set into matrix form [59],

$$\begin{pmatrix} \Gamma_{Ie} + \Gamma_{\mathrm{Re}} + \Gamma_{\mathrm{De}} + \Gamma_{eh,e} - i\Gamma_{Ee} & \Gamma_{eh,e} - \Gamma_{Ie} & -\Gamma_{\mathrm{die},e} \\ \Gamma_{eh,h} - \Gamma_{Ih} & \Gamma_{Ih} + \Gamma_{Rh} + \Gamma_{Dh} + \Gamma_{eh,h} + i\Gamma_{Eh} & \Gamma_{\mathrm{die},h} \\ i\Gamma_{Ee} - \Gamma_{\mathrm{De}} & i\Gamma_{Eh} + \Gamma_{Dh} & \Gamma_{\mathrm{die},e} + \Gamma_{\mathrm{die},h} \end{pmatrix}$$

$$\times \begin{pmatrix} n_1 \\ p_1 \\ N_{sc} \end{pmatrix} = I_1 \begin{pmatrix} \alpha + s_e(N_D - N_A - n_0 + p_0) \\ \alpha + s_h(N_A + n_0 - p_0) \\ 0 \end{pmatrix} \tag{3.14}$$

Several of the characteristic rates ($\Gamma_{Ie,h}$, $\Gamma_{eh,h}$, $\Gamma_{eh,e}$) are small compared to the others and can be neglected [59]; also, the photo- and thermal-generation rates from the defects can be neglected in comparison to the interband absorption. With this further approximation, the space-charge concentration is found to be

$$N_{sc} = -I_1\alpha \frac{\Gamma_{Rh}(i\Gamma_{Ee} - \Gamma_{\mathrm{De}}) + \Gamma_{\mathrm{Re}}(i\Gamma_{Eh} + \Gamma_{Dh})}{\Gamma_{\mathrm{Re}}\Gamma_{\mathrm{die},e}(i\Gamma_{Eh} + \Gamma_{Rh} + \Gamma_{Dh}) + \Gamma_{Rh}\Gamma_{\mathrm{die},h}(-i\Gamma_{Ee} + \Gamma_{\mathrm{Re}} + \Gamma_{De})} \tag{3.15}$$

In terms of the applied field E_0 and of the diffusion field $E_D = k_B T K/e$, the space-charge field is

$$E_1 = \frac{m}{\mu_e\tau_e\mu_h\tau_h} \frac{\mu_e\tau_e(E_0 + iE_D) + \mu_h\tau_h(E_0 - iE_D)}{2E_D K + (1/\mu_e\tau_e + 1/\mu_h\tau_h)} \tag{3.16}$$

where the electron and hole lifetimes are $\tau_{e(h)} = 1/\Gamma_{\mathrm{Re}(h)}$.

Because the externally applied electric fields are in general much larger than the diffusion field, the dependence of the space-charge field E_{sc} on the fringe spacing is given mainly by the denominator in Eq. (3.16). At small fringe spacings (large K) the space-charge field and therefore the photorefractive grating amplitude is reduced with respect to its $K \to 0$ value. One of the most important parameters for photorefractive devices is the device resolution, defined as the size of the smallest features that can be recorded by the photorefractive grating. This is given by the characteristic cut-off grating vector, for which the space-charge field is half of its maximum value,

$$K_c^2 = \frac{e}{k_B T\langle\mu\tau\rangle} = \frac{1}{\langle L_D\rangle^2} \tag{3.17}$$

where the bipolar $\mu\tau$ product is defined as

$$\frac{1}{\langle\mu\tau\rangle} = \frac{1}{2}\left(\frac{1}{\mu_e\tau_e} + \frac{1}{\mu_h\tau_h}\right) \tag{3.18}$$

and $\langle L_D \rangle^2 = k_B T \langle \mu\tau \rangle / e$ is the bipolar diffusion length. The cut-off fringe spacing $\Lambda_c = 2\pi \langle L_D \rangle$ depends only on the bipolar mobility-lifetime product for a given temperature. The carrier lifetime can be engineered via the defect density in the material, and can lead to cut-off fringe spacings of 1–10 μm, depending on the device structure.

Another key parameter for evaluating photorefractive performance is the device response time, which is essentially the grating formation time constant. If the ionic production and recombination rate can be neglected, as is usually the case [59], then the grating formation time constant is [60],

$$\tau = \tau_{\text{die}} \left[\left(1 + \frac{\Gamma_D}{\Gamma_R} \right) + \left(\frac{\Gamma_E}{\Gamma_R} \right)^2 \left(1 + \frac{\Gamma_D}{\Gamma_R} \right)^{-1} \right] \tag{3.19}$$

where only one type of carrier was assumed. It is apparent from Eq. (3.19) that the response time of the photorefractive device is limited by the dielectric relaxation time τ_{die}, as stated in Section 3.1.

4. OPTICAL PROPERTIES OF PHOTOREFRACTIVE MULTIPLE QUANTUM WELLS

This section describes the optical properties of semiconductor thin films and quantum wells that are relevant to the photorefractivity. Other optical properties are described in the chapters "Nonlinear Optical Properties of Nanostructures" and "Excitons in the Nanostructured Materials" of this handbook. The properties of quantum-confined excitons are discussed in Section 4.1, the transmission and absorption spectra are given in Section 4.2. In Section 4.3, we discuss the electroabsorption, on which the photorefractivity is based and then we reveal the Kramers–Kronig transform, which relates the electroabsorption to the electrorefraction.

4.1. Quantum-Confined Excitons

The exciton is a bound state of an electron and a hole by the Coulomb interaction, and is the principal electromagnetic two-body excitation of a semiconductor [61]. It is the lowest energy excitation of a perfect semiconductor crystal. In the AlGaAs system, excitons are analogous to positronium atoms, with the hole playing the role of the positron. The wavefunction is spread out over many lattice sites, and the Coulomb binding is modified by the relative dielectric constant ε of the material. At low temperatures the energy levels of the excitons near the bandgap E_g are

$$E_n = E_g - \frac{e^2}{2\varepsilon_0 \varepsilon n a_n} \tag{4.1}$$

with the Bohr radius,

$$a_n = n \frac{\hbar^2 \varepsilon_0 \varepsilon}{e^2 m^*} \tag{4.2}$$

where n is the principal quantum number, and the reduced effective mass of the exciton is given by $1/m^* = 1/m_e^* + 1/m_h^*$ with the effective mass of electrons m_e^* and holes m_h^*, respectively. Because of the large dielectric constant in semiconductors, the binding energy of an exciton is usually smaller than the thermal energy at room temperature. For example, this binding energy is about 4 meV for bulk GaAs. Therefore, excitons are usually thermally ionized at room temperature, making it difficult to see an exciton absorption feature in the transmission spectrum. At low temperatures, the exciton is clearly resolved [62].

The binding energy of an exciton is increased by spatial confinement in a quantum-well structure. In GaAs/AlGaAs systems both the holes and the electrons are confined within the wells. For well sizes less than the Bohr radius of the exciton (~120 Å in GaAs)

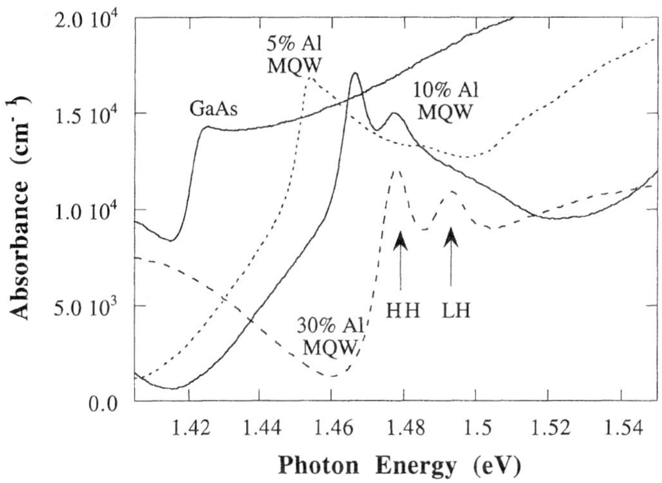

Fig. 15. Absorption spectra for different semiconductor samples at room temperature. The exciton features of the quantum wells become more pronounced as the binding energy increases, while there is no obvious exciton absorption feature for the semiconductor epilayer. HH and LH stand for heavy-holes and light-holes.

the excitons are confined in a quasi-two-dimensional potential, with a theoretical binding energy up to four times that of an unconfined exciton. The increased binding energy also increases the transition oscillator strength, producing a sharp resonant transition even at room temperature. Figure 15 shows the absorption spectra of GaAs/$Al_xGa_{1-x}As$ quantum wells with aluminum fraction of $x = 0.05$–0.3, in comparison with the spectrum of a GaAs epilayer. The absorbance in the figure is defined as

$$\alpha(h\nu) = -\frac{1}{L}\ln[T(h\nu)], \tag{4.3}$$

where T is the transmission without correction of Fresnel losses, L is the thickness of the sample, and $h\nu$ is the photon energy.

The quantum confinement lifts the degeneracy of the valence band, resulting in a heavy-hole (HH) exciton feature at lower energy and a light-hole (LH) exciton feature at higher energy. Both heavy-hole and light-hole excitonic transitions are clearly resolved in the quantum-well structures in Figure 15. As expected for room temperature, no exciton absorption feature is visible for the epilayer structure. The exciton features for the quantum wells become stronger as the barrier energy becomes larger, and therefore as the quantum confinement increases. Due to the surface reflections, Fabry–Pérot fringes are visible at photon energies below the bandgap in each sample. These fringes may be tuned to optimize various diffraction properties of the structures when desired, as will be discussed in Section 6.3. The lineshape of an excitonic absorption is often found experimentally to be Gaussian [63],

$$\alpha(h\nu) = \sqrt{\pi \ln 2}\,\frac{\hbar\omega_p^2}{cn}\frac{f}{\Gamma}\exp\left[-4\ln 2\frac{(h\nu - h\nu_0)^2}{\Gamma^2}\right] \tag{4.4}$$

where Γ is the full-width-at-half-maximum (FWHM), ω_p is the ionic plasma frequency, f is the oscillator strength, and ν_0 is the center frequency. For quantum-well structures with both heavy-hole and light-hole transitions, the absorption spectrum under an electric field E can be described as

$$\begin{aligned}\alpha(h\nu, E) = {} & \alpha_{HH}(E)\exp\left[-4\ln 2\frac{(h\nu - h\nu_{HH})^2}{\Gamma_{HH}^2(E)}\right] \\ & + \alpha_{LH}(E)\exp\left[-4\ln 2\frac{(h\nu - h\nu_{LH})^2}{\Gamma_{LH}^2(E)}\right] + \alpha_c(h\nu)\end{aligned} \tag{4.5}$$

with the FWHM and the center frequency of heavy- and light-holes Γ_{HH}, Γ_{LH}, and ν_{HH}, ν_{LH}, respectively. The contribution of the continuum is denoted as $\alpha_c(h\nu)$. Typical values for our QW-2 sample with 30% aluminum are: $h\nu_{HH} = 1.480$ eV, $\Gamma_{HH} = 7.8$ meV, $h\nu_{LH} = 1.496$ eV, $\Gamma_{LH} = 9.4$ meV.

4.2. Excitons in an Electric Field: Electroabsorption

The application of an electric field to a semiconductor changes the absorption of the material for photon energies near the bandgap. Depending on how the electric field is geometrically applied to the quantum wells, this change comes from different physical origins. When the electric field is applied parallel to the quantum-well layers (in the transverse or Franz–Keldysh geometry), the band edges are tilted, increasing the probability for carriers to tunnel into the continuum states. This effect is the Franz–Keldysh effect [64, 65], which can be described as exciton field ionization, where the field acts to reduce the lifetime of the exciton by lowering the relative Coulomb barrier [66]. The original calculations did not include the contribution of the exciton to the Franz–Keldysh effect, which greatly increases the magnitude of the absorption change [67].

Under an applied field in the transverse-field geometry, the excitonic absorption lines are broadened with the peaks reduced, but are not shifted. The broadening can be described as

$$\Gamma(E) = \Gamma(0) + \Delta\Gamma \tag{4.6}$$

The electroabsorption of an exciton can then be deduced from Eq. (4.2) when the applied field is not large,

$$\begin{aligned}\Delta\alpha(h\nu) \approx \frac{d\alpha}{d\Gamma} &= \sqrt{\pi \ln 2}\,\frac{\hbar\omega_p^2}{cn}\frac{f}{\Gamma^2}\Delta\Gamma\left(8\ln 2\frac{(h\nu - h\nu_0)^2}{\Gamma^2} - 1\right)\\ &\quad\times \exp\left[-4\ln 2\frac{(h\nu - h\nu_0)^2}{\Gamma^2}\right]\end{aligned} \tag{4.7}$$

with the maximum of the absolute electroabsorption at the center frequency ν_0,

$$|\Delta\alpha_{\text{Max}}| = \sqrt{\pi \ln 2}\,\frac{\hbar\omega_p^2}{cn}\frac{f}{\Gamma^2}\Delta\Gamma \tag{4.8}$$

Figure 16 shows the transmission spectra change $\Delta T(E)/T(0)$ of a GaAs/AlGaAs MQW sample with 30% Al in the barrier. At a field of $E = 14$ kV/cm, $\Delta T/T$ can exceed 85%. Experimentally, the electroabsorption can be extracted from the $\Delta T/T$ spectrum using the equation,

$$\Delta\alpha(E) = -\frac{1}{L}\ln\left[1 + \frac{\Delta T(E)}{T(0)}\right] \tag{4.9}$$

where the differential transmission ΔT is measured with a DC field modulated to several hundred Hz for lock-in detection. The transmission at zero-field is measured by chopping the source. The electroabsorption is insensitive to frequency from DC to many kHz.

Because the direction of the applied field should produce no difference in the effect, the change in absorption is quadratic in field to lowest order. At high fields, the energy of the exciton is shifted, but this is a much weaker effect than the lifetime broadening. Therefore, for small electric fields the electroabsorption has a quadratic dependence on field, which can be summarized by the use of effective quadratic electro-optic coefficients s_1 and s_2 [59],

$$\begin{aligned}\Delta n &= -\frac{1}{2}n^3 s_1 E^2\\ \Delta\alpha &= -\frac{2\pi}{\lambda}n^3 s_2 E^2\end{aligned} \tag{4.10}$$

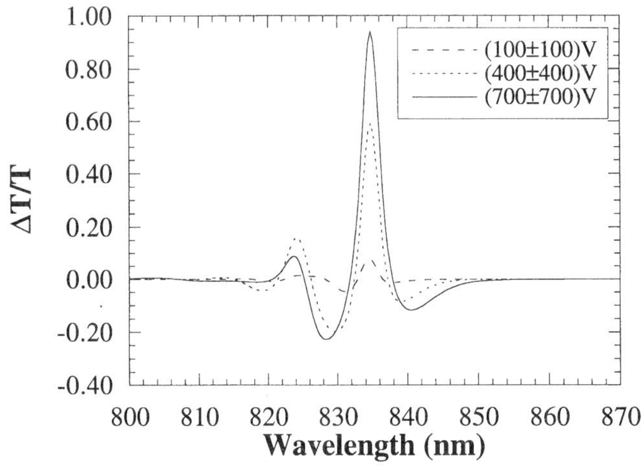

Fig. 16. Transmission change spectra $\Delta T(E)/T(0)$ of a GaAs/AlGaAs MQW sample with 30% aluminum at different applied voltages across the 1 mm electrode gap.

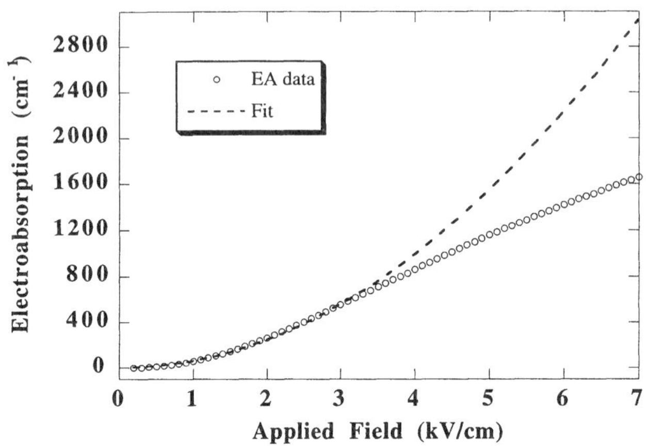

Fig. 17. Dependence of electroabsorption on the applied electric field for a 10% Al MQW sample at a photon energy near an electroabsorption peak, showing a quadratic function at fields lower than 3.5 kV/cm.

where n is the refractive index and E is the applied field. The dependence of electroabsorption on a field at a fixed wavelength is shown in Figure 17 for a 10% Al quantum-well sample. For fields below 3.5 kV/cm, the electroabsorption as well as electrorefraction is nearly quadratic.

When the field is applied perpendicularly to the quantum wells, the quantum-confined Stark effect (QCSE) [68, 69] occurs, changing the absorption behavior. The electric field adds a linear potential to the quantum-well potential which decreases the carrier confinement energy. This results in a red shift of the exciton transition, shown in Figure 18, that can be as large as 30 meV. The Stark shifting can be described as

$$\nu(E) = \nu(0) + \Delta\nu \tag{4.11}$$

The electroabsorption of an exciton due to the shift can then be deduced from Eq. (4.2),

$$\Delta\alpha(h\nu) \approx \frac{d\alpha}{d\nu}\Delta\nu = -\sqrt{\pi \ln 2}\,\frac{\hbar\omega_p^2}{cn}\frac{f}{\Gamma^2}\Delta\nu\left[8\ln 2\frac{(h\nu - h\nu_0)}{\Gamma}\right] \times \exp\left[-4\ln 2\frac{(h\nu - h\nu_0)^2}{\Gamma^2}\right] \tag{4.12}$$

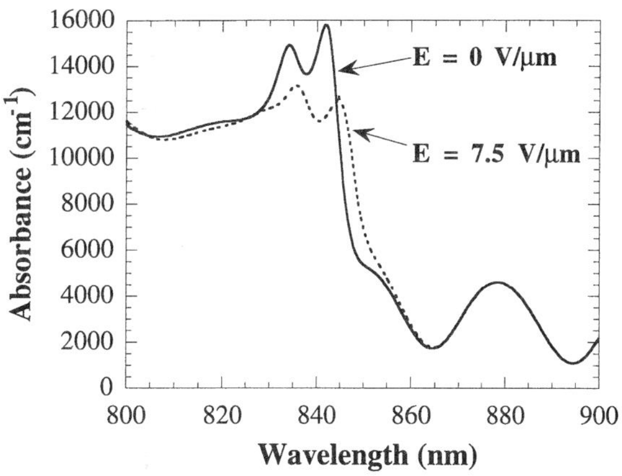

Fig. 18. Absorption spectra for a longitudinal device with and without an applied electric-field, showing the Stark shift of the excitonic transitions.

with the maximum of the absolute electroabsorption,

$$|\Delta\alpha_{\text{Max}}| = 2\ln 2\sqrt{\frac{2\pi}{e}}\frac{\hbar\omega_p^2}{cn}\frac{f}{\Gamma^2}h\Delta\nu \tag{4.13}$$

Both Eqs. (4.8) and (4.13) show that the maximum electroabsorption is inversely proportional to the square of the zero-field-linewidth, favoring narrower linewidth. Because of the small thickness of MQWs (typically 1 μm), a small applied voltage of several volts produces a very large electric field, leading to a large electroabsorption through the QCSE, which can exceed 8000 cm^{-1}.

4.3. Kramers–Kronig Relation

The Kramers–Kronig relation is an integral formula that relates a dispersive process to an absorption process [70, 71]. For an absorption spectrum $\alpha(\lambda)$ and the corresponding refractive index spectrum $n(\lambda)$ the Kramers–Kronig transformations are [72],

$$\begin{aligned} n(\lambda) &= 1 - \frac{\lambda^2}{2\pi^2}P\int_0^\infty \frac{\alpha(\lambda')}{\lambda'^2-\lambda^2}d\lambda' \\ \alpha(\lambda) &= 8P\int_0^\infty \frac{n(\lambda')}{\lambda'^2-\lambda^2}d\lambda' \end{aligned} \tag{4.14}$$

where P is the principal value, and the functions are given in terms of wavelength for convenience in analyzing data. When an electric field E is applied to the material, both the absorption and refractive index change,

$$\begin{aligned} \Delta\alpha(\lambda) &= \alpha(\lambda, E) - \alpha(\lambda, 0) \\ \Delta n(\lambda) &= n(\lambda, E) - n(\lambda, 0) \end{aligned} \tag{4.15}$$

The changes in absorption and refractive index are also related by

$$\begin{aligned} \Delta n(\lambda) &= -\frac{\lambda^2}{2\pi^2}P\int_0^\infty \frac{\Delta\alpha(\lambda')}{\lambda'^2-\lambda^2}d\lambda' \\ \Delta\alpha(\lambda) &= 8P\int_0^\infty \frac{\Delta n(\lambda')}{\lambda'^2-\lambda^2}d\lambda' \end{aligned} \tag{4.16}$$

which describe the electro-optic effect and are especially useful for photorefractivity. Because the changes $\Delta\alpha$ and Δn are only nonzero over a narrow range near the bandgap, it is easy to calculate the integrals.

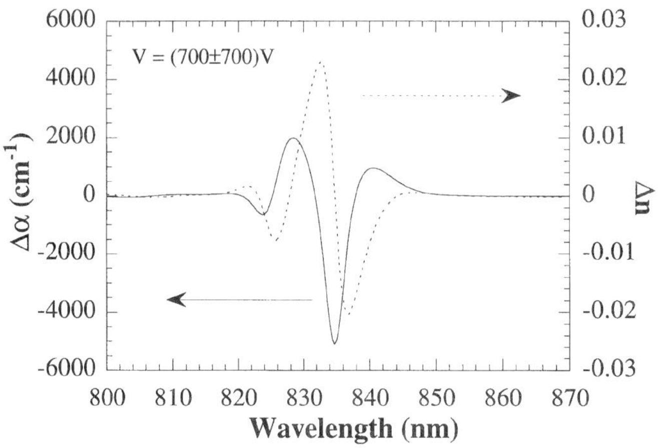

(a)

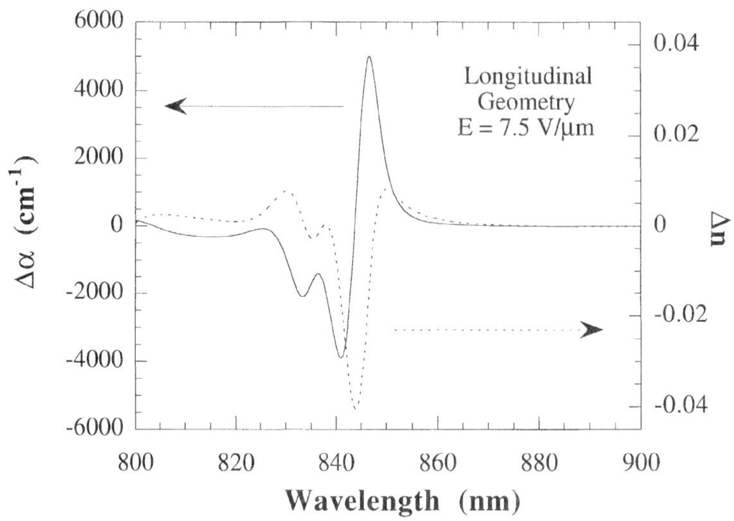

(b)

Fig. 19. Electro-optic properties of MQWs: (a) Electroabsorption and electrorefraction versus photon energy for a 30% Al at an applied field of 14 kV/cm with transverse geometry; (b) Electroabsorption and electrorefraction versus photon energy for a PIN quantum-well sample at an applied field of 75 kV/cm with longitudinal geometry.

The electrorefraction was calculated from the measured electroabsorption $\Delta\alpha$ using the Kramers–Kronig transformation in Eq. (4.16).Figure 19 shows the electroabsorption and the extracted electrorefraction for both transverse and longitudinal MQWs. For transverse devices, a typical absorption change is 5000 cm^{-1}, corresponding a refractive index change of 0.02 at an applied field of 14 kV/cm. For a longitudinal device, these values are 5000 cm^{-1} and 0.04, respectively, at an applied field of 75 kV/cm.

5. DIFFRACTION

5.1. Raman–Nath Diffraction

For gratings of finite extent, where there is an angular distribution of grating vectors, energy and momentum may be conserved for several scattering angles as seen in Figure 20. When

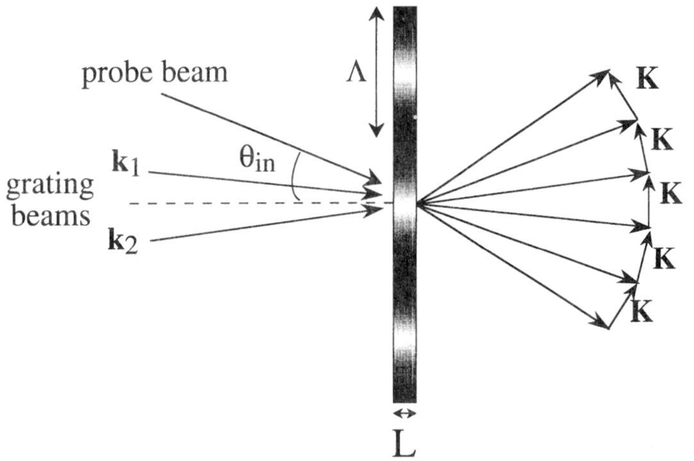

Fig. 20. Nondegenerate four-wave mixing in the Raman–Nath regime.

the angular distribution of the grating vector **K** is large compared to the Bragg angle θ_B, multiple diffraction orders are created. In MQW structures, the diffraction grating is finite in the z direction, perpendicular to the quantum-well planes, and the fringe spacing is typically larger than the film thickness. The resulting diffraction from the MQW is in the Raman–Nath [73] regime due to the resulting uncertainty in the z component of **K**. The value of the parameter Q, given by

$$Q = \frac{2\pi L\lambda}{n\Lambda^2} \tag{5.1}$$

determines the diffraction regime in which a particular device operates. Here L is the device thickness, λ is the wavelength, and Λ is the grating fringe spacing. For $Q > 1$ the device operates in the Bragg regime, although for $Q < 1$ it experiences Raman–Nath diffraction. For typical two- or four-wave mixing experiments with photorefractive GaAs-based MQWs, $L = 2\ \mu\text{m}$, $\lambda = 850$ nm, $n = 3.6$, $\Lambda = 20\ \mu\text{m}$. These values yield $Q = 7 \times 10^{-3}$ and therefore place the MQWs far into the Raman–Nath regime where no Bragg matching is required for diffraction to occur. Diffraction in this regime yields two- and four-wave mixing in photorefractive MQWs.

5.2. Nondegenerate Four-Wave Mixing

In the nondegenerate four-wave mixing geometry, depicted in Figure 20, two coherent pump lasers of energy at or above the bandgap of the MQW material are incident on the photorefractive material at equal angles, forming a sinusoidally varying interference pattern. A third probe beam, with a photon energy near the bandgap of the MQW, is incident at an angle θ_{in}, and diffracts off the grating to form the "fourth" wave. Multiple diffraction orders are present due to Raman–Nath diffraction.

The mth-order diffraction amplitude is calculated [5] using the Fraunhofer integral,

$$E_m = \frac{1}{\Lambda}\int_{-(\Lambda/2)}^{\Lambda/2} \left|E_t(x)\right| e^{-ikx(\sin\theta_d + \sin\theta_{\text{in}})}\, dx \tag{5.2}$$

where $E_t(x)$ is the transmitted field and the diffraction angles θ_d are given by

$$\sin\theta_d = -\sin\theta_{\text{in}} + m\frac{K}{k_{\text{vac}}} \tag{5.3}$$

The diffracted amplitude is given by

$$E_m = E_i e^{i(\delta_0 + m\phi + m\pi/2)} J_m(\delta_1) \tag{5.4}$$

where ϕ is the photorefractive phase shift and the complex phase factors δ_i are expressed as

$$\delta_i = \frac{2\pi n_i L}{\lambda \cos\theta'} + i\frac{\alpha_i L}{2\cos\theta'} \tag{5.5}$$

For small (<5%) diffraction efficiencies, the mth-order diffraction efficiency, η_m, in the Raman–Nath regime in a device where both absorption and index gratings are present is given by

$$\eta_m = \frac{1}{2^{2m}m!}\left[\left(\frac{2\pi\,\Delta n L}{\lambda\cos(\theta')}\right)^2 + \left(\frac{\Delta\alpha L}{2\cos(\theta')}\right)^2\right]^m \exp\left(-\frac{\alpha L}{\cos\theta'}\right) \tag{5.6}$$

where Δn and $\Delta\alpha$ are the index and absorption modulations, L is the thickness of the absorbing region, and $\sin\theta' = (1/n_0)\sin\theta_{\text{in}}$.

5.3. Two-Wave Mixing

For degenerate mixing, where the grating is written by two beams tuned near the bandgap absorption, coherent interference between the overlapping first-order diffraction from pump 1 and the pump 2 beam in the presence of a photorefractive phase shift ϕ creates nonreciprocal energy transfer, or two-wave mixing. This interference creates constructive interference in one beam, while creating destructive interference in the other. The two-wave mixing experimental configuration is shown in Figure 21.

Following Eq. (5.4) and the method in Ref. [5] the amplitude of beam 1 after passing through the grating is

$$E_1(L) = J_0(\delta_1)E_1(0)e^{i\delta_0} + J_1(\delta_1)E_2(0)e^{i(\delta_0+\phi+\pi/2)} \tag{5.7}$$

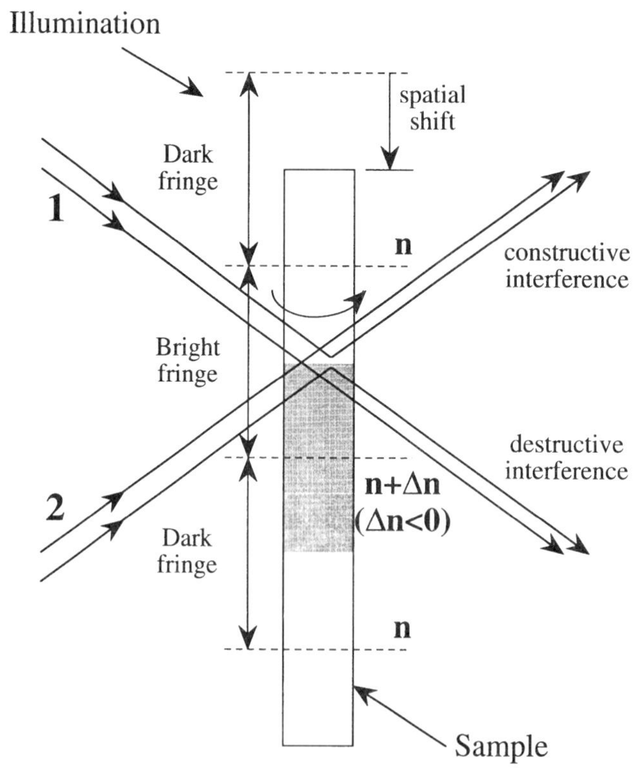

Fig. 21. Two-wave mixing experimental configuration. Nonreciprocal energy transfer occurs between the two pump beams.

and similarly for beam 2. The resulting photorefractive gain ratio, in terms of the beam intensities I_1 and I_2, is

$$\gamma = \frac{I_1(L)}{I_1(0)e^{-\alpha_0 L}} = 1 + \sqrt{\frac{I_2(0)}{I_1(0)}}\left[\frac{2\pi n_1 L}{\lambda \cos\theta'}\sin\phi + \frac{\alpha_1 L}{2\cos\theta'}\cos\phi\right] \tag{5.8}$$

6. PHOTOREFRACTIVE EFFECTS AND APPLICATIONS

6.1. Dynamics of the Stark Geometry

6.1.1. Introduction

The longitudinal-field Stark-geometry devices exhibit a rich variety of spatiotemporal dynamics because the transport is perpendicular to the grating vector. In this respect it differs from traditional photorefractive materials and is in many respects similar to self-electro-optic effect devices (SEED) [74–76] and other capacitive coupling structures such as spatial light modulators [77–81]. Photorefractive quantum wells [3, 5, 44] are a new class of optically addressed spatial light modulators that combine the advantages of large excitonic electroabsorption of quantum-confined excitons [69] with large carrier mobilities to produce high-sensitivity holographic devices operable at extremely low-light intensities [59].

The longitudinal-field Stark geometry [40, 41, 45, 47, 50], operates with photocarrier transport perpendicular to the grating vector, and requires insulating cladding layers to isolate the electro-optic quantum-well layer from the electrical contacts. The longitudinal-geometry operates only under transient electrical excitation, during which charge separates within the electro-optic layer (which is also the photoconductive layer) and moves to the interfaces between the electro-optic layer and the cladding layer.

6.1.2. Charge Transport

The photorefractive quantum-well longitudinal-field Stark geometry has the largest diffraction efficiency, operating at high field strengths using the quantum-confined Stark effect where the absorption changes approach 10,000 cm^{-1} [43], with index changes of several percent. The geometry of the longitudinal-field devices is shown in Figure 22. An electric field is applied perpendicular to the plane of the thin film. Charges are generated in the bright regions of the device which are trapped in the buffer layers. The trapped space charge sets up a space-charge field E_{sc}, which opposes the direction of the applied field E_0. Hence the device experiences alternating regions of high field (in the dark fringes) to low field (in the bright fringes) with the same spatial periodicity as the illuminating interference pattern. The optical properties of the device are modulated with the same periodicity as the modulated space charge through the electro-optic effect in the photoconductive electro-optic layer. The insulating buffer layers trap and store space charge and prevent charge leakage to the electrodes. Image charges are induced on the electrodes due to the trapped space charge. The complex interplay between the trapped space charge and the induced image charges has important consequences for the temporal response [82] and the spatial resolution [83, 84] of the device. A simple equivalent circuit model [85] has been developed which is useful to understand the basic aspects of the longitudinal-field devices such as speed and spatial resolution without the need to invoke lengthy material based calculations. These results are a simple consequence of the device geometry and the equivalent circuit.

In early designs, the applied field was screened by charge trapped by defects in the cladding layers, called buffer layers. The design of appropriate buffer layers with appropriate trap concentrations was considered a key feature of photorefractive longitudinal-

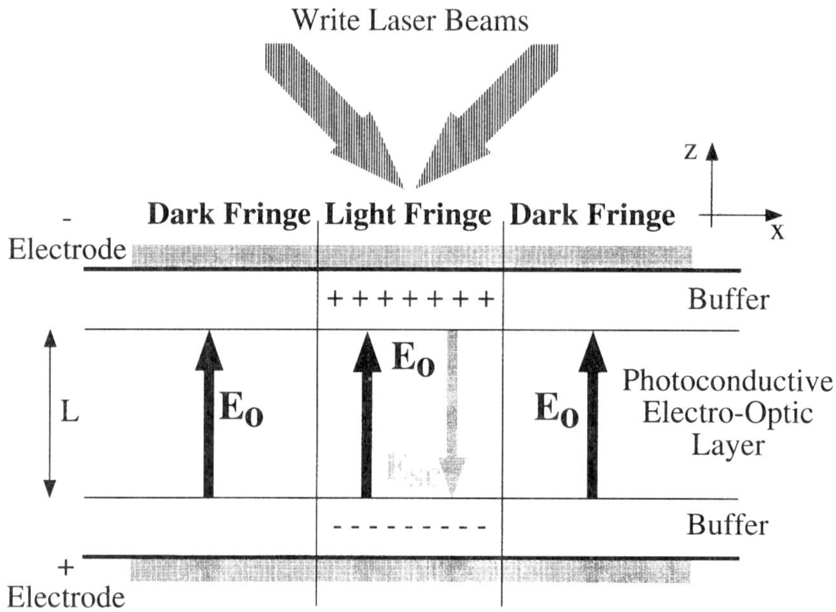

Fig. 22. The longitudinal-field Stark-effect device geometry. The photogenerated space-charge E_{sc} opposes the applied electric field E_0 which results in the modulation of the optical properties of the device with the same periodicity as the illuminating pattern.

field design to obtain optimum performance [40, 41, 45]. Some designs relied on non-semiconductor insulating buffers of SiO_2 and SiN to prevent charge transport to the contacts and to eliminate leakage currents in broad-area devices. In all previous device designs, buffer layers with large defect concentrations were used [41, 49, 50], based on the assumption that traps were required in these layers to efficiently trap and store space charge to screen the applied fields. However, large defect densities in the buffer layers produced short Debye screening lengths which prevented the electric field from penetrating uniformly through the device. Much of the voltage dropped across the buffer rather than dropping across the electro-optic layer. By removing the defects from the buffer layers, these layers act simply as stand-off layers with lower voltage drops, distributing more of the voltage across the electro-optic layer. Heavily doped p- and n-type contacts were proposed [38, 39] to replace the semitransparent contacts used in some previous designs [40, 41]. All-semiconductor photorefractive p-i-n quantum-well diodes were developed [47, 50] with low-temperature-grown (LTG) [42] multiple quantum wells which eliminate the need for Cr-doping [49] during growth or for any postgrowth processing. The LTG quantum wells have ultrafast lifetimes that extend down to picoseconds [31] and retain excellent excitonic properties [43]. The fundamental operation of the device is not affected when all defects are removed from the buffers in which the multiple quantum-well layer fulfills all three functions of photoconductive layer, electro-optic layer, and charge trapping layer [47].

6.1.3. Four-Wave Mixing

The equivalent circuit model [85] has demonstrated several different regimes of diffractive performance for the longitudinal-field photorefractive quantum-well geometry. It is possible to measure these three regimes of diffractive performance experimentally and this is illustrated in Figure 23. The figure shows a typical diffracted signal for two periods of the applied square-wave voltage recorded as a function of time on a digital storage oscilloscope. The transient time response is given by the largest diffracted intensity after

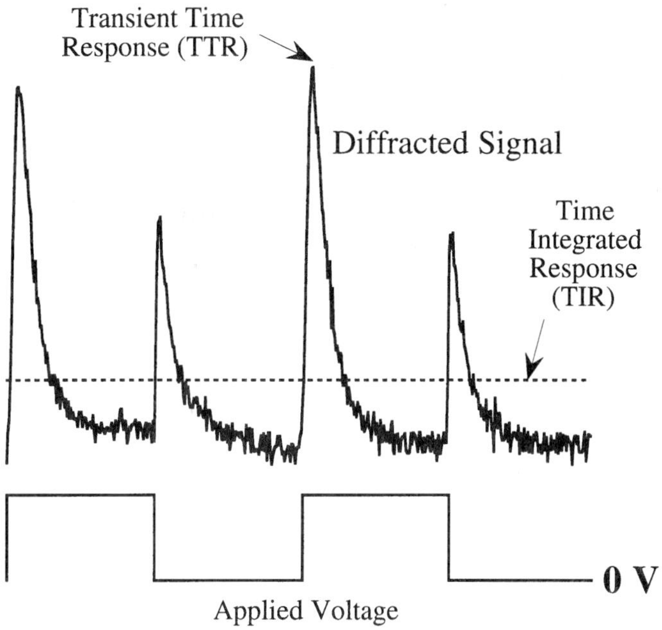

Fig. 23. The temporal response of the diffracted signal is shown for two cycles of the applied voltage. The time-integrated response (dashed line) can be considerably smaller than the peak transient time response.

the application of the electric field. The Fourier time response is obtained using lock-in detection, referenced to the frequency of the repetitive applied voltage. To obtain higher order Fourier components the lock-in can be referenced to the higher harmonics of the applied voltage. The Fourier diffraction efficiency obtained here experimentally is proportional to the first Fourier time coefficient (i.e., the second harmonic developed in the equivalent circuit model) of the diffracted signal. The time-integrated response is the average of the diffracted signal over one period of the applied modulated voltage. This is shown as the dashed line in Figure 23. The distinction between transient and time-integrated response is important, because the diffraction signal typically has a large and rapid diffracted signal spike shortly after the voltage is applied, but which quickly decays. For image processing applications, the time-integrated diffracted signal is the most important value, because it describes the average number of photons per second arriving at focal planes or charge-coupled device (CCD) arrays. An additional important issue for device performance in imaging applications is the fastest allowable repetition rate between pulses that defines the frame rate of the device. The optimum repetition rate for ac fields applied to the devices is considerably slower than the inverse rise time [50].

The devices used in our experiments consisted of low-temperature-growth (LTG) $Al_{0.10}Ga_{0.90}As/Al_{0.16}Ga_{0.84}As$ quantum wells grown by molecular beam epitaxy with an As_4 source. Contact and stop-etch layers of n-type materials were grown on an n^+ GaAs substrate at 600 °C. This was followed by the LTG (320 °C, [As] $\approx$ 0.2%) multiple quantum-well (MQW) layer consisting of a 100.5 period superlattice of 100 Å $Al_{0.10}Ga_{0.90}As$ wells and 100 Å $Al_{0.16}Ga_{0.84}As$ barriers. The shallow quantum wells provide high-mobility vertical transport, fast escape of photocarriers [86], and enhanced electro-optic sensitivity [87], which results in favorable performance during two-wave and four-wave mixing. We have previously shown that low-temperature growth of quantum wells can retain excellent electro-optic properties [43]. The LTG superlattice was sandwiched between 5000 Å $Al_{0.50}Ga_{0.50}As$ cladding layers. The lower and top cladding layers were grown at 600 and 450 °C, respectively. A 2000 Å p-$Al_{0.30}Ga_{0.70}As$ (1×10^{18} cm^{-3}) layer followed by a 1000 Å top p-GaAs (1×10^{19} cm^{-3}) layer were grown at 450 °C.

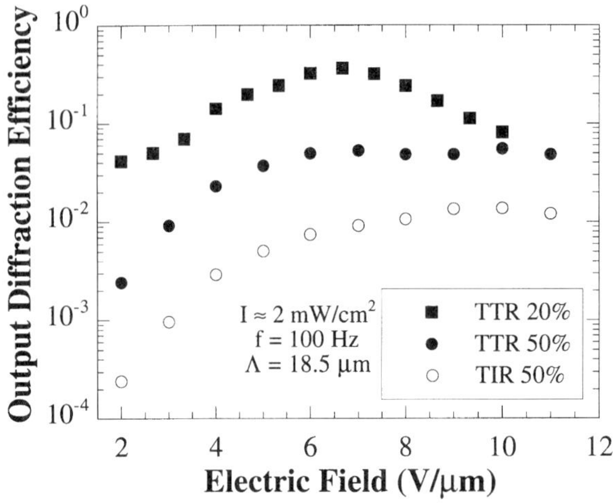

Fig. 24. The transient time response (TTR) and the time-integrated response (TIR) of the output diffraction efficiency of the photorefractive diode as a function of the applied electric field. The output diffraction efficiency approaches 35%. (Source: Reprinted with permission from [47]. © 1996 American Institute of Physics.)

The 450 °C growth temperature for the *p*-doped layers acts as a weak *in situ* anneal of the previously grown LTG layers. Hence the photoconductive electro-optic layer performs all the functions of photoconductivity, charge trapping, and electro-optics. The fabrication procedure of the final transmissive devices has been outlined in Section 2.3.2.

We performed degenerate four-wave mixing by writing a grating using a Ti:sapphire laser tuned near the heavy-hole exciton around 790 nm. The electric field was modulated using a single-sided square pulse. The devices used in the experiments were photorefractive diodes with intrinsic cladding layers. The transmitted zero-order and the diffracted first-order signals were detected using silicon photodiodes with 650 nm long pass filters. The diffracted signal was recorded as a function of time on a digital storage oscilloscope.

Figure 24 shows the transient time response (TTR) as well as the time-integrated response (TIR) of the diffraction efficiency as a function of the applied electric field. The devices were operated under reverse bias with a single-sided square pulse at a repetition rate of 100 Hz, an incident intensity of 2 mW/cm^2 and a fringe spacing of 18.5 μm. The transient output diffraction efficiency approaches 35% for a 20% duty cycle square wave, while it approaches 6% for a standard 50% duty cycle square pulse excitation. The enhanced diffraction at low duty cycle is likely caused by nonlinear dependence of the diffracted signal on the internal fields in the device. The transient input diffraction efficiency, which includes all losses, was measured to be 2.7%. The time-integrated response (TIR) of the diffraction efficiency is also shown for the same conditions with a 50% duty cycle square-wave electrical excitation. The diffracted signal in this case was measured to be half an order of magnitude smaller than the transient time response. In both cases of standard duty cycle operation the output diffraction efficiency follows a fourth-order power law at low electric fields. The deviation from fourth-power dependence at the higher electric fields may be caused by several mechanisms: lateral drift that reduces the electric-field modulation in the quantum wells [83], reduced quantum efficiency for carrier trapping at higher fields, and deviation of the electroabsorption from quadratic dependence on field. Because we use semiconductor buffers, the carriers can overcome the moderate potential barrier that separates the stand-off layers from the contacts. At higher fields, the electrons have higher ballistic velocities, which enable them to overcome the barrier more effectively, preventing sufficient time to trap the charge.

The frequency response of the device for a 5 V/μm reverse-biased single-sided square wave is shown in Figure 25. Again, the transient time response is consistently higher than

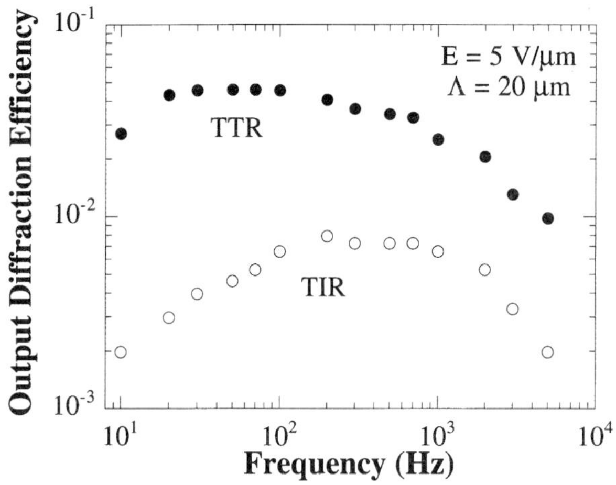

Fig. 25. The frequency response of the output diffraction efficiency of the photorefractive diode shows the fastest update rate of the device to be ≈1 kHz for a laser illumination of 2 mW/cm^2. The transient time response (TIR) is consistently greater than the time-integrated response (TIR). (Source: Reprinted with permission from [47]. © 1996 American Institute of Physics.)

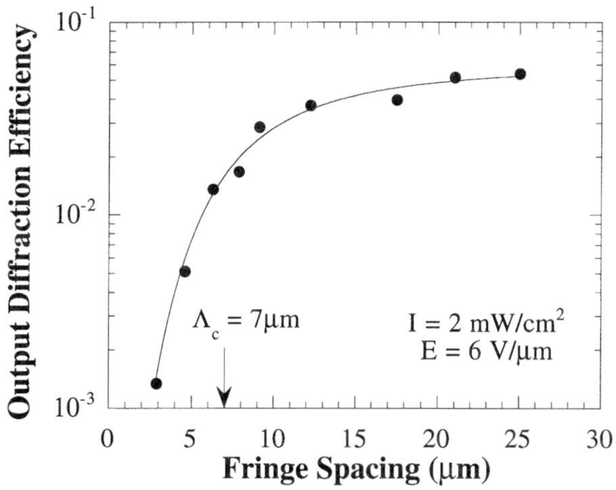

Fig. 26. The transient time response of the output diffraction efficiency as a function of the fringe spacing shows a spatial resolution of 7 μm. The curve is fitted to $(1 + (\Lambda_c/\Lambda)^2)^{-2}$. (Source: Reprinted with permission from [47]. © 1996 American Institute of Physics.)

the time-integrated response of the device. The roll-off frequency limits the fastest update rate for electrically addressing the device. This roll-off frequency can be tuned by changing the incident intensity (described earlier) on the device or by dynamically controlling the active trap density in the LTG MQW electro-optic layer [88]. Pre-illumination to control the active trap density of the carriers can ultimately lead to MHz repetition rate performance of these devices at low-light illumination levels of mW/cm^2.

The diffractive performance of the device is evaluated for its spatial resolution in Figure 26. The transient time response of the diffraction efficiency of the device is shown for an incident laser intensity of 2 mW/cm^2 while the device was reversed biased at 6 V/μm with a 50% duty cycle square wave at a repetition rate of 100 Hz. The device exhibits a smooth roll-off with decreasing fringe spacing, with a cut-off of approximately 7 μm. The diffraction efficiencies are fitted to be proportional to $(1 + (\Lambda_c/\Lambda)^2)^{-2}$.

The performance of these devices can be compared to other devices with different cladding layers [47, 50]. We find that although there are differences in the details of the device performance depending on the cladding layer properties, the general performances of the devices are similar and are relatively independent of the presence of defects in the cladding layer. The roll-off frequency in these devices with LTG quantum wells is independent of the cladding layer properties [47] and is dictated by the transport properties within the quantum wells. Larger diffraction efficiencies may be achieved from these devices by fabricating them as asymmetric Fabry–Pérot structures [51, 89].

6.1.4. Two-Wave Mixing

Two-wave mixing has been shown schematically in Figure 21, where each of the zero-order transmitted beams interferes with one of the first diffracted orders of the other beam. This interference can cause nonreciprocal energy transfer between the two laser beams if a spatial shift between the refractive index grating and the incident intensity pattern can be achieved. This has been well established in bulk photorefractive crystals [90, 91] as well as in thin films operating in the transverse Franz–Keldysh geometry [59, 92] with the electric field in the plane of the quantum wells. Energy transfer requires broken symmetry between the two mixing light fields to define an energy transfer direction. Nonreciprocal energy transfer is normally forbidden by symmetry in longitudinal Stark-effect photorefractive devices (electric field applied perpendicular to the quantum wells) because the carrier transport is perpendicular to the holographic grating vector, and all optical changes respond locally to the illumination. A spatial shift between the incident intensity pattern and the resulting space-charge formation can be achieved by using the well-known technique of moving gratings [93–96]. The moving gratings break the symmetry and make it possible to observe two-wave mixing in longitudinal Stark-geometry photorefractive p-i-n diode structures. Large photorefractive gains are one of the unique features of photorefractive phenomena that distinguish them from other optical nonlinearities and make them candidates for many optical applications.

The device structure uses low-temperature-grown (LTG) quantum wells as the active electro-optic layer, which retain good electro-optic properties [43] and have good diffraction efficiency during four-wave mixing [47, 50]. An applied electric field produces a space-charge field which is in phase with the intensity pattern. By introducing a running grating–intensity pattern which moves on the time scale of the formation of the photoinduced space charge, it is possible to achieve a condition in which the space charge lags behind the intensity pattern. Under optimal grating velocities, a $(\pi/2)$-phase shift can be introduced, which would result in large photorefractive gains [2].

The samples used in our experiments were photorefractive diodes grown by molecular beam epitaxy (MBE) using a flux of As_4. The structures were LTG MBE $Al_{0.3}Ga_{0.7}As$/GaAs multiple quantum wells (MQW) grown on n^+ GaAs substrates. Contact and stop-etch layers of n-type materials were grown on a n^+ GaAs substrate at 600 °C, followed by a LTG (320 °C) MQW layer consisting of a 120.5 period superlattice of 100 Å GaAs wells and 35 Å $Al_{0.3}Ga_{0.7}As$ barriers. The low-temperature growth at 320 °C results in approximately 0.1% excess arsenic in the MQW. The LTG superlattice was sandwiched between 5000 Å $Al_{0.5}Ga_{0.5}As$ cladding layers, grown at 320 °C. A 2000 Å p-$Al_{0.3}Ga_{0.7}As$ (1×10^{18} cm^{-3}) layer followed by a 2000 Å top p-GaAs (1×10^{19} cm^{-3}) layer were grown at 450 °C on top of the LTG layers. Gold contacts were deposited on the top p-GaAs layer. The samples were epoxied to glass and the substrate was removed using standard techniques. A final gold contact was made to the exposed n-$Al_{0.5}Ga_{0.5}As$ stop-etch layer after substrate removal.

We performed nearly degenerate two-wave and four-wave mixing using a CW Ti: sapphire laser tuned close to the low-energy zero crossing of the electroabsorption of the device, with a fringe–grating spacing in the device of $\Lambda = 30$ μm. Two acousto-optic modulators controlled the frequency difference, Ω, between the two equal-intensity writing laser beams. The moving interference pattern was created by keeping one of the acousto-optic modulators at a fixed frequency ($f = 80$ MHz) while varying the frequency of the other modulator, $f + \Omega$. The electric field across the photorefractive quantum well was modulated using a single-sided reverse-biased square pulse. The transmitted zero-order beams and the diffracted first-order signals were detected using silicon photodiodes with 650 nm long-pass filters. The transmitted and diffracted signals were recorded as a function of time on a digital storage oscilloscope.

The photorefractive gain is defined by the asymmetry in the transmitted intensity when both beams are copolarized. The gain is given by

$$\Gamma = \left(\frac{1}{2L}\right)\left[\left(\frac{I_1^{\parallel} - I_1^{\perp}}{I_1^0}\right) - \left(\frac{I_2^{\parallel} - I_2^{\perp}}{I_2^0}\right)\right] \tag{6.1.1}$$

where $I^{\parallel}$ is the transmitted intensity when the two writing beams are copolarized, $I^{\perp}$ is the transmitted intensity when the two writing beams are cross-polarized, I^0 is the transmitted intensity when the electric field is turned off, L is the active crystal length, and the subscripts 1 and 2 refer to the two writing beams. Diffraction efficiencies quoted are transient output diffraction efficiencies, defined as the ratio of the peak diffracted intensity to the zero-field transmitted intensity.

The transient response of the photorefractive diode is shown in Figure 27. The electric field was modulated at a repetition rate of 2 kHz using a 5 V/μm single-sided 100 μs square pulse. This electronic strobe produces a quasi-static grating in the photorefractive quantum-well diode. The electrical strobe "freezes" the intensity pattern in space. The change in transmitted intensity ΔI $(= I^{\parallel} - I^{\perp})$, caused by a moving intensity pattern for a detuning $\Omega = 19.6$ kHz is shown. The incident total laser intensity on the device was $\approx$10 mW/cm^2 with a wavelength of 850 nm. Nonreciprocal energy transfer is demonstrated as asymmetry between the two beams as one laser beam is amplified at the expense of the other. Oscillations persist after the electric field has been turned off, which is a consequence of the beating of the moving interference fringes against the quasi-static strobed grating created in the photorefractive diode by the electrical pulse. The direction of the energy exchange oscillates for a single period of the applied voltage.

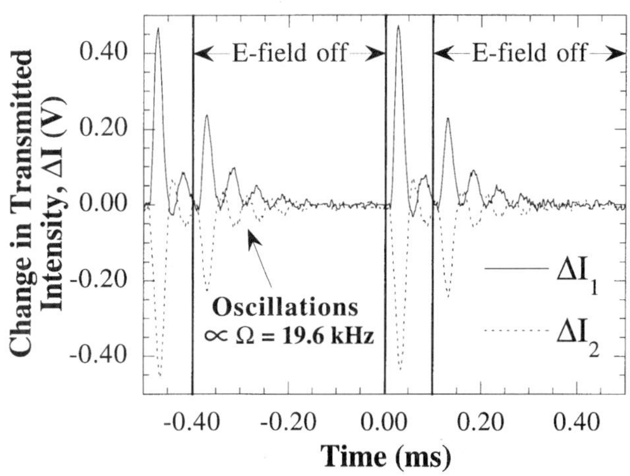

Fig. 27. The transient response of the photorefractive diode due to the formation of quasi-static gratings is shown. The change in transmitted intensity demonstrates nonreciprocal energy transfer where one laser beam is amplified at the expense of the other. (Source: Reprinted with permission from [99]. © 1998 American Institute of Physics.)

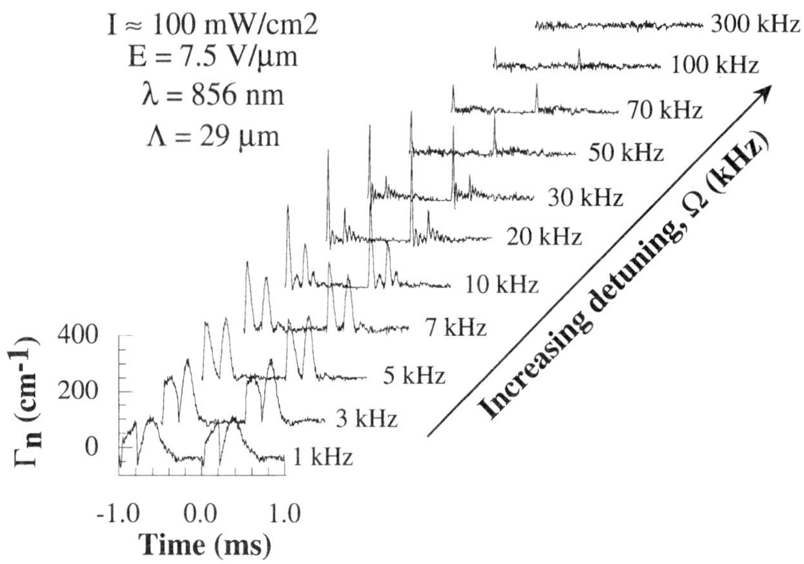

Fig. 28. The temporal response of the photorefractive gain for several Doppler detunings. (Source: Reprinted with permission from [97]. © 1996 American Institute of Physics.)

The temporal response of the photorefractive gain defined in Eq. (6.1.1) is shown in Figure 28 for varying frequency detunings Ω introduced by changing the frequency of the variable-frequency acousto-optic modulator [97]. The electric field applied to the device is a 7.5 V/μm single-sided 200 μs square pulse at a repetition rate of 1 kHz. The incident laser intensity is $\approx$100 mW/cm^2 with a fringe–grating spacing of 29 μm. The laser is tuned to 856 nm where there is negligible electroabsorption. Thus we rely almost exclusively on changes in index resulting from the quantum-confined Stark effect to induce unidirectional energy transfer in the photorefractive diodes. Fast response times ($\approx$20 μs) and large two-wave mixing gains are observed, induced by the broken symmetry using moving gratings. At low grating velocities (small detuning, $\Omega < 8$ kHz) we observe the oscillatory behavior of the photorefractive gain. This complex spatiotemporal behavior occurs when the detuning frequency is comparable to characteristic response times in the photorefractive quantum well. There are several different device response times, such as writing times or decay times, for a given average incident intensity. Therefore different processes in the grating formation sequence can "resonate" with different intensity pattern velocities. At higher detuning frequencies, the intensity pattern is changing too quickly for the space charge to track the intensity pattern throughout the voltage pulse, but a sharp gain transient is still observed at the voltage steps. At still higher detuning frequencies, the space charge fails to respond and the gain decays to zero.

The peak transient output diffraction efficiency and the peak transient gain are shown in Figure 29. The experimental conditions are identical during two-wave and four-wave mixing. The photorefractive gain approaches a maxima as the diffraction efficiency begins to roll off. The output diffraction efficiency is proportional to the magnitude of the space-charge field squared, $|E_{sc}(t)|^2$ which is constant up to the roll-off frequency, and then rapidly decreases because of the inability of the device to respond to the velocity of the moving grating. The photorefractive gain is proportional to the product of the magnitude of the space-charge field and the sine of the phase shift, $|E_{sc}(t)|\sin(\phi(t))$, which displays a resonant behavior with a peak that coincides with the detuning at which the output diffraction efficiency rapidly decreases. This detuning frequency coincides with the optimum delay between the moving intensity pattern and the lagging space charge.

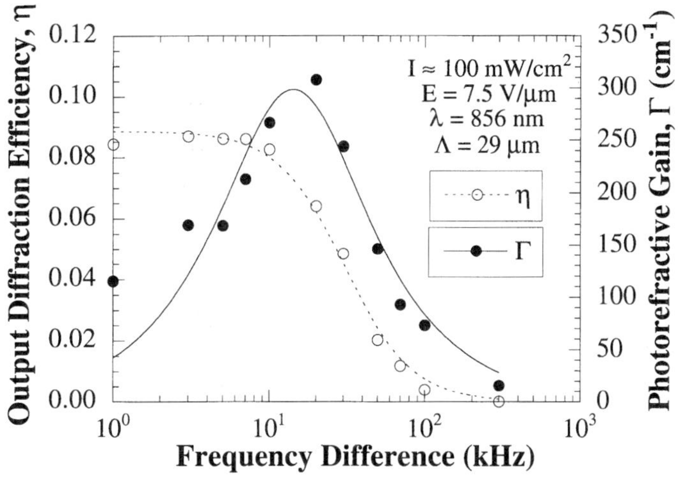

Fig. 29. Output diffraction efficiency and photorefractive gain as functions of the frequency detuning between the two laser beams. The curves are to guide the eye. (Source: Reprinted with permission from [97]. © 1996 American Institute of Physics.)

The dynamic interchange of energy between the two laser beams propagating in the sample perpendicular to the grating vector (in the direction of the applied field) can be described by a coupled-mode formalism for the intensities I_1 and I_2. The coupled-mode equations are given by [2, 98, 99],

$$\frac{d}{dx} I_1 = -\Gamma_n(t)\frac{I_1 I_2}{I_1 + I_2} - \Gamma_\alpha(t)\frac{I_1 I_2}{I_1 + I_2} - \alpha(t) I_1 \tag{6.1.2}$$

and

$$\frac{d}{dx} I_2 = \Gamma_n(t)\frac{I_1 I_2}{I_1 + I_2} - \Gamma_\alpha(t)\frac{I_1 I_2}{I_1 + I_2} - \alpha(t) I_2 \tag{6.1.3}$$

where

$$\Gamma_n(t) = \frac{4\pi n_m(t)}{\lambda \cos\theta'} \sin(\phi_0 + 2\pi\Omega t) \tag{6.1.4}$$

and

$$\Gamma_\alpha(t) = \frac{\alpha_m(t)}{\cos\theta'} \cos(\phi_0 + 2\pi\Omega t) \tag{6.1.5}$$

The coefficient $\Gamma_n(t)$ describes the photorefractive gain (caused by index gratings), $\Gamma_\alpha(t)$ describes the absorptive gain (caused by absorptive gratings), $\alpha(t)$ describes the absorption, and $n(t)$ describes the refractive index, all of which are time dependent. The time-dependent first Fourier coefficients of the absorption and index gratings are given by $\alpha_m(t)$ and $n_m(t)$, which have been normalized by the modulation index $m = 2\sqrt{I_1 I_2}/(I_1 + I_2)$. The coupling from the absorption gratings appears symmetrically in Eqs. (1) and (2), while the coupling from the index gratings is antisymmetric. Hence antisymmetric energy transfer is a result of the index gratings, where one laser beam is amplified at the expense of the other.

In Figure 27 the oscillations persist after the electric field has been turned off, which is a consequence of the beating of the moving interference fringes against the quasi-static strobed grating created in the photorefractive diode by the electrical pulse. These oscillations are caused by the time-varying argument of Eq. (6.1.4) describing the shift between the intensity pattern and the space-charge field. The decay of the oscillation amplitude is due to erasure of the index (and absorption) gratings $n_m(t)$ and $\alpha_m(t)$ in the photorefractive

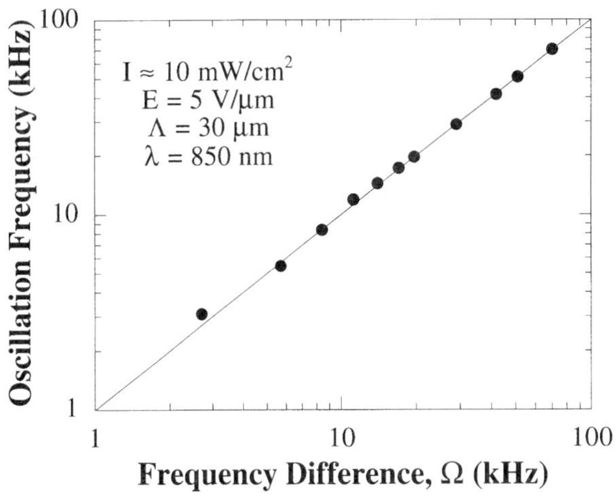

Fig. 30. The frequency of the oscillatory gain is found to be equal to the frequency difference Ω between the two laser beams. The line is $y = x$. (Source: Reprinted with permission from [99]. © 1998 American Institute of Physics.)

diode as the interference fringes move across the device, after the electric field is turned off. The decay is determined by the dielectric relaxation rate of the photorefractive diode.

The oscillations seen in Figure 27 after the electric field has been turned off have a characteristic frequency. This characteristic frequency is determined from the peak positions of the change in the transmitted intensity ΔI, which determine "optimum" conditions for nonreciprocal energy transfer between the two laser beams under specific operating conditions, such as the applied electric field, frequency offset, and illumination intensity. The applied electric field determines the strength of the index and absorption gratings [2] for a given intensity, while the frequency offset between the laser beams dictates the temporal dynamics. The frequency of the oscillatory gain is shown in Figure 30 as a function of the frequency difference, Ω, between the two laser beams. The oscillatory frequency is equal to the frequency difference between the two laser beams. The electrical strobe is caused by the transient voltage swing in the device. This sets up a displacement current which can be viewed as a capacitor in an equivalent circuit [85]. Hence in terms of the space-charge dynamics the change in voltage causes the strobe, while the pulse duration is of secondary importance.

Both the photorefractive gain, $\Gamma_n(t)$ and the absorptive gain, $\Gamma_\alpha(t)$ contribute to a complex gain defined by

$$\widetilde{\Gamma}(t) = \Gamma_n(t) + i\Gamma_\alpha(t) \qquad (6.1.6)$$

The photorefractive gain and the absorptive gain act in quadrature. Hence each component of the gain approaches a maxima at different times. Depending on the wavelength of the laser, and the electric field used, there is a finite contribution from both gains as a function of time. This is shown in Figure 31 as a function of the applied electric field for a detuning Ω of 7 kHz. The complex gain defined by Eq. (6.1.6) at our working wavelength of 852 nm is dominated by the index gratings at low fields. At high fields the absorption gratings dominate. This is probably due to a significant Stark shift at higher fields, which results in a larger contribution to the absorptive gratings. The photorefractive and absorptive gains were found to approach 1000 cm^{-1} independently, but at different times. In order to account for both contributions the temporal response of the magnitude of the complex gain as a function of the electric field is shown in Figure 32 for a frequency detuning of 7 kHz. The photorefractive diode was operated with 300 μs reverse-bias square pulses at a repetition rate of 1 kHz, with an incident laser intensity of $\approx$25 mW/cm^2 at a wavelength of 852 nm. The peak magnitude of the complex gain and its components (photorefractive

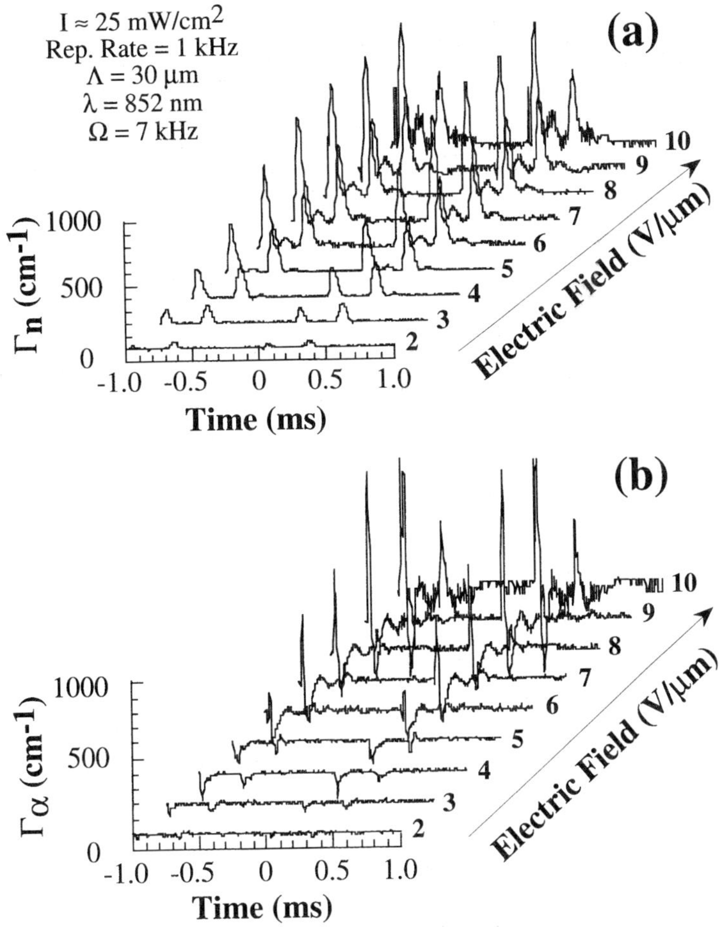

Fig. 31. The temporal response of the (a) photorefractive gain and (b) absorptive gain are shown for different applied electric fields to the photorefractive diode. Both gains approach 1000 cm^{-1}.

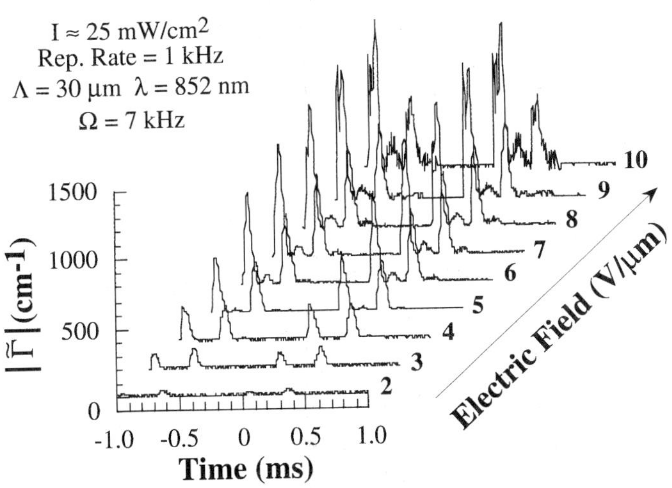

Fig. 32. The temporal response of the magnitude of the complex gain is shown for increasing electric fields. The magnitude of the complex gain approaches 1200 cm^{-1} even though its individual components approach 1000 cm^{-1}.

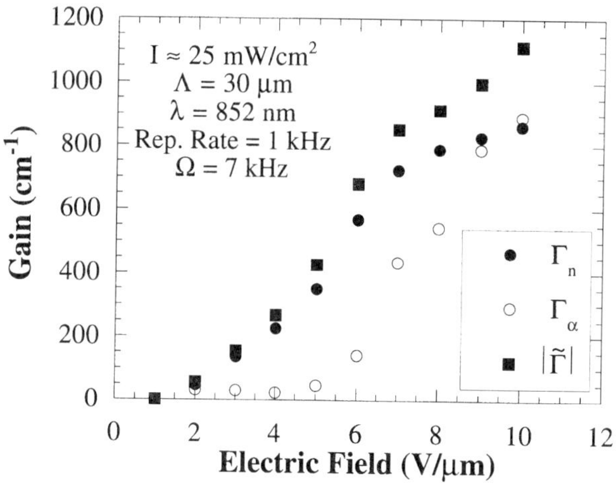

Fig. 33. The peak magnitude of the complex gain and its components are shown for increasing electric fields for a fixed detuning of 7 kHz. (Source: Reprinted with permission from [99]. © 1998 American Institute of Physics.)

gain and absorptive gain) is shown in Figure 33 for a detuning of 7 kHz. Effective gains approaching 1200 cm^{-1} are demonstrated at a fringe spacing $\Lambda = 30\ \mu\text{m}$. In Figure 33 at large electric fields (approaching 10 V/μm) the effective gain decreases because the space charge needs more time to screen the applied field, causing saturation and even a decrease in the grating amplitude. Higher detunings produce more pronounced decreases [99]. Thus for a given Doppler shift between the laser beams there is an optimum field that should be applied to the device for a particular laser intensity.

These photorefractive diodes can accurately measure Doppler shifts and hence can become useful tools for detecting surface imperfections and vibrations. This technique can be used with a simple laser based velocimeter [100] to measure a range of velocities. Earlier techniques of laser velocimetry [100] relied on photocurrents generated in photoconductive semiconductors, which was limited by the dielectric relaxation rate. The use of photorefractive diodes in the two-wave mixing configuration has advantages over other techniques for measuring velocities that require interferometric arrangements [101] needing precise alignment.

It is important to note that this technique is significantly different from standard square-law heterodyne detectors. The photorefractive quantum-well diode acts as an adaptive beamsplitter and compensates for any slowly varying changes (compared to the response time of the photorefractive diode). The dynamic hologram ensures that the phasefronts are exactly matched and thus the detection is not speckle-limited. This makes it possible to measure a wide range of surface displacements and vibrations by measuring the Doppler shifts in the laser beams in the presence of slowly varying changes caused by the environment. This technique does not require precise interferometric alignment and can operate when speckle would limit other techniques.

6.2. Asymmetric Fabry–Pérot and Microcavity Effects

6.2.1. Introduction

Photorefractive multiple quantum wells (MQWs) have the advantages of working at low incident light intensities and at faster speeds than bulk photorefractive materials. Operating in transmission they have one of the largest dynamic diffraction efficiencies per interaction length of known optical materials [4, 5, 59]. However, the device input diffraction efficiencies, defined as the ratio of first-order diffracted intensity to the reflected or transmitted intensity when no grating is present, which are limited by the trade-off between device absorption and electroabsorption, have not exceeded 3% [49]. Larger absorption at the exciton transitions produces larger electroabsorption and stronger gratings, but it also reduces the interaction length. This design incompatibility can partly be avoided by operating the photorefractive quantum wells in reflection as asymmetric Fabry–Pérot [102, 103] (ASFP) structures. Quantum-well ASFP modulators have generated the largest contrast ratios in quantum-well structures by utilizing multiple beam interference [104]. Careful placement of the Fabry–Pérot fringes can yield greater than an order of magnitude enhancement in the diffraction efficiency of the photorefractive MQW.

Condensed matter optical microresonators display a wide variety of interesting behaviors and are suited to many applications. For example, the modification of electromagnetic modes in microcavity resonators is the basis for the design of semiconductor lasers and devices which rely on strong cavity quantum electrodynamic effects. Incorporation of an absorbing material within a cavity bounded by mirrors leads to modifications in both the amplitude and phase of the optical wave modes. A small number of modes of the optical field can be isolated with careful control of the cavity geometry and size. Other microcavity devices such as electro-optic modulators, SEEDs, and photorefractive multiple quantum wells all utilize cavity effects, yet are not generally described in terms of microcavity resonators.

A multiple quantum-well microcavity sandwiched between unequal reflector interfaces can operate as a perfect absorber of light, as shown in Section 6.2.2, despite the fact that the thickness of the active region is only on the order of one absorption length. This is a consequence of microcavity effects which modify the optical modes and the density of states within the cavity such that the absorption coefficient increases to unity. Photorefractive-induced index and absorption gratings cause additional modifications in the propagating optical modes and subsequent enhancement in photorefractive four-wave mixing performance.

6.2.2. *Theory of Operation—Asymmetric Fabry–Pérot Modulator*

A classical Fabry–Pérot [105] consists of two plane parallel surfaces of equal reflectivity separated by a nonabsorbing spacer. Control of the relative phases of the reflected and transmitted beams is achieved by changing the thickness of the spacer. Asymmetric Fabry–Pérot quantum-well modulators take advantage of dual absorption and cavity resonances in a thin semiconductor film to balance the unequal intensities reflected from the front and back faces of the film. ASFP structures have been incorporated into the design of electro-optic modulators [106] and self-electro-optic devices (SEEDs) [103] to increase the effective interaction length and to provide large on–off contrast ratios. The basic structure of an ASFP is shown in Figure 34. The absorptive effects of the ASFP create a cavity with a low finesse, F, and low quality factor, Q, given by

$$F = \frac{\pi (R_1 R_2)^{1/4}}{1 - (R_1 R_2)^{1/2}} \tag{6.2.1}$$

and

$$Q = \frac{2n_1 L_c}{\lambda_0} F \tag{6.2.2}$$

Typical values for GaAs-based ASFP MQWs are $F \approx 3$ and $Q \approx 50$. The total reflectivity of the Fabry–Pérot, R_{FP} is given by

$$R_{\text{FP}} = \frac{|\sqrt{R_T} - \sqrt{R_B}\, e^{-\alpha L_c}|^2}{|1 - \sqrt{R_T R_B}\, e^{-\alpha L_c}|^2} \tag{6.2.3}$$

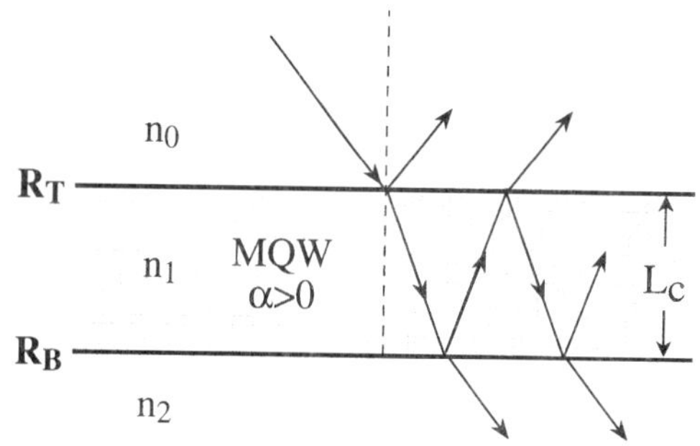

Fig. 34. Asymmetric Fabry–Pérot structure. The front (R_T) and back (R_B) reflectivities are unequal.

where R_T is the front interface reflectivity, R_B is the back interface reflectivity, α is the absorption, and L_c is the cavity thickness. The cavity absorption balances the reflectivities from the front and back interfaces when $R_B > R_T$. Most importantly, $R_{\text{FP}} = 0$ when

$$R_B e^{(-2\alpha L_c)} = R_T \tag{6.2.4}$$

One advantage of the Fabry–Pérot modulator over absorption modulators [107] is the π phase difference between the reflected waves off the front and back interfaces that can yield a net reflectivity of exactly zero.

Multiple beam interference within the cavity provides sensitive amplitude and phase control of the incident light. The reflected and transmitted field amplitudes from multiple layers are given by [108],

$$\begin{pmatrix} 1 \\ E_r \end{pmatrix} = \frac{C_1 C_2 \cdots C_{n+1}}{t_1 t_2 \cdots t_{n+1}} \begin{pmatrix} E_t \\ 0 \end{pmatrix} \tag{6.2.5}$$

where the C_m are the Fresnel matrices for each interface,

$$C_m = \begin{pmatrix} \exp(i\delta_{m-1}) & r_m \exp(i\delta_{m-1}) \\ r_m \exp(-i\delta_{m-1}) & \exp(-i\delta_{m-1}) \end{pmatrix} \tag{6.2.6}$$

The mth Fresnel reflection coefficient, r_m, and the mth phase, δ_m, are given by

$$r_m = \frac{n_{m-1} - n_m + i(\kappa_m - \kappa_{m-1})}{n_{m-1} + n_m + i(\kappa_m + \kappa_{m-1})} \tag{6.2.7}$$

and

$$\delta_m = \frac{2\pi}{\lambda} n_m \sum_{j=1}^{m-1} d_j \tag{6.2.8}$$

where d_j is the thickness of the jth layer. From the multilayer matrix product the field amplitudes E_r and E_t can be calculated and used in the Fraunhofer integral,

$$E_t^d = \frac{1}{\Lambda} \int_0^{\Lambda} \left| E_t(x) \right| \exp\left[i\phi_t(x) - iMKx \right] dx \tag{6.2.9}$$

to calculate the diffraction efficiencies η_{in} and η_{out} of the structure.

6.2.3. Simulations

Reflective, transmissive, electro-optic, and diffractive performance of both hybrid transmission–reflection and fully asymmetric Fabry–Pérot devices, depicted in Figure 35, can be computed following the previous procedure. Details of the simulation program, which has Kramers–Kronig transforms to accurately include spectral variation in the refractive index,

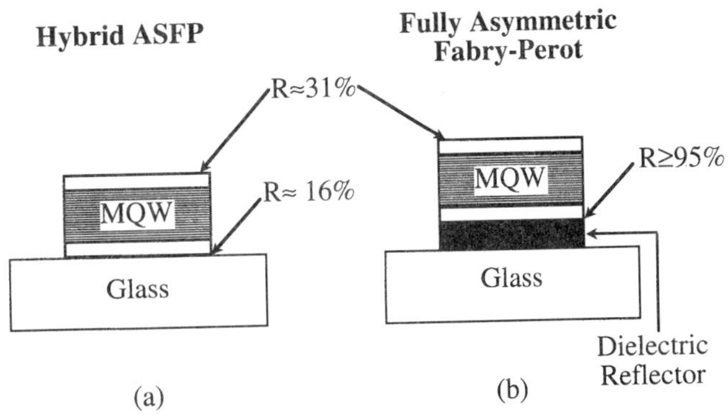

Fig. 35. ASFP MQW fabricated device configurations. (a) Hybrid transmission–reflection ASFP, (b) Fully asymmetric Fabry–Pérot with a highly reflective dielectric stack.

are given in Ref. [52]. The optical properties used in the simulation are described in Section 4. The device structures simulated here include a MQW region sandwiched between two nonabsorbing spacer layers of $Al_{0.1}Ga_{0.9}As$, used to vary the cavity Fabry–Pérot fringe placement. The incident layer may consist of air or glass although the final layer may consist of glass or air, depicted in Figure 35a, or a highly reflective coating, shown in Figure 35b. Reflectivities versus wavelength for both transverse- and longitudinal-field devices are shown in Figure 36. The reflected intensity can, in principle, be reduced to zero on the long-wavelength side of the excitonic transition. Even in the hybrid transmission–reflection structure, shown in Figure 36a, utilizing only the inherent reflectivities of the air–$Al_xGa_{1-x}As$ and $Al_xGa_{1-x}As$–glass interfaces, the reflected intensity therefore has large contrast ratios without the need for a high-reflectivity dielectric layer, as in the fully asymmetric Fabry–Pérots. For longer wavelengths, additional Fabry–Pérot fringes are clearly visible.

Best reflection diffractive performance for either field geometry of the photorefractive MQWs occurs for the fully asymmetric Fabry–Pérot configuration. Calculations for a transverse-field $Al_{0.1}Ga_{0.9}As/GaAs$ MQW with a thick layer of gold acting as the back reflector and a field-induced heavy-hole exciton absorption change of $\Delta\alpha_{max} = 2000\ cm^{-1}$ yield a maximum input diffraction efficiency of 1.5%. The diffraction efficiency of these devices operating in reflection is enhanced an order of magnitude above that obtained for antireflection-coated multiple quantum-well devices where Fabry–Pérot effects are not present [49].

A longitudinal-field geometry $GaAs/Al_{0.3}Ga_{0.7}As$ MQW p-i-n fully asymmetric Fabry–Pérot yields the best theoretical input diffraction for a thin-film quantum-well device [52]. Details of the fabrication, electro-optic properties, and operation of this type of photorefractive MQW are given in Sections 2.3.3, 4, and 6.2, respectively. Figure 36c shows the calculated reflectances with and without field for the spacer thicknesses which yield the maximum and minimum diffraction efficiencies. Maximum diffraction for the longitudinal-field geometry occurs when the Fabry–Pérot fringe is placed so that an antireflection condition occurs under the applied field. The relative intensities of the reflectance, with no field or grating present, and the zeroth- and first-order diffraction efficiencies are shown in Figure 37. Fabry–Pérot phase effects cause the zeroth-order diffraction to dip below the first-order [109]. A maximum input diffraction efficiency of 6.5% is calculated in this case.

Careful control of the cavity dimensions yields quenching of the intensity of the zeroth-order diffracted beam below that of the first-order beam. In principle, control of the relative magnitudes of the various diffracted orders may be achieved. The effect of the Fabry–Pérot fringe placement on the diffraction efficiency for both field geometries is shown in Figure 38. Optimum placement of the Fabry–Pérot fringe yields an order of magnitude enhancement in the diffraction efficiency.

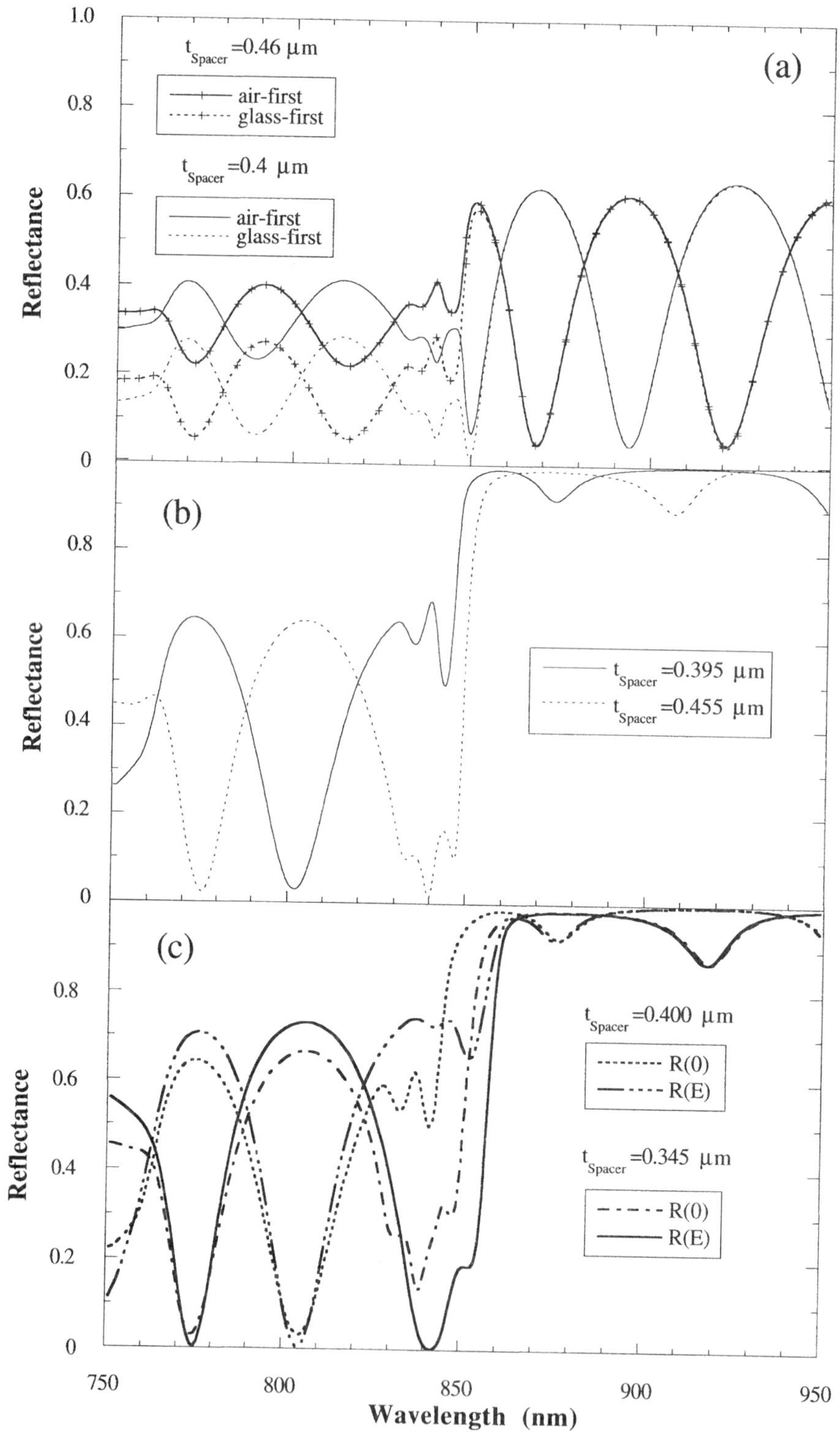

Fig. 36. Calculated reflectance and transmittance of transverse-field (a) and (b) and longitudinal-field (c) ASFPs for spacer thicknesses yielding minimum and maximum η_{in}. (a) Glass- and air-first incidence of a hybrid transmission–reflection geometry 1.5 μm thick GaAs/$Al_{0.1}Ga_{0.9}As$ MQW device for η_{min} (0.46 μm) and η_{max} (0.4 μm), (b) Spectra for a 1.0 μm GaAs/$Al_{0.1}Ga_{0.9}As$ fully ASFP for η_{min} (0.395 μm) and η_{max} (0.455 μm), (c) Reflectance at zero-field ($R(0)$) and 100 kV/cm field $R(E)$ for a 2.0 μm GaAs/$Al_{0.3}Ga_{0.7}As$ longitudinal-field fully ASFP for η_{min} (0.4 μm) and η_{max} (0.345 μm).

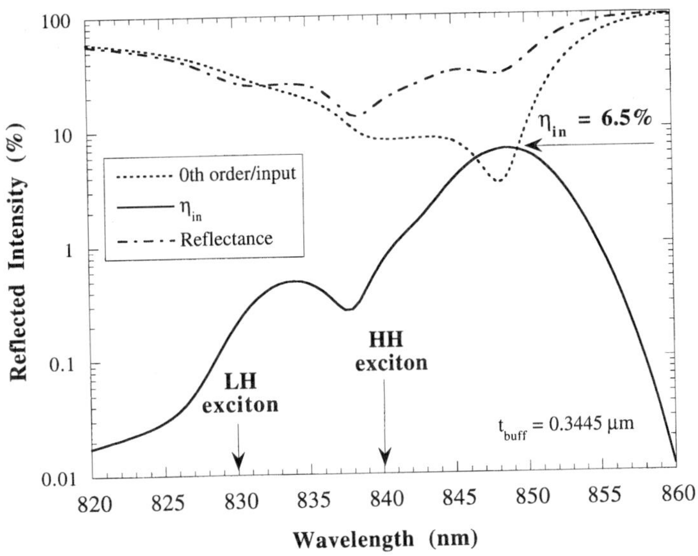

Fig. 37. Calculated longitudinal-field geometry fully ASFP performance.

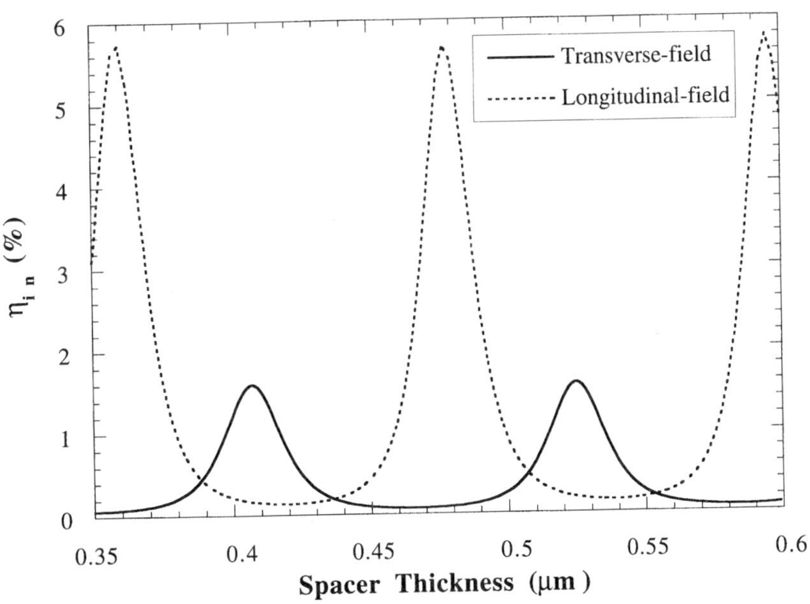

Fig. 38. Calculated maximum reflection diffraction efficiencies versus spacer thickness for the transverse- and longitudinal-field fully ASFP devices in Figure 36b and c.

6.2.4. Experimental Results

Hybrid Transmission–Reflection Partially Asymmetric Fabry–Pérot

Diffractive performance in MQW devices depends on the orientation of the cavity reflectors relative to the incident light beams. In a device consisting of a multiple quantum-well structure bonded to glass the diffraction efficiency can increase by an order of magnitude when the incident light direction is changed from semiconductor-first incidence to glass-first incidence [89]. In this case the quantum-well device is fabricated, following the procedure in Section 2.3.1, as a partially asymmetric Fabry–Pérot by adhering one side of the

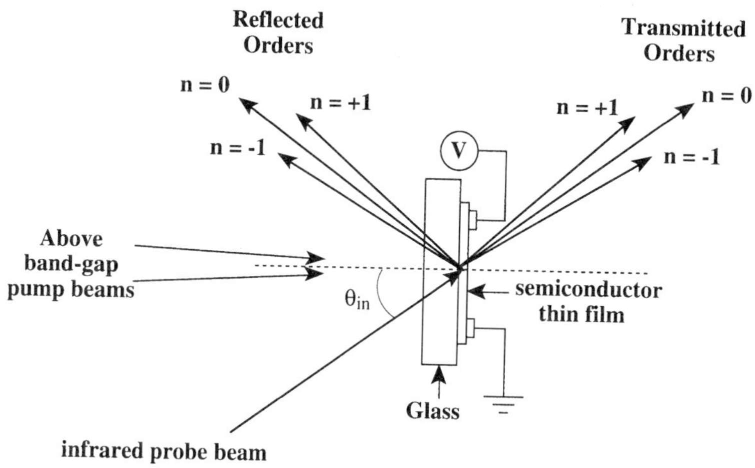

Fig. 39. Nondegenerate four-wave mixing experimental setup. (Source: Reprinted with permission from [89]. © 1994 American Institute of Physics.)

free-standing quantum-well structure to glass with a refractive index of approximately 1.5. The configuration of this hybrid transmission–reflection ASFP is shown in Figure 35a. The glass-semiconductor interface has a lower reflectivity, $R_T \sim 16\%$, than the semiconductor-air interface, $R_B \sim 31\%$. This front–back reflectivity ratio, combined with the intrinsic absorption within the semiconductor, makes it possible for a light beam incident from the glass side to satisfy an antireflection condition. The read beam must be incident on the glass-semiconductor interface to have the mismatch of interface reflections balanced, according to Eq. (6.2.2), by the intrinsic quantum-well absorption. Multiple beam interference occurs for both transmission and reflection because of partial reflections from the glass-semiconductor and the semiconductor-air interfaces.

In nondegenerate four-wave mixing experiments, shown in Figure 39, the diffraction grating is written by two laser beams that have a photon energy larger than the bandgap of the quantum-well structure; in this case HeNe beams [110] were used. The grating beams are incident at equal angles to produce a fringe spacing Λ of 26 μm. This interference pattern is turned into absorption and index gratings through the photorefractive process described in Section 3. To read the gratings written in the quantum wells, a read laser beam (with a laser wavelength tuned close to the exciton transition wavelength) interrogates the gratings and is diffracted into multiple diffraction orders. Here, an argon-ion-pumped Ti:sapphire laser, tunable through the exciton resonance near 845 nm, is used as a probe beam at an incident angle of 35°.

Reflection input diffraction efficiencies are shown in Figure 40 in a 75 Å GaAs/100 Å $Al_{0.1}Ga_{0.9}As$ MQW ASFP with $Al_{0.1}Ga_{0.9}As$ spacers grown at 600 °C, fabricated following the procedure in Section 2.3.1, for grating and probe beams incident from the glass-first interface compared with beams incident from the air-first interface. When the grating and probe beams are incident on the glass-semiconductor interface, the favorable ratio of reflectance from the front and back interfaces enhances the reflected diffraction efficiencies. On the other hand, when the beams are incident on the air-semiconductor interface, the front and back reflectivity ratios are unfavorable for interference effects and produce a diffraction efficiency that is an order of magnitude smaller.

Fully Asymmetric Fabry–Pérot

Peak transverse-field photorefractive ASFP performance relies on the optimization of the electroabsorption properties of the quantum wells, as well as of the Fabry–Pérot. The de-

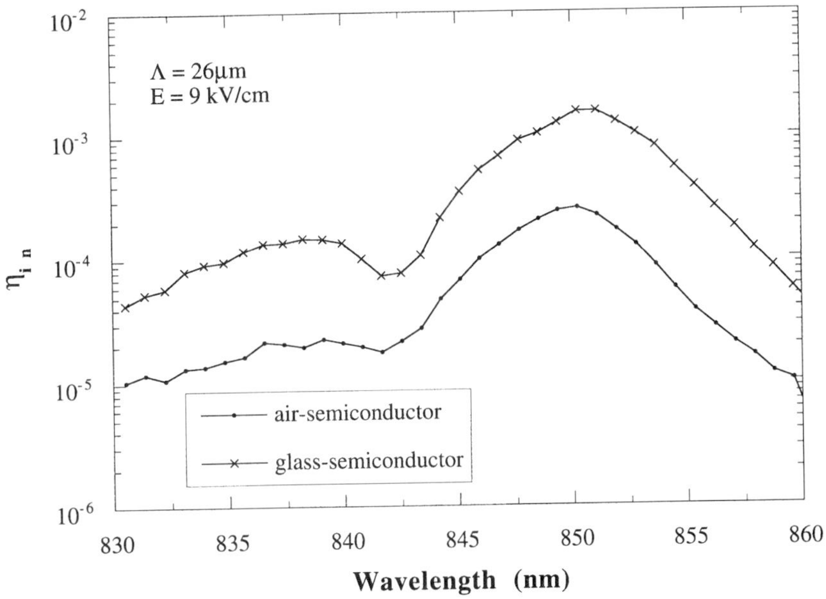

Fig. 40. Comparison of glass-first versus semiconductor-first for the incident interface. (Source: Reprinted with permission from [89]. © 1994 American Institute of Physics.)

sign of a standard temperature grown (STG) $Al_{0.1}Ga_{0.9}As$/GaAs ASFP was optimized using the computer simulations [52] described in Section 6.2.2. A 1.5 μm thick 100 Å GaAs/75 Å $Al_{0.1}Ga_{0.9}As$ MQW region provides the absorption necessary to balance the asymmetric front ($R = 31\%$) and back ($R > 95\%$) mirror reflectivities. The spacers in this structure consist of two $Al_{0.23}Ga_{0.77}As$ layers, one 1500 Å thick grown at the standard temperature of 600 °C and the other 0.75 μm thick grown at the low temperature of ~270 °C. The structure was made semi-insulating by annealing at 600 °C for 15 min for the low-temperature-growth (LTG) spacer and subsequent proton implantation after fabrication for the STG MQW region. The second structure is a LTG AlAs barrier MQW. The advantages of LTG AlAs/GaAs quantum wells include higher breakdown field and lower leakage currents, allowing the device to perform at higher fields than $Al_{0.1}Ga_{0.9}As$ barriers. LTG AlAs/GaAs is automatically semi-insulating when annealed [23], as described in Section 2.2.3. The LTG device structure consisted of STG-grown fabrication layers (described in Section 2.3.1) and LTG material consisting of a 2000 Å $Al_{0.23}Ga_{0.77}As$ spacer layer followed by 167 periods of 100 Å GaAs/20 Å AlAs MQWs and a 1000 Å $Al_{0.23}Ga_{0.77}As$ spacer, grown at 315, 305, and 305 °C, respectively, as measured in the Varian Gen II MBE system. A 30 s anneal at 600 °C made the material semi-insulating. The dielectric reflector stack was deposited on the topmost layer of both structures, the device was bonded to a glass slide, and the substrate was removed according to the procedure discussed in Section 2.3.1. For both structures, the nonabsorbing spacer was etched as outlined in Section 2.3.1 to set the cavity thickness to produce the best resonance condition for each device.

For the fully asymmetric Fabry–Pérots the device structure is shown in Figure 35b and the grating and probe beam configuration is the same as in Figure 39. Here, all beams are incident on the air-semiconductor interface and the reflector coating on the back of the device creates maximum diffraction efficiencies in reflection only. The reflection output diffraction efficiencies and the corresponding peak reflection input diffraction efficiencies of the two photorefractive ASFP structures are shown in Figure 41 as functions of wavelength. The diffraction peak of the STG 10% Al barrier device is shifted toward the

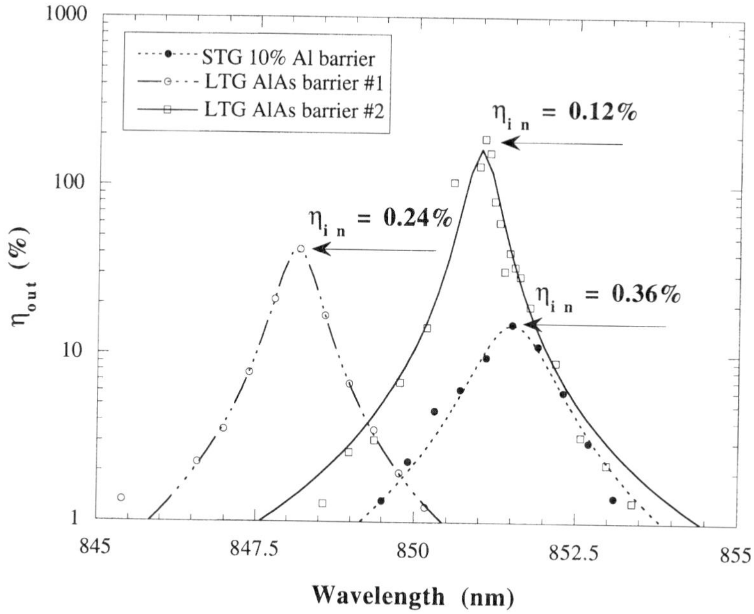

Fig. 41. Output and peak input reflection diffraction efficiencies for fully asymmetric Fabry–Pérot MQWs. The STG 10% Al barrier device was tested with a modulated DC field of 8 kV/cm at 142 Hz and an approximate total light intensity of 5.0 mW/cm^2. The two LTG AlAs barrier devices were tested with a modulated DC field of 15 kV/cm at 142 Hz and an approximate total light intensity of 155 mW/cm^2. (Source: Reprinted with permission from [51]. © 1995 American Institute of Physics.)

Fabry–Pérot fringe at 852.5 nm. An output diffraction efficiency as large as 15% has been obtained in this device, the corresponding input diffraction efficiency, taking into account all sources of insertion loss, is 0.34%. The diffraction spectra from two of the LTG devices are also shown. These two LTG devices differ from each other only in their buffer thickness, and therefore in their cavity resonance condition. The output diffraction efficiency reaches nearly 200% for a well-balanced cavity in LTG AlAs barrier device #2. A different buffer thickness reduces the output diffraction efficiency to 50% in LTG AlAs barrier device #1 as the cavity is moved away from resonance. The corresponding input diffraction efficiency is larger in LTG device #1 due to a Fabry–Pérot resonance closer to the heavy-hole exciton at 844 nm. No significant diffraction is seen at the exciton resonances in these devices. The largest contribution to the diffraction enhancement appears to be the Fabry–Pérot phase effects. A thicker MQW region and the additional background absorption from the incorporated arsenic precipitates [111] balance the cavity farther from the heavy-hole exciton energy and lower the overall input diffraction efficiencies for the LTG devices compared to the STG 10% Al barrier device. The maximum input diffraction efficiency from the LTG devices is 0.24%.

6.2.5. Microcavity Resonance Tuning

Several different methods to tune the cavity resonance with respect to the heavy-hole exciton resonance are discussed here. Interactions between cavity and light modes within the ASFP microcavities are also examined.

Angle Tuning

For off-normal incidence angles of the probe beam in the nondegenerate four-wave mixing configuration of Figure 39, a slight path length difference is created in the cavity, shifting

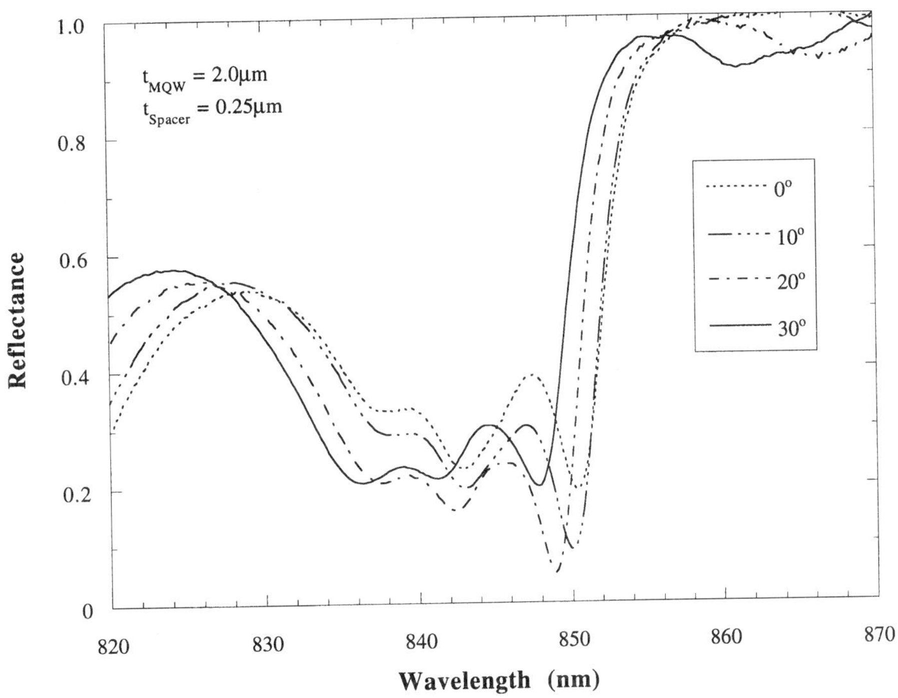

Fig. 42. Experimental reflectance spectra of a 2.0 μm thick GaAs/$Al_{0.1}Ga_{0.9}$As MQW device bonded to a dielectric mirror as a function of incidence angle.

the cavity resonance energy without modifying the heavy-hole exciton energy or linewidth. A shift of several nanometers in the cavity resonance positions can be achieved in a 2 μm thick fully asymmetric Fabry–Pérot transverse-field device, shown in Figure 42. This shift is enough to move the resonance located just to the low energy side of the heavy-hole from an energy condition which is too small at 0° incidence, through resonance near 20°, and out to an energy condition which is too high at 30°. The amount of shift is limited by the GaAs refractive index of ~3.6 and the thinness of the cavity.

Temperature-Tuning

The temperature-dependent bandgap shift of the multiple quantum wells can also be utilized to tune the excitonic resonance with respect to the cavity resonance. The effects of different resonance conditions on diffractive performance have been investigated experimentally in a transverse-field-geometry fully asymmetric Fabry–Pérot consisting of 2.0 μm-thick LTG GaAs/AlAs multiple quantum wells coated on one side with a highly reflective dielectric stack [112]. The reflectances versus wavelength for this device at two different temperatures are shown in Figure 43a. At 85 K the Fabry–Pérot appears to be at an on-resonant condition, while at 171 K it is detuned from resonance.

The results of nondegenerate four-wave mixing experiments conducted at 84 and 171 K are shown in Figure 43b, displaying nearly an order of magnitude difference in the diffraction efficiency at the two resonance conditions. While the diffraction spectra vary in magnitude, they have qualitatively similar shapes, suggesting that the mode-tuning is a weak function of the temperature shift. Factors influencing the magnitude of the diffraction efficiency include cavity-induced enhancement as a result of a shift from an on-resonance to an off-resonance condition, as well as thermal broadening of the excitonic transition. Diffraction enhancement is also seen at the low-energy cavity resonance 30–40 meV from

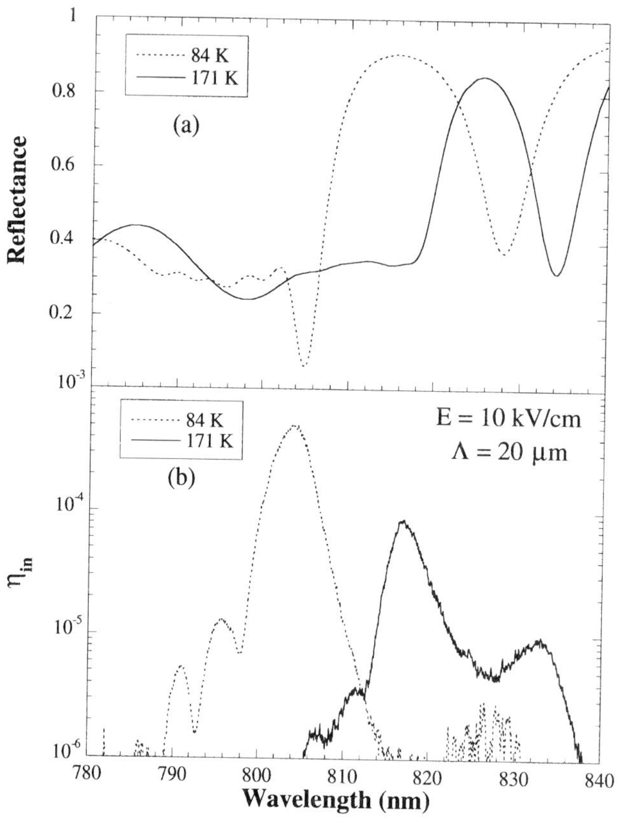

Fig. 43. (a) Reflectances versus wavelength versus temperature of a 2.0 μm LTG GaAs/AlAs multiple quantum-well fully asymmetric microcavity showing two different cavity resonance conditions. (b) Input diffraction efficiencies versus wavelength versus temperature. (Source: Reprinted with permission from [112].)

the main diffraction peak. A cavity with sharper room temperature excitons, as can be obtained with stoichiometric GaAs quantum wells, may allow better resolution.

Terraced Devices

An additional way to examine cavity effects is to create multiple terraces of varying thickness on a single device, thereby reducing device-to-device contact nonuniformity effects. An LTG MQW with the same structure as described previously was used. Terracing of the device was achieved before application of electrical contacts by etching strips on the surface to varying thicknesses. In Figure 44a, the reflectance spectrum for each terrace (A, B, C, D) of this fully asymmetric Fabry–Pérot-terraced MQW are shown. The difference in thickness between terraces A and D yields a shift of $\sim 1/2$ of the Fabry–Pérot fringe spacing, corresponding to a shift from an on-resonance (A) to an off-resonance (B) condition. The transmittance of the structure without a reflector coating highlights the exciton resonances. Cavity effects resulting from the differing terrace thicknesses create distinctly different diffraction spectra as shown in Figure 44b. For the thickness condition yielding the highest diffraction efficiency, the spectrum has a single main peak, while off-resonance condition spectra have several peaks of nearly the same magnitude. In the case of terrace D a peak is seen over 20 nm from the heavy-hole exciton position at 843 nm due to the presence of a secondary cavity resonance. The cavity shift from a near on-resonant to an off-resonant condition leads to an order of magnitude drop in diffraction efficiency, agreeing with the calculations in Section 6.2.2.

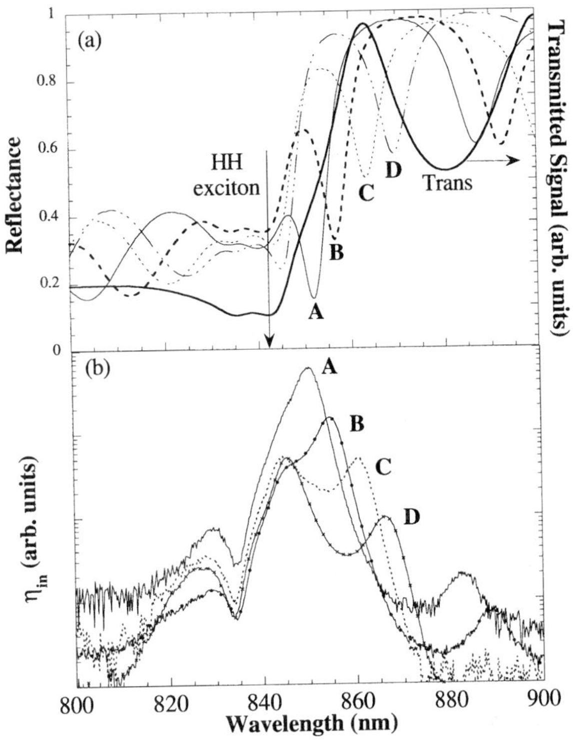

Fig. 44. (a) Transmittance of base LTG GaAs/AlGaAs thin-film structure and reflectance with added dielectric reflector coating for each terrace. (b) Reflection input diffraction efficiency spectra for each terrace for an applied field of 5 kV/cm, $\Lambda = 20\ \mu$m, $m = 1$, $\theta_{in} = 15°$. The relative magnitudes of the curves are normalized with respect to one another.

6.3. Novel Bandgap Engineering

Bandgap or bandstructure engineering is the practice of designing semiconductor heterostructure devices with controlled and well-defined electronic properties. Bandstructure engineering has been made possible by the advent of molecular beam epitaxy (MBE), a growth technique that allows the controlled deposition of atomically distinct layers on a substrate. In the usual meaning of the term, bandgap engineering entails designing the properties of a semiconductor heterostructure by varying its composition in the growth direction. This results in a material with controllable electronic properties in one dimension. Well-known examples of this concept are quantum wells and superlattices; the ubiquitous laser diode is also a commonplace example of bandgap engineering.

The particular implementation of bandstructure engineering depends on the desired application. For the design of electronic device properties, the one-dimensional control of electronic properties is often sufficient, and novel electronic devices (such as high-speed heterojunction transistors or floating gate memories) have been demonstrated [113]. For optoelectronic applications, where one desires to construct integrated waveguides and photonic switches, two-dimensional control of the semiconductor properties is necessary. Lateral variation can be achieved by patterning before growth, or by techniques such as strain-modulated epitaxy [114] or indium migration control [115]. In a more general sense of bandgap engineering, two-dimensional photonic bandgap structures have been demonstrated on GaAs [116, 117]; coupled with the "conventional" one-dimensional bandgap engineering in the growth direction, this can lead to structures where the electronic and/or photonic properties of the device are fully controlled in three dimensions.

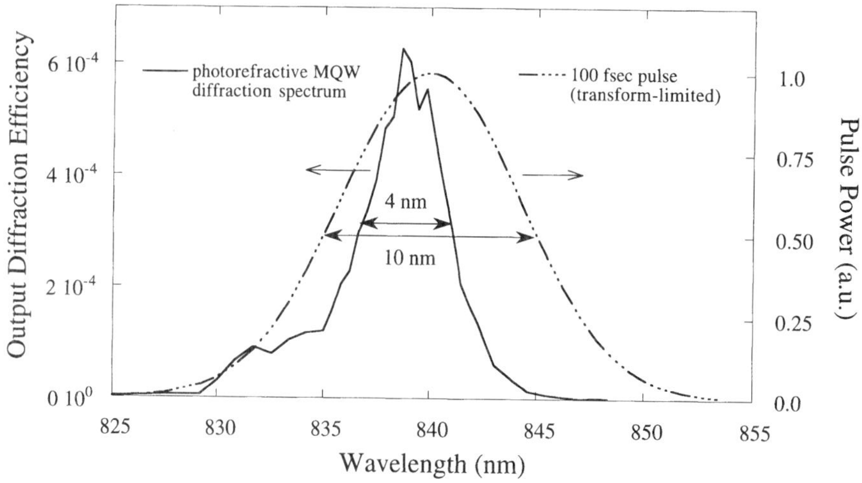

Fig. 45. The spectral dependence of the four-wave mixing diffraction efficiency for typical photorefractive quantum wells compared to the bandwidth of a transform-limited 100 fs pulse. The FWHM bandwidth of the quantum-well device is smaller by a factor of ≈ 3 than the bandwidth of the ultrafast pulse.

Our interest in bandgap engineering originates from the potential use of GaAs/AlGaAs thin-film devices for photorefractive applications in dynamic pulse shaping of ultrafast pulses and spectral holography [118]. Photorefractive quantum-well devices have advantageous characteristics for such applications (high-sensitivity and fast response times), but they suffer from the limited bandwidth of the excitonic transition, which is typically on the order of 4 nm. For pulse-shaping applications, a flat spectral dependence of the diffraction efficiency is desired over the entire spectral range of the femtosecond pulse, which is approximately 10 nm for a typical 100 fs pulse (Fig. 45). The diffractive bandwidth of a photorefractive multiple quantum-well device is mainly due to the phonon broadening of the excitonic transition at room temperature and to well thickness and well-barrier interface disordering.

The four-wave mixing diffraction spectrum of a quantum-well device depends quadratically on both the electroabsorption and the electrorefraction spectra. In order to achieve a desired bandwidth over a wide spectral range, it is necessary to control in a detailed manner the electroabsorption properties of the MQW device by selecting a range of excitonic transition energies. The strength and spectral dependence of the excitonic electroabsorption features are strongly influenced by the distribution of states available for transitions in the conduction and valence bands. Therefore, in order to control the bandwidth of the electroabsorption one needs to design the density of states available for the excitonic transitions. This leads to a new aspect of bandstructure engineering, namely, density-of-states engineering.

Within the framework of multiple quantum-well devices, there are two possible approaches to this problem (Fig. 46). The first consists in stacking isolated quantum wells with varying exciton transition energies in the vertical (growth) direction. In this case, the spectral properties of the MQW device will be determined by the convolution of the electroabsorption spectra over the individual isolated wells. Another approach would be to increase the coupling between the wells and design the superlattice to have a broad excitonic transition caused by multiple critical points within the wide superlattice miniband; this would constitute true density-of-states engineering. In all cases, due to the spreading of oscillator strength over a larger spectral range, the peak diffraction efficiency will be smaller than for a standard narrow-band device. Also, another characteristic property of MQW broadband devices arises due to the variation of the electric field at which the

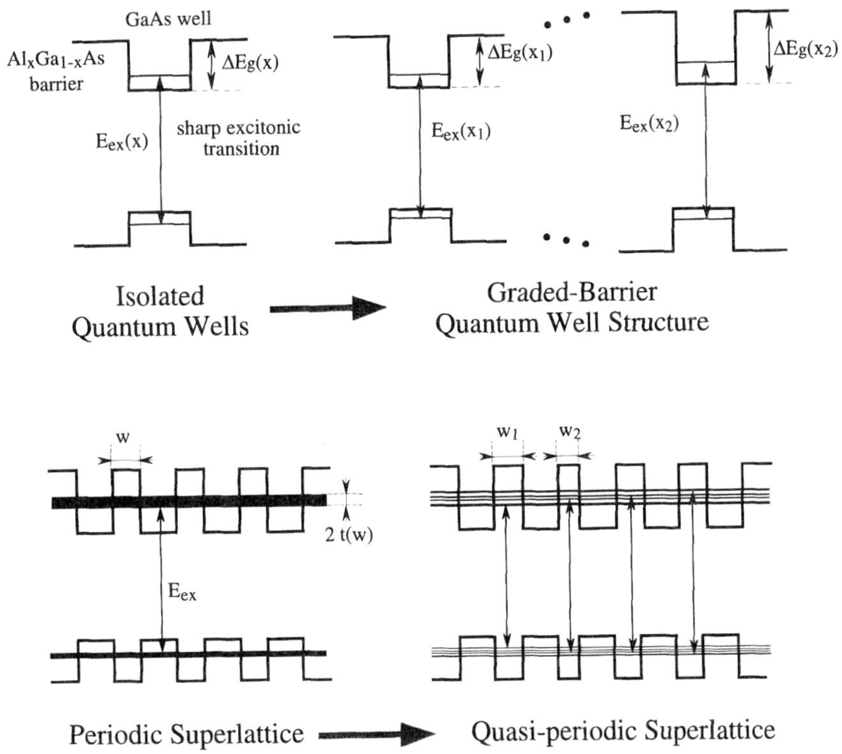

Fig. 46. Schematic of the strategies for the design of broadband photorefractive quantum wells: convoluting the exciton electroabsorption spectra of isolated quantum wells to obtain a graded-confinement quantum-well structure; or engineering the density of states in a superlattice, to obtain a dense set of critical points associated with excitonic transitions.

electroabsorption saturates across the electroabsorption spectral components; the shape of the electroabsorption spectrum and hence of the diffraction spectrum is then (weakly) dependent on the magnitude of the applied electric field.

6.3.1. Graded Confinement Structures

Graded Barriers

In a quantum well, the excitonic transition energy can be tuned by varying the carrier confinement energy, through the modification of either the well widths or barrier heights. Using the latter approach, of controlling the confinement potential in the individual wells by grading the composition of the AlGaAs barriers in the GaAs/AlGaAs multiple quantum-well structures, we have studied several devices. The electro-optic response of such a structure is given in the first approximation by a simple integration of the electroabsorption of each individual layer. A 1% variation in the Al fraction in the barriers induces approximately a 1 meV (0.6 nm) shift of the heavy-hole exciton energy. Therefore, a range of 15% Al fraction in the quantum-well barriers can achieve the necessary variation in exciton energy to cover the femtosecond pulse spectrum.

The structures we have studied are all standard transverse (Franz–Keldysh) geometry photorefractive quantum wells, differing only in the design of the electro-optically important MQW region. All GaAs wells were 75 Å wide. The double-layer MQW (DQW) consisted of 75 periods with $Al_{0.15}Ga_{0.85}As$ barriers followed by 75 periods with $Al_{0.30}Ga_{0.70}As$ barriers. The triple-layer device (TQW) consists of 65 periods with

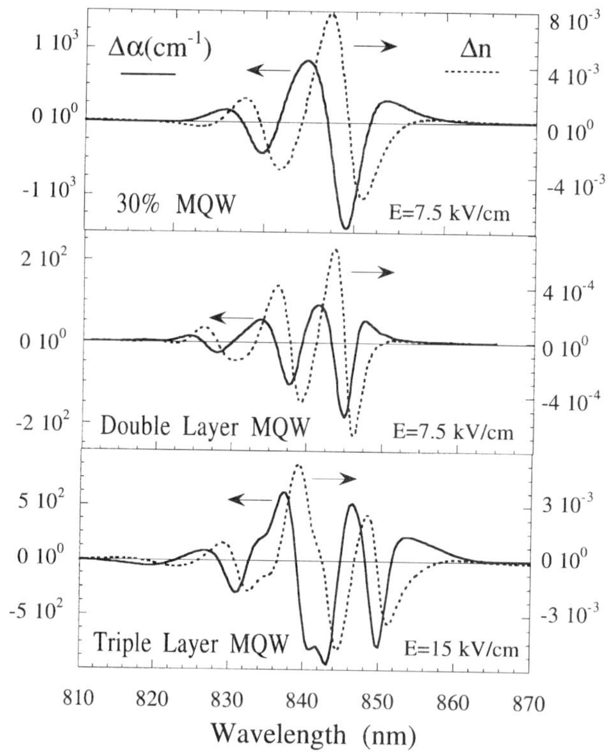

Fig. 47. Electroabsorption and electrorefraction spectra for the graded-barrier structures described in the text. The convolution of the multiple spectral components results in well-defined structure in the electroabsorption spectra.

$Al_{0.15}Ga_{0.85}As$ barriers, 70 periods with $Al_{0.225}Ga_{0.775}As$ barriers, and 75 periods with $Al_{0.30}Ga_{0.70}As$ barriers.

The response under applied electric field (electroabsorption) can be calculated for any such graded-barrier device provided relevant parameters are known: the heavy-hole and light-hole energies and oscillator strengths as functions of Al fraction in the barriers. The heavy-hole transition has higher oscillator strength and gives the main contribution to the diffraction spectrum; however, the electroabsorption associated with the light hole for low Al fraction can interfere with the electroabsorption corresponding to the high Al-fraction heavy hole, because the energy separation between the light- and heavy-hole energies is typically about 15 meV. The simulation parameters can be found from electroabsorption data for discrete Al fractions in the barriers. The simulations, as well as the experimental data, indicate that the spectral performance is a sensitive function of the structure parameters, because small changes in the electroabsorption lineshape can lead to large variations in the diffraction bandwidth. This is due to the quadratic dependence of the diffraction efficiency on the electroabsorption $\Delta\alpha$.

The electroabsorption and electrorefraction for the graded-barrier devices are shown in Figure 47 in comparison with a standard $GaAs/Al_{0.30}Ga_{0.70}As$ MQW device. The electrorefraction is calculated from the measured electroabsorption by means of a Kramers–Kronig transformation. The presence of multiple spectral components leads to well-defined peaks in the electroabsorption spectra. The wavelength dependence of the four-wave mixing diffraction efficiency can then be calculated using Eq. (5.6).

Photorefractive performance was quantified through degenerate four-wave mixing experiments in the transverse Franz–Keldysh geometry. The diffraction grating was written using a CW Ti:sapphire laser pumped by an argon ion laser. The output diffraction effi-

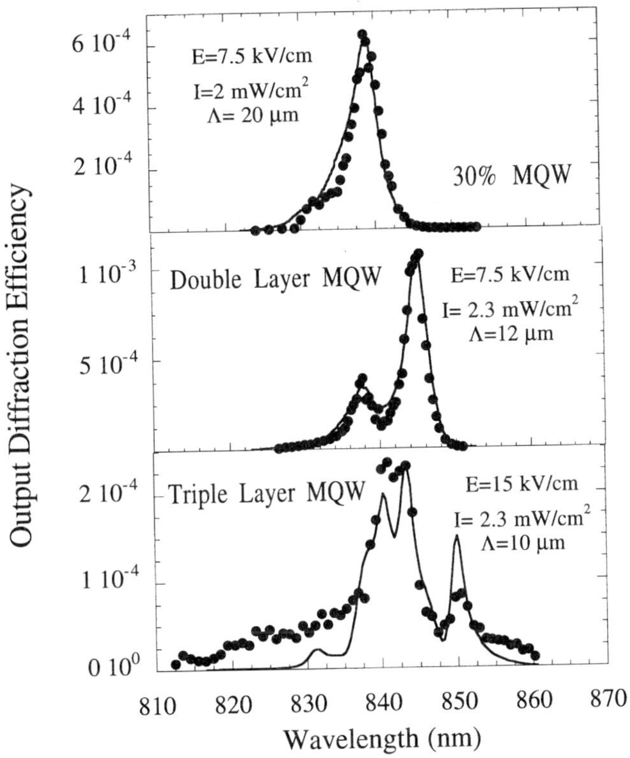

Fig. 48. Four-wave mixing output diffraction efficiency spectra for the graded-barrier structures. The points are experimental data and the continuous lines are the calculated curves using the measured electroabsorption and electrorefraction. The triple-layer MQW exhibits a FWHM bandwidth at the principal peak of about 8 nm.

ciency, defined as the ratio between the intensities of the first- and zeroth-order beams, was measured as a function of photon wavelength. The results of the four-wave mixing experiments are shown in Figure 48; the points indicate experimental data, while the continuous curve is the calculated diffraction efficiency, up to a multiplicative constant. There is a slight disagreement between the experimental and the calculated curves in the case of the triple-layer structures at photon energies above the gap, attributed to device heating. The bandwidth for the triple-layer quantum-well structure is approximately 8 nm in the central peak, almost double that of the standard MQW device.

Assuming oscillator strengths independent on Al fraction, simulations indicate that for a linearly graded structure the best electroabsorption spectrum (in terms of bandwidth and flatness) is achieved for an Al-fraction span of 15%, almost independent of the average Al fraction. However, the oscillator strength variation over the exciton energy range has to be compensated by relatively increasing the thickness of the high Al-fraction layers. This weighting of the high Al-fraction layers has been done in the triple-layer structure, resulting in a significantly higher bandwidth than for a standard MQW device.

The main disadvantage of the graded-barrier design is the difficulty of precisely controlling the Al fraction in the quantum-well barriers, coupled to the sensitivity of the diffraction spectrum on the electroabsorption lineshape. A more robust approach to exciton energy tuning is the variation of the quantum-well widths.

Graded Well Widths

In this case the carrier confinement and thus the exciton transition energies are controlled via the width of the potential well. Much better control over the spectral characteristics

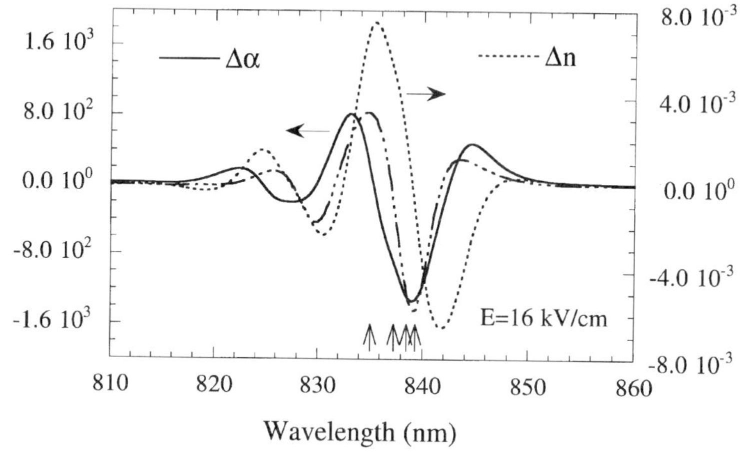

(a)

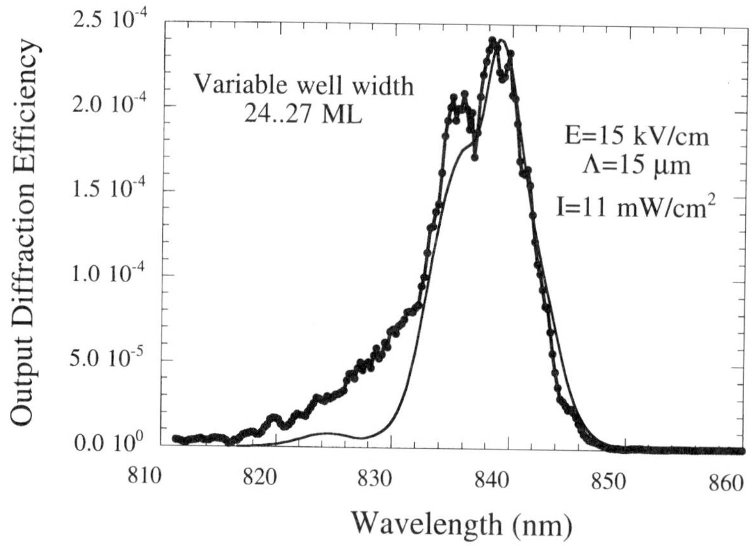

(b)

Fig. 49. (a) Room temperature electroabsorption and electrorefraction and (b) four-wave mixing output diffraction efficiency spectra for the variable well-width MQW device described in the text. The positions of the heavy-hole electroabsorption peaks for the four constant well-width layers ranging from 27 to 24 ML are indicated by arrows. The dot-dashed line is the electroabsorption corresponding to a 30% Al-fraction MQW, with well width equal to 75 Å.

of the device is allowed by the "quantization" of the quantum-well widths, inherent to the MBE growth process. Growth-to-growth variations occur then only in the confinement potential height, nominally kept fixed, and not in the well widths themselves.

Such a graded well-width design consisted in four layers of GaAs quantum wells, all with 100 Å $Al_{0.30}Ga_{0.70}As$ barriers. The well widths span from 24 to 27 monolayers (ML) of GaAs, and the device contains 6, 12, 18, and 16 periods for the well widths from 24 ML (67.9 Å) to 27 ML (76.4 Å), respectively. The number of periods for each layer has been chosen so as to give a relatively flat diffraction lineshape, approximately 9 nm wide. The electroabsorption spectrum is shown in Figure 49a. In this case the convolution of the various spectral components results in smooth broad electroabsorption peaks.

The diffraction spectrum for degenerate four-wave mixing is shown in Figure 49b; the diffraction linewidth FWHM is 9.2 nm, and the shape of the spectrum is nearly flat, consistent with the design. Again there is disagreement between the experimental and the calculated curves at photon energies above the gap.

In principle arbitrarily large bandwidths can be achieved either by grading the well widths or the barrier heights in the quantum-well devices. However, in practice, the graded well-width approach seems to be more robust in spite of the constraints in exciton transition energy imposed by the discrete thicknesses of the wells.

6.3.2. Fibonacci Superlattices

In a superlattice, excitonic transitions occur at the miniband critical points (miniband edges). Therefore, a venue toward designing a superlattice with broad excitonic transitions is engineering the density of states to exhibit multiple critical points within the minibands of interest. A family of systems with such a property are Fibonacci superlattices. Fibonacci superlattices are quasi-periodic structures built using two different constructive blocks that are arranged in a sequence which obeys the Fibonacci composition rule: $S_{n+1} = \{S_n S_{n-1}\}$, where S_n is the nth finite realization of the lattice. Fibonacci superlattices are one-dimensional quasi-crystals, having two spatial periods which are incommensurate. Fibonacci superlattices can be implemented in semiconductor heterostructures in various ways. The characteristic parameters are the interwell coupling and/or the carrier confinement, which can be made to vary quasi-periodically. One example is a superlattice with two types of AlGaAs barriers, long (L) and short (S) arranged in a Fibonacci sequence, following the pattern: LSLLSLSLLSLLS. . . .

The quasi-periodic interwell coupling is expected to induce qualitative changes in the envelope function of single-particle electronic states in the wells. For an infinite Fibonacci superlattice (FSL), the dense distribution of spatial periods are reflected in the self-similar structure of the energy spectrum and several quasi-bands emerge from the isolated quantum-well energy levels. The fractal structure of the one-electron density-of-states spectrum is broadband and might be expected to give rise to a broadband electroabsorption spectrum, because of the everywhere dense set of critical points associated with interband transitions.

We have investigated three GaAs/$Al_{0.3}Ga_{0.7}As$ 100-period Fibonacci superlattices for which the well coupling was varied by changing the barrier widths [119]. For all superlattices the well thickness was 75 Å. Growth FSL#1 had 35 and 25 Å barriers while the same thicknesses for FSL#2 were 25 and 18 Å, respectively. Growth FSL#3 had 18 and 25 Å barriers, and additionally the more numerous wells had an excess of $\approx 1\%$ Al to increase the bandgap. The transfer integrals characterizing interwell coupling for the shorter barriers in the conduction band were chosen to be comparable to the energy separation between the light-hole and the heavy-hole transitions in a GaAs/$Al_{0.3}Ga_{0.7}As$ MQW structure.

The measured electroabsorption and calculated electrorefraction for the structures are shown in Figure 50(a). The higher energy transitions within the minibands are manifested by the weaker electroabsorption features, and the miniband edges can be identified. Increasing interwell coupling and quasi-periodicity going from FSL#1 to FSL#2 is manifested by more prominent electroabsorption features in the high-energy photon range. The broadening due to room temperature phonon scattering and proton implantation damage obscures the self-similar structure within the quasi-bands, which can be nevertheless observed at low temperature (Fig. 51) in samples that have not been proton implanted [120].

The diffraction efficiency for the three structures as a function of wavelength is shown in Figure 50b. The higher energy spectral structure apparent in the electroabsorption of the Fibonacci superlattices disappears in the diffraction spectra. This shows the dramatic effect of the quadratic dependence of the output diffraction efficiency on the differential transmission: A flat broadband diffraction response requires that participating electroabsorption features have comparable magnitudes.

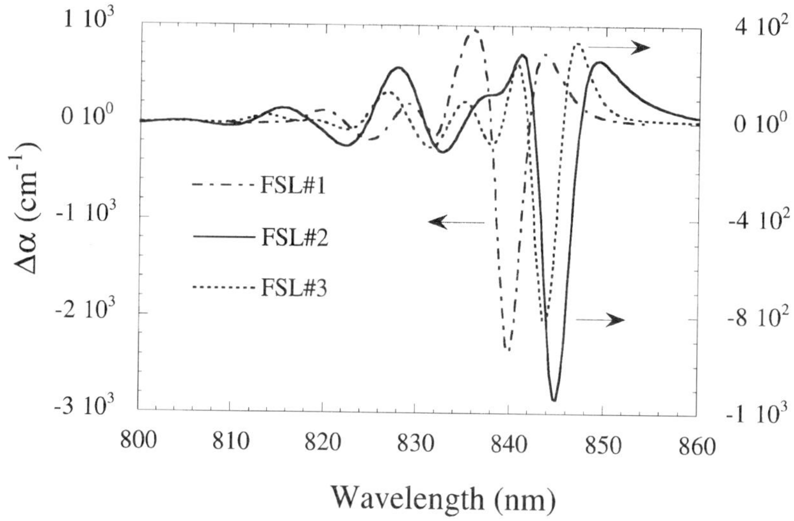

(a)

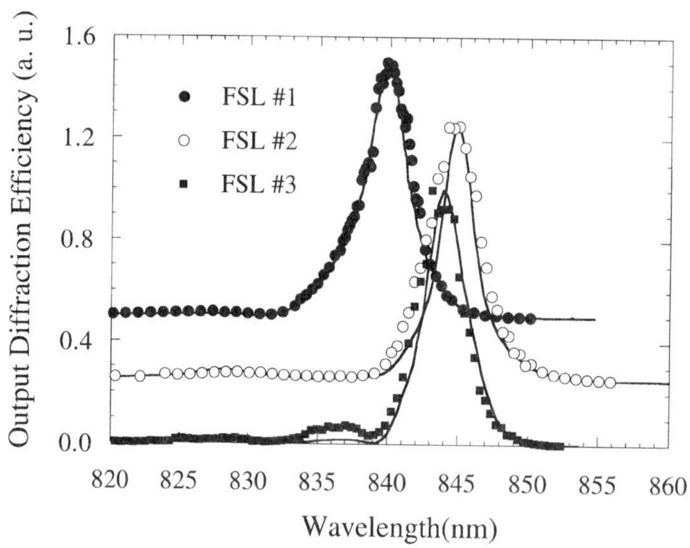

(b)

Fig. 50. (a) Room temperature electroabsorption spectra for the Fibonacci superlattices for an applied electric field $E = 15$ kV/cm (FSL#1 and FSL#2) and $E = 7.5$ kV/cm (FSL#3). (b) Four-wave mixing diffraction efficiency, normalized to the peak value, as a function of wavelength, for the Fibonacci superlattices. The experimental parameters were: $E = 15$ kV/cm, $\Lambda = 20$ μm, $I = 1$ mW/cm^2 (FSL#1); $E = 15$ kV/cm, $\Lambda = 20$ μm, $I = 2$ mW/cm^2 (FSL#2); $E = 7.5$ kV/cm, $\Lambda = 15$ μm, $I = 3.5$ mW/cm^2 (FSL#3). Due to the concentration of the oscillator strength at the bottom of the superlattice miniband, the higher energy electroabsorption features are weaker than the main heavy-hole peak, which leads to a diffractive bandwidth practically unmodified in comparison to a standard MQW device.

The smaller strength of the higher energy electroabsorption features is due to the settling of the excitonic oscillator strength to the bottom of the quasi-band [120]. This is a fundamental effect due to the Coulomb interaction which is also the origin of the strong excitonic transition at the band edge, and occurs in any structure where there is coupling between the quantum wells. Therefore, superlattices seem to offer little promise for practical broadband devices.

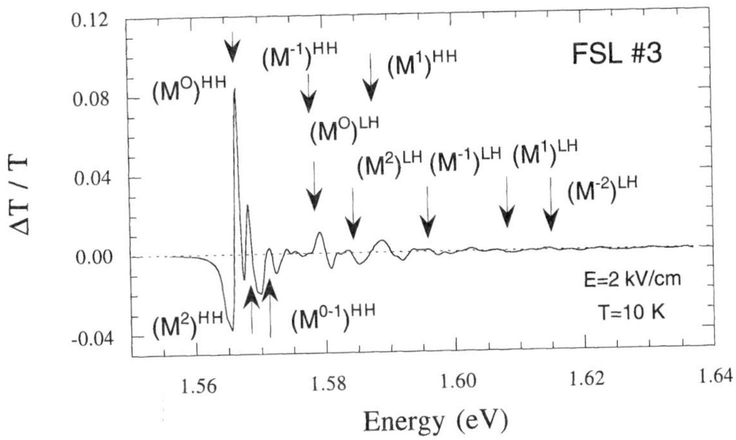

Fig. 51. Low-temperature electroabsorption spectrum of the Fibonacci structure FSL#3. The arrows indicate several identified transitions between major critical points in the quasi-bands, corresponding to both the heavy-hole and the light-hole excitons. The oscillator strength is concentrated at the lower energy transitions. (Source: Reprinted with permission from [120]. © 1997 American Physical Society.)

6.4. Applications of Photorefractive Quantum Wells in Ultrafast (Femtosecond) Optical Communications and Image Processing

As demonstrated by two- and four-wave mixing, photorefractive quantum wells (PRQWs) are excellent dynamic holographic media because of their high nonlinear sensitivities, small driving intensities, and fast response times in comparison with bulk photorefractive materials. In addition, holograms in such materials are thin and the diffraction is in the Raman–Nath regime with no Bragg-matching necessary, providing much larger space-bandwidth products than volume holograms in bulk materials. PRQWs also have great flexibility for design and fabrication using many possible material systems. For example, they can be designed by using bandgap engineering to match the wavelength of 1.55 μm for optical fiber communications [121]. These properties make PRQWs unique for holographic applications with femtosecond laser pulses containing a large bandwidth. In this section, we first discuss the diffraction of femtosecond laser pulses from gratings in a PRQW, which is the basis of most of the ultrafast applications that are described later, and then we give some examples of femtosecond applications of PRQWs, such as femtosecond pulse characterization of ultrafast lasers, all optical transmultiplexing and time-domain image processing for optical communications, and imaging through turbid media for optical coherence tomography.

6.4.1. *Femtosecond Diffraction from Photorefractive Quantum Wells*

When a femtosecond pulse with a given bandwidth enters a PRQW containing a static grating, it is diffracted (nondegenerate four-wave mixing). The electric field of the input pulse can be expressed as $e_{IN}(t) = E_0 f(t)\exp(-i\omega_c t)$, where the temporal shape is given by $f(t)$ and ω_c is the center frequency. For a Gaussian pulse,

$$f(t) = \exp\left(-2\ln 2\frac{t^2}{t_p^2}\right) \tag{6.4.1}$$

where t_p is the full-width-at-half-maximum (FWHM) of the intensity pulse shape. In the frequency domain, the electric field $E_{IN}(\omega)$ is the Fourier-transform of $e_{IN}(t)$. To discuss the diffraction, it is convenient to use the frequency domain. The transmitted field $E_T(\omega)$ from the PRQWs can be represented by a complex transmission function $F_T(\omega)$ in the frequency domain,

$$E_T(\omega) = E_{IN}(\omega) F_T(\omega) \tag{6.4.2}$$

Neglecting Fresnel reflection at the surfaces and the associated Fabry–Pérot effects, the transmission function is

$$F_T(\omega) = \exp\left[-\frac{\alpha(\omega)L}{2} + in(\omega)kL\right] \tag{6.4.3}$$

where k is the propagation constant in free space and L is the sample thickness. The transmission power spectrum is $T(\omega) = |F_T(\omega)|^2 = \exp[\alpha(\omega)L]$, which can be measured experimentally. From the discussion on four-wave mixing in Section 5.2, the first-order diffracted electric field is given by

$$E_D(\omega) = E_{IN}(\omega) F_T(\omega) F_D(\omega) \tag{6.4.4}$$

with the diffraction function,

$$F_D(\omega) = \frac{1}{2}\left(i\Delta n(\omega)kL - \Delta\alpha(\omega)\frac{L}{2}\right) \tag{6.4.5}$$

The power spectrum for the diffracted pulse is then

$$\left|E_D(\omega)\right|^2 = \left|E_{IN}(\omega)\right|^2 T(\omega)\frac{1}{4}\left[\left(\Delta n(\omega)kL\right)^2 + \left(\frac{\Delta\alpha(\omega)L}{2}\right)^2\right] \tag{6.4.6}$$

In the time domain, the electric field is the inverse Fourier-transform of the electric field in the frequency domain, i.e., $e_D(t) = F^{-1}\{E_D(\omega)\}$, and the envelope of the pulse is given by

$$I_D(t) = \frac{|e_D(t)|^2}{2Z} \tag{6.4.7}$$

where Z is the wave impedance.

From Eqs. (6.4.4) and (6.4.5), the shape of the diffracted pulse is strongly related to the electroabsorption $\Delta\alpha$ and the electrorefraction Δn. The striking aspect of the electroabsorption data in Figure 19 is the oscillatory behavior as a function of wavelength. The strong amplitude and phase dispersion enter directly into Eq. (6.4.6) for the electric field of the diffracted pulse. For many applications, such as dynamic Fourier manipulation and processing of femtosecond pulses, an ideal holographic material should have a flat amplitude and phase response over the bandwidth of the pulses. The diffraction from the hologram in such a material should not distort the probe pulse shape if the hologram itself is not intentionally modified by any masks. The oscillatory behavior as a function of wavelength might be expected to lead to severe distortion of the diffracted pulse shape.

To characterize femtosecond diffraction from the PRQWs, which is usually weak, linear techniques such as electric-field cross-correlation and spectral interferometry are used to measure the temporal shape and phase of the diffracted pulse. Electric-field cross-correlation is a technique suitable for weak ultrashort pulses [122, 123], where the field interferogram of the signal and a reference pulse with a known shape is recorded by scanning the time delay between the two pulses. In this technique, the characteristics of the reference pulse are important for determining the pulse shape of the signal pulse.

The femtosecond pulse source in the experiments described here was a mode-locked Ti-sapphire laser with 100 MHz repetition rate and a nominal pulse width of 100 fs. The electric-field and second harmonic autocorrelations of the reference pulse were measured and the time-bandwidth product was calculated directly from the spectrum using $t_p\Delta\nu = t_p c\Delta\lambda/\lambda^2$, where t_p is calculated from the second harmonic autocorrelation width (assuming Gaussian pulses). For the laser tuned at 850 nm, the pulse width was $t_p = 106$ fs, and spectra were measured using a spectrometer with a CCD camera. The time-bandwidth

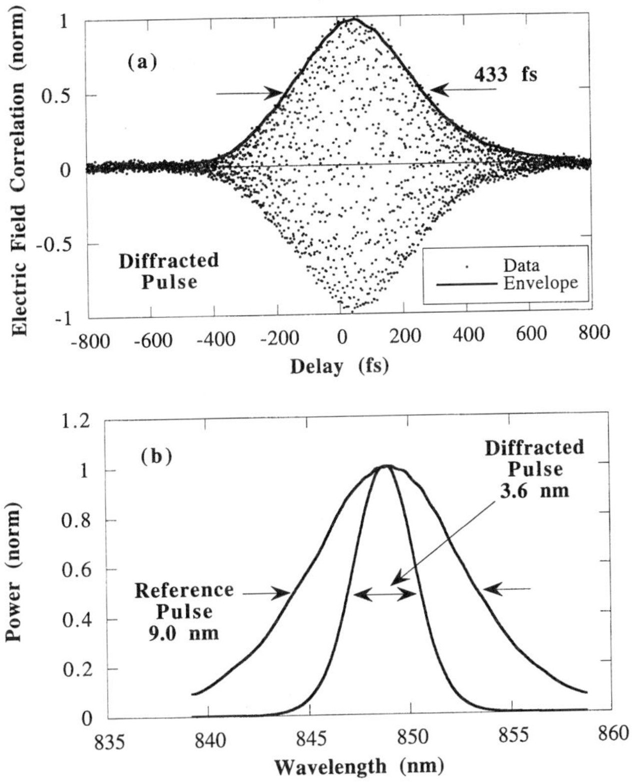

Fig. 52. First-order diffraction of a femtosecond pulse from holograms in PRQWs. (a) Electric-field cross-correlation data, (b) Power spectrum. For comparison the spectrum of the incident pulse is also given in (b). (Source: Reprinted with permission from [148]. © 1997 IEEE.)

product for the reference pulse was 0.41, which is very close to a transform-limited Gaussian pulse. A 685 nm diode laser with a photon energy above the bandgap of the quantum-well barriers was the source for writing the photorefractive grating. The fringe spacing of the interference pattern was 10 μm. The incident intensity from the two writing beams was a total of 2 mW with a beam ratio of unity. There was 2 mW in the probe pulsed beam, which resulted in a total intensity of approximately 40 mW/cm^2 on the GaAs/AlGaAs PRQWs with 10% Al. Over the duration of one 100 fs pulse, the photorefractive grating remains essentially static. To obtain diffraction, a dc voltage of 400 V was applied over the 1 mm gap between the electrodes. At this applied field and an intensity of 40 mW/cm^2, the grating response time is approximately 10 μs, which is the time for photocarriers to drift and trap.

The electric-field cross-correlation data of the diffracted pulse with the reference pulse are shown in Figure 52a. The envelope is nearly Gaussian with a FWHM of 433 fs. The estimated pulse width of the diffraction is 287 fs. The spectra of the reference and diffracted pulse are shown in Figure 52b. The diffraction spectrum appears also to be nearly Gaussian, although the FWHM is reduced from 9 to 3.6 nm due to the bandwidth limit of the excitonic absorption in the PRQWs. This spectrum and the estimate of the pulse length are used to calculate a time-bandwidth product of 0.43, which is very close to that for a Gaussian (0.44). To evaluate the possible contribution from second-order dispersion, spectral interferometry [124] is also used to analyze the diffracted pulses. It is a useful linear technique for determining both phase and shape of weak fs pulses, where the spectrum of a reference pulse interferes with the spectrum of the diffracted pulse with a chosen time delay. Using spectral interferometry, the phase of the diffracted pulse is measured [125], and is quite flat within the femtosecond pulse bandwidth, supporting the conjecture, and verifying that the pulse is nearly transform limited.

The explanation of why the diffracted pulse is nearly undistorted is based on two aspects: First, under an applied electric field, the excitonic absorption line is strongly broadened due to the field ionization in the transverse-field geometry. The field dependence of the transmission function $F_T(\omega)$ becomes very weak with a maximum modulation of 10^{-2} at fields of several kV/cm. Therefore, the phase of the diffracted field $E_D(\omega)$ is mainly determined by the phase of the diffraction function $F_D(\omega)$ in Eq. (6.4.5), which is

$$\Phi(\omega) = -\arctan\left(\frac{2k\Delta n(\omega)}{\Delta\alpha(\omega)}\right) \tag{6.4.8}$$

indicating that the phase is given by the ratio of the electrorefraction $2k\Delta n(\omega)$ and the electroabsorption $\Delta\alpha(\omega)$, which are related by the Kramers–Kronig relation. If the function $\Delta\alpha$ is a symmetric function around the center transition frequency ω_0, which is the case because of the Gaussian excitonic transition lineshape and the Franz–Keldysh effect, the index change Δn varies linearly with frequency over the range where the diffraction efficiency is largest. Because a linear phase with frequency represents only a delay, not a distortion, the strong dispersion of $\Delta\alpha$ and Δn do not distort the diffracted pulse. Both the phase calculated from Eq. (6.4.8) using electroabsorption $\Delta\alpha(\omega)$ and electrorefraction $\Delta n(\omega)$ data, and the phase measured directly by using spectral interferometry show a linear dependence on frequency, which does not introduce distortion.

In summary, the diffraction of 100 fs pulses from the static gratings of a photorefractive quantum well produces diffracted pulses that are nearly transform-limited, despite the strong dispersion near the quantum-confined excitonic transitions. The broadening is due to the limited bandwidth of the diffraction efficiency spectrum, which is determined by the electro-optic properties of the quantum wells. The ability to produce transform-limited pulses from photorefractive quantum wells indicates that photorefractive devices are useful for Fourier pulse shaping and other femtosecond applications.

6.4.2. Femtosecond Applications of Photorefractive Quantum Wells

Electric-Field Autocorrelation

Femtosecond pulses are too fast to be measured directly by electronic detectors. The temporal profile of the pulse is usually obtained by measuring the autocorrelation function of the pulses. Autocorrelation techniques fall broadly into two classes: intensity and field autocorrelation [126]. Intensity autocorrelation provides information on the instantaneous intensity of the pulse. Nonlinear optical effects, such as second harmonic generation, are typically used for this class of techniques. Field autocorrelation provides information on the electric field, including the phase, of the pulse. For many applications, field autocorrelation provides more useful information. It can be achieved through interferometric second harmonic generation [127], or through induced-grating techniques using multi-wave mixing in intensity-integrating materials [128]. Multiwave mixing of ultrafast pulses is attractive as a field-autocorrelation technique because of its relative ease of implementation.

As a dynamic holographic medium, PRQWs have the advantage of high sensitivity, requiring low-light intensities to generate large induced refractive index gratings. The low light intensities are a result of the relatively long response time of the materials, allowing low photon fluxes to be integrated over time. PRQWs can be used to find the electric-field correlation function of femtosecond pulses. This is accomplished both with direct interference [129], and interference in a Fourier-transform joint correlator geometry (time-to-space mapping) [130]. In both cases, PRQWs are used to characterize the shape of femtosecond pulses by obtaining the square of the electric-field correlation envelope.

Here we describe an autocorrelation measurement of femtosecond pulses by using direct interference of femtosecond pulses in a AlGaAs/GaAs PRQW sample. The setup is sim-

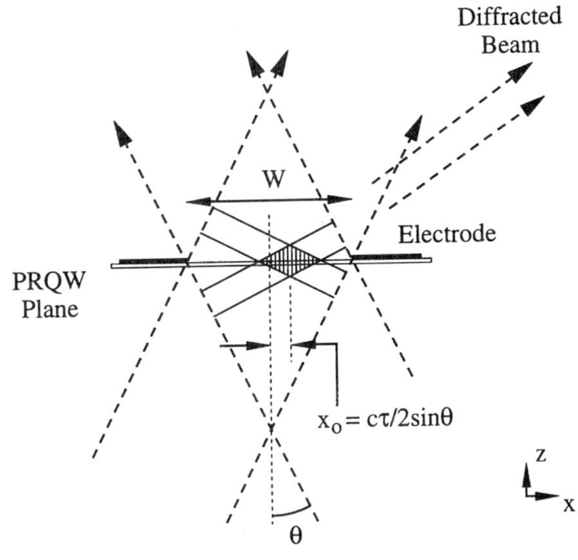

Fig. 53. Intersection volume and interaction area for two colliding pulses intersected by a thin film. The center position of the interaction area within the window defined by the electrodes depends on the relative time delay τ and on the crossing angle θ. (Source: Reprinted with permission from [129]. © 1994 Optical Society of America.)

ilar to degenerate four-wave mixing, where two laser beams intersect and the first-order diffraction is measured as a function of the delay of the two incident pulses. The intersection volume depends on the crossing angle and on the pulse duration t_p. A representative intersection volume is shown in Figure 53 for symmetric beams each incident by an angle θ. The delay of between the two beams is expressed as τ. The intersection volume intersects the plane of the PRQWs in a well-defined area within the aperture defined by the gold electrodes. The displacement of the center of the interaction area relative to the center of the aperture is given by

$$x_0(\tau) = \frac{c\tau}{2\sin\theta} \tag{6.4.9}$$

Within the coherence volumes of the pulses, the average light intensity is modulated by the interference between the two beams. The average light intensity is given by

$$I(x) = \overline{I}\left[1 + m(x,\tau)\cos(Kx + \phi)\right] \tag{6.4.10}$$

where x is the distance across the aperture, ϕ is an arbitrary phase, $\overline{I}$ is the average laser intensity, and $K = 2\pi/\Lambda$ is the grating wavevector of the fringe pattern with a fringe spacing Λ. All quantities are time averages. The modulation function $m(x,\tau)$ provides all the information concerning the coherence properties.

The modulated light intensity generates screening photocarriers that are trapped as space charge at deep defects. The PRQWs rely on the quadratic electroabsorption of the excitons, and the refractive index and absorption grating modulation are given by [59, 110]

$$\Delta n = -n^3 s_1 E_0^2 m(x,\tau) \quad \text{and} \quad \Delta\alpha = -\frac{4\pi}{\lambda} n^3 s_2 E_0^2 m(x,\tau) \tag{6.4.11}$$

with the applied electric-field E_0 and the effective quadratic electro-optic coefficients s_1 and s_2, where we have assumed full screening of the applied field.

The laser electric fields of the intersecting beams are

$$\vec{E}_{1,2}(\vec{x},t) = \hat{e}_{1,2}E_{1,2}\left(t - \tau_{1,2} - \hat{k}_{1,2}\cdot\frac{\vec{x}}{c}\right)\exp\left[i\left(\omega(t-\tau_{1,2}) - k_{1,2}\hat{k}_{1,2}\cdot\vec{x}\right)\right] \tag{6.4.12}$$

where $\hat{e}_{1,2}$, $E_{1,2}(t)$, $\tau_{1,2}$, and $\hat{k}_{1,2}$ are the polarization vector, the envelope function, the pulse delay, and the unit propagation vector for each beam, respectively, with $k_1 = k_2 = k = \omega/c$ for the center wavelength of the pulse. The x direction is across the aperture and the z direction is normal to the sample, giving $\hat{k}_{1,2} = \pm\sin\theta\hat{x} + \cos\theta\hat{z}$, where θ is the angle from normal. The beams are both polarized perpendicular to the plane formed by the beams, with $\hat{e}_{1,2} = \hat{y}$. Taking $\tau_1 = 0$ and $\tau_2 = \tau$, with $z = 0$ at the plane of the sample,

$$
\begin{aligned}
I(x,\tau) &= \frac{1}{T}\int_{-T/2}^{T/2} \frac{1}{2Z}\left|\vec{E}_1(x,t) + \vec{E}_2(x,t)\right|^2 dt \\
&= \bar{I}\left[1 + m(x,\tau)\cos\left(Kx - \omega\tau + \phi(x,\tau)\right)\right] \qquad (6.4.13)
\end{aligned}
$$

where ϕ is a phase that depends only weakly on x and τ. For $E(t) \equiv E_1(t) = E_2(t)$, the expression for the modulation index is

$$
\begin{aligned}
m(x,\tau) &\propto \left|\int_{-T/2}^{T/2} E\left(t - \frac{x\sin\theta}{c}\right)E\left(t - \tau + \frac{x\sin\theta}{c}\right)dt\right|^2 \\
&\propto \int_{-\infty}^{\infty} E\left(t + \tau - \frac{2x\sin\theta}{c}\right)E^*(t)\,dt = \left|\Gamma\left(\tau - \frac{2x\sin\theta}{c}\right)\right| \qquad (6.4.14)
\end{aligned}
$$

where $\Gamma(\tau)$ is the autocorrelation of the electric-field envelope function, also called the coherence function. This can be calculated directly using the Wiener–Khinchin theorem by taking the Fourier transform of the power spectrum.

For the case of a Gaussian pulse, the modulation index can easily be calculated. For equal beam intensities, it is

$$
m(x,\tau) = \exp\left[-\frac{8\ln 2}{c^2 t_p^2}\sin^2\theta\left(x - x_0(\tau)\right)^2\right] \qquad (6.4.15)
$$

which is a Gaussian function with a FWHM of

$$
\Delta x = \frac{ct_p}{\sqrt{2}\sin\theta} \qquad (6.4.16)
$$

Small interaction area widths result from large crossing angles. The angular control of the interaction area width provides control of the autocorrelator time resolution.

Because the diffraction power is proportional to the square of $\Delta\alpha$ and Δn, which depend linearly on $m(x,\tau)$, the diffracted signal is proportional to the square of the modulation from Eq. (6.4.15). The total diffracted intensity is proportional to

$$
\eta(\tau) \propto \frac{1}{W}\int_{-W/2}^{W/2} m^2(x,\tau)\,dx \qquad (6.4.17)
$$

where the integral is taken over the aperture of width W defined by the electrodes, and the transverse beam intensity profile is neglected. For a transform-limited Gaussian pulse, the integral in Eq. (6.4.17) yields

$$
\eta(\tau) \propto \operatorname{erf}\left(\frac{W/2 + x_0(\tau)}{\Delta x\sqrt{8\ln 2}}\right) + \operatorname{erf}\left(\frac{W/2 - x_0(\tau)}{\Delta x\sqrt{8\ln 2}}\right) \qquad (6.4.18)
$$

which can be evaluated numerically. The width $\Delta\tau$ (FWHM) of $\eta(\tau)$ cannot be given in closed form, but $\Delta\tau = \sqrt{2}t_p$ in the limit when W approaches zero.

The sensitivity of four-wave mixing for measuring the field autocorrelation depends on the ratio of the aperture width W to the interaction width Δx. When W is large compared to Δx, the diffraction does not change appreciably as x_0 moves over the interaction region. When W is small compared to Δx, then changes in x_0 cause changes in diffraction, resulting in good time resolution as τ is varied.

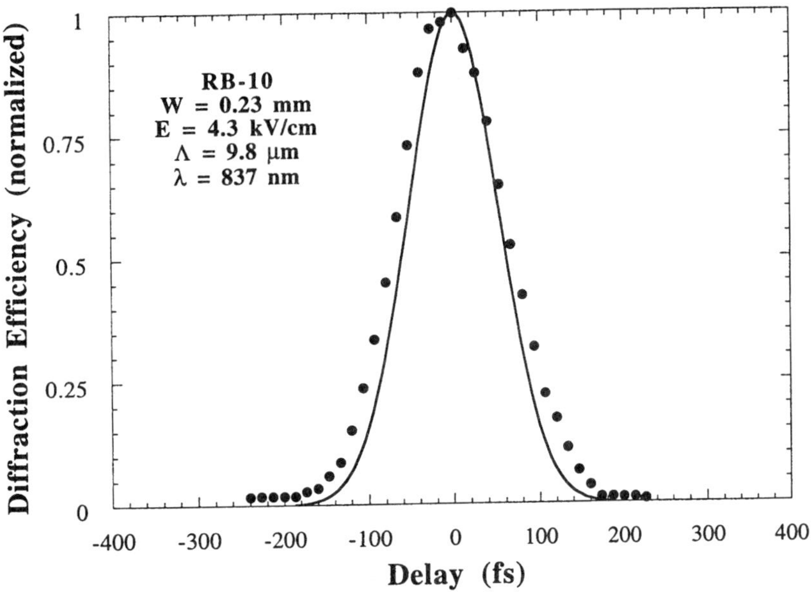

Fig. 54. Diffraction intensity versus time delay τ between the two incident pulses for a sample with an aperture width $W = 0.23$ mm. The data is the average of three runs (dots) and the theoretical curve is calculated from the Fourier transform of the power spectrum for pulse bandwidths $\Delta\lambda$ of 12.6 nm. (Source: Reprinted with permission from [129]. © 1994 Optical Society of America.)

Figure 54 shows the dependence of the diffracted intensity on relative pulse delay for sample RB-10 with an electrode spacing of $W = 0.23$ mm and an applied field of 4.3 kV/cm AC. The average intensity incident on the sample is 6 mW/cm^2, which is far above the saturation intensity [59]. The half-angle between the beams is $\theta = 2.44°$, and the laser wavelength is $\lambda = 837$ nm, resulting in a fringe spacing of $\Lambda = 9.8$ μm. The theoretical curve was calculated from the power spectrum data.

The difference between the data and the calculations in Figure 54 may be due to field inhomogeneity in the sample. The voltage applied to the sample through the gold electrodes was assumed to result in a uniform field across the aperture. A simple electrostatic calculation shows that the field near the electrodes should be larger than in the middle of the aperture [131]. Also, the gold electrodes may make nonohmic contact with the semi-insulating sample. Nonohmic contacts could lead to a depletion region near each contact, which would again lead to a stronger field near the electrodes.

A significant advantage of the photorefractive quantum wells that should be mentioned is the sensitivity of the autocorrelation schemes. The photorefractive quantum-well structures represent one of the highest sensitivity nonlinear optical materials. Under continuous wave (cw) operation, the hologram saturation intensity is only 10 μW/cm^2. Autocorrelation measurements on such weak pulses using conventional nonlinear crystals is extremely difficult without focusing.

Transmultiplexing in Optical Communications

Ultrafast pulses contain a significant bandwidth. For example, a typical 100 fs pulse has a bandwidth of 4 THz, corresponding to 10 nm at a center wavelength of 850 nm. The successful use of this bandwidth is desirable for high-rate communications and information access. Ultrashort pulse time-division multiplexing (TDM), wavelength-division multiplexing (WDM), and code-division multiple access (CDMA) are the most completely studied methods to explore the use of the large bandwidth of optics for optical communications.

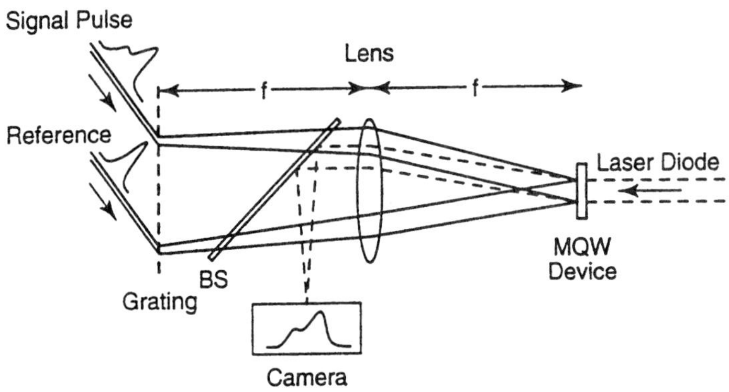

(a)

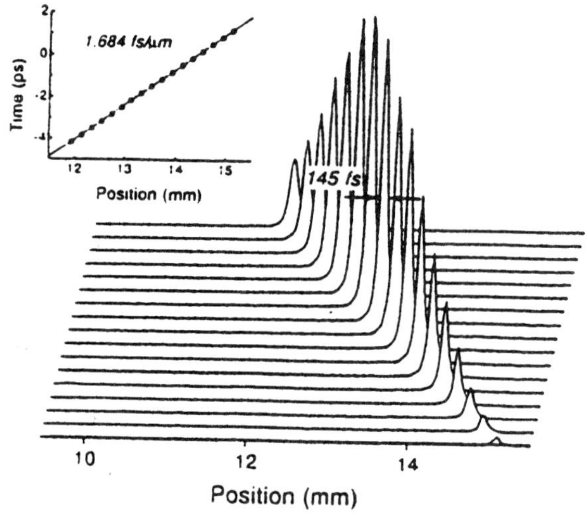

(b)

Fig. 55. Experimental setup for time-to-space conversion. Although a transmission grating is shown here for clarity, the experiment uses a 600 lines/mm reflection grating. The center wavelength of the femtosecond laser is 830 nm. The focal length of the Fourier lens for the writing beams is 200 mm and that of the one for the read-out beam (not shown) is 190 nm. (b) Autocorrelation traces in the space domain as recorded by an silicon array detector. The inset shows the position of the autocorrelation peaks as a function of the time delay between the two input pulses. (Source: Reprinted with permission from [130]. © 1994 American Institute of Physics.)

However, the need to interface to electronics for processing can impose a severe information bottleneck. All-optical interconnection between different multiplexing methods and between parallel electronics and serial optics, and ultrafast optical switches using time-slot interchangers can help lead to a high-performance information highway. Using holographic methods, transmultiplexing between time, space, and wavelength can be realized [118, 130, 132–136]. As an example, we discuss here time-to-space and space-to-time conversions using PRQWs as holographic media.

Time-to-Space Mapping. Figure 55a shows the principle for converting a time-domain signal to the space-domain, which is also called time-to-slot conversion or serial-to-parallel conversion [130]. As the signal pulse carrying information hits a dispersive element such as a grating, its spectrum is spatially spread. This spectrum interferes with the spectrum of a featureless reference pulse, which also diffracts from the grating, and forms a hologram in the PRQWs placed at the Fourier plane of the lens. The modulation depth of the hologram is given by

$$E_E^*(\omega)E_R(\omega) + E_S(\omega)E_R^*(\omega) \tag{6.4.19}$$

At the Fourier plane, the frequency information is converted to space information, i.e., each ω corresponds to a spatial coordinate x. In this stage, one cannot tell whether the hologram is written with the use of temporal or spatial information. When the hologram is read out by a plane wave, for example, from a cw diode laser working at the wavelength of 850 nm near the excitonic absorption of the PRQWs, either one of the two terms in Eq. (6.4.19) describing the correlations of the signal and the reference pulses is reconstructed in the space domain. The temporal signal is converted to a space-domain signal. The conversion calibration is linear [130] with a calibration factor of

$$\frac{\Delta t}{\Delta x} = \frac{\beta}{c}\frac{\lambda_w f_w}{\lambda_r f_r} \tag{6.4.20}$$

with the dispersion parameter of the grating $\beta = \lambda_w/d\cos\theta$, the speed of light c, the wavelength and focal length of the writing, and read-out beam, $\lambda_{w,r}$, and $f_{w,r}$, respectively, and θ is the incident angle of the beam to the grating.

Figure 55b shows the experimental results, which are the correlation peaks of the signal and reference pulses for different time delays between them. In this experiment, the signal and reference pulses are identical with a pulse width of 100 fs. The time delay between the successive traces is 333 fs. Other parameters of the experiment are given in the figure caption. As the delay changes, the position of the autocorrelation trace also changes. The dependence of this displacement on the time delay of the signal pulse is shown in the inset of Figure 55b, which gives a linear function with a slope of 1.684 fs/μm. This value is very close to the one obtained from Eq. (6.4.20) using the experimental parameter given in the figure caption. The FWHM of the correlation traces is 85 μm, corresponding to 145 fs. Assuming the pulses have a sech^2 shape, this is very close to the theoretical electric-field autocorrelation width of a 100 fs pulse. The envelope of the trace maxima in Figure 55b also shows that there is a limited range over which the input delay can be changed for obtaining an autocorrelation trace. This is due to the finite size of the beams at the grating. All temporal information projected onto the grating has to fit within the beam size without truncation.

Space-to-Time Conversion. Space-to-time conversion uses holographic diffraction from PRQWs [118], where a conventional image in the space domain is converted dynamically into the time domain. This is based on the Fourier synthesis of femtosecond pulses, which is shown in Figure 56. The setup contains a femtosecond pulse shaper [137] on the horizontal plane and a hologram-writing geometry in the vertical plane. As an ultrafast optical pulse with a wide spectrum is dispersed by the first grating, its spectral components are spatially spread. If these components are filtered and modified by an amplitude or phase mask, like the hologram written in PRQWs, and then recombined, the resultant temporal output, which is the Fourier transform of the modified spectrum, contains the information of the mask. If the amplitude–phase mask is itself the Fourier transform of a certain space-domain image, for example, a square aperture, the resultant temporal output envelope takes the same form as the space image.

The hologram is written in the PRQWs by two cw beams from either a diode laser operating at 685 nm or a He-Ne laser at 633 nm, whose photon energy is larger than the bandgap of the PRQWs. The power of each writing beam is 2–7 mW. In contrast to the conventional femtosecond pulse shaper, cylindrical lenses instead of spherical ones are used in the pulse shaper. In this case, the incident femtosecond beam at the Fourier plane remains unfocused in the vertical direction, so that its intensity is not so high as to erase the hologram, leading to an optimal photorefractive diffraction. The Fourier transforms

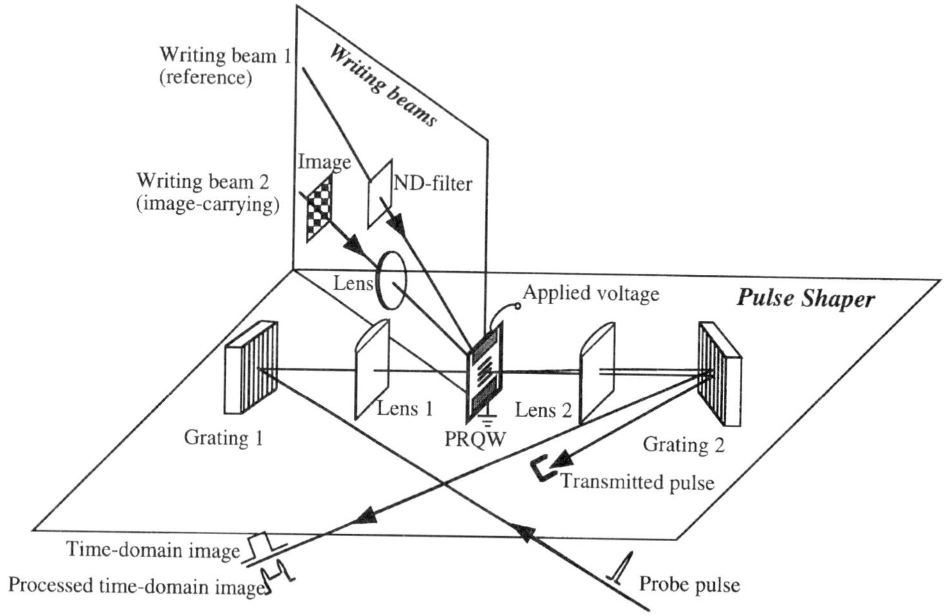

Fig. 56. Experimental setup for femtosecond space-to-time conversion and time-domain image processing using dynamic holograms written in photorefractive quantum wells. The focusing length of the cylindrical lenses in the pulse shaper is 10 cm and the two gratings have a groove density of 1800 lines/mm, leading to a spatial spectrum separation of 0.35 mm/nm. (Source: Reprinted with permission from [118]. © 1997 American Institute of Physics.)

of the fs probe beam and the diffracted beam by the cylindrical lenses are therefore one-dimensional transforms in the horizontal direction.

The femtosecond source is a mode-locked Ti-sapphire laser with a nominal pulse width of 100 fs and a central wavelength adjusted to the excitonic absorption line of the quantum wells. The fs probe beam with a power of 2.6 mW and a diameter of 1.8 mm is incident on the first grating of the pulse shaper, where it is spectrally dispersed. The frequency components are diffracted from the hologram in the PRQWs at the Fourier plane. The diffracted beam is traveling in the vertical plane determined by the two writing beams. The diffracted optical components are reassembled by the second cylindrical lens and the grating, forming the fs output carrying the information of the hologram. The spectrum of the femtosecond output is captured through a spectrometer with a CCD array, and the temporal shape of the pulse is measured by electric-field cross-correlation with an unshaped transform-limited reference pulse. In the hologram writing geometry, a lens is put into one of the writing beams, called the image-carrying beam, to project the Fourier transform of a mask or an image onto the PRQWs, whereas the reference beam is unaltered. The image Fourier plane is 22° off the Fourier plane of the pulse shaper, for technical simplicity. This has no detrimental effects on our results because of the long focal length of the lenses in the writing beams (100–200 mm). Neutral density filters are used to adjust the intensities of the two writing beams.

The pulse shaper is carefully aligned to be dispersion free with no PRQW in it, giving identical output pulses to the incident ones. Then a GaAs/AlGaAs PRQW sample is placed in the Fourier plane. The samples used for this experiment have 30% aluminum in the barriers, with the heavy-hole excitonic absorption centered at 836 nm. Titanium-gold contacts were evaporated, which have a 1 mm aperture, and across which a field of $\sim$7 kV/cm is applied in the plane of the quantum wells, perpendicular to the growth direction (transverse field geometry [5, 59]). The angle between the two writing beams is 3.3°.

The interference pattern of the two coherent writing beams has a modulation depth of

$$m(x) = \frac{2|E_1^* E_2(x)|}{|E_1|^2 + |E_2(x)|^2 + \delta|E_{\text{Probe}}|^2} = \frac{2\sqrt{I_1 I_2(x)}}{I_1 + I_2(x) + \delta I_{\text{Probe}}} \tag{6.4.21}$$

with field amplitude E_1 (intensity I_1) of the reference writing beam (spatially homogenous), those of the image-carrying beam E_2, $I_2(x)$, and of the fs probe beam E_{Probe}, I_{Probe}, respectively. δ is a factor describing the ratio of the photocarriers induced at the probe and writing wavelengths. Because the fs probe beam has a much longer wavelength (836 nm) than that of the writing beams (633 or 685 nm), it was shown that the erasure of the grating by the probe beam is much weaker than the effect of the writing beams [110] (i.e., $\delta \ll 1$) due to the difference in the photoconductivity and interaction length at the two wavelengths. From Eq. (6.4.21), m depends only on the intensity ratio of the two beams $I_2(x)/I_1$ and has the maximum value of unity when $I_2 = I_1$ (neglecting erasure). The hologram written in the PRQW is proportional to m as long as m is small. As m approaches unity, the amplitude of the photorefractive grating, and therefore the diffracted electric-field amplitude as well, become sublinear with m [138]. In holographic recording, the hologram pattern is a faithful replica of the original electric-field distribution only if the intensity of the image-carrying beam is much less than the intensity of the reference beam, and the material response is linear in intensity. In this case ($I_1 \gg I_2(x)$), $m(x) \sim E_2(x)$, and the hologram has the same distribution as the image-carrying beam $E_2(x)$.

As an example, a single slit with a width of 380 μm is used as a space-domain image and is put into the image-carrying beam. This image is Fourier transformed onto the PRQWs by a cylindrical lens with a focal length of $f = 150$ mm, performing the Fourier transform one-dimensionally. The intensity distribution at the Fourier plane takes the form of a squared sinc function when $x \gg f$, with the functional form

$$F = \frac{\sin^2\left(\frac{2\pi Dx}{\lambda f}\right)}{\left(\frac{2\pi Dx}{\lambda f}\right)^2} \tag{6.4.22}$$

for a slit width $2D$, writing wavelength λ, and the coordinate in the Fourier plane of the Fourier lens x.

The temporal shape of the diffracted (or reconstructed) femtosecond outputs, which are the Fourier transform of the electric-field spectra, corresponding to the power spectrum (Fig. 57a), is measured by electric-field cross-correlation. The results are shown in Figure 57b, which corresponds to the input space-domain image: a slit or a square-wave packet. The Fourier transform of a slit consists of an infinite number of Fourier components with different spatial frequencies. The lowest spatial component corresponds to the brightest spot in the Fourier plane. Several factors in the diffraction process prevent the second stage Fourier transform (Fourier transform from the frequency domain to the time domain) from transforming all the frequency components to the time domain. The finite spectral bandwidth of the material response, which is determined by the electroabsorption of the excitons in the PRQWs and has an approximately Gaussian shape, reduces the sharpness of the steps in the square-wave packet. The ratio of the spectrally dispersed femtosecond probe beam intensity to the writing intensity can influence the modulation in Eq. (6.4.21), resulting in small side lobes in the reconstructed image. Any asymmetry in the spectra due to the misalignment of the image Fourier lens will also introduce modulations in the time-domain image. Special care is taken in the experiments to minimize the influence of these factors to get an approximate square-wave packet. It is important to note that Figure 57b contains the electric-field cross-correlation data. The side lobes appear much smaller if the intensity rather than the field amplitude were plotted. Using this technique, any spatial information, such as a bit pattern, can be converted into the time domain, that is, a serial bit train with femtosecond accuracy.

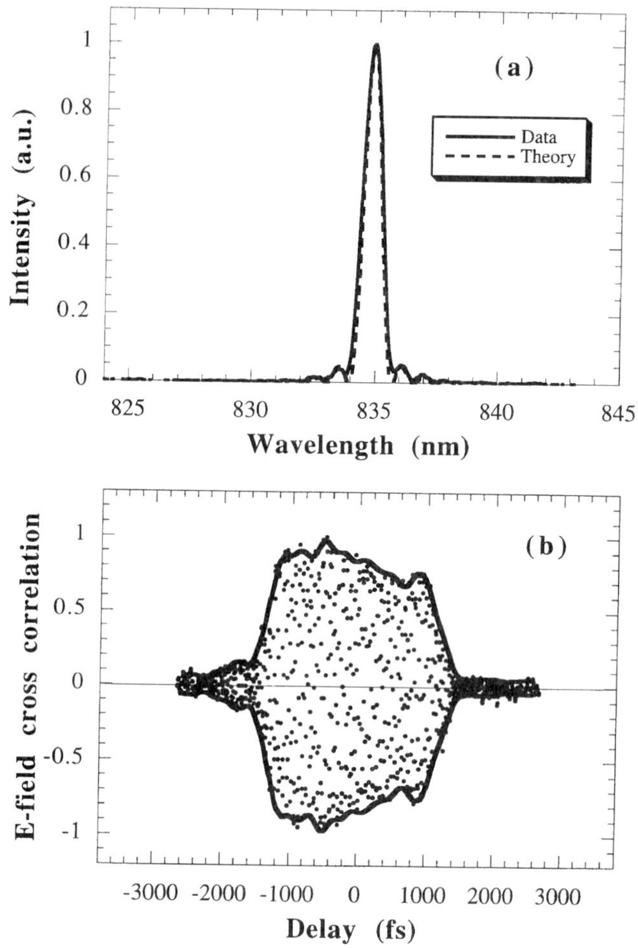

Fig. 57. Femtosecond output from the space-to-time conversion. (a) Spectrum, (b) Electric-field cross-correlation data and the extracted envelope of the diffracted temporal image. The input is the Fourier transform of a single. The reference beam intensity is higher than the image-carrying beam intensity. The corresponding theoretical calculation of the spectrum at $I_1/I_{2\,\mathrm{Max}} = 10$ is also given in (a). (Source: Reprinted with permission from [118]. © 1997 American Institute of Physics.)

Time-Domain Image Processing Using Dynamic Holography

Coded ultrafast optical pulses containing information can be treated as one-dimensional images in the time domain. The term "time-domain image" is most appropriate for cases when the shaped forms are obtained by converting space-domain images into the time domain. Processing time-domain images [118] (by analogy with processing conventional space-domain images) may open up new possibilities for applications in optical communication, signal processing, and data storage. Possible operations for time-domain image processing include (a) single-image processing, such as noise suppression, defect correction, edge enhancement, time reversal, and others, where there is only one image involved, and (b) multi-image processing, such as matched filtering, convolution or correlation, where two or more images are involved. By using spectral holography [139], some of these operations have been realized [140, 141]. Here we discuss image processing operations, such as edge enhancement, Fourier transform, and correlation, using dynamic holograms written in PRQWs. The images can be manipulated as they are converted from the space domain to the time domain by changing the hologram-writing intensities, writing geometries, and the relation between the input images, and so forth.

Edge Enhancement of a Time-Domain Image. Many applications require images with enhanced features, such as edges and noise, or require transformations of a image, such as a Fourier transform or time reversal. Using the experimental setup in Figure 56 edge enhancement can be performed on the converted time-domain image by choosing an appropriate hologram-writing intensity ratio. From Eq. (6.4.21) the hologram in the PRQW depends only on the intensity ratio of the two beams $I_2(x)/I_1$. For the hologram pattern to be a faithful replica of the original electric-field distribution, the reference beam intensity must be higher than the most intense spot of the Fourier transform of the image. When this condition is not satisfied, the hologram amplitude saturates and even decreases at positions where the reference beam intensity is weaker than the image-carrying beam. Such saturation and decrease can serve to enhance the high optical frequency components at the expense of the strong peak at the fundamental frequency. This produces a high-pass filter which can lead to edge enhancement in the reconstructed images. Such effects were demonstrated in the space domain [142, 143] and used to detect dust and other small defects in periodic photolithographic masks [144].

Experimentally, the same single slit used in the time-to-space conversion experiment is Fourier projected onto the PRQW sample, but with the intensity of the reference beam reduced by 2 orders of magnitude relative to the image-carrying beam. As a result, the interference of the two writing beams (the hologram) washes out the primary maximum in the image-carrying beam and extracts the higher spatial frequency components of the slit diffraction, which describe edges or other suddenly changing features. The diffracted spectrum from such a hologram is shown in Figure 58a, together with the theoretical calculation. By comparison with the diffraction spectrum in the space-to-time conversion in Figure 57a, the frequency side lobes are enhanced. The electric-field cross-correlation of the diffracted femtosecond output is given in Figure 58b, showing enhanced edges when compared with the time-domain image in Figure 57b. The side wings in Figure 58b are also expected from the theoretical Fourier transform of Eq. (6.4.21). Physically, the high-pass filtering operation in this experiment transforms the rising and falling edges of the input image into the observed derivativelike features.

Fourier Transform of a Time-Domain Image. It is known that the output time-domain image is the Fourier transform of the diffracted spectrum from the hologram. If the hologram corresponds to the space-domain image itself, then the resulting time-domain image is related to the Fourier transform of the space image. Experimentally, this is realized by directly projecting a mask (amplitude or phase) onto the PRQW sample in the pulse shaper using a 2f-2f geometry. The sample used in this experiment [125] is a PRQW with 10%-aluminum in the barriers with the heavy-hole exciton absorption centered at 847 nm. The mask used in the image-carrying beam is a phase mask containing a photolithographically etched phase step of π for a wavelength of about 620 nm. For the writing beams from the diode laser at around 685–690 nm for this experiment, the phase step σ has a value of about 0.9π. This mask is imaged onto the PRQW sample using a spherical lens with a focal length of 200 mm. The phase step of the mask changes the phase of one half of the writing beam by the amount of σ with respect to the other half, giving rise to a σ phase step in the hologram. The diffracted beam will then experience a phase function,

$$H(\omega) = \begin{cases} 1 & -\infty < \omega \leqslant \omega_c \\ e^{i\sigma\pi} & \omega_c \leqslant \omega < \infty \end{cases} \tag{6.4.23}$$

with the center frequency ω_c. The output electric field of Eq. (6.4.4) becomes

$$E_D(\omega) = E_{IN}(\omega) F_T(\omega) F_D(\omega) H(\omega) \tag{6.4.24}$$

The output pulse in the time domain is the Fourier transform of $E_D(\omega)$, which results in a double pulse. The experimental data of the electric-field cross-correlation and the envelope extracted from the data are shown in Figure 59.

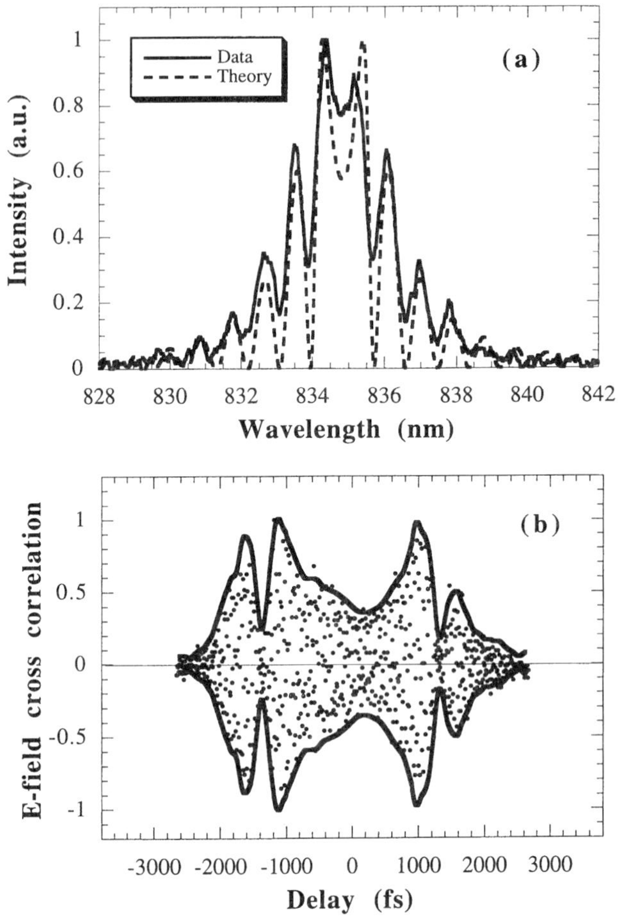

Fig. 58. Time-domain image (square-wave packet) edge enhancement. (a) Spectrum, (b) Electric-field cross-correlation data and the extracted envelope of the diffracted temporal image. The Fourier transform of a single slit is projected onto the PRQW and the reference beam intensity is lower than the maximum image-carrying beam intensity. The corresponding theoretical calculations of the spectrum at $I_1/I_{2\,\mathrm{Max}} = 0.01$ is also given in (a). (Source: Reprinted with permission from [118]. © 1997 American Institute of Physics.)

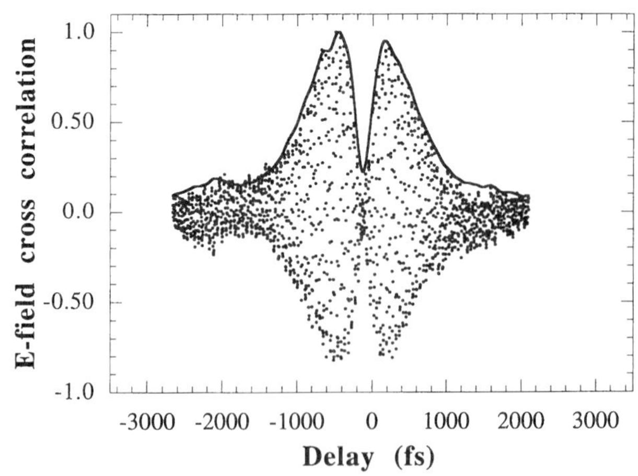

Fig. 59. Femtosecond output (a double pulse) showing the Fourier transform of a phase-step mask. (Source: Reprinted with permission from [125]. © 1997 American Institute of Physics.)

Correlation of Two Images. In optical processing, one of the most ubiquitous operations is the correlation, which is essentially a method for assessing the degree of similarity between two images. For two one-dimensional images $f(x)$ and $g(x)$, the correlation is defined as

$$f(x) * g(x) = \int_{-\infty}^{\infty} f(x')g(x + x')\,dx' \tag{6.4.25}$$

The relationship between Eq. (6.4.25) and the Fourier transform of the input images, known as the correlation theorem, is useful in image processing and makes it possible to correlate in the time domain by simple multiplication in the frequency domain. If $f(x)$ has the Fourier transform $F(\omega)$ and $g(x)$ has the Fourier transform $G(\omega)$, then the Fourier transform of the product $F(\omega)G^*(\omega)$ is the correlation of $f(x)$ and $g(x)$,

$$F\{F(\omega)G^*(\omega)\} \propto f(x) * g(x) \tag{6.4.26}$$

where $F\{\ \}$ represents the Fourier transform. Experimentally, the product of the Fourier transform of the first image and the phase conjugate of the Fourier transforms of the second image is obtained by projecting the Fourier transform of a space-domain image in each of the two writing beams onto the PRQW sample using Fourier lenses. The interference term of the intensity distribution on the PRQW ($I(x) = |E_1(x) + E_2(x)|^2$) is proportional to the product of the two fields $E_1(x)E_2^*(x)$, which gives the distribution or envelope of the hologram modulation, as long as the intensity response of the PRQWs remains linear, which was proven to be true in the experiments. The field diffraction from the hologram then is related to the product of the Fourier transform of the first image and the phase conjugate of the Fourier transform of the second image. The output time-domain image (the Fourier transform of this product) is then the correlation of the two input images.

The experiment of correlation is performed using an identical single slit as an input image in each of the two writing beams in the setup (Fig. 56), whose Fourier transform is a sinc function as given in Eq. (6.4.22). The diffracted power spectrum is then proportional to the fourth power of the sinc function, which is experimentally observed (Fig. 60a). The temporal output is the Fourier transform of this spectrum, resulting in a triangle function, corresponding to the correlation of two square-wave packets. Figure 60b shows the electric-field cross-correlation of such a processed time-domain image with a triangular envelope.

Biomedical Imaging

Coherence gating [35, 36, 145, 146], utilizing the short coherence length of fs-pulse lasers, is a promising technique for depth resolved biomedical imaging. This holographic method is insensitive to incoherent diffuse background light because only coherent ballistic photons participate. Tissue imaging with ballistic photons takes advantage of low skin absorption, $\mu_{abs} \sim 0.003$–$0.07\ \mathrm{mm}^{-1}$, in the wavelength range of 600–1000 nm where several photorefractive materials, including GaAs/AlGaAs MQWs, are suitable for use. This operating range also overlaps that of Ar-pumped Ti:sapphire and, possibly more importantly for applications, tunable diode pumped Cr:LiSAF lasers. Diode pumped lasers, in general, do not need the high operating voltage and current or water cooling that gas lasers require and are therefore more amenable to use in hospitals or clinics. Scattering in this wavelength regime is still high, however, limiting the maximum viewing depth to ~35 mean free paths (mfp), or ~4 mm of skin tissue [146].

The experimental setup used by Hyde et al. [36] for coherence gated imaging is shown in Figure 61. A grating is formed in the MQW by interfering an ~100 fs-pulse reference beam with one that has been reflected from a test object and passed through a scattering media. Only the nonscattered, or ballistic, photons participate in the grating formation. The hologram is then recorded on a CCD camera and can be downloaded to a computer for image processing. Depth resolution can be achieved by spatially translating one beam

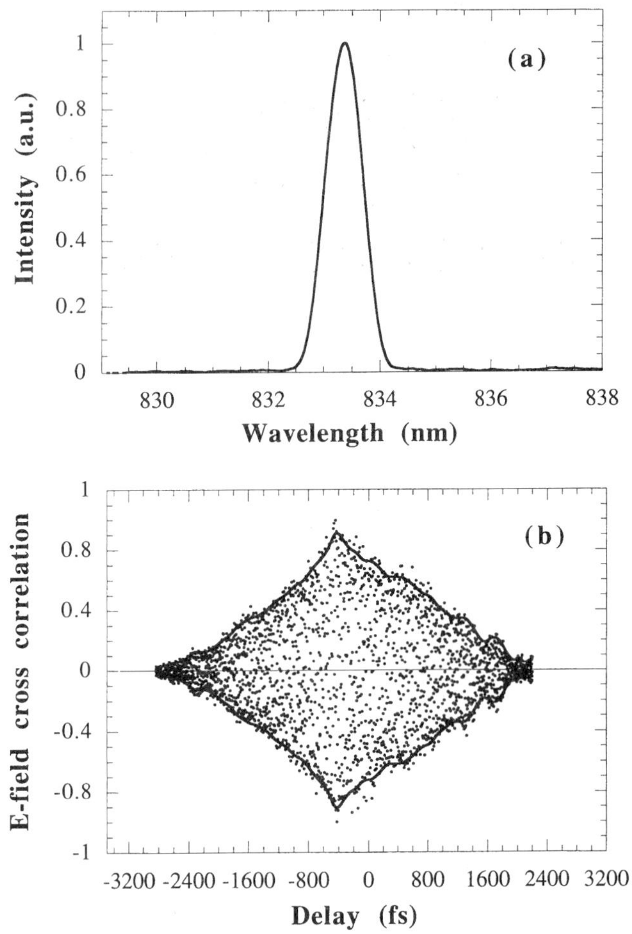

Fig. 60. Femtosecond output showing the correlation of two input images (square-wave packets). (a) Spectrum, (b) Electric-field cross-correlation data and the extracted envelope, showing a triangle function.

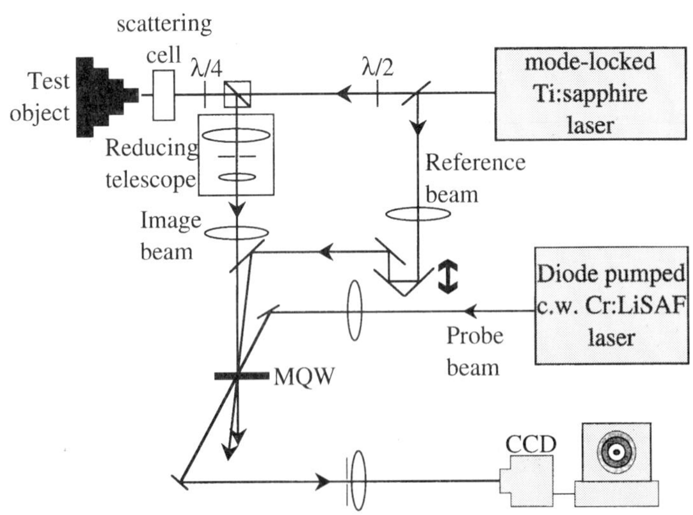

Fig. 61. Experimental configuration of the MQW holographic imaging system. (Source: Reprinted with permission from [36]. © 1997 American Institute of Physics.)

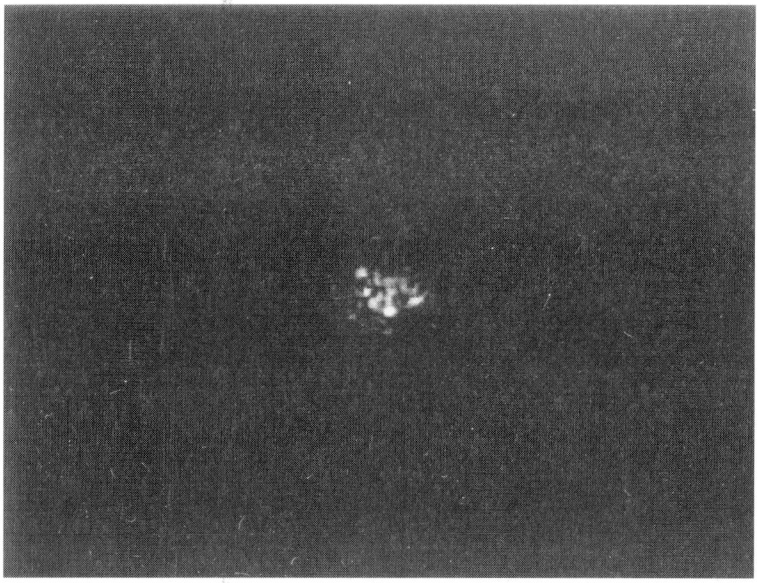

(a)

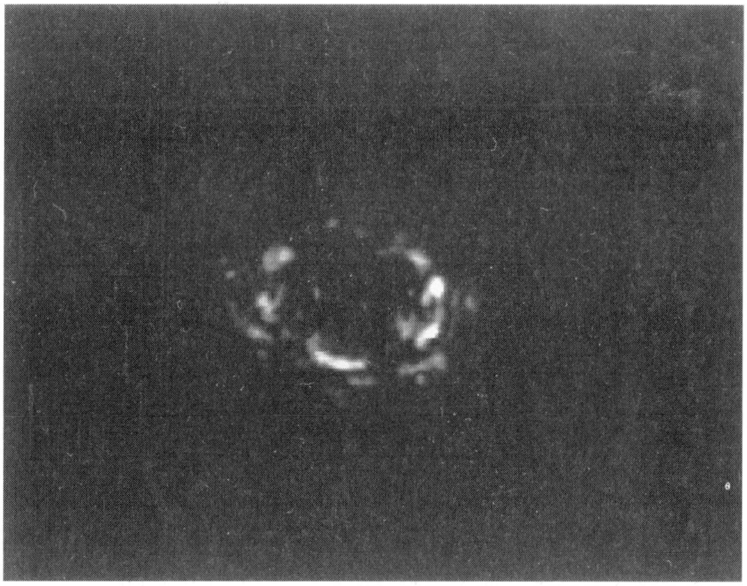

(b)

Fig. 62. Images of three-dimensional object through 8 mfp of scattering medium using MQW imaging system. (a)–(c) depth-resolved holographic images of three-dimensional object layers, (d) three-dimensional computer rendering from depth-resolved images. (Source: Reprinted with permission from [36]. © 1997 IEEE.)

path. In Figure 62 [36], a resolution of 50 μm is seen for a test object consisting of a set of decreasing radius cylinders with 100 μm high between-cylinder terraces.

The low-light intensity requirements of PRQWs are compatible with American National Standards Institute (ANSI) standards for laser intensities used on living tissue. In addition, the advantage of MQW operation in the Raman–Nath regime, described in Section 5.1, with fringe spacings of 20 μm or larger, eliminates degradation of depth resolution from

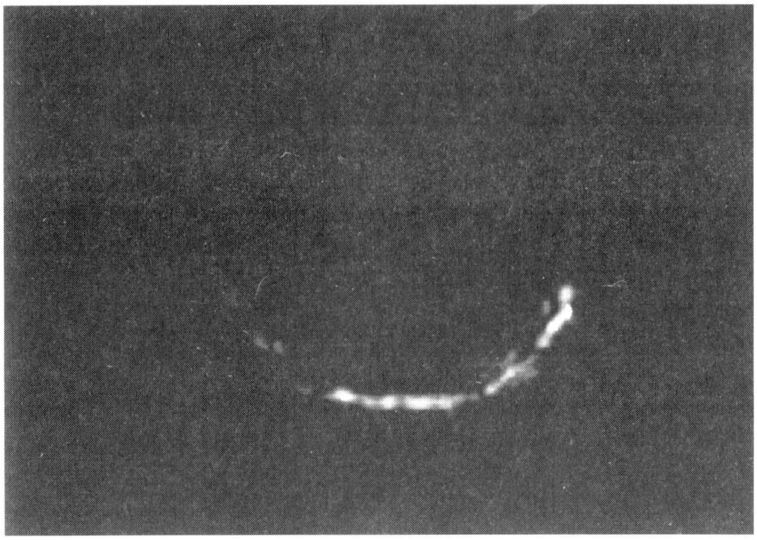

(c)

(d)

Fig. 62. (*Continued.*)

beam walkoff present in bulk crystals operating in the Bragg regime with small fringe spacings. Video-rate read out (<0.4 ms response time) of the image has been achieved by Hyde et al. [36] with PRQWs. With this coherence gating technique, "whole-field" imaging for *in vivo* applications allows the capture of an entire image at once, yielding greatly increased speed over line-by-line scanning methods. This holographic imaging technique may be useful for skin cancer detection or *in vivo* arterial imaging of hemoglobin oxygen levels or other blood compounds.

Acknowledgments

The authors gratefully acknowledge the support from the Department of Defense Focused Research Initiative through Air Force Office of Scientific Research Grant F49620-95-1-

0533 and F49620-96-1-0234A, the National Science Foundation Grant 9414800-ECS and the Rome Laboratory Grant F30602-96-2-0114.

References

1. Y. R. Shen, "The Principles of Nonlinear Optics," Wiley, New York, 1984.
2. D. D. Nolte, in "Photorefractive Effects and Materials, Electronic Materials: Science and Technology" (D. D. Nolte, ed.), pp. 1–66. Kluwer Academic, Dordrecht, 1995.
3. D. D. Nolte, D. H. Olson, G. E. Doran, W. H. Knox, and A. M. Glass, *J. Opt. Soc. Am. B*, 7, 2217 (1990).
4. D. D. Nolte and M. R. Melloch, *Mater. Res. Bull.* 19, 44 (1994).
5. D. D. Nolte and M. R. Melloch, in "Photorefractive Effects and Materials," (D. D. Nolte, ed.), pp. 373–452. Kluwer Academic, Dordrecht/Norwell, MA, 1995.
6. B. Schwartz, L. A. Koszi, P. J. Anthony, and R. L. Hartman, *J. Electrochem. Soc.* 131, 1703 (1984).
7. B. R. Pruniaux, J. C. North, and G. L. Miller, in "Ion Implantation in Semiconductors," Garmisch-Partenkirchen, 1971.
8. J. C. Dymet, J. C. North, and L. A. D'Asaro, *J. Appl. Phys.* 44, 207 (1973).
9. Y. Silverberg, P. W. Smith, D. A. B. Miller, B. Tell, A. C. Gossard, and W. Wiegmann, *Appl. Phys. Lett.* 46, 701 (1985).
10. M. B. Johnson, T. C. McGill, and N. G. Paulter, *Appl. Phys. Lett.* 54, 2424 (1989).
11. R. J. Matyi, M. R. Melloch, and J. M. Woodall, *J. Crystal Growth* 129, 719 (1993).
12. M. Kaminsak, E. R. Weber, Z. Liliental-Weber, R. Leon, and Z. U. Rek, *J. Vac. Sci. Technol. B* 7, 710 (1989).
13. D. E. Bliss, W. Walukiewicz, E. E. H. J. W. Ager, K. T. Chan, and S. Tanigawa, *J. Appl. Phys.* 71, 1699 (1992).
14. M. R. Melloch, N. Otsuka, J. M. Woodall, A. C. Warren, and J. L. Freeouf, *Appl. Phys. Lett.* 57, 1531 (1990).
15. M. R. Melloch, N. Otsuka, K. Mahalingam, A. C. Warren, J. M. Woodall, and P. D. Kirchner, in "Materials Research Society Symposium Proceedings," Vol. 241, 1992, pp. 113–1244.
16. M. R. Melloch, J. M. Woodall, N. Otsuka, K. Mahalingham, C. L. Chang, D. D. Nolte, and G. D. Pettit, *Mater. Sci. Eng. B* 22, 31 (1993).
17. K. Mahalingam, N. Otsuka, M. R. Melloch, J. M. Woodall, and A. C. Warren, *J. Vac. Sci. Technol. B* 9, 2328 (1991).
18. D. C. Look, G. D. Robinson, J. R. Sizelove, and C. E. Stutz, *J. Electron. Mater.* 22, 1425 (1993).
19. D. C. Look, D. C. Walters, M. O. Mansereh, J. R. Sizalove, C. E. Stutz, and K. R. Evans, *Phys. Rev. B* 42, 3578 (1990).
20. X. Liu, A. Prasad, A. Chen, W. M. Kurpiewski, A. Stoschek, Z. Liliental Weber, and E. R. Weber, *Appl. Phys. Lett.* 67, 1333 (1995).
21. A. C. Warren, J. M. Woodall, J. L. Freeouf, D. Grischowsky, D. T. McInturff, M. R. Melloch, and N. Otsuka, *Appl. Phys. Lett.* 57, 1331 (1990).
22. K. Mahalingam, N. Otsuka, M. R. Melloch, J. M. Woodall, and A. C. Warren, *J. Vac. Sci. Technol. B* 10, 812 (1992).
23. M. R. Melloch, J. M. Woodall, E. S. Harmon, N. Otsuka, F. Pollak, D. D. Nolte, R. M. Feenstra, and M. A. Lutz, in "Annual Review of Material Science" (B. W. Wessels, ed.), Vol. 25, Annual Reviews, Palo Alto, CA, 1995.
24. M. R. Melloch, D. D. Nolte, J. M. Woodall, J. C. P. Chang, D. B. Janes, and E. S. Harmon, *Crit. Rev. Solid State Mater. Sci.* 21, 189 (1996).
25. Y. Chen, S. Williamson, T. Brock, F. Smith, and A. R. Calawa, *Appl. Phys. Lett.* 59, 1984 (1991).
26. T. Motet, J. Nees, S. Williamson, and G. Mourou, *Mater. Res. Soc. Proc.* 241, 223 (1992).
27. A. C. Warren, N. Katzenellenbogen, D. Grischkowsky, J. M. Woodall, M. R. Melloch, and N. Otsuka, *Appl. Phys. Lett.* 58, 1512 (1991).
28. A. Rahman, D. Kralj, L. Carin, M. R. Melloch, and J. M. Woodall, *Appl. Phys. Lett.* 64, 2178 (1994).
29. R. H. M. Groeneveld, T. Rasing, L. M. F. Kaufmann, E. Smalbrugge, J. H. Wolter, M. R. Melloch, and H. v. Kempen, *Physica B Condensed Matter* 218, 294 (1996).
30. R. H. M. Groenveld, T. Rasing, L. M. F. Kaufmann, E. Smalbrugge, J. H. Wolter, M. R. Melloch, and H. v. Kempen, *J. Vac. Sci. Technol. B* 14, 861 (1996).
31. E. S. Harmon, M. R. Melloch, J. M. Woodall, D. D. Nolte, N. Otsuka, and C. L. Chang, *Appl. Phys. Lett.* 63, 2248 (1993).
32. A. J. Lochtefeld, M. R. Melloch, J. C. P. Chang, and E. S. Harmon, *Appl. Phys. Lett.* 69, 1465 (1996).
33. E. Yablonovitch, D. M. Hwang, T. J. Gmitter, L. T. Florez, and J. P. Harbison, *Appl. Phys. Lett.* 56, 2419 (1990).
34. E. Yablonovitch, T. Gmitter, J. P. Harbison, and R. Bhat, *Appl. Phys. Lett.* 51, 2222 (1987).
35. R. Jones, S. C. W. Hyde, M. J. Lynn, N. P. Barry, J. C. Dainty, P. M. W. French, K. M. Kwolek, D. D. Nolte, and M. R. Melloch, *Appl. Phys. Lett.* 69, 1837 (1996).

36. S. C. W. Hyde, R. Jones, N. P. Barry, J. C. Dainty, P. M. W. French, K. M. Kwolek, D. D. Nolte, and M. R. Melloch, *IEEE J. Select. Topics. Quantum Electron.* 2, 965 (1996).
37. H. Tanobe, T. Tamanuki, T. Uchida, F. Koyama, and K. Iga, *Jpn. J. Appl. Phys.* 31, 949 (1992).
38. N. T. Pelekanos, B. Deveaud, C. Guillemot, J. M. Gerard, P. Gravey, B. Lambert, A. L. Corre, and J. E. Vaillet, *Opt. Mater.* 4, 348 (1995).
39. A. Partovi, *Opt. Mater.* 4, 330 (1995).
40. A. Partovi, A. M. Glass, D. H. Olson, G. J. Zydzik, K. T. Short, R. D. Feldman, and R. F. Austin, *Appl. Phys. Lett.* 59, 1832 (1991).
41. W. S. Rabinovich, S. R. Bowman, D. S. Katzer, and C. S. Kyono, *Appl. Phys. Lett.* 66, 1044 (1995).
42. M. R. Melloch, N. Otsuka, K. Mahalingam, C. L. Chang, J. M. Woodall, G. D. Pettit, P. D. Kirchner, F. Cardone, A. C. Warren, and D. D. Nolte, *J. Appl. Phys.* 72, 3509 (1992).
43. I. Lahiri, D. D. Nolte, E. S. Harmon, M. R. Melloch, and J. M. Woodall, *Appl. Phys. Lett.* 66, 2519 (1995).
44. D. D. Nolte, M. R. Melloch, J. M. Woodall, and S. E. Ralph, *Appl. Phys. Lett.* 61, 3098 (1992).
45. C. S. Kyono, K. Ikossi-Anastasiou, W. S. Rabinovich, S. R. Bowman, and D. S. Katzer, *Appl. Phys. Lett.* 64, 2244 (1994).
46. P. Tayebati, K. Krishnaswami, D. D. Nolte, and M. R. Melloch, in "5th Top. Conf. Photoref. Mat. Eff. Dev.," Estes Park, CO, 1995.
47. I. Lahiri, M. Aguilar, D. D. Nolte, and M. R. Melloch, *Appl. Phys. Lett.* 68, 517 (1996).
48. F. W. Smith, A. R. Calawa, C.-L. Chen, M. J. Mantra, and L. J. Mahoney, *IEEE Electron Device Lett.* EDL-9, 77 (1988).
49. A. Partovi, A. M. Glass, D. H. Olson, G. J. Zydzik, H. M. O'Bryan, T. H. Chiu, and W. H. Knox, *Appl. Phys. Lett.* 62, 464 (1993).
50. I. Lahiri, K. M. Kwolek, D. D. Nolte, and M. R. Melloch, *Appl. Phys. Lett.* 67, 1408 (1995).
51. K. M. Kwolek, M. R. Melloch, and D. D. Nolte, *Appl. Phys. Lett.* 67, 736 (1995).
52. D. D. Nolte and K. M. Kwolek, *Opt. Commun.* 115, 606 (1995).
53. A. C. Warren, J. M. Woodall, P. D. Kirchner, X. Yin, F. Pollak, and M. R. Melloch, *Phys. Rev. B* 46, 4617 (1992).
54. D. D. Nolte, I. Lahiri, and M. R. Melloch, *Opt. Lett.* 21, 1888 (1996).
55. P. Yeh, "Introduction to Photorefractive Nonlinear Optics," Wiley, New York, 1993.
56. G. C. Valley and J. F. Lam, in "Photorefractive Materials and Their Applications" (P. Gunter and J.-P. Huignard, eds.), Vol. I, pp. 75–98. Springer-Verlag, Berlin/New York, 1988.
57. N. V. Kukhtarev, V. B. Markov, S. G. Odulov, M. S. Soskin, and V. L. Vinetskii, *Ferroelectrics* 22, 949 (1979).
58. R. M. Brubaker, Q. N. Wang, D. D. Nolte, and M. R. Melloch, *Phys. Rev. Lett.* 77, 4249 (1996).
59. Q. Wang, R. M. Brubaker, D. D. Nolte, and M. R. Melloch, *J. Opt. Soc. Am. B* 9, 1626 (1992).
60. J.-P. Huignard and P. Gunter, in "Photorefractive Materials and Their Applications II" (P. Gunter and J.-P. Huignard, eds.), p. 205, 1989.
61. N. W. Ashcroft and N. D. Mermin, "Solid State Physics," Saunders College, Philadelphia, PA, 1976.
62. M. D. Sturge, *Phys. Rev.* 127, 768 (1962).
63. D. S. Chemla, D. A. B. Miller, P. W. Smith, A. C. Gossard, and W. Wiegmann, *IEEE J. Quantum Electron.* QE-20, 265 (1984).
64. V. W. Franz, *Z. Naturforsch.* 13a, 484 (1958).
65. L. V. Keldysh, *Sov. Phys. JETP* 34, 788 (1958).
66. F. L. Lederman and J. D. Dow, *Phys. Rev. B* 13, 1633 (1976).
67. J. D. Dow and D. Redfield, *Phys. Rev. B* 1, 3358 (1970).
68. D. A. B. Miller, D. S. Chemla, T. C. Damen, A. C. Gossard, W. Wiegmann, T. H. Wood, and C. A. Burrus, *Phys. Rev. Lett.* 53, 2173 (1984).
69. D. A. B. Miller, D. S. Chemla, T. C. Damen, A. C. Gossard, W. Wiegmann, T. H. Wood, and C. A. Burrus, *Phys. Rev. B* 32, 1043 (1985).
70. H. A. Kramers, in "Atti del Congresso Internazionale dei Fisici, Como-Pavia-Roma," 1928.
71. R. d. L. Kronig, *J. Opt. Soc. Am.* 12, 547 (1926).
72. J. I. Pankove, "Optical Processes in Semiconductors," Dover, New York, 1971.
73. C. V. Raman and N. S. N. Nath, *Proc. Indian Acad. Sci.* 2, 406 (1935).
74. D. A. B. Miller, D. S. Chemla, T. C. Damen, A. C. Gossard, W. Wiegmann, T. H. Wood, and C. A. Burrus, *Appl. Phys. Lett.* 45, 13 (1984).
75. D. A. B. Miller, D. S. Chemla, T. C. Damen, T. H. Wood, C. A. Burrus Jr., A. C. Gossard, and W. Wiegmann, *IEEE J. Quantum Electron.* QE-21, 1462 (1985).
76. A. L. Lentine, D. A. B. Miller, J. E. Henry, J. E. Cunningham, and L. M. F. Chirovsky, *IEEE J. Quantum Electron.* 25, 1921 (1989).
77. S. L. Hou and D. S. Oliver, *Appl. Phys. Lett.* 18, 325 (1971).
78. J. Feinleib and D. S. Oliver, *Appl. Opt.* 11, 2752 (1972).
79. S. G. Lipson and P. Nisenson, *Appl. Opt.* 13, 2052 (1974).
80. R. A. Sprague, *J. Appl. Phys.* 46, 1673 (1975).
81. B. A. Horwitz and F. J. Corbett, *Opt. Eng.* 17, 353 (1978).

82. A. Partovi, A. M. Glass, D. H. Olson, G. J. Zydzik, K. T. Short, R. D. Feldman, and R. F. Austin, *Opt. Lett.* 17, 655 (1992).
83. D. D. Nolte, *Opt. Commun.* 92, 199 (1992).
84. S. L. Smith and L. Hesselink, *J. Opt. Soc. Am. B* 11, 1878 (1994).
85. D. D. Nolte, I. Lahiri, and M. Aguilar, *Opt. Commun.* 131, 119 (1996).
86. J. Feldmann, K. W. Goossen, D. A. B. Miller, A. M. Fox, and J. E. Cunningham, *Appl. Phys. Lett.* 59, 66 (1991).
87. K. W. Goossen, J. E. Cunningham, and W. Y. Jan, *Appl. Phys. Lett.* 59, 3622 (1991).
88. E. Canoglu, E. Garmire, I. Lahiri, D. D. Nolte, and M. R. Melloch, in "Ann. Meeting Opt. Soc. Am.," Rochester, NY, 1996.
89. K. M. Kwolek, M. R. Melloch, and D. D. Nolte, *Appl. Phys. Lett.* 65, 385 (1994).
90. J. P. Huignard and A. Marrakchi, *Opt. Commun.* 38, 249 (1981).
91. S. I. Stepanov and M. P. Petrov, *Opt. Commun.* 53, 292 (1985).
92. Q. N. Wang, D. D. Nolte, and M. R. Melloch, *Appl. Phys. Lett.* 59, 256 (1991).
93. S. I. Stepanov, V. V. Kulikov, and M. P. Petrov, *Opt. Commun.* 44, 19 (1982).
94. G. C. Valley, *J. Opt. Soc. Am. B* 1, 868 (1984).
95. B. Imbert, H. Rajbenbach, S. Mallick, J. P. Herriau, and J. P. Huignard, *Opt. Lett.* 13, 327 (1988).
96. A. Partovi, A. Kost, E. M. Garmire, G. C. Valley, and M. B. Klein, *Appl. Phys. Lett.* 56, 1089 (1990).
97. I. Lahiri, R. M. Brubaker, D. D. Nolte, and M. R. Melloch, *Appl. Phys. Lett.* 69, 3414 (1996).
98. A. Yariv, "Quantum Electronics," Wiley, New York, 1989.
99. I. Lahiri, D. D. Nolte, M. R. Melloch, and M. B. Klein, *Opt. Lett.* 23, 49 (1998).
100. C. C. Wang, F. Davidson, and S. Trivedi, *Appl. Opt.* 34, 6496 (1995).
101. S. I. Stepanov, I. A. Sokolov, G. S. Trofimov, V. I. Vlad, D. Popa, and I. Apostal, *Opt. Lett.* 15, 1239 (1990).
102. R. H. Yan, R. J. Simes, and L. A. Coldren, *IEEE Photon Technol. Lett.* 1, 273 (1989).
103. K.-K. Law, R. H. Yan, L. A. Coldren, and J. L. Merz, *Appl. Phys. Lett.* 57, 1345 (1990).
104. D. S. Gerber, R. Droopad, and G. N. Maracas, *IEEE Photon Technol. Lett.* 5, 55 (1993).
105. C. Fabry and A. Perot, *Ann. Chim. Phys.* 16, 115 (1899).
106. R.-H. Yan, R. J. Simes, and L. A. Coldren, *IEEE J. Quantum Electron.* 25, 2272 (1989).
107. T. H. Wood, *J. Lightwave Technol.* 6, 743 (1988).
108. O. S. Heavens, "Optical Properties of Thin Films," Dover, New York, 1991.
109. D. D. Nolte, *Opt. Lett.* 19, 819 (1994).
110. D. D. Nolte, Q. N. Wang, and M. R. Melloch, *Appl. Phys. Lett.* 58, 2067 (1991).
111. D. D. Nolte, *J. Appl. Phys.* 76, 3740 (1994).
112. K. M. Kwolek, M. R. Melloch, and D. D. Nolte, in "23rd International Conference on the Physics of Semiconductors," Berlin, 1996.
113. F. Capasso and A. Y. Cho, *Surf. Sci.* 299–300, 878 (1994).
114. C. Carter-Coman, A. S. Brown, R. Bicknell-Tassius, N. M. Joskert, and M. Allen, *Appl. Phys. Lett.* 69, 257 (1996).
115. A. R. Pratt, R. L. Williams, C. E. Norman, M. R. Fahy, A. Marinopoulou, and F. Chatenoud, *Appl. Phys. Lett.* 65, 1009 (1994).
116. T. Krauss, Y. P. Song, S. Thoms, C. D. W. Wilkinson, and R. M. DelaRue, *Electron. Lett.* 30, 1444 (1994).
117. Y. Chen, G. Faini, H. Launois, and J. Etrillard, *Superlattices Microstruct.* 22, 109 (1997).
118. Y. Ding, D. D. Nolte, M. R. Melloch, and A. M. Weiner, *Opt. Lett.* 22, 1101 (1997).
119. M. Dinu, M. R. Melloch, and D. D. Nolte, *J. Appl. Phys.* 79, 3787 (1996).
120. M. Dinu, D. D. Nolte, and M. R. Melloch, *Phys. Rev. B* 56, 1987 (1997).
121. C. D. Matos, A. L. Corre, H. L'Haridon, B. Lambert, S. Salaün, J. Pleumeekers, and S. Gosselin, *Appl. Phys. Lett.* 68, 517 (1996).
122. K. Naganuma, K. Mogi, and H. Yamada, *IEEE J. Quantum Electron.* 25, 1225 (1989).
123. K. G. Purchase, D. J. Brady, and K. Wagner, *Opt. Lett.* 18, 2129 (1993).
124. L. Lepetit, G. Chériaux, and M. Joffre, *J. Opt. Soc. Am. B* 12, 2467 (1995).
125. Y. Ding, R. M. Brubaker, D. D. Nolte, M. R. Melloch, and A. M. Weiner, *Opt. Lett.* 22, 718 (1997).
126. J.-C. Diels and W. Rudolph, "Ultrashort Laser Pulses Phenomena: Fundamentals, Techniques, and Applications on a Femtosecond Time Scale," Academic, San Diego, CA, 1996.
127. J.-C. Diels, E. W. V. Stryland, and D. God, in "Picosecond Phenomena," Springer-Verlag, Berlin, 1978.
128. H. J. Eichler, U. Kleinl, and D. Lanhans, *Appl. Phys.* 21, 215 (1980).
129. R. M. Brubaker, Q. N. Wang, D. D. Nolte, E. S. Harmon, and M. R. Melloch, *J. Opt. Soc. Am. B* 11, 1038 (1994).
130. M. C. Nuss, M. Li, T. H. Chiu, A. M. Weiner, and A. Partovi, *Opt. Lett.* 19, 664 (1994).
131. D. D. Nolte, N. P. Chen, M. R. Melloch, C. Montemagno, and N. M. Haegel, *Appl. Phys. Lett.* 68, 72 (1996).
132. K. Ema, M. Kuwata-Gonokami, and F. Shimizu, *Jpn. J. Appl. Phys.* 30, L2046 (1991).
133. Y. T. Mazurenko, S. E. Putilin, A. G. Spiro, A. G. Beliaev, V. E. Yashin, and S. A. Chizhov, *Opt. Lett.* 21, 1753 (1996).
134. P. C. Sun, Y. T. Mazurenko, W. S. C. Chang, P. K. L. Yu, and Y. Fainman, *Opt. Lett.* 20, 1728 (1995).
135. P. C. Sun, Y. T. Mazurenko, and Y. Fainman, *J. Opt. Soc. Am. A* 14, 1159 (1997).

136. M. C. Nuss and R. L. Morisson, *Opt. Lett.* 20, 740 (1995).
137. A. M. Weiner, J. P. Heritage, and E. M. Kirschner, *J. Opt. Soc. Am. B* 5, 1563 (1988).
138. P. Refregier, L. Solymar, H. Rajbenbach, and J. P. Huignard, *J. Appl. Phys.* 58, 45 (1985).
139. Y. T. Mazurenko, *Appl. Phys. B* 50, 101 (1990).
140. A. M. Weiner, D. E. Leaird, D. H. Reitze, and E. G. Paek, *IEEE J. Quantum Electron.* 28, 2251 (1992).
141. A. M. Weiner, D. E. Leaird, D. H. Reitze, and E. G. Paek, *Opt. Lett.* 17, 224 (1992).
142. J. P. Huignard and J. P. Herriau, *Appl. Opt.* 17, 2671 (1978).
143. J. Feinberg, *Opt. Lett.* 5, 330 (1980).
144. E. Ochoa, J. W. Goodman, and L. Hesselink, *Opt. Lett.* 10, 430 (1985).
145. H. Chen, Y. Chen, D. Dilworth, E. Leith, J. Lopez, and J. Valdmanis, *Opt. Lett.* 16, 487 (1991).
146. M. R. Hee, J. A. Izatt, J. M. Jacobson, J. G. Fujimoto, and W. Swanson, *Opt. Lett.* 18, 950 (1993).
147. P. Gunter and J.-P. Huignard, in "Photorefractive Materials and Their Applications" (P. Gunter and J.-P. Huignard, eds.), Vol. I, pp. 7–74. Springer-Verlag, Berlin/New York, 1988.
148. R. M. Brubaker, Y. Ding, D. D. Nolte, A. M. Weiner, and M. R. Melloch, *IEEE J. Quantum Electron.* 33, 2150 (1997).

Chapter 13

LINEAR AND NONLINEAR OPTICAL SPECTROSCOPY OF SEMICONDUCTOR NANOCRYSTALS

Victor I. Klimov

Chemical Science and Technology Division, CST-6, MS-J585, Los Alamos National Laboratory, Los Alamos, New Mexico, USA

Contents

1. INTRODUCTION

Artificially structured nanoscale materials have attracted widespread interest due to the potential for controlling their physical and chemical properties for a variety of applications. Semiconductor nanocrystals (NCs) represent a class of quasi-zero-dimensional objects, or

Handbook of Nanostructured Materials and Nanotechnology, edited by H.S. Nalwa
Volume 4: Optical Properties

quantum dots. Bulk crystalline structure is preserved in NCs, but due to quantum confinement NCs have molecularlike discrete energy spectra, which exhibit strong size dependence [1–4]. This provides an opportunity for a wide-range tailoring of their electronic and optical properties. For example, the energy gap in CdSe NCs can be tuned from $\sim$1.8 eV (red) to $\sim$3 eV (UV) by changing the NC diameter from 11.5 to 1.2 nm [5].

NCs bridge the gap between large molecules and bulk materials. The boundaries between molecular, NC, and bulk regimes are not well defined and are strongly material dependent. The NC regime occurs over a size range of approximately 10^2–10^4 atoms per cluster [3, 4]. The lower limit of this range is determined by the stability of the bulk crystalline structure with respect to isomerization into molecular structure. The upper limit corresponds to sizes for which the energy level spacing is approaching the thermal energy kT, meaning that carriers become mobile inside a NC.

Depending on the ratio between NC radius a and the bulk-exciton radius a_x, three different confinement regimes can occur. In the strong-confinement regime ($a/a_x \ll 1$), the Coulomb electron-hole (*e-h*) interaction is negligibly small in comparison with confinement energies, therefore an electron and a hole can be treated as independent particles [1]. In the regime of weak confinement ($a/a_x \gg 1$), the electron and hole motions are strongly correlated via the Coulomb interaction and NC energy spectra are determined by quantization of the motion of the exciton center of mass [1]. In the intermediate-confinement regime ($a/a_x \sim 1$), energy structures in NCs are determined by a complex interplay between quantum confinement and the Coulomb *e-h* interaction.

Molecular beam epitaxy has been a standard technique for making high quality two-dimensional quantum wells and superlattices [6]. Epitaxial techniques also allow the preparation of three-dimensionally-confined islands with lateral dimensions down to 10 nm [7], which, however is not sufficient to reach the interesting regime of strong confinement. The alternative approach to making high-quality three-dimensionally-confined nanostructures is a chemical synthesis based on the control of homogeneous nucleation. There are two main chemical routes for preparation of NCs: high-temperature inorganic precipitation in molten glasses, producing NCs embedded in transparent glass matrices [8, 9] and moderate-temperature organometallic reactions, resulting in colloidal NCs [3, 5]. Glass samples provide rigidity and environmental stability, but they have a broad NC-size distribution (typically greater than 15%) and a large number of surface defects. A much higher level of control is provided by colloidal NCs, which can be chemically manipulated in a variety of ways, including size-selective techniques (resulting in less than 5% size variations [5]), surface modification by exchanging the organic passivation layer [10], heterostructuring with formation of layered NCs [11–13], immobilization in sol–gel [14–16] and polymer [17–19] matrices, and self-assembly into three-dimensional superlattices [20].

Systematic research into NC materials began with the first observations of effects of quantum confinement in the optical absorption of semiconductor-doped glasses [21]. Since then, tremendous progress has been made in both the theory of three-dimensionally-confined systems and experimental studies. Theoretical descriptions of electronic structures in NCs have advanced from simple particle-in-the-box models [1] to much more sophisticated approaches, including finite-well effects [22–24], Coulomb interactions [22, 25–27], nonparabolicity of the conduction band [28, 29], and confinement-induced mixing of valence subbands [27, 29, 30]. A great deal of experimental work has been performed to detect and to carefully map the structure of electron and hole quantized states in different confinement regimes [29, 31–33], to study confined phonon modes [34–37] and the processes of electron-phonon [34, 35, 38–41] and many-particle Auger-type interactions [42–44], to investigate biexciton effects [45–47] and the effects of three-dimensional confinement and surface properties on carrier energy relaxation and recombination dynamics [31, 48–50]. New insight into the physics of three-dimensionally-confined structures has been provided by experiments on single NC spectroscopy [51–53].

Apart from exciting new physics and chemistry, NCs offer a number of device applications which can benefit from size-controlled spectral tunability, confinement-induced concentration of the oscillator strength, and ultrafast relaxation dynamics, which can be controlled by NC surface properties. Applications of NCs in laser emitters [54], electroluminescent devices [55], switching elements [56], and solar cells [57] have been reported.

This review summarizes some progress in NC research with a special emphasis on optical spectroscopy of II-VI NCs such as CdSe and CdS. Because of the highly developed preparation technology, these NCs have been prototype systems for studies of the effects of three-dimensional confinement on the electronic and optical properties of semiconductors. Due to a relatively large exciton Bohr radius ($\sim$5 nm in CdSe and $\sim$3 nm in CdS), II-VI NCs are convenient systems for studies of the strong- and intermediate-confinement regimes, the focus of this review. Experimental results are reviewed for two types of samples: NCs prepared by high-temperature precipitation in glass matrices (referred to as NC–glass composites or glass samples) and NCs made by colloidal synthesis (referred to as colloidal NCs).

The remainder of the chapter is organized into six sections entitled: 2. Energy states and optical transitions in semiconductor nanocrystals: Theoretical models; 3. Experimental studies of energy structures in semiconductor nanocrystals; 4. Fine structure of the lowest exciton state; 5. Effects of electron-phonon interactions on the optical spectra of semiconductor nanocrystals; 6. Band-edge optical nonlinearities in semiconductor nanocrystals; 7. Carrier dynamics in semiconductor nanocrystals.

In Section 2, energy structures in NCs are considered using a well-developed effective mass approach. The section begins with a description of the simple parabolic-band model, and then outlines the effects of valence-band mixing, Coulomb interactions, finite height of the confinement barrier, and nonparabolicity of the conduction band, and their impact on the energy structures and optical transitions in NCs. Section 3 contains some experimental results on the size dependence of the energy gap in NCs and the observations of discrete electron and hole energy states in NC-based structures. Section 4 gives a brief overview of theoretical and experimental studies of the fine structure of the lowest exciton state, associated with the effects of nonspherical NC shape, crystal-field splitting, and exchange interactions. Section 5 outlines the effects of electron-phonon interactions on the optical spectra of NCs. It starts with a description of the model of a "displaced" oscillator, which is then used to analyze the effects of carrier interactions with optical and acoustic phonons. This section also contains results of experimental studies of electron-phonon coupling performed by various optical techniques. Section 6 examines the results of studies of resonant optical nonlinearities in NCs observed at spectral energies near the band edge. It concentrates on several topics, including the role of state-filling and Coulomb many-particle interactions on the nonlinear optical response of NCs, the third-order nonlinear susceptibility of NC-based structures, optical nonlinearities in direct- and indirect-gap semiconductor NCs, and optical gain and lasing in NCs. Section 7 surveys some results on carrier dynamics in NCs, including those on intraband energy relaxation, radiative decay, carrier trapping, and nonradiative Auger recombination and associated ionization of NCs.

2. ENERGY STATES AND OPTICAL TRANSITIONS IN SEMICONDUCTOR NANOCRYSTALS: THEORETICAL MODELS

The effective mass approximation (EMA) has proven to be a powerful tool for describing energy structures in semiconductor NCs [1, 2, 22, 30]. Although it fails in the limit of very small NCs, it nevertheless allows a correct prediction of important trends in most of the NC sizes which have been synthesized and which have been studied experimentally. In the following text we briefly review several EMA models developed for strongly confined NCs.

2.1. Parabolic-Band Model

The first theoretical descriptions of NCs were based on the model of a spherical quantum well with an infinite potential barrier assuming a parabolic dispersion of electron and hole bands [1, 2]. However, even within this simple model, analytical solutions of the corresponding quantum mechanical problem are not possible if the Coulomb e-h interaction is explicitly taken into account. The Coulomb interaction scales with the NC radius as a^{-1}, whereas confinement energies are proportional to a^{-2}. Therefore, in the strong confinement regime ($a \ll a_x$), the Coulomb term in the Hamiltonian can be ignored which leads to a problem of independent quantization of electron and hole energies, having analytical solutions [1, 58]. In this approximation, the states are classified with respect to the orbital angular momentum l ($l = 0, 1, 2, \ldots,$ corresponding to states with S, P, $D, \ldots,$ symmetries), its projection m ($m = 0, \pm 1, \ldots, \pm l$) and the quantum number n ($n = 1, 2, 3, \ldots$) which counts the states of the same symmetry. The electron and hole wave functions are presented as a product of the envelope function ($\Phi^{e,h}_{n,l,m}$) and the Bloch function of the conduction (electrons) or valence (holes) bands. The envelope functions can be expressed in terms of the spherical Bessel (j_l) and the spherical harmonic ($Y_{l,m}$) functions [59],

$$\Phi^{e,h}_{n,l,m}(\mathbf{r}) = \sqrt{\frac{2}{a^3}} \frac{j_l(\phi_{l,n} r/a)}{j_{l+1}(\phi_{l,n})} Y_{l,m}(\Omega) \tag{1}$$

where $\phi_{l,n}$ is the nth root of the spherical Bessel function of the lth order and Ω is the solid angle of the vector $\mathbf{r}$. The envelope function of the ground state ($n = 1$, $l = 0$, $m = 0$) has a simple form,

$$\Phi^{e,h}_{1,0,0} = \frac{1}{\sqrt{2\pi a}} \frac{\sin(\pi r/a)}{r} \tag{2}$$

The quantized carrier energies are degenerate with respect to the quantum number m and they can be determined from the expression,

$$E^{e,h}_{l,n} = \frac{\hbar^2 \phi^2_{l,n}}{2 m_{e,h} a^2} \tag{3}$$

where $E^e_{l,n}$ ($E^h_{l,n}$) is the electron (hole) energy measured from the bottom of the conduction (valence) band, and $m_e(m_h)$ is the effective electron (hole) mass. The energy of the states increases in the order $1S, 1P, 1D, 2S \ldots,$ as illustrated in Figure 1. The first four roots $\phi_{l,n}$ are π ($l = 0$, $n = 1$), 4.49 ($l = 1$, $n = 1$), 5.76 ($l = 2$, $n = 1$), and 6.28 ($l = 0$, $n = 2$). Because in semiconductors the hole mass is typically much greater than that of the electron (in CdSe, e.g., $m_h/m_e \approx 6$), the electron levels are much more sparse than hole levels.

The envelope functions $\Phi_{n,l,m}$ with different n and/or l numbers are orthogonal ($\langle \Phi_{n',l',m'} \mid \Phi_{n,l,m} \rangle = \delta_{n,n'} \delta_{l,l'} \delta_{m,-m'}$), therefore, the optical transitions (shown by arrows in Fig. 1) are only allowed between the states of the same symmetry ($l = l'$, $n = n'$). The transition intensity is proportional to the degree of degeneracy of the states involved in this transition and is independent of the NC radius. As a result, the imaginary part of the optical susceptibility χ can be presented as follows,

$$\operatorname{Im} \chi(\hbar\omega) = \pi N_0 |d_{cv}|^2 \sum_{l,n} 2(2l+1) \delta\left(\hbar\omega - E_g - \frac{\hbar^2 \phi^2_{l,n}}{2 m_r a^2}\right) \tag{4}$$

where N_0 is a concentration of NCs in the matrix, d_{cv} is the matrix element of the interband transition, E_g is the bulk bandgap energy, and $m_r = m_e m_h/(m_e + m_h)$ is the reduced e-h mass. The absorption can be calculated using the relationship,

$$\alpha(\hbar\omega) = \frac{4\pi\omega}{n_0 c} \operatorname{Im} \chi(\hbar\omega) \tag{5}$$

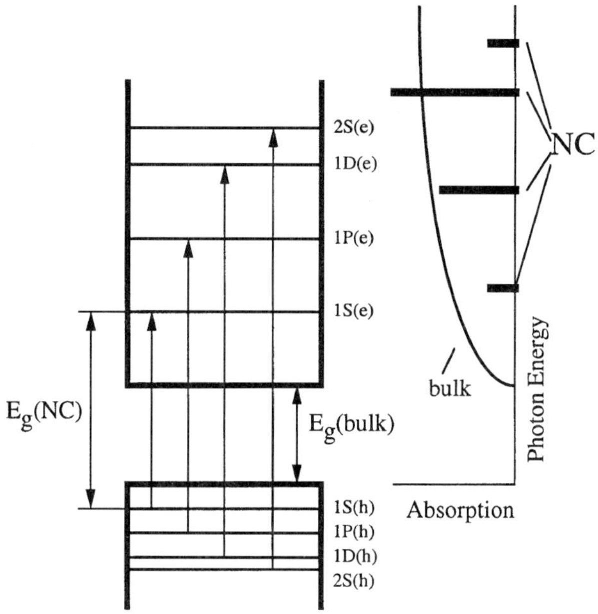

Fig. 1. Electron and hole quantized levels in the model of parabolic bands. Arrows indicate allowed optical transitions which couple electron and hole states of the same symmetry. Plot on the right is an illustration of the discrete structure of optical absorption in NCs which is compared with a continuous absorption spectrum of a bulk material.

where c is the light velocity and n_0 is a refractive index of the NC–matrix media. As seen from Eq. (4), the absorption spectrum of a NC consists of narrow peaks (Fig. 1) with spectral positions,

$$\hbar\omega_{ln} = E_g + \frac{\hbar^2 \phi_{l,n}^2}{2m_r a^2} \tag{6}$$

With increasing NC radius the spacing between absorption peaks decreases until they merge into continuous spectrum of the bulk material: $\alpha(\hbar\omega) \propto |d_{cv}|^2 \sqrt{\hbar\omega - E_g}$. The absorption edge in NCs is determined by the energy of the lowest $1S(e)$-$1S(h)$ transition,

$$\hbar\omega_{01} = E_g + \frac{\hbar^2 \pi^2}{2m_r a^2} \tag{7}$$

which can be treated as the NC energy gap. Due to quantum confinement, the gap increases as a^{-2} with decreasing NC size.

Expressing the NC concentration (N_0) in terms of the volume fraction of the semiconductor in the matrix ($\xi_0 = N_0 V_0 < 1$; V_0 is a volume of a single NC) we arrive at the expression $\alpha \propto \xi_0 / V_0$ which shows that absorption increases with decreasing NC volume, assuming that the volume fraction of the semiconductor is kept unchanged (this actually corresponds to increasing the NC concentration). This effect, referred to in Ref. [60] as a confinement-induced increase in the spectral weight of NC transitions, is a direct consequence of the fact that in the strong-confinement regime the probabilities of optical transitions are size independent; therefore, the absorption calculated per unit volume of the semiconductor scales as $1/V_0$.

2.2. Effects of Valence-Band Mixing

The simplified model described in the previous text gives a reasonable description of the NC conduction band. However, due to the complexity of the valence band of real semiconductors, the parabolic approximation fails to describe correctly hole energy structures. In semiconductors with diamond, zinc-blende, and wurtzite lattices, the lowest conduction minimum is formed from S-type orbitals and is twofold degenerate due to electron spin. The valence band formed from P-type wave functions is sixfold degenerate. Due to the spin-orbit interaction, the valence band is split into a fourfold degenerate band with a total unit-cell angular momentum $J = 3/2$ and a twofold degenerate band with $J = 1/2$ (Fig. 2). The energy separation between these bands (Δ_{so}) is determined by the strength of the spin-orbit interaction. In bulk semiconductors, the $J = 3/2$ band is further split into light- and heavy-hole subbands with J projections $M_J = \pm 1/2$ and $\pm 3/2$, respectively. At the Γ point of the Brillouin zone, this splitting is zero for diamond and zinc-blende lattices and nonzero (the "A-B splitting") for the wurtzite structure (Fig. 2).

The valence band can be described using three parameters γ_1, γ_2, and γ_3 introduced by Luttinger and Kohn [61]. In the case of the cubic crystal lattice, the number of parameters reduces to 2: γ_1 and $\gamma = \gamma_2 = \gamma_3$. Using these parameters, the masses of the heavy (m_{hh}), light (m_{lh}), and split-off (m_{so}) hole bands can be written in terms of the free electron mass (m_0) as follows,

$$m_{hh} = \frac{m_0}{(\gamma_1 - 2\gamma)} \qquad m_{lh} = \frac{m_0}{(\gamma_1 + 2\gamma)} \qquad m_{so} = \frac{m_0}{\gamma_1} \tag{8}$$

As was pointed out in Ref. [62], in the case of the spherical potential included into the hole Hamiltonian, the only true quantum number is the total momentum $\hat{\mathbf{F}}$ which is the sum of the Bloch-function angular momentum $\hat{\mathbf{J}}$ and the orbital momentum of the envelope function $\hat{\mathbf{L}}$: $\hat{\mathbf{F}} = \hat{\mathbf{J}} + \hat{\mathbf{L}}$. The hole wave functions, which are the eigenfunctions of the total momentum $\hat{\mathbf{F}}$, can be expanded using the eigenstates of the orbital momentum $\hat{\mathbf{L}}$, leading to mixing between different valence subbands.

The problem of quantization of hole states in a quantum sphere taking into account the band-mixing effects was theoretically addressed for both strong [27, 29, 30] and weak [26, 63] spin-orbit coupling. These limits are applicable to semiconductors with large (e.g.,

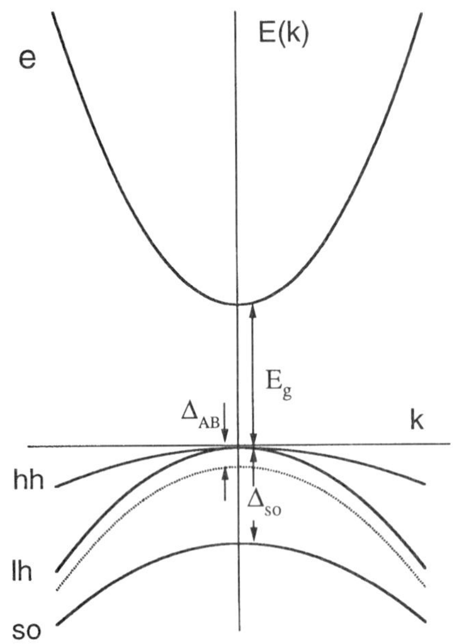

Fig. 2. Conduction and valence-band structures for diamond, zinc-blende and wurtzite semiconductors. Heavy, light, and spin-orbit split-off valence subbands are denoted as "hh," "lh," and "so," respectively. In the center of the Brillouin zone, the light–heavy hole splitting is zero in diamond and zinc-blende crystals (solid lines) and nonzero (the A-B splitting) in wurtzite semiconductors (dotted line).

CdSe) and small (e.g., CdS) spin-orbit splitting, respectively. To illustrate the main results of these models we briefly outline the approach which has been applied in the case of strong spin-orbit interaction ($\Delta_{so} \to \infty$) [30]. In this limit, a hole can be treated as a particle with spin 3/2 [62]. The spin-orbit term in the hole Hamiltonian with a spherical potential couples the hole states with orbital momenta L and $L+2$ [62]. This effect is commonly referred to as "S-D mixing." As a result, the hole envelope function has a general form

$$\Phi^h(L_F) = f(r)|L, J, F, M_F\rangle + g(r)|L+2, J, F, M_F\rangle \tag{9}$$

where M_F is a projection of the total momentum $\hat{\mathbf{F}}$, $f(r)$ and $g(r)$ are the radial wave functions, and $|L, J, F, M_F\rangle$ are eigenfunctions of the total angular momentum in the $L-J$ coupling scheme. The hole states are usually denoted as nL_F, where L is the lower of the two momenta involved in the wave function and n is the number of the state of given symmetry. The wave functions for $S_{3/2}$, $P_{1/2}$, and $P_{3/2}$ states can be written as follows,

$$\Phi^h(S_{3/2}) = f_0(r)|0, 3/2, 3/2, M_F\rangle + g_0(r)|2, 3/2, 3/2, M_F\rangle \tag{10}$$

$$\Phi^h(P_{1/2}) = f_1(r)|1, 3/2, 1/2, M_F\rangle \tag{11}$$

$$\Phi^h(P_{3/2}) = f_2(r)|1, 3/2, 3/2, M_F\rangle + g_2(r)|3, 3/2, 3/2, M_F\rangle \tag{12}$$

Because the total momentum $F = 1/2$ cannot be built from the states with $L = 3$ and $J = 3/2$, the $P_{1/2}$ wave function only has a contribution from the $L = 1$ state. As a result, the $P_{1/2}$ state is not affected by band mixing and has the same energy as in the parabolic model [Eq. (3)] with m_h replaced with m_{lh}. The wave functions of other states have contributions from two different $\hat{\mathbf{L}}$ eigenstates, which lead to the strong effect of band mixing on hole energies. The strength of band mixing can be characterized by the parameter,

$$\mu = \frac{2\gamma}{\gamma_1} \tag{13}$$

The effect of mixing light- and heavy-hole bands is illustrated in Figure 3 where the energies of the $S_{3/2}$ state are plotted versus μ [30]. For $\mu = 0$ (heavy and light holes have equal masses), this model gives the same results as the parabolic-band model. Without band mixing, the hole energies would have varied with increasing μ according to the linear relationships $(1+\mu)$ and $(1-\mu)$ for light and heavy holes, respectively. Band-mixing effects significantly modify these simple dependencies leading in particular to the avoided-crossing behavior of higher lying states (see Fig. 3). For small μ, the $3S_{3/2}$ state has a

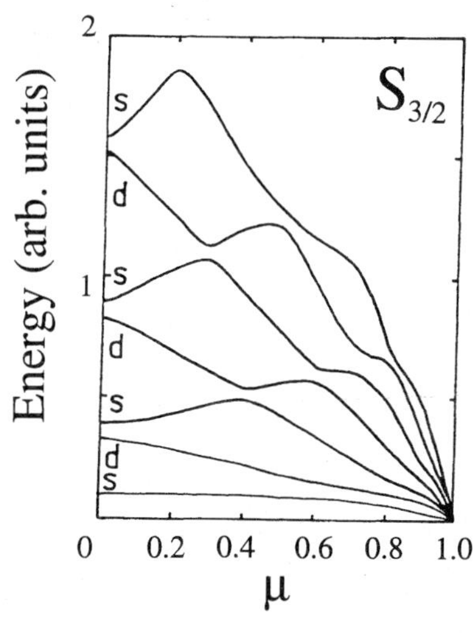

Fig. 3. Energies of the $S_{3/2}$ states as a function of μ calculated taking into account the valence-band mixing effects. (Source: Reprinted with permission from [30]. © 1989 American Physical Society.)

light-hole character (note the positive slope of the curve at $\mu < 0.4$) which switches to the heavy-hole character after an avoided crossing with the $4S_{3/2}$ term. The same behavior is also seen for other pairs of states: $4S_{3/2}$ and $5S_{3/2}$, $5S_{3/2}$ and $6S_{3/2}$, etc. Interestingly, for $\mu = 1$, the hole confinement energies drop to zero because the heavy-hole mass becomes infinite in this limit. However, even in the limit where $\mu = 1$, the energies of the $P_{1/2}$ states are nonzero as these states are solely due to light holes.

So far, we have been assuming that the spin-orbit splitting is infinitely large so that the effect of the split-off band ($J = 1/2$) on the hole energy spectra is negligibly small. The inclusion of a finite spin-orbit splitting leads to the additional mixing of existing states with the $J = 1/2$ band and to the appearance of two series of new states $nS_{1/2}$ and $nP_{1/2}^{so}$ [29]. As a result of the mixing with the split-off band, the $nS_{3/2}$ wave function, for instance, in addition to components $|L = 0,\ J = 3/2,\ F = 3/2,\ M_F\rangle$ and $|L = 2,\ J = 3/2,\ F = 3/2,\ M_F\rangle$ arising from the light-heavy hole subbands, has a component $|L = 2,\ J = 1/2,\ F = 3/2,\ M_F\rangle$ originating from the split-off band (compare with Eq. (10)).

Size-dependent hole energies in CdSe NCs calculated in Ref. [29], taking into account the mixing of all three valence subbands, are shown in Figure 4. According to these calculations, the four lowest hole states are $1S_{3/2}$, $1P_{3/2}$, $2S_{3/2}$, and $1P_{1/2}$, listed in the order of increasing energy. In the limit $m_{lh} \ll m_{hh}$ the hole ground state energy (the $1S_{3/2}$ state) can be found from the expression,

$$E(1S_{3/2}) = \frac{\hbar^2 \phi_{2,1}^2}{2m_{hh}a^2} \tag{14}$$

where $\phi_{2,1} \approx 5.76$ is the first root of the spherical Bessel function of the second order. An interesting result of the calculation is a prediction of the avoided crossing behavior for states $2S_{1/2}$ and $3S_{1/2}$ at $a \approx 3.5$ nm (see Fig. 4) which is a direct consequence of S-D mixing. Both, $2S_{1/2}$ and $3S_{1/2}$ states represent a mixture of the S- and D-type states from the $J = 1/2$ and $3/2$ subbands, respectively. In large NCs ($a > 3.5$ nm), the $2S_{1/2}$ state is mostly D-like (the $J = 3/2$ subband), whereas the $3S_{1/2}$ state is dominated by the S-type contribution from the split-off subband. After the avoided crossing (i.e., in small-size NCs) the character of the $2S_{1/2}$ and $3S_{1/2}$ states switches to S- and D-type, respectively.

Band-mixing effects not only modify hole energies but also significantly alter the selection rules for optical transitions. The intensity ($\Im$) of these transitions is determined by the overlap integral between electron and hole wave functions [29, 30],

$$\Im \propto \left| \int \Phi_{n',l,m}^{e}(\mathbf{r}) \Phi_{n,F,M_F}^{h}(\mathbf{r})\, d^3 r \right|^2 \tag{15}$$

where $\Phi_{n',l,m}^{e}$ is the electron wave function in the parabolic approximation (see the previous subsection), and Φ_{n,F,M_F}^{h} is the hole wave function calculated taking into account band-mixing effects.

In contrast to the parabolic model, the quantum numbers L and n are not strictly conserved for dipole-allowed transitions [29]. For example, transitions involving the $1S$ electron state are possible not only from the $1S_{3/2}$ and $1S_{1/2}$ states but also from the higher lying states $nS_{3/2}$ and $nS_{1/2}$ ($n > 1$) (see Fig. 5), because all of these states have a contribution from the spherical harmonic $Y_{0,0}$. Similarly, transitions to the $1P$ electron state are possible from $nP_{3/2}$, $nP_{1/2}$, and $nP_{5/2}$ hole states ($n \geqslant 1$). Due to S-D mixing, the transitions to the $1D$ electron level are possible from both S and D type hole states: $nS_{3/2}$, $nS_{1/2}$, $nD_{7/2}$, and $nD_{5/2}$. However, because the $nP_{1/2}$ and $nP_{1/2}^{so}$ hole states are not affected by band mixing, the selection rule $n = n'$ (n' is the electron-state quantum number) holds strictly for them as in the parabolic approximation. Transition probability calculations which take into account band-mixing effects can be found in Refs. [29, 30].

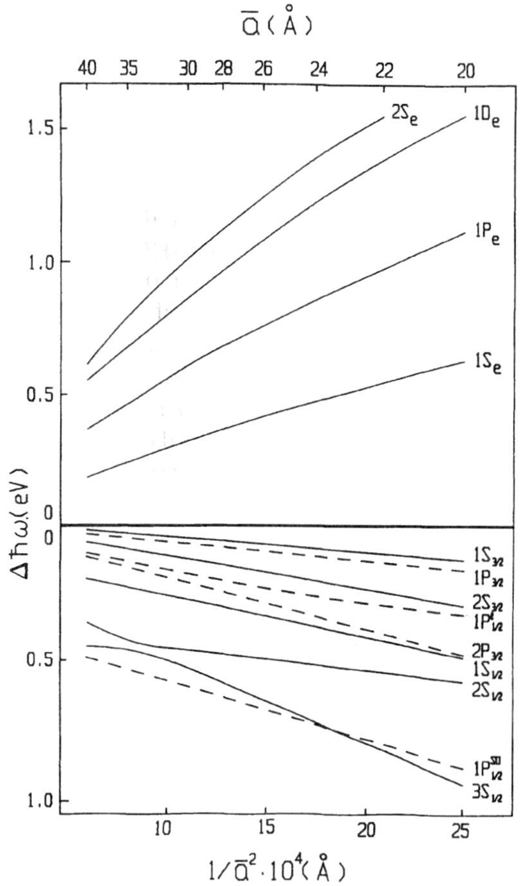

Fig. 4. Electron and hole energy spectra in CdSe NCs plotted versus the inverse-mean-square radius, as calculated in Ref. [29]. Hole energies are calculated taking into account valence band-mixing effects, and corrected for the Coulomb *e-h* interaction. Electron energies are calculated with taking into account nonparabolicity of the conduction band. The *S* and *P* hole states are shown by solid and dashed lines, respectively. (Source: Reprinted with permission from [29]. © 1993 Optical Society of America.)

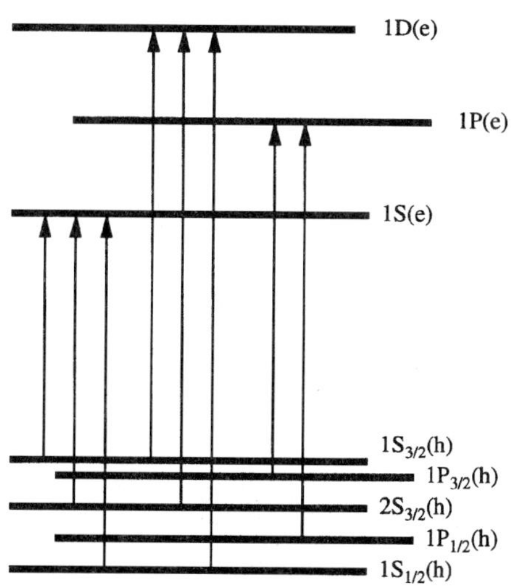

Fig. 5. Allowed optical transitions (arrows) between the several lowest electron and hole quantized states in the model accounting for valence-band mixing effects.

2.3. Coulomb Effects

In bulk materials with a dielectric constant ϵ_b, the Coulomb e-h interaction can be taken into account by the term,

$$V_{eh} = -\frac{e^2}{\epsilon_b |\mathbf{r}_e - \mathbf{r}_h|} \tag{16}$$

where $\mathbf{r}_e$ and $\mathbf{r}_h$ are electron and hole coordinates, respectively. This term gives rise to the e-h bound states or excitons [64]. The situation is more complicated in semiconductor NCs, because the dielectric constants of the semiconductor (ϵ_2) and the surrounding matrix (ϵ_1) can be significantly different, leading to image-charge effects [25, 65].

The e-h Coulomb interaction in a spherical NC can be represented as follows [25, 65],

$$V_{eh} = -\frac{e^2}{\epsilon_2 |\mathbf{r}_e - \mathbf{r}_h|} + P_e + P_h + P_{eh} \tag{17}$$

where $P_e(P_h)$ is the self-polarization energy arising from the electron (hole) interaction with an associated image charge (surface polarization charge), and P_{eh} is the mutual polarization energy due to interaction of one charge with the surface image of the other charge. Polarization energies can be expressed in terms of the Legendre polynomials (P_k) and coefficients $\alpha_n = (\epsilon - 1)(n+1)/(\epsilon n + n + 1)$ (ϵ is defined as ϵ_2/ϵ_1),

$$P_{e,h} = \frac{e^2}{2\epsilon_2 a} \sum_{n=0}^{\infty} \alpha_n \left(\frac{r_{e,h}}{a}\right)^{2n} \tag{18}$$

$$P_{eh} = -\frac{e^2}{\epsilon_2 a} \sum_{n=0}^{\infty} \alpha_n \left(\frac{r_e r_h}{a^2}\right)^{n} P_n(\cos\theta) \tag{19}$$

where θ is the angle between $\mathbf{r}_e$ and $\mathbf{r}_h$.

In the limit of strong confinement, the electron and hole motions can be separated using the adiabatic approximation [65]. Because m_h is typically much greater than m_e, the Coulomb potential applied to the hole (V_h) can be averaged over the electron wave function, leading to the expression,

$$V_h = P_h(r_h) - \frac{e^2}{\epsilon_2} \int d\mathbf{r}_e \frac{|\Phi^e_{n,l,m}(r_e)|^2}{|\mathbf{r}_e - \mathbf{r}_h|} + \int d\mathbf{r}_e \left|\Phi^e_{n,l,m}(r_e)\right|^2 P_{eh}(\mathbf{r}_e, \mathbf{r}_h) \tag{20}$$

where $\Phi^e_{n,l,m}(r_e)$ is the electron wave function in the parabolic approximation. The correction to the electron energy resulting from polarization-induced self-interaction (term P_e) can be found using a perturbation theory. Application of this approach within the limit of a strong spin-orbit interaction ($\Delta_{so} \to \infty$) leads to the following expression for the Coulomb correction (δE_c) to the energy of the lowest $1S(e)$-$1S_{3/2}(h)$ transition in strongly confined NCs [65],

$$\delta E_c = -1.765 \frac{e^2}{\epsilon_2 a} + \zeta(\epsilon) \frac{e^2}{\epsilon_2 a} \tag{21}$$

where ζ is a coefficient depending on the ratio of the semiconductor-to-matrix dielectric constants. This coefficient is zero for $\epsilon = 0$, and gradually increases up to 0.835 for $\epsilon = 10$. In the case of CdSe–glass samples, $\epsilon \approx 4$ which corresponds to $\zeta = 0.573$. The principal effect of the e-h Coulomb interaction is to red shift the optical transitions. Another consequence of this interaction is a spatial shift of the maximum of the hole charge density toward the NC center which occurs as a result of hole localization in the electron potential [25, 65].

2.4. Effects of the Finite Potential Barrier and Nonparabolicity of the Conduction Band

The particle-in-the-box model described in Section 2.1 is based on assumptions about the parabolic dispersion of the valence and conduction bands and the infinite height of the potential barrier at the NC surface. However, due to the small electron mass, the energies of electronic quantized levels can be large enough that the nonparabolicity of the conduction band and the leakage of the electron wave function into the potential barrier can lead to deviations from the energy structures calculated within the model of parabolic bands.

The nonparabolicity of the conduction band can be described in terms of the Kane model [66] by taking into account the interaction of the conduction band with valence bands [42, 67]. This approach leads to an energy-dependent electron mass which can be expressed as follows [29],

$$\frac{1}{m(E)} = \frac{1}{m_0}\left[1+2f+\frac{E_p}{3}\left(\frac{2}{E+E_g}+\frac{1}{E+E_g+\Delta_{so}}\right)\right] \tag{22}$$

where E is the electron energy measured from the bottom of the conduction band, f is a parameter accounting for the effect of higher lying bands [67], and E_p is a characteristic energy related to the Kane matrix element P_{cv} ($P_{cv} = p_{cv}/m_0$; p_{cv} is the momentum interband matrix element) by the expression $E_p = 2|P_{cv}|^2 m_0$. In the case of an infinite potential barrier, this approach leads to the following equation for determining the electronic quantized energies [29],

$$E^e_{ln} = \frac{\hbar^2\phi^2_{ln}}{2m(E^e_{l,n})a^2}, \tag{23}$$

which is a generalization of Eq. (3), obtained using a parabolic dispersion of the conduction band. The electron mass defined by Eq. (22) increases with increasing electron energy; therefore, the effects of nonparabolicity lead to a reduction of the electron energies which is particularly important for higher lying electron states.

In the case of an infinite potential barrier, the zero boundary conditions are used to derive the electron eigenenergies. If the height of the potential barrier is finite, the electron wave function penetrates into the barrier which modifies the electron energies. The trends resulting from this effect can be understood by analyzing a simple problem of a spherical quantum well with a potential barrier V_0 in the parabolic approximation [58] (see inset to Fig. 6). The effect of the penetration of the electron wave function into the barrier is taken into account by boundary conditions which require that both the wave function and its derivative are continuous at the NC surface. The S-type ($l = 0$) solutions inside ($r < a$) and outside ($r > a$) the NC have the form,

$$\Phi_S(r) = A\frac{\sin(kr)}{r} \qquad (r < a) \tag{24}$$

$$\Phi_S(r) = B\frac{e^{-\xi(r-a)}}{r} \qquad (r > a) \tag{25}$$

where

$$k^2 = \frac{2m_e E^e_{0,n}}{\hbar^2} \qquad \xi^2 = \frac{2m_e V_0}{\hbar^2} - k^2 \tag{26}$$

The spatial profile of the $1S$ function is shown in the inset to Figure 6. The electron energies $E^e_{0,n}$ can be found as solutions of the equation,

$$\tan(ka) = -\frac{k}{\sqrt{(2m_e V_0)/\hbar^2 - k^2}} \tag{27}$$

which is obtained by applying the boundary conditions to the wave functions given by Eqs. (24) and (25). Figure 6 shows the dependence of the first root of Eq. (27) (k_1) versus

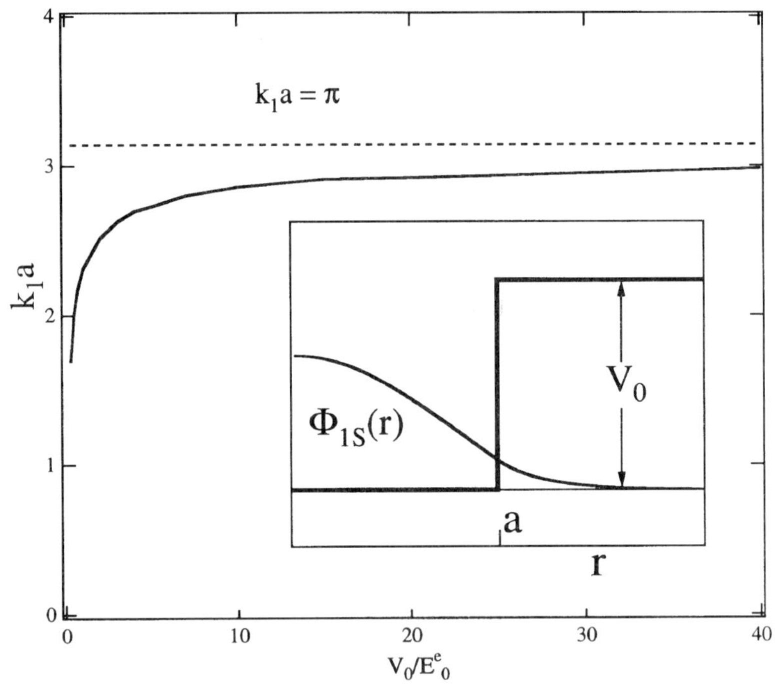

Fig. 6. Dependence of the first root of Eq. (27) (solid line) on the confinement potential in the model of the spherical quantum well with a finite potential barrier. The dashed line marks a first root of the same equation in case of the infinite confinement potential. Inset: illustration of the penetration of the $1S$ wave function into the potential barrier of the finite height V_0.

the height of the potential barrier presented in units of the $1S$ electron energy calculated for the infinite potential [$\eta = V_0/E_0^e$; $E_0^e = \pi^2\hbar^2/(2m_e a^2)$]. For large V_0, k_1 is close to π/a [a solution of Eq. (27) in the case of an infinite potential], deviating from this value by less than 10% in the range $\eta > 10$. The magnitude of k_1 (as well as the electron confinement energy) reduces sharply at $\eta < 2$–3. In the limit $\eta \rightarrow (2\pi)^{-1}$, k_1 approaches its minimum value of $\pi/(2a)$.

This simple model can be improved by including the Coulomb effects [22, 24] and interactions between conduction and valence bands [42] which allow one to account for the nonparabolicity of the electron energy spectrum.

3. EXPERIMENTAL STUDIES OF ENERGY STRUCTURES IN SEMICONDUCTOR NANOCRYSTALS

3.1. Energy Gap in Semiconductor Nanocrystals

Synthetic efforts have resulted in the development of procedures for preparation of high quality NCs with narrow size distributions using both high-temperature precipitation [8] and colloidal synthesis [5, 68]. The best results have been achieved using organometallic synthesis followed by a size-selective precipitation which allows isolation of samples with size dispersion <5% [5]. The preparation of NCs using high-temperature precipitation results in broader size distribution which is typically greater than 15%.

Due to the size dependence of carrier energies, optical spectra of NCs are strongly affected by inhomogeneous broadening which can smear out the discrete structure of optical transitions resulting from three-dimensional confinement. In the strong-confinement

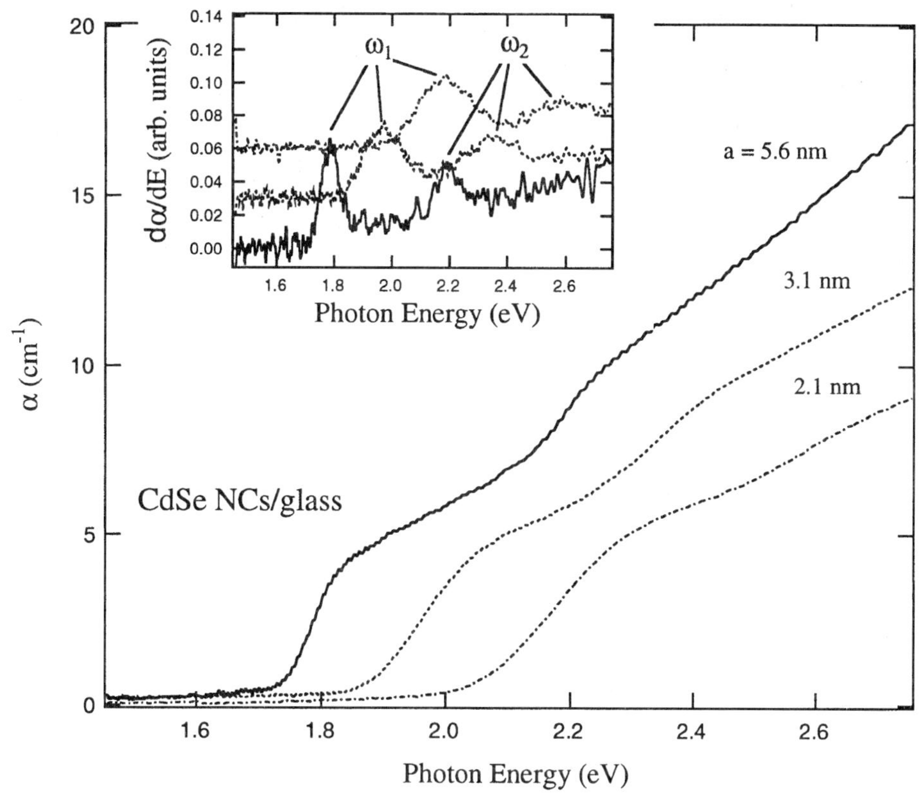

Fig. 7. Room-temperature absorption spectra of three CdSe NCs–glass samples with NC mean radii 2.1, 3.1, and 5.6 nm. Inset: first-derivative absorption spectra for the same samples.

regime, the size dependence of energy transitions in NCs is dominated by a term proportional to $1/a^2$. In this case, the broadening of transitions ($2\hbar\Gamma_{ih}$) arising from a size distribution can be estimated from the expression $2\hbar\Gamma_{ih} = (4\delta a/a)\Delta E$, where ΔE is the confinement-induced shift of the optical transition, and δa is the halfwidth of the size distribution. With $\delta a/a = 5\%$, typical for the best colloidal samples, and $\Delta E = 0.5$ eV ($a \approx 2$ nm in the case of CdSe NCs) the inhomogeneous linewidth is calculated to be extremely broad ($2\hbar\Gamma_{ih} = 100$ meV) even in this case of a very narrow size distribution.

Because of the broad size distribution, NC–glass samples typically do not exhibit sharp absorption features related to size quantization. However, even in samples with a relatively large size dispersion, the absorption edge shows a clear confinement-induced blue shift for NCs of smaller sizes as seen in Figure 7. Size-quantization-related features are better resolved in the first derivative spectra (inset to Fig. 7) which exhibit two peaks (ω_1 and ω_2), shifting synchronously to higher energies with decreasing NC sizes. The lowest of these peaks (ω_1) roughly gives the position of the lowest $1S(e)$-$1S_{3/2}(h)$ optical transition (NC energy gap). The spacing between the ω_1 and ω_2 features in all three samples is around 400 meV, which is close to the spin-orbit splitting in CdSe (420 meV). Therefore, the ω_2 feature can be assigned to the transition which couples the $1S$ electron state to the hole state $2S_{1/2}$ or $3S_{1/2}$ (depending on the NC radius), originating from the spin-orbit split-off valence subband.

Colloidal synthesis allows the fabrication of samples with narrow size distributions and a variety of closely space radii which provides an opportunity to map accurately the size dependence of the NC energy gap [5, 33]. Figure 8 shows a series of room-temperature absorption spectra of CdSe colloidal NCs with radii from 1.2 to 4.1 nm. Due to the high

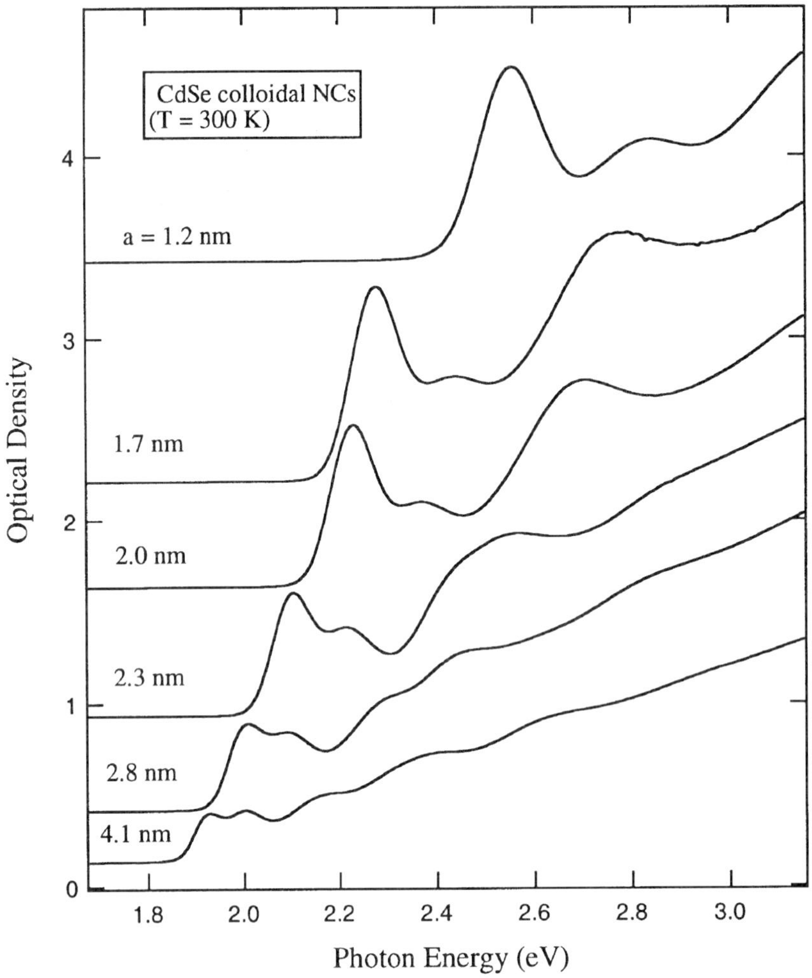

Fig. 8. Room-temperature absorption spectra of series of colloidal CdSe NCs with mean radii from the range 1.2–4.1 nm.

level of sample monodispersity, these spectra show discrete features associated with size quantization. Most of these features are due to transitions which couple the $1S$ electron state to various hole states which have an S-type component, as is discussed later. For all sizes, the peak associated with the lowest optical transition is well resolved, providing a simple means to determine the NC energy gap. The energies of the lowest transition derived from the spectra in Figure 8 are plotted in Figure 9 as E_g(NC) versus a, along with positions of the ω_1 peak from the spectra in Figure 7 obtained for NC–glass composites. The data for colloidal and glass samples are in good agreement with each other. These data are compared in Figure 9 with predictions of the parabolic-band model (see Section 2.1) corrected for Coulomb e-h interaction within a first-order perturbation theory [Eq. (21) in Section 2.3]. The following set of CdSe parameters was used in the calculation: $E_g(\text{bulk}) = 1.75$ eV ($T = 300$ K), $m_e = 0.13m_0$ (m_0 is a free electron mass), $m_h = 1.1m_0$, and $\epsilon = 9.4$. This simple model provides a good estimate of the energy of the lowest transition in NCs with radii greater than ~3 nm. However, it overestimates the confinement-induced energy shifts in NCs of smaller sizes, indicating the importance of such effects as penetration of the wave functions into the potential barrier [22–24] and nonparabolicity of the conduction band [28, 29]. Both effects tend to lower carrier energies as was discussed in Section 2.4.

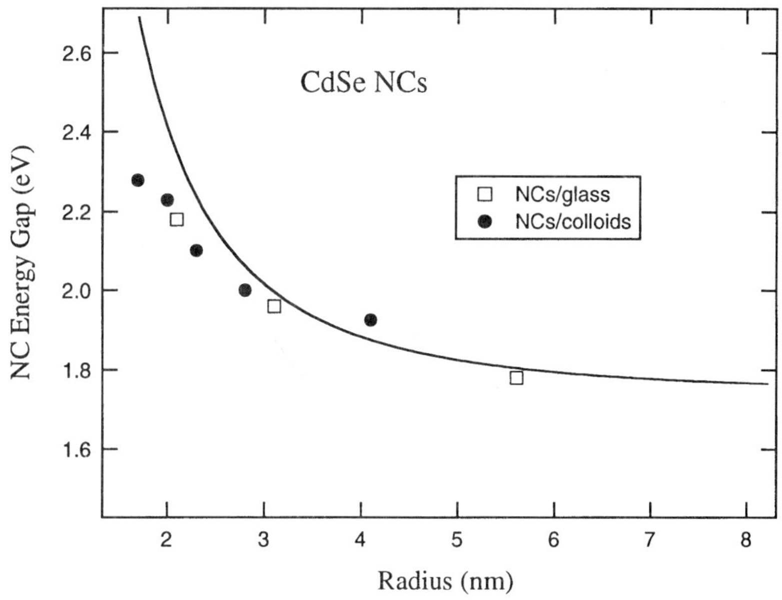

Fig. 9. The position of the lowest optical transition (NC energy gap) in CdSe NCs (room temperature) plotted versus NC mean radius. The data are derived from absorption spectra of colloidal (solid circles; Fig. 8) and glass (open squares; Fig. 7) samples, and compared with the bandgap energies (solid line) calculated within parabolic-band approximation, corrected for the Coulomb *e-h* interaction.

3.2. Observations of Electron Quantized States

As was mentioned, inhomogeneous broadening resulting from the size distribution masks discrete energy structures related to size quantization. Inhomogeneous broadening can be significantly suppressed using spectroscopic size-selective techniques such as fluorescence line narrowing (FLN) [31, 33, 69, 70], photoluminescence excitation (PLE) [31, 33], and spectral "hole burning" [31, 32, 71–73]. In these techniques, a narrow spectral window is used for resonance excitation (FLN and hole burning) or detection (PLE) of a small portion of NC sizes from the broad size distribution as is discussed in more detail in Sections 4.2 and 4.3 in the context of studies of electron-phonon interactions. Size-selective techniques are most efficient using excitation (detection) wavelengths on the red edge of the absorption spectrum (the lowest optical transition) which corresponds to the selective excitation of the largest NCs from the size distribution [33, 69]. However, excitation into higher lying states also provides some size-selective effect [74] which was used in Refs. [32, 54, 75] to resolve excited electron states in nonlinear differential transmission (DT).

DT gives a measure of pump-induced transmission changes and is defined as follows: $D = (T - T_0)/T_0 = \Delta T/T_0$, where T_0 and T are transmissions of the unexcited and excited samples, respectively. In the small-signal limit ($D \ll 1$), DT is proportional to the pump-induced absorption change $\Delta\alpha$: $D \approx -\Delta\alpha d$ (d is the sample thickness). In NCs, carrier-induced absorption changes are dominated by state filling (see Section 6.1) leading to pronounced bleaching bands located at the positions of allowed optical transitions. Due to the degeneracy of the valence band and the large difference between electron and hole masses (m_h is typically much greater than m_e), DT is dominated by electron populations [76] (see Section 6.1), although at low temperatures and/or in very small NCs, DT can also be affected by populations of hole states [50]. In addition to positions of optical transitions, time-resolved DT measured using picosecond (ps) or femtosecond (fs) pump-probe experiments (see Section 7.1) also provides information on the carrier population dynamics, which can be helpful in the assignment of observed energy structures.

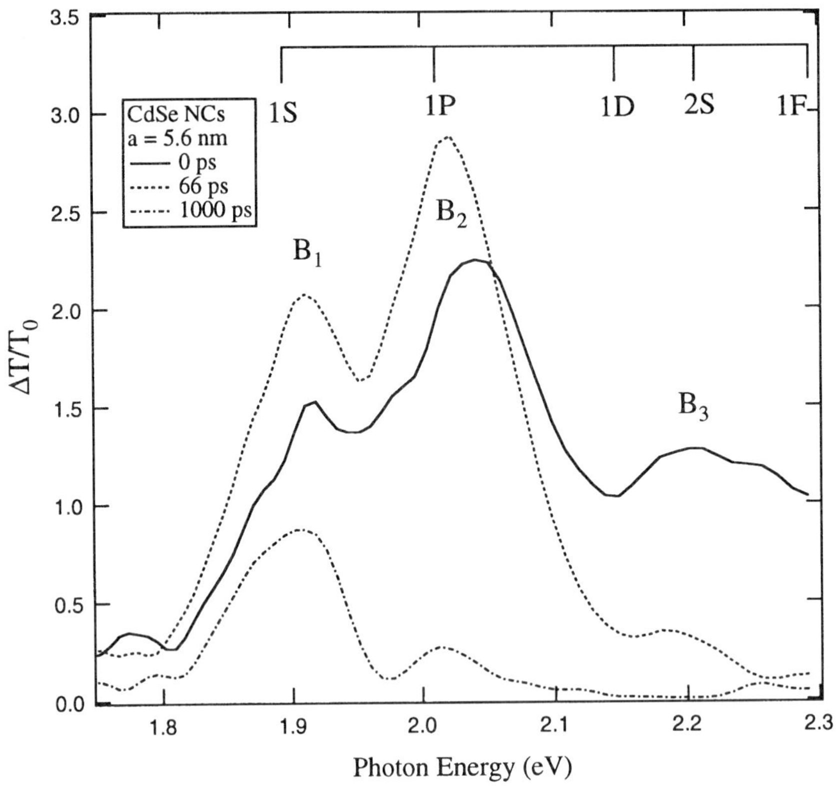

Fig. 10. Time-resolved differential transmission spectra of CdSe NC–glass sample ($a \approx 5.6$ nm) measured at three different delay times (0, 66, and 1000 ps) between pump and probe pulses. The sample temperature is 80 K. The positions of the bleaching features B_1, B_2, and B_3 are compared with spectral energies of transitions calculated within the parabolic-band model. (Source: Reprinted with permission from [32]. © 1994 American Institute of Physics.)

Figure 10 shows the ps DT spectra of CdSe NC–glass samples ($a \approx 5.6$ nm, $T = 80$ K) measured in Ref. [32]. These spectra were recorded at three different delay times (Δt) between pump and probe pulses (pump-photon energy was 2.33 eV). The linear absorption spectrum of the sample used in these measurements (shown by the solid line in Fig. 7) did not exhibit discrete features, and only the first-derivative spectrum (inset to Fig. 7; solid line) showed two broad resonances associated with transitions involving hole states from the $J = 3/2$ and $J = 1/2$ valence subbands. The pump intensity used in nonlinear-transmission measurements was sufficient to excite more than 10 *e-h* pair per NC, which corresponds to the initial position of the electron Fermi level above the $1P$ state [see Section 6.1]. In contrast to the structureless linear absorption, DT spectra exhibit well resolved discrete features at ~1.91 (B_1), ~2.02 (B_2), and ~2.18 eV (B_3). The positions of these bands were compared with the positions of allowed optical transitions calculated within the parabolic-band model taking into account the Coulomb corrections as given by Eq. (21). This comparison showed that B_1 and B_2 could be assigned to transitions involving the ground ($1S$) and the first excited ($1P$) electron states. The center of the B_3 band corresponded to the $2S$ transition, although two other spectrally close transitions ($1D$ and $1F$) could also contribute to the B_3 bleaching. These assignments are consistent with the observed DT dynamics. A fast decay of the B_3 bleaching accompanied by a complementary growth of the B_2 feature was attributed to a lowering down of the electron Fermi level during carrier cooling. The slower relaxation of the B_2 and B_1 bands was attributed to Auger recombination (see Section 7.3) followed by radiative decay.

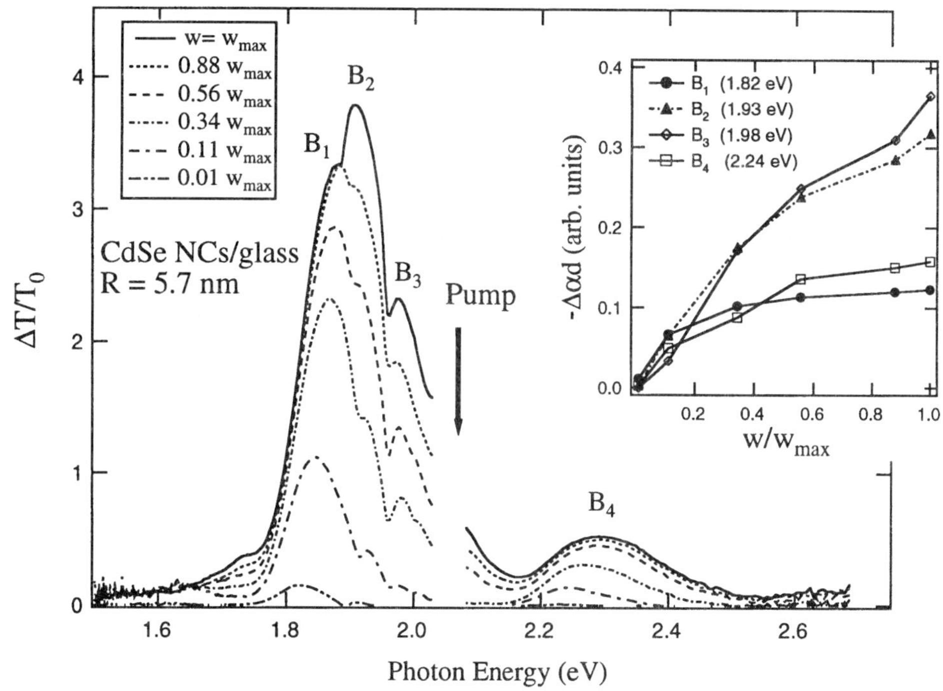

Fig. 11. Differential transmission spectra of CdSe NC–glass sample ($a \approx 5.7$ nm; room temperature) measured at 1 ps after excitation for different pump intensities. Inset shows pump-intensity dependence of bleaching features B_1–B_4. (Source: Reprinted with permission from [75].)

Excited electron states were also clearly manifested in pump-dependent fs DT spectra of CdSe NC–glass samples ($a \approx 5.6$ nm) studied in Ref. [75] (pump photon energy was 2.06 eV). These spectra (see Fig. 11) exhibited four well resolved features at ~1.85, ~1.93, ~1.98, and ~2.24 eV marked as B_1–B_4. The B_1 and B_4 bleachings were characterized by similar pump dependencies (inset to Fig. 11), and therefore, were attributed to transitions involving the same $1S$ electron but two different hole states: the $1S_{3/2}$ state (and likely unresolved $2S_{3/2}$ state) (B_1) from the $J = 3/2$ subband and the $3S_{1/2}$ state (B_2) from the split-off $J = 1/2$ subband. The pump dependence of the B_2 and B_3 features (inset to Fig. 11) was distinctly different from that of the B_1 and B_4 bands, indicating that the electron state involved in the transitions at 1.93 and 1.98 eV is different from $1S$. The B_1 and B_4 bleachings saturated at high-pump fluences, whereas a continuous increase was seen for the B_2 and B_3 bands (inset to Fig. 11). The saturation of the bleaching of the lowest transition can be explained by the fact that the population of the $1S$ electron state reaches its maximum (two particles), whereas the growth of the B_2 and B_3 features is due to the onset of the population of the first excited $1P$ electron state. The pump dependence of the DT spectra shown in Figure 11 is analyzed in more detail in Section 6.1.

3.3. Studies of Hole Energy Structures

Studies of valence-band energy structures are complicated by the high spectral density of hole states which results in a small spacing between quantized hole levels. Therefore, systematic analysis of hole energy spectra only became possible after the development of procedures for preparation of samples with very narrow size distributions exhibiting small inhomogeneous linewidths.

High-quality glass samples were used in Ref. [29] to study the linear absorption of CdSe NCs. The low-temperature absorption spectra of these samples (Fig. 12) showed

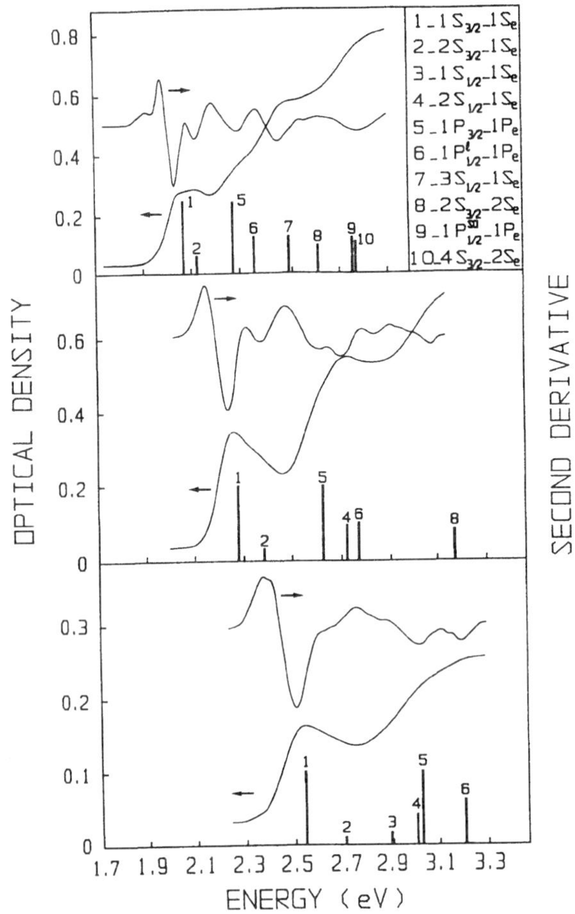

Fig. 12. Absorption and absorption-second-derivative spectra of CdSe NC–glass samples ($T = 77$ K) with mean NC radii 2.1, 2.6, and 3.8 nm (from the bottom to the top) in comparison with the positions of optical transitions calculated taking into account valence-band mixing effects. (Source: Reprinted with permission from [29]. © 1993 Optical Society of America.)

well resolved discrete features, indicating a high level of NC monodispersity. Discrete structures due to size quantization can be particularly well resolved in the second-derivative spectra shown in Figure 12 along with the linear absorption data. The effective-mass model, accounting for the valence-band mixing effects (see Section 2.2 and Fig. 4), was used to explain the positions of different absorption features. The positions and intensities of transitions predicted by this model are shown in Figure 12 by vertical bars. Several S- and P-type hole states were assigned by comparing experimentally detected features with the results of the modeling.

The lowest absorption peak was attributed to contributions from two transitions involving the $1S$ electron states and the hole states $1S_{3/2}$ and $2S_{3/2}$ (transitions 1 and 2 in Fig. 12, respectively). These transitions were well resolved in a sample with $a = 3.8$ nm and were merged into a single absorption band in samples with smaller NC sizes, though two separate features were still manifested in the second-derivative spectra (see Fig. 12) of the small-size NCs. The spacing between transitions involving $1S_{3/2}$ and $2S_{3/2}$ hole states as well as the relative strength of the first transition increased with decreasing NC sizes which was in agreement with theoretical predictions [29, 30]. The second-derivative spectra also exhibited features associated with the $F = 1/2$ hole states ($1S_{1/2}$, $2S_{1/2}$ $3S_{1/2}$; transitions 3, 4, and 7, respectively) and the higher lying $F = 3/2$ state $4S_{3/2}$ (transition 10).

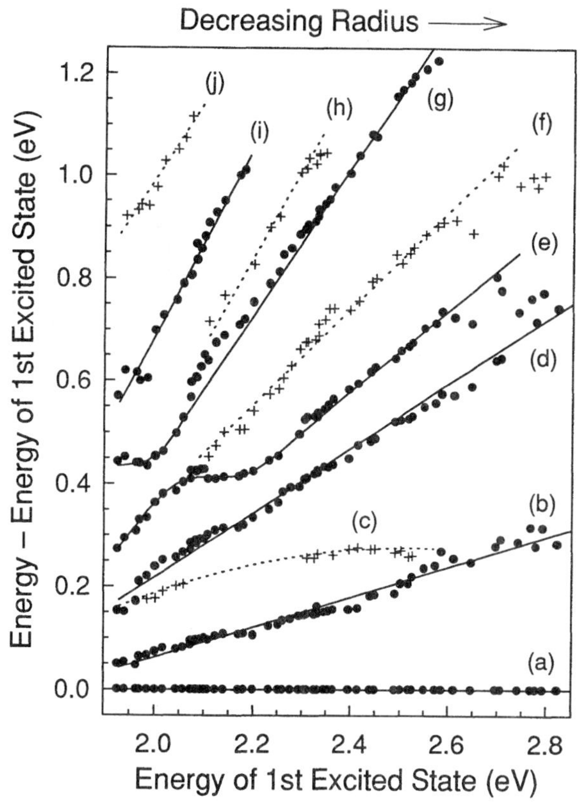

Fig. 13. Transition energies derived from PLE data of CdSe colloidal NCs of various sizes plotted versus the energy of the lowest $1S$ transition which gives a measure of the radius of NCs probed in PLE measurements. Strong (weak) transitions are denoted by circles (crosses). The lines are visual guides for transitions to clarify their size evolution. Based on comparison with calculated transition energies, the PLE features (a)–(g) were assigned as follows: (a) $1S(e)$-$1S_{3/2}(h)$, (b) $1S(e)$-$2S_{3/2}(h)$, (c) $1S(e)$-$1S_{1/2}(h)$, (d) $1P(e)$-$1P_{3/2}(h)$, (e) $1S(e)$-$2S_{1/2}(h)$, (f) $1P(e)$-$1P_{5/2}(h)$ and/or $1P(e)$-$1P_{1/2}(lh)$, (g) $1S(e)$-$3S_{1/2}(h)$. The features (h)–(j) were in the region with a high density of allowed optical transitions which complicated their accurate assignment. (Source: Reprinted with permission from [33]. © 1996 American Physical Society.)

Several hole states of P symmetry were also resolved in the spectra. A minimum in the second-derivative spectra at around 2.3 eV seen in a sample with $a = 3.8$ nm were assigned to the transition involving the $1P_{3/2}$ state (transition 5) with a possible contribution from the transition involving the $1P_{1/2}$ light-hole state (transition 6). These two P-type hole states were resolved as separate minima in the spectra of samples with $a = 2.6$ and 2.3 nm. The P state originating from the spin-orbit split-off subband ($1P^{so}$) was possibly seen as transition 9 in the spectrum of the sample with $a = 3.8$ nm.

Detailed studies of the size dependence of hole spectra were performed in Ref. [33] using a series of high-quality colloidal samples with mean radii from ~1.2 to ~5.3 nm and size dispersion $<5\%$ rms. To further reduce the effects of inhomogeneous broadening, the size-selective PLE technique was used to detect discrete absorption features, allowing resolution of up to eight bands in a single PLE spectrum. The results of size-dependent measurements of positions of these bands are plotted in Figure 13 as ΔE_i versus E_{1S}, where ΔE_i is the high-energy shift of the band number i with respect to the lowest $1S$ transition, and E_{1S} is the energy of the $1S$ transition. This choice of x axis allows one to minimize errors caused by uncertainty in the sizes of NCs actually probed in the PLE spectra. A comparison of the experimental data with calculations which take into account band-mixing effects [29, 30] allowed a confident assignment of the six lowest transitions involving five S-type ($1S_{3/2}$, $2S_{3/2}$, $1S_{1/2}$, $2S_{1/2}$, and $3S_{3/2}$) and one P-type ($1P_{3/2}$) hole states. Four more higher lying transitions were resolved in the spectra, but they were only tentatively assigned because of the high spectral density of allowed transitions at large confinement energies. The important result of these studies was the observation of the avoided-crossing behavior for states $2S_{3/2}$ and $3S_{3/2}$ [transitions (e) and (g) in Fig. 13] at $E_{1S} \approx 2$ eV, and for states $1S_{1/2}$ and $2S_{1/2}$ [transitions (c) and (e) in Fig. 13] at ~2.2 eV. This clearly indicates the important role of band-mixing effects in the valence-band energy spectra of NCs. The effects of the band mixing were also manifested in the size dependence of the oscillator strength of transitions involving $S_{1/2}$ states. These states represent a mixture of the D- and S-type wave functions from the $J = 3/2$ and $1/2$ subbands, respectively. The $1S_{1/2}$ state, for instance, in large NCs is dominated by the D-type wave function which is not coupled to the $1S$ electron level. For small sizes the dominant contribution to this state comes from the S-type component which is strongly coupled to the lowest electron state. This size-dependent behavior of the wave function should result in increasing oscillator strength for the $1S(e)$-$1S_{1/2}$ transition with decreasing NC radius, which was observed experimentally.

4. FINE STRUCTURE OF THE LOWEST EXCITON STATE

In spherical quantum dots with a cubic lattice the lowest e-h pair state $[1S_{3/2}(h)1S(e)]$ (the lowest exciton) is eightfold degenerate. However, the effects of the crystal field in the hexagonal lattice [77], nonspherical NC shape [78], and the electron-hole exchange interaction [79–81] lift this degeneracy, leading to fine structure of the lowest exciton. In the case of the exchange interaction, a good quantum number is the total momentum of the e-h pair $\hat{\mathbf{N}} = \hat{\mathbf{F}}_e + \hat{\mathbf{F}}_h$. When the anisotropy of the lattice and the nonspherical shape are included in the model, the only conserved quantity is a projection of the total exciton momentum N_m along a unique crystal axis.

The crystal field of the wurtzite lattice leads to a splitting of states with $M_h = \pm 1/2$ and $\pm 3/2$ (M_h is a projection fo the hole momentum $\hat{\mathbf{F}}_h$) (see Fig. 14a) which is analogous to the A-B splitting in bulk semiconductors (see Fig. 2). In the strong confinement regime, this splitting (δ_{AB}) is independent of the NC radius and is only determined by the ratio (β) between the light- and heavy-hole masses along the c axis ($\beta = m_{lh}/m_{hh}$) [77]: $\delta_{AB} = w(\beta)\Delta_{AB}$, where Δ_{AB} is the crystal-field-induced splitting in bulk materials. For $\beta = 0$, $w(0) \approx 0.2$, meaning that the A-B splitting in this case is $5\times$ smaller than in bulk semiconductors.

When the crystal-field-induced splitting is taken into account, the hole ground state is characterized by the momentum projection $M_h = \pm 3/2$. Holes in this state and $1S$ electrons can form e-h pair states (excitons) with total momentum projection $N_m = \pm 2$ or ± 1. The exciton with $|N_m| = 1$ is coupled to the ground state by an allowed dipole transition, whereas the $|N_m| = 2$ exciton is not optically active, because a single-photon emission (absorption) cannot account for the two-unit change of the angular momentum. The $|N_m| = 2$ excitons cannot be directly excited by light, but they can be created as a result of carrier relaxation after excitation into higher lying exciton states (associated, e.g., with the $M_h = \pm 1/2$ holes); this can significantly alter carrier recombination dynamics [77].

The A-B splitting is further affected by effects of the nonspherical shape [78]. The deviation from the spherical shape can be characterized by the parameter $\chi = a_1/a_2 - 1$, where a_1 and a_2 are the major and minor axes of a NC, respectively. χ is positive in prolate NCs and is negative in oblate ones. Due to the nonspherical NC shape, the hole states with $|M_h| = 3/2$ and $1/2$ shift in opposite directions. For the lowest $1S_{3/2}$ hole states, these shifts are [78],

$$\Delta E_{\pm 3/2} = -\Delta E_{\pm 1/2} = -\chi v(\beta) E(1S_{3/2}) \tag{28}$$

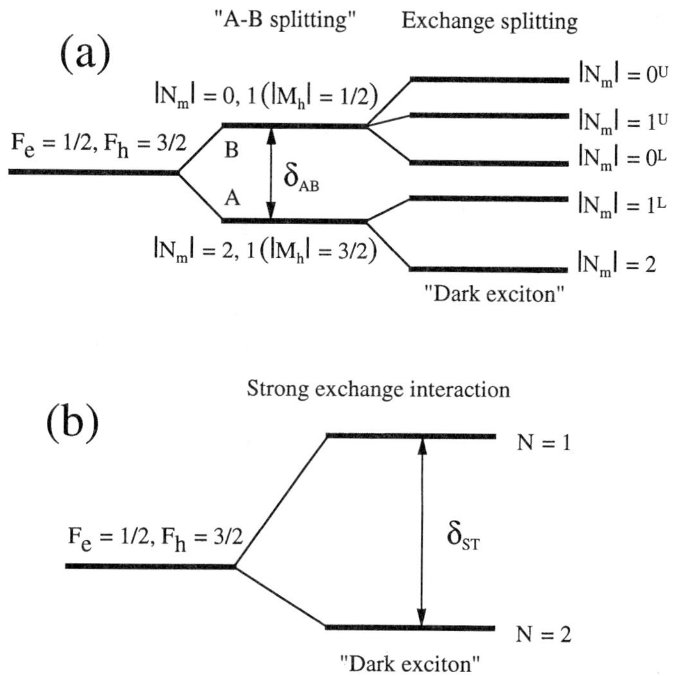

Fig. 14. (a) The splitting of the lowest exciton ($F_e = 1/2$, $F_h = 3/2$) due to the crystal field and/or the NC nonspherical shape (the A-B splitting). The A and B excitons are further split by the e-h exchange interaction into five sublevels (the limit of weak exchange interaction). (b) The splitting of the $N = 1$ and $N = 2$ excitons in the limit of strong exchange interaction which overwhelms the A-B splitting.

where $E(1S_{3/2})$ is the ground-state hole energy given by Eq. (14) and $v(\beta)$ is a parameter determined by the ratio of the light- to heavy-hole masses. With increasing β, $v(\beta)$ decreases from $4/15$ at $\beta = 0$ to $\sim(-0.26)$ at $\beta = 0.3$ with a crossover to positive v at $\beta \approx 0.14$; $v(\beta)$ increases in the range $\beta > 0.3$ until it reaches zero at $\beta = 1$ [78]. Due to the fact that energy corrections induced by the nonspherical shape are proportional to the hole energy, they show a size dependence, increasing as a^{-2} with decreasing NC radius. The sign of the energy shifts, which determines the alignment of the states with $|M_h| = 3/2$ and $1/2$, depends on the sign of χ and the magnitude of β. For example, in the case of CdSe ($\beta \approx 0.3$ [29]), the parameter v is negative ($v \approx -0.26$). Therefore, in prolate NCs ($\chi > 0$) (which is typical for CdSe colloids [82]), the effects of the nonspherical shape tend to reduce the A-B splitting resulting from the crystal field, which can ultimately lead to switching the ground hole state from one with $|M_h| = 3/2$ to that with $|M_h| = 1/2$.

The A and B exciton levels are further split by the e-h exchange interaction into five sublevels (Fig. 14a). These sublevels are labeled by the modulus of the projection of the total angular momentum $|N_m|$ with a superscript which denotes the upper (U) or the lower (L) states with the same $|N_m|$. The states with $|N_m| = 2$, 1^L, and $0^{U,L}$, 1^U originate from the A and B excitons, respectively. The exchange splitting is proportional to the overlap of the electron and hole wave functions and, therefore, is strongly enhanced by spatial confinement, following an a^{-3} dependence [80]. Because the crystal-field-induced contribution to the A-B splitting is size independent and the contribution due to nonspherical shape has the a^{-2} dependence, the exchange interaction overwhelms the A-B splitting effects in small NCs. In the limit of strong exchange interaction (Fig. 14b), the lowest exciton is split into two states with $N = 1$ (threefold degenerate) and $N = 2$ (fivefold degenerate). The state with $N = 1$ is strongly coupled to the ground state by dipole transitions, whereas the lowest state with $N = 2$ is optically forbidden. This can lead to the existence of so called "dark excitons" which are passive in optical absorption but can be detected in photoluminescence (PL) by their weak emission [80]. The splitting between the $N = 2$ and $N = 1$ states can be expressed in terms of the singlet–triplet splitting in bulk semiconductors (Δ_{ST}),

$$\delta_{ST} = 4\zeta \left(\frac{a_x}{a} \right)^3 \Delta_{ST} \tag{29}$$

where ζ is a constant on the order of unity. In bulk CdSe, for example, the exchange interaction leads to only a 0.13-meV difference in the energies of singlet and triplet excitons. However, due to quantum confinement, this splitting increases up to ~20 meV in NCs of ~1.4 nm radius [80].

The fine structure of the lowest exciton state was studied experimentally in both colloidal [80, 82] and glass [70, 83, 84] NC samples. These studies were performed using high-resolution PLE and FLN techniques. Low-temperature PLE spectra of CdSe colloidal NCs show fine structure in the band-edge absorption in which up to three features can be resolved [82]. The lowest of these features (α) was assigned to the $|N_m| = 1^L$ exciton, whereas the other two were attributed to states with $|N_m| = 1^U$ and 0^U (see Fig. 14a). In small NCs ($a \leqslant 3$ nm) two latter features merged into a single band (β). The $\alpha - \beta$ splitting increased with decreasing NC radius (up to ~50 meV for $a = 1.5$ nm), which was explained in terms of the confinement-enhanced exchange interaction [79–82].

The FLN studies of CdSe NCs indicate that the emitting state is different from the absorbing one [70, 80, 82–84] [see Fig. 15a], consistent with the predicted exchange-interaction induced splitting of the lowest exciton (A exciton) into the optically active state with $|N_m| = 1^L$ (absorbing state) and the optically passive "dark exciton" with $|N_m| = 2$ (emitting state). In agreement with theory [79, 80], experimental data indicate that the energy spacing between these states (the Stokes shift) increases with decreasing NC radius (see Fig. 15a and b) [70, 80, 82–84]. In CdSe colloids, the Stokes shift is only 1–2 meV

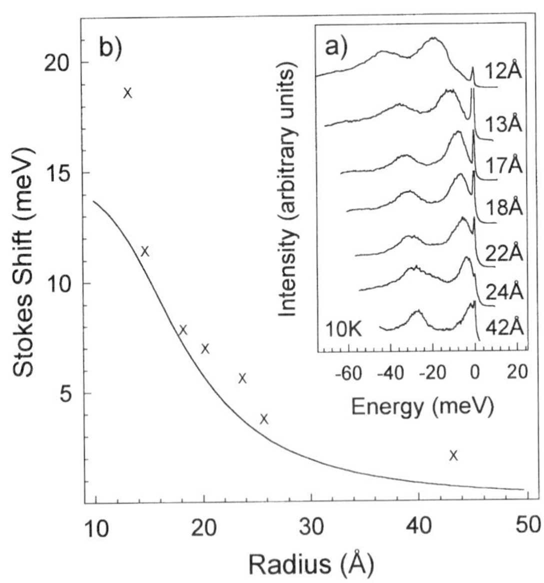

Fig. 15. (a) Normalized FLN spectra of CdSe colloidal NCs with radii from 12–42 Å. A small amount of the pump laser light is included for reference in each spectrum and set to zero energy for comparison. (b) Size dependence of the Stokes shift (crosses) between the pump-photon energy and the position of the zero-phonon line in the FLN spectra. The solid line is the calculated size dependence of the splitting between the $|N_m| = 2$ and 1^L states. (Source: Reprinted with permission from [80]. © 1995 American Physical Society.)

in large NCs ($a \approx 4$–5 nm), however, due to the confinement-induced enhancement in the exchange interaction, it increases up to about 20 meV in NCs of small radius ($a \approx 1.4$ nm) [80, 82]. The fact that two different states are involved in the absorption and emission processes is confirmed by PL polarization studies [70, 83, 84] which indicate that the polarization of the resonantly excited PL is different from that of the absorbed light.

The effect of the magnetic field on the PL dynamics support the assignment of the emitting state to the $|N_m| = 2$ dark exciton [80]. Without a magnetic field, the PL exhibits a long μs decay, characteristic of the optically passive "dark" exciton. The decay rate increases with increasing magnetic field which can be explained in terms of the magnetic-field induced mixing of the dark $|N_m| = 2$ state with the optically active $|N_m| = 1^L$ state.

5. EFFECTS OF ELECTRON-PHONON INTERACTIONS ON THE OPTICAL SPECTRA OF SEMICONDUCTOR NANOCRYSTALS

Electron-phonon interactions in semiconductor NCs are significantly modified with respect to those in bulk materials. Spatial confinement leads to quantization of the phonon modes. In spherical NCs, these modes are classified by their orbital momentum l and are $(2l + 1)$-fold degenerate with respect to momentum projection m [85, 86]. However, in contrast to the discrete electronic energy structures clearly manifested in optical spectra, the discrete structure of vibrational modes is difficult to resolve due to the close mode spacing resulting from the large ion masses. The electron-phonon interactions in NCs are strongly effected by the discrete structure of electronic energies, significantly reducing the efficiency of first-order processes involving a change in the electron (hole) state. These processes can only efficiently occur between states separated by integer numbers of phonons and are strongly hindered due to the scarcity of the electronic states [87].

5.1. The Model of a Displaced Oscillator

In strongly confined NCs with large energy-level separations, the mixing between electronic (exciton) states resulting from electron-phonon coupling can be ignored. Therefore, electron (exciton)-phonon interactions in these systems can be treated within the independent boson model [88], which has been traditionally used to describe a localized state coupled to lattice vibrations (the model of a displaced harmonic oscillator) [89, 90]. In the simple case of coupling to single-frequency phonons (Einstein model), the electron-phonon interaction can be characterized by a coupling parameter S (Huang–Rhys parameter). S is related to the dimensionless mode displacements Δ_k (k denotes the phonon mode) (see Fig. 16a) and further to the diagonal matrix element of the exciton-phonon interaction M_{jj}^k (j denotes the exciton state interacting with phonons) by the expression,

$$S = \sum_k \Delta_k^2 = \sum_k \frac{|M_{jj}^k|^2}{(\hbar\omega_0)^2} \tag{30}$$

where $\hbar\omega_0$ is the phonon-mode energy. In this model, the electron-phonon coupling gives rise to a series of equally spaced phonon sidebands in both emission and absorption spectra. These sidebands are associated with multiple phonon emission (absorption) processes accompanying the optical transition. The optical absorption spectrum taking into account the phonon progression can be presented as follows,

$$\alpha(\hbar\omega) \propto \sum_{-\infty}^{\infty} p_n \delta(\hbar\omega - E_0 - n\hbar\omega_0) \tag{31}$$

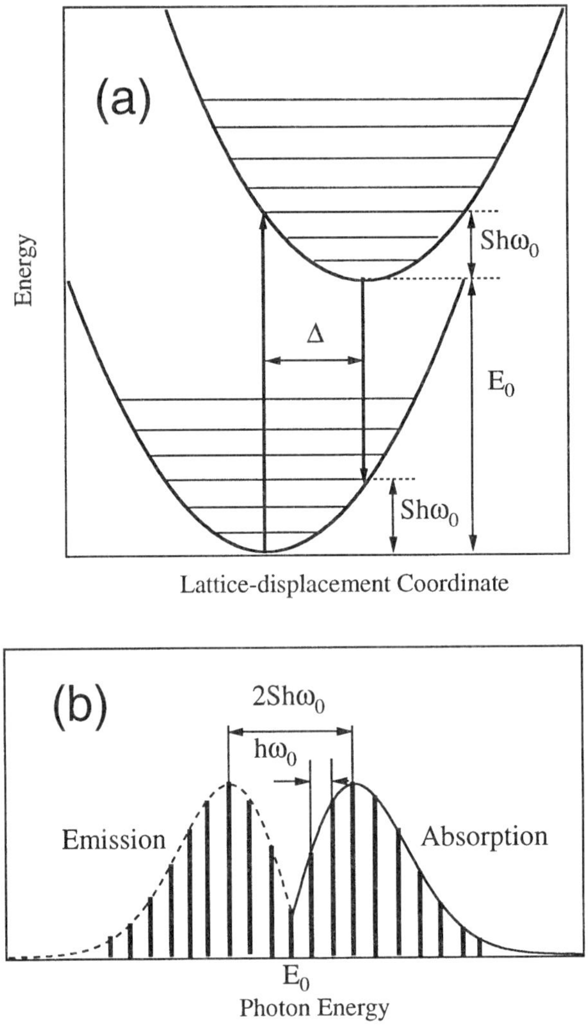

Fig. 16. (a) Configuration diagram representing the dependence of electronic energies of the ground and the first excited states on the lattice-displacement coordinate. Δ is the lattice relaxation accompanying the electronic excitation. Arrows show vertical transitions which determine the positions of the emission and absorption maxima. (b) Phonon progressions in absorption and emission; envelop functions for these two processes are shown by solid and dashed lines, respectively.

where E_0 is the energy of the zero-phonon transition (Fig. 16a). The corresponding emission spectrum is a mirror image of the absorption spectrum with respect to the energy E_0. At low temperatures when $kT \ll \hbar\omega_0$ the phonon absorption is suppressed and all phonon replicas in absorption (emission) spectra are located to the high- (low-) energy side from the zero-phonon line as shown in Figure 16b. The intensities of the phonon replicas can be found from the expression [89],

$$p_n = e^{-S(2N+1)} \left(\frac{N+1}{N} \right)^{n/2} I_n \left[2S\sqrt{N(N+1)} \right] \tag{32}$$

where $N = [\exp(\hbar\omega_0/kT) - 1]^{-1}$ is the phonon-mode occupation number density and I_n is the modified Bessel function. In the case of weak electron-phonon coupling ($S \ll 1$), the optical spectra are dominated by the zero-phonon line. In the opposite limit of strong coupling ($S \gg 1$), the zero-phonon line is suppressed due to an exponential factor $e^{-S(2N+1)}$

and the maximum of the absorption (emission) spectrum shifts upward (downward) in energy to the phonon replica with $n = S$, which corresponds to vertical transitions in the configuration diagram in Figure 16a. This results in a large Stokes shift ($\Delta_S = 2S\hbar\omega_0$) between absorption and emission maxima (Fig. 16b). In the low-temperature limit ($T \to 0$), the function in the right-hand side of Eq. (32) reduces to the Poisson distribution,

$$p_n = \frac{e^{-S} S^n}{n!} \tag{33}$$

which further reduces to a Gaussian distribution centered at $n = S$ in the limit $S \gg 1$.

The approach described in the previous text can be applied to both optical and acoustic modes. However, exciton interaction with these two types of lattice vibrations affects the optical spectra in different ways. Because the optical modes have relatively large frequencies and a narrow energy dispersion, the corresponding phonon side bands are spectrally well separated, leading to the appearance of the phonon progressions in absorption and emission spectra [31, 82, 91]. In contrast, acoustic modes have small energies and are characterized by almost continuous energy spectra. Therefore, the side bands associated with acoustic phonons tend to merge into a continuous band contributing to the homogeneous linewidth of optical transitions. Next we will examine in more detail the effects of carrier coupling to optical and acoustic phonons on optical spectra.

5.2. Electron-Optical Phonon Interactions

In NCs, electronic excitations are coupled to two different types of optical modes associated with longitudinal optical (LO) and surface optical (SO) vibrations [85, 86]. As shown by resonance Raman studies [34, 35], the LO-phonon frequencies in NCs are essentially identical to those in bulk materials. The frequencies of the SO modes corresponding to the orbital momentum l (ω_l) can be determined from the equation,

$$\frac{l+1}{l}\epsilon_m = \epsilon_\infty \frac{\omega_l^2 - \omega_{LO}^2}{\omega_l^2 - \omega_{TO}^2} \tag{34}$$

where ω_{LO} and ω_{TO} are longitudinal and transverse phonon modes, respectively, related by the expression $(\omega_{LO}/\omega_{TO})^2 = \epsilon_0/\epsilon_\infty$, ϵ_0 and ϵ_∞ are the high- and low-frequency dielectric constants of a NC, and ϵ_m is the dielectric-response function of the surrounding matrix which is usually assumed to be frequency independent. The frequencies of the SO modes are slightly smaller than those of the LO phonons. In the case of CdSe ($\omega_{LO} = 210\ \text{cm}^{-1}$), for instance, they are in the range 194–200 cm^{-1} [39].

In bulk semiconductors, the electron-optical phonon coupling is dominated by the polar Fröhlich interaction [92]. However, due to spatial confinement, this interaction is significantly suppressed in NCs. This suppression can be understood from the following simple arguments [60]. The polar coupling of the *e-h* pair (exciton) to optical modes is determined by the product of the exciton local charge density ρ_{eh} and the electric potential ϕ_{opt} associated with optical (LO and SO) vibrations,

$$M_{jj}^{\text{pol}} = \int_{\text{NC volume}} \phi_{\text{opt}}(\mathbf{r})\rho_{eh}^{j}(\mathbf{r})\,d\mathbf{r} \tag{35}$$

where $\rho_{eh}^{j}(\mathbf{r}) = e(|\Phi_e(\mathbf{r})|^2 - |\Phi_h(\mathbf{r})|^2)$, and Φ_e and Φ_h are the electron and hole wave functions corresponding to the $|j\rangle$ exciton (*e-h* pair) state. In the parabolic-band model applied to the spherical quantum well with an infinite barrier, the electron and hole ground-state wave functions have exactly the same shape, resulting in a zero local charge density in a NC. This immediately leads to the conclusion that the polar coupling dominating the electron-phonon interaction in bulk materials is completely suppressed in strongly confined NCs. However, the real situation is more complex, because any deviations in electron and/or hole wave functions which may result from such effects as penetration of the wave function into the potential barrier, mixing of valence subbands or the Coulomb e-h interaction lead to a nonzero local charge density with the associated restoration of the polar coupling.

The interaction of the ground-state exciton with optical vibrations in NCs is dominated by LO phonons with a minor contribution from coupling to the SO vibrations via the $l = 2$ component of the hole wave function [93] (in the case of purely $l = 0$ wave functions the exciton does not interact with the SO phonons [39, 94]). The results obtained for the absolute magnitudes of the exciton-LO-phonon coupling strength and its size dependence are strongly affected by the assumptions made regarding the electron and hole charge distributions. In calculations, electrons have been usually described by a simple parabolic-band model, whereas different assumptions have been made regarding the hole wave function. In the case of extremely high hole localization (δ-functionlike charge distribution), calculations predict a size independence for the coupling parameter S [39]. The use of more realistic hole wave functions which account for the band-mixing effects leads to the dependence $S \propto 1/a$ [93]. If the Coulomb e-h interaction is also taken into account, the coupling parameter shows a nonmonotonic size dependence with a minimum at ~7 nm in CdSe NCs [94]. This nonmonotonic behavior occurs because the parameter S increases due to increasing coupling to short-wavelength phonons [60] for smaller sizes, whereas in the range $a > 7$ nm, the coupling strength increases, approaching its bulk-exciton value due to increasing difference in the electron and hole wave functions caused by a large difference in the electron and hole masses.

Calculations using extended wave functions for both an electron and a hole result in small magnitudes for the coupling parameter S, typically below 0.01 [93, 94]. However, experimental studies indicate a much stronger coupling strength with S on the order of 0.1–1. The enhancement of the carrier-phonon polar interaction can be explained by carrier localization (due to, e.g., surface trapping or hole localization in the donorlike-exciton model [95, 96]) which leads to charge separation with an associated increase in the local charge density. The coupling strength for transitions involving localized states is much higher than for those involving extended states, being ~1 for the case of CdSe NCs [94, 96].

Experimental studies of the carrier LO-phonon coupling have been performed by PL [31, 82, 91, 97], absorption [31, 82], Raman [34, 35, 39, 41], and time-resolved degenerate four-wave mixing (FWM) [38, 98] spectroscopies. If the pump-photon energy significantly exceeds the energy of the lowest optical transition, NCs exhibit a broad structureless PL band, red shifted with respect to the absorption maximum, Figure 17a. This shift is typically several tens of meV in colloidal samples, and can be greater than 100 meV in glass samples. In early studies, particularly in those performed on NC–glass samples, the large PL band broadening and its low-energy shift were explained in terms of the strong coupling to LO phonons, with the coupling parameter estimated to be $S > 1$ [99–101]. However, the studies using spectroscopic size-selective techniques such as FLN and PLE [31, 69, 82], as well as time-resolved PL measurements [49, 91], indicate that the actual Stokes shift is relatively small. In CdSe NCs, it is in the range of 1–20 meV depending on the NC radius [80, 82] (see previous section). The large displacement of the PL maximum with respect to the absorption peak, seen without using size-selective techniques (in measurements of global PL and global absorption), can result from the effects of inhomogeneous broadening [69] and/or carrier trapping [91, 102], with the latter mechanism being particularly important in glass samples. In colloidal NCs, the global PL and the global absorption profiles as well as their energy shift can be modeled by summing contributions from "single-dot" LO-progression spectra of NCs of different radii, weighted using a size-distribution function [31, 82] (see illustration in Fig. 18).

In FLN experiments, a narrow-band laser with photon energy in the red part of the absorption peak is used to selectively excite only the largest NCs from the size distribution. This type of excitation results in a significant suppression of the inhomogeneous broadening and allows the observation of structured PL spectra in colloidal NCs which clearly

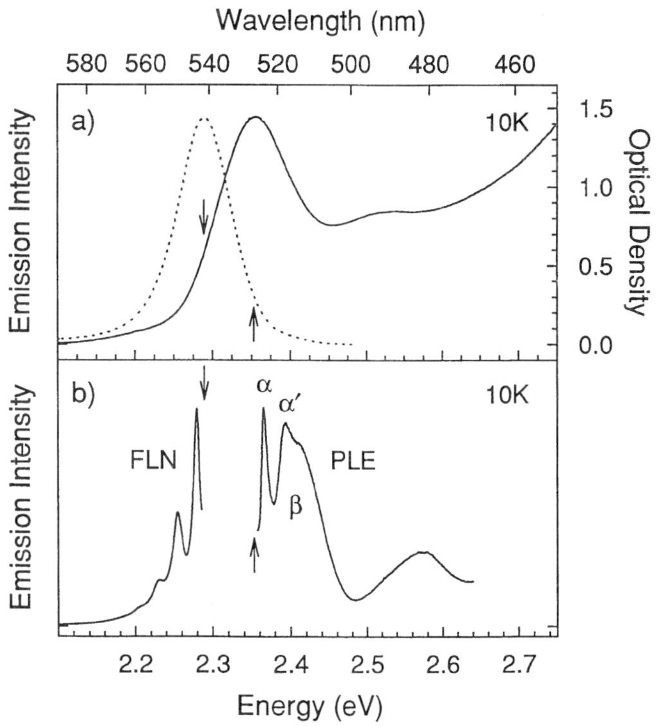

Fig. 17. Absorption (solid line) and PL (dotted line) spectra (a) of CdSe colloidal NCs ($a = 1.9$ nm; $T = 10$ K) in comparison to FLN and PLE spectra (b) of the same sample. The downward (upward) arrows denote the excitation (emission) positions for FLN (PLE). (Source: Reprinted with permission from [80]. © 1995 American Physical Society.)

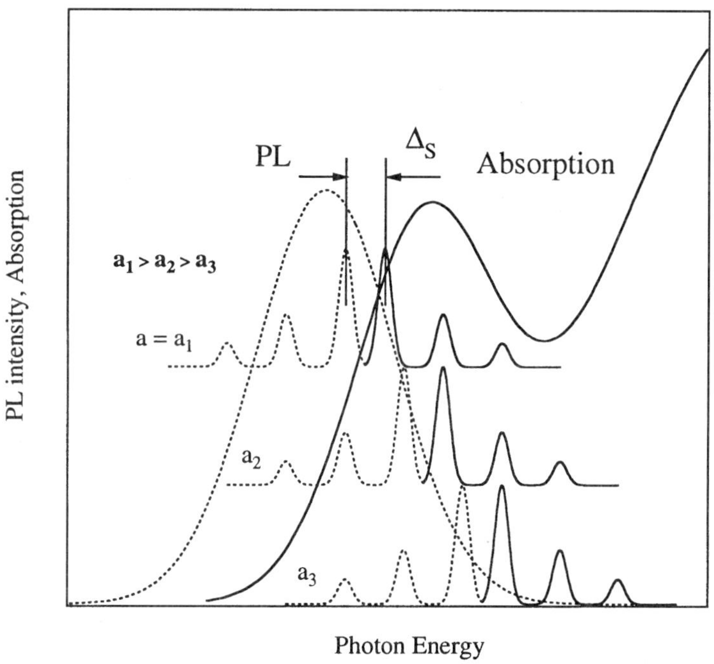

Fig. 18. Illustration of decomposition of the global PL and global absorption spectra of NC samples into single-dot spectra exhibiting phonon progressions. The actual Stokes shift between zero-phonon absorption and emission lines (Δ_S) is significantly smaller than the shift between the peaks in the global absorption and global emission spectra.

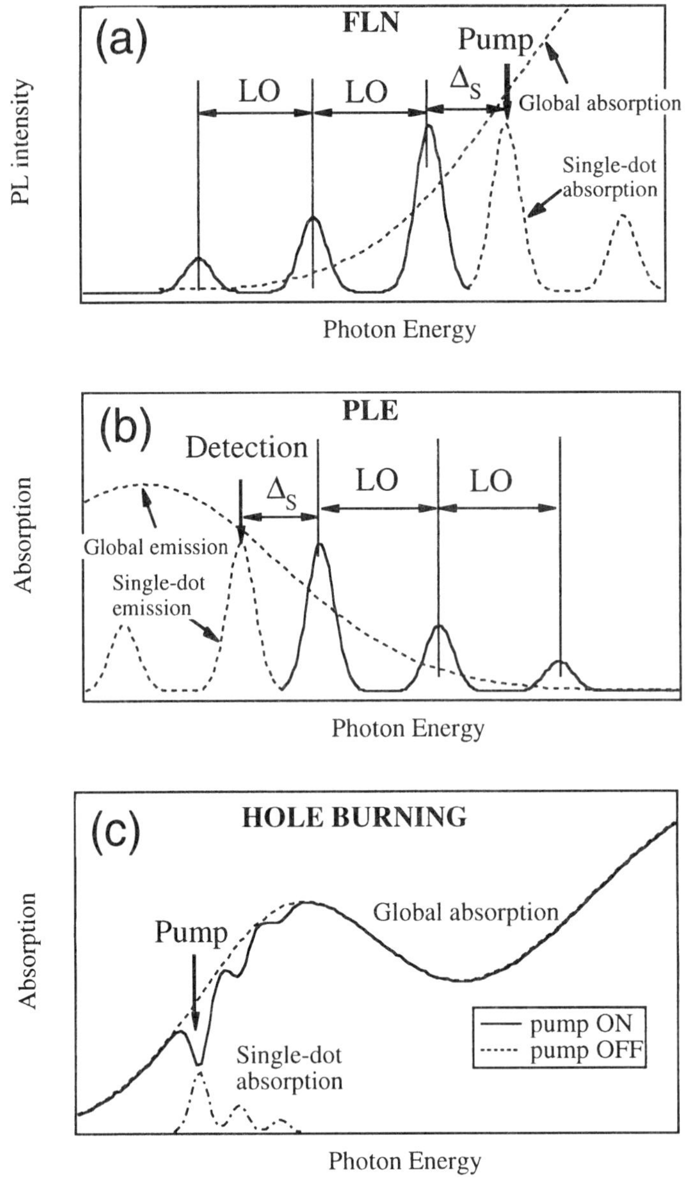

Fig. 19. Illustration of three different size-selective spectroscopic techniques: (a) fluorescence line narrowing (FLN), (b) photoluminescence excitation (PLE), and (c) spectral hole burning.

exhibit LO-phonon progressions [31, 69, 82] (see schematic in Figure 19a and the experimentally detected FLN spectrum of CdSe NCs in Figure 17b). The observation of phonon replicas is also possible in high-quality NC–glass composites [70, 83], although most of the glass samples exhibit structureless continuous wave (cw) emission dominated by trapping states. However, even in the samples which show efficient carrier trapping, the phonon progressions can be resolved using time-resolved techniques which allow one to access early stages of carrier relaxation before the carriers are removed from quantized states [91].

Inhomogeneous broadening is also significantly reduced in PLE spectra as illustrated in Figure 19b, allowing the observation of sharp features related to LO-phonon-assisted processes in absorption (compare the PLE spectrum in Figure 17b with the global absorption spectrum in Fig. 17a). The amplitudes of phonon replicas derived from the FLN and PLE spectra can be used to experimentally determine the strength of the electron-LO-phonon coupling. According to Eq. (33), the ratio between the spectrally integrated first- and zero-phonon lines provides a direct measure of the coupling parameter S. Interestingly, the magnitudes of S derived from FLN and PLE data are different from each other. In the case of CdSe colloids with radii from 1.5 to 5 nm, the coupling constant at $T = 10$ K is in the range 0.08–0.2 as derived from PLE spectra and in the range 0.36–0.7 as found from FLN measurements [82]. This difference is indicative of different electronic states involved in absorption and emission processes. Photoexcitation leads to the generation of carriers in extended quantized states which are relatively weakly coupled to lattice vibrations, as indicated by absorption spectra. Shortly after photoexcitation, one of the carriers (most likely a hole [69]) gets trapped into an almost isoenergetic surface state that leads to charge separation and the associated increase in the exciton-LO-phonon coupling, as detected by PL spectra. The role of carrier localization in the enhancement of the exciton-LO-phonon interaction is demonstrated by temperature dependent FLN studies [97], which indicate a decreasing coupling parameter S with increasing temperature. This can be explained by thermally activated hole delocalization.

The LO-phonon replica resulting from exciton-LO-phonon coupling have also been detected in resonant Raman spectra [34, 35, 39, 41]. In the low-temperature regime ($kT \ll \hbar\omega_0$), the Raman cross section (σ_n) for the nth-order line is determined by the expression [41, 103],

$$\sigma_n(\hbar\omega) \propto e^{-2S}\frac{S^n}{n!}\left|\sum_m \frac{L_m^{n-m}(S)}{E_0 + m\hbar\omega_0 - \hbar\omega + i\hbar\Gamma}\right|^2 \tag{36}$$

where E_0 is the electronic transition energy, $\hbar\Gamma$ is the homogeneous linewidth, m is the number of the intermediate vibrational level in the excited electronic state, and L_m^{n-m} is the associated Laguerre polynomial. Equation (36) can be used to derive the magnitude of the coupling parameter S from the measured ratio of the integral intensities of the second- and first-order Raman lines. This type of measurement also requires a knowledge of the homogeneous linewidth of optical transitions, which introduces some uncertainty in the results for the parameter S [35]. The Raman studies of CdSe NCs [104] indicate that the magnitude of the Huang–Rhys parameter is between 0.125 and 0.275 [104], consistent with results of PLE studies for the coupling strength seen in absorption [82]. Somewhat larger numbers (0.4–0.7) were obtained for CdS NCs [35].

The size dependence of the polar exciton-phonon coupling has been studied by using FLN, PLE, Raman, and degenerate FWM techniques [35, 39, 82, 98]. The Raman measurements performed on CdSe NC–glass samples yielded the ratio of the first- to second-order Raman line intensities of 2.7 ± 0.1 ($T = 77$ K), independent of the NC size for radii from 1.9 to 4 nm [39]. The corresponding magnitude of S was around 0.5, assuming that the broadening of the optical transition was 140 cm^{-1}. Raman studies of CdS NCs indicate a decrease in the Huang–Rhys parameter with decreasing NC radius [35], in contrast to the predicted enhancement of the electron-LO-phonon coupling for NCs of smaller sizes [60, 93, 94]. The same trend was also observed in the PLE data for CdSe NCs; in that data, the parameter S decreased from 0.2 to 0.1 as the NC radius was reduced from 5 to 1.5 nm [82].

5.3. Electron-Acoustic Phonon Interactions

As was mentioned, the phonon replicas associated with carrier-acoustic phonon interactions merge into a single band contributing to the homogeneous broadening of the optical transitions. To estimate the effect of carrier-acoustic phonon interactions on optical spectra, one can use the concept of an "acoustic-mode heat bath" [98]. The interaction of the exciton with this bath can be modeled by a coupling to a single low-frequency mode $\hbar\omega_a$, with a coupling parameter S_a. If the energy of this mode is much smaller than the thermal energy ($\hbar\omega_a \ll kT$) the distribution given by Eq. (32) takes a simple Gaussian form [90],

$$p_n = \sqrt{\frac{\hbar\omega_a}{4\pi S_a kT}} e^{-[(n-S)^2 \hbar\omega_a / 4kT S_a]} \tag{37}$$

with a width $\Delta_a \propto \sqrt{kTS_a}$. This shows that interaction with low-frequency acoustic phonons leads to a temperature-dependent contribution to the homogeneous broadening which scales as the square root of temperature in the single-frequency-mode model.

Let us consider the problem of homogeneous broadening in NCs in more detail. Studies of electronic dephasing demonstrate that there are three principal mechanisms which contribute to the homogeneous linewidth in NC systems [98]. These mechanisms are related to (1) a lifetime effect, associated with fast decay of the photoexcited state (due to, e.g., ultrafast trapping [49, 98]), (2) elastic scattering at impurities and defects, and (3) interactions with the low-energy acoustic phonons, as discussed earlier. The efficiency of the first two mechanisms is strongly dependent on the NC crystalline quality and properties of the NC surface; thus, they can be significantly reduced by synthetic means such as surface passivation with organic molecules [5] or building layered heterostructures [11]. Therefore, the ultimate linewidth achievable in NCs is determined by the phonon-related processes which are intrinsic to quantum-dot structures.

In bulk polar semiconductors, the carrier coupling to acoustic phonons is mediated by the deformation-potential (DP) and the piezoelectric (PZ) interactions [92]. In the case of NCs, these interactions exhibit size dependence, and scale with the NC radius roughly as $1/a^2$ and $1/a$ for the DP and PZ mechanisms, respectively, [105, 106]. This suggests an increasing relative contribution from DP scattering with decreasing NC size. Calculations performed for CdSe NCs show that the carrier-acoustic phonon interaction in the regime of strong confinment ($a < 7$ nm) is indeed dominated by DP processes [105, 106].

The experimental studies of homogeneous broadening have been performed in both the spectral and the time domains using spectroscopic size selective techniques [31, 69, 73, 82, 91], and nondegenerate and time-resolved degenerate FWM experiments [38, 98, 107]. Absorption and PL spectra of NCs recorded without using size-selective methods (see Figs. 17a and 18) exhibit broad linewidths dominated by inhomogeneous broadening, due to the NC size dispersion. Application of size-selective techniques such as PLE and FLN allows the detection of optical spectra in which the effects of size distribution are significantly suppressed [31, 69, 82, 91], as seen from FLN and PLE spectra in Figure 17b. However, even these spectra are still affected by residual inhomogeneous broadening which can be taken into account by deconvolving the experimental data based on the known size distribution of NCs [31, 97]. This method allows the construction of single-dot absorption and emission spectra which show narrow linewidths (typically several meV) at low temperatures [31, 82, 97].

Another method traditionally used to study systems with large inhomogeneous broadening is "spectral hole burning" in nonlinear absorption spectra [108]. In this experiment, a narrow-band laser line tuned within the inhomogeneous band selectively excites a narrow size distribution of NCs with a width determined by effects of homogeneous broadening. Photoexcited NCs can be detected due to carrier-induced bleaching (see Section 6.1) which leads to characteristic "spectral holes" in the nonlinear absorption as illustrated in Figure 19c. Hole-burning experiments were performed on both colloidal [31, 71] and glass [72, 73, 109] samples. Pristine and photodegraded forms of CdSe NC–glass samples behave differently in hole-burning measurements [73]. The pristine samples show broad spectral holes with a width typically exceeding 100 meV [72, 109]. However, the same samples after strong laser illumination (at pump fluence $\sim$100 kW/cm^2) exhibit much narrower resonances with a full width $\sim$20 meV at 20 K [73]. This difference was attributed to laser-exposure induced changes in the interface states affecting the polar electron-phonon coupling.

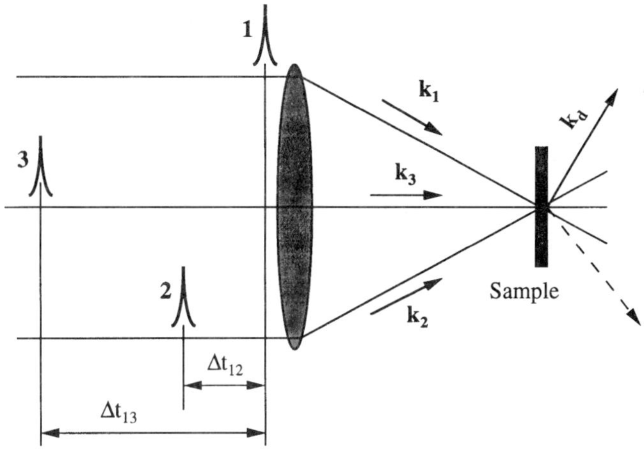

Fig. 20. Three-beam degenerate FWM configuration. Two ultrashort pulses with wave vectors $\mathbf{k}_1$ and $\mathbf{k}_2$ create a dynamical grating which is probed by a delayed pulse with wave vector $\mathbf{k}_3$. The diffracted beam is recorded in the background-free direction $\mathbf{k}_d = \mathbf{k}_2 - \mathbf{k}_1 + \mathbf{k}_3$.

In contrast to spectroscopic size-selective methods which do not allow a complete elimination of inhomogeneous broadening, time-resolved FWM techniques provide information on carrier dephasing and the homogeneous linewidth for an individual NC, averaged over the NC-size distribution. In time-resolved degenerate FWM experiments, two or three ultrashort (typically fs) pulses at the same frequency are used to probe carrier relaxation dynamics. In the three-pulse configuration shown in Figure 20, pulses 1 and 2 create a dynamical grating in the sample, which is probed by pulse 3. By monitoring the signal diffracted into the phase matching direction $\mathbf{k}_d = \mathbf{k}_2 - \mathbf{k}_1 + \mathbf{k}_3$, it is possible to study both the population relaxation (time constant T_1) and the electronic dephasing (time constant T_2). Within the model of a simple two-level system [110, 111] applied to a homogeneously broadened transition, the time-integrated signal measured in the $\mathbf{k}_d$ direction as a function of the delay between pulses 1 and 2 (Δt_{12}) decays with a time constant $T_2/2$, providing information on electronic dephasing. For the inhomogeneously broadened two-level system, the decay of the time-integrated signal measured as a function of Δt_{12} occurs faster (time constant $T_2/4$), but it is still determined by electronic dephasing processes. The same experimental configuration can also be used for measurements of population relaxation, which can be done by monitoring the time-integrated diffracted signal for the variable delay Δt_{13}. Within the two-level approximation the decay of this signal occurs with a time constant $T_1/2$.

FWM studies of NCs are complicated by the strong coupling of excitons to LO vibrations, leading to quantum-beat behavior in the diffracted signal which can distort the information on electronic relaxation, particularly that related to fast dephasing processes (T_2 constant) [38, 98]. In measurements of dephasing (variation of Δt_{12}), these quantum beats can be suppressed by adjusting the fixed delay Δt_{13} to an integer number of periods of the lattice vibrations, allowing recovery of the T_2 constant. The measurements of population dynamics (variation of Δt_{13}) undistorted by nuclear motion can be performed with Δt_{12} set to zero. If pulses 1 and 2 are separated in time ($\Delta t_{12} > 0$) the coherent signal recorded as a function of Δt_{13} shows pronounced oscillations related to lattice vibrations which can be used to measure the LO-phonon damping rate [98].

Detailed fs FWM studies of electronic dephasing in colloidal CdSe NCs were performed in Refs. [38, 98]. These studies revealed two different contributions to the dephasing process. The first contribution (temperature independent) was explained by a lifetime effect (T_1 process) and the elastic carrier scattering at impurities and defects (time constant T_{el}), whereas another one (almost linear in temperature) was attributed to interaction with acoustic phonons (time constant T_a) dominated by deformation-potential coupling. Taking into account these three processes, the total dephasing rate $[(T_2)^{-1}]$ which determines the homogeneous linewidth ($2\Gamma_h = 2\hbar/T_2$) can be presented as follows $(T_2)^{-1} = (T_{el})^{-1} + (T_a)^{-1} + (2T_1)^{-1}$. The FWM data of Refs. [38, 98] indicate that at low temperatures, dephasing is dominated by elastic scattering. With a temperature rise, the increasing contribution to the dephasing rate comes from the electron-acoustic phonon interaction, which dominates homogeneous broadening at room temperature. Studies of the dephasing size dependence showed an increasing relaxation rate for NCs of smaller radius. At $T = 15$ K, T_2 changed from 270 to 85 fs with decreasing NC diameter from ~4 to ~2 nm. This behavior was explained by the fact that all three mechanisms contributing to dephasing had an enhanced efficiency in NCs of smaller sizes.

6. BAND-EDGE OPTICAL NONLINEARITIES IN SEMICONDUCTOR NANOCRYSTALS

6.1. State-Filling and Optical Nonlinearities

This chapter concentrates on optical nonlinearities observed at energies near the band edge. These nonlinearities are dominated by real photoexcitations (excitons and/or free carriers) and are often referred to as resonant nonlinearities. The magnitudes of these nonlinearities are much greater than for nonresonant ones associated with virtual transitions and observed in the transparency range of optical materials. The increase in the magnitude of the nonlinearity in the resonant case occurs at the expense of the nonlinearity relaxation time. In contrast to the almost instantaneous response characteristic of nonresonant nonlinearities, the relaxation of resonant nonlinearities is determined by relatively slow decay of photoexcited carriers (excitons).

In bulk semiconductors at moderate pump levels, before the onset for exciton ionization (below the density for the Mott transition [112, 113]), the band-edge nonlinearities are dominated by excitonic effects. Within the limits of ideal boson behavior (ideal harmonic oscillator), the excitonic system does not exhibit optical nonlinearities. Nonlinear optical response arises only as a result of exciton–exciton interactions caused by such effects as Pauli exclusion (excitons are composite particles made from fermions), Coulomb, and exchange interactions [114–117]. Phenomenologically, excitonic optical nonlinearities can be described in terms of the pump-dependent oscillator strength of the excitonic transition, the broadening of this transition and the shift in the transition energy. Saturation of the excitonic transition (reduction of the exciton oscillator strength) at moderate carrier densities occurs primarily due to screening of the Coulomb electron-hole interaction and phase-space filling [114, 116]. Pump-dependent broadening arises from additional contribution to exciton dephasing due to exciton–exciton scattering [118, 119]. The transition shift is a result of Coulomb exciton–exciton interaction which can lead to formation of excitonic molecules or biexcitons if this interaction is attractive [120].

At pump levels exceeding the threshold for the exciton ionization due to screening effects, or at high temperatures where the excitons are ionized thermally, the band-edge nonlinearities in bulk materials are dominated by free-carrier effects. The blocking of interband optical transitions due to Pauli exclusion leads to the bleaching and ultimately to the optical gain at the band-edge spectral energies. This effect, referred to as the dynamic Burstein–Moss effect, can be explained in terms of band filling [121] and is manifested as a high-energy shift of the absorption edge. Screening of the *e-h* interaction by high density *e-h* plasmas leads to the renormalization of the energy gap in a semiconductor. This effect tends to reduce the energy gap and thus competes with band filling [114, 117].

Due to spatial confinement, the mechanisms for resonant optical nonlinearities in NCs are significantly different from those in bulk materials. In small-size NCs, hole and electron wave functions are almost indentical, resulting in a nearly zero local-charge density. Therefore, the role of Coulomb effects as a source of optical nonlinearities in this case is significantly reduced in comparison with bulk semiconductors [60]. The exchange interaction is also suppressed in three-dimensionally-confined systems [60]. Even at pump levels which correspond to optical gain (2 *e-h* pairs per NC), the exchange interaction between photoexcited electrons (holes) is zero because they have different spin projections.

In contrast to the Coulomb and exchange interactions which are suppressed in three-dimensionally-confined systems, the state filling resulting from Pauli exclusion is significantly enhanced in small-size NCs. This effect is analogous to the filling of continuous energy bands in bulk semiconductors. The absorption changes resulting from this effect are proportional to the sum of the electron and hole occupation numbers. In bulk materials, carrier populations are distributed over a large number of electron (hole) states from the quasi-continuous energy spectrum. Therefore, the single-state occupation numbers remain small even at very high pump levels. In NCs, energy level separation can greatly exceed thermal energies thus preventing thermal depopulation of the lowest quantized states. As a result the excitation of a single *e-h* per NC is sufficient to completely bleach the lowest optical transition [122].

The saturation of the lowest optical transition by state filling can be described using a two-level approximation taking into account spin degeneracy [60, 74, 122]. Using this model, the pump-induced absorption changes can be presented as follows,

$$\Delta\alpha(\omega) = -\gamma N_0 \frac{2(n^e + n^h)}{\Delta^2 + \hbar^2\Gamma_0^2} = -\alpha_0(\omega)\big(n^e + n^h\big) \tag{38}$$

where $\gamma = 4\pi\omega(\hbar\Gamma_0)\langle|d_{cv}|^2\rangle_{\text{pol}}(cn_0)^{-1}$, $\langle|d_{cv}|^2\rangle_{\text{pol}}$ is the square of the modulus of the interband matrix element averaged over light polarizations, $n^e(n^h)$ is the average occupation number of the electron (hole) state with one spin projection (spin up or spin down), $\Delta = \hbar\omega - E_0$, $E_0(\hbar\Gamma_0)$ is the energy (the broadening) of the lowest optical transition, and the factor of 2 in the numerator accounts for the spin degeneracy of the electron and hole states. Equation (38) shows that in this approximation state filling leads to the $\Delta\alpha$ spectra with a shape reproducing that of the liner absorption. The complete bleaching of the transition ($\Delta\alpha = -\alpha_0$) is achieved at densities corresponding to average number of *e-h* pairs per NC $N_{eh} = 1$ ($n^e = n^h = 1/2$). Further increase in carrier density can lead to the development of optical gain at the position of the lowest optical transition [54, 122].

Bleaching features resulting from saturation of optical transitions can be detected using nonlinear transmission measurements [32, 54, 72, 73, 109]. Figure 21a shows an example of room-temperature DT spectrum of CdSe colloidal NCs ($a = 2.3$ nm) taken at an excitation density of less than 1 *e-h* pair per NC on average. In this pump regime, all bleaching bands are due to the populating the lowest quantized states. The DT spectrum in Figure 21a exhibits two bleaching bands, one at 2.15 eV (ω_1) and another at 2.56 eV (ω_3). These bands are due to saturation of transitions involving the $1S$ electron state and two different hole states: $1S_{3/2}$ (ω_1) and $2S_{1/2}$ (ω_3). Additionally, the low-energy DT band shows a shoulder at ~2.21 eV (ω_2) which can be assigned to the intense transition involving the $1S$ electron and the $2S_{3/2}$ hole states. Due to a high spectral density of valence-band states, the occupation number of the $1S$ electron level is significantly higher than occupation numbers of hole states. Therefore, the amplitudes of all bleaching bands are dominated by $1S$ electrons which is clearly seen from single-wavelength DT dynamics (Fig. 21b), indicating the same relaxation behavior of the ω_1, ω_2, and ω_3 features.

Increasing the pump level it is possible to observe the bleaching of transitions involving higher lying electron states [32, 54, 75, 122]. Figures 10 and 11 in Section 3.1 show the DT spectra which in addition to the $1S$ bleaching clearly exhibit features due to filling of the $1P$ and possibly higher (feature B_3 in Fig. 10) electron states. State-filling models give a

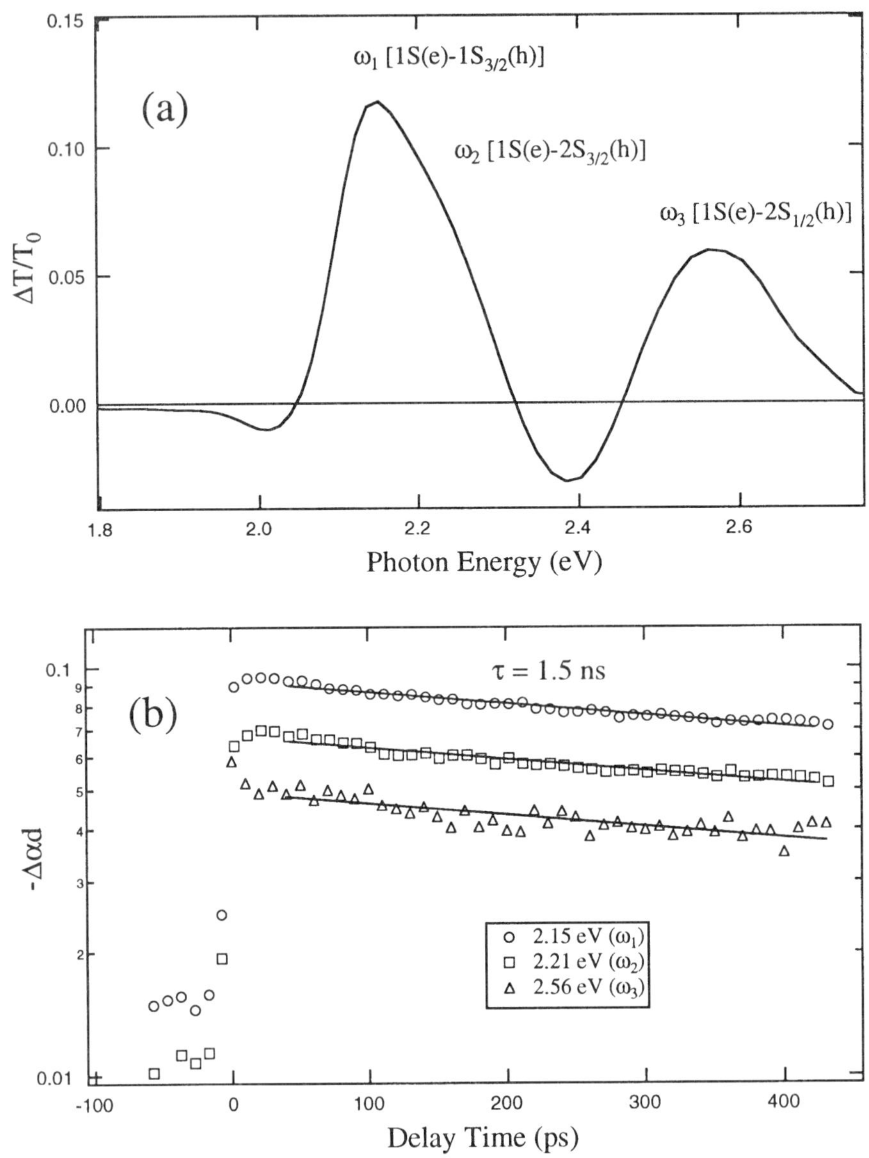

Fig. 21. (a) Room-temperature DT spectrum of CdSe colloidal NCs ($a = 2.3$ nm) recorded at 1 ps after excitation in the pump regime $N_{eh} < 1$. (b) DT time transients recorded for three bleaching features ω_1–ω_3; long-term dynamics of all bands are characterized by the same time constant of 1.5 ns.

fairly good description of density dependent DT spectra [32, 122]. In Ref. [122] nonlinear transmission of NCs was modeled using the parabolic-band approximation which led to the following expression for the absorption of the excited sample,

$$\alpha(\hbar\omega) \propto \sum_{iln} f_i(2l+1) \int_0^\infty \left(1 - n^e_{l,n} - n^h_{i,l,n}\right) g(\hbar\omega - E_{i,l,n}) P(a)\, da \tag{39}$$

where l and n are quantum numbers introduced in Section 2.1, i denotes different hole subbands, $E_{i,l,n}$ is the transition energy between the electron $|l, n\rangle$ and the hole $|i, l, n\rangle$ states, $g(\hbar\omega - E_{i,l,n})$ is the spectral factor accounting for the homogeneous broadening of optical transitions, f_i is the oscillator strength of the interband transition involving the i-valence subbands, averaged over polarizations, and $P(a)$ is the NC-size distribution function (taken in the Lifshitz–Slezov form [123]) accounting for inhomogeneous broadening. Calculations of nonlinear-absorption spectra were performed assuming that the inhomogeneous linewidth was much greater than the homogeneous one, hence allowing the replacement of the spectral factor $g(\hbar\omega - E_{i,l,n})$ with a δ-function. The occupation numbers were found under the assumption of a quasi-equilibrium Fermi distribution. The electron and hole chemical potentials were calculated using the equations,

$$\sum_{l,n} 2(2l+1)n^e_{l,n} = N_{eh} \qquad \sum_{i,l,n} 2(2l+1)n^h_{i,l,n} = N_{eh} \tag{40}$$

A similar model was used to fit pump-dependent DT spectra measured in Ref. [75] (see Fig. 11 in Section 3.1 and Fig. 22 in this section). These spectra exhibit four bleaching features at ~1.82 (B_1), ~1.93 (B_2), ~1.98 (B_3), and ~2.24 eV (B_4). As pointed out earlier, the room-temperature DT is dominated by population of electron states. The DT pump dependence (see inset to Fig. 11) indicates that in the pump regime studied the absorption changes can be explained by subsequent population of the two lowest electron states $1S$ and $1P$, as was discussed in Section 3.1. The population of each of these states gives rise to several bleaching bands associated with transitions involving different hole states of S and P symmetries. Matching pump dependencies of DT signals measured for the B_1 bleaching and the low-energy part of the B_4 band (see inset to Fig. 11) show that both signals are attributed to the population of the same $1S$ electron state. Similarly, the B_2 and B_3 bands can be explained by filling of the $1P$ state (see matching pump dependencies for the B_2 and B_3 features in inset to Fig. 11). The high-energy shift of the B_4 bleaching at high-pump levels indicates the unresolved contribution from a high-energy transition involving the $1P(e)$ state. Detailed spectral analysis allows one to distinguish six optical transitions [marked by vertical bars in Fig. 22a] three of which ($\omega_1^s = 1.83$ eV, $\omega_2^s = 1.875$ eV, and $\omega_3^s = 2.254$ eV) are associated with the $1S(e)$ state and three other ($\omega_1^p = 1.935$ eV, $\omega_2^p = 2.005$ eV, and $\omega_3^p = 2.359$ eV) are associated with the $1P(e)$ state. The hole states involved in these transitions can be tentatively assigned to $1S_{3/2}$ (ω_1^s), $2S_{3/2}$ (ω_2^s), $3S_{1/2}$ (ω_3^s), $1P_{3/2}$ (ω_1^p), $2P_{3/2}$, and/or $1P_{1/2}$ (light hole) (ω_2^p), and $1P^{so}_{1/2}$ (ω_1^p). The pump-dependent changes in DT spectra resulting from the state filling are illustrated in Figure 23a by spectra of absorption changes calculated for different numbers of e-h pairs per NC, varying from 1 to 8. In quasi-equilibrium, for these carrier densities the only populated electron states are $1S$ and $1P$; $N_{eh} = 8$ corresponds to the situation in which both the $1S$ and $1P$ electron states are fully occupied. These spectra were used to fit pump dependencies of spectrally integrated (Fig. 23b) and spectrally resolved (Fig. 22) absorption changes. The good agreement between the model and the experimental data seen from Figures 22 and 23b provides additional confirmation that resonant optical nonlinearities in NCs are dominated by the state-filling effect.

Simple reasoning within the two-level approximation shows that the threshold for the optical gain in NCs is determined by the condition $n^e + n^h > 1$, where n^e and n^h are the average occupation numbers of the lowest electron and hole quantized states. In the case of the twofold spin degeneracy for both an electron and a hole, this condition corresponds to excitation of one e-h pair per NC on average. Due to thermal depopulation of the lowest hole quantized state the gain threshold increases with increasing temperature. Experimentally the gain in NCs was first detected in low-temperature ($T = 80$ K) DT spectra of CdSe NC–glass samples [54, 124]. Time-resolved ps absorption spectra recorded in these experiments are shown in Figure 24. These spectra clearly show the negative absorption (gain) observed at band-edge spectral energies. Two gain bands ($\omega_1 \approx 1.9$ and $\omega_2 \approx 2.03$ eV) resolved in the spectra were assigned to transitions involving the $1S$ (ω_1) and $1P$ (ω_2) electron states. The gain within the ω_2 band was short lived and crossed over to the positive α at time ~100 ps. The ω_1-gain feature was characterized by a much longer relaxation time which allowed the observation of negative α up to 1 ns after excitation.

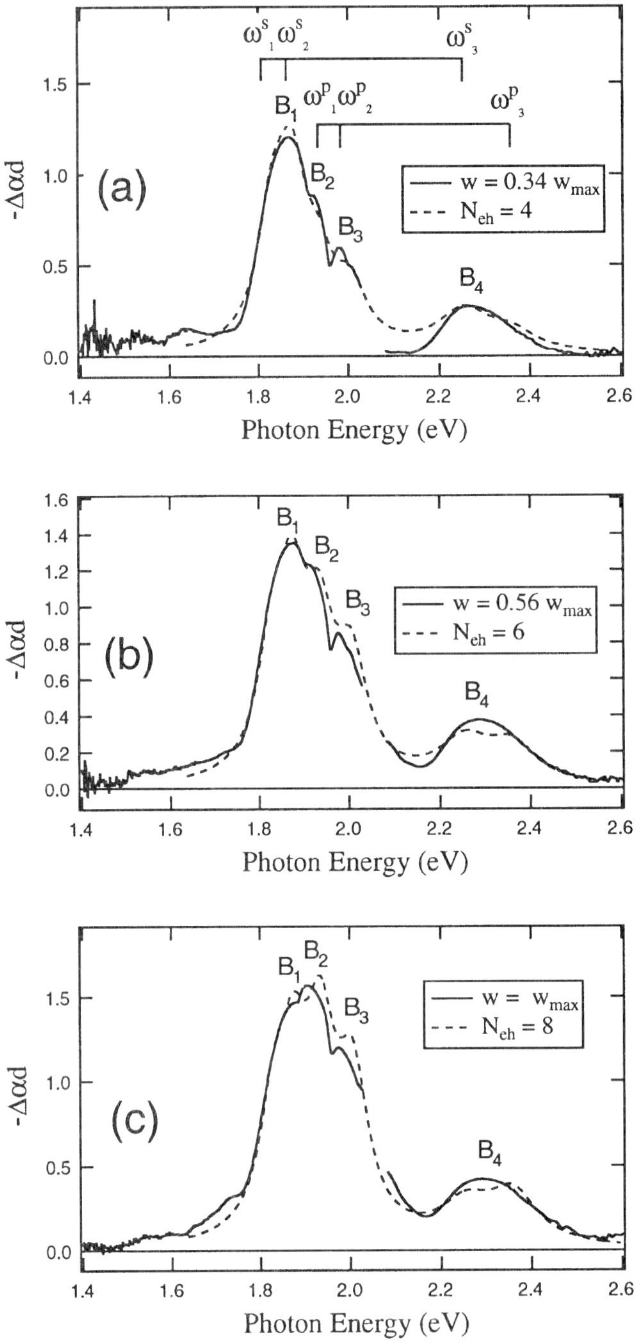

Fig. 22. The comparison of the room-temperature $\Delta\alpha d$ spectra (solid lines) of CdSe NC–glass sample ($a \approx 5.6$ nm) measured at different pump intensities [$w = 0.34w_{\max}$ (a), $0.56w_{\max}$ (b), $w_{\max}$ (c); $w_{\max} = 0.25$ mJ cm^{-2}] with spectra calculated within the state-filling model (dashed lines) assuming different numbers of e-h pairs in a NC [$N_{eh} = 4$ (a), 6 (b), 8 (c)].

The observed gain was used to demonstrate the lasing effect in NCs [54, 124]. The laser cavity was made by depositing dielectric coatings (95 and 100% reflection at 650 nm) on parallel faces of the 1-mm thick CdSe NC–glass sample. The sample was cooled down to 80 K and was pumped in the collinear configuration at 2.33 eV by 20-ps pulses from a

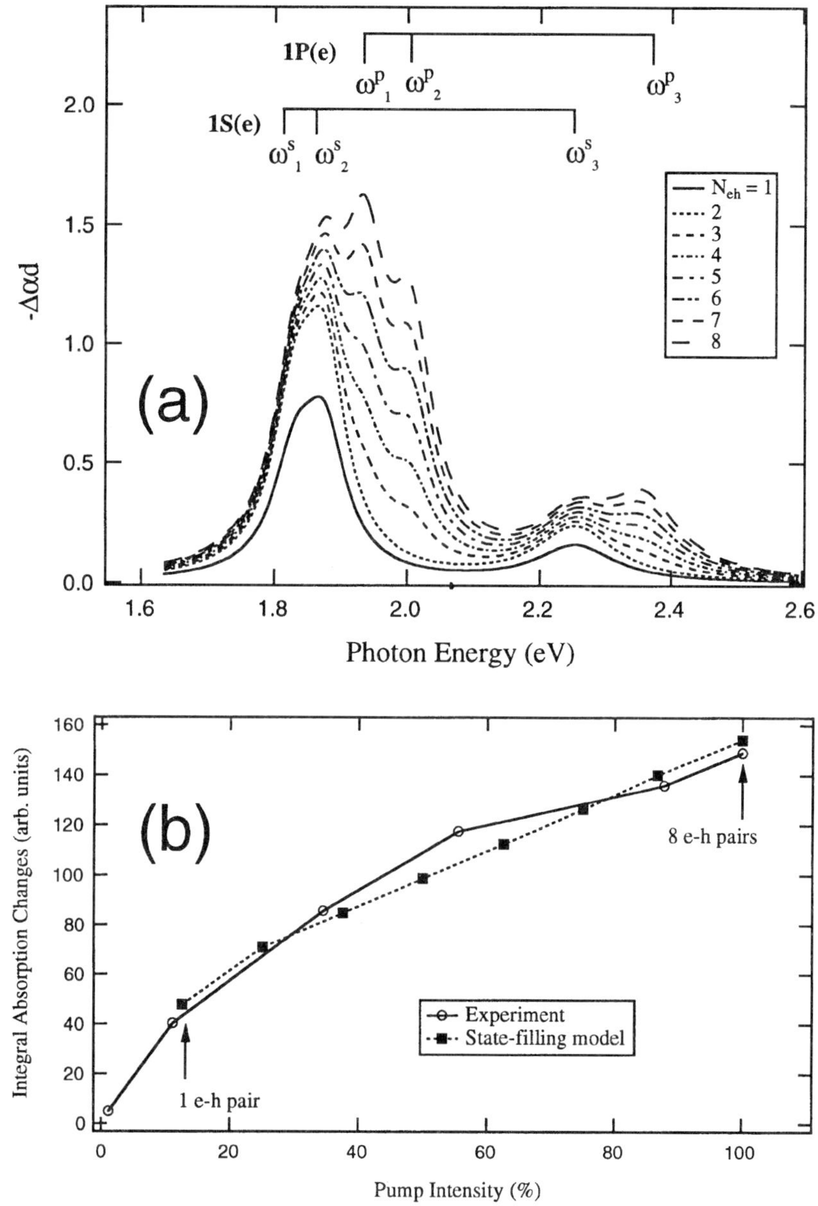

Fig. 23. (a) Spectra of absorption changes calculated for different numbers of *e-h* pairs excited in a NC assuming the state-filling mechanism for nonlinearity. Two series of transitions are taken into account (shown by vertical bars), one series arising from the $1S$ electron state $(\omega_1^s-\omega_3^s)$, and another originating from the $1P$ state $(\omega_1^s-\omega_3^s)$. (b) The comparison of the spectrally integrated absorption changes measured experimentally (open circles) with results obtained by integrating the calculated spectra shown in the panel (a).

mode-locked Nd-YAG laser. Figure 25 shows the comparison of the spontaneous-emission spectrum (line 1), recorded in the sample without coatings, with spectra of the stimulated emission (lines 2 and 3) detected in the coated sample in the direction perpendicular to the coated faces. The spontaneous emission was shifted to low energies from the lowest $1S$ transition, indicating that at least one of the states involved in the emission process was a localized state in the energy gap (see Section 7.2). In contrast to the spontaneous PL, the stimulated emission was located at exactly the position of the $1S$ transition which was consistent with the position of the maximum of the long-lived gain feature. The pump-intensity

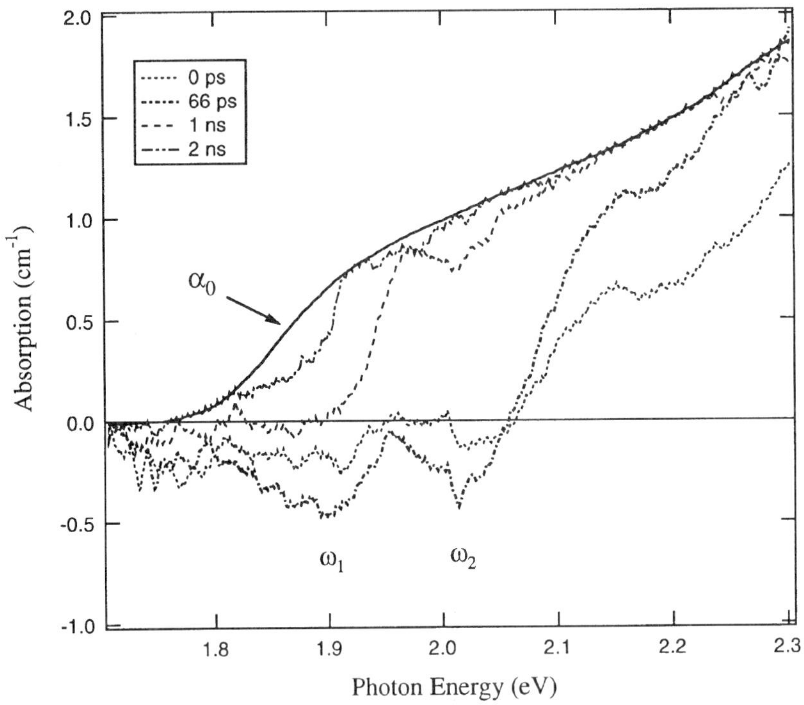

Fig. 24. Time-resolved absorption (gain) spectra of CdSe NC–glass sample ($a \approx 6$ nm) recorded at different delay times between pump and probe pulses, in comparison to the linear absorption spectrum. The sample temperature is 80 K. (Source: Reprinted with permission from [124]. © 1991 American Institute of Physics.)

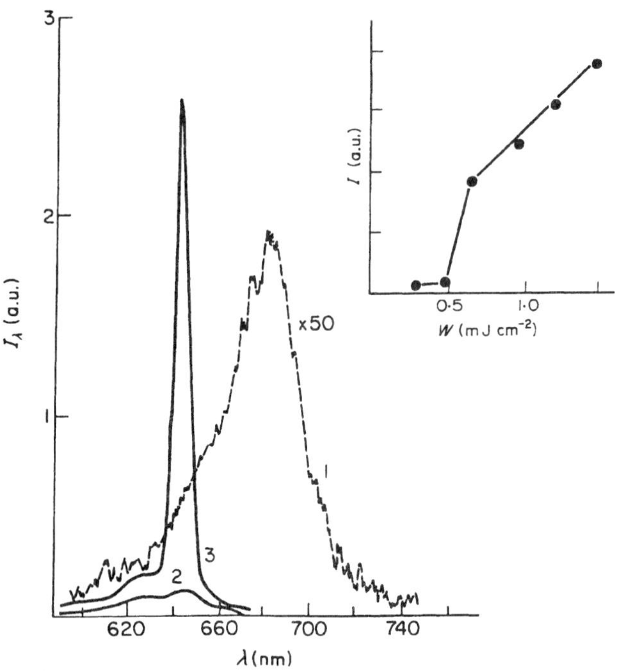

Fig. 25. Spectra of spontaneous (1) and stimulated (2, 3) emission of CdSe NC–glass sample ($a \approx 6$ nm, $T = 80$ K). The stimulated emission spectra were recorded using a sample with reflective coatings forming a Fabry–Perot resonator. Inset: pump-fluence dependence of the stimulated emission intensity indicating a lasing threshold of ~0.5 mJ cm^{-2}. (Source: Reprinted from [54], with permission from Elsevier Science.)

dependence of the stimulated emission (inset to Fig. 25) clearly shows the transition to the lasing regime observed as a sharp increase in the emission intensity above the lasing threshold ($\sim$0.5 mJ cm^{-2}) which was accompanied by the narrowing of the emission line.

6.2. Coulomb Interactions and Optical Nonlinearities

As stated above, in NCs, Coulomb interactions are of secondary importance to state filling effects. However, there are some special situations in which Coulomb interactions are enhanced, leading to new mechanisms for optical nonlinearities. One of the Coulomb-interaction-enhancing processes is carrier surface trapping which leads to charge separation and a buildup of local fields [44, 125]. Due to the dc-Stark effect, the local field leads to a low-energy shift of optical transitions which modifies the absorption spectra. Absorption changes resulting from the shift of the lowest optical transition due to trapped carriers can be presented as follows (we assume that both an electron and a hole are in localized states and that the lowest quantized states are not populated),

$$\Delta\alpha(\omega) \propto -\frac{\delta E(2\Delta + \delta E)}{(\Delta^2 + \hbar^2\Gamma_0^2)[(\Delta + \delta E)^2 + \hbar^2\Gamma_0^2]} \tag{41}$$

where δE is the local-field induced transition shift and the rest of the notation is the same as in Eq. (38). Instead of the single bleaching band seen in spectra of absorption changes for the state filling, the shift of the transition leads to a derivativelike feature with the bleaching at the original position of the transition and the induced absorption below this transition (see illustration in Fig. 26a). If only one of the carriers is localized and another one remains in the quantized state, the derivativelike feature is superimposed on the bleaching band due to state filling which might complicate the observation of the transition shift (see Fig. 26b).

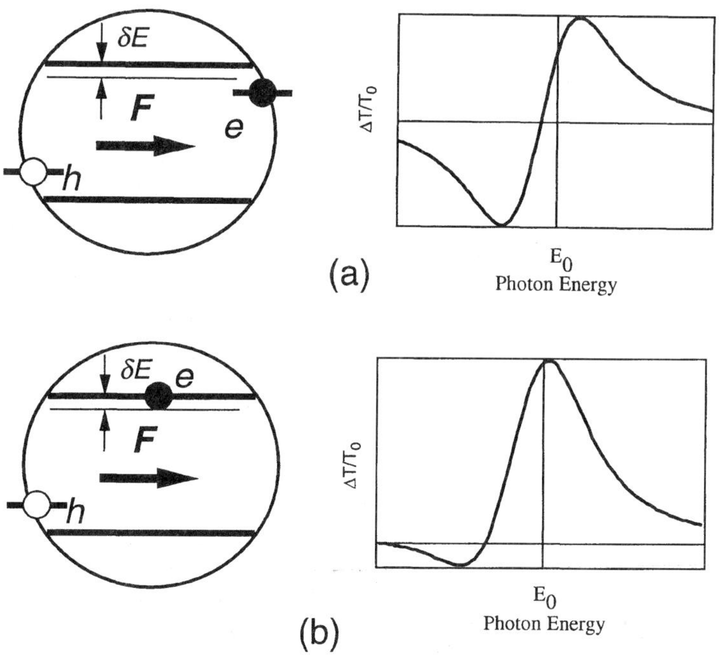

Fig. 26. Schematic of the surface-trapping-induced charge separation (on the left) leading to the generation of local fields and a resulting DT signal (on the right): (a) both carriers (an electron and a hole) are trapped at the surface (the lowest optical transition is only affected by the Stark effect); (b) the hole is trapped at the surface, but the electron remains in the lowest $1S$ quantized state (the lowest optical transition is affected by both state-filling and the Stark effect).

Experimentally, the nonlinearity induced by charge separation has been observed in both colloidal [125] and glass samples [44]. In colloids this effect was attributed to hole trapping at almost isoenergetic surface states [69, 125]. In glass samples efficient charge separation occurs only at high-pump fluences which can be explained in terms of Auger-assisted trapping at interface states [44]. The latter process will be discussed in more detail in Section 7.3.

The generation of nonzero local charge density with a resulting enhancement of Coulomb interactions can also occur in early stages of carrier relaxation when electrons and holes occupy states of different symmetries. This effect, referred to as a biexciton effect, leads to transient shifts of optical transitions which can be observed with fs pump-probe experiments [47]. The magnitude of the shift is determined by the two-e-h-pair interaction energy which can be treated as the biexciton binding energy [45, 126]. In the experiment reported in Ref. [47], the biexciton effect was observed in room-temperature DT spectra of CdSe NCs ($a \approx 5.7$ nm). The pump-photon energy ($\sim$4 eV) was intentionally chosen to significantly exceed the bandgap energy in order to extend the time of carrier relaxation to the lowest quantized states, and to facilitate in this way the observation of the transient shift of the lowest transition. As was mentioned earlier, this shift can be easily masked by the state-filling-induced bleaching even if only one of the states involved in the transition is populated. To avoid a fast nonlinear recombination, the pump intensity was below the level of excitation of one e-h pair per NC on average.

Figure 27a shows that the DT spectra recorded between 200 and 800 fs after excitation exhibit three well resolved features: two bleaching bands at $\omega_1 \approx 1.88$ and $\omega_2 \approx 1.94$ eV, and one band of induced absorption at $\omega_1' \approx 1.78$ eV. In Figure 27b the temporal evolutions of the changes of optical density, $-\Delta\alpha d$, derived from the DT spectra for different bands are shown. The bands ω_2 and ω_1' were characterized by extremely fast dynamics. Their amplitudes increased within the first 400–500 fs after excitation and then decayed during the next 500–600 fs. The band ω_1 reached a maximum at 800–900 fs, that is, after the decay of the bands ω_1' and ω_2. The exponential relaxation time of the ω_1 band was about 2.5 ps. The ω_1 feature was attributed to the bleaching of the lowest $1S$ transition, whereas the short-lived ω_2 band was explained by the bleaching of the $1P(e)$-$1P_{3/2}(h)$ transition ($1P$ transition). The DT dynamics were explained in terms of the competing contributions from the state-filling and the two-e-h-pair interaction. As was mentioned earlier, the state-filling-induced portion of the DT signal is dominated by electron populations. Because the pump-photon energy in the experiment discussed exceeded by far the energies of both $1S$ and $1P$ transitions, in the initial stage of the energy relaxation, $1S$ and $1P$ electron states were not occupied. During the energy relaxation the electrons reached initially the $1P$ state saturating the $1P$ transition. This was observed as the ω_2 bleaching band in the DT spectra. At the same time, the two-pair interaction resulted in a change of the energy (transition shift) for the creation of the second e-h pair including that in the ground state. This was manifested as a bleaching in the range of the lowest resonance (band ω_1) and the increased absorption below it (band ω_1'). The relaxation of the electron to the lowest $1S$ state was seen as a fast decay of the ω_2 and ω_1' bands, accompanied by the growth of the ω_1 band due to state filling. Because of the large linewidth of the optical transitions, the latter effect resulted in the suppression of the increased absorption below the lowest resonance.

Experimental data were modeled using the following expression which describes the changes in the optical susceptibility resulting from the state-filling and the Coulomb two-pair interaction in the presence of the electron-hole pair occupying the $|i_e', j_h'\rangle$ state,

$$\Delta\chi(\hbar\omega) \propto \sum_{i,j} f_{ij}\left[\frac{\delta_{ii'} + \delta_{jj'}}{\hbar\Delta\omega_{ij} + i\hbar\Gamma_{ij}} + \frac{(1-\delta_{ii'})(1-\delta_{jj'})\delta E_2(ij, i'j')}{(\hbar\Delta\omega_{ij} + i\hbar\Gamma_{ij})[\hbar\Delta\omega_{ij} + \delta E_2(ij, i'j') + i\hbar\Gamma_{ij}]}\right] \tag{42}$$

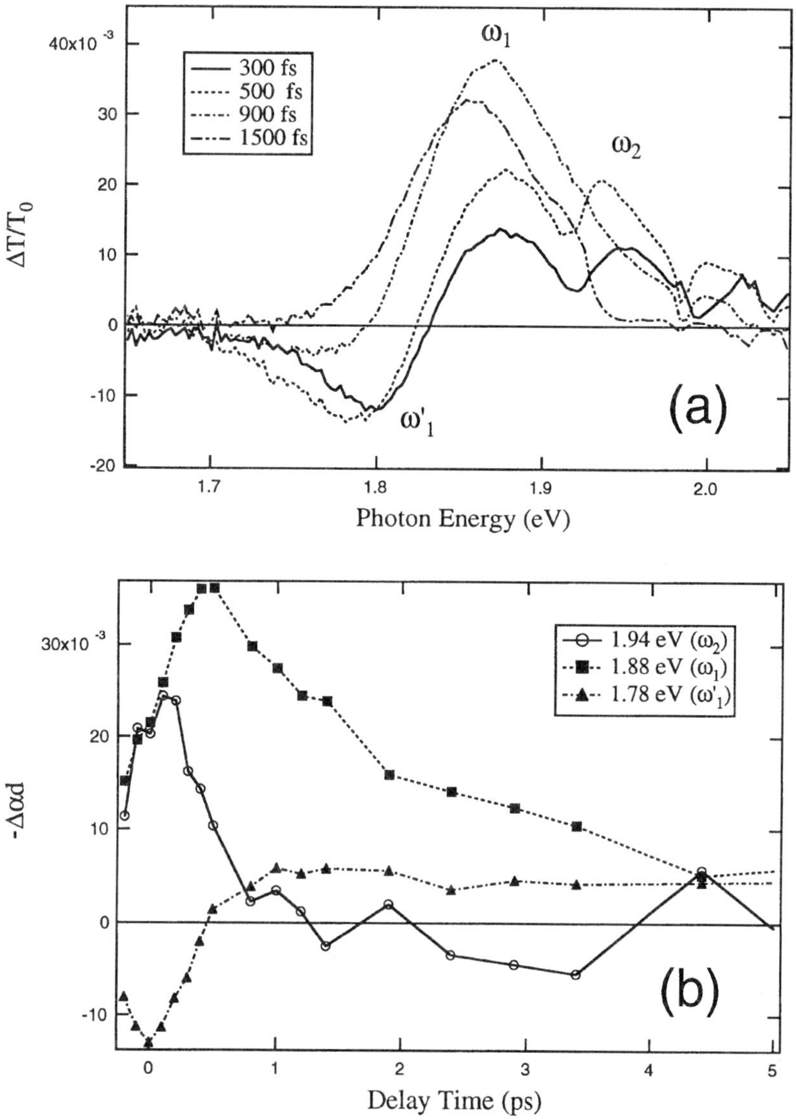

Fig. 27. (a) Room-temperature DT spectra of CdSe NC–glass sample ($a \approx 5.7$ nm) recorded at different delay times between pump and probe pulses. (b) Time dependence of the induced absorption changes at the maxima of the ω_1, ω_1', and ω_2 DT bands. (Source: Reprinted with permission from [47]. © 1994 American Physical Society.)

where $\hbar\Delta\omega_{ij} = \hbar\omega - E_{ij}$, E_{ij} and f_{ij} are the energy and oscillator strength of the optical transition resulting in the creation of the e-h pair in the $|i_e, j_h\rangle$ state, $\hbar\Gamma_{ij}$ is the broadening of this transition, $\delta E_2(ij, i'j')$ is the interaction energy of pairs in the $|i_e, j_h\rangle$ and $|i_e', j_h'\rangle$ states, and $\delta_{ii'}$ is the Kronecker symbol ($\delta_{ii'}$ is equal to 1 if $i = i'$ and to 0 otherwise). The DT spectra were calculated by taking into account the dipole-allowed transitions coupling the vacuum state, two one-pair states $|1s_e, 1s_h\rangle$ and $|1p_e, 1p_h\rangle$, and three two-pair states $|1s_e, 1s_h;\ 1s_e, 1s_h\rangle$, $|1s_e, 1s_h;\ 1p_e, 1p_h\rangle$, and $|1p_e, 1p_h;\ 1p_e, 1p_h\rangle$ which are schematically depicted in Figure 28 (we omit here the 3/2 index in the notation of the hole states). The calculations assuming the population of either $1P$ or $1S$ electron states were used to fit the DT spectra recorded at short ($\Delta t < 400$ fs) and long ($\Delta t > 1$ ps) times after excitation, respectively, as shown in Figure 29. As a result of the fitting procedure the value of 32 meV was estimated for the binding energy of the "hot biexciton" involving $1S$ and $1P$ electrons.

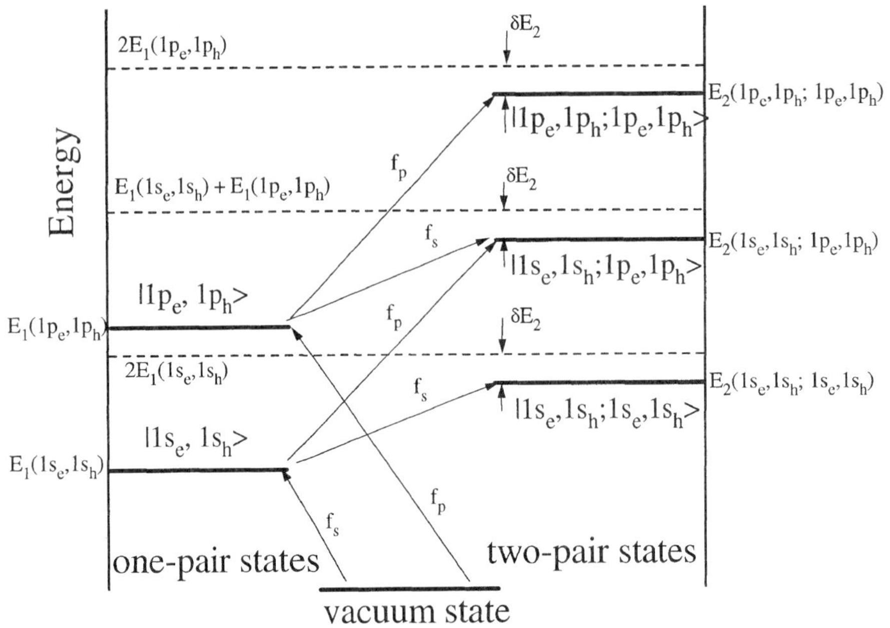

Fig. 28. The six-level model used to fit the pump-induced absorption spectra shown in Figure 27a. This model accounts for state-filling-induced bleaching of optical transitions and transition shifts due to the biexciton effect. The arrows show the six dipole-allowed transitions coupling vacuum, two one-*e*-*h*-pair and three two-*e*-*h*-pair states. (Source: Reprinted with permission from [47]. © 1994 American Physical Society.)

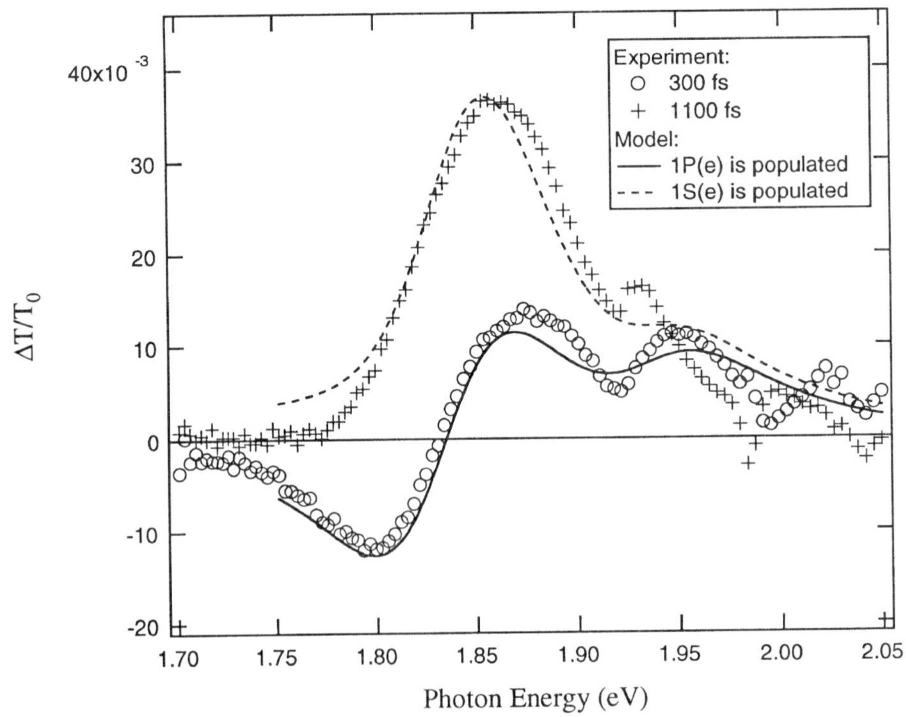

Fig. 29. Comparison of measured (symbols) and calculated (lines) DT spectra assuming the occupation of either $1S$ (solid line) of $1P$ (dashed line) electron states.

6.3. Third-Order Nonlinear Susceptibility

The quantitative analysis of the optical nonlinearities can be performed using nonlinear optical susceptibilities [127]. Due to the random orientation of NCs, the NC samples are optically isotropic. Therefore, the only nonvanishing nonlinear susceptibility for them are those of the odd order: $\chi^{(3)}$, $\chi^{(5)}$, etc. [127]. Many important phenomena such as those involving a light self-action are determined by so-called Kerr-type susceptibility $\chi^{(3)}(\omega; \omega, -\omega, \omega)$ [$\chi^{(3)}(\omega)$ in what follows] and other higher order susceptibilities with a set of frequencies constructed in a similar way. Using these nonlinear susceptibilities, the changes in $\chi(\omega)$ induced by a monochromatic field at a frequency ω can be presented as follows,

$$\Delta\chi(\hbar\omega) = \chi^{(3)}(\hbar\omega)|E_\omega|^2 + \chi^{(5)}(\hbar\omega)|E_\omega|^4 + \cdots \tag{43}$$

where E_ω is a field amplitude.

To derive expressions for nonlinear susceptibilities of NCs we will assume the dominant role of the state-filling effect. In this case, the carrier induced changes in the NC optical susceptibility can be presented as follows [74, 122],

$$\Delta\chi(\hbar\omega) = N_0 \sum_{j,k} \frac{\langle |d_{j,k}|^2 \rangle_{\text{pol}} (n_j^e + n_k^h)}{\hbar\omega - E_{j,k} + i\hbar\Gamma_{j,k}} \tag{44}$$

where $E_{j,k}$ and $\Gamma_{j,k}$ are, respectively, the energy and the broadening of the transition between $|j_e\rangle$ electron and $|k_h\rangle$ hole states, n_j^e and n_k^h are the occupation numbers of these states, and $\langle |d_{j,k}|^2 \rangle_{\text{pol}}$ is the square of the modulus of the matrix element of the k-to-j state transition averaged over light polarizations.

At moderate pump intensities, the nonlinear optical response is dominated by the third-order nonlinear susceptibility $\chi^{(3)}$. In NCs, this regime occurs when the average number of e-h pair per NC (N_{eh}) does not exceed 1 [122]. Next we consider the room-temperature case for which we can ignore the hole contribution to the state-filling-induced nonlinearity ($n_k^h \approx 0$). Due to a large spacing between electron levels, all electrons in the regime $N_{eh} < 1$ are in the lowest $1S$ state which corresponds to the occupation numbers of the electron $1S$ spin-up and spin-down states of $N_{eh}/2$. As a result of these simplifications Eq. (44) can be rewritten as follows,

$$\Delta\chi(\hbar\omega) = N_0 N_{eh} \sum_{k_S} \frac{\langle |d_{1S,k_S}|^2 \rangle_{\text{pol}}}{\hbar\omega - E_{1S,k_s} + i\hbar\Gamma_{1S,k_S}} \tag{45}$$

where summation only involves the S-type hole states ($|k_S\rangle$) coupled to the $1S$ electron state with one spin projection (spin up or spin down). The number of e-h pairs excited per NC can be found from the rate equation for carrier density which gives [74, 122],

$$N_{eh} = \frac{cn_0\tau_e}{2\pi\hbar\omega N_0}\alpha_0(\omega)|E_\omega|^2 \tag{46}$$

Combining Eqs. (45) and (46) we can see that in the pump-intensity regime studied, the changes in the optical susceptibility are proportional to $|E_\omega|^2$, meaning that they indeed can be described in terms of the third-order nonlinear susceptibility $\chi^{(3)}$. Using the definition $\chi^{(3)}(\omega) = \Delta\chi(\omega)/|E_\omega|^2$ we can obtain

$$\chi^{(3)}(\omega) = \frac{cn_0\tau_e\alpha_0(\omega)}{2\pi\hbar\omega} \sum_{j_S} \frac{|d_{1S,k_S}|^2}{\hbar\omega - E_{1S,k_s} + i\hbar\Gamma_{1S,k_S}} \tag{47}$$

In the vicinity of the lowest optical transition ($1S$ transition in what follows) Eq. (47) reduces to that for a simple two-level system [127],

$$\chi^{(3)}(\omega) = N_0 \frac{4|d_{1S}|^2\tau_e\Gamma_{1S}}{[(\hbar\omega - E_{1S})^2 + \hbar^2\Gamma_{1S}^2][(\hbar\omega - E_{1S}) + i\hbar\Gamma_{1S}]} \tag{48}$$

where E_{1S}, Γ, and d_{1S} are the energy, the broadening, and the matrix element of the $1S$ transition, respectively.

Experimentally, optical nonlinearities were studied in the highest detail in glass samples [73, 99, 109, 122, 128, 129]. In Refs. [54, 122], third-order nonlinearities in CdSe NC–glass composites were studied using ps DT measurements. DT spectra $[D(\omega)]$ recorded in this experiment provide a direct measure of pump-induced absorption changes. If the resonator effects are negligibly small,

$$\Delta\alpha(\omega) = -\frac{1}{d}\ln\big[D(\omega) + 1\big] \tag{49}$$

The spectra of absorption changes can be further used to derive the spectra of refraction changes via the Kramers–Kronig transformation,

$$\Delta n(\omega) = \frac{c}{\pi}\int_0^\infty \frac{\Delta\alpha(x) - \Delta\alpha(\omega)}{x^2 - \omega^2}\,dx \tag{50}$$

Because pump-induced absorption changes are located in the relatively narrow spectral range, the use of Eq. (50) does not involve uncertainties due to the infinite integration range which usually complicate the application of the Kramers–Kronig relationships in linear spectroscopy.

The pump-induced absorption and refraction changes are directly related to changes in the imaginary and the real parts of the optical susceptibility,

$$\Delta\alpha(\omega) = \frac{4\pi\omega}{cn_0}\,\mathrm{Im}\,\Delta\chi(\omega) \qquad \Delta n(\omega) = \frac{2\pi}{n_0}\,\mathrm{Re}\,\Delta\chi(\omega) \tag{51}$$

In the low-intensity regime, the changes in $\chi(\omega)$ are determined by the third-order nonlinearity and Eqs. (51) can be used to derive $\mathrm{Re}\,\chi^{(3)}(\omega)$ and $\mathrm{Im}\,\chi^{(3)}(\omega)$,

$$\mathrm{Re}\,\chi^{(3)}(\omega) = \frac{cn_0^2\Delta n(\omega)}{4\pi^2 I_\omega} \qquad \mathrm{Im}\,\chi^{(3)}(\omega) = \frac{c^2n_0^2\Delta\alpha(\omega)}{8\pi^2\omega I_\omega} \tag{52}$$

where $I_\omega = (cn_0/2\pi)|E_\omega|^2$ is the intensity of the resonant cw field applied to the system. In the case of nonresonant pulsed excitation, when the whole DT spectrum is recorded using a single pump wavelength, I_ω represents the effective spectrum of the cw resonant field which would result in the same DT signal as that measured in nonresonant pulsed experiments [74, 122]. Because the band-edge nonlinearity is dominated by carrier populations, the effective pump-intensity spectrum can be calculated using the rate equation, yielding

$$I_\omega = \frac{n_e\hbar\omega}{\tau_e}\alpha_0^{-1}(\omega) \tag{53}$$

where $n_e = N_{eh}N_0$ is the carrier density calculated for the unit volume of the NC–matrix media. Combining Eqs. (52) and (53), we arrive at the final expressions for the calculation of the third-order nonlinear susceptibilities using the results of time-resolved DT measurements,

$$\mathrm{Re}\,\chi^{(3)}(\omega) = \frac{cn_0^2}{4\pi^2\hbar\omega}\frac{\Delta n(\omega)\alpha_0(\omega)\tau_e}{n_e} \qquad \mathrm{Im}\,\chi^{(3)}(\omega) = \frac{c^2n_0^2}{8\pi^2\hbar\omega^2}\frac{\Delta\alpha(\omega)\alpha_0(\omega)\tau_e}{n_e} \tag{54}$$

It should be stressed that these expressions can be applied only in the case of low-pump intensities corresponding to excitation of less than 1 e-h pair per NC on average. Additionally, in the experiments with pulsed excitation, the $\Delta\alpha(\omega)$ and $\Delta n(\omega)$ spectra should be computed using the DT spectra recorded long after excitation when the nonlinear optical signal is not affected by energy relaxation processes.

The spectra of $\mathrm{Re}\,\chi^{(3)}$ and $\mathrm{Im}\,\chi^{(3)}$ of CdSe NC–glass sample ($a \approx 6$ nm, $T = 80$ K) derived from the DT data as outlined previously are shown in Figure 30 (symbols). In contrast to the two-level system, the $\chi^{(3)}$ spectra of NCs exhibit an asymmetric shape with a long

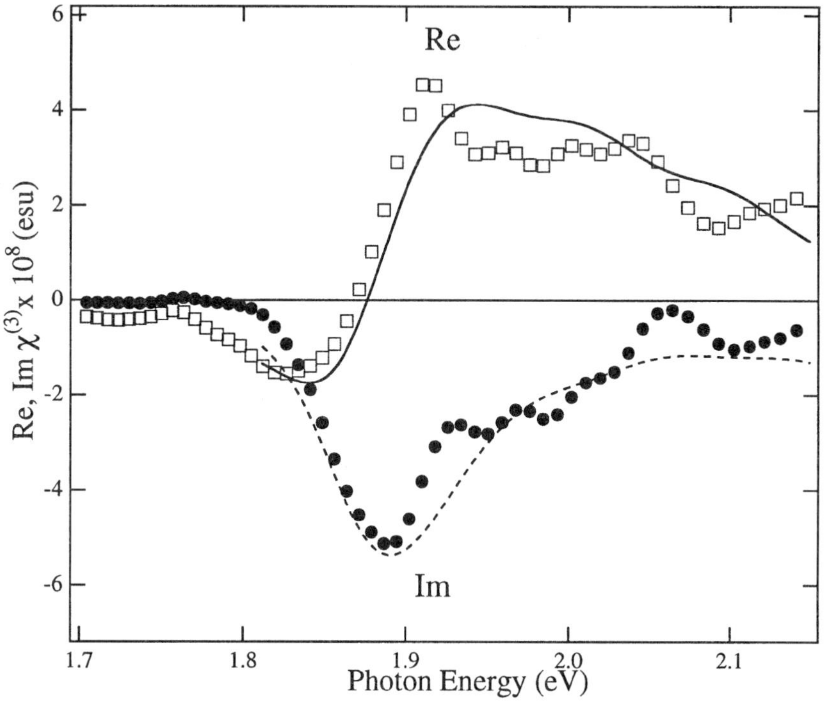

Fig. 30. Spectra of the real and imaginary parts of the third-order nonlinear susceptibility (symbols) of CdSe NC–glass sample ($a \approx 6$ nm, $T = 80$ K) derived from the measured DT spectra in comparison with calculations (lines) performed within the state-filling model. (Source: Reprinted with permission from [74]. © 1994 American Physical Society.)

high-energy tail due to contributions from higher lying hole states coupled to the $1S$ electron level. In Ref. [74] these spectra were described (lines in Fig. 30) within the parabolic-band model, assuming the state-filling-induced saturation of three transitions involving the $1S$ electron states and the $1S_A$, $1S_B$, and $1S_C$ hole states originating from the heavy, light, and split-off valence subbands, respectively. Nearly identical fits can be obtained using the improved model for the hole energy spectra which accounts for the band-mixing effects. Using this model the experimentally measured $\chi^{(3)}$ spectra can be explained assuming that the state-filling induced signal is dominating by transitions involving the hole $1S_{3/2}$, $2S_{3/2}$, and $3S_{1/2}$ states.

In Ref. [122], the values of the third-order nonlinearity measured for CdSe NCs (glass samples) at $T = 80$ K were compared with predictions of the state-filling model. Within this model the amplitude of $\mathrm{Im}\,\chi^{(3)}$ is determined by the expression [74, 122],

$$\left|\mathrm{Im}\,\chi^{(3)}\right|_{\max} = \frac{\xi}{V_0}\frac{8\langle|d_{1S}|^2\rangle_{\mathrm{pol}}\tau_e}{\hbar^3\Gamma_{1S}^2} \tag{55}$$

where $\xi = V_0 N_0$ is the semiconductor volume fraction in the matrix, first introduced in Section 2.1 [note a factor 8 in Eq. (55), as opposed to the factor 4 in Eq. (48), obtained for a room-temperature case; the factor of 2 difference accounts for a hole contribution to the low-temperature nonlinearity]. Estimations performed for a sample with $a \approx 6$ nm and $\xi = 0.1\%$, and using experimentally measured carrier lifetime of 1.7 ns yields $|\mathrm{Im}\,\chi^{(3)}|_{\max} \approx 3 \cdot 10^{-7}$ esu. This is a factor of 5 greater than the experimental value of $6 \cdot 10^{-8}$ esu. This discrepancy was explained in terms of a large difference between homogeneous and inhomogeneous linewidths, $\Gamma_{1S}/\Gamma_{ih} \ll 1$. Due to this difference, the volume concentration of actually excited ("working") NCs is smaller than the total semiconductor fraction ξ by approximately a factor of Γ_{ih}/Γ_{1S}.

Equation (55) predicts the increase in the nonlinearity with decreasing volume of a single NC for the case of a constant volume fraction of semiconductor, which in fact corresponds to increasing concentration of NCs in the matrix, as discussed in Section 2.1. However, experimental data indicate an opposite trend, namely, a strong decrease in the nonlinearity in NCs of smaller radii. As was demonstrated in Refs. [74, 122], this decrease was correlated with shortening of the carrier relaxation time which overwhelms the growth due to increasing concentration of NCs. The comparison of the data for samples with $a \approx 6$ and 3.5 nm indicated an almost 2 order of magnitude reduction of $\chi^{(3)}$ in NCs of smaller radius [122]. This decrease was mostly due to the strong shortening of the carrier lifetime which was 1.7 ns for the sample with $a \approx 6$ nm and only 60 ps in case of the sample with $a \approx 3.6$ nm. Time-resolved studies indicate a drastic shortening of carrier relaxation time with increasing temperature [32] which can be attributed to temperature-activated trapping. For instance, the CdSe NC–glass sample with $a \approx 6$ nm which showed a 1.7 ns relaxation constant at $T = 80$ K, had a carrier lifetime shorter than 40 ps at room temperature [32]. As a result of the enhanced carrier decay, the room-temperature nonlinearities in NC–glass samples are much smaller than those detected in the low-temperature regime, and in CdSe NCs are typically below 10^{-8}–10^{-9} esu [101].

In addition to DT measurements described previously, optical nonlinearities in NC–glass composites were studied by several other nonlinear optical techniques, such as optical phase conjugation in the degenerate FWM configurations [99, 128, 130, 131], intensity-dependent absorption saturation [129, 130, 132], and nonlinear interferometry [133]. The data of pump-dependent absorption and refraction measurements have been typically analyzed using expressions obtained in the two-level approximation [60, 127],

$$\Delta\alpha = \frac{\alpha_2 I}{1 + I/I_s} \qquad \Delta n = \frac{n_2 I}{1 + I/I_s} \tag{56}$$

where α_2 is the nonlinear absorption coefficient, n_2 is the nonlinear index of refraction, and I_s is the saturation intensity. In the case of the ideal two-level system α_2 and n_2 are connected with I_s by relationships $\alpha_2 = -\alpha_s/I_s$ and $n_2 = -n_s/I_s$, where α_s and n_s are saturable parts of the absorption coefficient and the refractive index, respectively. In contrast to the $\chi^{(3)}$ approach which is only valid in the low-pump intensity regime, application of Eqs. (56) allows one to describe also the saturation of the optical nonlinearity manifested as leveling off pump-induced absorption (refraction) changes at high-pump intensities [99, 133]. Using the first terms in expansions of Eqs. (56) in powers of I/I_s one can obtain the connection between $\mathrm{Re}\,\chi^{(3)}$ ($\mathrm{Im}\,\chi^{(3)}$) and $n_2(\alpha_2)$,

$$\mathrm{Re}\,\chi^{(3)} = \frac{cn_0^2}{4\pi^2} n_2 \qquad \mathrm{Im}\,\chi^{(3)} = \frac{c^2 n_0^2}{8\pi^2 \omega} \alpha_2 \tag{57}$$

The studies of CdSSe NC–glass composites indicate that pump-dependent absorption and refraction changes can be fairly well described by Eqs. (56). Depending on the samples, the reported saturation intensities vary in a broad range from tenths to hundreds MW cm^{-2} [132, 133]. In contrast to predictions of a simple two-level model, the saturation intensities derived from nonlinear absorption and nonlinear refraction measurements are different from each other, indicating a lower saturation threshold for absorption changes [133]. Some discrepancy is also observed regarding the correlations between $\alpha_2(n_2)$ and I_s.

6.4. Optical Nonlinearities in Direct- and Indirect-Gap Semiconductor Nanocrystals

In bulk indirect-gap semiconductors such Si and Ge the band-edge absorption is only allowed in the second order of the perturbation theory and occurs as a phonon-assisted process. As a result, the band-edge absorption in these materials is significantly weaker than in such direct-gap semiconductors as CdS and CdSe. The reduced efficiency of the interband absorption also affects the mechanisms for band-edge resonant nonlinearities. In indirect-gap materials, these nonlinearities are dominated by excited-state absorption (free-carrier absorption) due to reexcitation of electrons (holes) from the bottom of the conduction (valence) band to higher lying states within the same band [134, 135]. To satisfy wave-vector conservation, free-carrier absorption involves the emission (absorption) of phonons, as in the case of interband transition across the indirect energy gap.

It has been expected that in low-dimensional semiconductors, due to confinement-induced mixing of states, optical transitions across an originally indirect energy gap can be allowed in the first order of the perturbation theory with the associated increase in the intensity of the band-edge emission and absorption [136, 137]. This issue has been actively debated in the context of studies of efficient emission from porous silicon and Si NCs [138–143]. Because Si is an indirect-gap semiconductor, it does not exhibit efficient PL in the bulk form. However, highly porous Si structures prepared by electrochemical etching, exhibit bright emission in the range from red to blue which was initially explained in terms of quantum confinement in nanoscale objects existing in porous layers [138, 144]. This point of view is not universally accepted, and the origin of the visible PL from porous Si as well as from other Si nanostructures such Si NCs remains a matter of great controversy.

Discussions of the unusual luminescence properties of porous Si have raised an important issue concerning the effect of quantum confinement on optical properties of indirect-gap semiconductors. An interesting study tracing changes in optical properties of CdSe NCs during pressure-induced direct indirect-gap transformation is presented in Refs. [145, 146]. At atmospheric pressures, CdSe is a direct-gap semiconductor with a wurtzite crystalline structure. However, at high pressures ($\sim$2 Gpa [147]) it undergoes the structural transition to the rock-salt lattice with an accompanying transformation to the indirect-gap form. As shown by X-ray studies [145], the same type of transformation also occurs in CdSe NCs, although with a higher transition threshold which becomes NC-size dependent. The data from optical studies indicate that a structural transformation has a dramatic effect on the optical absorption of NCs [146]. At low pressures, CdSe NCs exhibit sharp features due to strongly allowed optical transitions coupling electron and hole quantized states. Above the threshold for structural reconstruction, these discrete features disappear although the absorption edge still exhibits a confinement-induced energy shift with respect to the bulk energy gap. The structureless band-edge absorption is a signature of indirect-gap optical transitions. These transitions occur with the participation of phonons which lead to the smearing out discrete absorption features. The size-dependent studies show that the ratio of oscillator strengths of the band-edge transition for direct- and indirect-gap forms of CdSe does not change with NC radius, indicating that the spatial confinement (at least in the size regime studied in Ref. [146]) does not lead to a significant enhancement of optical transition across the indirect energy gap.

An interesting material for comparative studies of the effect of three-dimensional confinement on optical properties of direct- and indirect-gap semiconductors is Cu_xS [148, 149]. Depending on the copper deficiency, the energy bandgap in copper sulfide varies from $\sim$1.2 eV for $x = 2$ (chalcosite) [150, 151] to $\sim$1.5 eV for $x = 1.8$ (digenite) [152], accompanied by a transformation of an originally indirect-gap semiconductor to a direct-gap one. Next we briefly review the linear and nonlinear absorption data obtained in Refs. [148, 149] for Cu_xS NCs in phases with $x = 1.8$ and 2 which exhibit, direct and indirect energy gaps in the bulk form, respectively.

Figure 31 shows linear absorption spectra for copper sulfide NCs ($a \approx 4$ nm) with $x = 1.8$ (a) and 2 (b). The NCs in the direct-gap phase ($x = 1.8$) show two well resolved peaks at 2.16 (ω_1) and 2.59 eV (ω_2) which can be explained as arising from transitions coupling the lowest electron and hole quantum confined states of S and P symmetries, respectively, [149]. In contrast to these observations, the absorption of indirect-gap Cu_2S

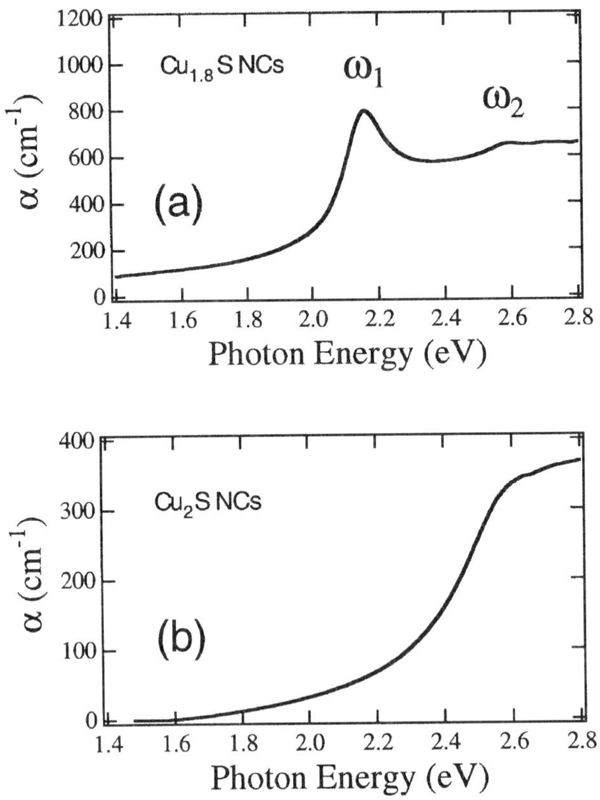

Fig. 31. Linear absorption spectra of Cu_xS NC–glass samples ($T = 300$ K) in phases with $x = 1.8$ (a) and $x = 2$ (b) which are direct- and indirect-gap in the bulk form, respectively. (Source: Reprinted with permission from [149]. © 1996 American Physical Society.)

NCs (Fig. 31b) does not show any discrete features and increases gradually with a photon energy starting at ~1.6 eV and showing a step around 2.6 eV. The absorption onset is blue shifted by ~0.4 eV with respect to that in the bulk material [150, 151] which is a clear signature of a three-dimensional confinement. However, the structureless absorption edge is indicative of the fact that the absorption remains a phonon-assisted process, three-dimensional confinement notwithstanding, which is consistent with results of pressure dependent studies of CdSe NCs [146].

The nonlinear optical response is also significantly different in direct- and indirect-gap copper sulfide NCs. Shown in Figure 32 are DT spectra of Cu_xS NCs with $x = 1.8$ (a) and 2 (b) recorded at 2 ps after excitation. The spectrum of direct-gap NCs shows two pronounced bleaching bands located at the positions of the discrete features seen in linear absorption (Fig. 31a). This is similar to DT spectra of direct-gap CdS and CdSe NCs [75, 153], and can be explained in terms of the state-filling-induced bleaching of strong optical transitions. In contrast to the direct-gap NCs, the NCs in the indirect-gap form with $x = 1.8$ do not show any bleaching but rather a broad band of photoinduced absorption peaked at around 2.6 eV. This induced absorption can be assigned to intraband transitions associated with reexcitation of carriers within the same energy band, analogous to free-carrier absorption in bulk indirect-gap materials. Due to confinement-induced relaxation in the wave-vector conservation, this process can be significantly enhanced in quasi-zero-dimensional structures [154], leading to a much higher probability of interband transitions than in bulk materials.

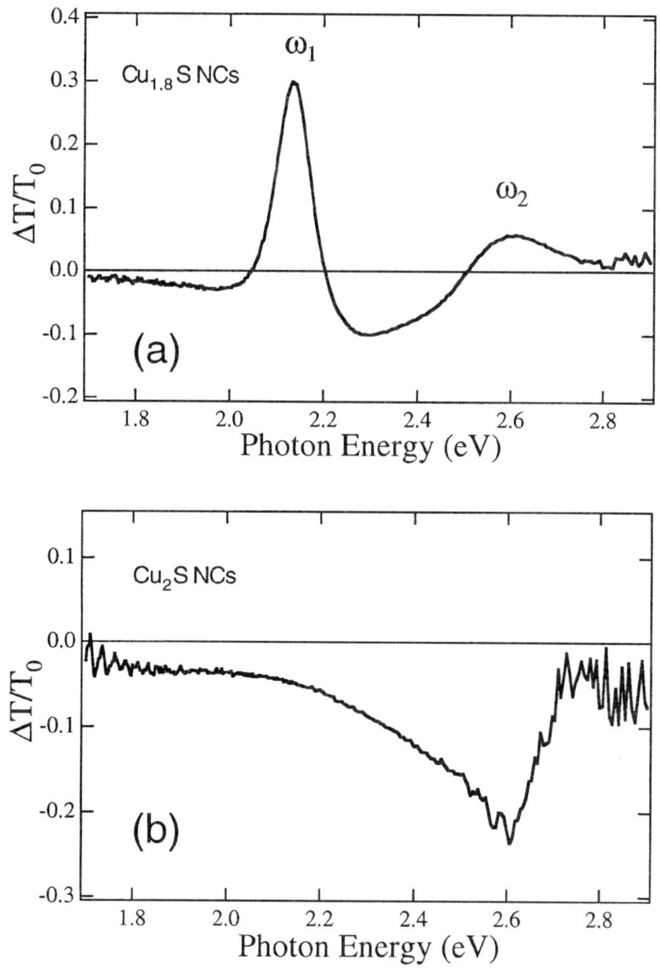

Fig. 32. DT spectra of Cu_xS NC–glass samples ($T = 300$ K) with $x = 1.8$ (a) and 2 (b). The spectra were taken at 2 ps after excitation at a pump fluence 9 mJ cm^{-2}. (Source: Reprinted with permission from [149]. © 1996 American Physical Society.)

7. CARRIER DYNAMICS IN SEMICONDUCTOR NANOCRYSTALS

Due to a three-dimensional carrier confinement and a large surface to volume ratio, energy relaxation and recombination dynamics in NCs are significantly different from those in bulk materials. In this section, we will address several issues related to different stages of carrier relaxation. We start our discussion with the effect of three-dimensional confinement on intraband energy relaxation. Next we consider depopulation dynamics of quantized states due to radiative decay and carrier trapping. Finally we will address the issue of the effect of multiparticle interactions on carrier relaxation behavior.

7.1. Intraband Energy Relaxation Dynamics

In bulk polar semiconductors energy relaxation is dominated by the Fröhlich interaction with LO phonons leading to fast (typically sub-ps) carrier cooling dynamics [92, 102]. In strongly confined NCs, the energy level spacing can greatly exceed typical phonon energies. Therefore, in the case of the phonon-dominated relaxation mechanism, energy losses

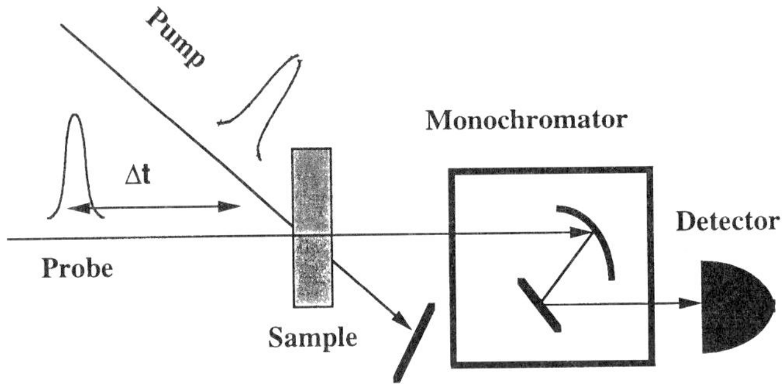

Fig. 33. Schematic of the fs pump-probe experiment applied in TA studies. The transmission of the sample excited with a pump pulse is probed by a variably delayed broad-band probe pulse.

can only occur via multiphonon processes. This should significantly inhibit carrier relaxation, an effect commonly referred to as a phonon bottleneck [87, 155, 156]. Calculations performed for GaAs NCs [87], show that the energy loss via two-phonon emission is orders of magnitude less efficient than for a single-phonon process, with a further reduction in the energy-relaxation efficiency for processes involving larger numbers of phonons. However, results from a number of experiments imply that carrier relaxation in strongly confined NCs is not significantly slower than in bulk materials. These results include the absence of hot-carrier emission in fs PL studies [49], the fast buildup of the nonlinear signal in transient absorption (TA) experiments [48, 50], and the sub-ps relaxation of TA features associated with the Coulomb two-pair interaction [47].

Direct studies of population–depopulation dynamics of the $1P$ and $1S$ electron states in CdSe NCs were performed in Ref. [157]. To evaluate the effect of surface (interface) states on the energy relaxation process, two different types of samples were investigated: NC–glass composites and colloidal NCs. Nonequilibrium carrier distributions were probed using high-sensitivity fs pump-probe experiment [158]. A typical configuration of this experiment is shown in Figure 33. The transmission of the sample excited with a short pump pulse is synchronously probed by delayed broad-band pulses of fs white-light continuum. The probe beam transmitted through the sample is dispersed with a spectrometer or a monochromator and is finally detected with either a charge-couple device (CCD) (which provides information on the entire DT spectrum) or a photodiode (single-wavelength detection). In the case of time-resolved sub-ps measurements, it is very important to account for temporal dispersion of the frequency (chirp) of the probe pulse. In typical fs experiments, the probe pulse is chirped such that the red portion of the pulse arrives at the sample before the blue portion. The chirp can be accounted for by postmeasurement numerical corrections employed usually in the case of CCD detection. Another approach, used in the measurements described in the following text, employs a single-wavelength phase-sensitive detection and a spectral scanning with simultaneous adjustment of the time delay between pump and probe pulses. This technique allows a direct measurement of chirp-free DT spectra with a high sensitivity up to 10^{-5} in DT [158].

Figure 34 shows the short-term dynamics of pump-induced absorption changes recorded in the chirp-free mode for glass [(a), $a = 4.2$ nm] and colloidal [(b), $a = 2.3$ nm] samples with $1S$-$1P$ energy separations of $\sim$6 and $\sim$14 LO-phonon energies, respectively. At probe delay times $\Delta t \geqslant 1$ ps, in both samples, the TA is dominated by two bleaching bands labeled as B_1 and B_2. B_1 marks the position of the lowest $1S(e)$-$1S_{3/2}(h)$ transition (the $1S$ transition) between electron and hole quantized states. The high-energy bleaching (B_2) is separated from B_1 by $\sim$400 meV which is close to the valence-band spin-orbit splitting in bulk CdSe (420 meV). This allows the assignment of the B_2 band to the transition

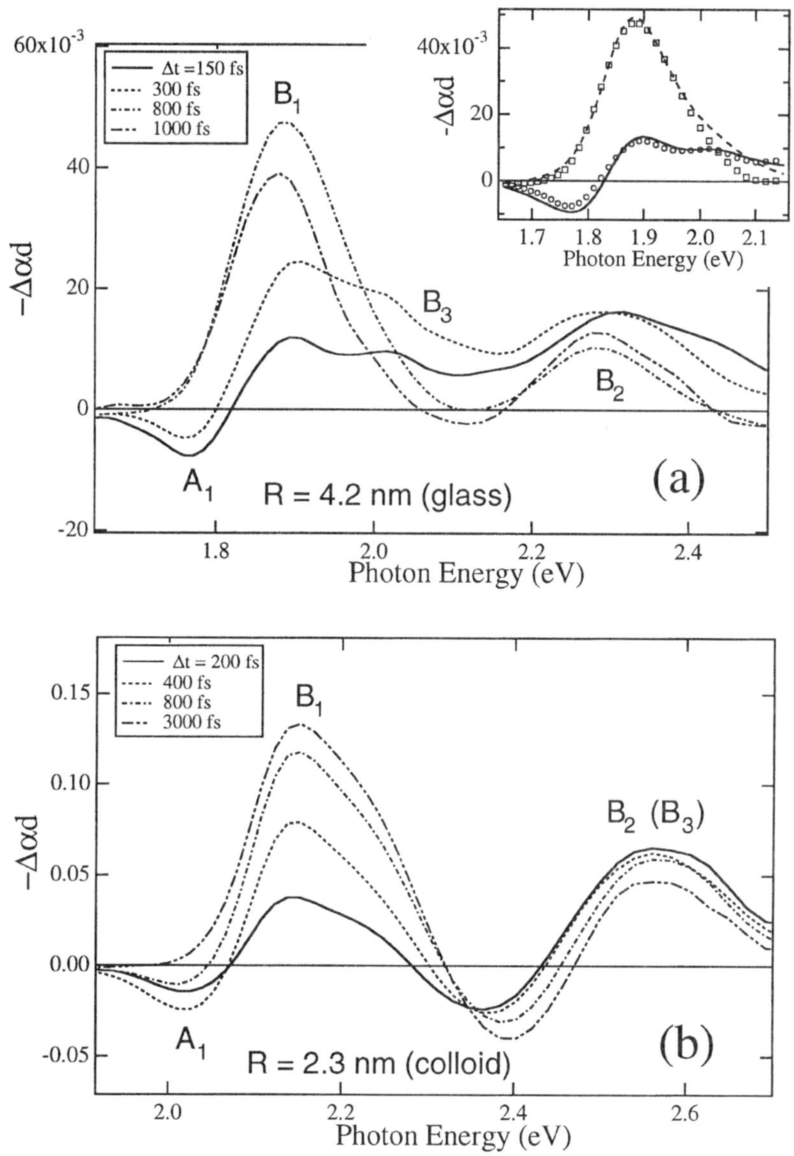

Fig. 34. Chirp-free TA spectra recorded at different delay times between pump and probe pulses for glass (a) and colloidal (b) samples with average radii 4.2 and 2.3 nm, respectively. Inset: TA spectra measured at 150 fs (circles) and 800 fs (squares) after excitation in comparison to calculations assuming that NCs are populated either in the $1P$ (solid line) or $1S$ (dashed line) states. (Source: Reprinted with permission from [157].

coupling the $1S$ electron state to the hole state $2S_{1/2}$ ($a = 2.3$ nm) or $3S_{1/2}$ ($a = 4.2$ nm) [33] originating from the spin-orbit split-off valence subband.

At sub-ps delay times between pump and probe pulses, both samples show a pronounced derivativelike feature at the position of the $1S$ transition, with an associated photoinduced absorption (PA) band labeled as A_1. The A_1 band decays rapidly on a sub-ps time scale, accompanied by a buildup of the B_1 bleaching. At $\Delta t < 500$ fs, TA spectra of the glass sample (Fig. 34a), along with the derivativelike TA features associated with the $1S$ transition, show an additional short-lived bleaching band (B_3) at ~2 eV. The position of this band is consistent with the energy of the $1P(e)$-$1P_{3/2}(h)$ transition (the $1P$ transition) [33].

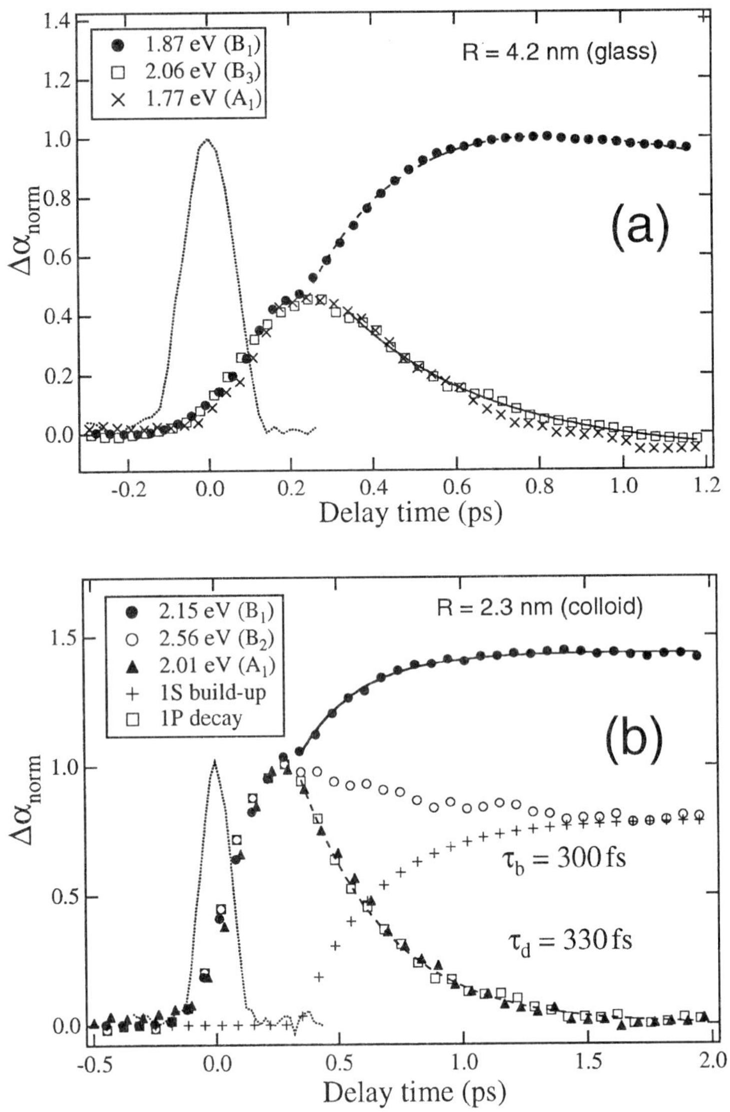

Fig. 35. (a) Normalized time transients recorded for the B_1 (solid circles) and B_3 (open squares) bleachings, and the A_1 photoinduced absorption (crosses) in the glass sample with $a = 4.2$ nm. Lines are fits assuming the $1P$-to-$1S$ relaxation time of 320 fs. (b) Normalized dynamics of the B_1 (solid circles), B_2 (open circles), and A_1 (solid triangles) bands in the colloidal sample with $a = 2.3$ nm along with exponential fits (solid line is a 300-fs exponential growth, dashed line is a 330-fs exponential decay). Open squares show the dynamics of the $1P$ electron state extracted from the B_2-band dynamics assuming an exponential population buildup of the $1S$ electron state with a time constant of 300 fs (crosses). Dotted line is a pump-probe cross-correlation. (Source: Reprinted with permission from [157]. © 1998 American Physical Society.)

Therefore, the B_3-band dynamics can be related to population changes of the $1P$ electron state. The decay of this band (Fig. 35a, open squares) is extremely fast and is characterized by a time constant of 320 fs.

The shape of the time transient recorded for the $1S$ transition (solid circles in Fig. 35a) indicates at least two different contributions. A fast initial rise is followed by a slower signal increase (time constant of ~300 fs) which is complementary to the B_3-band decay. The presence of two different contributions to the B_1 bleaching can be explained in terms of Coulomb two-pair interactions and state filling (see Section 6.2), which is confirmed by modeling of time-resolved TA spectra (see Eq. (42)) assuming carrier redistribution between the $1P$ and $1S$ electron states [47] (see inset to Fig. 34a). In the initial stage of hot-carrier relaxation, before there is significant population of the $1S$ level, the $1S$ transition shows a nonzero signal due to the Coulomb-interaction-induced transition shift. This mechanism for the fast rise of the B_1-band is confirmed by the fact that it matches that of the PA feature A_1 (crosses in Fig. 35a). The slower postrise of the B_1-band is due to the increasing population of the $1S$ state as a result of carrier relaxation, consistent with the complementary decay of the B_3-band associated with depopulation of the $1P$ state. The ~300-fs time constant measured for this process is in agreement with data of lower time resolution measurements in CdS_xSe_{1-x} NCs ($a = 4.6$ nm) [48] where the $1S$ buildup time was estimated to be below 700 fs.

Interestingly, the decay of the B_3-band matches closely that of the A_1-band, indicating that the dynamics of the PA feature are also related to carrier relaxation. This can be understood as occurring due to a large bandwidth of optical transitions, leading to the significant contribution of the $1S$ bleaching even at large detunings from the center of the B_1-band. Therefore, the increasing bleaching of the $1S$ transition accompanying carrier energy relaxation overwhelms the A_1 PA feature as the $1S$ state becomes occupied.

A similar analysis of TA can be performed for colloidal NCs as illustrated in Figure 35b for the sample with $a = 2.3$ nm. As in the NC–glass composites, in the colloidal sample the B_1 time transient shows a biexciton-effect induced fast rise (solid circles), followed by a postrise (time constant ~300 fs) due to increasing population of the $1S$ state. These dynamics are complementary to the decay of the PA feature A_1 (solid triangles) associated with the transition shift, as in the glass sample. However, TA spectra of the colloidal NCs (Fig. 34b) do not show the separate band associated with the $1P$ transition. For NCs of this size this transition is close to the $1S(e)$-$2S_{1/2}(h)$ transition [33]. Therefore, the high-energy bleaching at 2.56 eV is likely due to two contributions associated with populations of the $1S$ and $1P$ electron states. The dynamics of this band indeed show two different components: a fast and a slow one (Fig. 35b, open circles). Assuming that the slow component is due to the buildup of the $1S$ state, one can approximate it by an exponential growth with a time constant of 300 fs (crosses in Fig. 35b) as derived from the B_1 dynamics. Subtracting this component from the 2.56-eV signal, one can obtain the dynamics related entirely to population changes of the $1P$ state (open squares in Fig. 35b). As in the glass samples, these dynamics closely match those of the PA feature A_1 (solid triangles in Fig. 35b). A fit to the decay of the extracted signal gives a relaxation constant of 330 fs, which is close to the buildup time of the B_1 bleaching, as expected for the $1P$-to-$1S$ relaxation.

The data shown indicate that the initial TA dynamics are similar in both colloidal and glass samples, demonstrating the negligible role of surface effects and lattice imperfections on the initial energy relaxation process. The measured relaxation constants indicate a very high-energy loss rate: ~1.2 eV ps^{-1} in NCs with $a = 2.3$ nm. This value is almost a factor of 2 greater than the estimated rate for the unscreened polar interaction in bulk CdSe with a quasi-continuous energy spectrum [92], and many orders of magnitude higher than the relaxation rate due to multiphonon emission expected for strongly confined systems [87], indicating a dominant role of mechanisms other than phonon energy loss. As predicted in Refs. [159] and [160], one of the effects of spatial confinement is an enhancement of the Coulomb *e-h* interaction which can lead to an efficient energy transfer of electron excess energy to a hole, with subsequent fast hole energy relaxation through its nearly continuous spectrum of states. This process is analogous to nonradiative Auger recombination which leads to the efficient sub-ps carrier decay in II-VI NCs at high excitation densities [44]. According to calculations in Ref. [159], the $1P$-to-$1S$ relaxation time in CdSe NCs with radii from ~2 to ~8 nm is around 2 ps. This value requires a much higher energy-loss rate than possible for multiphonon emission, but is still nearly an order of magnitude larger than our experimental determination. This mismatch may be due to an underestimation of the hole-phonon interaction, which is not accurately known.

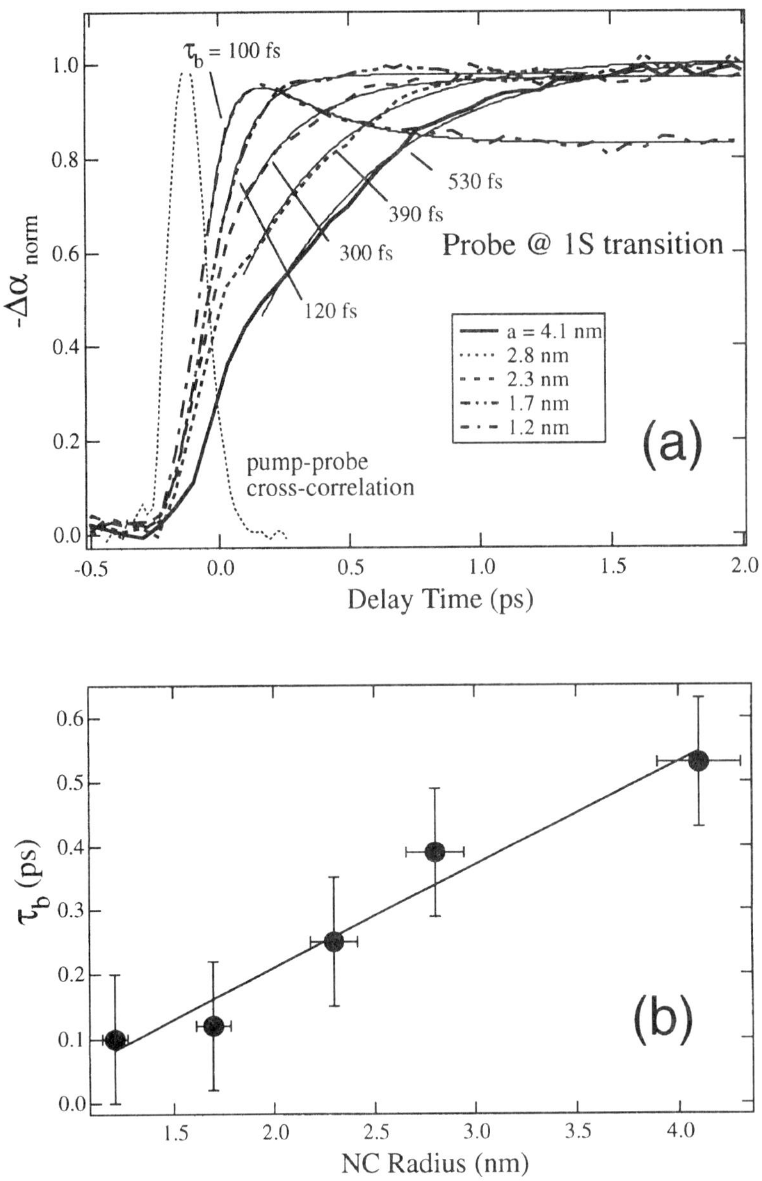

Fig. 36. (a) The 1S-bleaching build-up dynamics for colloidal CdSe NC's of different size (thick lines), fit to a single-exponential growth (thin solid lines). Thin dotted line is a pump-probe cross-correlation. (b) The dependence of the 1S bleaching build-up time on the NC radius (circles), fit to a linear size dependence.

The important role of confinement effects in the enhancement of the energy relaxation is confirmed by comparison of $1S$ state population dynamics in NCs of different radii (see Fig. 36a; colloidal samples). These data show a reduction in the $1S$ buildup time (τ_b) with a decreasing NC radius. This reduction is evident in Figure 36b where the relaxation data for five colloidal samples are plotted as τ_b versus a; τ_b shortens from 530 fs for $a = 4.1$ nm to 100 fs for $a = 1.2$ nm, roughly following a linear size dependence.

The issues of energy relaxation in quantum-dot systems remain largely unexplored from both experimental and theoretical points of view. The available data (see Fig. 36) strongly indicate that energy relaxation occurs via mechanisms other than electron-phonon interactions, and these mechanisms are enhanced with an increasing degree of confinement.

However, the detailed nature of energy-relaxation processes is still unclear and needs further investigation.

7.2. Carrier Recombination and Trapping Dynamics

Fast energy relaxation leads to establishing quasi-equilibrium populations of the lowest quantized states. These states can be depopulated by several processes. One of these processes is radiative e-h recombination leading to band-edge emission. Within the model of strong confinement, the constant of the radiative decay (τ_r) of the lowest e-h pair state (which we assume corresponds to the $|M| = 3/2$ hole) is independent of the NC radius and is given by the expression [77],

$$\frac{1}{\tau_r} = \frac{1}{\tau_0}\left|\int dr\, r^2 \Phi^e_{1,0,0}(r) R^h_0(r)\right|^2 \tag{58}$$

where $\Phi^e_{1,0,0}(r)$ is the radial wave function of the electron ground state (see Eq. (1)), $R^h_0(r)$ is the S-symmetry part of the hole ground-state wave function, and

$$\tau_0 = \left[\frac{4\omega n p^2_{cv}}{3 \times 137 m^2_0 c^2}\right]^{-1} \tag{59}$$

is the characteristic radiative-recombination time in the bulk material. The overlap integral in Eq. (58) is NC-size independent and is only determined by the ratio of the light-to-heavy hole masses (β). For CdSe ($\beta \approx 0.3$ [29]) the square of the overlap integral is close to 1 [77], meaning that $\tau_r \approx \tau_0$. If the parameters of CdSe are further used in Eq. (59), one deduces that $\tau_r \approx 1$ ns [77].

Experimentally, a slow ns decay attributed to radiative processes can be detected in colloidal NCs with well passivated surfaces [161]. However even in freshly prepared samples, the decay time exhibits a weak size dependence. Figure 37 shows long-term dynamics of the $1S$ bleaching measured at room temperature for colloidal CdSe NCs with six different radii from the range 1.2–4.1 nm. In NCs with the largest mean radius (4.1 nm) the relaxation constant is 2.55 ns. It gradually shortens with reducing NC size, being 1.42 ns in NCs with $a = 1.2$ nm. This shortening of the decay time can be due to the increasing role of nonradiative processes in NCs of small sizes resulting, e.g., from the enhanced surface trapping (see the discussion later). Slow ns dynamics which could be attributed to almost purely radiative decay were also observed in glass samples cooled down to liquid N_2 temperatures using time-resolved DT and PL measurements [122, 162]. Room-temperature carrier dynamics in glass samples are usually dominated by nonradiative processes as detailed in the following text.

In small-size NCs, carrier dynamics are strongly affected by surface and/or defect states which provide an efficient channel for nonradiative recombination [49, 153, 163–165]. The effect of trapping at localized defect (surface) states is particularly important in glass samples for which the methods for controlling NC surface properties have not been well developed. Due to weak oscillator strength, the transitions involving localized states are not pronounced in linear or nonlinear absorption and, therefore, are typically studied using cw and time-resolved PL spectroscopies. The cw-PL spectra of CdS, CdSe, and CdS_xSe_{1-x} NCs consist of two broad bands: one (high energy) at the band-edge spectral energies and the other (low energy) in the NC energy gap [69, 166–168]. Figure 38 shows an example of cw emission of CdS NCs ($a \approx 4$ nm) which is compared with a DT spectrum of the same sample recorded at 1 ps after excitation. The DT bleaching peak at ~2.62 marks the position of the lowest $1S$ transition between electron and hole quantized states. The high-energy PL band (band-edge emission) observed at 2.55 meV is red shifted from the $1S$ transition by ~70 meV, indicating that at least one of the carriers (an electron or a hole) involved in the emission process resides in the localized state. The low-energy band (referred to as deep-trap emission) for the sample shown in Figure 38 is located at ~1.98 eV

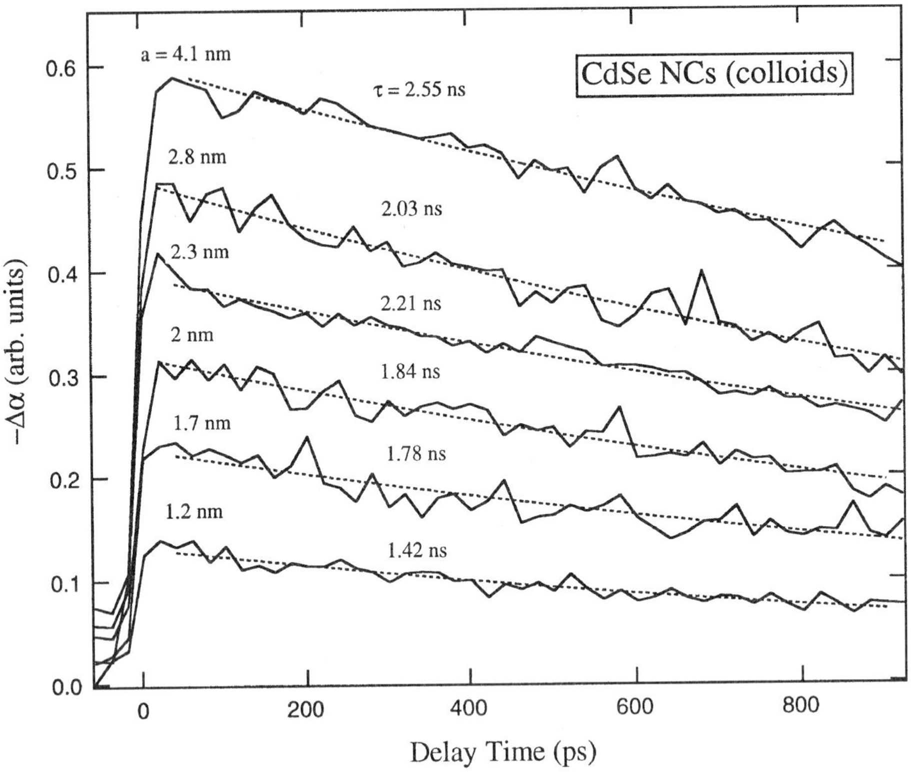

Fig. 37. Room-temperature dynamics of the 1S-bleaching in freshly prepared CdSe colloidal NCs of 6 different mean radii, fit to an exponential decay. NC radii and corresponding relaxation constants are indicated in the figure.

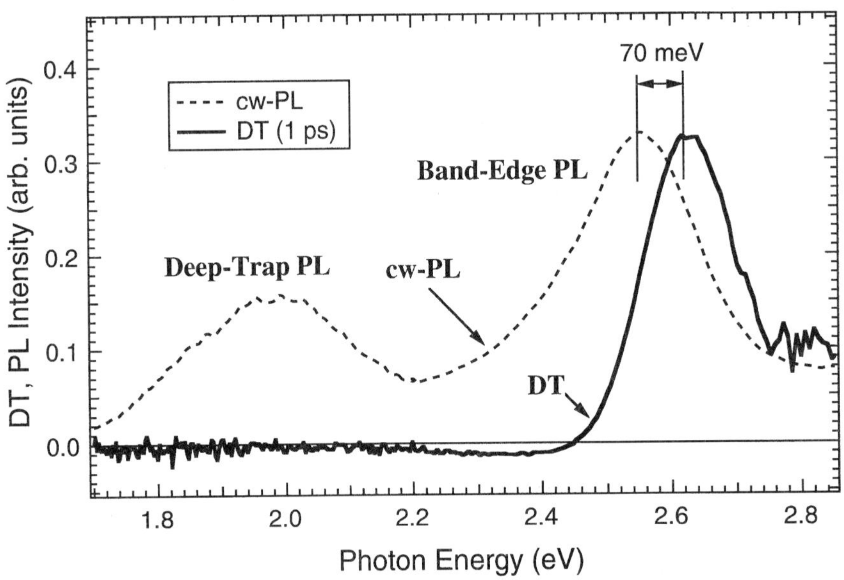

Fig. 38. The cw-PL spectrum of CdS NCs ($a \approx 4$ nm) in comparison to the DT spectrum taken at 1 ps after excitation ($T = 300$ K). The DT bleaching peak marks the position of the lowest 1S transition between electron and hole quantized states. The PL spectrum exhibits two bands usually referred to as the band-edge and the deep-trap emissions. (Source: Reprinted with permission from [153].)

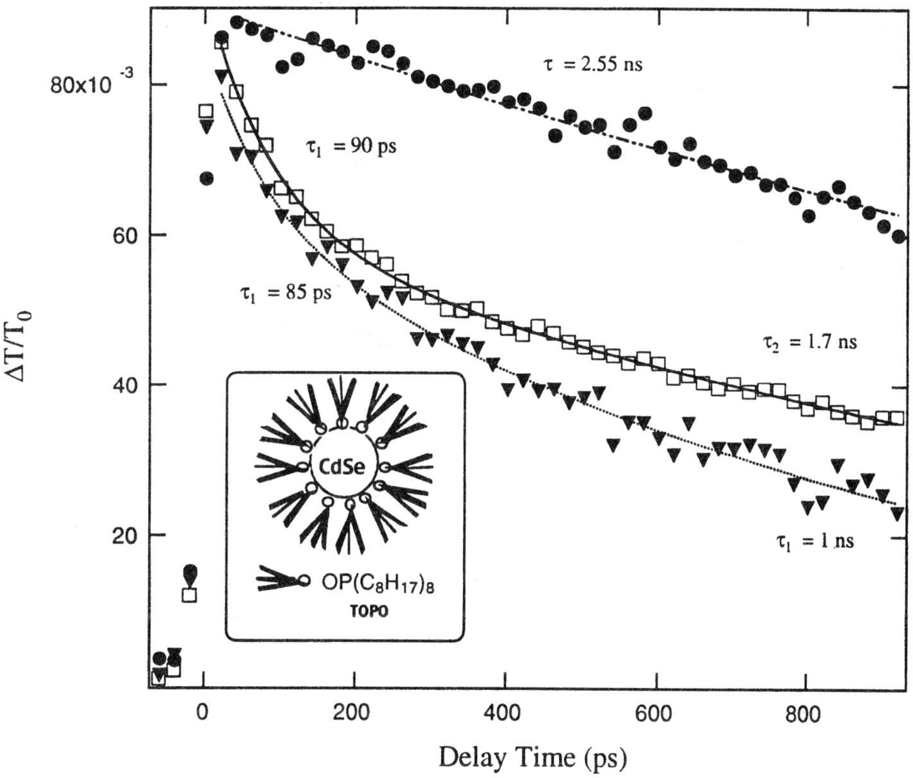

Fig. 39. The room-temperature $1S$-bleaching dynamics in CdSe colloidal sample ($a = 4.1$ nm) recorded immediately after preparation (solid circles), and 10 (open squares) and 15 (triangles) days after preparation. The freshly prepared sample shows an exponential decay with a time constant of 2.55 ns. The decay in the aged sample is strongly nonexponential with the initial fast component having a ~90-ps relaxation constant. Inset: schematic of a TOPO-capped CdSe NC.

which corresponds to the ~0.64-eV shift from the $1S$ optical transition. This shows the presence of the second group of trapping states located deep in the energy gap. It has been suggested that the deep-trap emission originates from the donor–acceptor recombination involving the electron trapped at the sulfur vacancy [163, 164]. The relative intensity of the low-energy PL band increases with decreasing NC radius [168], strongly suggesting that sulfur vacancies involved in the emission process are located at the surface. The fact that electron traps are surface related is confirmed by measurements of carrier dynamics in colloidal samples with different degree of surface passivation [161].

As prepared colloidal NCs are passivated with organic molecules which protect them from aggregation and render them soluble. One of the most developed synthetic routes is based on the organometallic reactions with tri-*n*-octylphosphine oxide (TOPO) as the surfactant [5]. Prepared in this way NCs have surfaces terminated via the metal ion with TOPO molecules (see inset to Fig. 39). As the sample ages, the organic surface layer is gradually destroyed, which can be used to study the effect of surface passivation on carrier relaxation dynamics [161]. Figure 39 shows the comparison of $1S$ bleaching time transients recorded in a fresh CdSe colloidal sample ($a = 4.1$ nm) with those recorded during the aging of the sample. The fresh sample in which the passivation layer is intact shows a slow exponential decay (circles in Fig. 39) with a time constant of 2.55 ns, presumbly corresponding to radiative *e-h* recombination. The loss of passivation accompanying sample aging leads to a dramatic change in $1S$ dynamics (squares and triangles in Fig. 39). The $1S$ decay becomes strongly nonexponential, revealing the presence of the additional fast component with a

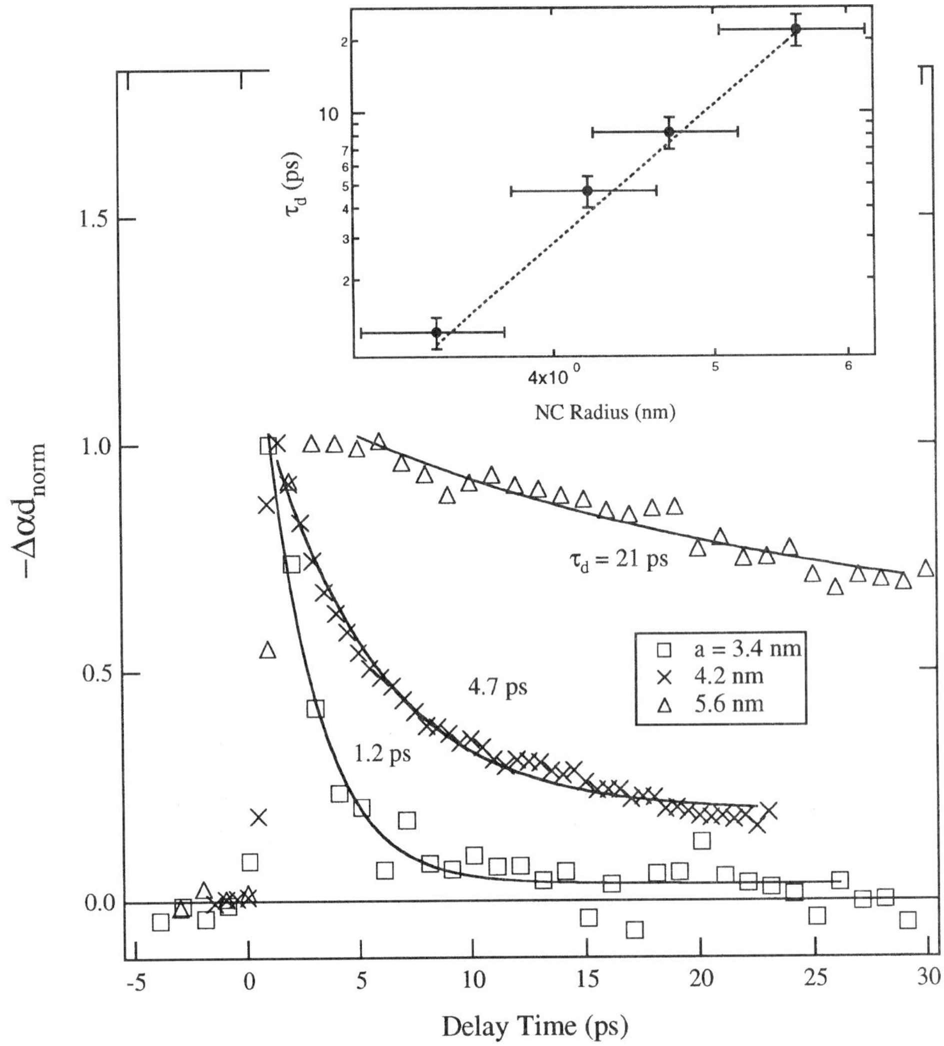

Fig. 40. The room-temperature $1S$-bleaching dynamics in CdSe NC–glass samples with NC radii 3.4 (open squares), 4.2 (crosses) and 5.6 nm (open triangles). Inset: size dependence of the $1S$ relaxation time (symbols), fit to the a^6 dependence (line).

time constant of about 80–90 ps. As was pointed out earlier, the $1S$ bleaching dynamics are dominated by electron populations. Therefore, the shortening of the $1S$-band relaxation time in the aged samples is indicative of the enhancement of electron trapping, leading to efficient depopulation of the $1S$ electron quantized state. The trapping sites responsible for this process are likely associated with surface Cd dangling bonds (sulfur vacancies) created as a result of the loss of passivation.

Electron trapping at deep states is a very efficient process in glass samples where it leads to very fast (picosecond) depopulation dynamics of the $1S$ electron state [49, 153, 157]. Figure 40 shows the comparison of the $1S$ time transients in CdSe–glass samples with NC mean radii of 3.4, 4.2, and 5.6 nm, recorded at room temperature. The observed dynamics are significantly faster than those in colloidal samples (compare with Fig. 37). The $1S$ decay time is 21 ps in the sample with $a = 5.6$ nm, and rapidly shortens for NCs of smaller radius (see inset to Fig. 40), being around 0.9 ps in the sample with $a = 3.4$ nm. The size dependence of the $1S$ relaxation time roughly follows an a^6 dependence (inset to Fig. 40).

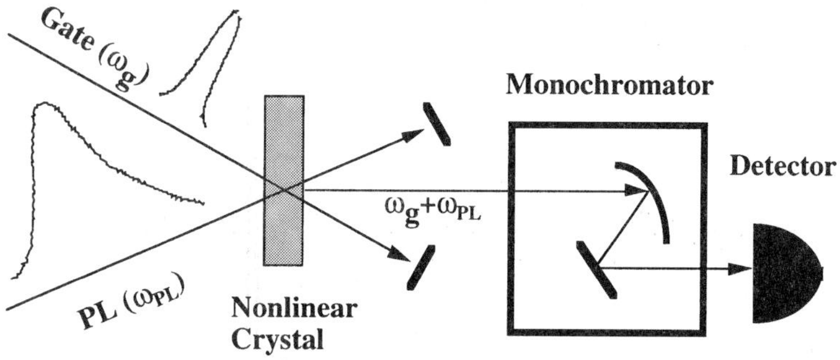

Fig. 41. Schematic of the PL fs up-conversion experiment. The PL pulse is mixed in the nonlinear crystal with a variably delayed short gating pulse to produce a sum-frequency signal which is detected with a monochromator coupled to a photodetector.

This can be explained in terms of the confinement-induced squeezing of the $1S$ electron wave function resulting in the enhancement of its overlap with the wave function of the deep-trap state [169].

At room temperatures, the $1S$ bleaching is dominated by the electron contribution. The hole dynamics are therefore practically unresolved in TA measurements. To probe hole dynamics one can apply time-resolved PL measurements [49, 153]. PL intensity is determined by a product of the electron and hole occupation numbers and therefore is significantly affected by the dynamics of both carriers. A comparative study of DT and PL dynamics of CdS NCs (glass samples), which allowed the separation of the electron and hole relaxation paths, was performed in Ref. [153]. Time-resolved DT transmission was measured using the fs pump-probe experiment (see Fig. 33). Time-resolved PL measurements were done using fs PL up conversion [170]. In these measurements, PL is mixed in a nonlinear crystal with a short gating pulse to produce a sum-frequency signal in blue-near-UV spectral range (see Fig. 41). Scanning the gating pulse along the PL pulse provides information on PL dynamics, whereas the spectral resolution is achieved using a monochromator to detect a sum-frequency signal.

Figure 42 shows time-resolved DT (a) and PL (b) spectra of CdS NCs ($a \approx 4$ nm; see cw emission spectrum in Fig. 38) recorded within the first 3 ps after excitation in the low-pump-intensity regime (pump fluence $w_p < 0.5\ \mathrm{mJ\,cm^{-2}}$), corresponding to excitation of less than one e-h pair per NC on average. This allows one to avoid the saturation of trap states and nonlinear recombination effects which might complicate the interpretation of the experimental data. DT spectra displayed in Figure 42a are dominated by the $1S$ bleaching at 2.62 eV which does not show significant spectral changes during the observation time and decays by only $\sim$20%. This is in a sharp contrast to PL dynamics which show more than a twofold reduction in PL intensity and a 70-meV red shift of its maximum (Fig. 42b). Initially (at $\Delta t < 0.5$ ps), PL is peaked at the position of the $1S(e)$-$1S_{3/2}(h)$ transition marked by a DT band. This rules out the explanation of the cw emission in terms of the strong electron-phonon coupling resulting in a large Stokes shift as was suggested in early stuidies of NC–glass composites [99–101]. At a later time, the PL maximum moves rapidly downward in energy and in 3 ps it reaches the position of the band-edge cw emission. A drastic difference in early-time PL and DT dynamics is well pronounced in single wavelength ($\hbar\omega = 2.6$ eV) time transients shown in Figure 43a (short terms dynamics) and (b) (long-term dynamics). Initial decay of PL occurs on a much faster time scale than that of DT. Double-exponential fit to the experimental data gives 1 and 11 ps relaxation times for the first stages of PL and DT dynamics, respectively, (Fig. 43a). A second slower stage of relaxation is almost identical for PL and DT, and is characterized by decay constants of 90 and 70 ps, respectively, (Fig. 43b).

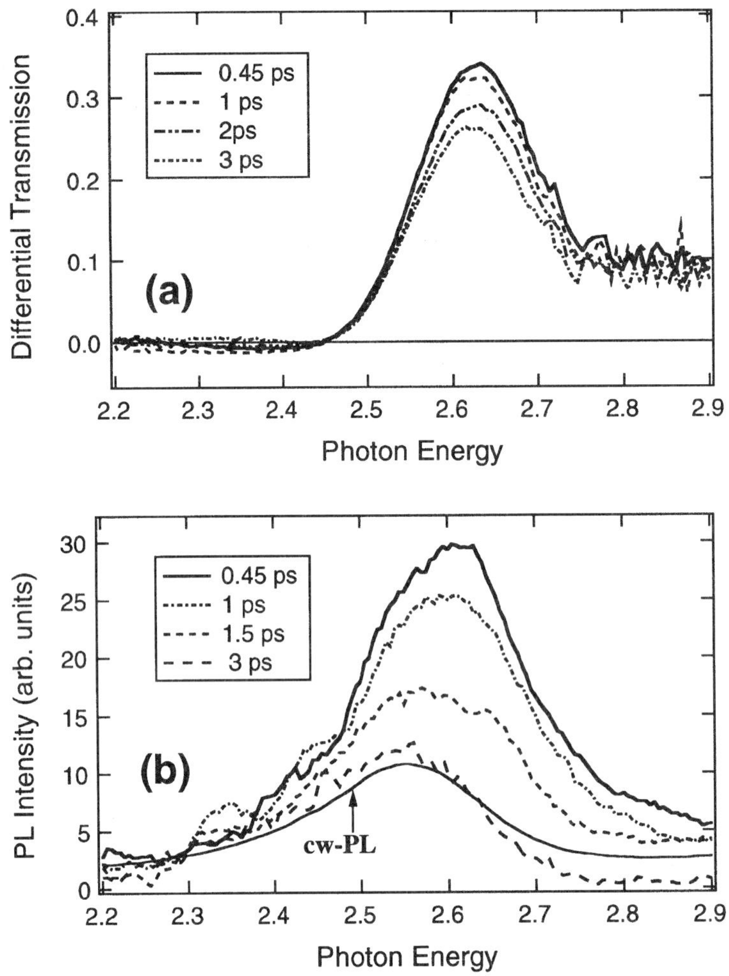

Fig. 42. Room-temperature short-term dynamics of the DT (a) and up-converted PL (b) spectra of CdS NCs ($a \approx 4$ nm; glass sample). For comparison, the cw-PL spectrum is also shown [(b), thin solid line]. (Source: Reprinted with permission from [153].)

A strong difference in the early-time DT and PL dynamics is a clear indication of the fact that the carrier decay is nongeminate and the electron and the hole have different relaxation paths. As discussed earlier, the $1S$ decay provides the measure of the electron trapping time at deep-trap states. According to time transients shown in Figure 43a this time is approximately 10 ps. Because PL dynamics are much faster than those of DT, they are almost entirely due to changes in hole occupation numbers. Therefore, the 1-ps relaxation of the PL intensity can be attributed to a hole trapping at localized states. The change in the position of the up-converted-PL maximum shows that the hole trap has a depth of around 70 meV. The transition involving this trap and the lowest electron quantized level dominates the cw band-edge emission. Almost the same dynamics observed for DT and PL on the longer time scale ($\Delta t > 10$ ps, Fig. 43b) indicate that the hole-trap state is long lived. Therefore, long-term DT and PL dynamics are governed by the same process: depopulation of the lowest conduction-band quantized state due to electron trapping.

The shallow hole trap with a 70-meV depth can also be detected in DT spectra taken at a long time after excitation [44]. As was pointed out previously, the DT spectral maximum does not show any noticeable shift within several picoseconds after excitation. However, on a longer time scale ($\Delta t > 10$ ps), the DT band starts to shift downward in energy until

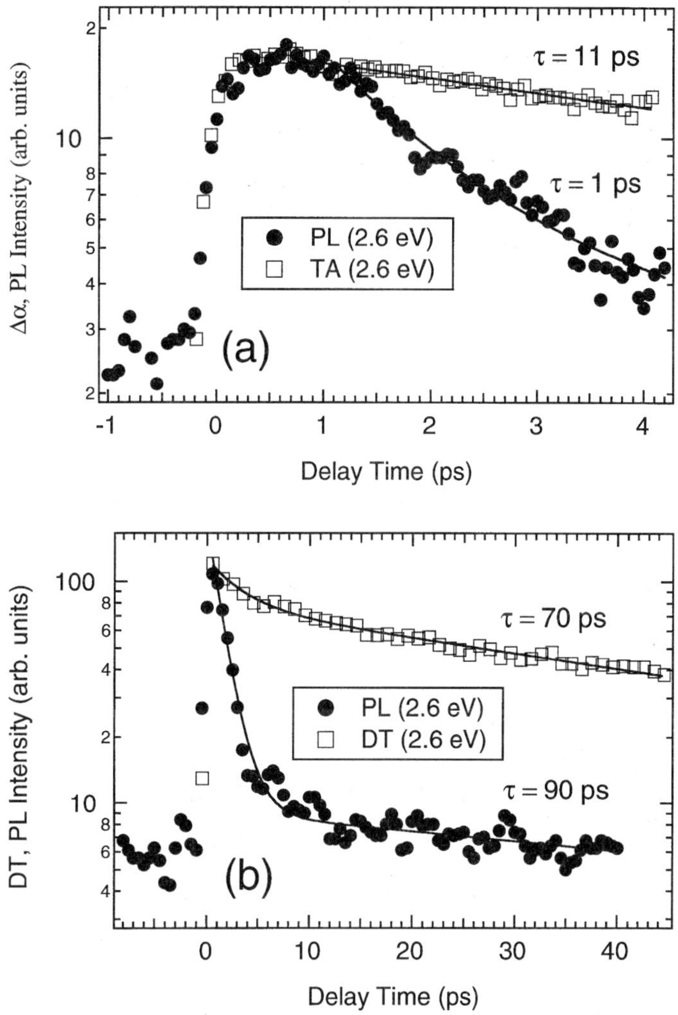

Fig. 43. Room-temperature single-wavelength [$\hbar\omega = 2.6$ eV (the $1S$ transition)] short- (a) and long-term (b) dynamics of PL (solid circles) and DT (open squares) in CdS NCs ($a \approx 4$ nm; glass sample). Solid lines show fits to a double-exponential decay. (Source: Reprinted with permission from [153].)

it finally reaches (at $\Delta t > 250$ ps) the position of the cw-PL maximum at 2.56 eV (see normalized spectra in Fig. 44b). This can be explained by the increasing relative contribution from the long-lived bleaching of the transition coupling the shallow hole trap to the lowest electron quantized state. At long delay times, after the electron state is completely depopulated, the bleaching of this transition is solely due to the trapped holes.

A detailed analysis of the up-converted PL spectra [49] (see Fig. 45) shows the presence of several discrete features (ω_1–ω_4) which are not resolved in cw emission. The data shown in the preceding text allows one to identify the high-energy band ω_1 with the $1S$ transition coupling the lowest electron and hole quantized states. The ω_2 band can be attributed to the transition involving the $1S$ electron and the hole-trap state. The low-energy bands ω_3 and ω_4 are red shifted from the $1S$ bleaching by ~170 and ~270 meV, respectively, indicating that they are due to transitions involving localized carriers. Interestingly, the low-energy portion of PL spectrum in the range of the ω_3–ω_4 bands has extremely fast buildup dynamics (time constant ~800 fs), essentially identical to those of the ω_1 feature (see inset to Fig. 45). These data can be understood under the assumption that the ω_3 and

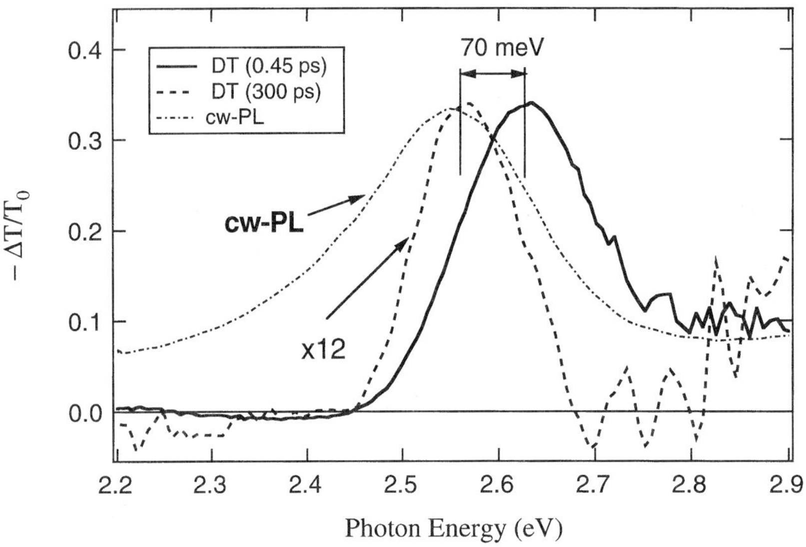

Fig. 44. Normalized DT spectra of CdS NCs ($a \approx 4$ nm; glass sample) measured at $\Delta t = 0.45$ ps (solid line) and 300 ps (dashed line) in comparison to the cw-PL spectrum ($T = 300$ K). (Source: Reprinted with permission from [44]. © 1997 American Physical Society.)

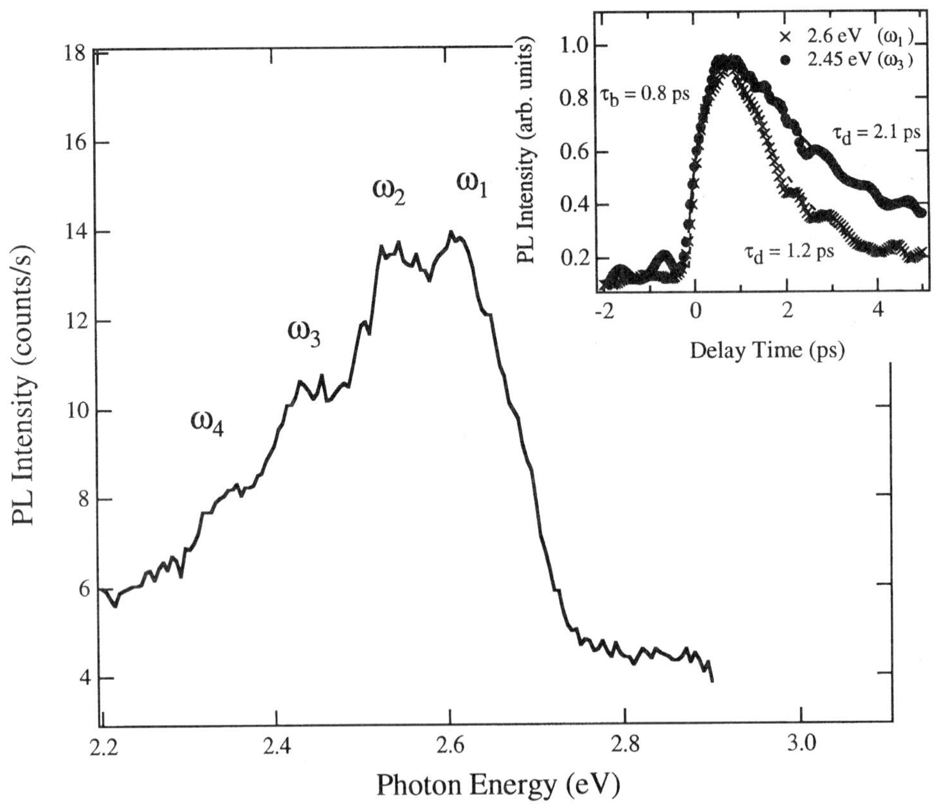

Fig. 45. The up-converted PL spectrum measured at 2 ps after excitation demonstrating four features labeled as ω_1–ω_4. Inset: Short-term PL dynamics measured at 2.6 (ω_1) and 2.45 eV (ω_3) along with fitting curves calculated for the exponential rise (time constant τ_b) followed by the exponential decay (time constant τ_d).

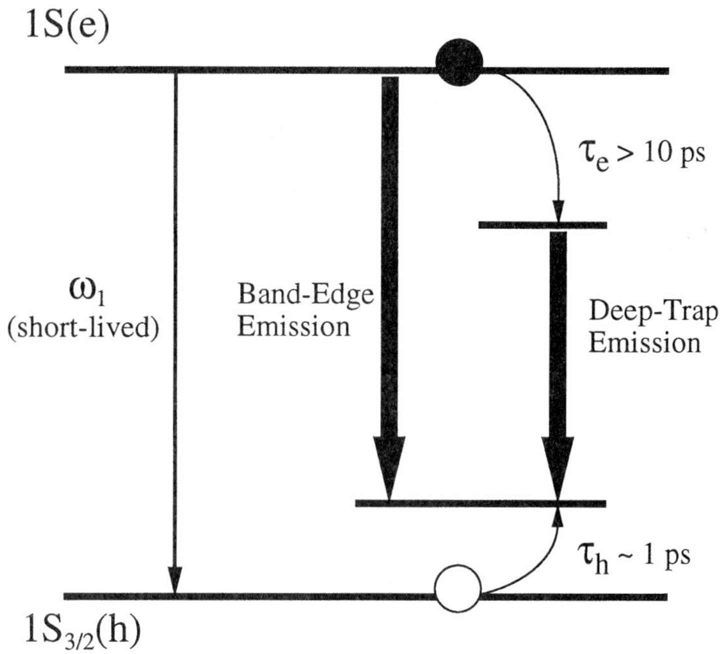

Fig. 46. Schematic of the trapping processes and optical transitions leading to principal emission features in CdS NCs. A short-lived ω_1 feature is located at the position of the $1S$ transition. It decays rapidly due to a fast 1-ps hole trapping (localization). The transition coupling the hole localized state and the $1S$ electron level dominates the cw band-edge emission. Electron gets trapped at a deep-trap state on the tens-of-ps time scale. The transition coupling this deep trap and the hole localized level gives rise to the low-energy deep-trap emission.

ω_4 PL bands are due to transitions coupling extended $1S$ electron state to localized hole states which are occupied in the absence of the pump. These hole states can be associated, with uncompensated acceptors [49].

In discussing the nature of the defect states contributing to the band-edge PL, it is natural to consider first surface located native defects such as sulfur or cadmium vacancies or add atoms. As was discussed in the previous text, the deep-trap emission in CdS NCs (band at 1.98 eV in Fig. 38) can be assigned to a surface sulfur vacancy with a level located ~0.7 eV below the conduction band [163]. A Cd vacancy can create two acceptor states in the energy band gap: One (singly ionized) above the valence band and the other (doubly ionized) below the conduction band [171, 172]. The other possible acceptor states are complexes associated with the Cd vacancy, such as a complex with a donor at the next lattice site [173]. Bulk CdS is an n-type material with a conductivity dominated by foreign donor atoms. However, one may expect a different situation in NCs (at least in a portion of them) with a large number of surface located native defects which can lead to the existence of uncompensated acceptors along with compensated ones. The presence of negatively charged compensated acceptors can explain a fast capture of the photoexcited holes resulting in a 1-ps decay of the ω_1 PL band associated with the $1S$ transition, whereas the processes involving the uncompensated neutral acceptors can account for a fast buildup of the emission in the NC energy gap. A schematic of the trapping processes and principal emission features in CdS NCs is given in Figure 46.

7.3. Auger Recombination in Semiconductor Nanocrystals

Three-dimensional carrier confinement leads to the enhancement of Auger-type processes and in particular, of Auger recombination [42, 160]. Auger recombination is a nonradiative process which involves three particles [174, 175]. Two particles (an electron and a hole)

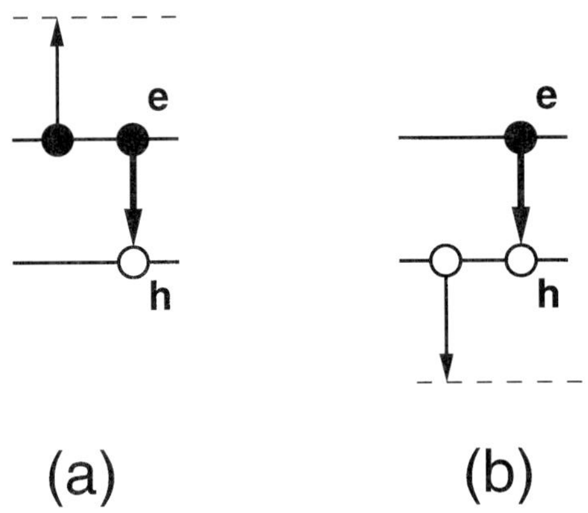

Fig. 47. Schematic of the Auger recombination accompanied by reexcitation of an electron (a) or a hole (b).

recombine, and the recombination energy is transferred to a third particle which can be either an electron or a hole (see Fig. 47). Auger recombination is mediated by the Coulomb electron–electron (hole) interactions which are much stronger than carrier interactions with the electromagnetic field. As a result, in strongly confined atomic and molecular systems, the Auger processes dominate carrier recombination in the case of multielectron excitations [160]. The efficiency of Auger processes is significantly reduced in bulk crystalline materials because of the kinematic restrictions imposed by the energy and momentum conservation [175]. These restrictions lead to the existence of the energy threshold for the Auger recombination and a temperature-activated behavior of the corresponding recombination rate $R_A \propto \exp(-(E_g/kT))n_{eh}^3$ (n_{eh} is the e-h pair concentration) [160].

In spherical NCs, the momentum conservation is replaced with a less-restrictive conservation of the angular momentum, which removes the energy threshold for the Auger process and results in a significant enhancement of the corresponding recombination rate [42, 160]. In NCs, Auger recombination is a confinement enhanced process. Calculations performed assuming that the confinement-induced energy shifts are significantly smaller than the confinement potential indicate a strong size dependence of the Auger-recombination rate which scales as $1/a^7$ [42, 160].

The first experimental indications of highly efficient Auger recombination in semiconductor NCs were obtained in pump-intensity-dependent studies of carrier decay in NC–glass composites [131, 162, 176–178]. In Refs. [162, 178] these studies were performed using streak-camera measurements of PL relaxation in CdSe NCs ($a \approx 3.5$, 6, and 6.5 nm) cooled down to the liquid-nitrogen temperature. In NCs with the largest mean radius ($a \approx 6.5$ nm), at low-pump intensities (pump fluence below ~0.5 mJ cm^{-2}) the PL decay was essentially exponential with a ~2.4-ns time constant. At pump fluences above ~0.5 mJ cm^{-2} the PL traces displayed strongly nonexponential behavior (see Fig. 48) indicating a significant role played by nonlinear recombination processes. The initial decay time (τ) rapidly shortened with increasing pump fluence (down to ~40 ps at 2.8 mJ cm^{-2}) following the w_p^{-2} dependence (see Fig. 49). Two other samples ($a \approx 6$ and 3.5 nm) showed a similar relaxation behavior at high-pump fluences, although the low-intensity PL dynamics for them were different from those in the sample with $a \approx 6.5$ nm. In the case of ps excitation used in Refs. [162, 178], the number of e-h pairs excited per NC (N_{eh}) was proportional to the pump fluence, therefore the pump dependence of the relaxation time observed at high-pump intensities could also be presented as $\tau \propto N_{eh}^{-2}$. This type of dependence indicates that the carrier decay at high-pump fluences is a cubic process governed by the rate equation $dN_{eh}/dt \propto -N_{eh}^3$, characteristic of Auger recombination. The difference in low-pump-intensity PL dynamics observed in samples with different radii was explained in terms of the competition between the radiative decay and carrier trapping. In the sample

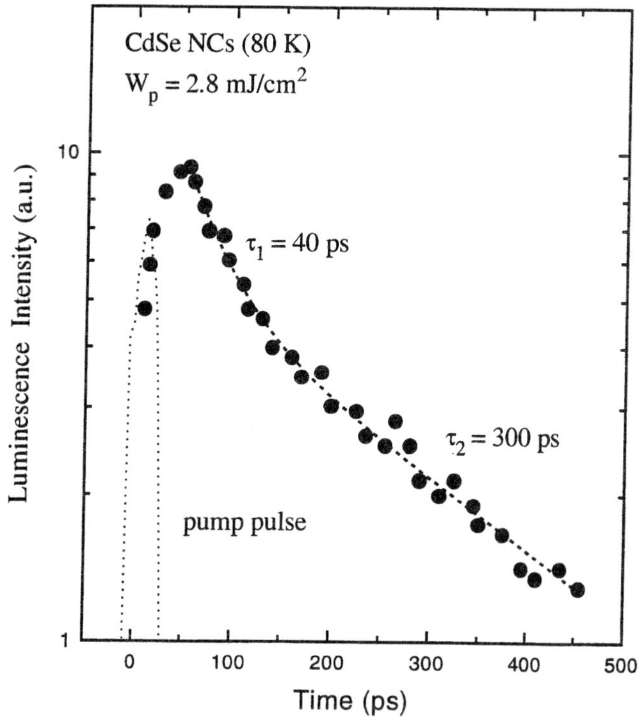

Fig. 48. PL dynamics (symbols) in CdSe NC–glass sample ($a \approx 6.5$ nm, $T = 80$ K) recorded at $w_p = 2.8$ mJ cm^{-2}. The PL relaxation is fit to a double-exponential decay (dashed line) with time constants of 40 and 300 ps. Dotted line shows a ps pump pulse.

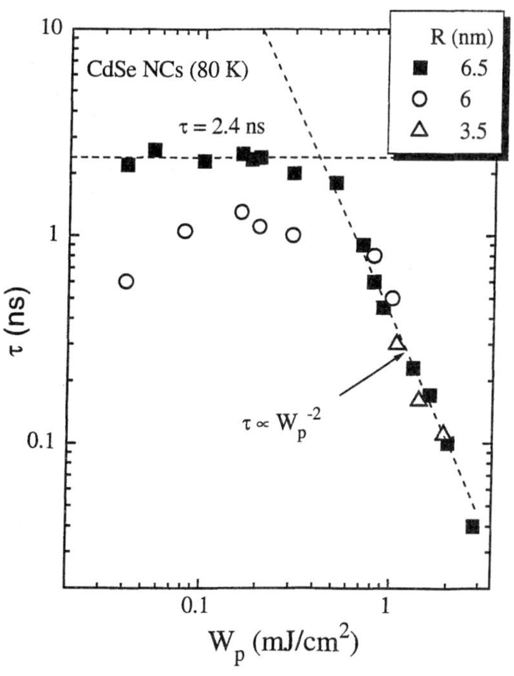

Fig. 49. The dependence of the PL initial relaxation time (τ) on the pump fluence (w_p) measured for CdSe NC–glass samples ($T = 80$ K) with $a \approx 3.5$ (open triangles), 6 (open circles), and 6.5 nm (solid squares). The high-pump-intensity data are fit to the w_p^{-2} dependence.

with the largest NCs, the PL decay was likely dominated by radiative processes. The increasing amount of surface defects in NCs of smaller sizes led to the enhancement of the surface trapping with an associated shortening of the PL relaxation time.

The electron (hole) reexcited as a result of the Auger process can have enough energy to overcome the NC surface barrier with resulting ionization of a NC. This process leads to the charge separation and the generation of local fields which can drastically modify NC optical properties. Auger-assisted NC ionization was observed in TA studies of CdS NC–glass samples [44]. The results of these studies are described in the following text.

DT spectra of CdS NCs ($a \approx 4$ nm, $T = 300$ K) recorded at low-pump intensities corresponding to $N_{eh} < 1$ (see Fig. 50a) are dominated by the $1S$ bleaching which decays almost

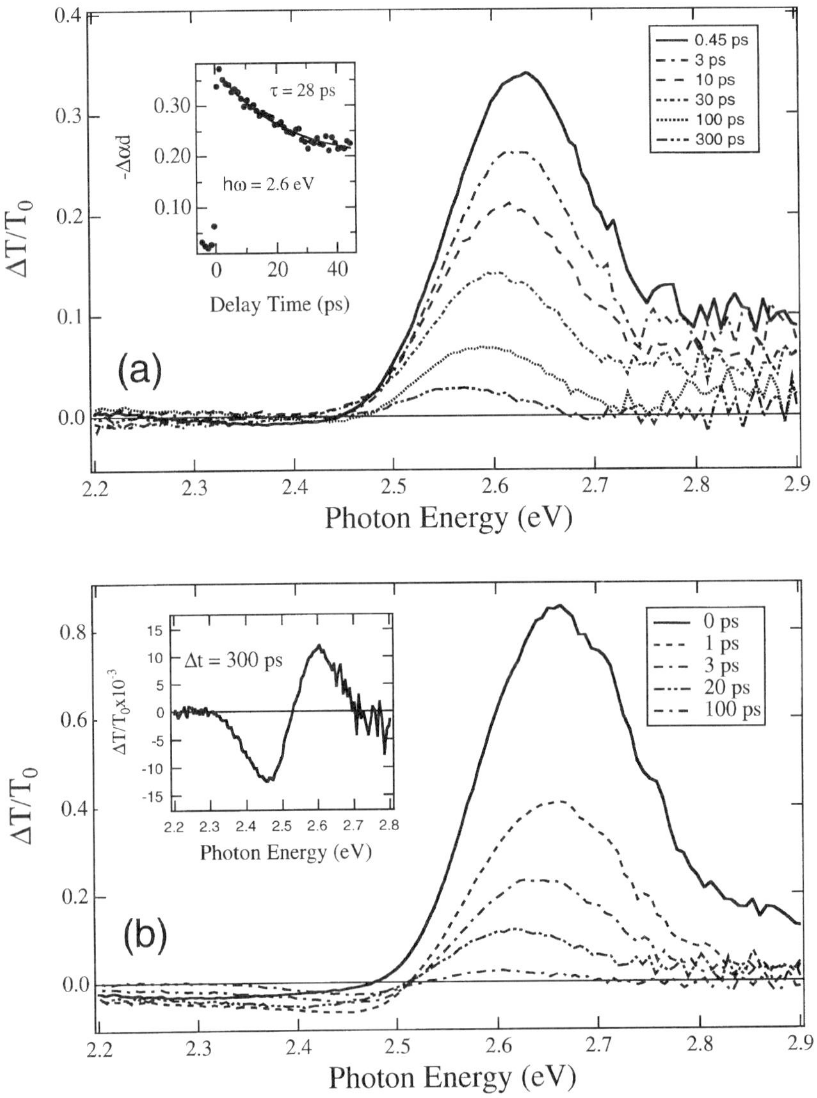

Fig. 50. (a) DT spectra of CdS NC–glass sample ($a \approx 4$ nm) recorded at different delay times after excitation at a pump level $w_p = 0.45$ mJ cm^{-2}, below the threshold for nonlinear recombination processes. Inset shows DT time evolution recorded at the same pump level at $\hbar\omega = 2.6$ eV (circles), along with a fit to an exponential decay (solid line). (b) DT spectra recorded at different delay times after excitation at a pump level $w_p = 10$ mJ cm^{-2}, above the threshold for nonlinear recombination processes. Inset shows an expanded view of the spectrum measured at $\Delta t = 300$ ps. (Source: Reprinted with permission from [44]. © 1997 American Physical Society.)

exponentially with a time constant of 28 ps (see inset to Fig. 50a). This decay can be explained by electron trapping at deep states, involved in a long-lived deep-trap emission (see the previous section). At pump levels above ~0.5 mJ cm^{-2}, the DT dynamics begin to show pump-intensity dependence, indicating the onset of the regime of the multiple *e*-*h*-pair excitation for which nonlinear decay processes start to play a significant role. The DT spectra recorded in this regime are shown in Figure 50b. In contrast to the low-pump-intensity data, the bleaching maximum at 2.63 eV exhibits a much faster initial decay on the sub-ps time scale. Another feature of the high-intensity DT is a new band of increased absorption located below the main bleaching band. The relative contribution from the induced absorption increases with increasing delay time until both features (bleaching and increased absorption) become essentially equivalent in magnitude (see inset to Fig. 50b). The shape of the DT spectra measured at high-pump levels and at long times after excitation is drastically different from that measured at low-pump fluences (see Fig. 51), although in both cases the average number of *e*-*h* pairs per NC is well below 1. Instead of a single bleaching band, the high-pump-intensity spectra show a derivativelike feature which is a clear signature of a dc-electric-field-induced shift of the optical transition (dc-Stark effect) (see Section 6.2). As was mentioned in the previous subsection, the low-pump-intensity DT observed at long times after excitation (after the electron has been trapped at a deep trap) is peaked at the position of the transition coupling the lowest 1S electron state and the hole shallow trap; and the bleaching of this transition is entirely due to the population of the hole trap. The shape of the high-intensity DT in Figure 51 clearly indicates that

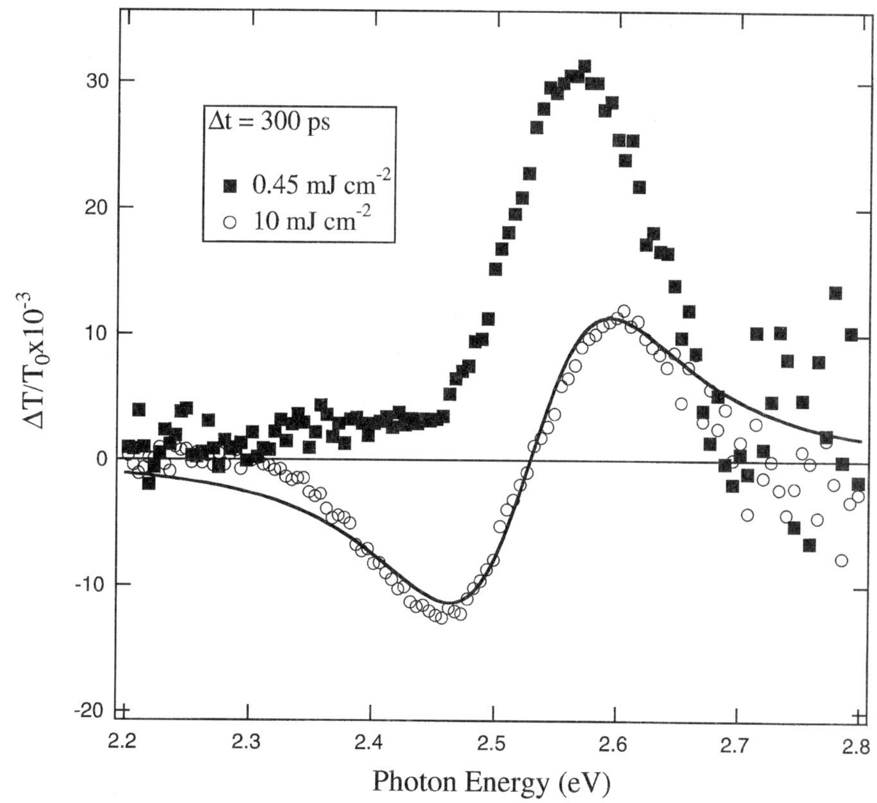

Fig. 51. Comparison of DT spectra of CdS NCs recorded at 300 ps after excitation at low (solid squares) and high (open circles) pump intensities. A solid line shows the fit of the high-pump-intensity spectrum to a model spectrum calculated for a dc-electric-field-induced shift of the optical transition centered at $E_0 = 2.546$ eV with a broadening factor $\Gamma = 110$ meV. (Source: Reprinted with permission from [44]. © 1997 American Physical Society.)

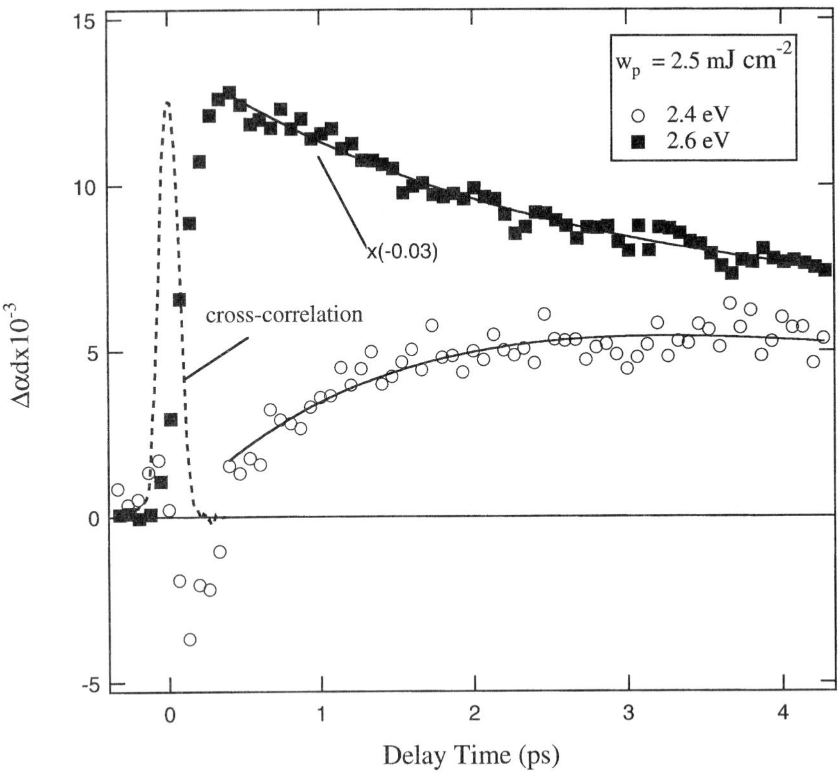

Fig. 52. Comparison of time evolution of DT of CdS NCs recorded at 2.4 eV (increased absorption; open circles) and 2.6 eV (bleaching; solid squares) at $w_p = 2.5$ mJ cm^{-2} (thin solid lines show fits to exponential decay and buildup functions with a same time constant of 2.5 ps). The dashed line shows the pump-probe cross-correlation function. (Source: Reprinted with permission from [44]. © 1997 American Physical Society.)

the transition responsible for the nonlinearity couples two unoccupied states (see Fig. 26a in Section 6.2), meaning that the hole relaxation path changes at high-pump intensities. Instead of being trapped at a shallow state with a relatively extended wave function, the hole most likely gets trapped at a state with much stronger localization. Due to the reduced overlap of the hole and the electron wave functions and/or a large low-energy shift of the corresponding transition (possibly below the observation range), this state is not marked by a pronounced feature in DT spectra. However, DT is affected by new trapped states via the dc-electric field generated as a result of the hole trapping. As the new hole relaxation channel is associated with an efficient charge separation, hole traps activated at high-pump fluences are most likely surface (interface) related.

Interesting information is provided by single-wavelength dynamics of bleaching (recorded at 2.6 eV) and induced absorption (recorded at 2.4 eV) (see Fig. 52, $w_p = 2.5$ mJ cm^2). A simple analysis of these dynamics using the exponential decay (rise) function shows that the buildup of the induced absorption is complementary to the decay of the bleaching feature; both are characterized by a time constant of 2.5 ps. The complementary behavior of the bleaching and the increased absorption is preserved at higher pump fluences (Fig. 53). Increasing pump intensity leads to a faster decay of the bleaching band (Fig. 53a) which is accompanied by faster buildup dynamics of the induced absorption (Fig. 53b). This indicates that the growth of the induced absorption is associated with a nonlinear decay of the bleaching band. Pump-intensity dependence of the bleaching and induced-absorption bands (Fig. 54) shows that the amplitude of the induced absorption grows roughly as the amplitude of the bleaching feature cubed. Combined with the

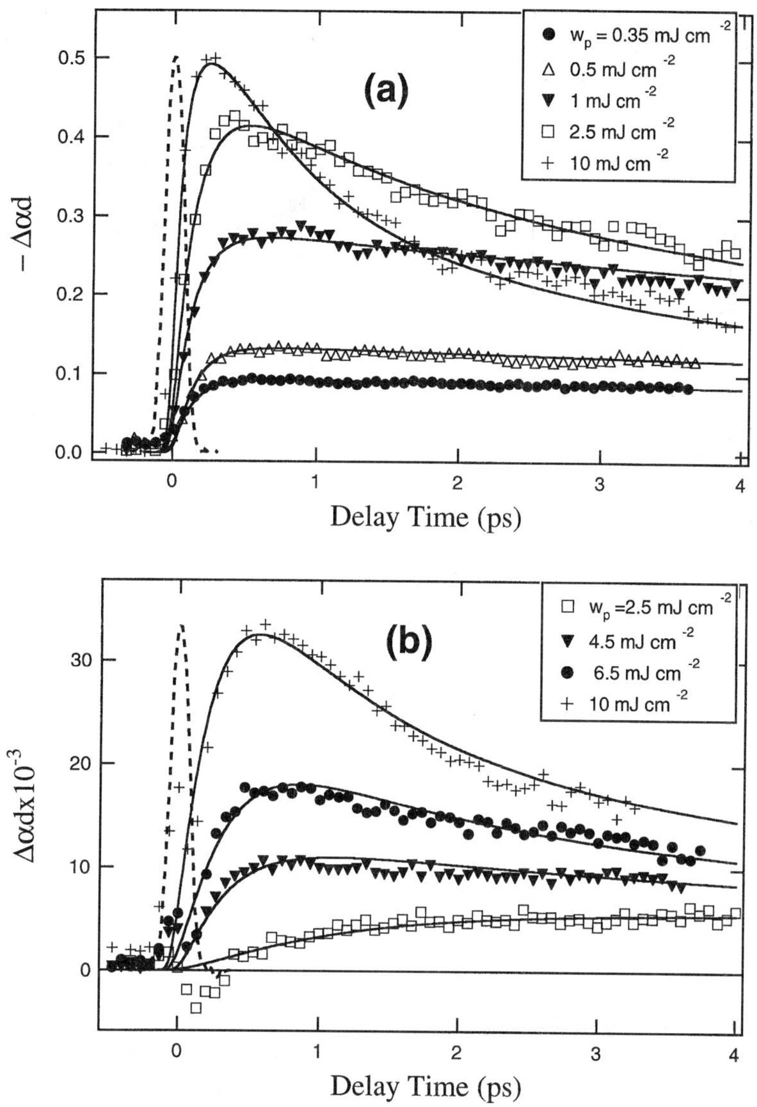

Fig. 53. Pump-dependent DT dynamics (symbols) of CdS NCs recorded at (a) 2.6 eV (bleaching) and (b) 2.4 eV (increased absorption) along with fits (lines) calculated using a model of Auger-process-assisted trapping (see text). (Source: Reprinted with permission from [44]. © 1997 American Physical Society.)

measured dynamics, this data strongly suggests that the charge separation leading to the DT induced-absorption feature is a result of the Auger process.

This conclusion is further confirmed by modeling DT time transients and DT pump dependence using the model schematically depicted in Figure 55. Laser pumping leads to a generation of carriers in the excited quantized states; concentrations of these carriers are denoted as n_x^e (electrons) and n_x^h (holes). With a characteristic constant of $\tau_r^{e,h}$ (intraband relaxation), the excited carriers relax to the lowest quantized states (corresponding carrier concentration is $n_l^{e,h}$). The numerical model includes one type of trapping states for electrons (localization constant τ_t^e, concentration of the trapped electrons n_t^e) and two types of different hole trapping states. One type ($t1$ in Fig. 55) is active at low-pump intensities and is accessible from the lowest hole quantized state (corresponding trapping constant τ_{t1}^h and concentration of the trapped holes n_{t1}^h). The other type of trapping states ($t2$), which

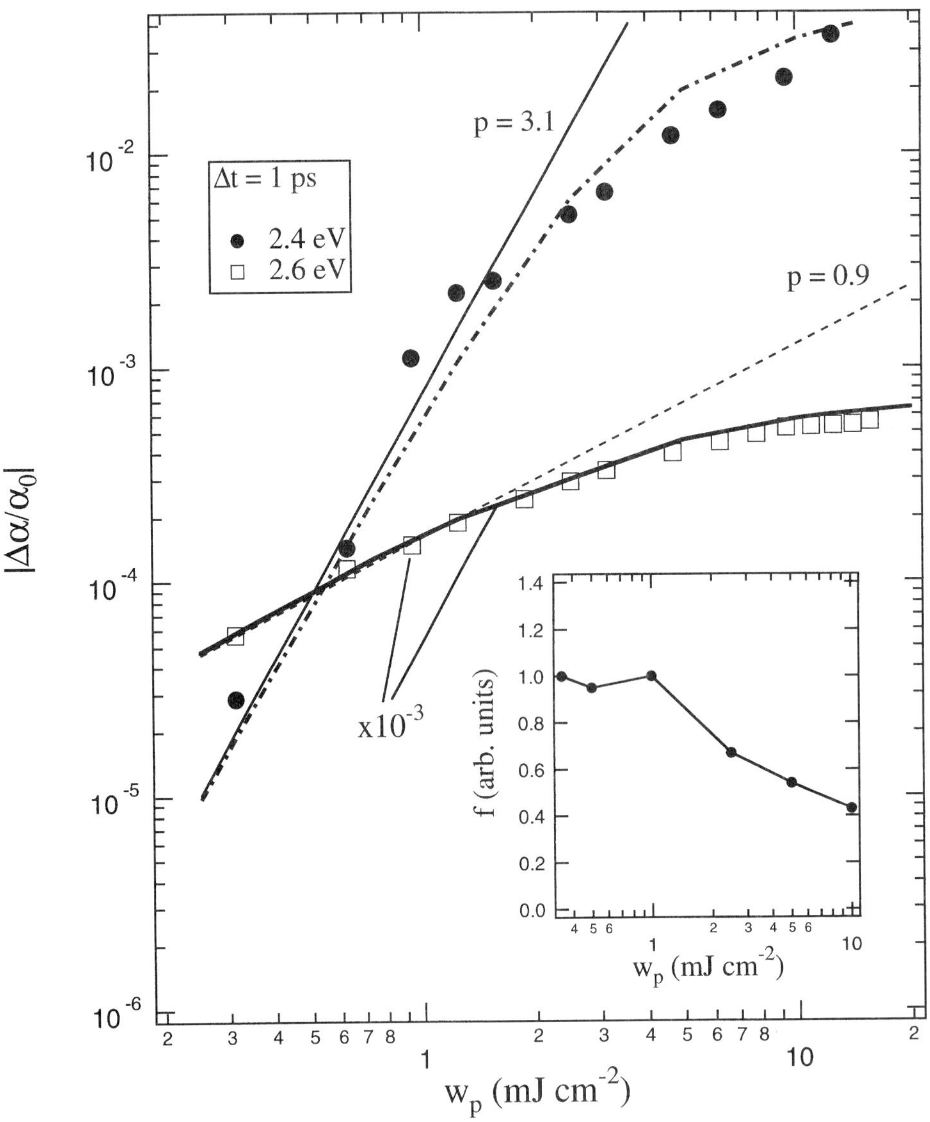

Fig. 54. Pump-intensity dependence of increased absorption at 2.4 eV (solid circles) and bleaching at 2.6 eV (open squares) in CdS NCs measured at 1 ps after excitation in comparison to modeling results shown by thick solid and dashed-dotted lines (see text). Dashed and thin solid lines show pump-power dependence with logarithmic slopes 0.9 and 3.1, respectively. Inset shows the pump dependence of the oscillator strength of the lowest optical transition between quantized states derived by fitting experimentally measured DT dynamics at 2.6 eV. (Source: Reprinted with permission from [44]. © 1997 American Physical Society.)

are presumably surface (interface) related, is separated from the hole quantized states by a potential barrier and can only be accessed after reexcitation. In this model, Auger process-initiated reexcitation occurs via an intermediate hole excited state (denoted as rx in Fig. 55; corresponding concentration n^h_{rx}) from which holes relax to the $t2$-trapped state with a time constant τ^h_{t2} (concentration of the holes trapped at this type of states is n^h_{t2}). The essential point of the model is that all eight types of carriers occupying different NC states interact with each other via the Auger process. The dynamics of photoexcited carriers in this system can be described by the following set of rate equations (shown are only five equations for occupations of hole states; the electron rate equations are similar to the first

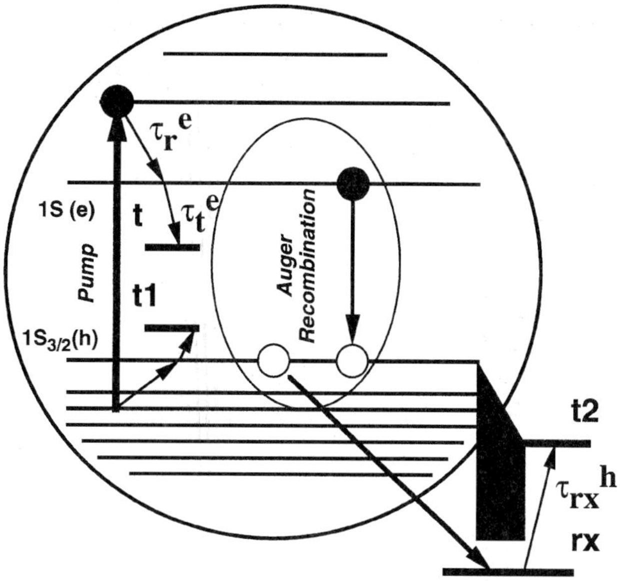

Fig. 55. Schematic illustrating the model of Auger-process-assisted hole trapping at surface–interface states. Laser pumping results in a population of the excited electron (hole) quantized states in a NC. With relaxation constants $\tau_r^{e,h}$ carriers relax to the lowest quantized states [$1S(e)$ and $1S_{3/2}(h)$]. From these states carriers can be trapped at t (electrons) or $t1$ (holes) localized states. Because of the potential barrier at a NC boundary, the hole trapping at the $t2$ surface–interface-related state can only occur after reexcitation via the Auger process. (Source: Reprinted with permission from [44]. © 1997 American Physical Society.)

three equations given in the following text),

$$\frac{dn_x^h}{dt} = g - \frac{n_x^h}{\tau_r^h} - n_x^h R_x^h \tag{60}$$

$$\frac{dn_l^h}{dt} = \frac{n_x^h}{\tau_r^h} - \frac{n_l^h}{\tau_{t1}^h} - n_l^h R_l^h \tag{61}$$

$$\frac{dn_{t1}^h}{dt} = \frac{n_l^h}{\tau_{t1}^h} - n_{t1}^h R_{t1}^h \tag{62}$$

$$\frac{dn_{rx}^h}{dt} = \frac{1}{2}\left(n_x^h R_x^h + n_l^h R_l^h + n_{t1}^h R_{t1}^h\right) - \frac{n_{rx}^h}{\tau_{rx}^h} - n_{rx}^h R_{rx}^h \tag{63}$$

$$\frac{dn_{t2}^h}{dt} = \frac{n_{rx}^h}{\tau_{rx}^h} - n_{t2}^h R_{t2}^h \tag{64}$$

where g is the pumping rate, $R_i^l = \sum_{j,k,m,n} C_{ijk}^{lmn} n_j^m n_k^n$ is the Auger-recombination term, C_{ijk}^{lmn} are the Auger constants, l, m, n indicate either electron (e) or hole (h) states [one of the carriers in the (l, m, n) combination is necessarily different from the other two] and i, j, k denote the type of these states (x, l, t for electrons and $x, l, t1, rex, t2$ for holes). The slow linear relaxation of the trapped states is ignored in Eqs. (60)–(64) because it does not affect fast ps carrier dynamics. Assuming a dominant role of the electron populations into the 2.6-eV bleaching, $\Delta\alpha(\hbar\omega = 2.6\text{ eV})$ can be modeled as being proportional to n_l^e: $\Delta\alpha \propto f n_l^e$, where f is the oscillator strength of the lowest optical transition (the contribution from the transition energy shift to the bleaching band is negligibly small on the time scale shown in Fig. 53). As the pump-induced absorption at 2.4 eV is associated with the dc-electric field induced by the surface-trapped holes, this signal can be modeled as being proportional to n_{t2}^h.

The application of the numerical model just described allows a close fit of experimentally measured time transients for both bleaching and induced absorption over the wide pump-intensity range (lines in Fig. 53). The same set of relaxation parameters as that derived by fitting the time transients provides also an excellent fit to the DT pump dependence measured experimentally (see Fig. 54). The Auger-recombination model explains well not only the slopes measured for the bleaching ($p \approx 1$) and the increased absorption ($p \approx 3$) at low-pump levels but also the saturation of both at high-pump intensities, demonstrating that the saturation observed experimentally is entirely due to the Auger recombination which effectively limits the occupation of the lowest quantized states at a level below the threshold for Fermi blocking.

Interesting behavior of bleaching time transients is observed at high intensities. It is manifested as intersections of traces recorded at different pump fluences (see Fig. 53a). This effect is entirely reversible and, therefore, cannot be attributed to a sample degradation (photodarkening) [99]. These observations can be explained assuming that the oscillator strength of the optical transition at 2.6 eV is pump dependent, decreasing gradually above the threshold pump level ($\sim$1 mJ cm^{-2}). This can be attributed to the influence of the electric field generated as a result of the hole surface trapping, leading to a modification of the symmetry of the electron and hole wave functions [179]. The pump dependence of the oscillator strength derived by fitting the amplitudes of time transients recorded at different pump intensities is shown in the inset to Figure 54.

8. CONCLUSIONS AND PROSPECTS

Since the first observations of quantum-size effect in NC–glass composites [21] and colloidal NCs [180], tremendous progress has been made in both synthesis, and understanding the physical and chemical properties of NCs. This review has concentrated on basic NC structures: NCs made by high-temperature precipitation in molten glasses, and colloidal NCs. Even these relatively simple systems provide a wide range of tunability of optical properties by varying NC sizes and surface passivation. A new level of tunability as well as an additional functionality can be achieved by using complex NC-based structures. The examples of such novel NC systems are layered quantum-dot heterojunctions [11–13], NC molecules (sandwich colloids) [68], and three-dimensional superlattices [20]. An important step toward integrating NCs into optical and electro-optical devices has been the development of methods for NC incorporation into sol–gel [14–16] and polymer [17–19] matrices. Studies of novel NC-based composite structures, which are presently just beginning, will certainly result in exciting new findings in the near future.

Intense theoretical and experimental work has resulted in much better understanding of the energy structures and optical transitions in NCs than that provided by a simple parabolic-band model. Advances in these studies are associated with theoretical predictions [29, 30] and experimental observations of the effect of valence-band mixing [29, 33, 125] and the fine structure of the lowest exciton state [80, 82]. Further work concentrates on evaluation of the effects of Coulomb interactions on energy structures, which are particularly important in the intermediate- and weak-confinement regimes. A significant effort also concentrates on the regime of very small sizes, which are beyond the effective-mass approximation.

A large amount of work has accumulated on electron-phonon interactions in three-dimensional-confined systems, which includes studies of electron-phonon coupling and electron dephasing performed by a variety of optical techniques, such as size-selective emission and absorption spectroscopies, Raman measurements, and degenerate and non-degenerate four-wave mixing. However, the issues concerning the strength of the electron-phonon coupling in NCs remain a matter of great controversy.

Significant progress has been made in understanding the effects of carrier dynamics on the optical properties of NCs. Modern methods of fs spectroscopy allow direct observation of ultrafast energy relaxation, recombination, and dephasing processes. Time-resolved studies indicate an important role of surface (interface) states in carrier relaxation, although the effects of controlled surface modification of carrier dynamics have not been systematically studied yet. Important issues about the effect of three-dimensional confinement on energy relaxation have been raised by theoretical [159] and experimental [157] studies of carrier intraband dynamics in strongly confined NCs. These studies indicate a significant role of energy-loss mechanisms which are different from electron-phonon interactions. However, the exact nature of these mechanisms needs further investigation. Experimental studies of pump-dependent carrier dynamics indicate a significant role of nonradiative Auger recombination at high-pump levels. In addition to very fast ps dynamics, this process can lead to efficient charge separation (NC ionization) with an associated dramatic change in NC optical properties [44, 160].

Important steps toward practical applications have been realized with studies of optical nonlinearities, optical gain, and lasing in NCs. Due to three-dimensional spatial confinement, the mechanisms for resonant band-edge optical nonlinearities in NCs are different from those in bulk materials. The dominant contribution to the nonlinear optical response in NCs is due to a trivial state filling [60], leading to pump-induced bleaching of NC samples spectrally concentrated at positions of optical transitions which couple populated energy states. Charge separation (due to, e.g., carrier surface trapping) leads to the enhancement of Coulomb effects, with an associated increased contribution to the optical nonlinearity by transition shifts (dc-Stark effect) [44, 125]. Transient nonlinearity resulting from the Coulomb two *e-h* pair interaction is also pronounced in the early stages of carrier relaxation, when an electron and a hole occupy states of different symmetries [47]. Due to relaxation of the requirement for momentum conservation, the excited-state absorption due to intraband transitions has been expected to be enhanced in three-dimensionally-confined systems [154]. However, this effect has not yet been studied experimentally in detail. After the first demonstrations of gain and lasing in CdSe NCs [124], some followup work has been done in this direction. However, these effects still remain a subject of future systematic studies.

Acknowledgments

I would like to acknowledge my colleagues and students I have been working with on many intriguing subjects of quantum-dot physics for the past several years. I am grateful to my colleagues at Los Alamos National Laboratory, D. McBranch, R. Kohlman, B. Kraabel, J. M. Robinson, and R. Oldenborg, for numerous discussions on the photophysics of NCs and related systems, for supporting a creative research atmosphere in our group, and for critical remarks concerning the manuscript. I am grateful to my colleagues and former students in Moscow, V. Dneprovskii, V. Karavanskii, M. Novikov, Yu. Vandyshev, and D. Okorokov, with whom I made the first steps toward understanding the quantum-dot systems. I am grateful to H. Kurz, and the members of his institute, P. Haring Bolivar and S. Hunsche, who helped me in my first studies of femtosecond phenomena in NCs. I thank A. Efros, A. Ekimov, M. Bawendi, P. Alivisatos, A. Nozik, and W. White for stimulating discussions. I am grateful to A. Ekimov and M. Bawendi who provided NC samples for several studies described in this chapter. I thank M. Nirmal, D. Norris, D. Ricard, and J.-B. Xia for kindly providing figures from their original publications. I also acknowledge financial support from Los Alamos Directed Research and Development Funds, under the auspices of the U.S. Department of Energy.

References

1. A. L. Efros and A. Efros, *Sov. Phys. Semicond.* 16, 772 (1982).
2. L. Brus, *J. Chem. Phys.* 79, 5566 (1983).
3. L. Brus, *Appl. Phys. A* 53, 465 (1991).
4. A. Alivisatos, *Science* 271, 933 (1996).
5. C. Murray, D. Norris, and M. Bawendi, *J. Am. Chem. Soc.* 115, 8706 (1993).
6. M. Herman, "Semiconductor Superlattices," Akademie Verlag, Berlin, 1986.
7. M. Grundmann, J. Christen, N. Ledentsov, J. Bohrer, D. Bimberg, S. Ruvimov, P. Werner, U. Richter, U. Gosele, J. Heydenreich, V. Ustinov, A. Egorov, A. Zhukov, P. Kopev, and Z. Alferov, *Phys. Rev. Lett.* 74, 4043 (1995).
8. A. Ekimov, *J. Lumin.* 70, 1 (1996).
9. N. Borrelli, D. Hall, H. Holland, and D. Smith, *J. Appl. Phys.* 61, 5399 (1987).
10. S. Majetich and A. Carter, *J. Phys. Chem.* 97, 8727 (1993).
11. A. Kortan, R. Hull, R. Opila, M. Bawendi, M. Steigerwald, P. Carroll, and L. E. Brus, *J. Am. Chem. Soc.* 112, 1327 (1990).
12. A. Eychmueller, A. Haesselbarth, and H. Weller, *J. Lumin.* 53, 113 (1992).
13. A. Mews, A. Kadavanich, U. Banin, and A. Alivisatos, *Phys. Rev. B* 53, 13,242 (1996).
14. T. Rajh, M. Vucemilovic, N. Dimitrijevic, O. Micic, and A. Nozik, *J. Phys. Chem. Lett.* 143, 305 (1988).
15. M. Nogami, K. Nagasaka, and M. Takata, *J. Non-Cryst. Solids* 122, 101 (1990).
16. G. Counio, S. Esnouf, T. Gacoin, and J. Boilot, *J. Phys. Chem.* 100, 20,021 (1996).
17. I. Akimov, I. Denisyuk, and A. Meshkov, *Opt. Spectrosc.* 72, 558 (1992).
18. H. Noglik and W. Pierto, *Chem. Mater.* 7, 1333 (1995).
19. N. Greenham, X. Peng, and A. Alivisatos, *Phys. Rev. B* 54, 17,628 (1996).
20. C. Murray, C. Kagan, and M. Bawendi, *Science* 270, 1335 (1995).
21. A. Ekimov, A. Onushenko, and V. Tzehomskii, *Sov. Phys. Chem. Glass* 6, 511 (1980).
22. Y. Z. Hu, S. W. Koch, and D. B. T. Thoai, *Mod. Phys. Lett. B* 4, 1009 (1990).
23. L. Banyai, P. Gilliot, Y. Hu, and S. Koch, *Phys. Rev. B* 45, 14,136 (1992).
24. D. B. T. Thoai, Y. Z. Hu, and S. W. Koch, *Phys. Rev. B* 42, 11,261 (1990).
25. L. Brus, *J. Chem. Phys.* 80, 4473 (1984).
26. A. L. Efros and A. Rodina, *Solid State Commun.* 72, 645 (1989).
27. S. W. Koch, Y. Z. Hu, B. Fluegel, and N. Peyghambarian, *J. Cryst. Growth* 117, 592 (1992).
28. S. Nomura and T. Kobayashi, *Solid State Commun.* 78, 677 (1991).
29. A. I. Ekimov, F. Hache, M. C. Schanne-Klein, D. Ricard, C. Flytzanis, I. A. Kudryavtsev, T. V. Yazeva, A. V. Rodina, and A. L. Efros, *J. Opt. Soc. Am. B* 10, 100 (1993).
30. J.-B. Xia, *Phys. Rev. B* 40, 8500 (1989).
31. M. Bawendi, W. Wilson, L. Rothberg, P. Carroll, T. Jedju, M. Steigerwald, and L. Brus, *Phys. Rev. Lett.* 65, 1623 (1990).
32. Y. Vandyshev, V. Dneprovskii, and V. Klimov, *JETP Lett.* 53, 314 (1991).
33. D. Norris and M. Bawendi, *Phys. Rev. B* 53, 16,338 (1996).
34. A. Alivisatos, T. Harris, P. Carrol, M. Steigerwald, and L. Brus, *J. Chem. Phys.* 90, 3463 (1989).
35. J. Shiang, S. Risbud, and A. Alivisatos, *J. Chem. Phys.* 98, 8432 (1993).
36. A. Mlayah, A. M. Brugman, R. Carles, J. Renucci, M. Valakh, and A. V. Pogorelov, *Solid State Commun.* 90, 567 (1994).
37. A. Tu and P. D. Persans, *Phys. Rev. B* 53, 16,338 (1991).
38. R. Shoenlein, D. Mittleman, J. Shiang, A. Alivisatos, and C. Shank, *Phys. Rev. Lett.* 70, 1014 (1993).
39. M. Klein, F. Hache, D. Ricard, and C. Flytzanis, *Phys. Rev. B* 42, 11,123 (1990).
40. T. Itoh, M. Nishijima, A. Ekimov, C. Gourdon, A. L. Efros, and M. Rosen, *Phys. Rev. Lett.* 74, 1645 (1995).
41. M. Silvestry and J. Schroeder, *Phys. Rev. B* 50, 15,108 (1994).
42. D. Chepic, A. L. Efros, A. Ekimov, M. Ivanov, V. A. Kharchenko, and I. Kudriavtsev, *J. Lumin.* 47, 113 (1990).
43. M. Ghanassi, M. Schanne-Klein, F. Hache, A. Ekimov, and D. Ricard, *Appl. Phys. Lett.* 62, 78 (1993).
44. V. Klimov and D. McBranch, *Phys. Rev. B* 55, 13,173 (1997).
45. Y. Hu, S. Koch, M. Linberg, N. Peyghambarian, E. Pollock, and F. Abraham, *Phys. Rev. Lett.* 64, 1805 (1990).
46. Y. Masumoto, S. Katayanagi, and T. Mishina, *Phys. Rev. B* 49, 10,782 (1994).
47. V. Klimov, S. Hunsche, and H. Kurz, *Phys. Rev. B* 50, 8110 (1994).
48. K. Shum, W. B. Wang, R. Alfano, and K. Jones, *Phys. Rev. Lett.* 68, 3904 (1992).
49. V. Klimov, P. H. Bolivar, and H. Kurz, *Phys. Rev. B* 53, 1463 (1996).
50. U. Woggon, H. Giessen, F. Gindele, O. Wind, B. Fluegel, and N. Peyghambarian, *Phys. Rev. B* 54, 17,681 (1996).
51. L. Brus and J. Trautman, *Philos. Trans. R. Soc. London A* 353, 313 (1995).
52. S. Empedocles, D. Norris, and M. Bawendi, *Phys. Rev. Lett.* 77, 3873 (1996).

53. S. Blanton, M. Hines, and P. Guyot-Sionnest, *Appl. Phys. Lett.* 69, 3905 (1996).
54. V. Dneprovskii, V. Klimov, D. Okorokov, and Y. Vandyshev, *Solid State Commun.* 81, 227 (1992).
55. V. Colvin, M. Schlamp, and A. Alivisatos, *Nature* 370, 354 (1994).
56. A. Sombra, *Solid State Commun.* 88, 305 (1993).
57. A. Hagfeldt and M. Graetzel, *Chem. Rev.* 95, 49 (1995).
58. S. Fluegge, "Practical Quantum Mechanics I," Vol. 1, Academic Press, Berlin, 1971.
59. M. Abramowitz and I. Stegun, "Handbook of Mathematical Functions," Dover, New York, 1972.
60. S. Schmitt-Rink, D. Miller, and D. Chemla, *Phys. Rev. B* 35, 8113 (1987).
61. J. Luttinger and W. Kohn, *Phys. Rev.* 102, 1030 (1955).
62. A. Baldereschi and N. O. Lipari, *Phys. Rev. B* 8, 2687 (1973).
63. J.-B. Xia, *J. Lumin.* 70, 120 (1996).
64. R. Knox, "Theory of Excitons," Academic Press, New York, 1963.
65. G. Grigoryan, A. Rodina, and A. L. Efros, *Sov. Phys. Solid State* 32, 2037 (1990).
66. E. Kane, *J. Phys. Chem.* 1, 249 (1956).
67. R. Aggarwal, in "Semiconductors and Semimetals" (R. Willardson and A. Beer, eds.), Vol. 9, Academic Press, New York, 1972.
68. H. Weller, *Angew. Chem. Int. Ed. Engl.* 32, 41 (1993).
69. M. Bawendi, P. Carroll, W. Wilson, and L. Brus, *J. Chem. Phys.* 96, 946 (1990).
70. M. Chamarro, C. Gourdon, P. Lavallard, O. Lublinskya, and A. Ekimov, *Phys. Rev. B* 53, 1336 (1996).
71. A. Alivisatos, A. Harris, N. Levinis, M. Steigervald, and L. Brus, *J. Chem. Phys.* 89, 4001 (1988).
72. N. Peyghambarian, B. Fluegel, D. Hulin, A. Migus, M. Joffre, A. Antonetti, S. Koch, and M. Lindberg, *IEEE J. Quantum Electron.* QE-25, 2516 (1989).
73. U. Woggon, S. Gaponenko, W. Langbein, A. Uhrig, and C. Klingshirn, *Phys. Rev. B* 47, 3684 (1993).
74. V. Klimov, *Sov. Phys. Solid State* 34, 1326 (1992).
75. V. Klimov, S. Hunsche, and H. Kurz, *Phys. Status Solidi B* 188, 259 (1994).
76. S. Hunsche, T. Dekorsy, V. Klimov, and H. Kurz, *Appl. Phys. B* 62, 3 (1996).
77. A. L. Efros, *Phys. Rev. B* 46, 7448 (1992).
78. A. L. Efros and A. Rodina, *Phys. Rev. B* 47, 10,005 (1993).
79. T. Takagahara, *Phys Rev. B* 47, 4569 (1993).
80. M. Nirmal, D. Norris, M. Kuno, M. Bawendi, A. L. Efros, and M. Rosen, *Phys. Rev. Lett.* 75, 3728 (1995).
81. K. Nash, L. C. P. D. Calcott, and R. Needs, *Phys. Rev. B* 51, 17,698 (1995).
82. D. Norris, A. L. Efros, M. Rosen, and M. Bawendi, *Phys. Rev. B* 53, 16,347 (1996).
83. M. Chamarro, C. Gourdon, and P. Lavallard, *J. Lumin.* 70, 222 (1996).
84. U. Woggon, F. Gindele, and O. W. C. Klingshirn, *Phys. Rev. B* 54, 1506 (1996).
85. R. Endlman and R. Ruppin, *J. Phys. C* 1, 614 (1968).
86. R. Ruppin, *J. Phys. C* 8, 1969 (1975).
87. T. Inoshita and H. Sakaki, *Phys. Rev. B* 46, 7260 (1992).
88. G. Maham, "Many-Particle Physics," Plenum, New York, 1981.
89. K. Huang and A. Rhys, *Proc. R. Soc. London Ser. A* 204, 406 (1951).
90. B. Ridley, "Quantum Processes in Semiconductors," Oxford Univ. Press, Oxford, U.K., 1988.
91. V. Jungnickel and F. Henneberger, *J. Lumin.* 70, 238 (1996).
92. E. Conwell, "High Field Transport in Semiconductors," Academic Press, New York, 1967.
93. A. L. Efros, A. Ekimov, F. Kozlovski, V. Petrova-Koch, H. Schmidbaur, and S. Shumilov, *Solid State Commun.* 78, 853 (1991).
94. S. Nomura and T. Kobayashi, *Phys. Rev. B* 45, 1305 (1992).
95. A. Ekimov, A. L. Efros, M. Ivanov, A. Onushenko, and S. Shumilov, *Solid State Commun.* 69, 565 (1989).
96. J. Marini, B. Stebe, and E. Kartheuser, *Phys. Rev. B* 50, 14,302 (1993).
97. M. Nirmal, C. Murray, and M. Bawendi, *Phys. Rev. B* 50, 2293 (1994).
98. D. Mittleman, R. Shoenlein, J. Shiang, V. L. Colvin, A. Alivisatos, and C. Shank, *Phys. Rev. B* 49, 14,435 (1994).
99. P. Roussignol, D. Ricard, J. Lukasik, and C. Flytzanis, *J. Opt. Soc. Am.* 4, 5 (1987).
100. M. Tomita, T. Matsumoto, and M. Matsuoka, *J. Opt. Soc. Am. B* 6, 165 (1989).
101. V. Dneprovskii, V. Karavanskii, and V. Klimov, *Phys. Solid State* 35, 1297 (1993).
102. V. Klimov, P. H. Bolivar, and H. Kurz, *Phys. Rev. B* 52, 4728 (1995).
103. P. Champion, G. Korenowski, and A. Albrecht, *Solid State Commun.* 32, 7 (1979).
104. J. Shiang, R. Grubbs, and A. Alivisatos, unpublished.
105. T. Takagahara, *Phys. Rev. Lett.* 71, 3577 (1993).
106. T. Takagahara, *J. Lumin.* 70, 129 (1996).
107. U. Woggon and M. Portuné, *Phys. Rev. B* 51, 4719 (1995).
108. W. Demtroeder, "Laser Spectroscopy: Basic Concepts and Instrumentation," Springer-Verlag, Berlin, 1982.
109. C. Spiegelberg, F. Henneberger, and J. Puls, *Superlattices Microstruct.* 9, 487 (1991).
110. T. Yajima and Y. Taira, *J. Phys. Soc. Jpn.* 47, 1620 (1979).
111. W. Zinth and W. Kaizer, in "Ultrashort Laser Pulses: Generation and Applications" (W. Kaiser, ed.), p. 235, Springer-Verlag, Berlin, 1993.

112. N. Mott, "Metal-Insulator Transitions," Taylor & Francis, London, 1979.
113. W. Ebeling, W. Kraeft, and D. Kremp, "Theory of Bound States and Ionization in Plasmas and Solids," Akademie Verlag, Berlin, 1976.
114. H. Haug and S. Schmitt-Rink, *J. Opt. Soc. Am. B* 2, 1135 (1985).
115. D. Chemla and D. Miller, *J. Opt. Soc. Am. B* 32, 1155 (1985).
116. S. Schmitt-Rink, D. Chemla, and D. Miller, *Phys. Rev. B* 32, 6601 (1985).
117. S. Koch, N. Peyghambarian, and M. Lindberg, *J. Phys. C* 21, 5229 (1988).
118. G. Manzke, K. Henneberger, and V. May, *Phys. Status Solidi* 130, 233 (1987).
119. V. Dneprovskii, V. Klimov, D. Okorokov, and Y. Vandyshev, *Opt. Comput. Proc.* 2, 47 (1992).
120. O. Akimoto and E. Hanamura, *Solid State Commun.* 10, 253 (1972).
121. D. Miller, C. Seaton, M. Prise, and S. Smith, *Phys. Rev. Lett.* 47, 197 (1981).
122. Y. Vandyshev, V. Dneprovskii, and V. Klimov, *Sov. Phys. JETP* 74, 144 (1992).
123. I. Lifshitz and V. Slesov, *Sov. Phys. JETP* 8, 331 (1959).
124. Y. Vandyshev, V. Dneprovskii, V. Klimov, and D. Okorokov, *JETP Lett.* 54, 442 (1991).
125. D. Norris, A. Sacra, C. Murray, and M. Bawendi, *Phys. Rev. Lett.* 72, 2612 (1994).
126. T. Takagahara, *Phys. Rev. B* 39, 10,206 (1989).
127. Y. Shen, "The Principles of Nonlinear Optics," Willey, New York, 1984.
128. R. Jain and R. Lind, *J. Opt. Soc. Am.* 73, 647 (1983).
129. S. Park, R. Morgan, Y. Hu, M. Lindberg, S. W. Koch, and N. Peyghambarian, *J. Opt. Soc. Am. B* 7, 2097 (1990).
130. A. Uhrig, A. Woerner, C. Klingshirn, L. Bayai, I. Lacis, N. Neuroth, B. Speit, and K. Remitz, *J. Cryst. Growth* 117, 598 (1992).
131. S. Yao, C. Karaguleff, A. Gabel, R. Fortenberry, C. Seaton, and G. I. Stegeman, *Appl. Phys. Lett.* 46, 801 (1985).
132. L. Zimin, S. Gaponenko, V. Leved, I. Malinovskii, I. Germanenko, E. Podorova, and V. Tsekhomskii, *Phys. Status Solidi* 159, 267 (1990).
133. J.-L. Coutaz and M. Kull, *J. Opt. Soc. Am.* 8, 95 (1991).
134. H. Fan, *Rep. Prog. Phys.* 19, 107 (1956).
135. E. Haga and H. Kimura, *J. Phys. Soc. Jpn.* 18, 777 (1963).
136. F. Buda, J. Kohanoff, and M. Parrinello, *Phys. Rev. Lett.* 69, 1272 (1992).
137. A. Read, R. Needs, K. Nash, L. Canham, P. Calcott, and A. Qteish, *Phys. Rev. Lett.* 69, 1232 (1992).
138. L. Canham, *Appl. Phys. Lett.* 57, 1046 (1990).
139. D. J. Lockwood, *Solid State Commun.* 92, 101 (1994).
140. F. Buda and J. Kohanoff, *Prog. Quantum Electron.* 18, 201 (1994).
141. L. Tsybeskov, K. Moore, D. Hall, and P. Fauchet, *Phys. Rev. B* 54, 8361 (1996).
142. L. Brus, P. Szajowski, W. Wilson, T. Harris, S. Schuppler, and P. Citrin, *J. Am. Chem. Soc.* 117, 2915 (1995).
143. C. White, J. Budai, S. Withrow, J. Zhu, S. Pennycook, R. Zuhr, D. Hembre, D. Henderson, R. Magruder, M. Yacaman, G. Mongragon, and S. Prawer, *Nucl. Instrum. Methods Phys. Res. Sect. B* 127, 545 (1997).
144. A. Cullis and L. Canham, *Nature* 353, 335 (1991).
145. S. Tolbert and A. Alivisatos, *Science* 265, 373 (1994).
146. S. Tolbert, A. Herhold, C. Johnson, and A. Alivisatos, *Phys. Rev. Lett.* 73, 3266 (1994).
147. A. Edwards and H. Dickamer, *Phys. Rev.* 122, 3196 (1962).
148. V. Klimov, P. H. Bolivar, H. Kurz, V. Karavanskii, V. Krasovskii, and Y. Korkishko, *Appl. Phys. Lett.* 67, 653 (1995).
149. V. I. Klimov and V. A. Karavanskii, *Phys. Rev. B* 54, 8087 (1996).
150. B. Mulder, *Phys. Status Solidi A* 15, 409 (1973).
151. P. McLeod, L. D. Partain, D. E. Sawyer, and T. M. Peterson, *Appl. Phys. Lett.* 45, 472 (1984).
152. B. Mulder, *Phys. Status Solidi A* 13, 79 (1972).
153. V. Klimov, P. H. Bolivar, H. Kurz, and V. Karavanskii, *Superlattices Microstruct.* 20, 395 (1996).
154. V. Milanović and Z. Ikonić, *Phys. Rev. B* 39, 7982 (1989).
155. H. Benisty, C. Sotomayor-Torres, and C. Weicbuch, *Phys. Rev. B* 44, 10,945 (1991).
156. T. Inoshita and H. Sakaki, *Physica B* 227, 373 (1996).
157. V. Klimov and D. McBranch, *Phys. Rev. Lett.* 80, 4028 (1998).
158. V. Klimov and D. McBranch, *Opt. Lett.* 23, 277 (1998).
159. A. L. Efros, V. A. Kharchenko, and M. Rosen, *Solid State Commun.* 93, 281 (1995).
160. V. A. Kharchenko and M. Rosen, *J. Lumin.* 70, 158 (1996).
161. V. Klimov, C. Schwarz, D. McBranch, C. Leatherdale, and M. Bawendi, submitted for publication.
162. V. Dneprovskii, A. Efros, A. Ekimov, V. Klimov, I. Kudriavtsev, and M. Novikov, *Solid State Commun.* 74, 555 (1990).
163. N. Chestnoy, T. Harris, R. Hull, and L. Brus, *J. Phys. Chem.* 90, 3393 (1986).
164. Y. Wang and N. Herron, *J. Phys. Chem.* 92, 4988 (1988).
165. J. Zhang, *Acc. Chem. Res.* 30, 423 (1997).
166. J. Zheng, L. Shi, F. Choa, P. Liu, and H. Kwok, *Appl. Phys. Lett.* 53, 643 (1988).
167. K. Misawa, H. Yao, T. Hayashi, and T. Kobayashi, *J. Cryst. Growth* 117, 617 (1992).

168. T. Inokuma, T. Arai, and M. Ishikawa, *Phys. Rev. B* 42, 11,093 (1990).
169. A. L. Efros, private communication.
170. J. Shah, *IEEE J. Quantum Electron.* 24, 276 (1988).
171. M. Lorenz and H. Woodbury, *Phys. Rev. Lett.* 10, 215 (1963).
172. M. Lorenz and B. Segal, *Phys. Lett.* 7, 18 (1963).
173. S. Shionoya, *J. Lumin.* 1, 17 (1970).
174. P. Landsberg, "Recombination in Semiconductors," Cambridge Univ. Press, Cambridge, U.K., 1991.
175. V. Abakumov, V. Perel, and I. Yasssievich, "Nonradiative Recombination in Semiconductors," North-Holland, Amsterdam, 1991.
176. K. Natterman, B. Daniezik, and D. V. der Linde, *Appl. Phys. A* 44, 111 (1987).
177. F. de Rougemont, R. Frey, D. Ricard, P. Roussignol, and C. Flytzanis, *Appl. Phys. Lett.* 50, 1619 (1987).
178. V. Dneprovskii, V. Klimov, M. Novikov, D. Okorokov, and Y. V. Vandyshev, *SPIE Proc.* 1841, 196 (1991).
179. Y. Hu, M. Lindberg, and S. Koch, *Phys. Rev. B* 42, 1713 (1990).
180. A. Henglein, Ber. Bunsenges. *Phys. Chem.* 88, 301 (1982).

Chapter 14

MOLECULAR AND SUPRAMOLECULAR NANOMACHINES

Marcos Gómez-López, J. Fraser Stoddart

Department of Chemistry and Biochemistry, University of California at Los Angeles, California, USA

Contents

1. INTRODUCTION

At the dawn of the third millennium, the prospect of information storage and processing on the nanoscale level has become an issue of some considerable importance. To date, lithographic techniques [1] have been employed to create "state-of-the-art" computers capable of handling massive amounts of information in very little space. However, lithography is rapidly reaching its limits, and the production of nano-sized silicon-based binary switches—which are the basic components of computers—is becoming more and more difficult as they become smaller and smaller. Increasingly, the challenge lies now with molecular engineers—i.e., chemists—to design and synthesize, from simple chemical blocks, more intricate molecular and supramolecular machines displaying switching

Handbook of Nanostructured Materials and Nanotechnology, edited by H.S. Nalwa
Volume 5: Organics, Polymers, and Biological Materials

properties on nanoscale dimensions. This strategy—the so-called bottom-up approach—was first proposed by the late Nobel laureate and visionary Richard Feynman [2] in 1960. It differs from the current lithographic techniques in so far as miniaturization—the "top-down" approach—is not invoked, because, in principle at least, the switching units are constructed from molecular subunits by synthetic procedures. Thus, the possibility now exists that chemists could be constructing artificial nanoscale-functioning entities with relative ease in the not too distant future [3].

Traditional synthetic approaches [4] rely upon the formation of covalent bonds. (The power of traditional synthetic organic chemistry cannot be underestimated. Its practice in the research laboratories of some of the most talented synthetic organic chemists has led to the controlled synthesis of molecular compounds composed of large molecules such as vitamin B_{12} and palytoxin. For notable examples, see [4].) Without exception, these approaches require stepwise, divergent strategies that are time-consuming and demand a great deal of human effort. It is difficult to envisage the controlled construction of large molecules—with the exception of the so-called dendrimers [5]—incorporating many functionalities, using what has been termed "covalent synthesis." The reason for this difficulty is that covalent synthesis is dependent upon a sequential approach to the construction of molecular compounds, and many of the incremental steps can only be achieved under kinetic control. Hence, overall yields are small and errors (i.e., side reactions) cannot be corrected. However, there is an alternative to covalent synthesis, which relies upon the much easier preparation of relatively small molecules that have stereoelectronically complementary binding sites judiciously located on their molecular surfaces, such that they can recognize each other as a consequence of matching noncovalent bonding interactions, resulting in well-defined nanosized superstructures. By relying upon these relatively weak noncovalent bonding interactions [6], aggregation processes can occur under thermodynamic control, and therefore errors can be corrected constantly.

Strict molecular self-assembly [7] can be defined as the process that leads to the construction of large and ordered superstructures by the use of weak, reversible noncovalent bonding interactions between complementary molecular structures. The final superstructure represents a thermodynamic minimum for the system. In addition, these self-assembly processes are error-checking—a feature which means that defective or incorrectly attached subunits can be eliminated from growing superstructures. The processes are also highly convergent, which permits much greater synthetic efficiencies to be attained than in the classical stepwise approaches to the syntheses of molecular compounds.

Supramolecular chemistry [8]—borrowing, as it does from Nature, the concept of self-assembly—has made it possible for chemists to start constructing unnatural molecular assemblies and supramolecular arrays that can mimic some of the functions found in living systems, such as substrate binding, transportation of ions [9], molecular sensing [10], and information storage [11].

Self-assembly with covalent modification [7] is another category of self-assembly. It relates to processes in which the formation of covalent linkages within an assembly is irrevocable and irreversible. One subclass of this type of self-assembly is the so-called post-assembly processing, in which an assembled superstructure undergoes covalent modification to attain functional competence. The main difference with strict self-assembly is that the final structure and/or superstructure is produced under kinetic control—involving the formation of a covalent bond—and not thermodynamic control. This strategy has been employed [12] successfully in the self-assembly of unnatural interlocked molecular compounds, such as catenanes and rotaxanes, which are now being utilized for the fabrication of molecular switching devices (see Section 4 and references therein).

A molecule or a supermolecule with the capacity to store information might be expected to behave like a molecular/supramolecular switch [13]. Such a switch can be envisaged as a molecule/supermolecule that can exist in two different states that can be interconverted (Fig. 1) reversibly by some external stimuli. The basic requirement for such a switch is

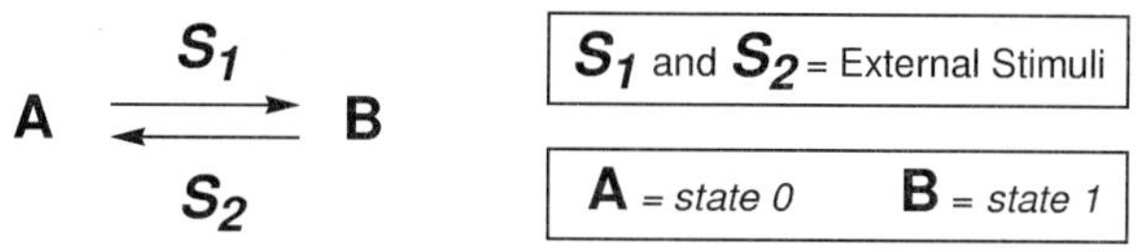

Fig. 1. Schematic representation of a molecular switch. Bistability is the basic requirement. The molecule must exit in two different forms, A and B, which can be interconverted by an external source, S_1 or S_2.

bistability. Therefore, any material, including a molecule/supermolecule, that possesses two independent states, which can be changed ("writing") one into the other by a perturbation from its environment—and that, in a given instant, has the property that both states can be identified ("reading")—may be used, in principle at least, as the basis for a memory element in a computer using binary logic.

Organic compounds are potentially a very interesting source of molecular switches. Moreover, the organic chemist possesses in his/her toolbox many easily accessible building blocks that have the ability to be switched from one state to another by an external source. The following features make organic compounds and complexes some of the most important choices for the possible fabrication of functioning nanoscale devices:

- The ease of fabrication
- The possibility of fine-tuning of the physical and chemical properties of the compounds and complexes by very small controlled changes in the molecular structures and supramolecular superstructures
- Two states that can be identified easily and differentiated with a number of spectroscopic techniques
- The fact that it is possible to synthesize simple isolated systems that can act as prototypes for much larger and more intricate systems
- The advent of supramolecular chemistry and the subsequent rapid development of self-assembly as a synthetic method. We can now design and fabricate, in a fast, efficient and controlled way, large molecular assemblies and supermolecular arrays. The topologies [14] of these structures and superstructures are controlled by weak noncovalent bonding interactions that can be easily perturbed by external stimuli. Such perturbations can disrupt the complementarity between the subunits, inducing a profound geometrical change within a molecular assembly, or even the total disintegration of a supramolecular array.

What now follows is the discussion of a selection of switching systems at the molecular and supramolecular levels, which have been described in the scientific literature in recent times. We have classified them first on the basis of the nature of the switch, that is, based on whether they are a traditional molecule, supramolecular arrays, or interlocked molecular assemblies, and, within molecular switching classes, according to the type of structural change that the molecule undergoes on being exposed to an external stimulus.

2. CONVENTIONAL MOLECULAR SYSTEMS

2.1. Conformational Change

2.1.1. A Molecular Brake

Kelly and co-workers [15] have designed a molecular compound **1**, consisting of a triptycene unit covalently linked to a 2,2′-bipyridine moiety, which they have described as a

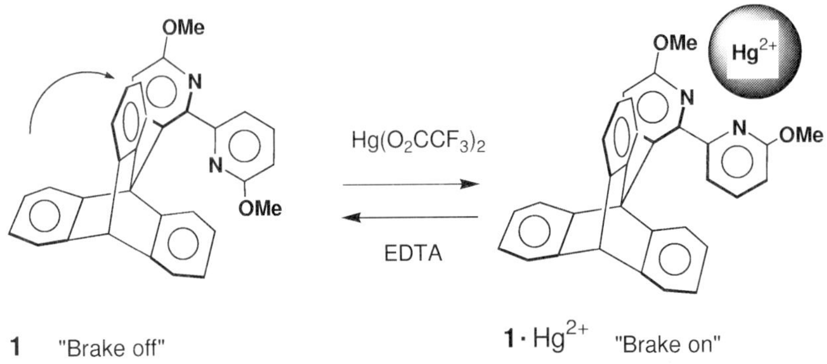

Fig. 2. A molecular brake. The rotation of the triptycene component in **1** can be hindered by the addition of an Hg^{2+} salt. The brake is released by treatment with EDTA.

molecular brake. In the absence of metal ions, the triptycene unit rotates (Fig. 2) spontaneously around the single bond that serves as a connector to the bipyridine moiety. ^{1}H NMR spectroscopic studies at 30 °C in CD_3COCD_3 revealed that the triptycene wheel spins rapidly [16], giving rise to a relatively simple spectrum. However, the addition of $Hg(O_2CCF_3)_2$ produces dramatic changes in the ^{1}H NMR spectrum. The presence of the metal ion in solution forces a conformational change upon the bipyridine moiety, which hinders the rotation of the adjacent triptycene unit such that the bipyridine unit appears to be functioning as a molecular brake. Treatment of the Hg^{2+} complex with EDTA removes the Hg^{2+} ions and releases the break, restoring the free rotation to the triptycene wheel and rendering the system reversible. The researchers concluded that they have created a molecular brake that operates by coordination of a metal to a remote site, which, in turn, imposes a conformational change that stops the rotation of the triptycene wheel.

2.2. Configurational Change

2.2.1. Bacteriorhodopsin

The *cis–trans* isomerization of a carbon-carbon double bond is the basis of the process of vision [17]. Upon irradiation of the 11-*cis* retinal **2a**—which is linked to the protein opsin via the amino acid lysine to form rhodopsin—*cis–trans* isomerization takes place (Fig. 3), giving rise to the 11-*trans* retinal **2b**. The isomerization of the retinal portion of rhodopsin induces a conformational change within the entire protein in the shape of cascade reactions. This change ultimately creates a nerve pulse responsible for the process of vision.

In the laboratory [18], bacteriorhodopsin has been used as the switching component in the development of an information storage device. It is readily available in macroscopic quantities and possesses photochemical activity when dissolved in polymer films. Its all-*trans* form can be converted into the corresponding *cis*-isomer upon irradiation at 568 nm. These two states could represent the states 0 and 1 of a binary molecular device. The

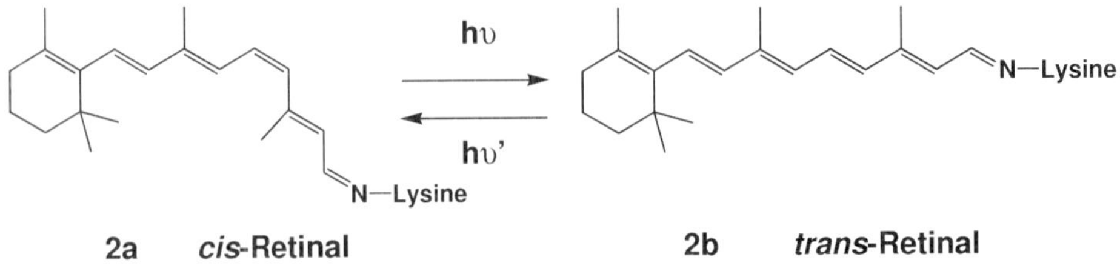

Fig. 3. The *cis–trans* photoisomerization of 11-retinal **2**.

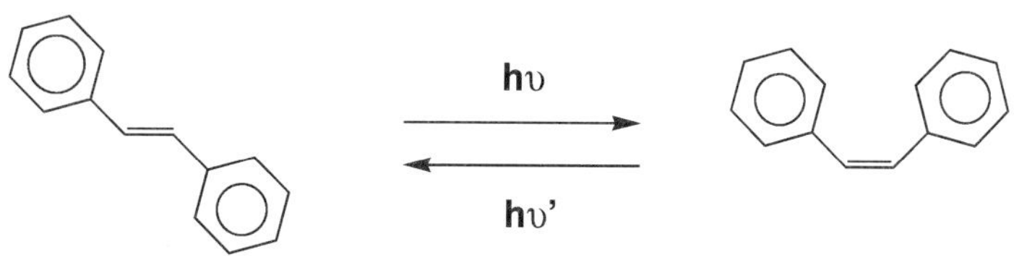

3a ***trans*-Stilbene** **3b** ***cis*-Stilbene**

Fig. 4. The *cis–trans* photoisomerization of stilbene **3**.

nondestructive reading of the information has been carried out using laser diodes. Starting with this naturally occurring molecular compound, it has been possible to store patterns of digital information.

2.2.2. *Stilbenes*

The *cis–trans* isomerization of stilbene **3** is a well-known [19] configurational change within a molecule. *Trans*-stilbene **3a** can be isomerized reversibly (Fig. 4) to the corresponding *cis*-stilbene **3b**. However, its application as a device for information storage has been hampered by the many possible side reactions. *Cis*-stilbene undergoes conrotatory ring closure to dihydrophenathrene upon irradiation with light, and this compound can be readily oxidized to phenanthrene. In addition, however, dimerization can also take place upon irradiation to form a cyclobutane derivative.

2.2.3. *Azobenzenes*

Azobenzenes **4** exist (Fig. 5) in two different isomeric forms—the (E)- and (Z)-isomers (i.e., usually equating with *trans* and *cis* forms, respectively), which can be recognized by the differences in their UV/vis absorption spectra. The photoisomerization of azobenzenes leads to a photostationary state, where the *cis–trans* ratio is dependent upon the wavelength used. In the case of azobenzene itself, irradiation at 313 nm of the (E)-isomer **4a** results in approximately 80% conversion to the (Z)-isomer **4b**. The azobenzenes can form the basis of molecular switches [20] (Fig. 5) with the application of light of different wavelengths, in the knowledge that an excess of one isomer can be detected by the change in the UV/vis absorption spectra. As a result, azobenzene units have been used, for example, (i) to induce structural variations in polymeric liquid crystals [21], (ii) to effect gelation by employing a cholesterol derivative functionalized with an azobenzene moiety [22], and (iii) to induce conductivity into a Langmuir–Blodgett film of a charge-transfer complex [23].

The main disadvantage that hinders the further development of azobenzenes into practical switching devices is the facile thermal reisomerization of the (Z)-isomer back to the

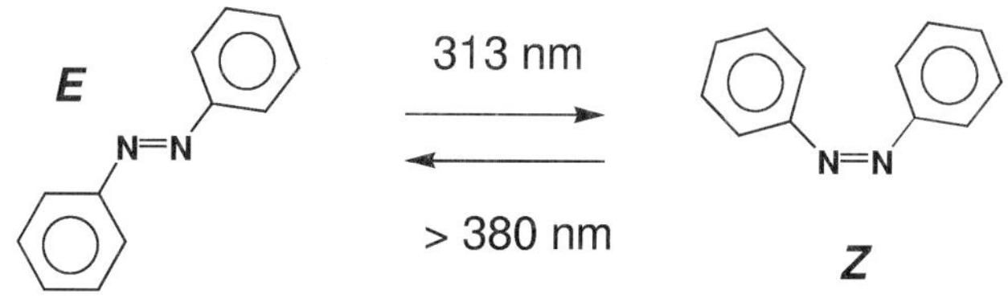

4a **4b**

Fig. 5. The *cis–trans* photoisomerization of azobenzene **4**.

more stable (E)-isomer. An elegant method for circumventing this problem was reported by Liu and co-workers [24], who used a selective reduction procedure (Fig. 6) to convert the (Z)-isomer to the corresponding hydrazobenzene. This storage system is based on a combination of two different types of reversible processes—namely, photochemical *cis–trans* isomerization and electrochemical oxidation and reduction. (Z)-Azobenzene can be reduced to hydrazobenzene at a more anodic potential than the (E)-isomer, and the hydrazobenzene can be oxidized exclusively to the (E)-isomer of azobenzene. Therefore, information can be first of all written (Fig. 6), then it can be stored by selective reduction of the (Z)-isomer, allowing for nondestructive readout of the information (Fig. 6), and then, finally, it can be erased (Fig. 6) by oxidizing the hydrazobenzene moiety back to the (E)-isomer. This reversible cycle can be repeated several hundred times without discernible fatigue.

2.2.4. *Chiroptical Molecular Switches*

A photoreversible system, based upon the interconversion of enantiomers (or pseudoenantiomers) of inherently dissymmetric molecules, can be used as an alternative approach to the creation of molecular switching systems. The two chiral forms [25] (Fig. 7) of **5** are denoted as P and M, for the right- and left-handed helical structures, respectively. The principle behind a chiroptical molecular switch using circularly polarized light (CPL) for writing is shown in Figure 7. Reading can be achieved with linearly polarized light (LPL)—by monitoring the change in optical rotation—and erasing with unpolarized light (UPL)—as a result of photochemical racemization. Such a system has been used recently for the reversible photochemical modulation of the helical pitch of a twisted nematic liquid crystalline phase and for the reversible transition between cholesteric and nematic phases [26].

2.3. Constitutional Change

2.3.1. *Fulgides*

The fulgides **6** are derivatives of dimethylene-succinic anhydrides, substituted with an aromatic ring. Their switchable properties [27] depend upon a reversible photochemical

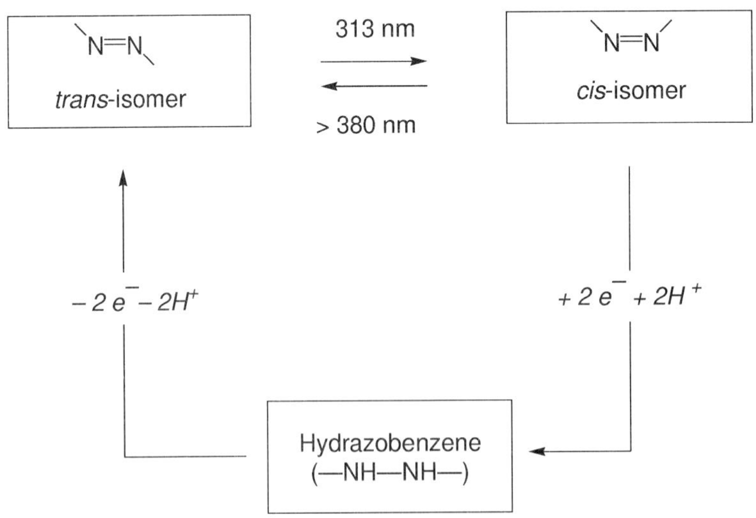

Fig. 6. Dual mode of controlling the azobenzene switching properties by employing photo- and electrochemistry.

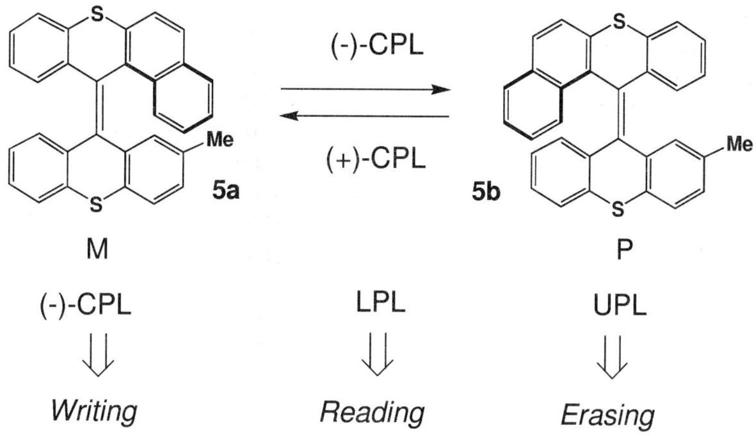

Fig. 7. A chiroptical molecular switch **5**. The interconversion between the two states is accomplished by circularly polarized light. CPL, Circularly polarized light; LPL, linearly polarized light; UPL, unpolarized light.

conrotatory electrocyclization. Upon irradiation with UV light, the colorless fulgide **6** is converted (Fig. 8) into the colored dihydronaphthalene derivative **7**. Thus, a photochromic switch has been created, inasmuch as the cycle can be made reversible by irradiation of **6** with visible light. The readout procedure for the two different states is based on the differences in the absorption spectra of the two isomers. The main problem is that this detection method can interfere with the "writing" of the data. An easily accessible, nondestructive method of "reading" the two states will have to be uncovered if fulgides are to play an important role in storing data at the molecular level.

2.3.2. Diarylalkenes

Irie and co-workers [28] first reported the preparation of thermally stable diarylalkenes by using furan or thiophene rings instead of phenyl rings. In this manner, the possible *cis–trans* isomerization of the central double bond was prevented. Upon UV irradiation at 405 nm, the yellow-colored open form **8** was converted (Fig. 9) into the brown-colored closed dihydro derivative **9**: compound **8** could be regenerated by exposure of **9** to visible light ($\lambda = 520$ nm).

Based upon this system, Lehn et al. [29] have devised (Fig. 10) a dual-mode molecular switching device. The open form **10** is electrochemically inert, whereas the photocyclized hydroquinoid **11** can be oxidized reversibly at +0.72 V in MeCN. The resulting cyclic quinoid **12** is not photochemically active and cannot be reversed to the open form, that is, it is locked in the closed state. In summary, information can be "written" photochemically, and the data can be safeguarded or "locked" by oxidation to the quinoid form **12,** which

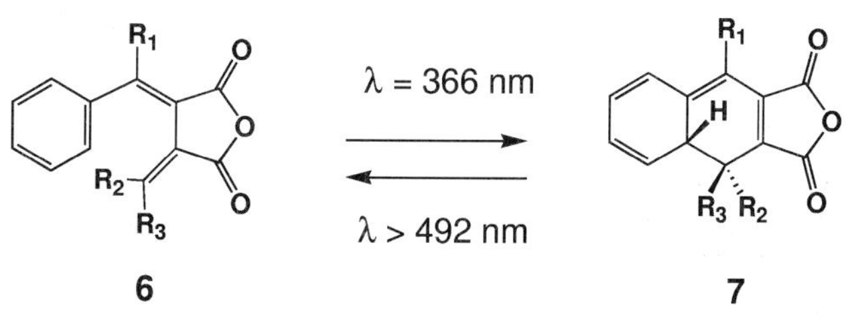

Fig. 8. The reversible photoinduced electrocyclization of the fulgide **6** into the dihydronaphthalene derivative **7**.

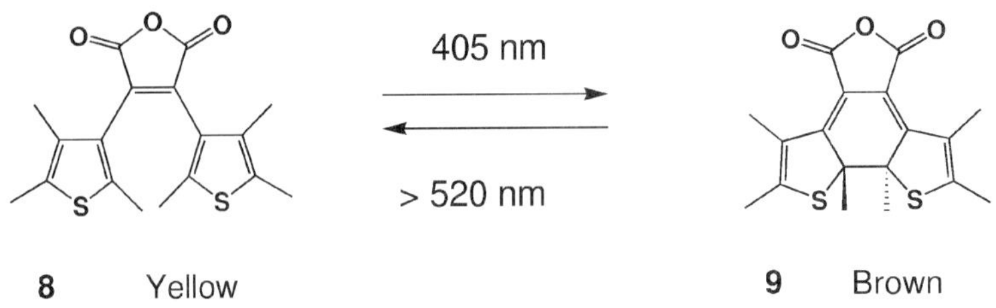

Fig. 9. The reversible photoinduced electrocyclization of the diarylalkene **8** into the dihydro species **9**.

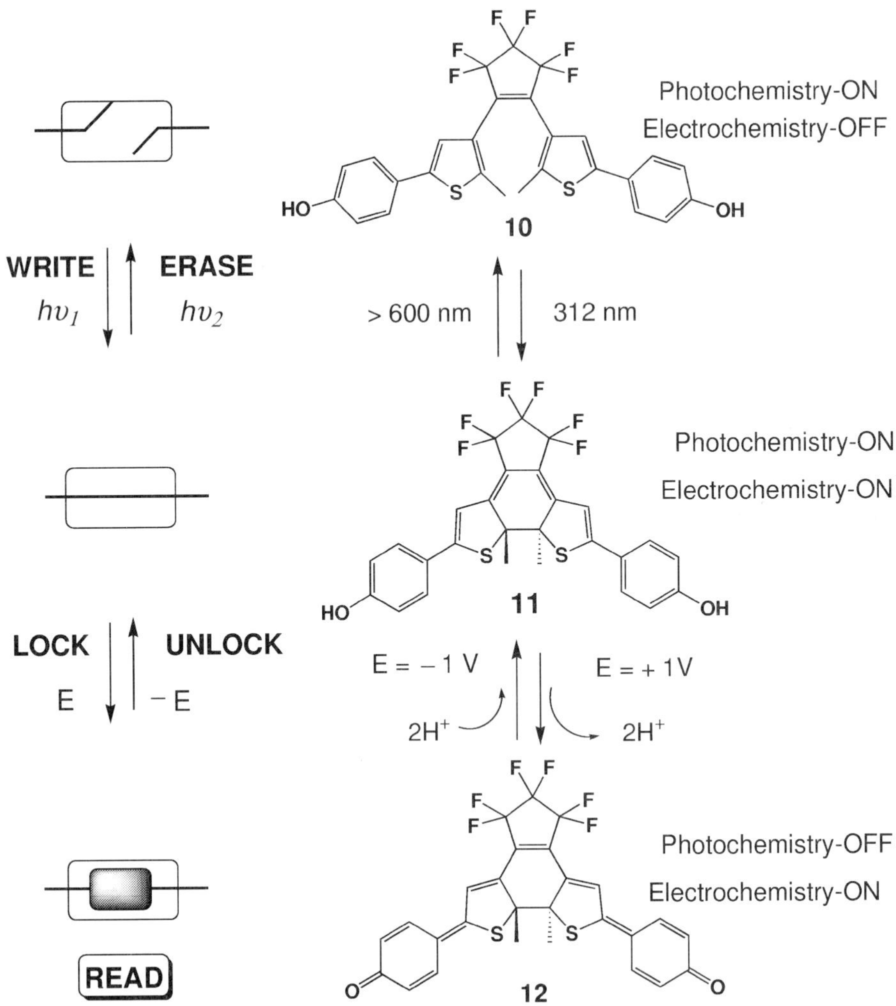

Fig. 10. Schematic representation of an erasable optical memory system with nondestructive readout capacity based on the dual mode—electro- and photochemical—of switching of compounds **10–12**.

allows for nondestructive "reading." A reduction process—back to the hydroquinoid **11**—can then be used to unlock the information and permit subsequent photochemical erasing—photochemical ring opening to **10**—thus rendering the cycle reversible.

At the microscopic level, the system clearly behaves as a dual-mode optical-electrical molecular switching device. On the macroscopic level, the spectral and redox properties are well suited to form the basis for a potential erasable data storage system with non-destructive readout capacity—as shown in the schematic representation on the right-hand side of Figure 10.

2.3.3. Spiropyrans

The photochromic behavior of spiropyrans [30] is based on the reversibility (Fig. 11) that exists between the colorless "closed" spiropyran form **13** and the colored open merocyanine dye **14** as a result of the heterolytic cleavage of the spiro carbon–oxygen bond. On ultraviolet irradiation of the spiro form **13**, the merocyanine **14** is formed. It can be made to revert, either by irradiation with visible light or thermally, to the spiro form **13**. The lifetime of the zwitterionic merocyanine was increased dramatically from seconds to hours by the introduction of electron withdrawing nitro substituents.

The photochromic behavior of the spiropyrans has been studied widely, and their conversion into the merocyanines have been exploited (i) to induce order into aggregated mesomorphic materials, (ii) to photocontract liquid spiropyran–merocyanine films [31], (iii) to assist the cross-linking of mesomorphic polymers [32], and (iv) to create organic materials with nonlinear optical responses, to cite but a few examples [33].

The most elegant use of the spiropyran–merocyanine switching cycle to date is one that combines surface chemistry, self-assembly processes, molecular recognition events, and biology. Willner and co-workers [34] have constructed (Fig. 12) a fascinating biomolecular device, in which the electron transfer communication between a cytochrome *c* (Cyt *c*) and an electrode is controlled reversibly by light. The electron transfer communication of the biomaterial Cyt *c* with electrodes is normally accomplished by the attachment of the pyridine substituents to the surface of the electrode, thus facilitating the recognition of the Cyt *c* by the pyridine substituents. The surface of the electrode can be modified by the introduction of the spiropyran component **15**. The spiropyran moiety can be photoconverted into the charged merocyanine unit when electrostatic repulsions perturb the associative interactions of Cyt *c* with the pyridine units. Hence, the electrical communication between the Cyt *c* and the electrode becomes blocked. This electronic communication can be reversibly switched back on again by transforming photochemically the merocyanine unit into the neutral spiropyran form, thus restoring the molecular recognition between the pyridine units and the Cyt *c*. Furthermore, the Cyt *c* acts as an electron mediator in a series of biological pathways, such as the reduction of molecular oxygen to water by Cyt *c* oxidase (COX). Therefore, the biocatalyzed reduction of oxygen to water can be switched (Fig. 12) off and back on again by photochemically impeding the electronic communication of the Cyt *c* with the gold electrode. These results are of considerable importance because they prove that the chemist can achieve the control of intricate biological pathways in a logical manner.

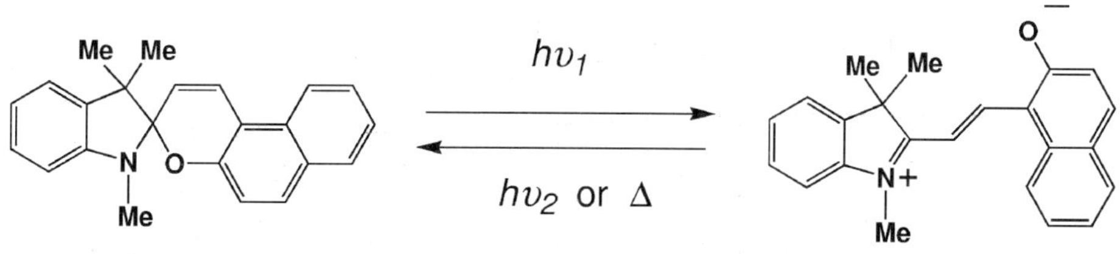

13 **14**

Fig. 11. The photochemical switching from the spiropyran **13** to the merocyanine **14**.

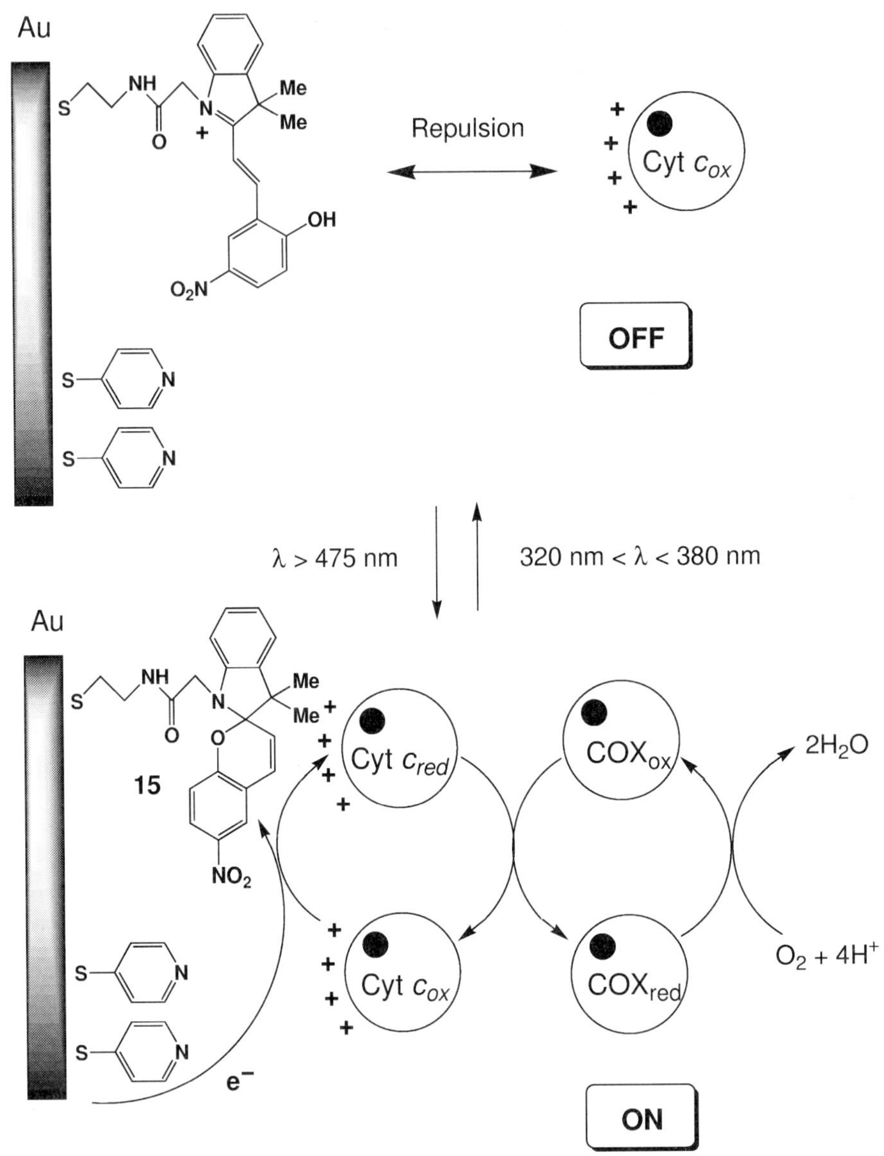

Fig. 12. A photochemically controlled biomolecular device in which the reduction of molecular oxygen to water can be switched on and off reversibly by photochemical means.

2.3.4. Redox Switching of the Luminescence of a Bipyridine Metal Complex

Lehn and co-workers [35] have designed and synthesized (Fig. 13) a molecular device that combines an electroactive component with a light emitting one. Compound **16** consists of a $Ru(bpy)_3^{2+}$ as the photoactive center and a quinone moiety linked together by an ethylene spacer. In the case of **16**, the luminescence of the $Ru(bipy)_3^{2+}$ complex at 610 nm is very efficiently quenched by the quinone unit. Electrochemical reduction of the quinoid **16** to the hydroquinoid **17** at −0.60 V results in a highly emitting solution, that is, the hydroquinone moiety does not quench the emission of the Ru complex. Compound **17** can be reversibly oxidized back to the quinone **16** at +1.10 V, regenerating a nonlumiscent solution. Hence, it is possible to control electrochemically the luminescence of a bipyridine metal complex

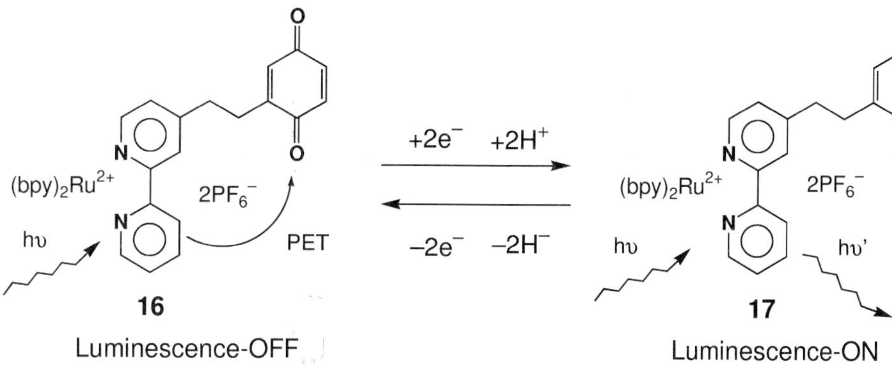

Fig. 13. Electrochemically controlled switching of the luminescence of the $Ru(bipy)_3^{2+}$.

in a reversible manner. Because both **16** and **17** are stable and isolable, they have the potential to function as an information storage device.

3. SUPRAMOLECULAR SYSTEMS

A supramolecular system is composed of two or more subunits that come together to form an ordered superstructure because of the stereoelectronical complementarity of its components. The forces that control the formation of the superstructure are noncovalent bonding interactions, which are weak and reversible, making them ideal candidates for the construction of switching systems at the nanoscale level. Because there are many examples of supramolecular switching systems, discussing all of the examples is beyond the scope of this chapter. In our discussion, we will concentrate on more recent results, as well as on those that have emerged from within our own research group.

3.1. Crown Ethers

Supramolecular chemistry came into being with the discovery of the so-called crown ethers by Pedersen [36] in 1967. It is well documented that crown ethers act as receptors for alkali metal cations and that the selectivity of this interaction depends mainly upon the size and conformation of the crown ether ring.

In 1979, Shinkai and co-workers [37] described how they could modify the binding properties of crown ethers toward alkali metal cations by introducing switchable azobenzene units into their rings. The (E)-isomer of the azacrown **18** shows no complexation ability for alkali metal cations as a result of the "tight" conformation adopted by the polyether chain. On irradiation, however, the azacrown **18** is converted (Fig. 14) into its (Z)-isomer, producing a crown ether with a much bigger cavity. The (Z)-azacrown shows very high affinities toward Na^+, K^+, and Rb^+ ions. Furthermore, thermal reisomerization was prevented by the formation of the complex. The bistability in this case is total, that is, the (E)-isomer shows no binding ability, whereas the (Z)-isomer binds strongly to the alkali metal cations. Thus, they can be defined as all-or-nothing switches.

Shinkai [38] also exploited the *cis–trans* isomerism of azobenzene and crown ether complexation in a different manner. He managed to photoregulate crown ether–metal ion complexation by competitive intramolecular ammonium ion binding. A derivative of benzo[18]crown[6] (**19**) incorporating an azobenzene unit with a primary alkylammonium ionic center at its terminus was synthesized. The (E)-isomer of crown ether **19** is able to extract alkali metal cations with no interference from the ammonium group, which is located far from the crown ether component. However, photoisomerization (Fig. 15) of

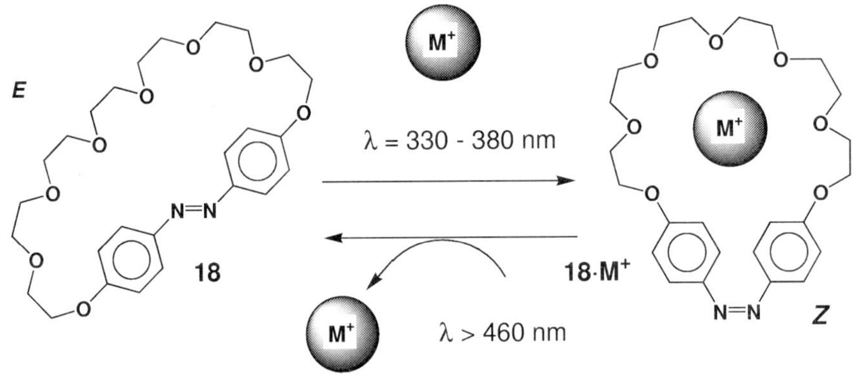

Fig. 14. Photochemically induced isomerization in the crown ether **18**. The complexation is effectively switched on and off by irradiation.

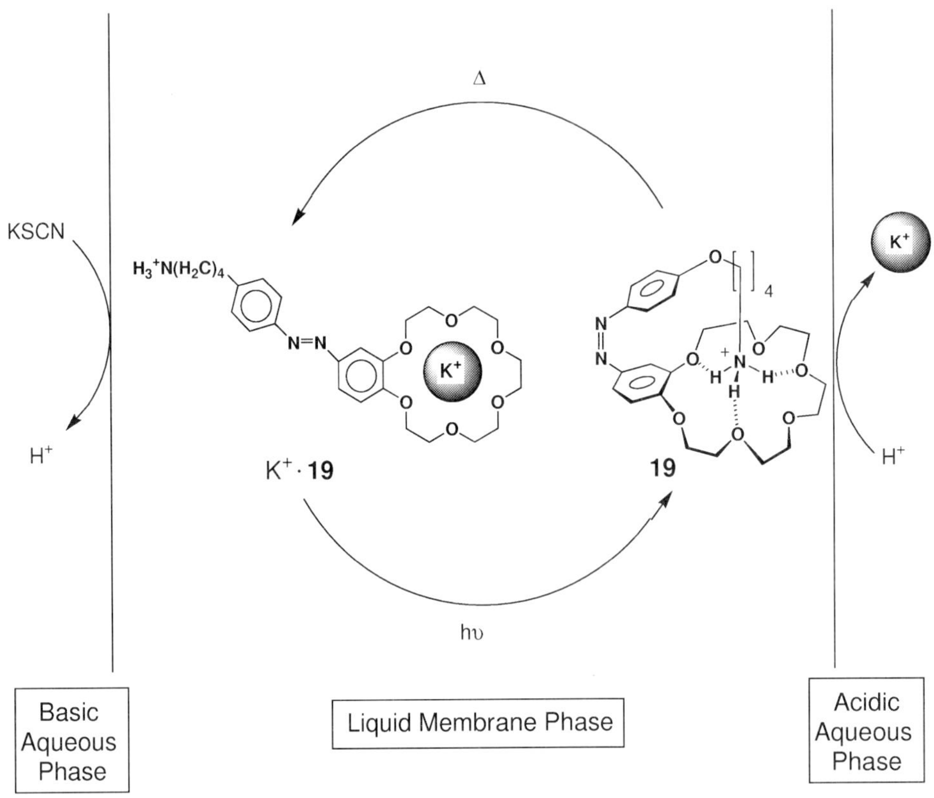

Fig. 15. Photochemically controlled transport of alkali metal ions across a liquid membrane.

the (*E*)-isomer to the (*Z*)-isomer forces the ammonium ion into close proximity with the crown ether component, so that the macrocyclic polyether can bind with the ammonium ion center in an intramolecular manner, creating a "tail-biting" molecule, thus hindering the complexation of alkali metal cations. Back-isomerization to the (*E*)-isomer of the crown ether regenerates the original geometry, making it possible to extract alkali metal cations across a liquid membrane phase.

Recently, Kimura and co-workers [39] have designed and prepared the malachite green derivative **20**. This bis-crown ether is able to form strong complexes with Na^+ and K^+

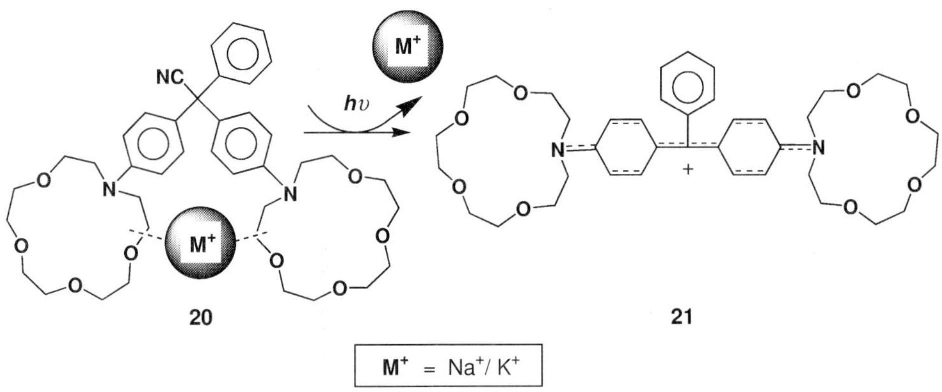

Fig. 16. Photochemically controlled complexation of alkali metal cations by the crown ether **20**.

ions, at least as indicated by mass spectrometric studies. When irradiated with UV light, the malachite green ionizes (Fig. 16) to its corresponding quinoid cation **21**, and then, because of electrostatic repulsion, almost all of the metal ions complexed by the crown ether moiety of **20** were ejected out of the cavity. Thus, this biscrown ether can undergo clear-cut switching from a powerful cation binder in the form of **20** to the perfect cation release species in the shape of **21**. This photoinduced switching behavior of cation binding is an all-or-nothing type of switch with potential device-like applications.

3.2. Fluorescent Signaling Systems

A fluorescent signaling system [40] is a switch and/or a sensor in which the fluorescence of one of the components of the molecule can be switched on or off by the presence of a external chemical species that can interact with a different part of the molecule, giving rise to a supermolecule whose photophysical properties differ from those of the original molecule. A fluorescent ion signaling system is one in which the fluorescence is controlled by the presence of an ionic species. The design of such a system requires the presence of a component with fluorescent properties and an ion receptor component, which are separated by a spacer unit. The choice of the components is critical, and even the spacer unit is of crucial importance, because it has to allow electronic interactions between the fluorescent group and the receptor component of the molecule. A schematic representation of how this system works appears in Figure 17. The "fluor-spacer-receptor" system can be chosen such that its fluorescence is switched off (Fig. 17a) by a photoinduced electron transfer process (PET). The PET process can be suppressed by the entry of a cation into the receptor, thus producing an increase in the oxidation potential of the receptor. The logic of the system can easily be reversed (Fig. 17b). In this case, the PET only occurs when a cation is bound and the fluorescence is switched off in the free receptor.

A very simple proton signaling system can be designed by combining an organic base (i.e., an amino function) with a common fluorescent unit (viz. anthracene), attached via a methylene spacer, as in compound **22**. PET occurs (Fig. 18) in this system in so far as the fluorescence of the anthracene unit is quenched by the basic amino function. However, protonation of the amine nitrogen raises its oxidation potential to $> +2.54$ eV, the PET is suppressed, and a strong fluorescence is observed. Treatment with base regenerates the free amino function, resulting in the disappearance of the fluorescence and thus creating a reversible system. One of the main advantages of this system is that it can be used as a fluorimetric pH indicator.

A special feature of modular systems is that their level of sophistication can be increased progressively by the designer. And so the next challenge was to extend the concept of

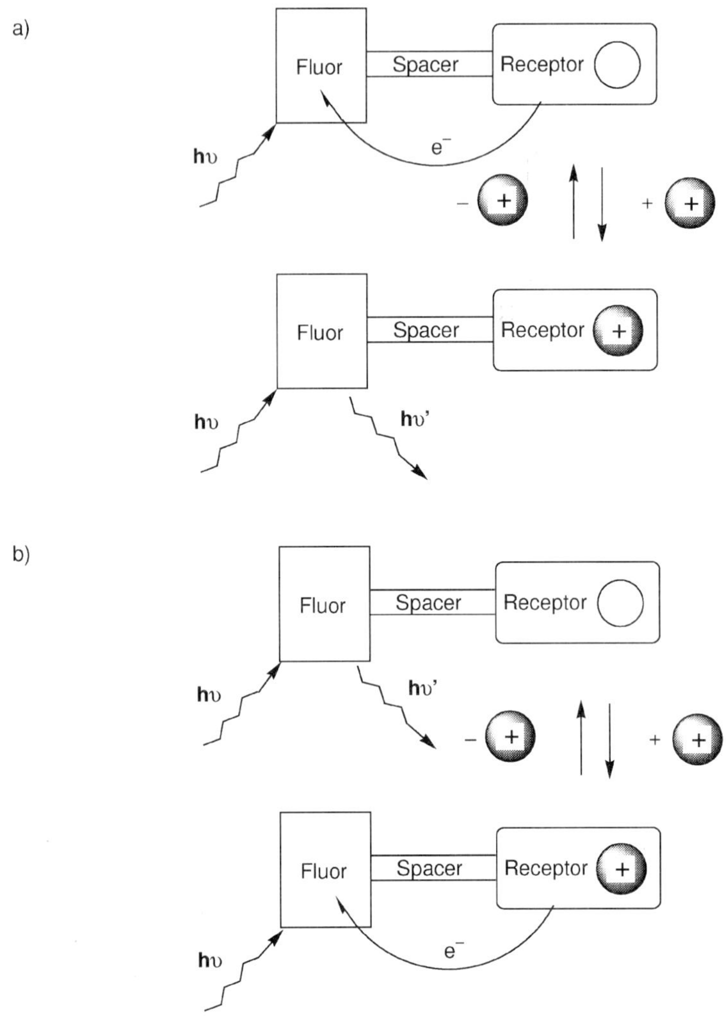

Fig. 17. Cartoon representation of the design logic behind a fluorescent signaling system. a) The fluorescence of the system is switched on by the complexation of a cation by the receptor. b) The fluorescence of the system is switched off by entry of a cation into the receptor.

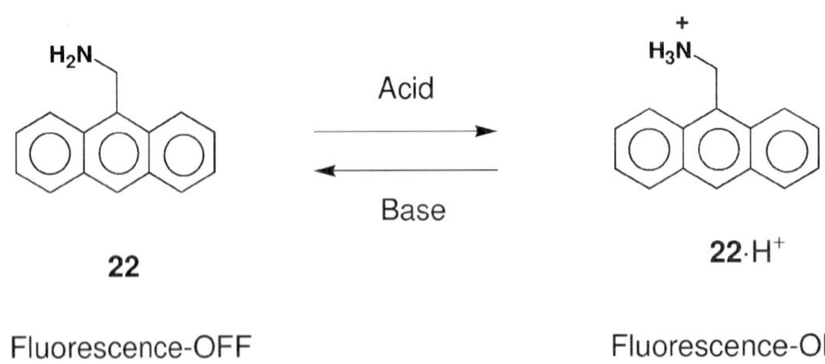

Fig. 18. The pH-controllable fluorescence of the anthracene-containing compound **22**.

fluorescent signaling to cations other than protons. Compounds **23–26** listed in Figure 19 all behave as expected. When they are free of cations, the fluorescence of the anthracene unit is quenched by a PET process. However, in the presence of metal cations, the forma-

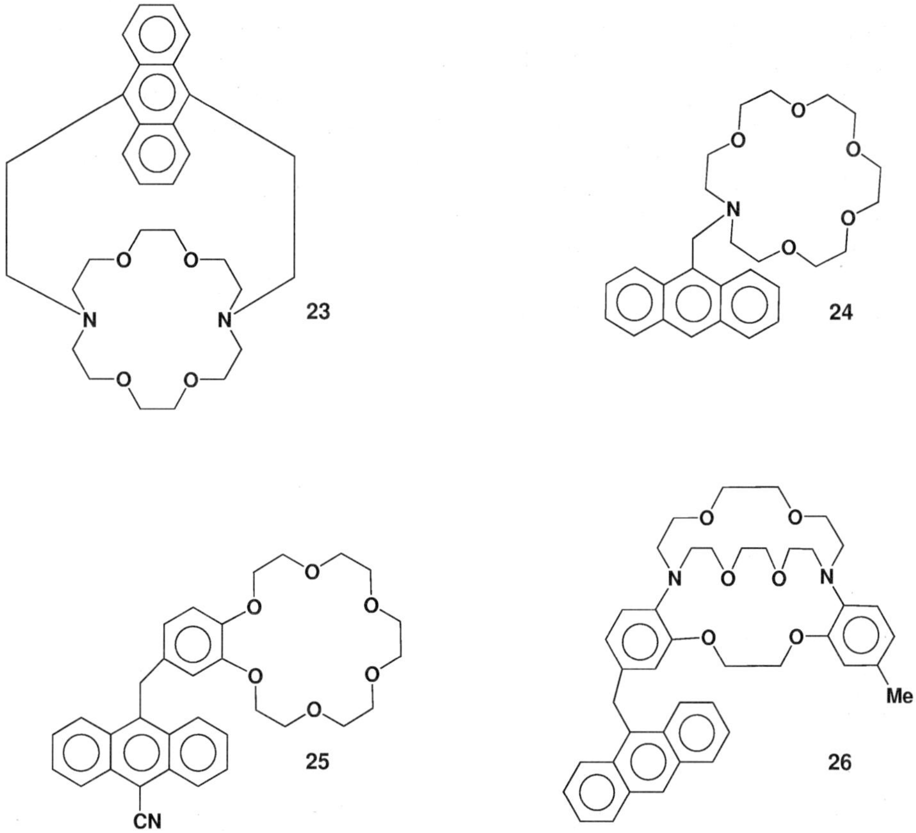

Fig. 19. Structural formulas of the fluorescent signaling systems **23**–**26**.

tion of the corresponding complex inhibits the PET pathway, resulting in the generation of a fluorescent signal. Compounds **23** and **24** exhibit remarkable responses toward Na^+ and K^+ ions, in addition to showing pH-responsive fluorescence. Compound **25** exhibits very good selectivity toward Na^+ ions in alcoholic solutions, and compound **26** gives a maximum fluorescence emission in the presence of Rb^+ ions.

In summary, the design of "fluor-spacer-receptor" signaling systems is simple in concept and quite general in scope and applicability. The very same concept can be applied in the construction of molecular logic gates (see Section 3.8).

3.3. Redox Switches by Ligand Exchange

This supramolecular switch is based upon a receptor molecule that possesses two sets of ion binding groups, one a hard set and the other a soft set of ligating groups, both of which are anchored onto a calix[4]arene framework in an alternating fashion [41]. The calixarene [42] torus was chosen because it can direct the two different binding sites to the same face of the ring and allows either set to converge as a result of a conformational change. The selected "hard" binding sites were hydroxamate groups, whereas the "soft" binding sites were 2,2′-bipyridyl rings. Both of these binding sites have been attached to a calix[4]arene ring in compound **27** shown in Figure 20. In the presence of Fe^{III}, the hydroxamates—the "hard" binding sites—converge to embrace the "hard" metal ion, whereas the "soft" binding sites diverge. Upon reduction to Fe^{II}, compound **27** rearranges so that the "soft" bipyridyl binding sites engulf Fe^{II}, whereas the hydroxamate groups separate. Subsequent oxidation reverses the process. The two states have different spectroscopic characteristics

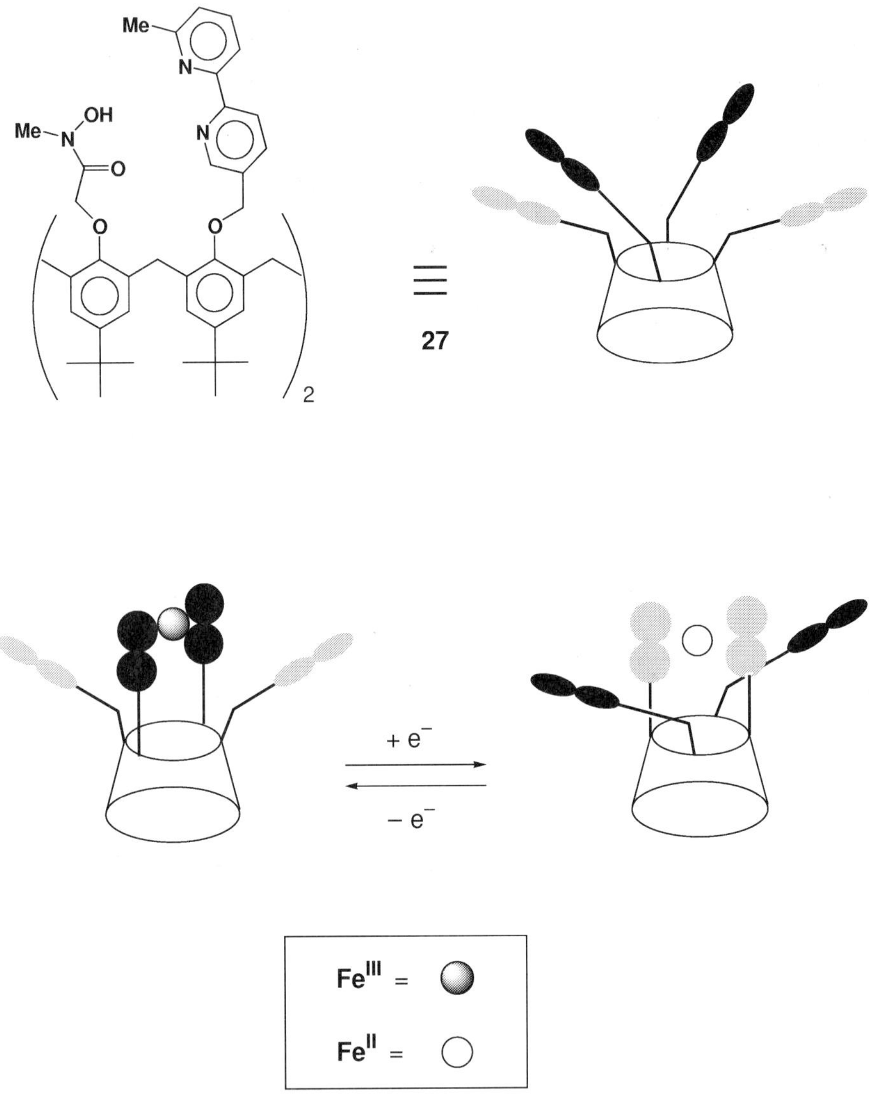

Fig. 20. Electrochemically controlled switching of the mode of complexation of an iron metal cation by the calix[4]arene **27**.

that enable nondestructive "reading" by UV/vis spectroscopy. Compound **27** behaves as a chromophoric switch: it changes from orange to pink upon reduction. As a result of the amphiphilicity of this switch, it is well suited for self-organization as a monolayer at air–water interfaces and for transfer into Langmuir–Blodgett films.

3.4. Translocation in Helical Complexes

Based upon the principle already outlined in Section 3.3, the triply branched compound **28** shown in Figure 21, containing hydroxamate "hard" binding sites (located proximal to the core) and "soft" 2,2′-bipyridyl binding sites (at the periphery of the dendrimer-like molecule), was synthesized [43]. Addition of Fe^{III} to **28** results in the formation of a Fe^{III}-hydroxamate complex possessing a left-handed triple-helix geometry. At this starting point, the single metal ion is located in the internal "hard" binding cavity: reduction (with ascorbic acid) of the guest ion from Fe^{III} to Fe^{II} induces (Fig. 21) translocation from the

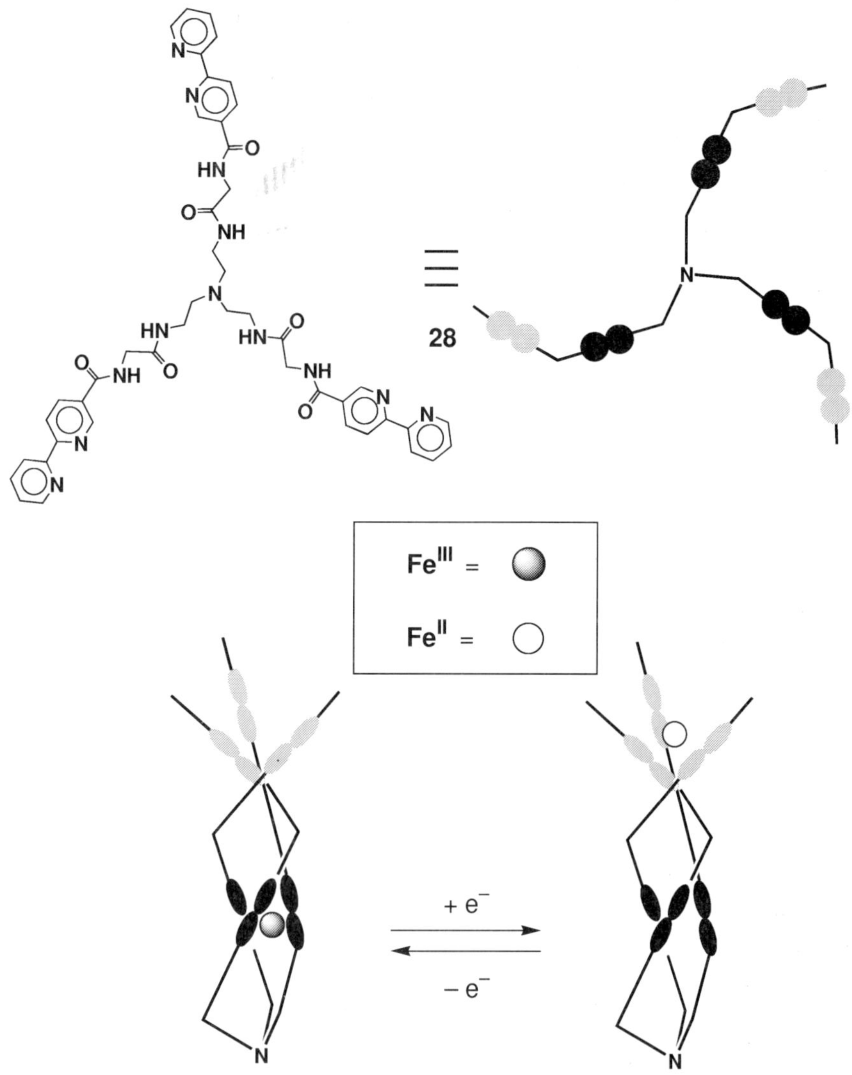

Fig. 21. Electrochemically controlled translocation of an iron metal cation along the triple-helical ligand **28**.

internal "hard" hydroxamate binding sites to the external "soft" bipyridyl binding sites. Simultaneously with the translocation of the binding site, a change in color from light brown (Fe^{III}-hydroxamate complexes) to deep purple (Fe^{II}-bipyridyl complexes) takes place, giving the switch chromophoric properties. Subsequent oxidation by ammonium persulfate of the Fe^{II} ions regenerates Fe^{III} ions, which then move back into the internal "hard" hydroxamate binding sites. One of the challenges that still has to be met is the immobilization of this supramolecular switch on a conducting surface.

3.5. Photoswitchable Complexation of Metalloporphyrins

The reversibly photocontrolled "on–off" complexation of zinc and magnesium phorphyrins with a well-designed stilbazole derivative can be induced by the *cis–trans* photoisomerization of the stilbazole [44]. Pyridine derivatives are well known to act as guests for metalloporphyrin hosts. Moreover, it has been reported that steric hindrance by porphyrin substituents—especially in the 2 position of the pyridine derivative—has a considerable influence on the binding constants. In the case of the zinc tetraarylmetalloporphyrin **29**

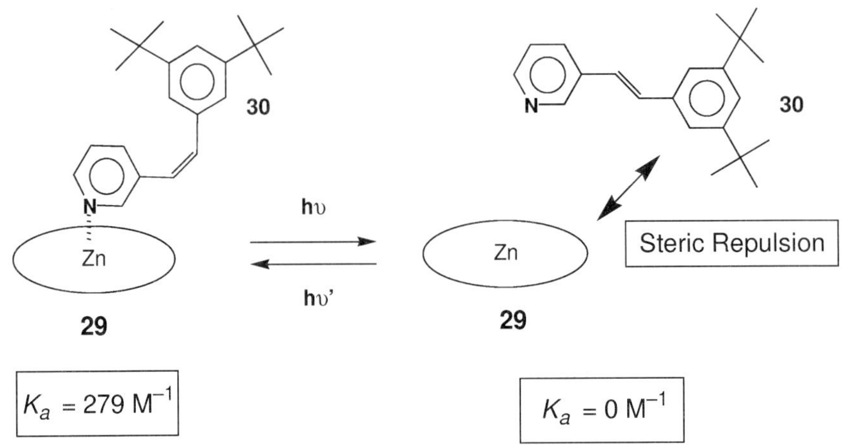

Fig. 22. Photoswitchable complexation of the stilbazole **30** with the 4-*tert*-butyltetraphenyl zinc porphyrin **30**. The K_a values were obtained in benzene at 23 °C.

with bulky *tert*-butyl substituents, the binding constant with the *trans*-stilbazole **30** is negligible because of the steric interactions (Fig. 22) between the phenyl ring of **30** and the *tert*-butyl substituents on the aryl groups of **29**. However, irradiation of the mixture causes the photoisomerization of **30** from the *trans*-isomer into the *cis* one. This change in geometry results in the formation of a 1:1 complex between the porphyrin **29** and the *cis*-isomer of **30**, the binding constant being 279 M^{-1}. The two states, complexed and uncomplexed, can be monitored by UV/vis and ^1H NMR spectroscopies. The study of the complexation of metalloporphyrins with bases is of particular importance, because it influences the reactivity of metalloporphyrins in heme enzymes.

3.6. Dendritic Boxes: Ships in a Bottle

Dendrimers can be defined as monodisperse, highly branched macromolecules with well-defined constitutions [5]. One of the main potential applications of dendrimers is their ability to encapsulate guest molecules within their globular structures. Meijer and coworkers [45] have recently reported that it is possible to imprison guest molecules in a molecular dendritic container with a diameter of 5 nm—the so-called dendritic box. These dendritic boxes are constructed from a flexible poly(propylene imine) dendrimer with 64 amino functions as end groups. These end groups can be capped with a *t*-BOC-protected-L-phenylalanine derivative, producing a 64 L-Phe box with a highly dense hydrogen-bonded shell that confers upon the dendritic box a solid-state character, as demonstrated by ^{13}C NMR relaxation data. The guest molecules present in solution, 3-carboxyproxyl, TCNQ, and Rose Bengal, are captured (Fig. 23) within the internal cavities of the globular dendritic boxes like ships in a bottle. When the *t*-BOC groups are hydrolyzed by formic acid, the guest molecules residing in the cavity of the dendritic box are released back into solution again.

Furthermore, it has been possible more recently to achieve the shape-selective liberation of guest molecules from the dendritic boxes [46]. Thus, Rose Bengal and *p*-nitrobenzoic acid were encapsulated within the dendritic box. Hydrolysis, followed by dialysis (5% water in acetone) resulted in the liberation of the guest *p*-nitrobenzoic acid molecules, whereas the Rose Bengal molecules remained in the cavities of the dendritic box. However, hydrolysis of the outer L-Phe shell (12 N HCl), followed by dialysis (100% water), brought about the liberation of the Rose Bengal molecules and the recovery of the starting poly(propylene imine) dendrimer in 50–70% yield.

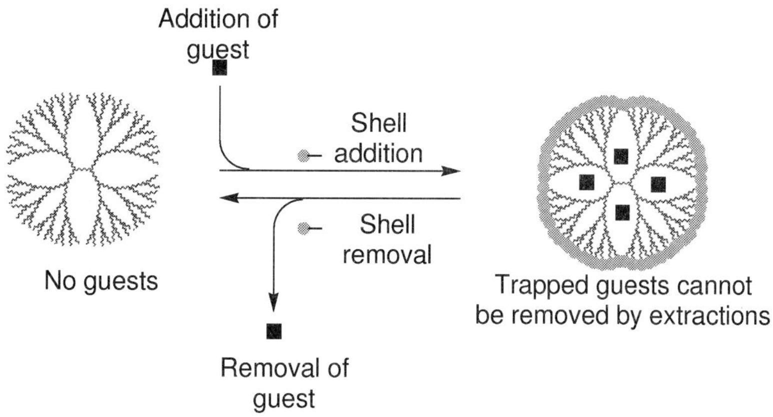

Fig. 23. Cartoon representation of the encapsulation and release of chemical species by a poly(propylene imine) dendrimer operating as a dendritic box.

In summary, a chemically driven supramolecular switch has been created. The fact that the shape-selective liberation of guests can be easily accomplished in a two-step process indicates that the fine-tuning of dendrimers is possible and opens up new opportunities for research in host–guest chemistry.

3.7. Complexation/Decomplexation of Pseudorotaxanes

Much effort has been invested in recent years within our research group in exploring the donor–acceptor interactions between π-systems. In the mid-1980s, it was discovered [47] that the crown ether bis-p-phenylene-34-crown-10 (**BPP34C10**) binds strongly ($K_a = 720\ M^{-1}$) with the herbicide Paraquat in acetone solution. The solid state crystal structure reveals the formation of a 1:1 complex with a pseudorotaxane-like geometry in which the Paraquat dication threads itself through the cavity of the crown ether. Closer inspection of this X-ray crystal structure shows that the driving forces for the formation of the complex are the noncovalent bonding interactions associated with [C–H···O] hydrogen bonding interactions involving both the methyl and α-bipyridinium methine hydrogen atoms on the Paraquat dication and the central oxygen atoms of the two polyether loops in **BPP34C10**, and π-π stacking, including charge transfer interactions between the π-electron deficient bipyridinium unit and the π-electron rich hydroquinone rings incorporated within the macrocyclic polyether. With these recognition features established, the design logic was inverted and the Paraquat units were incorporated [48] into a macroring in the shape of the tetracationic cyclophane, cyclobis(paraquat-p-phenylene). This compound, as its hexafluorophosphate salt, binds strongly with π-electron-rich aromatic substrates, such as 1,3- and 1,4-dioxybenzene [49], and 1,5-, 1,6-, 2,6-, and 2,7-dioxynaphthalene [50] derivatives in both acetone and acetonitrile solutions. The corresponding tetrachloride salt has been shown to bind biologically significant compounds, such as amino acids and neurotransmitters (e.g., phenylalanine, tryptophan, serotonin, etc.), in water [51].

Cyclobis(paraquat-p-phenylene) complexes the π-electron-rich template 1,4-bis-2-(2-(2-hydroxy)ethoxy)ethoxy)benzene. The solid-state structure of this 1:1 complex shows a familiar pseudorotaxane-like arrangement, with stabilization provided by cooperative noncovalent bonding interactions [52]. The oxygen atoms associated with the hydroxyl groups at the termini of the polyether chains of the linear compound are involved in [C–H···O] hydrogen bonds with two of the α-bipyridinium hydrogen atoms on the tetracationic cyclophane. This additional hydrogen bonding is reflected in a dramatic increase in the binding constant, relative to that observed for 1,4-dimethoxybenzene. In addition to this source of stabilization, [C–H$\cdots\pi$] interactions between two of the the aromatic hydrogen atoms of the hydroquinone ring of the diol and the π-cloud of the *p*-xylyl spacers in the tetracationic cyclophane also play a significant role in adding to the stability and defining the geometry of the complex.

Against this background of knowledge it became possible to construct large and ordered molecular assemblies and supramolecular arrays. The structures and superstructures of these compounds and complexes depend upon the nature of the noncovalent bonding interactions and the fact that small changes in the stereoelectronic characteristics of the subunits allow for extensive changes in the binding constants. The reversible nature of the noncovalent bonding interactions provides the framework for the self-assembly of various supramolecular switching systems with pseudorotaxane-like geometries.

3.7.1. A Photochemically Driven Supramolecular Machine

The self-assembly of the receptor **31**.4Cl and the naphthalene-containing substrate **32**, which gives rise to the pseudorotaxane **32·31**.4Cl (Fig. 24), is a result of donor–acceptor interactions between the π-electron-rich 1,5-dioxynaphthalene moiety of **32** and the π-electron–deficient bipyridinium units of **31**.4Cl, as well as of [C–H$\cdots$O] hydrogen bonding and [C–H$\cdots\pi$] interactions. When **32** is added to an aqueous solution of **31**.4Cl, the substrate threads spontaneously through the center of the receptor to produce the 1:1 complex with a pseudorotaxane-like geometry. The threading process is supported by absorption and emission spectroscopic data in addition to ^{1}H NMR spectroscopy. In a 6×10^{-5} M aqueous solution of **32** and **31**.4Cl, a charge transfer band ($\lambda_{max} = 520$ nm), appears and the intensity of the naphthalene fluorescence ($\lambda_{max} = 345$ nm) of **32** is quenched.

It is well known that intermolecular redox reactions can be driven with light by means of suitable photosensitizers. A deoxygenated solution containing 9-anthracenecarboxylic acid (**33**), [$E_{red}(\mathbf{33}^+/\mathbf{33}^*) = -0.88$ V vs. SCE], the complex **32·31**.4Cl, and a sufficient amount of a sacrificial reductant, triethanolamine, so that the oxidized **33**$^+$ species can be rapidly scavenged, was irradiated (Fig. 25) with 365-nm light to cause the reduction of the π-electron acceptor bipyridinium unit ($E_{red} = -0.35$ V for the "alongside" bipyridinium unit of an analogous [2]catenane).

On photoreduction, dethreading takes place [53]. Proof of the occurrence of the dethreading process is the increase in the fluorescence of the 1,5-dioxynaphthalene moiety, which can only originate from free **32**. After dethreading has occurred, if oxygen is allowed to enter the solution, the reduced cyclophane is promptly reoxidized, and **32** threads

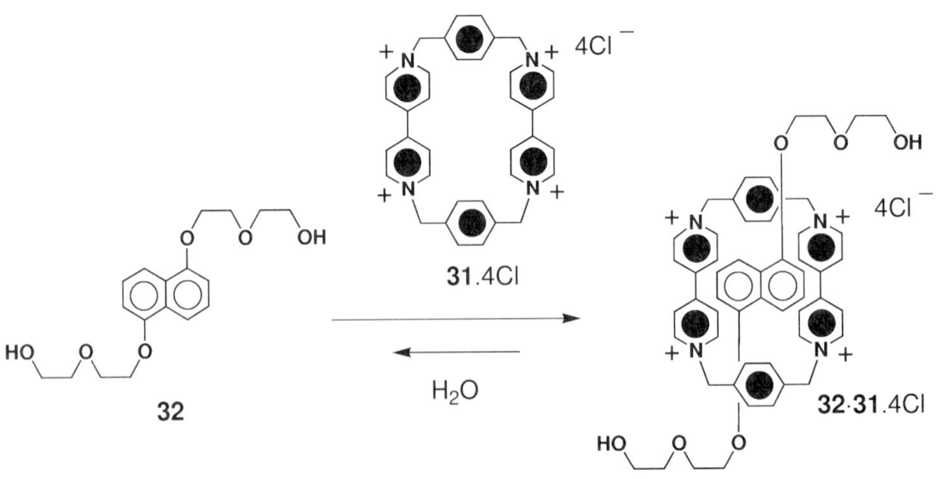

Fig. 24. The formation of the [2]pseudorotaxane **32·31**.4Cl.

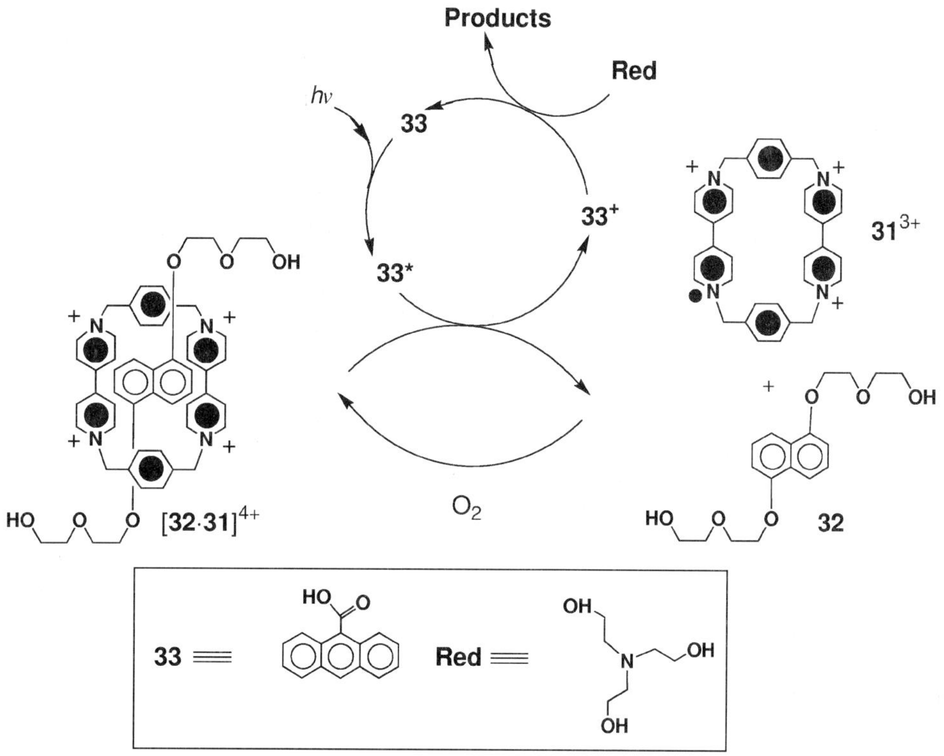

Fig. 25. A photochemically driven supramolecular machine. The dethreading of the [2]pseudorotaxane **32**·**31**.4Cl occurs after the photoinduced reduction of one of the bipyridiniun units in **31**.4Cl.

through it again, as shown by the decrease in the intensity of the fluorescence band and the recovery of the initial absorption spectrum. Thus, a simple supramolecular system, where complexation can be switched on and off photochemically in a reversible fashion, can be constructed rather easily.

3.7.2. An Optically Responsive Supramolecular Switch

It is possible to self-assemble [2]pseudorotaxanes from cyclobis(paraquat-*p*-phenylene) $\mathbf{31}^{4+}$ as the ring with dioxy-disubstituted arenes bearing macrocyclic polyethers at their termini as the linear thread-like components. These [2]pseudorotaxanes can be used [54] to trigger an optical response on metal binding.

The 18-crown-6 derivative **34**, containing a 1,5-dioxynaphthalene ring system, can act both as a host toward alkali metal cations and as a guest toward the tetracationic cyclophane **31**.4PF$_6$. The binding of the π-electron-rich substrate **34** by **31**.4PF$_6$ can be impaired significantly by the addition of alkali metal cations, such as K^+, which form complexes with 18-crown-6 derivatives. The resulting electrostatic repulsive forces induce dethreading of **34**, leaving the tetracationic cyclophane **31**.4PF$_6$ free to act as a receptor for the π-electron-rich cosubstrate **35**, which lacks a crown ether terminus. Mixing equimolar proportions of **34**, **35**, and **31**.4PF$_6$ in MeCN results in a deep purple color arising from a charge transfer band at 518 nm, indicating that the tetracationic cyclophane resides mainly around the 1,5-dioxynaphthalene residue of **34**. Upon the addition of KPF$_6$ to the solution, the purple color fades and the UV/vis spectrum reveals a new (red) charge transfer band at 475 nm, which is consistent with the 1,4-dioxybenzene ring of **35** residing within the cavity of the tetracationic cyclophane. It may be concluded that when KPF$_6$ is added to the solution, an unstable complex **34**·K·**31**.5PF$_6$ is formed (Fig. 26) in which electrostatic

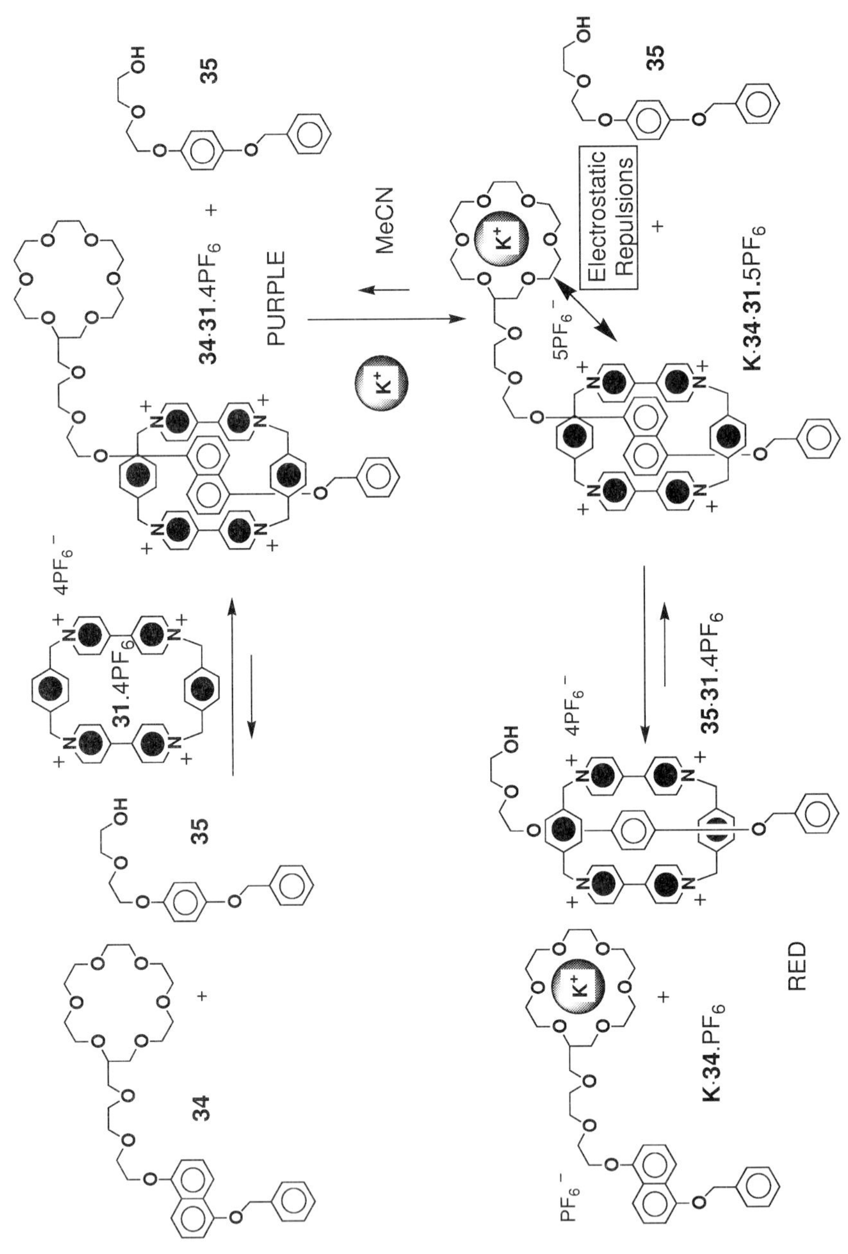

Fig. 26. The chemically controlled complexation of the linear threads **34** and **35** by the tetracationic cyclophane **31**.4PF$_6$. The change of complexation is accompanied by a change of color, rendering the system a chromophoric supramolecular switch.

repulsions between the bound K^+ ion within the 18-crown-6 complement of **34** and the tetracationic cyclophane **31**.4PF_6 force the disassembly of the complex **34**·K·**31**.5PF_6, producing the metalated derivative **34**·K.PF_6 and the free tetracationic cyclophane **31**.4PF_6, which can now form a complex with the neutral cosubstrate **35**. The development of self-assembling systems such as this one could lead to functioning supramolecular entities that can be switched by chemical stimuli to give optical responses.

3.7.3. An Electrochemically Driven Supramolecular Machine

Tetrathiafulvalene (TTF) **36** is a π-electron-rich building block widely used in supramolecular and materials chemistry [55]. Being a π-donating unit, TTF is complexed [56] by the tetracationic cyclophane **31**.4PF_6, forming a 1:1 complex in which TTF is threaded through the cavity of **31**.4PF_6. One attractive feature of TTF is that it is readily oxidized—in two, reversible, single-electron processes—to form sequentially the cation radical and dicationic species. An equimolar mixture of TTF and the tetracationic cyclophane in Me_2CO results [57] in the formation of a dark green solution, and the UV/vis spectrum reveals a charge transfer band at λ_{max} 854 nm associated with a 1:1 complex with a K_a value of 2600 M^{-1}. Electrochemical oxidation of the TTF unit produces (Fig. 27) the positively charged TTF radical cation ($TTF^{\bullet+}$), which is ejected from the cavity of the tetracationic cyclophane **31**.4PF_6 as a result of repulsive electrostatic interactions. This process can be monitored by UV/vis spectroscopy and by cyclic voltammetry. Electrochemical reduction back to the neutral TTF π-electron-rich unit regenerates the 1:1 complex **36·31**.4PF_6 with its associated charge transfer band, rendering the cycle reversible. This process leads to a distinct color change from dark green (when a complex is present) to pale brown (when the system is uncomplexed). Hence, this supramolecular system behaves as a chromophoric switch.

3.7.4. A Chemically Driven Supramolecular Machine

In the case of the [2]pseudorotaxane discussed in this section, we will be inverting the molecular recognition pattern employed in the last three examples. The π-electron-deficient 2,7-dibenzyldiazapyrenium salt **37**.2PF_6 will be the linear threadlike component, and 1,5-dinaphtho-38-crown-10 **1/5-DN38C10** (**38**), containing π-electron-rich 1,5-dioxynaphthalene ring systems, will be acting as the ring component. When **37**.2PF_6 is added to a solution of **38** in MeCN, the dication threads itself through the center of the macrocyclic polyether ring, giving rise to the 1:1 complex **37·38**.2PF_6 with a pseudorotaxane-like geometry and with a reported [58] K_a value of 3×10^5 M^{-1}. The dication $\mathbf{37}^{2+}$ also forms adducts with aliphatic amines, presumably as a consequence of charge transfer interactions between the electron-accepting dication and the electron-donating amines. Thus, advantage can be taken of the chemical affinity of $\mathbf{37}^{2+}$ for

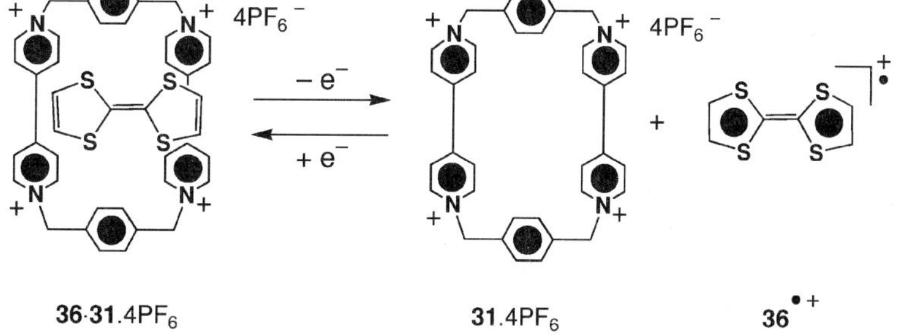

Fig. 27. Electrochemically controlled complexation of TTF by the tetracationic cyclophane **31**.4PF_6.

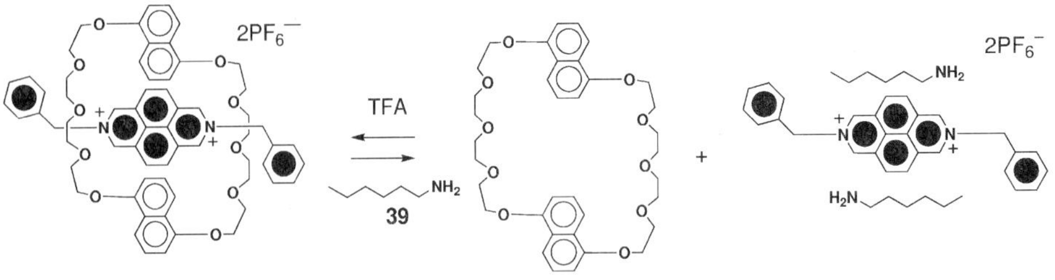

$\mathbf{37{\cdot}38}.2PF_6$ **38** $\mathbf{39_2{\cdot}37}.2PF_6$

Fig. 28. The chemically controlled complexation of the salt **37**.2PF$_6$ by the **1/5-DN38C10** macrocyclic polyether.

aliphatic amines, such as hexylamine (**39**), to induce (Fig. 28) the unthreading of the ring and threadlike components of the [2]pseudorotaxane **37·38**.2PF$_6$. Upon the addition of **39**, large spectral changes occur with the appearance of an absorption band, which can be assigned to a charge transfer band in the $\mathbf{39_2{\cdot}37}$.2PF$_6$ adduct, along with the recovery of the emission band for the naphthalene rings in **38**, this emission having been quenched when **38** is complexed by **37**.2PF$_6$. The unthreading process can be reversed quantitatively by the addition of TFA to the solution. Protonation of the nitrogen atom in **39** disrupts the $\mathbf{39_2{\cdot}37}$.2PF$_6$ adduct and liberates **37**.2PF$_6$, and the [2]pseudorotaxane **37·38**.2PF$_6$ is formed once again.

This system represents a prototype at the supramolecular level for the design and construction of a simple nanoscale machine in which the changes in the relationships between the components can be followed by differences in the absorption and luminescence spectra.

3.7.5. pH-Controllable Supramolecular Switching

Recently, it has been discovered [59] that secondary dialkylammonium cations form pseudorotaxane-like 1:1 complexes with suitably constituted crown ethers, such as dibenzo-24-crown-8 (**DB24C8**). When dibenzylammonium hexafluorophosphate is mixed in apolar solvents (e.g., $CHCl_3$ and CH_2Cl_2) with **DB24C8**, a strong 1:1 complex is formed that can be isolated in a crystalline form. X-ray crystallography reveals that the NH_2^+ center resides in the center of the cavity of the crown ether. The main noncovalent bonding interactions that stabilize this complex are [N^+–H···O] hydrogen bonds between the two hydrogen atoms on the nitrogen atom and some of the oxygen atoms in the polyether chains of the crown ether. They are augmented by weaker [C–H···O] hydrogen bonds involving some of the benzylic methylene hydrogen atoms in the cation and oxygen atoms in the crown ethers and π-π stacking interactions between one of the cation's phenyl groups and one of the catechol rings in the crown ethers. When dibenzylammonium hexafluorophosphate is cocrystallized with a crown ether possessing a bigger cavity, such as bisparaphenylene-34-crown-10 (**BPP34C10**), a 2:1 pseudorotaxane-like complex is formed in which two cations are threaded simultaneously through the cavity of the crown ether [60].

The secondary dialkylammonium salt **40**-H.PF$_6$ containing an anthracenylmethyl unit as one of its substituent groups [61] has been used as a photochemical probe in the complexation–decomplexation processes by Balzani's group in Bologna [62]. When one equivalent of **DB24C8** is added to **40**-H.PF$_6$, a 1:1 pseudorotaxane-like complex is formed between the two species. The [2]pseudorotaxane formation can be monitored by photophysical studies as a result of the presence of the anthracene ring system. If **DB24C8** and **40**-H.PF$_6$ are brought together in close proximity to each other, the luminescence properties of the catechol rings in **DB24C8** are quenched by energy transfer (ET) to the anthracene. However, after the addition of a suitable base—diisopropylethylamine—to the [**40**-H]·[**DB24C8**].PF$_6$ complex, deprotonation of the NH_2^+ center occurs, disrupting the

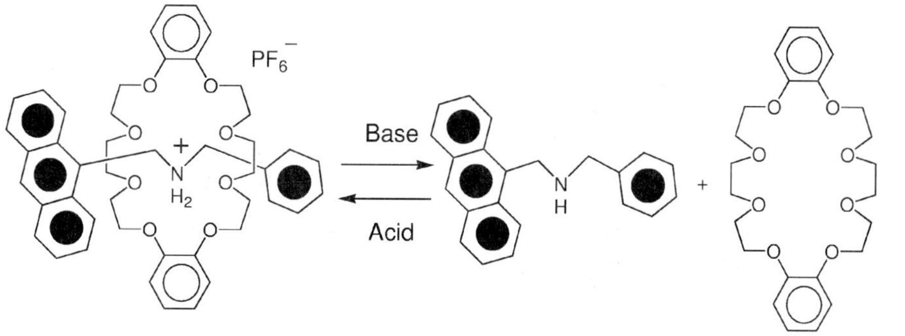

Fig. 29. The pH-controllable complexation of the linear secondary dialkylammonium salt **40**-H.PF_6 by the **DB24C8** macrocycle.

noncovalent bonding interactions responsible for the complex formation, such that unthreading of the cation from the crown ether takes place (Fig. 29). This process can be monitored by luminescence studies because the anthracene ring system and **DB24C8** are no longer in close proximity, with the consequence that the emission of the catechol rings in **DB24C8** is no longer quenched—the ET pathway is not operative—and can be observed. Unthreading can be reversed upon the addition of TFA to the solution. Protonation of the amino function takes place, regenerating the NH_2^+ center and reforming the 1:1 complex.

3.7.6. Electrochemically Controlled Complexation in Cyclodextrins

Cyclodextrins (CDs) are among the building blocks most widely used for the self-assembly of molecular and supramolecular entities [63]. They constitute a series of naturally occurring cyclic oligosaccharides, consisting of six or more α-1,4-linked D-glucopyranose rings. Their rigid hydrophobic cavities give rise to high binding affinities for guests of all types in aqueous solutions. The binding interactions in CD–guest complexes are most likely a summation of several relatively weak forces and effects—for example, van der Waals interactions, hydrophobic binding, and the release of high energy water.

Kaifer et al. [64] have studied the complexation of water-soluble viologen derivatives, such as **41**.2Cl, with β-CD and its heptakis(2,6-*O*-dimethyl) analogue DM-β-CD in aqueous solution. It was found that the interaction between the fully oxidized viologen and the CD is negligible, that is, it is too weak to be observed experimentally. However, cyclic voltammetric experiments indicate (Fig. 30) that, upon reduction of the viologen unit, its hydrophobic character is enhanced, thus increasing the stability of its CD inclusion complex. The neutral fully reduced viologen derivative is a much better guest, particularly for β-CD. This system constitutes an excellent example of redox control of the strengths of host–guest interactions. Furthermore, the researchers were able to discern the mechanism of the electron transfer process. The viologen guests do not engage directly in heterogeneous electron transfer reactions when they are included inside the cavities of CDs acting as hosts.

3.8. Logic Gates

Logic gates are switches whose output states (0 or 1) depends on the input conditions (0 or 1). The design and construction of molecular systems capable of performing complex logic functions is of considerable fundamental scientific interest, because it introduces new concepts into the field of chemistry [65]. YES and NOT single-input gates are the simplest logic devices. A YES gate operates without changing the input bits (input: 1, output: 1;

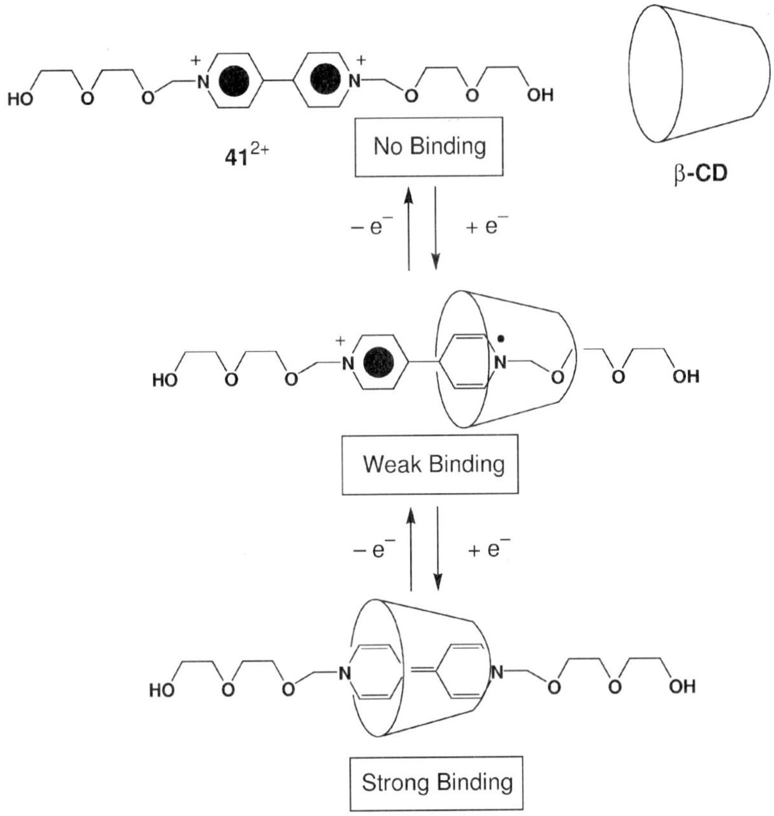

Fig. 30. Electrochemically controlled complexation of the viologen derivative **41**.$2PF_6$ by β-CD.

Table I. The Truth Table for an AND Logic Gate

X	Y	O
0	0	0
0	1	0
1	0	0
1	1	1

The output (O) peformed is a function of two chemical inputs (X and Y).

input: 0, output: 0). On the other hand, a NOT gate inverts the input data (input: 1, output: 0; input: 0, output: 1). More complex logic gates—AND, OR, and XOR—are those whose operations occur under the influence of two chemical inputs (X and Y). An AND gate performs (Table I) the logic product between the inputs—that is, the output signal is 1, provided all inputs are 1. An OR gate performs (Table II) the logic sum between inputs—i.e., the output is 1, provided at least one of the inputs is 1. An XOR gate (Table III) is a more complicated situation in which the truth table is the same as for the OR logic gate, except that the output is 0 if both of the inputs are 1.

The design and synthesis of molecules and supermolecules capable of performing multiple logic functions represents a major challenge to chemists. In this section, we will describe how the challenge is being met using the principles of supramolecular chemistry.

Table II. The Truth Table for an OR Logic Gate

X	Y	O
0	0	0
0	1	1
1	0	1
1	1	1

The output (O) peformed is a function of two chemical inputs (X and Y).

Table III. The Truth Table for a XOR Logic Gate

X	Y	O
0	0	0
0	1	1
1	0	1
1	1	0

The output (O) peformed is a function of two chemical inputs (X and Y).

3.8.1. Photonic AND Logic Gates

The logic gate that will be discussed here is based upon receptors of the type described in Section 3.2. De Silva and co-workers [66] showed that, by exploiting photoinduced electron transfer (PET), it was possible to develop chemical sensors based on fluorescent signaling. Compound **42** is a receptor that operates as a logic device with two input channels, that is, the fluorescence signal depends on the binding of the molecule with protons, Na^+ ions, or both. The receptor is an anthracene derivative, linked to a tertiary amine and a benzo crown ether. It is well known that anthracene derivatives undergo PET processes from both free tertiary amine nitrogen atoms and benzo crown ethers, hence switching off the fluorescence of the anthracene ring system. The PET process can be suppressed by the addition of protons and Na^+ ions simultaneously, causing the switching on of the fluorescence. The successful operation of **42** as an AND logic gate depends on the fact that tertiary amines and benzo crown ethers are very selective for protons and Na^+ ions, respectively. This selectivity allows the molecule to self-select its input ionic signals into the appropriate site. The fluorescence of **42** increases by a factor of 1.1 when only 10^{-2} M of Na^+ ions are present and by a factor of 1.7 when only 10^{-2} M of protons are present. On the other hand, when both inputs are simultaneously applied, the fluorescence output increases (Fig. 31) by a factor of 6.0. In conclusion, the fluorescent signaling is only high when both inputs, Na^+ ions and protons, are coincident in their occupation of **42** and remains low for the other three possible input signal conditions.

Receptor **42** not only acts as a photonic molecular device—a logic gate—but it also allows direct fluorescence monitoring of situations where two (or more) chemical species must come together to cause a vital process.

A constitutionally similar AND logic gate **43** has been reported [67] recently by the Belfast group. The main advantage of this system over the one described previously is that the "off" states are virtually nonfluorescent and the fluorescence efficiency in the "on"

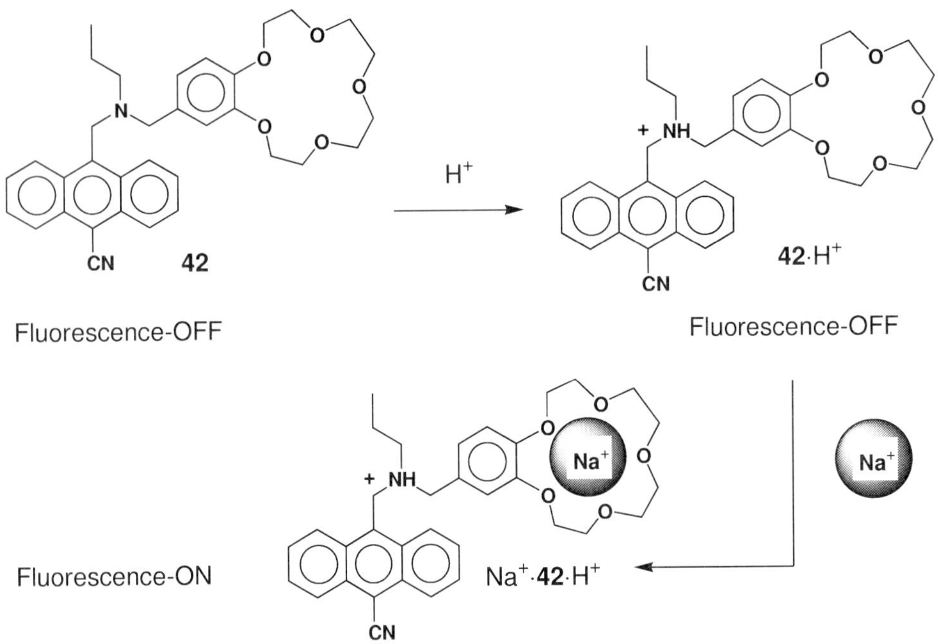

Fig. 31. The fluorescent signaling system **42**, behaving as a photonic AND logic gate. Note that the anthracene fluorescence is only observed in the presence of Na^+ and protons.

state is very high indeed. The fluorescence of **43** is unobservable unless both H^+ and Na^+ are present (Fig. 32) in suffuciently high concentrations, and when this happens a large fluorescence enhancement factor of >37 is observed. Another remarkable feature of these photoionic gates is that they operate in a wireless mode and are able to self-select the two ionic inputs with negligible cross-talk.

3.8.2. A Photonic OR Logic Gate

The design principle [68] behind the photonic OR logic gate **44** containing a fluorescent anthracene unit, which is attached via a methylene linker to London's receptor, relies upon the lack of selectivity of London's receptor toward Mg^{2+} and Ca^{2+} ions. The fluorescence of the anthracene unit in **44** is quenched by a PET from the 1-amino-2-alkoxybenzene unit. The binding of Ca^{2+} and Mg^{2+} ions to the receptor **44** causes an increase in its oxidation potential which is translated into a suppression of the PET process, resulting in fluorescence emission. For **44**, the fluorescent signal enhancement caused by Ca^{2+} and Mg^{2+} ions is very similar, and hence it behaves as a two-input photonic OR logic gate. Both Mg^{2+} and Ca^{2+} ions cause the amine electron pair to be directed orthogonally with respect to the plane of the π-orbitals on the aromatic ring upon complexation. Thus, fluorescence recovery (Fig. 33) by PET suppression is equally efficient for the two ion inputs. This situation represents, in chemical terms, a good quality truth table for the OR logic function and another triumph for de Silva's research group.

3.8.3. A XOR Gate Based upon Pseudorotaxane Formation

Attention has already been drawn to the fact that the 2,7-diazapyrenium unit is a good electron acceptor, capable of forming pseudorotaxane-like complexes with aromatic crown ethers. The π-electron accepting 2,7-dibenzyldiazapyrenium dication $\mathbf{37}^{2+}$ forms a pseudorotaxane-like complex with the crown ether **45** (which contains two 2,3-dioxy-naphthalene π-electron donating units) in which the diazapyrenium unit is sandwiched between the naphthalene rings [69]. Because of the donor–acceptor interactions, there

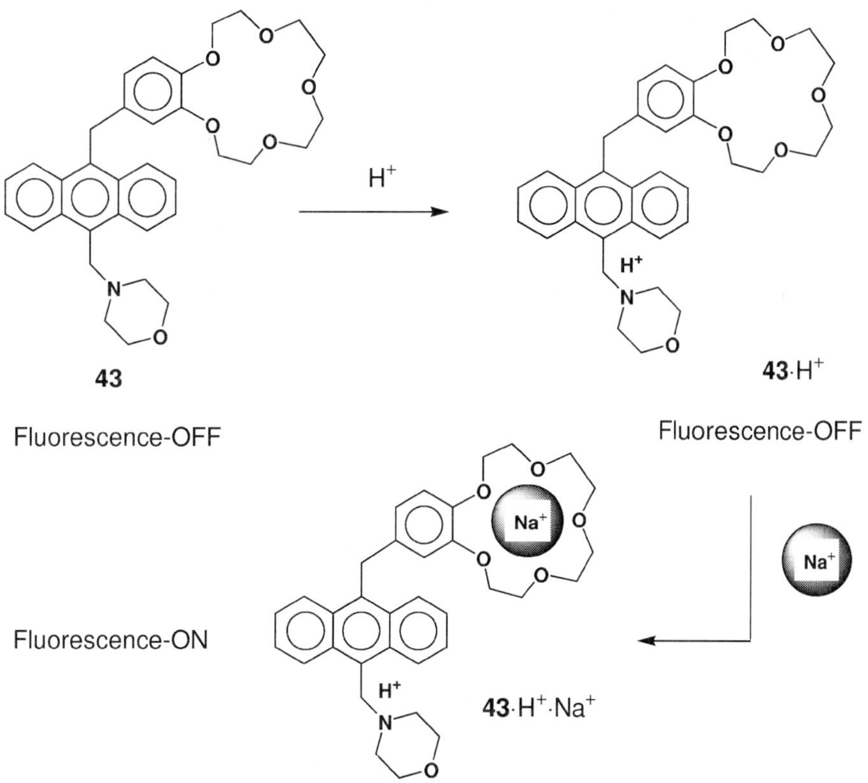

Fig. 32. The improved molecular AND logic gate **43** with greater fluorescence enhancement.

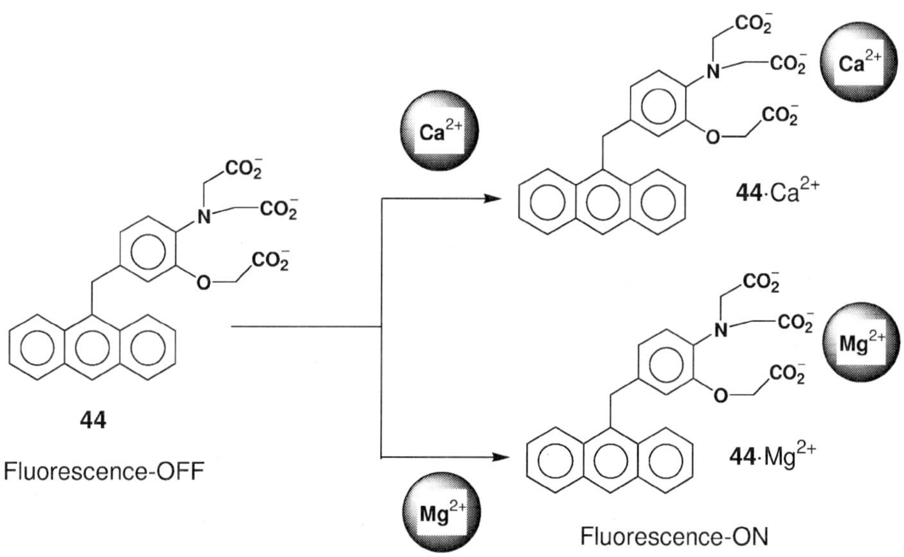

Fig. 33. The fluorescent signaling system **44**, behaving as a photonic OR logic gate. The fluorescence of the anthracene unit is observed when either Ca^{2+} or Mg^{2+} metal ions are present.

is a charge transfer band at λ_{max} around 400 nm. This charge transfer interaction is responsible for the disappearance of the strong fluorescence exhibited by the two separated components—a λ_{max} at 432 nm for **37**.2PF$_6$ and a λ_{max} at 343 nm for **45**. Upon the addition of tributylamine (**46**), a 1:2 complex is formed (Fig. 34) between the dication **37**$^{2+}$ and

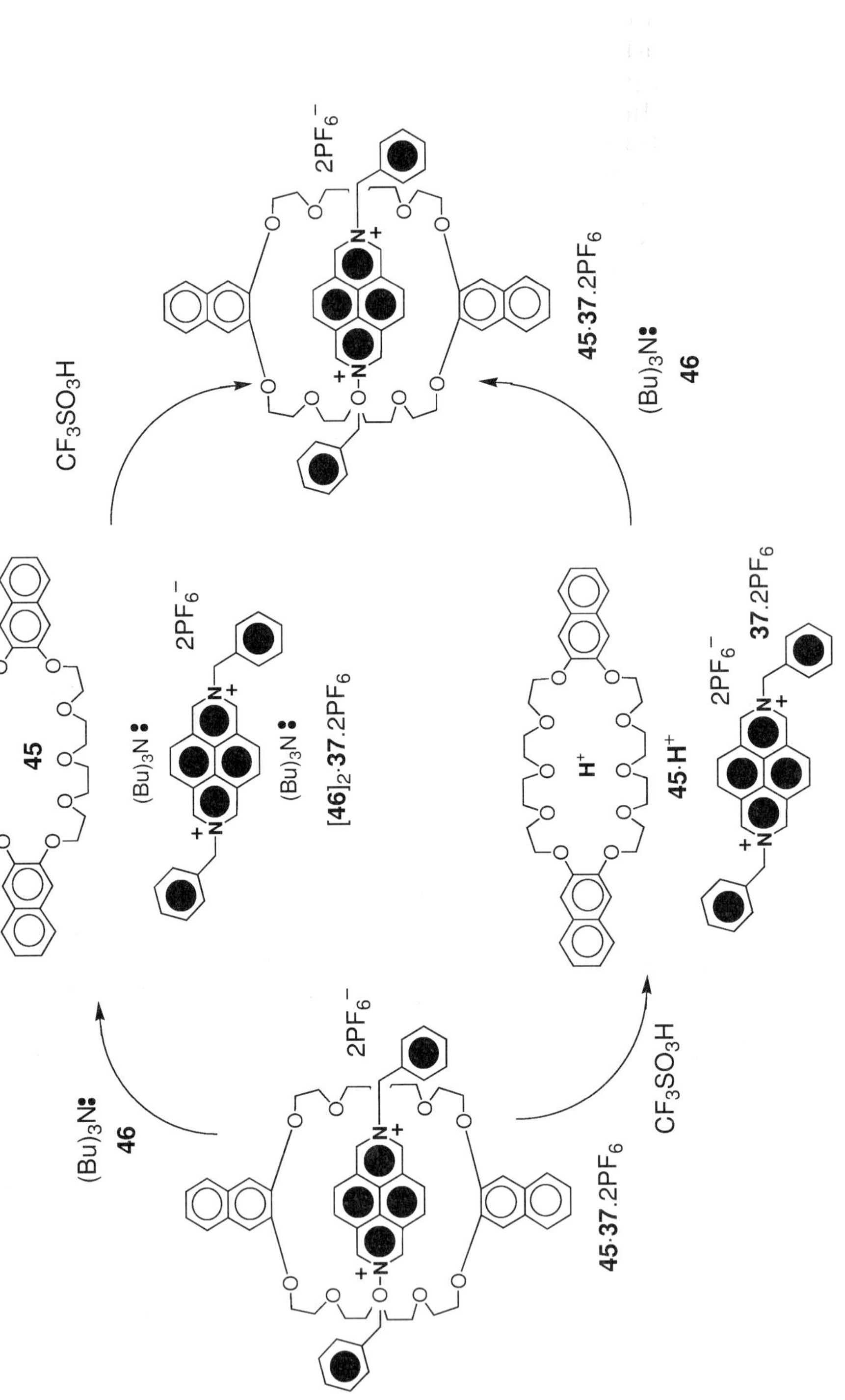

Fig. 34. A XOR logic gate based upon the complexation of the dication **37**.2PF$_6$ with the macrocyclic crown ether **45** and the amine **46**.

the amine **46**, with the consequent unthreading of the pseudorotaxane. This process causes spectral changes associated for the most part with the recovery of the emission of the free **45** and the appearance of a broad absorption band characteristic of the $[\mathbf{46}]_2 \cdot \mathbf{37}.2PF_6$ complex. Subsequent addition of a stoichiometric amount of trifluoromethanesulfonic acid unlocks $\mathbf{37}.2PF_6$ and allows for rethreading of $\mathbf{37}.2PF_6$ through the cavity of crown ether **45**. The unthreading–rethreading cycle can also be performed (Fig. 34) by reversing the order of two inputs. The addition of trifluoromethanesulfonic acid causes protonation of the crown ether **45**, with the subsequent unthreading of the pseudorotaxane, a process that is accompanied by large spectral changes, including the appearance of an emission band for $\mathbf{45} \cdot H^+$ (which is identical to that of **45**), the recovery of the emission band for $\mathbf{37}.2PF_6$, and the disappearance of a CT band between $\mathbf{37}.2PF_6$ and **45**. The addition of a stoichiometric amount of tributyl amine deprotonates the crown ether and permits the rethreading of $\mathbf{37}.2PF_6$ through the cavity of **45**. Both processes cause a strong increase in emission intensity at 343 nm, and when both inputs (acid and base) are present, the fluorescence output is switched off altogether. This supramolecular system exhibits the input–output relationship associated with the truth table of a XOR logic gate. The fluorescence at 343 nm only occurs when either amine **46** or H^+ are present: on the other hand, the fluorescent signal is absent when none or both of the inputs are present.

4. INTERLOCKED MOLECULAR SYSTEMS

Once the noncovalent bonding interactions, which control the formation of pseudorotaxane-like complexes, have been established, efficient template-directed syntheses of novel interlocked molecular compounds become a reality. The main families of interlocked molecular compounds that have been self-assembled so far include the so-called catenanes and rotaxanes.

4.1. Switching Properties in Catenanes

Catenanes are composed of two or more interlocked rings that are joined together, not by covalent bonds, but rather by mechanical bonds. The first catenane, which exploits the molecular recognition between π-electron rich and -deficient aromatic units, was self-assembled by our research group, which at the time (1989) was located in Sheffield [70]. The crown ether **BPP34C10** (**47**) was used as a template for the cyclization of cyclobis(paraquat-p-phenylene) around the π-electron rich hydroquinone rings of **BBP34C10**. This reaction produced (Fig. 35) the [2]catenane $\mathbf{48}.4PF_6$ in 70% yield when one molar equivalent of the salt $\mathbf{49}.2PF_6$ was reacted with 1,4-bis(bromomethyl)benzene in the presence of three molar equivalents of **BPP34C10**. This yield is remarkable when one considers that three molecules are being brought together in the self-assembly steps to form the [2]catenane $\mathbf{48}.4PF_6$. As a result of this observation, it became possible to construct a large new family of molecular assemblies.

The next challenge was to proceed to the construction of higher order [n]catenanes. To address this goal, the macrocyclic polyether and tetracationic cyclophane components were enlarged, and hence the multistep self-assembly of a [5]catenane [71] via a key [3]catenane intermediate became possible. This [5]catenane was baptized "Olympiadane" because of its topological similarity to the logo used by the Olympics movement. The next challenge lies in the production of polycatenated compounds, which may possess interesting properties as novel polymeric materials.

A vital feature for mechanical switching properties of the [2]catenanes is that they possess [52] dynamic properties associated with the circumrotation of the two macrocyclic components with respect to one another (Fig. 36). In the case of a desymmetrized [2]catenane, this relative motion gives rise to two different so-called translational isomers [72] (if

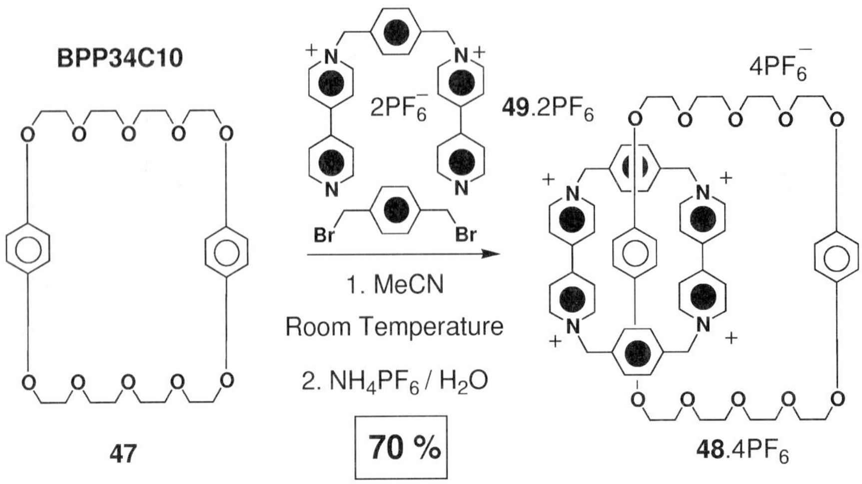

Fig. 35. The self-assembly in 70% yield of the [2]catenane **48**.$4PF_6$.

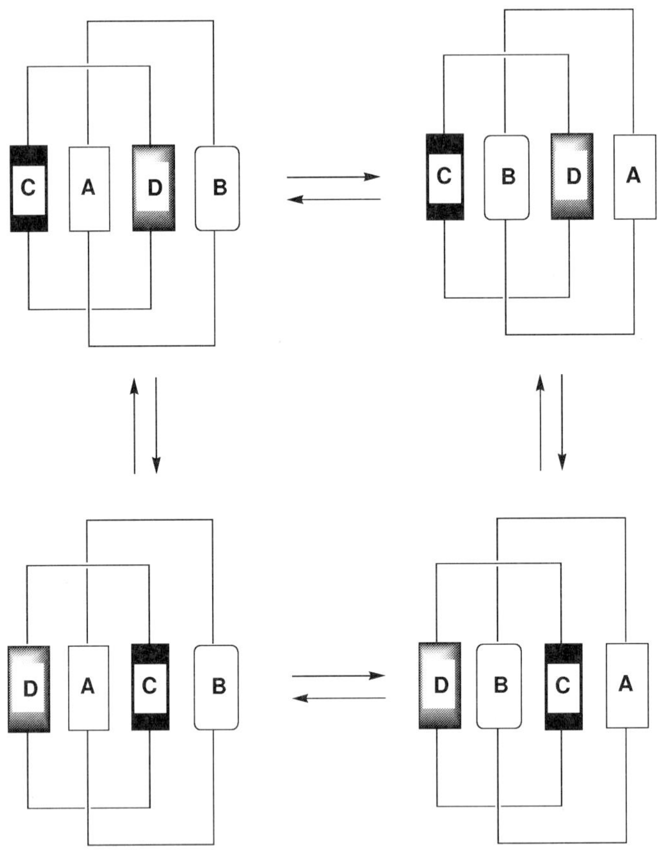

Fig. 36. Cartoon representation of the four possible isomers in a desymmetrized [2]catenane.

only one of the interlocked rings is desymmetrized) or four different translational isomers (if both ring components are desymmetrized). Recently the term "coconformation" [73] was coined to describe the different relative spatial arrangements of the components in molecular assemblies and the building blocks in supramolecular arrays. (Strictly speaking, the term "conformation" refers to those arrangements of atoms in space that differ only as a result of rotation about a single bond in a discrete molecular species. The term "coconformation" is used to designate the relative three-dimensional spatial arrangements of (i) the constituent parts (e.g., host and guest) in supramolecular systems, such as precatenanes and pseudorotaxanes, and of (ii) the components of interlocked molecular systems, such as catenanes and rotaxanes. See [73].) If the movement of the rings can be controlled by an appropiate external stimulus and the different states can be easily addressed by spectroscopic techniques, then this translational isomerism is certainly a phenomenon that can be exploited in the fabrication of molecular compounds with switching properties.

4.1.1. A Solvent-Dependent [2]Catenane

The unsymmetrical [2]catenane **50**.4PF$_6$ comprises [74] two different π-electron-rich units in its crown ether component—that is, a hydroquinone ring and a 1,5-dioxynaphthalene ring system (Fig. 37). In the supramolecular context, the cyclobis(paraquat-*p*-phenylene) exhibits a much larger binding affinity toward 1,5-dioxynaphthalene derivatives than toward 1,4-dioxybenzene derivatives. However, in many cases—and this one is no exception—it is not possible to extrapolate from the supramolecular world into the molecular one.

^{1}H NMR spectroscopic studies reveal that, in CD_3SOCD_3, the tetracationic cyclophane component encircles the 1,5-dioxynaphthalene unit preferentially in **50**.4PF$_6$. The ratio between the two translational isomers is 70:30 in favor of translational isomer 2. On the contrary, when the [2]catenane **50**.4PF$_6$ is dissolved in CD_3COCD_3, the cyclobis(paraquat-*p*-phenylene) tetracation encircles the hydroquinone ring preferentially—the isomer ratio being 70:30 in favor of translational isomer 1 this time (Fig. 37).

This behavior can be explained in the following way. In solvents with low dielectric constants, such as CD_3COCD_3, the need for the 1,5-dioxynaphthalene ring system to act as a shield "alongside" the positively charged bipyridinium units is more important than the fact that the tetracationic cyclophane possesses a much higher binding affinity for the 1,5-dioxynaphthalene ring system located "inside" than for the hydroquinone ring in the analogous pseudorotaxane-like complexes. In more polar solvents, such as CD_3SOCD_3, the 1,5-dioxynaphthalene unit is not required to act as a shield, because the bipyridinium unit is solvated by the highly polar solvent molecules and is therefore able to reside "inside" the cavity of the cyclobis(paraquat-*p*-phenylene) tetracation. In summary, the position of the tetracationic cyclophane component can be controlled by the polarity of the

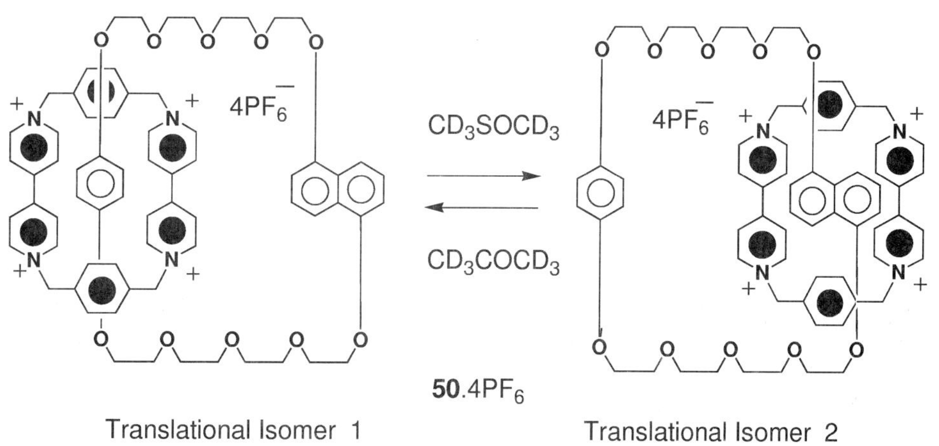

Fig. 37. The solvent-dependent translational isomerism in the [2]catenane **50**.4PF$_6$.

solvent, that is, the equilibrium proportions of the translational isomers are highly solvent dependent.

4.1.2. An Electrochemically Controlled [2]Catenane

Another way to exercise control over translational isomerism in [2]catenanes is by using π-extended viologen units as building blocks in the construction of the interlocked tetracationic cyclophane component. The [2]catenane **51**.4PF$_6$, in which the macrocyclic polyether component **BPP34C10** is encircled (Fig. 38) by an unsymmetrical tetracationic cyclophane containing one bipyridinium unit and one bis(pyridinium)ethylene unit, has been self-assembled [75]. ^{1}H NMR spectroscopic studies in CD_3COCD_3 at $-60\,°C$ revealed that the isomer in which the bipyridinium unit resides in the center of the neutral component becomes favored (Fig. 38) to the extent of 98:2. This selectivity is a consequence of the lower π-electron accepting ability of the π-extended viologen unit relative to that of a bipyridinium unit.

The [2]catenane **51**.4PF$_6$ shows four single-electron reduction waves [76]. The first reduction occurs at the bipyridinium unit with a potential very similar to that of the first reduction potential of the parent symmetrical [2]catenane **48**.4PF$_6$. This observation indicates that the bipyridinium unit resides "inside" the cavity of the crown ether component, as demonstrated by ^{1}H NMR spectroscopy. Next, there are two single-electron reductions that occur at the extended viologen unit—the bis(pyridinium)ethylene unit—before the final one reduces the bipyridinium radical cation. The first reduction wave associated with the extended viologen unit is very close to that observed in the analogous [2]catenane with two bis(bipyridinium)ethylene units present in its tetracationic cyclophane component. This result implies that, after a single-electron reduction, the translational isomerism is inverted in favor of the less π-electron deficient unit. The molecule is therefore switchable electrochemically. At the outset, the predominant isomer in solution is the one in which the bipyridinium unit is located within the cavity of the macrocyclic polyether. The first reduction occurs at the "inside" bipyridinium unit because of its lower reduction potential, converting it (Fig. 39) to a radical cation. After reduction, the "inside" radical cation is no longer a good π-electron acceptor, and so circumrotation of the cyclophane takes

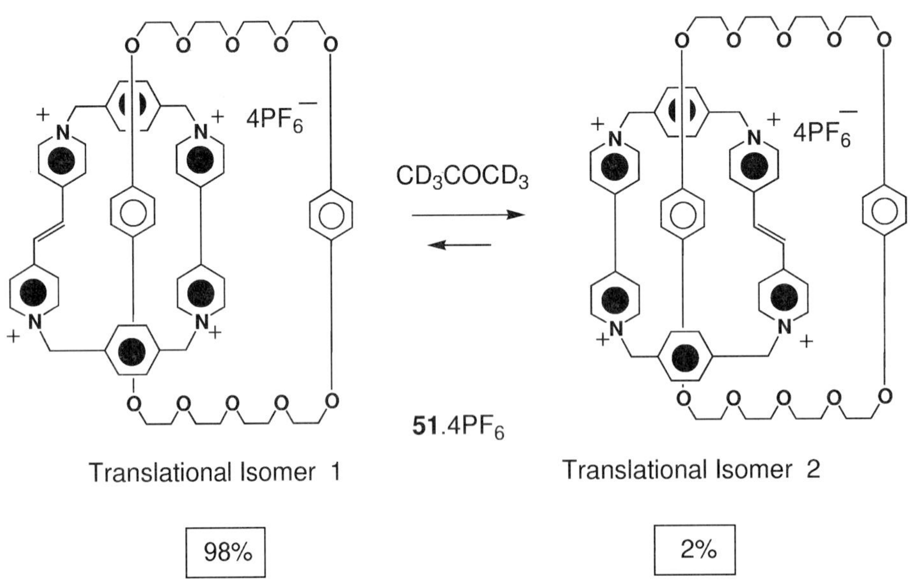

Fig. 38. The two translational isomers in the extended viologen-containing [2]rotaxane **51**.4PF$_6$.

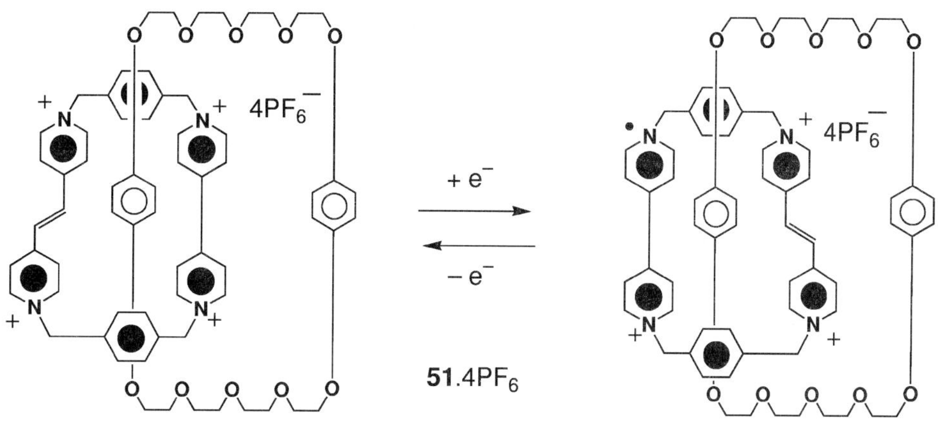

Fig. 39. Electrochemical switching properties of the [2]rotaxane **51**.$4PF_6$.

place (Fig. 39) to locate the extended viologen bis(pyridinium)ethylene unit "inside" the crown ether component. The predominant species is now related to translational isomer 2. The process is reversible. Electrochemical oxidation regenerates the bipyridinium unit and induces circumrotation of the tetractionic cyclophane to relocate the bipyridinium unit "inside" the neutral crown ether component. It may be concluded that, in this unsymmetrical [2]catenane, the components can be controlled electrochemically in a reversible fashion, conferring upon the [2]catenane **51**.$4PF_6$ molecular switching properties.

4.1.3. A Solvent-Dependent Amphiphilic [2]Catenane

The use of complementary hydrogen bonding interactions between benzylic amides to afford a new family of [2]catenanes was first described by Hunter [77]. This approach to catenane formation has been extended considerably by Vögtle [78] and Leigh [79] in recent years. Leigh et al. [80] have reported the solvent-dependent coconformational behavior of some amphiphilic benzylic amide [2]catenanes (Fig. 40). The homo[2]catenane **52** is composed of two identical interlocked rings that have a lipophilic linear alkyl chain portion and a benzylic diamide portion. In $CDCl_3$, the lipophilic sebacoyl chains are located on the surface of the molecule, and the hydrogen bonding groups are buried in the middle of the molecular structure and are involved in inter-ring intramolecular hydrogen bonding, a feature that causes a downfield shift in the signal for the amide protons by 1.5 ppm. The network of hydrogen bonds locks the aromatic rings in close proximity to each other (see translational isomer 1). On the other hand, in CD_3SOCD_3, a hydrogen-bond-accepting

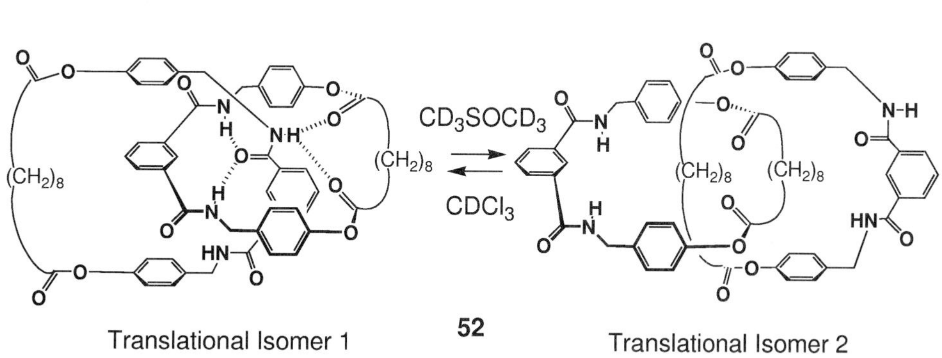

Fig. 40. The solvent-dependent translational isomerism in the [2]catenane **52**.

solvent, the ^{1}H NMR spectrum suggests that the aromatic rings spend very little time close to each other. Moreover, the fact that the protons on the sebacoyl chain are heavily shielded is consistent with a structure in which the lipophilic chains are buried in the center of the catenane, whereas the amide groups are exposed so that they interact with the polar solvent, that is, translational isomer 2 is the predominant isomer in CD_3SOCD_3 solution.

4.1.4. Electrochemically Controlled Circumrotation of a [2]Catenane

Sauvage and his group [81] were first to achieve the efficient synthesis of a [2]catenate by exploiting the tetrahedral coordination of Cu^+ with 1,10-phenanthroline ligands. Using this approach, the Strasbourg group has managed to construct an impressive series of catenanes, polycatenanes, and rotaxanes, as well as a topologically fascinating molecular trefoil knot [82].

The dissymmetric [2]catenane **53**.PF_6 consists [83] of a transition metal (Cu) complex whose organic components consist of two interlocked coordinating rings. The interconversion between the two forms of the complex can be triggered electrochemically and corresponds to a sliding motion of one ring with respect to the other ring. The fundamental reason for the process is the difference in coordination number (CN) for the two different redox states of the metal: CN = 4 for Cu(I) and CN = 5 for Cu(II). The organic backbone of the [2]catenane **53**.PF_6 has a 2,9-diphenyl-1,10-phenanthroline (dpp) bidentate chelate located in one ring, which is interlocked with another ring containing two different subunits—a dpp moiety and a tridentate ligand 2,2′,6,2″-terpyridine (terpy). Depending on the relative positions of the interlocked rings, the central metal atom can be tetrahedrally complexed (two dpps), or it can be 5-coordinated (dpp + terpy) . The interconversion between these two states, which results from the circumrotation of one ring with respect to the other, can be induced by an external stimulus.

At the outset, Cu(I) is 4-coordinate (two dpps), electrochemical oxidation to Cu(II) leads (Fig. 41) to an unstable 4-coordinate Cu(II) complex, which then rearranges to the more stable 5-coordinate state (dpp + terpy). The process can be reversed by reducing Cu(II) back to Cu(I). The reduction gives rise to a transient 5-coordinate Cu(I) species, before circumrotation of the rings takes place with regeneration of the 4-coordinate Cu(I). The switching process can be monitored by UV/vis spectroscopy, because all of the four possible species have different spectroscopic properties. Therefore, by exploiting the coordination properties of two different redox states of a transition metal within a catenate environment, a reversible switching system has been constructed.

4.1.5. A Switchable Hybrid [2]Catenane

A novel hybrid [2]catenane **54**.5PF_6, in which the cation complexing abilities of the dpp ligand and the noncovalent bonding interactions between π-electron-rich and -deficient aromatic units are combined, has been synthesized [84]. This hybrid interlocked molecular compound has two different modes of interaction that govern the relative geometries of the component macrocycles. These different modes can be interchanged by the inclusion or exclusion of a cation that is chelated by the catenane. If the self-assembly of the [2]catenate takes place in the presence of Cu(I)—employed as a metal ion template—the topology of **54**.5PF_6 is the one shown in Figure 42, in which the Cu(I) ion is complexed tetrahedrally between the two dpp units. However, upon the addition of KCN, the Cu(I) ion is released and circumrotation of the rings in **54**.4PF_6 induces a change in topology with the π-electron-rich 1,5-dioxynaphthalene unit sandwiched between the two bipyridinium units of the other ring. The topology of the molecule is now controlled by the noncovalent bonding interactions between the 1,5-dioxynaphthalene unit and the bipyridinium units. Addition of $LiBF_4$ to the solution causes the disappearance of the charge transfer band associated with the noncovalent bonding interactions between the π-electron-rich and

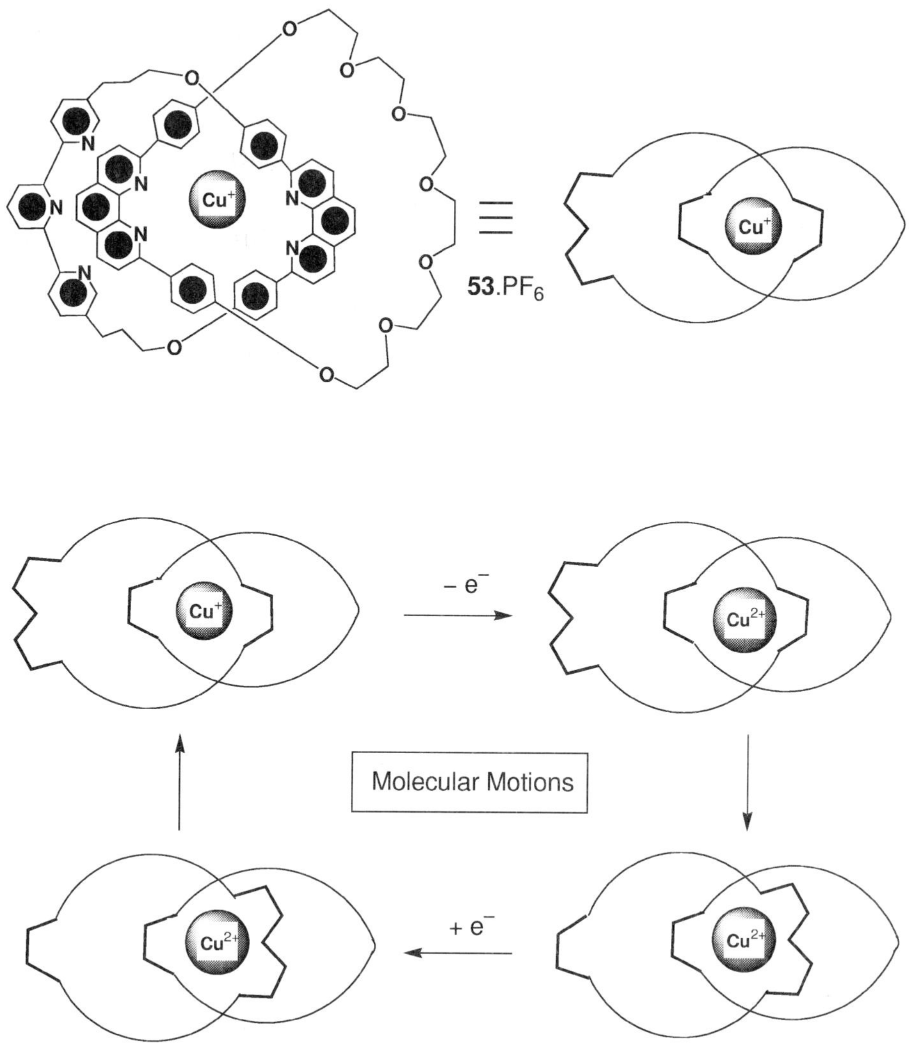

Fig. 41. The electrochemically driven circumrotation in the Cu(I)-containing [2]catenane **53**.PF_6.

-deficient aromatic units. The Li^+ ion can form a tetrahedral complex with the dpp ligands, inducing a coconformational change within the [2]catenane. Unfortunately, the Li^+ ion could not be removed from the catenane. Hence, although it was possible to switch from the π-electron donor–acceptor mode of complexation to the metal ion mode, the reverse process is not easy to control in this case.

Despite this failure, the behavior of catenane **54**.4PF_6 was studied in acidic and basic media. In the beginning, **54**.4PF_6 possesses a topology in which the π-electron donor–acceptor interactions dominate (Fig. 43). The addition of TFA to a MeCN solution induces large changes in the ^{1}H NMR spectrum, which are consistent with the disappearance of the π-electron donor–acceptor mode of complexation, which is accompanied by the suppression of the charge transfer band. The topology of the catenane corresponds now to one in which an H^+ ion is surrounded by the two dpp ligands. In this respect, the H^+ ion is functioning in a manner very similar to that of a transition metal ion. Upon the addition of a base, such as pyridine, the original ^{1}H NMR and UV/vis spectra are reconstructed, indicating complete reversion (Fig. 43) to the π-electron donor–acceptor mode of complexation. It is clear that the [2]catenane **54**.4PF_6, which was first synthesized by a metal

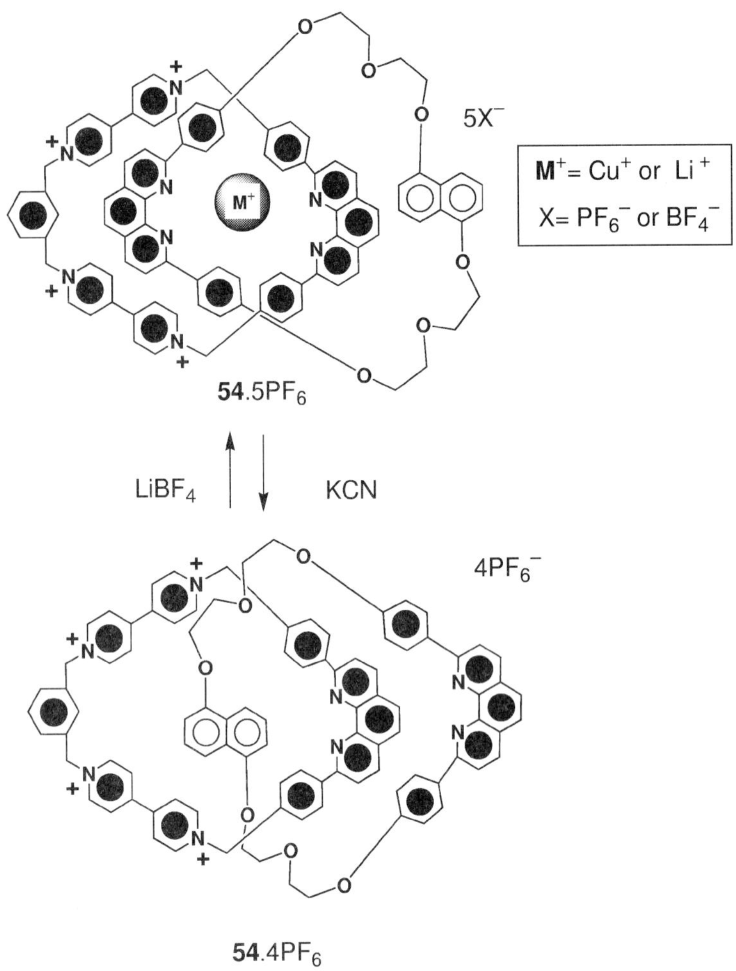

Fig. 42. The chemically induced change of geometry in the hybrid [2]catenane **54**.$4PF_6$.

ion templated strategy, behaves as a pH-dependent molecular switch in which the system changes from a proton catenate to a π-electron donor–acceptor mode of complexation in the [2]catenate.

4.1.6. A Chemically and Electrochemically-Switchable [2]Catenane

The [2]catenane **55**.$4PF_6$ is composed of a crown ether component comprising two π-electron-rich units—an isomeric *cis–trans* mixture of tetrathiafulvalene (TTF) and a 1,5-dioxynaphthalene ring system—and the π-electron deficient cyclobis(paraquat-*p*-phenylene) cyclophane tetracation. X-ray crystallography in solid state and NMR spectroscopy and UV/vis spectroscopy in solution revealed that the TTF unit is positioned (Fig. 44) "inside" the cavity of the tetracationic cyclophane, with the 1,5-dioxynaphthalene ring system lying "alongside" one of the bipyridinium units. Chemical oxidation of the TTF unit with $Fe(ClO_4)_3$ causes dramatic changes in both the ^1H NMR and UV/vis spectra. After oxidation, the resonances of the H-4 and H-8 protons of the 1,5-dioxynaphthalene ring are shifted to higher fields (δ 2.45), indicating that the 1,5-dioxynaphthalene moiety is now encircled by the tetracationic cyclophane. Reduction with $Na_2S_2O_5$ regenerated the original NMR spectrum, thus rendering the cycle reversible. Similarly, UV/vis experiments show a dramatic shift in the maximum of absorbance of the charge transfer band from

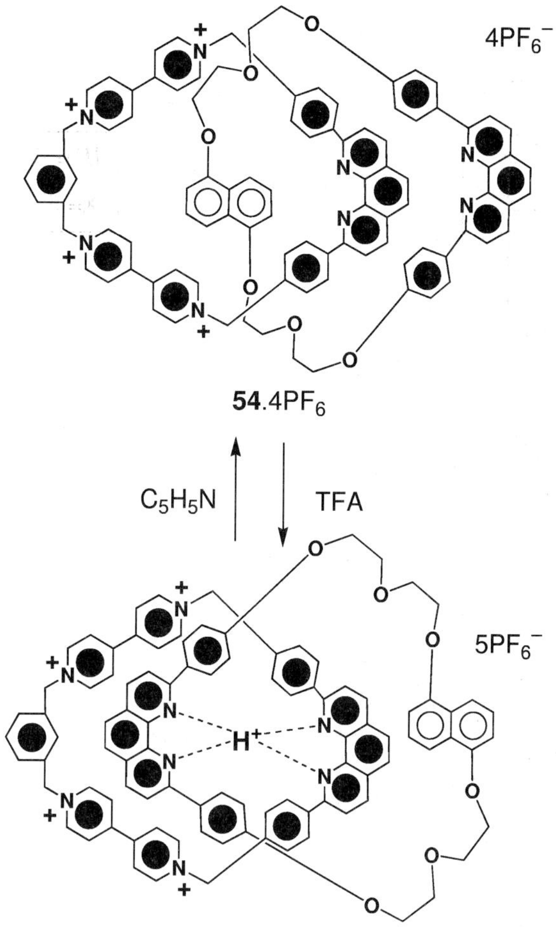

54.4PF$_6$

54.5PF$_6$

Fig. 43. A pH-controllable molecular switch.

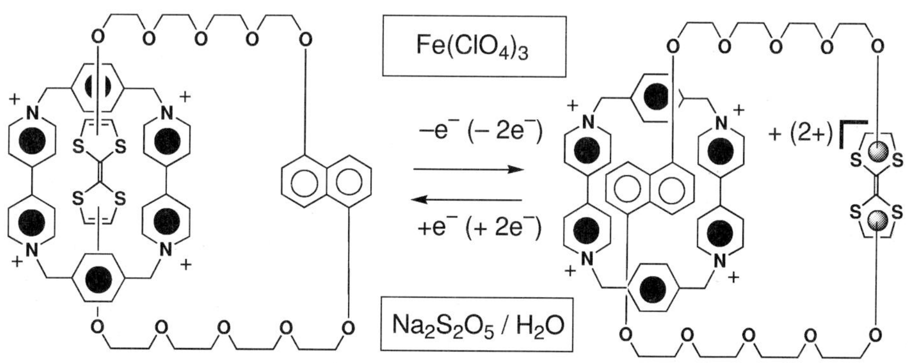

55^{4+} Dark Green

55^{5+} (**55**$^{6+}$) Maroon

Fig. 44. The TTF-containing [2]catenane **55**.4PF$_6$. Note that it behaves as a chemically and electrochemically driven molecular switch.

$\lambda_{max} = 835$ nm (when the TTF unit is located inside the cyclophane) to $\lambda_{max} = 515$ nm (when the 1,5-dioxynaphthalene ring system is located inside the cyclophane).

The switching can also be achieved by electrochemical means. One-electron electrochemical oxidation of the TTF unit forces the positively charged $TTF^{+\bullet}$ unit out of the cavity of the cyclophane, compelling the circumrotation of the macrocyclic polyether with respect to the tetracationic cyclophane so that the 1,5-dioxynaphthalene ring is located inside the cyclophane. This process can be monitored by cyclic voltammetry, and it can be made reversible.

4.2. An Electrochemically Controlled Self-Complexing Macrocycle

Recently we self-assembled [85] a series of molecules that display self-complexing geometries. These types of molecules consist of a linear component comprising a π-electron-rich unit—that is, 1,5-dioxynaphthalene, which is linked covalently to the π-electron-deficient tetracationic cyclophane by means of an ester linkage. The π-electron-rich aromatic unit lies inside the cavity of the cyclophane, giving rise to a molecule that exhibits an exotic self-complexing topology. One example of this type of molecule is the self-complexing macrocycle **56**.4PF_6, which is also a molecular analogue of the supramolecular system already discussed in Section 3.8.1. Therefore, one can expect **56**.4PF_6 to behave as a compound that exhibits switching properties at the molecular level.

A photochemical experiment similar to that performed on **32·31**.4PF_6 was carried out on **56**.4PF_6. However, after photoinduced reduction of the bipyridinium units, no luminescence could be observed emanating from the free 1,5-dioxynaphthalene ring system. This lack of emission does not rule out the possibility that decomplexation has occurred. It is well known that the emission from a 1,5-dioxynaphthalene ring system can be quenched by the presence of a nearby bipyridinium unit. In the case of the supramolecular system **32·31**.4PF_6, once decomplexation has taken place, the 1,5-dioxynaphthalene-containing thread **32** can diffuse away from the bipyridinium-containing tetracationic cyclophane. On the other hand, such a motion is not possible in **56**.4PF_6 because the 1,5-dioxynaphthalene- and the bipyridinium-containing components are linked together covalently.

Nevertheless, the self-complexing macrocycle **56**.4PF_6 acts as an electrochemically driven molecular machine. After the first two-electron reduction of **56**.4PF_6 to the corresponding diradical dication, the noncovalent bonding interactions responsible for the self-complexing topology are disrupted, and the 1,5-dioxynaphthalene moiety is ejected (Fig. 45) from within the cavity of the cyclophane. This fact is supported by the value of the second two-electron reduction potential of the bipyridinium units in **56**.4PF_6, which is

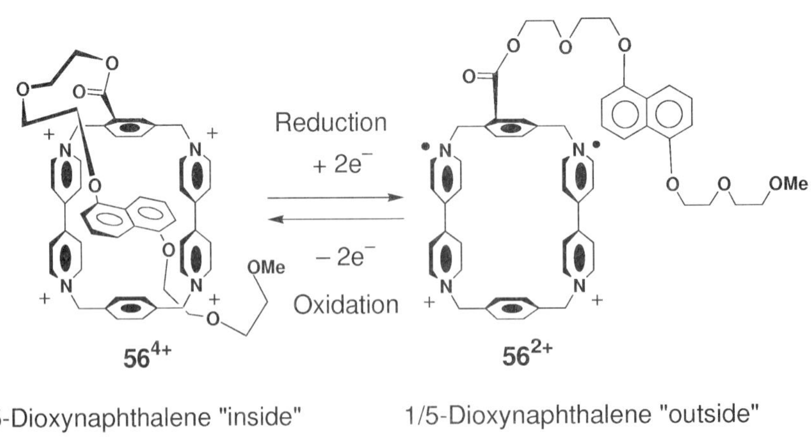

Fig. 45. Electrochemically controlled self-complexation in the self-complexing macrocycle **56**.4PF_6.

the same as that for the last reduction potential for the "empty" parent tetracationic cyclophane **31**.4PF$_6$, indicating that, in this instance, the 1,5-dioxynaphthalene does not reside inside the cavity of the cyclophane. The process can be reversed by electrochemical or chemical oxidation to the tetracationic cyclophane.

4.3. Rotaxanes: From Molecular Shuttles to Molecular Switches

Another class of mechanically interlocked molecules, which may be self-assembled by employing the molecular recognition that exists between the π-electron-rich and -deficient components, are the so-called rotaxanes [86]. A rotaxane consists of a ring component (Fig. 46), which can be clipped, slipped, or threaded onto a dumbbell-shaped component—and a rod with large stoppers at both its termini that prevent the ring component from slipping off the dumbbell-shaped component. In many cases, this component is a linear polyether chain interrupted by π-electron-rich aromatic units, and the ring component is the cyclobis(paraquat-p-phenylene) tetracation.

Just like the catenanes, rotaxanes also exhibit translational isomerism. Not surprisingly perhaps, rotaxanes, as a result of their abacus-like geometries, have become one of the most popular starting points in the search for molecular switches.

4.3.1. A Molecular Shuttle

In 1991, the first molecular shuttle, **57**.4PF$_6$, was reported [87]. It is a [2]rotaxane (Fig. 47) consisting of a dumbbell-shaped component possessing two hydroquinone rings separated by polyether chains (a linear analogue of **BPP34C10**) and terminated by two bulky triisopropylsilyl groups as stoppers and encircled by the cyclobis(paraquat-p-phenylene) tetracation. Dynamic ^{1}H NMR spectroscopic studies in CD_3COCD_3 revealed that the tetracationic cyclophane component shuttles back and forth (Fig. 47) between the two degenerated hydroquinone rings approximately 500 times a second at room temperature.

Because the movement of the cyclic component cannot be controlled, the molecular shuttle **57**.4PF$_6$ does not possess molecular switching properties. However, it prepares the way for the construction of abacus-like controllable molecular shuttles (Fig. 48). By providing two different nondegenerate, π-electron-rich recognition sites on the dumbbell-shaped component in such a manner that they can be addressed selectively by an external stimulus, which can be chemical, electrochemical, or photochemical in nature, the resulting translational isomerism can be exploited in a molecular switching sense.

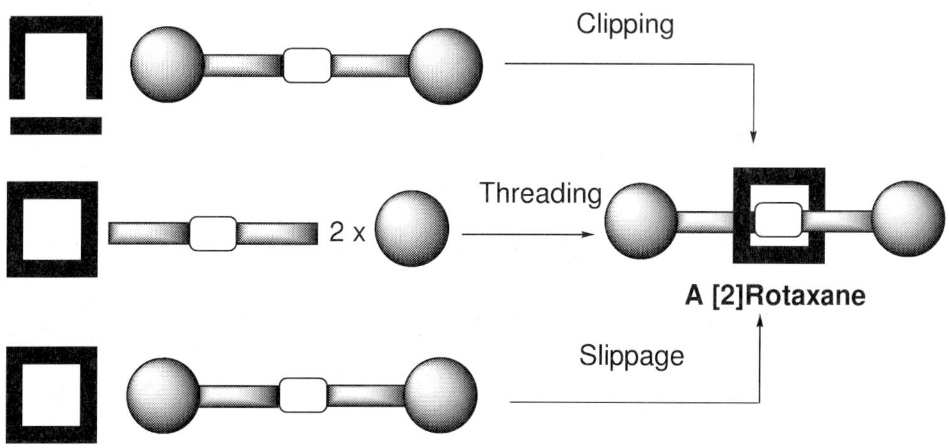

Fig. 46. Cartoon representation of the different modes of self-assembling a [2]rotaxane.

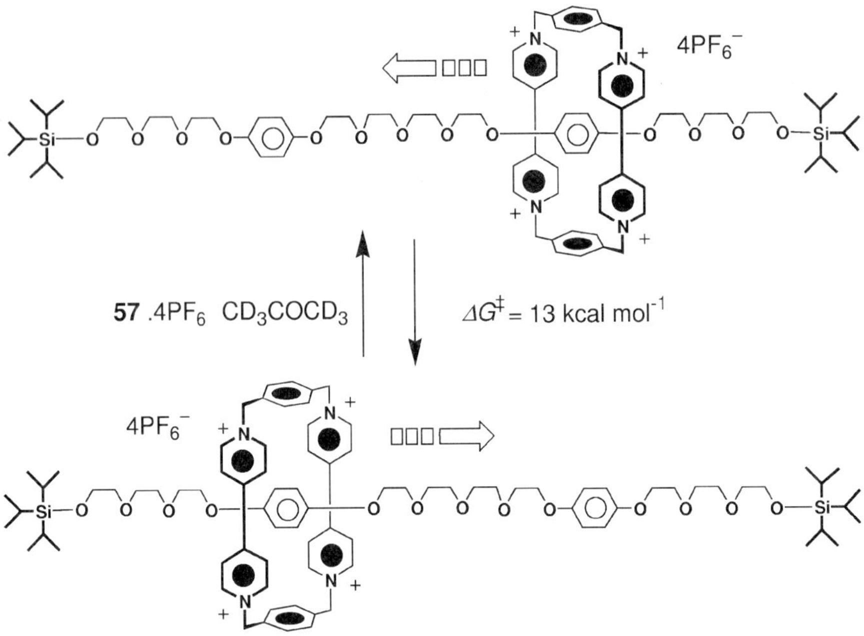

Fig. 47. The original molecular shuttle, **57**.$4PF_6$.

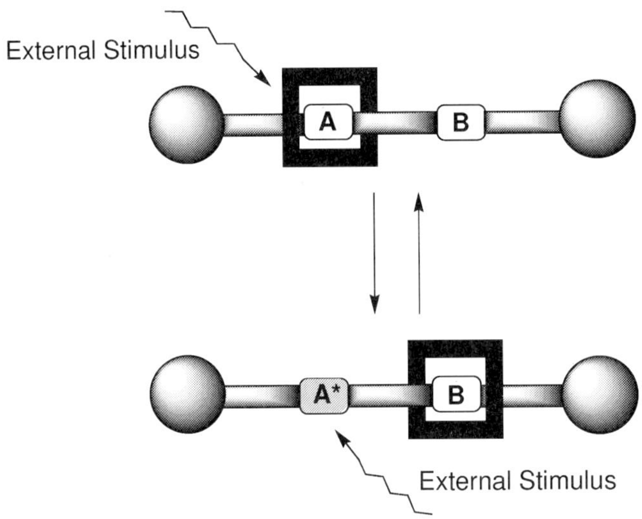

Fig. 48. A cartoon representation of the design logic for a desymmetrized [2]rotaxane with binary switching properties.

4.3.2. Nondegenerate Molecular Shuttles

A series of desymmetrized molecular shuttles were designed and self-assembled that comprise two nondegenerate π-electron rich recognition sites—one always being a hydroquinone ring—in the dumbbell-shaped component.

The [2]rotaxane **58**.$4PF_6$ possesses [88] a p-xylene unit and a hydroquinone ring as its recognition sites in the dumbbell-shaped component. The ^1H NMR spectrum of the molecular shuttle **58**.$4PF_6$ in CD_3COCD_3 reveals (Fig. 49) that the tetracationic cyclophane component distributes itself between the two recognition sites with 70% occupancy

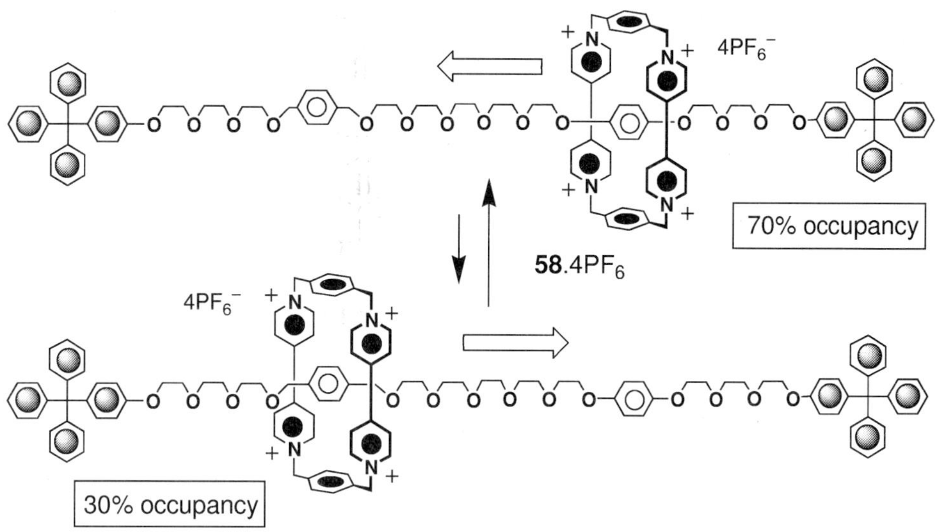

Fig. 49. The desymmetrized molecular shuttle **58**.4PF$_6$ comprising two nondegenerate stations, a hydroquinone ring and a *p*-xylene moiety.

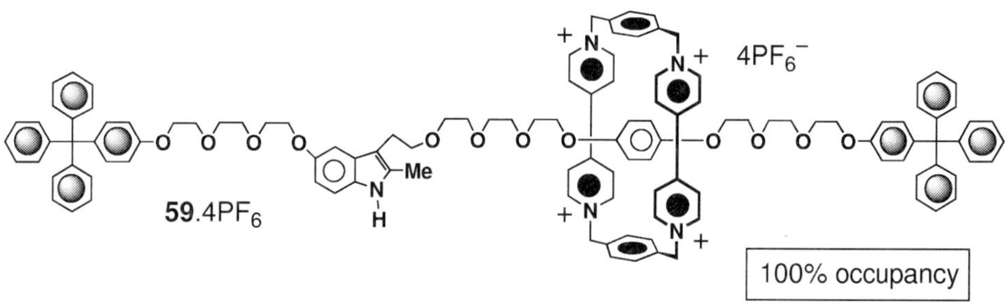

Fig. 50. The desymmetrized [2]rotaxane **59**.4PF$_6$ comprising two nondegenerate stations, an indole ring and a hydroquinone moiety. The tetracationic cyclophane resides exclusively on the hydroquinone station.

at the more π-electron-donating hydroquinone ring and 30% occupancy at the less π-electron-donating *p*-xylene unit.

The [2]rotaxane **59**.4PF$_6$ comprises [89] (Fig. 50) an indole residue and a hydroquinone ring as its recognition sites in the dumbbell-shaped component. However, in this case, the design logic failed and ^{1}H NMR spectroscopy showed that the tetracationic cyclophane component is located exclusively at one recognition site—namely, the hydroquinone ring—and so it is unable to exhibit molecular shuttling properties. Although the indole residue is more π-electron-rich than the hydroquinone ring, its inclusion inside the cavity of the cyclophane is much more sterically demanding. Thus, the hydroquinone ring is included preferentially within the cavity of the tetracationic cyclophane.

4.3.3. A Solvent-Controllable [2]Rotaxane

Another molecular shuttle that was investigated by us [90] is the [2]rotaxane **60**.4PF$_6$. It comprises (Fig. 51) two degenerate hydroquinone rings and a tetrathiafulvalene (TTF) unit as the π-electron-rich recognition sites present in the dumbbell-shaped component. This [2]rotaxane adopts different coconformations, depending on the nature of the solvent. ^{1}H NMR spectroscopic studies show that, in CD_3COCD_3, the tetracationic cyclophane

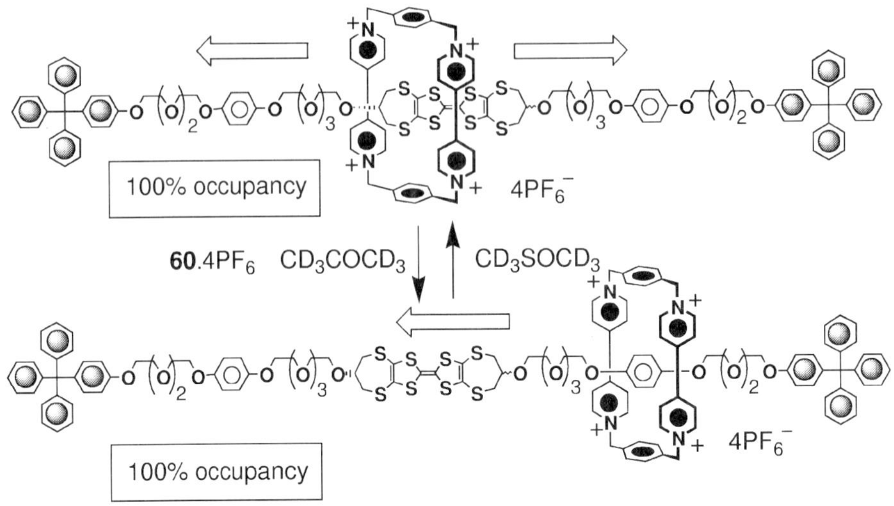

Fig. 51. The desymmetrized molecular shuttle **60**.4PF$_6$ containing a TTF unit sandwiched by two hydroquinone stations. This molecular shuttle possesses solvent-dependent switching properties.

occupies exclusively the two degenerate hydroquinone rings, whereas, in CD_3SOCD_3, the cyclophane is located exclusively on the TTF unit.

In principle, the proportions of the translational isomers populated in **60**.4PF$_6$ can be controlled by changing the nature of the solvent. However, none of the [2]rotaxanes described in this section have the potential of becoming useful as materials that will exhibit molecular switching properties, because the external stimulus that might control the translational isomerism is a change of solvent, which implies evaporation of one solvent followed by the addition of another one.

4.3.4. A Chemically and Electrochemically Controlled Molecular Shuttle

In view of the less than satisfactory results obtained with the [2]rotaxanes **58**–**60**.4PF$_6$, the design logic was changed [91] to produce **61**.4PF$_6$, which has turned out to be the most successful π-acceptor-donor rotaxane so far in terms of its molecular switching properties. The dumbbell-shaped component in **61**.4PF$_6$ consists (Fig. 52) of two non-degenerate π-electron-rich recognition sites: a benzidine unit and a biphenol one. Dynamic ^{1}H NMR spectroscopic studies in CD_3CN show that the tetracationic cyclophane resides predominantly on the more π-electron-rich benzidine residue. Nonetheless, this preference can be addressed (Fig. 52) by one of two external stimuli as follows: (i) The benzidine ring can be oxidized electrochemically and reversibly to the corresponding radical cation, whereupon the positively charged tetracationic cyclophane is forced to migrate to the neutral π-electron-rich biphenol ring system as a result of the repulsive electrostatic interactions from the benzidine radical cation. When the benzidine radical cation is reduced back to the neutral benzidine unit, the tetracationic cyclophane forsakes the biphenol ring system to occupy the more π-electron-donating benzidine unit once again, rendering the cycle reversible. (ii) Alternatively, the basic nitrogen atoms of the benzidine unit can be protonated by treatment with acid, resulting in the expulsion of the tetracationic cyclophane from the fully protonated benzidine unit, thus obliging the cyclophane to reside on the biphenol unit. Because, treatment with base regenerates the neutral π-electron-donating benzidine unit, the molecular switching process is reversible. It may be concluded that the [2]rotaxane **61**.4PF$_6$ behaves as a prototype binary molecular switch. This step represents the first in

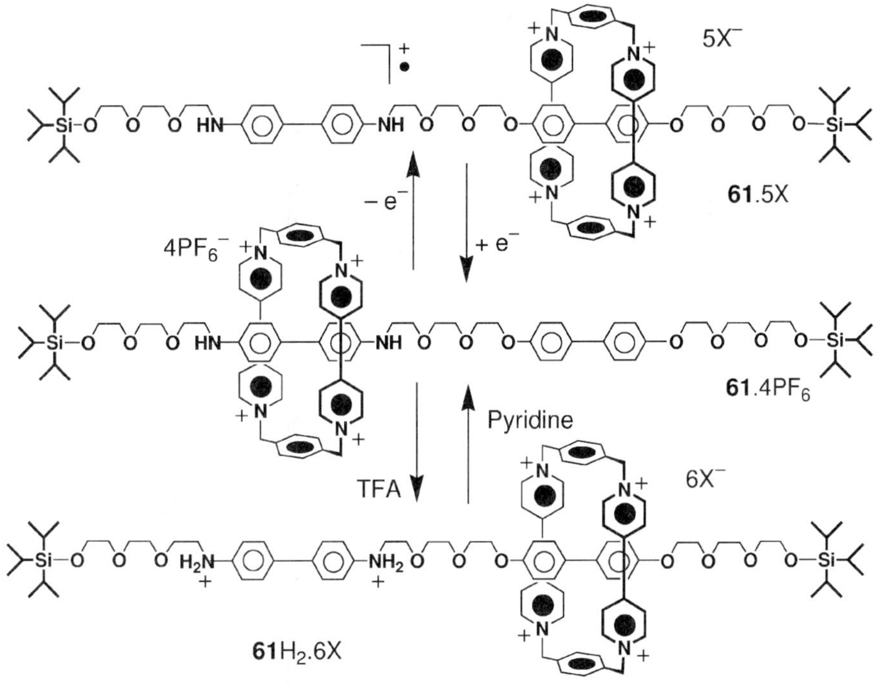

Fig. 52. The chemically and electrochemically controllable molecular shuttle **61**.4PF$_6$, comprising benzidine and biphenol stations.

the construction of more intricate molecular assemblies with the ability to receive, store, transfer, and transmit information in a controllable manner.

4.3.5. A pH-Controllable Molecular Shuttle

By exploiting the molecular recognition between secondary dialkylammonium ions and suitably sized crown ethers, it has been possible [92] to design and self-assemble the [2]rotaxane **62**-H.3PF$_6$. This [2]rotaxane is composed of a dumbbell-shaped component, comprising a secondary NH_2^+ center and a π-electron-deficient bipyridinium unit, together with a crown ether, **DB24C8**, as the ring component (Fig. 53). It is well known that **DB24C8** has a considerable affinity for secondary dialkylammonium ions: on the other hand, it enters into only rather weak interactions with bipyridinium units. Hence, at the outset, the crown ether ring resides exclusively on the NH_2^+ center, as indicated by ^{1}H NMR spectroscopy. Addition of a suitable base (the choice of base was not trivial, because bipyridinium units are very labile to nucleophilic bases), Hünig's base, deprotonates the NH_2^+ center to give rise to the amine-containing [2]rotaxane **62**.2PF$_6$. Thus, the noncovalent interactions that brought together the crown ether and the NH_2^+ center have been destroyed, and the crown ether moves toward the bipyridinium unit, forming a weak association with it. The appearance of a new charge transfer interaction can be monitored by the change in color of the solution from colorless to weak orange. The switching of the macrocycle from the NH_2^+ center to the bipyridinium unit can be also monitored by the dramatic changes evident in the ^{1}H NMR spectrum. The process is reversible. Upon the addition of TFA, the acid reprotonates the amine group, forcing the **DB24C8** to return to the NH_2^+ center, thus providing **62**-H.3PF$_6$ with its initial topology.

This molecular shuttle behaves as a pH-dependent molecular switch. The main advantage of this system is that, at both low and high pHs, only one state is detected by ^{1}H NMR spectroscopic techniques. The shuttling process is 100% controllable in both directions: it is an on–off switch.

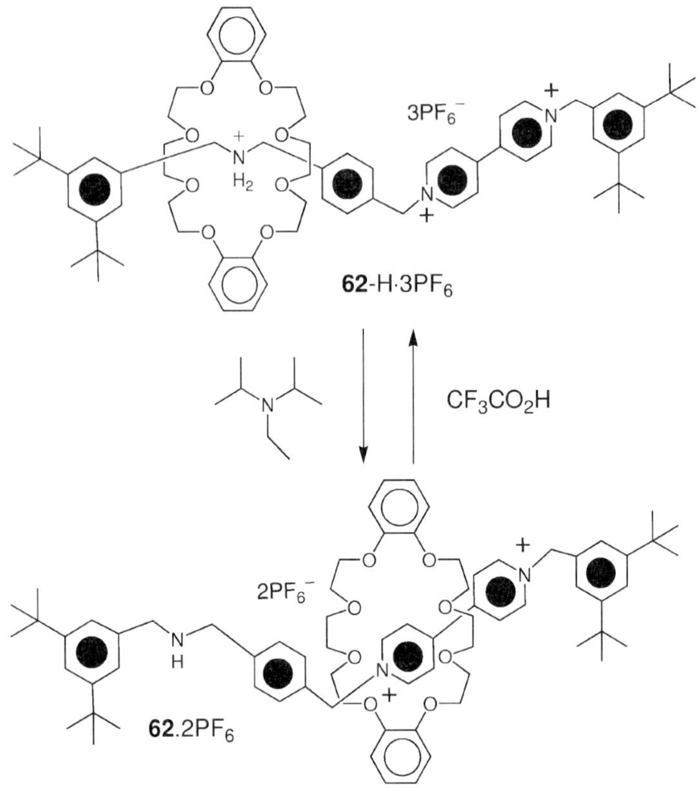

Fig. 53. The pH-controllable molecular shuttle **62**-H.3PF$_6$. The switching process is 100% controllable.

4.3.6. A Light-Driven Molecular Shuttle

Very recently, Nakashima and co-workers [93] designed and self-assembled a [2]rotaxane **63**$^{4+}$ in which a dumbbell-shaped component, composed of a photoisomerizable azobenzene unit and two bulky 2,4-dinitrobenzene terminal stoppering groups, is threaded through α-CD as the ring component. ^{1}H NMR spectroscopy indicates that the α-CD ring component is located on the azobenzene unit. Irradiation ($\lambda = 360$ nm) causes (Fig. 54) the azobenzene unit to undergo photoisomerization from the *trans* to the *cis* configuration. Nuclear Overhauser effect experiments show that, after UV iradiation, the α-CD resides on the bismethylene spacers. Upon irradiation at $\lambda = 430$ nm, the photostationary state reverted to that of the *trans* configuration, and the original situation with the α-CD encircling the azobenzene unit was restored, rendering the cycle reversible.

5. CONCLUSIONS AND REFLECTIONS

In this review chapter, we trust that we have demonstrated that the synthetic chemist possesses in his/her toolbox numerous simple building blocks that exhibit bistability and thus have molecular switching properties when brought under the influence of an external source. These molecular switches are becoming an attractive alternative for the fabrication of devices capable of writing, reading, and storing information at the nanoscale level—the so-called bottom-up approach.

Although the switching properties of some of these compounds have been known for many years, it is only in the last decade or so that they have attracted the attention they deserve. In addition, with the development of supramolecular science and self-assembling

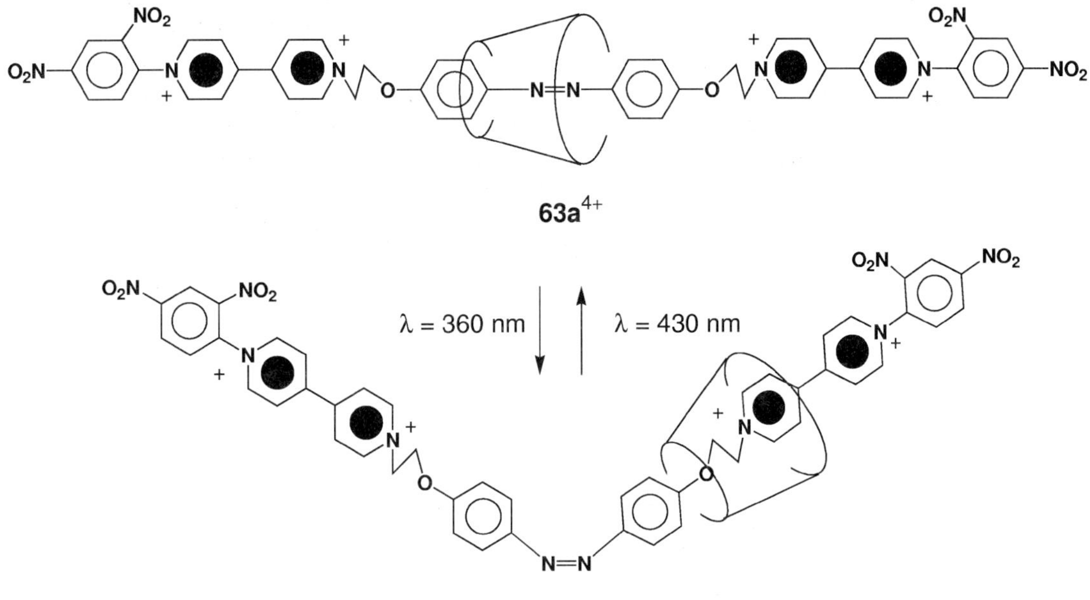

Fig. 54. The photochemically controlled molecular shuttle $\mathbf{63}^{4+}$. This is first molecular shuttle in which the switching process is controlled by light.

synthetic methodologies, we are now able to construct relatively large, well-defined, and highly ordered superstructures. The order of these superstructures is controlled by weak and reversible noncovalent bonding interactions with small changes in the environment or in the stereoelectronic complementarity of the building blocks, producing dramatic changes in the shapes of the superstructures, sometimes destroying them. These changes are reversible upon regeneration of the original conditions, because strict self-assembly is a thermodynamically controlled process. These particular properties open up new avenues for the construction of switching systems at the supramolecular level. Sometimes one of the previously known simple building blocks—susceptible to change under the right conditions—can be incorporated into one of the precursors of the superstructure, thus conferring molecular switching properties on the supermolecule.

Even more, thanks to the increasing availability of more complex self-assembling techniques, it is possible to design and synthesize, in the research laboratory, novel molecular compounds in which the components are not covalently linked but are instead mechanically interlocked, for example, the catenanes and rotaxanes. These molecular assemblies display fascinating geometries and are not static compounds; they also show molecular motions of one component with respect to the remainder, such as translational isomerism. These motions are ultimately regulated by the noncovalent bonding interactions that hold the components together in the first place. Thus, small changes produced by external stimuli can perturb the interactions, inducing a change in the molecular topology. Hence, catenanes and rotaxanes are especially well suited for the study of switching phenomena at the molecular level.

The versatility of the approach is appreciated when it is realized that the molecular motions present in the supramolecules and molecules can be addressed by photochemical, electrochemical, and/or chemical means.

We have shown numerous examples of molecular and supramolecular switching systems in the solution state. However, molecules swimming around in solution behave in a fairly incoherent manner. For molecular and supramolecular species that are able to support device-like properties at a macroscopic level, the self-organization of molecular compounds into larger macroscopic superstructures will be necessary. One of the initial approaches to addressing this problem has been to self-organize these structures into two-dimensional Langmuir films at an air–water interface and then, subsequently, as Langmuir–Blodgett multilayers on solid supports [94]. The next logical step is to study the ordering of molecular structures possessing molecular switching properties onto solid supports, thus permitting the all important investigation of such properties at the macroscopic level.

It is hoped that this series of research papers will eventually culminate in the long term in the inclusion of molecular components with switching characteristics into electronic circuitry that will display real device-like functionality at the macroscopic level. This prospect still lies in the domain of science fiction—like a trip to the moon used to.

References

1. A. L. Robinson, *Science* 223, 267 (1984).
2. R. P. Feynman, *Sat. Rev.* 43, 45 (1960).
3. G. M. Whitesides, J. P. Mathias, and C. T. Seto, *Science* 254, 1312 (1991).
4. (a) R. B. Woodward, *Pure Appl. Chem.* 33, 145 (1973).
   (b) A. Eschenmoser and C. E. Wintner, *Science* 196, 1410 (1977).
   (c) R. W. Armstrong, J.-M. Beau, S. H. Cheon, W. J. Christ, H. Fujioka, H. Fujioka, W. H. Ham, L. D. Hawkins, H. Jin, S. H. Kang, Y. Kishi, M. J. Martinelli, W. W. McWhorter, M. Mizuno, M. Nakata, A. E. Stutz, F. X. Talamas, M. Taniguchi, J. A. Tino, K. Ueda, J. Uenishi, J. B. White, and M. Yonaga, *J. Am. Chem. Soc.* 101, 7525 (1989).
   (d) K. C. Nicolaou and E. J. Sorensen, "Classics in Total Synthesis." VCH, Weinheim, 1996.
5. (a) C. J. Hawker and J. M. J. Fréchet, *J. Am. Chem. Soc.* 112, 7638 (1990).
   (b) D. A. Tomalia, A. M. Waylor, and W. A. Goddard, III, *Angew. Chem. Int. Ed. Engl.* 29, 138 (1990).
   (c) J. M. J. Fréchet, C. J. Hawker, and K. L. Wooley, *Pure Appl. Chem.* 11, 1627 (1994).
   (d) S. C. Zimmerman, F. W. Zeng, D. E. C. Reichert, and S. V. Kolotuchin, *Science* 271, 1095 (1996).
   (e) F. W. Zeng and S. C. Zimmerman, *Chem. Rev.* 97, 1681 (1997).
6. (a) S. K. Burley and G. A. Petsko, *Science* 229, 2 (1985).
   (b) G. R. Desiraju, *Acc. Chem. Res.* 24, 290 (1991).
   (c) C. B. Aakeröy and K. R. Seddon, *Chem. Soc. Rev.* 20, 397 (1993).
   (d) S. Paliwell, S. Geib, and C. S. Wilcox, *J. Am. Chem. Soc.* 116, 4497 (1994).
   (e) M. Nishio, Y. Umezawa, M. Hirota, and Y. Takeuchi, *Tetrahedron* 51, 8665 (1995).
   (f) G. R. Desiraju, *Angew. Chem. Int. Ed. Engl.* 35, 2311 (1995).
   (g) C. A. Hunter and J. K. M. Sanders, *J. Am. Chem. Soc.* 112, 5525 (1990).
   (h) W. L. Jorgensen and D. L. Severance, *J. Am. Chem. Soc.* 112, 4768 (1990).
   (i) C. A. Hunter, *Angew. Chem. Int. Ed. Engl.* 32, 1584 (1993).
   (j) G. R. Desiraju and C. V. Krishnamoham-Sharma, "The Crystal as a Supramolecular Entity," pp. 31–61. Wiley, New York, 1996.
   (k) G. R. Desiraju, *Chem. Comm.* 1475 (1997).
7. (a) J. S. Lindsey, *New J. Chem.* 15, 153 (1991).
   (b) G. M. Whitesides, E. R. Simanek, J. P. Mathias, C. T. Seto, D. N. Chin, M. Mammen, and D. M. Goedon, *Acc. Chem. Res.* 28, 37 (1995).
   (c) D. S. Lawrence, T. Jiang, and M. Levett, *Chem. Rev.* 95, 2229 (1995).
   (d) D. Philp and J. F. Stoddart, *Angew. Chem. Int. Ed. Engl.* 35, 1154 (1996).
   (e) M. C. T. Fyfe and J. F. Stoddart, *Acc. Chem. Res.* 30, 393 (1997).
8. (a) J.-M. Lehn, *Angew. Chem. Int. Ed. Engl.* 27, 89 (1988).
   (b) D. J. Cram, *Angew. Chem. Int. Ed.* 27, 1009 (1988).
   (c) C. J. Pedersen, *Angew. Chem. Int. Ed. Engl.* 27, 1021 (1988).
   (d) J.-M. Lehn, "Supramolecular Chemistry—Concepts and Perspectives." VCH, Weinheim, 1995.
9. R. Langer, *Science* 249, 1527 (1990).
10. (a) T. D. James, K. R. A. S. Sandanayake, and S. Shinkai, *Nature* 374, 345 (1995).
    (b) M. J. Marsella, P. J. Carroll, and T.M. Swager, *J. Am. Chem. Soc.* 117, 9832 (1995).
11. (a) V. Balzani, L. Moggi, and F. Scandola, in "Supramolecular Photochemistry" (V. Balzani, ed.). Reidel, Dordrecht, The Netherlands, 1987.
    (b) H. Kuhn, in "Molecular Electronics" (P. I. Lazarev, ed.), Chap. 12. Kluwer Academic, Dordrecht, The Netherlands, 1991.
12. G. Schill, "Catenanes, Rotaxanes and Knots." Academic Press, New York, 1971.

13. (a) B. L. Feringa, W. F. Jager, and B. de Lange, *Tetrahedron* 49, 8267 (1993).
(b) M. D. Ward, *Chem. Soc. Rev.* 24, 121 (1995).
(c) M. D. Ward, *Chem. Ind. (London)* 16, 640 (1997).

14. (a) D. M. Walba, *Tetrahedron* 41, 3161 (1985).
(b) C. O. Dietrich-Buchecker and J.-P. Sauvage, *Chem. Rev.* 87, 795 (1987).
(c) J-.C. Chambron, C. O. Dietrich-Buchecker, and J.-P. Sauvage, *Top. Curr. Chem.* 165, 131 (1993).
(d) D. B. Amabilino and J. F. Stoddart, *Chem Rev.* 95, 2725 (1995).

15. T. R. Kelly, M. C. Bowyer, K. V. Bhaskar, D. Bebbington, A. Garcia, F. Lang, M. H. Kim, and M. P. Jette, *J. Am. Chem. Soc.* 116, 3657 (1994).

16. A. Guenzi, C. A. Johnson, F. Cozzi, and K. Mislow, *J. Am. Chem. Soc.* 105, 1438 (1983).

17. R. R. Birge, L. A. Findsen, G. C. Fox, E. W. Hyde, and J. R. Tallent, *J. Am. Chem. Soc.* 114, 1581 (1992).

18. R. R. Birge, *Sci. Am.* March, 66 (1995).

19. C. Brächle and N. Hampp, in "Photochromism, Molecules and Systems," Studies in Organic Chemistry 40 (H. Dürr and H. L. Bouas, eds.), Chap. 3, p. 64. Elsevier, Amsterdam, 1990.

20. G. S. Hartley, *Nature* 140, 281 (1937).

21. S. Tazuke, S. Oriuchi, T. Ikeda, D. B. Karanjit, and S. Kurihara, *Macromolecules* 23, 36 (1990).

22. S. Shinkai, M. Aoki, A. Ikeda, K. Murata, and T. Nishi, *J. Chem. Soc., Chem. Commun.* 1715 (1991).

23. H. Tachibana, Y. Kawabata, H. Kmomizu, E. Manda, M. Matsumoto, T. Nakamura, H. Niino, and A. Yabe, *J. Am. Chem. Soc.* 111, 3080 (1989).

24. Z. F. Liu, K. Hashimoto, and A. Fujishima, *Nature* 347, 658 (1990).

25. (a) B. L. Feringa, W. F. Jager, and B. de Lange, *J. Am. Chem. Soc.* 113, 5468 (1991).
(b) B. L. Feringa, W. F. Jager, and B. de Lange, *J. Chem. Soc., Chem. Commun.* 288, (1993).
(c) A. M. Schoevaars, W. Kruizinga, R. W. Zijlstra, N. Veldman, A. L. Speck, and B. L. Feringa, *J. Org. Chem.* 62, 4943 (1997).

26. B. L. Feringa, N. P. M. Huck, and H. A. van Doren, *J. Am. Chem. Soc.* 117, 9929 (1995).

27. H. G. Heller and R. M. Megit, *J. Chem. Soc., Perkin Trans.* 1, 923 (1974).

28. (a) M. Irie and M. Mohri, *J. Org. Chem.* 53, 803 (1988).
(b) M. Irie and S. Nakamura, *J. Org. Chem.* 53, 6136 (1988).

29. (a) S. L. Gilat, S. H. Kawai, and J.-M. Lehn, *Chem. Eur. J.* 1, 275 (1995).
(b) S. H. Kawai, S. L. Gilat, R. Ponsinet, and J.-M. Lehn, *Chem. Eur. J.* 1, 285 (1995).

30. Y. Hirshberg, *J. Am. Chem. Soc.* 78, 2304 (1956).

31. I. Cabrera and V. Krongauz, *Nature* 326, 582 (1987).

32. I. Cabrera, F. Shvartsman, O. Veinberg, and V. Kronagauz, *Science* 220, 341 (1984).

33. I. Cabrera and V. Krongauz, *Macromolecules* 20, 2713 (1987).

34. (a) M. Lion-Dagan, E. Katz, and I. Willner, *J. Chem. Soc., Chem. Commun.* 2741 (1994).
(b) I. Willner and S. Rubin, *Angew. Chem. Int. Ed. Engl.* 35, 367 (1996).
(c) I. Willner, *Acc. Chem. Res.* 30, 347 (1997).

35. V. Gfoulle, A. Harriman, and J.-M. Lehn, *J. Chem. Soc., Chem. Commun.* 1034 (1993).

36. C. J. Pedersen, *J. Am. Chem. Soc.* 89, 7017 (1967).

37. (a) S. Shinkai, Y. Kusano, O. Manabe, T. Nakaji, and T. Ogawa, *Tetrahedron Lett.* 20, 4569 (1979).
(b) S. Shinkai, Y. Kusano, O. Manabe, T. Nakaji, Y. Nishida, and T. Ogawa, *J. Am. Chem. Soc.* 102, 5860 (1980).
(c) S. Shinkai, T. Minami, Y. Kusano, and O. Manabe, *J. Am. Chem. Soc.* 105, 1851 (1983).

38. S. Shinkai, M. Ishira, K. Ueda, and O. Manabe, *J. Chem. Soc., Perkin Trans. II* 511 (1985).

39. K. Kimura, R. Mizutani, M. Yokoyama, R. Arakawa, G. Matsubayashi, M. Okamoto, and H. Doe, *J. Am. Chem. Soc.* 119, 2062 (1997).

40. R. A. Bissell, A. P. de Silva, H. Q. M. Gunaratne, P. L. M. Lynch, G. E. M. Maguire, and K. R. A. S. Sandanayake, *Chem. Soc. Rev.* 21, 187 (1992).

41. C. Canevet, J. Libman, and A. Shanzer, *Angew. Chem. Int. Ed. Engl.* 35, 2657 (1996).

42. C. D. Gutsche, "Calixarenes, Monographs in Supramolecular Chemistry" (J. F. Stoddart, ed.). The Royal Society of Chemistry, Cambridge, 1989.

43. L. Zelikovich, J. Libman, and A. Shanzer, *Nature* 374, 790 (1995).

44. Y. Iseki and S. Inoue, *J. Chem. Soc., Chem. Commun.* 2577 (1994).

45. (a) J. F. G. A. Jansen, E. M. M. de Brabander-van den Berg, and E. W. Meijer, *Science* 266, 1226 (1994).
(b) S. Stevelmans, J. C. M. van Hest, J. F. G. A. Jansen, D. A. F. J. van Boxtel, E. M. M. de Brabander-van den Berg, and E. W. Meijer, *J. Am. Chem. Soc.* 118, 7398 (1996).

46. J. F. G. A. Jansen and E. W. Meijer, *J. Am. Chem. Soc.* 117, 4417 (1995).

47. (a) B. L. Allwood, N. Spencer, H. Shahriari-Zavareh, J. F. Stoddart, and D. J. Williams, *J. Chem. Soc., Chem. Commun.* 1064 (1987).
(b) P. R. Ashton, A. M. Z. Slawin, N. Spencer, J. F. Stoddart, and D. J. Williams, *J. Chem. Soc., Chem. Commun.* 1066 (1987).
(c) A. M. Z. Slawin, N. Spencer, J. F. Stoddart, and D. J. Williams, *J. Chem. Soc., Chem. Commun.* 1070 (1987).

48. B. Odell, M. V. Reddington, A. M. Z. Slawin, N. Spencer, J. F. Stoddart, and D. J. Williams, *Angew. Chem. Int. Ed. Engl.* 27, 1550 (1988).
49. (a) D. B. Amabilino, P. R. Ashton, G. R. Brown, W. Hayes, J. F. Stoddart, M. S. Tolley, and D. J. Williams, *J. Chem. Soc., Chem. Commun.* 2479 (1994).
(b) D. B. Amabilino, P. L. Anelli, P. R. Ashton, G. R. Brown, E. Córdova, L. A. Godínez, W. Hayes, A. E. Kaifer, D. Philp, A. M. Z. Slawin, N. Spencer, J. F. Stoddart, M. S. Tolley, and D. J. Williams, *J. Am. Chem. Soc.* 117, 11142 (1995).
50. M. Asakawa, P. R. Ashton, S. E. Boyd, C. L. Brown, R. E. Gillard, O. Kocian, F. M. Raymo, J. F. Stoddart, M. S. Tolley, A. J. P. White, and D. J. Williams, *J. Org. Chem.* 62, 26 (1997).
51. (a) T. T. Goodnow, M. V. Reddington, J. F. Stoddart, and A. E. Kaifer, *J. Am. Chem. Soc.* 113, 4335 (1991).
(b) A. R. Bernardo, J. F. Stoddart, and A. E. Kaifer, *J. Am. Chem. Soc.* 114, 10624 (1992).
52. P. L. Anelli, P. R. Ashton, R. Ballardini, V. Balzani, M. Delgado, M. T. Gandolfi, T. T. Goodnow, A. E. Kaifer, D. Philp, M. Pietraszkiewicz, L. Prodi, M. V. Reddington, A. M. Z. Slawin, N. Spencer, J. F. Stoddart, C. Vicent, and D. J. Willimas, *J. Am. Chem. Soc.* 114, 193 (1992).
53. R. Ballardini, V. Balzani, M. T. Gandolfi, L. Prodi, M. Venturi, D. Philp, H. G. Ricketts, and J. F. Stoddart, *Angew. Chem. Int. Ed. Engl.* 32, 1301 (1993).
54. M. Asakawa, S. Iqbal, J. F. Stoddart, and N. D. Tinker, *Angew. Chem. Int. Ed. Engl.* 35, 976 (1996).
55. T. Jørgensen, T. K. Hansen, and J. Becher, *Chem. Soc. Rev.* 23, 41 (1994).
56. D. Philp, A. M. Z. Slawin, N. Spencer, J. F. Stoddart, and D. J. Williams, *J. Chem. Soc., Chem. Commun.* 1584 (1991).
57. W. Devonport, M. A. Blower, M. R. Bryce, and L. M. Goldenberg, *J. Org. Chem.* 62, 885 (1997).
58. (a) R. Ballardini, V. Balzani, A. Credi, M. T. Gandolfi, S. J. Langford, S. Menzer, L. Prodi, J. F. Stoddart, M. Venturi, and D. J. Williams, *Angew. Chem. Int. Ed. Engl.* 35, 978 (1996).
(b) P. R. Ashton, E. J. T. Chrystal, P. T. Glink, S. Menzer, C. Schiavo, N. Spencer, J. F. Stoddart, P. A. Tasker, A. J. P. White, and D. J. Williams, *Chem. Eur. J.* 2, 709 (1996).
(c) P. T. Glink, C. Schiavo, J. F. Stoddart, and D. J. Williams, *Chem. Commun.* 1843 (1996).
59. P. R. Ashton, P. J. Campbell, E. J. T. Chrystal, P. T. Glink, S. Menzer, D. Philp, N. Spencer, J. F. Stoddart, P. A. Tasker, and D. J. Williams, *Angew. Chem. Int. Ed. Engl.* 34, 1865 (1994).
60. P. R. Ashton, E. J. T. Chrystal, P. T. Glink, S. Menzer, C. Schiavo, J. F. Stoddart, P. A. Tasker, and D. J. Williams, *Angew. Chem. Int. Ed. Engl.* 34, 1869 (1995).
61. P. R. Ashton, R. Ballardini, V. Balzani, I. Baxter, A. Credi, M. C. T. Fyfe, M. T. Gandolfi, M. Gómez-López, M.-V. Martínez-Díaz, A. Piersanti, N. Spencer, J. F. Stoddart, M. Venturi, A. J. P. White, and D. J. Williams, *J. Am. Chem. Soc.* 120, 11932 (1998).
62. M. Montalti, R. Ballardini, L. Prodi, and V. Balzani, *Chem. Commun.* 2011 (1996).
63. G. Wenz, *Angew. Chem. Int. Ed. Engl.* 33, 803 (1994).
64. A. Mirzoian and A. E. Kaifer, *Chem. Eur. J.* 3, 1052 (1997).
65. A. Aviram, *J. Am. Chem. Soc.* 110, 5687 (1988).
66. A. P. de Silva, H. Q. N. Gunaratne, and C. P. McCoy, *Nature* 364, 42 (1993).
67. A. P. de Silva, H. Q. N. Gunaratne, and C. P. McCoy, *J. Am. Chem. Soc.* 119, 7891 (1997).
68. A. P. de Silva, H. Q. N. Gunaratne, and G. E. M. Maguire, *J. Chem. Soc., Chem. Commun.* 1213 (1994).
69. A. Credi, V. Balzani, S. J. Langford, and J. F. Stoddart, *J. Am. Chem. Soc.* 119, 2679 (1997).
70. P. R. Ashton, T. T. Goodnow, A. E. Kaifer, M. V. Reddington, A. M. Z. Slawin, N. Spencer, J. F. Stoddart, C. Vicent, and D. J. Williams, *Angew. Chem. Int. Ed. Engl.* 28, 1396 (1989).
71. (a) D. B. Amabilino, P. R. Ashton, A. S. Reder, N. Spencer, and J. F. Stoddart, *Angew. Chem. Int. Ed. Engl.* 33, 1286 (1994).
(b) D. B. Amabilino, P. R. Ashton, S. E. Boyd, J. Y. Lee, S. Menzer, J. F. Stoddart, and D. J. Williams, *Angew. Chem. Int. Ed. Engl.* 36, 2070 (1997).
72. G. Schill, K. Rissler, H. Fritz, and W. Vetter, *Angew. Chem. Int. Ed. Engl.* 20, 187 (1981).
73. P. R. Ashton, A. Collins, M. C. T. Fyfe, P. T. Glink, S. Menzer, J. F. Stoddart, and D. J. Williams, *Angew. Chem. Int. Ed. Engl.* 36, 735 (1997).
74. P. R. Ashton, M. Blower, D. Philp, N. Spencer, J. F. Stoddart, M. S. Tolley, R. Ballardini, M. Ciano, V. Balzani, M. T. Gandolfi, L. Prodi, and C. H. McLean, *New J. Chem.* 17, 689 (1993).
75. P. R. Ashton, L. Pérez-García, J. F. Stoddart, A. J. P. White, and D. J. Williams, *Angew. Chem. Int. Ed. Engl.* 34, 571 (1995).
76. P. R. Ashton, R. Ballardini, V. Balzani, A. Credi, M. T. Gandolfi, S. Menzer, L. Pérez-Gracía, L. Prodi, J. F. Stoddart, M. Venturi, A. J. P. White, and D. J. Williams, *J. Am. Chem. Soc.* 117, 11171 (1995).
77. C. A. Hunter, *J. Am. Chem. Soc.* 114, 5303 (1992).
78. F. Vögtle, S. Meier, and R. Hoss, *Angew. Chem. Int. Ed. Engl.* 31, 1619 (1992).
79. (a) A. G. Johnston, D. A. Leigh, R. J. Pritchard, and M. D. Deegan, *Angew. Chem. Int. Ed. Engl.* 34, 1212 (1995).
(b) A. G. Johnston, D. A. Leigh, L. Nezhat, J. P. Smart, and M. D. Deegan, *Angew. Chem. Int. Ed. Engl.* 34, 1209 (1995).
80. D. A. Leigh, K. Moody, J. P. Smart, K. J. Watson, and A. M. Z. Slawin, *Angew. Chem. Int. Ed. Engl.* 35, 306 (1996).

81. C. O. Dietrich-Buchecker, J.-P. Sauvage, and J. M. Kern, *J. Am. Chem. Soc.* 106, 3043 (1984).
82. (a) C. O. Dietrich-Buchecker and J.-P. Sauvage, *Angew. Chem. Int. Ed. Engl.* 27, 1095 (1988).
(b) F. Bistch, C. O. Dietrich-Buchecker, A.-K. Khémis, J.-P. Sauvage, and A. van Dorsselaer, *J. Am. Chem. Soc.* 113, 4023 (1991).
(c) J.-C. Chambron, V. Heitz, and J.-P. Sauvage, *J. Chem. Soc., Chem. Commun.* 1131 (1992).
(d) J.-C. Chambron, C. O. Dietrich-Buchecker, J.-F. Nierengarten, and J.-P. Sauvage, *J. Chem. Soc., Chem. Commun.* 801 (1993).
(e) J.-C. Chambron, C.O. Dietrich-Buchecker, J.-F. Nierengarten, J.-P. Sauvage, N. Solladie, A. M. Albrecht-Gary, and M. Meyer, *New J. Chem.* 19, 409 (1995).
83. A. Livoreil, C. O. Dietrich-Buchecker, and J.-P. Sauvage, *J. Am. Chem. Soc.* 116, 9399 (1994).
84. D. B. Amabilino, C. O. Dietrich-Buchecker, A. Livoreil, L. Pérez-García, J.-P. Sauvage, and J. F. Stoddart, *J. Am. Chem. Soc.* 118, 3905 (1996).
85. (a) P. R. Ashton, V. Balzani, S. E. Boyd, M. Gómez-López, S. Iqbal, D. Philp, J. A. Preece, H. G. Ricketts, J. F. Stoddart, M. S. Tolley, A. J. P. White, and D. J. Williams, *Chem. Eur. J.* 3, 146 (1997).
(b) P. R. Ashton, M. Gómez-López, S. Iqbal, J. A. Preece, and J. F. Stoddart, *Tetrahedron Lett.* 38, 3635 (1997).
86. (a) C. Wu, P. R. Lecavalier, Y. X. Shen, and H. W. Gibson, *Chem. Mater.* 3, 569 (1991).
(b) A. C. Benniston and A. Harriman, *J. Am. Chem. Soc.* 116, 11531 (1994).
(c) A. G. Kolchinski, D. H. Busch, and N. W. Alcock, *J. Chem. Soc., Chem. Commun.* 1289 (1995).
(d) P. R. Ashton, P. T. Glink, J. F. Stoddart, P. A. Tasker, A. J. P. White, and D. J. Williams, *Chem. Eur. J.* 2, 729 (1996).
(e) S. Anderson and H. Anderson, *Angew. Chem. Int. Ed. Engl.* 35, 1956 (1996).
(f) D. A. Leigh, A. Murphy, J. P. Smart, and A. M. Z. Slawin, *Angew. Chem. Int. Ed. Engl.* 36, 728 (1997).
(g) R. Jäger and F. Vögtle, *Angew. Chem. Int. Ed. Engl.* 36, 930 (1997).
87. (a) P. L. Anelli, N. Spencer, and J. F. Stoddart, *J. Am. Chem. Soc.* 113, 5131 (1991).
(b) P.-L. Anelli, M. Asakawa, P. R. Ashton, R. A. Bissell, G. Clavier, R. Górski, A. E. Kaifer, S. J. Langford, G. Mattersteig, S. Menzer, D. Philp, A. M. Z. Slawin, N. Spencer, J. F. Stoddart, M. S. Tolley, and D. J. Williams, *Chem. Eur. J.* 3, 1113 (1997).
88. P. R. Ashton, R. A. Bissell, N. Spencer, J. F. Stoddart, and M. S. Tolley, *Synlett* 914 (1992).
89. P. R. Ashton, R. A. Bissell, R. Görski, D. Philp, N. Spencer, J. F. Stoddart, and M. S. Tolley, *Synlett* 919 (1992).
90. P. R. Ashton, R. A. Bissell, N. Spencer, J. F. Stoddart, and M. S. Tolley, *Synlett* 923 (1992).
91. R. A. Bissell, E. Córdova, A. E. Kaifer, and J. F. Stoddart, *Nature* 369, 133 (1994).
92. M. V. Martínez-Díaz, N. Spencer, and J. F. Stoddart, *Angew. Chem. Int. Ed. Engl.* 36, 1904 (1997).
93. H. Murakami, A. Kawabuchi, K. Kotoo, M. Kunitake, and N. Nakashima, *J. Am. Chem. Soc.* 119, 7605 (1997).
94. R. C. Ahuja, P.-L. Caruso, D. Móbius, D. Philp, J. A. Preece, H. Ringsdorf, J. F. Stoddart, and G. Wildburg, *Langmuir* 9, 1534 (1993).

Chapter 15

FUNCTIONAL NANOSTRUCTURES INCORPORATING RESPONSIVE MODULES

Andrew C. Benniston, Philip R. Mackie

Chemistry Department, Glasgow University, Glasgow G12 8QQ, Scotland

Contents

1. INTRODUCTION

Nanotechnology, as the name suggests, is concerned with the practical development of functional materials (e.g., devices), the workings of which can be considered to be operating at or around the nanometer scale. The idea of nanoscale molecular devices is not particularly new, and has been around since the earlier ideas of Feynman [1, 2], who addressed the issue of quantum mechanical computers. As Feynman concluded, "at any rate, it seems that the laws of physics present no barrier to reducing the size of computers until bits are the size of atoms, and quantum behavior holds dominant sway"—a large task, considering the small size that is wanted!

Handbook of Nanostructured Materials and Nanotechnology, edited by H.S. Nalwa
Volume 5: Organics, Polymers, and Biological Materials

However, it is clearly evident from the increase in publications over recent years [3] that chemists and material scientists have taken up the scientific challenges, especially on the development side of nanotechnology [4]. The main impetus for this increased interest is the argument that the miniaturization of conventional silicon components below the current 0.5-μm level will, at the present rate, soon run into both practical manufacturing and operating difficulties [5, 6]. Accordingly, it is the need to develop new miniaturization processes that is the main practical obstacle to real progress in the field of nanotechnology. The conventional synthesis used in today's manufacture of silicon chips [7] is appropriately termed the "bottom-down" engineering approach and relies on smaller and smaller components being "chiseled" to the desired size by lithographic techniques [8]. The main inadequacy of this method is that the limit to the final component's dimension is primarily governed by the wavelength of the radiation used [9]. Of course, other restrictions stem from quantum mechanical restrictions (tunneling currents), although this may not be a hindrance, as the development of quantum dots specifically relies on quantum phenomena [10]. On the other hand, chemists are attempting to prepare nanoscale molecular systems by first constructing small molecular pieces and allowing them, using in-built self-recognition processes, to assemble themselves into a final structure; the structural, chemical, or physical properties of which may significantly differ from those of the individual building blocks. Because of the very nature of starting at the molecular level and growing outward from a central core, this methodology has been accordingly given the title of the "bottom-up" approach (Fig. 1). As a consequence, the development side of the bottom-up approach is heavily tied into the ever expanding field of supramolecular chemistry [11]; the science that uses weak intermolecular forces to construct large architectures.

It is clear that each method offers unique advantages over the other, but both are fraught with their own practical problems. Indeed, it is difficult to envisage the bottom-down approach being able to construct a nanostructured system that contains, for example, a single photoactive center that has both specificity in its orientation and positional order. Alterna-

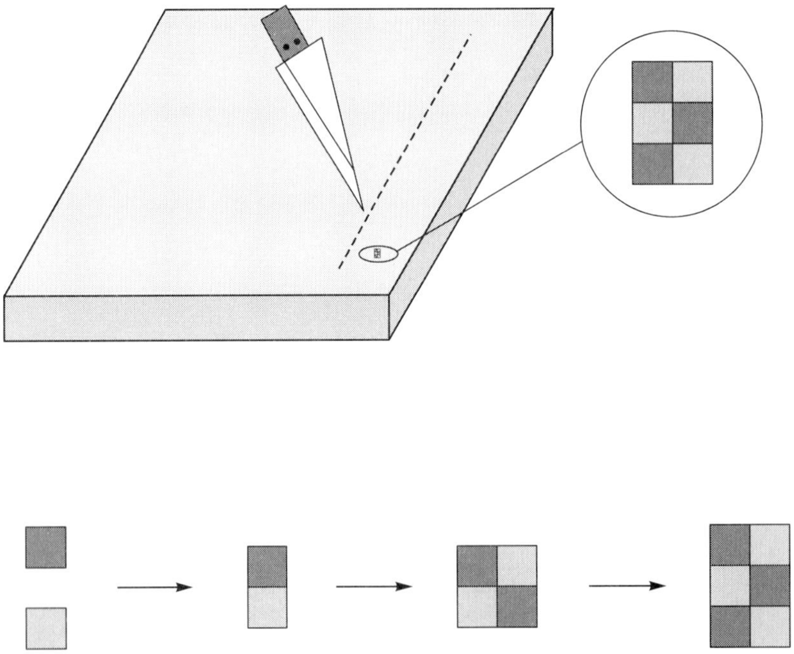

Fig. 1. A conceptual approach to manufacturing nanoscale devices, where the top picture depicts the conventional "cutting" of a large piece of material, and the lower diagram depicts the "self-assembly" of small units by self-recognition.

tively, constructing a device that is able to communicate with the "outside world" via real wires is just as much a task for builders using the bottom-up technique.

It is worth stressing that manufacturing aspects of nanotechnology are not the only important issue, which leads to the relevant question, of what use is this technology and what are the final goals, considering all the current research effort? For many researchers working in the nanotechnology field, particularly from a molecular-based approach, the main aim is to tailor-make atomic-scale structures (nanostructures), whose control by electronic, optical, mechanical, or chemical means will have applications in future electronic, optronic, photonic, and magnetic molecular-based devices. Even within this narrow realm of nanotechnology, subdivisions into fields concerned with property, characterization, structuralization, preparation, theory, order, etc. are all important aspects, which themselves can be further subdivided.

This difficulty in tying down nanotechnology has somewhat to do with the interdisciplinary and multidisciplinary nature of the area and to some extent reflects its importance and the current optimism shown by researchers worldwide as the new century approaches.

1.1. Functional Molecular Structures: General Definition

In a sense the definition of a molecular system that can be termed functional is somewhat arbitrary, considering that a vast majority of organic and inorganic compounds contain chromophoric or redoxactive centers as part of their structure. For instance, a module containing benzene units will absorb strongly in the UV spectral region and thus is active toward photons, yet it would be incorrect to term the assembly photoactive if the benzene unit serves no purpose. Accordingly, in the context of this review, a functional molecular structure is defined as *an assemblage in which responsive modules can be made to perform specific operations by appropriate activation from external reagents such a photons, electrons, protons, and ions, and in which the structure plays an important role in the operation of that system.* With this definition in mind we can imagine that external triggering of a molecular structure could, in principle, lead to a number of outcomes associated more with bulk material changes, such as refractivity, magnetic, dielectric, or even thermal properties. However, for the sake of simplicity, only molecular-scale operations will be considered; namely, electron transfer reactions and large-scale structural alterations (Scheme I).

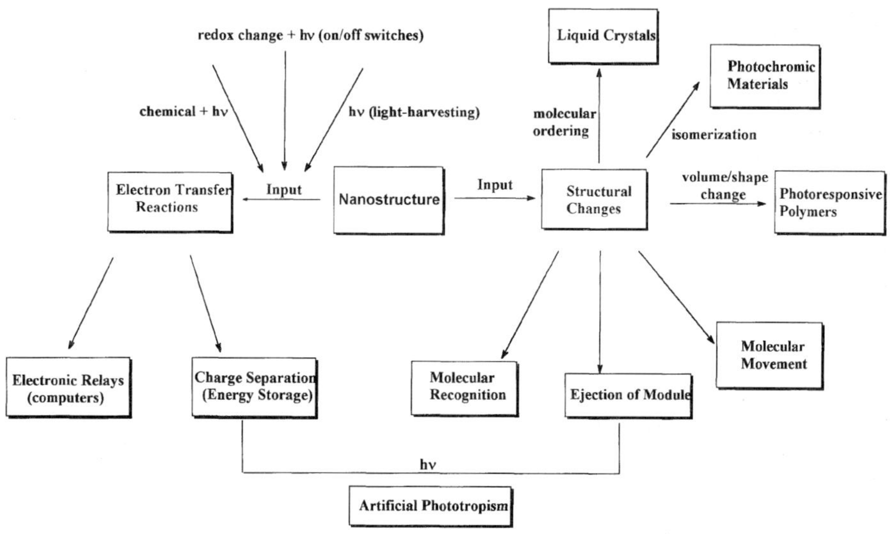

Scheme I. Possible outcomes and uses of nanostructures following input from an external source.

The processes in Scheme I demonstrate possible roles of functional nanostructures and represent typical outcomes, reactions, and potential uses of the structures following external input. It should be stressed that the depicted subdivision, however, is not clear-cut, as many of the individual sections rely on more than one process and can quite easily be exchanged. For instance, if we imagine a molecular system in which a photon is able to generate a pair of redox ions by their mutual separation along a unique axis (structural change), it could be said that the overall process is crudely reminiscent of phototropism in plants (specific light-induced motion). Alternatively, the separated ions are also equivalent to charge separation that is consequential to energy storage in light-harvesting bacteria (electron transfer).

Within the above general definition of functional nanostructures, if input is by way of photons, the assembly is deemed *photoactive*, and conversely, if induction is via redox changes, the system is *redoxactive*. Other forms of external agents (i.e., protons, metal ions) will consequently render a system *ionactive*.

1.2. Scope and Context of Review

Over the years the volume of research into photoactive, redoxactive and ionactive molecular systems has increased dramatically, particularly in line with improvements in both synthetic and spectroscopic techniques (i.e., ultrafast laser flash photolysis). The growth in the use of molecular-scale systems based on crown ethers such as 24C8 (1,4,7,10,17,20,23,26,29-octaoxa[10,10]paracyclophane) (**1**) [12] and other cyclophane-derived structures (**2**) [13] has been noticeable, especially in light of their strong binding affinity toward π-deficient/rich aromatics and metal ions; the associated equilibrium is susceptible to perturbation by controllable outside factors (i.e., excitation via the charge transfer (CT) band) [14] (Fig. 2).

In recent times, topologically intriguing molecules such as rotaxanes and catenanes have emerged as real contenders for functional nanostructures, especially for use as photoactive and redoxactive devices. Accordingly, the aim of this chapter is to focus on these new and intriguing molecules, as well as on cyclophane-derived systems such as cryptands, crowns, etc. Particular emphasis will be placed on those photoactive systems that mimic biological processes, such as charge separation and molecular switching (Section 2).

In an attempt to place this work in a broader prospective, other photoactive, redoxactive, etc. architectures will also be briefly reviewed in the final section. It is hoped that similarities will become apparent and pave the way for the possible future development of functional nanostructures.

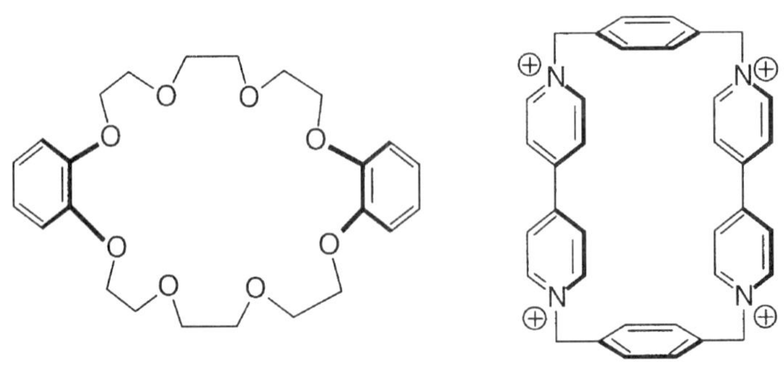

1 **2**

Fig. 2. Two examples of simple molecular building blocks.

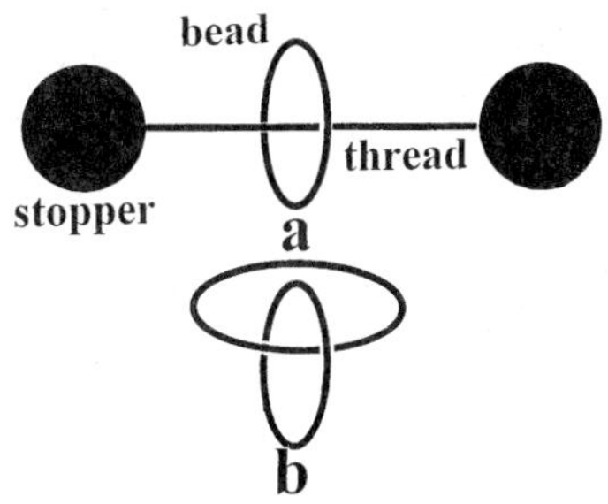

Fig. 3. Simple cartoons of (a) a [2]rotaxane and (b) a [2]catenane and common nomenclature. (Reproduced by permission of The Royal Society of Chemistry.)

1.3. Rotaxanes and Catenanes: Nomenclature and General Synthetic Methods

Considering the importance placed on rotaxanes and catenanes, not only in this chapter, but as a whole in supramolecular science [15], it is perhaps pertinent to acquaint readers with terms often used to describe these systems and the general synthetic methodologies used in their preparation. It should be pointed out, however, that a comprehensive insight into rotaxane and catenane syntheses is not the intent of this chapter, as this will be more than adequately covered elsewhere. The two simple cartoons in Figure 3 attempt to illustrate the similarity between rotaxane and catenane assemblies and contain the typical nomenclature used to refer to these structures.

It should be noticeable that the "bead" used to refer to rotaxanes is somewhat analogous to that found on a child's abacus, with the "stopper" groups restricting the unit from simply flying off the end. With removal of the stopper units we can imagine the bead being able to simply slip on and off the thread, thus forming a structure that has been termed a *pseudorotaxane*.

To be strictly correct, the numerical prefix used when naming rotaxane/catenane structures is related to molecular graph theory and corresponds to the total number of crossing points in a structure's molecular graph. To this end—and as pointed out by Sauvage in one of his early papers [16]—only a bead threaded on an infinitely long thread can be truly termed a rotaxane, because by simply reducing the size of the terminal stoppers, or alternatively growing the bead, dismantling into the individual constituents occurs; these separated units can be drawn as planar molecular graphs. No such uncertainty occurs in the catenane cases, however, because by virtue of the interlocked rings, a nonplanar molecular graph always arises. To be specific, in rotaxane systems the numerical prefix is not directly related to the number of beads/stoppers present or, as discussed in the following section, the stations on a thread, but represents the number of components. If we therefore regard the ball-and-stick examples depicted in Figure 4a as simple cartoon representations of rotaxanes, the incremental increases in the number of beads from one to three corresponds to the naming [2]rotaxane, [3]rotaxane, and [4]rotaxane, respectively. In the same vein, it is hoped that the simple interlocked chains in Figure 4b and their corresponding nomenclature, [2]catenane, [3]catenane, and [4]catenane, are self-evident.

To limit the number of examples, only [2]catenane and [2]rotaxane assemblies will be considered in this review.

1.3.1. Stations and Shuttling

A particularly relevant concept in rotaxane chemistry (as will become more evident in Section 3) is the incorporation of elements onto a thread, where the bead will prefer to spend significantly more time. For instance, one can imagine that a protuberance on a thread,

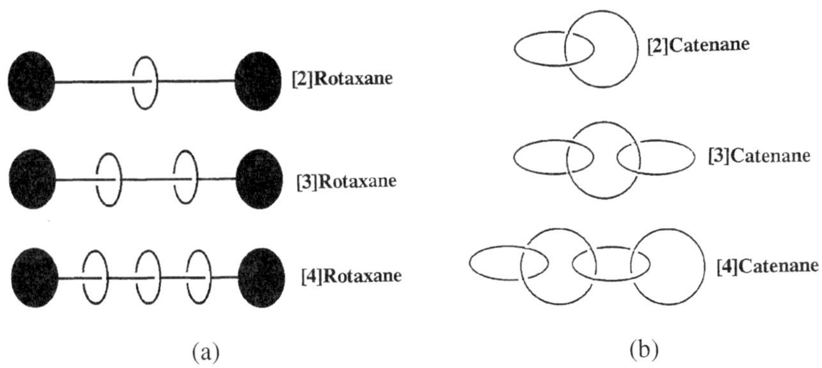

Fig. 4. Pictorial representations of (a) rotaxanes and (b) catenanes and their appropriate nomenclature.

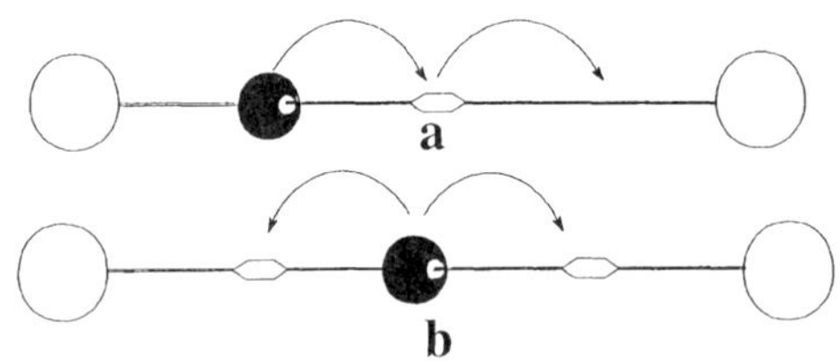

Fig. 5. A simple representation of (a) a station on a thread and (b) the shuttling of a bead between stations.

the size of which is marginally smaller than a bead's hole, will restrict motions by way of excessive friction (Fig. 5a). Moreover, when the size of the restriction exactly matches or even exceeds the hole in the bead, "sticking" at that given site results, or more importantly, motion ceases in a specific direction. As a consequence, an element somewhere on a thread where a bead will predominantly reside is often termed a *station*. Extending the analogy even further, there now exists the possibility of having more than one station and the bead being able to "shuttle" back and forth between sites, or even spend significantly more time at one station, owing to an idiosyncrasy in the two sites (Fig. 5b). It is not difficult to imagine that the control of shuttling by restriction of a bead at given site by the use of an external agent is the input-structural change required for a functional nanostructure (molecular switch).

1.3.2. General Synthesis

In the real world, construction of rotaxane/catenane structures is easily accomplished by simply linking two loops together and welding them shut, or threading a bead onto a string and tying a sufficiently large knot. At the molecular level this is not so easy to carry out, as synthetic tools are required that are far more intricate, and as a consequence the task has fallen to the ingenuity and synthetic skills of chemists. In terms of simplicity the synthetic strategy easiest to envisage in rotaxane/catenane production is a statistical approach in which two species encounter, become interwoven, and react before dethreading has time to take place. Without the existence of any substantial binding between components, at any given time only a very small percentage of necessary organized complex is formed, and hence the vast majority of reactions lead to noninterwoven products (Fig. 6). For this reason, it is not too surprising that the earliest reported examples of rotaxanes and catenanes prepared by such random-reaction procedures were in very low yield [17].

The reliance on such chance reactions is evidently rather limited and is a particularly inappropriate method for the generation of workable quantities of material. Fortunately, more sophisticated synthetic methods have been developed whereby prior to catenation or

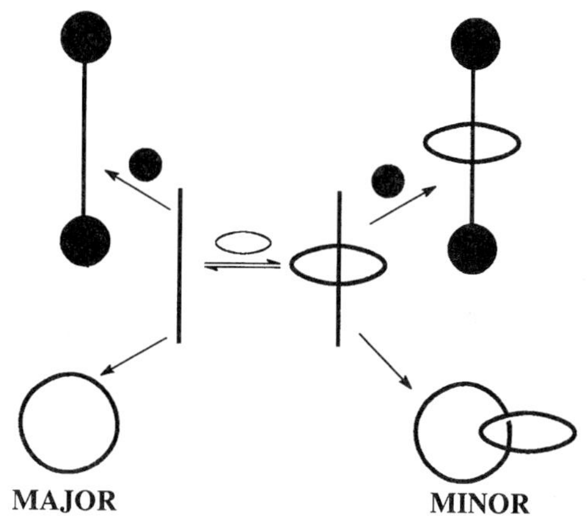

Fig. 6. Chance reaction procedures for generation of rotaxanes and catenanes.

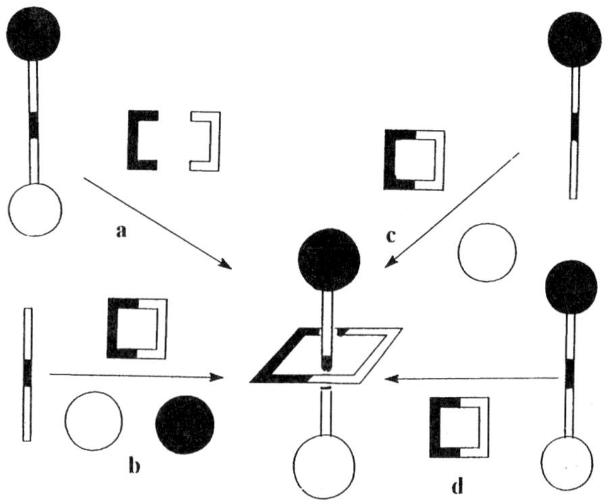

Fig. 7. Illustrations of the four synthetic approaches that can be used in the manufacture of a [2]rotaxane.

stoppering, individual modules are preassembled into a more stable entity; the process is regarded as analogous to shifting the equilibrium in Figure 5 further to the right. Clearly, the generation of preorganized assemblies requires reacting species to be able to perform self-recognition through appropriate "smart" portions or fashion themselves around suitable templates [18]. The understanding and practical development of this "self-assembly" procedure are where the field of supramolecular chemistry and nanotechnology cross over and common goals arise.

1.3.3. Rotaxane Synthesis

Without going into actual chemical examples, the ball-and-stick cartoons of Figure 7 attempt to conceptually illustrate four possible (but by no means the only) methods of rotaxane production. For instance, Figure 7a depicts how a presynthesized stoppered thread can be used as a template in generation of a trapped bead by clipping together two complementary segments. Alternatively, the need for bead production during rotaxane formation is completely eliminated if the bead can be separately preconstructed and simply acts as the templating agent for encapsulation of an open-ended thread that is finally stoppered with suitably sized blocking groups (Fig. 7b). Clearly, this method is limited to the manufacture

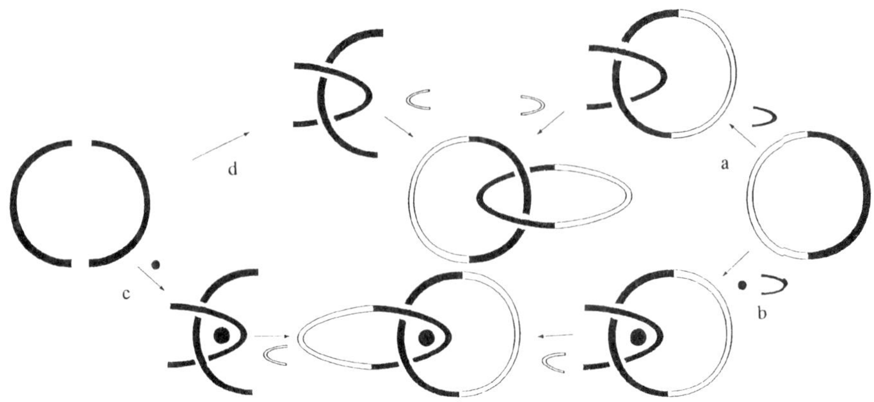

Fig. 8. Cartoons representing the ring closure reactions that can be used to generate a [2]catenane. Note: The black circle represents an external templating agent, which, if necessary, can be removed at the end of the reaction.

of symmetric-stoppered rotaxanes as a statistical mixture of products would certainly arise under the conditions outlined in Figure 7b. However, incorporation of a monostoppered thread allows functionalization with a different end group (Fig. 7c), although synthesis of a suitable thread is still necessary for success. A more recently conceived synthetic approach is one in which a bead is forced onto an already stoppered thread, in what has been termed "slippage" [19]. As the name suggests, the bulk of the stoppers is just too large for a bead to simply slip onto the thread. Elevating the reaction mixture temperature results in increased molecular motions and possibly slight bead expansion, which allows the bead to propel itself over the stopper. The thermodynamic "trap" caused by interaction of the bead and thread retards dethreading, and upon cooling the rotaxane is formed (Fig. 7d).

For the sake of simplicity, only one-station rotaxane examples have been discussed, but clearly by introduction of more than one station and by use of careful reaction conditions, two-station or even three-station rotaxanes can be constructed along the same lines as those discussed above.

1.3.4. Catenane Synthesis

It is not to difficult to imagine that slight adaptation of the synthetic methodologies outlined for a rotaxane production lend themselves to the construction of catenanes (Fig. 8). Indeed, pursuing a line similar to rotaxane stoppering, Figure 8a and b shows the catenation procedures developed using preorganization around a template [20] or via self-recognition [21] of a macrocyclic ring and reaction with a complementary linker unit. As catenation reaction procedures have matured, it is now not uncommon to perform two ring closures by using two preorganized half-units around an external reagent [22] and more recently by self-recognition of two modules [23] (Fig. 8c, d).

As well as enhancing yields of rotaxane/catenane assemblies, the above-mentioned techniques have lent themselves rather nicely to building into the structure's further functionality. For instance, it is possible to make one of the linker units or stopper groups a redoxactive moiety that can extract an electron from the bead segment or play a role in electron transfer (ET) from one stopper to the other (Section 3.3).

2. LEARNING FROM NATURE: BIOACTIVE MODULES

As in many aspects of science, Nature has already paved the way with "biological machines" that receive, process, and transmit energy/information to the various parts, in many cases, of large protein structures. A common mechanistic feature of many biological systems is light-triggered processes in which a chomophoric center is either (i) activated and alters in shape (e.g., isomerizes), with the form change transduced to the surrounding biomembrane or protein structure, or (ii) photoexcited, and relays an electron to a series of acceptors, with charge separation as the final result. At the biological level in a sense, a cell is the ultimate molecular-based device, which not only performs many functions, but is also able to self-program by adjusting to environmental factors (e.g., phototaxis) [24]. It is perhaps not too difficult to see why aspects of biological systems are deemed important, considering their evolutionarily optimized advantages such as small size, speed, sensitivity, selectivity, and stability. For this reason, the mimicking of biological functions has been particularly linked to the development of electronic devices (Section 3), especially sensory and information storage/retrieval applications [25]. The following biological examples are highlighted as relevant biological functional assemblies that reveal insights into the necessary active groups, design criteria, and reactions (charge separation (CS), switching) that are required in the development of artificial systems.

2.1. Light-Harvesting Antenna

Photosynthetic membranes or reaction centers (RCs) are light energy transducers that transform photon energy into electrochemical potential, which is further used to drive chemical reactions, including oxidative splitting of water into oxygen, conversion of ADP to ATP, and assimilation of CO_2 by the $NADP^+$/NADPH cycle. The underlying process common to light-harvesting systems is the efficient separation of charge across a cytoplasmic membrane, using a series of directional and effective ET reactions. It is this control of long-range directional ET in the RC that has been focused on as an essential property in the design of, for instance, molecular-scale electronic devices. It is also interesting to note that CS occurs across distances on the order of ~30–70 Å (3–7 nm), which is precisely within the realm of nanotechnological systems.

2.1.1. Photosystem of Purple Bacteria

The structure of the RC for the purple bacteria *Rhodopseudomonas viridis* [26] is depicted in Figure 9, showing the specific positioning, orientations, and intermolecular distances of the prosthetic membrane-bound pigments. The workings of the RC incorporate a "special pair" of bacteriochlorophyll molecules (P), which act as the primary electron donor, with the ancillary bacteriochlorophyll (B) mediating ET to the primary electron acceptor bacteriopheophytin (H). The final redox equivalent resting place by ET is the ubiquinone residue (Q_B). The hole left behind at the oxidized P unit is filled by reduction, using a heme group from a nearby (23 Å) cytochrome residue. The efficiency of charge separation approaches unity, and it is clear to see why the RC center works so well, as forward ET rates are several orders of magnitude greater than the inherent energy-wasting reverse ET processes (Fig. 9). It would seem, therefore, that if one is to develop an artificial system capable of comparable CS efficiency, then rates of forward electron transfer processes would need to greatly exceed those of any charge recombination reactions.

One intriguing aspect of electron transfer within the RC is the observed unidirectionality along the L-branch at the expense of the near-identical M-branch (as shown). It is worth noting that the two sets of redoxactive cofactors are placed symmetrically about a transmembrane rotational axis and that the relative distances and orientations of these units make the two pathways very similar. The reason for unidirectionality has been attributed to a higher effective dielectric constant on the L-side, which effectively lowers the energy of the key CS states because of stronger ion pair solvation [27]. The mimicking of this process and its potential importance in the development of photoactive systems are only just being realized. Aspects of the necessary design criteria for artificial systems will accordingly be discussed in Section 3.2, particularly in relation to functionalized catenanes.

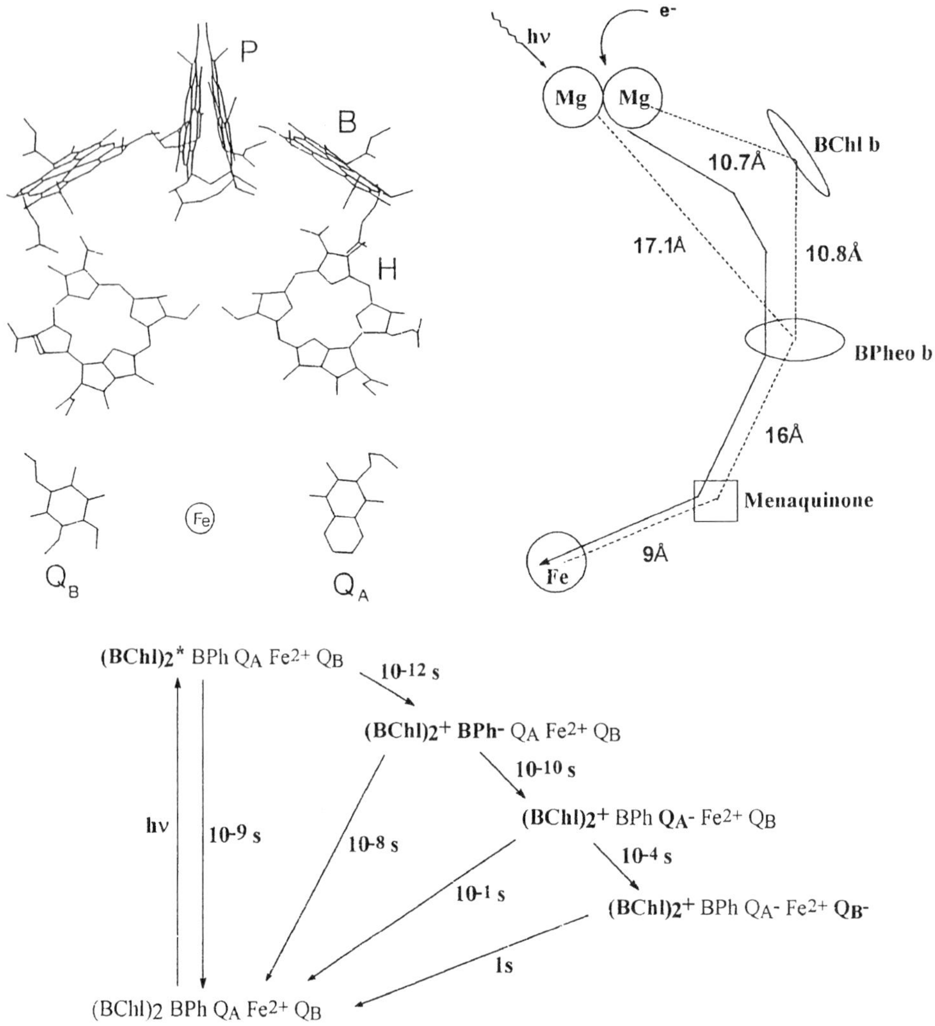

Fig. 9. X-ray structure of the photosynthetic reaction center for the purple bacteria *Rhodopseudomonas viridis*. Also shown are approximate distances between the redox partners and the relative rates of electron transfer that accompany light activation of the "special pair." (Reproduced in part by permission of The Royal Society of Chemistry.)

2.1.2. Photosystem II

In common with the RC of purple bacteria is the oxygen-generating center used in plant photosynthesis (photosystem II), which consists of at least 10 protein subunits [28]. Within photosystem II, electron migration processes take place via prosthetic groups associated with two particular proteins designated D1 and D2 (Fig. 10). The bound redox- and photoactive centers involved in light-induced charge separation include a special pair of chlorophylls (P_{680}), pheophytin, and two quinones, Q_A and Q_B, that are separated by a non-heme iron residue; these subunits are collectively called the "acceptor side" and the corresponding "donor side" associated with redoxactive tyrosine residues. As depicted in Figure 10, the reaction center P_{680} is orientated to the lumen face of the thylakoid membrane, and the reducible plastoquinones (Q_A and Q_B) are located in the membrane that faces the chloroplast space. In common with RCs of purple bacteria, photoexcitation of P_{680} leads to the rapid movement of an electron through a distance of $\sim$30 Å, leaving an oxidized P_{680} residue that in this instance is utilized in the water-splitting reaction.

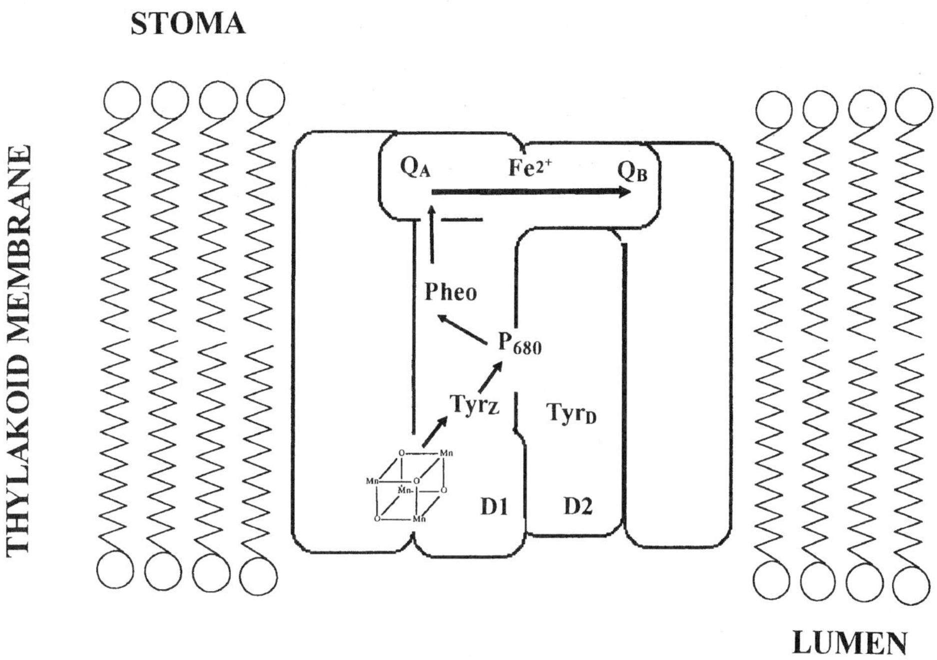

Fig. 10. A simplified picture of the workings inside photosystem II, with dark arrows marking the pathway of electron flow. Note the similarity with the reaction center of purple bacteria.

The common feature of RCs in both photosystem II and purple bacteria is that primary (fast) electron transfer reactions occur between expanded macrocyclic rings and are accompanied by slower processes involving quinone moieties. It is also noted that control of individual electron transfer steps is imparted through both the protein hydrophobic environment and the specific orientation of cofactors in the lipid bilayer.

2.1.3. Antenna Complex LH2

Although charge separation is associated with RCs of light-harvesting bacteria, the primary photon capture is performed in membrane-bound antenna complexes (LH1 and LH2). The recent X-ray structure [29] of the complex LH2 from the bacteria *Rhodopseudomonas acidophila* (Fig. 11) containing two "rings of porphyrins" represents a breakthrough that opens up a new dimension of the field of artificial photosynthetic models.

The crystal structure reveals that the active assembly consists of two concentric cylinders of helical protein units whose purpose is to enclose a collection of pigment molecules. Situated between the two protein helices is an overlapping ring of 18 bacteriochlorophyll-*a* (Bchl *a*) molecules with a further nine positioned between the outer helices and perpendicular to the transmembrane helix axis. The ring of 18 bacteriochlorophyll (Bchl *a*) molecules absorb around 850 nm and are strongly exciton-coupled, whereas the smaller ring of nine Bchl *a* molecules behaves more like monomers and have an absorption maximum at 800 nm; the observed differences are a consequence of the different ring environments. Energy transfer from the B800 to the B850 molecules is an efficient process and occurs on a time scale of 700 fs, and is followed by a rapid transfer among the pigment residues. The ring of B850 Bchl *a* molecules acts like a "storage ring," with the excited-state energy rapidly delocalized over a large area. This observed large-scale delocalization is facilitated by the highly hydrophobic environment, which tends to reduce the local dielectric constant and thus promotes exciton coupling.

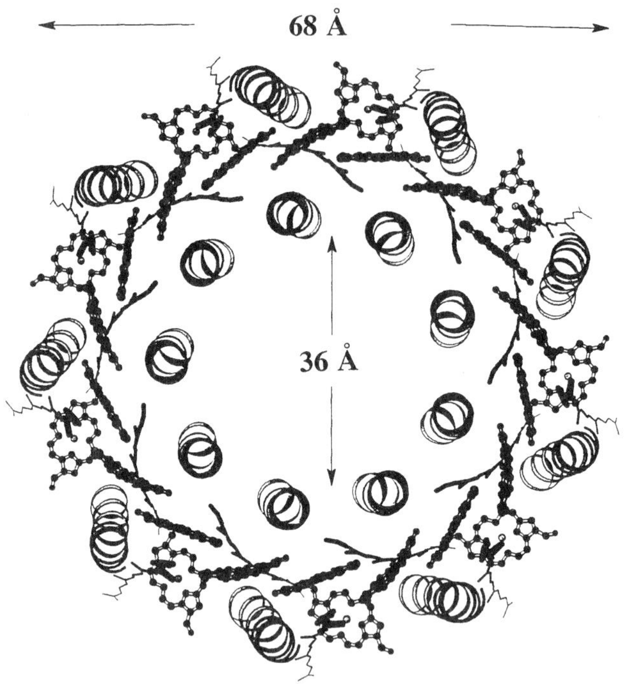

Fig. 11. The X-ray structure, looking down the protein helix of the light-harvesting antenna complex LH2 from the bacteria *Rhodopseudomonas acidophila*.

The dimensions of each complex are only 68 Å × 16.5 Å, leaving a relatively small volume within which a total of 27 light-harvesting pigments are judiciously positioned.

2.2. Light-Activated Biological Switches

In common with photosynthesis, the optical switching of a chromophore within a biomembrane is used to convert light energy into chemical energy via the synthesis of ATP [30]. Unlike RCs, however, there is no reliance on vectorial electron transfer processes, but there is instead a repeated cycling of structural changes in a chromophoric center; the bioevent is switched on by an optical signal and off in the dark. It should be noted that photoreversible transformations need not only lead to more localized processes, but after initial sensing coupled with signal amplification and output can give a large-scale biological response (phototropism) [31].

2.2.1. Proton Pumps: Bacteriorhodopsin

Bacteriorhodopsin (bR) is a light-driven proton translocase (26 kDa) found in the purple membrane of the extremely halophilic bacterium *Halobacterium halobium* [32]. The function of the bR is to eject protons from cells of halobacteria and to create an energy gradient across the cell membrane; the energy is provided by direct sunlight. Light-energy transduction relies on a fast cyclic photoreaction in a retinal center, which is found to be a protonated Schiff base and is bound to a lysine residue from the surrounding polypeptide. Although several intermediates have been characterized in the photocycle, the key process is a simple isomerization about the retinal C13–C14 double bond, which initiates a deprotonation of the Schiff base residue. The resetting of the system is a simple reprotonation of the Schiff base from the cytoplasmic side, followed by thermal reisomerization (Fig. 12).

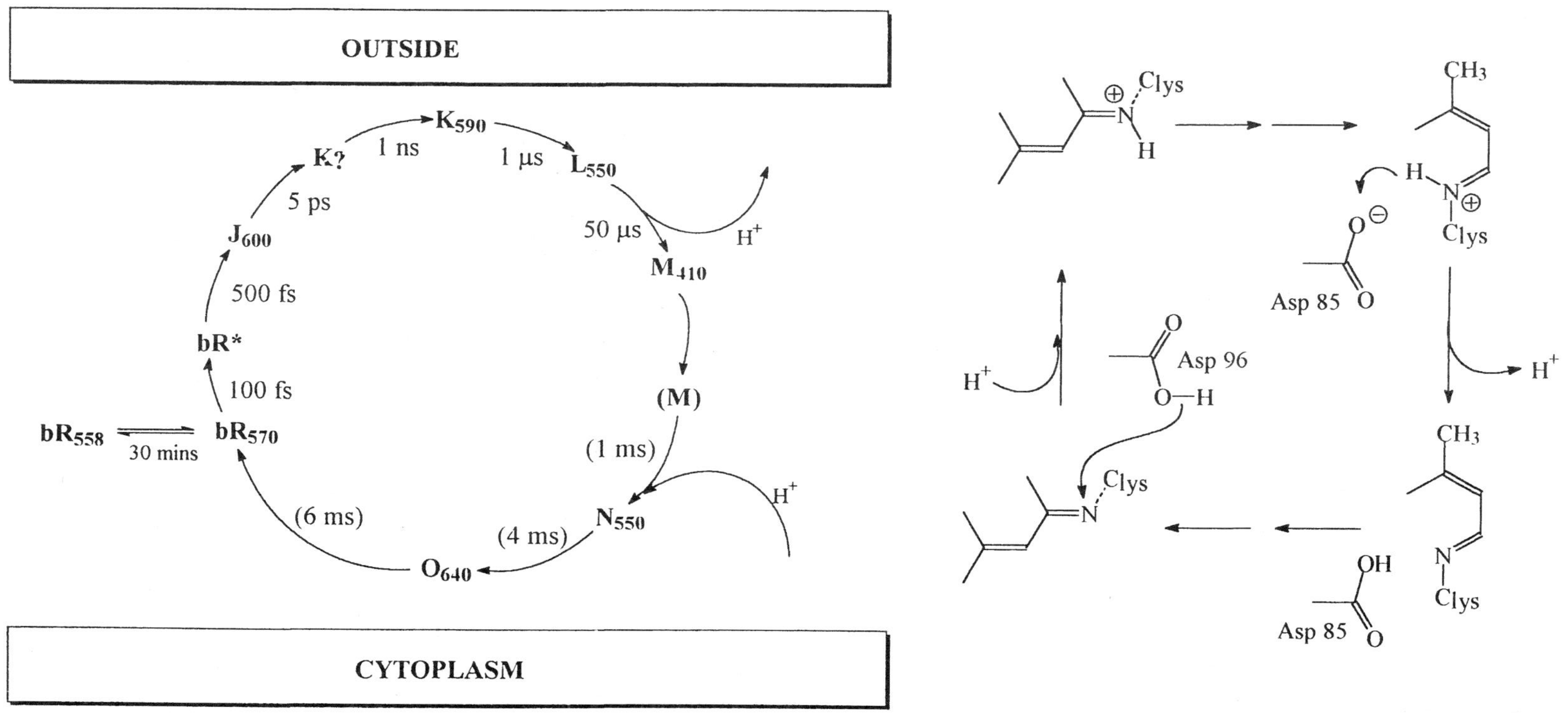

Fig. 12. The events following excitation of *Bacteriorhodopsin*, and a pictorial representation of the structural changes that accompany isomerization. (Reproduced in part by permission of Plenum Publishing.)

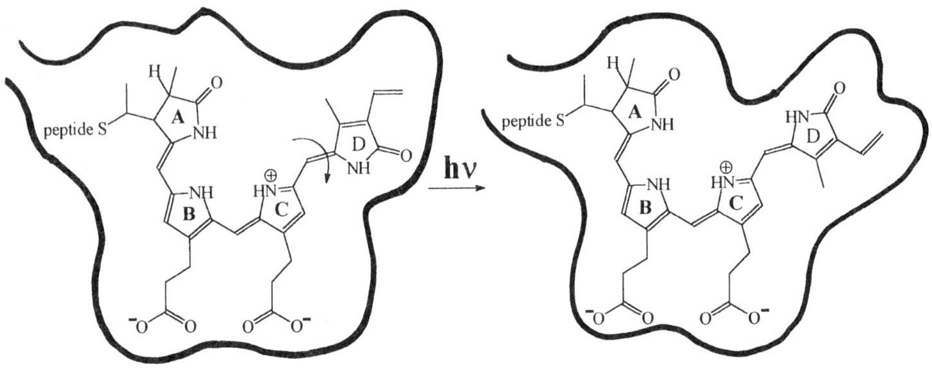

Fig. 13. The proposed isomerization unit of the chromoprotein phytochrome. (Reproduced by permission of Plenum Publishing.)

2.2.2. *Photophobic Responses*

The highly sensitive photoperception to red light coupled with, for example, photophobic reactions in higher plants is linked to the initial ultrafast photoprocesses in the tetrapyrrolic chromoprotein phytochrome [33]. Absorption of light (660 nm) by the Pr form of the photoreceptor protein is accompanied by phototransformation to the physiologically active Pfr form, which can be reverted back to the Pr form by by far red light (730 nm).

$$\mathrm{Pr} \underset{730\ \mathrm{nm}}{\overset{660\ \mathrm{nm}}{\rightleftharpoons}} \mathrm{Pfr} \rightarrow\rightarrow\rightarrow \text{biological response}$$

From amino acid sequence studies in oat phytochrome, it has been shown that the tetrapyrrole chromophore is found tethered via a thioether linkage to a cysteine residue. A significantly hydrophobic region is found adjacent to the chomophore linkage site, suggesting that it has an important role to play in the working of the chromophore. The Pr-to-Pfr alteration of phytochrome occurs in the Q_y excited singlet state, is complete in a few picoseconds, and is ascribed to a C=C double bond isomerization between rings C and D of the tetrapyrrolic segment (Fig. 13).

At the macroscopic scale, the photoreceptor phytochrome has been linked to the phototropic behavior of certain moss and fern protonema. In such cases it is assumed that the axis of maximum absorption of the phytochrome turns by 90° during transition from the Pr to the Pfr form, leading to an orientation of the Pfr molecules perpendicular to the cell surface. The time scale for the phototropic response in caulonema of *Ceratodon purpureus* is on the order of 20–30 min [34]. Despite the primary event being an elementary isomerization process, repeated amplification results in an overall bulk displacement that far exceeds the primary molecular-scale event. As such, the specific light-induced movement of a molecular module along a specific trajectory can be considered molecular-based artificial phototropism.

2.3. Overview

Using the features of the previous biological examples, a number of key design points/reactions can be extracted that may either be incorporated into artificial nanosystems or serve as processes to be mimicked at the nanoscale. In summary, Nature is adept at (i) using light to initiate reactions (i.e., ion movement), (ii) instigating fast unidirectional electron transfer processes, (iii) utilizing structural changes in reaction centers that impart information through proteins, (iv) managing relatively simple processes that have a significant impact on large structures, (v) using expanded macrocyclic rings and quinone groups,

(vi) handling chromophores that possess two different accessible states that are readily interchanged (switched), (vii) using highly sensitive groups, and (viii) utilizing the relative positioning and orientation of redoxactive groups to control processes.

3. ARTIFICIAL SYSTEMS: APPLICATIONS AND EXAMPLES

The development of artificial systems that operate specifically in the nanoscale domain is synonymous with the areas of research of "molecular electronics" (information techniques with molecular functional units) [35] and "molecular devices" (a system in which the physical properties of individual molecules are used, as opposed to group properties resulting from their interaction) [36]. For the latter case, it should not be forgotten that although the individual functioning of a molecular unit is of overriding importance, for a working device to be controllable, a significant number of molecules would be required to obtain a statistical behavior. Within the realm of molecular devices one can imagine such systems acting as sensors [37], wires [38], switches [39], shift registers [40], memory storage modules [41], and ultimately, molecular computers [42]. A detailed account of the aspects and physicochemical requirements needed in the development of such systems is too vast to cover, but the following key points are noted as issues that certainly have to be considered:

(i) *Speed of Process*. The kind of molecular device required clearly governs how quickly a process needs to be initiated and the results processed. The current interest in molecular-based optical storage devices having read, write, and erasable memory serves as a relevant example. In this instance, a molecule is required that contains two distinctly different states that can be interchanged by external means, where the transformation is detectable macroscopically (i.e., absorption/fluorescence changes). For a fast on/off memory system, the external stimulation mode is the rate determining step, as reading is carried out at the speed of light. In this case slow diffusional processes seem inappropriate, and use of photons (as used by Nature) to procure structural or transient absorption changes is more reasonable.

(ii) *Selectivity*. Considering that molecules are inherently built up from many different components with specific redox potentials and absorption and fluorescent properties, the ability to specifically activate one component is challenging. The selective excitation of specific chromophores with ultrashort laser pulses is feasible and offers a definite advantage over conventional techniques in the initiation of a photochemical process.

(iii) *Compound Degradation*. The most significant problem facing the use of molecular materials is their stability to repeated sampling from external sources; this is specifically true for reversible cycling systems using light activation. For instance, an optical switch operating at 100 MHz that degrades by only 1% for every 10^5 cycles would degrade in less than a second [36].

In the end, the ultimate challenge for molecular devices will of course come from their ability to successfully compete with the existing technologies, which have much longer and deeply rooted development histories. Although integration of molecular-scale devices into real workable machines may seem far off, the growing research effort is surely indicative of future optimism. The molecular systems outlined in the following sections are intended to exemplify one small part of the research into molecular-scale functional assemblies.

3.1. Artificial Molecular Systems Based on Rotaxanes, Catenanes, and Cyclophanes

Previously (Section 1) the topologically interesting catenanes and rotaxanes were introduced as prototypes for functional molecular systems. The potential reactivity and biological mimicry of these assemblies are linked to their unique structural attributes, which are otherwise unavailable to simple "open-ended" molecular structures. We will begin, however, with more simple cyclophane-based architectures that can be viewed as counterparts to the more complex rotaxane/catenane systems, but still incorporate unique features that can be utilized for beneficial purposes.

3.1.1. Light-Activated Artificial Phototropic Assemblies

Molecular structures, especially those constructed by the interlocking of redoxactive host-guest species, offer the possibility of devising artificial phototropic assemblies. The main feature of such units is that light-induced triggering of a large-scale conformational change along a preferred trajectory allows the reversible cycling of energy and information in the superstructure (Fig. 14). Particularly adept systems are those in which primary binding involves electron donor–acceptor (EDA) interactions, because the molecular association can be readily perturbed selectively and spontaneously by laser excitation.

The ubiquitous feature that facilitates the selective triggering of common EDA systems is that molecular interaction is characterized by the appearance of a colored Gaussian-shaped CT absorption band. This CT band is also beneficial, as it permits calculation of a binding constant (K) for the donor-acceptor complex by standard Benesi–Hildebrand analysis [43]. Furthermore, curve-fitting the Gaussian bandshape gives parameters that are used in the determination of the corresponding electronic coupling matrix (H_{DA}) (Eq. 1) using Hush theory [44], which in turn gives an insight into the electron transfer properties of such assemblies:

$$H_{DA} = \frac{0.0206}{R_d}\sqrt{\varepsilon_{max}\Delta\nu_{1/2}E_{op}} \tag{1}$$

Here, ε_{max} is the molar absorption coefficient (M^{-1} cm^{-1}), E_{op} ($= 1/\lambda_{max}$) represents the optical energy (in cm^{-1}) of the absorption transition, $\Delta\nu_{1/2}$ is the half-width maximum (in cm^{-1}) of the CT band, and R_d equals the separation (in Å) of the donor/acceptor moieties.

The direct excitation via the CT absorption bands of EDA complexes results in electron transfer from donor to acceptor and formation of a compact radical ion pair (RIP). Provided that energy-wasting charge recombination (k_{cr}) is not too fast, the mutual electrostatic repulsion between the molecular partners facilitates dissociation (with a rate constant k_{sep}) of the RIP to form solvated radical ions [45]. For efficient photoinitiated separation of the interlocked species, k_{sep} must be able to compete with k_{cr}, with the quantum efficiency (Φ_{sep}) given by

$$\Phi_{sep} = \frac{k_{sep}}{k_{sep} + k_{cr}} \tag{2}$$

It is clear from the simple cartoon in Fig. 14 that one can construct artificial phototropic systems by constraining either electron donors or acceptors into a cyclophane ring, thus enabling the incarceration of their counterpart aromatics.

3.1.1.1. Electron-Donating Crown Ether Complexes

It is well documented that the electron-accepting cationic molecule $\mathbf{PQT^{2+}}$ readily binds inside the central void of electron-rich dibenzocrown ethers (**CE7–CE12**) when the two are mixed in a polar solvent such as acetonitrile (Fig. 15) [46]. The magnitudes of the binding

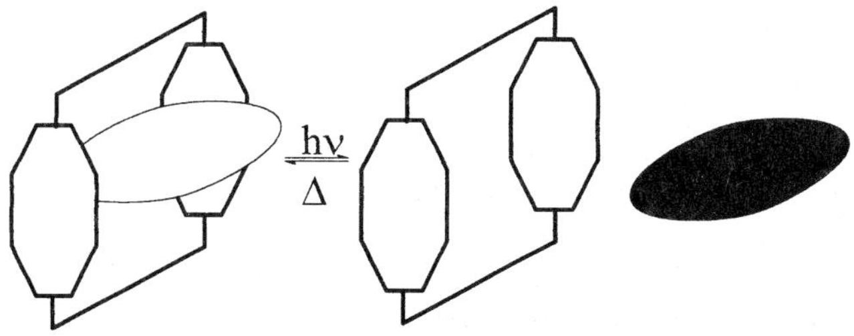

Fig. 14. A cartoon representation of artificial phototropism, showing the ejection of a molecular unit along a specific trajectory following light activation.

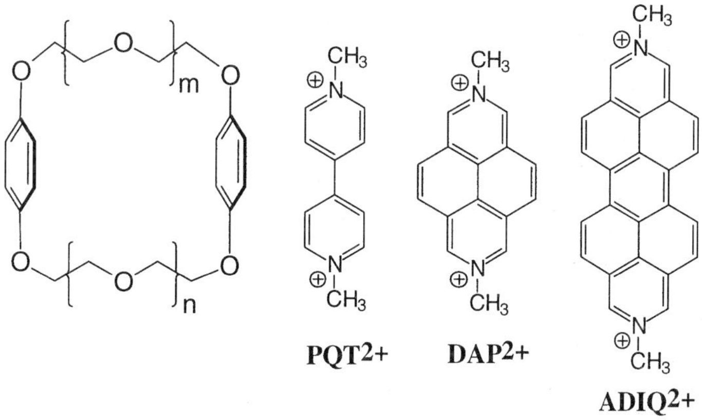

CE7 m = 2, n = 1; **CE8** m = 2, n = 2; **CE9** m = 3, n = 2;
CE10 m = 3, n = 3; **CE11** m = 4, n = 3; **CE12** m = 4, n = 4

Fig. 15. Potential EDA units based on electron-deficient planar aromatics and electron-rich crown ether cyclophanes.

Table I. Calculated Parameters for Charge-Transfer Complexes with PQT^{2+} in Acetonitrile

Cmpd	K (M^{-1})	λ_{max} (nm)	H_{DA} (cm^{-1})	K_{CR} (10^{10} s^{-1})
CE(7)	0.3	383	2065	4.00
CE(8)	12.0	419	1220	1.03
CE(9)	22.3	432	1400	1.23
CE(10)	48.7	440	5500	1.82
CE(11)	25.7	422	1445	1.35
CE(12)	15.5	415	1270	0.95

Values are from [47].

constant (K), wavelength maximum (λ_{max}), and H_{DA} (Table I) have been shown to vary with the size of the cyclophanes, reaching a maximum at **CE10** (1,4,7,10,20,23,26,28,32-decaoxa[13.13]paracyclophane), corresponding to the "best fit" situation.

The dynamics of charge recombination within the various donor–acceptor complexes have also been shown to mirror closely the host–guest size matching. In all examples there is no sign of the photogenerated ions separating into their solvated species, as charge recombination is far too efficient (Table I) [47]. This work carried out by Harriman and co-workers has also determined that for the complex **CE(10).PQT^{2+}** solvent polarity has a pronounced effect on the H_{DA} values, suggesting that the intimate structure of the CT complex may alter somewhat.

The practical implication from this work is that although a selective triggering mode is available (i.e., laser excitation of CT band), the imparted energy is wasted, transmitted to the surrounding solvent as heat, with no phototropic behavior. However, work carried out by the same workers has identified ways in which to "slow-down" the charge recombina-

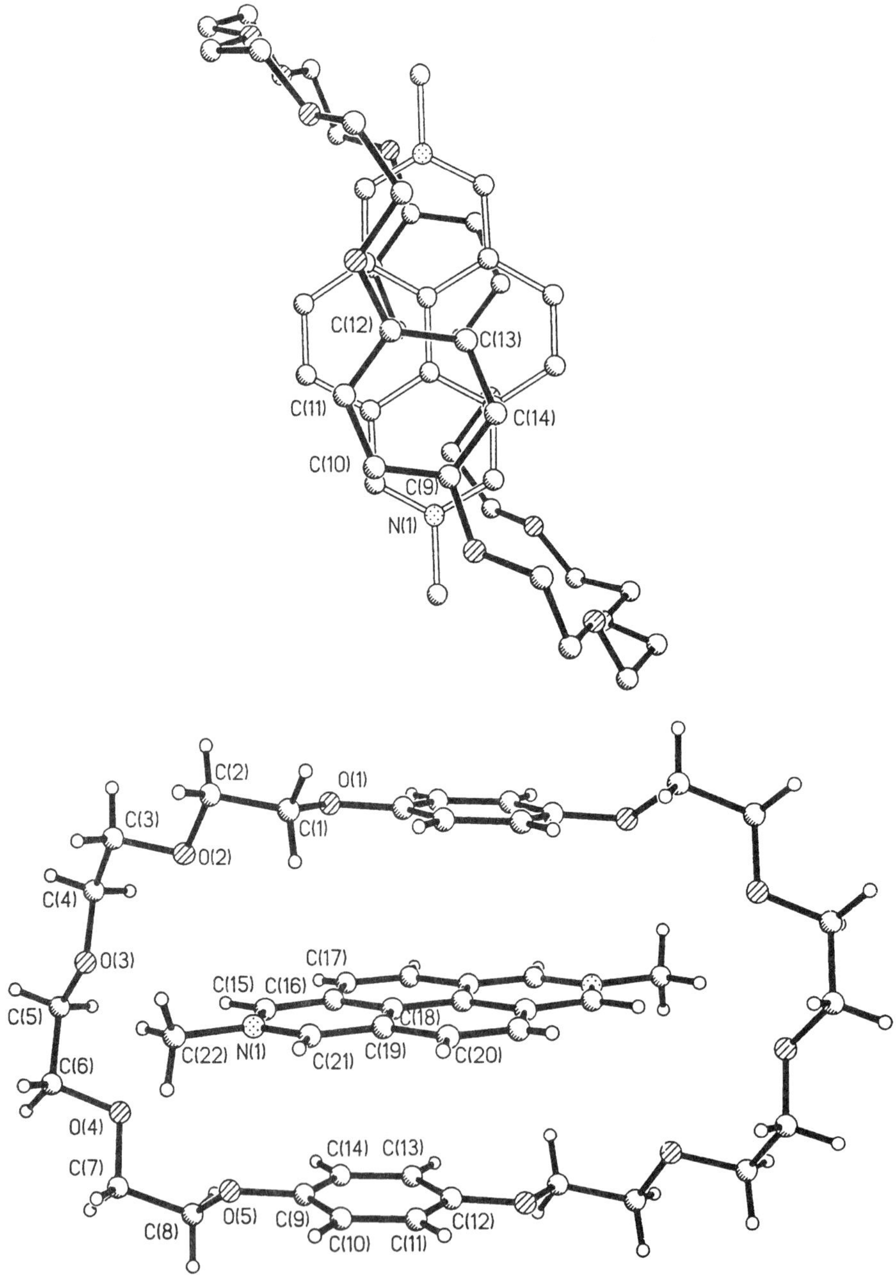

Fig. 16. X-ray crystal structure of **DAP^{2+}.CE10**, showing the top and side views of the molecular complex.

tion step and hence induce phototropic behavior through the use of larger planar aromatic acceptors; namely, **DAP^{2+}** (*N,N′*-dimethylanthra[2,1,9-def: 6,5,10-d′e′f′]diisoquinoline dication) and **ADIQ^{2+}** (*N,N′*-dimethyldiazapyridinium dication) [48]. Both of these extended π-stacking moieties predominantly form 1:1 complexes with **CE10**, the binding constants (3900 M^{-1} (DAP^{2+}) and 425 M^{-1} (ADIQ^{2+})) being significantly greater than for **PQT^{2+}** (4,4′-dimethyl-4,4′-bipyridinium dication; methyl viologen or paraquat). From the X-ray structure of the complex formed between **DAP^{2+}** and **CE10** (Fig. 16), the dication fits within the central void created by the macrocyclic ring and aligns itself with the two hydroquinol units. From the side view, it is clear that the dication sits almost centrally in the cavity, with the two hydroquinol units of the crown interacting with each half of the **DAP^{2+}** unit. The slight twisting of the **DAP^{2+}** unit within the cavity (top view) ensures maximum interaction between the host–guest molecules.

Laser excitation of the complex **DAP^{2+}.CE10** in acetonitrile results in the rapid establishment of the RIP from the initially formed Frank–Condon state. Unlike in the previous case, deactivation of the RIP occurs on a slower time scale, with $k_{obs} \approx 7.7 \times 10^9\ s^{-1}$. Moreover, some 10% of the initially produced ions separate into their solvated π-radical cations, indicating that k_{sep} is able to compete, although still to a rather limited extent, with back electron transfer. It has been proposed that the difference between the diazapyrene and viologen CT complexes is linked to the existence of low-lying π-π^* states that exist in the **DAP^{2+}.CE10** complex, which are otherwise absent in **PQT^{2+}.CE10**, and mix with the CT state. The introduction of π-π^* character into the CT state therefore stabilizes the donor–acceptor complex and decreases the rate of charge recombination.

Unlike the viologen dication, the diazapyrene residue is highly fluorescent ($\Phi_F = 0.67$) in its uncomplexed form but fully quenched when residing in the crown ether cavity. If one considers the photoprocesses in Figure 17, it is possible to envisage utilizing the cycling process as a fast molecular switch. In this instance, the starting complex represents a nonfluorescent species, which when triggered results in the generation of two redox ions by dissociation. The return electron transfer by bimolecular collisions of these two ions consequently regenerates the **DAP^{2+}** residue, whose fluorescence can be detected before formation of the starting complex by its reinsertion into the crown.

Although far from ideal, the system represents a move forward in the manufacture of artificial phototropic materials that have the potential to act as fast optical switches.

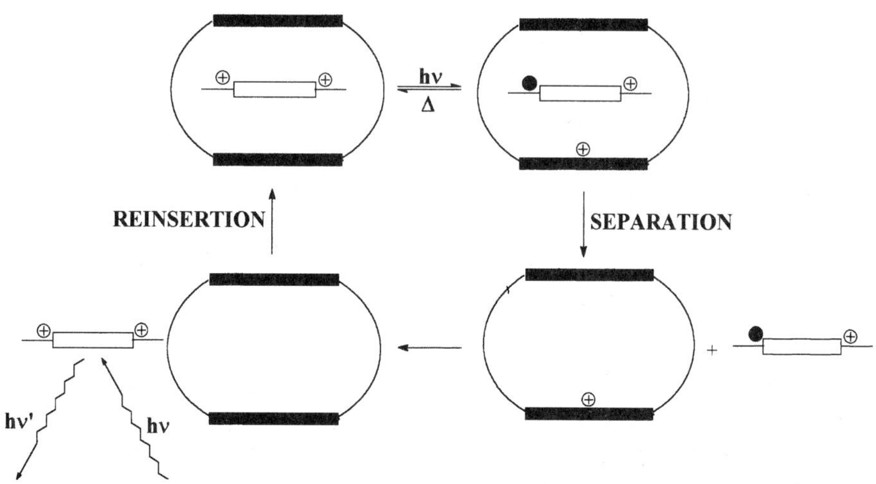

Fig. 17. The reversible photocycle that is proposed to occur in the supramolecular complex **DAP^{2+}.CE10**. Note: Before reinsertion of the **DAP^{2+}** unit, the fluorescence emitted by the molecule is detectable and hence acts as an optical read-out probe.

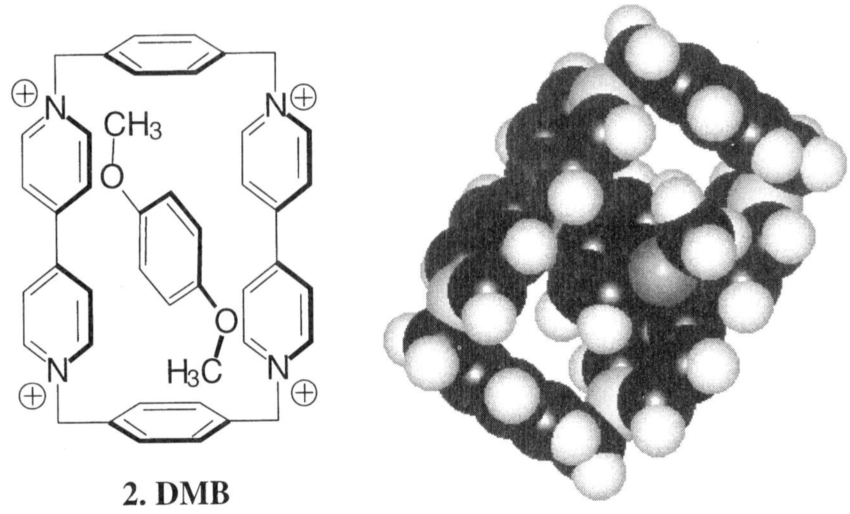

2. DMB

Fig. 18. The complex **2.DMB** and a space-fill computer-generated picture of the complex formed, showing the compact structure.

3.1.1.2. Electron-Accepting Cyclophanes

Instead of constraining the electron donor to a macrocyclic ring, formation of an electron-accepting cyclophane means that electron rich aromatics can be bound to form highly compact donor-acceptor complexes. The most significant contribution to this area has been by Stoddart and co-workers in their construction of a viologen-based box (**2**) (Fig. 2), the dimensions of which ($\sim$6.8 $\times$ 10.3 Å) are optimal for encapsulation of varying electron-rich aromatics such a 1,4-dimethoxybenzene (DMB) (Fig. 18).

As in the crown-based systems, the CT interaction can be destabilized by laser excitation of the resultant EDA complex, with the dynamic processes that follow reminiscent of those shown in Figure 17. Listed in Table II are some of the results obtained by a study carried out on such a series of EDA complexes.

The most striking feature of these systems is that the λ_{max} can be readily tuned from the near infrared to deep into the UV region by simply altering the redox potential of the bound guest. As found previously in crown-based assemblies, the charge recombination process in the majority of the EDA complexes is highly efficient, and no separation of the RIP into solvated ions occurs. It is only toward the lower end of the table that examples arise in which any appreciable molecular dissociation is possible, even though for the best case the quantum efficiency is only a modest 24%. For these host–guest systems, the photonic energy imparted to the RIP has to dissipate through appropriate skeletal vibrational modes, and as a consequence, energy loss mechanisms effectively control the rate of charge recombination. The more energy the RIP has to dissipate, the slower the charge recombination process, although the actual rate energy dependency is rather shallow and reminiscent of a simple energy gap law:

$$k_{CR} = A \exp[\delta \Delta G^{\circ}] \tag{3}$$

where A and δ are constants and ΔG° is the free energy change for charge recombination.

From this work it is clear that if photodissociation of EDA complexes based on cyclophane **2** is to be both achievable and useful, then donors with high oxidation potentials are required. The down side of this finding is that the CT absorption band of such an EDA complex would fall deep into the UV region and make selective excitation difficult. In the following section, however, we address ways in which these problems have been by-passed, and ways in which the module **2** is used as a bead in one- and two-station [2]rotaxanes.

Table II. Parameters Measured for EDA Complexes Formed between Cyclophane **2** and Various Donors in Acetonitrile

Donor	K (M^{-1})	λ_{max} (nm)	k_{CR} (10^9 s^{-1})	k_{sep} (10^8 s^{-1})
TMPD[a]	42	820	248	—
PD[b]	112	655	230	—
MTOL[c]	15	601	145	—
BHQ[d]	18	473	56	—
MEOPH[e]	24	469	65	—
DMB[f]	18	475	46	—
Xylene	16	355	9.3	2.0
tBBENZ[g]	5	333	7.3	10
FTOL[h]	10	320	6.0	4.0
Toluene	16	319	5.5	4.0
DCB[i]	3	312	5.1	3.0
Benzene	17	290	2.6	8.0

Data are from [14].

[a] *N,N,N′,N′*-Tetramethyl-1,4-diaminobenzene.

[b] 1,4-Diaminobenzene.

[c] *N,N*-Dimethylaminobenzene.

[d] 1,4-Benzohydroquinone.

[e] 4-Methyloxyphenol.

[f] 1,4-Dimethoxybenzene.

[g] *tert*-Butylbenzene.

[h] 4-Fluorotoluene.

[i] 1,4-Dichlorobenzene.

3.1.1.3. Light-Induced Molecular Shuttles Based on Rotaxane Assemblies

To alter the position of a bead on the thread in a [2]rotaxane, it is clear that any associative interaction between the bead and its station has to be destabilized long enough for molecular motion to take place. In principle, this can be achieved by a number of methods, but if photochemical activation is considered, then three main methods are available, which are graphically illustrated in Figure 19.

In the first case scenario (A), destabilization of the bead–station interaction is brought about by direct excitation of the resultant CT complex in a way analogous to that described in Section 3.1.1. As discussed previously, there is a problem with this method, in that the resultant deactivation photoprocess that accompanies excitation is fast, and accordingly, any molecular motion of the bead can be ruled out. However, a main advantage of this approach is that the assembly is self-contained and only a single photoevent is required as the trigger. The two methods B and C are similar in that an external reagent is used as a photoreductant, which itself is cycled with a secondary agent to remove the return electron transfer process. In case B the photoreagent is added together with the rotaxane, whereas in case C the stopper is designed to act as the light-activated agent. The main disadvantage of these two methods is that agents are consumed and must be replenished to keep the whole system cycling. Relevant examples of assemblies based on cyclophane 2 and using the above-outlined methods are described in the following sections.

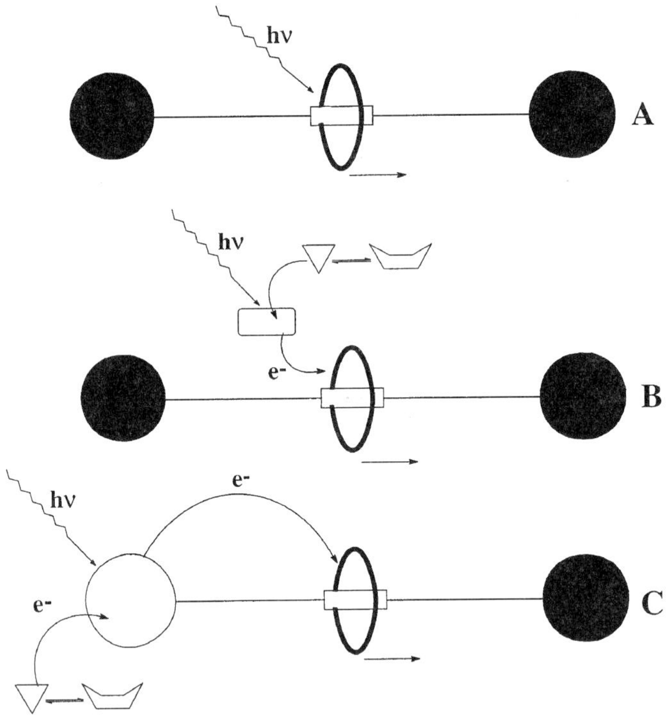

Fig. 19. Three possible methods of initiating molecular motion of a bead along a thread. In the diagrams the triangle represents an externally added reagent that physically changes (i.e., protonates) during the cycling and is required for the system to work.

3.1.1.3.1. Direct Excitation (Method A). A series of one- and two-station [2]rotaxanes developed by Harriman and co-workers [49] for direct activation via the CT absorption band are shown in Figure 20. Using the photophysical results obtained for simple dimethoxybenzene-based EDA complexes of **2** (Table II), it was surmised that to compete with rapid charge recombination, a secondary ET process would be required that shifted charge away from the primary RIP. In an attempt to achieve this goal, potential electron donor groups were incorporated directly into the rotaxane assembly via the terminal stopper groups.

For the rotaxanes **3–7**, the lifetime of the RIPs generated from photoexcitation using the CT band fall within the range of 14–30 ps, which is consistent with lifetime data obtained for the simple donor–acceptor complex **2.DMB**. In the anthracene-stoppered rotaxanes **6** and **7**, oxidation of an anthracene stopper by the photogenerated dialkoxybenzene π-radical cation was not observed because the charge-shift process is thermodynamically unfavorable ($\Delta G^\circ \approx +0.3$ eV) [50]. This was not, however, the case for the ferrocene-stoppered rotaxanes **3–5**, in which oxidation of a ferrocene unit ($\Delta G^\circ \approx -1.0$ eV) was observed to generate a charge transfer state (CTS); the rate constant for this process (k_{cs}) is $\sim 1.5 \times 10^{10}$ s^{-1} [51]. Although the quantum yields (ϕ_{cts}) for generation of the CTS appear low (~25% for **3**), the observed rate of oxidation for the ferrocene is far too high for a simple through-space process involving widely spaced molecular partners. However, the X-ray structures of rotaxanes **3** and **5** clearly demonstrate why the secondary oxidative stopper-based process is able to compete with rapid charge recombination (Fig. 21). In both structures the stopper units are seen to wrap around and form a secondary π-stack with a viologen unit of the cyclophane. This leads to a more closed and compact assembly, which, from ^{1}H NMR nuclear Overhauser enhancement spectroscopy (NOESY) experimental results, persists to a varying extent, even in the solution phase.

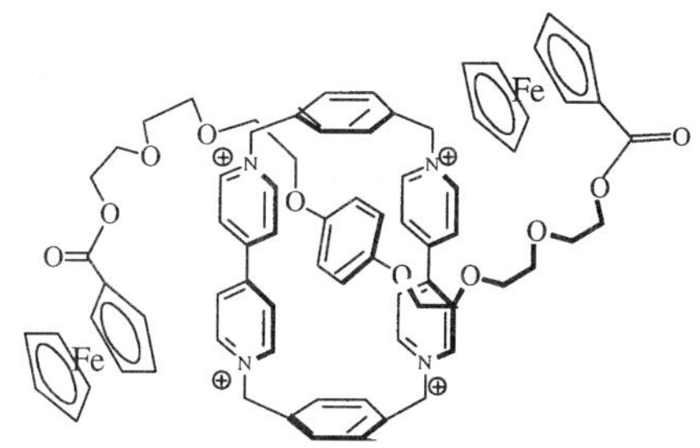

3

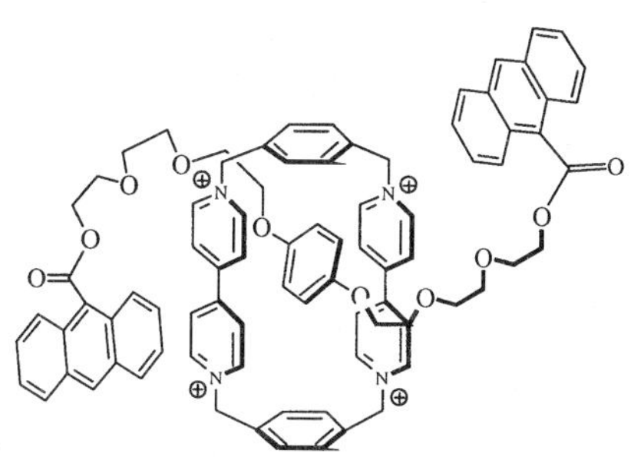

6

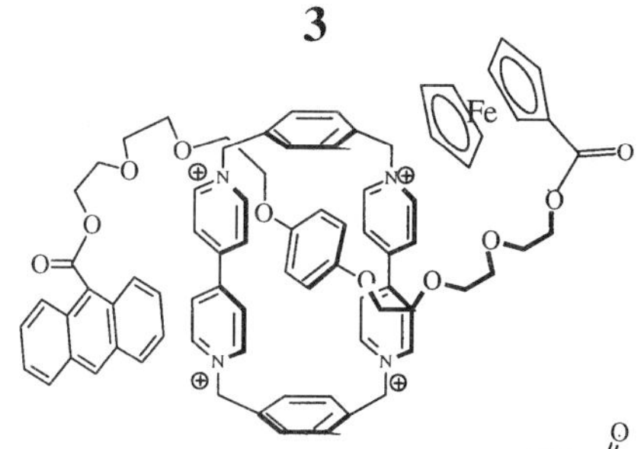

4

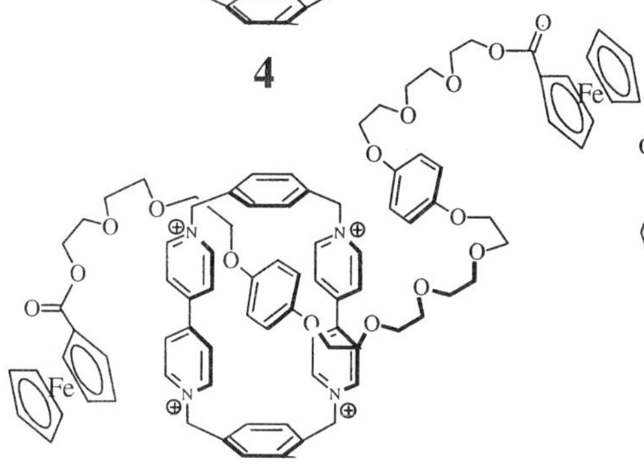

5

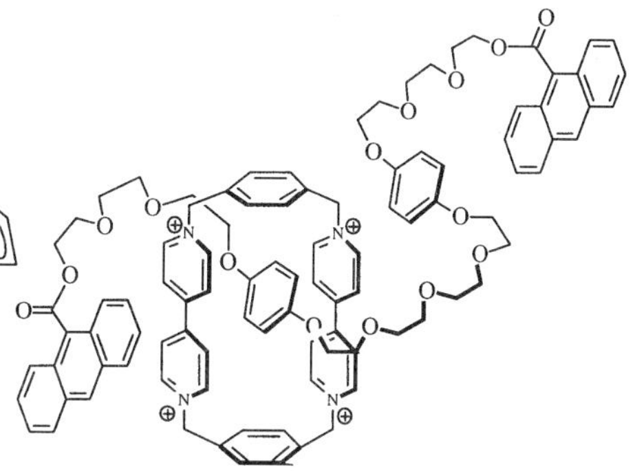

7

Fig. 20. Examples of ferrocene- and anthracene-stoppered [2]rotaxanes produced for light activation via the charge transfer absorption band.

The lifetimes of generated CTSs are relatively long (0.5–1 μs), suggesting an increase in separation between the ferricinium and reduced cyclophane groups. A picture of the overall processes occurring in the [2]rotaxane assemblies is illustrated in Figure 22. In solution the rotaxanes exist in two extreme conformations in which the thread is either fully extended or has the stopper units π-stacking to produce a compact assembly. The competitive electron transfer processes occur exclusively in the closed conformation, with the generated electrostatic repulsion between stopper and bead forcing open the conformer and driving the bead to the opposite end. The relatively long-lived CTS lifetimes can be explained by the large distances created between the generated redox ions. It is interesting to note that despite the long-lived CTS for two-station rotaxane **5**, the bead does not shuttle to the opposing station before return electron transfer takes place. In this instance, the highly mobile thread allows intramolecular diffusional encounter of the redox ions, which suggests that changes in the flexibility of the thread could be used to control the rate of return electron transfer and hence charge separation.

More recently, a light-driven molecular shuttle based on the cyclodextrin derived rotaxane **8** (Fig. 23) has been reported by Nakashima and co-workers [52]. Unlike in the previous rotaxane molecular shuttles, the bead segment (alpha-cyclodextrin, $\alpha-$CD) is made to move by the isomerization of an azobenzene moiety that resides in the thread.

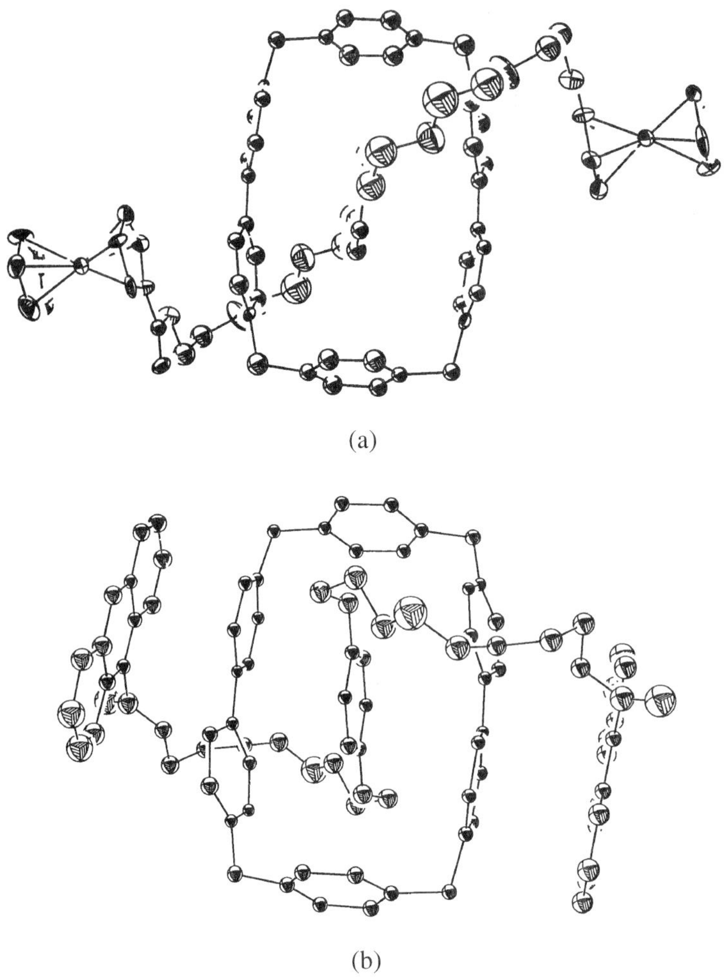

Fig. 21. The X-ray crystal structures of (a) [2]rotaxane **3** and (b) [2]rotaxane **6**, showing the stoppers π-stacking to the bipyridinium cyclophane.

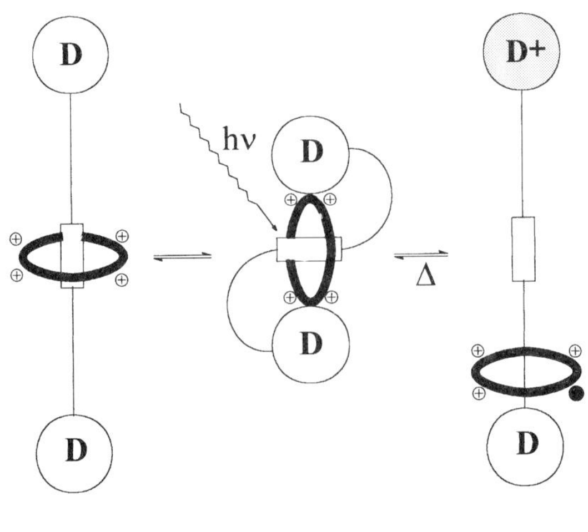

Fig. 22. The dynamic and photophysical processes that are proposed to occur in the ferrocene-stoppered [2]rotaxanes **3–5**.

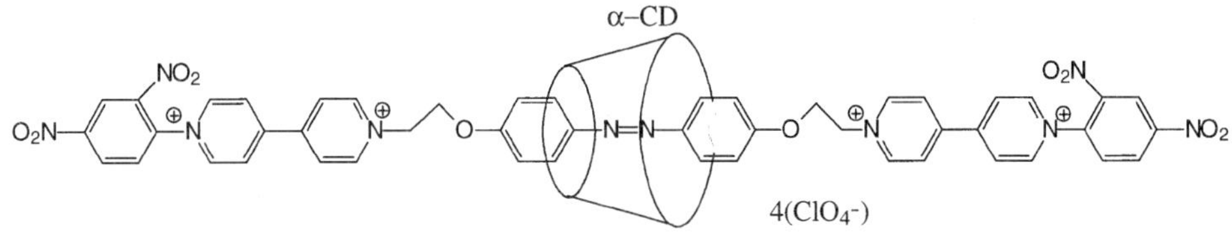

8

Fig. 23. The light-induced molecular shuttle prepared by Nakashima et al.

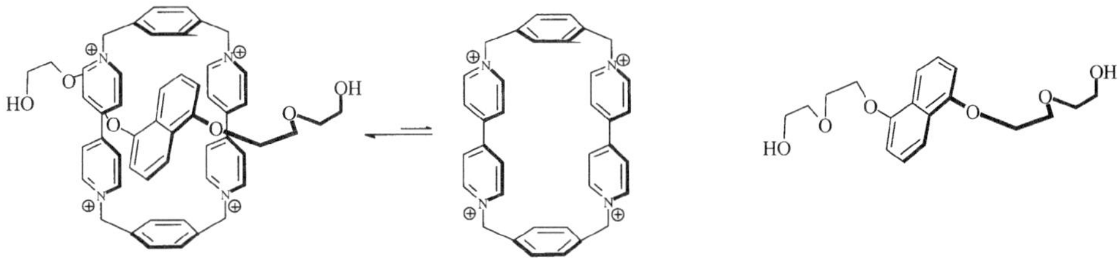

9

Fig. 24. The pseudorotaxane studied by Balzani and co-workers.

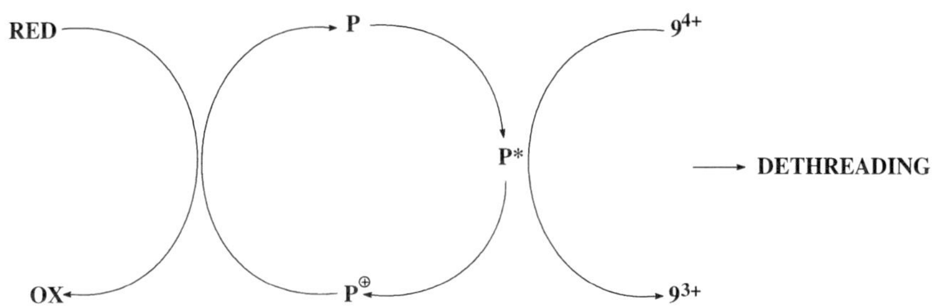

Fig. 25. Overall processes occurring in the dethreading of pseudorotaxane **9**. Abbreviations are defined in the main text.

3.1.1.3.2. Indirect Photoactivation (Methods B and C). The preliminary work on the indirect method of CT destabilization (as favored by Balzani and co-workers) was carried out on pseudorotaxane **9** (Fig. 24) incorporating the naphthoxy-based thread [53].

Because there are no stopper units on the assembly, the bead is free to slip off the thread, although in aqueous solution at an equimolar concentration of components (6.0×10^{-5} M), a vast proportion (85%) of solution species is pseudorotaxane. To perturb the host–guest equilibrium, the CT interaction is attenuated by the reduction of a bipyridinium unit, using an external photosensitizer (P). In this case, the highly favorable reducing ability of 9-anthracenecarboxylic acid ($E_{red}(P^+/P^*) = -0.88$ V vs. SCE) is used to transfer an electron to the cyclophane and disrupt molecular association. Although this reduction of a viologen unit leads to reduced binding, the pseudorotaxane does not become dethreaded, as the fast intermolecular return electron transfer process restores the ground state. To remove the species P^+, a large excess of triethanolamine(~200-fold) is added to the solution to act as a sacrificial reducing agent. As a consequence, turning off the wasteful return electron transfer reaction means that the cyclophane remains reduced and dethreading can take place. Even under these conditions, the extent of dissociation is extremely inefficient and slow. Under the conditions employed by Balzani, in 25 min 25% of the pseudorotaxane becomes unthreaded. A pictorial representation of these outlined processes appears in Figure 25.

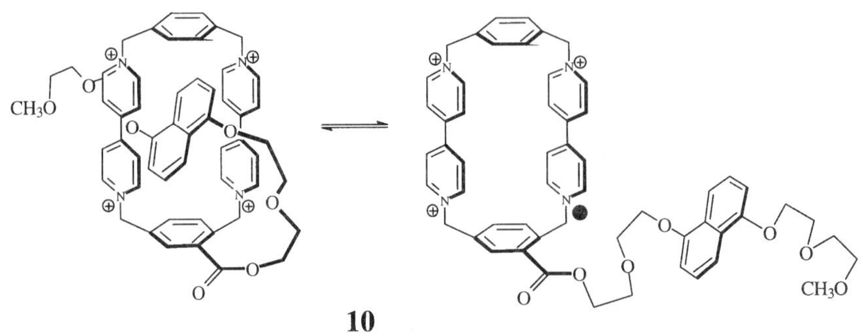

10

Fig. 26. Threading and dethreading observed in the functionalized pseudorotaxane **10** following reduction of the bipyridinium unit. In this case the naphthoxy unit is unable to totally escape and become completely detached from the molecular structure.

It is apparent that the outlined indirect method is not conducive to fast conformational switching, which is the consequence and advantage of the direct excitation strategy. The use of such systems would seem to be more in line with miniaturized long-term data storage modules rather than fast optical switching devices.

Despite the apparent problems with pseudorotaxane, more recently the same workers have taken this work further by tethering the thread directly to the cyclophane (Fig. 26). Irradiation for 15 min of an aqueous solution of **10** (6×10^{-5} M), 9-anthracenecarboxylic acid (5×10^{-6} M), and $EDTANa_2$ (1×10^{-2} M) as sacrificial reductant resulted in changes in the absorption spectrum that were consistent with dethreading [54].

To conclude this section, two further examples are noted that possess a sensitizer unit linked via flexible chains to electron donor groups that enter into EDA complexes with cyclophane **2** (Fig. 27). The work published on these systems does not directly address the issue of driving the cyclophane off the assembly, but gives an insight into the photoinduced electron transfer processes occurring within the molecular system.

In the first example, **11**, binding of **2** readily occurs at the ends of the ethoxy "arms" with one cyclophane associated with two of the alkoxybenzene groups [55]. The overall photophysical picture associated with this system is best described using the scheme shown in Figure 28.

In the absence of any quencher, the luminescence lifetime (τ) of **11** is 704 ns, but is reduced upon the addition of **2**, in accordance with quenching by photoinduced electron transfer. Using nanosecond time-resolved techniques, the appropriate quenching rates for the bimolecular process (k_{dq}) and intramolecular rate (k_{iq}) have been determined to be 3.4×10^8 M^{-1} s^{-1} and 2.9×10^8 s^{-1}, respectively.

More recently, Kimura and co-workers [56] have produced the pthalocycanine-based binding moiety **12**, which has been shown to form a 1:4 complex with cyclophane **2**. The steady-state fluorescence data obtained for **11** upon the addition of **2** are consistent with static quenching, where the calculated Stern–Volmer K_{SV} is 820,000 dm^3 mol^{-1}.

3.1.1.4. Light-Activated [2]Catenanes

The removal or reduction of the CT interaction between donor–acceptor groups of [2]catenanes possessing matching units as their counterpart [2]rotaxanes could, in principle, induce mutual ring rotation. Alternatively, an increase in the CT interaction by changes in the internal volume of a [2]catenane could be used to reduce ring rotation. Attempts to instigate such molecular motions have included direct excitation of the molecular assembly, in one instance via the CT absorption band and in the other by using a photoisomerizable group built into the catenane structure. Rather limited success has been achieved but

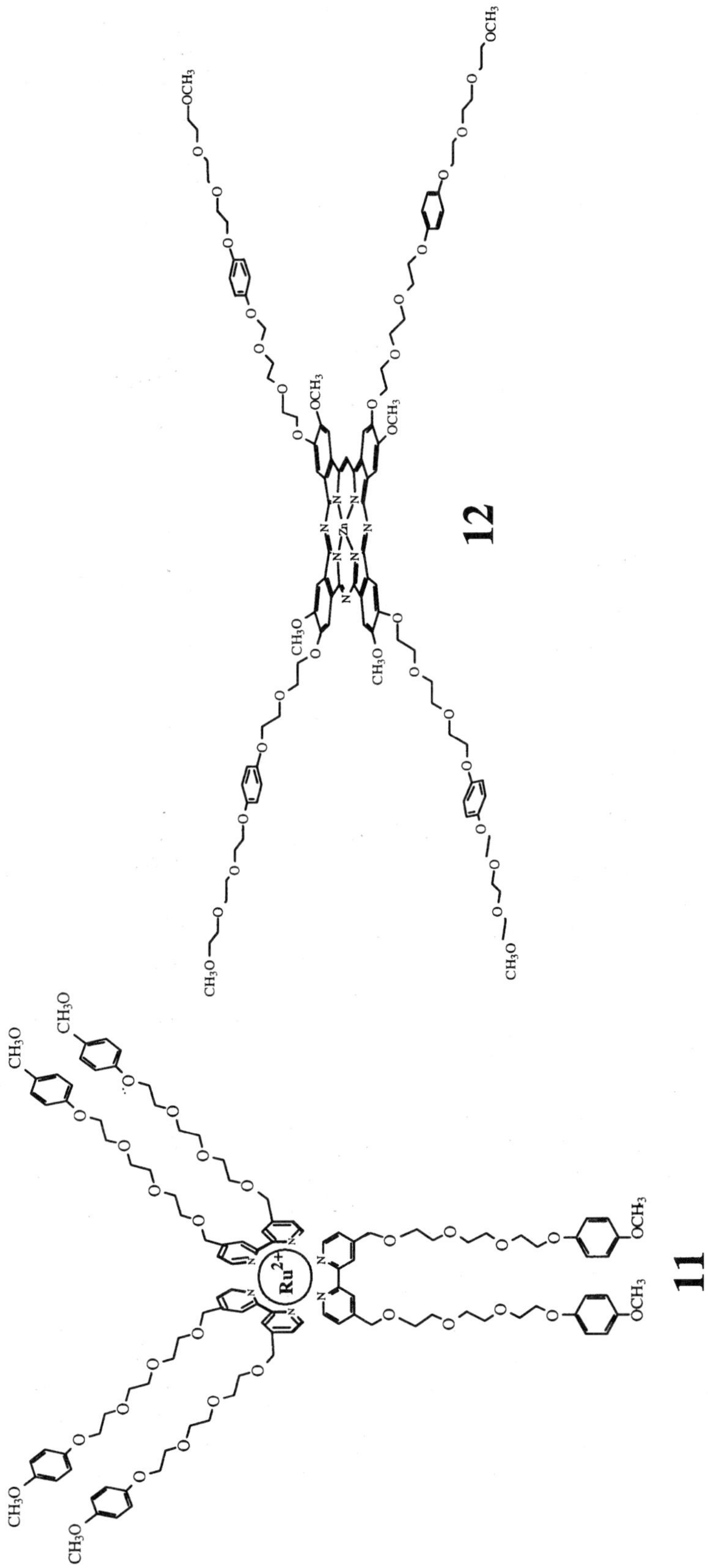

Fig. 27. Photoactive threads incorporating sensitizer units and binding sites for cyclophane **2**.

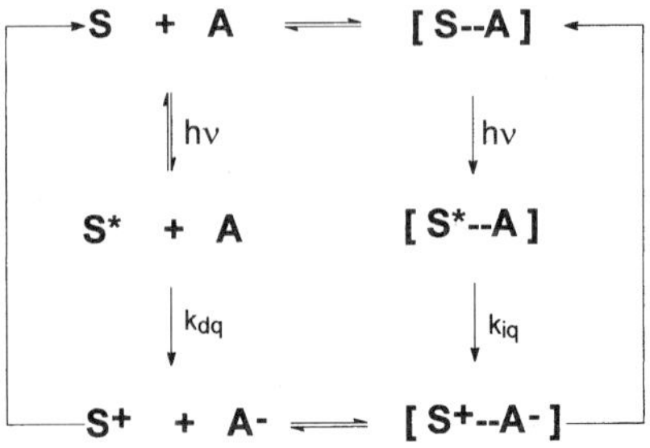

Fig. 28. Overall photoprocesses that are possible in the deactivation of a sensitizer (S) and electron acceptor (A). Here, k_{dq} represents the bimolecular rate constant for excited state deactivation, and k_{iq} is the first-order rate constant for quenching within the supramolecular complex.

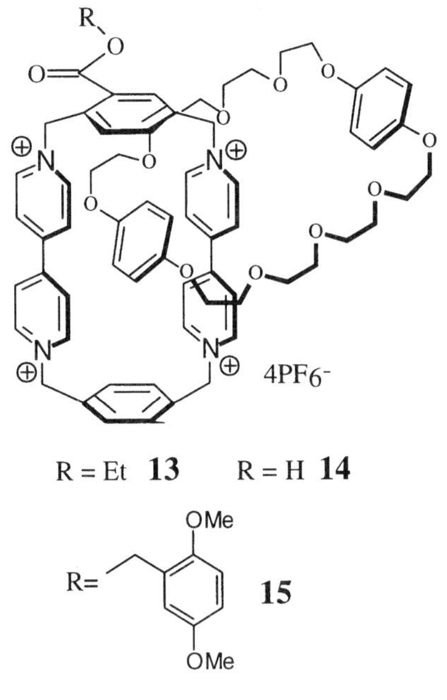

Fig. 29. Examples of [2]catenanes containing functional groups tethered to the cationic cyclophane.

in one instance has given insight into the potential usefulness of catenane assemblies as light-activated charge carriers.

3.1.1.4.1. Miniaturized π-conductors. The [2]catenanes shown in Figure 29 are examples of systems developed for activation by illumination of the CT absorption band [57]. Unlike the rotaxanes, in which the thread contains a secondary donor, the design strategy used here allows functionalization directly onto the π-acceptor cyclophane [58]. Specifically, the carboxylate group attached to the xylyl unit in one ring is readily altered and allows the tethering of a dimethoxybenzene secondary electron donor close to the CT reaction center.

The photochemistry that accompanies selective laser excitation into the CT absorption band of catenanes **13–15** is similar to that observed in the related EDA complexes and rotaxanes. In the cases of **13** and **14**, lifetimes of the primary RIP are on the order of 50–60 ps, which are slightly longer than those found in the rotaxane counterparts, but still not long enough to allow ring rotation to take place. The increased RIP lifetimes for the catenanes are credited to the increase in the total number of skeletal stretching modes, which help dissipate the photonic energy as vibronic energy.

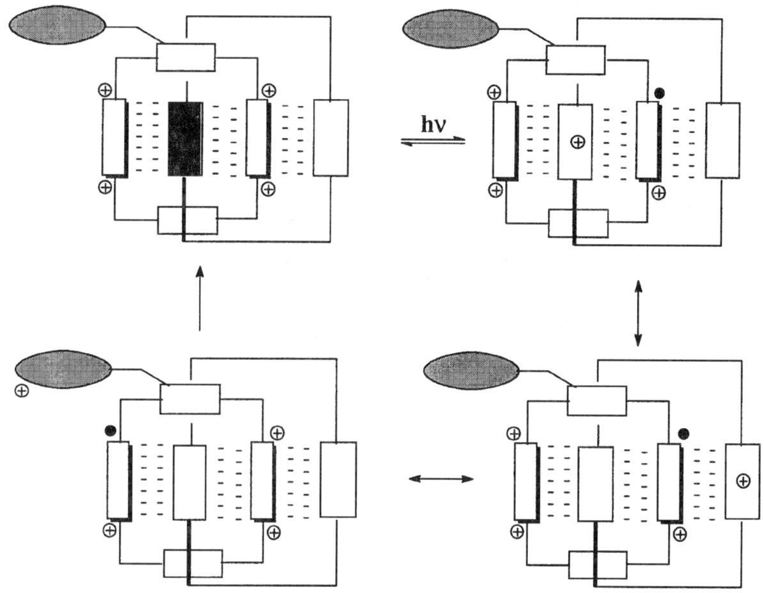

Fig. 30. Proposed hole migration in the functionalized [2]catenane **15**, which accounts for the more complex kinetics that result after light excitation.

In contrast to catenanes **13** and **14**, the photophysics of **15** was shown to be somewhat different, in that decay of the photogenerated RIP did not obey a simple first-order rate law. In this case the decay kinetics are more consistent with a fast process ($\tau \approx 50$ ps), corresponding to 70% of the profile, and a slightly slower ($\tau \approx 90$ ps) step making up the final 30%. It has been suggested that these kinetic data are consistent with fast decay corresponding to charge recombination in the intimate RIP, where the slower step is return electron transfer between the oxidised appendage and the reduced cyclophane. The prolonged lifetime observed for the RIP is, accordingly, a result of extended hole delocalization throughout a π-system that is made possible by a combination of π-stacking of the attached group to the cyclophane and by efficient hole migration across an aromatic nucleus (Fig. 30). It should be noted that no long-lived species were detected that could be attributed to a CTS in which the tethered p-dimethoxybenzene unit "swung away" from the initially produced RIP. Although conformational switching within these [2]catenanes is not possible, the delocalization of charge across a one-dimensional π-stack is in itself interesting. In particular, the possibility exists of further developing these systems and using them as light-activated miniaturized π-conductors [59].

3.1.1.4.2. Photoactive switches. To modify the overall form of a [2]catenane, a unit must be incorporated into the structure that undergoes a large shape change upon illumination. A well-established photoswitchable unit is azobenzene, which isomerizes upon UV irradiation (300 nm) from the more favored E-to-Z configuration. Irradiation at 430 nm alternatively drives the reisomerization process, although in the dark thermal Z-to-E conversion is possible, where the half-life is ~25 days at room temperature (Fig. 31). More importantly, the change at the N=N bond results in a large structural change in the appended phenylene rings. Specifically, the distance between the carbon atoms at the *para* position shortens from ~9.0 to 6.2 Å, and the phenyl groups are no longer coplanar.

Examples of photoswitchable [2]catenanes incorporating azobenzene units produced by Vögtle's group are illustrated in Figure 32 [60]. In systems **16–18**, the volume in which the interlocked crown has to move is significantly varied, starting from the relatively loose assembly **17** to the highly compressed catenane **18**.

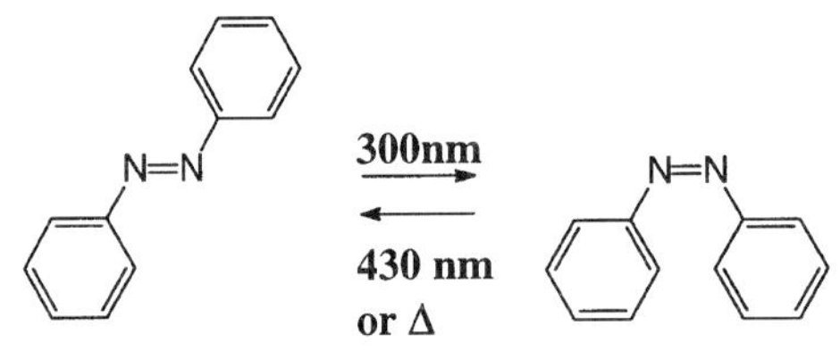

Fig. 31. The *cis* and *trans* isomerization of an azobenzene group.

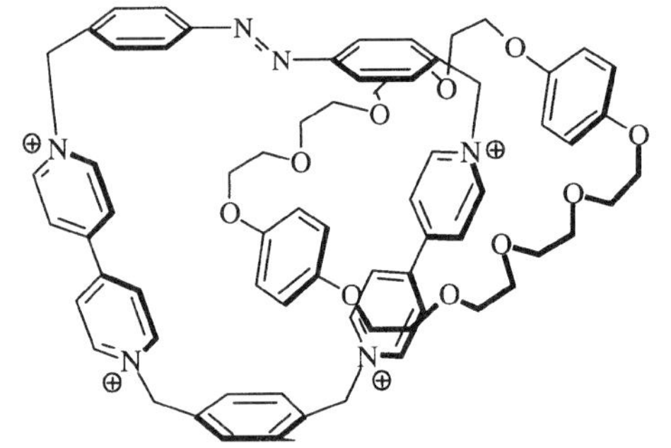

16

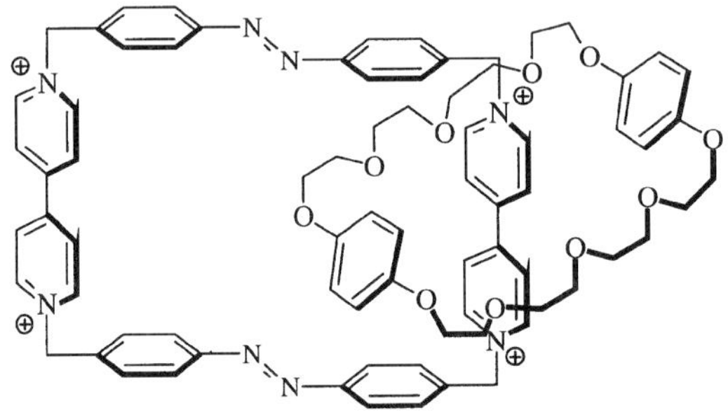

17

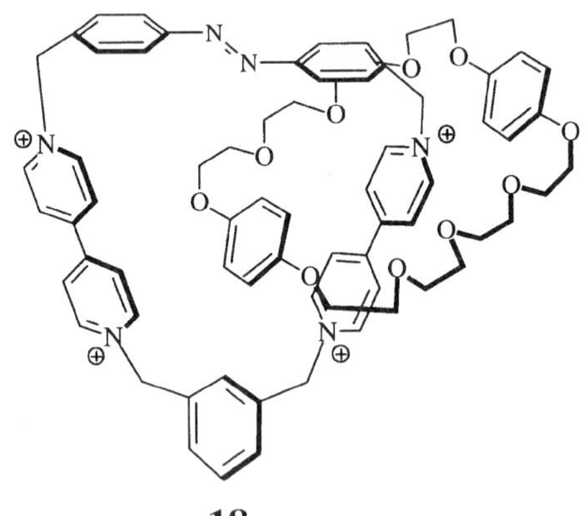

18

Fig. 32. Photoswitchable [2]catenanes based on crown ethers and tetracationic cyclophanes.

The irradiation of catenane **16** with UV light at 300 nm resulted in the production of a photostationary equilibrium consisting of ~40% of the Z isomer. It has been proposed that this mechanical constriction leads to increased "friction" between the rotating components

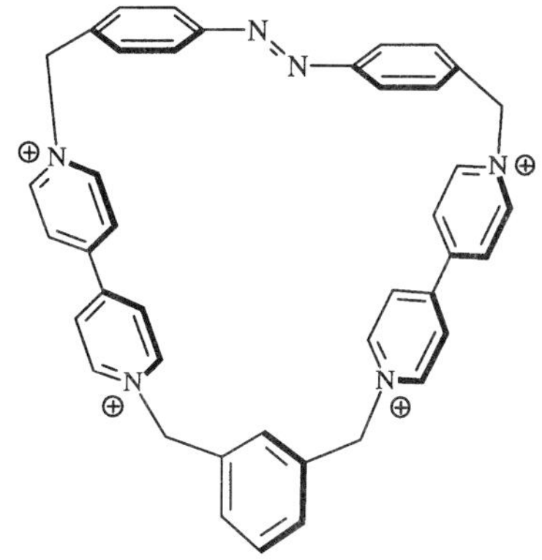

19

Fig. 33. A tetracationic cyclophane incorporating an azobenzene moiety as a light-activated switchable section.

and hence acts as a brake. A similar photoisomerization is also observed in catenane **17**, but because there are two azobenzene units, both (E,Z) and (Z-Z) conformations are formed. The situation, however, is somewhat simplified because of the faster thermal reisomerization of (Z-Z) compared to the (E,Z) isomer. As might be expected, the rate of thermal (Z) to (E) isomerization of the azobenzene unit is dependent on the size of the macrocyclic cavity. The half-life of Z-**16** in acetonitrile at room temperature is ~12.5 h, whereas reformation of E,E-**17** from the Z,Z configuration and mixed isomers occurs with a half-life of 12 days. The most striking steric influence on photoisomerization is that displayed by catenane **18**, containing the metasubstituted phenyl spacer. This substitution change results in a diminished distance between the bipyridinium groups, which severely hinders the translational rotation of the crown ether. The compressed nature of the catenane also means that isomerization of the azobenzene unit is not possible. In contrast, cyclophane **19** (Fig. 33), which represents the uncatenated version of **18**, readily isomerizes to the Z isomer [61]. These observations have been rationalized in terms of a higher induced steric "pressure" that would result from an E-to-Z isomerization in **18**.

3.1.2. Redoxactive Molecular Switches

In this section the perturbation of molecular systems by electrochemical means is examined. Unlike in the previous sections, the addition or removal of electrons is performed electrochemically, but as before, it is used to create either unfavorable intramolecular electrostatic repulsion or geometries that initiate conformational changes.

3.1.2.1. π-π Stacked Molecular Assemblies

The affinity of the cationic cyclophane **2** for extended electron-rich aromatics such as benzidine (**20**) and 4,4′-biphenol (**21**) has formed the basis of the study by Kaifer and co-workers of redoxactive systems (Fig. 34) [62, 63]. As found with the simple 1,4-disubstituted benzene derivatives (Table I), these planar aromatics form 1:1 complexes with **2**, although the corresponding binding constants are greater as a result of increased π-overlap. It is interesting to note that although the two aromatics **20** and **21** are similar in size, the 1:1 geometries formed with **2** in CH_3CN have been shown by steady-state NOE experiments to be somewhat different. In the benzidine case, the aromatic is buried deep in the cyclophane cavity and forms a centrosymmetric complex, whereas for **21** only one aromatic ring seems to be associated with the bipyridinium units.

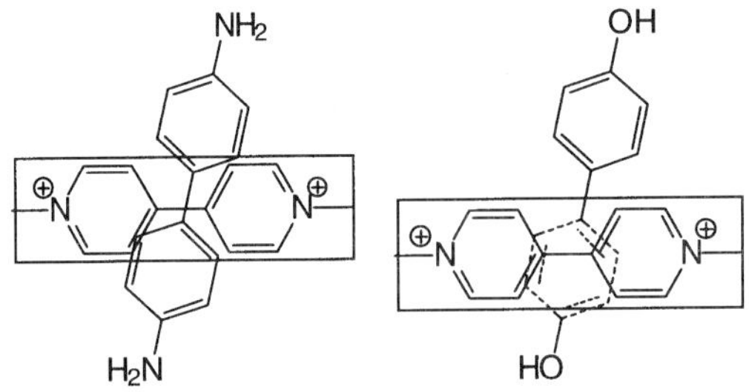

20 **21**

Fig. 34. A side view of cyclophane **2** showing the difference in geometry of the benzidine and biphenol CT complex.

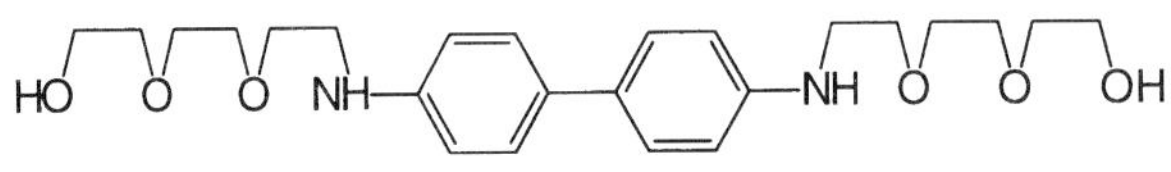

22

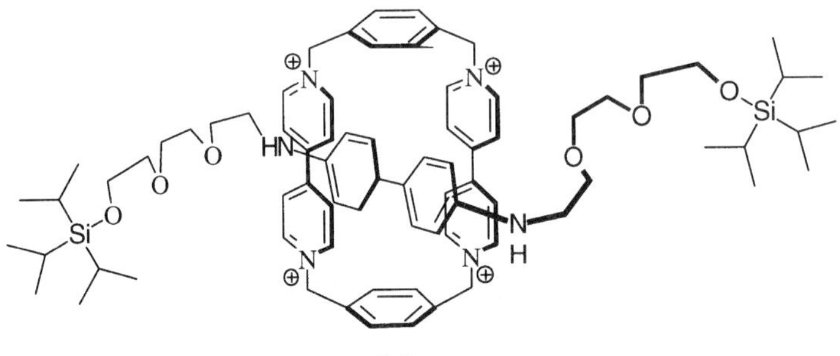

23

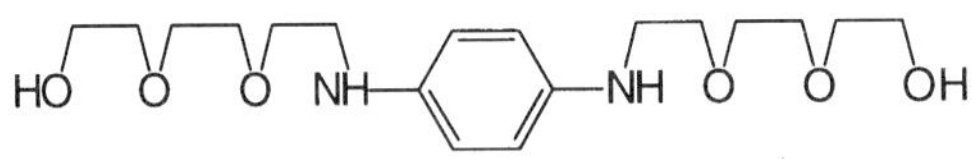

24

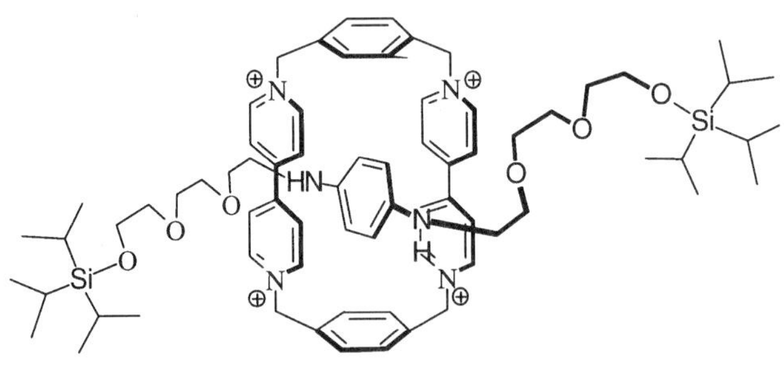

25

Fig. 35. Examples of switchable [2]rotaxanes based on benzidine and phenylenediamine groups and their corresponding thread units.

Expanding on this simple host–guest binding work, the same workers have incorporated extended aromatics into rotaxane structures (Fig. 35). The particular interest in the systems built around threads **22** and **24** is the highly reversible and readily accessible anodic elec-

Table III. Anodic Electrochemical Data for [2]Rotaxanes and Their Precursors

Compound	$E^{1}_{1/2}$ (V)[a]	$E^{2}_{1/2}$ (V)[a]
PD[b] (unit)	0.324	0.871
22 (thread)	0.203	0.732
23 (rotaxane)	0.463	1.505
20 (unit)	0.591	0.800
24 (thread)	0.529	0.714
25 (rotaxane)	0.695	0.988

Data are from [62].

[a] Referenced vs. Ag/AgCl.

[b] 1,4-Diaminobenzene.

trochemistry exhibited by the 1,4-diaminobenzene and benzidine units. The anodic electrochemistry can be summarized as follows:

$$\mathrm{S} - \mathrm{e} \rightleftharpoons \mathrm{S}^{\bullet+}$$
$$\mathrm{S}^{\bullet+} - \mathrm{e} \rightleftharpoons \mathrm{S}^{2+}$$

where S represents the aromatic residue. Some of the relevant electrochemical data obtained on these compounds are collected in Table III. It is clear that the incorporation of units **22** and **24** into a rotaxane structure notably affects their anodic electrochemistry. Specifically, both of the half-wave potentials are shifted to more anodic potentials, which is consistant with the increased difficulty of generating additional positive charge in close proximity to the tetracationic cyclophane. It is worth noting that the anodic shift is far more pronounced for rotaxane **23** compared to **25** and is likely a result of the larger π-area in the benzidine case, which allows the positive charge to be spread out over a larger area.

Although the electrostatic repulsion generated within rotaxanes **23** and **25** is unfavorable, the limited distance the cyclophane bead is allowed to move to alleviate the condition is very much restricted. Even though a simple extension of the thread in rotaxane would lead, in principle, to enhanced displacement of the cyclophane, to make the system useful another station is required, so that upon shuttling the bead becomes trapped at the second site. The incorporation of the benzidine and biphenol units into the two-station rotaxane **26** (Fig. 36) goes some way toward addressing this issue [64]. In CD_3CN at room temperature, the bead shuttles between the two stations of rotaxane **26**, although low-temperature two-dimensional correlation spectroscopy (2D-COSY) experiments confirm that the bead prefers residence at the benzidine (84%) rather than the biphenol (16%) station. This result is understandable, because benzidine has a greater binding affinity for cyclophane **2** when compared to biphenol. The electrochemical behavior of **26** is significantly different from the counterpart "tighter" rotaxane **23** in that the second oxidation wave, corresponding to removal of a second electron from the benzidine unit, occurs at a potential that is almost identical to that found for the simple thread **22**. The reason for this similarity in electrochemical behavior is easily explained if one takes into account what occurs after the first oxidation process. In this instance, the generation of the benzidine cation radical creates enough electrostatic repulsion to drive the bead off the station and over to the opposing side. With the cyclophane bead well separated from the benzidine unit, removal of the second electron to generate the cation is relatively easy, and consequently the potential for this process is similar to that for the uncomplexed thread. Maintaining the benzidine unit

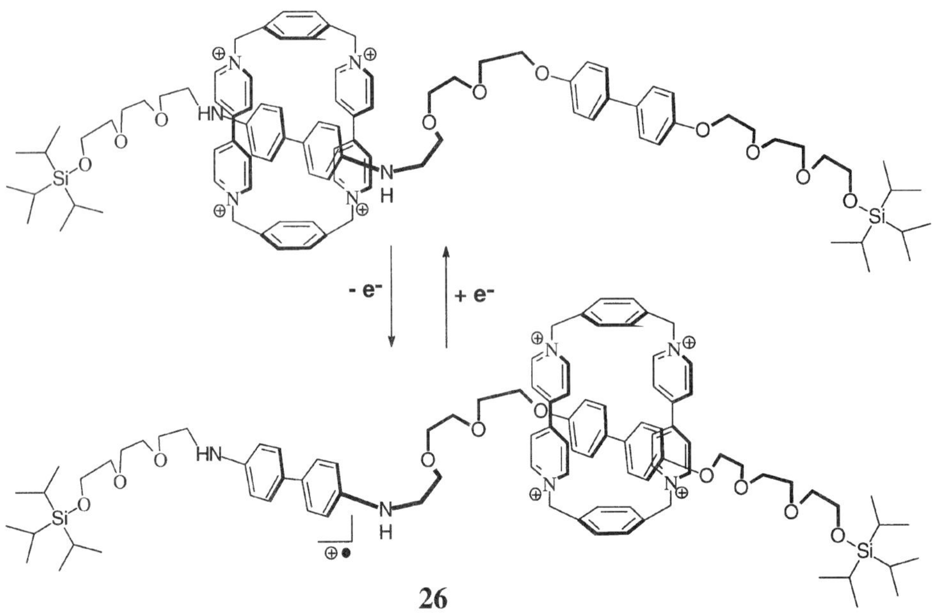

26

Fig. 36. An electrochemically activated two-station [2]rotaxane showing the relative bead positioning after oxidation and reduction of the benzidine unit on the thread.

in an oxidized state forces the bead to sit at the biphenol station and hence controls the conformation of the rotaxane.

In contrast to using the oxidation of the encapsulated moiety, the control of catenane conformation has primarily focused on using the preferential reduction of asymmetric cyclophanes [65]. Illustrated in Figure 37 is an example of a catenane, which although primarily produced for light activation via the bis(pyridinium)ethylene subunit, is better suited for electrochemical activation. Because of the higher binding affinity, the preferred translational isomer for **27** is as shown, with the bipyridinium unit "sandwiched" between the two hydroquinol groups of the crown. Upon preferential reduction of this bipyridinium unit, the CT interaction is significantly diminished and permits circumrotation of the cyclophane and insertion of the bis(bipyridinium)ethylene unit. As reoxidation of the molecular assembly forms the nonpreferred isomer, a further rotation of the cyclophane is required to regenerate the starting conformation. Beginning in the top left-hand corner, these overall processes are depicted in the simple cartoon in Figure 37.

3.1.2.2. *Cation Chelating Structures*

It is well established that the oxidation state of first-row transition metal ions and the coordination number of their complexes are somewhat interrelated. In particular, the tetrahedral geometry requirement of copper(I) is well known and explains the preference of the metal ion to bind two bidentate chelators and form tetrahedral complexes. The use of such specificity in binding geometry has been extensively employed by Sauvage and co-workers in the manufacture of catenane and rotaxane assemblies built around copper(I) chelates [66]. More recently, the same workers have begun to use changes in the copper center oxidation state to induce large-scale conformational changes using systems containing multiple binding sites. The pseudorotaxane **28** [67] represents an elegant example in which two different chelators form part of the thread, with a phenanthroline chelator built directly into the bead segment (Fig. 38).

The starting copper(I) complex **28** in CH_2Cl_2 is deep red ($\lambda_{max} = 442$ nm), and its cyclic voltammogram displays a reversible wave at $+0.84$ V vs. the saturated calomel electrode

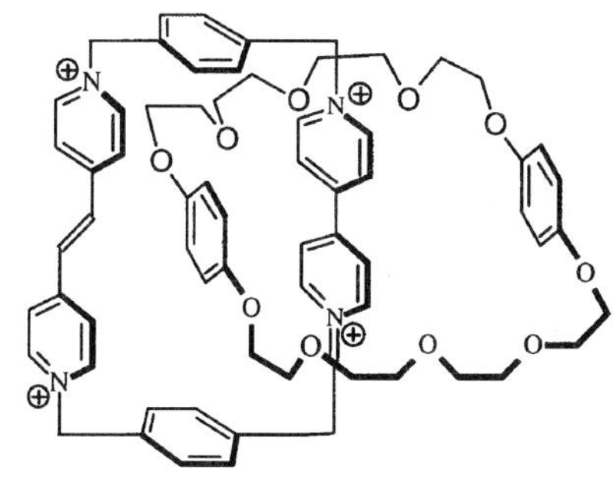

27

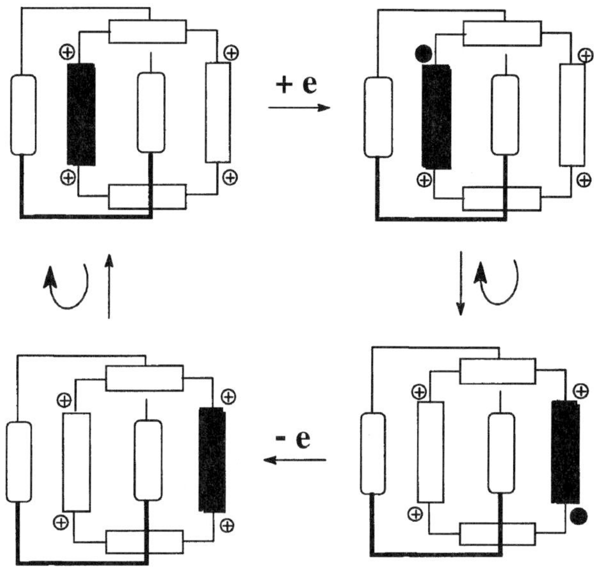

Fig. 37. An illustration of a switchable [2]catenane and the corresponding conformational changes that occur upon electrochemical reduction of the bipyridinium moiety.

(SCE), which is characteristic of a tetrahedral copper(I) species. Upon controlled electrolysis ($E_{app} = +1.1$ V), however, the solution turns light green ($\lambda_{max} = 658$ nm), and a new wave appears at +0.1 V vs. SCE, which is consistent with a copper(II) species in a five-coordinate environment. The starting complex can be readily restored by reductive electrolysis ($E_{app} = -0.6$ V), as evidenced by the reappearance of the initial red color. A slight problem with this system is that in CH_3CN, in addition to the switching of stations by the bead, appreciable dethreading occurs. In an attempt to solve this problem the same workers have recently constructed the stoppered rotaxane **29** (Fig. 39) [68]. A detailed comparison of the electrochemical data for these systems has been reported [69]. In the case of the rotaxane **29**, a perfect reversibility of motions of the bead along the thread has been demonstrated.

Of course, the problem of dethreading does not exist for [2]catenanes, the only restriction being the limited distance by which molecular partners can become separated. In the catenane case, instead of shuttling a bead along a thread, any redox changes induce swinging of one of the rings. This principle has been successfully demonstrated by Sauvage and co-workers using the prototype catenane **30** (Fig. 40), which represents the ring closed equivalent of pseudorotaxane **28**.

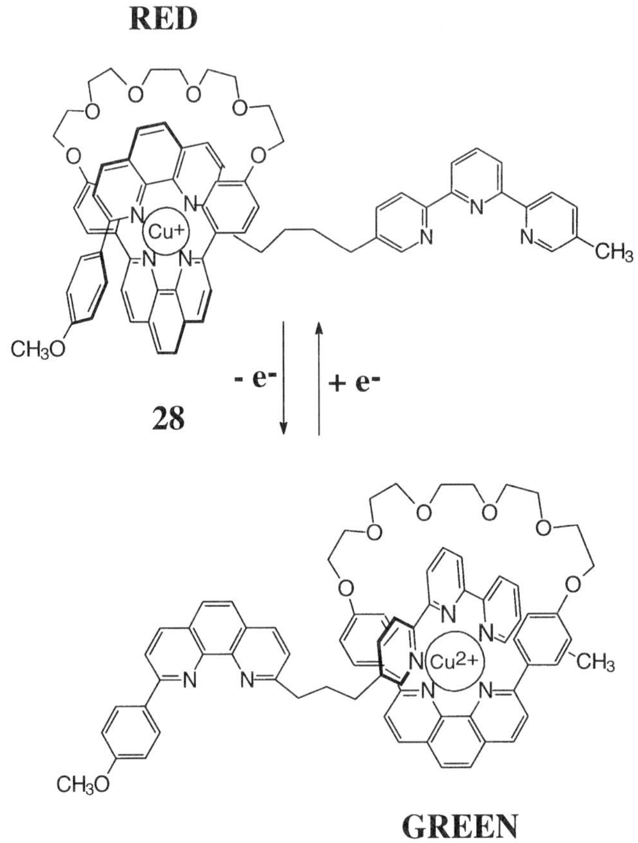

Fig. 38. A phenanthroline-based pseudorotaxane displaying the bead positioning and accompanying color changes that occur upon reduction and oxidation.

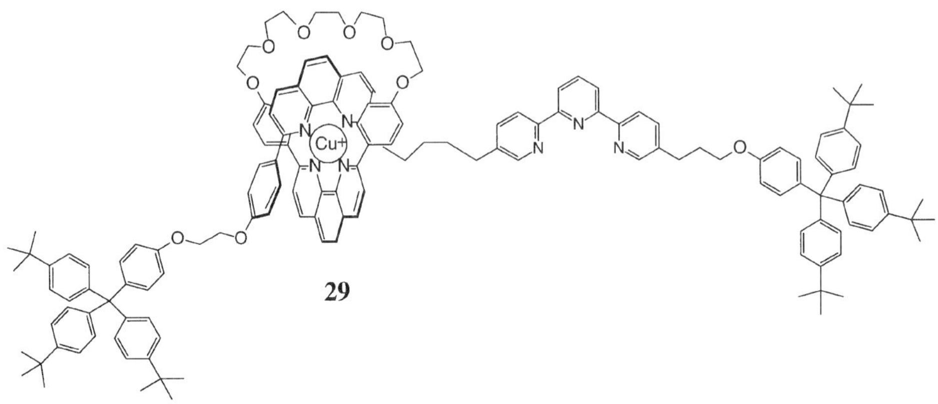

Fig. 39. A stoppered [2]rotaxane incorporating a phenanthroline-type bead and two stations based on phenanthroline and terpyridine.

As observed in the pseudorotaxane, the more stable copper(I) complex is the one with the two phenanthroline ligands bound and encapsulating the metal in a tetrahedral environment. Upon oxidation of **30** in CH_3CN, the characteristic red color disappears and is replaced by a deep green ($\lambda_{max} = 670$ nm), which represents at first the copper(II) complex. Over a period of a few days, the green color is seen to disappear and is replaced by a

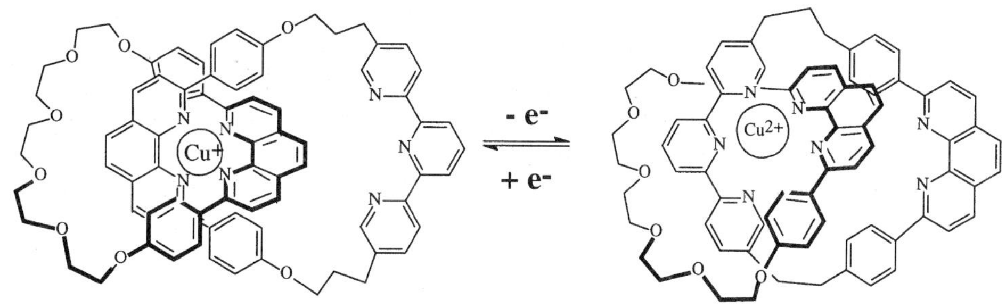

30

Fig. 40. A redoxactive [2]catenane prepared by Sauvage's group, showing swinging of the terpyridine subunit to bind copper(II) ion after oxidation.

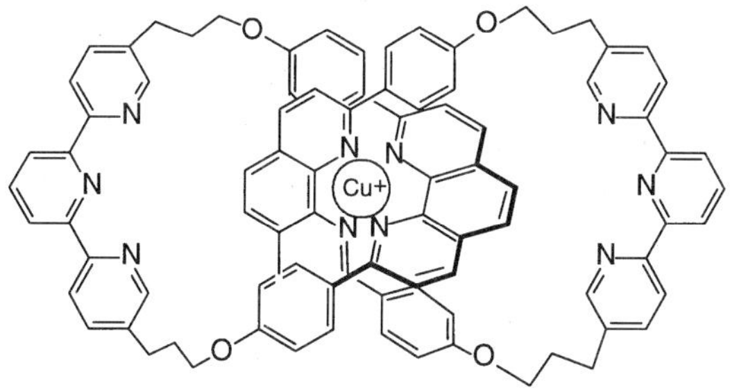

31

Fig. 41. A [2]catenane containing four chelating units prepared by the group of Sauvage.

yellow-green coloration, which is consistent with a five-coordinate complex in which the terpyridine (2,2′:6′,2″-terpyridine) ligand is bound. It is interesting that whereas formation of the pentacoordinate copper(II) species is very slow, the changeover process after reduction back to copper(I) occurs in a matter of seconds [70].

As part of the development of metal-based catenanes, the incorporation of four metal binding sites into the structure has recently been addressed by Sauvage et al. In the catenane **31** (Fig. 41), an extra terpyridine chelator has been introduced into the overall structure [71].

Once again starting from the red copper(I) complex, chemical oxidation ($NO^+BF_4^-$) generates a green species ($\lambda_{max} = 670$ nm) that is formulated as the four coordinate copper(II) species. Over a period of time, the intensity of this absorption decreases, a fact that is attributed to slow ring rotation in the catenane accompanied by generation of the hexacoordinate copper(II) complex. This ring motion is very solvent dependent and has been shown to be retarded by the presence of water and accelerated when DMF is present. As observed with catenane **30**, the regeneration of the copper(I) species results in a much faster reformation of the starting material geometry, but not before passing through the pentacoordinate copper(I) complex. These overall processes are illustrated in Figure 42, starting with the preferred conformation in the top left-hand corner.

It is clear that with metal-based systems, rates of ring gliding processes can be controlled very effectively by the oxidation state of the metal center. Although molecular changes in the outlined catenanes are too slow for fast optical switching purposes—especially after Cu(II) generation—the two distinctly different colored states are useful for optical reading purposes.

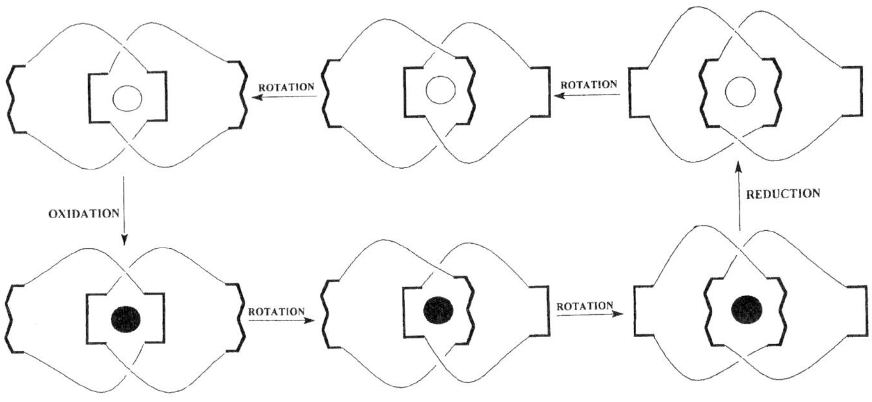

Fig. 42. A simple cartoon representing the possible conformations of catenane **31**. The open circles represent Cu(I) and the filled circles Cu(II); phenanthroline is indicated by half-open squares.

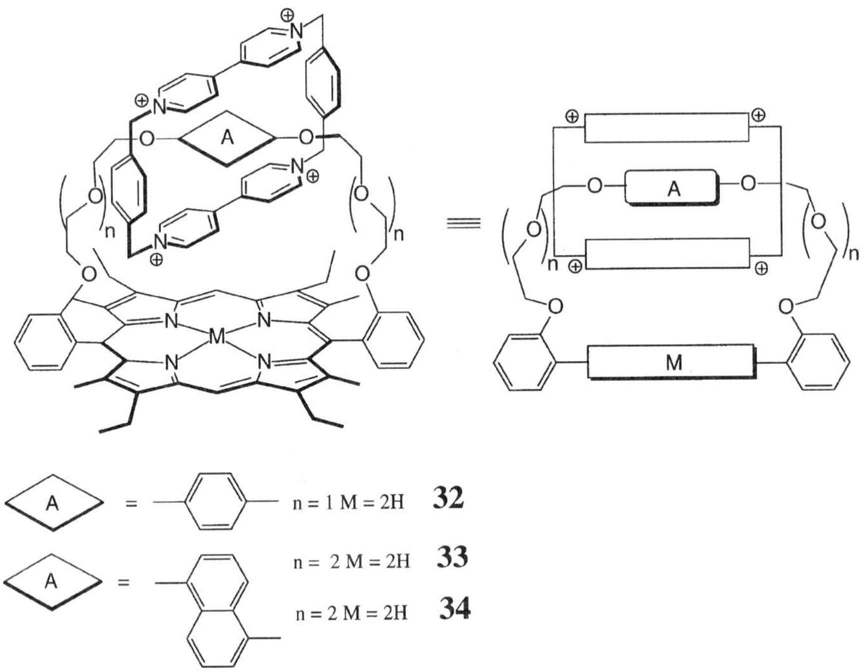

Fig. 43. Illustrations of porphyrin-based [2]catenanes and a simple cartoon representing their face-to-face arrangement.

3.1.3. Ion Active Molecular Assemblies

A common theme in the control of molecular conformation is the generation of unfavorable interactions between adjacent modules that force these units apart. In the case of photo- and redox-driven processes, the reduction or oxidation of a single subunit is enough to trigger the switching process. In the following examples, chemical stimuli are in the form of protons and monovalent metal ions.

3.1.3.1. Catenanes

The versatility of cyclophane **2** is once again demonstrated in the work carried out by Gunter's group on a series of porphyrin-based catenanes (Fig. 43) [72–74].

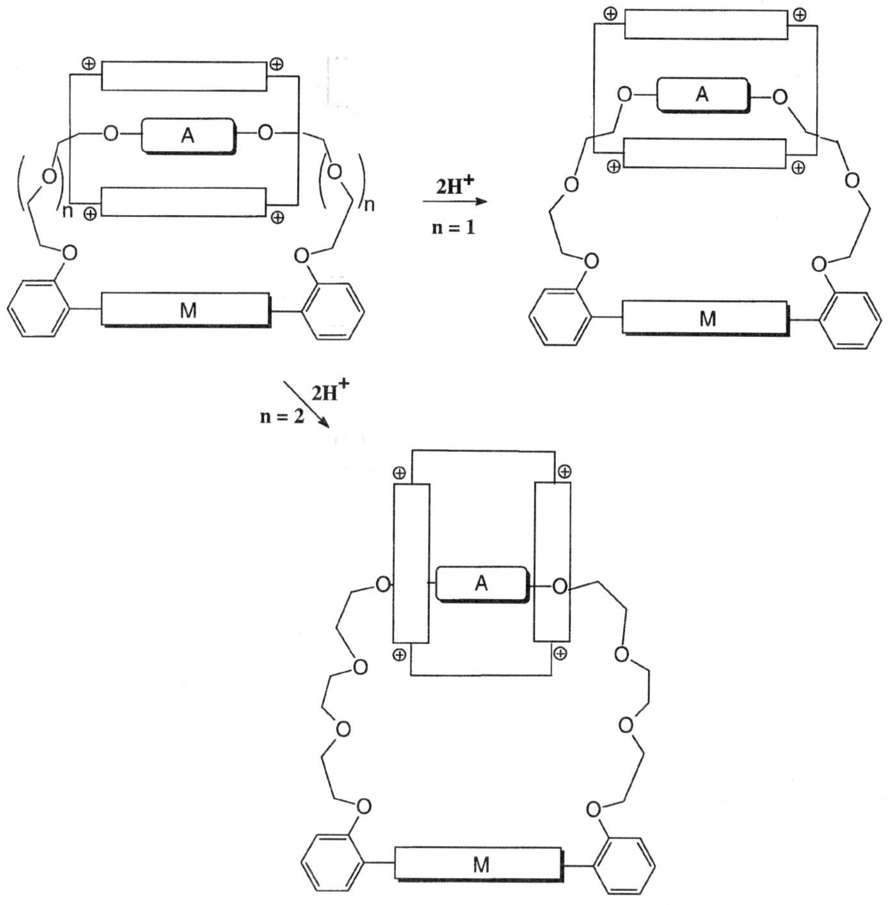

Fig. 44. Conformational changes that occur in [2]catenanes **32–34** following protonation of the porphyrin segment.

In the catenanes **32–34**, the bipyridinium unit of the cyclophane sits in a face-to-face arrangement with the porphyrin ring. Because of this close proximity, the porphyrin Soret band in **32** appears at 416 nm, which is some 18 nm blue shifted with respect to the uncatenated porphyrin. In the slightly looser systems, **33**, **34**, this shift is less pronounced (~8 nm). With protonation of the porphyrin unit in **32**, the increase in electrostatic repulsion causes separation of the two molecular components. From ^{1}H NMR experiments in CD_3COCD_3/trifluoroacetic acid, however, this increase in separation is not accompanied by a rotation of the cyclophane, as the tight ethoxy strap is too restrictive. In contrast, protonation of the more unconstrained cyclophanes, **33**, **34**, is accompanied by both separation of molecular partners and a 90° rotation of the cyclophane unit. These processes are illustrated in the simplified cartoons in Figure 44.

The hybrid [2]catenane **35** (Fig. 45) is rather unique in that both metal binding and donor–acceptor sites are incorporated into the structure. The relative positioning of the two rings can be readily altered through the introduction/removal of protons, with the switching detected by the appearance of the CT absorption band [75].

3.1.3.2. Pseudorotaxanes

The perturbation of an equilibrium through the use of ions in pseudorotaxane structures is examined in this final section. Illustrated in Figure 46 is an example of an assembly in

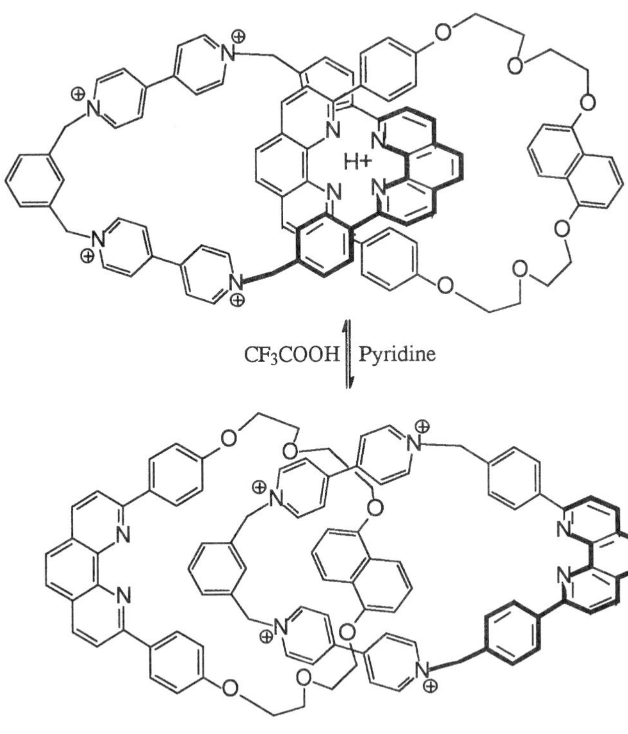

35

Fig. 45. An example of a switchable [2]catenane incorporating metal ion and donor–acceptor binding sites.

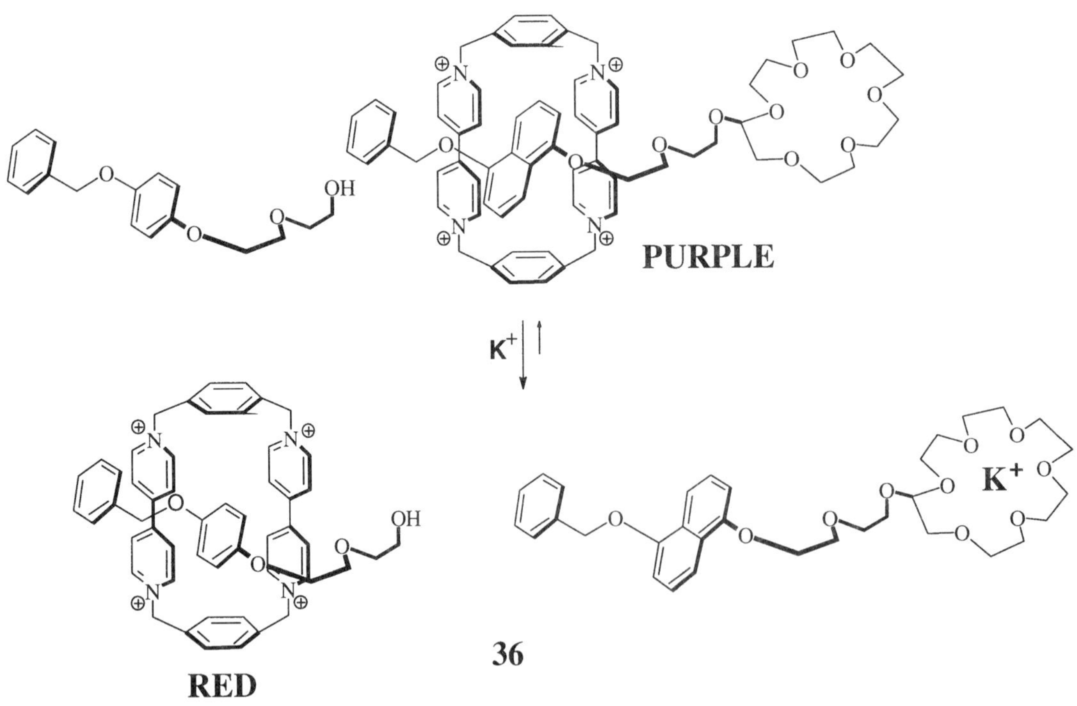

Fig. 46. A crown-functionalized pseudorotaxane and alkoxybenzene thread (top view) and the corresponding unthreading and color change that takes place upon the addition of excess KPF_6.

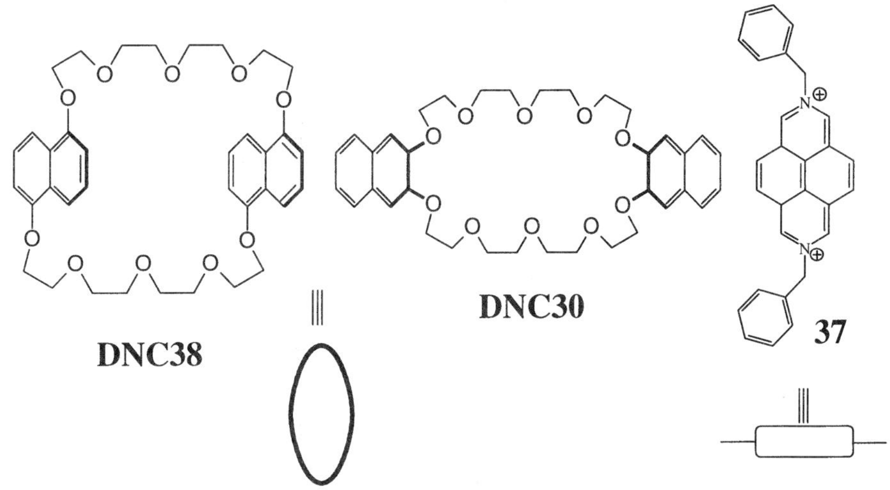

Fig. 47. Molecular components that self-assemble to produce simple pseudorotaxane assemblies that can be chemically activated.

which the thread contains a crown ether moiety that binds potassium in its central cavity [76].

In acetonitrile, and under normal conditions, the top trace structures as shown in Figure 46 dominate the solution mixture, as expected because of the greater binding affinity of the naphthalene-based thread for cyclophane **2**. The addition of a 10-fold excess of KPF_6 to a solution of **36** results in a considerable blue-shift in the absorption maximum indicating expulsion of the naphthoxy thread and insertion of the more weakly binding alkoxybenzene moiety. This observation has been rationalized in terms of the increased electrostatic repulsion between **2** and the K^+ ion bound in the crown cavity, which destabilizes the resultant complex. This negative allosteric effect is potentially useful in that an optical response to the addition of potassium is observed, although reversible cycling of the system is difficult, because it would require removal of the potassium from the crown cavity.

In section 3.1 the phototropic behavior of **DAP^{2+}.CE10** molecular systems was discussed in which direct destabilization caused large-scale molecular changes. Examples of similarly based systems are shown in Figure 47, but in which the dethreading is performed by the addition and protonation/deprotonation of amines [77].

In acetonitrile the diazapyrene-based thread **37** binds in the central cavity of the crown ether **DNC38** (1,4,7,10,13,24,27,30,33,36-decaoxa[13.13]-[1,5]-naphthophane), forming a 1:1 complex; the structure resembles that shown in Figure 16. The dethreading of this structure can be carried out by the addition of a suitable amine B (hexylamine) to a solution of the assembly; the driving force is the generation of a CT complex between **37** and the added amine. Protonation of the amine through the addition of a suitable acid (trifluoroacetic acid, TFA) destabilizes the **37.2B** complex and regenerates the pseudorotaxane (Fig. 48). This simple cycling threading/dethreading process is readily followed by both absorption and fluorescence spectroscopy. First, the crown and diazapyrene adducts when free in solution both display characteristic emission spectra, which disappear upon formation of the donor–acceptor complex **37.DNC38**. Second, the absorption band for the amine adduct **37.2B** is easily distinguished from the CT profile of the pseudorotaxane complex, as it appears well toward the red end of the visible region.

The ability to turn on/off particular pathways and selectively monitor conformational changes has also been studied in the analogous pseudorotaxane structure **37.DNC30**. Consistent with nanotechnology, the usefulness of such a system as a molecular machine in logic gate switching has recently been proposed by Balzani and co-workers. In their paper

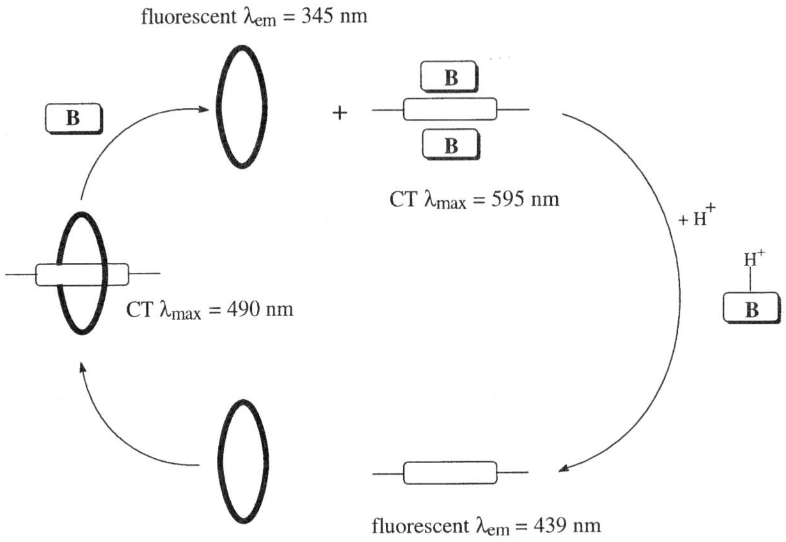

Fig. 48. Conformational changes and absorption/fluorescence maxima of individual components and the resultant CT complexes. B = amine.

it has been suggested that the input (chemical)/output (fluorescence) characteristics of the molecular assembly correspond to an XOR (exclusive OR) logic gate [78].

3.2. Photosynthetic Reaction Center Mimics

In this final section the use of rotaxane and catenane assemblies is reviewed as a model for the study of ET processes in relation to photosynthetic RCs. It should be noted that as photosynthetic models, rotaxane/catenane structures are merely a small part of a much larger field that has been comprehensively reviewed by Wasielewski [79].

3.2.1. Charge Separation Models Using [2]Rotaxanes

If one considers the incorporation into a [2]rotaxane structure of donor and acceptor units as stoppers, then depending on the rigidity of the thread, it is possible to envisage photoinduced electron transfer processes occurring between these two subunits. These processes are conceptually illustrated in Figure 49, in which pathway (a) represents through-space ET from donor to acceptor, and (b) involves a superexchange mechanism using molecular orbitals residing on the bead portion. The collapse of the CTS back to the ground state (step c) can occur by a simple solvent process or involve a real redox intermediate associated with the bead segment.

Contributions to the field of artificial photosynthetic models based on [2]rotaxanes have mainly been made by Sauvage and Harriman and have stemmed from their earlier work on oblique porphyrin-based assemblies [80, 83]. In the rotaxane examples, **38–40**, the phenanthroline functionalized thread is stoppered with gold(III) and zinc(II) porphyrinic subunits, and the bead section is made from a phenanthroline-based macrocycle (Fig. 50). The relevant portion of the natural RC structure is also shown to help illustrate the structural similarity these rotaxanes have to the porphyrin cofactors in which primary ET takes place. Specifically, the stopper units represent the porphyrins marked P and H, and the role of the phenanthroline moiety is similar to that of the bacteriochlorophyll B, in possibly facilitating electron transfer [84]. In this instance, the highest occupied molecular orbitals/lowest occupied molecular orbitals (HOMO/LUMO) of the phenanthroline spacer are expected to

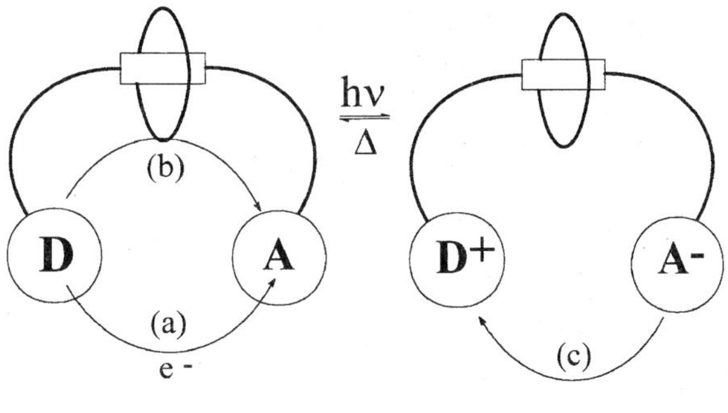

Fig. 49. A simple representation of the forward electron transfer (a and b) and return electron transfer processes (c) in porphyrin-stoppered [2]rotaxanes. Here, process (a) represents electron transfer through solvent, and (b) is a superexchange mechanism involving molecular orbitals on the bead portion. Pathway (c) is a through-solvent return electron transfer process. D = donor, A = acceptor.

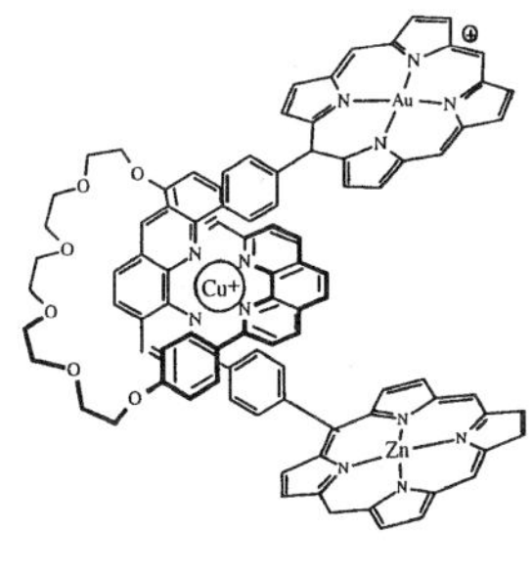

38

39

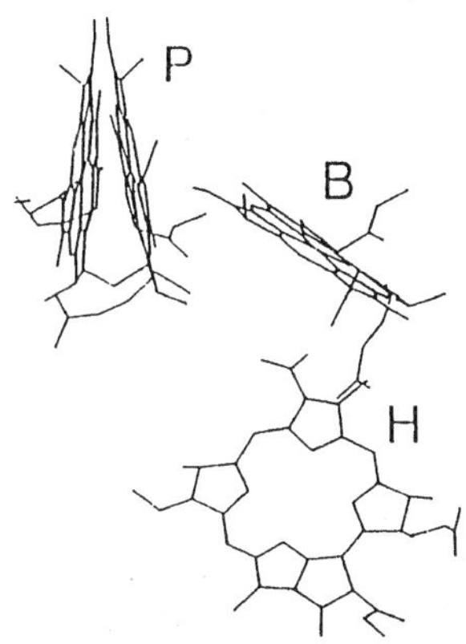

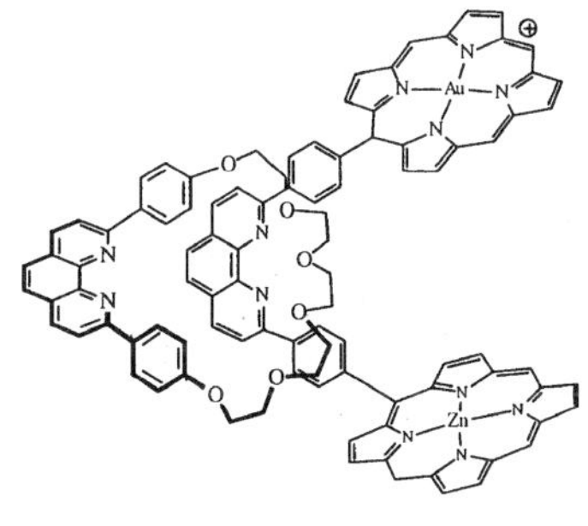

40

Fig. 50. Examples of porphyrin-stoppered [2]rotaxanes prepared of Sauvage's group, and the corresponding section of the photosynthetic RC structure that the rotaxanes mimic. Note: Substituents appended to the rotaxanes' porphyrin groups have been omitted for clarity. For full structures see [83].

play a role in promoting forward electron transfer, where observed differences in rates are a consequence of modulation in energy of these orbitals by changes of cation [85].

The selective laser excitation of **38–40** by way of the zinc porphyrin unit causes rapid singlet excited state ET to the adjacent gold porphyrin with a concomitant generation of a CTS:

$$(\mathrm{ZnP})^{*}\text{-----Spacer-----}(\mathrm{AuP}^{+}) \Rightarrow (\mathrm{ZnP}^{+\bullet})\text{-----Spacer-----}(\mathrm{AuP}^{\bullet}) \qquad (4)$$

For the rotaxanes **39**, **40**, the lifetimes of the singlet state (τ_S) and CTSs are significantly shorter than those measured for **38** (Table IV), suggesting that the copper(I) central moiety strongly facilitates electron transfer.

Table IV. Photophysical Properties Measured for Rotaxanes **38–40** in Butyronitrile at 25 °C

Compound	τ_s (ps)	τ_t (ps)	τ_{cts} (ps)
38	1.7	17	22
39	28	56	360
40	36	71	540

Data are from [85].

Alternative excitation of the rotaxanes via the gold(III) porphyrin segment produces the triplet state, which again rapidly collapses to generate a CTS:

$$(\text{ZnP})\text{- - - - -Spacer- - - - -}(\text{AuP}^{+})^{*} \Rightarrow (\text{ZnP}^{+\bullet})\text{- - - - -Spacer- - - - -}(\text{AuP}^{\bullet}) \tag{5}$$

A similar trend in the triplet lifetimes (τ_t) is also seen in rotaxanes **38–40**, where τ_t of the copper(I)-based system is considerably shorter than those measured for the other two rotaxanes (Table IV). These observed differences in lifetimes of the rotaxane assemblies can be explained, to a first approximation, by using a superexchange mechanism. In this instance, the rate constants (k) for the electron transfer processes can be equated to the expression $k \propto (1/\delta E_{AB})$, where δE_{AB} represents the energy gap between the spacer LUMO or HOMO and the donating or accepting orbitals of the appropriate porphyrin subunits. It is seen for rotaxane **38** that the δE_{AB} value is small, and accordingly rate constants are at their peak value. It is interesting to note that whereas the rate of formation of the CTS for rotaxane **38** is very similar to that observed in the natural RC the rate of charge recombination is very much faster. This result is as expected, because CR in the model system falls close to the apex of a Marcus-type rate vs. driving-force profile, whereas in the natural RC it is well within the inverted region. It is clear that systematic modulation of spacer energy by variation of the metal center in rotaxane structures permits fine-tuning and maximization of the efficacy of photoinduced electron transfer processes [86].

3.2.2. Unidirectional Electron Transfer Mimics

The unidirectionality of electron transfer in the natural RC is an intriguing phenomenon, inasmuch as discrimination between the two sets of cofactors is merely a result of differences in their local environment. The illustration in Figure 51 attempts to demonstrate how a similar difference in environment can be achieved in an artificial system. In this instance, the two chemically identical electron acceptors are connected to a primary donor, and by wrapping an inert sheath around one of the units, shield it from the surrounding solvent.

The next step of incorporating the outlined design idea into a real chemical system has recently been put into practice by the construction of ruthenium-based [2]catenane and cyclophane assemblies, as shown in Figure 52 [87]. Specifically, the building of the 2,2′-bipyridyl unit into the [2]catenane **41** has allowed the attachment of a photoactive group in close proximity to the electron-accepting cationic bipyridinium subunits. In structure **42** the wrapping of the electron-donating crown ether around one of the viologen units effectively shields it from the surrounding solvent and decreases its electron affinity by way of charge-transfer interactions. As a consequence, the reduction potentials of the two units are significantly different (~130 mV), meaning there exists a difference in the thermodynamic driving force for photoinduced electron transfer from the ruthenium chromophore to the electron acceptors. The removal of redox asymmetry in cyclophane **43** has meant that a better interpretation of the electron transfer processes occurring in these metal-based systems has been possible [88].

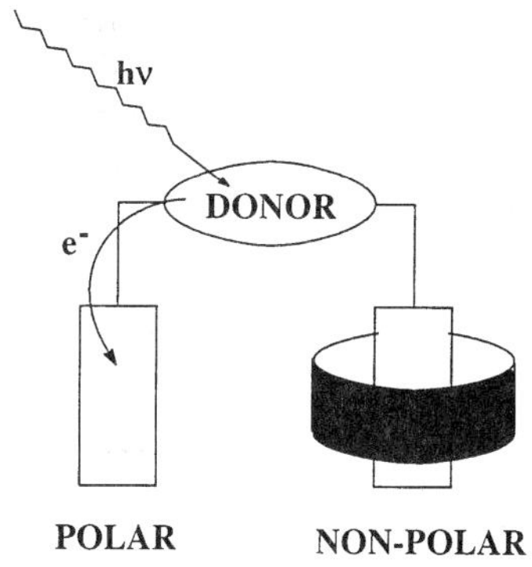

Fig. 51. A cartoon representation of preferential electron transfer from a primary donor to an electron acceptor, using an environmentally induced process. The cylinder represents an inert sheath that protects the acceptor unit from surrounding solvent.

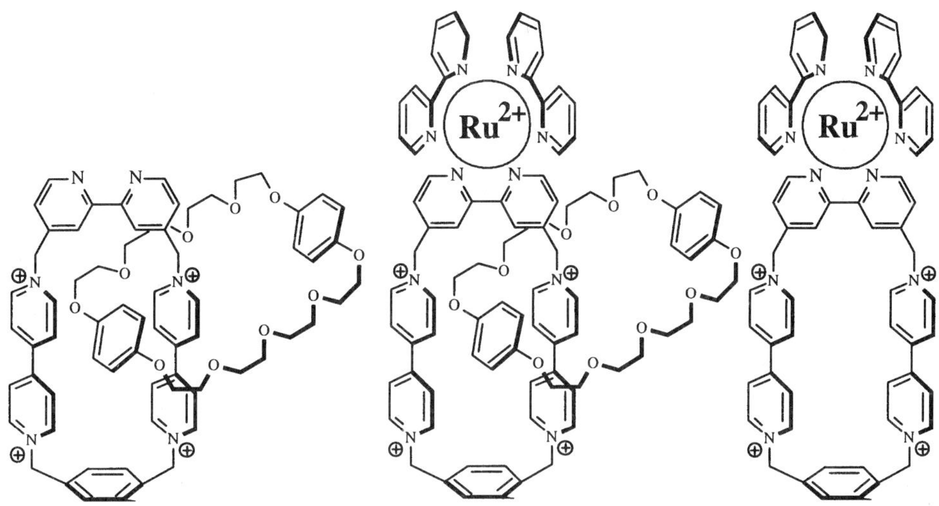

41 **42** **43**

Fig. 52. Illustrations of prototype [2]catenane and cyclophane assemblies for studying preferentially induced electron transfer.

By the use of fast kinetic techniques, the electron transfer processes occurring in **42** and **43** have been studied in varying solvents and at a range of temperatures (Fig. 53). For both systems the rate of forward electron transfer is extremely fast ($k_{CS} \approx 6 \times 10^{10}\ s^{-1}$) and is accompanied by a slower charge recombination process ($k_{CR} \approx 2 \times 10^{10}\ s^{-1}$). More importantly, it has been established that k_{CS} increases with the driving force, suggesting that a difference of 130 mV in catenane **42** could mean that electron transfer proceeds preferentially to the more "open" bipyridinium unit. Indeed, from simple calculations this selectivity has been deduced to be on the order of 85%, although direct observation of this effect will require an alternative approach because the viologens cannot be distinguished spectroscopically.

The usefulness of molecular systems that display unidirectional electron transfer is perhaps not immediately obvious, particularly because in solution all molecules are randomly

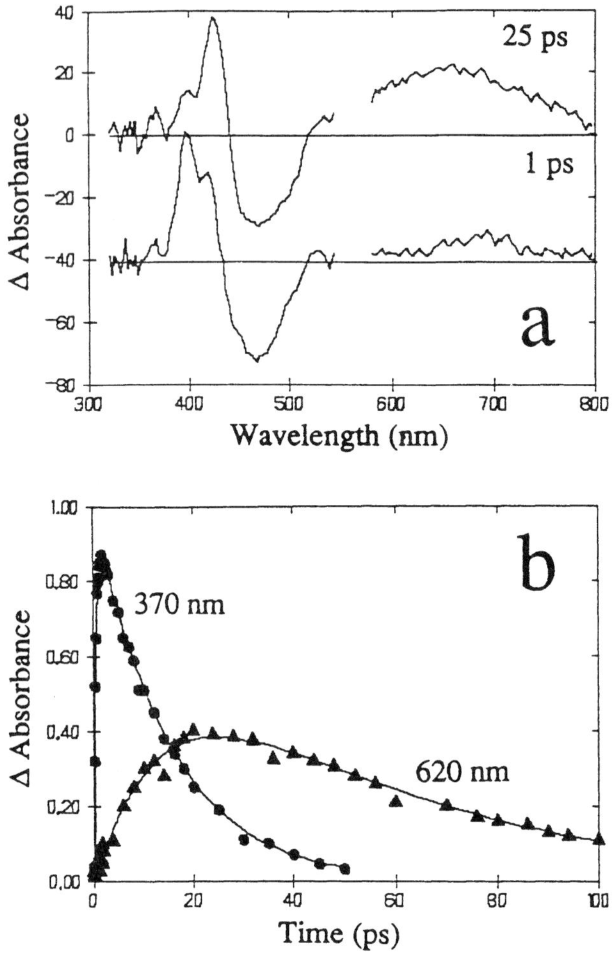

Fig. 53. (a) Transient differential absorption spectra recorded following laser excitation of **42** in acetonitrile at 20 °C at varying delay times. (b) Decay profiles showing formation and decay of the triplet state of the ruthenium(II) complex at 370 nm and of the radical ion pair at 620 nm.

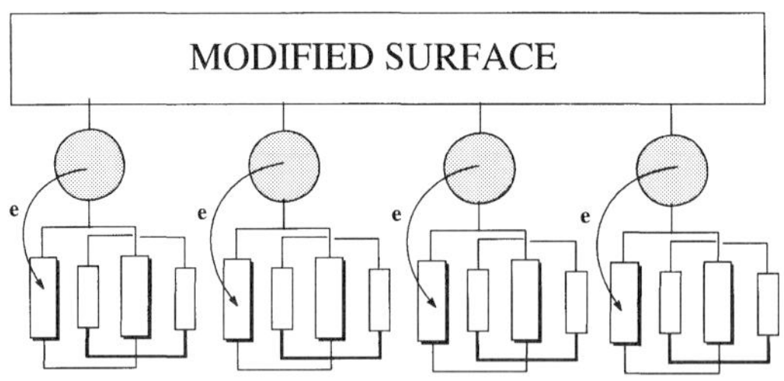

Fig. 54. A conceptual approach for the use of preferential electron transfer in a one-dimensional array made from [2]catenanes. Note: Return electron transfer is not shown, but for a practical working system this would have to be curtailed.

arranged. However, it is not too difficult to imagine that it would be possible by synthetic means to assemble, on an appropriate surface (i.e., an electrode), a one-dimensional array (Fig. 54). This is not unreasonable, inasmuch as recently Kaifer and co-workers have shown that surface-modified catenanes can be readily prepared, albeit with rather low surface coverage [89]. Furthermore, the necessary alignment of all of the molecules is also equitable, considering that solid-state structures of [2]catenanes often exhibit an extensive one-dimensional π-stack [90]. Of course, whether all of this is possible is yet to be seen, and before commencing it will be far more productive to optimize the efficiency of the simple molecular catenane unit.

3.3. Overview

Molecular-scale structures based on rotaxanes, catenanes, and simpler cyclophanes display a wide range of interesting properties. These include phototropic responses, light-activated molecular shuttling, regulation of molecular structure, switching of components, extended π-conduction, and photoinduced electron transfer reactions. The operation of outlined systems is to some extent reminiscent of biological processes, but still very crude and far from practical applications. It should also be noted that in a majority of cases the beneficial traits of rotaxane/catenane structures often stem from their unique molecular topology. As synthetic methodologies improve, the synthesis of these more elaborate molecular architectures is no longer an arduous task, and molecules that were once mere curiosities may become real, viable working devices.

4. MISCELLANEOUS EXAMPLES

The control of large-scale molecular motions is in no way restricted to topological exotica, and so to conclude this review, a brief description of other functional systems will be given. The following list of examples, however, is in no way exhaustive and merely attempts to give an insight into the general field commonly termed "supramolecular devices" (e.g., chemical sensors) [91].

4.1. Ion Expulsion from Crown-Based Assemblies

The expulsion of divalent cations (e.g., calcium, magnesium) upon photocleavage of the functionalized cryptate **44** (Fig. 55) represents a large-scale irreversible conformational change [92]. The release of the bound ion is an extremely efficient process because the cryptate is a strong binder, however, a large change in ion affinity occurs upon cleavage.

The discrimination in cation affinity of a binder upon photoactivation has more recently been put into practice by using the malachite green containing azacrown binder **45** (Fig. 56) [93]. Here, the photoionization of **45** switches the molecule from a powerful

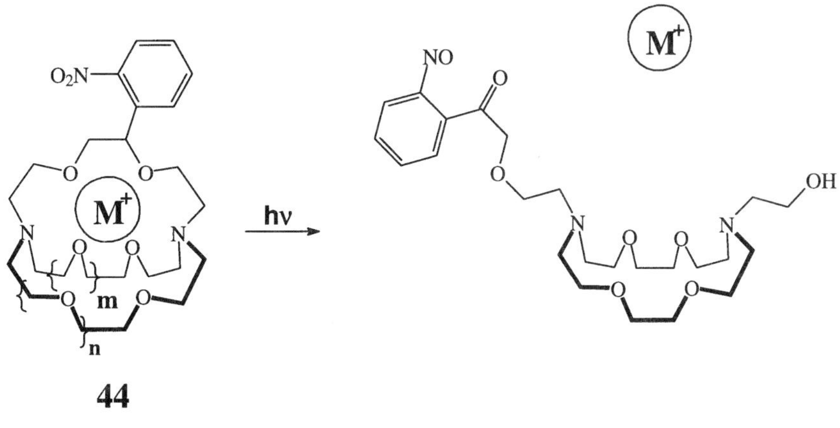

44

Fig. 55. An example of a photoactive cryptand that ejects alkali metal ions upon light activation.

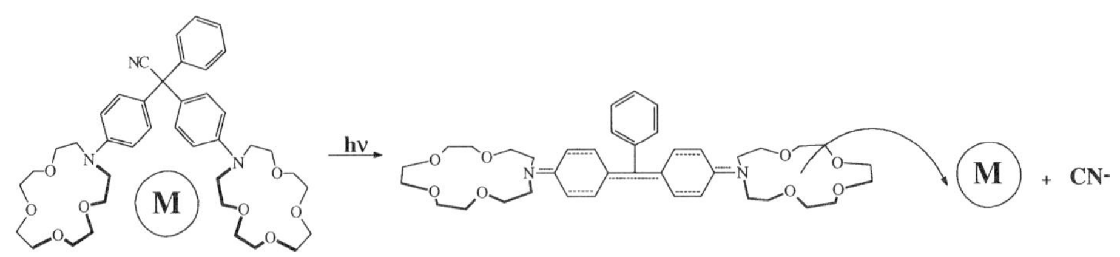

45

Fig. 56. Metal ion ejection from a bis-crown assembly and the resultant conformational change that accompanies ion release.

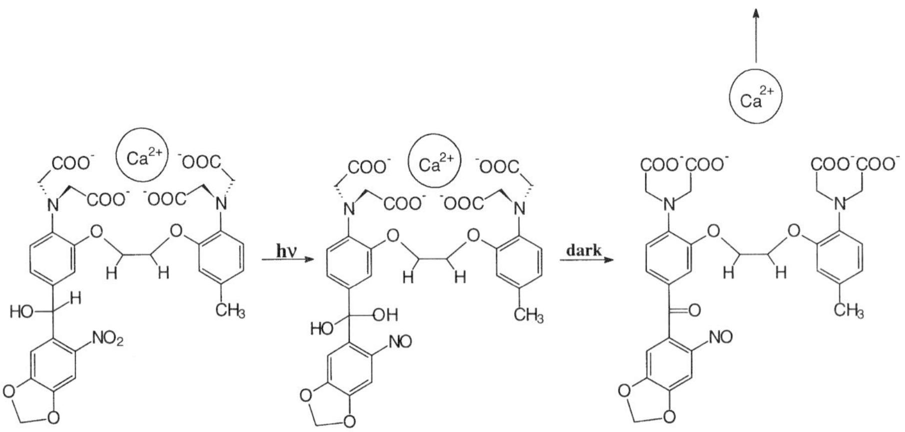

46

Fig. 57. Photoinitiated Ca^{2+} release from a carboxylate-derivatized binder.

cation binder to a near-perfect cation-releasing species. Strong supporting evidence has been obtained by both ^1H NMR spectroscopy and electrospray mass spectrometry. In the dark, and in the presence of $KClO_4$, a significant peak at $m/z = 742$ is consistent with a K^+ adduct of **45**. Irradiation of this solution for just 30 s with UV radiation is accompanied by a complete loss of this peak and the appearance of a new peak at $m/z = 677$, which is consistent with the quinoid cation of **45** (Fig. 56). In common with the previous cryptand, the light-activated alteration of molecular structure is the impetus required for ion release.

In the quest to create ion ejective systems that are more biologically useful, the carboxylate functionalized chelators shown in Figure 57 are worthy of mention [94]. The series of Ca^{2+}-selective binders (**46**) incorporates varying photosensitive *o*-nitrobenzylhydryl subunits, which upon irradiation at 365 nm are converted to *o*-nitrobenzophenones and release the bound ion. These photoactive chelators show potential as artificial agents for the creation of fast controlled jumps in intracellular calcium concentrations.

4.2. Molecular Capture by Conformational Switching

The modification of a structure to aid the encapsulation of a specific molecular guest is illustrated by the photochromic saccharide "tweezer" **47** (Fig. 58). The photoswitch within this molecule is the diarylethene unit, which is known to be both fatigue resistant and thermally irreversibly opened. In the open-ring form there are two rapidly exchanging species (i.e., parallel and antiparallel), but only the parallel conformer has the correct geometry to allow binding. It should be noted that the closed-ring form **47** is not able to act as a binder.

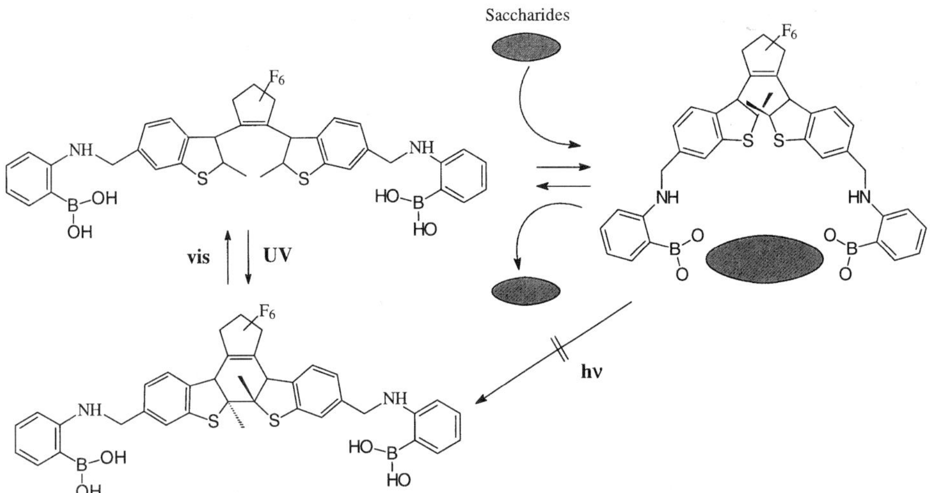

Fig. 58. An example of a photoresponsive molecular "tweezer" for encapsulating saccharides.

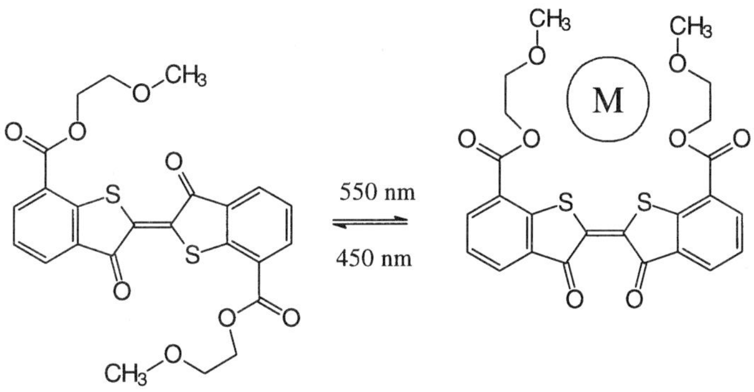

Fig. 59. A photoresponsive tweezer based on thioindigo used for regulating the capture and release of ions.

The addition of D-glucose to the open-ring form of **47** (EtOH, Tris-HCl buffer) is accompanied by changes in the circular dichroism spectrum in line with host–guest binding [95].

One of the simplest ways in which to alter the conformation of a molecular system is through the *trans–cis* isomerization of an ethene subunit. Under light activation the removal of the double-bond character allows bond rotation, with in many cases the *cis* and *trans* forms displaying distinctly different colors (e.g., photochromic materials) [96]. The large-scale structural change that accompanies isomerization can be put to use by the appropriate functionalization of the isomerizing group. This is demonstrated in molecule **48** (Fig. 59), which consists of a photolabile thioindigo portion and ethylenedioxy side groups for metal ion binding [97]. The binding of metal ions by the *trans* form of **48** is essentially zero, whereas K^+ and Rb^+ and Na^+ are readily bound by the photogenerated *cis* conformer. The use of this conformational selectivity toward ions has been successfully used in the transportation of Ag^+ ions across a liquid membrane. By a systematic alteration of the length of the ethylenedioxy units, optimization of the performance of these ion transportation devices has been made possible [98].

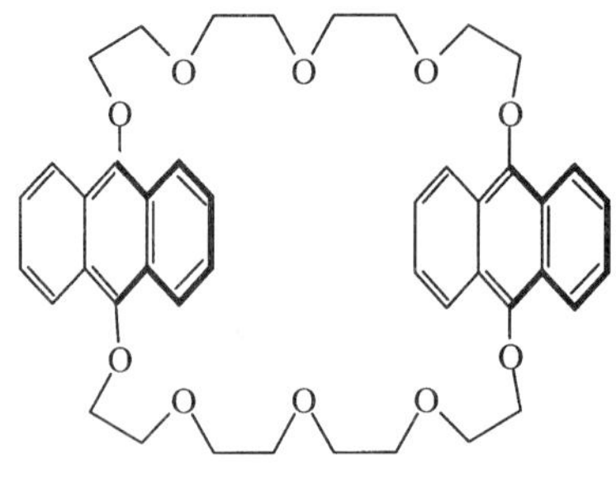

49

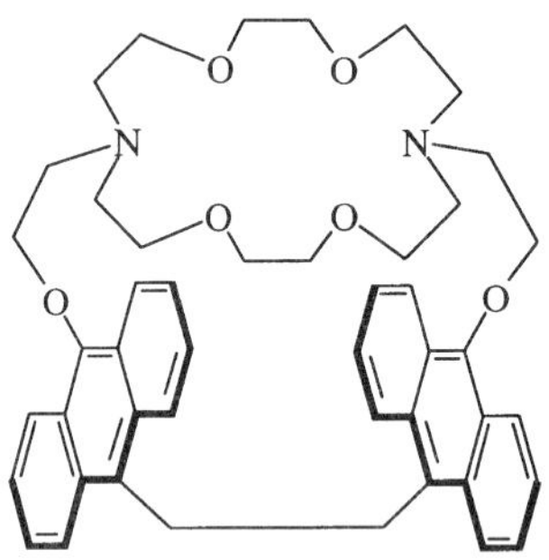

50

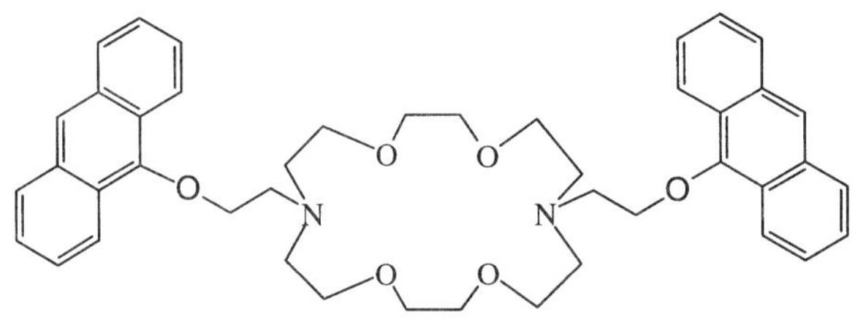

51

Fig. 60. Crown ether-based macrocycles incorporating photoactive anthracene moieties.

4.3. Structural Modification by Ion Binding

The incorporation of ions into molecular architectures can affect their physicochemical properties in a number of ways, such as forcing together of units, enhancing chemical reactions, and making gross structural changes [99]. Figure 60 represents a small cross section of such systems, which incorporate photoactive anthracene groups that are tethered to a macrocyclic ligand. The anthracene units within these molecules are important because they display intense fluorescence emission in the visible region, including excimer contributions that strongly depend on structural factors. Moreover, anthracenes can be readily dimerized, resulting in substantial shifts in their electronic absorption spectra.

The addition of excess $NaClO_4$ to a solution of **49** in methanol leads to a deformation of the structure and a concomitant change in the fluorescence profile. That the fluorescence spectrum for **49** shifts to emit at 570 nm is consistent with excimer formation and suggests that two Na^+ ions bind to the polyether chains and bring the anthracene units within a van der Waals distance. The extent of π-overlap between two anthracene groups is more effectively controlled by the addition of a crown binding site (**50**) (Fig. 61). The modulation of the rotation angle about the ethano bridge upon cation complexation (e.g., Na^+, K^+, Rb^+, NH_4^+, Ag^+, Tl^+) results in spectral changes in the UV-vis region [100]. More recently, the

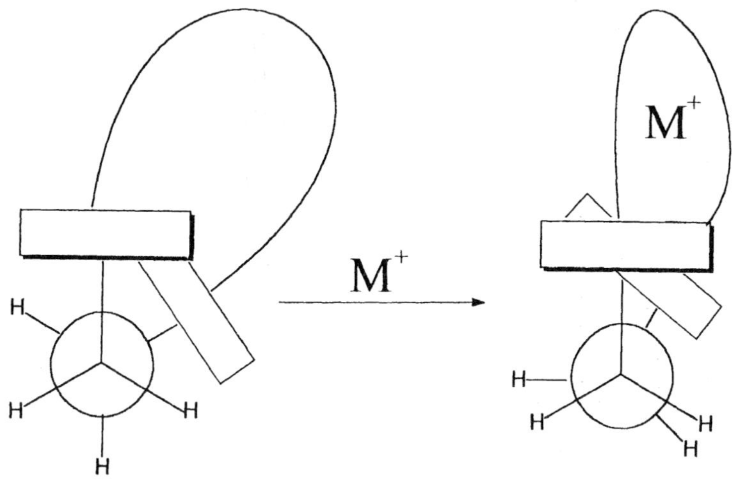

Fig. 61. The proposed overlap between the anthracene groups in **50** upon metal ion complexation.

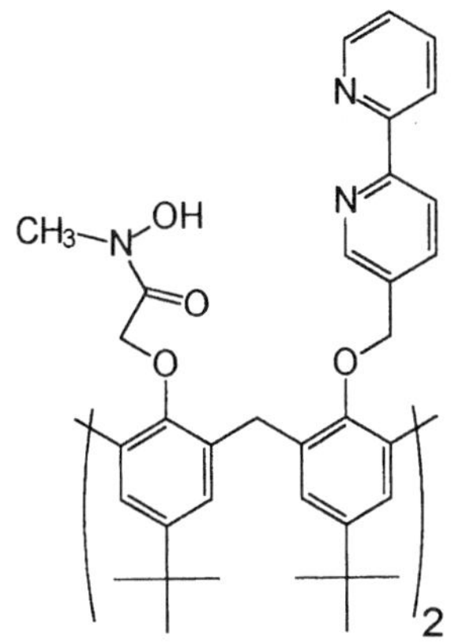

52

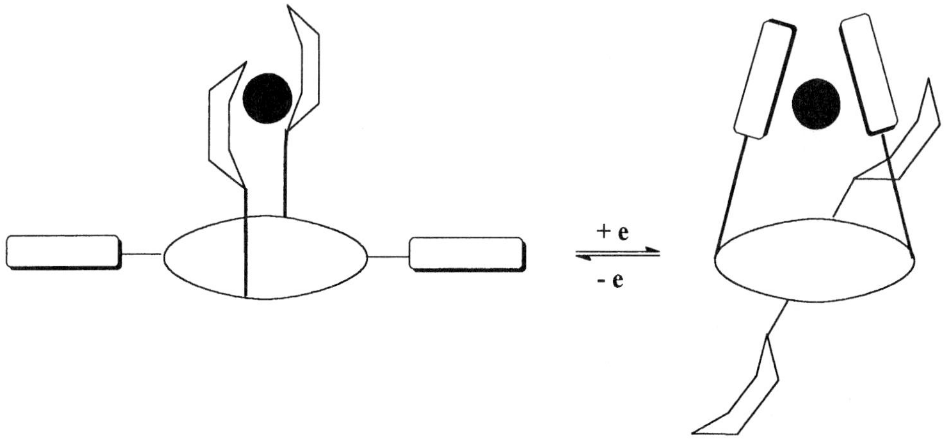

Fig. 62. An electrochemically switchable supramolecular cage and a cartoon representing the changes that occur upon reduction/oxidation of the metal ion.

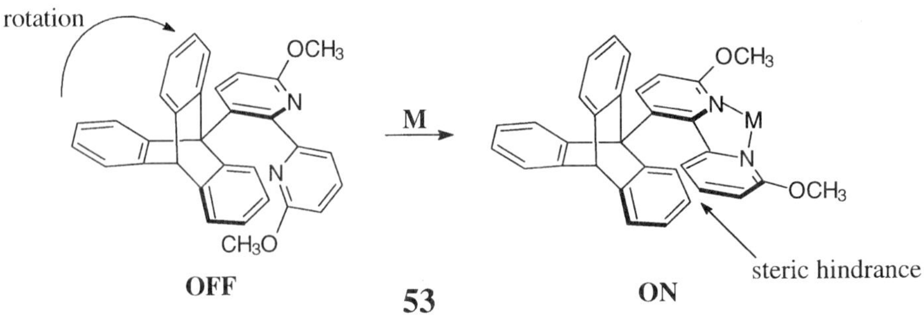

Fig. 63. An illustration of an ion-active molecular brake showing the on/off positions.

regulatory binding of **51** has been reported whereby the molecule reversibly switches between the crown and cryptand form. It has been shown that photocylization quantum yields and rates of thermal ring opening are dependent on the nature of the bound guest [101].

The electrochemical switching of a structure is exemplified in the calix[4]arene functionalized ligand **52** (Fig. 62) recently reported by Shanzer and co-workers [102]. Here, the molecule possesses two sets of binding groups; one for the binding of "soft" metal centers (2,2′-bipyridyl, bipy), and the other for more conventional "hard" metal ions (hydroxamates). The loading of Fe(III) to **52** is accompanied by the hydroxamates swinging up to engulf the metal ion, and the soft binders lying away. The reduction of the metal center to Fe(II) results in a reverse in the binding mode, with the bipy ligands wrapping up the ion. The switching off process is reversible because oxidation of the metal center back to Fe(III) regenerates the starting conformer.

The retardation of motion in a system, such as the rotation of a large group on its molecular axis, constitutes a molecular brake [103]. In the system developed by Kelly and co-workers, the introduction of a bipy unit into the "spokes" of **53** (Fig. 63) has been shown to reduce triptycene rotation. In the absence of metal ions, the ^{1}H NMR of **53** at 30 °C is consistent with a rapidly spinning triptycene moiety, suggesting that the bipy unit in no way interferes with the spinning of the molecular wheel. Upon the addition of Hg^{2+} to **53**, however, the ^{1}H NMR spectrum changes dramatically with the broadening of the triptycene resonances and a large shift observed in the methoxy signal.

5. CONCLUDING REMARKS

Clearly, from the simple molecular systems outlined in this chapter, the "bottom-up" approach has a long way to go before practical working systems start to emerge, let alone compete with well-established methodologies. A great many problems are immediately obvious and mainly concern how molecular-scale systems are to be addressed without the need for solution-phase probing and reliance on highly energy wasting and irreversible processes. In addition to these problems, the "wiring" of molecular systems into macroscopic devices represents a major task, but, it is hoped, not an insurmountable problem. Indeed, it should not be forgotten that the first computers filled an entire room and used valves, but have nevertheless progressed very quickly to now only occupy a mere fraction of that space. One could speculate that in the same time span a similar reduction in size is not too unreasonable and may utilize the properties of molecular-based functional assemblies.

Acknowledgments

This review has been possible by financial support from the University of Glasgow (ACB) and the Engineering and Physical Sciences Research Council (PRM). Pictures by

Dr. A. Freer are also gratefully appreciated. I also thank all of the authors of the papers reviewed for their inspirational work. This review is dedicated to the memory of Kim Wesson.

References

1. J. Mehra, "The Beat of a Different Drum. The Life and Science of Richard Feynman," p. 441. Clarendon Press, Oxford, 1994.
2. R. P. Feynman, *Found. Phys.* 16, 507 (1986).
3. T. Bein and G. D. Stucky, *Chem. Mater.* 8, 1569 (1996).
4. A. S. Moffat, *MOSAIC* 21, 30 (1990).
5. M. J. Little and J. Grinberg, *Byte* 13, 311 (1988).
6. A. L. Robinson, *Science* 223, 267 (1984).
7. D. V. Morgan and K. Board, "An Introduction to Semiconductor Microtechnology." Wiley, New York, 1991.
8. S. A. Campbell, "The Science and Engineering of Microelectronic Fabrication," p. 151. Oxford Univ. Press, London, 1996.
9. B. Martin, *GEC J. Technol.* 14, 2 (1997).
10. N. F. Johnson, *J. Phys. Condens. Matter.* 7, 965 (1995).
11. J-M. Lehn, "Supramolecular Chemistry: Concepts and Perspectives." VCH, Weinheim/New York, 1995.
12. P. R. Ashton, S. J. Langford, N. Spencer, J. F. Stoddart, A. J. P. White, and D. J. Williams, *J. Chem. Soc., Chem. Commun.* 1387 (1996).
13. A. R. Bernardo, J. F. Stoddart, and A. E. Kaifer, *J. Am. Chem. Soc.* 114, 10624 (1992).
14. A. C. Benniston, A. Harriman, D. Philp, and J. F. Stoddart, *J. Am. Chem. Soc.* 115, 5298 (1993).
15. A. C. Benniston, *Chem. Soc. Rev.* 25, 427 (1996).
16. J-C. Chambron, V. Heitz, and J-P. Sauvage, *J. Am. Chem. Soc.* 115, 12378 (1993).
17. G. Schill, "Catenanes, Rotaxanes and Knots." Academic Press, New York, 1971.
18. J-P. Sauvage, *Acc. Chem. Res.* 23, 319 (1990).
19. P. R. Aston, M. Bêlohradský, D. Philp, and J. F. Stoddart, *J. Chem. Soc., Chem. Commun.* 1269 (1993).
20. J-C. Chambron, D. K. Mitchell, and J-P. Sauvage, *J. Am. Chem. Soc.* 114, 4625 (1992).
21. D. Philp and J. F. Stoddart, *Synlett* 7, 445 (1991).
22. C. O. Dietrich-Buchecker and J-P. Sauvage, *J. Am. Chem. Soc.* 106, 3043 (1984).
23. A. G. Johnston, D. A. Leigh, R. J. Pritchard, and M. D. Deegan, *Angew. Chem. Int. Ed. Engl.* 34, 1209 (1996).
24. W. Haupt, *Philos. Trans. R. Soc. London B. Biol. Sci.* 303, 467 (1983).
25. H. Shichi, "Molecular Electronics: Biosensors and Biocomputers" (F. T. Hong, ed.), p. 165. Plenum Press, New York, 1990.
26. J. Deisenhofer and H. Michel, *Angew. Chem. Int. Ed. Engl.* 28, 829 (1989).
27. M. A. Steffen, K. Lao, and S. G. Boxer, *Science* 264, 810 (1994).
28. R. Manchanda, G. W. Brudvig, and R. H. Crabtree, *Coord. Chem. Rev.* 144, 1 (1995).
29. G. McDermott, S. M. Prince, A. A. Freer, A. M. Hawthornwaite-Lawless, M. Z. Papiz, R. J. Cogdell, and N. W. Isaacs, *Nature* 374, 517 (1995).
30. W. Draber, K. Tietjen, J. F. Kluth, and A. Trebst, *Angew. Chem. Int. Ed. Engl.* 30, 1621 (1991).
31. O. Totland, *Am. J. Botany* 83, 452 (1996).
32. M. B. Sporn, A. B. Roberts, and D. S. Goodman, eds., "The Retinoids," Vol. 2, p. 130. Academic Press, San Diego, 1984.
33. P-S. Song, "Molecular Electronics: Biosensors and Biocomputers" (F. T. Hong, ed.), p. 173. Plenum Press, New York, 1990.
34. V. Meske, V. Ruppert, and E. Hartmann, *Protoplasma* 192, 189 (1996).
35. A. Näubauer, P. Berg, and I. Ruge, "Nanostructures Based on Molecular Materials" (W. Gopel and Ch. Ziegler, eds.), p. 73. VCH, Weinheim/New York, 1992.
36. T. W. Ebbesen, *New J. Chem.* 15, 191 (1991).
37. A. W. Czarnik, "Fluorescent Chemosensors for Ion and Molecular Recognition" (A. W. Czarnik, ed.), p. 104, 1993.
38. A. Harriman and R. Zeissel, *J. Chem. Soc., Chem. Commun.* 15, 1707 (1996).
39. J. R. Barker, *Hybrid Circuits* 14, 19 (1987).
40. J. J. Hopfield, J. N. Onuchic, and D. N. Beratan, *J. Phys. Chem.* 93, 6350 (1989).
41. B. L. Feringa, W. F. Jager, and B. de Lange, *Tetrahedron* 49, 8267 (1993).
42. R. C. Haddon, *Proc. Natl. Acad. Sci. USA* 82, 1874 (1985).
43. H. A. Benesi and J. H. Hildebrand, *J. Am. Chem. Soc.* 71, 2703 (1949).
44. N. S. Hush, *Prog. Inorg. Chem.* 8, 391 (1967).
45. I. A. Gould and S. Farid, *Acc. Chem. Res.* 29, 522 (1996).

46. P. R. Ashton, A. M. Z. Slawin, N. Spencer, J. F. Stoddart, and D. J. Williams, *J. Chem. Soc., Chem. Commun.* 1064 (1987).
47. A. C. Benniston and A. Harriman, "NATO ARW Physical Supramolecular Chemistry" (A. E. Kaifer and L. Echegoyen, eds.), p. 179. VCH, Weinheim/New York, 1996.
48. A. C. Benniston, A. Harriman, and D. S. Yufit, *Angew. Chem. Int. Ed. Engl.* 36, 2356 (1997).
49. A. C. Benniston, A. Harriman, and V. M. Lynch, *J. Am. Chem. Soc.* 117, 5275 (1995).
50. A. C. Benniston, A. Harriman, and V. M. Lynch, *Tetrahedron Lett.* 35, 1473 (1994).
51. A. C. Benniston and A. Harriman, *Angew. Chem. Int. Ed. Engl.* 32, 1459 (1993).
52. H. Murakami, A. Kawabuchi, K. Kotoo, M. Kunitake, and N. Nakashima, *J. Am. Chem. Soc.* 119, 7605 (1997).
53. R. Ballardini, V. Balzani, M. T. Gandolfi, L. Prodi, M. Venturi, D. Philp, H. G. Ricketts, and J. F. Stoddart, *Angew. Chem. Int. Ed. Engl.* 32, 1301 (1993).
54. P. R. Ashton, R. Ballardini, V. Balzani, S. E. Boyd, A. Credi, M. T. Gandolfi, M. Gómez-López, S. Iqbal, D. Philp, J. A. Preece, L. Prodi, H. G. Ricketts, J. F. Stoddart, M. S. Tolley, M. Venturi, A. J. P. White, and D. J. Williams, *Chem. Eur. J.* 3, 152 (1997).
55. M. Seiler, H. Dürr, I. Willner, E. Joselevich, A. Doron, and J. F. Stoddart, *J. Am. Chem. Soc.* 116, 3399 (1994).
56. M. Kimura, Y. Misawa, Y. Yamaguchi, K. Hanabusa, and H. Shirai, *J. Chem Soc., Chem. Commun.* 2785 (1996).
57. A. C. Benniston, S. Gardner, L. J. Farrugia, and A. Harriman, unpublished results.
58. A. C. Benniston and A. Harriman, *Synlett* 3, 223 (1993).
59. A. C. Benniston, V. Grosshenny, A. Harriman, and R. Ziessel, *Angew. Chem. Int. Ed. Engl.* 33, 1884 (1994).
60. F. Vögtle, W. M. Müller, U. Müller, M. Bauer, and K. Rissanen, *Angew. Chem. Int. Ed. Engl.* 32, 1295 (1993).
61. M. Bauer, W. M. Müller, U. Müller, K. Rissanen, and F. Vögtle, *Leibigs Ann.* 649 (1995).
62. R. A. Bissell, E. Córdova, and A. E. Kaifer, *J. Org. Chem.* 60, 1033 (1995).
63. E. Cördova, R. A. Bissell, N. Spencer, P. R. Ashton, J. F. Stoddart, and A. E. Kaifer, *J. Org. Chem.* 58, 6550 (1993).
64. R. A. Bissell, E. Córdova, A. E. Kaifer, and J. F. Stoddart, *Nature* 369, 133 (1994).
65. P. R. Ashton, R. Ballardini, V. Balzani, A. Credi, M. T. Gandolfi, S. Menzer, L. Pérez-García, L. Prodi, J. F. Stoddart, M. Venturi, A. J. P. White, and D. J. Williams, *J. Am. Chem. Soc.* 117, 11171 (1995).
66. C. O. Dietrich-Buchecker and J-P. Sauvage, *Tetrahedron* 46, 503 (1990).
67. J-P. Collin, P. Gaviña, and J-P. Sauvage, *J. Chem. Soc., Chem. Commun.* 17, 2005 (1996).
68. P. Gaviña and J-P. Sauvage, *Tetrahedron Lett.* 20, 3521 (1997).
69. J-P. Collin, P. Gaviña, and J-P. Sauvage, *New. J. Chem.* 21, 525 (1997).
70. A. Livoreil, C. O. Dietrich-Buchecker, and J-P. Sauvage, *J. Am. Chem. Soc.* 116, 9399 (1994).
71. D. J. Cárdenas, A. Livoreil, and J-P. Sauvage, *J. Am. Chem. Soc.* 118, 11980 (1996).
72. M. J. Gunter and M. R. Johnston, *J. Chem. Soc., Chem. Commun.* 1163 (1992).
73. M. J. Gunter, D. C. Hockless, M. R. Johnston, B. W. Skelton, and A. H. White, *J. Am. Chem. Soc.* 116, 4810 (1994).
74. M. J. Gunter and M. R. Johnston, *J. Chem. Soc., Chem. Commun.* 829 (1994).
75. D. B. Amabilino, C. O. Dietrich-Buchecker, A. Livoreil, L. Pérez-García, J-P. Sauvage, and J. F. Stoddart, *J. Am. Chem. Soc.* 118, 3905 (1996).
76. M. Asakawa, S. Iqbal, J. F. Stoddart, and N. D. Tinker, *Angew. Chem. Int. Ed. Engl.* 35, 976 (1996).
77. R. Ballardini, V. Balzani, A. Credi, M. T. Gandolfi, S. J. Langford, S. Menzer, L. Prodi, J. F. Stoddart, M. Venturi, and D. J. Williams, *Angew. Chem. Int. Ed. Engl.* 35, 978 (1996).
78. A. Credi, V. Balzani, S. J. Langford, and J. F. Stoddart, *J. Am. Chem. Soc.* 119, 2679 (1997).
79. M. R. Wasieleski, *Chem. Rev.* 92, 435 (1992).
80. A. M. Brun, S. J. Atherton, A. Harriman, V. Heitz, and J-P. Sauvage, *J. Am. Chem. Soc.* 114, 4632 (1992).
81. A. M. Brun, A. Harriman, V. Heitz, and J-P. Sauvage, *J. Am. Chem. Soc.* 113, 8657 (1991).
82. A. Harriman, V. Heitz, and J-P. Sauvage, *J. Phys. Chem.* 97, 5940 (1993).
83. J-C. Chambron, S. Chardon-Noblat, A. Harriman, V. Heitz, and J-P. Sauvage, *Pure Appl. Chem.* 65, 2343 (1993).
84. J-P. Sauvage and A. Harriman, *Chem. Soc. Rev.* 35, 41 (1996).
85. J-C. Chambron, A. Harriman, V. Heitz, and J-P. Sauvage, *J. Am. Chem. Soc.* 115, 7419 (1993).
86. J-C. Chambron, A. Harriman, V. Heitz, and J-P. Sauvage, *Coord. Chem. Rev.* 132, 229 (1994).
87. A. C. Benniston, P. R. Mackie, and A. Harriman, *Tetrahedron Lett.* 38, 3577 (1997).
88. A. C. Benniston, P. R. Mackie, and A. Harriman, *Angew. Chem. Int. Ed. Engl.* 37, 354 (1998).
89. T. Lu, L. Zhang, G. W. Gokel, and A. E. Kaifer, *J. Am. Chem. Soc.* 115, 2542 (1993).
90. P. L. Anelli, P. R. Ashton, R. Ballardini, V. Balzani, M. Delgado, M. T. Gandolfi, T. T. Goodnow, A. E. Kaifer, D. Philp, M. Pietraszkiewicz, L. Prodi, M. V. Reddington, A. M. Z. Slawin, N. Spencer, J. F. Stoddart, C. Vicent, and D. J. Williams, *J. Am. Chem. Soc.* 114, 193 (1992).
91. A. W. Czarnik, *Acc. Chem. Res.* 27, 302 (1994).
92. R. Warmuth, J-M. Lehn, J. W. Bats, and G. Quinkert, *Helv. Chim. Acta* 74, 671 (1991).

93. K. Kimura, R. Mizutani, M. Yokoyama, R. Arakawa, G. Matsubayashi, M. Okamoto, and H. Doe, *J. Am. Chem. Soc.* 119, 2062 (1997).
94. S. R. Adams, J. P. Kao, G. Grynkiewicz, A. Minta, and R. Y. Tsien, *J. Am. Chem. Soc.* 110, 3212 (1988).
95. M. Takeshita, K. Uchida, and M. Irie, *J. Chem. Soc., Chem. Commun.* 1807 (1996).
96. M. Irie, "Advances in Polymer Science," p. 28. Springer-Verlag, Berlin/New York, 1990.
97. M. Irie and M. Kato, *J. Am. Chem. Soc.* 107, 1024 (1985).
98. M. F. Rahman, K. Fukunishi, M. Kuwabara, H. Yamanaka, and M. Nomura, *Bull. Chem. Soc. Jpn.* 66, 1461 (1993).
99. J-P. Desvergne, F. Fages, H. Bouas-Laurent, and P. Marsau, *Pure Appl. Chem.* 64, 1231 (1992).
100. J-P. Desvergne, J. Lauret, H. Bouas-Laurent, P. Marsau, N. Lahrahar, H. Andrianatoandro, and M. Cotrait, *Recl. Trav. Chim. Pays-Bas.* 114, 504 (1995).
101. J. H. R. Tucker, H. Bouas-Laurent, P. Marsau, S. W. Riley, and J-P. Desvergne, *J. Chem. Soc., Chem. Commun.* 13, 1165 (1997).
102. C. Canevet, J. Libman, and A. Shanzer, *Angew. Chem. Int. Ed. Engl.* 35, 2657 (1996).
103. T. R. Kelly, M. C. Bowyer, K. V. Bhaskar, D. Bebbington, A. Garcia, F. Lang, M. H. Kim, and M. P. Jette, *J. Am. Chem. Soc.* 116, 3657 (1994).

Chapter 16

STRUCTURE, BEHAVIOR, AND MANIPULATION OF NANOSCALE BIOLOGICAL ASSEMBLIES

Timothy Bayburt, Joseph Carlson, Bruce Godfrey, Mary Shank-Retzlaff, Stephen G. Sligar

Beckman Institute, Urbana, Illinois, USA

Contents

1. BIOLOGICAL MOLECULES AS NANOSTRUCTURED MATERIALS

Several billion years of evolution have led to the generation of an astonishing variety of biological organisms. These organisms vary in many ways, but their fundamental building blocks remain the chemical structures of nucleic acids, proteins, and lipids that make

Handbook of Nanostructured Materials and Nanotechnology, edited by H.S. Nalwa
Volume 5: Organics, Polymers, and Biological Materials

life possible. The hierarchical assembly of these fundamental building blocks into working biochemical machinery is the foundation of cellular function. Many of these assembly processes, however, occur at a size regime in which visualization and characterization are difficult. Detailed structural characterization is an essential goal, not only for understanding these processes, but also for harnessing this technology for later technological application.

Each of the biological materials mentioned—nucleic acids, proteins, and lipids—is nanostructured in different ways. Nucleic acids, through base stacking and specific hydrogen bond base pairing, are the informational media for the genetic code. Furthermore, nucleic acids can fold into complex structures that are chemically active. Nucleoprotein complexes add an additional level of functional diversity to these assemblies, allowing the manipulation and expression of genetically encoded information. By the same token, amphiphilic phospholipids can self-assemble to form cellular compartments, which are essential for life. Phospholipid bilayers are the most common interface upon which biological processes occur, and yet the exact role of phospholipids in these processes is still being determined. This natural interface is a logical starting point for the construction of synthetic interfaces for biotechnology. Proteins are responsible for the majority of functions in biological systems. Enzymes capable of catalyzing an amazing diversity of reactions, channels that control the flow of ions and other molecules from compartment to compartment, electron transport proteins, and receptors, which translate the binding of signal molecules such as steroids into a biochemical response, are examples of this functional diversity.

The structure of biological materials is a result of both their chemical composition and their environment. Chemical composition, such as the polypeptide sequence of a protein or the sequence of bases in a nucleic acid strand, is a key determinant of structure. Environment affects the structure of these materials in much more subtle and deterministic ways than most man-made materials. The components found in an aqueous solution, as well as their concentrations, will determine the detailed structure of a biomolecule.

The behavior of these biomaterials is varied and often goes beyond a simple assessment of their function. For example, although an enzyme's function may be to convert a substrate into a product, its behavior in a given system may involve not only that catalytic process, but also such processes as diffusive motions, regulatory controls, conformational changes, and environmental factors. These behaviors are important, for they dictate what is actually occurring with the enzyme in the broadest sense; they are not limited, for example, to a simple chemical transformation. As biological systems are utilized more and more for technological applications, their behavior in unique, nonphysiological environments becomes more critical.

Manipulation of these biomolecular assemblies and the nanomaterials they are used in involves engineering and control at the nanometer-scale. This often means ordering a material by crystallization or confining it to nanometer scale regions through the use of patterning or lithography. The ability to change a system at the nanometer scale to construct new, scientifically interesting systems or to suit new design requirements is one of the most exciting aspects of this field and is broadly termed nanostructured materials research.

This review will focus on three different aspects of biologically derived nanostructured materials. Each of these three aspects is an area of focus for our laboratory, and each is reviewed in a comprehensive fashion, with the goal of imparting to the reader the current state of these subfields. First, the use of scanning probe microscopy in studying biological assemblies, specifically supported biological membranes, nucleic acids, and nucleoprotein complexes will be described. Second, the use of more general protein phospholipid assemblies as nanostructured materials for the study of protein function at interfaces and the development of operational molecular devices is discussed. Third, the use of general surface immobilization techniques to engineer nanostructured materials from functional proteins is presented.

2. SCANNING PROBE MICROSCOPY OF NANOSCALE BIOLOGICAL ASSEMBLIES

2.1. Scanning Probe Microscopy

Scanning probe microscopy (SPM) has begun, in recent years, to have an impact on the study of biological membranes and their associated proteins. SPM is an umbrella term for any microscopic technique that utilizes the scanning principles first pioneered in the scanning tunneling microscope (STM). These microscopes can vary so much that they are best discussed in terms of their guiding central principle. SPM combines independent mobility in all three spatial directions (scanning) with a detection system capable of detecting some characteristic of the surface (probing). The various surface characteristics that can be probed (conductivity, surface forces, compressibility, capacitance, magnetic field, fluorescence emission) demonstrate the wealth of information that can be obtained.

The scanning used in SPM is achieved through the use of voltage-sensitive piezo crystal actuators to move either the sample or the scanning probe in angstrom increments in the x, y, and z directions. The fine approach of the probe to the surface being imaged is controlled by an electronic feedback loop operated by a digital computer interface. Coarse approach of the probe to the surface is accomplished with a digitally controlled stepper motor and micrometer screws. The microscope scanner must be installed with some form of vibration isolation if it is to achieve nanometer or subnanometer resolution. These principles are depicted in a schematic form in Figure 1.

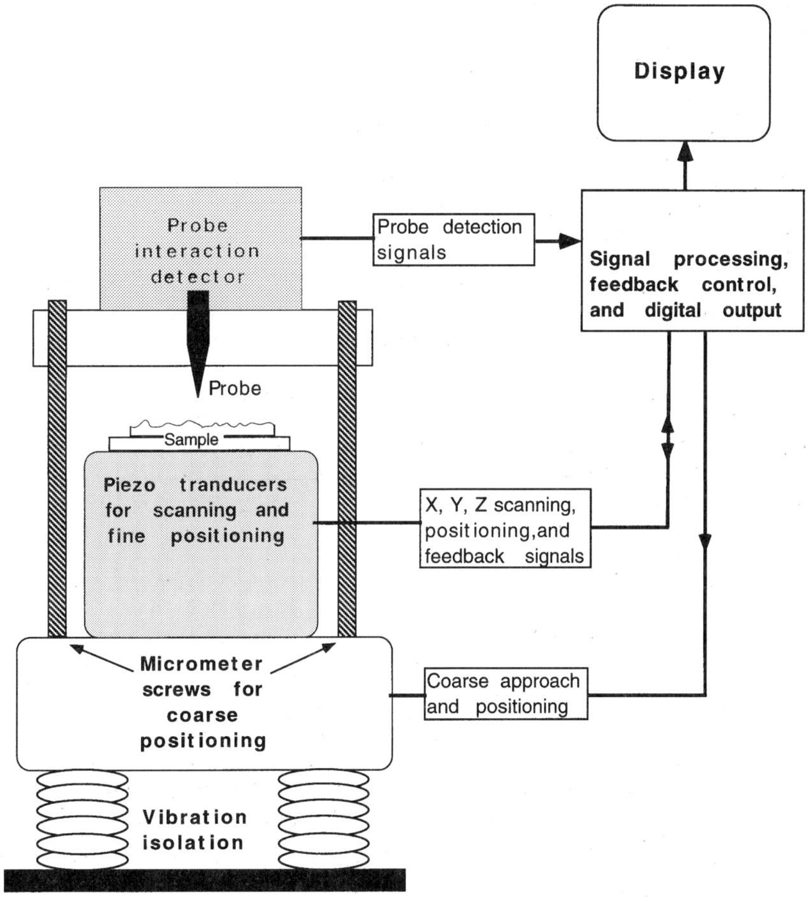

Fig. 1. General schematic of optical lever detection in the scanning force microscope.

Most piezo scanners used today are tube-type scanners, consisting of a vertically positioned hollow tube divided lengthwise into quadrants. Each quadrant has an individual electrode and ground, so that it can be expanded or contracted lengthwise by an applied voltage. The voltage is supplied by a controller circuit/amplifier that is directed by a digital signal processor in the computer. The expansion of one quadrant in coordination with the contraction of the opposite face causes the tube to bend and displaces the tube's end laterally. This motion is used to raster scan the sample under the tip. The motion produced is primarily lateral, but it is accompanied by a slight vertical shift as well. This vertical shift is corrected by the controlling software. Piezo elements have an inherently nonlinear response to applied voltage. The nonlinearity can be handled either through software correction or through a physical measurement of scanner motion provided by a strain gauge or similar means. The motion of the sample in the z direction can be accomplished by adding a baseline voltage to all four quadrants simultaneously or by incorporating a separate piezo section to the quadrant tube for independent vertical motion control. These issues are discussed in depth in several books on scanning probe microscopy, including [1, 2], and [3], which contains an up-to-date and simplified treatment of the operation of currently available commercial instruments.

The different forms of SPM data are distinguished by the type of interaction the probe is detecting, by the use of this signal to control the feedback loop, and by the type of data plotted against scan position to form the image. For example, a scanning tunneling microscope probe detects tunneling current between tip and sample. The feedback loop responds to the current and adjusts the tip–sample separation continuously to maintain a constant tunneling current. A scanning force microscope probe is mounted on a flexible cantilever, and the feedback maintains a constant deflection of this cantilever. In either case, an image is formed by plotting the z axis corrections made by the feedback loop at each point in the scan. Other forms of data from the scan can also be plotted, such as the actual output signal from the probe. The various imaging and data display modes available will be discussed in more detail in the specific sections below.

2.1.1. Scanning Tunneling Microscopy

The STM was invented in 1982 by Gerd Binnig [4]. The STM scans a surface with a conductive tip made of a metal wire etched to a fine point, and the microscopes' electronics detects the tunneling current passing between the tip and surface, typically from a distance of about 10 Å. The tunneling current varies exponentially with distance. This sensitive distance dependence allows atomic resolution to achieved, but only on conductive or semiconductive surfaces [2]. The most common modes of STM operation are constant current mode and constant height mode. In constant current mode the feedback loop maintains a constant height of the probe tip above the sample, assuming a constant tunneling barrier height over the surface, by maintaining the tunneling current at a constant value. Artifacts in the height image created in this mode can arise from variations in the tunneling barrier strength caused by changes in the composition of the surface. In constant height mode, the feedback loop is used to bring the tip within tunneling range of the surface, and then it is turned off and the tip is scanned at that height while the variations in tunneling current are recorded over the surface. This mode only gives usable results for extremely flat surfaces. These modes are depicted schematically in Figure 2.

Scanning tunneling microscopy can be performed under aqueous solution if the wire probe tip is fully insulated over its length except at the apex [5]. In general, however, this approach has not yielded images of biomolecules such as DNA with a resolution as good as that which scanning force miroscopy can give on similar samples. It is also possible to image DNA on a mica surface in humid air by detecting the tens or hundreds of picoamps that flow through the thin layer of adsorbed water that is always present on the surface [6]. This type of imaging is performed using the constant current mode to produce a topographical image.

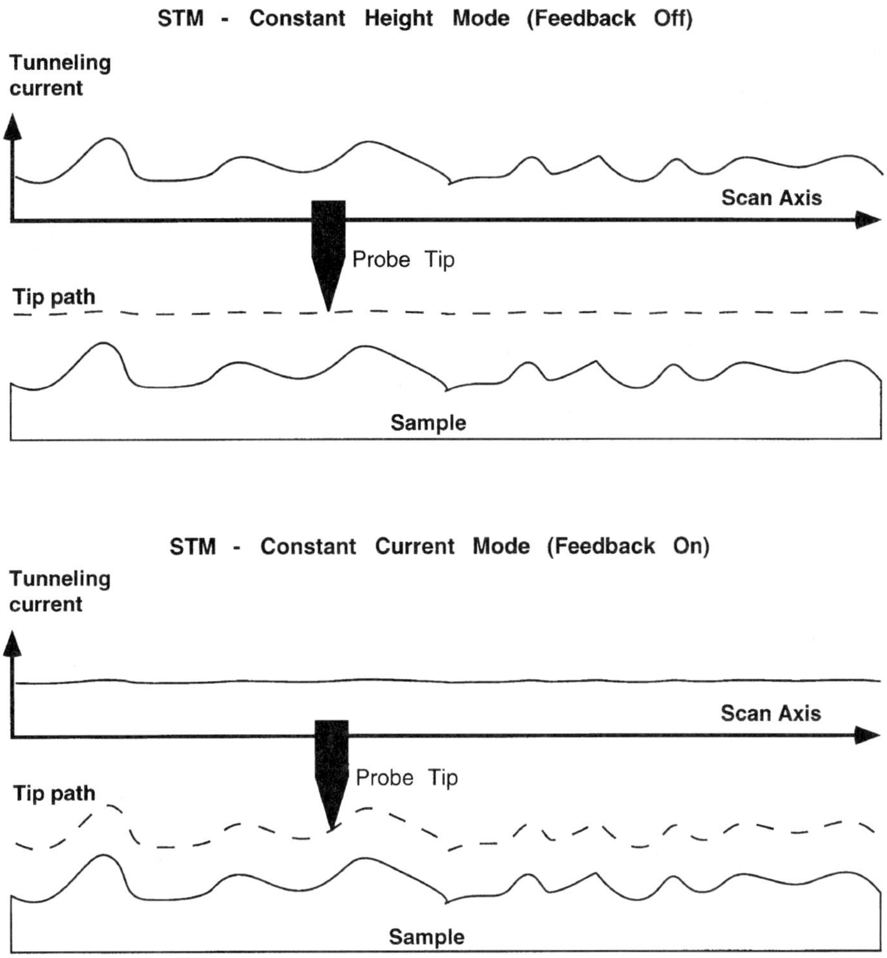

Fig. 2. Schematic cross section of a high coverage rHDL surface.

2.1.2. Scanning Force Microscopy

The scanning force (also called the atomic force) microscope (SFM) was invented in 1986 by Binnig, Quate, and Gerber [7] as a means of surface imaging with resolution similar to that obtained from a STM but on nonconducting surfaces [7]. The scanning force microscope works by raster scanning a sharp tip that is mounted on a flexible cantilever over the sample surface and detecting the deflection of the cantilever as surface features are encountered by the tip. The basic components of a scanning force microscope are the scanner, the cantilever and tip, and a means of detecting the motion of the cantilever. Cantilevers with integral tips are microfabricated on silicon wafers [8]. The detection of cantilever motion is accomplished with an optical lever and a split photodiode on most commercial instruments, illustrated in Figure 3. This makes the SFM a very versatile instrument, capable of acquiring images and force data in several different modalities [1], three of which are described in Figure 3. These imaging techniques can be used in air, in a vacuum, and under fluid. The last capability has made scanning force microscopy the most important scanning probe technique in the biological sciences.

2.1.2.1. Contact Mode

The first mode of operation developed was the contact mode, in which a sharp tip on the end of a microfabricated cantilever is brought into contact with the sample surface at a

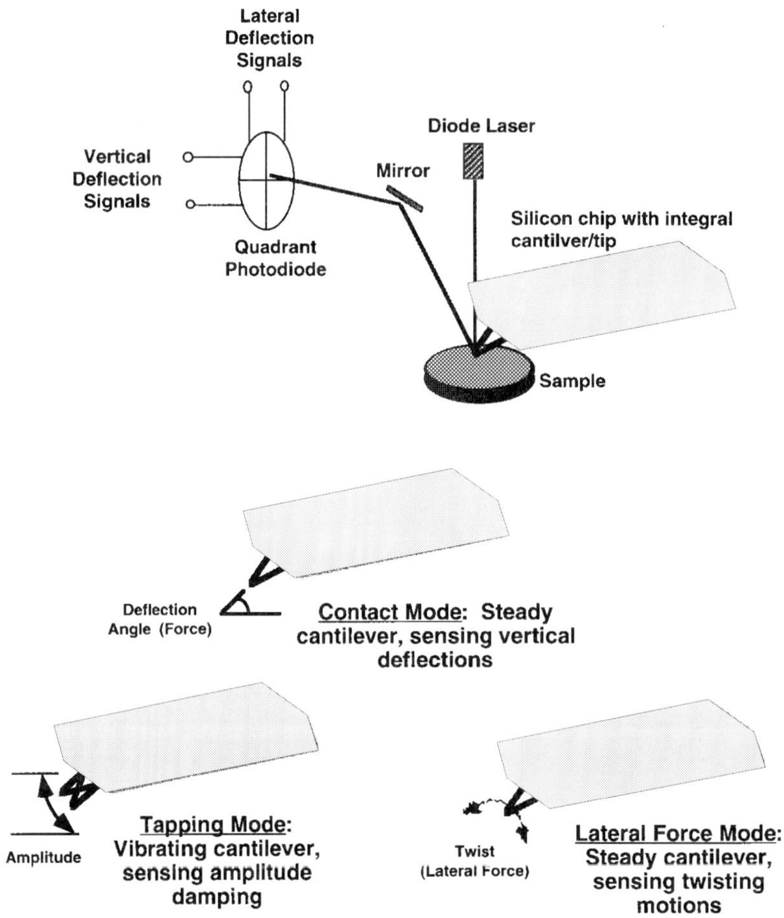

Fig. 3. General schematic of a scanning probe microscope.

set force level and scanned while the feedback loop maintains the force at the desired setpoint. As was mentioned above, the image can be displayed as height, where the data point plotted is the feedback correction at each point, or as the instantaneous deflection of the cantilever at each point. Because the feedback loop quickly corrects for any deflection of the tip, the edges of features are accentuated in deflection images, but the apparent heights are obscured. A further type of image that can be acquired from a contact mode scan is a friction or lateral force image. As the tip scans, the cantilever can twist laterally as features or areas with different coefficients of friction on the surface are encountered. The twisting can be detected by the quadrant photodiode and can be displayed as a lateral force or friction image.

Contact mode SFM can provide atomic resolution on hard surfaces like mineral crystals. Achieving such high resolution requires an exceptionally sharp tip, and the scan must be conducted at a relatively high force level, in the tens of nanonewtons range. The resolution attainable on biomolecules is lower because they are often deformed or damaged at such high tip forces and because it is generally difficult to bind them to a surface sufficiently rigidly to withstand high scan forces. The best imaging of biomolecules requires the use of soft cantilevers that allow scanning at lower forces, in the nanonewton to subnanonewton range. Such soft cantilevers easily respond to the van der Waals attraction as they approach a surface and can jump into contact. Furthermore, the layer of adsorbed water that is always present on a hydrophilic surface exposed to the atmosphere can trap the tip by its capillary forces and pull it to the surface strongly enough to result in an applied force much higher than is desirable for imaging biomolecules.

This problem is generally overcome either by imaging in a low-humidity chamber or by imaging under a fluid such as an aqueous buffer or an alcohol, which eliminates the capillary forces. Imaging under an aqueous buffer, in which biomolecules in general and nucleic acids in particular are highly soluble, makes greater demands on specimen preparation techniques. It is often more difficult to achieve tight binding of a soluble biomolecule to a surface under buffer than under air or vacuum. Since SFM was invented, much effort has been expended on developing methods of binding nucleic acids to flat surfaces such as mica well enough for imaging under aqueous buffer solutions.

2.1.2.2. Oscillating Tip Scanning Modes

SFM images can also be acquired by using a probe that is oscillating perpendicular to the surface and thus only interacts with the surface strongly at the bottom of each oscillation cycle. When operating in this mode, the photodiode detects the amplitude of the oscillation and the feedback loop maintains a constant value for the amplitude at a set point below the free resonant amplitude of the cantilever. This type of operation is called intermittent contact mode (or tapping mode). As in contact mode, the height image displays the amount of feedback correction needed to maintain the constant interaction with the surface at each point. There are also several other ways of acquiring data from an oscillating probe. The instantaneous response of the cantilever can be displayed as an amplitude image, analogous to the deflection image in contact mode, with the same limitations on interpretation of the image. Moreover, a phase change in the cantilevers' response with respect to the driving signal can be displayed as a phase image, and the related shift in frequency caused by coupling with surface features can be displayed in a frequency image.

In intermittent contact mode the phase difference between the oscillation driving the tip and the actual motion of the tip is sensitive to the viscoelasticity of the surface [9, 10]. It has recently been shown that molecular adhesion can also induce phase changes [11, 12]. For example, Finot and McDermott produced images with reproducible phase contrast using a bare silicon tip, which is hydrophilic, when imaging a gold surface patterned with carboxyl- and methyl-terminated thiols.

Intermittent contact mode can be used in air and under fluids. The cantilevers used for operation in air are stiff enough to have resonant frequencies typically in the 200-kHz range, which gives the oscillating tip enough energy to overcome the capillary forces encountered in ambient air. The major advantage of intermittent contact is that it greatly reduces the lateral forces applied to the sample. Soft samples that would deform and weakly bound samples that can be swept aside in contact mode can therefore be imaged.

In fluid intermittent contact mode, the cantilever is damped by the fluid and generally is driven at a frequency that corresponds to an acoustic mode of the fluid cell rather than at the cantilever's free air resonance. A means of driving the cantilever without involvement of the fluid may allow a greater range of choices for the drive frequency. Two direct-drive methods of operating commercial cantilevers in intermittent contact mode have been explored, electrostatic [13] and magnetic [14], and improvements in noise levels have been reported. Electrostatically driven intermittent contact has so far only been demonstrated for operation in air. This approach has the advantage that a commercial tip can be used without further modification. Driving a tip magnetically has been shown to give excellent results in fluid imaging of DNA with low noise levels [15]. This direct-drive method requires that a thin magnetic layer be evaporated onto the cantilever, which can easily be done during the manufacturing process or by the experimenter with access to standard electron microscopy specimen preparation equipment. In addition, a simple solenoid coil has to be added to the microscope to drive the magnetically coated cantilever [14, 15]. Significant noise reduction in contact mode operation has also been achieved using resonance contact mode, where a low-amplitude, high-frequency oscillation is imparted to a cantilever scanning in contact mode [16].

Decreasing the force applied to a soft sample, such as a biological macromolecule, helps to increase resolution by reducing deformation of the molecule. An alternative to this approach is to stiffen the molecule. This can be done by cooling the specimen and scanning apparatus to liquid nitrogen temperatures. Cryo-scanning force microscopy has been pioneered by the Shao group at the University of Virginia. They have found that stiffer frozen samples do allow higher resolution scans, with excellent results reported for single-stranded DNA specimens. The cryo-SFM work on biological specimens has been reviewed recently, and images of many different types of biomolecules, including single-stranded DNA, have been presented [17].

Fluid intermittent contact mode imaging has become very important in biological research because of the ability to image soft samples. Developments in intermittent contact mode SFM operation are ongoing, and improvements in resolution, reduction in the forces applied to the sample, and reduction in background noise levels are being pursued by instrument manufacturers and users.

2.1.3. Near-Field Scanning Optical Microscopy

Near-field scanning optical microscopy is a technique that is reaching the stage of development where it is beginning to become useful [18]. Very few of these instruments are sold commercially, and most of the publications containing images of biological specimens have been made on custom built or modified commercial near-field scanning optical microscopes (NSOMs). The original NSOM design uses a probe tip that consists of an optical fiber pulled to a sharp point, 30–50 nm in diameter, coated with aluminum. This tip is vibrated laterally to detect proximity to the surface for feedback operation. This is referred to as shear force detection, a variation of noncontact SFM. This type of NSOM is usually used for dry imaging, but it is also being developed for use under fluid [19]. Near-field optical images of chromosomes labeled with fluorescent oligonucleotide probes by in situ hybridization have been reported [20]. These images, scanned on dry samples, have a lateral resolution of about 100 nm, substantially better than that produced by standard or confocal scanning optical microscopy.

Another version of NSOM has been developed that shows promise for imaging in fluid: the SNOAM, or scanning near-field optical atomic force microscope [21]. It uses a standard NSOM optical fiber, which is bent at the tip so that the fiber can be mounted nearly parallel with the sample surface and act as an SFM cantilever and optical probe simultaneously. Motion of this optical cantilever is detected by the standard SFM optical lever system, and the cantilever is vibrated vertically as in intermittent contact mode SFM. This has been found to work in a noncontact regime in fluid and in intermittent contact regime in air. This apparatus was used to make near-field optical images of a nucleoprotein complex, metaphase chromosomes, labeled with an intercalating fluorescent dye [22]. The resolution appears to be comparable to that reported for the shear force feedback NSOM. Another approach to combining the functions of SFM and NSOM, and possibly STM as well, has been experimentally demonstrated using luminescent porous silicon particles fabricated on SFM and STM tips as light-emitting probes for NSOM [23]. This technique has not yet achieved the same resolution as the more conventional NSOM techniques, but it may offer one of the simplest means of mass-producing NSOM probes.

The ultimate advantage of NSOM techniques is their ability to use the optical spectra of regions of the sample to identify and the molecular compositions. This would allow more flexible chemical mapping than scanning with modified SFM probes. It would be possible, for instance, to map chromosomes marked with numerous FISH labels simultaneously at high resolution [20]. It is not clear at present whether resolution sufficiently fine to study individual nucleic acid–protein complexes, typically less than 50 nm across, can be obtained by NSOM [18]. Apertureless NSOM instruments like the photon scanning tunneling microscope described in [18, 24] may eventually achieve sub-50-nm resolution with continued development.

2.1.4. Chemical Force Microscopy

An SFM image is a high-resolution topographical map of a surface. Objects in the map are typically identified with the help of their unique shapes and sizes. The ability to obtain image contrast based on the chemical composition of the object adds a great deal to the interpretation of SFM images. An SFM probe tip surface treated to create a thin molecular layer with a particular chemical property can "sense" an area of the sample that has an attractive or repulsive interaction with the treated tip [25]. In earlier work investigating this mode of operation, the tip was scanned in constant height contact mode, and lateral force data were displayed as the image [26, 27]. Regions that have an attractive interaction with the tip increased the measured friction as the tip passed over them.

A similar data set, an adhesion force map, can be generated by taking force curves at each point in the scan. This type of operation is much slower than intermittent contact mode and results in the accumulation of a large amount of digital data, but more detailed information can be extracted [27]. Image contrast based on the chemical nature of the surface [28] and on specific molecular recognition [29] has been demonstrated by using force-curve maps. For example, in the study of molecular recognition, mapping force–distance curves were taken at each point in the scan to detect the specific molecular interaction between molecules. A streptavidin pattern on a gold surface was easily detected by a biotinylated probe tip. The specificity of the interaction was demonstrated by blocking the contrast in the adhesion image by treating the surface with free biotin. Chemical mapping of hydrophobic self-assembled thiols patterned on gold [28] was accomplished in buffer with bare hydrophilic silicon nitride tips.

2.1.5. Electrostatic Force Microscopy

The first use of electrostatic forces to generate image contrast, described in 1994 [30], used a specially fabricated cantilever that had a conductive tungsten tip. The tip was oscillated while scanning a few nanometers above the surface, and a voltage bias was applied between the tip and sample. This noncontact mode of operation gave a resolution of about 30 nm, with frequency shift detection of electrostatic effects on the vibrating cantilevers. These experiments were performed in air, which may be a source of ambiguities in biological samples as a result of buffer salts drying onto the sample. Another approach was taken by Xu and Arnsdorf [31], who coated standard silicon nitride tips with charged lipids and used these to measure the changes in adhesion force against different charged surfaces under buffer. These authors did not use a scanned array of force curves as was described in the previous section, but the possibility of that imaging mode with the use of charge-modified tips is clearly implied by their finding of large changes in adhesion forces based on surface charge differences. Given the results described for chemical contrast using intermittent contact mode with phase detection, based on adhesion forces, this approach would also be expected to work for charged-modified tips interacting with charges on the sample surface. This type of imaging of a biological specimen under buffer has not yet been reported.

2.1.6. Tip Technology

Another area of technology in which improvements would be welcomed, and which could immediately be taken advantage of by users of current instruments, is in tip sharpness and composition. The best commercially available tips have radii of curvature of about 5–40 nm. Advantages have been reported for electron beam-deposited (EBD) carbon "super-tips" that are due to their sharpness and chemical inertness [32–35]. Even though the EBD tips are not significantly sharper than the best commercially available silicon tips, they seem more resistant to contamination from the sample and exhibit low adhesion to mica substrates, so they often produce better imaging results. It has been reported that simply carbon-coating a commercial tip can provide much the same benefit [36].

There is another type of carbon tip that can provide still further improvement on this. Multiwalled carbon nanotubes can be used as SPM probes, but they must be custom fabricated by the user. A method has been published for making multiwalled nanotube probes by gluing nanotubes onto the tip of a commercial cantilever, and excellent results have been reported [37]. These tips have the advantages of greatly reduced capillary adhesion, a predictable force at which point reversible buckling occurs, and a very high aspect ratio. They are also conductive enough to use as STM tips. A single-walled nanotube has a diameter of 1 nm, and these are claimed to be quite stiff. Such a nanotube may make the ultimate SFM tip, if a way can be found to produce them and attach them to cantilevers reliably and cheaply [37].

If probe tips that exhibit very low adhesive forces against common imaging substrates become readily available, it will become possible to image routinely at lower forces than are now possible. For instance, intermittent contact in air with cantilevers having a spring constant of only 0.01 N/m was reported to work with carbon nanotube probes [37], whereas commercial silicon probes for intermittent contact in air have spring constants of 30 N/m or more to overcome the capillary adhesion forces. Softer cantilevers are normally used in contact mode for imaging biological specimens under fluid, where adhesion forces are very low. A sufficiently soft cantilever may allow high-resolution noncontact imaging, which has the potential to completely avoid tip damage of soft specimens. A possible means of using cantilevers as soft as 0.0001 N/m in contact mode has been demonstrated recently [38]. The method uses laser radiation pressure to exert feedback positioning on the cantilever.

2.1.7. Image Analysis

The main limitation on interpreting the image data provided by SFM is the effect of tip convolution, distorting primarily the lateral dimensions of scanned objects, but also their overall shape [39]. It has been shown that it is possible to reconstruct the tip shape by imaging a sharp point [40] or a sphere with a radius comparable to that of the apex of the tip. Several algorithms have been written to subtract the known shape of a tip from images taken with that tip. Only one of the publications on nucleic acid imaging, which is a subject of this review, has taken advantage of these techniques for calibrating a tip and deconvoluting its contribution to the images obtained [41]. Several obstacles to reliable deconvolution exist. One is the fact that the shape and sharpness of a tip often change while it is in use, because of wear and/or contamination. Therefore, it is important to use a standard on the same specimen, preferably in the same imaging field, as the subjects being scanned. Having to change specimens between a standard and the subject greatly increases the risk of damaging or contaminating the tip during the exchange and makes calibration considerably less practical.

Colloidal gold spheres, which are incompressible and commercially available with diameters in the 5–50-nm range, have been shown to be a suitable imaging standard. However, colloidal gold does not adhere to mica under fluid, and so it can only be used to calibrate the tip for air imaging experiments. Another potential source of uncertainty is that although a standard such as colloidal gold is incompressible in the force regime normally used to image biological specimens, the specimens themselves are not. Therefore, even after subtraction of the tip shape from an image, one may still not have the image of a subject in a natural or relaxed state [41]. These effects depend on the sample, and the imaging conditions used and are not necessarily an insurmountable problem. For instance, some samples have been found to behave as if they are quite stiff under forces applied at high frequency for very short durations, as is the case in intermittent contact mode scans [42]. Fixation of samples can increase their stiffness somewhat, and some sample preparations reagents, like uranyl salts [43, 44], also may increase the effective stiffness of biological molecules. For now, many researchers quote the widths of features at half-maximum height as a reasonable compromise for the lateral dimensions of an object in an uncalibrated image.

There is much room for improvement in SPM data handling software, and a number of ideas for making needed improvements have been proposed. A method of restoring the true surface from an image that does not require an imaging standard to calibrate the tip has been developed: the blind restoration method [45, 46]. This technique uses the mathematical morphology operations of erosion and dilation to calculate the tip radius from the image data and restore the image. Blind restoration has some advantages but is still subject to the same uncertainties about sample compliance and possible changes in tip shape during a scan. Even with blind restoration, there may be some merit in imaging a standard object at the beginning and end of a scan to confirm that no serious changes in tip shape occurred during acquisition of the image.

All of the methods for reconstructing the true surface of a sample ultimately rely, at least to some extent, on an accurate determination of the background slope of the sample substrate. Determination of this slope with the software provided with commercial SPMs has been found to be a large source of uncertainty [41]. An improved algorithm for extracting the true background slope from the image data has claimed to provide results up to 70% more accurate than the manufacturers' software [47]. Other improvements in image processing of SPM data have also been proposed, which could allow automation of data filtering [48]. For images of repeating structures such as 2-D arrays, correlation averaging is often employed to extract an enhanced image of the repeat unit. This algorithm has been found to contain a bias that could lead to overestimation of sample heights when applied to SPM data, a problem of little concern for data processing of electron microscopy images [49]. These researchers have proposed an improvement in this algorithm for SPM data analysis that minimizes such bias.

2.2. Scanning Probe Microscopy of Supported Biological Membranes

2.2.1. Supported Biological Membranes as Nanostructured Materials

Thin films and membranes allow the functionalization of solids in both scientifically and technologically interesting ways. They allow the incorporation of unique optical and electronic properties, and they can be the foundation for unique chemical capabilities. From a biological standpoint, they provide an important scaffold for biofunctionalization.

Biological systems possess an enormous diversity of capabilities. The versatility and specificty of antibodies and receptors, the catalytic ability of enzymes, and the charge transfer properties of electron transport proteins are clear examples. Harnessing these capabilities in functional technological devices has led to the formation of an enormous industry in biotechnology and bioengineering. From a scientific standpoint, supported biological membranes are a logical choice for maintaining a protein in its native environment. These membranes are therefore becoming surface laboratories for the study of protein structure and function, often providing model systems where the important parameters governing protein structure and function can be controlled.

Nanostructured supported biological membranes (SBMs) can be prepared in a number of ways. Phospholipids and fatty acids can be transferred to a solid support by using the Langmuir–Blodgett technique. Furthermore, membrane vesicles will spontaneously fuse with a hydrophilic solid such as mica or quartz to generate a supported bilayer. Polymer films can also be used with vesicles to generate hybrid layers. Layers consisting entirely of nonbiological polymers can also be used to attach biological molecules. This section of our review will deal with layers generated from purely biological building blocks, either entirely or in a hybrid form. The general issues associated with immobilizing biomolecules to solid supports, particularly those substrates that are nonbiological, will be covered in a separate section on protein immobilization.

The structure of SBMs can be investigated in a number of ways. X-ray diffraction, transmission electron microscopy (TEM), Fourier transform infrared spectroscopy, and far-field epifluorescence microscopy (FFM) are among the most common structural methods. These structural methods have several limitations, however. First, in the case of crystallographic methods, the structure determined may not be representative of the molecule's noncrystalline structure. Crystallographic structures are often unhydrated, which is not the typical state of a biomolecule. On the other hand, noncrystallographic methods do not have sufficient resolution to image nanometer-scale film behavior. The resolution of FFM is the optical diffraction limit, roughly several hundred nanometers. Furthermore, for FFM to be useful, a fluorophore must be present, and this can greatly affect the structures observed [50]. The resolution of TEM on SBMs is limited by the absolute electron density, which means that averaging techniques or exogenously applied stains must be used. Thus structural information available on SBMs before the advent of SPM was limited to an averaged structure over several square microns.

Studying the functional behavior of SBMs involves many of the same limitations. Average behaviors over micron-scale dimensions can easily be characterized. That is not to imply that only large-scale motions can be described. Clearly, a great deal has been learned about chain motion, hydration of phospholipids, and structural and hydration forces, but localizing and examining the behavior of SBMs at the nanometer scale has been extremely difficult.

Manipulation of SBMs has typically been dependent on the deposition technique chosen. Well-packed monolayers of protein can be generated through the use of protein-containing liposomes or through Langmuir–Blodgett techniques. These protein films can often be coaxed into forming two-dimensional crystals. More recently, SBMs have been engineered through lithographic modification of surface properties, which has allowed the generation of micron-scale phospholipid bilayer domains. SPM is a clear choice for the manipulation and control of SBMs at the nanometer scale.

For the purposes of this review section, discussion of SBMs will be limited to biologically derived molecules, such as phospholipids or fatty acids. Polymer films or other nonbiological chemical functionalization will be described subsequently, and the reader is also referred to a recent review of these fields [51]. Supported biological membranes have also been used extensively as substructures for the high-resolution imaging of membrane proteins. This aspect of SPM, however, has been reviewed frequently in the past and so will not be covered here [52]. Rather, this review will focus on the use of scanning probe microscopy to understand the structure and behavior of these films, as well as to manipulate these films. The goal will be an appreciation of attempts to control the architecture of these films at the nanometer scale.

2.2.2. Supported Biological Membranes

2.2.2.1. Near-Field Scanning Optical Microscopy

Hwang and co-workers recently used NSOM to shed light on the structure of supported phospholipid bilayers at the submicron scale [53]. Using the Langmuir–Blodgett technique, bilayers of dipalmitoylphosphatidylcholine (DPPC) supported by quartz were generated at three different pressures, yielding liquid expanded (LE), mixed LE and liquid condensed (LE/LC), and liquid condensed dominant (LCD) membranes. Bodipy-PC, present at 0.5 mol %, was used to evaluate the domain structure of the membranes. An exogenous fluorescent probe partitions into the LE regions in an LE/LC film. Although far-field epifluorescence microscopy (FFM) can provide information about domain structure, its resolution is diffraction limited. NSOM was capable of resolving features of these regions too fine to distinguish using FFM. The distribution of Bodipy-PC in the LE regions was found to be uneven, and such features as gradual domain boundaries and fine web structures were observed. Cholesterol, added at 1 mol %, reduced the line tension between the LE and LC domains, so that the LC domains became long and thin. NSOM revealed that large LC domains were often connected by thinner LC domains that had gone undetected by FFM. NSOM was also capable of quantitating the composition of the domains, revealing that the domain boundaries were gradual and not sharp.

Tamm and co-workers used NSOM and SFM to reveal structural details of dipalmitoylphosphoethanolamine (DPPE) and DPPC supported bilayers [54]. Transfer of DPPC bilayers (containing a fluorescent probe) at high surface pressures resulted in weblike structures of apparently crystalline hexagonal domains from which the fluorophore was excluded. Transfer of DPPE bilayers resulted in a much different partitioning of the fluorophore. The solid domains in DPPE were much larger, and the fluorophore was excluded into large patches with no sharp angles. Bilayers containing 10 mol % of the ganglioside G_{D1a} were also constructed, and the binding of influenza virus was monitored by total internal reflection fluorescence and SFM.

2.2.2.2. *Scanning Force Microscopy*

Meyer and co-workers used the SFM to obtain molecular-resolution images of cadmium arachidate on amorphous silicon, revealing that although the silicon is disordered, the forces regulating molecular self-assembly of the film are sufficient to drive ordering [55]. Weisenhorn and co-workers obtained molecular-resolution images of LB films of dimyristoylphosphatidylethanolamine (DMPE), dimyristoylphosphoglycerol (DMPG), dimethyldioctadecyl ammonium bromide (DODAB) and 1:1 DODAB/DPPG on mica [56]. They used the known lattice spacing of mica and data from X-ray crystallography to interpret the periodicity appearing in their images. Singh and co-workers used SFM to image vesicles of choline DPPC and DPPE in air [57]. Vesicles of dimyristoylphosphatidylcholine (DMPC) were also imaged in aqueous solution, and the ability to resolve individual headgroups was reported [58].

The detailed structure of supported biological membranes has been investigated repeatedly. Zasadzinski and co-workers imaged the headgroups of DMPE bilayers on mica and were able to determine the location of individual headgroups by visualizing the lipids' lattice structure [59]. Their results indicate that the DMPE molecules in the bilayer have good long-range orientational order, but only short-range positional order. Luo and co-workers used an SFM operating at $-25\,^{\circ}\mathrm{C}$ to image solid-phase vesicles deposited on mica [60] and found the samples more robust to repeated scanning at $-25\,^{\circ}\mathrm{C}$ than at room temperature. They also made initial attempts to use a simplified freeze-fracture technique but found that even at $-25\,^{\circ}\mathrm{C}$, the tip appeared to become stuck to the icy sample. More recent work by Woodward and co-workers to develop a freeze-fracture technique for STM will be discussed below.

Garnaes and co-workers used SFM to assess the presence of domain boundaries in crystalline layers of cadmium arachidate [61]. In multilayer films, a subtle buckling structure was observed for the first time. This buckling, which frequently had a characteristic wavelength of 2.3 ± 0.3 nm, was seen with two substrates, mica and amorphous silicon, indicating that it was an inherent film characteristic. They proposed that the buckling is related to the ripple phases seen in phospholipid systems, which is a result of differences in packing between headgroups and acyl chains.

Li and co-workers investigated the fine domain structure of phase-separated fatty acid LB films [50]. Their images revealed fine structures too small to be seen with conventional far-field microscopy. Several unanticipated behaviors were also observed. After several hours, domains were found to change shape after deposition of the layer on the surface. Films with an added fluorescent dye had a different domain structure, in which various "star-like" domains were replaced with "fiber-like" domains.

Egger and co-workers used supported phospholipid bilayers of DMPG to obtain images of hydrated phospholipid headgroups [62]. The measured mean molecular area of 45 ± 19 $Å^2$ compares well with dimensions determined from the crystal structure, 44 $Å^2$. They also imaged a synthetic hybrid molecule, consisting of a Fab' fragment of a monoclonal antibody linked to a phospholipid via a short spacer. This hybrid molecule was found to form stable, well-defined monolayers.

In general, the SFM technique allows more than structural information to be obtained. It also allows the behavior of membrane films under various conditions to be examined. Mou and co-workers found that the commonly used buffer compound, tris (hydroxymethyl)aminomethane, induces a ripple phase [63]. This ripple phase had previously been observed only in mulilamellar supported bilayers, but never in single bilayers, and it had never been induced by a chemical agent. Mou and co-workers also studied the interdigitation of phosphatidylcholines induced by alcohol [64]. The SFM was very sensitive to the appearance of interdigitated domains, with a value of 2.5 mol % being reported as the threshold for interdigitation. Czajkowsky and co-workers found that asymmetric supported bilayers, where the lower leaflet was a neutral and saturated phospholipid, and the upper leaflet was negatively charged and unsaturated, formed a ripple phase [65]. This behavior was contrary to expectations based on previous work, in which neither of the phospholipids used was in a ripple phase at room temperature. This observed ripple phase required the presence of sodium and phosphate.

The capability of the SFM to manipulate phospholipid bilayers was noted very early in the imaging of these films. For example, Radmacher and co-workers estimated a worst-case value for the force required to introduce disorder into a crystalline Langmuir–Blodgett film of 0.3 nN [66]. Here, the tip could be used to remove phospholipids from a supported bilayer revealing the surface beneath, a technique that has become a standard method for measuring film thickness [64, 66].

Brandow and co-workers used SFM to manipulate self-assembled lipid tubules adsorbed onto mica [67]. The lipid tubules, composed of a diacetylene lipid, form a supported bilayer after adsorption. By scanning repeatedly across the bilayer and steadily increasing the applied force, a damage threshold was found. By maintaining the force and continuing to scan, typically 2500 times, they could selectively slice the rectangular area into two regions. If the phospholipids were polymerized, repeated scanning at higher forces had no effect. If the slice was scanned in a perpendicular direction at a force below the damage threshold value, lipid was pushed back into the gap, closing it. This process could also occur spontaneously over a longer time scale.

Work in our laboratory has focused on reconstituted high-density lipoproteins (rHDLs) [68]. There are many important differences between rHDL particles and other forms of phospholipid aggregates, that is, liposomes and supported bilayers. The most commonly used model membranes are unilamellar liposomes, composed of a spherical bilayer enclosing an aqueous interior. Liposomes are soluble particles with diameters in the range of tens of nanometers to micrometers, whose concentration in solution can be varied. The spherical nature of liposomes can present a problem in that not both sides of the bilayer are accessible to components added to a liposome solution. Furthermore, the small-diameter liposomes have highly curved bilayers and are often unstable. We have found that planar rHDL bilayer particles in solution are a possible alternative to spherical liposomes. Their structure makes them exciting candidates for a new generation of supported bilayer structures, in which lateral control of a bilayer is possible at the nanometer scale. Proteins with functional domains on both sides of the membrane in such a particle would be simultaneously accessible to ligands (e.g., hormones or substrates in the case of receptors). The

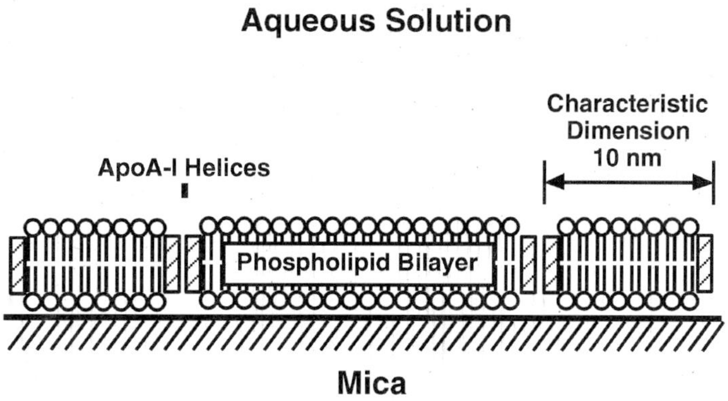

Fig. 4. Modes of scanning tunneling microscope operation.

low stoichiometry of phospholipid per particle (200–600) compared to liposomes (5000 or more) may be advantageous in some situations. For example, for studying protein association with membranes, the rHDL particle may in effect present two localized sites of defined composition, rather than many delocalized sites whose composition may change as more protein binds.

The size, shape, and presence of a protein matrix at the edges of the bilayer and its propensity to sit flat on surfaces suggest several additional advantages over existing supported bilayer systems. Membrane proteins incorporated into the bilayer domain of rHDL can be adsorbed onto a surface with a known orientation. The protein portion at the perimeter of the phospholipid bilayer can potentially be genetically engineered and used to manipulate or immobilize the particles, possibly in oriented arrays, without disturbing the phospholipid domain. They are also useful as single membrane protein corrals, serving to isolate single membrane proteins for individual study. High-resolution imaging in aqueous solution makes the study of the particles completely feasible.

rHDLs are prepared through the incubation of phospholipids (primarily phosphatidylcholines), detergent (typically cholic acid), and protein (apolipoprotein A-I, although other exchangeable apolipoproteins will form discoidal particles), followed by removal of the detergent through dialysis [69]. The resulting solution will contain either a monodisperse preparation of disks or various sizes that can be separated later through high-resolution chromatography, depending on the phospholipids used and the stoichiometry of the incubation mix [70]. Disk sizes are easily characterized by native gradient gel electrophoresis (GGE).

In the human body, apoA-I plays a central role in lipoprotein metabolism and contributes a great deal to the structure of the rHDL produced. The perimeter of the disks formed is related directly to the stoichiometry of apoA-I on the disk, dictating discrete particle size distributions [71, 72]. Furthermore, subclasses of size are formed that are consistent with the addition or removal of single α-helices from the disk edge. This combination of conformational flexibility and specificity in phospholipid interactions makes these exchangeable lipoproteins unique.

Incubation of rHDL with mica results in creation of a monolayer of rHDL. This monolayer is equivalent to a supported bilayer divided into nanodomains separated by apolipoprotein A-I (Fig. 4). Varying concentrations of disks can be used to follow the formation of this monolayer, but once a monolayer is formed, additional increases in concentration result in no further rHDL adsorption (as detected by SFM imaging). Low coverage samples have provided direct evidence of the rHDL's bilayer structure. The movement of a single rHDL under the influence of the SFM tip can be followed using simultaneous detection of topographical and lateral force data (Fig. 5).

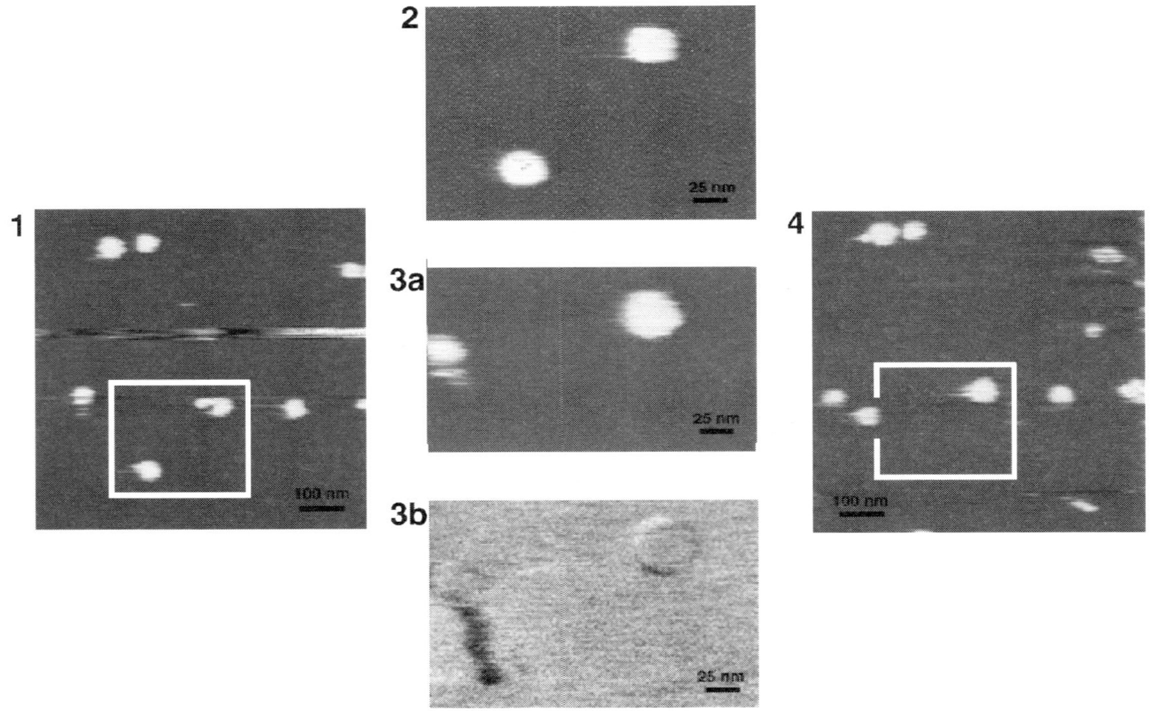

Fig. 5. Movement of single rHDL by the SFM tip. The sequential images (1–4) show the step-by-step movement process. Frame 1: Initial image showing a number of rHDL on the mica surface (scale bar = 100 nm). Frame 2: Zoomed-in image, where the enlargement is indicated in the initial image by the white box. Both rHDL are stable (scale bar = 25 nm). Frame 3a: Height image showing the movement of the lower left rHDL in frame 2 to a new position up and to the left. The SFM tip is moving from the bottom to the top of the frame (scale bar = 25 nm). Frame 3b: Simultaneously obtained lateral deflection image showing the motion of the cantilever as it nudges the rHDL along the surface, which is detected as the black line in lower left of the frame (scale bar = 25 nm). Frame 4: Zoomed-out image, showing the original rHDL for reference, and the new position of the moved rHDL (scale bar = 100 nm).

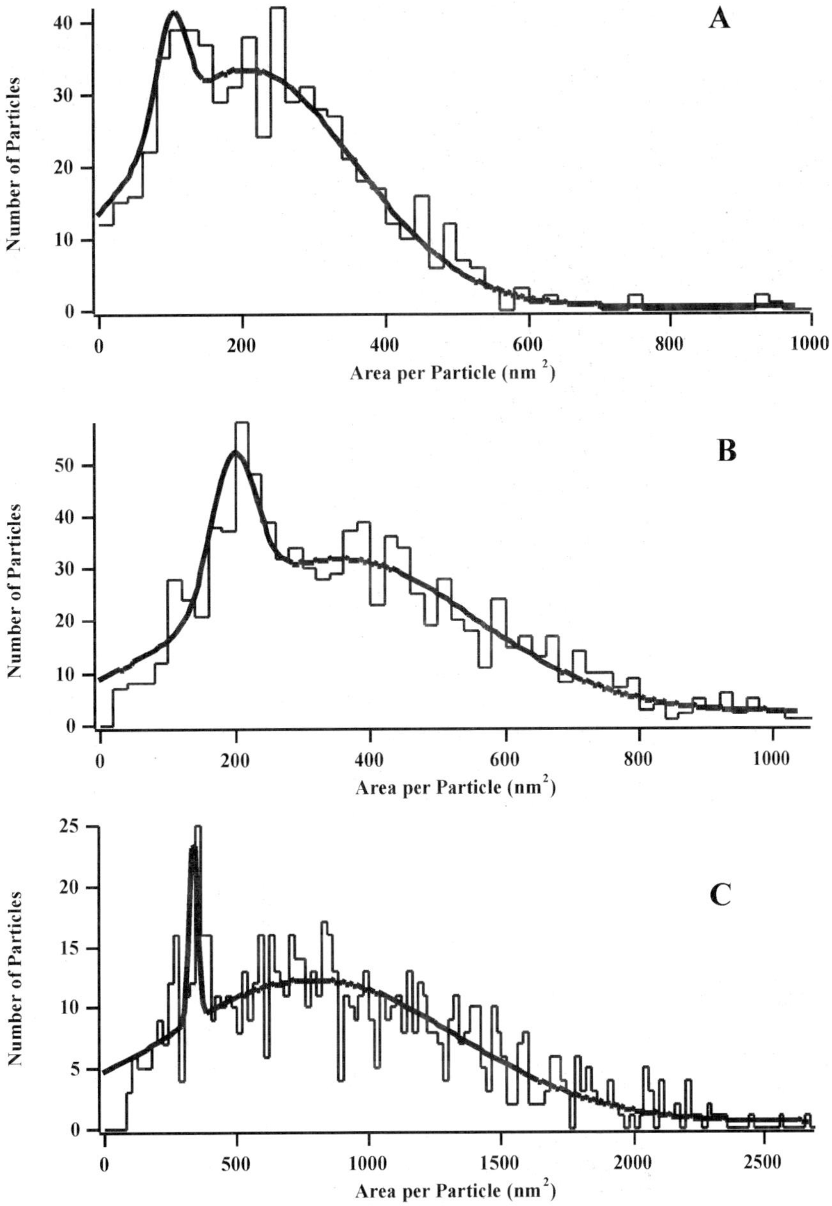

Fig. 6. Histograms of area per particle for each sample. The solid line corresponds to a fit to two distributed populations. (A) Lp2 rHDL ($n = 635$). (B) Lp3 rHDL ($n = 953$). (C) Lp4 rHDL ($n = 835$). (Source: Reprinted with permission from [68].)

In addition to static structural information, SFM can reveal the behavior of these lipoprotein-covered surfaces as well. SFM imaging of full coverage samples shows that these nanodomains spontaneously fuse, increasing their size. This fusion process occurs over minutes to hours. Measurements of the area per particle revealed a bimodal distribution consisting of the initial population of disks prepared in solution and a new population, believed to be the result of interdisk fusion (Fig. 6). This fusion process can be followed in real time with the SFM. Images indicate that the process begins with an initial connection between the two particles. This connection then widens until the particles are linked. This widening is presumably accompanied by either the departure of the protein into solution or a change in its binding mode from disk edge to disk top.

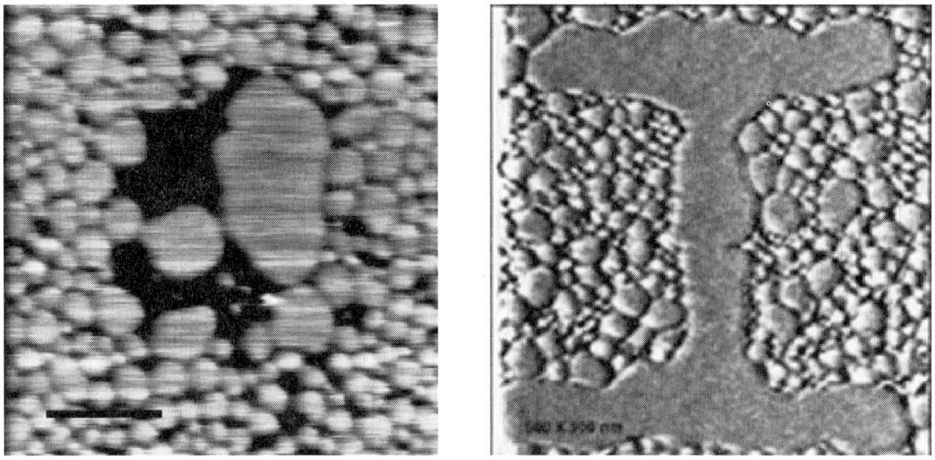

(A) (B)

Fig. 7. SFM images of manipulated rHDL. (A) Example of digging into a high-coverage sample (scale bar = 100 nm). (B) Example of patterning a high coverage sample (image size = 500 × 500 nm). (Source: Reprinted with permission from [68].)

We have discovered that this normally spontaneous fusion process can be directed with the SFM tip under appropriate conditions. To direct this fusion, the applied force is increased, the scan rate is increased, and the gains are decreased, thus reducing the ability of the microscope to feed back and causing the tip to "plow" through the surface-attached lipids. This is continued for 30 s, and then the parameters are reset to those suitable for imaging. The resulting manipulated portion of the bilayer will show signs of "digging" or patterning. Digging results in a hole with variable depth (Fig. 7). Occasionally the hole will extend entirely to the mica surface below the disks, or alternatively, it can extend only through the first sheet of the bilayer. Patterning of the monolayer can result from this directed fusion (Fig. 7). Whether patterning or digging is the result of a given manipulation is not dependent on the applied force, scan speed, or feedback gains, but rather appears to be a complex function of tip and sample condition.

This directed patterning allows creation of nanometer-scale phospholipid bilayer domains. These domains may be useful for molecular devices, in which a bilayer is divided into various lateral regions with resultant heterofunctionality. These domains may also be useful for the controlled interaction of membrane proteins, where two proteins are only allowed to interact when their corrals are opened. The interaction could then be imaged from their initial contact.

Initial attempts in our laboratory to incorporate membrane proteins into rHDL focused on NADH cytochrome P450 reductase (P450 reductase) and cytochrome b_5 from the hepatic endoplasmic reticulum. NADH cytochrome P450 reductase is a microsomal protein involved in the reduction of cytochrome P450, which is responsible for the breakdown of a large number of pharmaceutical compounds. Cytochrome b_5 is a small electron transfer protein whose membrane binding topology and redox properties have been evaluated by a number of other techniques. Addition of these proteins during detergent-based reconstitution allows rHDL to be created with an associated membrane protein, as determined by native GGE. The behavior of the two protein upon incorporation into the rHDL is different. Attempts to image cytochrome b_5-rHDL results in the formation of multilayers, images of which are noisy and unstable. This is often indicative of a soft, changing surface, which may have an interaction with the tip. P450 reductase-rHDL, in contrast, can be adsorbed to the mica surface in a monolayer similar to that seen for rHDL, and single molecules of P450 reductase can be imaged reproducibly (Fig. 8).

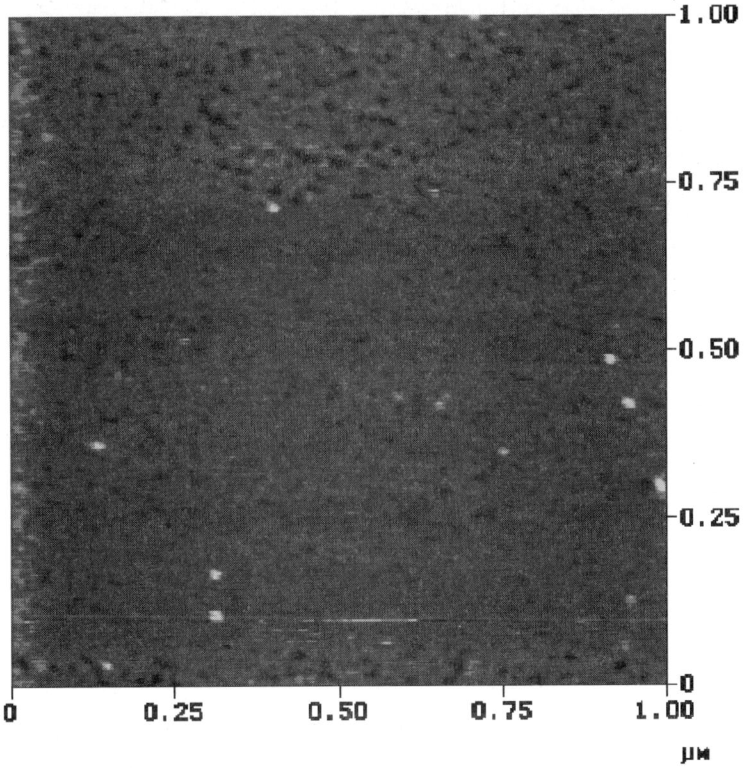

Fig. 8. SFM images of cytochrome P450 reductase-rHDL (image size = 1.0 μm × 1.0 μm). (Source: Reprinted with permission from [73].) See also Plate 8.

The SFM has been used to characterize these single protein molecules incorporated into rHDL disks, providing reproducible height measurements. In low force contact mode, the mean height of P450 reductase was determined to be 1.7 ± 0.9 nm, whereas in intermittent contact mode the mean height was determined to be 2.5 ± 1.1 nm. Increasing the force in contact mode leads to the disappearance of the P450 reductase molecules, which can be reversed by returning the applied force to the original low value. This indicates that although the molecules are not imaged by the SFM at high forces, they remain in their original spatial location, that is, the SFM is not removing the molecules from the surface, but rather is deforming them and/or pushing them into the bilayer.

Investigating these unique nanostructured materials has shed light on the structure of rHDL, on the behavior of lipids, and on the structure and behavior of membrane proteins. They indicate what is an important new paradigm in the control of supported phospholipid bilayers: control not only of the layer, but also of its lateral homogeneity. This control can be achieved in many ways, for example, through patterning the substrate to change the adsorption of the lipids. But for manipulation at the nanometer scale, a system such as rHDL is of obvious benefit.

2.2.2.3. Scanning Tunneling Microscopy

STM has been used many times to image Langmuir–Blodgett films of fatty acids and phospholipids [73–75]. Organic bilayers are essentially nonconductive, however, so interpretation of the images was difficult. Gregory and co-workers used an octadecanethiol monolayer on gold to support a monolayer of dipalmitoylphosphoglycerol (DPPG) [76]. Their images indicate multiple hexagonal domains in the DPPG/ODT bilayer. Further evidence for this structure comes from infrared spectroscopy measurements. The domains are

7–10 nm in diameter, and domain boundaries have an interior angle of $119 \pm 2°$; molecular resolution of the headgroups was achieved. Observed step heights of 2.5–3.0 nm were consistent with a predicted monolayer thickness of 2.72 nm, but step heights between neighboring domains indicated that surface roughness may play a role in domain extent (i.e., only smooth regions can support the hexagonally packed domains). Their results indicated a smaller headgroup area than expected and little or no chain tilt, which may indicate a reduction in the expected electrostatic repulsion between the DPPG headgroups. This repulsion could be absent for a number of reasons, including hydrogen bonding, headgroup protonation, or the presence of counterions. Manipulation of these monolayers by STM was reported as a loss of definition of the domain edges, typically after only 5 min of imaging.

Heim and co-workers used the lateral conductivity of adsorbed waters to investigate headgroup hydration of supported phospholipid bilayers [77, 78]. They used bilayers of DPPE or DPPE/DPPC and found a humidity-dependent conductivity. The strength of this current depended on the hydrophilicity of the headgroup. The water layers involved are very thin; for example, adsorption of one water molecule per headgroup is sufficient to conduct a current of 0.1 pA. The currents supported were found to be two- to threefold stronger than those that can be supported by bulk water. It was proposed that a proton transfer mechanism is responsible for the conduction process.

Although not on a supported bilayer system, recent freeze-fracture results on phospholipid vesicles reported by Woodward and co-workers are of note. In their work, STM, in conjunction with metal replication and freeze fracture, was used to image fully hydrated ripple-phase bilayers [79]. These samples were flash frozen from aqueous solution, precluding the necessity of supporting the bilayer on solid substrate. After fracture and coating with Pt/C, STM images revealed the ripple phase, showing its asymmetric nature and revealing an unexpected temperature dependence to the ripple amplitude. Correlation averaging was used to obtain resolution values of <1 nm laterally and 0.3 nm vertically (better than the roughly 2-nm granularity of the metal coating).

2.2.2.4. Other SPM Imaging Techniques

Phase imaging is a combination of oscillating the imaging cantilever (called tapping mode imaging, ac mode, or intermittent contact imaging) and monitoring the phase shift between the driving oscillation and the cantilever's actual oscillation. This phase shift is the result of damping the cantilever's oscillation, either through the sample's force curve or through mechanical properties of the sample (such as variations in sample elasticity). It is this sensitivity to the sample's mechanical properties that makes phase imaging very interesting for the imaging of supported phospholipid bilayers. Where a topographical image may not be sensitive to domains of different composition, phase imaging may detect changes in the domain's characteristics.

Radmacher and co-workers demonstrated the capability of phase imaging to resolve crystalline domains of a polymeric diacetylene film from a monomeric fluid phospholipid (DMPC) [80]. Their images, as well as their discussion of the mechanism of image formation, demonstrated that phase shifts could be explained by using the cantilever's spring constant and the effective stiffness of the surface. When the tip elastically indents the sample, the forces can be described by a Hertzian model.

Inoue and co-workers used Maxwell stress microscopy to examine the electrostatic surface potential distribution of phase-separated LB phospholipid monolayers deposited on metal [81]. Scanning Maxwell stress microscopy uses harmonic analysis of the oscillation of the cantilever off-resonance to simultaneously obtain topographical information, the surface charge and potential, the dielectric constant, and its high-frequency dispersion. Supported phospholipid bilayers of DPPC on aluminum were transferred at a pressure that resulted in the formation of a mixed liquid expanded/liquid condensed state. Images revealed regions of varying surface potential, and the regions of high surface potential corresponded well with the liquid condensed regions (examined using fluorescence imaging). An increase in both molecular density and the perpendicular component of the dipole moment per molecule were sufficient to explain the observed potential.

2.3. DNA Imaging

The first SPM images of DNA were made with an STM. Some of these images were made of DNA mounted on highly ordered pyrolytic graphite (HOPG) and are now considered problematic because of the large number of artifacts that can be seen on this substrate [82, 83]. However, images taken on epitaxial gold [111] surfaces clearly reveal DNA structures [84]. The physical basis of STM imaging of insulating macromolecules on conducting surfaces is still not well understood. A model for STM image formation for short oligonucleotides on a gold surface has been proposed [85, 86]. However, an STM tip scanning over a conductive surface under tunneling current-controlled feedback would be expected to crash into an object that is insulating. The images obtained thus far for large DNA molecules appear to have been formed by a process with a mechanical component as well as an electrical component [87, 88], whereas STM of conducting and semiconducting surfaces is purely electrical, that is, it is mediated by true electron tunneling.

Recently a new mode of STM imaging that can be conducted on the atomically flat but insulating mica surface that is widely used for SFM imaging has been developed [89]. It relies on the molecularly thin water layer that forms on mica once it is exposed to ambient humid air. This layer of water is capable of conducting a few picoamps of current, so it is possible for an STM tip to scan under feedback control, if sufficiently sensitive current detection is available. Some nice images of DNA have been obtained by this method [6]. So far, however, the SPM images of DNA that show the best submolecular detail, such as a distinct periodicity for the helical turns, have been produced by SFM, even though tip convolution adds to the widths of objects in these SFM images. DNA has served as a test case for developing methods generally applicable to SFM imaging of biological macromolecules because of its easily recognizable appearance, well-understood structure, and good chemical stability.

2.3.1. Specimen Preparation Methods

The majority of the material published on the subject of DNA imaging by SFM through about 1993 concerned various means of adsorbing or mounting DNA to flat surfaces so that it could be imaged reliably. Some of the methods developed are intended to provide the best submolecular resolution of the double-helix structure of the DNA itself, and others are intended to provide acceptable images under conditions in which DNA–protein complexes could be observed in a native state, but with the compromise of some loss of detail. To date, these two goals have not been achieved simultaneously on the same specimen.

The probe tip of the scanning force microscope operating in contact mode makes repulsive contact with the surface being scanned and applies a force normal to the surface that is typically in the 0.1–100-nanonewton range when imaging biological samples. Such forces can deform biological macromolecules or push them around on or wipe them from the surface. It is therefore critical that the molecules to be imaged are solidly fixed to the surface. Beyond the sharpness of the tip, the firmness of surface attachment is one of the most important factors controlling the attainable resolution. No matter how good the tip, if the object moves under the tip, the image will appear smeared.

The surface to which the DNA is attached should have background roughness much smaller than the size of DNA, so that the DNA can easily be distinguished from the background. A number of different mineral and glass surfaces have been tried [90], but ruby mica remains the most widely used substrate. Mica cleaves easily and reproducibly to give atomically flat areas square microns in size. It is quite hydrophilic and carries a negative surface charge. The surface structure of mica, at the atomic scale, consists of hexagonal rings of oxygen atoms, so the surface is polar and rich in hydrogen bond acceptors, which explains the hydrophilicity [91].

Because both the DNA and the mica are negatively charged, DNA applied to mica in solution does not adhere well. Mono- and divalent cations are often used to neutralize the charges or act as bridges between the like-charged molecules and surface [92, 32]. Chemical modifications such as treatment of the mica with silanes bearing amine or quaternary amine groups have been used successfully to make the surface positively charged, with resultant DNA adsorption [93]. Mica has also been coated with cationic lipid bilayers to create a flat and positively charged surface [94]. The cationic lipid bilayer can be annealed by warming it to a temperature above the transition temperature for the lipid and thus allowing attached DNA strands to equilibrate into a close-packed state. Contact mode scans of DNA samples prepared in this way have yielded some of the highest resolution SFM images of DNA reported to date [94].

The first attempts to image DNA by SFM were made by contact mode in ambient air. The DNA was applied to a mica surface as a dilute solution in water or a salt-containing aqueous buffer and allowed to dry [95, 96], or it was applied as a solution containing a divalent cation, generally magnesium, allowed to bind for a short time, then rinsed with water and dried [32]. It was found that the drying step was necessary for good adhesion. It was discovered that the apparent height and width of DNA samples prepared by these methods varied with changes in the ambient humidity [95, 97]. Researchers concluded that the capillary forces acting between sample surface and tip were increased greatly by the water layer that exists under humid conditions. The excess force between the tip and sample surface would cause compression of the DNA in the vertical dimension [98], and could even cause damage to the strands [99, 100]. The humidity also appeared to cause the salt adsorbate layer, which may be partially responsible for binding the DNA to the mica, to swell and thus reduce height contrast between the DNA and surrounding surface [101, 96].

Other methods of spreading DNA on mica have been adapted from electron microscopy. One of these is the cytochrome *c* spreading method. This involves evaporating a layer of carbon onto the mica, to which the cytochrome *c*-coated DNA adheres [102]. This method has yielded nice images, and these preparations were found to be less sensitive to humidity than specimens with a bare mica surface as the background [99]. Another technique makes use of dilute quaternary ammonium salts to prevent aggregation of the DNA [103]. In this method, the DNA is bound to mica without the need for divalent cations, so it allows greater flexibility in buffer choices.

Imaging under fluid was found to eliminate the capillary effect and shield-long range van der Waals forces, which also contribute to the total adhesion force, but it is difficult to achieve tight binding of the DNA to mica under aqueous buffers [92]. The use of alcohols, particularly propanol, was found to enhance DNA binding while greatly reducing the capillary and van der Waals adhesion forces [32]. Elegant images of isolated DNA molecules have been obtained by scanning under propanol [43]. Alcohols, although they may enhance binding and imaging of DNA strands, are not a native environment for either DNA or DNA–protein complexes.

A systematic study of the effects of different divalent cations on DNA binding to mica revealed that those ions that have radii compatible with the oxygen lattice on the mica surface give the best DNA binding efficiency [104]. These ions, Co^{2+}, Zn^{2+}, and Ni^{2+}, were found to provide DNA binding strong enough for contact imaging under water. The use of HEPES buffer in place of Tris buffers has been reported improve the efficiency of DNA binding to mica, in the presence of magnesium ions, by a factor of 10–100 [43]. The treatment of mica with amine-terminated silanes such as aminopropyltriethoxy silane (APTES) also gives binding tight enough for fluid imaging, although it causes a small increase in surface roughness [93]. APTES-treated mica is also claimed to retain binding activity for long storage periods [105]. Recently, an improved preparation method for APTES-treated mica has been reported that avoids the increase in surface roughness associated with the original dilute vapor phase reaction [106]. The cationic lipid bilayer deposited on mica also allows imaging of DNA under aqueous buffers [94].

Another promising technique for preparing specimens of DNA on mica makes use of cationic detergents and is a modification of a technique originally developed for electron microscopy, briefly mentioned earlier [103]. Benzyldimethylammonium chloride is diluted into the DNA solution, and this mixture is then applied to glow-discharge-treated mica for a few minutes. The specimen is then gently washed with water and dried in air and can be scanned in contact mode under ambient conditions. This preparation, like the APTES treatment, allows flexibility in the choice of mono- or divalent cations used in the buffer. This technique has been further improved by the substitution of an alkyl-terminated cationic detergent and by imaging in *n*-propanol [107]. It has not been reported that these specimens are compatible with imaging under aqueous fluid, however. Cationic detergents have also been used to spread DNA on gold [108].

A means of producing flat gold surfaces, and of covalently attaching DNA, has been developed by Hegner and co-workers [109]. For this approach to work, modified nucleotides containing a thiol on an alkyl spacer arm must be incorporated into the DNA fragment of interest. This was done by using the modified nucleotides to synthesize oligonucleotides, which were then used as polymerase chain reaction (PCR) primers to produce the DNA chains. Gold was evaporated onto mica, and then the mica was peeled away to expose the flat gold surface, to which the thiol-modified DNA fragments were bound. The DNA fragments were scanned in contact mode under 10 mM ammonium actetate and under propanol. The fragments imaged well near the ends, where they were covalently attached to the surface, but appeared fuzzy in the central regions. Presumably improved results for either SFM or STM could be obtained with other covalent linker approaches by using this means of preparing very flat gold surfaces. Moreover, as the authors suggested, anchoring the DNA only at the ends might prove advantageous for imaging dynamic events such as enzymatic activities on the fragments.

Reversible binding of DNA to mica under aqueous buffer has been studied with the same goal in mind of observing dynamic events. Thomson and co-workers [110] developed a method to image DNA on mica in a fluid cell equipped for gravity-fed buffer exchanges in which the DNA molecules could be held tightly for good imaging or released to allow interaction with other molecules. Their method made use of a buffer containing Zn(II) to achieve tight binding and one in which the Zn(II) was omitted to allow the release of the DNA from the mica surface. Repeated binding/release cycles for individual DNA strands was demonstrated using stalled *Escherichia coli* RNA polymerase (RNAP) complexes. The RNAP remained tightly bound to mica in both buffers used, so the bound DNA strand could be observed to appear and disappear as the buffer was exchanged.

All of the mica specimen preparation techniques discussed above involve some sort of modification of the mica surface to enhance DNA binding, and most involve the introduction of positive charges. Some researchers have taken a different approach to making molecules, including DNA, stick to mica by using a light coating to hold the molecule down. Plasma-polymerized methane has been used successfully to make sturdy phage specimen preparations for imaging in ambient conditions [111]. The sample has to be dry for the polymer deposition, so this type of approach may not be useful for imaging complexes in their hydrated state, but it could be appropriate for studies of nucleic acid structures. Thin (0.5 nm) carbon/Pt shadowing has also been used to mount DNA on mica [112]. This method provided good contact mode images in ambient air that were not affected by humidity. Precoating the mica and then binding the DNA to the C/Pt surface worked as well [112]. In either case, adhesion forces would be reduced by the inertness of the C/Pt surface, which does not collect moisture.

2.3.2. DNA Imaging Methods

DNA and RNA have been imaged by SPM for numerous reasons, including visualizing nucleic acid structures such as bends, supercoils, loops, and triple helices, measuring effects of intercalating agents and DNA-binding drugs, mapping specific sequences, and developing labels for SFM imaging of biomolecules. Force measurements have also been made on DNA by SFM, to estimate elasticity or hydrogen bond strengths. Some of the studies described in this review have been aimed at understanding the nature of the parameters that influence imaging of biomolecules under various conditions and have made use of DNA as a test-case biomolecule.

In one of the earliest reports of DNA imaging by SFM, Weisenhorn and co-workers imaged a single-stranded 25mer oligonucleotide that was either covalently attached to a polymerized Langmuir–Blodgett lipid film or ionically bound to mica with aluminum chloride [113]. Both images were taken under water or in aqueous buffer in contact mode. The DNA on the mica surface was found to be mobile, and the images were not stable. The DNA on the LB film, however, appeared to be firmly attached. The single-stranded DNA in both cases appeared to be in an extended state, which is not unreasonable, considering it was imaged in water without any counterions present, although more globular structures have been reported for similar DNA samples by other groups. A much longer single-stranded molecule was imaged by Hansma and co-workers, the ϕX174 virion DNA of 5386 bases [32]. They used 0.5% formaldehyde to extend the molecule and still obtained images that appeared to contain sections of extended strand interrupted by lumps of secondary structure. They suggested that 2% formaldehyde may be necessary to fully extend DNA strands. This imaging was done in contact mode under propanol or butanol. In the images of a double-stranded plasmid, attached to magnesium-treated mica and scanned in contact mode under propanol, some periodic substructure was also seen.

Images of supercoiled pUC9 plasmid were published by Zenhausern and co-workers, which were prepared by a method adapted from electron microscopy [96]. The DNA solution in a buffer containing triethanolamine-HCl, $MgCl_2$, and glutaraldehyde was evaporated onto mica. This resulted in plasmids embedded in a salt layer, which could easily be imaged by contact mode in air, although the samples were sensitive to ambient humidity. An improved method of spreading circular plasmid DNA was reported by Li and co-workers, who found that heating the DNA solution at 35 °C for 10 min before depositing on magnesium-treated mica gave good coverage with only 2–10 ng of plasmid [33]. The scans were made in contact mode under propanol and showed the substructure seen in previous DNA images made under these conditions.

Other work published at about the same time included investigations of the effect of humidity on DNA samples prepared by air-drying droplets of DNA solution in NH_4OAc on mica for imaging in air [95]. Calculations indicated that capillary forces as high as 100 nN could be caused by the water layer on mica in ambient air, and thus this was proposed as an explanation for negative contrast images obtained from scanning in humid air. Yang and co-workers showed that excellent resolution could be obtained from the cytochrome *c* spreading technique, as indicated by the narrow apparent width (6 nm) of the plasmid strands in their images [102]. However, no substructure on the order of helix turns was observed. Yang and co-workers also made images of single-stranded DNA from M13 phage, in which the measured length agreed well with the known size of 7100 bases. The presence of 50% formamide in the preparation in addition to the abundant cytochrome *c* probably caused the single-stranded DNA to fully extend. These authors also noted that DNA could be cut by scanning a small area overlapping a plasmid strand at high force. In another study specifically aimed at the nanodissection of DNA with the SFM tip, Henderson found that hydrated plasmids could be cut easily and that fragments of 100–150 bp could be excised [114]. Dehydrated samples proved much more difficult to dissect in the SFM.

To investigate the role of surface hydration and residual salt adsorbates on images of DNA made in air, Vesenka and co-workers investigated the effects of humidity on DNA samples labeled with colloidal gold through the biotin–avidin interaction [101]. This effort was based on the expectation that the gold would remain incompressible regardless of humidity, and thus it would act as an internal standard for the height and width of the DNA and surrounding adsorbates. As the relative humidity was increased from 10%, it was possible to reproducibly measure the height of the DNA until the humidity reached 55%. The gold particles could be monitored until the surface became saturated with moisture at 64% relative humidity, at which point the gold particles would be wiped from the surface by the tip. The height of the gold particles was found to decrease as the humidity increased from 10% to about 35% and then leveled off. The heights of the DNA changed by a smaller factor, but followed the same pattern. To explain these data, the authors proposed a model in which the layer of salt adsorbates swells as humidity increases, partially elevating the DNA along with it, while the larger gold particles remain in contact with the mica surface and thus appear to decrease in height relative to the background.

Two studies published in 1993 addressed the effect of probe force, particularly the adhesion force, on resolution in acquired images of DNA. Each investigation used different specimen preparation techniques, and they reached the same general conclusion that minimizing adhesion forces by any of several means increases the resolution and the reproducibility in images of DNA, as well as other biological samples. Yang and Shao made measurements of adhesion forces on the background and over DNA molecules on samples prepared by the cytochrome *c* spreading technique [99]. They found a strong correlation between low-adhesion forces and good images from contact mode scans in air at controlled moderate humidity, as judged by the apparent width of DNA, observed to be 6 nm or less. This resolution was routinely achieved using commercial cantilevers, so they concluded that the adhesion force actually plays a greater role in SFM resolution than tip radius. Scanning the same specimens under propanol or butanol, which totally eliminates the capillary force, also produced images in which the apparent width of the DNA was typically in the 4–6-nm range.

Lyubchenko and co-workersmade a similar study of adhesion forces on resolution, using APTES-treated mica as the substrate for DNA and filamentous phage specimens [98]. They found that in addition to a correlation between low adhesion forces and small apparent widths of DNA, there was a correlation between low adhesion forces and greater heights of the DNA in their images. They also found that the measured heights were not greatly affected by the probe force applied or the spring constant of the cantilevers. These results, taken together, led the authors to conclude that tip contamination was the main source of excessive adhesion forces on these substrates. They also suggested that the chemical composition of the tip may be as important a factor as tip geometry in determining the resolution in SFM images.

The most detailed images of DNA published so far have been made using a cationic lipid bilayer deposited on mica as the imaging substrate [94]. The densely cationic surface readily binds DNA, but it is the controllable fluidity of the bilayer that ultimately makes the high resolution possible. When bilayers with a high coverage of DNA were warmed above the phase transition temperature of the lipid for an hour in a buffer containing EDTA, the DNA chains form close-packed arrays resembling fingerprints. These arrays were scanned in contact mode under aqueous buffers to give images in which the helical repeat distance and right-handedness of the helix are clearly discernible.

A further improvement in fluid imaging came about as the result of the development of the intermittent contact mode SFM operation for fluid imaging [115, 42]. The reduction of lateral forces while scanning in intermittent contact mode allows samples that weakly adhere to mica to be imaged in fluid. This development has increased the researcher's flexibility in choices of buffer and surface preparation for fluid imaging, and it has been found to be quite useful for imaging DNA–protein complexes in native conditions in which buffer choices are limited by the requirements of the sample [116].

Another area of research that is aimed at enabling future biological investigations is the development of site-specific labels for DNA that are detectable by SFM. Two methods have been published so far, and both make use of the high binding constant interaction between biotin and streptavidin. Nucleotide triphosphates bearing biotin on a polar spacer arm are commercially available, so DNA strands can be modified with biotin by incorporating biotinylated bases with a polymerase-based methodology, or DNA can be synthesized with biotin attached. Avidin is a relatively small protein (about 3.5 nm in diameter), so the labeling strategies use either colloidal gold conjugated to avidin or a fusion between avidin and another protein to create a bulky complex that will be easier to recognize in scanning probe images. An advantage of using gold particles is that they are spherical and incompressible, so they can serve as internal tip characterization standards as well as labels. Murray and co-workers used an avidin–protein A fusion with a mass of 125.6 kDa and could clearly distinguish this protein from the DNA chains in their images [117]. They imaged by contact mode in propanol. Shaiu and co-workers used streptavidin-gold particles and imaged by contact mode in air [118]. They reported the effects of humidity on the stability of images scanned in air, showing that at high humidity neither gold nor the DNA remained fixed to the mica well enough to produce clear images. So far, there is apparently no reliable method of labeling DNA for imaging under aqueous buffers.

2.3.2.1. *Scanning Tunneling Microscopy*

The requirement of STM for a conductive specimen limits STM imaging to conductive substrates. The most common substrates are highly ordered pyrolytic graphite (HOPG) and gold films evaporated onto mica and subsequently annealed. Neither substrate, however, can be routinely prepared to give atomically flat terraces as large as those obtained with mica. The nature of sample binding is also quite different. HOPG presents a hydrophobic surface, and molecules physisorb to this surface primarily by van der Waals interaction. On the other hand, covalent chemistry based on the thiolate linkage is the usual means of attaching molecules to gold. The earliest DNA imaging experiments with the STM using HOPG substrates will not be discussed, because those results were later shown to be questionable, in as much as it was subsequently demonstrated that defects in the HOPG surface itself could mimic DNA structures [82, 83]. DNA is easily distinguished from gold, and most of the more recent work has made use of this substrate. The individual nucleotide bases, however, can be cleanly imaged on HOPG because they form a monolayer, apparently by a hydrophobic interaction between their aromatic ring faces and those of the graphite surface [119]. Allen and co-workers showed that it was possible to distinguish between the purine and pyrimidine heterocycles of adenine and thymine on HOPG in STM images taken in air [120].

In 1992 Allison and co-workers imaged plasmid DNA ionically bound to amine-modified gold by scanning in air [121]. The apparent width of the DNA observed was 3–5 nm, which is closer to the known width of 2 nm than the numbers typically measured from SFM images. However, most of the molecules showed negative height contrast in their images, appearing to be up to 1.6 nm below the surface. No submolecular detail on the order of helical turns was seen. Lindsay and co-workers also published STM images of DNA deposited on gold potentiostatically from phosphate buffer and scanned under water with an insulated tip [122]. They measured widths very close to 2 nm on these preparations, but no submolecular detail was observable. These studies point to a lacuna in the understanding of STM image formation on large insulating molecules.

Nevernov and co-workers explored the origins of the negative contrast that is sometimes seen in STM images of insulating molecules on a conducting surface by imaging 10.7-nm-diameter dendrimer microspheres [87]. The dendrimers gave negative contrast in constant current images, where they appear as depressions. This was interpreted as a combination of mechanical and electronic interactions of the tip with the substrate and the adsorbate. Modeling the imaging process, with the dendrimers represented as elastic spheres, gave good correspondence with actual images. This type of imaging mechanism could explain why STM can give high resolution of the conducting background and produce images of DNA where the widths are very close to known values and yet not reveal any intramolecular detail.

The mechanical aspect of the STM imaging process for DNA would seem to place requirements on specimen preparation similar to those for SFM. The DNA would have to be bound tightly enough to the surface to avoid being moved or deformed by the tip. The potentiostatically bound and ionically bound specimens mentioned above satisfy this requirement. It has proved more difficult to image short oligonucleotides than long DNA molecules on mica by SFM, presumably because the shorter chains present fewer sites for ionic interaction than the longer chains, and so the tip more easily displaces these molecules during scanning.

Lindsay and co-workers demonstrated the efficacy of a potentiostatic deposition method for attaching DNA oligonucleotides to gold to produce specimens for scanning in humidity-controlled nitrogen [122]. They obtained images of double- and single-stranded oligonucleotides that contain repetitive features that correlate well with the helical rise of B-DNA. They proposed a simple tunneling model to explain the image contrast observed. The same group also showed that this imaging method could distinguish between several 20-base oligonucleotide samples with different structures in blind trials [85]. These experiments support their interpretation of contrast generation in this imaging mode and demonstrate the reproducibility of this specimen preparation method. Rekesh and co-workers have shown that short oligonucleotides can be covalently bound to gold by incorporating a thiol-bearing spacer arm into both ends of the molecules [123]. The measured lengths of the oligonucleotides in their images were in good agreement with calculated values and shared some of the repetitive features with the potentiostatically deposited samples discussed above.

Another approach to imaging DNA with the STM has been pioneered by Guckenberger and co-workers [6]. They have shown that it is possible to image DNA on insulating substrates such as glass and mica by taking advantage of the conductivity of the thin water layer that is present on these hydrophilic surfaces at high relative humidity. This approach allows the use of the various specimen preparation techniques developed for SFM on mica and allows researchers to take advantage of the long-range atomic-scale flatness of mica to image long DNA molecules without interfering background artifacts. It is not entirely clear if the physical basis of this type of STM imaging involves conduction from the direct contact of the tip with the adsorbed water layer (sometimes referred to as electrochemical imaging), or if the mechanism is tunneling [124, 125]. This type of imaging is now referred to as hydration scanning tunneling microscopy [89].

Heim and co-workers [89] have recently published a study in which hydration scanning tunneling microscopy was carried out at different tip bias voltages, and concluded that at high bias voltage, image formation is the result of field emission or tunneling, whereas at low voltages a water meniscus between the tip and sample is involved. The images of plasmid DNA in this report are about 3 nm wide for either imaging regime, so there is very little contribution of tip size to the image. The images contained periodic features along the DNA strands with a repeat of about 5 nm. Although these submolecular features may not be a direct reflection of the B-form helical repeat, the images do have a resolution higher than has been obtained by SFM on isolated DNA molecules to date. Hydration scanning tunneling microscopy of DNA has also been demonstrated in the Hansma laboratory [105]. This high-resolution scanning probe imaging method appears very promising and may become a routine type of STM imaging for biological samples in the future.

2.3.3. DNA Structure Studies

A body of literature is being built up in the area of SFM visualization of DNA structure, partially because of the relative quickness and ease of use of the scanning force microscope, and taking advantage of the precise height measurements that can be made directly from images taken with this instrument. Several studies of the structural effects of the binding of intercalating agents or drugs have been made. Rampino [126] examined DNA fibers produced by spinning DNA from aqueous solution into ethanol via fine spinerettes, the fiber preparation technique used for fiber X-ray diffraction analysis. He showed that the fibers are composed of twisted assemblages of DNA with an average width of 76 nm by 20 nm thick, a characteristic vertical undulation of 10 nm, and a right-handed twist. It was also shown that the drug cisplatin alters DNA structure such that the mean width of assemblages increases and kinks are introduced.

Bending effects on circular and linear DNA caused by interaction with small-molecule DNA-binding ligands have been observed and quantified by SFM. The standard means of assessing the extent of bending induced by a ligand is to record changes in electrophoretic gel mobility [127], and this technique was used to confirm the SFM results in the experiments described below. A synthetic mimic of the naturally occurring antibiotic distamycin, called MGT-6b, was shown to decrease the curvature of bent kinetoplast circular DNA and to increase the curvature of a linear 400-bp molecular weight standard [128]. The difference is attributed to different base compositions of the two DNAs and the sequence preference of distamycin.

Coury and co-workers were able to mark the position of an intercalated small molecule by using a biotinylated psoralen, which intercalates into DNA and can be cross-linked by exposure to UV radiation [129]. Psoralen photo-cross-linked plasmids were labeled with streptavidin gold and deposited on mica. The authors hope to use this approach to explore the relationship between DNA structure and intercalator binding. The same group has also shown that SFM can be used to distinguish between intercalative and nonintercalative modes of DNA binding by uncharacterized ligands [130]. They made length measurements on the DNA fragments after treatment with various concentrations of the ligand. An intercalating ligand would cause a concentration-dependent increase in fragment length. This was not observed for the ligand in question, 2,5-bis(4-amidinophenyl)furan, whereas ligands known to intercalate, like ethidium bromide, do cause an increase in length. Exclusion numbers and cooperativity parameters were obtained for the intercalating ligands from their data.

As soon as DNA imaging techniques were established, researchers began to explore the possibilities of obtaining data about novel non-B-form DNA structures. Hansma and co-workers published images of triple-stranded oligos, distinguished by their height and width being greater than double-stranded oligos [131]. Thundat and co-workers obtained evidence that the DNA helix unwinds in a small percentage of molecules during the course of preparing samples of DNA strands bound to mica and subsequent drying [132]. Conditions that affect the formation of kinks in certain sequences have been explored by using DNA minicircles that contain repeats of sequences dA5 and dGGGCCC [15]. These sequences are known to bend DNA. The minicircles were formed by ligating linear oligo duplexes without introducing supercoils, and they were imaged using intermittent contact mode in fluid. When the minicircles were imaged in 1 mM $MgCl_2$ they were smooth, but introduction of 1 mM $ZnBr_2$ caused them to form abrupt kinks. Similar investigations of kinking in minicircles have been made by Larquet and co-workers, using a combination of transmission electron microscopy (TEM) and SFM, to provide complementary data [133].

Henderson [134] has imaged guanosine monophosphate gels, which are known to contain ordered arrays displaying structural parameters very similar to those of covalently linked guanosine tetrads. Guanosine tetrads can take on a family of structural forms in a dynamic equilibrium sensitive to the chemical environment known as G4 DNA or G4 RNA. Guanosine tetrads occur in the telomere and may have a role in telomere function. Images of GMP gels that were dried onto mica revealed the presence of fibers of varying sizes from 1.3 to 40 nm across.

In later work by the same group, Marsh and co-workers studied the self-assembly behavior of a 10-bp oligonucleotide based on the telomeric G-strand repeat sequence dGGGGTTGGGG [135]. The G-quartet is expected to form a structure termed a G-helix, which is roughly tubular with a large central cavity into which cations and water may fit. The small structures were found to coalesce into long polymeric strands that were stable to drying, as the SFM images were acquired on dry samples in low-humidity air. The long polymers have been termed G-wires. G-quartets have a diameter of 28 Å, from X-ray crystallography measurements. The G-wires had measured heights in the SFM images of about 23 Å. In contrast, B-DNA normally appears about half as tall in SFM images as its known diameter. Vesenka and co-workers subsequently used G-wires as an imaging subject to test the effectiveness of an image reconstruction algorithm [41]. The G-wire DNA was used because of its inherent rigidity, which makes it possible to reconstruct the true shape of the subject molecule. The image reconstruction algorithm uses imaging of colloidal gold spheres to reconstruct the tip shape and then subtracts this shape from the raw images of the subject molecule acquired with the same tip.

Another novel conformation of DNA has been observed by TEM and SFM in preparations of phage l DNA that were boiled and quenched [136]. This form of DNA is resistant to DNase I and appears as thick curved rods in the images. It has distinctive circular dichroism spectra quite different from those of B-form or heat-denatured DNA. The authors speculate that it may be a triple-helical form, but no evidence to support this conjecture was presented.

One of the issues that needs to be considered to make meaningful interpretations of macromolecular structures visualized in the SFM is the nature of structural equilibration that may or may not occur after the molecule binds to the imaging substrate. This question, with regard to linear DNA, has been investigated by the Bustamante group in some depth. Rivetti and co-workers made measurements to determine the persistence length of DNA molecules bound to mica in a way that allows structural equilibration on the surface [137]. The measured values corresponded well with theoretical calculations based on a statistical polymer chain analysis. They were able to define those binding methods that allowed equilibration and those that cause kinetic trapping of structural states on the mica surface. Mica that was equilibrated in water before binding of the DNA samples and mica that was glow-discharge-treated before DNA binding caused kinetically trapped DNA forms to predominate on the surface.

Recently research has focused on the use of SFM to visualize and quantify supercoiling and the solution conditions that influence the degree of supercoiling in circular DNA. Lyubchenko and Shlyakhtenko imaged a 3.76-kb and a 688-bp plasmid in fluid by intermittent contact mode, using APTES-treated mica as the substrate [138]. They found that at high salt concentrations, the supercoiled plasmids had a tightly twisted geometry, yet at low ionic strength the plasmids formed large twisted loops and rosette structures. The results obtained for specimens prepared in the same ionic conditions, but where the samples were dried and imaged in air, were substantially the same, so it appears that DNA can withstand drying with little structural alteration.

Rippe and co-workers also took advantage of the ability of the SFM to image under buffer in their study of supercoiling in an 1868-bp plasmid [139]. They used the conditions defined by Rivetti and co-workers [137] as allowing structural equilibration on the mica surface and imaged the specimens in intermittent contact mode. They also compared images obtained in air, calculating the writhing number and the twist for both cases. The writhing number and twist were slightly higher in air, but there was no significant difference in the number of nodes per molecule, the superhelical radius, or the contour length. The length of a superhelical turn was shown to be easily obtained from the distance between height maxima along the molecule. Taken together, these studies point out the possibilities of studying the effects of DNA-processing proteins on the superhelicity of DNA, and vice versa.

Because it is feasible to image DNA strands and measure their lengths in images, it should be possible to use SFM to map long DNA molecules. However, a way of labeling a specific sequence is needed. There are many highly sequence-specific DNA-binding proteins that could fill this need, although a majority of them are restriction endonucleases. However, Allison and co-workers have demonstrated the feasibility of using a restriction enzyme as a sequence-specific label [140]. They used a mutant form of *Eco*RI, which lacks nucleolytic activity but retains sequence specificity, to mark the *Eco*RI sites with a distance measurement accuracy of roughly 50–100 bases. The same group subsequently used the mutant *Eco*RI protein to map the *Eco*RI sites on a 35-kb cosmid [141]. They calibrated the technique on λ-DNA and were able to demonstrate an accuracy for the distances between sites of better than 1%.

An interesting proposal for using SFM or STM to rapidly sequence long strands of DNA has been put forth by Kelson and Nussinov [142]. They propose that it should be possible to detect where radioactive decay breaks a DNA strand at the site of incorporation of a radioactive base into the chain. Then chains with the four different bases radioactively labeled would be imaged in parallel, and the order of bases would be determined by comparison. This idea, although theoretically feasible, requires the ability to reliably prepare specimens of single-stranded DNA with totally extended chains. This type of specimen preparation has not yet been demonstrated.

Developing medical technology is driving structural research on nucleic acid complexes capable of delivering DNA sequences into cells by using cationic liposomes. These liposomes are made up of a mixture of a cytofectin, a synthetic lipid analog with a long-chain hydrocarbon tail and a polar headgroup bearing a positive charge, and neutral zwitterionic lipids. Mixing DNA with such cationic liposomes causes aggregation of the liposomes and leads to ordering of the DNA. This ordered DNA has been studied by high-resolution X-ray scattering, and it has been found that DNA condenses into a highly ordered phase similar to a smectic liquid crystal [143]. A similar condensed phase of DNA on a lipid bilayer has recently been imaged by SFM, and the interaxial spacing was in agreement with that seen by X-ray scattering [144]. In addition, the spacing was found to increase with increasing salt concentration. It is not clear what chemical property of cytofectins is responsible for their ability to deliver DNA into cells. Wheeler and co-workers [145] investigated the effect of changing an alcohol group to a primary amine on the polar headgroup of a cytofectin, adjacent to the quaternary amine, which is also present on the headgroup. The complexes of DNA with the new cytofectin were characterized by transfection assays to measure their efficiency for DNA delivery and by SFM to image the structure of the complexes. They reported difficulties with reproducible SFM specimen preparation, but they were able to conclude that liposome produced with the amine-modified cytofectin and DNA produced rods and branched structures, whereas the liposomes formed with the original cytofectin and DNA appeared as globules of varying sizes.

A related approach to the problem of introducing DNA into mammalian cells makes use of DNA–polyelectrolyte complexes. SFM was used as the primary method for the structural characterization of these complexes by Wolfert and co-workers [146]. Complexes formed between low-molecular-weight poly(L)lysine and DNA were found to be more homogeneous in size and to have lower cytotoxicity than complexes formed with high-molecular-weight poly(L)lysine. All of the complexes had a globular appearance. The same group also studied similar self-assembled structures formed with block copolymers containing polycationic blocks alternating with neutral hydrophilic polymeric blocks [147]. Most of these structures were also globular in appearance, except for the complexes made using the polyethylene glycol-poly(L)lysine copolymer. The latter complexes formed curved rodlike and toroidal structures. The presence of the hydrophilic blocks appeared to reduce aggregation, as the sizes of the globular structures produced by the block copolymers were more homogeneous than were those produced by the polycationic polymers alone.

2.3.3.1. *Intermolecular Force Measurements*

DNA has also served as a test-case molecule for demonstrating the detection of forces involved in molecular recognition. Two groups [148, 149] have made experimental measurements of the forces between complementary DNA bases in solution. In one case, 20mer oligonucleotides with different length stretches of complementary bases were attached to the SFM tip, and complementary oligos were attached to the surface [148]. Interactions forces were measured after bringing the tip into contact with the surface and allowing the complementary strands to form base pairs. Specific binding forces were not detected if the oligonucleotide sequences on the tip and surface were not complementary, and the interaction could be blocked by supplying one of the complementary oligos in solution. A different approach was taken by Boland and Ratner [149], who took advantage of the ability of three of the deoxynucleotide bases to form self-assembled monolayer arrays on a gold surface. These authors were able to measure the interactions between the individual bases in monolayers formed on a gold-coated tip and those on a gold-coated mica surface. The interaction forces between as few as two individual pairs of bases could be detected, and the overall interaction could be blocked by supplying a complementary base in solution. Again, the forces were not detected if the monolayers on the tip and sample were not complementary.

These experiments demonstrate the ability to detect molecular recognition forces between small numbers of molecules and suggest the possibility of using SFM to produce images with contrast based on specific molecular recognition with nanometer-range spatial resolution.

2.3.4. *RNA Imaging*

After more than 5 years of development, SFM imaging of DNA and DNA–protein complexes has become relatively routine. In contrast, at this time there are only a small handful of published images of RNA molecules, and only one of a single-stranded RNA, a poly-A oligomer. Single-stranded RNA (ss-RNA) has proved to be more difficult to work with than DNA, and there are still no publications that discuss ss-RNA sample preparation methods for SFM. Single-stranded RNA has great conformational flexibility, and the structures it forms under various solution conditions are not easily predictable. In addition, the extra handling precautions required when dealing with RNA make it unnecessarily tedious for use in developing basic imaging methods. There is much that could potentially be learned about folding of single-stranded RNAs, by using a technique such as SFM that allows imaging in native buffer conditions and by observing changes induced by altering buffer conditions in pseudo-real time. There are also many RNA–protein complexes of great importance to biology, including mRNA splicing complexes, complexes of translational regulators with RNA secondary structures, and transcription termination complexes. This list includes many complexes that are unlikely to be crystallized or formed into 2-D arrays for X-ray or electron diffraction structure analysis because of the conformational flexibility of the RNA portion of the complex. Thus RNA–protein complexes provide both special opportunities and special challenges for study by SFM techniques.

The first SPM images of RNA molecules published were of a double-stranded RNA isolated from an RNA virus [93]. The objective of that investigation was simply to demonstrate the efficacy of the amine-terminated, silane-modified mica (aminopropyltriethoxy silane or APTES) surface for preparing SFM specimens of nucleic acids. The RNA molecules appeared essentially identical to DNA. The double-stranded genomic RNA from Reovirus, which has a multisegment genome, was used to demonstrate that SFM can be an attractive alternative to EM for routine nucleic acid length measurements. Lyubchenko and co-workers [150] used APTES-treated mica as the imaging substrate for their measurements, which were made by contact mode in air. They reported a statistical analysis of the length distribution for the retrovirus RNA with three clear maxima, which were in good agreement with previous electron microscopy data.

The only images of isolated single-stranded RNA molecules that have been published to date are of polyriboadenosine (poly(A)), although images of crystallized phenylalanine transfer RNA have been used to study RNA crystal growth [151], and images of single-stranded viral RNAs have been presented [152]. The poly(A) RNA homopolymer was imaged by depositing it on mica and scanning in air by intermittent contact mode [131]. The RNA appeared to organize into a series of globular domains separated by stretches of extended strand, with a wide range of apparent lengths. The complex formed between the riboadenosine homopolymer and poly(A) binding protein (PABP) has also been imaged by SFM. Smith and co-workers [153] formed complexes in solution and then deposited them on mica, or first dried the poly(A) on mica and imaged it to confirm its presence and then added a solution of PABP to the mica, allowing the complexes to form for 5–10 min. This was followed by a gentle water rinse and drying in air. Images were acquired by intermittent contact mode in ambient air. Knobby or globular structures were observed on all of the strands of poly(A), and consistently at the ends of the strands. Images of poly(G) prepared in the same way from a solution in water were also presented, and the poly(G) appeared as randomly sized globulat objects, probably aggregates, with no obvious structure.

Images of poly(A) binding protein deposited on mica showed fields of objects 1 nm tall by 9 nm wide, which are presumed to represent the monomeric state of the protein. The PABP–poly(A) complexes also appeared as a range of sizes, but the knoblike structures were larger than those observed on the poly(A) RNA alone and were presumed to be bound PABP molecules, probably clusters of several molecules. The height of the RNA between the knoblike structures was reduced in the complexes relative to that seen in poly(A) alone, so they concluded that binding of the protein altered the RNA structure in some way. Smith and co-workers [153] also reported that attempts to image single-stranded RNA under buffer met with no success. They were not able to achieve tight binding of RNA to mica, and secondary structure formation or aggregation of the RNAs made image interpretation difficult. Similar difficulties were reported by Kasas [154].

A systematic investigation of pre-mRNA splicing complexes, using SFM to image the complexes in a native buffer environment, is ongoing in our laboratory. The splicing complexes are formed on a pre-mRNA that contains a single short intron ($\sim$100 nucleotides) with exonic sequences at both ends. We first determined how to prepare specimens of this single-stranded pre-mRNA molecule bound to mica such that it can be imaged in buffer. This work is discussed in detail by Godfrey et al. [155] and will be briefly summarized here. Nickel-modified mica was used as the substrate, as it was found to provide the strongest binding of DNA to mica by Hansma and Laney [156]. Acid pretreatment was used in combination with nickel treatment to further improve binding. This surface preparation would be expected to cause kinetic trapping of DNA due to the strength of the binding interaction. We have not yet been able to obtain satisfactory images of this RNA by contact mode under buffer, even though these conditions allowed easy contact imaging of DNA under buffer [156]. To obtain fields of RNA molecules that all have a similar appearance, the RNA is annealed by heating to 70 °C and allowing it to cool slowly to room temperature. The conformation of the RNA was found to vary, depending primarily on the concentration of magnesium ions present during the annealing process and to a lesser extent on the concentration of monovalent salt. At 50 mM monovalent salt concentration, the RNA molecules appear as large globular objects or as smaller globular domains separated by stretches of extended strand. At 10 mM monovalent salt, a variety of lobed structures appears in samples if they were previously annealed in the presence of 10 mM magnesium. If the samples are annealed at 1 mM magnesium and imaged at low monovalent salt concentration, globular structures are seen that appear larger than those seen in 50 mM monovalent salt and which are more variable in size and form (Figs. 9 and 10).

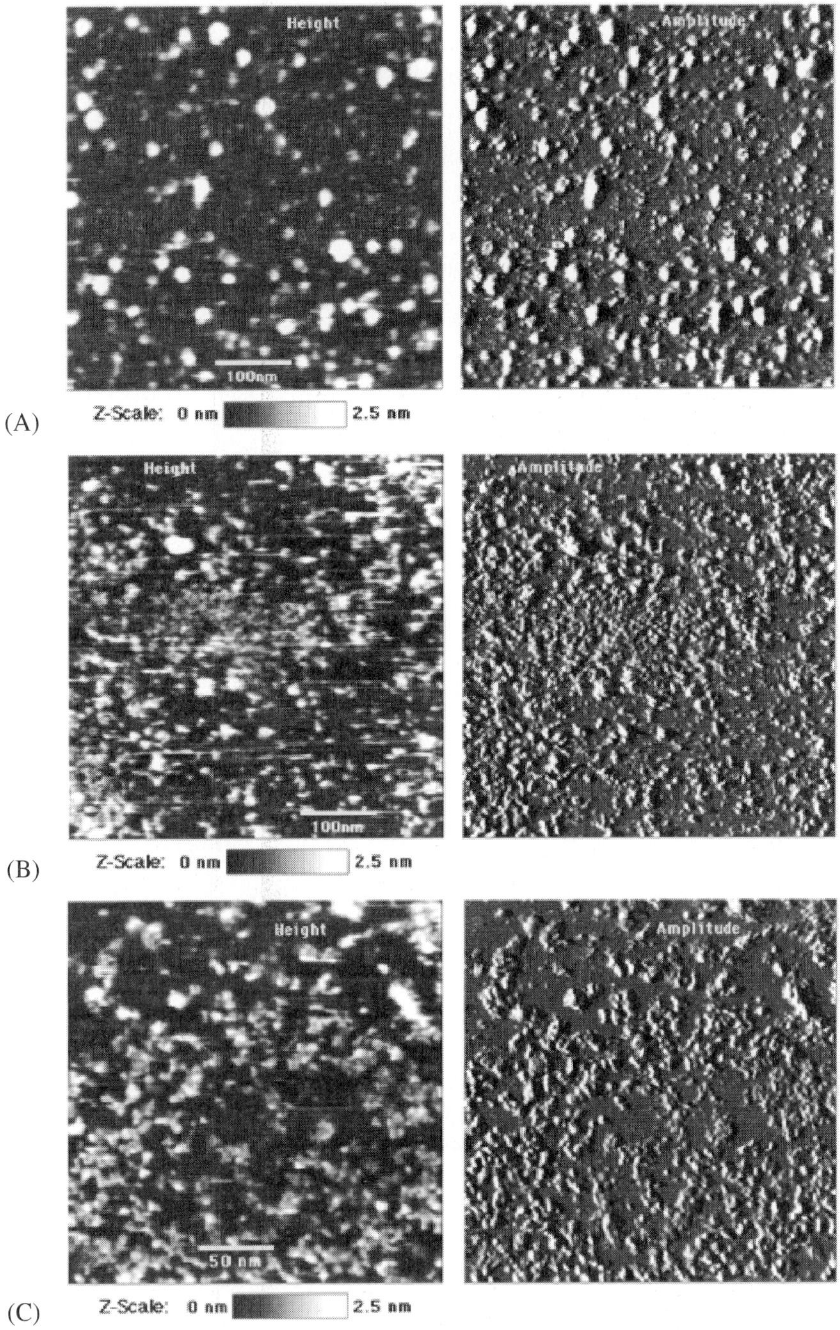

Fig. 9. Major-late adenovirus 2 RNAs annealed and imaged under different ionic strength buffers. (A) Major-late adenovirus 2 pre-mRNA fragment, 416 bases, hybridized with a biotinylated 20mer oligonucleotide. The sample was annealed in the presence of 1 mM $MgCl_2$, diluted $1{:}10^6$ in a buffer containing 1 mM $MgCl_2$, 50 mM KCl, and applied to nickel-treated mica. Imaging was carried out in tapping mode under the dilution buffer. Height and amplitude images are shown. The RNA takes on a largely globular form under these conditions, with some short stretches of extended strands between globular features. (B) The same RNA sample described above, but the imaging buffer was changed to 1 mM $MgCl_2$, 10 mM KCl, while the specimen was in the fluid cell. The RNAs are refolding into forms with fewer large globular domains. The streaking artifacts are caused by the loosening of the binding interaction with the mica surface during the conformational transition caused by the change in ionic strength. (C) MLAd2/oligo hybrid annealed in the presence of 10 mM $MgCl_2$ and then diluted $1{:}10^6$ in a buffer containing 1 mM $MgCl_2$, 10 mM KCl before application to the mica substrate. This preparation results in the formation of stable structures that bind tightly and can be repeatedly scanned with no apparent changes for many frames. There are some streaking artifacts in this image that appear to be associated with larger aggregates that may not be in good contact with the mica surface. This is probably caused by the high coverage on this specimen.

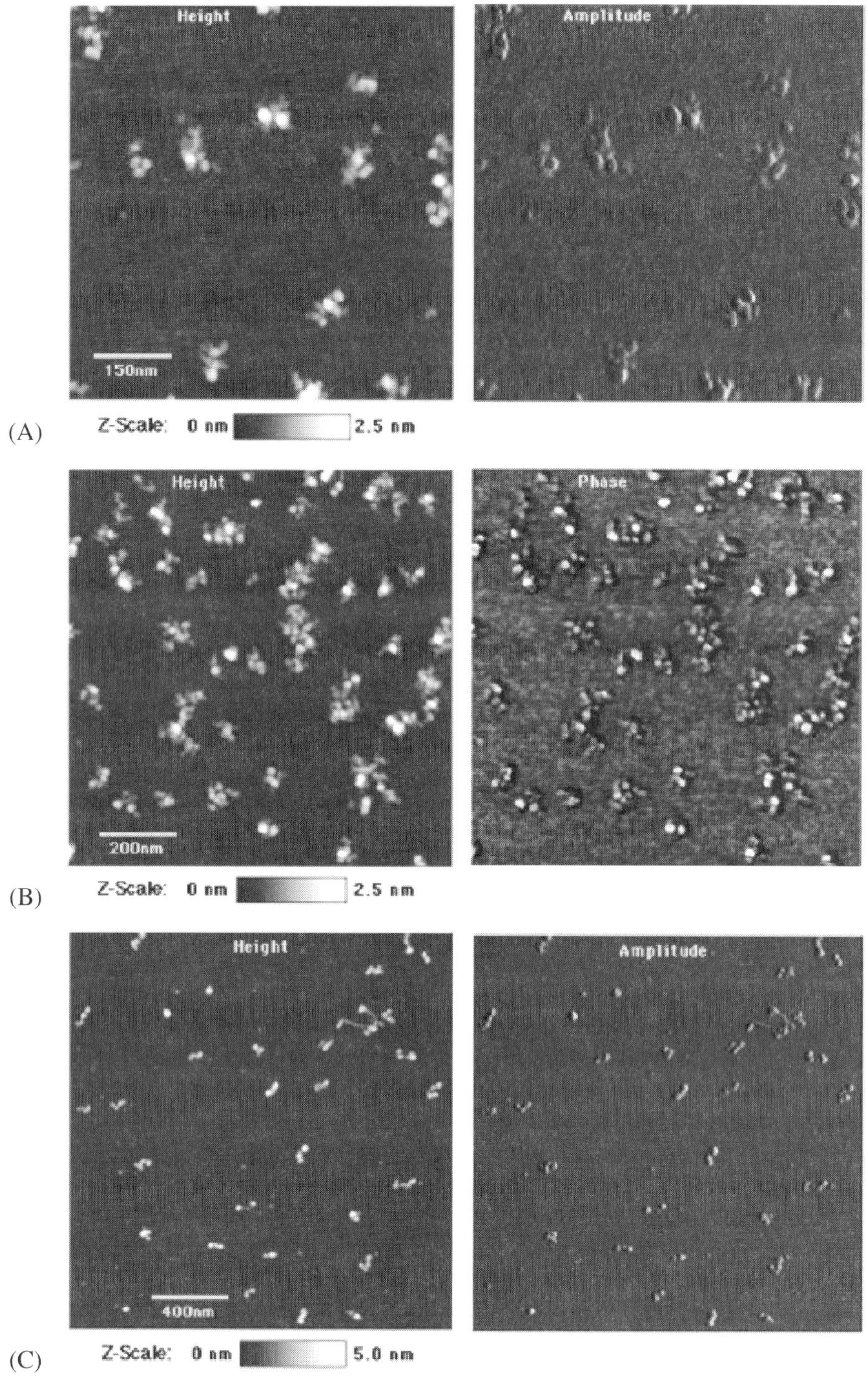

Fig. 10. 23S ribosomal RNA imaged in air by tapping mode. (A) 23S ribosomal RNA (rRNA) was transcribed in vitro and diluted into 10 mM $NiCl_2$, applied to mica, and rinsed with water. Height and amplitude images of low-coverage specimens were contributed by W. Blaine Stine of Abbot Laboratories. This RNA folds into a discrete structure as part of the ribosome. The interesting secondary structures are clearly visible in these images. (B) A larger field of the same 23S ribosomal RNA preparation. Height and phase images are presented. The phase image provides better contrast on this specimen than the amplitude image. (C) Height and amplitude images were made with a freshly transcribed and diluted 23S RNA preparation. The structure appears more compact than in the previous images. This could be due to the shorter exposure of this rRNA sample to the low-ionic-strength 10 mM $NiCl_2$ buffer.

2.4. Scanning Probe Microscopy of Nucleoprotein Complexes

2.4.1. Protein-Induced DNA Bending

A number of transcription factors isolated from prokaryotes and eukaryotes have been found to induce bends in DNA when bound to their cognate DNA binding sites. In many cases transcription factors bind to their specific DNA sites as dimers, and this seems to correlate with bending activity. For example, the prokaryotic helix-turn-helix protein CAP binds as a dimer and bends DNA, whereas the eukaryotic homeodomain proteins that also contain the helix-turn-helix DNA binding motif and bind as monomers do not bend the DNA. There is also some evidence that the bendability of a DNA sequence is important in determining the affinity of a protein for a specific binding site. The role of DNA bending in transcriptional control could take the form of providing proper orientation of DNA-binding proteins relative to one another, aiding in site recognition, or providing a conformational energy linkage that could facilitate maturation of initiation complexes [157].

DNA bending is usually detected and measured by comparative gel electrophoresis. For example, the location of a bend in sequence can be mapped by moving the suspected sequence closer to the end of a fragment, the basis for the circular permutation electrophoretic assay, and the relative orientation of two bends can be mapped by varying the number of bases and, thus, the phase between them [127]. Cyclization analysis, rotational relaxation, and electron microscopy [158] have also been used to characterize DNA bends. The primary advantage of using high-resolution microscopy is that one can measure the distribution of bend angles in a population of complexes. Observing the distribution of angles can give researchers insight into the flexibility of a complex and may reveal the existence of different populations in solution. Scanning force microscopy has found increasing application in observing and measuring bend angles of DNA–protein complexes because of the simplicity of the technique and the similarity of the common specimen preparations to solution conditions. One caveat, however, is that the possibility of kinetic trapping has to be considered when interpreting the data, as was discussed above. Furthermore, the possibility that a three-dimensional complex is formed, such as a combination of two noncoplanar bends like that in the *Eco*RI crystal structure, must be considered [159]. Finally, consideration must also be given to the possibility that the surface effects the three dimensional nucleic acid topology.

A number of imaging studies of prokaryotic transcription factors and transcription complexes have been published, including *E. coli* RNA polymerase (RNAP) bound to λ-phage promoters, an ampicillin resistance gene promoter, and complexes of Cro protein with the O_R region of λ. The degree of DNA bending in open and closed transcription complexes was measured on the λ-P_L promoter (lP_L) by Rees and co-workers [34], using contact mode SFM in low-humidity air with electron beam-deposited (EBD) carbon tips. The specific RNAP/lP_L open promoter complexes were identified by the position of RNAP binding on the DNA fragment. Closed promoter complexes were prepared by providing three of the four ribonucleotides for the polymerase, which allows synthesis of the first 15 bases of the transcript. Statistical analysis of 120 images of each type of promoter complex was performed. The mean bend angle was 54° for the open complexes and 92° for the closed complexes. Bending of the DNA template in transcription complexes had been inferred from electrophoresis experiments, and it has been observed in EM images before but not analyzed quantitatively.

The same group made a study of DNA bending by another λ transcription factor, the Cro protein [160]. Cro, a 14.7-kDa protein, binds as a dimer to three operator sites in the O_R region between two divergent promoters, lP_R and lP_{RM}, where it acts as a transcriptional repressor. Cro is known to bend DNA when engaged in specific complexes with the operator sites and was chosen as a system in which to study the effects of DNA bending on the process of target localization, binding specificity, and the mechanism of specific site recognition. As was done for RNAP, the images were acquired in contact mode in low-humidity air with EBD-modified tips. The O_R sites were two-fifths from one end of the fragment to facilitate identification of specifically bound Cro dimers. The distribution of bend angles determined for specifically bound complexes was narrow and centered at $69^\circ \pm 11^\circ$. The average angle for the nonspecific complexes was nearly the same, 62°, but the spread was broader at $\pm 23^\circ$. One possible interpretation of these data, discussed by the authors, is that the specific Cro binding site is more bendable than most random sequences of similar length and that this flexibility helps to define the site. Several other examples of DNA bending by prokaryotic transcription factors are also contained in reference [161].

Another case of bending induced by a transcriptional regulator protein has been investigated by scanning tunneling microscopy [162]. The repressor activator protein 1 (or RAP 1 protein) in yeast binds to a 13-bp consensus sequence found in transcriptional regulatory regions, silencer elements, and telomeric repeats. Electrophoretic mobility assays suggest that RAP 1 bends DNA when it binds, with the vertex of the bend is located 5′ of the consensus sequence. The authors demonstrated, using the circular permutation assay, that removal of the N-terminal 230 amino acid reduces the aberrant mobility and that removal of the C-terminal domain increases the aberrance. STM and TEM were used in parallel by preparing high-angle metal-shadowed specimens on mica with the same procedures for both microscopies. Both techniques show a bend angle of more than 50° for complexes with the full-length wild-type protein, and only 29° for complexes with the minimum DNA binding domain alone.

2.4.2. *DNA Looping Induced by Complex Formation*

Another structural modification of DNA that can be induced by binding of transcription factors is loop formation. Formation of loops between proteins bound to specific binding sites more than 50 bp apart has been imaged by electron microscopy in several cases [163 and references therein]. The ability to prepare and image samples in native buffers makes investigation of looping in DNA–protein complexes by SFM an attractive alternative to EM. SFM has the additional advantage of excellent height resolution, which has aided in the determination of the number of molecules bound to DNA in complexes in several cases reviewed here. Wyman and co-workers imaged human heat shock transcription factor 2 (HSF 2) as it binds to a 2.1-kb DNA fragment that contains two heat shock elements (HSEs) separated by about 880 bp [164]. HSF 2 binds DNA as a trimer, and biochemical evidence suggests that the trimers bound to distant HSEs interact. Complexes with HSF 2 and the DNA fragment were prepared and imaged by electron microscopy and scanning force microscopy. In both cases more looped complexes were formed than complexes in which two HSF trimers bound separately to both HSEs, which is consistent with a cooperative association in forming loops. The SFM imaging was carried out using intermittent contact mode in ambient air. They were able to resolve a bilobed structure at the base of the loops, which had a size consistent with the presence of two or more individual HSF 2 trimers, based on size comparisons of various areas of the same specimen with the same tip.

2.4.2.1. *Role of Jun in the Formation of Transcriptional Preinitiation Complexes*

The functioning of another eukaryotic transcriptional regulator, the Jun protein, has been studied by SFM. It is thought that bending induced by transcription factors facilitates the formation of looped complexes containing the transcription factors and the preinitation complex. Jun is a member of the bZIP family of basic region leucine zipper DNA-binding proteins and forms homodimers. The DNA binding domain of bZIP proteins undergoes a transition to a form with high α-helical content upon binding to its cognate DNA sequence. This suggests that a complementary change in conformation of the DNA is likely upon Jun binding. However, the crystal structures available to date for bZIP DNA-binding domains and DNA do not show any bending in the DNA.

Becker applied SFM to image complexes of Jun with a linearized plasmid containing an AP-1 binding site to determine if Jun binding bends the DNA [165]. They prepared transcriptional preinitiation complexes (PICs) isolated from nuclear extracts that had been depleted of naturally occurring Jun and then reconstituted the PICs with recombinant Jun dimers and the plasmid DNA. Transcriptional activity was confirmed by in vitro transcription assays for the nuclear extract preparations used in the experiments. The DNA–protein complexes were deposited on mica, dried, and imaged in contact mode under humidity control. They found a mean bend angle of 36° for Jun/DNA complexes. Loops were observed in complexes containing both the PIC and Jun, and not in complexes where either nuclear extract or Jun was omitted, or where the DNA lacked an AP-1 site.

2.4.2.2. Mode of Interaction of Ku Protein with DNA

Ku is an abundant nuclear DNA-binding protein. It binds to nicks, gaps, ends, and transitions between double- and single-stranded structures. Ku is also a subunit of the DNA-dependent protein kinase (DNA-PK), which is thought to play a role in DNA repair, but the exact function of Ku in this process remains unclear. Cary and co-workers investigated the association of the Ku subunit of the DNA-PK with double-strand breaks [166]. They performed binding reactions between Ku and a 660-bp DNA fragment or *Sma*I-linearized pUC19 and imaged the resulting complexes by SFM and TEM. The SFM specimens were fixed with glutaraldehyde after binding to mica and imaged in air by intermittent contact mode. They found that though Ku does not self-associate in water, it does associate once it binds to a DNA fragment. A significant fraction of the fragments were looped, and some were formed into linear multimers, presumably by the interaction between Ku molecules bound at their ends. The size of the protein bump at such sites of interaction was consistent with the presence of two or more Ku molecules. Addition of the catalytic subunit of DNA-PK neither inhibited nor enhanced loop or multimer formation.

Pang and co-workers used a somewhat different approach to study of the Ku protein [167]. They formed complexes of Ku with circular pUC19 plasmid DNA and with plasmid that had been gamma-irradiated to produce breaks. They imaged unfixed specimens bound to mica by SFM in intermittent contact mode in ambient air. Ku did not bind circular plasmids but did bind to the ends of plasmids linearized by radiation exposure. Complexes of Ku with 400-bp linear DNA fragments contained strands of lengths varying in increments of 132 nm, with internally bound Ku. Pang et al. concluded that Ku functions to bring the ends of fragments together for ligation.

2.4.2.3. GalR and HU Interaction with the gal Operon Promoter

Lyubchenko and co-workers [168] have investigated a case of suspected DNA looping in the regulation of the *E. coli gal* operon. Biochemical evidence suggested that repression of the *gal* operon required binding of GalR dimers to two operator sites and formation of a loop, but loop formation had not been observed with purified GalR protein. It was recently discovered that the histone-like protein HU, in addition to GalR, is required for repression of *gal* promoters [169]. This system was investigated by SFM of GalR–*gal* promoter complexes. A 688-bp circle containing the wild-type *gal* promoter (called a *minicircle*) was used as the DNA substrate. With a distance of 113 bp between the two operator sites, a loop formed between GalR dimers would appear to be about one-fifth the size of the minicircle, making an asymmetric figure eight shape. Specimens were prepared on APTES-treated mica, rinsed with water, and blown dry with argon. Imaging was by intermittent contact mode in ambient air. As expected, little loop formation was observed with GalR alone. With the addition of histone-like protein HU, numerous loops were formed.

Lyubchenko et al. also investigated whether the role of HU in facilitating loop formation was in promoting supercoiling of the minicircle or in promoting tetramerization of the GalR protein. A linear 3.7-kb DNA fragment containing the *gal* promoter was used to form GalR complexes with or without HU present. The GalR bound to the linear DNA specifically and nonspecifically, appearing as globular features of uniform size distributed along the strands. In the presence of HU, the globular features were not of uniform size and shape, indicating that GalR had aggregated in several positions on each strand. This result suggests that HU may facilitate tetramerization of the GalR.

Most of the studies of DNA bending and looping have included estimates of the stoichiometry of the protein complex formed on the DNA. Nettikadan and co-workers addressed this point alone in their comparison of the complexes formed by the interaction of two eukaryotic transcription factors with their respective binding sites [170]. They analyzed SFM images of the mammalian transcriptional regulators AP-2, a helix-span-helix protein, and SP-1, a zinc-finger protein, bound to their respective cognate DNA binding sites. They were able to show that AP-2 probably binds as a dimer, whereas SP-1 binds as a monomer. The size analysis of the complexes made use of the apparent width of the DNA as a calibration for the radius of curvature of the tip, and they used samples of 50 images for statistical analysis of the sizes of the isolated proteins bound to mica as well as in complexes with DNA.

2.4.3. *Other Sequence-Specific DNA–Protein Complexes*

2.4.3.1. SV40 T-Antigen Complex with the SV40 Origin

The SV40 T-antigen binds as a double hexamer to the viral origin of replication, a 27-bp palindrome, and interacts with numerous proteins of the eukaryotic host cell as well as the viral replication proteins. T-antigen destabilizes the DNA helix at the SV40 origin in an ATP-dependent step during replication. Mastrangelo and co-workers [44] imaged the complexes of SV40 T-antigen on the SV40 origin palindrome sequence to confirm the structure suggested by earlier scanning transmission electron microscope (STEM) images. Complexes were formed between recombinant T-antigen and a plasmid containing the SV40 origin palindrome sequence and fixed with glutaraldehyde. They were then applied to mica, rinsed, and dried, or positively stained with uranyl acetate before being rinsed and dried. Scans were made in contact mode in air. The uranyl acetate was found to increase the resistance of the protein to damage by the tip in small area scans and to increase the apparent height of the DNA strand. The cleft between the two hexamers, formed with ATP present, was resolved, and they were able to image binding of monomers in the absence of ATP. The sizes of the monomers could reasonably account for the size of the hexamer, and the size of the hexamers was in good agreement with previous measurements from STEM images.

2.4.3.2. Interaction of a Prokaryotic Histone-Like Protein with DNA

Following infection of mammalian cells, the prokaryote *Chlamidia trachomatis* goes through a life-cycle stage during which the chromosome becomes condensed. The organism contains a lysine-rich histone-like protein, called Hc1, which is expressed during that stage and which is thought to be involved in compaction of the DNA. Kaul and co-workers [171] found that the Hc1 protein also appears to have specific binding sites in its gene upstream of its ORF. A 24-bp region there was found to bind Hc1 specifically with high affinity and was found to be $>75\%$ unique to *Chlamydia trachomatis*. A similar specific site was found in a 7.5-kb Chlamidial plasmid. Complexes of recombinant Hc1 and linearized Chlamydial plasmid DNA were imaged by contact mode SFM at low humidity. The Hc1 was consistently seen bound about one-third of the way from one end of the DNA, where the specific binding site is located. Binding was accompanied by coiling, shrinking, and aggregation of the DNA. Even if DNA was first bound to mica and then incubated with Hc1, the protein bound and produced dense foci. Thus Hc1 may serve as a sequence-specific gene regulation factor in addition to its non-sequence-specific function, which appears to be similar to that of histone H1 in eukaryotes.

2.4.3.3. Mode of Binding of Fur to Iron-Regulated Promoters

Le Cam and co-workers [172] applied two types of TEM specimen preparations together with SFM to a study of the binding of the *E. coli* Fur iron(II)-regulated repressor protein to a Fur-binding DNA sequence. Dark-field TEM of positively stained complexes and SFM were used to visualize Fur interaction with regulatory sequences of aerobactin and hemolysin operons. Fur binding to an aerobactin promoter DNA fragment resulted in partial to total coverage, depending on the ratio of DNA to protein used. This was interpreted to mean that the protein initiates polymerization along the DNA strand at specific sites. TEM of negatively stained complexes suggested a helical arrangement of the protein on the DNA strand. The areas of high Fur coverage were always straighter than naked DNA, so Fur polymerization appears to cause a rigidification of the DNA strand. The authors suggest that the rigidification resulting from polymerization could be involved in the repression process. There was no change in the length of DNA with Fur bound, so the protein does not affect the helical twist.

SFM images of Fur complexes prepared at an intermediate ratio of protein to DNA showed a constriction between two areas of Fur coverage of the aerobactin promoter DNA fragment that correspond to two previously defined promoter elements. The SFM, even though it overestimates widths, was found to give a better indication of complex volumes because it gives more accurate heights. The authors pointed out that SFM and EM measurements can be combined to give even higher precision measurements, with EM providing width information and SFM providing height information.

2.4.3.4. Formation of Cruciforms in an HSP70 Promoter Element

Ohta and co-workers [173] investigated the complex formed between the promoter of the *Staphylococcus aureus* HSP 70 gene and the protein product of the first ORF of the HSP 70 operon, ORF37. The HSP 70 promoter contains an inverted repeat that has also been identified in other heat shock genes and is referred to as the *controlling inverted repeat of chaperone expression* or CIRCE. This sequence appears to be capable of forming a stem-loop structure, which may be important to its function. A 1200-bp fragment containing the promoter of the HSP 70 operon was used as the DNA substrate and combined with recombinant ORF37. The complexes were bound to mica and imaged in air by intermittent contact mode. Ten percent of the strands had a feature at one end that protruded vertically. Binding of the protein ORF 37 was observed at the position where the vertical protrusion sometimes appears, within error. These data led the authors to propose that the inverted repeat can fold into a cruciform structure in which the two stem-loops lie flat against each other and point away from the mica surface to produce the vertical protrusion seen in the SFM images. This is a very interesting system in which a nonstandard DNA structure may form the binding site for a regulatory protein. However, it is hard to understand why dual stem-loops would stably protrude away from the surface when they are also negatively charged and would be expected to bind flat to mica as DNA normally does. No length measurements for strands with and without protruding structures were given, so it is not possible to determine how many base pairs are actually involved in the formation of these structures. This study points out an area where the ability of SFM to operate in ambient air and buffer may be key to investigating structure–function relationships. DNA structures that may exist in equilibrium with B-DNA and may be sensitive to ionic conditions can be formed and imaged in buffers favoring one conformation or the other, and protein binding could be correlated with structure formation.

2.4.3.5. Translocation of TATA Binding Protein

The last case covered here does not directly concern a DNA–protein complex, but rather involves a question concerning the nuclear translocation of a sequence-specific DNA-binding protein and serves to demonstrate the potential usefulness of fluid imaging to studies of protein complexes involved in gene regulation. The TATA box binding protein, or TBP, is translated in the cytoplasm and must be imported through the nuclear pore complex to the nucleus, where it serves a vital role in gene expression. Oberleithner and co-workers [174] sought to determine the role ATP plays in the process of nuclear import of TBP. They observed aggregates of 14 TBP molecules in solution with no ATP present. In the presence of ATP, the aggregates primarily dissociate into tetramers, which can be transported into the nucleus. The apparent dissociation constant was found to be 10 μM for ATP. Transport of TBP was assayed by electrophysiological measurement of nuclear pore complex open states. The 14mer aggregates do not transport through the nuclear pore unless ATP is supplied. Thus ATP works in conjunction with the nuclear pore complex to regulate the access of TBP to its DNA-binding sites.

2.4.4. Complexes with Non-Sequence-Specific DNA-Binding Proteins

A number of complexes involving non-sequence specific nucleic acid binding proteins and single-stranded binding proteins have also been imaged by scanning probe microscopies. The body of literature on chromatin structure investigated by SPM is building rapidly, primarily because of the ability to apply SPM techniques to hydrated specimens. Some structural questions about large-scale protein–DNA complexes such as chromatin, which are sensitive to many environmental conditions, may be more accessible to SFM study than other current structural techniques. This rapidly growing field deserves a dedicated review and will not be covered in depth here because of space limitations. For a recent review of chromatin structure studies emphasizing the contribution of scanning probe microscopies, see [175].

2.4.4.1. T4 Gene 32 Single-Strand Binding Protein

Hansma and co-workers have published images of the complex of the phage T4 gene 32 single-strand-binding protein with ϕX-174 single-stranded DNA circles. Excess gene 32 protein was incubated with ϕX-174 DNA and fixed with glutaraldehyde before dilution and binding to mica [176]. This procedure is similar to the preparation originally used for TEM characterization of this protein [177], except that no stains or cytochrome *c* thickness enhancements for the single-stranded DNA were used on the SFM specimens. As would be expected, the lateral dimensions of the complex as compared to those measured in the TEM images are enlarged by tip convolution (22 nm wide vs. 6 nm by TEM). However, more substructure is visible in the SFM images, where the complex appears as lumps spaced about 22 nm apart along a strand.

The gene 32 protein is thought to bind DNA in a cooperative manner, but it is not clear how the cooperativity comes about if there is no direct protein–protein contact in the resulting complexes, as these images seem to imply. This could be an indication that once the protein has initially attached to the strand, possibly with the aid of an already bound molecule, it becomes free to diffuse linearly along the strand, with the entire collection of gene 32 proteins bound to a strand reaching an equilibrium in which they are nearly uniformly spaced.

2.4.4.2. E. coli RecA Protein

One of the most interesting nonspecific DNA-binding proteins currently under study is the *E. coli* RecA protein. RecA is part of the bacterial recombination system, where it aids in the alignment of single-stranded DNA with complementary sequences in double-stranded molecules. Briefly, RecA binds to double stranded DNA and polymerizes into nucleoprotein filaments in which the DNA is extended to about 1.5 times the length of B-DNA, with a rise of 5.1 Å/bp. RecA also binds to single-stranded DNA, and the resulting complexes are similarly extended to about 5.1 Å/bp. A RecA single-stranded DNA filament can bind to double-stranded DNA to form a RecA-coated filament with both the double- and single-stranded DNA in its center. The single-stranded DNA in this complex can slide along the duplex until it finds a complementary sequence. At this point strand exchange occurs, which is followed by ATP hydrolysis and release of the RecA from the DNA [178].

Two groups have published SFM images of RecA–DNA complexes with both single- and double-stranded DNAs. Hansma and co-workers [179] formed complexes of RecA with a mixture of single-stranded ϕX-174 circles and double-stranded fragments of ϕX-174 *Hae*III restriction digest, so that the single- and double-stranded DNAs share sequence. The complexes were formed in the presence of nonhydrolyzable ATP[γS] to trap the RecA on the DNA. The complexes were scanned in intermittent contact mode, some dry under helium and some under propanol. The DNA molecules were extended in length by the factor of 1.5, as expected, and some junctions between molecules were observed.

Lyubchenko imaged complexes of RecA with a 1200-bp circular plasmid that was adsorbed to APTES-treated mica and scanned in air in contact mode [180]. The 1200-bp circles appeared in nearly perfect circular form, which is reasonable in view of the rigidity imparted to DNA by binding RecA, and with the expected length. Features with a periodicity of 20–30 nm appeared in these images, which could not easily be explained, as a 10-nm periodicity was previously observed in EM and STM studies. One possibility, suggested by the authors in the discussion with reviewers, is that the RecA protein may be mobile on the DNA and may spread out along the strand when it binds to mica from solution. They noted that no periodicity was seen in images of the same complexes scanned under propanol, even though they had the same heights and lengths as the complexes scanned in air.

2.4.4.3. Bacteriophage Connector Complex

The ϕ29 bacteriophage connector, or portal, protein is involved in the process of packaging DNA into viral particles. The connector complex is located in the head-to-tail connecting region of the virion. It has been proposed that the connector complex forms a channel through which linear DNA is translocated into the viral particle. Isolated connectors generally bind DNA as multimeric aggregates of 12 or 13 molecules [181 and references therein]. Valle and co-workers [181] analyzed complexes between the connector and either linear or circular DNA by SFM and TEM. TEM showed a specific binding with the free ends of linear DNA. SFM of these samples by intermittent contact mode in air showed that with circular DNA there is a 132° angle between entering and leaving DNA strands. When the connector was incubated with linear DNA and later circularized, a greater bend of 168° was seen. These results are consistent with the DNA passing through an inner channel of the complex, as the complexes formed on circular DNA have a different bend and appear to bind on the side. These data are not enough in themselves to positively determine the structural details of the interaction between the connector complex and DNA. However, they do support a distinction between the mode of association of the connector complex with linear and circular DNA, which is consistent with the natural substrate being linear duplex viral DNA. This group intends to persue two-dimensional crystallization of the connector–DNA complex to obtain high-resolution structural data.

2.4.4.4. Immunolabeling Z-DNA

Probing for local conformational changes in DNA using an antibody that recognizes a particular structural state would be a potentially useful tool for studies of the effects of topological changes on processes such as transcription and replication. Pietrasanta and co-workers have investigated this possibility with SFM by using an anti-Z-DNA antibody to form complexes with plasmid DNA that contains an inserted sequence known to convert from the B to Z conformation under certain conditions [182]. The insert consisted of a

$(GC)_{11}$ sequence that assumes a Z conformation in a superhelical background, provided by the circular plasmid. Antibody binding to circular and linearized plasmid was assessed by gel electrophoresis, and binding of the antibody to only the superhelical form was confirmed. The position of binding for the antibody to the DNA in the images was as expected, assuming B-form for the DNA outside of the insert. The sizes of the bound protein in the images are consistent with one IgG bound per insert. The authors also stated that the DNA appears to bend to accommodate the interaction with both Fab domains of the antibody, but no attempt was made to quantify the bending.

2.4.4.5. HMG1 and Slippage Synthesis of Dinucleotide Repeats

Gibb and co-workers studied the effect of binding HMG1 to a repeat sequence, $(dTG)_n$, on the elongation of that sequence by the Klenow fragment of *E. coli* DNA polymerase in an in vitro assay [183]. The HMG proteins are ubiquitous basic nuclear proteins in eukaryotes and the most abundant proteins in the nucleus besides the histones. The authors sought evidence that these basic nuclear proteins could explain the unstable nature of dinucleotide repeats in the genome, possibly by influencing the replication of these sequences by the "slippage" mechanism. This work is a good example of the use of relatively easily obtained SFM data to bolster the conclusions from biochemical experiments that provided the bulk of the data discussed in this publication. HMG1 was expected to bind to the repeat sequence because the sequence should be able fold into cruciform structures based on the G-T wobble base pair, and HMG1 is known to bind preferentially to branches.

The appearance of up to five different bands in gel electrophoresis experiments using the DNA fragments alone and the differential effects of HMG1 binding on the mobilities of those bands support this expectation. In biochemical experiments HMG1 was shown to inhibit the elongation of the dinucleotide repeat sequence at 37 °C, but to enhance it at 46 °C. In contrast, polylysine, a nonspecific basic protein, inhibited elongation at both temperatures. SFM images were made of HMG1 protein alone and decorating DNA strands, to determine if the HMG1 protein bound only to the ends, or if it would bind at numerous sites in the middle of the sequences. The latter result would be consistent with binding to internal branched structures. The protein was seen binding in the center of the strands and at the ends, providing further evidence for the presence of cruciforms. The SFM tips used were treated with dichlorodimethyl silane to make them hydrophobic, and the scans were done in contact mode under dry nitrogen. These scanning conditions were claimed to give more accurate representation of heights than would be obtained with untreated commercial tips, which are inherently polar.

Lyubchenko and co-workers published images of a complex between the p25 protein from *Vaccinia* virus with the double-stranded RNA [180] from retrovirus that this group has worked with in the past [184]. Double-stranded RNA is produced during most viral infections, and it induces synthesis of interferon. The *Vaccinia* p25 protein is an inhibitor of the protein kinase PKR, which is involved in the induction of the antiviral state in mammalian cells in response to interferon. PKR activation requires binding to double-stranded RNA, and p25 presumably acts as a competitor for the dsRNA substrate. Complexes for imaging were made using an approximately 3:2 molar ratio of p25 to dsRNA. What was observed was a mixture of naked dsRNA and completely coated strands of RNA, which were shorter than the naked molecules and had a thickness several times that of naked dsRNA. The thick, short filaments could be aggregates induced by p25 binding, but further structural studies will be necessary to confirm this suspicion. These results indicate that the binding of p25 to dsRNA is highly cooperative. No examples of a single p25 molecule bound to RNA were seen. The images were acquired in contact mode in dry air and under propanol.

2.4.4.6. *Polymerases and Nucleases*

Several imaging studies of DNase and DNA polymerase complexes have been undertaken. One of these studies reperesents the first example of imaging a dynamic process in "pseudo-real time" will be covered in more depth in the next section. Li and co-workers imaged the thermostable polymerase from *Bacillus stearothermophilus* bound to mica with plasmid DNA and were able to resolve the bilobed form of the enzyme [185]. Sequenase, a variant of *E. coli* DNA polymerase, has also been imaged in the process of replicating single-stranded cirlces [186], and this reference also contained images of DNase I digesting a short linear DNA fragment, imaged under fluid. The degradation of DNA by DNase I was covered in greater depth by Bezanilla and co-workers [187], who were able to image the same DNA strands before and after several cleavage events had occurred.

2.4.4.7. *Time Lapse of RNAP Transcription*

Kasas and co-workers [154] recently published the first dynamic imaging study of an enzymatic activity, containing a time-lapse series of images of transcription by *E. coli* RNA polymerase. This work made use of Zn^{2+} in a reversible DNA binding technique to image molecules of RNAP with their DNA templates tightly or loosely bound. A 373-bp template with no terminator and a 1047-bp template with a terminator were used, and the transcription reactions were run with dilute rNTPs to slow the reaction enough for observation at the SFM image acquisition time scale. Their conditions give RNA polymerization rates of about 2 bp/s. The activity could be observed in time lapse only during the transition from the Zn-containing buffer, in which the DNA was bound to mica firmly enough to appear clearly in the images, to the transcription buffer without Zn^{2+}, where the DNA was mobile enough to allow enzymatic processing. The RNA transcript could not be seen directly in these images, because of the low affinity of RNA for mica. All of the images were made in intermittent contact mode, using electron beam deposited carbon tips. Intermittent contact mode operation was critical in this application, as it allows imaging of relatively weakly bound molecules under buffer because of the low lateral forces applied to the sample.

To confirm the activity of the RNAP bound to mica under these conditions, they also used a rolling circle template with a sequence consisting mostly of C and T bases to produce a transcript with no intramolecular base pairing potential. This template could be transcribed indefinitely, given an adequate supply of nucleotides, so that the transcripts produced would be long enough to be clearly visible. Even when transcripts roughly 2 μm long were made, they could still only be imaged after the specimens on which transcription had taken place were rinsed with Zn^{2+}-containing buffer and dried for imaging in air by intermittent contact mode. These experiments demonstrate that SFM can be used to follow a dynamic reaction in a time lapse fashion if the circumstances can be engineered appropriately. As SFM technology improves to allow faster high-resolution scanning of samples in fluid, more routine dynamic imaging will become practical.

3. PROTEIN–PHOSPHOLIPID STRUCTURES

3.1. Protein–Lipid Complexes

The primary form of phospholipid aggregate used to study membrane proteins and processes that occur in membranes is the unilamellar liposome or vesicle. Upon hydration, phospholipids spontaneously self-assemble into bilayer structures. Multilamellar liposomes, consisting of concentric bilayer spheres, can be formed by mechanical agitation of the hydrated bilayers. Unilamellar liposomes, consisting of a single bilayer surrounding an aqueous interior, can be formed from multilamellar liposomes by sonication or extrusion through membranes with defined pore size. Alternatively, unilamellar liposomes can be formed by reverse-phase evaporation, in which organic solvent is removed from a reverse-phase emulsion of lipid and solvent in water, or by dissolution of phospholipid in detergent followed by detergent removal. Unilamellar liposomes have nanometer scale to micrometer scale diameters with a bilayer thickness of approximately 5–10 nm. Incorporation of proteins into unilamellar liposomes has been used as a way to study the function of purified membrane proteins in a state that mimics the native membrane-bound state in phospholipid bilayers with a defined and variable lipid composition. Besides their use in basic research, liposomes have found practical applications such as delivery of drugs, DNA, and protein, and in liposome-based immunoassays.

The ability to form supported bilayer membranes has enhanced the study of protein–membrane interactions and membrane protein function through the use of surface analysis techniques such as scanning probe microscopy, surface plasmon resonance, and fluorescence microscopy. Supported bilayers are typically formed over large surface areas, however, some work has been done to form microscale or nanoscale phospholipid bilayers through patterning [188]. Research in our laboratory is being done on the uses of new types of nanoparticle containing a phospholipid bilayer into which membrane proteins can be incorporated. The potential for studying single membrane-bound proteins in nanoscale bilayers and construction of biosensor or optoelectronic chip-based devices based on functional protein–phospholipid complexes now exists.

3.2. Morphology and Function of Native Membranes

Natural cell membranes are complex bilayer structures composed of a variety of phospholipids, fatty acids, cholesterol, and proteins. The weight ratio of protein to lipid varies from 20% to 70% [189], however, it is the phospholipid component that gives the membrane the gross morphology of a closed bilayer. The early simplified Singer and Nicolson model of cell membranes [190] is of a fluid lipid bilayer formed through hydrophobic and hydrophilic interactions of the amphiphilic phospholipids, with integral membrane proteins embedded within the lipid phase to maximize hydrophobic interactions between protein hydrophobic surfaces and the lipid hydrocarbon domain. Peripheral membrane proteins were proposed to interact electrostatically with the polar headgroups. Proteins are free to diffuse laterally in the lipid milieu.

Phospholipids, in many instances, are necessary for the stabilization and function of native membrane-bound proteins [191, 192]. Lipid activation of proteins is believed to arise from solvation of membrane proteins by boundary lipids surrounding the membrane-inserted domain of the protein [191]. The phospholipid components of natural cell membranes include phosphatidylcholine, phosphatidylethanolamine, phosphatidylserine, phosphatidylinositol phosphates, and glycolipids, each differing in charge, acyl chain composition, and physical properties. Such complexity may be attributed in part to the functions of cellular membranes. Membrane functions will be discussed in the following sections, because an understanding of them may be helpful in discerning what is necessary or beneficial for the successful application of phospholipid–protein complexes to the developing areas of nanotechnology.

3.2.1. *Molecular Self-Assembly and Compartmentalization*

Phospholipids are amphiphilic molecules that spontaneously self-assemble into predominantly closed bilayer structures and thus impose the general structure of cellular membranes. For a review of phospholipid bilayer properties, see Cevc and Marsh [193]. Phospholipid bilayers are rather impermeable to ionic molecules and polymers and so have high electrical resistances and can retain cellular contents. Cells and membrane-enclosed organelles, such as the mitochondria, utilize the permeability properties of the membrane as barriers for the storage of energy through formation of electrochemical gradients. Active, passive, and facilitated transport mechanisms, as well as signal transduction across the membrane, are mostly protein-mediated and allow the cell to communicate with and interact with its environment. It must be stressed that proteins functionalize the membrane. Protein components of the cell membrane increase the organizational complexity of the cell membrane through bilayer asymmetry of protein and lipid species and by lateral self-assembly of protein–lipid domains and other supramolecular structures [194, 195]. Advantage can be taken of the self-assembling nature of artificial phospholipid bilayers and their properties in the construction of electrodes, compartmentalization of components in nanoscale regions, and production of self-assembling biomaterials.

3.2.2. Localization and Orientation of Membrane Proteins

The cell membrane presents a surface on which proteins can be concentrated and oriented in the presence of their oriented lipidic or protein targets. This function can be useful or even necessary for building nanoscale supramolecular structures, providing a biological surface on which to compartmentalize and orient proteins, especially those that are membrane-bound in their native states. Fluid bilayers allow diffusion and recognition in protein systems, for example, formation of enzyme–substrate complexes and multiprotein aggregates. During some signal transduction events, proteins actually translocate to or from membranes by various switching mechanisms to interact with functional membrane-bound targets. The function of membrane proteins, in turn, depends on the physical characteristics of the cell membrane, such as fluidity, charge, and chemical composition. These properties are dynamic and depend of the state of the cell. Thus, for exploitation of natural membrane proteins and even artificial membrane proteins, the chemical composition and physical properties of the artificial bilayer must be considered [196].

3.2.3. Source of Chemical Signals

Eukaryotic cellular phospholipids can be broken down by lipolytic enzymes activated through signal transduction mechanisms to give second-messenger signaling molecules such as arachidonic acid, lysophospholipids, diacylglycerol, and inositol triphosphate, leading to a cascade of signal transduction events within the cell. The chemical composition of artificial bilayers can be simplified drastically compared to natural membranes. Fluidity, charge, and stability are the major concerns when artificial membrane systems are constructed.

3.3. Protein–Lipid Interactions

Knowledge of how proteins interact with phospholipid bilayers is critical to understanding the structure and organization of artificial supramolecular assemblies and may lead to ideas for the construction of new nanostructured biomaterials. Membrane proteins were originally placed into two classes defined by ease of extraction from the membrane [190]. Peripheral membrane proteins could be extracted with a high ionic strength or metal chelator, whereas integral proteins required detergent for solubilization. These two classes of membrane proteins will be presented, followed by examples of engineered artificial membrane proteins with desirable functions. Figure 11 depicts several types of membrane-bound proteins.

3.3.1. Integral Membrane Proteins

Typically, integral membrane proteins have one or more membrane-spanning α-helices, each composed of approximately 22 hydrophobic amino acids, to generate a length sufficient to span the hydrophobic core of the bilayer. The carbonyl groups of the phospholipid

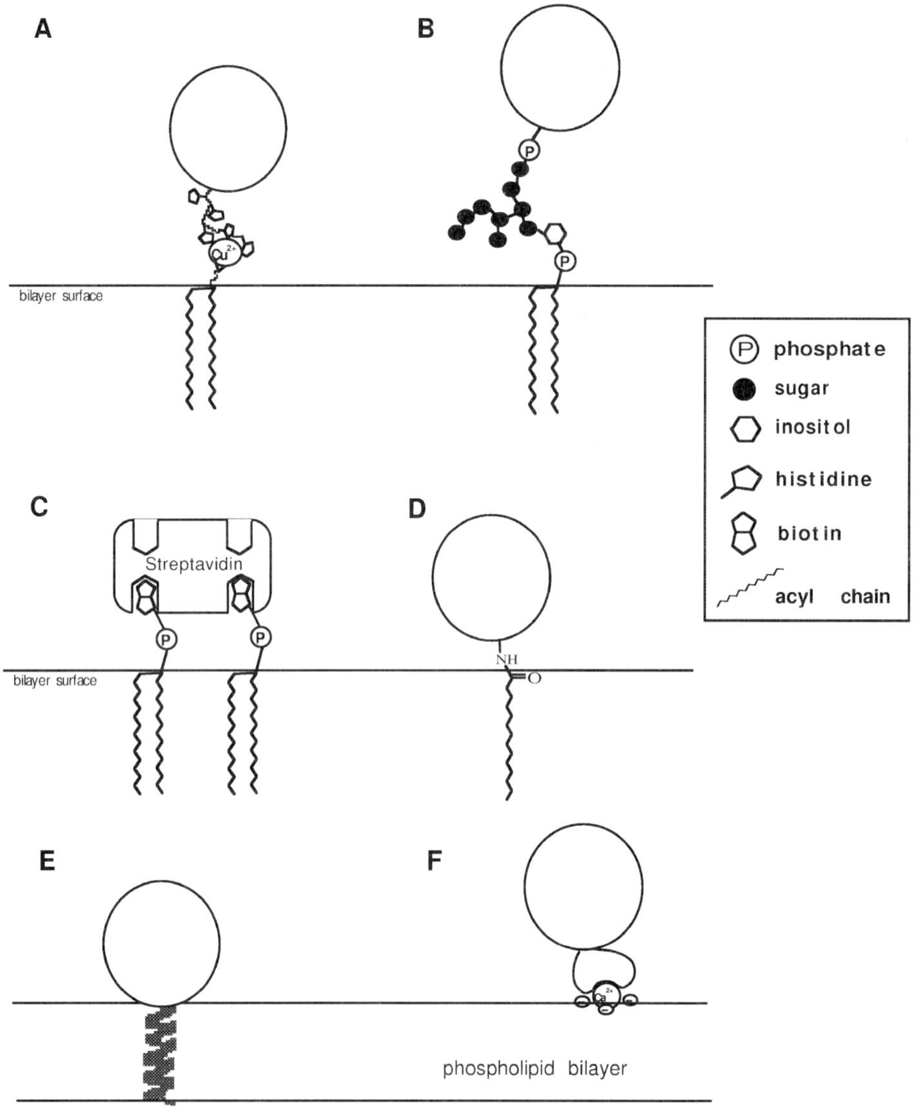

Fig. 11. Representations of membrane-bound proteins. (A) Histidine-tagged protein bound to the bilayer interface by chelation of copper through protein histidine residues and a chelator lipid. (B) GPI-anchored protein. (C) Streptavidin bound to a bilayer surface through biotinylated phospholipid. (D) Protein with a N-linked myristoyl group. (E) Integral membrane protein with a single membrane-spanning helix. (F) Protein with a calcium-binding domain interacting with negatively charged phospholipid headgroups through a calcium-bridging mechanism.

acyl chains mark the boundary of the hydrocarbon region and the polar headgroup region, as determined by neutron and X-ray diffraction of fluid dioleoylphosphatidylcholine bilayers [197]. In addition to the hydrophobic interaction of the hydrophobic membrane-spanning helix, positively charged residues are typically found flanking the hydrophobic domain [198, 199]. An asymmetry in phospholipid distribution exists across cell membranes, with negatively charged phospholipids more often found facing the cytoplasm. The basic residues flanking hydrophobic membrane anchors are also found on the cytoplasmic side of the membrane and probably interact with these acidic phospholipids.

The mattress model of lipid–protein interactions [200, 201] provides a thermodynamic treatment of the phase behavior of mixtures of proteins and phospholipids in terms of hydrophobic mismatch, defined as the difference in hydrophobic length of a particular protein and the hydrophobic thickness of the bilayer. As a result of hydrophobic mismatch, the protein can alter the phase behavior of the phospholipid, causing lateral phase separation, in which the protein partitions preferentially into either the fluid or gel phase. Segregation of protein within a single lipid phase can also arise. Protein segregation within a single phase is theoretically favored in gel-phase lipids (proteins have low solubility in the gel phase) and in thick phospholipid bilayers. Experimental evidence supports these considerations [200]. Hydrophobic mismatch may not exist in vivo, but can be produced in artificial bilayers. A conceptual knowledge of these phase-related phenomena should be used when an appropriate phospholipid component for artificial systems is chosen.

Electron paramagnetic resonance (EPR) measurements of spin-labeled lipids have been a popular method for measuring interactions of integral membrane proteins with phospholipids in artificial biomembranes [202]. Briefly, phospholipids associated with membrane-inserted domains of proteins are more immobilized relative to the bulk phospholipid in the bilayer, giving rise to a characteristic EPR spectrum. The ratio of free to motionally restricted spin label can be quantitated from the EPR spectrum, and the average relative association constant of all sites for spin-labeled lipid versus nonlabeled lipid can be calculated. Measured specificity values of different phospholipids with respect to phosphatidylcholine for several proteins range from 1 (no specificity with respect to phosphatidylcholine) to about 10 (preference by a factor of 10 over phosphatidylcholine), with a trend toward a preference for negative lipid species [202].

The assumption that integral membrane proteins are irreversibly associated with a bilayer is not always true. The ability of proteins to dissociate from artificial phospholipid matrices could have the consequence of loss of protein or an inability to image proteins by scanning probe microscopy due to the dynamics of proteins leaving and inserting into the interface on short time scales. Cytochrome b_5, cytochrome *c* reductase, NADPH-cytochrome P-450 reductase, and cytochrome P-450$_{scc}$, for example, are believed to exchange between membranes through the aqueous phase [203–206]. For cytochrome b_5 and NADPH-cytochrome P-450 reductase, the ability to exchange between membranes is dependent on the method of reconstitution of the protein in phospholipid membranes.

3.3.2. Peripheral Membrane Proteins

Peripheral membrane proteins were once believed to interact solely through electrostatic interaction with membranes. However, many proteins are now known to have post-translational modifications, such as an amide-linked myristate on an N-terminal glycine, palmitic acid linked to cysteines via thioester linkages, thioether-linked isoprenyl groups attached to C-terminal cysteine residues, and glycosyl-phosphatidylinositol conjugation of the C-terminal carboxyl group. N-Myristoylation [207, 208], acylation [209], prenylation [210, 211], and glycosyl-phosphatidylinositol anchors [212] have been reviewed. These types of protein modifications present a powerful tool for easily attaching proteins to membranes that have been genetically engineered to have signal sequences for modification either in vitro by purified enzymes or in cultured eukaryotic cells or bacteria. Proteins associated through lipid anchors are most likely to be highly mobile in supported bilayers, in contrast to many membrane-spanning proteins. For example, Groves and coworkers [213] have produced glycosyl-phosphatidylinositol anchored proteins, incorporated them into supported bilayers, and verified high lateral mobility.

Three common protein domains that bind to phospholipids are presented briefly. These include annexins, which bind to negatively charged membranes in a calcium-dependent manner; the so-called C2 domain of protein kinase C and other diverse proteins that bind to negatively charged membranes in the presence of calcium; and the pleckstrin homology domain found in a diverse group of proteins that have a specificity for phosphatidylinositol triphosphate. It should be noted that many proteins with these types of domains are involved in cellular signal transduction and can bind to membranes reversibly: the domains act as switches sensitive to chemical signals.

Annexin V requires calcium and negatively charged phospholipids such as phosphatidylserine for membrane binding and has four similar domains, of which at least three may be involved in calcium-dependent binding. The structures from 3-D crystals as well as 2-D crystal structures on phospholipid layers are known [214, 215], and annexin V is one of the few peripheral membrane proteins whose membrane-bound orientation is known decisively. The specific interactions between Ca^{2+} and glycerophosphoserine and glycerophosphoethanolamine have been determined by determining the co-crystal of annexin V with these phospholipid mimics [216]. The complexed calcium ions interact with phosphate oxygens and the carboxylate moiety of phosphoserine, while other hydrogen-bonding interactions occur between the amines of phosphoserine and phosphoethanolamine and the protein.

The C2 domain is a calcium-dependent membrane-binding domain of about 100 amino acids originally found in the classic protein kinase C isoforms (for a review, see [217]). This domain is found in proteins involved in vesicular transport, GTPase regulation, lipid modification, and phosphorylation. Crystal structures of the domains found in synaptotagmin 1 and phosphatidylinositol-specific phospholipase C $\delta 1$ have been determined. The calcium is believed to bridge the protein and phospholipid headgroups in membranes, as is the case for annexin V as described above. Different C2 domains have different specificities: some bind to phosphatidylcholine liposomes, some require phosphatidylserine or phosphatidylinositol for binding, and some lack acidic residues needed to complex the calcium ion and show no calcium-dependent binding to membranes. As far as is known, the C2 domain has only been found in membrane-binding proteins.

The 100 amino acid pleckstrin homology domain (for a review, see Shaw [218]) is represented by a family of more than 100 genes. The domain binds with high affinity to membranes containing phosphatidylinositol triphosphate. NMR solution structures [219–221] and crystal structures of pleckstrin domains [222, 223] have been reported. The mode of interaction with phosphatidylinositol triphosphate is not known; however, it is believed that binding occurs in a region of loops present on the domain whose NMR spectra are perturbed with the binding of inositol triphosphate.

In summary, proteins can have direct ionic interactions with negatively charged lipids or use a bridging mechanism such as annexin and protein kinase C through a calcium ion. Many peripheral membrane proteins require negatively charged headgroups for membrane binding and function; these include protein kinase C, cytochrome *c*, annexin, phospholipases, and coagulation factors. Specificity for headgroups also exists, as protein kinase C appears to have a stereospecific interaction with phosphatidylserine, and proteins containing the pleckstrin homology domain have a specificity for phosphatidylinositol. Of the three types of domains mentioned above, the pleckstrin and C2 domains are relatively small domains found in a variety of different proteins, fold into active structures independently, and are therefore likely to be generally useful for constructing genetically engineered membrane-binding proteins.

3.4. Nonnative Interactions

3.4.1. Chelator Lipids

Affinity metal chelation chromatography of genetically engineered proteins containing six histidine residues at the polypeptide terminus has become a popular way of easily purifying proteins of interest. The incorporated histidines can complex metal ions held on chromatographic supports by chelating groups. Recently, researchers have begun to investigate binding of histidine-tagged proteins to metal chelator lipids present in monolayers. Nitrilotriacetic acid–dioctadecylamine (NTA-DODA) [224–226] and lipid with an iminodiacetic acid headgroup (IDA-lipid) [227] have been used to attach proteins to phospholipid monolayers and bilayers. Protein binding to chelator lipid is reversible upon removal of metal ion. The IDA-lipid was developed for the purpose of attaching proteins and the formation of template-directed patterning of metal ions in lipid structures. The IDA-lipid was used to bind horse heart myoglobin, which contains at least four histidines that can bind Cu^{2+}IDA complex, to monolayers of pure IDA-lipid and to monolayers and liposomes containing a small percentage of IDA-lipid. Fluorescent peptides containing a 5-His tag were shown to specifically associate with phosphatidylcholine/NTA-DODA monolayers in the presence of Ni^{2+} and to form arrays of NTA-DODA domains approximately 20 μm across containing bound peptide surrounded by a fluid domain of phosphatidylcholine [224]. It was suggested that such domains could be used to study molecular recognition processes in membranes and for the 2-D crystallization of proteins. NTA-DODA has also been used to immobilize a DNA binding protein containing a C-terminal 6-His tag. Preservation of protein function was shown through a specificity for DNA binding to condensed domains of the chelator lipid in the presence of Ni^{2+}. Lipid bilayers containing NTA-DODA have also been formed on solid substrate by vesicle fusion and Langmuir–Blodgett techniques, and a monolayer containing arrays of chelator lipid domains could be transferred to a solid support and were shown to bind a histidine tagged peptide [226].

3.4.2. Biotin–Streptavidin Interactions

The strong specific tetravalent binding of biotin by streptavidin has been exploited to form protein–phospholipid complexes. Streptavidin can form 2-D crystals when bound to a monolayer of biotinylated lipids [228]. An interesting concept is the use of such a well-ordered self-assembling system as a template for the self-assembly of more complex structures. Darst and co-workers [228] proposed using a crystalline monolayer of streptavidin as a self-assembling template for 2-D crystallization of other proteins. By linking a monolayer of streptavidin bound to a monolayer of biotinyolated lipid via a biotinylated helical peptide spacer, a second monolayer of streptavidin can be formed that also has crystalline form [229]. However, a problem lay in the instability of the helical structure of the peptide linker, such that both ends of the peptide could bind to the first monolayer, resulting in incomplete second monolayer formation, as evidenced by surface plasmon resonance spectroscopy of the solid-supported structure. This problem points to the fact that although streptavidin is a good adapter for constructing membrane-bound structures, in many instances a bifunctional adapter would be even more versatile for the formation of self-assembling structures.

3.5. Reconstitution of Integral Membrane Proteins

The ability to reconstitute integral membrane proteins in defined structures is one of the most important concerns in studying protein–lipid complexes. Because reconstitution of peripheral membrane proteins is a relatively straightforward matter, only reconstitution of integral membrane proteins will be covered here. Reconstitution of integral membrane proteins can be acheived by four general strategies [230]: from mixtures in organic solvent, by mechanical means, by direct incorporation into preformed liposomes, and by detergent-mediated reconstitution. The first method involves the use of organic solvent (such as ether or pentane) for dissolution of protein and lipid followed by solvent removal and rehydration. Another organic solvent-mediated reconstitution method is reverse-phase evaporation in which solvent is removed from an emulsion of lipid, protein, and solvent in water [231]. The proteins, however, must be stable in the presence of organic solvent. Mechanical incorporation of proteins can be achieved by sonication of a suspension of protein and lipid. Direct incorporation of lipid-free protein into liposomes has the advantage of asymmetric incorporation of protein. The most widely used, gentlest, and probably most successful reconstitution method is to solubilize both protein and phospholipid in the presence of detergent, followed by detergent removal by dialysis, gel filtration, or adsorption of detergent by lipophilic beads (Biobeads SM2 or Amberlite XAD-2). The choice of method for detergent removal depends upon the detergent used for reconstitution. Only detergents with high critical micelle concentration (cmc), such as cholate, octylglucoside, and 3-[(3-cholamidopropyl)dimethylammonio]-1-propanesulfonate (CHAPS), can be removed by dialysis or gel filtration. The use of adsorptive beads removes dialyzable detergents as well as detergents that have a low cmc, such as Triton X-100.

A systematic approach to the study of reconstitution of membrane proteins in the presence of detergent has been undertaken [231–233] and is based upon the assumption that the process of solubilization of liposomes by detergent is the reverse of the process of formation of liposomes from mixed micelles of detergent and phospholipid. Three stages of liposome solubilization have been proposed. In the first stage, detergent partitions into liposomal bilayers with resultant destabilization. Upon saturation, increasing detergent concentration induces the formation of mixed micelles in coexistence with detergent-saturated liposomes. Further increases result in complete solubilization. The efficiency of incorporation of active enzymes at the different stages of solubilization was found to depend on the type of detergent used and the properties of the protein. Incorporation of proteins into detergent-destabilized liposomes was found to result in highly asymmetric incorporation into liposomal bilayers (70–95%).

Supported bilayer systems are useful for the study of membrane protein structures by surface-sensitive techniques. The Langmuir–Blodgett technique of supported bilayer formation [234] is not well suited to incorporation of protein because the protein must withstand incorporation into a monolayer at the air–water interface and subsequent transfer onto a solid support, and requires substantial quantities of the protein because of the large subphase of Langmuir–Blodgett troughs. Proteoliposomes formed by the methods that have been described can be used for the formation of supported bilayers containing protein. Fusion of proteoliposomes directly onto a hydrophilic surface has been used to form supported bilayers, but complications due to the bidirectional orientation of proteins in the liposome bilayers and interaction of lipids and proteins with the hydrophilic support are cause for concern [235, 236]. Liposome fusion to a supported phospholipid monolayer has been found to result in bilayers with high lateral lipid mobility, as determined by fluorescence recovery after photobleaching (FRAP) and total internal reflection fluorescence microscopy [236, 237]. A proposed mechanism for fusion of liposomes with the monolayer is compatible with formation of supported bilayers having hydrophilic domains of proteins extending from only the outside leaflet of the supported bilayer [236]. Bacterial membrane-insertion signal peptides [238], cytochrome b_5 [236], phospholamban [239], bacteriorhodopsin, acetylcholinesterase, cytochrome oxidase, nicotinic acetylcholine receptor [240], and ATPases CF_0CF_1 and EF_0F_1 have been successfully incorporated using this method.

The lateral mobility of integral membrane proteins on supported bilayers has been the subject of much research [241–244]. In most studies it is assumed that a water layer of only approximately 10 Å separates the phospholipid bilayer from the hydrophilic support [243]. Integral membrane proteins with hydrophilic regions extending from the membrane could therefore significantly interact with the support. It has been shown that lipids within the bilayer may also couple to the support, resulting in lateral immobilization. The use of thin hydrophilic polymer films has been suggested as a way to overcome this problem [241–244]. In one study, polyacrylamide films 30–40 μm thick were formed on glass by silanizing with 3-methacryl-oxypropyl-trimethoxysilane followed by polymerizing acrylamide monomer in the presence of bisacrylamide cross-linker. Deposition of DMPC bilayers was performed by the Langmuir–Blodgett technique. Apparently homogeneous bilayers could be formed, although the bilayers allowed penetration of carboxyfluorescein, a membrane-impermeable dye, into the gel layer, indicating holes or defects in the membrane structure. FRAP measurements revealed that fluorescent lipids in such bilayers show nearly complete fluorescence recovery with diffusion coefficients in the range of 6–10 $\mu m^2/s$ at 30 °C and exhibit a thermal transition at 22.5 °C, which is similar to the behavior of multibilayer DMPC systems. Diffusion measurements of a 20-amino acid lipopeptide showed greater recoveries and diffusion coefficients with decreasing cross-linker concentration, indicating that immobilization of peptide and frictional coupling to the gel layer can be modulated by altering the gel composition. In another scheme, a reactive polymer cushion was used to support and covalently attach phosphatidylcholine bilayers through reactive lipid groups present in the bilayer. Covalent attachment to 20 mol% *N*-succinimidomyristidic ester present in the bilayer resulted in apparently homogeneous bilayers. However, the diffusion coefficient for fluorescently labeled lipids was considerably lower than that of noncovalently tethered bilayers, possibly because of the formation of fixed diffusion barriers to the dye lipid upon coupling of the thiol lipid to the support [244]. In yet another scheme, thin dry dextran films with controlled thicknesses in the range of 2–30 Å were covalently attached to oxidized silicon wafers [241]. DMPC bilayers deposited on hydrated films by the Langmuir–Blodgett technique showed budded liposomes, whereas incorporation of 20 mol% cholesterol, which tends to stabilize bilayers, resulted in apparently homogeneous membranes with high fluorescence recovery and a diffusion coefficient of 2.8 $\mu m^2/s$ at 30 °C (similar to that of DMPC/cholesterol liposomes, 3.2 $\mu m^2/s$).

3.6. Practical Applications of Protein–Lipid Complexes

3.6.1. Biomaterials/Templates

An interesting use of the biotin–streptavidin interaction has been explored for the formation of tubular biomaterials [245]. The biotinylated lipid DODA-EO_2-biotin forms hollow tubules 27 nm in diameter. The addition of streptavidin results in the formation of helical arrays of streptavidin on the surface of the tubules with a crystalline order up to 1.5 nm. Addition of liposomes containing biotinylated phospholipid and monosialoganglioside G_{M1} resulted in the formation of liposome-wrapped tubules to which cholera toxin could bind through ligation with G_{M1}. These experiments are examples of how complex nanostructures may be manufactured through molecular self-assembly and molecular recognition processes. A suggested use for the tubular structure was helical 2-D crystallization of proteins, which may have advantages over planar 2-D crystals for structural determination [245]. The tubules could act as a template for directing the shape of the 2-D crystal through specific ligand interaction at the tubule surface and for building more complex functional structures through the use of biotin–streptavidin interaction with functional biotinylated molecules.

Another biomaterial with implications for the development of supramolecular assemblies also takes advantage of the specific molecular recognition of the biotin–streptavidin system [246, 247]. In this scheme, biotin-bearing liposomes can be aggregated by the tetravalent streptavidin molecule to form higher order supramolecular structures composed of liposomes. Proposed uses of such structures were the investigation of cellular recognition and development of artificial tissues and other biomaterials.

Walker and co-workers [248] produced vesicle-encapsulated liposomes termed "vesosomes." Phosphatidylserine liposomes containing a small amount of biotinylated phospholipid were transformed into cochleate cylinders by the addition of calcium. The liposomes to be encapsulated were cross-linked through the use of biotinylated lipid and streptavidin to form vesicle aggregates. The aggregates could be sized by extrusion through track-etched filters with a 1-μm pore size. The cylinders and vesicular aggregates were complexed through streptavidin linkers by simply mixing the two. Upon depletion of calcium, vesosomes were formed by unrolling the cylinder to make a vesicle 500–5000 nm in diameter containing aggregates of cross-linked 100-nm-diameter liposomes. These authors proposed that the vesosomes could be used for functional compartmentalization of components within the interior of the vesosome with application in drug delivery.

3.6.2. Biosensors

The development of phospholipid membrane-based biosensors has recently begun to show promise. Examples of biosensors are given in Table I. These include ion channels, a membrane-bound receptor, membrane-bound enzymes (a native membrane protein, acetylcholinesterase, and streptavidin-coupled xanthine and glucose oxidase), and soluble enzyme (glucose oxidase) encapsulated within liposomes attached to a solid support. Ions are detected by measuring current between two compartments separated by the ion-channel-containing bilayer. Single-channel conductances can be measured, however, the aqueous compartments must be large to support current flow. The device of Nikolelis and Siontorou [249] used a bilayer supported on a silver electrode and had no obvious compartment on the inner side of the membrane. Current flow was believed to be due to reaction of the ammonium ion with the silver electrode. The principle of detection of analyte by glucose and xanthine oxidase is well known and involves the measurement of peroxide formed during the enzymatic reaction. The photometric method of detection used by Klee and co-workers [250] to detect specific binding of fluorescein-labeled streptavidin to a biotinylated membrane on a quartz fiber was an experimental test of the "evanescent wave sensor" and is an example of a promising detection technique. Kinnear and Monbouguette [251] used a gold electrode modified with alkyl mercaptan and short-chain disulfides to immobilize fructose dehydrogenase along with phospholipid and coenzyme Q by detergent dialysis for the amperometric measurement of fructose. Three of the more notable devices were constructed by Krull and co-workers [252], Cornell and co-workers [253], and Eray and co-workers [254, 255], which will be described next in more detail.

Krull and co-workers [252] could detect fluorescence enhancement of a fluorescent phospholipid present at 1–2 mol% in phospholipid liposomes containing acetylcholine receptor upon binding of carbamylcholine or α-bungarotoxin to the receptor. The enhancement appeared to be due to specific binding, because it was not seen with denatured receptor. Moreover, binding of one ligand precluded fluorescence enhancement by the

Table I. Examples of Phospholipid Membrane Biosensors

Analyte	Receptor	Detection	Reference
K^+, Rb^+, Cs^+, Na^+	Valinomycin	Amperometric	[272]
Glucose	Glucose oxidase	Amperometric	[273, 274]
Acetylcholine, carbamylcholine, α-bungarotoxin	Acetylcholinesterase, acetylcholine receptor	Photometric	[252]
Ammonium ion	Gramicidin D	Amperometric	[249]
Antigens	Modified nicotinic acetylcholine receptor	Amperometric	[254, 255]
Fructose	Fructose dehydrogenase	Amperometric	[251]
Antigens	Modified gramicidin	Amperometric	[253]
Streptavidin	Biotinylated lipids	Photometric	[250]
Glucose	Glucose oxidase	Amperometric	[275, 276]
Xanthine	Xanthine oxidase		

other ligand. Acetylcholinesterase incorporated into fluorescent liposomes showed time-dependent fluorescence changes upon the addition of the substrate acetylcholine. The authors ascribed the fluorescence change to a localized pH change in the vicinity of the liposome membrane leading to fluorescence enhancement. Fluorescence microscopy images of acetylcholinesterase monolayers suggested pH-dependent formation of domains that excluded the fluorescent phospholipid. The mechanism of fluorescence enhancement, however, is unclear. These studies indicate that it may be possible to use the phase behavior of membranes for the detection of ligand binding events. In the one case, acetylcholine receptor would represent a protein that alters the structure of the phospholipid bilayer upon ligand binding. The other case could be generalized to enzymes whose products can alter the bilayer structure, as exemplified by esterases whose carboxylic acid product dissociates to release hydrogen ions.

Eray and co-workers [254, 255] have developed a sensing concept based on the acetylcholine receptor functionalized with two bispecific antibodies. The receptor was incorporated into a phospholipid bilayer spanning a 40-μm polyimide aperture on a silicon wafer. After binding of bispecific antibody, channels remained capable of opening and closing. Antigen, in this case IgG, caused the closing of channels, presumably by cross-linking the two bispecific antibodies. Such an approach may be generalized to other antigens but requires that the bispecific antibodies can both bind to the antigen at the same time. The presence of a cross-linked analyte used in a competitive immunoassay format has been suggested to overcome this difficulty [256]. It might be envisioned that ion channels genetically engineered with a ligand-binding site or an antibody attachment site in the proper orientation could be used in a similar manner.

Cornell and co-workers [253] have developed a more complex biosensor device, the transduction element of which can be generalized to almost any analyte. The device is based on gramicidin, a protein that dimerizes to form ion channels. The ion channel is formed when two monomers present in opposing leafs of the bilayer line up to make a channel through the bilayer. A membrane-bound gramicidin monomer in the leaf of the bilayer nearest to a gold support was fixed to the gold by a spacer. Freely diffusing biotinylated gramicidin monomers present in the opposite leaflet of the bilayer were coupled to a biotinylated antibody fragment containing a single antigen-binding site (Fab fragment) via streptavidin. Another biotinylated Fab fragment in the outer leaflet is fixed in position (via a streptavidin linker) by a membrane-spanning biotinylated phospholipid that is coupled to the gold surface. The bilayer itself is fixed to the gold surface through spacers ($\sim$4 nm in length) attached to lipids in the bilayer, creating a reservoir between the bilayer and the gold surface. Diffusing gramicidin/Fab monomers in the outer leaflet line up with partner gramicidin monomers fixed in the inner leaflet to form conducting channels. When antigen with two identical epitopes are bound by both a fixed Fab and a freely diffusing Fab/gramicidin monomer, there is a loss of ability to form a complete channel because gramicidin monomers in both leaflets are fixed in position. In another scheme, a competitive-type assay is used. Here, antigen is covalently linked to the freely diffusing gramicidin monomer and hence is bound by the gold-coupled Fab fragment, resulting in a loss of ability to form complete channels. Free antigen (analyte) can compete with the gramicidin-coupled antigen, resulting in release of the gramicidin-coupled antigen from the Fab anchor and an increase in membrane conductance upon channel formation. This ingenious device also addresses the problem of membrane stability, as it includes membrane-spanning and diphytanoyl-containing lipids that stabilize the membrane and is reported to be stable for months.

3.6.3. Liposome-Based Immunoassays

Liposome-based immunoassays have been extensively reviewed (for example, see [257].) In general, the roles of liposome in these assays are as carriers of labels that can be detected after binding of antigen. The liposome provides a vessel ~100 nm in diameter. Labels are usually encapsulated within the aqueous interior and are fluorescent dyes or enzymes typically used in enzyme-linked immunoassays (alkaline phosphatase or horseradish peroxidase). Antibodies are conjugated to the liposome surface by one of a variety of techniques. Apart from this, most techniques do not differ drastically from the usual ELISA (enzyme-linked immunosorbent assay).

3.6.4. Drug Delivery

The use of engineered nanoparticles for the delivery of therapeutic drugs, proteins, and DNA has been a major focus of research [258]. For many useful applications, the particle must function in ways other than as a simple carrier of therapeutic agents. The particle might have to recognize a specific tissue or cell type, fuse with or be taken up by the cell (akin to viruses), and transfer of its contents directly to the correct cellular compartment (e.g., DNA delivery to the nucleus for gene therapy). An ideal liposome-based complex would be essentially a cell-like device with multiple functionalities needed to accomplish its task.

An obstacle to the use of liposomes for drug delivery is short circulation time in the bloodstream. Particles become coated with plasma constituents, a process called *opsonization*, followed by phagocytosis by macrophages. Uptake can be minimized by using nanoparticles coated with hydrophilic polymer [258]. The "stealth liposome" is an example of such a coated particle designed for long circulation times in the bloodstream. For a review of the stealth liposome see Lasic and Needham [259]. The use of polymer coatings raises the question of obstruction of recognition by targeted liposomes. To overcome the problem of steric hindrance of association of the targeted antibody with its target caused by a polyethylene glycol coating on liposomes, Maruyama and co-workers [260] synthesized a polyethylene glycol lipid with a terminal carboxylic acid for the attachment of antibody. The lipid monosialoganglioside G_{M1} also increases circulation times. Peptides bound to liposomes containing monosialoganglioside G_{M1} lipid were shown to have a prolonged peptide circulation time [261].

Targeting of liposomes to specific cells, such as neoplasias, can be achieved through molecular recognition at the surface of the cell. A typical scheme is to attach an antibody raised against a specific cellular epitope. Progress is also being made in targeting to cells through specific receptors. Grimaldi and co-workers [262] have demonstrated that Epstein–Barr virus envelopes could be used to deliver a fluorescent dye to hepatocarcinoma cells in vitro, presumably through receptor recognition, and that proteoliposomes bearing reconstituted band 3 protein from red blood cells increase phagocytosis by macrophages, which remove senescent red blood cells. Bloemen and co-workers [263] have proposed adhesion molecules as targets for targeting immunoliposomes to sites of inflammation, because the expression of some adhesion molecules is enhanced in areas of inflammation. They showed specific targeting of immunoliposomes to cultured cells that express intercellular adhesion molecule 1 (ICAM-1). Heeremans and co-workers [264] have coupled gluplasminogen to liposomes as a potential way of targeting liposomes to fibrin clots to deliver thrombolytic agents. Lee and Low [265] demonstrated specific folate receptor-mediated uptake of folic acid-conjugated liposomes by cultured cells. Lundberg and co-workers [266] conjugated apolipoprotein B to liposomes in an attempt to target cells through low-density lipoprotein receptors, which are present at abnormally high concentrations in some cancer cell lines.

Once liposomes are targeted to cells, they must exert some beneficial action. One scheme is a modification of antibody-directed enzyme prodrug therapy, in which liposomes containing an enzyme are targeted to tumor cells through a conjugated antibody [267]. This allows large quantities of enzyme to be delivered, compared to antibody–enzyme conjugates. The enzyme, for example β-glucuronidase, coupled to the surface of the liposome, activates the prodrug epirubicin-glucuronide, generating the drug in the vicinity of tumors.

The same group has encapsulated diphtheria toxin, a potential anticancer agent, in immunoliposomes with the idea that encapsulation would protect the toxin from inactivation by antibodies that are induced in patients treated with the toxin [268]. Similarly, McIntosh and Heath [269] used liposomes to deliver another inhibitor of protein synthesis, gelonin, to cells in culture. Immunomagnetoliposomes [270] have been proposed for cancer therapy. Liposomes 80 nm in diameter containing 10-nm magnetite particles were formed by sonication and coupled to a protooncogene antibody. The magnetoliposomes were shown to be taken up by cells in culture. The rationale for such a system is that tumor cells would take up the particles, and a high-frequency magnetic field could be used to generate heat to destroy the tumor.

Liposome-fusion methods must be developed to efficiently transfer the therapeutic agent into cells. Transfer of proteins from proteoliposomes to cell membranes is usually performed (in vitro) by simple incubation of the cells with liposomes. To study the mechanism of complement attack of Gram-negative bacteria, Tomlinson and co-workers [271] have developed a system for the transfer of preformed terminal attack complexes (C5b-9), which are transmembrane channels responsible for killing, from liposomes into the envelope of bacteria. They found that they could deliver lipidic components to bacterial membranes, liposome-encapsulated horseradish peroxidase to the periplasmic space, and reconstituted liposome-associated C5b-9 complexes to the outer membrane. Fusion of liposomes with bacteria was mediated by high calcium concentrations. More active fusion techniques are of great interest, such as the use of peptides or lipids that promote fusion [258]. Development of suitable triggers of fusion are also needed to increase the specificity of targeting. Natural vesicle fusion machinery, such as those that function in synaptic transmission, are currently being elucidated. Progress in this area of research could lead to better, more complex liposome-based devices for the delivery of therapeutic agents.

Phospholipid membranes in the form of liposomes and supported bilayers provide a way of immobilizing membrane proteins that are either natural or genetically engineered. They also provide a biocompatible surface containing phospholipid, which in many cases is needed for the stability and function of membrane proteins. Fluid membranes provide a 2-D surface in which proteins and their targets can be at least partially oriented, diffuse, and interact. The chemical composition and physical properties affect the way in which proteins interact with the bilayer surface and with each other at the surface. Methods exist for the reconstitution of membrane proteins and formation of supported bilayers containing functional protein systems of interest. The use of such supported bilayer systems and advances in the field of scanning probe microscopy are making the study of single membrane-bound protein molecules a possibility. Much work remains to be done in the area of constructing useful nanostructured biomaterials and devices composed of lipids and protein, as most of the work has yielded mainly proof of principle in self-assembly of nanoscale supramolecular structures, and in detecting analytes using crude large-scale devices such as electrodes. Last, the role of membranes and membrane-bound assemblies in nature can provide examples of the best uses of artificial phospholipid–protein complexes as opposed to simply attaching proteins to a solid support.

4. SURFACE-IMMOBILIZED PROTEIN NANOSTRUCTURES

The ability of proteins to undergo spectroscopic, electronic, and conformational changes in response to subtle changes in their environments may make them interesting as possible components of microcircuits and sensors. However, before the proteins can be incorporated into functional devices, methods for manufacturing nanoscale, defect-free assemblies must be developed. The goal of this section is to describe the application of surface-immobilized protein nanostructures, the methods currently available for making and characterizeing these nanostructures, and the challenges that remain.

The applications of surface-immobilized proteins are as diverse as the proteins immobilized. Surface immobilization of proteins was recognized in the first decade of the twentieth century as a powerful tool for reclaiming expensive industrial catalysts [277]. In the last two decades, surface-immobilized proteins have become extremely important as affinity chromatography resins [278, 279], and research on immobilization techniques for catalysis and chromatography has led to the development of a variety of effective immobilization techniques, including histidine tags and the chemical cross-linking of protein A and G to immobilize antibodies. In addition to the commercial applications, these techniques can be used to study protein–protein interactions, nonspecific adsorption, and the interactions between proteins and their substrates or ligands [280, 281]. Protein nanostructures are also being used to develop sensors for a variety of chemical species, including glucose, antibodies, and cocaine [166, 176, 282–284].

Several factors must be considered when protein nanostructures are being constructed, including the orientation of the protein, surface patterning, and prevention of nonspecifc interactions. There are currently a handful of techniques available for immobilizing proteins in an oriented fashion, many of which are applicable to a wide variety of substrates. The development of photolithographic techniques has led to the development of several surface patterning methods, and biomaterial researchers are making strides toward preventing nonspecific adsorption. However, universal methods of preventing nonspecific adsorption and means of simultaneously patterning and orienting immobilized proteins have still not been developed. Furthermore, many questions remain about the chemical and physical effects of both the substrates and other components of the nanostructures on the activity of the immobilized proteins. The challenge in the next decade will be to integrate the knowledge already gained and develop the control needed to produce defect-free protein nanostructures.

4.1. Oriented Protein Arrays

4.1.1. The Effect of Orientation on the Activity of Immobilized Proteins

Research into protein immobilization methods did not truly boom until the development of affinity chromatography. Researchers using antibodies for affinity purification realized that if antibodies are bound with the antigen-binding sites in contact with the surface, the antigen will be sterically excluded and the antibodies will be inactive (Fig. 12). On the other hand, antibodies that are bound with the antigen-binding sites exposed to solution will readily bind antigens. Despite the relationship between function and orientation, the majority of immobilization techniques used today, including nonspecific adsorption and chemical cross-linking of carboxyl and amino acid side chains, produce randomly oriented samples with limited activity. A variety of methods do exist, however, that will produce oriented arrays of proteins, including the use of protein A or G, the Langmuir–Blodgett

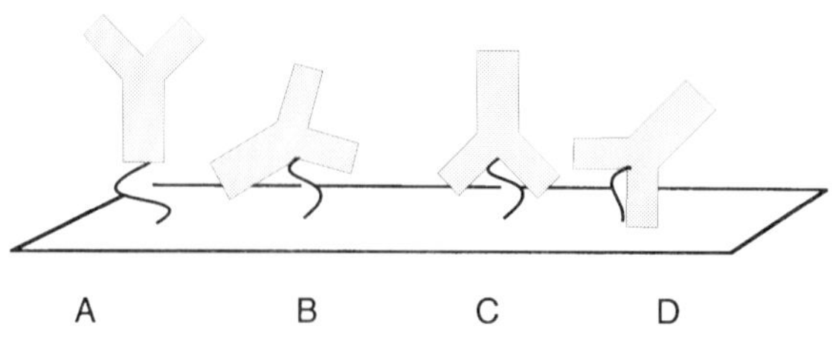

Fig. 12. The effect of random orientation on the ability of an antibody to bind antigen. (A) Full functional antibody oriented with the antigen binding site exposed to solution. (B and D) Semifunctional antibody with antigen binding sites partially buried, occluding antigen. (C) Inactive antibody with buried antigen binding sites.

technique, and the use of self-assembled monolayers (SAMs). These methods will be discussed individually below.

The importance of orientation in the activity of immobilized proteins has been demonstrated by several groups. Attili and Suleiman showed that immunosensors with oriented antibodies were more sensitive, more stable, and more reliable than sensors built with randomly oriented protein [282]. The standard deviation in the response of oriented IgG sensors was only 10% as compared to nearly 40% for sensors with randomly oriented protein. Oriented antibody fab fragments immobilized on derivitized silica were found to have 2.7 times the antigen-binding activity of antibodies immobilized by glutaraldehyde cross-linking, a method known to result in nonoriented samples [176]. And, as will be discussed later, cytochrome b_5 molecules that were specifically oriented to a chromatographic resin with their cytochrome *c*-binding site exposed to solution were found to bind cytochrome *c* more tightly than those oriented with an occluded binding site.

4.1.2. Immobilization of IgG by Proteins A and G

One of the first methods developed to immobilize proteins in an oriented fashion was the use of proteins A and G to immobilize antibodies. In 1966, Forsgren and Sjoquist discovered that the *Staphylococcus aureus* protein, known as protein A, binds to the Fc domain of IgG. Each protein A molecule contains four IgG-binding sites. Regardless of the orientation of protein A on a surface, two or more IgG-binding sites remain exposed to solution and available for coupling. Because the orientation of protein A has little effect on its ability to bind IgG, it can be attached to a surface by any immobilization technique. This allows for a great deal of flexibility in the choice of substrates and immobilization chemistries. Protein A has been used to couple IgG to silanized quartz, silicon, and gold through both physisorpotion and chemical cross-linking of the protein A layer. Despite the fact that the protein A sublayer is randomly oriented, the IgG will be partially oriented by its interaction with the protein A because only the IgG Fc domain can bind to the protein A layer. Furthermore, the antibody can only bind to those protein A sites that are exposed to solution (Fig. 13). To increase the stability of the system and prevent loss of antibody at high ionic strength or low pH, the IgG molecules can be chemically cross-linked to the protein A layer [285].

Protein G, like protein A, binds to the Fc domain of IgG and can be used to immobilize protein. Because protein G binds more IgG isoforms than protein A and does not cross-react with other immunoglobins such as IgM, it is both more selective and more universal [175]. Unexpectedly, however, a study comparing protein A immobilization to protein G showed that antibodies oriented with protein A had a greater antigen-binding enhancement compared with that of nonoriented samples than those oriented with protein G [286]. This result is curious because protein G immobilizes more IgG per square centimeter than protein A. The explanation for the anomalous results is that the orientation of IgG on protein G monolayers is not optimal for antigen binding, demonstrating that the ability to orient antibodies with their antigen-binding site exposed to solution is not necessarily sufficient to maximize activity. Another possibility that will be discussed later is that at high surface coverages, the steric interactions between neighboring proteins may decrease binding activity.

An application of the protein A and protein G immobilization technique was the development of a piezoelectric sensor for a cocaine metabolite [282]. In this device, either protein A or protein G was physisorbed or chemically cross-linked to the gold electrodes of a piezoelectric crystal. A solution of anti-benzoylecgonine IgG was added to the crystal and allowed to dry. After rinsing to remove excess antibody, the crystal was exposed to a solution of benzoylecgonine, and the shift in resonant frequency was measured. When the sensitivity and reliability of sensors made with different cross-linking methods were compared, the protein A/G immobilization method was found to be superior to physisorption and cross-linking with polyethyleneimine of the IgG directly to the electrode.

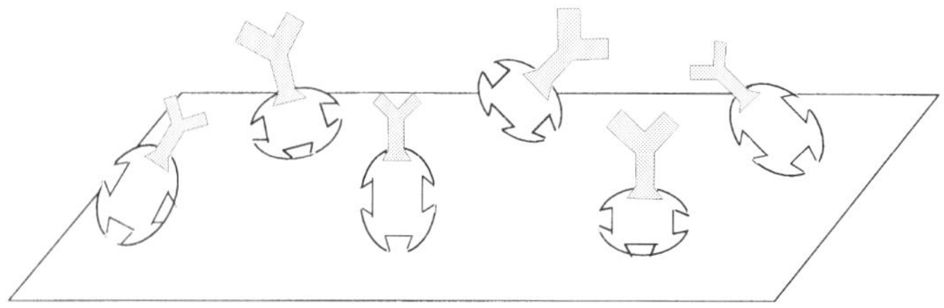

Fig. 13. The effect of coupling IgG to a surface through protein A or protein G. The IgG molecules are semioriented, and each has its antigen binding site exposed to solution.

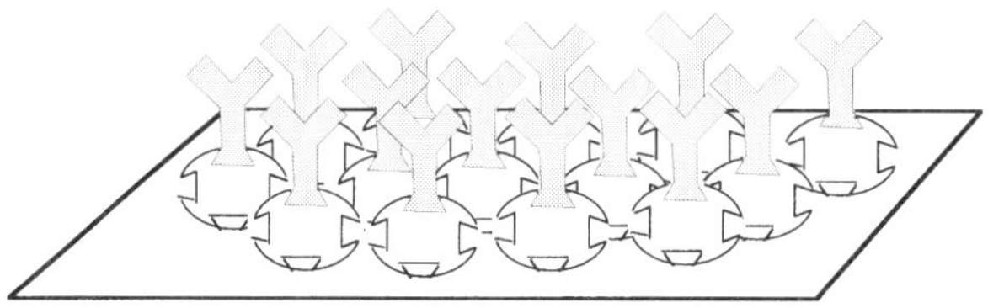

Fig. 14. The use of Langmuir–Blodgett monolayers of protein A, protein G, and the A/G chimera to orient IgG monolayers. The IgG orientation is enhanced relative to IgG bound to a randomly oriented protein A or G layer.

4.1.3. Orientation through Intrinsic Electric Dipole

An interesting technique that has seen limited use is the immobilization of proteins by their electric dipole moment. When a molecule with a naturally occurring dipole is exposed to a charged surface, it will adsorb in an orientation determined by electrostatics. Either the applied electric potential or the electric field of the double layer serves as the orienting force, and by reversing the electric field, the orientation of the protein can be inverted. Once oriented, the protein can be immobilized by electrostatic interactions or by chemical cross-linking. The strong dipole moment of cytochrome *c* has been used to orient it on glassy carbon [287]. As would be expected, the authors found that the ability of cytochrome *c* to transfer electrons to the electrode depended on its orientation. The photosynthetic reaction center has been vectorially oriented through electrostatic interactions with cytochrome *c* covalently immobilizd on Ge-Si [288].

4.1.4. Langmuir–Blodgett Monolayers

When amphiphilic molecules are spread at the air–water interface, the molecules spread to form a monolayer. The molecules can be compressed by the placement of an impermeable barrier at the interface. As the molecules are compressed, they will orient themselves with their hydrophobic ends away from the aquous subphase and the hydrophilic end toward the subphase. The monolayers can be transferred to solid substrates and are known as Langmuir–Blodgett films. Langmuir–Blodgett films have a variety of applications from nonlinear optical coatings [289] to biosensors [290]. Like other amphiphilic molecules, proteins spread at the air–water interface, and Langmuir–Blodgett films have been used to make monolayers of antibodies and enzymes and to study the properties of membrane proteins [291, 292].

To improve the orientation of IgG molecules oriented by the protein A/G method, the Langmuir technique has been used to make oriented arrays of protein A, protein G, and a recombinant protein A–protein G chimera protein [293] (Fig. 14). The protein contains four Fc-binding sites from protein A and the two Fc-binding sites of protein G. Two protein G albumin-binding sites were eliminated. The Langmuir–Blodgett technique was used to make a monolayer of the chimeric protein that was transferred to quartz resonators, silicon, or fused quartz. Despite the enhanced orientation of the A/G layer, the Langmuir–Blodgett protein A, G, and A/G immobilized IgG showed no enhancement of antigen binding over randomly oriented samples and actually had increased detection threshholds, as measured by fluorescence assays and microgravimetry. However, up to 90% of the IgG was nonspecifically adsorbed, which may have accounted for the unexpected results.

Despite problems with nonspecific adsorption, Langmuir–Blodgett monolayers of protein A can be used for optical immunosensing with relatively low detection limits. An optical immunsensor was constructed using anti-human IgG antibodies self-assembled on Langmuir–Blodgett monolayers of protein A deposited on alkylsilanized hydrophobic synthetic quartz plates. The device was capable of detecting human IgG at concentrations of 10^{-4} to 10^{-7} g/ml. No comparisons were made, however, between the detection limits of the Langmuir–Blodgett monolayer method for depositing the protein A layer and methods that result in a randomly oriented protein A layer.

In another approach, mixed Langmuir–Blodgett monolayers of phospholipids and lipid-tagged antibodies were used to make oriented arrays of IgG [168]. Single-chain antibodies of anti-2-phenyloaxazolone IgG1 were fused to the major lipoprotein of *E. coli*. Mixed monolayers of the antibody fusion protein 1,2-DMPC and 1,2-DMPE were made at the air–water interface and transferred to glass or gold substrates for SFM and surface plasmon resonance. The antibody retained its antigenic activity, but SFM studies indicated that the antibody was not homogeneously distributed, but sequestered in protein-rich domains.

Cytochrome *c* has also been immobilized on Langmuir–Blodgett multilayers [288]. Three to six monolayers of arachidic acid were deposited on an alkylated glass slide. Cytochrome *c* was found to bind to the polar slides containing an odd number of fatty acid layers. The system was developed as a tool to study the interaction between cytochrome *c* and other membrane proteins, particularly the reaction center. X-ray diffraction studies were performed to localize the cytochrome *c* position in the nanostructure.

The Langmuir–Blodgett technique has also been used to make oriented monolayers of cytochrome P450scc, IgG, and rhodopsin [293–295]. Furthermore, several groups have used fatty acid and lipid Langmuir–Blodgett monolayers to derivitize substrates and modify the chemical and physical properties of the surface before immobilizing the protein. The surface modifications are designed to enhance protein immobilization and minimize denaturation.

Like each of the other techniques discussed so far, there are inherent limitations with the use of Langmuir–Blodgett films to make nanostructures. Neither the protein nor the Langmuir–Blodgett monolayer is covalently attached to the substrate, so exposure to organic solvent, high ionic strength, extremes of pH, and sometimes air can destroy the systems. Furthermore, although one can make mixed monolayer systems, it is extremely difficult to pattern the nanostructures in the x-y plane. However, by manipulating the phase behavior of film components, one might be able to take advantage of domain formation to introduce controlled heterogeneity in the distribution of molecules across the solid surface [168, 170].

4.1.5. Self-Assembled Monolayers

One of the difficulties with making self-assembled nanostructures is that the chemical nature of the surface can affect the structure of the final nanostructure. For instance, silver and gold surfaces exhibit extremely high surface energies and denature many proteins upon contact. Quartz and glass are negatively charged and will repel negatively charged proteins and often will allow positively charged proteins to irreversibly adsorb in a nonspecific manner. Because the nature of the substrate is so important for signal transduction, one would like to be able to manipulate the chemical and physical properties of the interface at will. Langmuir–Blodgett films can be used to modulated the surface properties, but a more effective method is through the use of SAMs.

The discovery that organic thiols and silanes self-assemble on gold and glass, respectively, to form well-ordered, close-packed monolayers has brought about a revolution in surface science [296, 297]. Self-assembly has been recognized as an immensely powerful tool for modifying the chemical and physical properties of a substrate surface. By controlling the nature of the terminal group and chain length, one can modify the hydrophobicity, charge, and reactivity of the interface, thereby controlling the interaction of the surface and the protein. In some instances covalent immobilization of a protein increases its stability. Myoglobin resists denaturation by ethanol and urea more effectively when immobilized on a SAM on quartz than it does in solution [298].

A common method for immobilizing proteins on SAMs is to immobilize the protein's substrate or cofactor. One of the requirements for making highly ordered protein nanostructures is that the protein must interact with the surface in a highly specific manner. Few chemical interactions in nature are as specific as those between a protein and its substrate or cofactor. Cofactor or substrate immobilization is an excellent means of immobilizing proteins when the nanostructure application does not require substrate binding for its function. The general strategy is to immobilize a cofactor, ligand, or substrate on a solid surface using a SAM and allow the protein to bind. If the protein has only a single binding site for the immobilized cofactor, the interaction will orient the protein on the surface. Examples of this class of nanostructures are relatively diverse. The major challenge is to synthesize a thiol or silane with the cofactor as the terminal group and to determine the length of the spacer required, so that the cofactor has access to a deep substrate pocket if needed.

Katz and co-workers used a chemisorbed monolayer of pyroloquionoline quinone (PQQ) to immobilize glucose dehydrogenase (GD) [284]. The goal was to develop a glucose sensor that requires no reagent in addition to those that were immobilized on the electrode surface. The challenge with building these devices is to develop efficient electrical communication between the enzyme and electrode. Katz immobilized a monolayer of PQQ on a gold electrode and allowed GD to reconstitute the immobilized PQQ. The reconstitution not only immobilized the protein, but oriented it as well. Although the immobilized protein was catalytically active, no electron transfer between the immobilized redox site inside the holoenzyme and the electrode was achieved, regardless of the length of the spacer chain.

Heme proteins have also been immobilized through their cofactor. The common feature of all heme proteins is a large aromatic metal-chelating cofactor known as a porphyrin. Heme proteins have potential applications in a wide variety of sensors and catalytic systems because substrate and ligand binding affect both the electronic and spectroscopic characteristics of the proteins. Heme proteins are also useful models for studying protein immobilization because the porphyrin is a ring oscillator that can be used to determine the orientation of the immobilized protein. Liang-Hong Guo and co-workers have synthesized long-chain thiol derivatives of protoporphyrin IX [299]. The protoporphyrin thiols self-assemble on gold to form densely packed monolayers. Using cyclic voltammetry, Guo showed that the addition of myoglobin to the self-assembled monolayers resulted in electrochemical and spectroscopic changes similar to those expected for reconstitution of myoglobin (Mb) in solution. A mixture of long-chain porphyrin thiols to bind the protein without steric hindrance and short-chain thiols to prevent nonspecific protein adsorption was used. The possibility of nonspecific immobilization was ruled out by incubating the SAM-derivatized substrates with a solution of the holoprotein. No Mb was detected optically after a 24-h incubation. However, Mb denatures on some SAMs, resulting in the loss of the heme and the characteristic optical signature [300]. It is not clear whether denatured protein was present on the surface.

Nonspecific adsorption is a serious problem in nanostructure construction because it not only limits the net orientation of the proteins, but it can affect surface properties such as hydrophobicity. To truly achieve oriented samples, one must use substrates that resist nonspecific adsorption. Despite the millions of dollars spent on biomaterials research annually, the forces that govern adsorption are currently not well understood. However, several protein-resistant surfaces have been developed. The most promising method for preventing nonspecific adsorption to surfaces is derivatization of the surface with amphiphilic polymers. PEG, PEO, and dextran, with their large exclusion volumes, high mobility, and high hydration all effectively limit nonspecific adsorption [301, 302]. SAMs that resist adsorption include those presenting ethylene glycol oligomers and propylene sulfoxide groups [303–305]. By mixing thiols, one can make a surface that specifically binds protein only. Thiols with choline and hydroxyl headgroups can be mixed to provide a surface that resists nonspecific interactions but can specifically bind the target protein [306]. Ensuring a homogenous distribution of all components on the resultant surface can be difficult, and molecules that will not hydrolyze or cross-react must be chosen.

Carbonic anhydrase (CA) has been immobilized through its substrate on SAMs of benzenesulfonamide-terminated alkane thiols. Mrksich and co-workers used the high binding affinity of bovine CA for its ligand, para-benzenesulfonamide, to specifically immobilize and orient the protein on a SAM on gold [305]. To produce an inert background that would resist nonspecific binding of the CA, a mixed monolayer system of 0–0.08 mole fraction benzenesulfonamide and a tri-(ethylene glycol)-terminated alkanethiolate was used.

The common method of using biotin and streptavidin to immobilize proteins is another example of the cofactor technique. The interaction between biotin and streptavidin is believed to be one of the strongest noncovalent interactions in nature. Streptavidin monolayers can be formed by immobilization onto a self-assembled biotin monolayer or by forming a Langmuir–Blodgett film of lipid-tagged biotin. Because streptavidin contains four symmetric biotin-binding sites, after streptavidin immobilization, the remaining two biotin sites can be used to immobilize an additional layer of biotinylated enzymes or proteins [307, 308]. If the biotin is attached to the immobilized enzyme at a unique site (a surface-exposed cysteine, for example), the enzyme will be oriented by the biotin–streptavidin interaction. Besides an extremely tight binding interaction, the biotin–streptavidin immobilization method does not involve the substrate pocket, leaving the enzymes free to bind substrate and turn over product.

4.1.6. Immobilization of Genetic Fusion Constructs

The recent explosion of molecular biology techniques and the relative ease of genetic engineering have led several groups to explore the possibility of using genetic fusion constructs to immobilize proteins. With this technique, researchers genetically engineer a fusion protein between the protein of interest and a second protein or peptide with a highly specific and tight binding affinity for an easily immobilized reagent. Any protein whose substrate can easily be immobilized can be used as the fusion domain.

A common choice is *S*-glutathione transferase, because SAMs of its substrate glutathione are easy to assemble [309]. The most common choice for the development of nanostructures is perhaps the multihistidine domain described earlier in this chapter [278, 310, 311]. By engineering a series of two to six histidine residues at the tail end of the protein, molecular biologists can target a protein to a Zn^{2+}-chelated surface and immobilize the protein. The use of His domains is very popular because it only requires the addition of a few amino acids, minimizing the chance that the fused sequence will disrupt the tertiary structure of the fusion protein. Several metal chelates have been designed, including a nitrilotriacetic acid chelator lipid for Ni^{2+} immobilization and a nitrilotriacetic acid thiol [167, 312, 313]. The use of a metal chelator for immobilization is highly reversible, which may be advantageous, depending on the application.

4.1.7. *Covalent Immobilization through Surface Cysteines*

The issue of covalent versus noncovalent binding in nanostructures is an important one. The nature of the interaction that immobilizes the protein on the surface determines the range of experimental conditions in which the nanostructure can be submitted utilized. Whereas this is not a problem for the biotin–streptavidin interaction, many of the noncovalent systems described are relatively unstable and can be used only over a narrow pH and ionic strength range. In many cases, a covalent interaction would be far more preferable. However, to produce an oriented protein array, we must identify a unique chemical functionality on the protein that can be used to link the protein to the substrate. Surface-exposed cysteines are an excellent choice because they are are reactive toward thiols, maleimides, and halide-terminated alkyls and are relatively rare in nature. Genetic engineering can be used to add or remove cysteine residues to produce a unique attachment site.

Yeast cytochrome *c* contains a single surface exposed cysteine. Pachence and Blasie immobilized cytochrome *c* on an 11-siloxyundecanethiol on a Si-Ge substrate. The reaction center was then allowed to bind to the cytochrome *c* layer electrostatically, and the system was characterized by optical spectroscopy and X-ray diffraction [281]. The ability to covalently bind protein to a surface is important when the interactions of cytochrome *c* with other membrane proteins are studied. The covalent systems can be exposed to the detergents required to solubilize membrane proteins, allowing one to characterize the interaction of cytochrome *c* with the reaction center and electron transfer ligands.

With cysteine immobilization, not only can the protein be oriented, but unlike in the other methods discussed, the orientation of the protein can be altered by changing the position of the cysteine mutant. The ability to select protein orientation through genetic engineering was demonstrated with cytochrome b_5. The interaction between cytochrome b_5 and cytochrome *c* has been used as a model for electron transfer in proteins. Stayton and co-workers used site-directed mutagenesis to make two b_5 mutants, T65C and T8C, each containing a single cysteine residue on the surface [314]. Residue 65 was chosen because it is relatively close to the heme-binding pocket, but when it is derivatized with a bulky substituent, binding to cytochrome *c* is not impaired. Residue 8, on the other hand, is on the side of the protein opposite the exposed heme edge. The cysteine thiol was used to form a disulfide linkage between the protein and the Ti:Zn glass microscope slide (Fig. 15). Using

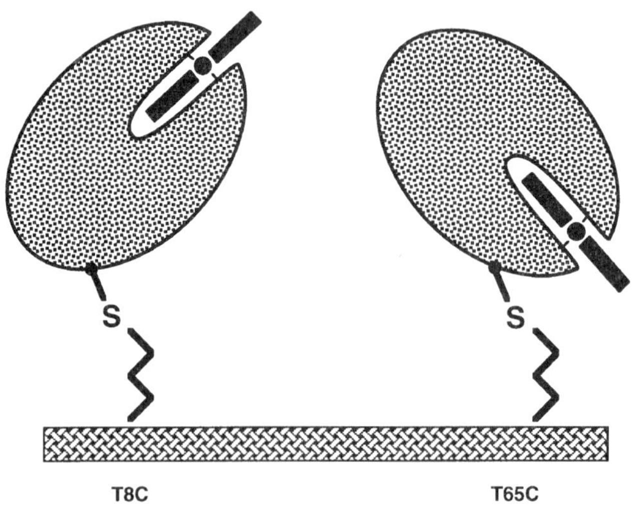

Fig. 15. The use of genetic engineering to control the orientation of covalently immobilized cytochrome b_5. Surface exposed cysteine residues were used to covalently immobilize cytochrome b_5 mutants on a microscope slide.

optical linear dichroism, the authors showed that changing the position of the cysteine mutation changes the orientation of the heme dramatically.

The T65C and T8C mutants were also used to examine the effect of protein orientation on protein–protein interactions [161]. The two proteins have complementary charged amino acids around the heme binding pocket that can form a salt bridge upon binding to cytochrome *c*. By measuring the retention time of cytochrome *c* on columns of cytochrome b_5 T8C- and T65C-linked affinity resins, McLean was able to compare the relative binding affinities. The isotherms clearly show that cytochrome b_5 T8C, with its cytochrome *c*-binding site accessible to solution, binds cytochrome *c* more tightly than the T65C mutant, with its buried binding site.

Myoglobin (Mb) is another heme protein that has been successfully oriented through a surface cysteine. Like other heme proteins, the optical properties of Mb are sensitive to small molecules in its environment, and it therefore has potential as a component of novel biomaterials and biosensors. Hong and co-workers successfully immobilized Mb on quartz, using a two-step process involving a heterobifunctional chemical cross-linker [315]. The quartz surface was first derivatized with a silane SAM. The SAM contained an amino-terminal functional group that was reactive toward the succinimidyl end of a bifunctional cross-linker. The remaining end of the cross-linker contained a maleimide that could react with the Mb thiol. This method overcame several problems with Mb immobilization on quartz, including the hydrolytic stability of the Si-O bond and the low reactivity of simple nucleophilic silanes. Most importantly, surface-immobilized Mb retains its secondary structure and its ability to bind CO, as determined by IR and UV/vis optical spectroscopy [298].

4.2. Azimuthal Orientation

Despite the fact that multiple methods have been developed to orient the protein with respect to the plane of the substrate, little work has been done to try to orient the protein azimuthally or around the surface normal. In many instances, the azimuthal orientation of the protein will have no effect on the nanostructure function. However, in some cases this orientation may be critical. To date there is only one immobilization technique that produces an azimuthally oriented sample, and that is the formation of 2-D protein crystals. Several proteins, including the bacterial S layer and streptavidin, will crystallize in two dimensions and, as will be discussed later, are being used for the development of nanostructures [180, 316].

Streptavidin forms 2-D crystals when bound to biotinylated lipid monolayers. Streptavidin does not distribute uniformly across the surface, but is sequestered in protein-rich domains, as evidenced by fluorescence microscopy [317]. Electron and X-ray crystallography indicates that the crystallized domains can reach dimensions of 200 μm across and contain two free biotin-binding sites per streptavidin molecule [318]. The free biotin-binding sites on the streptavidin have been used to build multilayer systems and immobilize additional protein layers of antibody Fab fragments and ferritin [308].

Antibodies can also be forced to crystallized at the air–water interface. When lipid-linked hapten monolayers are formed at the air–water interface, the antibody binds to the haptens, and lateral diffusion of the lipids facilitates crystallization of the antibodies [319]. Anti-*N*-dinitrophenyl phosphatidylethaolamine (DNP-PE) IgG antibodies have been crystallized on a DNP-PE Langmuir–Blodgett monolayer and transferred to an electron microscopy carbon grid. The antibodies crystallized into both linear and hexagonal domains and poled on the electric field.

4.3. Surface Patterning

The impetus for making patterned nanostructures comes from the field of electronics. In the world of modern electronics, the phrase "bigger is better" has been supplanted by "smaller is better." In the computer world, power is defined by the number of transistors that can accurately be written on a single silicon chip. The smaller the individual transistor, the more circuits that can be built in a small area. Biochemical researchers hope to use the power of the smaller electronics to make ever smaller biosensors that can be implanted in vivo. On the electronics side, the amazing sensitivity and selectivity of biological molecules make them interesting possibilities for components of electronic devices. However, before protein nanostructures can replace current technologies, circuit designers must be able to target a protein to the correct position on the chip. To obtain localization, the surface can be patterned into regions of high protein affinity and those that resist protein adsorption.

4.3.1. Two-Dimensional Protein Crystals as Nanostructuring Substrates

Most of the substrates that we have discussed so far can be described as bulk substrates that are homogeneous in both the x and y directions. Until now, we have not considered the possibility of nanostructuring the substrate itself. One way to make a 2-D nanostructured substrate is through the use of 2-D protein crystals. For example, it is well known that the bacterial S-layer protein forms a 2-D crystal when applied to a solid–liquid interface [267].

An example of a 2-D protein crystal is the bacterial S-layer coat protein. The S-layers have been used as an immobilization matrix for dipstick-style immunoassays [283]. In addition, they have been used as templates for making Ta/W oxide films with regular lattice pores [180]. The films can be made on a carbon grid and used to adsorb protein. Because proteins have a high affinity for carbon, they specifically adsorb to the porous regions. A 2-D pattern of ferritin was made this way. Unfortunately, electron microscopy of the substrate revealed that the system was far from defect free and contained a large number of unoccupied sites. Further work must be performed to develop more uniform structures and to develop a means of orienting the patterned protein.

4.3.2. Light Addressable Surface Patterning

Streptavidin monolayers and multilayers are not only used to immobilize other proteins; they can also be used to pattern surfaces [183]. Photobiotin is a photoactive biotin analogue that contains an aryl azide that can be activated by exposure to UV light. Upon activation, a nitrene is formed that can be used to immobilize additional proteins. By using multiple masks, several different proteins can be patterned onto a single surface, and by adding urea or guanidine, the surface can be cleaned and repatterned. Not only can these surfaces be repatterned, but the method is applicable to both SiO_2 and gold surfaces, making it extremely flexible. In a similar system the terminal thiol of an organosilane was photooxidized to produce an adsorption-resistant surface [320]. By using masks to selectively protect regions from the UV light, proteins can be patterned onto borosilicate glass coverslips. Photopatterning of the thiol-terminated silane not only allows for surface patterning, but when combined with the use of genetic engineering, it can provide a means of orienting the immobilized protein (although only one protein can be patterned at a time and the surfaces are not reusable). The lateral resolution of most optical patterning techniques is limited by diffraction and the ability to make small masks. The development of near-field optical scanning microscopy should push the lateral resolution of these techniques down to the nanometer regime.

4.3.3. The Use of SPM to Pattern Surfaces

Because of their high lateral resolution, several groups are exploring the use of SPMs to pattern surfaces. Lyding and co-workers have developed a method of patterning hydrogen-passivated silicon (100), using a STM in ultrahigh vacuum [321]. Because the chemistry of the surface is altered by the depassivation reaction, the STM can be used to make nanometer-sized domains of organic molecules, which can further be used to attach proteins and other biological molecules [322].

4.4. Three-Dimensional Protein Nanostructures: Protein Whiskers

In addition to patterning proteins in two dimensions, there is also interest in making 3-D protein nanostructures. Examples of 3-D nanostructures include heteromultilayer structures such as the protein A/G assemblies as discussed earlier. However, a unique 3-D structure has been built on nanometer-scale protein coated whiskers. Nanometer-scale whiskers can be formed by the rearrangement of some organic thin films, including *N*,*N*-di(3,5-xyly)perylene-3,4:9,10bis(dicarboxymide), also known as perlene red or PR149. Upon annealing at 250 °C, PR149 forms a dense array of crystalline whiskers, as seen by electron microscopy. The whiskers can be coated with metal and used to adsorb protein. Although the protein is not oriented, up to 70 times more protein can be adsorbed on the whiskers than on a flat surface of similar chemical composition.

4.5. Additional Factors Affecting Nanostructure Architecture

Obviously, the most important factors in the development of protein nanostructures are the nature of the substrate and the choice of immobilization techniques. Proteins that are immobilized through nonspecific interaction, such as physisorbing, glutaraldehyde cross-linking, and chemical reactions that use common functional groups, will be poorly oriented. But other factors, including surface roughness, linker length, and nonspecific adsorption, can greatly affect the ability to make oriented nanostructures.

When proteins are chemically cross-linked to the surface, the length of the linker chain can affect protein orientation and activity. Firestone and co-workers examined the orientation of 4-(dimethylamino)pyrindinium and myoglobin chemically bonded to quartz with chloropropylsilanes of odd and even lengths. Because the alkane chains assume a largely *trans* configuration in monolayers, the addition of each methyl group alternates the orientation of the final C-S bond and changes the orientation of the dithiol bond that will ultimately tie the protein to the surface. In addition, as the length of a silane or thiol alkyl chain increases, the hydrophobicity of the surface also increases. Myoglobin, which is stable for several hours on short alkyl chains, rapidly denatures on silanes with eight or more methylenes.

Surface coverage has also been found to effect protein activity. Intuitively, one would expect that the higher the surface coverage of immobilized enzymes, the higher the overall activity. With proteins that require only protons, photons, electrons, or small ligands for activity and do not require substantial conformational changes as part of their reaction cycle, this may be true. However, proteins such as IgG that interact with large analytes may require some movement of the protein domains for activity. Unfortunately, at high surface coverage, the immobilized proteins sterically hinder the mobility of neighboring proteins and block the interaction with ligands and analytes in solution [323]. A large antigen may bind to one IgG molecule but in the process sterically block the binding site of one or more additional antibodies. In the case of biotin SAMs, high packing densities have been shown to reduce the binding affinity of the biotin [324]. The density of binding sites on the surface must also be considered. When the density of protein-binding sites is significantly higher than the number required to bind a close-packed monolayer, multipoint attachment can occur. Multipoint attachment limits the conformational flexibility of the protein and can affect the kinetics of protein adsorption [325].

4.6. Characterization of Surface-Immobilized Nanostructures

A variety of techniques have been applied to the characterization of protein nanostructures, as it is crucial to be able to study the systems once they are built and to transduce signals from the nanostructure to the macroscopic world. The primary difficulty in studying nanostructure proteins is discriminating between signals from the interface and those from the bulk. Applicable methods include the scanning probe methodologies already discussed in this chapter, optical techniques, electrochemistry, and mass determination by microgravimetry. Optical techniques that are available for studying protein nanostructure include surface plasmon resonance, ellipsometry, and total internal reflection techniques.

The quartz crystal microbalance (QCM) technique can be used to measure the amount of adsorbed protein [293]. The shift in frequence is directly proportional to the mass of the adsorbed material. QCM is extremely sensitive to mass but is not selective and cannot discriminate between adsorbed protein and interferrents. It is therfore used in sensor development and characterization of immobilization methods, but is rarely used to study protein–protein interactions.

Ellipsometry allows one to measure film thickness or the index of refraction of a sample optically. In ellipsometry, the change in phase of the electric field polarization as light is reflected from an interface is used to determine the film thickness [326]. Ellipsometry has a height resolution of an angstrom or less, but ellipsometry and surface plasmon resonance require knowledge of the film thickness or the index of refraction of the material. Furthermore, both techniques provide only an average optical thickness. Neither can differentiate between a full monolayer of a species that is 5 nm high and a half monolayer coverage of a species that is 10 nm high. Therefore, knowledge of the protein dimensions is extremely useful.

Surface plasmon resonance spectroscopy is, in many ways, more powerful than ellipsometry, despite the fact that it must be performed on gold or silver, because it allows you to measure in situ kinetics [308, 324, 327]. The ability to study kinetics is particularly useful when studying adsorption phenomena and protein–protein interactions [303]. Close attention must be paid when using SPR to determine the binding constant and rates as the geometry of the cell can affect measured values [184].

Along with SPR and ellipsometry, total internal reflection (TIR) techniques are also extremely sensitive to surface structure [328, 329]. Examples of TIR methods that have been applied to protein nanostructures are fluorescence and Fourier transform infrared (FTIR) spectroscopy [298, 330]. In the case of fluorescent proteins, TIRF can be used to determine whether the protein is in its active state. TIR-FTIR can be used to probe the general conformation of a protein and the structure of underlying self-assembled monolayers [298, 300].

X-ray reflectivity also provides information about the thickness of protein nanostructure layers. X-ray reflectivity can provide additional information, such as depth profiling of a multilayer sample, and does not require knowledge of the index of refraction. The drawback of X-ray reflectivity is the fact that it can be highly destructive to the sample. As discussed earlier, X-ray diffraction has been applied to the characterization of cytochrome *c* immobilized on both SAM and Langmuir–Blodgett monolayers on silica [288]. Because diffraction from the inorganic components of biological nanostructures dominates the scanning patterns, X-ray diffraction is limited, however, to systems with ordered substrates and coupling layers. X-ray reflectivity has also been applied to the study of nonspecific adsorption. One approach to minimizing nonspecific adsorption is to derivatize interfaces with self-assembled monolayers and preadsorb human serum albumin (HSA), which has been shown to have some thromobogenic prevention character. X-ray reflectivity has been used to study the effect of monolayer packing on the tenacity of HSA adsorption (Petrash).

For proteins with strong optical transition moments, like the heme proteins, linear dichroism can be a very powerful tool for measuring the average orientation angle. The technique is based on the fact that molecules with their dipole moments aligned with the electric field of incoming light absorb strongly. Dipoles that are normal to the electric field cannot interact with the electric field and therefore cannot absorb the light. Experimentally, the absorption of s and p polarized light is measured independently as a function of the incident angle of the light. Equations describing the electric field can then be used to fit the data and determine the orientation angle [331, 332]. Linear dichroism has been widely used to study the orientation of heme proteins on quartz [300, 314, 330]. There are two major limitations to the application of linear dichroism. First, the protein of interest must have a strong linear or circular transition moment that does not overlap another transition. Furthermore, linear dichroism only allows one to calculate the average orientation angle. To measure the distribution of angles around the average, a second technique such as EPR must be used in connection with linear dichroism [330].

For nanostructures that exhibit crystalline domains, electron microscopy (EM) is an extremely powerful characterization technique [333]. Darst and co-workers used EM to examine streptavidin crystal grown on biotinylated lipids [318]. Stahl et al. used scanning electron microscopy to examine the 3-D whiskers discussed earlier [174]. Two-dimensional protein crystals of bacterial HPI layer and S-layers have also been subjected to EM [287, 334]. Electrochemical signal transduction and sample characterization of conducting substrates have been performes by a variety of electrochemical means, including cyclic voltammetry [284, 299, 306], amperometry [335], and nonlinear impendence measurements [336].

As described in more detail earlier in this chapter, scanning probe microscopies can be powerful tools in the study of biological nanostructures. SPMs have been applied to characterize a variety of biological nanostructures, including SAMs [300], transferred Langmuir–Blodgett films of antibodies [168], and 2-D bacterial coat proteins [334]. The advantage of SFM over other techniques lies in its high lateral resolution. Optical techniques (with the exception of NSOM) have a lateral resolution that is limited by the wavelength of light. The lateral resolution of most optical techniques is on the order of microns, although the z resolution of some optical techniques, particularly ellipsometry and surface plasmon resonance spectroscopy, can be much higher. SFM has a z resolution of angstroms and a lateral resolution that is limited only by the radius of curvature of the tip. On biological samples, the x-y resolution is generally in the range of tens of nanometers.

5. FUTURE DIRECTIONS

Although much progress has been made in the area of surface-immobilized nanostructures, there is much work to do. No method has yet been developed that will allow researchers to pattern proteins and simultaneously control their orientation. However, a judicious combination of the techniques described may provide the solution. Furthermore, additional research into the causes of nonspecific adsorption is crucial. Means of preventing nonspecific interactions must be found if defect-free nanostructures are ever to be achieved. Finally, additional methods must be developed to characterize these systems on the nanometer and angstrom levels if design methods are to be perfected. It is not enough to be able to measure the average orientation of immobilized molecules over a region of several hundred microns. One must be able to characterize groups of a few thousand or even a few hundred molecules to truly understand how to exert molecular-level control.

Acknowledgments

This work was supported by National Institutes of Health grant PHSGM31756.

References

1. D. Sarid, "Scanning Force Microscopy with Applications to Electric, Magnetic, and Atomic Forces." Oxford Univ. Press, New York, 1994.
2. C. J. Chen, "Introduction to Scanning Tunneling Microscopy." Oxford Univ. Press, New York, 1993.

3. R. Howland and L. Benatar, "A Practical Guide to Scanning Probe Microscopy." Park Scientific Instruments, Sunnyvale, CA, 1996.
4. G. Binnig and H. Rohrer, *Helv. Phys. Acta* 55 (1982).
5. S. M. Lindsay, L. A. Nagahara, T. Thundat, U. Knipping, R. L. Rill, B. Drake, C. B. Prater, A. L. Weisenhorn, S. A. C. Gould, and P. K. Hansma, *J. Biomol. Struct. Dyn.* 7, 279 (1989).
6. R. Guckenberger, M. Heim, G. Cevc, H. R. Knapp, W. Wiengrabe, and A. Hillebrand, *Science* 266, 1538 (1994).
7. G. Binnig, C. F. Quate, and C. Gerber, *Phys. Rev. Lett.* 56, 930 (1986).
8. M. Tortonese, *IEEE Eng. Med. Biol.* 16, 28 (1997).
9. A. J. Howard, R. R. Rye, and J. E. Houston, *J. Appl. Phys.* 79, 1885 (1996).
10. G. Bar, Y. Thomann, R. Brandsch, and H.-J. Cantow, *Langmuir* 13, 3807 (1997).
11. J. Tamayo and R. Garcia, *Langmuir* 12, 4430 (1996).
12. M. O. Finot and M. T. McDermott, *J. Am. Chem. Soc.* 119, 8564 (1997).
13. J. W. Hong, Z. G. Khim, A. S. Hou, and S.-i. Park, *Appl. Phys. Lett.* 69, 2831 (1996).
14. M. A. Lantz, S. J. O'Shea, and M. E. Weiland, *Appl. Phys. Lett.* 65, 409 (1994).
15. W. Han, S. M. Linsay, M. Dlakic, and R. E. Harrington, *Nature* 386, 563 (1997).
16. S. D. O'Connor, R. C. Gamble, R. K. Eby, and J. D. Baldeschwieler, *Rev. Sci. Instrum.* 67, 393 (1996).
17. Z. Shao and Y. Yang, *Ultramicroscopy* 66, 141 (1996).
18. N. F. v. Hulst and M. H. P. Moers, *IEEE Eng. Med. Biol.* 15, 51 (1996).
19. P. J. Moyer and S. B. Kamer, *Appl. Phys. Lett.* 68, 3380 (1996).
20. M. H. P. Moers, A. G. T. Ruiter, A. Jalocha, and N. F. v. Hulst, *Ultramicroscopy* 61, 279 (1995).
21. H. Muramatsu, N. Chiba, T. Umemoto, K. Homma, K. Nakajima, T. Ataka, S. Ohta, A. Kusumi, and M. Fukihira, *Ultramicroscopy* 61, 265 (1995).
22. S. Iwabuchi, H. Muramatsu, N. Chiba, Y. Kinjo, Y. Murakami, T. Sakaguchi, K. Yokoyama, and E. Tamiya, *Nucleic Acids Res.* 25, 1662 (1997).
23. H. Gottlich and W. M. Heckl, *Ultramicroscopy* 61, 145 (1995).
24. K. Fukuzawa and Y. Tanaka, *Appl. Phys. Lett.* 71, 169 (1997).
25. S. K. Sinniah, A. B. Steel, C. J. Miller, and J. E. Reutt-Robey, *J. Am. Chem. Soc.* 118, 8925 (1996).
26. C. D. Frisbie, L. F. Rozsnyai, A. Noy, M. S. Wrighton, and C. M. Lieber, *Science* 265, 2071 (1994).
27. J. L. Wilbur, H. A. Biebuyck, J. C. MacDonald, and G. M. Whitesides, *Langmuir* 11, 825 (1995).
28. M. Radmacher, J. P. Cleveland, M. Fritz, H. G. Hansma, and P. K. Hansma, *Biophys. J.* 66, 2159 (1994).
29. M. Ludwig, W. Dettmann, and H. E. Gaub, *Biophys. J.* 72, 445 (1997).
30. Y. Leng and C. C. Williams, *Colloids Surfaces A Physicochem. Eng. Aspects* 93, 335 (1994).
31. S. Xu and M. F. Arsndorf, *Proc. Natl. Acad. Sci. USA* 92, 10384 (1995).
32. H. G. Hansma, R. L. Sinsheimer, M.-Q. Li, and P. K. Hansma, *Nucleic Acids Res.* 20, 3585 (1992).
33. M.-Q. Li, H. G. Hansma, J. Vesenka, G. Kelderman, and P. K. Hansma, *J. Biomol. Struct. Dyn.* 10, 607 (1992).
34. W. A. Rees, R. W. Keller, J. P. Vesenka, G. Yang, and C. Bustamante, *Science* 60, 1646 (1993).
35. C. Bustamante, G. Zuccheri, S. H. Leuba, G. Yang, and B. Samori, *Methods* 12, 73 (1997).
36. B. B. Doris and R. I. Hegde, *Appl. Phys. Lett.* 67, 3816 (1995).
37. H. Dai, J. H. Hafner, A. G. Rinzler, D. T. Colbert, and R. E. Smalley, *Nature* 384, 147 (1996).
38. M. Tokunaga, T. Aoki, M. Hiroshima, K. Kitamura, and T. Yanagida, *Biochem. Biophys. Res. Commun.* 231, 566 (1997).
39. C. Bustamante and C. Rivetti, *Annu. Rev. Biophys. Biomol. Struct.* 25, 395 (1996).
40. F. Atamny and A. Baiker, *Surface Sci.* 323, L314 (1995).
41. J. Vesenka, T. Marsh, R. Miller, and E. Henderson, *J. Vac. Sci. Technol. B* 14, 1413 (1996).
42. C. A. J. Putman, K. O. V. d. Werf, B. G. D. Grooth, N. F. V. Hulst, and J. Greve, *Appl. Phys. Lett.* 64, 2454 (1994).
43. H. G. Hansma, R. L. Sinsheimer, J. Groppe, T. C. Bruice, V. Eilings, G. Gurley, M. Bezanilla, I. A. Mastrangelo, P. V. C. Hough, and P. K. Hansma, *Scanning* 15, 296 (1993).
44. I. A. Mastrangelo, M. Bezanilla, P. K. Hansma, P. V. C. Hough, and H. G. Hansma, *Biophys. J.* 66, 293 (1994).
45. S. Dongmo, M. Troyon, P. Vautrot, E. Delain, and N. Bonnet, *J. Vac. Sci. Technol. B* 14, 1552 (1996).
46. P. M. Williams, K. M. Shakesheff, M. C. Davies, D. E. Jackson, C. J. Roberts, and S. J. B. Tendler, *Langmuir* 12, 3468 (1996).
47. J. P. P. Starink and T. P. Jovin, *Surface Sci.* 359, 291 (1996).
48. F. K. Urban, P. Romine, and M. S. Islam, *Thin Solid Films* 253, 318 (1994).
49. J. T. Woodward, C. Kono, L. L. Madsen, and J. A. Zasadzinski, *J. Microsc.* 178, 86 (1995).
50. L. F. Chi, M. Anders, H. Fuchs, R. R. Johnson, and H. Ringsdorf, *Science* 259, 213 (1993).
51. E. Sackmann, *Science* 271, 43 (1996).
52. J. B. Heymann, D. J. Müller, K. Mitsuoka, and A. Engel, *Curr. Opin. Struct. Biol.* 7, 543 (1997).
53. J. Hwang, L. K. Tamm, C. Böhm, T. S. Ramalingam, E. Betzig, and M. Edidin, *Science* 270, 610 (1995).
54. L. K. Tamm, C. Böhm, J. Yang, Z. Shao, J. Hwang, M. Edidin, and E. Betzig, *Thin Solid Films* 284–285, 813 (1996).

55. E. Meyer, L. Howald, R. M. Overney, H. Heinzelmann, J. Frommer, H.-J. Güntherodt, T. Wagner, H. Schier, and S. Roth, *Nature* 349, 398 (1991).
56. A. L. Weisenhorn, M. Egger, F. Ohnesorge, S. A. C. Gould, S.-P. Heyn, H. G. Hansma, R. L. Sinsheimer, H. E. Gaub, and P. K. Hansma, *Langmuir* 7, 8 (1991).
57. S. Singh and D. J. Keller, *Biophys. J.* 60, 1401 (1991).
58. R. Lal, H. Kim, R. M. Garavito, and M. F. Arnsdorf, *Am. J. Physiol.* 265, C851 (1993).
59. J. A. N. Zasadzinski, C. A. Helm, M. L. Longo, A. L. Weisenhorn, S. A. C. Gould, and P. K. Hansma, *Biophys. J.* 59, 755 (1991).
60. K. Luo, M. Yorgancioglu, and D. Keller, *Ultramicroscopy* 50, 147 (1993).
61. J. Garnaes, D. K. Schwartz, R. Viswanathan, and J. A. N. Zasadzinski, *Nature* 357, 54 (1992).
62. M. Egger, F. Ohnesorge, A. L. Weisenhorn, S. P. Heyn, B. Drake, C. B. Prater, S. A. C. Gould, P. K. Hansma, and H. E. Gaub, *J. Struct. Biol.* 103, 89 (1990).
63. J. Mou, J. Yang, and Z. Shao, *Biochemistry* 33, 4439 (1994).
64. J. Mou, J. Yang, C. Huang, and Z. Shao, *Biochemistry* 33, 9981 (1994).
65. D. M. Czajowsky, C. Huang, and Z. Shao, *Biochemistry* 34, 12501 (1995).
66. M. Radmacher, R. M. Zimmerman, and H. E. Gaub, in "The Structure and Conformation of Amphiphilic Membranes" (R. Lipowsky, D. Richter, and K. Kremer, eds.), Vol. 66, pp. 25–29. Springer-Verlag, Berlin, 1992.
67. S. L. Brandow, D. C. Turner, B. R. Ratna, and B. P. Gaber, *Biophys. J.* 64, 898 (1993).
68. J. W. Carlson, A. Jonas, and S. G. Sligar, *Biophys. J.* 73, 1184 (1997).
69. A. Jonas, *Methods Enzymol.* 128, 553 (1986).
70. J. H. Wald, E. S. Krul, and A. Jonas, *J. Biol. Chem.* 265, 20037 (1990).
71. L. Calabresi, G. Vecchio, F. Frigerio, L. Vavassori, C. Sirtori, and G. Franceschini, *Biochemistry* 36, 12428 (1997).
72. A. Jonas, in "Structure and Function of Apolipoproteins" (M. Rosseneu, ed.), pp. 217–250. CRC Press, London, 1989.
73. (a) T. H. Bayburt, J. W. Carlson, and S. G. Sligar, *J. Struct. Biol.* 123, 37 (1998). (b) C. A. Lang, J. K. H. Hörber, T. W. Hänsch, W. M. Heckl, and H. Möhwald, *J. Vac. Sci. Technol. A* 6, 368 (1988).
74. W. Mizutani, M. Shigeno, K. Saito, K. Watanabe, M. Sugi, M. Ono, and K. Kajimura, *Jpn. J. Appl. Phys.* 27, 1803 (1988).
75. H. Fuchs, *Phys. Scr.* 38, 264 (1988).
76. B. W. Gregory, R. A. Dluhy, and L. A. Bottomley, *J. Phys. Chem.* 98, 1010 (1994).
77. R. Guckenberger, M. Heim, G. Cevc, H. F. Knapp, W. Wiegräbe, and A. Hillebrand, *Science* 266, 1538 (1994).
78. M. Heim, G. Cevc, R. Guckenberger, H. F. Knapp, and W. Wiegräbe, *Biophys. J.* 69, 489 (1995).
79. J. T. Woodward and J. A. Zasadzinski, *Biophys. J.* 72, 964 (1997).
80. M. Radmacher, R. W. Tillmann, M. Fritz, and H. E. Gaub, *Science* 257, 1900 (1992).
81. T. Inoue and H. Yokoyama, *J. Vac. Sci. Technol. B* 12, 1569 (1994).
82. W. M. Heckl and G. Binnig, *Ultramicroscopy* 42–44, 1073 (1992).
83. C. R. Clemmer and T. P. Beebe, *Science* 251, 640 (1991).
84. L. A. Bottomley, J. N. Hassletine, D. P. Allison, R. J. Warmack, T. Thundat, R. A. Sachleben, G. M. Brown, R. P. Woychik, K. B. Jacobson, and T. L. Ferrel, *J. Vac. Sci. Technol. A* 10, 591 (1992).
85. A. M. Jeffrey, T. W. Jing, J. A. DeRose, A. Vaught, D. Rekesh, F.-X. Lu, and S. M. Lindsay, *Nucleic Acids Res.* 21, 5896 (1993).
86. T. W. Jing, A. M. Jeffrey, J. A. DeRose, Y. L. Lyubchenko, L. S. Shlyaktenko, R. E. Harrington, E. Appella, J. Larsen, A. Vaught, D. Rekesh, F.-X. Lu, and S. M. Lindsay, *Proc. Natl. Acad. Sci. USA* 90, 8934 (1993).
87. I. Nevernov, I. Kurnikov, and C. Nicolini, *Ultramicroscopy* 58, 269 (1995).
88. D. D. Dunlap, R. Garcia, E. Schabtach, and C. Bustamante, *Proc. Natl. Acad. Sci. USA* 90, 7652 (1993).
89. M. Heim, R. Steigerwald, and R. Guckenberger, *J. Struct. Biol.* 19, 212 (1997).
90. M. Bezanilla, S. Manne, D. E. Laney, Y. Y. Lyubchenko, and H. G. Hansma, *Langmuir* 11, 655 (1995).
91. S. W. Bailey, "Micas," Vol. 13. Mineralogical Society of America, Washington, DC, 1984.
92. T. Thundat, D. P. Allison, R. J. Warmack, G. M. Brown, K. B. Jacobson, J. J. Schrick, and T. L. Ferrell, *Scanning Microsc.* 6, 911 (1992).
93. Y. L. Lyubchenko, A. A. Gall, L. S. Shlyatenko, R. E. Harrington, B. L. Jacobs, P. I. Oden, and S. M. Lindsay, *J. Biomol. Struct. Dyn.* 10, 589 (1992).
94. J. Mou, D. M. Czajkowsky, Y. Zhang, and Z. Shao, *FEBS Lett.* 371, 270 (1995).
95. T. Thundat, D. P. Allison, R. J. Warmack, and T. L. Ferrell, *Ultramicroscopy* 42–44 (part B), 1101 (1992).
96. F. Zenhausern, M. Adrian, B. T. Heggler-Bordner, R. Emch, M. Jobin, M. Taborelli, and P. Descouts, *J. Struct. Biol.* 108, 69 (1992).
97. C. Bustamante, J. Vesenka, C. L. Tang, W. Rees, M. Guthold, and R. Keller, *Biochemistry* 31, 22 (1992).
98. Y. L. Lyubchenko, P. I. Oden, D. Lampner, S. M. Linsay, and K. A. Dunker, *Nucleic Acids Res.* 21, 1117 (1993).
99. J. Yang and Z. Shao, *Ultramicroscopy* 50, 157 (1993).
100. G. Yang, J. P. Vesenka, and C. J. Bustamante, *Scanning* 18, 344 (1996).

101. J. Vesenka, S. Manne, G. Yang, C. J. Bustamante, and E. Henderson, *Scanning Microsc.* 7, 781 (1993).
102. J. Yang, K. Takeyasu, and Z. Shao, *FEBS Lett.* 301, 173 (1992).
103. A. Schaper, L. I. Pietrasanta, and T. M. Jovin, *Nucleic Acids Res.* 21, 6004 (1993).
104. H. G. Hansma and D. E. Laney, *Biophys. J.* 70, 1933 (1996).
105. Y. L. Lyubchenko, B. L. Jacobs, S. M. Lindsay, and A. Stasiuk, *Scanning Microsc.* 9, 705 (1995).
106. J. Hu, M. Wang, H.-U. G. Weier, P. Frantz, W. Kolbe, D. F. Ogletree, and M. Salmeron, *Langmuir* 12, 1697 (1996).
107. A. Schaper, J. P. P. Starink, and T. M. Jovin, *FEBS Lett.* 355, 91 (1994).
108. A. Criscenti, S. Selci, G. Chiarotta, and F. Amaldi, *J. Vac. Sci. Technol. B* 9, 1285 (1991).
109. M. Hegner, P. Wagner, and G. Semenza, *FEBS Lett.* 336, 452 (1993).
110. N. H. Thomson, S. Kasas, B. Smith, H. G. Hansma, and P. K. Hansma, *Langmuir* 12, 5905 (1996).
111. A. Umeda, K. Amako, A. Takade, and A. Tanaka, *J. Electron Microsc.* 44, 408 (1995).
112. I. V. Tuzov, E. V. Yurkova, D. V. Klinov, and V. V. Demin, *Russ. J. Bioorg. Chem.* 21, 269 (1995).
113. A. L. Weisenhorn, H. E. Gaub, H. G. Hansma, R. L. Sinsheimer, G. L. Kelderman, and P. K. Hansma, *Scanning Microsc.* 4, 511 (1990).
114. E. Henderson, *Nucleic Acids Res.* 20, 445 (1992).
115. P. K. Hansma, J. P. Cleveland, M. Radmacher, D. A. Walters, P. E. Hillner, M. Bezanilla, M. Fritz, D. Vie, H. G. Hansma, C. B. Prater, J. Massie, L. Fukunaga, J. Gurley, and V. Eilings, *Appl. Phys. Lett.* 64, 1738 (1994).
116. C. Bustamante, C. Rivetti, and D. J. Keller, *Curr. Opin. Struct. Biol.* 7, 709 (1997).
117. M. N. Murray, H. G. Hansma, M. Bezanilla, T. Sano, D. F. Ogletree, W. Kolbe, C. L. Smith, C. R. Cantor, S. Spengler, P. K. Hansma, and M. Salmeron, *Proc. Natl. Acad. Sci. USA* 90, 3811 (1993).
118. W.-L. Shaiu, D. D. Larson, J. Vesenka, and E. Henderson, *Nucleic Acids Res.* 21, 99 (1993).
119. W. M. Heckl, D. P. E. Smith, G. Binnig, H. Klagges, T. W. Hansch, and J. Maddocks, *Proc. Natl. Acad. Sci. USA* 88, 8003 (1991).
120. M. J. Allen, M. Balooch, S. Subbiah, R. J. Tench, W. Siekhaus, and R. Balhorn, *Scanning Microsc.* 5, 625 (1991).
121. D. P. Allison, L. A. Bottomley, T. Thundat, G. M. Brown, R. P. Woychik, J. J. Schrik, K. B. Jacobson, and R. J. Warmack, *Proc. Natl. Acad. Sci. USA* 89, 10129 (1992).
122. S. M. Lindsay, N. J. Tao, J. A. DeRose, P. I. Oden, Y. L. Lyubchenko, R. E. Harrington, and L. Shlyakhtenko, *Biophys. J.* 61, 1570 (1992).
123. D. Rekesh, Y. Lyubchenko, L. S. Shlyakhtenko, and S. M. Lindsay, *Biophys. J.* 71, 1079 (1996).
124. F. R. F. Fan and A. J. Bard, *Science* 270, 1849 (1995).
125. R. Guckenberger and M. Heim, *Science* 270, 1851 (1995).
126. N. J. Rampino, *Biochem. Biophys. Res. Commun.* 182, 201 (1992).
127. D. M. Crothers, M. R. Gartenberg, and T. E. Schrader, *Methods Enzymol.* 208, 118 (1991).
128. H. G. Hansma, K. A. Browne, M. Bezanilla, and T. C. Bruice, *Biochemistry* 33, 8436 (1994).
129. J. E. Coury, L. McFail-Isom, S. Presnell, L. D. Williams, and L. A. Bottomly, *J. Vac. Sci. Technol. A* 13, 1746 (1995).
130. J. E. Coury, L. McFail-Isom, L. D. Williams, and L. A. Bottomley, *Proc. Natl. Acad. Sci. USA* 93, 12283 (1996).
131. H. G. Hansma, I. Revenko, and D. E. Laney, *Nucleic Acids Res.* 24, 713 (1996).
132. T. Thundat, D. P. Allison, and R. J. Warmack, *Nucleic Acids Res.* 22, 4224 (1994).
133. E. Larquet, E. L. Cam, A. Fourcade, F. Culard, P. Furrer, and E. Delain, *C. R. Acad. Sci. Ser. III* 319, 461 (1996).
134. E. Henderson, *J. Microsc.* 167, 77 (1992).
135. T. C. Marsh, J. Vesenka, and E. Henderson, *Nucleic Acids Res.* 23, 696 (1995).
136. Y. Fang, C. Bai, Y. Wei, P. Zhang, Y. Tang, E. Gao, and Y. He, *Sci. China, Ser. B* 38, 1166 (1995).
137. C. Rivetti, M. Guthold, and C. Bustamante, *J. Mol. Biol.* 264, 919 (1996).
138. Y. L. Lyubchenko and L. S. Shlyakhtenko, *Proc. Natl. Acad. Sci. USA* 94, 496 (1997).
139. K. Rippe, N. Mucke, and J. Langowski, *Nucleic Acids Res.* 25, 1736 (1997).
140. D. P. Allison, P. S. Kerper, M. J. Doktycz, J. A. Spain, P. Modrich, F. W. Larimer, T. Thundat, and R. J. Warmack, *Proc. Natl. Acad. Sci. USA* 93, 8826 (1996).
141. D. P. Allison, P. S. Kerper, M. J. Doktycz, T. Thundat, P. Modrich, F. W. Larimer, D. K. Johnson, P. R. Hoyt, M. L. Mucenski, and R. J. Warmack, *Genomics* 41, 379 (1997).
142. I. Kelson and S. Nussinov, *Proc. Natl. Acad. Sci.* 91, 6963 (1994).
143. M. S. Spector and J. M. Schur, *Science* 275, 791 (1997).
144. Y. Fang and J. Yang, *J. Phys. Chem.* (to appear).
145. C. J. Wheeler, L. Sukhu, G. Yang, Y. Tsai, C. Bustamante, P. Felgner, J. Norman, and M. Manthorpe, *Biochim. Biophys. Acta* 1280, 1 (1996).
146. M. A. Wolfert and L. W. Seymour, *Gene Ther.* 3, 269 (1996).
147. M. Wolfert, E. H. Schacht, V. Toncheva, K. Ulbrich, O. Nazarova, and L. W. Seymour, *Hum. Gene Ther.* 7, 2123 (1996).
148. G. U. Lee, L. A. Chrisey, and R. J. Colton, *Science* 266, 771 (1994).

149. T. Boland and B. D. Ratner, *Proc. Natl. Acad. Sci. USA* 92, 5297 (1995).
150. Y. L. Lyubchenko, B. L. Jacobs, and S. M. Lindsay, *Nucleic Acids Res.* 20, 3983 (1992).
151. J. D. Ng, Y. G. Kuznetsov, A. J. Malkin, G. Keith, R. Giege, and A. McPherson, *Nucleic Acids Res.* 25, 2582 (1997).
152. B. Felden, Y. G. Kuznetsov, A. J. Malkin, A. Greenwood, and A. McPherson, Presented at RNA '96, Madison, Wisconsin.
153. B. L. Smith, D. R. Gallie, H. Le, and P. K. Hansma, *J. Struct. Biol.* 119, 109 (1997).
154. S. Kasas, N. H. Thompson, B. L. Smith, H. G. Hanmsa, X. Zhu, M. Guthold, C. Bustamante, E. T. Kool, M. Kashlev, and P. K. Hansma, *Biochemistry* 36, 461 (1997).
155. B. J. Godfrey, M. A. Schuler, and S. G. Sligar, submitted.
156. M. Suzuki, M. Shinkai, M. Kamihira, and T. Kobayashi, *Biotechnol. Appl. Biochem.* 21, 335 (1995).
157. P. C. v. d. Vliet and C. P. Verrizer, *Bioessays* 15, 25 (1993).
158. J. D. Griffith, A. Mahkov, L. Zawel, and D. Reinberg, *J. Mol. Biol.* 246, 576 (1995).
159. Y. C. Kim, J. C. Grable, R. Love, P. J. Greene, and J. M. Rosenborn, *Science* 249, 1307 (1990).
160. D. Erie, G. Yang, H. C. Schultz, and C. Bustamante, *Science* 266, 1562 (1994).
161. M. A. McLean, P. S. Stayton, and S. G. Sligar, *Anal. Chem.* 65, 2676 (1993).
162. T. Muller, E. Gilson, R. Schmidt, R. Giraldo, J. Sogo, H. Gross, and S. M. Gasser, *J. Struct. Biol.* 113, 1 (1994).
163. K. S. Mathewes, *Microbiol. Rev.* 56, 123 (1992).
164. C. Wyman, E. Grotkoop, C. Bustamante, and H. C. M. Nelson, *EMBO J.* 14, 117 (1995).
165. J. C. Becker, A. Nikroo, T. Brabletz, and R. Reisfeld, *Proc. Natl. Acad. Sci. USA* 92, 9727 (1995).
166. K. Owaku, M. Goto, Y. Ikariyama, and M. Aizawa, *Anal. Chem.* 67, 1613 (1995).
167. T. A. Keller, C. Duschl, D. Kroger, A.-F. Sevin-Landais, H. Vogel, S. E. Cervigni, and P. Dumy, *Supramol. Sci.* 2, 155 (1995).
168. I. Vikholm and J. Peltonen, *Thin Solid Films* 284–285, 924 (1996).
169. Y. L. Lyubchenko, L. S. Shlyakhtenko, T. Aki, and S. Adhya, *Nucleic Acids Res.* 25, 873 (1997).
170. M. Florsheimer and H. Mohwald, *Chem. Phys. Lipids* 49, 231 (1989).
171. R. Kaul, M. Allen, E. M. Bradbury, and W. M. Wenman, *Nucleic Acids Res.* 24 (1996).
172. E. L. Cam, D. Frechon, M. Barray, A. Fourcade, and E. Delain, *Proc. Natl. Acad. Sci. USA* 91, 11816 (1994).
173. C. Freij-Larsson, T. Nylander, P. Jannasch, and B. Wesslen, *Biomaterials* 17, 2199 (1996).
174. J. B. Stahl, M. K. Debe, and P. L. Coleman, *J. Vac. Sci. Technol. A* 14, 1761 (1996).
175. M. Elliasson, R. Andersson, A. Olsson, H. Wigzell, and M. Uhlen, *J. Immunol.* 142, 575 (1989).
176. B. Lu, J. Xie, C. Lu, C. Wu, and Y. Wei, *Anal. Chem.* 67, 83 (1995).
177. H. Delius, N. J. Mantell, and B. Alberts, *J. Mol. Biol.* 67, 341 (1972).
178. S. C. West, *Annu. Rev. Biochem.* 61, 603 (1992).
179. H. G. Hansma, D. E. Laney, M. Bezanilla, R. L. Sinsheimer, and P. K. Hansma, *Biophys. J.* 68, 1672 (1995).
180. K. Douglas, N. A. Clark, and K. J. Rothschild, *Mater. Res. Symp. Proc.* 174, 151 (1990).
181. D. M. Gersten and J. J. Marchalonis, *J. Immunol. Methods* 24, 305 (1978).
182. L. I. Pietrasanta, A. Schaper, and T. M. Jovin, *Nucleic Acids Res.* 22, 3288 (1994).
183. D. J. Pritchard, H. Morgan, and J. M. Cooper, *Anal. Chem.* 67, 3605 (1995).
184. J. Davies and L. Skelton, in "Surface Analytical Techniques for Probing Biomaterial Processes" (J. Davies, ed.), pp. 89–104. CRC Press, New York, 1996.
185. C.-Y. Wang, S. Hitz, J. D. Andrade, and R. J. Stewart, *Anal. Biochem.* 246, 133 (1997).
186. H. Muramatsu, N. Chiba, K. Homma, K. Nakajima, A. Tatsuaki, S. Ohta, A. Kusumi, and M. Fujihira, *Thin Solid Films* 273, 335 (1996).
187. M. Zachariou, I. Traverso, L. Spiccia, and M. T. W. Hearn, *Anal. Chem.* 69, 813 (1997).
188. J. T. Groves, N. Ulman, and S. G. Boxer, *Science* 275, 651 (1997).
189. M. K. Jain, "The Bimolecular Lipid Membrane: A System." Van Nostrand Reinhold, New York, 1972.
190. S. J. Singer and G. L. Nicholson, *Science* 175, 720 (1972).
191. H. Sandermann, *Trends Biochem. Sci.* 8, 408 (1983).
192. H. Sandermann, *Biochim. Biophys. Acta* 515, 209 (1978).
193. G. Cevc and D. Marsh, "Phospholipid Bilayers. Physical Principles and Models." Wiley-Interscience, New York, 1987.
194. P. K. J. Kinnunen, A. Koiv, J. Y. A. Lehtonen, M. Rytömaa, and P. Mustonen, *Chem. Phys. Lipids* 73, 181 (1994).
195. J.-F. Tocanne, L. Cézanne, A. Lopez, B. Piknova, V. Schram, J.-F. Tournier, and M. Welby, *Chem. Phys. Lipids* 73, 139 (1994).
196. R. B. Gennis, "Biomembranes: Molecular Structure and Function." Springer-Verlag, New York, 1989.
197. M. C. Wiener and S. H. White, *Biophys. J.* 61, 434 (1992).
198. G. von Heijne, *EMBO J.* 5, 3021 (1986).
199. E. Hartmann, T. A. Rapoport, and H. F. Lodish, *Proc. Natl. Acad. Sci. USA* 86, 5786 (1989).
200. O. G. Mouritsen and M. Bloom, *Biophys. J.* 46, 141 (1984).
201. M. M. Sperotto, J. H. Ipsen, and O. G. Mouritsen, *Cell Biophys.* 14, 79 (1989).

202. D. Marsh, in "Progress in Protein–Lipid Interactions" (W. a. D. Pont, ed.), pp. 143–172. Elsevier Science, New York, 1985.
203. T. L. Leto, M. A. Roseman, and P. W. Holloway, *Biochemistry* 19, 1911 (1980).
204. J. Poensjen and V. Ullrich, *Biochim. Biophys. Acta* 596, 248 (1980).
205. S. Kominami, S. Ikushiro, and S. Takemori, *Biochim. Biophys. Acta* 905, 143 (1987).
206. S. Kominami, M. Onizuka, C. Tasaka-Marumoto, and S. Takemori, *Biochemistry* 34, 4839 (1995).
207. J. A. Boutin, *Cell. Signaling* 9, 15 (1997).
208. J. I. Gordon, R. J. Duronio, D. A. Rudnick, S. P. Adams, and G. W. Gokel, *J. Biol. Chem.* 266, 8647 (1991).
209. G. James and E. N. Olson, *Biochemistry* 29, 2623 (1990).
210. M. Sinensky and R. J. Lutz, *Bioessays* 14, 25 (1992).
211. F. L. Zhang and P. J. Casey, *Annu. Rev. Biochem.* 65, 241 (1996).
212. M. G. Low, *Biochim. Biophys. Acta* 988, 427 (1989).
213. J. T. Groves, C. Wülfing, and S. G. Boxer, *Biophys. J.* 71, 2716 (1996).
214. R. Huber, J. Römisch, and E. P. Paques, *EMBO J.* 9, 3867 (1990).
215. A. Olofsson, V. Mallouh, and A. Brisson, *J. Struct. Biol.* 113, 199 (1994).
216. M. A. Swairjo, N. O. Concha, M. A. Kaetzel, J. R. Dedman, and B. A. Seaton, *Nature Struct. Biol.* 2, 968 (1995).
217. E. A. Nalefski and J. J. Falke, *Protein Sci.* 5, 2375 (1996).
218. G. Shaw, *Bioessays* 18, 35 (1996).
219. H. S. Yoon, P. J. Hajduk, A. M. Petros, E. T. Olejniczak, R. P. Meadows, and S. W. Fesik, *Nature* 369, 672 (1994).
220. M. J. Macias, A. Musacchio, H. Ponstingl, M. Nilges, M. Saraste, and H. Oschkinat, *Nature* 369, 675 (1994).
221. A. K. Downing, P. C. Driscoll, I. Gout, K. Salim, M. J. Zvelebil, and M. D. Waterfield, *Curr. Biol.* 4, 884 (1994).
222. K. M. Ferguson, M. A. Lemmon, J. Schlessinger, and P. B. Sigler, *Cell* 79, 199 (1994).
223. D. Timm, K. Salim, I. Gout, L. Guruprasad, and M. a. B. T. Waterfield, *Nature Struct. Biol.* 1, 782 (1994).
224. C. Dietrich, L. Schmitt, and R. Tampé, *Proc. Natl. Acad. Sci. USA* 92, 9014 (1995).
225. C. Dietrich, O. Boscheinen, K.-D. Scharf, L. Schitt, and R. Tampé, *Biochemistry* 35, 1100 (1996).
226. S. Gritsch, K. Neumaier, L. Schmitt, and R. Tampé, *Biosens. Bioelectron.* 10, 805 (1995).
227. D. R. Shnek, D. W. Pack, D. Y. Sasaki, and F. H. Arnold, *Langmuir* 10, 2382 (1994).
228. S. A. Darst, M. Ahlers, P. H. Meller, E. W. Kubalek, R. Blankenburg, H. O. Ribi, H. Ringsdorf, and R. D. Kornberg, *Biophys. J.* 59, 387 (1991).
229. K. Fujita, S. Kimura, Y. Imanishi, E. Rump, J. van Esch, and H. Ringsdorf, *J. Am. Chem. Soc.* 116, 5479 (1994).
230. J.-L. Rigaud and B. Pitard, in "Liposomes as Tools in Basic Research and Industry" (J. R. Philippot and F. Schuber, eds.), pp. 71–88. CRC Press, Ann Arbor, MI, 1995.
231. J. L. Rigaud, A. Bluzat, and S. Büschlen, *Biochem. Biophys. Res. Commun.* 111, 373 (1983).
232. M. T. Paternostre, M. Roux, and J. L. Rigaud, *Biochemistry* 27, 2668 (1988).
233. J. L. Rigaud, M. T. Paternostre, and A. Bluzat, *Biochemistry* 27, 2677 (1988).
234. H. Kuhn, *Thin Organic Films* 178, 1 (1989).
235. J. Salafsky, J. T. Groves, and S. G. Boxer, *Biochemistry* 35, 14773 (1996).
236. E. Kalb, S. Frey, and L. K. Tamm, *Biochim. Biophys. Acta* 1103, 307 (1992).
237. R. Naumann, A. Jonczyk, R. Kopp, J. van Esch, H. Ringsdorf, W. Knoll, and P. Gräber, *Angew. Chem. Int. Ed. Engl.* 34, 2056 (1995).
238. L. K. Tamm and S. A. Tatulian, *Biochemistry* 32, 7720 (1993).
239. S. A. Tatulian, L. R. Jones, L. G. Reddy, D. L. Stokes, and L. K. Tamm, *Biochemistry* 34, 4448 (1995).
240. G. Puu, I. Gustafson, E. Artursson, and P.-Å. Ohlsson, *Biosens. Bioelectron.* 10, 463 (1995).
241. G. Elender, M. Kühner, and E. Sackmann, *Biosens. Bioelectron.* 11, 565 (1996).
242. M. Kühner, R. Tampé, and E. Sackmann, *Biophys. J.* 67, 217 (1994).
243. E. Sackmann, *Science* 271, 43 (1996).
244. D. Beyer, G. Elender, W. Knoll, M. Kühner, S. Maus, H. Ringsdorf, and E. Sackmann, *Angew. Chem. Int. Ed. Engl.* 35, 1682 (1996).
245. P. Ringler, W. Müller, H. Rinsdorf, and A. Brisson, *Chem. Eur. J.* 3, 620 (1997).
246. H. Ringsdorf and J. Simon, *Nature* 371, 284 (1994).
247. S. Chiruvolu, S. Walker, J. Israelachvili, F. J. Schmitt, D. Leckband, and J. A. Zasadzinski, *Science* 264, 1753 (1994).
248. S. A. Walker, M. T. Kennedy, and J. A. Zasadzinski, *Nature* 387, 61 (1997).
249. D. P. Nikolelis and C. G. Siontorou, *Anal. Chem.* 68, 1735 (1996).
250. B. Klee, G. L. Duveneck, P. Oroszlan, M. Ehrat, and H. M. Widmer, *Sens. Actuators, B* 29, 307 (1995).
251. K. T. Kinnear and H. G. Monbouquette, *Anal. Chem.* 69, 1771 (1997).
252. U. J. Krull, J. D. Brennan, R. S. Brown, S. Hosein, B. D. Hougham, and E. T. Vandenberg, *Analyst* 115, 147 (1990).

253. B. A. Cornell, V. L. B. Braach-Maksvytis, L. G. King, P. D. J. Osman, B. Raguse, L. Wieczorek, and R. J. Pace, *Nature* 387, 580 (1997).
254. M. Eray, N. S. Dogan, S. R. Reiken, H. Sutisna, B. J. Van Wie, A. R. Koch, D. F. Moffett, M. Silber, and W. C. Davis, *BioSystems* 35, 183 (1995).
255. S. R. Reiken, B. J. Van Wie, H. Sutisna, D. F. Moffett, A. R. Koch, M. Silber, and W. C. Davis, *Biosens. Bioelectron.* 11, 91 (1996).
256. Y.-Y. Yu, B. J. Van Wie, A. R. Koch, D. F. Moffett, and W. C. Davis, *Biosens. Bioelectron.* 12, 135 (1997).
257. H. A. Rongen, A. Bult, and W. P. van Bennekom, *J. Immunol. Methods* 204, 105 (1997).
258. S. S. Davis, *Trends Biotechnol.* 15, 217 (1997).
259. D. D. Lasic and D. Needham, *Chem. Rev.* 95, 2601 (1995).
260. K. Maruyama, T. Takizawa, T. Yuda, S. J. Kennel, L. Huang, and M. Iwatsuru, *Biochim. Biophys. Acta* 1234, 74 (1995).
261. K. Maruyama, A. Mori, S. Bhadra, M. T. Subbiah, and L. Huang, *Biochim. Biophys. Acta* 1070, 246 (1991).
262. S. Grimaldi, A. Giuliani, A. Giuliani, L. Ferroni, A. Lisi, N. Santoro, and D. Pozzi, *Res. Virol.* 146, 289 (1995).
263. P. G. M. Bloemen, P. A. J. Henricks, L. van Bloois, M. C. van den Tweel, A. C. Bloem, F. P. Jijkamp, D. J. A. Crommelin, and G. Storm, *FEBS Lett.* 357, 140 (1995).
264. J. L. Heeremans, J. J. Kraaijenga, P. Los, C. Kluft, and D. J. Crommelin, *Biochim. Biophys. Acta* 1117, 258 (1992).
265. R. J. Lee and P. S. Low, *J. Biol. Chem.* 269, 3198 (1994).
266. B. Lundberg, K. Hong, and D. Papahadjopoulos, *Biochim. Biophys. Acta* 1149, 305 (1993).
267. M. H. Vingerhoeds, H. J. Haisma, M. van Muijen, R. B. J. van de Rijt, D. J. A. Crommelin, and G. Storm, *FEBS Lett.* 336, 485 (1993).
268. M. H. Vingerhoeds, P. A. Steerenberg, J. J. G. W. Hendriks, D. J. A. Drommelin, and G. Storm, *FEBS Lett.* 395, 245 (1996).
269. D. P. McIntosh and T. D. Heath, *Biochim. Biophys. Acta* 690, 224 (1982).
270. M. Shinkai, M. Suzuki, S. Iijima, and T. Kobayashi, *Biotechnol. Appl. Biochem.* 21, 125 (1995).
271. S. Tomlinson, P. W. Taylor, and J. P. Luzio, *Biochemistry* 28, 8303 (1989).
272. M. Uto, M. Araki, T. Taniguchi, S. Hoshi, and S. Inoue, *Anal. Sci.* 10, 943 (1994).
273. M. A. Taylor, M. N. Jones, P. M. Vadgama, and S. P. J. Higson, *Biosens. Bioelectron.* 10, 251 (1995).
274. M. F. Rosenberg, M. N. Jones, and P. M. Vadgama, *Biochim. Biophys. Acta* 1115, 157 (1991).
275. M. Snejdarkova, M. Rehak, and M. Otto, *Anal. Chem.* 65, 665 (1993).
276. M. Rehak, M. Snejdarkova, and M. Otto, *Biosens. Bioelectron.* 9, 337 (1994).
277. J. Michaelis, *Biochem. J.* 10, 283 (1908).
278. M. C. Smith, T. C. Furman, T. D. Ingolia, and C. Pidgeon, *J. Biol. Chem.* 263, 7211 (1988).
279. H. Hjelm, K. Hjelm, and J. Sjoquist, *FEBS Lett.* 28, 73 (1972).
280. O. Guryev, V. Erokhin, S. Usanov, and C. Nicolini, *Biochem. Mol. Biol. Int.* 39, 205 (1996).
281. J. M. Pachence and J. K. Blasie, *J. Biophys. Soc.* 52, 735 (1987).
282. B. S. Attili and A. A. Suleiman, *Microchem. J.* 54, 174 (1996).
283. A. Breitwieser, S. Kupcu, S. Howorka, S. Weigert, C. Langer, K. Hoffmann-Sommergruber, O. Scheiner, U. B. Sleytr, and M. Sara, *BioTechniques* 21, 918 (1996).
284. E. Katz, D. D. Sclereth, and H.-L. Schmidt, 368 (1994).
285. T. H. Sisson and C. W. Castor, *J. Immunol. Methods* 127, 215 (1990).
286. B. Lu, M. R. Smyth, and R. O'Kennedy, *Anal. Chim. Acta* 331, 97 (1996).
287. B. A. Kuznetsov, N. A. Byzova, and G. P. Shumakovich, *J. Electroanal. Chem.* 371, 85 (1994).
288. S. M. Amador, J. M. Pachence, R. Fischetti, J. P. McCauley, A. B. Smith, and J. K. Blasie, *Langmuir* 9, 812 (1993).
289. A. Ulman, "An Introduction to Ulthrathin Organic Films: From Langmuir–Blodgett to Self-Assembly." Academic Press, San Diego, 1991.
290. F. MacRitchie, "Chemistry at Interfaces." Academic Press, San Diego, 1990.
291. L. K. James and L. G. Augenstein.
292. R. Verger and F. P., *Chem. Phys. Lipids* 30, 189 (1982).
293. T. Dubrovsky, A. Tronin, S. Dubrovskaya, O. Guryev, and N. Claudio, *Thin Solid Films* 284–285, 698 (1996).
294. O. Guryev, T. Dubrovsky, A. Chernogolov, S. Dubrovskaya, S. Usanov, and C. Nicolini, *Langmuir* 13 (1997).
295. C. Salesse, D. Ducharme, R. M. Leblanc, and F. Boucher, *Biochemistry* 29, 4567 (1990).
296. R. G. Nuzzo and D. L. Allara, *J. Am. Chem. Soc.* 105, 4481 (1983).
297. J. Sagiv, *J. Am. Chem. Soc.* 102, 92 (1980).
298. M. Jiang, B. Nolting, P. S. Stayton, and S. G. Sligar, *Langmuir* 12, 1278 (1996).
299. L.-H. Guo, G. McLendon, H. Razafitrimo, and Y. Gao, *J. Mater. Chem.* 6, 369 (1996).
300. M. A. Firestone, M. L. Shank, S. G. Sligar, and P. W. Bohn, *J. Am. Chem. Soc.* 118, 9033 (1996).
301. E. Osterberg, K. Bergstrom, T. P. Schuman, J. A. Riggs, N. L. Burns, J. M. Van Alstine, and J. M. Harris, *J. Biomed. Res.* 29, 741 (1995).

302. J. H. Lee, B. J. Jeong, and H. B. Lee, *J. Biomed. Mater. Res.* 34, 105 (1997).
303. L. Deng, M. Mrksich, and G. M. Whitesides, *J. Am. Chem. Soc.* 118, 5136 (1996).
304. K. L. Prime and G. M. Whitesides, *J. Am. Chem. Soc.* 115, 10714 (1993).
305. M. Mrksich, J. R. Grunwell, and G. M. Whitesides, *J. Am. Chem. Soc.* 117, 12009 (1995).
306. J. Madoz, B. A. Kuznetzoz, F. J. Medrano, J. L. Garcia, and V. M. Fernandez, *J. Am. Chem. Soc.* 119, 1043 (1997).
307. J. Spinke, M. Liley, F.-J. Schmitt, H.-J. Guder, L. Angermaier, and W. Knoll.
308. J. Spinke, M. Liley, H.-J. Gruder, L. Angermaier, and W. Knoll, *Langmuir* 9, 1821 (1993).
309. N. Damrongchai, K. Yun, E. Kobatake, and M. Aizawa, *J. Biotechnol.* 55, 125 (1997).
310. F. H. Arnold and B. L. Haymore, *Science* 252, 1796 (1991).
311. J. Carlsson, K. Mosbach, and L. Bulow, *Biotechnol. Bioeng.* 51, 221 (1996).
312. C. Dietrich, L. Schmitt, and R. Tampe, *Proc. Natl. Acad. Sci.* 92, 9014 (1995).
313. G. B. Sigal, C. Bamdad, A. Barberis, J. Strominger, and G. M. Whitesides, *Anal. Chem.* 68, 490 (1996).
314. P. S. Stayton, J. M. Olinger, M. Jiang, P. W. Bohn, and S. G. Sligar, *J. Am. Chem. Soc.* 114, 9298 (1992).
315. H.-G. Hong, M. Jiang, S. G. Sligar, and P. W. Bohn, *J. Am. Chem. Soc.* 10, 153 (1994).
316. K. Douglas and N. A. Clark, *Appl. Phys. Lett.* 48, 676 (1986).
317. R. Blankenburg, P. Meller, H. Ringsdorf, and C. Salesse, *Biochemistry* 28, 8215 (1989).
318. S. A. Darst, M. Ahlers, P. H. Meller, E. W. Kubalek, R. Blankenberg, H. O. Ribi, H. Ringsdorf, and R. D. Kornberg, *Biophys. J.* 59, 387 (1991).
319. E. E. Uzgiris and R. D. Kornberg, *Nature* 301, 125 (1983).
320. S. K. Bhatia, J. L. Teixeira, M. Anderson, L. C. Shriver-Lake, J. M. Calvert, J. H. Georger, J. J. Hickman, C. S. Dulcey, P. E. Schoen, and F. S. Ligler, *Anal. Biochem.* 208, 197 (1993).
321. J. W. Lyding, T.-C. Shen, J. R. Hubacek, and G. C. Abeln, *Appl. Phys. Lett.* 64, 2010 (1994).
322. G. C. Ablen, J. W. Lyding, D.-S. Thompson, and J. S. Moore, *Appl. Phys. Lett.* 70, 2747 (1997).
323. R. L. Wimalasena and G. S. Wilson, *J. Chromatogr.* 572, 85 (1991).
324. L. Haussling and H. Ringsdorf, *Langmuir* 7, 1837 (1991).
325. R. D. Johnson, Z.-G. Wang, and F. H. Arnold, *J. Phys. Chem.* 100, 5134 (1996).
326. J. Davies and C. S. Nunnerley, in "Surface Analytical Techniques for Probing Biomaterial Processes" (J. Davies, ed.), pp. 131–152. CRC Press, Boca Raton, FL, 1996.
327. J. Davies and I. Faulkner, in "Surface Analytical Techniques for Probing Biomaterial Processes" (J. Davies, ed.), pp. 67–88. CRC Press, Boca Raton, FL, 1996.
328. A. J. Mello, in "Surface Analytical Techniques for Probing Biomaterial Processes" (J. Davies, ed.), pp. 1–42. CRC Press, Boca Raton, FL, 1996.
329. D. Axelrod, *Annu. Rev. Biophys. Bioeng.* 13, 247 (1984).
330. L. L. Wood, S.-S. Cheng, P. L. Edmiston, and S. S. Saavedra, *J. Am. Chem. Soc.* 119, 571 (1997).
331. M. Yoneyama, M. Suugi, M. Saito, K. Ikegami, S.-i. Kuroda, and S. Iizima, *Jpn. J. Appl. Phys.* 25, 961 (1986).
332. D. M. Cropek and P. W. Bohn, *J. Phys. Chem.* 94, 6452 (1990).
333. L. H. Radzilowski, B. O. Carragher, and S. I. Stupp, *Macromolecules* 30, 2110 (1997).
334. W. Wiegrabe, M. Nonnenmacher, R. Guckenberger, and O. Wolter, *J. Miscrosc.* 163, 79 (1991).
335. C. E. Hall, D. Datta, and E. A. H. Hall, *Anal. Chim. Acta* 323 (1996).
336. S. Nakata, N. Kido, M. Hayashi, M. Hara, H. Sasabe, T. Sugawara, and T. Matsuda, *Biophys. Chem.* 62, 63 (1996).

Index

K

L

T